TRANSACTIONS
OF THE
INTERNATIONAL
ASTRONOMICAL UNION

VOLUME XIXB — PROCEEDINGS

INTERNATIONAL ASTRONOMICAL UNION
UNION ASTRONOMIQUE INTERNATIONALE

TRANSACTIONS
OF THE
INTERNATIONAL ASTRONOMICAL UNION
VOLUME XIXB

PROCEEDINGS OF THE NINETEENTH GENERAL ASSEMBLY

DELHI 1985

Edited by

JEAN-PIERRE SWINGS
General Secretary of the Union

D. REIDEL PUBLISHING COMPANY

A MEMBER OF THE KLUWER ACADEMIC PUBLISHERS GROUP

DORDRECHT / BOSTON / LANCASTER / TOKYO

Library of Congress Cataloging in Publication Data

International Astronomical Union. General Assembly (19th : 1985 : Delhi, India)
  Proceedings of the nineteenth General Assembly, Delhi, 1985.

  (Transactions of the International Astronomical Union; v. 19B)
  1. Astronomy—Congresses. 2. International Astronomical Union—Congresses. 3. Astronomy—Societies, etc.—Directories. 4. International Astronomical Union—Directories.   I. Swings, J.-P.   II. Title.   III. Series.
QB1.I6   vol. 19B   520 s   [520]   86-17661
ISBN 90-277-2321-4

Published on behalf of
the International Astronomical Union
by
D. Reidel Publishing Company, P. O. Box 17, 3300 AA Dordrecht, Holland

All Rights Reserved
© *1986 by the International Astronomical Union*

Sold and distributed in the U.S.A. and Canada
by Kluwer Academic Publishers,
101 Philip Drive, Assinippi Park, Norwell, MA 02061, U.S.A.

In all other countries, sold and distributed
by Kluwer Academic Publishers Group,
P. O. Box 322, 3300 AH Dordrecht, Holland

No part of the material protected by this copyright notice may be reproduced or utilized in any form or by any means, electronic or mechanical, including photocopying, recording or by any information storage and retrieval system, without written permission from the publisher

Printed in The Netherlands

PRESIDENT DE L'UNION ASTRONOMIQUE INTERNATIONALE

ROBERT HANBURY BROWN

PRESIDENT OF THE INTERNATIONAL ASTRONOMICAL UNION

1982–1985

Preface

The XIXth General Assembly of the International Astronomical Union was held in New Delhi, India, from November 19 to 28, 1985. It was dedicated to the memory of a former IAU President, Professor M.K.V. Bappu, who tragically passed away on August 19, 1982. On the occasion of the Delhi General Assembly, the IAU Minor Planet Center announced that Minor Planet (asteroid) No. 2596 henceforth will carry the name Vainu Bappu. The full text of the announcement reads :

"(2596) VAINU BAPPU = 1979 KN (diameter about 8 kilometers, period 5 years 4 months, mean distance from the Sun around 450 million kilometers)

Discovered 1979, May 19, by R.M. West at the European Southern Observatory.

Named in memory of Manali Kallat Vainu Bappu (1927-1982), famous Indian Astrophysicist and a dear friend of the discoverer. Educated at Harvard and Caltech, he established, under difficult circumstances, the first modern Indian observatory at Naini Tal during the 1950s. Appointed director of the Kodaikanal Observatory in 1960, he subsequently founded and directed the Indian Institute of Astrophysics in Bangalore. He was the initiator and driving force of many projects, among them the 2.3-m Kavalur telescope, entirely designed and built in India. In the multiple roles of brilliant scientist, teacher and administrator, he contributed decisively to the high level of astronomy and astrophysics in India today. He served as Vice-President (1967-1973) and President (1979-1982) of the IAU and as chairman of the editorial board of the Indian Journal of Astronomy and Astrophysics".

During the Inaugural Ceremony, on November 19, a very special date since it corresponded to the birthday of the late Mrs. Indira Gandhi, a one-rupee stamp representing Halley's comet was issued in the presence of India's Prime Minister, Rajiv Gandhi, and of the Minister of Posts and Telegraphs Sri Ram Niwas Mirdha.

The scientific programme of excellent quality, as can be seen from the reports included in this book and from the 900 pages of the "Highlights of Astronomy", Volume 7, was organized by the Presidents of the 40 IAU Commissions and coordinated by the IAU General Secretary (1982-1985), Dr. R.M. West. The local arrangements were taken care of by numerous hard workers from Delhi and from Bangalore, under the supervision of Prof. M.G.K. Menon (Chairman of the National Organizing Committee), and of Prof. A.P. Mitra (Chairman of the Local Organizing Committee).

The present volume, IAU Transactions XIX B, summarizes the work of the XIXth General Assembly. The discourses given during the Inaugural Ceremony, held at Siri Fort Auditorium, are reproduced in Chapter I. The proceedings of the two sessions of the General Assembly are found in Chapter II, which includes the Resolutions and other aspects of the administration of the Union : they provide the official record of the business of the General Assembly, and, together with the Executive Committee report (Chapter III), constitute the permanent record for the Union in the period 1982-1985.

In addition, this volume contains the Commission Reports from Delhi, compiled by the Presidents of the Commissions (Chapter IV). Pending re-edition of a complete Astronomer's Handbook, which is in preparation, Chapter V gives some information about a few activities of the Union, as well as its Statutes and By laws. Finally, Chapter VI, "Membership", contains the list of countries adhering to the Union, member lists of IAU Commissions, and also an alphabetical list of the more than 6000 individual members of the IAU (with addresses, phone and telex numbers, affiliation to commission(s)). This list will also be published separately and updated every year.

In order to make the information contained in this volume available at the earliest possible date, a tight production schedule was imposed. This would not have been possible without the efficient cooperation of the members of the Resolutions' Committee, and in particular of Drs. McCarthy and Morton, and of the Commission Presidents who provided the camera-ready manuscripts for Chapter IV. I am in addition most grateful to Mrs. B. Manning and to Miss D. Lours from the Paris IAU Secretariat for their expert help in typing, preparing and assembling the documents for publication : this was indeed a tremendous job. Furthermore I acknowledge with thanks Mrs. Manning, Miss Lours, Dr. R. Ferlet, and some French speaking Commission Presidents or Vice-Presidents for their help in putting together a (hopefully) correct English to French translation of the numerous Resolutions.

IAU Secretariat
61, avenue de l'Observatoire
F-75014 Paris
France

Jean-Pierre Swings
General Secretary, IAU
May 15, 1986

CONTENTS

	Page
Preface	v

I. INAUGURAL CEREMONY - 19 November 1985 1

- Address by Prof. M.G.K. Menon, President of the National Organizing Committee 1

- Address by Prof. R. Hanbury Brown, President of the International Astronomical Union 3

- Address by Shri Ram Niwas Mirdha, Minister of Posts and Telegraphs 4

- Address by the Prime Minister of India, Shri Rajiv Gandhi 7

- Address by Dr. A.P. Mitra, Chairman of the Local Organizing Committee 8

II. REPORT ON THE XIXth GENERAL ASSEMBLY 9

Agenda 9

First session (19 November 1985) 11

- Opening of the 1st Session, Presidential Address 11

    1. Formal Opening 12
       - Representatives of other Organizations 13
       - Message from ICSU 13
       - Members deceased since the last General Assembly 14

    2. Appointment of Official Interpreters 15

    3. Report of the Executive Committee 1982-84 15

    4. Report of the General Secretary (January-November 1985) 16
       a) IAU Membership nominations 16
       b) Report by the General Secretary 16

    5. Report on the Work of the Special Nominating Committee 17

CONTENTS

6.	Announcement of	
	a) Official Representatives of Adhering Organizations	18
	b) Representatives to vote on the Nominating Committee	18
	c) Acting Presidents of Commissions	19
7.	Appointment of the Finance Committee	20
8.	Appointment of the Resolutions Committee	21
9.	Revisions of the Statutes and By-Laws	21
10.	Resolutions submitted by Adhering Bodies	21
11.	Resolutions submitted by Commissions or by Associated Inter-Union Commissions	21

Second session (28 November 1985) — 23

12.	Report of the Finance Committee	23
13.	Financial Resolutions of the Executive Committee	25
14.	Resolutions submitted by the Executive Committee Resolutions proposées par le Comité Exécutif	25
15.	Resolutions proposed by the Resolutions Committee Résolutions proposées par le Comité des Résolutions	26
	B1 : Responsibility for Time Responsabilité de l'Heure	27
	B2 : Reference Frames Systèmes de Référence	28
	B3 : CCIR Actions Actions du CCIR	30
	B4 : Radio Frequency Transmissions from Space Transmissions Radioélectriques à partir de l'Espace	32
	B5 : VLBI Coordination Coordination Relative à l'Interférométrie à Très Grande Base (VLBI)	34
	B6 : Protection of Observatory Sites Protection des Sites d'Observatoires	35
	B7 : Danger of the Contamination of Space Danger de la Contamination de l'Espace	36
	B8 : Tycho's Observatories Les Observatoires de Tycho Brahé	37
	B9 : Endorsement of Commission Resolutions Soutien des Résolutions des Commissions	37

CONTENTS

<u>Resolutions proposed by Commissions</u>		38
Résolutions proposées par les Commissions		
C1 :	Astronomical Constants Constantes Astronomiques	38
C2 :	Reference Systems Systèmes de Référence	39
C3 :	Astronomical Designations Désignations Astronomiques	40
C4 :	Space Astrometry Astrométrie Spatiale	44
C5 :	Data Handling for Solar Research Distribution des Données de Recherche Solaire	45
C6 :	Carrington Reference System Système de Référence de Carrington	45
C7 :	Adequate Network of Solar Observatories Réseau Adéquat d'Observatoires Solaires	46
C8 :	Working Group on Planetary Surveys Groupe de Travail sur des Etudes Planétaires Systématiques	47
C9 :	Polarimetry and Large Telescopes Polarimétrie et Grands Télescopes	47
C10 :	Designation of Supernovae Affectation de Noms aux Supernovae	48
C11 :	Stellar and Solar Research at Mount Wilson Recherche Stellaire et Solaire au Mont Wilson	49
C12 :	Recommendations of Radio Source Nomenclature Recommandations sur la Nomenclature des Radiosources	49
C13 :	Faint Standard Stars Studies Etudes d'Etoiles Standard Faibles	51
C14 :	Future High-Energy Space Missions Futures Missions Spatiales en Astrophysique des Hautes Energies	52

16. Appointment of the Special Nominating Committee — 53

17. Nomination of new Members of the Union — 53

18. Application for IAU Membership — 53

19. Changes in Commissions — 54
    a) Presidents and Vice-Presidents of Commissions (1985-1988) — 54
    b) Organizing Committees of Commissions — 56
    c) Changes in Names of Commissions — 56

20. Place and Date of the XXth General Assembly — 56

21.	Election to the Union of a President, six Vice-Presidents, a General Secretary and an Assistant General Secretary		56
22.	Address by Prof. R. Hanbury Brown, President 1982-85		58
23.	Address by Prof. J. Sahade, President 1985-88		59
24.	Closing Ceremonies		61
	- Address on behalf of the Registered Guests by Mrs. T. West		61
	- Address by the retiring General Secretary, Dr. R.M. West		62
	- Address by the newly elected General Secretary, Dr. J.-P. Swings.		63

III. REPORT OF THE EXECUTIVE COMMITTEE (1982-1984) — 65

IV. REPORTS OF MEETINGS OF COMMISSIONS — 91
COMPTES RENDUS DES SEANCES DES COMMISSIONS

No.	Commission		
4	Ephemerides	Ephémérides	93
5	Documentation and Astronomical Data	Documentations et Données Astronomiques	97
6	Astronomical Telegrams	Télégrammes Astronomiques	101
7	Celestial Mechanics	Mécanique Céleste	103
8	Positional Astronomy	Astronomie de Position	111
9	Instruments and Techniques	Instruments et Techniques	123
10	Solar Activity	Activité Solaire	131
12	Radiation and Structure of the Solar Atmosphere	Radiation et Structure de l'Atmosphère Solaire	133
14	Atomic and Molecular Data	Données Atomiques et Moléculaires	141
15	Physical Study of Comets, Minor Planets and Meteorites	Etude Physique des Comètes, des Petites Planètes et des Météorites	159
16	Physical Study of Planets and Satellites	Etude Physique des Planètes et des Satellites	163
19	Rotation of the Earth	Rotation de la Terre	165

CONTENTS

20	Positions and Motions of Minor Planets, Comets and Satellites	Positions et Mouvements des Petites Planètes, des Comètes et des Satellites	177
21	Light of the Night Sky	Lumière du Ciel Nocturne	187
22	Meteors and Interplanetary Dust	Météores et Poussière Interplanétaire	193
24	Photographic Astrometry	Astrométrie Photographique	197
25	Stellar Photometry and Polarimetry	Photométrie et Polarimétrie Stellaires	205
26	Double and Multiple Stars	Etoiles Doubles et Multiples	211
27	Variable Stars	Etoiles Variables	217
28	Galaxies	Galaxies	223
29	Stellar Spectra	Spectres Stellaires	235
30	Radial Velocities	Vitesses Radiales	237
31	Time	L'Heure	245
33	Structure and Dynamics of the Galactic System	Structure et Dynamique du Système Galactique	251
34	Interstellar Matter	Matière Interstellaire	255
35	Stellar Constitution	Constitution des Etoiles	259
36	The Theory of Stellar Atmospheres	La Théorie des Atmosphères Stellaires	265
37	Star Clusters and Associations	Amas Stellaires et Associations	275
38	Exchange of Astronomers	Echanges d'Astronomes	279
40	Radio Astronomy	Radioastronomie	281
41	History of Astronomy	Histoire de l'Astronomie	285
42	Close Binary Stars	Etoiles Doubles Serrées	287
44	Astronomy from Space	L'Astronomie à partir de l'Espace	291
45	Stellar Classification	Classification Stellaire	301
46	Teaching of Astronomy	Enseignement de l'Astronomie	305

	47	Cosmology	Cosmologie	311
	48	High-Energy Astrophysics	Astrophysique des Hautes Energies	319
	49	The Interplanetary Plasma and the Heliosphere	Plasma Interplanétaire et Heliosphère	329
	50	Protection of Existing and Potential Observatory Sites	Protection des Sites d'Observatoires Existants et Potentiels	331
	51	Bioastronomy : Search for Extraterrestrial Life	Bioastronomie : Recherche de la Vie dans l'Univers	337
		Working Group for Planetary System Nomenclature	Groupe de Travail sur la Nomenclature du Système Planétaire	339

V. <u>FUNCTIONS AND STATUTES OF THE UNION</u>   355

VI. <u>MEMBERSHIP</u>   381

  1. List of Adhering Countries
  2. Membership of Commissions   389
  3. Alphabetical List of Members   467

CHAPTER I

INAUGURAL CEREMONY

19th November 1985

The Inaugural Ceremony was held at Siri Fort Auditorium, Delhi, with distinguished representatives from the Government of India, the City of Delhi, the University of Delhi, and the Indian National Science Academy.

Prime Minister Rajiv Gandhi was the Chief Guest. Shri Ram Niwas Mirdha represented the Ministry of Posts and Telegraphs of India, Prof. C.N.R. Rao represented the Indian National Science Academy, Prof. M.G.K. Menon was the representative of the National Organizing Committee.

The chair was taken by Prof. A.P. Mitra, Chairman of the Local Organizing Committee, Director of the National Physical Laboratory, in Delhi.

The musical interludes were performed by Gandharva Mahavidyalaya. The group consisted of the Principal of Shri Vinaya Chandra Maudgalya College and of his students. They sang the Indian National Anthem and one invocation song.

All the addresses were taped during the Inaugural Ceremony, and typed in Paris. The editor and the IAU staff apologize for the errors or misinterpretations that might have occured during this process ; Mr. Sahu nevertheless provided quite a useful input in deciphering the tape.

After some words of welcome from Prof. Rao, Prof. Mitra introduced Prof. M.G.K. Menon, Chairman of the National Organizing Committee.

Address by Prof. M.G.K. Menon, President of the
National Organizing Committee

Professor R.H. Brown was born in the Nilgiris at Aruvankadu and he was just mentioning standing outside, that his mother used to play golf where the large cylindrical radio telescope built by the Tata Institute of Fundamental Research is located at Ootacamund. Of course, Hanbury Brown has also been a visitor to India particularly as a Raman Professor of the Indian Academy of Sciences. A very special welcome to you Hanbury. The International Astronomical Union has a special character and perhaps it is worth just mentioning it briefly. It is a Union in which there is individual membership. That is very important. It is also a Union which supports on a significant basis young scientists. And indeed there are about three hundred young astronomers attending this meeting including about a hundred from India ; and this is part of the programme of IAU and it is certainly a programme which augures well for the future.

Mr. Prime Minister, you have practically in every address you have given, since you took over as Prime Minister a little over a year ago, referred to the modernization of India, referred to the great role of science and technology, and we have been aware of the fullest support which science and technology have received in India as Prof. C.N.R. Rao mentioned. First, at the hands of our first Prime Minister Pandit Jawaharlal Nehru, and thereafter in the hands of Srmt. Indira Gandhi, our Prime Minister for a period of almost eighteen years. This particular General Assembly takes place today on the 19th of November, the birth anniversary of Indiraji. In fact, we had quite intentionally kept it so. She had graciously consented, when I had spoken to her, to inaugurate this General Assembly. We had all hoped to welcome her here on a very auspicious day in her life, her birthday. But, alas, that was not to be, and I certainly hope that this very large meeting of scientists coming from all parts of the world, in an area which received the fullest support from her : international understanding, cooperation, and inter-dependence that exist in the world today, that this meeting, will indeed be a fitting tribute to the memory of that great lady who stood for peace, secularism, social justice, human understanding and the broadest holistic concept of culture to which very brief reference was made this morning, Mr. Prime Minister, in the function when the Indira Gandhi National Center was opened by you.

India has had a long and distinguished tradition in science from its earliest days. And in this tradition were recorded the great contributions of Indian astronomy covering mathematical aspects such as the decimal place value system, the Āryabhaṭiya written by the great astronomer mathematician Ārybhata, the trigonometric system which is characteristic of Hindu astronomy best known through the Sūryasiddhanta, and much else. But, in the in between period, prior to the British period, there has been a decline ; we of course have the last of the great efforts in observational astronomy, the masonry structures, the Jantar Mantars, of which those who are attending this conference will be able to see in Delhi itself, but which there are other specimens in the country. And indeed, it is from that we have the logo of this particular meeting.

We, of course, had during the British period, and prior to independence, a great deal of work arising through the setting up of the first Madras observatory, the Kodaikanal observatory, the great work of Megh Nad Saha. And indeed some of the great work on solar eclipses was done in this country, including the first observation of the element helium done in the plains of Andhra Pradesh. However, it is since independence and particularly over the past thirty years that we have had a renaissance in astronomy, with a large number of institutions, a fairly large community of astronomers, astrophysicists and cosmologists in the country, working over practically the entire spectrum. This covers new instruments such as several large optical telescopes, the latest being the 234-cm telescope which is now installed at Kavalur in the south, the large cylindrical radio telescope at Ootacamund, which I referred to, the proposal, Mr. Prime Minister, to take up in this plan a giant metre-wavelength radio telescope which will fill an important gap and which will be one of the major instruments when completed, the millimeter-wave telescope of the Raman Research Institute, the 1.2 meter infrared telescope at Gurushikar, of the Physical Research Laboratory in the Indian Space Research Organisation. There is, in addition, a considerable amount of ground-based, rocket-based, balloon-based and satellite-based astronomy, covering infrared, gamma-ray astronomy and a great deal of cosmic ray physics. And it is thus that we are now at a stage where one can say there are major new instruments that exist which offer opportunities for our scientists, particularly of the younger generation, and with the most modern techniques and technologies and electronics and so on, to work at the frontiers of this field. And, as I mentioned earlier, we have a large group here, about three hundred Indian delegates in what could be roughly about 1400 delegates in all as registered during the course of the conference, and of these, a large number are young astronomers. And we certainly, with these facilities, hope to work in this major international cooperative entreprise with our colleagues from round the world and in that sense, this particular meeting will be a milestone in the development of Indian astronomy.

We are honoured, Mr. Prime Minister, that in spite of the very heavy schedule which you have, you arrived early this morning from the Middle East, and since 6:30 or thereabouts, you have had a non-stop schedule of commitments. In spite of that, you have agreed to come here to inaugurate this meeting which your mother had so graciously agreed to earlier. We are very happy to have you with us and would like to thank you for this interest in science, in fundamental research, and particularly, in this field which has been with man since the beginnings of human civilization, an area where we are dealing with distances, with energies, with phenomena on a scale unimaginable compared to anything that one experiences on Earth, where all of our current knowledge has been derived not by actual experimentation, as one does for example in physics, but purely through observation, analysis, building of a systematic picture which is truly magnificent. And here we see the great powers of science, of observation, analysis and building up brick by brick these capabilities, this understanding, we find these powers of science at its very best. And astronomy, as a challenging field, is one indeed through which one can develop interest in science and appreciation of the scientific methods, not only in terms of challenges and excitement that exist in it, but also in terms of the universality of this particular field, and we therefore are very happy to have this conference in India, and on behalf of the National and Local Organizing Committees, once again, a very warm welcome. We hope all of you will enjoy this meeting, and not only this meeting, but also the country you have come to, where we hope to receive you over the next ten days with hospitality, with friendship and in the great traditions of Indian culture.

Thank you very much.

Address by Prof. R. Hanbury Brown, President of the

International Astronomical Union

On behalf of the International Astronomical Union, may I say how very honoured we are that you, Mr. Prime Minister, have consented to be with us today, on such a busy day for you. We greatly appreciate your presence here as a mark of your government's interest in science. May I also thank you, Professor Rao for your welcome.

I would also like to convey to all those concerned our very real gratitude for the invitation to meet in this country and for the generous facilities which you have provided for our meeting. We are indeed happy to be here - happy to be with people who are so welcoming and to meet in a city which is so handsome and so historic.

Our Union has been bringing its members together in General Assemblies ever since 1922 ; in fact it is one of the oldest of the Scientific Unions. Its broad aim is to develop astronomy through international cooperation and these General Assemblies are intended to serve that aim in three main ways. Firstly, they have the straightforward scientific function of exchanging and reviewing the latest scientific results and of planning the international cooperation in research which is so essential to astronomy - no matter what our nationality may be, we all work in the same sky.

Secondly, these Assemblies help to make individuals feel that they are part of a real, live, world-wide community of astronomers. For a short while they make visible the invisible community to which all true scientists belong, the invisible college of science. To know that they are part of that great community is particularly valuable to young astronomers who may have little, if any, opportunity of attending international meetings. I am glad to say that in planning this meeting considerable efforts have been made to help young astronomers to attend.

Thirdly, it has been our experience that these Assemblies help to promote a public interest in astronomy in the country in which they are held ; I hope that it will prove to be true in India.

Many members of our Union, will, I feel sure already, know something about India. If they were lucky they could have learnt it from our late President Vainu Bappu whose death was such a grievous loss to astronomy, especially to Indian astronomy and to our Union. He was an excellent ambassador for Indian science - courteous, charming and, moreover, extremely good at the task which he had undertaken - the modernising of Indian optical astronomy. We all miss him greatly at this meeting ; I know how much he valued the prospect of our Union meeting in India.

Many members will, no doubt, have learnt something about India from those guidebooks which we all buy and promise ourselves that we will read, and often end up reading on the plane. From those books we can learn quite a lot about Indian history and culture. The word culture as it is commonly used, includes literature, architecture, painting, music, dancing, sculpture, religion and so on ; but it never includes science. Astronomy is an integral part of science and the pursuit of science should be an integral part of any worthwhile conception of modern culture and vision of progress. Maybe the more serious guidebooks have something to say about India's extensive and distinguished scientific past - perhaps they tell us something about India's contributions to algebra or to astronomy in the 5th century, or more likely they tell us about the magnificent medieval observatories such as the one at Jaipur. But what they do not tell us, unlike Professor Bappu, is anything much about what Indian science is like today. To take a very few examples from my own experience of this great country, I could tell you about the fine optical observatory at Kavalur, the impressive radioastronomical installation at Ootacamund or about the excellent work on radioastronomy at the Raman Research Institute in Bangalore. But I hope that some of you will see these things for yourselves.

One of the best ways of getting to know an unfamiliar country in a short time is to meet your opposite numbers in that country. I hope that you will meet some of the very many Indian astronomers and other Indian scientists at this General Assembly. If you do, you will discover something which cannot be discovered from a guidebook, that the tradition of scientific excellence which we find in Indian history is still very much alive today.

Shri Ram Niwas Mirdha, Minister of Posts and Telegraphs, was then asked to request the Prime Minister to release the one rupee commemorative stamp, depicting Comet Halley, which was issued at the occasion of the XIXth General Assembly of the IAUThe Minister then delivered the following speech :

सीरीफोर्ट सम्मेलनकक्ष में 19-11-85 को हुई खगोलीय संघ की 19वीं आम सभा में स्मारक डाक टिकट जारी करने के अवसर पर संचार मंत्री श्री राम निवास मिर्धा द्वारा दिया गया भाषण ।

माननीय प्रधानमंत्री श्री राजीव गांधी, प्रो० हेनरी ब्राउन, अध्यक्ष, अंतरराष्ट्रीय खगोलीय संघ, प्रो० सी० एन० राव, अध्यक्ष, भारतीय राष्ट्रीय विज्ञान अकादमी, प्रो० एम० जी० के० मेनन, अध्यक्ष, राष्ट्रीय संगठन समिति तथा विशिष्ट अतिथिगण ।

हमारे लिये यह वास्तव में गौरव का विषय है कि अंतरराष्ट्रीय खगोलीय संघ की 19वीं आम सभा का आयोजन यहाँ हो रहा है। ऐसा पहली बार हुआ है कि स्व० डा० के वेणु बाप्पु के विशेष प्रयत्नों के फलस्वरूप भारत में संघ की आम सभा का आयोजन किया गया है। हमें इस बात की प्रसन्नता है कि सौर-निकाय से लेकर ब्रह्माण्ड की उत्पत्ति तक के विषय के अध्ययनों पर चर्चा करने के लिए संसार भर के अनेक वैज्ञानिक इस सभा में भाग ले रहे हैं। इसके द्वारा विश्व भर की सहकारिता पर चल रही परियोजनाओं की समीक्षा की जायेगी तथा नये कार्यक्रमों पर सहयोग द्वारा कार्य आरम्भ होगा। सन् 1946 से भारत अंतरराष्ट्रीय खगोलीय संघ का एक सदस्य है। पिछले कई वर्षों में खगोलविज्ञान तथा खगोल-भौतिकी के प्रायोगिक पहलू पर भारत ने महत्वपूर्ण योगदान दिया है।

भारत में खगोलविज्ञान को सदैव सम्मानजनक स्थान दिया गया है यद्यपि प्राचीन भारत में खगोलविज्ञान और ज्योतिष का परस्पर विलय हो गया था। उदाहरण के लिए वाराहमिश्र तथा ब्रह्मगुप्त दोनों को ही यह भलीभांति ज्ञात था कि चन्द्रग्रहण पृथ्वी की छाया तथा सूर्यग्रहण चन्द्रमा के कारण होता है।

आधुनिक भारत में हमारे वैज्ञानिकों ने खगोल विज्ञान तथा खगोल-भौतिकी के क्षेत्र में बहुत उपलब्धियाँ प्राप्त की हैं तथा विश्वभर में उनके कार्य को सराहा गया है।

खगोलज्ञों की श्रेणी में मेघनाथ साहा का विशिष्ट स्थान है। डा० सुब्रह्मण्यम शेखर ने भी खगोल विज्ञान के क्षेत्र में बहुत योगदान दिया है।

खगोल भौतिकी के प्रयोगों में भारत ने असाधारण प्रगति की है, तमिलनाडु में केवलूर, ऊटी में उधागमन्डलम् तथा कोडाईकनाल में स्थित विशेष दूरबीनों से हमारे वैज्ञानिकों को कार्य करने की उपयुक्त सुविधाएँ तथा वैज्ञानिक वातावरण उपलब्ध हो गया है।

खगोल विज्ञान तथा खगोल भौतिकी केवल वैज्ञानिकों का ही विषय नही हैं बल्कि सभी युगों में इसके द्वारा स्त्री - पुरूष रोमांचित होते रहे हैं । विशेषकर इस समय यह रोमांचक स्थिति नई ऊंचाइयों को छू रही है । हैली धूमकेतु 76 वर्षों के बाद पुन: सूर्य तथा पृथ्वी के निकट आ रहा है तथा सन् 1985 के अंत और सन् 1986 के आरम्भ तक निकट बना रहेगा । 27 नवम्बर को यह पृथ्वी से 9 करोड़ 30 लाख मील की दूरी पर होगा । हम डाक विभाग के सदस्य ब्रह्माँड की उत्पत्ति में किये जा रहे मानव-प्रयासों में भागीदार होना चाहते हैं। इस ऐतिहासिक घटना के अवसर पर हमने हैली धूमकेतु दर्शाते हुए एक डाक टिकट को स्मारक के रूप में जारी किया है । हम यह भी आशा करते हैं अंतरराष्ट्रीय खगोलीय संघ को इस आम सभा के आयोजन से उत्पन्न वैज्ञानिक वातावरण से भारत की खगोल भौतिकी के क्षेत्र को नया जीवन मिलेगा ।

The English translation as provided by the Indian hosts is given below :

"Respected Prime Minister Shri Rajiv Gandhi, Prof. Hanbury Brown, President of the International Astronomical Union, Prof. C.N.R. Rao, President of the Indian National Science Academy, Prof. M.G.K. Menon, Chairman of the National Organizing Committee, and distinguished guests,

It is indeed a matter of great honour for us that the XIXth General Assembly of the International Astronomical Union is being held here. This is the first time that a General Assembly of the Union is held in India mainly as a result of the efforts of late Dr. M.K. Vainu Bappu. We are glad that the General Assembly is being attended by a large number of scientists from all over the world to discuss a wide range of subjects ranging from the study of the solar system to the origin of the Universe. The progress in on-going international cooperative projects will be reviewed and new collaborative programmes will be instituted. India has been a member of the International Astronomical Union from 1946. Over the years, it has made significant contributions to the practical aspects of astronomy and astrophysics.

India has always had a place of honour in astronomy, though invariably in ancient India, astronomy got merged with astrology. For example, Varahmishra and Brahmagupta knew perfectly well that the Moon is eclipsed by the shadow of the Earth and the Sun is eclipsed by the Moon.

In modern India, we have scientists who have achieved a great deal in the field of astronomy and astrophysics and have been acclaimed all over the world.

Meghnath Saha is one bright name in the galaxy of astronomers. Dr. Subramanian Chandrasekhar has also contributed significantly to the world of astronomy.

India has also made remarkable progress in experimentation in atrophysics and now with specially equipped telescopes in Kavalur, in Tamil Nadu, in Udhyamandalam, in Ooty and in Kodaikanal, our scientists have proper facilities and scientific climate to work in.

Astronomy and astrophysics do not belong to the scientists alone, but have for long fascinated men and women of all ages. At this particular juncture, the fascination is at a new height. The Halley's comet is coming back towards the sun and the earth after 76 years, and will pass near them in late 1985 and early 1986. On November 27 it will be 93 million kms from the Earth. We, in the Department of Posts like to share the human endeavour in understanding the origin of Universe. On this historic occasion we have brought out a commemorative postage stamp depicting the Halley's comet. We also hope that the field of Indian astrophysics will get a new life in the scientific climate generated by the XIXth General Assembly of the International Astronomical Union."

Address by the Prime Minister of India, Shri Rajiv Gandhi

Sri Mirdha, Professor Hanbury Brown, Professor C.N.R. Rao, Professor Menon, Professor Mitra, distinguished astronomers, friends, let me first welcome you to India and I hope that you do have not just a good meeting which I am sure you will have, but also a good time in India and get to see our country just a little bit. Professor Brown, I believe, is connected much more closely to India than we were told by Professor Menon. Not only his mother, but his father and grandfather as well were born in India. And well, we welcome him back.

Our traditions with astronomy are very old. Man has always looked up at the stars in wonder and India was no different. Our star gazers developed a science in our very early stages of development. Some of it as you mentioned is still visible in Jaipur and in Delhi at the Jantar Mantars which, as I was reminded, are still serving a very useful purpose in bringing tourists to India. And that set me wondering, I was thinking, how would people, maybe three or four hundred years hence, look at what we are setting up today, the big radiotelescopes, the other equipment. I wonder if they would be coming back to look at it as tourists. Indian astronomers have been in touch with their counterparts in all parts of the world, for a long time. And it has helped our sciences to develop themselves. When India became independent, we had various choices in front of us, and although I was not very old at that time, I still remember the heated arguments that used to go on, the discussions, the controversies in the newspapers, about whether we should try and develop our own science and technology, or we should just try to give work to our people, limit ourselves to rural industries. And we are fortunate that our leaders had the vision to look ahead, look at science and technology for answers to our day-to-day problems. Jawaharlal Nehru gave a thrust for not just applied science but also for fundamental research in India at that time, and it is the fruits of that vision that we are having today. India, unlike many other developing countries, set out to develop its own basic sciences. And we have produced results, we have produced eminent personalities, many have mentioned Professor Vainu Bappu, but he is only one amongst a long chain of scientists that have come out of our system. It is a continuation in the same theme that we are trying to do today. We are looking to science for most of our answers. We have found that we have only succeeded where the latest technology, the newest technology, in that particular field, has been applied to the problem. If it was agriculture, today we are self-sufficient, we have vast stocks of food grains, because the most modern genetic technology was applied to the seeds, the best methods for irrigation, fertilizers, equipments were applied and the results are in front of us. Wherever we have hesitated, we have not achieved such success. And the incredible thing is that the average Indian farmer may be illiterate, not educated, but he has the capacity and capability to absorb the technology when it is put in front of him. He is able to cope. This has been perhaps our biggest achievement in post-independence India.

We set out to use science and technology as a tool to remove poverty. One of the primary goals that was put in front of us by Panditji was to develop what he called a scientific temper. Professor Brown, you have mentioned the word "culture" and what one pictures when one sees culture. Is it just architecture, paintings, dance, music ? Or is it something more than that ? We have always looked at culture slightly differently, in a much wider perspective and science and technology have been very much part of that perspective. We have felt that if our country is to progress, then this scientific temper must be built into our culture ; not just in pockets, at the highest levels of development, where we might have top class scientists and technologists, but the pyramid must start from the bottom. It must have a wide base which will support these small groups right at the top. And this is the direction that we would like to take.

Ultimately, when we are talking of development in a developing country, we are not talking just of major projects, whether they are large industries or steel plants or dams, we are not really talking about anti-poverty programmes. What we are really talking about is the development of the human being, how are we going to make a better human being out of, well in India, our 750 million people. The root has to be from our heritage, from our traditions, but today that is coming under tremendous pressure with modern technological development, with modern sciences, which we ourselves are striving to develop. And it is this balance that we are trying to build, balance between science and technology, the 20th century, and the values, the spirituality, the inner strength that we have inherited from our ancient civilizations. It is such questions that need answers today. We, in India, are striving for these answers, striving to see that we are able to produce just such a balance.

In today's world, we have perhaps as the biggest problem, barriers that have set up, maybe I should say political barriers that we have set up, in between us. Fortunately, there are many areas where we cut across all such barriers : science and scientists are definitely one such area. It does not matter what the political problems are, it does not matter what other complications there are, but invariably, we find that when it comes to scientific knowledge, there is a sharing among scientists. There is a brotherhood, a oneness, which we would like to spread to the rest of our society, to other fields, in modern day life. I hope that this seminar, this annual meeting, will lead to much more such comradery, much more friendship among scientists, not just scientists from abroad and India, but also among scientists from all the different countries that are here today. I wish you all the best and, once again, I hope you have a nice time in India.

Thank you.

Address by Dr. A.P. Mitra, Chairman of the Local Organizing Committee

Honourable Prime Minister, Sri Mirdha, Professor Hanbury Brown, Professor C.N.R. Rao, Professor Menon, distinguished delegates and guests, for us in the Organizing Committee, it has been a labour of love, to have brought together a galaxy of distinguished astronomers on the occasion of this inauguration, many distinguished diplomats, administrators, and intellectuals. As Professor Menon mentioned, in spite of the distance and the costs involved, the total number of registered participants we expect will exceed 1400, of which about 300 are Indian astronomers. A very satisfying feature is the large number of young astronomers present here, we would like to welcome them. I would like to thank the distinguished delegates and invitees who are present here for coming to this inaugural function and encouraging us. The Organizing Committee would like to thank the Prime Minister for finding time to inaugurate this General Assembly on a very busy day. We are happy that he came here and encourage us at the beginning at of this important scientific conference.

Thank you.

CHAPTER II

REPORT ON THE XIXth GENERAL ASSEMBLY OF THE INTERNATIONAL ASTRONOMICAL UNION

AGENDA
Session 1 - November 19, 1985

1. Opening of the General Assembly by the President

2. Appointment of Official Interpreters

3. Report of the Excecutive Commitee, 1982-1984

4. Report of the General Secretary January November 1985

   a) IAU Membership nominations

   b) Report of the General Secretary

5. **Report by the President on the work of the Special Nominating Committee 1982-1985**

   The names proposed by the Special Nominating Committee to the XIXth General Assembly at Delhi as members of the IAU Executive Committee for the term 1985-88 are reproduced in the proceedings of the First Session of the XIXth General Assembly, on p.17.

6. **Announcement of :**

   a) The names of Official Representatives of Adhering Organizations

   b) The names of Representatives to serve on the Nominating Committee

   c) The names of Acting Presidents of Commissions

7. **Appointment of the Finance Committee**

8. **Appointment of the Resolutions Committee**

9. **Revisions of the Statutes and By-Laws**

   The Executive Committee proposes that By-Law 17 be changed as follows :

   "The Executive Committee appoints the Union's Representative to the International Council of Scientific Unions ; if not already an elected member of the Executive Committee, this representative will become its adviser."

   "Le Comité Exécutif nomme le représentant de l'Union qui doit siéger au sein du Conseil International des Unions Scientifiques ; si ce représentant n'est pas déjà un membre élu du Comité Exécutif, il devient conseiller."

   The change consists in deleting the words "... on the Executive Committee..." and reflects a change in the ICSU Structure.

10. Resolutions submitted by Adhering Bodies

11. Resolutions submitted by Commissions or by associated Inter-Union Commissions

    Session 2, 28th November 1985.

12. Report of the Finance Committee

    To be presented by the Chairman of the Subcommittee of Finance Committee. The report will include recommendations about what concerns Item 13.

13. Financial Resolutions of the Executive Committee

    a) Proposed Budget 1985
    b) Unit of Contribution to the IAU, 1986-1988
       The Executive Committee proposes that the unit of contribution be fixed at SFr 1 885 (1986), 1 945 (1987) and 2 005 (1988).
    c) Proposed Budget 1986-1988
       The Executive Committee proposes a budget of Swiss Francs 1 722 000 for this triennium.
    d) Other financial resolutions
       Not included in the proposed budget.

14. Resolutions submitted by the Executive Committee

15. Recommendations of the Resolutions' Committee on the resolutions submitted by Commissions

16. Appointment of the Special Nominating Committee

17. Nominations of the new Members of the Union

18. Application for IAU Membership

19. Changes in Commissions

    a) Election of Presidents and Vice-Presidents of Commissions for the period 1985-1988
    b) Organizing Committees of Commissions
    c) Names of Commissions
       The following change of name has been proposed for Commission 51 :
       "Search for Extraterrestrial Life", to be renamed : "Bioastronomy : Search for Extraterrestrial Life".

20. The place and date of the XXth General Assembly

21. Election to the Union of a President, six Vice-Presidents, a General Secretary and an Assistant General Secretary

22. Address by the President 1982-1985

23. Address by the President 1985-1988

24. Closing Ceremonies

PROCEEDINGS

XIXth GENERAL ASSEMBLY
FIRST SESSION

Held in the Siri Fort Auditorium
19th November 1985, 15:30
Prof. R. Hanbury Brown, President in the Chair

Opening of the First Session

Before we go any further I want to say just a few words to the many people in this Hall for whom these proceedings are their first experience of an IAU meeting. I am concerned that much of the Agenda may seem to them not only boring but unnecessary. I shall not try to persuade them that these proceedings are entertaining, but only that they are necessary. The IAU, thanks to a lot of hard work by few dedicated people, can justifiably claim to have the least possible administrative machinery at the least possible cost. The agenda of this session represents the essential nuts and bolts of that machinery.

I have in mind an awful picture of some newcomers to our Union trying to find out what the IAU is all about by reading our Transactions Volume B in their library before setting out for the General Assembly. Our Statutes would tell them the purposes of this Union, but only in the most general terms ; however if they were lucky enough to come across the recent Report of the Subcommittee on IAU Commission Structure, to which I shall refer later, then they would find the functions of this Union spelled out more in terms of what it actually tries to do.

Very briefly the principal aim of the Union, a very worthy aim, is to develop astronomy through international cooperation. In pursuit of that aim it sponsors international scientific meetings, international scientific programmes, international scientific services which are needed by astronomers, and it represents the interests of astronomy on a host of international bodies. If you cannot always see how these functions are related to what is going on in this meeting, then there is one major function of our Union which you can appreciate quite simply by looking aroud the hall. I refer, of course, to the fact that the IAU promotes international cooperation in astronomy by fostering an international community and by encouraging an awareness of that community. In a world which is divided by nationalism the IAU aims to preserve the unity of astronomy, and to preserve the unity of astronomy with the other sciences. One of the principal ways in which it does so, is to create a community of people with a common interest in the one sky which we all share.

Let me remind you of just two of the many ways in which it does this : firstly, unlike many other international Unions, it enlists individuals as members : secondly, it spends a great deal of time, trouble and money in bringing these individuals together every three years. And so I ask you to remember that this General Assembly, with all the inevitable difficulties which attend any large meeting, is not to be seen simply as a number of smaller meetings of specialised Commissions, but as an important step towards one of the principal objectives of our Union, the making of a real, live, international community of astronomers.

One of the first things I did when I was elected President was to try to find out whether or not there was a need to reorganise the IAU either to meet its present obligations or to meet those of the future. I asked the opinion of a great many people, Presidents of Commissions, past Officers of the Union and so on. To my surprise and dismay, very few people were prepared to make serious criticisms of the Union ; all they had to offer were minor improvements. It seemed that either the Union was perfect or, perhaps, that it was beyond repair !

Finding both these conclusions hard to accept I persuaded, with the help of the Executive Committee, that faithful friend of the IAU, Patrick Wayman, to chair a sub-committee on IAU Commission Structure : this was one limited aspect of the IAU which we thought should be examined in detail. As I expected, the sub-committee did not find the IAU to be perfect ; they made a number of constructive criticisms - nothing very radical but nevertheless valuable. I hope you will read their report ; a summary was published in the IAU Bulletin No. 53. To my mind one of the most important sections of that report is where they draw our attention to the importance of the part played by the Presidents and Vice-Presidents of our Commissions and suggest how the procedure for electing them might be improved.

Looking to the future let me draw your attention to one or two points ; firstly I invite you to note the presence at this General Assembly of a large contingent of astronomers from the People's Republic of China - the Executive Committee has done everything in its power to encourage their attendance and we are delighted to see them here. We look forward to the part which I am sure they will play in the life of the Union.

There are, I should add, a number of countries in which astronomy is practised but which do not, as yet, adhere to this Union. The Executive, largely thanks to the personal efforts of the General Secretary, is exploring whether or not the IAU can play any useful part in the development of astronomy in those countries.

May I also invite you to note from the Draft Budget for 1986-1988 put forward by the Executive, that we propose to increase substantially the fraction of our budget which is spent on the support of international symposia and colloquia ; this proposal is put forward in the simple conviction that these meetings represent one of the most worthwhile activities of this Union.

Finally let me say that my own experience of the last three years suggests that, although the IAU is not perfect, nevertheless it is meeting its present obligations very well and, as far as I can see, it is in good shape to meet the future.

1. Formal Opening

The President welcomed the members of the Union, Invited Participants and Registered Guests. He emphasized the value of individual membership, and extended a special welcome to those of the members who have belonged to the Union for fifty years or more, especially those who were present at the XIXth General Assembly :

>Prof. W.J. Luyten
>Prof. K.A. Strand
>Prof. Z. Kopal.

He proposed to send a message of greetings to the others, not present :

>Prof. Y. Oehman
>Prof. J.H. Oort
>Prof. S. Plakidis
>Dr. H.B. Sawyer-Hogg
>Dr. F. Whipple.

This suggestion was approved by acclamation.

A special welcome was extended to members who in the past have served on the Executive Committee of this Union, and who were attending the General Assembly :

>Prof. W. Iwanowska
>Prof. E. Müller
>Prof. J.-C. Pecker
>Prof. J. Sahade
>Prof. S. van den Bergh

and to the official representatives of the Adhering Organizations which support this Union.

Furthermore, the President extended a hearty welcome to the official representatives of ICSU and sister Unions as followed :

ICSU	E.A. Müller
UNESCO	M. Derkatch
ESA	R. Bonnet
IUPAC	C.N.R. Rao
IUHPS	O. Pedersen
BIPM	P. Giacomo
BIH	B. Guinot, M. Feissel
IGU	M. Shafi
FAGS	E.A. Tandberg Hanssen
NASA	C. J. Pellerin
COSTED	S. Radhakrishna
IUTAM	R. Narasimha
IAF	L. Perek
IUPAP	G. Swarup
IAGA	R.G. Rastoji
URSI	V. Radhakrishnan
CTS	D. McNally
CCIR	Mr. Nalbandian
IAG	I.I. Mueller
IUGG	P. Pâquet
INTERCOSMOS	A.G. Massevitch
ISWA	J. Cornell

The representative of ICSU, Prof. Edith Müller read a message from ICSU :

"The President of ICSU has asked me to express his regrets that he is unable to be with you on the occasion of the XIXth General Assembly of the International Astronomical Union. He has asked me to extend to you his best wishes for a successful Assembly.

As one of the founder organizations of ICSU's predecessor, the International Research Council, the International Astronomical Union has played and continues to play an important role in ICSU and in its development. This applies particularly at the present moment when your Secretary General, Dr. Richard West, is a member of the Executive Board of ICSU.

The development of science in the last forty years has been considerable. ICSU has made a number of changes in its organization, statutes and methods of work in an endeavour to adapt to such changes. Such modifications have been beneficial but it was recently decided to have a more thorough study of the organization of international science and ICSU's role therein, and a conference was organised recently which brought together about 50 people from within and outside the ICSU family. A wide range of suggestions were put forward and these are being reviewed by a small group which will suggest short and longer term follow-up actions to the Executive Board, which will then put its proposals to the ICSU General Assembly in September 1986.

One of the points that came out strongly in the conference discussions was the importance of ICSU's role in basic science and particularly in the preparation and organization of programmes of basic science studies, such as that proposed on Global Change, not being undertaken by other organisations.

Another point that came out strongly was the need for ICSU and the members of the ICSU family to increase their visibility. In this respect it is a pleasure to note the care with which arrangements for the Press have been made during this Assembly.

Mr. President, distinguished participants, you have a busy period in front of you and I would not wish to diminish the time available to discuss science. Let me close, therefore, in reiterating the best wishes and to say how pleased I am to see once again so many friends and colleagues."

The President then asked those present to stand while the General Secretary would read the names of the members who had died since the XVIIIth General Assembly. The General Secretary then read the following list :

ABELL, George O.	DIECKVOSS, Wilhelm K.E.
ABETTI, G.	DOAN, H.H.
ARAKELIAN, M.A.	DOUGLAS, A.E.
BAEHR, U.	DUNHAM, Théodore
BAPPU, M.K.V.	EBBIGHAUSEN, E.G.
BERTAUD, Charles	FATCHIKIN, Nikolaj
BEYER, Max	FILLIOZAT, Jean
BIGAY, J.H.	FLORENSKY, K.P.
BINNENDIJK, Leendert	FORBES, Eric C.
BOGORODSKY, A.F.	FRESA, Alfonso
BOK, Bart J.	GAPOSHKIN, S.I.
BOTELHEIRO, A.	GAVRILOV, I.V.
BULLEN, K.E.	GEMANTSEV, German G.
BURCH, Cecil	GIOVANELLI, Ronald
CANNON, Chris	GRIGOREVSKY, V.M.
CECCARELLI, M.	HARTNER, W.R.
COX, John	HATTORI, Akira
D'AZAMBUJA, M.L.	HECKMANN, Otto
DE MARCUS, W.C.	HEMENWAY, Curtis L.
DECAUX, Bernard	HOHL, Frank
DEMENKO, A.A.	HOUTGAST, J.

JOHNSON, Martin
JUNKES, Josef
KASTLER, Alfred
KATASEV, L.A.
KIPPER A.
KIZILIRMAK, Abdullah
KOBRIN, M.M.
KOROLKOV, D.V.
KOZYREV, Nikolaj A.
KRAT, Vladimir A.
KRINOV, E.L.
KUZMENKO, K.N.
LABORDE, Georges
LAMBRECHT, Hermann
LAPAZ, Lincoln
LI, Zhong Ryul
LITTLE, Alec
LINFOOT, E.H.
LINK, François
LOHMANN, W.
MALMQUIST, K.G.
MAMEDBEJLI, G.D.
MARCUS, Ella
MCKINLEY, Donald
MEFFROY, Jean F.
MEHLTRETTER, J.P.
MELNIKOV, O.A.
MERTON, Gerard
MICHEL, Karl-Wolfgang
MIKHAILOV, A.A.
MÜLLER, R.
NICOLINI, Tito
NIKOLSKY, G.M.
O'CONNELL, D.J.K.
OKUDA, Toyozo
PIOTROWSKI, Stefan
PLAUT, Lucas
POPOV, N.A.
PRICE DE SOLLA, Derek J.

PRZYBYLSKI, Antoni
PURGATHOFER, Alois
REINHARDT, Michael
ROSSELAND, Svein
RYLE, Martin
SAMAHA, A.H.M.
SCHOMBERT, John
SHCHEGLOV, V.P.
SHEN, Zee
SHKLOVSKY, Joseph
SIEGEL, C.L.
SIMONENKO, A.N.
SOCHER, Hermann
SCHUERMAN, D.W.
SHANE, Donald C.
SLAUCITAJS, S.
SMITH, Henry J.
SOCHER, H.
STARYTSIN, Gennadij
STEARNS, C.L.
STERNBERK, B.
SWINGS, Pol
TANAKA, Haruo
TAVASTSHERNA, K.N.
TEMESVARY, Stefan
TUVE, Merle
VANDERLINDEN, H.L.
VARSAVSKY, Carlos
VASILEV, Vladimir
VELTMANN, Ulo Ilmar K.
VON SOCHER, Hermann
VSEKHSVYATSKIJ, Sergius K.
WAGMAN, N.E.
WIDORN, Thomas
WILCOX, John M.
WOOD, Harley Weston
ZHAO, Que-Min
ZHONGOLOVICH I.D.

2. Appointment of Official Interpreters

The General Secretary appointed J. Rountree as official interpreter from French to English and C. Cesarsky from English to French.

3. Report of the Executive Committee 1982-84

The President invited discussion about the Report of the Executive Committee for the 3-year interval 1982-84, as an extract was presented in IAU Information Bulletin No. 54.

He mentioned that it had been given consideration by the official representatives of

Adhering Organizations, who, with the important reservation that the financial section of the report would come under scrutiny of the Finance Committee of the General Assembly, were prepared to recommend adoption of the report, subject to further discussion here.

There were no points raised from the audience, and the General Assembly unanimously approved the Report of the Executive Committee (1982-1984), subject to a relevant report by the Finance Committee (the full report is printed in this volume, see Chapter III).

4. Report of the General Secretary (January-November 1985)

a) IAU Membership Nominations

The President asked the General Secretary to inform the General Assembly about proposals of individuals to IAU Membership.

The General Secretary indicated that he had received almost 900 proposals of individuals to IAU membership. A further 43 were received from Presidents and members of the Executive Committee.

b) Report by the General Secretary

First of all, I should like to thank most heartily the President and the Assistant General Secretary, as well as the other members of the Executive Committee for their very efficient support during the recent period of great activity. Moreover, I believe that a few of you realize how much work has been done during the past three years by the two secretaries, Mrs. B. Manning and Miss D. Lours, as well as by my secretary in Munich, Mrs. E. Völk, and by Mrs. D. Fraipont, secretary to the Assistant General Secretary in Liège. We are privileged to have been helped by them during our period of office and I should like to express my sincere thanks to them.

The main effort during the past ten months has been directed towards the organization of the XIXth IAU General Assembly here in Delhi. The work involved the arrangement of the scientific programme and all the practical details of travel and lodging. As you undoubtedly already noticed, our Indian colleagues have established a most impressive organisation around this meeting which is running as smoothly as anybody could expect. It is a particular pleasure to thank the members of the National and Local Organizing Committees and the Chairmen, Profs. M.G.K. Menon and A.P. Mitra. They have managed to gain access to Vigyan Bhavan for us. It is the Prime Minister's conference complex, and I am told that very few scientific organizations have ever been given this great privilege.

We are also thankful for the extensive financial support which has been given to this General Assembly and which has allowed the IAU and the Local Organisers to support no less than 250 scientists. Most of these are young and many come from countries where astronomy has not yet been fully developed. I am sure that they and their colleagues from other countries who come for the first time to an IAU General Assembly will find this a most useful and scientifically beneficial experience.

The programme in front of us is very full, but you should be able to keep track of what is going on through the daily issues of the newspaper "Mandakini" and by reading regularly the notice boards. I should also like to say that the organisers and the IAU Executive Committee have placed a special emphasis on the press relations at this meeting. There will be several press conferences and I ask you kindly to be available for interviews with scientific reporters.

Another important item during the past months has been the now completed computerization of the entire IAU member file. Telephone, telex, commission membership, etc., have been added and this will greatly facilitate interaction among IAU members in the future. It is expected that the updated lists, including the new members, who will be admitted at this General Assembly, will become available in early 1986. As a special service to Commission Presidents, the Secretariat will be able to produce address labels for commission members. This will facilitate the work of the Commission Presidents and hopefully increase the frequency with which they are in contact with their members through circular letters etc.

There have been many other activities, and you will hear about most of them during this General Assembly. I wish you a profitable meeting and look forward to have the opportunity to meet many of you in person during the coming days.

The General Secretary's report was received with acclamation.

5. Report on the Work of the Special Nominating Committee

The President informed the Assembly that the Special Nominating Committee had selected the following IAU members to be proposed as members of the Executive Committee from November 28, 1985 :

As President :   Professor J. Sahade (Argentina)

As Vice-Presidents continuing from the former period :

Professor R.P. Kraft (USA)
Professor M. Peimbert (Mexico)
Professor Ya. S. Yatskiv (USSR)

As Vice-Presidents to be newly proposed :

Professor A. Batten (Canada)
Professor R. Kippenhahn (F.R. Germany)
Professor P.-O. Lindblad (Sweden)

As General Secretary :   Dr. J.-P. Swings (Belgium)

As Assistant General Secretary :

Dr. D. McNally (UK)

As Advisers to the Executive Committee :

Professor R. Hanbury Brown (Australia)
Dr. R.M. West (Denmark)

The President informed the Assembly that the formal election would take place at the Final Session of the General Assembly.

6. Announcement of

    a) <u>Official Representatives of Adhering Organizations</u>

and

    b) <u>Representatives to vote on the Nominating Committee.</u>

As requested by the President, the General Secretary announced the following names :

COUNTRY	CAT/UNITS	NATL. REPRESENTATIVE	NOMINATING COMM. REP.
Arab Rep. Egypt	III/4	A. Aiad	M. Khairy Aly
Argentina	II/2	J.L. Sersic	A. Ringuelet
Australia	III/4	K.C. Freeman	K.C. Freeman
Austria	I/1	H.F. Haupt	H.F. Haupt
Belgium	IV/6	P. Smeyers/ J. Dommanget	J. Dommanget
Brazil	II/2	S. M. Viegas Aldrovandi	S. M. Viegas Aldrovandi
Bulgaria	I/1	Z. Kovachev	M.K. Tsvetkov
Canada	VI/14	G. Michaud	J. R. Percy
Chile	I/1	F. Noël	F. Noël
China (Nanjing)	III/4	Wang Shou-guan	Qu Qin-yue
China (Taipei)	I/1	Typhoon Lee	Typhoon Lee
Colombia	I/1	------	------
Cuba	I/1	------	------
Czechoslovakia	III/4	V. Bumba/ J. Sykora	L. Kresak
Denmark	II/2	O. Pedersen	O. Pedersen
Finland	I/1	K. Mattila	K. Lumme
France	VII/20	J.C. Pecker	P. Lantos
German Dem. Rep.	II/2	H. Lorenz	H. Lorenz
German Fed. Rep.	VII/20	R. Kippenhahn	R. Wielebinski
Greece	II/2	------	------
Hungary	II/2	I. Almar	B. Szeidl
India	III/4	J.C. Bhattacharyya	V.R. Venugopal
Indonesia	I/1	B. Hidayat	W. Sutantyo
Iran	I/1	------	------
Iraq	I/1	H.M. Al-Naimy	T.H. Kadouri
Ireland	I/1	I. Elliott	P.A. Wayman
Israel	II/2	------	------

Italy	V/10	V. Castellani	V. Castellani
Japan	VI/14	Y. Kozai/ K. Kodaira	K. Kodaira
Korea Dem. People	I/1	Kim Yong Hyok	Kim Yong Hyok
Korea, Rep. of	I/1	Tu Hwan Kim	Tu Hwan Kim
Mexico	II/2	S. Torres-Peimbert	D. Dultzin
Netherlands	IV/6	H. van Woerden	W.B. Burton
New Zealand	I/1	E. Budding	E. Budding
Norway	I/1	J. Solheim	J. Solheim
Poland	III/4	W. Iwanowska	B. Kolaczek
Portugal	II/2	J. Pereira Osorio	J. Pereira Osorio
Roumania	II/2	------	------
South Africa	III/4	E.E. Baart	J. Menzies
Spain	II/2	A. Orte	M. De Pascual
Sweden	III/4	A. Sandqvist/ B. Gustafsson	A. Hjalmarsson
Switzerland	III/4	E.A. Müller	B. Hauck
Turkey	I/1	------	------
United Kingdom	VII/20	A. Hewish	F.G. Smith
U.S.A.	VIII/30	F. Drake	H. Smith
U.S.S.R.	V/10	V. Straizys	A.V. Tutukov
Uruguay	I/1	A. Benavidez	A. Benavidez
Vatican State City	I/1	M. McCarthy	M. McCarthy
Venezuela	I/1	J. Stock	J. Stock
Yugoslavia	II/2	S. Ninkovic	S. Ninkovic

c)   Acting Presidents of Commissions

The General Secretary announced that the Executive Committee had appointed the following persons to act for Presidents of Commissions unable to attend the General Assembly :

Commission	Acting President
15	J. Rahe
31	D. McCarthy
46	C. Iwaniszewska

7. Appointment of the Finance Committee

In accord with Statute 18(a), the General Assembly appointed the following Finance Committee consisting of one representative from each Adhering Body:

COUNTRY	CAT/UNITS	FINANCE REP.	DEPUTY
Arab Rep. Egypt	III/4	A. Aiad	
Argentina	II/2	E. Brandi	
Australia	III/4	K. Freeman	
Austria	I/1	H.F. Haupt	
Belgium	IV/6	P. Smeyers	
Brazil	II/2	S. M. Viegas Aldrovandi	
Bulgaria	I/1	V.G. Shkodrov	
Canada	VI/14	S. van den Bergh	
Chile	I/1	L. Campusano	
China (Nanjing)	III/4	Ye Shu-hua	
China (Taipei)	I/1	Typhoon Lee	
Colombia	I/1	------	
Cuba	I/1	------	
Czechoslovakia	III/4	J. Sykora	V. Bumba
Denmark	II/2	J. Andersen	
Finland	I/1	V. Piirola	
France	VII/20	G. Wlérick	
German Dem. Rep.	II/2	H. Lorenz	
German Fed. Rep.	VII/20	M. Grewing	
Greece	II/2	------	
Hungary	II/2	I. Almar	
India	III/4	M.B.K. Sharma	
Indonesia	I/1	Mr. Darsa	
Iran	I/1	------	
Iraq	I/1	S.A.A. Abdulla	T.H. Kadouri
Ireland	I/1	I. Elliott	
Israel	II/2	A. Serban	
Italy	V/10	G. Tofani	
Japan	VI/14	Y. Kozai	K. Kodaira
Korea Dem. People	I/1	Li Son Jae	

Korea, Rep. of	I/1	------
Mexico	II/2	D. Dultzin
Netherlands	IV/6	E.P.J. van den Heuvel
New Zealand	I/1	E. Budding
Norway	I/1	R. Stabell
Poland	III/4	S. Grzedzielski
Portugal	II/2	J. Pereira Osorio
Roumania	II/2	------
South Africa	III/4	E. Baart
Spain	II/2	A. Orte
Sweden	III/4	A. Ardeberg
Switzerland	III/4	E.A. Müller
Turkey	I/1	------
United Kingdom	VII/20	F.G. Smith
U.S.A.	VIII/30	P. Boyce
U.S.S.R.	V/10	V.K. Abalakin
Uruguay	I/1	A. Benavidez
Vatican State City	I/1	M. McCarthy
Venezuela	I/1	J. Stock
Yugoslavia	II/2	S. Ninkovic

8. Appointment of the Resolutions Committee

The President informed the Assembly that the Executive Committee proposed a Resolutions Committee to be established under the chairmanship of Prof. M. McCarthy, and with Drs. D. Morton and M. Kitamura as members. The General Assembly unanimously agreed to this composition of the Resolutions Committee.

9. Revisions of the Statutes and By-Laws

The change in By-Lay 17 such as mentioned in § 9 of the Agenda, is approved.

10. Resolutions submitted by Adhering Bodies

No resolutions were proposed to the XIXth General Assembly by Adhering Bodies.

11. Resolutions submitted by Commissions or by Associated Inter-Union Commissions

The President mentioned that six motions proposed by Commissions had been notified to Adhering Organizations for inclusion in the Agenda. However, the Executive Committee proposed that some of these Resolutions be referred to the Resolutions Committee and be taken under Item 5 of the Agenda of the Second Session of the XIXth General Assembly.

The President then formally adjourned the meeting to Thursday, November 28, 1985 at 10:00 and closed the Session with a word of thanks to the participants.

REPORT ON THE XIXTH GENERAL ASSEMBLY

SECOND SESSION

Held in the Main Auditorium

Vygyan Bhavan

Thursday, November 28, 1985, 10:00 to 12:00

Prof. R. Hanbury Brown, President in the Chair

The president declared the session open. Messages of good wishes were sent to absent former IAU Presidents and General Secretaries : V. Ambartsumian, A. Blaauw, G. Contopoulos, C. de Jager, L. Goldberg, J.H. Oort, L. Perek, D. Sadler, and B. Strömgren.

Before passing to the Agenda, the General Secretary, called upon by the President, established the quorum and found eight Adhering Countries not represented. The General Assembly then appointed G. Westerhout and O. Gingerich as Tellers.

12. Report of the Finance Committee

The General Secretary mentioned that copies of the financial report were in the hands of the official representatives and he thus invited the Chairperson of the Finance Committee, Prof. Ye Shu-hua, to read the report. Mrs. Ye reported as follows :

"The Finance Subcommittee has inspected the accounts of the IAU for the period 1982-84 and finds everything to be in order. We commend the General Secretary for making a fundamental advance in the modernization of the operations of the IAU Secretariat. We particularly applaud the installation of the IBM PC-XT for keeping the membership files, the installation of a telex machine for rapid communication and for revising the format and presentation of the fiscal accounts. The new format with explanatory notes is much clearer than the previous, which not only makes our job easier, but will also result in better fiscal management.

Along this line, we commend the General Secretary for continuing the effort started by his predecessor of maintaining as large a fraction of the IAU funds in interest bearing accounts as possible as well as taking advantage of favorable exchange rates whenever possible.

The Finance Subcommittee notes with satisfaction the low administrative costs of the IAU. We acknowledge the debt owed by the IAU and its members to the Officers, particularly the General Secretary and Assistant General Secretary, and to their institutions for the generous support which makes this possible.

We have examined both the proposed 1985 budget and the income and expenses for the first three quarters of the year. We have assured ourselves that the proposed budget will be met very closely, provided that all countries make their payments before the end of the year.

The Finance Subcommittee discussed the proposed 1986-1988 budget at some length. First, the proposed annual increase in the unit of contribution received particular attention. We are all aware that the costs of operating all organizations are rising nearly everywhere. In the IAU, the problem is compounded because the inflation rates differ in different countries and the exchange rates between countries also vary substantially. We endorse the approach taken by the General Secretary to form an average range of inflation around the world and then to set the increase of the unit at, slightly below, the low end of this range. The Committee realizes that such an approach will, from time to time, cause hardships for individual countries, but, over the course of time, such problems will average out. Therefore, realizing that we must continue to pay for the operations of the Union in the face of rising costs, we strongly endorse the proposed value of the unit of contribution.

As a corollary, we note that the payment of the 1985 contributions are running somewhat late. We urge that the adhering bodies take note of the importance to the IAU of receiving the contributions early in the year, and we urge all countries to pay their contributions as rapidly as possible. We gratefully thank those countries which actually pay in advance.

With regards to the actual budgets proposed for 1986-1988 we make the following comments. We note with enthusiasm the increasing emphasis upon the scientific programmes of the IAU, but this increase can only be obtained by squeezing the rest of the budget. We feel that we must carefully examine all the IAU programmes and put our resources into areas which are best, or uniquely, done by the IAU. The Finance Subcommittee unanimously agrees that the IAU is unique in fostering international exchange of, and contact between, astronomers. This is particularly important for developing countries, and those with foreign currency exchange problems where the IAU funding often acts as important seed money for maintaining international contacts.

Therefore, the Finance Committee recommends increasing the funding for the exchange of astronomers by 3 333.- SFr. per year, making up this amount by reducing support for the Regional Astronomy Meetings.

There are three resolutions pending before the General Assembly which call for increasing funding for support of services. The Finance Committee feels that such resolutions should be brought to the attention of the Executive Committee in time to receive careful attention when the draft budget is prepared. It seems unwise to increase the projected deficit.

As final comment, we note that the IAU membership is growing rapidly and the operations are becoming more complex. We feel that the Union should anticipate that the time will arise in the future when the General Secretary's job cannot be carried out effectively by one or two scientists who volunteer their time and services to the Union, and that the IAU Secretariat staff may have to be enlarged. In this case, the expenses will rise and additional sources of funding may have to be developed. We realize the important role that the IAU plays for astronomy, and it may be time to investigate the balance between the services rendered and the charges assessed. It is clear that many IAU functions cannot be self-supporting. Therefore, unless the adhering countries can afford to pay increased assessments, the IAU will have to look to those activities which produce income to carry the burden for the rest of the programmes.

We recommend that the Executive Committee looks into this matter in the near future in consultation with people experienced in the problems of funding non-profit, scientific societies. The IAU operations are now approaching in size and complexity those of a small corporation and deserve expert attention by experienced people."

The President thanks Prof. Ye and puts the report of the Finance Committee to the vote, each item separately.

1) The proposed budget 1985 was unanimously voted.
2) The increase of the unit of contribution was voted with 135 votes in favour, 15 against, and 7 abstentions.
3) The proposed budget 1986-1988 was voted by 151 votes in favour, none against, and 4 abstentions.

13. Financial Resolutions of the Executive Committee

None.

14. Resolutions submitted by the Executive Committee

The following resolutions were submitted by the Executive Committee :

1) The XIXth General Assembly of the International Astronomical Union,

    <u>resolves</u>   that the name of Commission 51 shall henceforth be : "Bioastronomy : Search for Extraterrestrial Life".

1) La XIXème Assemblée Générale de l'Union Astronomique Internationale,

    <u>décide</u>   que l'intitulé de la Commission 51 sera dorénavant : "Bioastronomie : Recherche de la Vie dans l'Univers".

2) The XIXth General Assembly of the International Astronomical Union,

    <u>recalling</u>   Resolution 13 adopted by the International Council of Scientific Unions (ICSU) in Cambridge (September 1982) which expressed great concern about the refusal of exit visas to scientists, and noting that this is a serious obstacle to international cooperation in science,

    <u>resolves</u>   that the IAU, as an Adhering Body of ICSU, do its utmost, in the spirit of the aforementioned Resolution 13, to urge the responsible authorities in all member countries to resolve as speedily as possible all cases referred to in that resolution.

2) La XIXème Assemblée Générale de l'Union Astronomique Internationale,

    <u>rappelant</u>   la Résolution 13 adoptée par le Conseil International des Unions Scientifiques (ICSU) à Cambridge (Septembre 1982), qui exprimait de sérieuses inquiétudes concernant le refus de visas de sortie à des scientifiques, et considérant que ceci est un obstacle serieux à la coopération scientifique internationale,

    <u>décide</u>   que l'UAI, en tant qu'Organisation Adhérente de l'ICSU, fasse tout son possible, dans l'esprit de la résolution 13 mentionnée ci-dessus, pour que les autorités responsables dans tous les pays membres résolvent aussi rapidement que possible tous les cas auxquels il est fait allusion dans cette résolution.

3) The XIXth General Assembly of the International Astronomical Union,

    <u>recalling</u>   Resolution 9 adopted by ICSU in Ottawa (1984) which expressed great concern about violations of ICSU's principles of universality, for example in the granting of entry visas to scientists wishing to attend scientific conferences,

<u>resolves</u>   that the IAU, as an Adhering Body to ICSU, do its utmost, in the spirit of the aforementioned Resolution 9, to urge the responsible authorities in all countries to adhere to the guidelines given by the ICSU Standing Committee on the Free Circulation of Scientists and embodied in its publication "Advice to Organizers of International Scientific Meetings".

3) La XIXème Assemblée Générale de l'Union Astronomique Internationale,

<u>rappelant</u>   la Résolution 9 adoptée par ICSU à Ottawa (1984), qui exprimait de sérieuses inquiétudes au sujet de la violation des principes d'ICSU sur l'universalité, par exemple, concernant la délivrance de visas d'entrée à des scientifiques désireux de participer à des conférences scientifiques,

<u>décide</u>   que l'UAI, en tant qu'Organisation Adhérente de l'ICSU, fasse tout son possible, dans l'esprit de la Résolution 9 mentionnée ci-dessus, pour que les autorités responsables dans tous les pays, adoptent les directives du Comité Permanent de l'ICSU sur la Libre Circulation des Scientifiques, qui sont contenues dans sa publication "Conseils aux Organisateurs de Réunions Scientifiques Internationales".

were approved unanimously.

15. Resolutions proposed by the Resolutions Committee

At the request of the President, Dr. M. McCarthy, Chairman of the Resolutions Committee, reported about the work of the Committee. Nine resolutions were put forward by the Commissions (French translations are also included). Prior to reading the resolutions, Dr. McCarthy made the following announcement :

"Mes chers Collègues, Mesdames et Messieurs, je suis très content de vous présenter le texte des résolutions de la XIXème Assemblée Générale de l'Union Astronomique Internationale. Pour mes collègues membres du Comité des Résolutions, les Professeurs Kitamura et Morton et pour moi-même, ces travaux ont duré trois jours. Mais pour les Commissions et les membres de l'Union, ils représentent plusieurs mois et même plusieurs années d'efforts. Ces travaux - et particulièrement ceux qui traitent de l'établissement du temps atomique international et des effets néfastes et désastreux de l'interférence avec les observations radioastronomiques et optiques - sont vraiment un 'magna carta' et une 'rosetta stone' pour l'avenir de notre science. Je tiens à remercier par dessus tout Mesdames Elisabeth Völk et Danielle Lours, ainsi que la rédaction de notre journal anglais Mandakini, pour les efforts dignes d'Hercule qu'ils ont fournis. Maintenant, nous allons commencer.

Now in English : today in New England it is Thanksgiving day and we wish, my colleagues Professors Kitamura and Morton and I, to express our thanks to all of you in the Commissions of the Union for your works of months and years in preparing the material and ordering it for the resolutions. We owe a special debt of thanks to the officials of the General Secretariat. I wish to thank especially Elizabeth Völk and Danielle Lours who shared our work and concern and to the producers of Mandakini, their editors, typesetters and printers. The official text will be in Transactions IAU XIXB. Please note that Resolution C9 as printed in Mandakini is withdrawn from List C and has just been returned to List D. Proper adjustments will be made before publication in French for Transactions XIXB."

Resolution B1 : Responsibility for Time

The International Astronomical Union,

<dl>
<dt>recalling</dt>
</dl>

1) that the establishment of International Atomic Time (TAI) and of Coordinated Universal Time (UTC) is one of the present tasks of the Bureau International de l'Heure (BIH), and

2) that the IAU is the main parent scientific Union of the BIH, the other parent unions being the International Union of Geodesy and Geophysics (IUGG) and the International Union of Radio Science (URSI), and

considering

1) that the atomic time scales, originally used mainly in astronomy, have now a much wider use, including numerous and important technical and public applications,

2) that TAI is based solely on physical measurements independent of astronomy,

3) that there exists an inter-governmental organization of which the Bureau International des Poids et Mesures (BIPM) is the Executive Body in charge of the unification of measurement of the major physical quantities,

4) that UTC is based both on TAI and on the astronomical time scale designated as Universal Time (UT1), and

5) the URSI recommendation A-1, 1984, relative to the transfer of TAI to the BIPM,

approves of TAI being taken over entirely by the Bureau International des Poids et Mesures, under the responsibility of the International Committee of Weights and Measures (CIPM) and of the General Conference of Weights and Measures,

recommends

1) that the function of determining and announcing the leap seconds of the UTC system, as well as the function of determining and announcing the $\Delta UT1$ corrections, be given to the new International Earth Rotation Service entrusted by the IAU and IUGG with the evaluation of the Earth rotation parameters, and

2) that a permanent committee, where the IAU will be represented, be created, under the sponsorship of CIPM in order to take care of the interest of TAI users, and

extends to the Paris Observatory its thanks for the service provided to the international community by supporting the BIH.

Résolution B1 : Responsabilité de l'Heure

L'Union Astronomique Internationale,

rappelant 1) que l'établissement du Temps Atomique International (TAI) et du Temps Universel Coordonné est une des tâches actuelles du Bureau International de l'Heure (BIH), et

                      2) que l'UAI est la principale Union scientifique mère du BIH, les autres unions apparentées étant l'Union Internationale de Géodésie et de Géophysique (UGGI) et l'Union Internationale des Sciences Radio (URSI), et

<u>considérant</u>   1) que les échelles de temps atomiques, originellement utilisées principalement en astronomie, ont maintenant acquis une diffusion plus large comprenant de nombreuses et importantes applications techniques et publiques,

                      2) que le TAI est basé uniquement sur des mesures physiques indépendantes de l'astronomie,

                      3) qu'il existe une organisation inter-gouvernementale dont le Bureau International des Poids et Mesures (BIPM) est le Corps Exécutif, ayant la charge de l'unification de la mesure des quantités physiques importantes,

                      4) que l'UTC est basé à la fois sur le TAI et sur l'échelle de temps astronomique connue sous le nom de Temps Universel, (UT1), et

                      5) la recommandation de l'URSI A-1, 1984, relative au transfert du TAI au BIPM

<u>approuve</u>   la prise en charge complète du TAI par le Bureau International des Poids et Mesures, sous la responsabilité du Comité International des Poids et Mesures (CIPM) et de la Conférence Générale des Poids et Mesures,

<u>recommande</u>   1) que la fonction de détermination et d'annonce des secondes intercalaires du système UTC, comme celle de déterminer et d'annoncer les corrections ΔUT1 soient confiées, en plus de l'évaluation des paramètres de rotation de la Terre, au nouveau service international "Rotation de la Terre", à qui l'UAI et l'UGGI donnent leur entière confiance,

                      2) qu'un comité permanent, dans lequel l'UAI sera représentée, soit créé, sous le parrainage du CIPM, dans le but de prendre soin des intérêts des utilisateurs du TAI, et

<u>adresse</u>   à l'Observatoire de Paris ses remerciements pour le service rendu à la communauté internationale par le support apporté au BIH.

Resolution B2 : Reference Frames

The International Astronomical Union,

<u>recognizing</u>   the highly significant improvement in the determination of the orientation of the Earth in space as a consequence of the MERIT/COTES* programmes of observation and analysis, and

                      the importance for scientific research and operational purposes of regular earth-orientation monitoring and of the establishment and maintenance of a new Conventional Terrestrial Reference Frame,

thanks	all the organizations and individuals who have contributed to the development and implementation of the MERIT and COTES programmes and to the operations of the International Polar Motion Service (IPMS) and the Bureau International de l'Heure (BIH),
endorses	the final report and recommendations of the MERIT and COTES Joint Working Groups,
decides	1) to establish in consultation with the IUGG a new International Earth Rotation Service within the Federation of Astronomical and Geophysical Services (FAGS) replacing both the IPMS and the BIH from 1988 January 1, for monitoring earth rotation and for the maintenance of the Conventional Terrestrial Reference System,
	2) to extend the MERIT/COTES programmes of observation, analysis, intercomparison and distribution of results until the new service is in operation,
	3) to recommend that an optical astrometric network be maintained for the rapid determination of UT1 for so long as this is recognized to be useful, and
	4) to set up a Provisional Directing Board to submit recommendations on the terms of reference, structure and composition of the new service, and to serve as the steering committee for the extended MERIT/COTES programmes, and
invites	National Committees for Astronomy and for Geodesy and Geophysics to submit proposals for the hosting of individual components of the new service by national organisations and observatories, and
urges	the participants in Project MERIT to continue to determine high-precision data on earth rotation and reference systems and to make the results available to the BIH until the new service is in operation.
*MERIT =	A programme of international collaboration to monitor Earth-rotation and intercompare the techniques of observation and analysis.
*COTES =	A programme of international collaboration to establish and maintain a new conventional terrestrial reference system.
*FAGS =	Federation of Astronomical and Geophysical Services.
*IPMS =	International Polar Motion Service.

Résolution B2 : Systèmes de Référence

L'Union Astronomique Internationale,

> reconnaissant les améliorations très significatives apportées à la détermination de l'orientation de la Terre dans l'Espace, suite au programme MERIT/COTES* d'observation et d'analyse, ainsi que
>
>> l'importance pour la recherche scientifique et pour des buts opérationnels de la surveillance de l'orientation de la Terre et de l'établissement et du maintien à jour d'un système de référence conventionnel terrestre,

remercie	toutes les organisations et les individus qui ont contribué au développement et à la mise en place des programmes MERIT et COTES et aux opérations du service du mouvement du pôle (IPMS*) ainsi que le Bureau International de l'Heure,
adopte	le rapport final et les recommandations des groupes de travail MERIT et COTES,
décide	1) d'établir, en consultation avec l'Union Internationale de Géodésie et de Géophysique (UGGI), un nouveau service international au sein de FAGS*, pour la surveillance de la rotation terrestre et la mise à jour du système de référence conventionnel terrestre, devant remplacer à la fois IPMS et le BIH à partir du 1er Janvier 1988,
	2) d'élargir les programmes MERIT/COTES d'observations, d'analyse, d'inter-comparaison et de distribution des résultats jusqu'à ce que le nouveau service soit opérationnel,
	3) de recommander qu'un réseau d'astrométrie optique soit maintenu pour la détermination rapide de UT1 aussi longtemps que cela sera jugé utile,
	4) de former un Bureau Directeur Provisoire, chargé de soumettre des recommandations sur les termes de références, la structure et la composition du nouveau service, et de servir de Comité Directeur pour l'extension des programmes MERIT/COTES,
invite	les Comités Nationaux d'Astronomie, de Géodésie et de Géophysique, à faire des propositions pour l'hébergement des composants individuels du nouveau service par des organisations nationales et/ou des observatoires, et
insiste	pour que les participants du projet MERIT continuent de déterminer des données de haute précision sur la rotation de la Terre et sur les systèmes de référence, et en communiquent les résultats au BIH jusqu'à ce que le nouveau système soit opérationnel.

Resolution B3 : CCIR * Actions

The International Astronomical Union,

recalling	the considerations (a) to (d) of IAU Resolution No. 3 passed at the XVIIth General Assembly in 1979 concerning harmful interference to radio astronomy observations, and
noting	a) that the IAU, URSI *, and COSPAR * have collaborated over many years in the Inter-Union Commission on the Allocation of Frequencies for Radio Astronomy and Space Science (IUCAF) in obtaining such bands by international agreement,
	b) that certain experiments have begun in which transmissions take place from space in one of these bands, and that these transmissions may interfere with observations of OH emission from Halley's Comet,
	c) that proposals for revision of Recommendation 314 of the CCIR*, for consideration by its XVIth Plenary Assembly, reflect the interests of astronomers, and

d) that additions to CCIR Reports 224 and 697, and a draft new Recommendation (DOC. 2/196) emphasize the concern of radio astronomers regarding the possible effects of spurious emissions from space stations, especially those which are geostationary,

<u>resolves</u>   1) that the documentation of Study Group 2 of the CCIR, regarding revisions to Recommendation 314, Reports 224 and 697 and draft Recommendation (Doc. 2/196) is welcomed by astronomers as contributions to the XVIth Plenary Assembly of CCIR, and

2) that in respect to draft Recommendation (Doc. 2/196), astronomers should heed the likely limitations on observations within 5° of a geostationary satellite orbit from any single observatory, and of the need to reduce the side-lobe gains of their antennae to the greatest practicable extent, and

<u>recommends</u>   in view of the particular danger of interference to radio astronomy from space-based radio transmissions, that all those concerned in the design of systems requiring radio transmissions from space should consult with IUCAF at the planning stage to ensure that sensitive passive radio observations are not jeopardized in the future.

* CCIR = International Radio Consultative Committee
* URSI = Union Radio Scientifique Internationale
* COSPAR = Committee on Space Research

Résolution B3 : Actions du CCIR *

L'Union Astronomique Internationale,

<u>rappelant</u>   les considérations (a) à (d) de la Résolution No. 3 votée à la XVIIème Assemblée Générale de l'UAI en 1979 concernant l'interférence nuisible sur des observations radioastronomiques, et,

<u>notant</u>   a) que l'UAI, l'URSI (2) et le COSPAR (3) ont collaboré durant de nombreuses années au sein de la Commission Inter-Union pour l'Attribution des Fréquences pour la Radioastronomie et les sciences spatiales (IUCAF) à l'obtention, par accord international, de telles bandes,

b) que certaines expériences ont été entreprises dans lesquelles des transmissions ont lieu depuis l'espace dans une de ces bandes, et que ces transmissions peuvent interférer avec les observations de l'émission OH de la Comète de Halley,

c) que des propositions de révision de la Recommandation 314 du CCIR, dans le contexte de sa XVIème Assemblée Plénière, reflètent l'intérêt des astronomes,

d) que des additions aux rapports 224 et 697 du CCIR*, et un projet de nouvelle Recommandation (Doc. 2/196) soulignent le souci des radio-astronomes en ce qui concerne les effets possibles d'émissions intempestives depuis des stations spatiales, en particulier les stations géostationnaires,

<u>décide que</u> 1) la documentation du Groupe d'étude 2 du CCIR, concernant les révisions de la Recommandation 314, des rapports 224 et 697 et du projet de recommandation (DOC 2/196) est accueillie favorablement, par les astronomes, comme contribution à la XIXème Assemblée Plenière du CCIR,

2) en ce qui concerne le projet de Recommandation (Doc. 2/196), les astronomes devront veiller aux limitations probables sur les observations à moins de 5° de l'orbite d'un satellite géostationnaire à partir de tout observatoire, et à la nécéssité de réduire les gains de lobe de leurs antennes au maximum des possibilités pratiques, et

<u>recommande</u> pour parer au danger particulier d'interférence avec la radioastronomie des transmissions radio à partir des stations spatiales, tous ceux qui sont concernés par la définition de systèmes utilisant des transmissions radio depuis l'espace se mettent en contact avec l'IUCAF dès la phase de définition de leur projet afin de garantir que des observations radio-astronomiques passives et sensibles ne soient pas compromises dans le futur.

Resolution B4 : Radio Frequency Transmission from Space

The International Astronomical Union,

<u>considering</u> a) that certain frequency bands in the range 1300-1800 MHz are very important to the science of radio astronomy, in particular the allocated bands 1330-1427 MHz, 1610.6-1613.8 MHz, 1660-1670 MHz and 1718.8-1722.2 MHz,

b) that radio astronomy observatories are particularly vulnerable to interference from transmitters located on aircraft and spacecraft,

c) that the frequency range 1300-1800 MHz is also the object of considerable attention for satellite systems in a number of countries for navigation, position location, and communications,

d) that certain modulation methods are coming into more common usage in Space Radio Services, such as spread spectrum techniques which may cause interference to radio astronomy, not only in frequency bands adjacent to transmission bands, but also at frequencies far removed from bands allocated to space services,

e) that the International Telecommunications Union (ITU) World Administrative Radio Conference (WARC) for Mobile Services, which is scheduled for 1987, may allocate frequencies in the band 1300-1800 MHz in order to accomodate satellite services, and

f) that the Mobile Service WARC in 1987 and the WARC on the Use of the Geostationary-Satellite Orbit and the Planning of the Space Services Utilizing It, which is scheduled for 1988, may establish technical standards governing unwanted emissions from the transmitters in the space services,

<u>urges</u>   a) that administrations avoid, whenever practicable, planning space systems with transmitters on spacecraft or aircraft which operate in the frequency bands listed in consideration (a) above,

b) that administrations take into account the current allocations to the radio astronomy service and its vulnerability to air and space transmissions when preparing proposals for the 1987 WARC for the Mobile Services and the 1988 Space WARC,

c) that administrations devise and adopt technical standards governing unwanted transmissions from transmitters in the space services both nationally and through the Radio Regulations of the UTI, and

d) that administrations coordinate those satellite systems which may impact radio astronomy through the Interunion Commission on the Allocation of Frequencies for Radio Astronomy and Space Science (IUCAF) with sufficient lead time in the planning phase for an effective exchange of concerns to take place.

IUCAF * = <u>I</u>nter-<u>U</u>nion <u>C</u>ommission on <u>F</u>requency <u>A</u>llocations for Radio Astronomy and Space Science

Résolution B4 : Transmissions Radio-électriques à partir de l'Espace

L'Union Astronomique Internationale,

<u>considérant</u> a) que certaines bandes de fréquences dans le domaine 1300-1800 MHz sont très importantes pour la radioastronomie, en particulier les bandes allouées 1330-1427 MHz, 1610,6-1613,8 MHz, 1660-1670 MHz et 1718,8-1722,2 MHz,

b) que les observatoires radioastronomiques sont particulièrement vulnérables aux interférences provoquées par les émissions issues d'avions ou de satellites,

c) que le domaine de fréquences 1300-1800 MHz est aussi d'un intérêt considérable pour les systèmes de satellites dans nombre de pays pour la navigation, la localisation de la position, et les communications,

d) que certaines techniques de modulation sont de plus en plus couramment employées dans les services de radio spatiale, telles par exemple les techniques de "spread spectrum", qui peuvent causer des interférences pour la radioastronomie, non seulement dans des bandes de fréquences adjacentes aux bandes de transmission, mais aussi à des fréquences très éloignées des bandes allouées aux services spatiaux,

e) que la Conférence Administrative Mondiale sur la Radio (WARC), pour les services mobiles de l'Union Internationale des Télécommunications qui est prévue pour 1987, pourra décider d'attribuer des fréquences dans la bande 1300-1800 MHz en vue d'installer des services de satellites artificiels, et

f) que les Services Mobiles WARC en 1987 et le WARC prévu en 1988 pour l'Utilisation de l'Orbite Géostationnaire et la Planification des Services Spatiaux qui l'utilisent, pourra décider de l'établissement de standards techniques relatifs à des émissions non voulues à partir d'émetteurs dans des services spatiaux,

<dl>
<dt>insiste pour</dt>
<dd>

a) que les administrations évitent, chaque fois qu'il est possible, de proposer des projets de systèmes spatiaux comprenant des émetteurs sur satellites ou avions opérant dans les bandes de fréquences dont la liste figure ci-dessus (considération a),

b) que les administrations prennent en compte les allocations actuelles au service de la radioastronomie et sa vulnérabilité aux transmissions aériennes et spatiales lors de la préparation des propositions pour la WARC 1987 pour les Services Mobiles, et pour la WARC spatiale de 1988,

c) que, au plan national, ou à partir des règlements radio de l'UIT, les administrations mettent au point et adoptent les standards techniques régissant les émissions intempestives à partir d'émetteurs spatiaux, et

d) que les administrations coordonnent les systèmes satellisés qui peuvent interagir avec la radioastronomie, par l'intermédiaire de IUCAF*, en tenant compte d'un délai suffisant au niveau de la planification pour qu'un échange de vues puisse avoir lieu.

</dd>
</dl>

Resolution B5 : VLBI Coordination

The International Astronomical Union,

recognizing
1) that the well-established international collaboration in ground-based VLBI has resulted in high-angular-resolution radio imaging,

2) that ground-based VLBI images have demonstrated the need for even higher resolution which can be achieved by the combination of ground arrays and future space-based antennae,

3) that the feasibility of launching space-based VLBI elements into Earth orbit is under investigation by space agencies around the world,

4) that the full scientific benefits of VLBI will result only from observations obtained through the combined and simultaneous use of all space-based antennae with existing ground facilities, and

5) that COSPAR has established an Ad-Hoc Committee to examine the requirements for coordinated space and ground-based VLBI activities,

recommends that the appropriate national and international authorities concerned with space and ground-based VLBI make every effort to coordinate in a timely way the contributions to this important international programme.

* COSPAR = Committee on Space Research

Résolution B5 : Coordination Relative à l'Interférométrie à Très Grande Base (VLBI)

L'Union Astronomique Internationale,

<u>reconnaissant</u> 1) que la collaboration internationale dans le domaine de la VLBI au sol conduit à de l'imagerie à haute résolution angulaire,

2) que les images VLBI au sol ont prouvé la nécessité d'une résolution encore plus grande qui peut être atteinte par la combinaison de réseaux d'observatoires au sol et de futures antennes placées dans l'espace,

3) que la faisabilité de placer sur orbite des éléments de VLBI est à l'étude dans différentes agences spatiales du monde,

4) que le maximum de retombées scientifiques du VLBI ne pourra résulter que d'observations à la fois combinées et simultanées de toutes les antennes en orbite avec les installations existant au sol,

5) que le COSPAR* a constitué un Comité ad hoc pour examiner les conditions nécessaires aux activités VLBI coordonnées entre l'espace et le sol,

<u>recommande</u> que les autorités compétentes nationales et internationales concernées par les VLBI dans l'espace et au sol consacrent tous leurs efforts à coordonner les contributions à cet important programme international.

Resolution B6 : Protection of Observatory Sites

The International Astronomical Union,

<u>reaffirms</u> the importance of resolutions, adopted by previous General Assemblies, which relate to the protection of observatory sites and observing techniques, including

(1961) Resolutions Nos. 1 and 2, Transactions IAU XI,

(1964) Resolutions Nos. 3 and 5, Transactions IAU XIIB,

(1967) Resolution No. 2, Transactions IAU XIIIB,

(1970) Resolution No. 10, Transactions IAU XIVB,

(1976) Resolutions Nos. 8 and 9, Transactions IAU XVIB, and

(1979) Resolution No. 3, Transactions IAU XVIIB, and

<u>requests</u> that astronomers urge civil authorities to make all possible efforts to preserve the quality of observing conditions at the remaining excellent sites on this planet.

Résolution B6 : Protection des Sites d'Observatoires

L'Union Astronomique Internationale,

> réaffirme   l'importance des résolutions adoptées lors de précédentes Assemblées Générales, qui concernent la protection des sites d'observatoires et des techniques d'observation. Celles-ci comprennent :
>
> > (1961)   les Résolutions Nos. 1 et 2, (Transactions IAU XI),
> >
> > (1964)   les Résolutions Nos. 3 et 5, (Transactions IAU XIIB),
> >
> > (1967)   la Résolution No. 2, (Transactions IAU XIIIB),
> >
> > (1970)   la Résolution No. 10, (Transactions IAU XIVB),
> >
> > (1976)   les Résolutions Nos. 8 et 9, (Transactions IAU XVIB), et
> >
> > (1979)   la Résolution No. 3, (Transactions IAU XVIIB),
>
> et,
>
> demande   que les astronomes agissent auprès des autorités civiles afin que ces dernières fassent le maximum d'efforts pour préserver la qualité des conditions d'observation dans les excellents sites restant sur cette planète.

Resolution B7 : Danger of the Contamination of Space

The International Astronomical Union,

> noting with grave concern   the dramatically increasing uses of space for scientific and other purposes and the accompanying contamination of space that adversely affects astronomical observations from the ground and from space,
>
> re-affirms   its previous resolutions bearing on the uses of space,
>
> maintains   that no group has the right to change the Earth's environment in any significant way without full international study and agreement, and
>
> urges   that all national representatives bring this concern to the notice of adhering organizations and space agencies in their countries.

Résolution B7 : Danger de la Contamination de l'Espace

L'Union Astronomique Internationale,

> constatant avec de graves inquiétudes   l'augmentation dramatique des utilisations de l'espace dans des buts scientifiques et autres et par là même la contamination de l'espace qui affecte dangereusement les observations astronomiques depuis le sol et l'espace,
>
> réaffirme   ses résolutions précédentes concernant les utilisations de l'espace,

maintient qu'aucun groupe n'a le droit de changer l'environnement de la Terre en aucune façon sans qu'une étude internationale complète n'ait lieu, et qu'un accord en découle, et

insiste auprès de tous les représentants nationaux afin qu'ils fassent part de leur souci aux organisations adhérentes et aux agences spatiales dans leurs pays respectifs.

Resolution B8 : Tycho's Observatories

The International Astronomical Union,

noting with great satisfaction that action has been initiated in both Sweden and Denmark which aims at an improvement of the state of the remains of Tycho Brahe's observatories on the Island of Ven, a site of unique significance in the history of astronomy,

urges the relevant authorities to make every possible effort to preserve the ruins and to keep the site as a whole in a condition worthy of its past.

Résolution B8 : Les Observatoires de Tycho Brahe

L'Union Astronomique Internationale,

constatant avec grande satisfaction qu'une procédure a été engagée en Suède et au Danemark pour la restauration des restes des observatoires de Tycho Brahe sur l'île de Ven, site d'une importance unique dans l'histoire de l'astronomie,

insiste auprès des autorités compétentes afin qu'elles fassent un maximum d'efforts pour préserver les ruines et pour maintenir le site dans un état digne de son passé.

Resolution B9 : Endorsement of Commission Resolutions

The XIXth General Assembly of the International Astronomical Union,

having full confidence in its Commissions,

endorses the other Resolutions submitted by them to the Resolutions Committee for publication in the official languages of the Union, French and English, in Transactions IAU XIXB.

Résolution B9 : Soutien des Résolutions des Commissions

La XIXème Assemblée Générale de l'Union Astronomique Internationale,

accordant son entière confiance à ses Commissions,

souscrit aux autres résolutions qu'elles ont soumises au Comité des Résolutions, pour être publiées dans les deux langues officielles de l'Union, le français et l'anglais, dans les Transactions de l'UAI XIXB.

The full texts of the 14 Resolutions by IAU Commissions (C1-C14) are as follows :

Resolutions Proposed by Commissions

Resolution C1 : Astronomical Constants

Commissions 4, 7, 8, 19 and 31

<u>recognizing</u>   the importance of ensuring that the IAU system of astronomical constants is rigorously defined and is well suited to current applications,

<u>invites</u>   the Presidents of IAU Commissions 4, 7, 8, 19 and 31 to form a Working Group to serve in collaboration with the appropriate special study group of the International Association of Geodesy which will

1) review current determinations of astronomical and geodetic constants,

2) provide for informational purposes the current, best estimates of the values, accuracies and sources of these constants,

3) propose appropriate changes in the relevant definitions and values of the constants of the IAU system,

4) urge all authors to specify completely the values and accuracies, as well as the sources, of the constants used in their work, and

5) submit a preliminary report in 1987.

Résolution C1 : Constantes Astronomiques

Les Commissions 4, 7, 8, 19 et 31

<u>reconnaissant</u>   l'importance de garantir que le système de constantes astronomiques de l'UAI est défini rigoureusement et qu'il est bien adapté aux applications courantes,

<u>invite</u>   les Présidents des commissions 4, 7, 8, 19 et 31 de l'UAI à former un Groupe de Travail qui devra, en collaboration avec le groupe spécial d'études approprié de l'Association Internationale de Géodésie,

1) passer en revue les déterminations des constantes astronomiques et géodésiques,

2) fournir dans un but d'information les meilleures estimations de ces constantes ayant cours : valeur, précision et source,

3) proposer des changements appropriés dans la définition et les valeurs des constantes du système de l'UAI,

4) inciter tous les auteurs à spécifier complètement la valeur, la précision et la source des constantes utilisées dans leur travail, et

5) soumettre un rapport préliminaire en 1987.

Resolution C2 : Reference Systems

Commissions 4, 7, 8, 19, 20, 24, 31, and 33

<u>recognizing</u>   1) the existence of inconsistent reference systems based upon different theories and modes of observations,

2) the significant improvement in the accuracy of observations using new techniques, and

3) the importance of a space-fixed reference system, independent of the mode of observation, for use in astronomy and geodesy and satisfying the requirements of relativistic theories,

<u>invites</u>   the Presidents of interested IAU Commissions (for example, 4, 7, 8, 19, 20, 24, 31, 33 and 40) to form an IAU Working Group, with appropriate sub-groups devoted to specialized topics, under the overall chairmanship of the Chairman of the Joint Discussion on Reference Frames, which will report to the XXth General Assembly in 1988 with recommendations for

1) the definition of the Conventional Terrestrial and Conventional Celestial Reference Systems,

2) ways of specifying practical realizations of these systems,

3) methods of determining the relationships between these realizations, and

4) a revision of the definitions of dynamical and atomic time to ensure their consistency with appropriate relativistic theories, and

<u>invites</u>   the President of the International Association of Geodesy to appoint a representative to the Working Group for appropriate coordination on matters relevant to Geodesy.

Résolution C2 : Systèmes de Référence

Les Commissions 4, 7, 8, 19, 20, 24, 31 et 33

<u>reconnaissant</u> 1) l'existence de systèmes de référence non cohérents les uns avec les autres, fondés sur des théories et des modes d'observation différents,

2) l'amélioration significative de l'exactitude des observations par les nouvelles techniques, et

3) l'importance de disposer pour l'astronomie et la géodésie d'un système de référence fixe dans l'espace indépendant du mode d'observation et répondant aux exigences des théories relativistes,

<u>invite</u>   les Présidents des Commissions de l'UAI concernées (par exemple 4, 7, 8, 19, 20, 24, 31, 33 et 40) à former un Groupe de Travail de l'UAI comprenant des sous-groupes chargés d'aspects particuliers, sous la responsabilité générale du Président de la Discussion Conjointe sur les Systèmes de Référence ; ce Groupe de Travail préparera pour la XXème Assemblée Générale de 1988 des recommandations pour

1) la définition des Systèmes de Référence Conventionnels Terrestre et Céleste,

2) les façons de spécifier les réalisations pratiques de ces systèmes,

3) les méthodes pour déterminer les relations entre ces réalisations, et

4) une révision des définitions du temps dynamique et du temps atomique qui garantissent leur cohérence dans le cadre des théories relativistes appropriées, et

<u>invite</u>   le Président de l'Association Internationale de Géodésie à désigner un représentant à ce Groupe de Travail afin d'exercer la coordination nécessaire en ce qui concerne les problèmes intéressant la géodésie.

Resolution C3 : Astronomical Designations

Commission 5, with a view to avoiding confusion,

<u>recommends strongly</u> that IAU resolutions which concern the designation of astronomical objects outside the solar system be forwarded to the Working Group of Commission 5 on Designations for its advice before being passed on to the General Assembly.

Furthermore, Commission 5,

<u>recognizing</u>   the many benefits that would follow from the clear and unambiguous identification of all astronomical objects to which reference is made in astronomical journals and other sources of data,

<u>strongly urges</u> 1) that all astronomers follow the IAU recommendations on the designation of objects outside the solar system that were adopted by Commission 5 in 1979 (IAU Transactions XVIIB, pp. 87-88) and the supplementary precepts that are given below in the Memorandum on Designations adopted in 1985,

2) that the editors of astronomical journals draw the attention of authors to these recommendations, preferably by providing a summary, and request referees to refer back any papers or tabulations that do not provide satisfactory designations, and

3) that the Space Telescope Science Institute adopt these recommendations for objects discovered with the Space Telescope.

REPORT ON THE XIXTH GENERAL ASSEMBLY

Memorandum on Designations - New Delhi 1985

1) The IAU Style Book, 1986 Revision, will provide rules on designations to be used for constellations, stars and other astronomical objects.

2) Astronomers should consult "The First Dictionary of the Nomenclature of Celestial Objects" by Fernandez, Lortet and Spite and its supplement (1983 A. & A. Suppl. 52, No. 4, and ibid 1986 in press) for the designations of various types of astronomical objects already in the literature and to avoid duplication when proposing designations of new objects.

3) The following precepts are now added :

   (i) The IAU approved three-letter abbreviations for the constellations together with LMC and SMC for the Magellanic Clouds should be used. These abbreviations should not occur with any other meanings.

   (ii) Abbreviation of abbreviations (e.g. 'N' for 'NGC') should never be used.

   (iii) Personal names such as 'Gum Nebula' should be preserved as in the First Dictionary.

   (iv) New acronyms for abbreviating catalogues, types of objects, authors' names, observatories, etc. should have at least two letters.

   (v) The list of types of objects (e.g. GCL, SNR, etc.) given in the First Dictionary should be followed closely.

   (vi) Specific references are needed for acronyms that do not appear in the First Dictionary and for those that appear in the First Dictionary with classification E (explain) or Z (avoid) or those with S (systematic) which could be ambiguous, (e.g. OH).

   (vii) For designations based on coordinates

   - use truncated coordinates, not rounded up ones ;
   - use explicit leading zeroes and the declination sign ;
   - use decimal points if appropriate ;
   - adopt the EINSTEIN extended format if possible
     (e.g. acronym HHMMSS $\pm$ DDMMSS
     or acronym HHMMSS.SS etc. $\pm$ DDMMSS.S etc.) ;
   - when necessary to distinguish old names based on Besselian 1950.0 coordinates from new names based on Julian 2000, precede the right ascension with a J in the latter case ;
   - adopt for galactic coordinates the prefix G
     (e.g. acronym GLLL.LL $\pm$ BB.BB) ; and
   - if a coordinate designation includes the catalogue name, rather than the type of object, do not change the designation when the coordinates are improved.

   (viii) The recommended form for designation of individual objects inside a larger object is
   (e.g. LARGE : Acronym Number).

(ix)   When objects are designated on finding charts, the coordinate equinox, the scale, and the N-S and E-W orientations should be indicated clearly.

4) Astronomers may obtain further advice, if necessary, from representatives of the Working Group of Commission 5 on Designations. The current (November 1985) representatives are :

Name	Address	Tel./Tlx.
C. Jaschek	Observatoire de Strasbourg 11, rue de l'Université F-67000 Strasbourg, France	88-35-43-00 890506 STAROBS F
J.M. Mead, or W.H. Warren, Jr.	Astronomical Data Center, Code 633 NASA Goddard Space Flight Center Greenbelt, MD 20771, USA	(301) 344-8310 89675 NASCOM GBLT
M.-C. Lortet	Obs. de Paris, Section de Meudon, DAPHE F-92195 Meudon Principal Cedex, France	(1) 45-34-75-70 201571 LAM F
H.R. Dickel	Astronomy Department University of Illinois 1011 West Springfield Avenue Urbana, IL 61801-3000, USA	(217) 333-5602 910-245-2434 AST

Résolution C3 : Désignations Astronomiques

La Commission 5,

afin d'éviter toute confusion,

<u>recommande fermement</u> que les résolutions de l'UAI concernant la désignation d'objets astronomiques en dehors du système solaire soient transmises au Groupe de Travail de la Commission 5 sur les Désignations pour tenir compte de son avis avant d'être soumises à l'Assemblée Générale.

Par ailleurs, la Commission 5,

<u>tenant compte</u> des avantages nombreux qui découleraient d'une identification nette et sans ambiguïté de tout objet astronomique mentionné dans les journaux astronomiques et autres sources de données,

<u>invite instamment</u>   1) tous les astronomes à suivre les recommandations de l'UAI sur la désignation d'objets en dehors du système solaire qui ont été adoptées par la Commission 5 en 1979 (Transactions UAI XVIIB, pp. 87-88) ainsi que les préceptes supplémentaires contenus dans le Memorandum sur les Désignations (voir ci-dessous) adopté en 1985,

2) les éditeurs de journaux astronomiques à attirer l'attention des auteurs sur ces recommandations en leur fournissant un résumé, et en demandant aux "referees" de refuser tout article ou table qui ne présente pas les désignations d'une façon satisfaisante, et

3) le Space Telescope Science Institute à adopter ces recommandations pour les objets découverts par le Space Telescope.

Memorandum sur les Désignations - La Nouvelle Delhi 1985

1) Le Manuel de Rédaction, version 1986, indiquera les règles concernant les désignations à utiliser pour les constellations, les étoiles et autres objets astronomiques.

2) Les astronomes doivent consulter "The First Dictionary of the Nomenclature of Celestial Objects" par Fernandez, Lortet et Spite, ainsi que son supplément (1983 A. & A. Suppl. 52, No. 4, et ibid 1986 (sous presse)) pour se conformer aux désignations du "First Dictionary", dans le cas des objets astronomiques de types divers déjà mentionnés dans la littérature et aussi avant de proposer des désignations pour des objets nouveaux, ceci afin d'éviter la duplication.

3) Les préceptes suivants sont maintenant ajoutés :

(i) Les abréviations de constellations en trois lettres approuvées par l'UAI, ainsi que LMC et SMC pour les Nuages de Magellan, doivent être employées. Ces abréviations ne doivent jamais être employées avec d'autres significations.

(ii) Une abréviation ne doit <u>jamais</u> être abrégée elle-même (p. ex. "N" pour "NGC").

(iii) Les désignations contenant des noms propres, comme "Gum Nebula" doivent être préservées conformément au "First Dictionary".

(iv) Les acronymes nouveaux pour l'abréviation de catalogues, types d'objets, noms d'auteurs, observatoires, etc. doivent être composés d'au moins deux lettres.

(v) La liste des abréviations des types d'objets (p. ex. GCL, SNR, etc) donnée dans le First Dictionary doit être suivie de près.

(vi) Des références spécifiques sont requises pour les acronymes qui ne figurent pas dans le First Dictionary ainsi que pour ceux qui sont dans le First Dictionary avec la classification E (expliquer) ou Z (éviter) ainsi que ceux avec la classification S (systématique) quand ils pourraient être ambigus, p. ex. OH.

(vii) Pour les désignations basées sur les coordonnées :

- utiliser les coordonnées <u>tronquées</u> au lieu des coordonnées arrondies ;
- indiquer explicitement les zéros qui se trouvent en première position, ainsi que le signe de la déclinaison ;
- mettre les points décimaux dans les cas appropriés ;
- adopter si possible le format EINSTEIN
 (p. ex. acronyme HHMMSS $\pm$ DDMMSS
 ou acronyme HHMMSS.SS etc. $\pm$ DDMMSS.S etc.) ;
- lorsqu'il y a lieu de faire une distinction entre les anciennes désignations basées sur les coordonnées de Bessel 1950.0 et les nouvelles désignations basées sur les coordonnées Juliennes 2000, faire précéder l'ascension droite d'un J dans ce dernier cas.
- le préfixe G sera adopté pour les coordonnées galactiques
 (p. ex. acronyme GLLL.LL $\pm$ BB.BB) ; et

- si une désignation basée sur les coordonnées comprend le nom d'un catalogue, plutôt que le type d'objet, la désignation ne doit pas être changée lorsque les coordonnées seront améliorées.

(viii) La forme recommandée pour la désignation d'objets à l'intérieur d'un objet plus grand est la suivante
(LARGE : No. Acronyme).

(ix) Lorsque les objets sont désignés sur les cartes d'identification, l'équinoxe des coordonnées, l'échelle, et les orientations N-S et E-O doivent être clairement indiquées.

4) Les astronomes peuvent s'adresser aux représentants suivants du Groupe de Travail de la Commission 5 sur les Désignations pour tout autre renseignement ou conseil :

Nom	Adresse	Tél./Tlx.
C. Jaschek	Observatoire de Strasbourg 11, rue de l'Université F-67000 Strasbourg, France	88-35-43-00 890506 STAROBS F
J.M. Mead, ou W.H. Warren, Jr.	Astronomical Data Center, Code 633 NASA Goddard Space Flight Center Greenbelt, MD 20771, USA	(301) 344-8310 89675 NASCOM GBLT
M.-C. Lortet	Obs. de Paris, Section de Meudon, DAPHE F-92195 Meudon Principal Cedex, France	(1) 45-34-75-70 201571 LAM F
H.R. Dickel	Astronomy Department University of Illinois 1011 West Springfield Avenue Urbana, IL 61801-3000, USA	(217) 333-5602 910-245-2434 AST

Resolution C4 : Space Astrometry

Commission 8,

<u>noting</u> the resolution of IAU Symposium 114 (Leningrad, May 1985), and

<u>considering</u> 1) that new technologies become most useful for astrometry when the observations are made from space, and

2) that accuracies of 10^{-4} to 10^{-6} second of arc appear to be possible,

<u>requests</u> appropriate agencies

1) to consider the importance of the scientific results obtainable by space astrometry,

2) to develop advanced astrometric instruments, particularly interferometers operating in the optical, UV and IR domains, and

3) to promote programs utilizing such instruments.

Résolution C4 : Astrométrie Spatiale

La Commission 8,

<u>notant</u> la Résolution du Symposium 114 de l'UAI (Léningrad, mai 1985), et

<u>considérant</u>
1) que les nouvelles technologies deviennent extrêmement utiles pour l'astrométrie quand les observations sont faites de l'espace, et

2) que des précisions de 10^{-4} à 10^{-6} seconde d'arc apparaîssent possibles,

<u>demande</u> aux agences compétentes

1) de considérer l'importance des résultats scientifiques qu'il est possible d'obtenir par l'astrométrie spatiale,

2) de développer des instruments astrométriques de pointe, en particulier des interféromètres opérant dans les domaines optiques, UV er IR, et

3) de promouvoir des programmes utilisant de tels instruments.

Resolution C5 : Data Handling for Solar Research

Commission 10,

<u>considering</u> the large amount of solar information published in the Quarterly Bulletin of Solar Activity (QBSA), and the bulkiness and cost of disseminating hard copies of the data versus the increasing use of magnetic tape storage,

<u>recommends</u> that judicious choices be made in order to reduce the use of hard copying and promote the use of magnetic tapes in data distribution for the QBSA.

Résolution C5 : Distribution des Données de Recherche Solaire

La Commission 10,

<u>considérant</u> la quantité importante d'informations solaires publiées dans le "Quarterly Bulletin of Solar Activity" (QBSA), le volume et le coût financier de la distribution de copies et d'autre part l'utilisation croissante du stockage de données sur bandes magnétiques,

<u>recommande</u> que des choix judicieux soient effectués de façon à diminuer l'utilisation des copies et de promouvoir l'utilisation de bandes magnétiques dans la distribution des données du QBSA.

Resolution C6 : Carrington Reference System

Commission 10,

<u>considering</u> the importance of the Carrington reference system for statistical studies of solar phenomena from long time-series observations, and the importance of continuity between more than 100 years of past solar data and future

data of similar type, and the difficulty at the present time to improve significantly the determination of the position of the solar axis,

    <u>recommends</u>   that the Carrington reference system continue to be used generally.

Résolution C6 : Système de Référence de Carrington

La Commission 10,

    <u>considérant</u>   l'importance du système de référence de Carrington pour les études statistiques des phénomènes solaires basés sur de longues séries d'observations et l'importance de la continuité entre plus de 100 ans de données solaires passées et les données futures de type similaire,

    <u>recommande</u>   que le système de référence de Carrington continue à être utilisé en général.

Resolution C7 : Adequate Network of Solar Observatories

Commissions 10 and 12,

    <u>considering</u>   the importance of ground-based optical and radio solar observatories for obtaining data which are critical to our understanding of the Sun, which are complementary to solar space projects, and which have scientific importance beyond solar physics, and

    that many ground-based solar observatories around the world are threatened by closure,

    <u>recommends</u>   that the appropriate organizations within the countries involved cooperate to ensure that an adequate network of observatories be maintained to study the Sun, taking account of the need for proper longitude coverage as well as the special contributions of some unique instruments.

Résolution C7 : Réseau Adéquat d'Observatoires Solaires

Les Commissions 10 et 12,

    <u>considérant</u>   l'importance des observatoires optiques et radio au sol pour l'obtention des données essentielles pour notre compréhension du Soleil, qui sont complémentaires des programmes spatiaux solaires et qui ont une portée scientifique qui dépasse la seule physique solaire, et

    que de nombreux observatoires au sol répartis dans le monde sont menacés de fermeture,

    <u>recommandent</u>   que les organismes appropriés dans les pays concernés, coopèrent pour garantir qu'un réseau suffisant d'observatoires soit maintenu pour l'étude du soleil, en tenant compte d'une répartition adéquate en longitude ainsi que des contributions particulières de quelques instruments uniques.

Resolution C8 : Working Group on Planetary Surveys

Commission 16,

<u>proposes</u>   the formation of a special Working Group on Planetary Surveys to coordinate ground-based and space-based observations of variable phenomena on the surfaces and in the atmospheres of the planets and satellites. These observations must be regular and are meant to serve in planning future space missions and to complement spacecraft encounter data. They will also contribute to the understanding of possible correlations of solar activity and planetary phenomena.

Résolution C8 : Groupe de Travail sur des Etudes Planétaires Systématiques

La Commission 16,

<u>propose</u>   la formation d'un groupe de travail spécial consacré à des examens systématiques de planètes, afin de coordonner les observations, à partir du sol et de l'espace, de phénomènes variables sur les surfaces ou dans les atmosphères des planètes et des satellites. Ces observations doivent avoir lieu à intervalles réguliers, et sont destinées à planifier de futures missions spatiales ainsi qu'à servir de compléments à des données obtenues à partir de véhicules spatiaux atteignant des planètes. Elles contribueront également à la compréhension de corrélations possibles entre l'activité solaire et des phénomènes planétaires.

Resolution C9 : Polarimetry and Large Telescopes

Commissions 25 and 9,

<u>considering</u>   that certain properties of astronomical objects are revealed best through measures of their polarized radiation, which is generally quite small, and

<u>noting</u>   that relatively large telescopes are often needed to provide the necessary high signal-to-noise ratio,

<u>recommends</u>   that, in achieving the compromises involved in the design of the very largest telescopes, due weight be given to the need to avoid instrumental polarization as far as possible.

<u>Note</u> : The more detailed proposal for resolution, submitted at the time of the General Assembly is in Commission 25 report on p. 207.

Résolution C9 : Polarimétrie et Grands Télescopes

Les Commissions 9 et 25,

<u>considérant</u>   que certaines propriétés d'objets astronomiques sont mieux mises en évidence en mesurant la polarisation de leurs radiations, qui est en général faible, et

<u>notant</u>   qu'il faut souvent utiliser des télescopes relativement grands pour obtenir le rapport signal sur bruit nécessaire,

recommandent que les compromis indispensables à l'élaboration des projets de très grands télescopes tiennent suffisamment compte de la nécessité d'éviter autant que possible la polarisation instrumentale.

Remarque : lorsqu'elle a été soumise, la formulation de cette résolution était plus détaillée et nuancée. Cette première version est publiée dans le rapport de la Commission 25 p. 207 du présent volume.

Resolution C10 : Designation of Supernovae

Commission 28,

recognizing the need both for an immediate assignment of designations to supernovae as they are discovered, so that researchers can refer to them unambiguously, and for a permanent list of confirmed events for archival purposes,

recommends a) that the IAU Central Telegram Bureau, which has provided temporary designations since January 1985, continue to do so,

b) that the events in order of discovery be designated SN1985A, SN1985B,...., SN1985Z, and as required SN1985aa, SN1985ab,..., SN1985az, SN1985ba, SN1985bb,..., SN1985bz,...,..., SN1985za,..., SN1985zz, a system which provides essential continuity with that established many years ago by F. Zwicky, while permitting easy computer sorting, because the lower case letters will automatically follow the upper case ones, and

c) that the archival list continue to be maintained at the California Institute of Technology for the present.

Résolution C10 : Affectation de Noms aux Supernovae

La Commission 28,

reconnaissant la double nécessité d'attribuer une désignation à des supernovae dès leur découverte, afin que les chercheurs puissent y référer d'une manière non ambigüe, et d'établir une liste actualisée d'événements confirmés pour des besoins d'archivage,

recommande a) que le Bureau Central des Télégrammes de l'UAI, qui a attribué des désignations temporaires depuis janvier 1985, poursuive cette activité,

b) que les événements, dans l'ordre chronologique de leurs découvertes soient appelés SN1985A, SN1985B,...., SN1985Z, si nécessaire, SN1985aa, SN1985ab,..., SN1985az, SN1985ba, SN1985bb,..., SN1985bz,...,..., SN1985za,..., SN1985zz, suivant un système qui procure une continuité avec celui établi, il y a de nombreuses années, par F. Zwicky, tout en permettant une classification informatisée, facilitée par la séquence des lettres utilisées,

c) que, dans un premier temps, la liste des archives continue d'être maintenue à jour au California Institute of Technology.

Resolution C11 : Stellar and Solar Research at Mount Wilson

Commissions 29, 36, and 12,

<u>recognizing</u>   the continuing excellence of the facilities of the Mount Wilson Observatory for solar, stellar and interstellar research,

<u>encourage</u>   efforts to ensure continuity of research at this observatory.

Résolution C11 : Recherche Stellaire et Solaire au Mont Wilson

Les Commissions 12, 29 et 36

<u>reconnaissant</u> que les installations de l'Observatoire du Mont Wilson continuent d'être d'une excellente qualité pour des recherches solaires, stellaires et interstellaires

<u>encouragent</u> les efforts qui permettront de poursuivre des recherches dans cet observatoire.

Resolution C12 : Recommendations of Radio Source Nomenclature

Commission 40, with the support of Commissions 5 and 48,

<u>considering</u>   1) that the IAU adopted a new standard epoch designated J2000 that uses definitions of time and earth motion that are superior to former conventions,

              2) that the new standard epoch and astronomical constants adopted by the IAU have been used for astronomical ephemerides since January 1984, and

              3) that the increased use of computers for data handling, data archiving, and telescope control requires unambiguous easily readable formats for radio source names,

<u>recommends</u>   1) that observers should use the approved catalogue designations with source names (1983 <u>A. & A. Suppl.</u> 52, No. 4 ; 1986 <u>ibid</u>, in press),

              Other abbreviations should not be used. Since catalogue parameters such as epoch, wavelength, and naming convention are implied by the catalogue name, they will not have to be displayed explicitly. In naming new catalogues, the approved updated list of designations should be consulted to avoid duplication, and the new catalogue name should be registered promptly. If reference is made to a catalogue that has been listed, the appropriate literature citation should be given.

              2) that the use of J2000 for source names based on equatorial coordinates start as soon as possible with the format

                      Catalogue name HHMMSS.SS etc. $\pm$ DDMMSS.S etc.,

The leading zeroes should be included. When fewer digits are needed, the coordinates should be truncated, not rounded. If more digits are required, they can be added to the decimal fraction of a second. Note that old catalogue names need not be changed, since the prefixed catalogue name implies the epoch. The source name is a convention, and is not necessarily a statement of an accurate position. However, when there is any possibility of confusion between designations based on B1950.0 and J2000, the prefix B or J should be used in front of the right ascension.

3) that if galactic coordinates are used, the name should be prefixed by G (without a space), and the digits should be truncated as in 2) with the format

$$\text{Acronym GLLL.LL etc. } \pm \text{ BB.BB etc.,}$$

4) that pulsar designations should retain the format

$$\text{PSR HHMMSS.SS etc. } \pm \text{ DDMMSS.S etc.}$$

and J2000 should be used as soon as possible, in view of the astrometric precision needed for some pulsar-related problems,

5) that observers should include the epoch of observation in all catalogues and, when proper motion is important, the relevant supplementary publications concerning J2000 should be consulted, and

6) that observers should note that the relevant IAU resolutions concerning J2000 and the new astronomical constants are summarized in USNO Circular No. 183 with illustrative examples.

Résolution C12 : Recommandations sur la Nomenclature des Radiosources

La Commission 40, avec le soutien des Commissions 5 et 48

considérant
1) que l'UAI a adopté une nouvelle époque d'origine, dite J2000, qui utilise des définitions de temps et de mouvement terrestre qui sont supérieures aux conventions précédentes,

2) que la nouvelle époque d'origine et les nouvelles constantes astronomiques adoptées par l'UAI ont servi à établir les éphémérides astronomiques depuis Janvier 1984, et

3) que l'utilisation accrue d'ordinateurs pour le traitement de données, l'archivage des données, et la commande des télescopes demande une lecture des formats de nomenclature de radiosource qui soit à la fois facile et non-ambigüe,

recommande
1) que les observateurs utilisent les désignations approuvées de catalogue avec noms de sources (1983, A. & A. Suppl. 52, No. 4 ; ibid, sous presse),

L'emploi d'autres abréviations est à proscrire. Il n'y a pas besoin de citer les paramètres du catalogue tels qu'époque, longueur d'onde, etc. puisque ceux-ci sont implicitement compris dans le nom du catalogue. Lors d'appellation de nouveaux catalogues, la liste approuvée et mise à jour des désignations doit être consultée afin d'éviter toute duplication, et le nouveau nom du catalogue doit être enregistré sans délai. Une référence à la littérature doit être donnée en cas de mention de catalogue listé.

2) que l'emploi de la désignation J2000 pour les noms de sources basés sur les coordonnées équatoriales commence aussitôt que possible avec le format suivant :

Nom de catalogue HHMMSS.SS etc. ± DDMMSS.S etc.,

Tous les zéros, même en tête, doivent être inclus. Les coordonnées seront tronquées au lieu d'être arrondies si le nombre de chiffres doit être limité. Si davantage de chiffres sont nécessaires, ils seront ajoutés à la fraction de seconde. Noter qu'il n'y a pas besoin de changer les noms d'anciens catalogues, puisque l'époque est sous-entendue par le préfixe du nom du catalogue. Le nom de la source est une convention plutôt qu'une affirmation de position précise. Cependant, lors de possibilité de confusion entre les désignations B1950.0 et J2000, il faudrait ajouter le préfixe B ou J devant l'ascension droite,

3) que, dans le cas des coordonnées galactiques, on ajoute le préfixe G (sans espace) avant le nom, en tronquant les chiffres comme en 2) avec le format

Acronyme GLLL.LL etc. ± BB.BB etc.,

4) que les désignations de pulsar gardent le format

PSR HHMMSS.SS etc. ± DDMMSS.S etc.

avec l'utilisation de J2000 aussitôt que possible en vue de la précision astrométrique requise pour certains problèmes ayant rapport aux pulsars,

5) que les observateurs indiquent l'époque des observations dans tous les catalogues et que, lorsque le mouvement propre est important, les publications supplémentaires appropriées concernant J2000 soient consultées, et

6) que les observateurs prennent note que les résolutions de l'UAI qui se rapportent à J2000 ainsi qu'aux nouvelles constantes astronomiques, se trouvent résumées dans la circulaire USNO No. 183 et avec des exemples à l'appui.

Resolution C13 : Faint Standard Star Studies

Commission 45,

realizing   the importance for all astronomy of obtaining a network of standard stars for spectral, photometric, radial velocity, and astrometric research, and

recognizing   that the progress of our science continues to bring new knowledge by exploring to ever-fainter magnitude limits, and noting   the increase in the number of large telescopes with efficient detectors both on the ground and in space,

recommends   that the necessary time on large telescopes be devoted to fundamental work of establishing a network of faint standard stars around the sky for situations in which it is not appropriate to use neutral-density filters.

Résolution C13 : Etude d'Etoiles Standard Faibles

La Commission 45,

prenant conscience   de l'importance pour toute l'astronomie de disposer d'un réseau d'étoiles standard pour les recherches spectroscopiques, photométriques, de vitesses radiales, et astrométriques, et

reconnaissant   que le progrès de notre science continue d'apporter de nouvelles connaissances en explorant jusqu'aux limites de magnitudes les plus extrêmes, et

remarquant   que s'accroit, tant au sol que dans l'espace, le nombre de grands télescopes équipés de détecteurs sensibles et efficaces,

recommande   que soit accordé, sur les grands télescopes, le temps nécessaire au travail fondamental d'établir un réseau d'étoiles faibles dans tout le ciel pour tous les cas où il n'est pas possible d'utiliser des filtres neutres.

Resolution C14 : Future High-Energy Space Missions

Commission 48,

notes with approval   the existing and planned studies of a number of space missions of great relevance to its work, specifically the Advanced X-ray Astronomy Facility, the X-ray Multimirror Facility and the Superconducting Magnet Facility on the Space Station. The Commission looks forward to the early construction of these facilities which can contribute significantly to international cooperation in high-energy astrophysics.

Résolution C14 : Futures missions spatiales en Astrophysique des Hautes Energies

La Commission 48,

note avec satisfaction   les études en cours ou en projet de missions spatiales telles que l'"Advanced X-ray Astronomy Facility", le "X-ray Multi-mirror Facility", et le "Superconducting Magnet" sur la Station Spatiale. La Commission espère que ces missions, qui peuvent contribuer de façon importante à la coopération internationale en astrophysique des hautes énergies, seront réalisées à court terme.

16. Appointment of the Special Nominating Committee

The President asked the General Secretary to report on the names of the members proposed for appointment by the General Assembly to the Special Nominating Committee 1985-1988. These persons will be convened by the President of the IAU for the purpose of proposing names to the XXth General Assembly (1988) for the IAU Executive Committee membership (1988-1991). The five persons appointed are :

> V. Abalakin (USSR)
> K. Kodaira (Japan)
> G. Swarup (India)
> F. Pacini (Italy)
> B. Pagel (UK).

The appointment was unanimously confirmed by the General Assembly.

17. Nomination of new Members of the Union

The General Secretary announced that the Executive Committee had, on the proposal of the Adhering Bodies and with the advice of the Nominating Committee, admitted around 935 new members to the Union. The names of the new members were displayed in a prominent place in the course of the General Assembly. The General Secretary informed that the names will be incorporated in the alphabetical list of IAU Members to appear in Transactions Vol. XIXB.

18. Application for IAU Membership

The President invited Mr. Chidi E. Akujor, representative of Nigeria, to present an application for membership from Nigeria :

"Mr. President, Members of the Executive Committee of the International Astronomical Union, fellow astronomers, distinguished ladies and gentlemen, I address you on behalf of Professor Sam Okoye, Chairman of the Nigerian National Committee on Astronomy, and Secretary (Physical Sciences), Nigerian Academy of Sciences.

May I convey the greetings of the President and Fellows of our National Academy of Sciences on the occasion of the XIXth General Assembly of the International Astronomical Union, taking place in New Delhi, India.

Mr. President, I wish to formally request the admission of my dear country, Nigeria, as an adhering member of the International Astronomical Union.

The Nigerian Academy of Sciences, as an adhering member of ICSU as well as of a number of other scientific unions within the ICSU family, believes that the modern development of any nation, especially the developing countries, must be seen beyond the issue of solving social and economic problems. It is for this reason that our Academy has decided to provide auspices for the future development in Nigeria for Astronomy - a science that continues to remain at the esoteric level in many a developing country.

Mr. President, as you may recall, the 1978 International School for Young Astronomers co-sponsored by the IAU was hosted by Nigeria, and this triggered the teaching of Astronomy routinely as an optional subject for the physics degree in many Nigerian Universities. Now, astronomy programmes at the level of BSc, and PhD are available in the University of Nigeria, Nsukka. To date one PhD and eight MScs have been trained, while three PhDs and three MScs are in the pipeline.

You may also recall that Nigeria has maintained regular presence at the IAU General Assemblies since 1976. Of recent, the IAU, through the General Secretary, Dr. Richard West, has been instrumental to the installation and commissioning of a 10-metre radio-telescope at Nsukka, which it is hoped will be useful in low frequency radio astronomy, especially in VLBI observations. We are indeed grateful.

It is our recognition of the importance of Astronomy as a science, as well as of these pioneering efforts, that the Nigerian Academy of Science hopes, in the truest tradition of science, that these international cooperative efforts and ventures in science involving Nigeria can be consolidated and expanded. We also believe that time has come for Nigeria to fully identify with the International Astronomical Union.

Once again, Mr. President, accept on behalf of the International Astronomical Union, the esteem and best wishes of the Nigerian Academy of Sciences."

The General Secretary informed the General Assembly that the application from Nigeria had been carefully examined by the Executive Committee and it was found that the degree of astronomical development in that country met the standards for IAU membership. The General Secretary therefore moved that the General Assembly admit Nigeria as a new Adhering Country. This motion was adopted unanimously.

19 Changes in Commissions

a)   The General Secretary read the proposals of the Executive Committee for Commission Presidents and Vice Presidents for the period 1985-88

	Commission	Presidents, Vice-President(s)
4	Ephemerides	**B. Morando** (France), K. Seidelman (USA)
5	Documentation and Astronomical Data	**G. Wilkins** (UK), B. Hauck (Switzerland)
6	Astronomical Telegrams	**A. Mrkos** (Czechoslovakia), E. Roemer (USA)
7	Celestial Mechanics	**V.A. Brumberg** (USSR), J. Henrard (Belgium)
8	Positional Astronomy	**Y. Requième** (France), M. Miyamoto (Japan)
9	Instruments and Techniques	**C.M. Humphries** (Australia), J. Davis (Australia)
10	Solar Activity	**M. Pick** (France), E.R. Priest (UK)
12	Radiation and Structure of the Solar Atmosphere	**M. Kuperus** (Netherlands), John W. Harvey (USA)
14	Atomic and Molecular Data	**R.W. Nicholls** (Canada), S. Sahal-Bréchot (France)

15	Physical Study of Comets, Minor Planets and Meteorites	L. Kresak (Czechoslovakia), J. Rahe (Germany, FR)
16	Physical Study of Planets and Satellites	G.E. Hunt (UK), A. Brahic (France) and D. Morrison (USA)
19	Rotation of the Earth	W. Klepczynski (USA), M. Feissel (France) and B. Kolaczek (Poland)
20	Positions and Motions of Minor Planets, Comets and Satellites	Y. Kozai (Japan), J.V. Batrakov (USSR)
21	Light of the Night Sky	K. Mattila (Finland), A.-C. Levasseur-Regourd (France)
22	Meteors and Interplanetary Dust	P.B. Babadzhanov (USSR), C.S.L. Keay (Australia)
24	Photographic Astrometry	A. Upgren (USA), W.F. van Altena (USA)
25	Stellar Photometry and Polarimetry	F. Rufener (Switzerland), I.S. McLean (UK)
26	Double and Multiple Stars	K. Rakos (Austria), H. McAlister (USA)
27	Variable Stars	B. Szeidl (Hungary), M. Breger (Austria)
28	Galaxies	P. van der Kruit (Netherlands), G.A. Tammann (Switzerland)
29	Stellar Spectra	G. Cayrel de Strobel (France), P. Conti (USA)
30	Radial Velocities	J. Andersen (Denmark), D. Latham (USA)
31	Time	D.D. McCarthy (USA), P. Pâquet (Belgium)
33	Structure and Dynamics of the Galactic System	W.B. Burton (Netherlands), M. Mayor (Switzerland)
34	Interstellar Matter	J. Lequeux (France), J.S. Mathis (USA)
35	Stellar Constitution	D. Sugimoto (Japan), A. Maeder (Switzerland)
36	Theory of Stellar Atmospheres	K. Kodaira (Japan), D.F. Gray (Canada)
37	Star Clusters and Associations	D.C. Heggie (UK), G.L.H. Harris (Canada)
38	Exchange of Astronomers	E. Müller (Switzerland), F.G. Smith (UK)
40	Radio Astronomy	J.E. Baldwin (UK), P. Mezger (Germany, FR)
41	History of Astronomy	J.A. Eddy (USA), J.D. North (Netherlands)

42	Close Binary Stars	**J.I. Smak** (Poland), R.H. Koch (USA)
44	Astronomy from Space	**Y. Kondo** (USA), K.A. Pounds (UK)
45	Stellar Classification	**R.F. Garrison** (Canada), M. Golay (Switzerland)
46	Teaching of Astronomy	**C. Iwaniszewska** (Poland), A. Sandqvist (Sweden)
47	Cosmology	**G. Setti** (Italy), K. Sato (Japan)
48	High Energy Astrophysics	**C. Cesarsky** (France), R.A. Sunyaev (USSR)
49	The Interplanetary Plasma and the Heliosphere	**S. Grzedzielski** (Poland), L.F. Burlaga (USA)
50	Protection of Existing and Potential Observatory Sites	**S. van den Bergh** Observatory (Canada), D.L. Crawford (USA)
51	Bioastronomy : Search for Extraterrestrial Life	**F.D. Drake** (USA), G. Marx (Hungary)
	Working Group for Planetary System Nomenclature	**H. Masursky** (USA)

This proposal was received by acclamation by the General Assembly.

b) <u>Organizing Committees of Commissions</u>

The General Secretary informed that, in order to save time, it had been decided not to present the lists of members in Organizing Committees of Commissions, but that the lists were available at the IAU Secretariat for inspection. They will be printed in Transactions XIXB (see Chapter VI).

c) <u>Change in Names of Commissions</u>

The General Secretary informed the General Assembly that the Executive Committee had accepted the change of name of Commission 51 : "Search for Extraterrestrial Life" into "Bioastronomy : Search for Extraterrestrial Life".

20. Place and Date of the XXth General Assembly.

The President called upon Prof. F.D. Drake, who presented an invitation from the National Academy of Sciences of the USA, to hold the XXth General Assembly of the IAU in Baltimore, USA, from August 2-11, 1988. The General Assembly accepted this invitation with acclamation and the President asked Prof. F.D. Drake to convey the acceptation and the gratidude of the Union to the President of the National Academy of Sciences of the United States.

21. Election to the Union of a President, six Vice-Presidents, a General Secretary, and an Assistant General Secretary

The General Assembly approved by acclamation the proposal of the President that Prof. J. Sahade be elected the new President of the Union, for the term 1985-1988.

REPORT ON THE XIXTH GENERAL ASSEMBLY

The President then moved that Profs. A. Batten, R. Kippenhahn and P.-O. Lindblad be elected Vice-Presidents for the term 1985-1991. This motion was also approved by acclamation.

The President finally proposed that Dr. J.-P. Swings be elected General Secretary of the Union, and Dr. D. McNally Assistant General Secretary of the Union, for the term 1985-1988. This proposal was also approved by acclamation.

The President then invited Prof. J. Sahade, Profs. Batten, Kippenhahn and Lindblad, and Dr. McNally to join the Executive Committee on the platform.

Following these eletions, the IAU Executive Committee for the period 1985-1988 will thus be as follows :

President : Prof. J. Sahade
Casilla de Correo 677
Observatorio Astronomico
Universidad Nacional de La Plata
1900 La Plata (Bs. As.)
Argentina
Tel. : 54 (21) 249790
Tlx. : 31151 BULAP AR

General Secretary :
Dr. J.-P. Swings
IAU Secretariat
61, avenue de l'Observatoire
F-75014 Paris
France
Tel. : (33-1) 43 25 83 58
Tlx. : 205671 IAU F

Home Institute :

Institut d'Astrophysique
5, avenue de Cointe
B-4200 Cointe-Ougrée
Belgium
Tel. : 41-52-99-80
Tlx. : 41264 ASTRL B

Assistant General Secretary :

Dr. D. McNally
University of London Observatory
University College London
Mill Hill Park
London NW7 2QS, UK
Tel. : 1-959-7367
Tlx. : 28722 UCPHYS G

(Colloquia, Symposia,
Regional Meetings,
Co-sponsorship)

Vice-Presidents :

Dr. A. Batten
Herzberg Institute of Astrophysics
Dominion Astrophysical Observatory
5071 West Saanich Road
Victoria BC, V8X 4M6, Canada

Prof. R.P. Kraft
Lick Observatory
University of California
Santa Cruz, CA 95064, USA

Prof. M. Peimbert
Instituto de Astronomia
Apdo. Postal 70264
Mexico 04510 DF
Mexico

Prof. R. Kippenhahn
MPI für Physik & Astrophysik
Karl-Schwarzschild-Strasse 1
D-8046 Garching bei München
Germany, FR

Prof. P.-O. Lindblad
Stockholm Observatory
S-133 00 Saltsjöbaden
Sweden

Prof. Ya. S. Yatskiv
Main Astronomical Institute
Ukrainian Academy of Sciences
SU-252127 Kiev
USSR

Advisers:

Prof. R. Hanbury Brown
School of Physics
University of Sidney
Sidney, NSW 2006
Australia

Dr. R.M. West
European Southern Observatory
Karl-Schwarzschild-Strasse 2
D-8046 Garching bei München
Germany, FR

22. Address by the President 1982-85

Prof. Hanbury Brown spoke as follows:

"Before I ask you, next President to speak, I would like to say a few words of thanks to a few of the many people who have served our Union during the last few years. I will leave the pleasant task of thanking our hosts to a later speaker, but, even so, I must express the sincere thanks of this Union to the National Organizing Committee whose Chairman is Professor M.G.K. Menon and to the Local Organizing Committee, whose Chairman is Dr. A.P. Mitra, for all they have done to make this meeting a success.

First of all I thank all the members of the Executive for their careful work and for attending all the meetings. I particularly wish to express our thanks to the three retiring Vice-Presidents, Michael Feast, Lubor Kresak and Robert Wilson and to the retiring Adviser, Patrick Wayman. Let us assure them that we appreciate what they have done for the Union.

I am sure you would also like to join with me in thanking the Editors of Mandakini, Archie Roy and Mr. Ratnakar for their really excellent newspaper which has added significantly to the success of this meeting.

This brings me to the secretaries of the Union, the Assistant General Secretary, Jean-Pierre Swings, let me tell you, has handled his job with all the skill and tact which it demands. We can look forward to his term of office as General Secretary with complete confidence. We are also grateful to the Belgian Ministry of National Education for their help in his work for our Union.

We welcome Dr. Derek McNally as the next Assistant General Secretary.

Richard West, the General Secretary, told us in his speech at Patras when he was invited to be a Secretary of the IAU, he saw the invitation as an opportunity to contribute in a new way to the furthering of astronomy. I am sure you will all agree with me that he has made full use of that opportunity. With the help of the admirable services of Brigitte Manning and Danielle Lours in Paris, and of Elisabeth Völk in Munich, he has done more for this Union, and so for the furthering of international Astronomy, than we have any right to expect. Vainu Bappu told me that would be so and he was right. I would also like to thank the European Southern Observatory for their support.

Richard West is a man who values the purposes - the ideals - for which this Union was originally formed and for which it still exists, and he has been prepared to devote his considerable talent for administration, a great deal of his time and prodigeous energy to serving those purposes. This Union has been blessed with a series of remarkably able and industrious General Secretaries - Richard West has sustained that tradition with distinction. Please join me in thanking him."

23. Address by the President 1985-1988

Prof. J. Sahade addressed the General Assembly as follows :

"Professor Hanbury Brown, members of the Union, Ladies and Gentlemen, mes chers collègues,

Habituellement, afin de rassurer l'audience que les langues officielles de l'Union Astronomique Internationale sont deux, l'anglais et le français, le Président qui vient d'être élu s'adresse à l'Assemblée Générale en utilisant chaque langue alternativement pour les différents paragraphes de son allocution.

J'ai trouvé très difficile de faire cela de façon équitable et efficace, et, en conséquence, j'ai préféré utiliser aujourd'hui une seule des deux langues, celle qui est la plus proche de la tradition du pays amphitryon.

Alors, avec votre permission, je m'adresserai à vous en anglais.

Mais avant de commencer, je voudrais rendre hommage à la mémoire de mon très cher ami Vainu Bappu qui a été avec nous en esprit pendant toute cette Assemblée Générale. Un de ses rêves les plus chers, celui de réunir l'Union Astronomique Internationale en Inde vient de se concrétiser et j'en suis, comme vous tous, particulièrement heureux et satisfait.

I do not need to say how deeply honored, and moved and thankful I feel for having been so kindly raised to the highest position in the International Astronomical Union, that is, to the highest position an astronomer could dream of in his career.

I am conscious of the fact that the honor that has been bestowed upon me goes well beyond my merits, and, as I wrote to Professor Hanbury Brown in my letter of acceptance of the nomination, "when I think of the eminent astronomers who have been Presidents of the IAU before me, I cannot refrain from feeling a little uneasy"... even more than a little uneasy, I would say now.

I have, however, accepted because I felt that my nomination actually meant a recognition of the importance and of the level Astronomy has reached in Latin America, in general, and in Argentina, in particular, and because I also felt that my three years in office might perhaps encourage support to and induce the growth of our science in Latin America at large and, likewise, in countries from other geographical areas, where no or little Astronomy is being done at the present time. And this might have been taken into consideration by the Special Nominating Committee on account of the IAU's concern for spreading Astronomy teaching and research among countries that are not members of the Union.

May I also say that some prospective facts added up to help me overcome my initial reluctance to accept the nomination. I was perfectly aware that during my term of office I would enjoy the experienced and friendly advice of Robert Hanbury Brown and Richard West, the help of two active and enthusiastic General Secretaries, Jean-Pierre Swings and Derek Mc Nally, and the warm cooperation of the six Vice-Presidents, most of whom I had known for quite some time now. Lastly, I did likewise take into consideration that the IAU Secretariat is indeed very efficient and... is located in Paris... and I must confess that I do like to come to Paris !

Naturally I also counted on the cooperation of all of you, members of the Union. I hope that you will let me have your thoughts, your suggestions, your comments,... and indeed that you will let me hear from you as often as you have ideas that you might like to share with me. Please do feel strongly that **you** are the Union.

The International Astronomical Union was established sixty six years ago. In the meantime, and, particularly in the last decade or so, Astronomy has grown tremendously. The number of researchers and of research papers, the number of meetings (symposia and colloquia and the like, IAU - and non IAU - sponsored) have increased extraordinarily and is continuously increasing at a pace which is difficult, perhaps I should say impossible to keep up with. The utilization of new technologies dominates the scene and the emphasis of the research subjects shifts dynamically. A number of new branches of Astronomy, especially since the beginning of the space technology era, have acquired stature and are bringing about extraordinary progress in our knowledge and understanding of the Universe. Space experiments like those connected with the Halley Comet, the flyby of the Voyager near planet Uranus, the placing in orbit of the Space Telescope and of the Hipparcos mission satellite, are some of the most, should we say spectacular ?, forthcoming sources of new information and new advances that will most probably change, in more or less drastic way, the conclusion from previous knowledge.

The rapid growth of our science - which also reflects in the rapid growth of the membership of the Union - is the **fact of life** for us, astronomers ; but it is not the only one. Another fact of life arises from the realization that we are immersed in a world, a large fraction of which faces socio-economic problems that bring about, as a consequence, a curtailment of the funds devoted to science in general, and to Astronomy, in particular. Some may argue that this is of no concern to us, because actually brains, which cost nothing, are the most important assets for research. I fully agree with this, but brains alone may not amount to much. An up-to-date library is the **sine qua non** complement. And if you are an observational astronomer, you would need well-equiped instruments and, perhaps, also access to astronomical satellite information.

The yearly cost of journal subscriptions and the prices of books have increased so much that not all libraries can keep up with their, until rather recently, normal purchases. On the other hand, modern astronomical facilities and the application of advanced technology imply expenditure of such an order of magnitude that only very very few countries can afford them.

Another feature of our science today is that non-conventional information is being secured at such a rate that it is impossible for the relatively not too large a number of scientists who are involved, to undertake an exhaustive analysis, and, as a result, data banks are storing a truly scientific treasure, not completely exploited, which is at the disposal of the astronomical community.

As the person who is going to preside the International Astronomical Union for the next three years, I cannot but try to build up in my mind a picture of the state of Astronomy worldwide and try to identify areas, if any, in which we should attempt to take action.

It may be that we should try to find ways to actually create and/or increase contacts and cooperation between astronomers, particularly involving young astronomers, and between different groups of the astronomical community. It may be that we should try to encourage some efforts on a regional basis, following the example set by ESO or ESA, which have proved to be so successful and rewarding in Europe. We should engage in some careful thinking to best decide in which directions our endeavors ought to go, one question that readily comes up has certainly been asked each time by everyone of my predecessors. Is the present structure and operational procedure of the IAU the best to cope with today's and near future Astronomy ? In the life of the Union, no large changes seem to have been, so far, necessary. Is such still the case. In trying to find the best way to steer the Union and preserve its unity and integrity during the next three years I hope that we shall have opportunities to develop strong interaction with the members, individually and through the different commissions.

May I finish these words by wishing you all the best in your astronomical, as well as in your personal activities, throughout the next triennium. Thank you much for your attention."

24. Closing Ceremonies

The President then invited **Dr. P. Shaver** to propose a vote of thanks to all the organizers on behalf of all participants. In a few sentences, Dr. Shaver thanked all those who so efficiently contributed to the great success of the XIXth General Assembly. On behalf of the registered guests, **Mrs. T. West** then expressed her thanks as follows :

"Ladies and Gentlemen,

On behalf of the registered guests, let me with a few words express our great thanks and gratitude to all those who made our stay in India so wonderful and unforgettable. Thanks to the Local Organizing Committee and in particular to those who organized and prepared the programme for the registered guests : to Mrs Saxena and Mrs Chatterjee. They did an excellent job.

The so-called "Ladies' Programme" was just marvelous and we greatly enjoyed it while our dear husbands worked very, very hard.

Thanks to those who organized the wonderful trip to Agra, thanks to the hotel management, to the photographers and to the volunteers who took part in the preparation of the programme for us. Thanks to all those who organized concerts. I would like to congratulate you on having such a great talent as Master Srinivasan who, I am sure, will be a good ambassador of your country.

We want you to know that from the very first day of our arrival, and until now, everything has been perfectly organized.

We really admired your hospitality, food, your fabulous country, elegant ladies in their saris, and kind Indian people.

India, which has always been a fairy tale and a far, far away country, has all of a sudden become very close to me, and I will tell you why. This is not only because of the beauty, but also because of the language. Here I discovered the words which are similar in my mother tongue, Georgian. The Georgians also say "puri" for bread, "patara" for small, "magali" for big, "dukan" for shop, and so on. I will not bother you too much with these similarities - I just want to say that for me, as for the linguist, a whole new world opened. I am sure that all of the registered guests, and particularly me, wish to come back to India and discover not only linguistic phenomena, but also more of the history, beauty, and meet again the wonderful Indian people with smiles on their faces.

Thank you."

The President then called upon the retiring General Secretary, **Dr. R.M. West**:

"Ladies and Gentlemen,
Friends and Colleagues,

During the past ten days, I must have spoken thousands and thousands of words - and since I became General Secretary 39 months ago, I must have dictated a million words to the poor secretaries. You have all suffered at the receiving end, so let me try to be brief today before I return to my telescope.

In the nordic sagas, the individuals are described as links in a universal chain of characters, who come and go - and leave behind only the memory of their doings and deeds in the minds of their successors. Maybe it is because I was born in Viking territory that I consider myself very much in the same way, that is as a link between past and future IAU General Secretaries.

Before me were the warriors who laid the foundations for the Union. And I see behind me the future fighters lining up, eagerly awaiting to test their youthly strength and endurance. To my predecessors, to my successors and all others in this hall, I should like to say some words about the work of a General Secretary.

I could tell you about happy moments and difficult hours - about travels to meetings in distant countries and encounters with marvelous people. About waiting in airports and thoughts about the family back home. About talks in foreign languages and discussions of how the IAU can help. How I have felt wonderfully elated when problems were successfully solved, and also how there were glimpses of despair when some of my letters went unanswered.

I can tell you of coming to Paris, that pearl of a city, without time to see its beauty, but only the bleak Metro labyrinths. To arrive early in the morning at the IAU Secretariat, so small and insignificant when seen from the outside and yet big enough to embrace the entire astronomical world within its ancient walls.

You climb the squeaking stairs to your office and sit down on the chair that always rolls off to the right because the old oak-floor is no longer horizontal, and you watch with unbelieving eyes the mountain of mail in front of you.

You hear the noises of doves on the roof, the honking of horns from the boulevard. The clattering of the telex below and the loud voices of the secretaries when they have a bad intercontinental telephone connection. And when they prepare lunch, there are the undecipherable smells of real French cuisine.

But let me now come back to realities. Those in the past who thought up the schedule of IAU General Secretaries were clever indeed. Three years training as Assistant General Secretary, three years at maximum momentum as General Secretary and three years as Adviser is as much as a normal person can do. A degree in Astronomy does not necessarily prepare you for the intricacies of international administration and diplomacy. However, with the right help, everything is possible.

Thank you, Hanbury, for all your support and advice. I have indeed been exceedingly fortunate that your term coincided with mine. You and the other members of the Executive Committee have given me courage and strength ever since Patras. And I have great respect for you, Jean-Pierre, who have dealt so well with all matters of IAU meetings, be assured, dear members of the Union, that you will be in very good hands.

REPORT ON THE XIXTH GENERAL ASSEMBLY

Let me also cordially thank my secretaries in Paris, Mrs. Manning and Miss Lours, and Mrs. Völk in Munich for your never failing patience and dedication. Without your support, the administration of the IAU would never have been as efficient as I think it is today.

And you, my Indian friends, how happy I am to see your incredible efforts crowned with success. On top of all the common features of an IAU General Assembly, which you have managed so well; where did we ever get a freshly printed Sunday issue of the Assembly Newspaper, as we here were given "Mandakini", in the buses half way to Agra. Now that is organizational talent !

Il me semble que j'avais promis d'être bref. Eh bien, je termine ici. Permettez-moi de proposer l'introduction d'une petite cérémonie qui, peut-être, fera tradition.

Mon cher successeur, voici les clefs du secrétariat à Paris. Bonne chance, Jean-Pierre, et merci pour votre attention."

Finally the President invited the newly elected General Secretary, **Dr. J.-P. Swings** to address the General Assembly :

"Being the last one to speak today is either a good or a bad thing, since most of what I wanted to say has been said already ! Anyhow, let me pronounce a few brief words before a horrible burden falls on me.

It is both with pride and with fear that I stand here today : it is definitely a great honor to have been chosen by the Special Nominating Committee, and I am most grateful to the Union for this, but, also, I am aware that a lot of responsibilities are linked to dealing with matters concerning more than six thousand IAU members.

Je demanderai donc beaucoup de patience et d'indulgence à tous les membres, tout en étant certain qu'ils m'aideront à mener à bien le mandat de Secrétaire Général. For those who did not realize, this was a sentence in French asking the IAU members to be patient with me during my term !

In the past three years I relied appreciably on the experience of my predecessor, Richard West, and I am afraid (for him) that I will have to continue. He has helped me a great deal already, and he has contributed to make the Secretariat in Paris more and more efficient, by acquiring modern equipment, so that Brigitte Manning and Danielle Lours can now "play" with a word processor, a computer, and a telex, for the true benefit of the whole Union. Thank you Richard for all you did and thanks for your help in the future.

During the last three years I had the opportunity to meet several times the past President of the IAU, Professor R. Hanbury Brown : this has been a very nice experience because Hanbury is a delightful person with a great sense of humor, and also a very efficient chairman. His advice in the coming three years will be very useful too, I am sure.

Let us now come to the future, the very near future I am afraid to say. The officers of the IAU work as a team, of course, and I am sure that Derek McNally will do a great job as Assistant General Secretary. During several years the newly elected President, Professor Jorge Sahade and I, have had similar astrophysical interest concerning those weird objects called Symbiotic Stars. In fact I hope that the two of us will form a good and productive symbiosis, and that it will be the main goal of the IAU to continue on promoting astronomy in all geographical areas, and on sponsoring as many scientific activities as possible : colloquia, symposia, regional meetings, etc. Mailing travel grants to young and less young astronomers, or sending money to meeting organizers provides the Assistant General Secretary or the General Secretary, I am sure, with the nice feeling of doing something concrete in the promotion of our discipline. It should be said that all this happens while trying to keep the administrative costs to a strict minimum, which has been and still is the wish of the IAU Executive Committee, and has been noted with satisfaction by the Finance Committee.

Ayant vu mon prédécesseur en action depuis 1982, je me rends compte que la période entre aujourd'hui et août 1988 sera très occupée et que ma femme et mes deux fils ne me verront pas beaucoup à la maison. Je leur demande d'être indulgents..., comme ils le sont déjà maintenant. Je ne voudrais pas poursuivre cette petite allocution sans remercier Mr. Deloz, du Ministère de l'Education Nationale Belge pour son soutien financier à l'antenne de Liège de l'UAI et à Denise Fraipont qui s'occupe si bien de celle-ci.

I believe I am a "porte-parole" of all of you when I say that I was impressed by the quality of the opening ceremony : all speeches were very nice, but, of course, we have all been moved by the friendly words of the Prime Minister of India Honorable Rajiv Gandhi. His smile, his attitude that was "décontractée", the fact that he definitely enjoyed the company of his fellow countryman Hanbury Brown, all this made the ceremony enjoyable, and something that we shall all undoubtedly remember and tell our children about. So once again, after the past President, the past General Secretary, Peter Shaver and Tamriko West, let me thank the National Organizing Committee, as well as the Local Organizing Committee, that is both teams from Bangalore and from Delhi, for all they did for us. There were many Indian astronomers attending this General Assembly, about three hundred I believe : I thus sincerely hope that this gathering has been successful in promoting Astronomy in this country (I am coming back to one of the aims of the IAU I mentioned earlier), that many contacts have been made between Indian and foreign astronomers, and, perhaps and hopefully, that a few collaborations have started on this occasion.

I actually wish to express my gratitude to all those who came here to Delhi and to tell them that I hope to see them in Baltimore in 1988. The US National Organizing Committee and Local Organizing Committee have seen a beautiful example here in Delhi and I am sure they will do their best to set up a very successful General Assembly. In the meanwhile, this is a message for all of you astronomers, members or not of the IAU, do not hesitate to call upon me for help and/or especially for constructive suggestions : the life of the Union relies on all of you, definitely not only on the officers !

Merci pour votre patience et votre attention, et "Au revoir", à Baltimore en août 1988.

CHAPTER III

REPORT OF THE EXECUTIVE COMMITTEE

1982-1984

President :

Prof. R. Hanbury Brown, School of Physics, University of Sidney, Sidney, NSW 2006, Australia.

Vice-Presidents :

Dr. M.W. Feast, South African Astronomical Observatory, P.O. Box 9, Observatory, Cape 7935, South Africa.

Prof. R.P. Kraft, Lick Observatory, University of California, Santa Cruz, CA 95064, USA.

Dr. L. Kresak, Slovak Academy of Sciences, Dubravska Cesta 5, 84228 Bratislava, Czechoslovakia.

Prof. M. Peimbert, Instituto de Astronomia, Apartado Postal 70264, Mexico 04510 DF, Mexico.

Prof. R. Wilson Department of Physics and Astronomy, University College London, Gower Street, London WC1E 6BT, UK.

Prof. Ya. S. Yatskiv, Main Astronomical Observatory, Ukranian Academy of Sciences, SU-252127 Kiev, USSR.

General Secretary :

Dr. R.M. West, European Southern Observatory, Karl-Schwarzschild-Strasse 2, D-8046 Garching bei München, F.R. Germany.

Assistant General Secretary :

Dr. J.-P. Swings, Institut d'Astrophysique, 5 avenue de Cointe, B-4200 Cointe-Ougrée, Belgium.

Adviser :

Prof. P.A. Wayman, Dunsink Observatory, Dublin 15, Ireland.

I. INTRODUCTION

The present report covers the period from 1st January 1982 to 31 December 1984. This period includes the first eight months (to 26 August) of 1982, which was part of the period in office of the previous Executive Committee ; it does not include the first eleven months of 1985, still the responsibility of the present Executive Committee (to 28 November 1985).

The address of the General Secretary at the XIXth General Assembly will include a report for the period 1st January-19 November 1985.

The report includes a summarized financial report for the calendar years 1982, 1983 and 1984 and a budget proposal for the calendar year 1985, during the 3-year budgetary period 1983-1985.

II. ADMINISTRATION

1. Executive Committee

The Executive Committee held the following meetings :

49th Meeting : Patras, Greece, 15, 16 and 24 August 1982.

50th Meeting : Patras, Greece, 26 August 1982.

51st Meeting : Munich, F.R. Germany, 12-15 September 1983.

52nd Meeting : Ile d'Yeu, France, 10-13 September 1984.

The 49th and 50th meetings were held in the shadow of the illness and subsequent death on 19 August 1982, of the President of the Union (1979-1982), Prof. M.K.V. Bappu. The participants in the the XVIIIth General Assembly paid a moving tribute to the memory of Prof. Bappu on 23 August 1982, cf. IAU Transactions XVIIIB, page 1.

In the absence of the President, the 49th Meeting was chaired by Vice-President, Dr. D.S. Heeschen ; all other members were present. Prof. R. Hanbury Brown, President-elect, and Dr. J.-P. Swings, Assistant General Secretary elect, participated also in this meeting.

The 50th, 51st and 52nd meetings were held under the chairmanship of Prof. R. Hanbury Brown, President of the IAU. All other members participated. It was decided that the position as adviser to the Executive Committee, which is normally taken by the Past President during the three years after a General Assembly, would remain unfilled after Prof. Bappu's death.

2. Officers' Meetings

The President, the General Secretary and the Assistant General Secretary met as follows :

Patras, Greece, 17-18 February 1982   Munich, F.R. Germany, 11 September 1983

Patras, Greece, 14 August 1982   Delhi, India, 22-24 January 1984

Paris, France, 20 January 1983   Paris, France, 9 September 1984

3. IAU Secretariat

The IAU Secretariat continued to function at its permanent seat in Paris according to an agreement signed with the Paris Observatory on 7 December 1981 ; the address is :

I A U - U A I
61, avenue de l'Observatoire
F-75014 Paris

Telephone : 33 (1) 43-25-83-58   Telex : 205671 IAU F

A Xerox 860 word processor was acquired in November 1982 and a Canon NP 120 photocopier in April 1983. The telex connection was installed in May 1983. For the IAU computerized member file, an IBM XT minicomputer with a 10 Mbyte hard disc and an ink-jet printer was acquired in December 1984. Various minor improvements (e.g. floor, furniture) were made.

4. IAU Staff

Mrs. B. Manning continued as IAU Secretary. Ms. P. Smiley, Typist, left in November 1983. Miss D. Lours was employed as IAU Assistant Secretary from 15 March 1984.

5. Adhering Countries

Chinese membership of IAU was ratified by the XVIIIth General Assembly (Resolution A1 ; cf. Transactions XVIII B, page 25).

6. Members of the IAU

There were about 4497 members on 1 January 1982. 795 new members were elected at the XVIIIth General Assembly. 91 names were deleted by resignation and decease up to December 1982. Since that date there have been recorded 55 further names for deletion. The number of members at 31 December 1984 is therefore estimated to be around 5146. The approximate numbers of IAU members, residing within the geographical areas of IAU Adhering Bodies, are indicated in Section VIII of this report. In addition, IAU members reside in Libya (1), Nigeria (2), Lebanon (1), Philippines (1), Saudi Arabia (2), Singapore (2), Trinidad and Tobago (1).

7. Consultants to IAU Commissions

The 40 IAU Commissions had a total of approximately 188 consultants at 31 December 1982.

III. COMMISSIONS OF THE IAU

By resolution of the XVIIIth General Assembly, Commission 51 "Search for Extraterrestrial Life" was created. The names of Commission 26 and 34 were changed to "Double and Multiple Stars" and "Interstellar Matter", respectively.

Commission Reports, covering the period 1 July 1981 - 30 June 1984 have been published in early 1985 (IAU Transactions Vol. XIX A, 725 pages).

The following remarks concern some of the many Commission activities not otherwise included in subsequent sections of this report :

Commission 5 : Work continued on the new IAU Style Book, which is expected to be ready before the XIXth General Assembly.

Commission 6 : The IAU Central Bureau for Astronomical Telegrams (Director : Dr. B.G. Marsden) issued 369 Circulars and 170 "Telegram Books" in this period.

Commission 15 : The XVIIIth General Assembly decided to support the International Halley Watch (IHW) as coordinating authority for observations of Comet P/Halley during its 1985-86 return.

Commissions 19 and 31 : The MERIT Main Campaign took place from 1 September 1983 - 31 October 1984. Reports about the work of the MERIT and COTES groups appeared in successive issues of the Information Bulletin (e.g. 53, pp. 16-17).

Commission 20 : Together with Commission 6 and the IAU Working Group for Planetary System Nomenclature (WGPSN), procedures were established for assigning names and designations to various classes of newly-discovered bodies in the solar system. The IAU Minor Planet Center (Director : Dr. B.G. Marsden) issued more than 2600 MPC's and numbered no less than 642 new minor planets during 1982-84.

Commission 22 : An IAU Meteor Data Center was created in Lund, Sweden, under the directorship of Prof. B.A. Lindblad.

Commission 38 : Under the guidance of its President, Prof. F.B. Wood, this Commission continued to give assistance to (young) astronomers under agreed rules (cf. IB 50, pp. 11-13).

Commission 46 : The Visiting Lecturers' Programme Subcommittee (Chairman : Prof. D.G. Wentzel) decided to recommend that the VLP's be initiated in Lima, Peru and Nsukka, Nigeria.

IV. SCIENTIFIC MEETINGS

1. Symposia and Colloquia

The following IAU meetings were held :

IAU Symposium No. 101 : "Supernovae Remnants and their X-ray Emission", Venice, Italy, 30 August - 2 September 1982.

IAU Symposium No. 104 : "The Early Evolution of the Universe and its Present Structure", Kolymbari, Crete, Greece, 30 August - 2 September 1982.

IAU Symposium No. 105 : "Observational Tests of the Stellar Evolution Theory", Geneva, Switzerland, 12-16 September 1983.

IAU Symposium No. 106 : "The Milky Way Galaxy", Groningen, Netherlands, 30 May - 3 June 1983.

IAU Symposium No. 107 : "Unstable Current Systems and Plasma Instabilities in Astrophysics", College Park, MD, USA, 8-11 August 1983.

IAU Symposium No. 108 : "Structure and Evolution of the Magellanic Clouds", Tübingen, FRG, 5-8 September 1983.

IAU Symposium No. 109 : "Astrometric Techniques", University of Florida, USA, 9-12 January 1984.

IAU Symposium No. 110 : "VLBI and Compact Radio Sources", Bologna, Italy, 27 June - 1 July 1983.

IAU Symposium No. 111 : "Calibration of Fundamental Stellar Quantities", Como, Italy, 24-29 May 1984.

IAU Symposium No. 112 : "The Search for Extraterrestrial Life : Recent Developments", Boston, USA, 18-21 June 1984.

IAU Symposium No. 113 : "Dynamics of Star Clusters", Princeton, USA, 29 May - 1 June 1984.

IAU Colloquium No. 73 : "Ultraviolet and X-ray Spectroscopy of Astrophysical and Laboratory Plasmas", Dublin, Ireland, 30 August - 1 September 1982.

IAU Colloquium No. 74 : "Dynamical Trapping and Evolution in the Solar System", Chalkidiki, Greece, 30 August - 2 September 1982.

IAU Colloquium No. 75 : "Planetary Rings", Toulouse, France, 30 August - 2 September 1982.

IAU Colloquium No. 76 : "The Nearby Stars and the Stellar Luminosity Function", Middletown, CT, USA, 12-13 June 1983.

IAU Colloquium No. 77 : "Natural Satellites", New York, USA, 5-9 July 1983.

IAU Colloquium No. 78 : "Astronomy with Schmidt-type Telescopes", Asiago, Italy, 30 August - 2 September 1983.

IAU Colloquium No. 79 : "Very Large Telescopes, their Instrumentation and Programs", Garching bei München, F.R. Germany, 9-12 April 1984.

IAU Colloquium No. 80 : "Double Stars - Physical Properties and Generic Relations", Bandung, Indonesia, June 3-7, 1983.

IAU Colloquium No. 81 : "Local Interstellar Medium", Madison, WI, USA, 4-6 June 1984.

IAU Colloquium No. 82 : "Cepheids : Theory and Observations", Toronto, Canada, 28 May - 1 June 1984.

IAU Colloquium No. 83 : "Dynamics of Comets : Their Origin and Evolution", Rome, Italy, 11-14 June 1984.

IAU Colloquium No. 84 : "Longitude Zero", Greenwich, UK, 9-13 July 1984.

IAU Colloquium No. 85 : "Properties and Interactions of Interplanetary Dust", Marseille, France, 9-12 July 1984.

IAU Colloquium No. 86 : "Eighth International Colloquium on Ultraviolet and X-ray Spectroscopy of Astrophysical and Laboratory Plasmas", Washington, USA, 27-29 August 1984.

2. Cosponsored Meetings

IAU cosponsored the following meetings :

COSPAR/IAU Symposium on "Gamma-Ray Astronomy in Perspective of Future Space Experiments", Ottawa, Canada, 18-19 May 1982.

COSPAR/IAU/IAMAP/IUTAM/URSI Symposium on "Giant Planets and their Satellites", Ottawa, Canada, 18-21 May 1982.

COSPAR/IAU Symposium on "Advanced Space Instrumentation in Astronomy", Ottawa, Canada, 20-22 May 1982.

COSPAR/IUTAM/IAU/IUGS Symposium on "Impact Processes of Solid Bodies", Ottawa, Canada, 21-22 May 1982.

COSPAR/IAU/SCOSTEP Symposium on the "Solar Maximum Year", Ottawa, Canada, 25-27 May 1982.

IUPAP/IAU "International Conference on Plasma Physics", Gothenburg, Sweden, 9-15 June 1982.

EPS/IAU : "Fourth CESRA Workshop on Solar Noise Storms", Trieste, Italy, 9-13 August 1982.

IAU/UN International Seminar on Astronomy (at the time of UNISPACE 82), Vienna, Austria, 12 August 1982.

COSPAR/IAU Symposium on "Advances of High Energy Astrophysics and Cosmology", Pamporovo, Bulgaria, 18-22 July 1983.

URSI/IAU Symposium "Measurement and Processing for Indirect Imaging", Sydney, Australia, 30 August - 3 September 1983.

COSPAR/IAU Symposium on "High Energy Astrophysical Sources", August 1983.

COSPAR/IAU Workshop on "Venus International Reference Atmosphere", Hamburg, FRG, 24-25 August 1983.

IUGG/IAU Symposium on "Cosmic Dust in Planetary Atmosphere", Hamburg, FRG, 17 August 1983.

URSI/IAU Workshop on "Solar Terrestrial Prediction", Meudon, France, 18-22 June 1984.

SCOSTEP/IAU Symposium on "Solar Maximum Analysis", Graz, Austria, 25-27 June 1984.

COSPAR/IAU Symposium on "Planetology of Venus, Mars, and Satellites of Outer Planets", Graz, Austria, 26-28 June 1984.

COSPAR/IAU Symposium on "Nucleosynthesis and Acceleration of Cosmic Rays", Graz, Austria, 27-30 June 1984.

COSPAR/IAU Workshop on "Venus Atmosphere", Graz, Austria, 25-30 June 1984.

3. Regional Meetings

Third **Latin-American** Regional Astronomy Meeting, Buenos Aires, Argentina, 28 November - 3 December 1983.

Seventh **European** Regional Astronomy Meeting, Florence, Italy, 12-16 December 1983.

Eighth **European** Regional Meeting, Toulouse, France, 17-21 September 1984.

Third **Asian-Pacific** Regional Astronomy Meeting, Kyoto, Japan, 1-5 October 1984.

Fourth **Latin-American** Regional Astronomy Meeting, Rio de Janeiro, Brazil, 19-24 November 1984.

4. IAU/UNESCO Young Astronomers' Schools

IAU Commission 46 "Teaching of Astronomy" organized the **XIIIth School** at Lembang, Indonesia, 16 May - 2 June 1983. It brought together 21 students from India, Indonesia, Malaysia, Philippines and Thailand. The curriculum included Solar Physics and Galactic Structure.

5. XVIIIth General Assembly, Patras, Greece, 17-26 August 1982

IAU Transactions, Volume XVIIIB (1983), contains a report of the Proceedings of the General Assembly of the Union, including the resolutions passed at the Assembly. It also contains the reports of Commission meetings during the period of the General Assembly.

The IAU Statutes, By-laws and Working Rules, as updated at the XVIIIth General Assembly, are also included in Vol. XVIIIB, page 397 ff.

V. PUBLICATIONS

1. Publisher

In November 1984, the contract with D. Reidel Publishing Co. was renewed for a six-year period, from 1985-1991.

2. Sales

Number of copies sold (hard,soft) by D. Reidel Publishing Company in the period 1982-1984.

Transactions : XIIIA (4), XIIIB (21), XIVA (8), XIVB (5), XVA (7), XVB (4), XVIA (3), XVIB (6), XVIIA part 1 (22), XVIIA part 2 (21), XVIIA part 3 (23), XVIIB (90), XVIIIA (493), XVIIIB (517).

Highlights : I (8), II (15), III (18), IV 1 (13, 5), IV 2 (10, 3), V (40, 55), VI (335, 188).

Symposia Proceedings : 32 (17), 33 (0), 34 (32), 35 (22), 36 (0), 37 (7), 38 (44), 39 (26), 40 (18), 41 (1), 42 (17), 43 (26), 44 (39), 45 (32), 46 (33), 47 (15), 48 (32), 49 (17, 56), 50 (12, 38), 51 (7, 58), 52 (14, 57), 53 (14, 37), 54 (15, 34), 55 (12, 27), 56 (4, 4), 57 (8, 30), 58 (6, 53), 59 (17, 54), 60 (13, 36), 61 (8, 25), 62 (8, 36), 63 (22, 68), 64 (18, 48), 65 (2, 15), 66 (15, 54), 67 (28, 74), 68 (19, 32), 69 (15, 35), 70 (7, 6), 71 (10, 10), (11, 29), 80 (7, 25), 81 (37, 23), 82 (24, 27), 83 (11, 18), 84 (59, 71), 85 (38, 77), 86 (44, 25), 87 (62, 144), 88 (34, 45), 89 (23, 14), 90 (35, 26), 91 (41, 32), 92 (51, 82), 93 (82, 115), 94 (116, 86), 95 (108, 110), 96 (76, 115), 97 (473, 424), 98 (323, 195), 99 (384, 224), 100 (443, 344), 101 (403, 230), 102 (377, 209), 103 (390, 252), 104 (385, 221), 105 (342, 154), 107 (110, 37), 108 (359, 131), 110 (344, 268).

3. Information Bulletin

Starting with issue No. 49 (January 1983), the IAU Information Bulletin was prepared and composed on the Xerox 860 Word Processor by the IAU Secretariat, and printed and distributed by Reidel. It appeared twice a year and was sent, free of charge, to IAU members, consultants, scientific institutions, and selected international organizations. Each print-run was 6300 copies.

VI. RELATIONS TO OTHER ORGANIZATIONS

IAU was represented by the General Secretary at the following meetings of the International Council of Scientific Unions (ICSU) : 15/16th Meetings of the General Committee and 19th General Assembly (Cambridge, UK, September 1982) ; 17th Meeting of the General Committee (Warsaw, Poland, August 1983) ; 18/19th Meetings of General Committee and 20th General Assembly (Ottawa, Canada, September 1984). The IAU General Secretary was elected Ordinary Member of the ICSU Executive Board (1982-1986).

IAU was also represented at meetings of the ICSU Committee for Space Research (COSPAR) : XXIVth Plenary Meeting (Ottawa, Canada, May 1982) - P.A. Wayman and R.M. West ; XXVth Plenary Meeting (Graz, Austria, June-July 1984) - R.M. West, L. Kresak and R. Wilson.

The following IAU Representatives to other ICSU and international bodies were active during the period 1982-84 :

CCDS	Consultative Comm. for the Definition of the Second	W. Markowitz
		J. Benavente
CCDM	Consultative Comm. for the Definition of the Metre	A.H. Cook
CCIR	International Radio Consultative Committee	
	Study Group 2	J. Whiteoak
		L.H. Doherty
	Study Group 7	H.M. Smith
		J. Pilkington
CCMP	Coordination Committee for Moon and Planets	P.M. Millman
		M. Ya. Marov
CODATA	Committee on Data for Science and Technology	B. Hauck
COSTED	Comm. on Science and Technology in Developing Countries	G. Swarup
CTS	Committee on the Teaching of Science	D. McNally
EPS	European Physical Society - Conference Committee	J.-P. Swings

FAGS	Federation of Astronomical & Geophysical Services	E. Tandberg-Hanssen
- BIH	Bureau International de l'Heure	S. Iijima
		P. Pâquet
- IPMS	International Polar Motion Service	B. Kolaczek
- IUWD	International Ursigram & World Day Services	H. Coffey
- QBSA	Quarterly Bulletin on Solar Activity	F. Moriyama
		E. Hiei
IAF	International Astronautical Federation	M. Papagiannis
ICSTI	International Council for Scientific & Technical Information	W. Heintz
		G.A. Wilkins
IUAA	International Union of Amateur Astronomers	J.-P. Swings
		M. Rigutti
IUCAF	Inter-Union Commission on Frequency Allocation for Radio Astronomy and Space Science	R. Schilizzi
		G. Swarup
IUCS	Inter-Union Commission on Spectroscopy	B. Edlen
		J.G. Phillips
		M.J. Seaton
SCOPE	Scientific Committee on Problems of Environment	F.G. Smith
SCOSTEP	Scientific Committee on Solar-Terrestrial Physics	M.R. Kundu

CCMP was discontinued in 1984.

UNESCO was regularly kept informed about relevant IAU activities. The IAU/UNESCO GERT project was delayed due to the necessity of a technical revaluation. A site in Sumatra, Indonesia, was found and a geological study was made.

A joint IAU/UN International Astronomy Seminar was held on 12 August, 1982 in Vienna, Austria at the time of the UNISPACE 82 Conference.

VII. FINANCIAL MATTERS

1. Summarized Account of Receipts and Payments for the 3-year Period 1982-84

The 3-year summarized account shown in Table I, has been certified by the IAU Auditor, R. Bacle, H.E.C., Paris. The certified "Vérification des Comptes" for 1982, 1983 and 1984 will be available for examination by the Finance Committee of the XIXth General Assembly.

2. Balance 1982-1984

		Swiss Francs	Swiss Francs
Bank Balance : 31.12.1981			663 991.14
Re/Devaluation :	1982	- 4 745.28	
	1983	+ 1 903.40	
	1984	+ 31 284.86	+ 28 442.98
Result (Income - Expenditure)	1982	- 6 804.27	
	1983	+ 29 133.05	
	1984	+ 111 380.14	+ 133 708.92
Bank Balance : 31.12.1984			+ 826 143.04

3. Comments to Account and Balance 1982-84

The ICSU mean conversion rates to Swiss Francs for the years 1982, 1983 and 1984 were :

	1 US$	1 Dfl	1 FF	1 DM
1982	2.0125	0.7589	0.31025	---
1983	2.0858	0.7421	0.2784	0.8288
1984	2.3275	0.7342	0.2692	0.8263

The <u>gains on transfers</u> arise because of the difference between the conversion rate <u>at the time of a transfer</u> from one currency to another, and the above <u>yearly mean</u> rates used for establishment of the annual and 3-year accounts.

The <u>re/devaluation</u> figures represent the difference in the bank balances on 31.12. and on 1.1. as expressed in Swiss Francs ; e.g. the conversion from other currencies to Swiss Francs on 31.12.82 is done according to the mean conversion rate of 1982, and on 1.1.1983 according to the mean conversion rate of 1983.

As was the case for the (1979-81) accounts, no distinction is made between "Savings" and "Current" accounts.

Income

<u>ad 1</u> The XVIIIth General Assembly (1982) decided to increase the annual unit of contribution to 1 760 Swiss Francs. Two countries, India and South Africa, asked to enter a higher category, so that the total number of units were 229 from 1982. Since 1982, some Adhering Organizations have paid their annual dues in advance. On 1.6.1985, eight Adhering Organizations were in arrears (up to and including 1984 dues) with a total of 26 units.

<u>ad 2</u> Royalties from Reidel. A new six-year contract (1986-1991), signed in November 1984, guarantees the IAU a minimum of 75 000 DFl/year in royalties.

ad 3   When calculated straightforward from the interest and the mean of the opening (1.1.) and closing (31.12) balances, the mean interest rates were 6.36 % (1982), 6.39 % (1983) and 5.34 % (1984), reflecting the 1984 drop in interest rates, mainly on US $ accounts. By currency, the IAU money was held as follows (in percent):

	SFr	US$	FF	DFl
31.12.1981	33.2	41.1	5.9	19.8
31.12.1982	30.9	41.9	7.9	19.3
31.12.1983	41.2	41.4	1.2	16.2
31.12.1984	37.0	47.3	0.8	14.9

By cash-flow management, a minimum of money (normally about 15 %) was held in current, low-interest accounts.

ad 4   A basic allocation and support for specific projects was received from ICSU (UNESCO subvention or ICSU grant) as follows (in US$):

	1982	1983	1984	(1982-1984)
Basic allocation	13 991	13 495	12 634	40 120
Project MERIT		4 000	5 000	9 000
Commission 38 Travel Grants		3 000	3 000	6 000
IAU VLP		--	4 000	4 000
IAU ISYA	4 000	--	--	4 000
TOTAL	17 991	20 495	24 634	63 120

ad 5   Includes FF 45 000.- from Ministère des Relations Extérieures (Paris) in support of the 52nd Executive Committee Meeting (Ile d'Yeu, France), which is herewith gratefully acknowledged.

Expenditure

ad 1   Including equipment (Xerox 860 Word Processor, Canon Photocopier, telex machine and IBM XT 256 computer), charges for building to Paris Observatory (electricity, heating, overheads), operation costs (stationery, postal charges, etc.), salaries and social charges for the IAU Secretary and Assistant Secretary, General Secretary expenses (travel and subsistence) and Audit Fees. Major contributions to the functioning of IAU were made by the home institution of the General Secretary (ESO, Garching) and Assistant General Secretary (Institut d'Astrophysique, Liège) in the form of administrative assistance. The Assistant General Secretary was also supported by grants from the Belgian Ministry of Education.

ad 2   Contribution to ICSU is 2.5 % of the contribution from Adhering Organizations during the preceding year.

ad 3   Includes payment of the ICSU contributions to Project MERIT (to BIH and in support of MERIT meetings) and IAU support to Meteor Data Center, as decided by the XVIIIth General Assembly, and for which no provisions were made in the adopted (1983-85) budget.

ad 4   Details in Table II. Note that the extra 10 000 Swiss Francs, allocated to Commission 38 (cf. IB 48, p. 39) were not used. Due to delays in the implementation of the

Lima and Nsukka VLP's, spending only started in 1984. The Commission 27 money, which is handled within the IAU accounts, was drawn upon in 1982 in connection with the General Assembly. This sum was erroneously debited to Item 5. On 31.12.84, the credit to Commission 27 was 11 548.03 Swiss Francs.

ad 5   Details about the spending in 1982 will be found in the "Vérification des Comptes 1982". The additional amount paid in 1983 to the Local Organizing Committee in Patras was authorized by the Executive Committee during its 51st meeting (1983).

ad 6   Each issue of the Information Bulletin was printed in 6300 copies. The cost in Dutch Guilders, including postage which depends on the number of pages was : IB 47 (9 694.88) ; IB 48 (12 381.70) ; IB 49 (19 616.00, including covers for IB 49-54) ; IB 50 (11 067.00) ; IB 51 (13 410.35) and IB 52 (12 240.25). Due to increasing postage rates in the Netherlands, it was decided to make use of "air-speeded" mail-service to North America.

ad 7   The costs were significantly reduced by distributing soft-cover copies only. Some institutes were added to the distribution list.

ad 8   The added transportation cost to the 52nd meeting was covered by a grant (see above). The Executive Committee acknowledges with gratitude the hospitality offered by ESO (1983) and by Prof. J.-C. Pecker and the Ile d'Yeu authorities (1984).

ad 9   Comparatively large expense in 1984 because of the Officers' Meeting in Delhi, as part of preparations for General Assembly in 1985.

ad 10   In support of 30 IAU-sponsored and 18 co-sponsored meetings (cf. Section IV 1) :

	Symposia	Colloquia	Co-Sponsored Meetings
1982	5	5	8
1983	5	4	5
1984	4	7	5

ad 11   Contribution to CTS (US$ 400 per year), FAGS (SFr 2 000 in 1982, US$ 2000 in 1983 and 1984), ICSTI (US$ 360 per year), IUCAF (US$ 1250 per year), and SCOSTEP (US$ 1 000 in 1982).

ad 12   "First Dictionary of the Nomenclature of Celestial Objects" (delayed from 1982 ; SFr 11 148 paid in 1983) ; Working Group on Photographic Problems (SFr 1 384.95 in 1984) ; AAVSO Publication (SFr 6 000 in 1984).

ad 13   Part of cost of IAU Representation to meetings of FAGS, CCIR, IAF, ICSTI, EPS, COSPAR, SCOSTEP, SCOPE, COSTED and URSI.

ad 15   Late payments for 1980 Hvar School (US$ 110) and 1981 Cairo School (US$ 1 250) in 1982. Payment to Bandung School in 1983. No School in 1984.

ad 16   Extra travel support to Kyoto (1984) meeting, since this was the first major IAU-sponsored meeting with participants from P.R. China.

ad 17   No SNC meetings were held. Communication costs only.

ad 18   Including Swiss anticipated tax (35 % of interest on current accounts in SFr) and NRA on US$-account in 1983 (the type of account was changed and no NRA tax was paid in 1984).

Conclusion

The total balance on 31.12.1984 was 162 151.90 Swiss Francs higher than on 31.12.1981. As explained above, this 24 % increase is mainly due to the increased SFr-value of US$ holdings, following the revaluation of the US$ against all other currencies in late 1984. Moreover, some Adhering Bodies paid their annual 1985 dues already in 1984 (altogether 32 units, i.e. SFr 56 320).

4. Residual Budget for the year 1985 and Budget Proposal 1985

The figures in the accounts for 1983 and 1984, added together for each item, are subtracted from the figures for 1983-85, as approved by the General Assembly in 1982. In this way a Residual Budget 1985 is obtained. Negative figures mean that the 1983-85 Budget figure has been exceeded. Based upon careful examination of trends and figures actually achieved, the Executive Committee approved some changes during its 52nd Meeting in September 1984. Two sets of figures are given in Table III :

1. Residual Budget 1985, as explained above.

2. Budget Proposal 1985, to be referred to the Finance Committee at the XIXth General Assembly for presentation to the Assembly, according to IAU Statute 18(b).

5. Explanatory Notes to Budget Proposal 1985

The changes proposed by the Executive Committee are explained as follows :

Income

ad 1   No change. Arrears and advance payments are expected to cancel out.

ad 2   Same level of royalties as in earlier years.

ad 3   Conservative estimate. More funds must be mobilized in low-interest accounts during a General Assembly year. Lower interest on US$ accounts due to general fall in rates may be expected in 1985.

ad 4   US$ 27 989.- was allocated by ICSU in September 1984 to IAU for transfer in 1985. However, the payment is dependent upon the receipt by ICSU, in mid-1985, of the foreseen UNESCO subvention. Hence the cautious figure.

Expenditure

ad 1   Same level of expenditure as in 1983 and 1984. Although no purchase of major equipment is planned, there are additional expenses in connection with the preparations for the General Assembly and the verification of the computerized membership file (postage, stationery).

ad 2   Paid in April 1985.

ad 3   Payment of ICSU 1985 contribution to MERIT and IAU contribution to Meteor Data Center, plus minor expenses.

ad 4   Expenses now foreseen as follows :

Item 4.1	Commission 38	Exchange of Astronomers	25 000
Item 4.2	Commission 46	Visiting Lecturers' Programme	12 000
		Newsletter etc.	1 800
Item 4.3	Commission 6	Telegram Bureau	3 100
	Commission 20	Minor Planet Center	3 100
	Commission 27	Variable Stars	10 000
			55 000

The Commission 38 Travel Grant Scheme is supplemented with funds received from ICSU. Due to the late start, less funds are needed for the Lima and Nsukka VLP's.

ad 5   SFr 30 000 added. The reasons are the unforeseen payment in 1983 to the Patras LOC and the wish of the EC to help the largest possible number of IAU members to participate in the XIXth General Assembly.

ad 6   Information Bulletin Nos. 53 and 54. Free distribution to Commission Presidents of IAU Transactions Vol. XIX A (1985), purchased at a very reduced rate from Reidel for this purpose.

ad 7   A few institutes in developing countries have been added to the distribution list.

ad 8   No expenses in 1985.

ad 9   Meeting at Paris Secretariat in June 1985.

ad 10   SFr 25 000 added in order to increase number of meetings around General Assembly which can be supported. For 1985, 7 Symposia and 4 Colloquia have been planned.

ad 11   SFr 12 000 less than foreseen in (1983-85) budget.

ad 12   No major expenses expected in 1985.

ad 15   ISYA in Punjab (October - November 1985) was postponed.

ad 16   No Regional Astronomy Meeting in 1985.

ad 17   No expenses expected in 1985.

VIII. LIST OF ADHERING ORGANIZATIONS

(Not reproduced here, see p. 383).

IX. LIST OF DECEASED MEMBERS

(See updated list on page 14).

X. LIST OF IAU PUBLICATIONS

1. Transactions and Highlights

Transactions XVIIIA : (Patras, 1982) pp. viii + 669, D. Reidel Publishing Company, 1982.

Transactions XVIIIB : (Patras, 1982) pp. x + 604, D. Reidel Publishing Company, 1983.

Highlights of Astronomy, Volume 6, as presented at the XVIIIth General Assembly of the IAU, 1982, pp. viii + 818, D. Reidel Publishing Company, 1983.

2. IAU Symposia Volumes

97 -Extragalactic Radio Sources, D.S. Heeschen & C.M. Wade, pp. xviii + 490, Reidel, 1982.

98 -Be Stars, M. Jaschek & H.G. Groth, pp. xvi + 523, Reidel, 1981.

99 -Wolf-Rayet Stars : Observations, Physics, Evolution, C.W.H. de Loore & A.J. Willis, pp. xx + 618, Reidel, 1982.

100 -Internal Kinematics & Dynamics of Galaxies, E. Athanassoula, pp. xvi + 432, Reidel, 1982.

101 -Supernova Remnants and their X-Ray Emission, J. Danziger and P. Gorenstein, pp. xvii + 614, Reidel, 1983.

102 -Solar and Stellar Magnetic Fields : Origins and Coronal Effects, J.O. Stenflo, pp. x + 564, Reidel, 1983.

103 -Planetary Nebulae, D.R. Flower, pp. xxi + 554, Reidel, 1983.

104 -Early Evolution of the Universe and its Present Structure, G.O. Abell and G. Chincarini, pp. xxi + 536, Reidel, 1983.

105 -Observational Tests of the Stellar Evolution Theory, A. Maeder and A. Renzini, pp. xxiv + 590, Reidel, 1984.

107 -Unstable Current Systems and Plasma Instabilities in Astrophysics, M.R. Kundu and G.D. Holman, pp. xxii + 566, Reidel, 1984.

108 -Structure and Evolution of the Magellanic Clouds, S.van den Bergh and K.S. de Boer, pp. xvii + 425, Reidel, 1984.

110 -VLBI and Compact Radio Sources, R. Fanti, K. Kellermann, and G. Setti, pp. xx + 489, Reidel, 1984.

3. IAU Colloquia Volumes

60 -Uranus & the Outer Planets, G. Hunt, Cambridge University Press, pp. ix + 307, 1982.

61   -Comets, Gases, Ices, Grains & Plasmas, L.L. Wilkening, review papers will be published by the University of Arizona Press : "Comets", edited by J.P. Burns and contributed papers in Icarus, issue dedicated to comets.

62   -Current Techniques in Double & Multiple Star Research, R.S. Harrington and O.G. Franz, Lowell Observatory Bulletin No. 167, vol. IX No. 1, 1983, Lowell Observatory, PO Box 1269, Flagstaff, AZ 86002, USA.

63   -High-Precision Earth Rotation & Earth-Moon Dynamics, O. Calame, Astrophysics & Space Science Library 94, pp. 376, Reidel, 1982.

64   -Automated Data Retrieval in Astronomy, C. Jaschek & W.D. Heintz, Astrophysics & Space Science Library 97, pp. xx + 324, Reidel, 1982.

66   -Problems of Solar & Stellar Oscillations, D.O. Gough, pp. 494, reprinted from "Solar Physics", vol. 82, Nos 1 and 2, Reidel, 1983.

67   -Instrumentation for Astronomy with Large Optical Telescopes, C. Humphries, Astrophysics & Space Science Library, 92, pp. xvii + 321, Reidel, 1982.

68   -Astrophysical Parameters for Globular Clusters, A.G. Davis Philip, L. Davis Press, Inc., 1125 Oxford Place, Schenectady NY 12308, USA, pp. 630, 1982.

69   -Binary & Multiple Stars as Tracers of Stellar Evolution, Z. Kopal & J. Rahe, Astrophysics & Space Science Library, 98, pp. xxx + 503, Reidel, 1982.

70   -The Nature of Symbiotic Stars, M. Friedjung & R. Viotti, Astrophysics & Space Science Library 95, pp. xix + 310, Reidel, 1982.

71   -Activity in Red-Dwarf Stars, P.B. Byrne & M. Rodono, Astrophysics & Space Science Library 102, xxvi + 669 pp., Reidel, 1983.

72   -Cataclysmic Variables and Related Objects, M. Livio & G. Shaviv, Astrophysics & Space Science Library 101, pp. xii + 351, Reidel, 1983.

73   -7th International Colloquium on Ultraviolet and X-ray Spectroscopy of Astrophysical and Laboratory Plasmas, **no proceedings,** but a limited number of copies of the Programme and Abstract booklet are available from Professor P.K. Carroll, Physics Dept, University College, Dublin, Ireland.

74   -Dynamical Trapping and Evolution in the Solar System, V.V. Markellos and Y. Kozai, Astrophysics & Space Science Library 106, xvi + 424 pp., Reidel, 1983.

76   -The Nearby Stars and the Stellar Luminosity Function, A.G. Davis Philip and A. Upgren, Van Vleck Observatory Contribution No. 1, pp. xviii + 487, L. Davis Press, Inc., Schenectady, NY, USA, 1983.

78   -Astronomy with Schmidt-type Telescopes, M. Capaccioli, Astrophysics & Space Science Library 110, pp. xii + 620, Reidel, 1984.

80   -Double Stars, Physical Properties and Generic Relations, B. Hidayat, Z. Kopal, and J. Rahe, reprinted from Astrophysics and Space Science 99, Nos. 1-2, pp. vi + 412, Reidel, 1984.

4. Proceedings of Regional Astronomy Meetings

Fifth European
Liège, Belgium, 28 July - 1 August 1980,
Variability in Stars and Galaxies, P. Ledoux, xix + 456 pp., 1980. Sold and distributed by :
Institut d'Astrophysique, avenue de Cointe 5, B-4200 Cointe-Ougrée, Belgium.

Sixth European
Dubrovnik, Yugoslavia, 19-23 October 1981,
Sun and Planetary System, W. Fricke and G. Teleki, xiii + 538 pp., Astrophysics & Space Science Library 96, Reidel, 1982.

Second Asian-Pacific
Bandung, Indonesia, 24-29 August 1981,
M. Feast & B. Hidayat, University Press, ITB, Bandung, Indonesia.

5. Miscellaneous Publications

Commission 5 : The First Dictionary of the Nomenclature of Celestial Objects.
    By A. Fernandez, M.C. Lortet and F. Spite, Astronomy & Astrophysics Supplement,52, 4, June 1983.

Commission 6 : IAU (Telegram Bureau) Circulars.
    Issued by the IAU Central Bureau for Astronomical Telegrams, Smithsonian Astrophysical Observatory, 60 Garden Street, Cambridge, Massachusetts 02138, USA. Applications should be sent to B.G. Marsden at the above address.

Commission 10 : Cartes Synoptiques de la Chromosphère solaire et Catalogue des Filaments et des Centres d'Activité.
    A complete list of the volumes of the "Cartes Synoptiques" which have been published may be obtained from the Observatoire de Paris, Section d'Astrophysique de Meudon, DASOP, 92190 Meudon, France.

Commission 10 : Quarterly Bulletin on Solar Activity.
    Published at the Tokyo Astronomical Observatory, Attn. Dr. F. Moriyama, Mitaka, Tokyo 181, Japan.

Commission 10 : Sunspot Bulletin.
    Published by the Sunspot Index Data Center (S.I.D.C.), Dr. André Koeckelenbergh, 3 Avenue Circulaire, B-1180 Bruxelles, Belgium.

Commission 19 : Circulaires du Bureau International de l'Heure.
    A monthly publication of Bureau International de l'Heure, 61, Avenue de l'Observatoire, 75014 Paris, France.

Commission 19 : Monthly Notes of the International Polar Motion Service.
    Prepared and distributed by the Central Bureau of the International Polar Motion Service, International Latitude Observatory of Mizusawa-shi, Iwate-ken, Japan.

Commission 20 : Minor Planet Circulars.
    Issued by the Minor Planet Center, Attn. Dr. B.G. Marsden, Center for Astrophysics, 60 Garden Street, Cambridge, MA 02138, U.S.A.

Commission 21 : Newsletter.

Commission 26 : Circulaires d'Information de la Commission 26 (Double Stars).
    Prepared and distributed by P. Couteau, Observatoire de Nice, B.P. 139, F-06003 Nice, France.

Commission 27 : Information Bulletin on Variable Stars.
    Prepared and distributed by the Konkoly Observatory of the Hungarian Academy of Sciences, 1525 Budapest XII, Box 67, Hungary.

Commission 27 : General Catalogue of Variable Stars
    Catalogue of Suspected Variable Stars.
    Name-lists of Variable Stars.
    Distributed by Publishing House Nauka, Moscow, U.S.S.R.

Commission 29 : International Register of Stellar Spectroscopists, Edition 2, March 1982.
    Edited by W.K. Bonsack, University of Hawaii and A. Slettebak, Ohio State University.

Commissions 29, 30 and 45 : Newsletter on Be Stars

Commission 41 : Newsletter.

Commission 42 : Standard Stars Newsletter

Commission 46 : Newsletter on the Teaching of Astronomy.

Commission 46 : Astronomy Education Materials.

Commission 51 : Newsletter.

IAU Information Bulletin Nos. 47 - 52.

MERIT Monthly Circular 1-15

MERIT Newsletter 1-6

6. Summary of Accounts 1982-1984 (Swiss Francs)

	1982	1983	1984	1982-84
INCOME				
1. Contributions	378 911.90	370 756.44	438 529.98	1 188 198.32
2. Publications	18 407.79	17 641.79	19 305.08	55 354.66
3. Interest	41 701.49	36 810.77	41 170.39	119 682.65
4. UNESCO/ICSU	35 786.00	42 716.10	57 335.64	135 837.74
5. Miscellaneous	-	835.20	12 114.26	12 949.46
Gain on Transfers	3 119.39	1 808.71	204.32	5 132.42
	477 926.57	470 569.01	568 659.67	1 517 155.25
EXPENDITURE				
1. Administration	165 659.46	160 411.70	161 631.26	487 702.42
2. ICSU Contribution	7 724.00	9 817.86	9 268.91	26 810.77
3. Commission Expenses	-	7 014.50	15 353.75	22 368.25
4. Commission Projects				
4.1. Exchange	8 468.11	15 754.67	30 889.38	55 112.16
4.2. Teaching	3 622.50	2 010.27	10 200.84	15 833.61
4.3. Other	6 237.75	6 685.24	7 047.68	19 970.67
5. General Assembly	151 825.24	16 827.19	-	168 652.43
6. IAU Publications	53 736.94	25 361.27	19 330.04	98 428.25
7. Dev. Count. + EC	2 466.04	6 866.11	5 031.29	14 363.44
8. EC Meetings	-	32 933.93	37 841.04	70 774.97
9. Officers' Meetings	7 134.35	4 747.46	13 420.70	25 302.51
10. Symposia & Colloquia	52 954.71	85 119.46	79 218.67	217 292.84
11. Inter-Unions Comm.	7 333.12	8 397.15	4 678.28	20 408.55
12. E.C. Projects	8 439.71	11 148.00	7 384.95	26 972.66
13. Representation	3 422.91	2 901.80	14 172.08	20 496.79
14. Bank Charges	1 199.24	1 020.78	1 467.27	3 687.29
15. Young Ast. Schools	2 737.00	20 881.52	-	23 618.52
16. Regional Meetings	1 006.25	20 000.00	39 560.33	60 566.58
17. SNC	-	-	412.30	412.30
18. Miscellaneous	763.51	3 537.05	370.76	4 671.32
	484 730.84	441 435.96	457 279.53	1 383 446.33
BALANCE	- 6 804.27	29 133.05	111 380.14	133 708.92

7. Projects of IAU Commissions 1982-1984 (Swiss Francs)

Item	Comm.	Project	Allocation 1982	Paid 1982	Allocation 1983-1985	Paid 1983	Paid 1984	Residual 1985
4.1.	38	Exchange of Astronomers	18 500.00	8 468.11	58 000.00	15 754.67	30 889.28	11 355.95
4.2.	46	VLP	--	--	35 000.00	0.00	8 975.54	26 024.46
		Newsletter etc.	2 800.00	3 622.50	5 000.00	2 010.27	1 225.30	1 764.43
4.3.		Other projects						
	6	IAU Telegram Bureau	2 000.00	1 972.25	10 000.00	3 342.62	3 523.83	3 133.55
	16	Planetary Documentation	1 770.00	1 770.00	--	--	--	--
	20	Minor Planet Center	2 500.00	2 495.50	10 000.00	3 342.62	3 523.83	3 133.55
	27	Catalogue of Variable Stars	9 100.00	8 214.00	10 000.00	0.00	0.00	10 000.00

8. 1985 Residual and Proposed Budget (Swiss Francs)

		Residual	EC Proposed
RECEIPTS			
1.	Contributions	378 714	379 000
2.	Publications	14 054	17 000
3.	Interest	47 019	32 000
4.	UNESCO/ICSU	-32 052	50 000
5.	Miscellaneous	-12 949	0
		394 786	478 000
	Excess of Payments over Receipts	140 200	141 100
		534 986	618 100
PAYMENTS			
1.1.	Administration	177 958	158 000
1.2.	Secretarial Help	10 000	6 000
2.	ICSU Contribution	10 613	11 000
3.	Commission Expenses	-7 369	15 600
4.	Commission Projects	55 412	55 000
5.	General Assembly	183 173	213 000
6.	IAU Publications	15 309	23 000
7.	Developing Countries & EC	8 103	10 000
8.	EC Meetings	-775	0
9.	Officers' Meetings	1 832	7 000
10.	Symposia and Colloquia	60 662	95 000
11.	Inter-Unions Comm.	26 925	15 000
12.	EC Projects	-4 533	1 000
13.	Representation	12 926	7 000
14.	Bank Charges	1 488	1 500
15.	Young Astronomers' Schools	15 118	0
16.	Regional Meetings	-29 560	0
17.	SNC	4 588	0
18.	Miscellaneous	-3 908	0
		534 986	618 100

Budget (1986-1988)
All amounts in Swiss Francs

	Triennial (1986-88)	1986	Annual 1987	1988
INCOME				
1. Contributions				
1.1 Adhering Organisations (226 units)	1 318 710	426 010	439 570	453 130
1.2 ICSU/UNESCO	105 000	35 000	35 000	35 000
1.2.1 ICSU Allocations				
1.2.2 ICSU Grants				
1.2.3 UNESCO Contracts				
2. Publications	165 000	55 000	55 000	55 000
2.1 Royalties from IAU Publisher				
2.2 Other				
3. Interests etc.	105 000	35 000	35 000	35 000
3.1 Current Accounts				
3.2 Savings Accounts				
3.3 Gain on Exchange				
4. Other Receipts	—	—	—	—
4.1 Special Contributions				
4.2 Grants				
4.3 Refunds				
4.4 Other				
TOTAL INCOME	1 693 710	551 010	564 570	578 130

EXPENSES

	Triennial (1986-88)	1986	Annual 1987	1988
1. Executive Committee etc.				
1.1 EC Meetings	80 000	40 000	40 000	-
1.2 Officers' Meetings	24 000	8 000	8 000	8 000
1.3 SNC Expenses	5 000	-	5 000	-
1.4 Other	-	-	-	-
TOTAL 1	109 000	48 000	53 000	8 000
2. Publications				
2.1 Information Bulletin	42 000	13 000	14 000	15 000
2.2 Other Publications	-	-	-	-
2.3 Distribution to Developing Countries, etc.	36 000	12 000	12 000	12 000
TOTAL 2	78 000	25 000	26 000	27 000
3. Scientific and Related Activities				
3.1 Meetings				
3.1.1 General Assemblies	240 000	-	-	240 000
3.1.2 Symposia and Colloquia	365 000	120 000	120 000	125 000
3.1.3 Regional Astronomy Meetings	65 000	32 000	33 000	-
3.1.4 Young Astronomers' Schools	50 000	25 000	25 000	-
3.1.5 Visiting Lecturers' Programmes	72 000	24 000	24 000	24 000
3.1.6 Other	-	-	-	-
TOTAL 3.1	792 000	201 000	202 000	389 000

	Triennial (1986-88)	Annual 1986	1987	1988
3.2 Commission Activities				
3.2.1 Commission Expenses	12 000	4 000	4 000	4 000
3.2.2 Commission Projects				
3.2.2.1 Exchange of Astronomers	70 000	23 000	23 000	24 000
3.2.2.2 IAU Telegram Bureau	10 500	3 500	3 500	3 500
3.2.2.3 IAU Minor Planet Center	10 500	3 500	3 500	3 500
3.2.2.4 Variable Star Catalogue	10 500	3 500	3 500	3 500
3.2.2.5 IAU Meteor Data Center	3 000	1 000	1 000	1 000
3.2.2.6 Other	7 000	2 000	2 000	3 000
TOTAL 3.2	123 500	40 500	40 500	42 500
3.3 Relations with Other Organisations				
3.3.1 Dues to ICSU	33 000	10 700	11 000	11 300
3.3.2 Support of Inter-Union Commissions	27 000	9 000	9 000	9 000
3.3.3 Representation to Other Organisations	24 000	8 000	8 000	8 000
TOTAL 3.3	84 000	27 700	28 000	28 300
3.4 Other Activities				
3.4.1 Executive Committee Projects	15 000	5 000	5 000	5 000
3.4.2 Other Projects	-	-	-	-
TOTAL 3.4	15 000	5 000	5 000	5 000
TOTAL 3	1 014 500	276 200	275 500	464 800

4. Administration
  4.1 Secretariat

	Triennial (1986-88)	1986	Annual 1987	1988
4.1.1 Salaries and Social Charges, etc.	315 000	100 000	105 000	110 000
4.1.2 Office Expenses	150 000	45 000	50 000	55 000
4.1.2.1 Rent, Heating, Light, Cleaning, etc.	-	-	-	-
4.1.2.2 Communication (PTT)	-	-	-	-
4.1.2.3 Equipment (Major)	-	-	-	-
4.1.2.4 Office Supplies	-	-	-	-
4.1.3 General Secretary Expenses	30 000	9 000	10 000	11 000
4.1.3.1 Travel, Subsistence	-	-	-	-
4.1.3.2 Representation	-	-	-	-
4.1.4 President's Expenses	12 000	4 000	4 000	4 000
4.1.5 Assistant General Secretary's Expenses	6 000	2 000	2 000	2 000
TOTAL 4.1.	513 000	160 000	171 000	182 000

REPORT OF THE EXECUTIVE COMMITTEE

	Triennial (1986-88)	1986	Annual 1987	1988
4.2 Other				
4.2.1 Bank Charges	3 000	1 000	1 000	1 000
4.2.2 Audit Fees, Legal Advice	4 500	1 500	1 500	1 500
4.2.3 Loss on Exchange	-	-	-	-
4.2.4 Miscellaneous	-	-	-	-
TOTAL 4.2	7 500	2 500	2 500	2 500
TOTAL 4	520 500	162 500	173 500	184 500
TOTAL EXPENDITURE	**1 722 000**	**509 700**	**528 000**	**684 300**
INCOME – EXPENDITURE	**-28 290**	**+37 310**	**+34 570**	**-100 170**

CHAPTER IV

REPORTS OF MEETINGS OF COMMISSIONS

COMMISSION 4: ÉPHÉMÉRIDES (EPHEMERIDES)

Report of Meeting 1985 November 21

PRESIDENT: T Lederle   SECRETARIES: H Schwan
                                           B D Yallop

1. Organization and Membership

The President's proposals for the Officers of the Commission for the next three years were adopted as follows:

President: B L Morando

Vice-President: P K Seidelmann

Organizing Committee: V K Abalakin, S Aoki, J Chapront, R L Duncombe, T Lederle, J H Lieske, B D Yallop, A Yamazaki

New Members of the Commission: J-E Arlot, V A Brumberg, N Capitaine, M Catalan, A de Castro, X-h Di, J O Dickey, A M Fominov, M Garcia de Polavieja, M L Gi, H L Hyok, M Iliyas, G A Krasinsky, Y Kubo, B-l Liu, M P Romero Perez, G Rosselo, J L Simon, D-zh Xian, X-z Zhao

Consultants: H-J Felber, X X Newhall, K G Steinert

2. Reports of the Commissions for 1983 to 1985

The President pointed out a correction in the report of the Commission in Trans. IAU 19A: line 8 on page 5 should read "the ratio of the mean solar day to the mean sidereal day". The President stated that he had asked for reports from the ephemeris offices, but the changes in the Almanacs would be discussed at the meeting of the Commissions on Tuesday, November 26.

3. Resolutions

At the Joint Discussion on Reference Systems on Wednesday, November 20, which was sponsored by Commissions 4, 7, 8, 19, 20, 24, 31, 33 and 40, two resolutions were adopted, one on astronomical constants, the other on reference systems. The Chairman of the Joint Discussion, J A Hughes, had referred these two resolutions back to the individual Commissions to obtain their separate approval since not enough time had been available for a full discussion at the Joint Discussion. After considerable discussion it was decided that the resolution on astronomical constants would be acceptable to Commission 4 if the words "physical and geophysical" were replaced by "geodetic", if the word "publish" was replaced by "provide for information purposes" and if the rider "5. to submit a preliminary report in 1987" was added at the end of the resolution.

There was also a long discussion on the resolution on reference systems. There was general agreement that item 4 in the resolution, which dealt with the organizational structure instead of the scientific aim should be deleted.

In addition P K Seidelmann commented that it would be appropriate if Commission 4 was given the responsibility for reference frames and he suggested that the Presidents of the interested Commissions should consider the matter.

It was agreed that the President should draw these proposed changes to the attention of the Resolutions Committee.

Report of Meeting 1985 November 26

PRESIDENT: T Lederle

1. The New Japanese Ephemeris

T Fukushima (Japanese Hydrographic Department) described the characteristics of the new Japanese Ephemeris which had been introduced in 1985. The main characteristics are that it uses the IAU 1976 system of astronomical constants, general relativity, a new planetary and lunar ephemeris and an "FK5-like" star catalogue. There are no apparent changes in the style of the published almanac except for the introduction of a rotation matrix to replace the Besselian day numbers.

The Einstein Infeld Hoffmann metric is used to derive the equations of motion which are applied to the Sun, Moon, eight major planets and four minor planets. The finite body interactions of the Sun, Earth and Moon and the effect of the Earth Moon tidal dissipation are also included in the new Japanese fundamental ephemeris (referred to here as JFE). In addition an analytical lunar libration theory has been used in the computations.

The initial conditions for the JFE are found by fitting over the period 1969 to 1981 the data from the JPL fundamental ephemeris DE118/LE62, which is equivalent to DE200/LE200 except that it is on the FK4 system, and the ephemeris of Duncombe for the minor planets.

The main difference between JFE and DE200/LE200 are that DE200/LE200 uses its own self-consistent set of planetary masses, it includes two more minor planets and it uses a numerical lunar libration theory. A comparison of JFE with DE200/LE200 shows differences of 1 mas or less. There were very small differences between the Japanese fundamental ephemeris (JFE) derived using the IAU 1976 planetary masses and JFE' derived the same way as JFE but using the DE200/LE200 planetary masses.

One conclusion that could be drawn was that the three new ephemerides published in the Japanese Ephemeris (JFE), the Astronomical Almanac (DE200/LE200) and the Connaissance des Temps (VSOP82/ELP-2000) are all equivalent to the precision printed.

The JFE has been extended to cover the period 1950 to 2050 and is available for export in machine readable form with software written in FORTRAN 77. The JFE has also been enlarged to include more planets and satellites and to contain physical information.

J O Dickey: Your comparison of the longitude of the Moon between JFE and DE200/LE200 shows a run-off at the ends which indicates that the wrong masses have been used.

P K Seidelmann: That type of characteristic could be obtained if libration were out of synchronization.

T Fukushima: We did synchronize lunar libration and in the comparison with JFE' the effect does not appear.

2. New developments with the JPL Ephemeris

J O Dickey (Jet Propulsion Laboratory) described the new developments at JPL on the DE125/LE125 ephemeris which is required for the Voyager 2 encounter with Uranus and the launch ephemeris for the Galileo mission to Jupiter. It is not intended as an export ephemeris. Other colleagues involved with the project but who are unable to be present are Standish, Williams, Newhall and Lieske. In addition to the observational data that were used to produce DE200/LE200 it incorporates new observations and a new integration. The new observations include transit circle observations from Herstmonceux and La Palma. Certain data were not included, but were used instead for verification. The optical navigation of Voyager demonstrates that it is dead on target.

T Lederle: JPL has the task and the need to compute the best possible ephemeris, but I am not sure that the ephemeris published in almanacs needs to be changed so frequently.

G A Wilkins: The published almanacs are consistent to the precision given.

T Lederle: The users are concerned at not knowing the basis of the ephemeris.

B Guinot: This matter will be considered by the new Working Group. The tendency is to impose too many rules on ephemeris producers when there should be more freedom.

S V Debarbat: We also need an ephemeris that we can use for long-term research, not one that lasts for three years.

The discussion turned to more general considerations about the publication of ephemerides and other information to meet the needs of observers. Wilkins raised the point that perhaps we had jumped to J2000.0 too soon and left others behind. Seidelmann suggested that the main problem was the lack of star catalogues for J2000.0. There were also problems with the transformation of star places from one equinox to another, and USNO would shortly produce a paper helping to solve this problem.

3. The Connaissance des Temps

B Morando discussed current ephemeris work at the Bureau des Longitudes. From 1984 onwards new ephemerides were introduced in the Connaissance des Temps. The Bureau is now working on analytical theories that are valid for long periods of time. Some of the results will be published shortly in Astronomy and Astrophysics. In particular the ephemerides have been used in climatic studies. New ephemerides of the satellite systems of Jupiter, Saturn and Uranus have been developed in a compressed form which combines Fourier series and Chebyshev polynomials. Ephemerides of minor planets brighter than magnitude 12 are available in machine readable form for observers who need them.

G A Wilkins: Sinclair and Taylor at the Royal Greenwich Observatory have done quite extensive work on the satellites of Saturn and more recently on those of Uranus.

4. The USSR Astronomical Yearbook

V Abalakin talked about the Yearbook which may be traced back to 1919 when it was a much smaller booklet. The basis of the 1986 volume changed with the introduction of the 1976 IAU system of astronomical constants and the use of DE200/LE200 which was supplied by JPL and USNO. The only noticeable change in appearance was the replacement of the Moon's hourly ephemeris with daily Chebyshev coefficients. He was grateful to Seidelmann, Lieske, Standish and Lederle for their help in introducing the changes. About 2,000 copies of the Yearbook are published and about 250 copies are distributed to other institutes. Explanations in English and in German are being produced. Numerical research into the libration of the Moon using Euler's coordinates is being undertaken at the Institute of Theoretical Astronomy.

5. The Apparent Places of Fundamental Stars (APFS)

T Lederle reviewed the history of APFS. It was Sadler and Clemence who first conceived the idea of an international almanac with USNO, Washington and HMNAO, Herstmonceux responsible for the Sun, Moon and planets, ARI, Heidelberg for the stars and ITA, Leningrad for the minor planets.

At present APFS is based on the FK4. All other changes recommended by the IAU were introduced with the 1984 edition. The 1988 volume will be based on the FK5 and the corrections FK5-FK4 will be included for the years 1984 to 1987. The introduction will be rewritten, but perhaps some of the translations into other languages will disappear. Comrie first introduced the multi-language versions in 1940 and this may be a suitable time to break the tradition.

C A Smith: APFS is invaluable for checking computer software results and it was remarkable how many users were disclosed when the publication was late to appear in the USA.

A Bandyopadhyay: We have not been able to compute our own fundamental ephemeris. I would be glad to have the opportunity for some of my staff to be able to work at other ephemeris offices.

6. Lunar Occultations

T Fukushima presented the first report of the International Lunar Occultation Centre (ILOC). The ILOC group now comes under the Maritime Safety Agency, Hydrographic Department, Geodesics and Geophysics Division. The Head of the Division is K Sugimoto. The ILOC will supply grazing predictions and limit line maps. Reduction was changed to the new reference system in 1985 but because of problems in convergence the (O-C)'s have not been returned to the observers yet. The number of observations has fallen from over 15000 in 1982 to just over 5000 in 1984 but the number of photoelectric observations has risen above 1000, mainly due to Japanese observers. A preliminary analysis of $\Delta\lambda$ and $\Delta\beta$ using the photoelectric observations shows a large systematic difference of between $0''\!.2$ and $0''\!.5$ in longitude and $-0''\!.17$ and $0''\!.25$ in latitude. This could possibly be due to an error in Watt's charts which are fitted to the centre of figure and not the barycentre.

7. Physical Ephemeris of the Sun

P K Seidelmann announced that Commission 10 had passed a resolution that Carrington's rotation elements for the Sun should be used. The Astronomical Almanac will return to using Carrington's values instead of those given by the working group on cartographic coordinates and the rotational elements of the planets.

COMMISSION 5: DOCUMENTATION AND ASTRONOMICAL DATA
DOCUMENTATIONS ET DONNEES ASTRONOMIQUES

Report of Meetings 21 and 27 November 1985

CHAIRMAN: W.D.Heintz

Activities of the Commission, its Working Groups, and the representation in CODATA and ICSTI were reported and discussed. The Organising Committee has approved G.A.Wilkins and B.Hauck, respectively, as president and vice-president; P.Wayman and G.Westerhout will replace Hauck and L.Schmadel on the OC, and nine new members raise the membership tally to 67. The major part of discussions was occupied by efforts toward a reasonably comprehensive resolution text on nomenclature (see "Designations", below, and IAU Resolution C3).

Late contributions supplementing the Commission Report: The Space Science Dictionary (J.Kleczek) is nearing completion. The body of the volumes, 5000 pages, has largely been typed by Dr. and Mrs.Kleczek; the alphabetical indexes are being currently prepared, and negotiations to arrange for publication are under way.

The International Colloquium on "Stellar Catalogues: Data Compilation, Analysis, and Scientific Results" convened in Tbilissi, September 10 - 15, 1984, organised by the Soviet Astronomical Datacenter, Abastumani Observatory, the Georgian Academy of Sciences, E.K.Kharadze (LOC Chair), and C.Jaschek (SOC Chair). Problems of catalogue compilation, datacenter activity and development plans were covered. It was appreciatively acknowledged that datacenter representatives of different countries do sometimes need to strengthen their scientific bonds by meeting in person and not only online. Publication of the proceedings in Bull. Astrophys.Obs. Abastumani 59 is expected in late 1985   (O.B.Dlushnevskaya).

A first report on vocabulary activity in China was presented by Miau Fu-He: Astronomy vocabulary defined as a long-term task (1922); translation committee established (1930); a five-language dictionary appeared in 1934, and new English/Russian/Chinese editions in the 1950's. English/Chinese glossaries (1974-85), including an edition by T.Kiang (Dublin), have now increased to 16 000 astronomical terms and seven appendices. The committee on astronomical terminology (founded in 1983, Chair Y.C.Chang) studies translation of terms, also from ancient Chinese books, and of references.

B.Corbin (USNO, Washington), "Commission 5 and Astronomy Libraries: A Cooperative Venture", commented on the importance of the "Astronomy and Astrophysics Abstracts", the bibliographic description of IAU publications, and the preservation of historical material. Plans are under way to digitally store 19th-century observatory publications on optical disks. B.Corbin also described the "Union List of Astronomy Serials", published in 1983. Toward the preparation of a 2nd edition, librarians worldwide are urged to supply additions and corrections. A meeting of astronomy librarians (the first since Hamburg 1964) will be arranged - at the occasion of the IAU GA in 1988 - between USNO, the Special Libraries Association, and R.Shobbrook (Anglo-Australian Obs., Epping) who points out that (1) the frequently isolated work of librarians needs more communication, (2) methods and procedures in libraries, particularly where applications of new technology are concerned, are to be discussed, (3) improvements in the collection, organisation, and dissemination of information, as fostered by the contacts between librarians and scientists through Commission 5, will be benefits for research.

COMMISSION 5

Working Group on Astronomical Data (IAU Commission 5)

Report of Meeting 26 November 1985

CHAIRMAN: B. Hauck   SECRETARY: A. Heck

Chairman's Report: B. Hauck summarised the WG activity over the past three years, especially in relation with CODATA work. The proceedings of the Trieste School on "Data Handling in Astronomy" (1984) have been published in Mem.Soc.Astr. Ital. 56 (1985). Subsequent discussion emphasized the two-way learning process of the relationship with CODATA.

Officers and Members: The new Organising Committee has been elected as follows: O.B.Dlushnevskaya, B.Hauck (ex officio), A.Heck, C.Jaschek, H.Jenkner, G. Lynga (Vice-Chairman), S.Nishimura, A.G.D.Philip, W.Warren, G.Westerhout (Chairman), and G.A.Wilkins (ex officio). - Continuing members are W.P.Bidelman, R.Duncombe, P.Grosbol, W.D.Heintz, J.Mead, L.E.Pasinetti, F.Spite, Y.Terashita.

The following were proposed and accepted as new members: V.K.Abalakin (Pulkovo), L.Benacchio (Padova), M.Bessel (Mt.Stromlo), H.R.Dickel (Los Alamos), K.F. Hartley (RGO), H.Jenkner (STScI), J.C.Mermilliod (Lausanne), A.Pamyatnikh (Moscow) G.Sedmak (Trieste), A.Uesugi (Kyoto), D.C.Wells (NRAO).

The meeting expressed its gratitude to the retiring chairman B.Hauck for his dynamic and efficient leadership over the past six years.

Microfiche on Standard Stars: A.G.D.Philip informed the assembly that the first version of the microfiche has been included in the proceedings of IAU Symposium 111, just published by Reidel Co. The second edition is planned for March 1986 and will be published by CDS.

Miscellaneous: Priorities for the next three years were discussed, in particular the integration of radio data into the data banks.

SCIENTIFIC MEETING A: Problems due to the large amount of data from Space Astronomy.

CHAIRMAN: G. Westerhout.

J.Mead: Coping with the IUE and IRAS data from the user's point of view. Data products and software routines developed to handle the large amounts of data from the International Ultraviolet Explorer (IUE) satellite and Infrared Astronomy Satellite (IRAS) were described. These products include cross identifications of infrared objects, special atlases of ultraviolet objects, and bibliographical indices of UV and IR objects from a search of the literature.

H.Jenkner: Comments on astronomical data related to the Hubble Space Telescope. Space Telescope data -- ranging from proposal, engineering, and spacecraft data to the various kinds of scientific data -- will amount to 2.5 Gbyte per day or 13.5 Tbyte over the ST lifetime. Other ST related databases already under development are the Guide Star Catalogue (about 30 million objects), the Guide Star Photometry Catalogue (9000 stars), and the Guide Star Image Archive (1500 plates, 14 000 x 14 000 each). - In the following discussion A.Heck informed the meeting that a conference will be held in autumn of 1987 at the ST-ECF in Garching (FGR) on "Astronomy around large data bases: Scientific objectives and methodological approaches", dealing essentially with the statistical methodology applicable to reduced data.

SCIENTIFIC MEETING B: Astronomical Catalogues - Transfer and Archiving.

CHAIRMAN: B.Hauck.

P.Grosbol: Extension of FITS to catalogues. During the last three years the Data Exchange Task Force has carried out a project of an extension of FITS (Flexible Image Transfer System) to tables and catalogs. The detailed proposal is published in the proceedings of the Trieste Seminar (Mem.Soc.Astr.Ital. 56, 437,1985). This extended format has been verified by exchange of catalogues between NASA WDC-A, NRAO, and ESO.

W.Warren: Can we use FITS for catalogues? The FITS format is suitable for distributing machine readable catalogs on magnetic tapes, although the format introduces certain constraints in the flexibility of tape characteristics and documentation. We intend to prepare FITS headers for major catalogs as time permits, but will retain the capability to distribute catalogs as simple character coded files as at present.

W.Warren: General philosophy concerning modification and archiving of astronomical catalogs. We believe that the primary roles of astronomical data centers are to acquire catalogs as produced, to prepare machine versions, to quality assure them for completeness, homogeneity, and standard presentation of data, to add adequate documentation (in particular when an original catalog is modified), correction of errors in consultation with authors, error-alert previous recipients of the material, and to arrange dissemination with adequate accompanying information. - In a later discussion, Warren specified archiving conditions as: easy-to-use character form, uniform formatting of the data body, availability of documentation on revised and corrected data, and (a FITS requirement) separation of text from data. Further guidelines on the cooperation between authors and data operators are deemed desirable.

G.Westerhout: Data networks and communications. The most common method for communication is through a modem using ordinary voice networks. Data exchanges through packet-switching networks encounter still difficulties to cross borders. Special astronomical networks are available in some countries. Remote control of telescopes is also implemented. Various examples were cited, also in fields other than astronomy.

A.Heck: Strasbourg meeting on a European astronomical data network. In November 1985 a highly successful meeting was held at CDS (Strasbourg) to review the various achievements and plans with respect to establishing national and international astronomical data networks (see Proceedings in CDS Bull. 30). An "International Directory of Professional Astronomical Institutions" with listing of institutional computer access identifications is being prepared at CDS.

TASK FORCE ON ABSTRACTING GUIDELINES.

The Commission met briefly on 23 November to consider the preliminary report by L.Schmadel (absent) as published in Bull. CDS 28, 95 (1985), which will be considered for the new IAU Manual. R.Wielen (Heidelberg) reported on the current status of the Astronomy and Astrophysics Abstracts (AAA), and W.Lueck (Karlsruhe) on the services and near-future plans of the Fachinformationszentrum Physik. P.Lantos has his vocabulary version completed, save for some more concordance with the AAA version.

WORKING GROUP ON DESIGNATIONS (IAU Commission 5)

Report of Meeting 19 November 1985

CHAIRMAN: F. Spite

Contributions and discussions are being prepared for publication in a forthcoming CDS Bulletin. C.Jaschek had previously contacted the IAU Commissions concerning the problems encountered in the proliferation of celestial-object designations. The major part of the meeting was devoted to drafting designation guidelines and caveats. Through appreciable efforts particularly by F.Spite and by H.Dickel, on behalf of the WG Nomenclature of Commission 34, the draft went through several subsequent, informal meetings (with members of Commissions 28, 37, and 40 participating) and several modified versions, until finalised by Commission 5.

JOINT SESSION OF COMMISSIONS 5, 38, 46: DEVELOPING ASTRONOMICAL INSTITUTIONS

Report of Meeting 25 November 1985

CHAIRMAN: J.-C. Pecker

The IAU EC has suggested that the problems be considered, with which new centers of research (particularly, though not exclusively, in "developing countries") are confronted.

The major needs, and burdens on the budgets, are past and present books, serials, and abstract publications. Microfilm copies are of limited availability. It was noted that reduced "individual" serials prices are sometimes already granted to third-world libraries; further contacts through organisations (ICSU) with publishers are suggested. The continuation and updating of the Serials List was encouraged, and so was the formation of national bibliographic centers; existing centers have already shown their usefulness through providing literature documentation and access. Long mail delays of serials shipments (sometimes inevitable due to geographic distances, but sometimes surprising and excessive) are an aggravating impediment. Information about preprints (as in Astrophys.J.) is thus appreciated.

The shortage of instruments makes an international trading post for (unused but available) equipment feasible; the IAU will be approached to make the IAU Bulletin accessible to "Wanted" ads. Travel support is more difficult to get for the instrument technicians than for researchers. Attention was called to the Newsletter of IAU Commission 46 on educational materials.

The services provided by scientific and bibliographical databanks often encounter shortcomings at the receiving end, i.e., through the lack of linking datacenters in many countries, and through the incompatibility of computers and tapes. A Working Group within IUGG is already considering this problem. - The agenda being unfinished, the topic - with review of possible progress, and exploration of administrative avenues - is intended to be resumed at a future meeting.

COMMISSION 6 : ASTRONOMICAL TELEGRAMS
(TELEGRAMMES ASTRONOMIQUES)

Report of Meeting : 25 November 1985

PRESIDENT : M.P. Candy

SECRETARY : B.G. Marsden

The President announced the nominations of A. Mrkos and E. Roemer as President and Vice-President for the triennium 1985-1988, and these were confirmed by the members present. Although the Commission has traditionally not had an Organizing Committee, it was felt appropriate that B.G. Marsden, as Director of the Central Bureau for Astronomical Telegrammes, should now serve in this capacity. S. Nakano was elected a new member of the Commission.

Supplementing the information in the report in Transactions XIXA, Marsden affirmed that, after reaching a low point in 1983, there appeared to have been some improvement in the speed with which Circulars were received by subscribers ; it had been necessary to increase the 'regular' and 'special' subscription rates in late 1984 to 65c and 39c, respectively. On the other hand, the introduction of the 'dial-in' computer service had been in many respects the most significant event of the past triennium, and to encourage further use of this service there had been a reduction in the daily charges from 30c to 20c in late 1984. Following consultation with the Finance Committee, it had been agreed to restrict the request for IAU financial support during the upcoming triennium at the level of the preceding one, and this request was approved by the Executive Committee.

The Commission considered and supported a proposal by former President P. Simon that the IAU cosponsor a symposium on "The Contribution of Amateur Astronomers to the Discovery of the Universe", to be held in Paris in June or August 1987.

Nakano gave an account of the manner in which he makes use of the IAU Computer Service and how this led to his development of a similar computer service for the Oriental Astronomical Association. The O.A.A. service helps coordinate the observational and computational activities of many amateur astronomers in Japan, and it also provides immediate linkages to the Tokyo Observatory and back to the IAU Central Bureau.

Marsden gave a general description of the IAU Computer Service, which also involves Commission 20 and the Minor Planet Center. Noting that Nakano was indeed one of its most frequent users, he remarked that the resulting multi-way interaction with the Japanese astronomers was an excellent example of how astronomers ought to be conducting business in the 1980's. There exists the unfortunate problem of the general incompability of modem protocol in different countries, and frequent international telephone calls can soon of course become a prohibitive expense. P.B. Boyce suggested that use might be made of a linkage to GTE Telenet via the American Astronomical Society and the American Institute of Physics. On the other hand, while this linkage, as well as others of its type, can be very useul if only an electronic-mail system is required, the IAU Computer Service also involves logging in to a computer and actually performing computations on it. By making the computer a Telenet node the problem of international incompatibility could be removed, but that of expense clearly remained.

COMMISSION 7 : CELESTIAL MECHANICS (MECANIQUE CELESTE)

Reports of meetings on November 22 and 25

PRESIDENT: J. Kovalevsky   SECRETARY: J. Henrard

BUSINESS SESSION (November 22)

The President opened the session by paying tribute to the memory of two deceased members of the Commission: Professor C.L. Siegel and Doctor J. Meffroy. He also greeted Prof. B. Garfinkel who has just celebrated his 81-st birthday.

OFFICERS AND MEMBERSHIP OF THE COMMISSION
  After discussion and some proposals from the floor, the following list was moved and unanimously approved :
- President : V.A. Brumberg, the retiring Vice-President.
- Vice-President : J. Henrard.
- Organizing Committee : Ju.V. Batrakov, K.B. Bhatnagar, J. Chapront, A. Deprit, S. Ferraz-Mello, J.D. Hadjidemetriou, H. Kinoshita, J. Kovalevsky, H. Scholl, P.K. Seidelmann, A. Sinclair and Z.H. Yi.

  Three members of the IAU were appointed members of the commission between the general assemblies: M.F. He (China), G.A. Krasinsky (USSR) and A.M. Nobili (Italy). The following members of the IAU were proposed to become members of Commission 7 and were approved: B.D. Jovalovic (Yugoslavia), J. Osorio (Portugal), A. Pal (Romania), M.J. Valtonen (Finland) and M. Yuasa (Japan).

  The following new members of the IAU have requested to become members of Commission 7 and were approved after a short presentation: M. Ahmed (Egypt), B. Barberis (Italy), F. Boigey (France), N. Caranicolas (Greece), Z. Chen (China), K.H. Choi (Korea Rep.), C.F. Cui (China), H. Din (China), A. Drozyner (Poland), A. Elipe-Sanchez (Spain), S.M. Fernandez (Argentina), S. Ferrer Martinez (Spain), C.G. Fong (China), D. Galletto (Italy), A. Gonzalez Camacho (Spain), A. Hanslmeier (Austria) T.Y. Huang (China), A. Journet (France), A. Lemaitre (Belgium), B.K. Lu (China),S. Mikkola (Finland), C.D. Murray (U.K.), Y.S. Sun (China), C. Veillet (France), J-J. Walch (France), I.W. Walker (U.K.), E. Wnuk (Poland) L.D. Wu (China), P.X. Xu (China), H. Yoshida (Japan), S.P. Zhang (China), J.Q. Zheng (China), X.T. Zheng (China), H.N. Zhou (China) and W.Y. Zhu (China).

  The 1982-85 list of consultants was reviewed. For the new term, eight consultants were presented and approved. They are: V.R. Bond (USA), P.P. Hallan (India), J. Moser (Switzerland), D. Saari (USA), S.K. Shrivastava (India), C. Simo (Spain), M. Soffel (Fed.rep. of Germany) and J. Waldvogel (Switzerland).

IAU COLLOQUIUM
  The IAU executive committee has approved a colloquium by Commission 7 together with commissions 20, 26, 33 and 37. The title will be "The few body problem". It will be organized in June 1987 in Turku (Finland) by P. Valtonen. The chairman of the scientific committee is V.A. Brumberg.

RESOLUTIONS
  The joint discussion on reference frames has voted two resolutions. The first resolution sets up a working group on reference frames. As it is of the interest of Commission 7 to be represented in such a working group, V.A. Brumberg and J. Kovalevsky were proposed to become members. The second resolution calls for a wor-

king group on the best current values of various constants. Commission 7 nominated P.K. Seidelman to be its representative in the working group. In addition, it is remarked that the working group should not concern itself about physical and geophysical constants, but only on astronomical and geodetic constants.

REPORTS ON MEETINGS

V.A. Brumberg reported on the IAU Symposium 114 "Relativity in Celestial Mechanics and Astrometry" held in Leningrad on May 28-31, 1985. It was the first IAU meeting devoted exclusively to relativistic problems of Celestial Mechanics and Astrometry, even if some such problems were partly treated in other meetings. Since the relativistic treatment of the motion of celestial bodies is useful only if the related Newtonian contributions are discussed with the necessary high-level accuracy, high precision dynamical theories were also included in the program.

There were 80 soviet and 58 foreign participants including astronomers engaged in relativistic problems of Celestial Mechanics and Astrometry, specialists of classical but very high precision approaches to these sciences and physicists interested in applying their methods to practical problems. The proceedings will be published by D. Reidel Company in the IAU Symposium series. They will appear in February 1986.

V.G. Szebehely reported on the first meeting dealing with Celestial Mechanics held in India: the workshop was held in Delhi just before the IAU General Assembly. There were 130 indian and 27 foreign participants; 48 papers were presented during 8 sessions. A great variety of topics were addressed ranging from qualitative to quantitative problems and from artifical satellite mission analysis to regularization. The proceedings will be published by D. Reidel Company.

REPORT ON CELESTIAL MECHANICS IN CHINA

Yi Zhao-Hua reported on the contribution of chinese astronomers in Celestial Mechanics during the last four years.

1. <u>Numerical studies of dynamical systems</u>. Sun Yi-Sui continued his studies in this field and obtained a series of results on dynamical systems with even and odd dimensions. In particular new results were obtained in the case of odd dimensions.
2. <u>General and restricted three body problem</u>. Sun Yi-Sui together with C. Marchal and J. Yoshida obtained a new test for escape in the general three body problem. Huang Tian-Yi and his collaborators dealt with stability region of the planar three body problem and applied the results to the solar system. Yi Zhaohuc and J.H. Jefferys used the Liapounoff characteristic numbers to study the region of stability of the restricted three body problem. Ding Hua and Tong Fu found periodic orbits of the first kind for the restricted three body problem with an ellipsoid.
3. <u>Perturbation and resonance theory</u>. Liu Lin, Zhang Sheng-pan and Zhao De-Zhi presented an interesting method to study high order perturbations that can be used for artificial satellites. Recently, together with F.A. Inanen, he established a new theory of resonance different from those of Garfinkel and Schubart and applied it in the case of the solar system.
4. <u>Dynamics of the solar system</u>. Tong Fu derived analytically the precise post-Newtonian terms for all major planets. Zhang Jia-xiang and Li Gang-yui gave the secular perturbations of many asteroids. Liu Lin studied the resonance phenomena of asteroids and of satellites of Neptune.
5. <u>Other subjects</u>. Huang Tian-yi, K.L. Innanen and Zhou Qing-lin obtained some interesting results by using integral invariants to check numerical computations. Zhan Hong-nan et al. used numerical methods to study the evolution of galaxies and close binaries, while Zheng Xue-tang et al. studied the velocity dispersion of galaxies.

CELESTIAL MECHANICS

SCIENTIFIC SESSION (November 22, chaired by J. Kovalevsky)

V.G. Szebehely presented a paper on binary asteroids. Several theoretical and observational approaches have been reported in the literature since 1966 to the present time. One of the best references is F. Whipple's paper presented in 1982 at the IAU meeting in Patras.

The theoretical models are the restricted problem of three and many bodies. The actual dynamical system consists of the Sun, Jupiter, Asteroid and its Satellite. If the effect of Jupiter is neglected the limiting distance between the asteroid and its satellite is given by:

$$d = (\mu/81)^{1/3}$$

where μ=(Mass)asteroid/(Mass)Sun and d is in nondimensional units to be multiplied by the distance of the asteroid from the Sun. The above equation for the limiting distance for stability may also be written as:

$$d < 30 \, R \, l$$

where R is the radius of the asteroid in km and l is the distance of the asteroid to the Sun in a.u. The many assumptions made in the derivation of the above equation restrict seriously its applicability, therefore, better models are presently being developed. These models are known as the restricted problem of many bodies. Analytical solutions are not known but the generalization and the extension of the Jacobian integral has been established. Regarding the questionable observational results, we mention Lucina and Herculina which might have satellites inside the stability limit given by the above equation.

- A. Deprit presented a paper entitled "The critical inclination in artificial satellite theory" co-authored with S.L. Coffey and B.R. Miller.

After a Delaunay normalization to average the Hamiltonian over the mean anomaly, invariance with respect to the group of rotation about the polar axis is used to assimilate the flow on the manifold of constant L and constant H to a flow on a two-dimensional sphere in a three-dimensional space (ξ,η,ζ). Expressed in the averaged Delaunay elements (G,H,L,g,h,l) the coordinates are:

$$\xi = LGe \, \sin l \, \cos g, \quad \eta = LGe \, \sin l \, \sin g, \quad \text{and} \quad \zeta = G^2 - \frac{1}{2}(L^2+H^2);$$

the sphere's equation is: $\xi^2 + \zeta^2 + \eta^2 = \frac{1}{4}(L^2-H^2)$.

At any point in the interval $0<H<L$ the flow on the sphere presents two isolated singularities: S_0 at the North pole ($\xi=\eta=0, \zeta>0$) corresponding to the circular orbits, and S_5 at the South pole ($\xi=\eta=0, \zeta<0$) corresponding to the equatorial orbits. In addition, when only the terms in J_2 are retained, the flow admits, as long as $0<H<L/\sqrt{5}$, a continuous manifold of equilibria lying on a small circle of latitude representing the orbits with stationary perigee at critical inclination.

However, the second order perturbation in the reduced Hamiltonian removes the degeneracy to leave only four isolated orbits with stationary perigees: two of them, say S_1 and S_3, such that the average perigee lies on the average line of nodes, the other two, S_2 and S_4, keeping their perigee at right angle with the line of nodes. The orbits S_1 and S_3 are stable whereas S_2 and S_4 are unstable.

A detailed analysis reveals that, as H approaches $L/\sqrt{5}$ from below, all four equilibria S_1, S_2, S_3 and S_4 move toward S_0. A bifurcation occurs at a value $H=L/\sqrt{5}-O(J_2)$ when the merging of S_1 and S_3 with S_0 results in S_0 regaining its stability. Extensive numerical integrations of the differential equations including first and second terms in the perturbations fully confirm the mathematical deductions.

CELESTIAL MECHANICS IN INDIA (November 22, chaired by K.B. Bhatnagar)

Five papers, representative of the indian achievements in Celestial Mechanics were presented during this session prepared by K.B. Bhatnagar.

- "The effect of perturbations in Coriolis and centrifugal forces on the non-linear stability of equilibrium points in the restricted problem of three bodies" by K.B. Bhatnagar and P.P. Hallan.

The non-linear stability of the equilibrium points in the restricted problem has been studied when small perturbations ϵ and ϵ' are given to the Coriolis and the centrifugal forces respectively. It is found that the collinear points are unstable and the triangular points are stable for all mass ratios in the range of linear stability except for three mass ratios.

- "Stability of the particular solutions for the restricted as well as unrestricted problem of three bodies" by R.K. Choudhry.

Under the joint authorship with Smt. Nanju Kumari, we studied the stability of the triangular points of libration for the restricted problem of three bodies when the effect of solar radiation pressure is also taken into account. We have investigated the stability for all orders and all times. We took into consideration the resonance case also. For the unrestricted problem of three bodies we have taken the equations of motion in Whittaker's form and we have examined the stability of Lagrange's particular solutions in full detail taking into consideration the effect of light pressure as well. Under the joint authorship of Dr D.N. Singh, we have investigated the stability of the libration points for the generalized restricted problem of three bodies when the shape of the infinitesimal mass is an oblate spheroid. Here we have considered the stability only of the first order. We have also found the existence of such points of three bodies when one of the finite bodies is taken to be a radiating body. Here we have not considered the stability of libration points. In the latter problem we get coplanar libration points in addition to the triangular and the colinear libration points.

- "The existence and stability of the equilibrium points of a triaxial rigid body moving around another triaxial rigid body" by K.B. Bhatnagar and Mrs Usha Gupta.

The motion of two mutually attracting triaxial rigid bodies has been considered. Thirty six particular solutions corresponding to the libration points analogous to the points Spoke, Arrow and Float (Duboshin, 1959) have been found. The stability of these libration points has been discussed in two categories of cases. In the first category, different shapes of the bodies have been taken and in the second category, the mass and the linear dimensions of one of the bodies have been taken small in comparison to the other.

- "Influence of planets on the Sun" by S.D. Verma.

A fundamental and basic model is developed in which the influence of planets, mainly Jupiter, Venus, Earth and Mercury on the Sun is evaluated. There is observational support for the model, which spans from few months to hundred of years. The simplicity and the beauty of the model is a point to discuss together with the convincing experimental observations. In turn these effects on the "Sun", cause changes in the solar atmosphere which easily extend to the orbit of Earth. They produce an influence on the Earth's environment (ionized and neutral atmosphere) especially in the Northern and Southern polar regions of the Earth.

- "Attitude dynamics and control of artificial satellites" by Shashi K. Shrivastava.

The motion of a spacecraft presents two dynamical aspects of interest: the trajectory of its center of mass, and the rotational motion about its center of mass, commonly referred to as libration or attitude motion. There are numerous situations of practical importance such as communications, scanning of cloud coverage, Earth resources survey, astronomical observations, etc... where it is necessary to maintain a fixed orientation with respect to the Earth or a given direction in space. There are many sources of environmental and internal perturbations which tend to disturb the attitude of the satellites. To overcome these disturbances a wide range of attitude stabilization and control concepts has been proposed over the years and several concepts have found practical applications. Broadly these may be classified as active, passive and semi-passive procedures.

Over the past 30 years, several hundred papers have appeared which deal with various aspects of attitude dynamics and control of satellites. Currently most of the attention is focussed on dynamics and control of large flexible spacecrafts. This paper presents an overview of studies on attitude dynamics and control of satellites. Sources of environmental and internal disturbances and the concepts behind various methods of attitude control are discussed. The presentation also includes a brief discussion on dynamics and control of large flexible spacecrafts along with some results of the related studies by the author and his associates. A few important and challenging areas needing researcher's attention are identified.

UNSOLVED PROBLEMS IN CELESTIAL MECHANICS
(November 25, chaired by A. Deprit and U. Gupta)

The objective of this session was to list a certain number of problems in various aspects of Celestial Mechanics and related fields (such as reference frames and observations) that are not solved and to discuss their relative importance as an incentive to think about areas that really need attention.

- P.K. Seidelmann discussed the problems that are connected with the study of the solar system and grouped them in six general topics.
1. <u>Discrepancies between observations and ephemerides</u>. The present Uranus predicted ephemerides deviate from observations by about $0\overset{"}{.}5$. A systematic difference exists also in Saturn declination and there exists a seasonal effect in solar right ascension. There still exist systematic errors in correcting planetary phases and satellite separations. For faint objects, a good reference catalogue of faint stars is badly needed. Finally, a possible error of 1" per century in the solar observations can have several causes (equinox motion, constant of precession, longitude of the Sun, etc.). It still remains an unsolved problem.
2. <u>Formation and stability of the solar system</u>. Many problems remain unsolved or the solutions proposed are not satisfactory. The history of the Earth-Moon system or of the Neptune-Pluto system and more generally the very long time period motions; the non-gravitational effects on comets, the interactions between minor bodies of the solar system and more generally the explanation of the distribution of these bodies; the problems connected with the existence, the evolution of planetary rings and their interactions with satellites. One may add to these problems, the relation between chaotic motion theory and the reality which may lead to a partial long term unpredictability of the motions.
3. <u>Resonances</u>. They are largely present in the solar system. Why these resonances exist, or do not exist? How they were built up. In connection with resonances are the problems of Kirkwood gaps, families of minor planets and rings of Saturn.
4. <u>Theory</u>. A number of types of representation of general theories (Fourier series, Chebyshev polynomials, rational functions, elliptic functions, etc..) exist but not all are sufficiently well studied to assess their usefulness.

Important problems are the construction of general theories to a designed accuracy, of compact very long period theories and of very accurate theories with the relativistic effects.
5. <u>Reference systems</u>. The present important problems refer to the practical realization of ideal systems and the relationship between such realizations (in particular the space fixed versus inertial references). Several definitions are to be clarified (equinox, obliquity, inertial, mean, proper versus dynamical time).
6. <u>Artificial satellites</u>. An accurate modelling of all forces acting on a satellite is still a problem. It governs the long time period accurate prediction and the determination of various physical parameters.

- J. Kovalevsky presented a paper co-authored with V.A. Brumberg. Every problem is considered under four different aspects: its physical bases, the mathematical aspects, the computational techniques and the astronomical objectives. Some of these aspects might be trivial and others might present major difficulties or different importance and oldness. Three main fields were considered.
1. <u>Relativistic Celestial Mechanics</u>. The application of the PPN formalism to all actual motions is the most important present problem. The tools have been developed by physicists, but their applications to the solar system are far from being completed. The main theoretically and practically unresolved problem is the general rotational-translational motion of a finite body. Two mathematical problems deserve attention: the problem of relativistic resonance (solution developed in fractional powers of the relativistic parameters) and building up perturbation theory in general relativity. Finally, an important astronomical aspect is the construction of reference frames in astronomy and geodynamics in a general relativistic context.
2. <u>Newtonian Celestial Mechanics</u>. The main problem is evidently the N-body problem. The qualitative behaviour is still unknown for N 3. Many questions are to be answered concerning the structure of the phase space, the ergodicity of some regions, stability, the existence of families of periodic orbits, the strange attractors, etc... A whole field of separating the predictable results in overall non deterministic long term trajectories is opening now. Many of these problems are to be dealt within simplified cases. At this point, the importance of the problem in Celestial Mechanics may be questionned and should rather refer to the study of dynamical system. Difficult numerical problems exist in long term integration, the construction of litteral solution and all the studies dealing with the dynamics of complex stellar systems.
3. <u>Evolutionary problems</u>. Very modern, difficult and unsufficiently studied problems concern the evolution of the solar system, of clusters and galaxies. Non-gravitational effects have to be introduced (tidal effects, interaction in planetary rings, cometary dynamics, mass loss and depletion in clusters). Even the physical background of these problems is still to be studied and the mathematical aspects are not satisfactorily dealt with. Problems like capture into resonance, very long term evolution of a planetary or satellite system or the escape and multiple star formation in clusters present a number of unsolved difficulties. Other problems like the evolution of double stars exchanging matter or with strong magnetic interaction or other wild force fields are studied only by astrophysicists: they could be the source of major and useful activities for celestial mechanicists - who have built up very powerful tools that are unsufficiently used on problems of astronomical interest.

During the discussion, A. Deprit suggested a few other fields of interest to Celestial Mechanics: autonomous navigation on satellites including real time reduction, and the use of the possibilities offered by the new generation computers.

- J. Henrard presented a paper on the Emerging Problems in Resonance Theory and concentrated on two specific problems.
1. <u>Planetary rings</u>. It is certainly an important problem in our understanding of the evolution of the solar system and its dynamical aspects are numerous and

varied. The main questions connected with resonance are at present: formation of the Cassini division, density waves, sharp edges and the theory of shepherd satellites. The first difficulty about dynamical aspects of planetary rings is that it is not clear how to model the various forces acting on them. A ring can be considered in a first approximation as a collection of particles or as a continuous medium leading to very different models. The theory of shepherd satellites leads to perturbations of a type quite unusual in Celestial Mechanics. Two mathematical models are presently proposed to deal with this aspect: Hill's problem (Henon, Waldvogel) and the stream-line (Borderies, Goldreich, Tremaine).

2. <u>Chaotic motions and the double resonance problems</u>. The fact that chaotic motions can be of importance in Celestial Mechanics was clearly revealed by the paper of Wisdom, Peale and Mignard about the rotation of Hyperion. Chaotic motions is also at the base of one of the possible explanation for the formation of Kirkwood gaps. Wisdom has shown numerically that the 3/1 resonance gap associates with chaotic motions and has proposed a mechanism explaining the appearance of such motion: the periodic crossing of the critical curve of the circular averaged restricted three body problems forced by the perturbation due to the eccentricity of Jupiter. Such motion was also described by Froeschlé and Scholl.

This problem is related to the two resonance problem which can be described by the Hamiltonian function :

$$H = H_0(P,Q) + \varepsilon H_1(P,Q,p) + \varepsilon' H_2(P,Q,p,q)$$

If we assume that $\partial H/\partial P$ vanishes along a curve $f(P,Q)=0$, we say that we have a resonance problem. When $\partial H_0/\partial Q$ does not vanish the angular variable q can be averaged out and the problem is approximated by a one degree of freedom problem But when $\partial H_0/\partial Q$ vanishes also in the domain of interest we say that we have a double resonance problem. This is the case in the elliptic restricted three body problems. When ε' is much smaller than ε, one may consider introducing action angle variables (J,ψ) for the $H_0+\varepsilon H_1$ problem and the Hamiltonian H then reduces to :

$$K = K_0(J,Q) + \varepsilon' K_2(J,Q,\psi,q)$$

Of course the action angle variables are singular along the critical curves of the one degree of freedom $H_0+\varepsilon H_1$ problem. This will lead to chaotic motion if the orbit of K crosses periodically one of these critical curves. If certain conditions are met, the Hamiltonian K can be studied by perturbation methods. Hopefully the approximations so obtained will describe meaningfully the very long time behaviour of the system.

A more general classification of double resonance classes of problems is needed to describe and predict the onset of chaotic motion. This kind of problem has already received much attention in the fields of dynamical systems, plasma physics and particle physics. Celestial Mechanics may yet trace a new path in the study of these phenomena.

<u>SCIENTIFIC SESSION (November 25, chaired by J. Kovalevsky)</u>

- P. Farinella "Numerical simulations of asteroid families", co-authored with Ch. Froeschlé, Cl. Froeschlé, R. Gonczi, M. Carpino, P. Paolicchi and V. Zappalà.

The accuracy and reliability of the proper orbital elements used to define asteroid families are investigated by a new method, that is by simulating numerically the dynamical evolution of families assumed to arise from the "explosion" of a single object. The equations of motion of the simulated family asteroids are integrated in the frame of the elliptic restricted three body problem Sun-Jupiter-asteroid, over times of the order of the circulation periods of perihelia and nodes. By filtering out short-periodic perturbations, the behaviour of the proper eccentricities and inclinations is monitored. Significant long-period variations have been found especially for families having non-negligible eccentricities and (or) inclinations (like the Eos family), and strong disturbances due to the proximity of commensurabilities have been evidenced (e.g. the Themis family). We conclude that further dynamical studies on the collisional origin of families and on the properties of their parent bodies must probably await the development of more accurate and reliable secular perturbation theories.

- Yi Zhao-hua: "On the qualitative study of the N-body problem", co-authored with Wong Qiou-dong.

A transformation of the phase space in the N-body problem has been established. It is suited to study the cases when time tends to infinity or to any collision moment. The following results were obtained.
a) The existence of a global solution departing from any initial condition which corresponds to any ordinary point in the original space was proved. The solution is analytic in all axes.
b) An extension of McGehee's total collision manifold of the N-body problem was given, which includes all the cases of k-tuple collisions as well as the total collisions. We showed that in order to understand the character of motions near a collision is equivalent to understand the characters of motion in the case h=0 of the transformed system.
c) This transformation was used to discuss qualitative properties of the isoceles three body and trapezoidal four body problems. The singular manifold obtained is compact and contains all singular motions corresponding to escape, super hyperbolic or tending to a collision or a libration point.

- D. Galleto: "On the collapse of a homogeneous incoherent cloud of matter".

It is observed that if an isolated homogeneous incoherent cloud of matter is radially collapsing towards one of its elements and if the homogeneity is preserved during the collapse, the cloud is also radially collapsing towards any other element of the cloud. The shape of the cloud is necessarily spherical and the forces at distance are necessarily those expressed by the Newton's law of gravitation or those of an elastic type. Furthermore, a principle of impotence follows.

THE JOURNAL "CELESTIAL MECHANICS" (November 25, chaired by J. Kovalevsky)

M.S. Davis, the new executive editor of the journal "Celestial Mechanics" presented some of the problems related to the journal. They are: the length of time between submission and publication of a paper, the low rate of rejection which may be due to the fact that some editors feel compelled to help the authors rewrite their papers, the new typography, the editing of grammar and spelling for non english speaking authors, the need of an Index, a possible publications of letters.

Some of the problems may be alleviated soon. M.S. Davis talked with the publishers: the typography will be improved soon, Reidel is willing to take charge of the editing concerning the english grammar and spelling. The executive director will send a memorandum to editors and prepare instructions and a questionnaire for reviewers should be sufficient. As he is in the process of computerizing the files of Celestial Mechanics, an Index composed of key words provided by the authors should be forthcoming. The publication of letters does not seem to correspond to a real need.

COMMISSION 8. POSITIONAL ASTRONOMY (ASTRONOMIE DE POSITION)

PRESIDENT: J.A. Hughes   VICE PRESIDENT: Y. Requieme

ORGANIZING COMMITTEE: G. Billaud, G. Carrasco, W. Fricke, V.S. Gubanov, E. Høg, B.L. Klock, J.A. Lopez, Luo Ding-jiang, M. Miyamoto, L.V. Morrison, I. Nikoloff, G. Teleki, Y.S. Yatskiv

INTRODUCTION

The activities of the commission during the XIX GA are chronicled below. The Joint Meeting on HIPPARCOS (7, 8, 24, 25, 33 and 37) and those on Earth Rotation Parameters (7, 8, 19 and 31) are covered elsewhere. The reporter (and now past-president) wishes to thank the commission members who made these activities very worthwhile, and to especially acknowledge the various chairmen and secretaries. In the latter category, L.V. Morrison and C.A. Smith are particularly noteworthy.

Business Meeting, 21 November 1985, 0900-1030
Chairman: J. Hughes

The results of the commission elections were certified by the president and accepted by the members. The new officers of the commission are: President: Y. Requieme; Vice President: M. Miyamoto; Organizing Committee: P. Benevides-Soares, D.P. Duma, L. Helmer, J. Hughes, L. Lindegren, Luo Ding-jiang, F. Noel, G.I. Pinigin, L. Quijano, S. Sadzakov, H. Schwan, C.A. Smith and M. Yoshizawa.

In the absence of the chairman of the SRS committee, W. Fricke, the report of that group was read by committee member J. Hughes. The report consisted of the text of the agreements concluded at a meeting held at the Pulkovo Observatory in June, 1985. Additional members include T. Lederle, D.D. Polozhentsev, V.A. Fomin and C.A. Smith. The cited text follows.

The SRS Committee recommends:

1. That the Pulkovo Observatory and the U.S. Naval Observatory should produce a single catalog of SRS positions which should be both completed and published jointly within a reasonable time in the FK5 system.
2. That the preliminary catalog of SRS positions should be referred to the system of the FK4 catalog and later transferred to the FK5 system.
3. That work on the computation of proper motions of the SRS, already in progress at the U.S. Naval Observatory, should be completed as soon as possible after the completion of the SRS catalog.

Recommendations with regard to future work:

1. That second-epoch observations of IRS (AGK3R + SRS) should receive high priority, and that an international effort to reobserve the IRS differentially and fundamentally should be organized.
2. That observations in the southern hemisphere should be strongly encouraged.
3. That observations of IRS should be referred to the FK5 system.
4. That an *IRS Committee* should be formed.
5. That a *Supplementary List* of IRS stars be made available to observers in order to improve both the homogeneity of the areal distribution of the IRS and its spectral type distribution in selected areas of the sky.

The report of the Astrolabe Working Group (AWG) was given by its chairman, G. Billaud. Some 22 astrolabes participated in the MERIT program. The use of laser ranging and VLBI for EOP measurements will decrease the astrolabe contribution to this work although efforts should continue, at least for the present, for rapid prediction services. Future work is likely to center on catalog observations. Such work is currently underway at twelve stations. It is believed that improved EOP observations will indirectly benefit the astrolabes by allowing much better evaluations of local effects thus improving results for star positions. Chollet has shown that humidity can cause errors of $0\overset{"}{.}003/^{\circ}C$, while star colors can contribute a few $0\overset{"}{.}01$'s. Visual astrolabes have established a precision of $0\overset{s}{.}005$ and $0\overset{"}{.}07$ in time and latitude, while new, automatic instruments lower these numbers, and are intended for stellar observations (China, France and Japan) or solar observations (Brazil and France). The latter have a monthly precision of $0\overset{"}{.}06$. The equinox difference, FK5-FK4, given by Fricke is confirmed to within $0\overset{s}{.}002$. Journet found $E=0\overset{s}{.}026 \pm 0\overset{s}{.}005$. Studies of changes in the solar diameter showed a period of 975 days, not correlated with Wolf's number, and of interest to astrophysicists. The South American Group (Natal, Rio, Rio Grande, San Juan, San Martin and Valhinos) plans a catalog covering $0°$ to $-75°$. A similar northern program involving the Chinese astrolabes, plus Potsdam and Boroviec would be very desirable. Natal and Quito could link the hemispheres. The planets, Jupiter, Saturn and Uranus are observed at Santiago and San Juan; radio stars at Paris, Potsdam, San Juan and Santiago.

The report of the Working Group on Astronomical Refraction (WGAR) was given by its chairman, G. Teleki. A *Workshop on Refraction in Optical and Radio Astrometry* was held at the Pulkovo Observatory from 3 to 5 June, 1985. The LOC was chaired by I.I. Kanaev, and the SOC by Teleki. The latter will edit the proceedings which will be published as the *Belgrade Astronomical Observatory Publication, No. 35* during the first half of 1986. Some 60 participants from 7 countries together with about 50 attendees from the USSR heard 42 papers. The appearance of the *Pulkovo Tables, Fifth Edition* is of particular interest. The new treatment of chromatic effects in this work is of special interest.

A meeting of the members of the WGAR present at Leningrad took place after the Workshop. Attending were, D. Currie, B. Garfinkel, Huang Kun-yi, I.G. Kolchinskiy, A.I. Nefed'eva, V.I. Sergienko and G. Teleki. It was agreed that closer connections between the researchers and institutions working in the field would be very useful. Specific areas of investigation and the most interested institutions were identified as: Local anomalous refraction, Irkutsk, Nanjing, Tokyo, Tomsk and Washington; Atmospheric modeling, Belgrade, Irkutsk, Leningrad, Kazan, Lvov, Moscow, Nanjing, Tokyo, Tomsk and Washington; Experimental determination of refraction effects, Irkutsk, Kazan, Kiev, Lvov; Terminology and Standards, Belgrade, Kiev. Dispersion methods were strongly endorsed. The meeting agreed that it would be advisable to continue the existing groups within the IAU and IAG. An inter-union group including astronomers, geodecists, meteorologists, physicists, et cetera, was endorsed. Finally, the year 1987 and the location of Belgrade were mentioned in the context of a special workshop on the basic questions of refraction.

The continuation of the above Working Groups and the Study Group on Horizontal Meridian Circles was approved by the commission without dissent.

Membership in the commission has become very popular. In all, 30 new members were approved. Arranged by country, they are:

     Australia: D. Harwood
     Brazil: L.B.F. Clauzet
     Chile: E. Costa

China: (Nanjing)	Hua Yingmin, Jiang Chong Guo, Li Zhifang, Li Zhigang, Lu Chun Lin, Mao Wei, Miao Yongkuon, Quian Zhi Han, Quin Zhi Han, Shen Kaizian, Shi Guagchen, Xia Yi Fei, Xie Liangyun, Xu Bangxin
Denmark:	K. Fabritius
FRG:	R. Wielen
France:	F. Crifo, A. Journet
Italy:	G. Chuimiento
Korea DPR:	Du Jin Cha
UK:	M.A.C. Perryman, C. Thoburn
USA:	S.J. Dick
Yugoslavia:	D. Durovic, N. Solaric, D. Saletic

The commission approved four consultants. They are: M. Dachich, I. Pakvor (Yugoslavia); T. Rafferty (USA); K.G. Steinert (DDR).

A discussion centering on the need for unified star lists took place. Actually, L. Morrison had broached the subject at the *Joint Discussion on Reference Frames* held on 20 November where he, together with C.A. Murray, pointed out the need for defined, faint fundamental lists, perhaps to 13th magnitude. Other lists should also be standardized or at least made known. For example, V. Abalakin proposed a list including some 42,000 IRS, 5,000 BS, 3,000 High Luminosity Stars, 2,500 Doubles and 2,000 Radio Source Reference Stars. In view of the universal need for and interest in such lists, a new *Working Group on Star Lists* was formed. Its members are: G. Carrasco, T. Corbin, L. Helmer, M. Miyamoto, D.D. Polozentsev and Y. Requieme. The latter, in his capacity as president-elect, agreed to serve as chairman of the group, at least provisionally. Additional members may be approved by the Chairman. The report of the SRS committee should be read in connection with this Working Group.

The possibility of a conflict between "astrometric" meetings proposed for 1987 in Paris and in Belgrade was discussed. It was agreed to leave the matter in the hands of the principles involved until the joint meeting with Commission 24. Some discussion of that forthcoming meeting, particularly involving joint interests, took place, but the business meeting adjourned without further action.

General Space Astrometry, 21 November 1985, 1100-1230
Chairman: R. L. Duncombe

A. and M. Meinel (read by J. Hughes) reported on their plans for *Astrometry on a Thousand Astronomical Unit (TAU) Voyage*. The technology to send a 1.5m telescope on such a long baseline parallax voyage has been examined at JPL. With an initial mass of 25,000 kg the TAU spacecraft will be deployed into LEO by the shuttle. Ion propulsion using a 1Mw nuclear electric module and 13,000 kg of Hg can attain a velocity of 100 km/sec upon leaving the planetary neighborhood, and a distance of 1,000 AU is reached in 50 years. Optical communication at 20 kb/sec can return data by using a 5w laser and a 1m communication telescope. Objects at 20th magnitude could be reached in a reasonable integration time, and it appears that centroiding diffraction limited CCD images to 30 microarcsconds is possible. An object exhibiting a parallax of 200 microarcseconds at a baseline of 200 AU is at a distance of a megaparsec, thus members of the local group are accessible. Since TAU cannot separate parallax and proper motion, a second 1.5 m telescope in LEO must provide the latter. A conference is planned for summer 1986 and Commission 8 members involved in new astrometric techniques are especially invited.

R.D. Reasenberg gave a presentation on *Optical Interferometry in Space* pointing out that although interferometers are more complex than telescopes, they are useful for astrometry since, for instruments of comparable size, an interferometer is orders of magnitude faster for most scientific objectives. POINTS (Precision Optical INTerferometry in Space) is a design concept for an articulated, space-based, dual astrometric interferometer having two 2 m baselines and four 25 cm telescopes capable of measuring the angular separation (of about $90°$) of a pair of tenth magnitude stars to 5 μas in about 10 minutes. This would allow about 60 measurements per day. The problem of systematic error is addressed by using stable materials and by means of realtime metrology and post analysis of data including bias estimation. The latter, depending upon $360°$ closure, has been investigated by covariance studies which show that no special observing schedule is required and several biases per day can be estimated with only a small increase in the statistical uncertainty in star coordinates. Other covariance studies show that with redundant observations, the separation of all pairs of stars (including those pairs not directly observed) becomes well determined; the grid becomes rigid. A set of 300 stars and 5 QSO's could be redundantly observed in a month. If such a sequence were repeated 40 times in 10 years, positions, proper motions, and parallaxes typically would be determined to 0.6 μas, 0.4 μas/year and 0.4 μas, respectively. As a result of aberration, these coordinates would be known to better than a milliarcsec in the frame of the earth's orbit. Additional objects would then be observed with respect to these 300 stars thus thousands of additional objects could be investigated with accuracies only slightly reduced. A POINTS mission would support a diverse set of scientific objectives. These include a second-order deflection test of general relativity and a deep search for other planetary systems. However, the most important results of such a mission are likely to come from lines of investigation not yet considered including serendipitous discoveries.

W. Van Altena reported that the Space Telescope Astrometry Team plans to observe Hyades cluster members and field subdwarfs with the Fine Guidance Sensors to determine the stars' trigonometric parallaxes and improve the Pop I and II distance scales. They expect that both distance scales will be defined to approximately 0.01 -0.02 magnitude in the distance modulus. Observations with the Planetary Camera are also planned for six globular clusters to determine their internal velocity dispersions.

L.W. Fredricks indicated that the Hubble Space Telescope Astrometry Team has drawn up a list of 19 objects without parallaxes and hopefully within 200 parsecs of the sun. Redundancy was desired so the list concentrates on five planetary nebulae, 2 dwarf novae, four old novae, Feige 24, and seven T Tauri stars. The latter will yield distances to their nebulae as well. Special filters have been inserted. The passband starts above the OII lines and cuts off before the H-alpha line.

P.K. Seidelmann talked about the Hubble Space Telescope, which is to be launched in 1986, and carry a Widefield/Planetary Camera, which will have fields of view of 2.′63 and 68″. For astrometric observations, methods of centroiding, plate solutions, calibration procedures and achievable accuracies must be considered. The Widefield/Planetary Camera Investigation Definition Team plans to include in its guaranteed time observations for low-mass companion studies, searches for and observation of faint satellites of the outer planets, search for satellites of asteroids, studies of the rings and shepherd satellites, study of the orbit and colors of Pluto and Charon, and motions within planetary nebulae.

T.B.H. Kuiper described work carried out with S.P. Synnott and E.F. Tubbs. Synthesis of large telescope apertures in space at submillimeter wavelengths is feasible with current or near-term technology if the instantaneous vector between two spacecraft can be established in a stable coordinate system. A technique by Kuiper et al., *Radio Science* in press, 1985 has been developed in which clustered star trackers can be used to determine the direction of any visible point source with respect to the stars of the HIPPARCOS catalogue. The accuracy of the determination will be limited by the catalogue accuracy (approx. 2 millarcsec), degrading with time due to the uncertainties in the proper motions (approx. 2 milliarcsec/year). A reflight of the HIPPARCOS mission is thus of great importance to our ability to produce high resolution submillimeter images of astronomical sources. Several submillimeter missions are planned between 1995 and 2010, including FIRST (Far InfraRed Space Telescope, by ESA) and LDR (Large Deployable Reflector, by NASA), as well as possible Spartan and Explorer class precursors. This approach to interferometry should be evaluated for the possibility of combining two or more of these spacecraft, after they have fulfilled their primary missions, to perform aperture synthesis. Such a mission would yield valuable experience for future missions, such as SAMSE and TRIO. The design of currently planned missions would be affected by an intent to do interferometry and this should therefore be determined in the next few years.

J. Hughes described an Astrometry Satellite proposed by a group headed by D. York (Chicago) of which the former is a member. Based upon a 1 to 1.5m class, compound, reflecting telescope, the primary mission of this satellite is the determination of relative positions to 10^{-4} arcsecond with a limiting magnitude of 19.5. This, together with a field of about 20 arcminutes in diameter, ensures that over most of the sky at least one QSO appears in each field. If the roll angle can be monitored by precision gyros even a single QSO could suffice for accurately referring the measures to the rest frame defined by such objects, although in many, or most, cases multiple QSO's will be in each field. It was remarked parenthetically that such gyros could also generate a global reference system, albeit at a somewhat reduced accuracy, say 0".01. The detector is a combination of a Ronchi ruling and a CCD. The parallaxes and inertial proper motions to be obtained from these measurements open up many, many scientific possibilities including much improved distance scale calibrations, the determination of the relative motions of globular clusters, and the measurement of the motions of the galaxies in the local group.

Observing Programs I, 22 November, 0900 - 1030
Chairman: L. Helmer

M. Miyamoto reported on the observing programs planned for the Tokyo Photoelectric Meridian Circle. With a scan time of about 30s per star and an average of 150 clear nights per year, it is possible to carry out 350,000 observations of stars brighter than 11.0 magnitude in 5 years. With 3 observations per star, this will give positions of about 100,000 stars to an accuracy of around 0".1. The Tokyo PMC program will include 100,000 stars in the AGK3 and Yale catalogues in the declination range $-30°$ to $+90°$. This will include about 20,000 IRS stars which will be observed with higher weight. Proper motions will be determined with an accuracy of 0".003/yr. The program also includes observations of the Sun, major and minor planets.

M. Yoshizawa reported on solar observations made with the Tokyo PMC since December 1984. The first and second limbs of the Sun are scanned with 3 pinholes 3" in diameter. The goodness-of-fit to the observations varies from around 2".0 in winter to 1".3 in summer, of which the image motion contributes about 1". Fourteen scans with 3 pinholes, taking 200s, gives an expected accuracy of 0".3. In fact, an accuracy of 0".4 was obtained, with a mean correction of -3".7 to the solar radius. The definition of the solar limb is, of course, dependent on wavelength. The standard deviations in RA, DEC and radius were $0^s.022$, 0".62 and 0".38, respectively.

G. Teleki reviewed present and future programs in Belgrade. The Meridian Circle (Askania 190/2578 mm, visual) presently observes a program of double stars, stars around radio sources, and the Sun, Mercury, Venus and Mars (daytime observations). The future program may also include IRS stars. The Transit Instrument (Askania 190/2578 mm, vacuum meridian marks) is undergoing a program of modernization. The future program will include the Sun and planets. The Vertical Circle (Askania 190/2578 mm, visual) carries out observations of Mars (rms error 0".55), Jupiter (0".65), Saturn (0".57), Uranus (0".73) and Neptune (0".64). The future program will include the Sun and inner planets. The Zenith Telescope (Askania 110/1287 mm, visual) continues its program of star observations.

S. Debarbat described the Paris astrolabe program. The long series of observations of fundamental stars continues. Among these stars is the radio star β Persei which has been observed on a regular basis for 10 years. These observations showed a systematic drift with respect to the FK4 proper motion. The observations will be continued and compared with the radio positions. The correction for the "phase effect" of planets, calculated by Chollet, has been used at the Bureau des Longitudes for the HIPPARCOS asteroid model, and will be employed at JPL. Chollet started making observations of the Sun with a modified astrolabe (in 1984), following the results obtained with the CERGA astrolabe which appear to show a fluctuation in the diameter of 0".5 over 3 years.

V. Abalakin (for D. Polozhentsev) reported on the organization of meridian observations of the IRS in the USSR. Meridian circles at Pulkovo, Odessa and Nikolaev have been equipped with photoelectric micrometers and some have been moved to high altitude sites. About 5 or 6 instruments will begin observing the IRS between $-20°$ and $+90°$ in 1986. An IAU Resolution stressing the importance of the IRS is desirable.

C. Smith reported on the programs of the Flagstaff Transit Circle. In the period June 1983 to 1985, the positions of 362 reference stars for Halley's comet with m_v between 5.9 and 9.0, and spectral types B to M, were observed. The internal mean errors in RA and DEC were 0".08. Future programs will include radio stars (those measured with the VLA will have high priority), AGK3R stars within $6°$ of selected quasars, and non-AGK3R stars within $0°.5$ of quasars.

Observing Programs II, 22 November 1985, 1100 - 1230
Chairman: H. Schwan

R. Stone reported on the USNO New Zealand program. The USNO 7-inch automated transit circle has been relocated to New Zealand and has become operational. The USNO double 8-inch astrograph has completed its northern hemisphere program and is currently enroute to New Zealand. In July 1985 the transit circle started a comprehensive observational program including an evaluation of its errors and an investigation of various instrumental effects. Preliminary results from collated data give standard deviations of $0^s.014 \cos \delta$, $0".17$, and $0^m.06$ in α, δ and V respectively. These are averages from all zenith distances. The limiting magnitudes for night and daytime stellar observations are about $11^m.5$ and $4^m.5$, and the instrumental constants are found to be very stable with respect to time. Solar observations are beginning.

Y. Requieme considered the question of faint star lists for automatic meridian circles. The latter are now able to determine positions of faint stars up to B=13 practically without magnitude error. It would now be possible to construct a secondary reference frame with 11<B<13 for the needs of photographic astrometry. Other valuable programmes concern local frames and extragalactic objects and selected areas for galactic research. Furthermore, about 200,000 stars have been proposed for the HIPPARCOS mission, but 100,000 will not be included in the input catalogue. The faintest stars of these latter stars, of high astrophysical interest, could make up a priority complementary program for meridian circles, but the HIPPARCOS Input Catalogue Consortium plans to release the list only by the end of 1989.

Hu Ningsheng described work by Mao Wei and Guo Xinyian of the Yunnan Observatory concerning the use of a CCD to set up an inertial coordinate system. The method involves the measurement of the absolute proper motions of fundamental reference stars with respect to extragalactic systems. The residual rotation of the fundamental system is then calculated from these motions. Since the field observed by a CCD is quite small, overlapping techniques were described to link adjacent areas of the sky. The method compares very favorably with photographic techniques.

Meridian observations of the Sun and planets at the Kislovodsk Station of the Pulkovo Observatory were described by K.G. Gnevysheva, A.V. Devyatkin and G.S. Kossin in a report read by Ya. Yatskiv. The Kislovodsk location ($\phi=44°$, h=2100m) has proved to be an excellent site for a daytime observing program. The Struve-Ertel vertical circle started regular observations of the declinations of the Sun, Mercury and Venus in 1984. The standard deviations of one observation of these objects are, 0".51, 0".31 and 0".35 respectively. For day stars the figure is 0".35.

A paper by A. S. Karin, N.F. Minyajlo and V.L. Voronkevich of the Main Astronomical Observatory at Goloseevo, Kiev (also read by Yatskiv) concerns the accuracy of observations of the Sun and major planets. A data bank exists involving some 18 years of observations and three optical methods. Extensive analyses were carried out to determine random, systematic and total error estimates. Comparisons were made of six series of observations each from the Nikolaev Transit Instrument and Vertical Circle and the Washington Transit Circle. Extensive estimations of the errors of recent observations (published 1981-84) were also given.

A progress report on the *Belgrade Catalog of Absolute RA's of Bright Polar Stars* was given by I. Pakvor, Belgrade Observatory, Yugoslavia. He indiated that the observations of absolute RA's of bright polar stars from the list of 308 stars in the declination zone $+65°$ to $+90°$ with the Large Transit Instrument (LTI) of the Belgrade Observatory were finished early in 1983. This is the first absolute RA catalog observed by the LTI of Belgrade Observatory. These observations involve the first use of vacuum meridian marks for determining the values of differential azimuth and collimation errors. The annual variations of these constants were given. Reductions are in progress and the chain method is in use.

Common Interests of Commissions 8 & 24, 25 November 1985, 1100 - 1230
Chairmen: W. Gliese and J. Hughes

The chairmen made brief introductory remarks emphasizing the need for close cooperation between the two commissions and pointing out that indeed many astronomers have dual memberships.

The discussion regarding the two "astrometric" meetings proposed for 1987 was continued. As a result of communications between the proposers, namely S. Debarbat (Paris) and G. Teleki (Belgrade), it was determined that the subject matter for these two meetings is distinct. The Paris meeting will have an Historical/Scientific thrust, concentrating on the heritage of the past including large scale catalogs as well as future directions. The Belgrade meeting, commemorating the 100th anniversary of the Belgrade Observatory, will concentrate on Fundamentals of Astrometry and indeed is so named. Current plans call for a spring 1987 meeting in Paris sponsored by Commissions 24 and 41; and a fall 1987 meeting in Belgrade sponsored by Commissions 8 and 19. Given the many new developments and initiatives in astrometry, it was held that these meetings are each appropriate, distinct and desirable.

The members of Commisssion 8 present voted without dissent to accept the two *Joint Discussion* I resolutions as written, and to forward them to the Resolutions Committee for submission to the GA. It was noted that Commissions 4 and 20 had deleted the section of the resolution on reference frames dealing with the duties of commissions, but Commission 8 members present did not accept this deletion. Members of Commission 24 had previously adopted the same position as Commission 8 in the matter. A resolution, proposed to Commission 8 by J. Kovalevsky and R. Reasenberg, concerning space and interferometric observations, was approved and forwarded to the Resolutions Committee. (**NB** See the list of resolutions as adopted by the GA.)

A discussion took place regarding the names and "duties" of the Commissions (8 and 24). It was generally agreed that both commissions should change their names. It was also apparent that it was good to discuss the matter jointly. Beyond this, however, a concensus did not emerge. Several suggestions and comments were made, many meeting with general approval, but no single idea was received with enthusiasm. It was finally noted by one of the chairmen (JH) that if the JDI resolution was adopted by the GA as written, i.e., as approved by Commissions 7, 8, 19, 24, 31, 33 and 40, and not as amended by commissions 4 and 20, then a forum would exist to discuss the matter in a much broader context. On this note the meeting adjourned.

Instrumentation and Techniques, 27 November 1985, 0900 - 1030
Chairman: C.A. Smith

A new Slit Micrometer for the CAMC on La Palma was described by L. Helmer. Its purpose is to achieve higher accuracy, a fainter limiting magnitude and to require less service. The slit plate is equipped with a linear encoder having a resolution, in x, of 1/4 micron, and the mounting is more rigid. Efficiency is maximized and a cooled, fixed Ga-As photocathode is introduced. Accuracy is expected to be better than 0."15 in both coordinates and the limiting magnitude should reach 14.5 to 15.

M. Yoshizawa of the Tokyo PMC reported that regular measurements of the instrumental constants have been made since May 1983. In addition to measuring the constants each day at midnight (24^h JST), several programs of more intensive measurements were also performed. Each program consists of continuous measurements lasting 2 to 5 days with a 0.5 to 1 hour interval between successive measurements. The standard deviation of a single measurement is typically about 0."05. The diurnal and annual variations of the constants are studied as a function of temperature or its rate of change with time. It was found that an optimal procedure to obtain the invidividual constants during a tour of observations is to (1) measure the relative azimuth and the zero point of the divided circle once per 2 hours (or once per 1°K change), (2) measure the level and collimation errors at least once per 6 hours (or once per 3°K change), (3) measure the flexure at least once per day, and (4) measure the seasonal variation of the pivots and division errors of the circle. In the daytime, an increase in the frequency of measurement is strongly recommended.

Miyamoto described an annual variation of the graduation error of the Tokyo PMC. He stated that the fully automated meridian instrument enabled the determination of the graduation error of the circle within a few days. Frequent measurements of the error shows its annual change. The amplitude of the change in the circle error amounts to about 0."05 (the corresponding value in the diameter error is about 0."3), which may cause, in compiling an absolute catalogue, a systematic declination error depending on right ascension. In the classical graduated circle, the amount of the change is expected to be much larger, in general.

V. Abalakin read several reports starting with a description of the first version of the Axial Meridian Circle (AMC) being developed at Pulkovo by G.I. Pinigin, A.V. Sergeev and O.E. Shornikov. This instrument consists of a single, fixed horizontal telescope with the tube lying East/West. An optical device mounted in front of the telescope objective transfers the image of a transiting star into the telescope and permits the observation of a distant prime vertical mark. Thus one can directly measure the distance between the observed star and the mark. Provision is also made for auto-collimation measurements with the eyepiece micrometer. Among the advantages of such an instrument is freedom from errors caused by variation of the position of the eyepiece micrometer due to, e.g., pivot errors, temperature deformations, tube flexure, etc. The horizontal light path is minimized and circle errors do not double. It is believed that by careful use of appropriate materials the systematic errors will be about 0."02 - 0."03 in both coordinates.

A *Catalog of Absolute Declinations of Stars* as compiled from PVC Observations in the Southern Hemisphere by V.A. Naumov and A.A. Naumova was discussed by Abalakin. It contains 698 FK4, 691 PFKSZ and 35 GC stars, and is the result of observations made at the Cerro Calan Observatory (Santiago) by five observers using the Zverev photographic vertical circle (PVC) during 1965-66. The reductions and scale derivation were described. Corrections for flexure, diameter errors and a dependence of zenith distance on photographic emulsion were determined and applied. The standard deviation is given by $(S.D.)^2 = (0\rlap{.}{''}27)^2 + (0\rlap{.}{''}20 \tan z)^2$.

The first observational results with the Sukharev Horizontal Meridian Circle (HMC) at Pulkovo were described in a paper writted by R.I. Gumerov, V.B. Kapkov, T.R. Kirian and G.I. Pinigin. Extensive measurements of circle diameter errors were made giving corrections with an accuracy of $0\rlap{.}{''}02$. The circle is monitored by special rosette measurements of $6°$ diameter errors. The reduction of 1300 observations (1981-83) of the declinations of 300 FK4 stars is completed and the error of one observation is $0\rlap{.}{''}20 \sec z$. The instrument appears stable with time and temperature and measures of the same star using the north and south tubes indicate that mirror deformations are small.

J.A. Hughes reported on the progress being made in the construction of the Mark III optical, phase coherent interferometer. A group, led by M. Shao, and involving SAO, MIT, NRL and USNO, is engaged in this effort. Construction on a site at Mt. Wilson Observatory is essentially complete. The instrument will use a 20m N/S baseline and somewhat shorter N/E and S/E baselines. The optical delay lines were designed and constructed at SAO and MIT, the siderostats at USNO. Tests of these and other components together with the associated software are underway. It is anticipated that first fringe tracking will occur in mid-1986. Preliminary stellar data from the Mark II instrument have been reduced and give a precision of about 1 arcsecond in large angle measurements. This was strictly a proof of principal effort however, and the refinements being incorporated in the Mark III will improve results by orders of magnitude.

T.J. Rafferty talked on the subject of *The Circle Scanning Systems and Glass Circles of the U.S. Naval Observatory*. Since 1970 the observatory has operated four different types of electronic circle scanning devices to determine the pointing position of its three transit circles. Most of the systems used a moving slide micrometer with a photoelectric diode as the detector. (Currently a CCD scanner, with no moving parts, is being tested.) During the same period, the observatory has replaced its engraved metal circles with glass circles. The improved precision of these devices, besides decreasing the error of measurements made with the transit circles, has revealed new areas of concern. Short term changes in the circles, apparently caused by temperature, have been found. (One of the circles shows such a large change over a short period that modelling is being used to apply diameter corrections.) Changes to a circle when it is rotated have also appeared and a new approach for monitoring and determining the diameter corrections has been developed.

Catalogs I, 27 November 1985, 1400 - 1530
Chairman: Y. Requieme

H. Schwan gave a survey concerning the process of improving a fundamental catalogue. Completed tasks are the derivation of corrections to the constants of general precession and the determination of the FK5 equinox and equator. The selection of new fundamental stars and the improvement of the individual and systematic accuracy are in progress. A description of the observational material and the new methods developed for the improvement of the FK4 and the preliminary FK5 system was presented. The basic FK5, providing improved positions and proper motions of the traditional fundamental stars, will be available in the course of 1986.

L. Helmer reported that the *Carlsberg Meridian Catalogue No. 1* has now been published. It contains the positions, in the FK4 system for the equinox J2000.0, and the proper motions and magnitudes, of 5,292 stars north of $-45°$. These include 2,369 AGK3R, 1,296 SRS, 227 PZT and 838 faint reference stars around radio sources. The zenith

mean errors are: 0″.193, 0″.184, and 0^m.054 in α, δ and V respectively. The mean
error of a proper motion is typically 0″.003/year and the limiting magnitude is
13.5. The catalogue also contains 857 observations of solar system objects. The
total number of observations is 35,100 plus planets and FK4 stars (12,000 obs.),
all obtained in the first eight months of 1984.

First results from the Tokyo PMC were described by M. Yoshizawa and M. Miyamoto.
More than 10,000 observations of FK4 and other stars have been made with the Tokyo
PMC in the period Nov. 1983 to Sep. 1985. The scanning time for one observation
is 2 min. The reduction of part of those observations (Dec. 1984 to Sep. 1985)
was made relative to the FK4 system. The mean error of a single 2 min. observation
is given by: $\epsilon(\alpha)$ cos δ = 0^s.009 (sec z ** β1) and $\epsilon(\delta)$ = 0″.14 (sec z ** β2), where
β1 and β2 vary between 0.5 and 1.0. Preliminary results of an analysis of the
declination dependence of the (O-C's) was also presented.

The *General Catalog of Faint Fundamental Stars* (FKSZ) has been compiled, according
to a report by A.N. Kuryanova, D.D. Polozhentsev, A.D. Polozhentsev, Ya.S. Yatskiv
and M.S. Zverev, as part of the KSZ-program (Zverev, 1951). Positions and proper
motions are given for epochs near 1962 and 1947, respectively, of 931 faint stars
(7.3< m <8.4) for both hemispheres. Seventeen catalogues in RA and 15 catalogues
in DEC as well as the AGK3R and SRS programs contributed. The system is the FK4,
the equinox and epoch 1950.0. The mean error of the mean positions is 0″.050 in RA
and 0″.074 in DEC. The mean errors of the proper motions are 0″.26 and 0″.32 per
century, respectively. The method of improving the proper motions and how the
PFKSZ catalogue for the southern hemisphere was compiled were described.

A program of observations of 3 to 5 reference stars in the magnitude range of 7 to
9 in the vicinity of 87 radio sources was described by A.S. Kharin, P.F. Lazorenko
and I.I. Kumkova and read by Ya. Yatskiv. This list contains 315 stars from +90°
to -40°. Observations to -15° have been made at Goloseevo (VC) and at Kiev State
University and Beograd (MC's). Up to 10 meridian instruments (Soviet Union) will
be used in this program which will be referred to the system of FK4 (FK5). Meridian
circles at Washington, Greenwich, Bordeaux, Tokyo, Uccle, Perth, and the Canary
Islands are invited to participate. Star lists are available.

A *Belgrade Catalogue of 308 Polar Star Declinations* by M. Mijatou, G. Teleki and
D.J. Bozhichkovich from observations with the Askania 190/2578 mm vertical circle
was described by Teleki. A quasi-absolute method was used with observations from
both instrumental positions. Each star was observed at least four times at both
upper and lower culmination. The total number of observations per star is about
10. The internal accuracy was 0″.33 and 0″.39 for upper and lower culminations,
respectively. Results will be sent to the Astronomisches Rechen-Institut and the
Kiev University Observatory in December 1985, and will be published in 1986.

A paper on *The Problem of Classifying Modern Methods for the Determination of Right
Ascensions* by S.A. Tolchelnikova-Murri, Pulkovo Observatory, was also read by
Teleki. Several methods intended for the improvement of the right ascension system
of FK4 were discussed. Suggestions were made to overcome difficulties in two
commonly used methods of determining the absolute azimuth of a meridian circle.

The *Declinations and Proper Motions of 36 Belgrade Zenith Stars* as determined by
R. Grujich and G. Teleki of the Astronomical Observatory, Belgrade, Yugoslavia,
using the 111/1287 mm Askania ZT from 1960 to 1982 were reported by the latter
author. Stars were observed from 18 to 189 times. Declinations and proper motions
were computed using three models of the variation of latitude. The most accurate
results were obtained using unsmoothed measurements of the variation of latitude
made in Belgrade. The mean error in DEC is 0″.27, and in the proper motion, 0″.006
per year. Due to instrumental modifications and procedural changes a considerable
improvement in accuracy has occurred. In the period 1960.0 to 1968.5, the mean

error was 0".294, while for the period 1969 to 1982, the mean error was 0".220.

The *Preliminary Results of Observations of Double Stars and Stars Near Radio Sources* made by S. Sadzakov and M. Dachich with the Belgrade Meridian Circle were described. Observations involve a list of 2,322 double stars in four zones from +60° to -20° with FK4 stars. The standard deviation in RA is 0^s.021, and in DEC is 0".31. Observations of 285 stars in the vicinity of 78 radio sources are also in progress in the zone +90° to -30°. There are 3 to 5 stars for each source. This program is about 60% completed; 2,452 transits registered. The double star program is 90% completed; 7,622 transits. The double star program should be completed in 1986 and the radio source program in 1987. Kustner series are regularly observed. About 20 series are observed per year, so that instrumental parameters are well known.

An analysis of systematic deviations of the well known *GCLS and AGK3 Catalogues* from 20 catalogues obtained from latitude observations was carried out by S. Sadzakov, Belgrade. Comparisons were made between the GCLS and AGK3 declinations and those in the individual latitude catalogues, at the epoch of the individual catalogs, with the GCLS transferred to the system of FK4. The differences were represented by 3rd order Fourier expansions in the right ascension. Tables of results were given.

Catalogs II, 27 November 1985, 1600 - 1730
Chairman: M. Miyamoto

C.A. Smith (Washington) reported on behalf of himself and L. Yagudin (Pulkovo) on the excellent progess made in the collaboration of the U.S. Naval and Pulkovo observatories on the compilation of the SRS catalog. The complete data base of SRS observations became available for the first time in 1984. About 482,000 observations of approximately 20,000 SRS and 900 FK4 stars from 12 transit circles are included. In many cases a further reduction to the system of the FK4 catalog beyond that already achieved by the individual participants was necessary. Preliminary mean errors of the mean positions in each coordinate are about 0.1 arcsec at the mean epoch of observation. An exchange of data bases between the U.S. Naval and Pulkovo observatories has been effected. Intercomparison of the two data bases has been completed, and has shown deficiencies in both which have since been overcome. The final stages of the collaboration will involve comparisons of outliers, of reductions to the FK4 system, of the two preliminary unweighted systems, and of the two weighted systems. If both the agreement between the Pulkovo and U.S. Naval observatories' systems and the distribution of differences for individual SRS are satisfactory, it is anticipated that an unweighted mean of the results from the two compilations will be recommended for general use.

G. Billaud described the results obtained with the CERGA Photoelectric Astrolabe while J. Kovalevsky described the work in progress at CERGA with the solar astrolabe by F. Laclare and A. Journet. Since 1983, the instrument observes the Sun at 8 different zenith distances between 30° and 60°. Normal points over one month of observation give an rms of 0".05 on the radius of the Sun. Corrections to the equinox that can be derived from these observations have an rms of 0".01.

F. Noel, on behalf of L.B.F. Clauzet (Brasil), described the basis for cooperative efforts towards a southern hemisphere astrolabe catalog involving instruments at Natal, Rio de Janeiro and Valinhos, in Brasil; San Juan, San Martin and Rio Grande, in Argentina; and Santiago, Chile. The catalog will cover 0° to -75° with Natal allowing a connection with the northern hemisphere. Recommendations agreed to at the *IV Regional Latin American Astronomical Meeting*, (Rio de Janeiro 12/84) were given regarding observation and reduction procedures, star lists and constants. Also, stations must ultimately be prepared to re-reduce data to the FK5 system, and observe planets and radio stars. Solar observations at Valinhos and Rio de

Janeiro must be continued. Agreements are sought to facilitate exchanges of personnel, and a working meeting is envisioned for 1986 in Brasil to finalize the procedures for the compilation of the general catalog which will contain some 1,000 stars.

A paper by Luo Ding-jiang (Beijing) concerning planned Photoelectric Astrolabe Observations was read. The international character of star catalogues was stressed, and determinations of the systematic errors of the FK4 by various instruments were compared. The main characteristics of the Mark III Photoelectric Astrolabe were listed including its 260/5000 mm optical system, the on-line PDP 11/23, and a photon counting technique reaching to 11-12 mag. Major planets are observable and almucantars at zenith distances of 30, 45 and/or 55 degrees are available. The many contributions which such an instrument can make were detailed including the improvement of the fundamental systems, extension of the same to fainter stars and various programs involving the optical counterparts of radio sources, radio stars and so on. Preliminary study has shown that two stations at $+40°$ and $+25°$ plus one or more in the southern hemisphere (including an assumed site near $-42°$) could produce 10,000 star positions as the first step of a complete program.

The work involved in determining *The Positions and Proper Motions of 4949 Geodetical Stars* pole to pole by E.V. Khrutskaya was described by V. Abalakin. The compilation of this catalog involved material from 17 catalogs in RA and 12 in DEC. The mean epochs are respectively 1967.74 and 1968.16. The number of catalogs contributing to a compiled position varies from 3 to 11. Weighted values of the observationally determined zonal systematic errors of the FK4 were determined. The catalog is given in the system of the FK4 and in the system of the FK4 as corrected by the weighted systematic error $\Delta\alpha(\delta)$. This will facilitate its use pending the arrival of the FK5. Errors are $0\overset{s}{.}003$ and $0\overset{''}{.}10$ in RA and DEC, while the corresponding errors in the centennial proper motions are $0\overset{s}{.}008$ and $0\overset{''}{.}11$ respectively. The new proper motions for the bright stars were used in statistical investigations which gave:

$$\Delta p = 1\overset{''}{.}10 \pm 0\overset{''}{.}07; \quad \Delta E = 1\overset{''}{.}21 \pm 0\overset{''}{.}07;$$

$$A/47.4 = 0\overset{''}{.}34 \pm 0\overset{''}{.}03; \quad B/47.4 = -0\overset{''}{.}20 \pm 0\overset{''}{.}02;$$

and a solar apex at:
$$A = 271\overset{°}{.}2 \pm 3\overset{°}{.}8, \quad D = 33\overset{°}{.}8 \pm 3\overset{°}{.}2$$

J. Dickey reported on work by her colleagues, O.J. Sovers, J.L. Fanselow, R.N. Treuhaft, K.M. Liewer, A.E. Niell and C.J. Jacobs on the *JPL 1985-1 Celestial Radio Reference Frame*. Dual frequency, VLBI observations during the past seven years have yielded an extragalactic radio source catalog containing 137 sources uniformly distributed north of $-40°$ declination. The positional uncertainties are of the order of a few milliarcseconds, and 38 sources have declination uncertainties smaller than 1 mas. Positions of sources that are also monitored by the East Coast VLBI group agree well with their results. The VLBI solution also indicates a need for slight revisions of the IAU 1980 precession and nutation constants.

J. A. HUGHES

COMMISSION 9:   INSTRUMENTS AND TECHNIQUES (INSTRUMENTS ET TECHNIQUES)

PRESIDENT: W. Livingston

Commission Meeting, 26 November 1985

Dr. Colin M. Humphries was elected the new President. For Vice-President there were two nominations with John Davis having the winning vote. The following members of the Organizing Committee were slated and approved: J. C. Bhattacharyya, J. Davis, O. Engvold, B. Fort N. Hu, W. Livingston (Retiring President), D. Malin, N. Steshenko, R. Tull, G. Walker, W. L. Wilcock, and G. Wlerick. A list of new applications for membership in the Commission was submitted and approved.

Brief summaries of business in the Working Groups were presented by their Chairman. It was noted that Dr. Elizabeth M. Sims retired as Chairperson of Astronomical Photography WG, being replaced by Dr. David Malin. Dr. John Davis continues as Chairman of the High Angular Resolution Interferometry WG, and Dr. Gordon Walker has accepted Chairmanship of the Detector WG.

Dr. J. Tinbergen, President of Commission 25, presented a Resolution on Instrumental Polarization for our consideration. Calling attention to the fact that several proposed designs for large telescopes in the future incorporate severe oblique reflections, he pointed out that this feature will virtually exclude precision polarimetry. He urges designers to avoid such optical schemes where possible. Our membership unanimously approved Tinbergen's resolution although it was too late in the General Assembly for official recognition of this vote.

Special Science Report: First Light at India's 2.3m Telescope

Professor J. C. Bhattacharyya, India Institute of Astrophysics, beamed as he circulated first-light pictures taken at the prime focus of the 2.34m telescope at Kavalur. The star images proved first rate and India can well be proud of this achievement by indigenous skills. Mr. A. P. Jayarajan, project optican, and Mr. A. Tapde, mechanical engineer, described the history and special features of the telescope.

This project was the brainchild and dream of one individual, the late Dr. M. K. Vainu Bappu. It was he who visualized and practically designed the entire instrument. No detail was beyond his concern. Some milestones include: Funding provided 1973, mirror blank received and engineering started 1974, housing completed 1978, mirror grinding began 1979, dome installed 1981, mechanical mount shop tested 1984, and 'first light' October 31, 1985.

Four foci are available: the prime focus at f/3.25, the cassegrain at f/13, and coudé at f/43.25. A 'Zerodur' blank was obtained from Schott which, after grinding, has a thickness diameter ratio of 1/6. During figuring two tests proved both convenient and precise. The 'axial wire test' was emplyed daily complemented by the 'Abe Offner null corrector test' especially towards the final stages of figuring. Some local zonal errors persist but the accuracy of the entire surface is better than a tenth of a wave. Cut-off measurements by a Foucault knife edge indicate that 90% of the imaged light falls within 0.6 arc sec.

The telescope housing is ample: 23m in diameter with the declination axis 18.5m above ground. The telescope itself has an equatorial horse-shoe yoke configuration with hydrostatic bearings. Drives, both RA and Dec, employ 2.8-m bull gears with two pinions for antibacklash. The final cost was equivalent to about US $6 million which turns out to be 2.34 (or the aperture in meters!) times the original estimate.

Other Reports

Mr. Parn-an Shen, Nanjing Astronomical Instruments Factory, presented a summary concerning 16 major new facilities in China. These include a 1.26-m infrared telescope and a 2.16-m reflector for Beijing Observatory; a 1.56-m astrometric telescope for Shanghai; a 0.35-m solar magnetic field telescope for the Huairou Reservoir Station, Beijing; and a new solar fine-structure telescope for Yunnan Observatory. Several radio installations and a balloon-borne telescope were described.

Mr. Chang-xin Cao, also at NAIF, considered the optimization of optical systems especially for large telescopes. Designs proceeding from collaborations with A. Meinel and R. N. Wilson were mentioned.

B. N. Karkera at Bhabhe Atomic Research in Bombay proposed a novel and simplified equatorial mount which consists of a cylindrical horse shoe with 5 oil pad bearings. Less than 50% as massive as conventional designs the advantage is in cost, stiffness, and ease of fabrication.

Albert Betz, University of California Space Sciences, described a far-infrared (150 μ to 500 μ) laser heterodyne spectrometer for airborne astronomy. Operating above the tropopause, the instrument has been used to detect fine structure of CI in molecular clouds.

Multi-object Spectroscopy - 25 November
(Organizer: W. L. Wilcock; Chairman: C. M. Humphries)

Richard Green (KPNO): "Multi-object Spectroscopy with the Kitt Peak Cryogenic Camera and 4-meter Telescope." A computer planned photographic mask in the focal plane serves as entrance apertures (i.e., slits). Followed by a field lens, doublet collimator and a grating prism, resolutions of 8A to 30A are provided between 4900A and 8000A. Detection is by a Texas Instrument 3-phase 800x800 thinned CCD after a demagnification of 8.4. Imagery is at f/1.0. Typical format consists of 10-12 slits, each 10-15 arcseconds. Major limitation comes in sky subtraction because of focus variations arising from ripples in the thinned chip.

Donald Morton (AAT) "The Multi-object Fiber Spectroscopy System at the AAO."

Bernard Fort (Toulouse): "Multi-object Observations of Faint Objects at the CFHT with an On-line System, PUMA1."

Fred Watson (UKSTU): "The UK Schmidt Telescope 'FLAIR' Multi-object Spectroscopy System." 45 μ fibers, glued to a positive copy glass plate, pick up objects over the 40 square degree FOV. These fibers feed a simple spectrograph with photographic film as the detector. Useful spectra are obtained to V ~ 14. With a CCD detector the system should go to V ~ 17.

Chris Impey (Caltech): "Radio Galaxies and Active Galaxies in Clusters" Operating at the Cass focus of the 3.9-m AAT, 50 fibers of 200 μ diameter are divided between objects and sky positions specified to 0.3 arcseconds. Centering and alignment takes about 5 minutes at the telescope. Detection is by the IPCS with a 2048x200 format corresponding to 0.88A per channel. During a three-night run, over 65% of the time was spent in target integration. Results for clustered radio galaxies were presented.

Keith Taylor (RGO): "A Low Dispersion Survey Spectrograph for the La Palma 4.2-m Telescope and The AAT."

Dave Carter and Peter Teague (Mt. Stromlo): "Velocity Structure of Clusters of Galaxies."

R. Foy (CERGA), A. Baranne (Marseille), and F. Thevenin (Nice): "The Large Multi-slit Spectrograph 'SFM'." An array of custom entrance slits are prepared on aluminized plates coated with photoresist. Matching the CFHT 15 arcminute FOV of the Cass focus, a resolution ~ 0.5A from 3500 to 5500A should be achieved. Guiding is by two field stars. Radial velocity is derived with aid of two comparison spectra and by inserting a 'lame en toit' spectrum divider. Aim is V ~ 20 in 5 hours for a S/N = 10.

H Lorenz (GDR): "Multi-object Spectroscopy with the USSR 6-m Telescope." A multi-slit system is described which photographically records 200-250 spectra per the 14 arcminute field. Recently an electronographic camera and a SIT TV have been used to record the spectra.

WG On Photoelectronic Image Devices - 22 November
(Chairman: Bernard Fort)

Gordon Walker (Univ. B. C.): "On Choosing a Detector." There is still no all-purpose detector. Choice must reflect the application and available funds. A full system consisting of detector, control electronics, data recording and display will cost about US $75,000 whether built commercially or in the lab. Requirements for the differing applications were summarized. Present practical detector specifications were reviewed.

Richard Green (KPNO): "Overview of the New Tektronix CCDs." Tektronix is the successor to TI technology. Experimental 27 μ pixel chips of size 512x512 (13.8x13.8mm) and 2048x2048 (55.3x55.3mm) have been frabricated. Each chip has two serial registers: one for low-noise slow-scan and one for fast reading. A read noise of less than 10e is promised. Many specifications and results of tests were given but at this stage everything is preliminary. Such a large format will affect future instrument planning significantly.

R. Griffiths (STScI), et al.: "An Advanced Radial Camera for the Hubble Space Telescope." The promise of Tektronix 2048x2048 arrays with 27 μ pixels, low read noise and large full well capacity (7×10^5e) suggest pursuing a second generation of cameras for the HST.

D. Thorne (RGO): "Experiments with the GEC/AAO 1500x1500 CCD."

D. Thorne (RGO): "Microchannel Plate Development at the RGO."

M. Cullum (ESO): "Photon-counting Systems vs. CCD's." It is pointed out that even with the availability of low read noise CCDs there is a domain of applications such as UV spectroscopy, speckle interferometers, and imaging Fabry-Perots where photon-counting wins out.

Richard Green (KPNO): "The Kitt Peak Photon-Counting Array." This is a more-or-less conventional system with a rapid scan detector which records intensified events and centrodes to a fraction of a pixel referred to the photocathode. First stage is a Carnegie 2-stage intensifier followed by an ITT dual microchannel plate tube. This is coupled to a Fairchild 222 CCD clocked at 20 mHz yielding a 6 ms full frame time for a 3040x976x16 bit accumulator. An option is dithering the first stage to suppress a centroiding fixed pattern.

A. Blezit and R. Foy (CERGA): "The "Photon-counting Camera CP40." Under development is a 2000x2000 pixel PCD for speckle and multislit spectroscopy. First stage is a 40mm VARO intensifier followed by a 50mm microchannel plate RTC intensifier. Four Thompson 288x385 CCDs are imaged in quadrants by a custom fiber-optic coupler. The accumulator size is then 3080x2304.

Craig McCreight (Ames Res. Cent.): "Infrared Astronomical Applications of Two-Dimensional Detector Arrays."

I. S. McLean (R. O. Edinburgh): "IR Imaging at the UKIRT." R. O. Edinburgh is developing imaging systems in the 1-5 μ region for use on the 3.8-m UKIRT. Capable of operating a variety of devices requiring different temperatures and control voltages, the first detector will be a SBRC 58x62 InSb direct readout device.

R. Puetter (San Diego): "Work on IR Array Detectors at the U.C.S.D." Two Si:Bi detectors in the thermal IR (5-18 μ) are described. The first is a 128 linear array of 20x6 mil pixels on 8 mil centers. Each pixel is connected to a 100pF capacitor so full well size of 10^9e accommodates long exposures.

F. Lacombe, P. Lena, D. Rouan, and F. Sibille (Obs. de Paris et Lyon): "Ground-Based Imaging in the Near IR (1-5 μ): Progress Report on a Charge-Transfer Array." (Presented by C. Cesarsky) A 32x32 InSb CID array has been manufactured for us by the French company SAT. The array has 100 μ x 100 μ pixels with a fill factor of 80%. Quantum efficiency in the middle of the 2-5 μ region is ~50% and integrations of up to 100 seconds have been achieved at 4K. The readout noise is measured as 1200 e^- and the device does not exhibit much time lag or slow response. The CID array has already been used to obtain astronomical images at Pic du Midi. Jupiter and Saturn were imaged, both in broad-bands and in a methane absorption feature, with a spatial resolution of 2".

J. W. V. Storey, J. Karianski, S. T. Shanahan, M. A. Green, and U. Theden (U. New South Wales): "IR CCD Development at UNSW." Metals-silicide Schottky arrays for 1 to 6 μ with CCD readouts are being tested on the AAT.

G. Wlerich (Obs. de Paris): "Observations avec la Camera Electronique Grand Champ au CFHT." In operation since December 1982, at the f/8 focus, the large (80mm) field of effectively 22×10^6 pixels permits full use of the excellent resolution as allowed by the telescope and site. Exposures up to 4 hours are effective. Application includes crowded field photometry and the study of small extended objects against foreground stars.

Abhijit Saha (KPNO): "A High Speed Plate Scanner with CCD Imaging and On-line Data Processing."

Harold Ables (USNO Flagstaff): "CCD Astrometry."

WG on Astronmical Photography - 22 November
Chairman: Elizabeth M. Sims

BUSINESS: Most of the discussion was based on the contents of a telex received from Eastman Kodak in response to questions from the WG Chairman about the supply, cost and availability of materials.

AVAILABILITY OF SPECTROSCOPIC EMULSIONS ON FILM: Technical Pan 2415 emulsion may be available by the end of 1986 coated on to thick (0.7 mil) Estar base film, in large formats, although the minimum order quantity is likely to be about 750 square feet. The production of other spectroscopic emulsions on thick base large format film is under consideration but there are as yet no plans to begin production. Many potential users welcomed the availability of spectroscopic emulsion on film, and especially the price advantage over the same emulsion on glass, but it was also clear that a significant number of users still require glass plates for more critical applications.

COST OF PHOTOGRAPHIC PLATES: The large increase in the cost of photographic material in recent years, made worse by the strong US dollar, has created grave problems for many users of spectroscopic plates. Small departments often find that they cannot afford even the modest number of plates they need. Unfortunately by commercial standards the astronomers are an extremely small fraction of the photographic market, and is only marginally profitable. Two approaches to this problem were agreed: (i) The incoming Chairman will write to Eastman Kodak Company expressing concern at the recent price increases and the consequent effects on astronomical research facilities; (ii) Users of spectroscopic materials are encouraged to collaborate in ordering materials, especially where several small orders can be combined into one larger one, thereby overcoming the minimum order constraints and costs. If timing orders to meet specified production dates, possibly twice a year, would help to keep costs down, many users seemed willing to fit in with such a time scheme.

ORDERING AND DELIVERY PROBLEMS: Many local Kodak agents do not know about spectroscopic materials, and consequently users experience difficulty in obtaining information, placing orders, and receiving shipments. In these circumstances help may be forthcoming from Nanci Snyder, of the International Division of Eastman Kodak Company, Rochester. In some circumstances, and with the agreement of the appropriate national Kodak agent, it may even be possible for users to place orders directly with Eastman Kodak International Division.

The new officers and Organizing Committee for the Working Group were agreed as follows: Chairman: D. F. Malin (Australia), Secretary: S. Marx (GDR); Organizing Committee: O. Dokuchaeva (USSR), J-L. Heudier (France), B. Hidayat (Indonesia), K. Ishida (Japan), A. G. Millikan (USA). W. E. Schoening (USA), J. G. Schumann (FRG), M. E. Sim (UK), K. R. Sivaraman (India), A. G. Smith (USA), M. K. Tsvetkiov (Bulgaria), R. M. West (FRG), and O. Zichova (Czechoslovakia). It was agreed that the name of the Working Group should become the "Working Group on Astronomical Photography." There is a proposal to hold another meeting of the Working Group, probably in Jena, GDR, in 1987.

Reports

J-L. Heudier, "Calibration and Sensitometry at TESCA." A Kitt Peak-style sensitometer is used to project a stepwedge onto each plate. After processing, the stepwedge density measures are used in a BASIC program to provide H + D curve, DQE and S:N curves for each plate automatically.

A. R. Good and M. E. Sim, "Microspots - A Progress Report." Further inspection of the plate collection reveals a higher than expected number of IV-N plates developing microspots.

J. Quebatte, B. Dumoulin, and R. M. West, "Grid Processing of Large Photographic Plates." This new method of developing large format plates gives more uniform and efficient results than the rotating rocker.

M. F. McCarthy & V. M. Blanco, "Detection of Carbon Stars - IIIa-J And IV-N Plates Compared."

Syuzo Isobe and Hiroshi Maseda (Tokyo), "An Image Detection System Developed For Schmidt Plates With Large Numbers of Stars."

G. Wlerick, G. Lelievre, B. Servan, L. Renard, D. Horville and M. Poinse, "Emploi de la Camera Electronique A Grand Champ Avec Le Telescope CFH."

M. Capaccioli, "Calibration of Photographs." The quality of calibrations provided on photographic plates is sometimes reasonable, occasionally good, but usually very poor. Managers of telescope facilities are encouraged to make good-quality calibrations available more often.

It is still the policy of the Working Group to encourage authors to publish papers relating to astronomical photography in the AAS Photobulletin.

WG on High Angular Resolution Interferometry - 25 November
(Chairman: J. Davis)

P. Nisenson, C. Papaliolios, S. Ebstein and R. Stachniki (CFA), "Center for Astrophysics Speckle Imaging Program." The program is built around two cameras: the PAPA (precision analog photon address) detector camera for low light level experiments and a Photometrics Inc., RCA-chip-based CCD camera for solar surface imaging. The CCD camera can be used in intensified mode for narrow spectral line work. The PAPA camera is a 512x512 element sensor which records photons in sequential format, with arrival times, at mHz rates. Decoding is done with the same electronics upon return to the laboratory. The Knox-Thompson procedure is used for image recovery. Recent results include detection of a 15th mag. companion to 10th mag. T Tau and a 5th mag. companion to α Ori (separation: 0".05).

J. E. Baldwin (Cambridge), "First Measurements of Closure Phase for High-resolution Optical Imaging." As a first step in applying phase closure techniques developed for radio VLBI to optical wavelengths, short exposure CCD images have been recorded with the University of Hawaii 88 inch telescope through aperture masks in the pupil plane. Each mask contained 3 or 4 holes corresponding to sizes in the aperture plane in the range 13-45cm. Fringes of high visibility were observed for single stars for exposures t < 50ms. Closure phases were derived for a variety of masks and exposure times. The results suggest that closure phases accurate to < 1° should be possible after a few-minute observation. Further observations using photon-counting detectors have been made and are expected to give diffraction limited images of simple objects. The limiting magnitude of the technique is expected to be at least +10.5 and perhaps +15 depending on atmospheric conditions.

P. Venkatakrishnan & S. Chatterjeel (IIA), "On the Saturation of the Refractive Index Structure Function at Large Separations: Enhanced Hopes for Long Baseline Optical Interferometry." The conventional "2/3 power law" for the refractive index structure function is an approximation which is valid only in the inertial subrange of turbulence. For separations larger than the outer scale the structure function will tend to a constant value. A form for the structure function including saturation has been assumed to rederive the phase structure function following Roddier. This function tends to a constant at large separations and follows a power law with index 1.37 at small separations. Such a behavior of the structure function does not require the indefinite narrowing of bandwidth with increasing baseline and thus enhances the hopes for long baseline optical interferometry.

F. Vakili, Y. Rabbia & L. Koechlin (CERGA, Saint Vallier, France), "Recent Results With I2T in the Visible." The stellar interferometer at CERGA is currently working with a spectral range of 0.4-0.7 µ and baselines from 6-67m which provide angular resolution from 15-2 mas. Fringes are found in dispersed light using either low dispersion of 2nm/pixel with a total bandwidth of 200nm or high dispersion of 0.5nm/pixel with a 30nm bandwidth. Stellar images are superposed to better than 1". The guiding limit is +8 mag using a photon-counting camera. Methods of fringe processing have been developed for the determination of angular diameters and for the automatic detection of fringes at low light levels (used to find high dispersion fringes on the +3.7 mag star λ Lyr). The envelope around the shell star γ Cas has recently been resolved using this device.

W. J. Tango and J. Davis (University of Sydney), "The University of Sydney Prototype Stellar Interferometer." A prototype stellar interferometer has been constructed as an essential step in the development of a very high angular resolution instrument. Except for a single baseline of 11.4m and limited range optical delay lines, the prototype has all the essential features of the large interferometer. All the optics, including the fixed siderostats, are mounted on massive concrete supports to provide mechanical stability and active optical delay control is used. The effects of atmospheric turbulence are minimized by using apertures ~r_o, active wavefront tilt correction and rapid data sampling. Interference was observed with starlight in October on the first observing night after alignment of the instrument was complete. The visibility maximum has been tracked for a number of stars (optical bandwith 0.3nm) allowing the baseline coordinates to be determined to ~0".2. The servo bandwidths, etc. are being optimized to maximize the detected fringe visibility and measurements on "standard" stars are planned to commence soon.

J. Davis & W. J. Tango (University of Sydney), "The University of Sydney New High Angular Resolution Stellar Interferometer." Based on experience in developing the prototype interferometer, a new high angular resolution stellar interferometer has been designed. The new instrument will have a N-S linear array of baselines covering the range 5-640m which will permit angular sizes in the range 2-0.05 mas to be measured. Siderostats with 0.2m diameter mirrors will give a primary beam diameter of 0.14m and the limiting magnitude is expected to be ~+8.5. The critical beam-combining optics from the prototype interferometer will be used in the new instrument. A proposal for funding has been submitted to the Australian Federal Government.

J. Hughes; (USNO), "The Mark III Astrometric Interferometer." Following the successful test program with Shao's Mark II astrometric interferometer the Mark III instrument is being constructed on Mt. Wilson. The new instrument features 10m and 20m N-S baselines and 7m and 14m NE-SE baselines. Piers for the siderostats and the path compensation system have been poured and a temperature-controlled instrument laboratory has been constructed. Siderostats have been built at USNO with apertures of 0.25m. Corner cubes have been included in the back of the siderostat flats, as close to the front reflecting surface as possible, to permit their positions relative to the piers to be monitored with laser interferometers. The pier separation will also be monitored by laser interferometry. The project is a collaborative effort between SAO, MIT, USNO and USNRL.

W. Traub, M. Shao & P. Nisenson (CFA), "The 40 Metre Imaging Stellar Interferometer on Mt. Wilson." It is proposed that an Imaging Stellar Interferometer (ISI) be built on Mt. Wilson with a 40m N-S variable baseline. It would have its own two ~0.6m diameter collecting telescope, focal plane optics and control software but would time-share the variable delay line and star-tracker system from Shao's Mark III astrometric interferometer. The ISI would operate in three modes: (1) single r_o to measure visibility amplitudes to 9th mag; (2) as for (1) but with phase also measured; and (3) multi-r_o to measure visibility amplitudes in photon-starved mode to about 13th mag. The focal plane optics (including photon-counting camera) are being assembled and tested and the detailed design of the siderostats, variable baseline, etc. will commence soon.

R. Stachnik & W. Traub (CFA), "Report on the October 1985, Cambridge Space Imaging Interferometry Workshop." This meeting, convened by the astronomy subcommittee of a U.S. National Academy panel, developed recommendations to NASA on scientific and technological drivers for space imaging interferometry in the interval 1995-2015. Seven recommendations were formulated including: early development of critical technologies identified in the report, support for a precursor experiment, encouragement of ground-based efforts, long-term planning for a major observatory-class instrument, formation of a U.S. working group analogous to the one in Europe and encouragement of the international efforts characterizing work in this area to date.

R. Stachnik & M. Faucherre (CFA), "SAMSI, A Multiple Spacecraft Interferometer." SAMSI is a proposed space Michelson interferometer consisting of three spacecraft: two light collectors and a central beam-combiner. Detailed calculations indicate certain orbits are extremely favorable for efficient one-dimensional surveys of angular diameters to 0.01 mas corresponding to a baseline of 10km. In a second observing mode, synthesis of filled apertures hundreds of metres in diameter would be possible on a timescale of about a day using 1m collectors and 0.5N thrusters. Numerical simulations have been undertaken to demonstrate the feasibility of image construction using a self-referencing scheme which relies on approximate determination of Fourier plane phase from information contained in the dispersed fringes and on optimal error distribution across the highly-overdetermined u-v plane.

M. Faucherre, R. Stachnik & W. Traub (CFA), "SPI, Space Platform Interferometer." The large and growing number of proposed space interferometry projects suggest the need for a relatively modest precursor experiment to test concepts and produce unique science. SPI would be a 10m class baseline imaging device which might fly on either the Eureca or U.S. co-orbiting platform. Emphasis would be on measurements in the ultra-violet and on testing different approaches to image fourier transform phase recovery. A highly desirable scientific goal would be synoptic imaging of several supergiants at 5 to 10 resolution elements across the surface over a wide range of wavelengths.

W. Traub, M. Lacasse & N. Carleton (CFA), "COSMIC, A High Angular Resolution Imaging Array in Space." COSMIC is a proposed dilute-aperture direct-imaging telescope in space. Engineering studies at NASA-Marshall Space Flight Center and the Perkin-Elmer Corp. have shown COSMIC to be a good candidate for an early space interferometer, in part because it utilizes technology which is close to that currently available. COSMIC produces images using several identical afocal collecting telescopes, a central beam-combining telescope, a rigid support structure, internal path compensation, and a one or two-dimensional layout. Baselines on the order of 35m are planned. Direct imaging has the advantage of a wide field of view and very faint limiting magnitudes. In its fail-safe mode, COSMIC can operate as a speckle imager. Current activities include laboratory simulations with six separator mirrors.

R. D. Reasenberg (CFA), "POINTS, A Small Astrometric Optical Interferometer in Space."

Business: Matters discussed included future meetings and methods of improving communication between members of the Working Group. The Working Group's Organizing Committee for 1985-8 is J. Davis (Chairman), G. P. Di Benedetto, R. Foy, R. D. Reasenberg, R. V. Stachnik & C. H. Townes.

COMMISSION 10 : SOLAR ACTIVITY (ACTIVITE SOLAIRE)

Report of Business Meeting

PRESIDENT : E. Tandberg-Hanssen   SECRETARY : R. Howard

A business meeting was held on November 20. E. Tandberg-Hanssen opened the meeting with a few introductory remarks about the present level of solar activity, research activity of the members, and new applications for membership. He then introduced the new secretary, S.R. Kane, and proposed the following persons for the positions of Commission President and Vice-President for 1985-89 :

President : M. Pick (France)

Vice President : E. Priest (Scotland/U.K.)
The members approved by applause.

Next, the following members were proposed and approved for the Organizing Committee : C. Alissandrakis (Greece), R. Falciani (Italy), T. Hirayama (Japan), S.R. Kane (U.S.A.), V. Krishan (India), G.V. Kuklin (U.S.S.R.), M. Machado (Argentina), B.Valnicek (Czechoslovakia), Z.D. Zhang (China). Because Machado is not a U.A.I. Member, the name of P. Kaufman (Brazil) has been proposed after the General Assembly.

Reports of standing committees (FAGS, IUWDS, QSBA, SIDC and SCOSTEP) were then presented. Except SCOSTEP, standing committee representatives will continue in their present positions for the next 4 years. The new representative for SCOSTEP will be F. Wu (USA).

Special reports on the Max '91 Study Plan and the Status of Solar-Terrestrial Monitoring Programs were presented by S. Jordan and H. Coffey, respectively. Other planned observations in the U.S.S.R., Europe and the U.S.A. for the 1991 solar maximum were also presented.

The importance of ground-based observations was stressed by J.C. Pecker. It was supported by M. Kundu, R.W. Noyes and M. Pick. A resolution has been formulated to express this need for solar activity observations.

Changes in the heliographic Coordinate system were proposed by Commission 4 (Ephemerides). The members did not agree with these changes and approved the following resolution :

Considering the importance of the Carrington reference system for statistical studies of solar phenomena from long time-series observations, and the importance of continuity between more than 100 years of past solar data and future data of similar type; and
Considering the difficulty at the present time to significantly improve the determination of the position of the solar axis;
Commision 10 recommends that the Carrington reference system continue to be generally used.

Future colloquia on the Contribution of Amateur Astronomers to the Discovery of the Universe and on Solar Prominences were endorsed by the membership. Also, the suggestion that some of the solar activity data in Solar Geophysical Data be made available on a magnetic tape (instead of the "yellow book") was supported by the following resolution :

Considering the amount of solar data processed by the QBSA (Quarterly Bulletin of Solar Activity), and

Considering the cost and bulkiness of providing hard copies of all data versus the increasing use of magnetic tape storage,

Commission 10 recommends that judicious choises be made in order to reduce the use of hardcopying and promote the use of magnetic tapes in data handling for QBSA.

SCIENTIFIC SESSIONS

1. Magnetic Reconnection in Astrophysical Plasmas (E.R. Priest), November 21

   Joint Commission (10, 12) Meeting (half-day)
   Programme : 5 invited talks.
   T.G. Forbes (Durham, U.S.A.) : Numerical Reconnection Experiments
   R.S. Steinolfson (Texas, U.S.A.) : Tearing Mode Instability
   Y. Uchida (Tokyo, Japan) : Reconnection and Stellar Activity
   A.O. Benz (Zurich, Switzerland) : Reconnection in Solar Flares
   M. Dubois (Fontenay, France) : Reconnection In Laboratory Machines

2. Coronal Activity and Interplanetary Disturbances

   Joint Commission (10, 12, 49) Meeting, (M.R. Kundu) November 26 (half-day).

   Altogether 10 papers were presented at this meeting. Coronal activity at radio wavelengths was discussed by M.R. Kundu and P. Lantos. Kundu used Clark-Lake 2-dimensional radioheliograph pictures (obtained with time resolution of 0.6 sec - 1 min) in the frequency range 15-25 MHz to discuss large scale structures of the upper corona, synoptic charts at 50 MHz, their comparison with white light synoptic charts and He 10830 A spectroheliograms and microbursts at meter-decameter wavelengths. Lantos used Nançay 169 MHz two dimensional synthesis maps to discuss their relationships with white light synoptic charts and coronal holes as well as Clark Lake low frequency maps obtained at nearly the same time. M. Pick discussed the multifrequency use of the Nançay radioheliograph. R.C. Stone discussed kilometer wavelength radiobursts and their use in remote sensing of the interplanetary medium. In particular, Stone discussed three dimensional trajectories, type III bursts and the three dimensional large scale magnetic field structures along which type III electron streams propagate, and the properties of interplanetary shock structure derived from type II bursts. E. Antonucci discussed the coronal response to the energy relased in flares, using the Solar Maximum Mission (SMM) soft X-ray spectroscopic data. She emphasized the role of the onset of turbulence a few minutes prior to the flare onset. B.C. Low reviewed the properties of coronal transient phenomena, as observed with the SMM coronograph polarimeter (C/P) experiment and reviewed several models to interpret these phenomena. A. Hewish, using the radio scintillation observations of a grid of 3000 sources per day, discussed the properties of heliospheric disturbances and traced their sources to high speed solar wind streams originating from coronal holes. In particular, Hewish emphasized that the role of flares in producing such disturbances was minimal or purely by chance coincidence. Y. Smolkov described the Sibizmir (Siberian) Solar Radio Telescope and discussed some observations of the slowly varying component at 3cm wavelength. X. Liu discussed the role of sheared coronal magnetic field in the production of flares and R. Shelke discussed differential rotation of coronal holes.

COMMISSION 12: RADIATION AND STRUCTURE OF THE SOLAR ATMOSPHERE
(RADIATION ET STRUCTURE DE L'ATMOSPHERE SOLAIRE)

Report of Meetings Held in Conjunction with the 19th General Assembly

PRESIDENT: Robert W. Noyes

I. BUSINESS MEETING OF COMMISSION 12

A business session was held on November 20, chaired by R. Noyes. A second brief meeting was held on November 26. Items addressed were:

1. Election of Organizing Committee.

The Commission elected the following slate of officers and members of the Organizing Committee for the term 1985 to 1988:

President	Max Kuperus
Vice President	John W. Harvey
Organizing Committee:	G.E. Brueckner
	J. Christensen-Dalsgaard
	L.E. Cram
	E.A. Gurtovenko
	E. Hiei
	V.A. Kotov
	K.R. Sivaraman
	J.O. Stenflo
	R.W. Noyes (Past President)

2. Election of New Members of the Commission.

Applications for membership were made by 72 individuals, of whom some were new members of the IAU, others old IAU members who had indicated their interest in joining the Commission, and still others whose membership had not formally been entered in the IAU records. All were welcomed by acclamation.

3. Commission Resolutions.

After discussion of the threatened closure of a number of optical and radio astronomy observatories around the world, the Commission co-sponsored the following Commission Resolution:

Commission 10 and 12,

considering the importance of ground-based optical and radio solar observatories for obtaining data which are critical to our understanding of the Sun, which are complementary to solar space projects, and which have scientific importance beyond solar physics, and

that many ground-based solar observatories aorund the world are threatened by closure, *recommends* that the appropriate organizations within the countries involved cooperate to ensure that an adequate network of observatories be maintained to study the Sun, taking account of the need for proper longitude coverage as well as the special contributions of some unique instruments.

Also, because the progress of both solar and solar-related stellar research at Mt. Wilson Observatory (which is also threatened with closure) are central to the activities of Commission 12, the following joint Commission Resolution was passed:

Commission 29, 36, and 12,

recognizing the continuing excellence of the facilities of the Mount Wilson Observatory for solar, stellar and interstellar research,

encourage efforts to ensure continuity of research at this observatory.

Both of these resolutions were subsequently endorsed by the General Assembly.

<u>4. Working Group on Eclipses.</u>

Dr. Fiala reported that the Working Group on Eclipses had not been asked to submit a report, but that it was still active. The Commission reconfirmed the importance of this Working Group and voted that a report of its activities should be presented at the next General Assembly.

<u>5. Discussion of Scientific Scope of Commission 12.</u>

Dr. Noyes reviewed the discussions of the relation of commissions 12 and 10, and the question whether they should be combined. It was pointed out that their activities are complementary but not redundant. Whereas Commission 10 emphasizes solar activity both as a physical phenomenon and with reference to its effect on the heliosphere and geophysics, Commission 12 emphasizes the Sun as a star, with particular reference to astrophysics. The purview of Commission 12 has broadened in recent years beyond the study of the solar atmosphere by itself, due to the development of observational and theoretical studies of the solar interior. There was discussion of changing the name of the Commission to reflect this broadened purview; a name such as "Solar Structure" was generally felt to be more appropriate to the present situation. However, it was decided to postpone decision on such an important matter, pending widespread discussion within the Commission and a possible formal vote in connection with the next General Assembly.

II. SCIENTIFIC MEETINGS SPONSORED BY COMMISSION 12

<u>1. Small-scale Structure and Dynamics of the Solar Atmosphere</u>

This meeting, co-organized by R. W. Noyes and Y. Uchida (Chairman), was held on November 20. The purpose was to discuss observations and theories of small-scale motions and magnetic fields in the photosphere and chromosphere, and their effects on coronal structure and energetics.

R. Muller discussed "Spatial Properties of the Solar Granulation", reviewing the morphological, evolution, and brightness properties of the granulation, as well as its vertical temperature and velocity structure. Of particular interest was the question of the penetration of granules, according to their size, into the stable photospheric layers. The origin and properties of the smallest granules (smaller than one arcsec) were discussed.

J. Leibacher, on behalf of the Solar Optical Universal Polarimeter (SOUP) team of investigators, discussed "Solar Granulation Time Series Obtained from the Space Shuttle". White light distortion-free images of the photospheric granulation were obtained by the SOUP instrument on Spacelab in August 1985. The data were obtained during approximately 12 hours of observing time, with a spatial resolution of about 0.4 arcsec; image stabilization in the seeing-free space environment permitted obtaining excellent data strings of a sunspot, an active region, and the quiet photosphere.

A. Dollfus described the use of the FPSS (Filtre Polarisant Solaire Selectif) instrument for high angular resolution two-dimensional imagery of solar line profile parameters. These include doppler shift (longitudinal velocity), circular polarization (longitudinal magnetic field), depth, strength, and halfwidths of spectral lines. Vector polarization maps may be obtained, and transverse magnetic field measurements are possible. A detailed description is given in *Astron. Astrophys.*, **151**, 253 (1985).

N. Weiss discussed "Theory of Interaction of Magnetic Fields and convection," stressing that the structure of photospheric magnetic fields is determined by the interaction of granular convection and isolated flux tubes. He summarized current understanding of compressible convection and magnetoconvection, and related this to the formation and location of the photospheric network of strong magnetic fields.

P. Venkatakrishnan discussed "Inhibition of Convective Collapse of Magnetic Flux Tubes by Radiative Diffusion." Using simplified models, he discussed the role played by convection in the formation of thin magnetic flux tubes. The inclusion of heat transport by radiative diffusion reduces the efficiency of collapse for the formation of strong fields. The twin effects of radiative diffusion and external convection on magnetic flux concentration leads to the prediction of three classes of flux tubes: (1) thick ($>$ 300 km) and strong (1400 G); thin ($<$ 100 Km) and weak (700 G); and intermediate size, oscillating between strong and weak phases. This result has interesting testable observational consequences, given sufficient angular resolution.

G. E. Brueckner presented "New Results of the High Resolution Telescope and Spectrograph (HRTS) Instrument on Spacelab". This instrument consists of a 30 cm Gregorian telescope with a stigmatic Tandem Wadsworth Spectrograph for the uv wavelength region 1200 to 1700 A, plus a 50A bandwith heliograph centered at 1550A and a 0.8A bandwidth mica interference filter centered on Hα. Spectra can be obtained either in a single position on the solar disk or as a raster. Time resolution can be as short as 0.9 seconds. The rasters allow construction of images of intensity, doppler shift, and linewidth. Data were presented showing observations of impulsive energy release, apparently with enough energy dissipation to heat the corona.

John Cook, with co-authors J.-D. F. Bartoe, G. Brueckner, K. P. Dere, and D. G. Socker, described "HRTS Spacelab 2 Observations of Spicular Emission at the Solar Limb". These observations were taken in all three of the wavelength bands mentioned above. The data show evolution of EUV spicules and transient bursts in chromospheric and transition region line emission above the limb over approximately 15 minutes of observations with relatively stable pointing.

Jagdev Singh, S. K. Jain, P. Venkatakrishnan, F. Recely, and W. C. Livingston discussed "Temporal Variations of HeI 10830A in the Solar Chromosphere". Very preliminary measurements of the line depth at one location on the Sun indicate periodicities near 2 min, 5 min, and 10 min. These periodicities could have their origin in either coronal dynamics, or chromospheric dynamics, or both. However, a clearer picture will emerge only after complete analysis of the data at all the 512 spatial locations that were observed.

Y. Uchida discussed "Formation of Fine Structures in the Solar Chromosphere and Corona". The origin of fibrils and spicules in the chromosphere, and fine loops in the corona above active regions, were considered in terms of a magnetodynamic picture. The mere presence of a potential magnetic field does not by itself produce activity, or even any inhomogeneity as long as the field lines have their footpoints in an iso-enthalpy potential level in a static photosphere, although it defines the shape of the structure to be produced. The coupling of the magnetic field with the motion in the high-β ($=p_g/p_m$) sub-photospheric layers, however, can introduce conspicuous inhomogeneity or activity in the low-β overlying layers. Emphasis was given to a mode of energization of the mean magnetic field by magnetic twists coming out from subphotospheric layers. This process can solve the problem of the mass supply as well as the energization of these structures. It was suggested that the vorticity in the large scale convection can be the source of the helicity to produce mass-filled hot structures in active regions, while that in the granular motion may give rise to the mass-filled structures in the quieter regions.

V. Krishan discussed "Collisions and Coalescence of Solar Coronal Loops". The corona is observed to be threaded by loop-like structures, which have their footpoints in the convection zone. The footpoints are continuously jostled by the eddies in the convection zone, as a result of which the coronal loops move in a random manner in the atmosphere in a plane perpendicular to the direction of gravity. The author studied the binary collisions between these loops and what happens at the contact surface when the loops collide. Depending upon the collision velocity, the coronal loops may coalesce and stay that way or may separate after colliding. Coronal loops with the same sense of twist have oppositely directed azimuthal components of the magnetic field, which undergoes reconnection during the collision. If the kinetic energy associated with the relative motion of the loops is more than the energy associated with the diffused or reconnnected magentic field, the loops separate after the collision; otherwise they stay bound together. An estimate of the critical collision velocity was made.

P. Sturrock discussed "Small-scale Energetic Phenomena in the Solar Atmosphere". He investigated the possibility that solar activity, known to occur on a wide range of scales of length, time, and energy, extends down to the scale characteristic of solar granulation, and what the possible observational consequences might be. He suggested that spicules are the small-scale analogues of surges, and proposed an MHD process that is responsible for both spicules and surges. He proposed that coronal heating may be attributed to the small-scale analogue of the same mechanism that produces hot plasma in solar flares.

S. Jordan described "The Solar Optical Telescope". This is the premier NASA facility for solar studies; its purpose is to obtain high-resolution data for studying the dynamics of magnetic fields at the scales where the basic physical processes occur. The SOT telescope and focal plane payload were described, as well as the major scientific objectives of the SOT mission.

Finally, two papers were read by title only, and the text made available to the attendees at the meeting. The first, by V. N. Karpinsky, discussed "Some Results of Photospheric Fine Structure Investigations at Pulkovo Observatory". Observations with the Soviet Stratospheric Observatory and Pamirs telescope lead to the conclusions that (a) the photosphere is strongly inhomogenous, with excitation temperature differences as large as 1000 K over small distances; (b) the spatial power spectrum of granulation may be represented by two power functions K^n with n=1.2 and n=-5 for the ascending and descending branches respectively; (c) the photospheric velocity field may be represented by homogenous vertical columns, abruptly terminating in a very thin (20-50 km) transition layer about 300 km above $\tau_5=1$; (d) the idea of convective overshoot is not a constructive explanation for the major observed facts concerning the granulation. The second paper, by A. B. Delone, E. Makarova, and G. V. Yakunina, dealt with "The Interpretation of Profiles of the Coronal Red and Green Emission lines for the Eclipses of 1965, '68, '70, '72, and '81: Interferometric Observations". Many of the observed profiles of both emission lines have a complicated shape. This is interpreted as a result of small moving structures in the corona, even though they were undetected by previous authors. It is suggested that earlier authors failed to detect such moving structures because of insufficient spatial resolution.

2. Convection and the Solar Radiative Output

This meeting, co-organized by W. C. Livingston and K. R. Sivaraman (chair), was held on November 26. The purpose was to discuss how convection influences the global radiation from the Sun, and by extension similar stars, as well as other aspects of the global solar radiation.

D. Dravins delivered a paper "Solar and Stellar Granulation". He noted that the study of granulation now has impact on several fields of astrophysics, for example: (a) the physics of stellar convection (possible replacement of "mixing length" concepts), (b) mechanisms of spectral line broadening (possible replacement of "turbulence" concepts), (c) interpretation of stellar radial velocity variations, such as required in the search for extra-solar system planets, (d) constraints on magnetic flux concentrations in stellar photospheres. Promising areas for study include: (a) differences between active and quiet region granulation, and between different epochs in the solar cycle, (b) differences between line profiles for lines with different conditions of formation, such as Fe I and Fe II, (c) stellar granulation, inferred through the photospheric line asymmetry signature, (d) numerical simulations of the hydrodynamics of stellar granular convection.

R. Muller discussed "Variation of the Solar Granulation". He noted that the mean intergranular distance decreases with increasing activity, while the number of granules per unit area increases; this was derived from measurements directly from photographs. Muller and Roudier have numerically processed Pic du Midi Observatory granulation photographs, obtained between 1977 and 1984. This work confirms that the number of granules per surface area decreases with decreasing activity, from 47 per 10" x 10" area in 1979 to 38 in 1984. The image processing shows that the granulation has a critical size at which changes of granule properties occur, both in the power spectrum of the area distribution, and in the fractal dimension. In addition granules of a *critical* size are the main contributors to the granule area and radiation. This critical size varies over the solar cycle from 1.15" in 1979 (solar maximum) to 1.30" in 1984. The fact that the number of granules increases with increasing global magnetic flux (active plus quiet) might indicate that the variation of granulation scale is due to the interaction between convection and global magnetic flux. On the other hand the number of granules is a decreasing function of the number of network bright points in the quiet Sun; this might indicate that the interaction occurs locally. Additional observations are needed to determine at which scale the interaction occurs.

D. Gray discussed "Stellar Granulation as Seen in Spectral Line Bisectors". Asymmetries in stellar spectral lines have been known for some time (Gray, *Ap.J.*, **235**, 508; **251**, 583). Line bisectors have been used in interpretation of these asymmetries in terms of stellar granulation (Dravins *et al.*, *Astr. Ap.*, **96**, 345; Gray, *Ap. J.*, **255**, 200; Gray and Toner, *Publ.Astr.Soc.Pac.*, **97**, 543). Granulation is observed to be more vigorous in stars of higher luminosity and higher effective temperature. Two separate effects in line bisectors lead to these conclusions. First, the velocity span of the bisectors is greater for hotter and for more luminous stars. Second, the blueward displacements of the bisectors continues right into the cores of the lines for these stars, indicating substantially greater height penetration than seen with the solar granulation. Numerical simulations involving disk integrations of two-stream models can reproduce the observed stellar line asymmetries. Disk-integration effects are very important and so introduce complications not experienced by solar observers. One prediction of the numerical simulations is a rotation effect, which was subsequently looked for and found (Gray, *Publ.Astr.Soc.Pac.*, 1986) and which allows a measure of the mean velocity of rise of the granules that is nearly independent of the other parameters of the simulation. Values of 1.5 to 2.0 km/s are deduced.

K. R. Sivaraman, R. Kariyappa, and W. C. Livingston presented a review paper "Solar Line Bisectors as a Function of Disk Position". Wavelength shifts and line asymmetries are an important diagnostic tool to infer atmospheric inhomogeneities and convective motions in solar and stellar atmospheres. After correction for sun-earth motion and gravitational redshift, typical photospheric lines show a blueshift of 300-400 m/sec, decreasing toward the limb. In addition, they show asymmetry, which also varies toward the limb. The asymmetries and their center-to-limb variation are different for different lines, depending most strongly on the height of formation. Observations made by the authors of 17 Fe I lines were analyzed, and suggest the existence of a poleward meridional motion; this result is in qualitative agreement with earlier conclusions from the migration of Hα filaments, and also Mt. Wilson and Stanford velocity data.

W. Livingston discussed "Activity Cycle Dependence of Line Bisectors in the Solar Irradiance Spectrum". Two pieces of evidence were presented that suggest a granulation component to irradiance variability. With the advent of minimum activity a slight increase of bisector amplitude is noted, which is the converse of the weakening noted at solar maximum. Secondly, his line strength archives indicate that Mn 5394 is more variable than Fe 5250, in agreement with high resolution granulation samples. In plage the opposite is true, so it is inferred that there is a significant granulation signature in the variability of the Sun observed as a star.

Jagdev Singh and T. P. Prabhu discussed "Variations in the Solar Rotation Rate Derived from Ca II Plage Areas". Daily calcium plage areas for 1951-1981 were used to derive the solar rotation period in four latitude belts. The rotation period was found to change with time quasi-periodically, with time scales ranging from the solar activity cycle period down to about 2 years. Variations in adjacent latitude belts are in phase, whereas those in different hemispheres are uncorrelated. Rotation rates from sunspot numbers have a similar behavior, although a variation on the timescale of the solar cycle is not very apparent. The total plage area, integrated over the disk, shows a dominant periodicity of 7 years in rotation rate.

H. Neckel discussed "Radiative Output of the Sun from 330 to 1250 nm". A detailed analysis of the absolute energy distributions of the Sun, of Vega, and of three "solar analog" stars yielded the results: (a) The magnitude differences between "solar analog" and the Sun can be described by a gradient proportional to $1/\lambda$, on which is superimposed a wavy pattern with amplitude of about 0.02 mag. (b)This wavy pattern is not real but is caused by systematic errors of order 0.01 mag, which affect the energy distributions of both Sun and Vega. (c) The relative gradients, spectral types, and nearly all results of detailed spectrocopic studies yield about the same position of the Sun among the stars: somewhere between 16 Cyg A and 16 Cyg B. (d)The UBV- data for the Sun are: B-V = 0.650 ± 0.005, U-B = 0.195 ± 0.005, V = -26.75 ± 0.025.

J.P.Goutail, D. Labs, H. Neckel, P.C.Simon, and G. Thuillier presented a paper "Solar Spectral Irradiance Measured from Spacelab 1". The presentation gave a short description of the experiment conception, the calibration procedure, the problems experienced in orbit and on the ground, and some results. A full length paper will be published in *Solar Physics*.

G. Brueckner described "Solar Irradiance Measurements in the Ultraviolet". A series of Solar Ultraviolet Spectral Irradiance Monitors (SUSIM) has been developed at the Naval Research Laboratory. Their purpose is to improve solar UV output data in order to define the magnitude of UV variability over a solar activity cycle. The instruments carry several spectrometers and in-flight calibration sources to distinguish between solar UV output changes and variation of instrument sensitivity. The first SUSIM was flown on Spacelab-2. From the calibration data obtained before, during, and after this flight it should be possible to derive absolute solar UV intensities which have error not greater than a few percent. Parallel to the instrument development an extensive calibration program has been carried out, using the Synchrotron Storage Ring at the National Bureau of Standards, which will be used to track the instrument sensitivity over the next 11 years.

III. OTHER SCIENTIFIC MEETINGS CO-SPONSORED BY COMMISSION 12

Two scientific meetings were sponsored by Commission 10 and co- sponsored by Commission 12. The first, on November 21, was *Reconnection in Astrophysical Plasmas*. The second, on November 26, was *Coronal Activity and Interplanetary Disturbances*. These are described in the Report of Commission 10.

In addition, Commission 12 was a co-sponsor of two Joint Discussions. The first was *Solar and Stellar Non-Radial Oscillations* (November 22). The second was *Stellar Activity: Rotation and Magnetic Fields* (November 25). The texts of these two Joint Discussions will appear in **Highlights of Astronomy**.

IV. REPORTS OF OBSERVATORIES

Two observatories submitted brief reports to Commission 12:

a) Uttar Pradesh State Observatory, Naini Tal, India (M.C. Pande)

Equivalent widths of molecular absorption lines of C O, C_2, C H, Mg H, N H, O H, and Si II have been calculated for different photospheric and facular models. The calculations indicate that observations of these lines can be useful to test homogeneous and inhomogeneous models of these regions (Tripathi *et al.*, *Bull.Astron.Soc.India*, **10**, 150, 334). The observed center-to-limb behavior of rotational temperatures of the 0-0 band of the C_2 Swan system has been compared with model calculations, accounting for saturation effects. The modeled excitation temperatures near the limb are slightly lower than observations (Sinha, *Bull.Astron.Soc.India*, **12**, 172). Line profiles of CH at disk center and near the limb yield a photospheric turbulent velocity of 4.0 km/sec with a very slight increase toward the limb (Punetha and Joshi, *Bull.Astron.Soc.India*, **12**, 249). The oscillator strength of C_2 Phillips systems was determined from observations (Sinha, *Bull.Astr.Soc.India*, **12**, 45) and found to be in good agreement with other results. Several unidentified lines in the photospheric spectrum (4800 to 5200A) were identified as lines of the C_2 Swan system (Sinha, *Bull.Astr.Soc.India*, in press).

b) Purple Mountain Observatory, Nanjing, People's Republic of China (Chen Biao).

This report stressed mainly work in China on the area of solar activity. A memorable event was the international workshop on solar physics and interplanetary travelling phenomena held at Yunnan Observatory in the fall of 1983. This was the first international meeting on such subjects ever held in China. There were comprehensive reviews of observational data from the previous solar maximum, and its interpretation. Following this meeting, a project was undertaken to organize scientific studies of the coming solar maximum. Through a series of workshops it was decided to emphasize:

1. Observations and studies on solar magnetic fields and velocity fields.

2. Observations and studies on pre- and post-flare phenomena, concentrating upon the real-time sequence high-time resolution radio and optical (10830A included) observations. Special attention should be paid to the impulsive phase of solar flares.

3. Solar activity prediction. Since Chinese astronomers are not prepared to start systematic research on fine structures on the Sun within the Chinese boundary, they have decided to join efforts with their European colleagues in the LEST Project. The Yunnan Observatory has applied for a membership in LEST Foundation and the Nanjing Astronomical Instrument Factory has also committed itself to some research efforts for this project. In a joint effort with HAO, the possibility will be explored of the installation of a K-coronagraph in the western parts of China. It is hoped that there will be increased activity on the theoretical study of solar seismology and internal motion in the solar interior.

COMMISSION 14: ATOMIC AND MOLECULAR DATA
DONNÉES ATOMIQUES ET MOLÉCULAIRES

Report of Meetings 21 and 26 November 1985

PRESIDENT: A. H. Gabriel   SECRETARY: W. L. Wiese

Business Session, 21 November 1985

1. Subject Matter of the Commission

   A useful discussion was held on this topic following the debate which started at the Patras meeting. The principal question concerns whether the Commission should broaden its area to include the provision to astronomers of data in fields other than atomic and molecular physics, such as elementary particles, thermonuclear reactions, etc. Dr. A. H. Gabriel and a small committee had been charged with a task at Patras of reviewing this question. He reported the outcome of various consultations which resulted in the following conclusions and recommendations:-

   a) The expansion to include these view fields of interest would lead to a substantial increase in the scale of the total activity and would result in a large diverse Commission, having few common interests.

   b) Only certain sub-disciplines are suited to a critical compilation of data aimed at providing a service to astronomers. These generally have already a specialized medium available for dissemination. In other areas, the Commission works best when in a collaborative science role, in which each side participates a little in the other's scientific activity.

   c) There is no evidence of pressure from the astronomy community for the Commission to move into these new areas.

   d) Until or unless such pressure is evident, the Commission should continue to concentrate on the field of atomic and molecular data.

   Following a discussion of this recommendation it was decided that for the present time the Commission should limit its field to atomic and molecular physics.

   There followed a critical examination of the Working Group structure which resulted in the following changes.

   Working Group 1 formerly included two district areas. The first concerns the primary standard of wavelength. It was felt that this problem had now moved into the field of metrology and was not of sufficient concern to astronomers. It should be dropped from the Commission's work. The remaining area concerns wavelength standards for practical spectroscopy. This should be incorporated in the Working Group which deals with Atomic spectra.

   Working Group 3 formerly concerned collision cross-sections and line broadening. This wide topic had always been split in two for preparing the Report, and it was agreed that it should now be dealt with by two Working Groups. The name of the first should be modified to Collision Processes in order to cover rate coefficients, etc.

Working Group 5 on Molecular Spectra should be expanded to include other molecular data, such as transition probabilities and dipole moments but not collision data. Its name should be changed to Molecular Structure and Transition Data.

A discussion was held on the manner in which the Commission carries out its work. There was strong support for a collaborative mode. It was felt that at the General Assembly meeting, joint commission meetings were most valuable and should be encouraged, with the astronomers taking a lead in formulating the programme.

2. Appointment of offices for the period 1985-8

The following committee was approved:-

President: R.W. Nicholls
Vice President: S. Sahal
Organizing Committee:
    A. H. Gabriel
    T. Kato
    F. J. Lovas
    S. L. Mandelstam
    H. Nussbaumer
    W. H. Parkinson
    Z. B. Rudzikas
    W. L. Wiese

Following the earlier discussion on reorganisation, the Working Groups and their Chairmen were approved as follows:-

1. Atomic Spectra and Wavelength Standards (excluding primary standards), W. C. Martin

2. Atomic Transition Probabilities, W. L. Wiese

3. Collision Processes, A. Dalgarne

4. Line Broadening, (A. H. Gabriel to discuss possible Chairman)

5. Molecular Structure and Transition Data (R. W. Nicholls to recommend Chairman)

The Commission approved the acceptance of 11 names from list A as new members of the Commission

Scientific Session, 26 November 1985

1. Photo Rate Coefficient

W. F. Huebner
T-4
Los Alamos National Laboratory
Los Alamos
NM 87545

The original list of species for which we calculated photo rate coefficients for dissociation, ionization, and dissociative ionization was limited to those of interest in comet coma studies. This list has now been expanded to include species of interest to planetary atmospheres, polutants in the earth atmosphere, and interstellar clouds. Rate coefficients are defined by

$$k \equiv \int_0^{\lambda_{th}} \tau(\lambda)\, e^{-\tau(\lambda)}\, F(\lambda)\, d\lambda,$$

where $\tau(\lambda)$ is the cross-section for the appropriate process taking into account the branching ratio, $\tau(\lambda)$ is the optical depth as measured from the top of the atmosphere or interstellar cloud, $F(\lambda)$ is the photon number flux, and λ_{th} is the wavelength at threshold. We also calculate the mean excess energy of the photolysis products

$$E_X \equiv ch \int_0^{\lambda_{th}} \left(\frac{1}{\lambda} - \frac{1}{\lambda_{th}}\right) \tau(\lambda)\, e^{-\tau(\lambda)}\, F(\lambda)\, d\lambda\, /\, k.$$

We use detailed, wavelength dependent cross-sections for all process for which such data are available, including predissociation and autoionization. But since the measured solar (or interstellar) flux is only available for finite wavelength bins, the above integrals are replaced by summations of the appropriate values in these bins.

Rate coefficients and mean excess energies have been calculated for about 90 atomic and molecular species using a detailed unattenuated ($\tau = 0$) solar spectrum for the quiet sun at 1 AU heliocentric distance. Similar calculations are being done for the active sun and for the interstellar radiation field. Several more atomic and molecular species will be added.

Cross-sections averaged over larger wavelength bins will be made available on magnetic tape so that users can calculate effective photo rate coefficients with suitable optical depth effects. These data will be too crude if Doppler shifted coincidences of atomic or molecular spectral lines with solar spectral features are important in the determination of the rate coefficients.

IMPROVEMENTS IN ATOMIC TRANSITION PROBABILITY DATA
_____

W. L. Wiese
National Bureau of Standards
Gaithersburg
MD 20899
USA

Atomic transition probability data for many atomic spectra have been improved recently, especially for the ions and neutral atoms of lighter elements. The overall distribution of new literature papers containing numerical data for the three-year period 1982-1984 is presented in Figure 1.

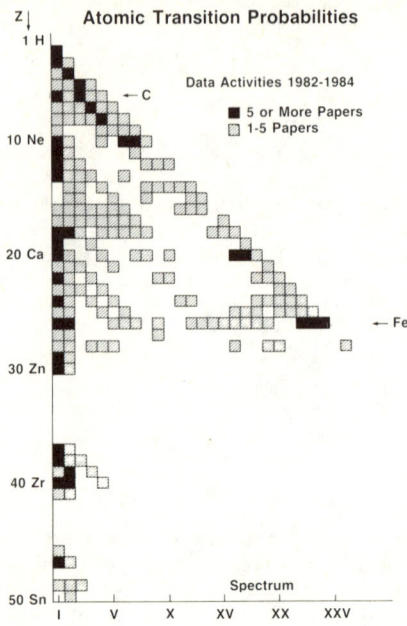

Fig. 1. Distribution of literature references on atomic transition probabilities for the time period 1982 to 1984.

This overview shows clearly that a large part of the new work is concentrated on neutral and singly ionized species. Furthermore, among the more highly ionized atoms, recent data determinations have concentrated primarily on ions of the helium-, lithium-, and beryllium-like isoelectronic sequences, which are represented by the longest diagonal lines in the figure.

From a detailed study of the literature, four major conclusions emerge:

(1) For neutral atoms of nuclear charges Z smaller than 20, atomic structure theory has advanced to a stage where it is capable of delivering quite accurate data. An important atomic structure problem is the mutual interaction between the atomic electrons, leading to the electron-correlation problem. Thus, atomic eigenstates are best described in terms of combinations of configuration state functions, each of which is a product of single-electron functions (orbitals). Such multi-configuration or superposition-of-configuration treatments, carried out within the self-consistent field or Hartree-Fock model, are the most powerful theoretical approaches which show in many cases excellent agreement with accurate experimental data, typically at the 10% level.

(2) Turning to elements of the iron group (Z = 21-28) with their increasingly complex valence shells, one observes that theorectical methods are still inadequate for the accurate determination of atomic transition probabilities while experimental approaches have made significant progress. Thus measurements are the major source of accurate data and numerous accurate results have become available on both neutral atoms and singly ionized ions in recent years. A very productive and successful approach is to combine two quite different experimental techniques:

(a) Determinations of atomic lifetimes for those levels which interact directly with the ground state via electric dipole transitions (resonance transitions).

(b) Relative emission (or absorption) measurements, including "branching ratio" measurements, for all transitions from (or to) such upper levels, as well as neighbouring levels.

Thus the strengths of the lifetime technique and emission or absorption methods are combined, i.e., the lifetime determinations yield accurate absolute scales for the relative emission or absorption line strengths covering the individual lines involved with each level. The resulting data are estimated to be quite accurate, typically in the 10-25% accuracy range, because atomic level lifetimes can be reliably determined with delayed coincidence methods using selective tunable laser excitation; furthermore, the emission or absorption measurements, being restricted to relative measurements from or to the same atomic level are independent of the properties of the source and no diagonostic measurements have to be made. When the measurements are extended to neighbouring levels, some source properties must be known, but they do not enter sensitively into the results. With the combination of these two techniques, data for the spectra of neutral and singly ionized iron, cobalt, and nickel have been drastically improved in recent years.

(3) For the heavier elements beyond the iron group many recent experiments have been concerned with the determination of atomic lifetimes. This is due to the advent of the above-cited laser excitation technique which allows a wide range of atomic species to be studied. These lifetime determinations yield the transition probabilities of a few resonance lines but do not provide a comprehensive description of the spectrum. Emission, absorption or anomalous dispersion methods involving many transitions similar to the above-described case for the iron group elements are necessary to complete the task.

(4) For highly stripped ions, only a few electrons remain and the situation for theoretical work becomes similar to that of light neutral atoms (case (1)). However, an important difference is that the nuclear charge is strongly increased, and thus relativistic effects will become noticeable. Thus relativistic multi-configuration techniques have been developed and applied to many light-element isoelectronic sequences. For highly stripped ions of Fe-group and heavier elements, relavistic effects become increasingly important.

This brief summary thus shows that newly developed theoretical and experimental techniques as well as the combination of the strong features of such techniques have brought about significant improvements in atomic transition probability data.

Reference

1. W. L. Wiese, in "Reports on Astronomy 1985," Comm. 14, D. Reidel Publ. Co., Dordrecht, Holland.

REALISTIC NUMERICAL SYNTHESIS OF MOLECULAR SPECTRA

R. W. Nicholls and M. W. P. Cann
Centre for Research in Experimental Space Science
York University
4700 Keele St.
North York.
Ontario
Canada M3J 1P3

INTRODUCTION

Spectrocopy is the classical diagnostic tool of astrophysics. Intensities and line shapes of well identified emission and/or absorption atomic and molecular features are used to provide information on species concentrations, and degree of excitation, from which gas kinetic, rotational, vibrational, electronic and excitation "temperatures" can be inferred when LTE conditions exist. Departures from LTE can also be determined spectroscopically. Diagnostic interpretation of spectra in optically thin circumstances is fairly straightforward. However, in optically thick conditions when the photon mean free path is very much less than the geometrical path, the emission spectrum is controlled by the absorption coefficient (Armstrong and Nicholls, 1972), (see equation 4a).

In high temperature and high pressure circumstances, thermal and collision broadening effects often cause strong overlap between neighbouring lines. In such cases it is very useful to be able to compare an observed spectrum profile with a realistically calculated synthetic profile whose form is controlled by relevant input data on species concentrations, temperature and pressure conditions, and basic atomic and molecular structure and transition probability constants. Over the past three decades, spectral synthesis methods of increasing power and flexibility have been developed for the diagnostic interpretation of molecular spectra which occur in astrophysics, atmospheric physics and laboratory circumstances. It is the purpose of this paper to provide a brief review of the methods and to give an example of one of them.

THE PRINCIPLE OF SPECTRAL SYNTHESIS

Spectrum synthesis is based on the well known equations (1a and 3a) for emission and absorption intensities. The integrated power of an emission spectral feature arising from transitions between an U(pper) and a L(ower) level is:

$$I_{UL} = K\, N_U\, A_{UL}\, E_{UL} \qquad (1a)$$

where k is a constant which takes account of units, optical collection geometry and instrumental response, N_U is the species number density in the upper state, A_{UL} is the Einstein coefficient or transition probability per particle per second, and E_{UL} is the energy quantum involved.

For diatomic molecular spectra, the terms of equation (1a) are:

$$E_{UL} = hc/\lambda_{UL} = hc\nu_{UL} \qquad (1b)$$

$$A_{UL} = (64\Pi^4/3hc^3)\, \nu_{UL}^3\, R_e^2\, (\bar{r}_{v'v''})\, q_{v'v''}\, S_{J'J''} \qquad (1c)$$

$$N_U = (N/Q)\, \exp(-E_U/kT) \qquad (1d)$$

where $\nu_{UL} = \nu_{oo} + \Delta G_v + \Delta F_J \qquad (1e)$

is the wavenumber of the spectral feature. It is determined by spectral structure constants of the vibrational ($G(v)$) and rotational ($F(J)$) eigenvalues (Herzberg, 1950). R_e is the electronic transition moment of the transition, $\bar{r}_{v'v''}$ is the r-centroid and $q_{v'v''}$ is the Franck-Condon factor of the band in question, and $S_{J'J''}$ is the Hönl-London factor of the line (Nicholls, 1969). When LTE conditions obtain, the Boltzman equation (1d) (in which Q is the partition function, N is the total species number density, and E_U is the upper level energy) can be used to predict species population and also to determine an appropriate temperature T. If the species of interest are formed in chemical reactions, chemical kinetics are used to estimate species concentration N and its variation with T.

Equation (1a) is recast in equation (2) as a function of ν to show how the contributions from the profiles of a number of lines contribute to the emission intensity at wavenumber ν.

$$I(\nu) = K \sum N_U R_e^2(\bar{r}_{v'v''}) \nu^4_{v'J'v''J''} q_{v'v''} S_{J'J''} b(\nu) \qquad (2)$$

Here $b(\nu)$ is the line profile function and Σ is the summation over every feature which contributes to the profile at wavenumber ν. Numerical spectral synthesis methods use this equation to calculate $I(\nu)$ at each of a number of closely spaced grid points.

Similarly, under optically thin conditions, the spectral transmission of an absorbing slab is given by:

$$I/I_o = \exp(-\tau) \qquad (3a)$$

where I_o is the incident flux and I is the emergent flux. τ is the optical depth, which when integrated over a spectral feature is:

$$\tau = N \sigma X \qquad (3b)$$

where N is the number density of absorbers, τ is the cross section per absorber, and X is the path length. When N_i absorbers of type i contribute to the extinction at wavenumber ν, equation (3b) becomes:

$$\tau = X \sum_i N_i \sigma_i(\nu) \qquad (3c)$$

and $\sigma_i(\nu) = (8\pi^3/3hc) \nu R_e^2(\bar{r}_{v'v''}) q_{v'v''} E_{v'v''J'J''} S_{J'J''} b(\nu) \qquad (3d).$

Numerical synthesis methods for absorption spectra are based on these equations.

In optically thick gases the absorption cross section $\sigma(\nu)$ controls the emission spectrum through equations (4a,b,c). Spectrum synthesis of $\sigma(\nu)$ is thus of great importance in both emission and absorption applications.

$$I(\nu) = B(\nu,T) (1-\exp(-k'(\nu,T) X)) \qquad (4a)$$

where $B(\nu,T)$ is the Planck-Function and $k'(\nu)$ is the absorption coefficient with stimulated emission taken into account.

$$k'(\nu,T) = k(\nu) (1-\exp(-h\nu/kT)) \qquad (4b)$$

and

$$k(\nu) = \sum_i N_i \sigma_i(\nu) \qquad (4c)$$

These equations have also been used to predict radiation flux through radiometer pass bands in shock tube radiometry (Arnold and Nicholls, 1973; Cooper and Nicholls, 1975).

To use equations (2), (3) and (4) for spectrum synthesis it is necessary to:
i) Determine all spectral feature line centres.
ii) Place a line profile function $b(\nu)$ appropriate to each line at its centre.

iii) Adjust the amplitude of the profile function to take account of transition probability and population factors.
   iv) Select a closely spaced grid of ν-values, and at each grid point add the contributions from each line whose profile has a significant value at the point.

Use of these procedures enables one to plot a realistic high resolution synthetic spectrum. The line profile function $b(\nu)$ derives from combined effects of collision (Lorentz) and thermal (Gaussian) line shapes. It is a Voigt function, algorithms for which exists (Armstrong, 1967; Whiting, 1968). To produce a synthetic spectrum plot for comparison with observed spectra it is also necessary to degrade the high resolution spectrum by convolution with an instrumental resolution profile. A normalized triangle function with suitable half-width can often be used for this purpose.

To perform the above procedures the following information is needed:

a) Line centre locations: They can either be determined from "look-up" tables of line centres, or from equation (1e) using molecular structure constants. Molecular structure data are not always known accurately enough, particularly for transitions between high quantum states, to locate line centres with sufficient precision for very high resolution spectra. The important compilation and assessment of Huber and Herzberg (1979) provides authoritative information on the reliability of available molecular data. Many of these data were derived from low temperature excitation sources, or in absorption.

b) The transition probability constants: Electronic transition moment $R_e(r)$, Franck-Condon factors $q_{v'v''}$ and r-centroids $r_{v'v''}$, Hönl-London factors $S_{J'J''}$. Many measurements of electronic transition moments and band strengths $S_{v'v''} = R_e^2 q$ have been reported in the literature for numerous band systems of astrophysical importance (Nicholls, 1977). Tables of Franck-Condon factors and r-centroids have been published for many band systems, based on a number of models of molecular potentials, including realistic potentials (Jarmain, 1971; Jarmain and McCallum, 1970). Approximate formulae for Franck-Condon factors are available for cases where insufficient molecular data, required for more complete calculations, exist, (Nicholls, 1981,1982). Extensive information on Honl-London factors is available (Tatum, 1967; Kovacs, 1969; Whiting and Nicholls, 1974; Whiting, Paterson, Nicholls and Kovacs, 1973). None of these data should be accepted uncritically.

c) Effective half-widths of the Gaussian and Lorentzian line profiles for evaluation of the effective Voigt profile: These depend upon temperature, pressure and molecular constants (see equations (5a), (5b) below).

MOLECULAR SPECTRAL SYNTHESIS METHODS

The concept of Band Models was the earliest method of spectral synthesis to be developed. It often adopts a statistical approach to the distribution of lines in a band and their profiles (Goody, 1974; Penner, 1959; Plass, 1958; Kidd and King, 1971). This method is widely used to model intensity distributions of polyatomic molecular spectral features.

In the 1960's a number of laboratories developed dedicated spectral synthesis codes to calculate low resolution spectral absorption coefficient data of heated air (1000 to 200,000°) from soft X-ray to IR spectral regions. The SACHA (spectral absorption coefficient of heated air) code developed at the Lockheed Palo Alto laboratories is an example of such work (Armstrong and Nicholls, 1972; Churchill and Meyerott, 1965; Landhoff and Magee, 1969).

The Atmospheric Optics Laboratory of the Air Force Geophysics Laboratory has developed a most useful set of synthetic spectrum programmes for calculation of the spectral transmittance of and radiance through the earth's atmosphere. AFGL has also developed a massive compilation of line parameters; locations: strengths, and widths for 120,000 absorption lines of atmospheric molecules (Rothman et al., 1983). The best known of these programmes is LOWTRAN (Kneizys et al., 1983) for low spectral resolution calculations based on a band model method. For the synthesis of high resolution spectra FASCOD1 which calls on the AFGL line data set has been developed (Clough et al., 1981). HITRAN is referred to in various AFGL reports for precision line-by-line calculations in small spectral regions, for example around laser frequencies (McClatchey et al.,178). All of these programmes are for applications to the terrestrial atmosphere.

One of the first truly flexible programmes for calculation of intensity distributions of molecular spectra on a line-by-line basis was developed at the NASA Ames Laboratory by Arnold, Whiting and Lyle (1969). It treats emission and absorption spectra from electronic transitions from diatomic molecules, and some atoms, on a line-by-line basis in the spirit of equations (3,4,5). The capability of this programme has been extended in our laboratories over the past decade and several new programmes have been written for application to shock tube spectra (Shin and Nicholls, 1978) discharge spectra (Danylewych and Nicholls, 1978), astrophysical spectra (Danylewych et al., 1978), atmospheric absorption studies in the UV (Cann et al., 1979), Red (Cann et al., 1982), IR (Cann et al., 1985a) and microwave regions of the spectrum (Cann and Nicholls, 1985).

HIGH PRESSURE HIGH TEMPERATURE ABSORPTION
SPECTRUM OF OXYGEN BETWEEN 180 AND 300 NM

To illustrate the numerical synthesis of molecular spectra, a brief description is given here of the results of a recent study made in our laboratories of the absorption coefficient of O_2 between 180 and 300 nm for temperatures between 300 and 3000°K, and for pressures between 1 and 50 atm. (Cann et al., 1984). While the work was done for applications to combustion and explosions, it does illustrate how the detailed character of molecular spectra are very sensitive to the molecular environment. High pressures and high temperatures are found in numerous astrophysical situations. A brief review of the principal trends is given below. Full details are given in Cann et al. (1984).

The principal contributors to the absorption coefficient of O_2 in the waveband are the O_2 Herzberg I ($X^3\Sigma - A^3\Sigma$) and Schumann-Runge ($X^3\Sigma - B^3\Sigma$) band systems and associated photodissociation contunua.

Input data for the calculations were as follows:

a) Molecular Structure Constants: The constants from which line locations were calculated were taken from the literature (Creek and Nicholls, 1975; Cann et al., 1979).

b) Transition Probability Data: These were adopted from the recommendations of a critical review of the literature (Cann et al., 1979). Franck-Condon factors for the bands, and Franck-Condon Densities for the photodissociation continua were determined by use of the TRAPB programme of Jarmain and McCallum, 1970 for realistic potentials. Hönl-London factors for effective singlet states were used, (Herzberg, 1950).

c) Line Profiles: The Voigt function profile $b(\nu)$ was generated by Whiting's (1968) approximation. The profile is specified by effective Gaussian and Lorentzian half widths W_G and W_L given by:

$$W_G = 2\nu(2kT \ln2/mc^2)^{1/2} \text{ cm}^{-1} \qquad (5a)$$

$$W_L = aP(273.2/T)^s \text{ cm}^{-1} \qquad (5b)$$

and their ratio. P is the pressure in atmospheres, T is the temperature in degrees Kelvin, a is 0.3 and s is 0.7. Predissociation broadening was added to W_L as appropriate.

d) Thermodynamic Constants: Total number densities N of O_2 molecules were interpolated from thermochemical tables (Gray, 1972) and the relative populations in thermally excited vibration levels of $X^3\Sigma$ were evaluated from equation (1d).

To minimise computing, before making calculations of high resolution spectra, quantitative assessments were made in each case of which bands make major contributions to the absorption coefficient. Lines of these bands were incorporated in the calculations. High resolution spectra included about 7000 lines (at 300°) and 32,000 lines (at 3000°). The computation increment is typically 5×10^{-5} nm and a total number of points computed for each spectrum was about 5×10^5. For comparison with typical observational spectra the high resolution spectra were degraded numerically by convolution with a triangular "slit" function of full width at half maximum of 0.05nm which as stepped across the spectrum in 0.0125nm increments.

Examples of the many spectra which were calculated are displayed in Figures 1 and 2. They respectively display spectra at 300°K (1 and 50 atm) and at 3000°K (1 and 50 atm). In each case the lower plot is 1 atm and the upper is 50 atm. As temperature and pressure increase, spectral detail is quickly lost. At lower temperatures, the absorption coefficient decreases by many decades with increase in wavelength across the spectrum. The similar decrease at high temperatures is only a few decades. This is due to the effects of thermal and pressure broadening.

REFERENCES

Armstrong, B.H. (1967), Journ. Quant. Spect. Rad. Transf. 7, 61-88.
Armstrong, B.H. and Nicholls, R.W. (1972), "Emission, Absorption and Transfer of Radiation in Heated Atmospheres", Pergamon Press, Oxford.
Arnold, J.O. and Nicholls, R.W. (1973), Journ. Quant. Spect. Rad. Transf. 13, 115-173.
Arnold, J.O., Whiting, E.E. and Lyle, G.C. (1969), Journ. Quant. Spect. Rad. Transf. 9, 775-798.
Cann, M.W.P., Nicholls, R.W., Evans, W.F.J., Kohl, J.L., Kurucz, R., Parkinson, W.H. and Reeves, E.M. (1979), Applied Optics 24, 1374-1384.
Cann, M.W.P., Miller, J.R., Nicholls, R.W. and Peterson, R.N. (1982), Chapter 2 "The Effects of the Atmosphere", pp. 2-1 to 2-39 of "Investigation of the Feasibility of Mapping Chlorophyl Fluorescence from Space", Final Report, DSS Contract, Serial OS681-00229.
Cann, M.W.P., Nicholls, R.W., Roney, P.L., Blanchard, A. and Findlay, F.D. (1985), Applied Optics 24, 1374-1384.
Cann, M.W.P. and Nicholls, R.W. (1985), "Theoretical Computations of the Microwave Transmittance and Emittance of the Earth's Atmosphere: Atmospheric Slant Path Calculations",11th Quarterly Report DSS/AES Contract, Serial OES81-00214.
Cann, M.W.P., Shin, J.B. and Nicholls, R.W. (1984), Can. J. Phys. 62, 1738-1751.
Churchill, D.R. and Meyerott, R.E. (1965), Journ. Quant. Spect. Rad. Transf. 5, 69-86.
Clough, S.A., Kneizys, F.X., Rothman, L.S. and Gallery, W.O. (1981), S.P.I.E. 277, 153-166.
Cooper, D.M. and Nicholls, R.W. (1975), Journ. Quant. Spect. Rad. Transf. 15, 139-150.
Creek, D.M. and Nicholls, R.W. (1975), Proc. Roy. Soc. Lond. 341A, 517-536.
Danylewych, L.L. and Nicholls, R.W. (1978), Proc. Roy. Soc. Lond. 360A, 557-573.
Danylewych, L.L., Nicholls, R.W., Neff, J.S. and Tatum, J.B. (1978), Icarus 35, 112-120.
Goody, R.M. (1964), "Atmospheric Radiation", Oxford.
Gray, D.E. (ed) (1972), American Institute of Physics, Handbook, 3rd edn. McGraw Hill.
Herzberg, G. (1950), "Spectra of Diatomic Molecules", Van Nostrand.
Huber, K.P. and Herzberg, G. (1979), "Constants of Diatomic Molecules", Van Nostrand.
Jarmain, W.R. (1971), Journ. Quant. Spect. Rad. Transf. 11, 421-492.
Jarmain, W.R. and McCallum, J.C. (1970), "TRAPRB, A Computer Programme for Molecular Transitions", Department of Physics, University of Western Ontario.
Kidd, K.G. and King, G.W. (1971), J. Mol. Spect. 40, 461-472.
Kneizys, F.X., Shuttle, E.P., Gallery, W.O., Chetwynd Jr., J.H., Abreu, L.W., Selby, J.E.A., Clough, S.A. and Fenn, R.W. (1983), "Atmospheric Transmittance/ Radiance Computer Code LOWTRAN 6", Air Force Geophysics Laboratory, AFGL-TR-83 0187.
Kovacs, I. (1969), "Rotational Structure in the Spectra of Diatomic Molecules", Adam Hilger.

McClatchey, R.A. and D'Agati, A.P. (1978), "Atmospheric Transmission of Laser Radiation: Computer Code LASER", Air Force Geophysics Laboratory, AFGL-TR-78-0029.

Landshoff, R.K.M. and Magee, J.L. (eds) (1969), "Thermal Radiation Phenomena: Vol. 1 Radiative Properties of Air, Vol. 2: Excitation and Non-Equilibrium Phenomena in Air", Plenum Press.

Nicholls, R.W. (1969), "Electronic Spectra of Diatomic Molecules", Chap. 7, Electronic Structure of Atoms and Molecules, (ed. H. Eyring, D. Henderson and W. Jost), Academic Press.

Nicholls, R.W. (1977), Ann. Rev. Astron. and Astrophys. 15, 197-233.

Nicholls, R.W. (1981), Astrophys. J. Supp. 47, 279.

Nicholls, R.W. (1982), Journ. Quant. Spect. Rad. Transf. 22, 481-492.

Penner, S.S. (1959), "Quantitative Molecular Spectroscopy and Gas Emissivities", Addison Wesley.

Plass, G.N. (1958), Journ. Opt. Soc. Amer. 48, 490-702.

Rothman, L.S., Gamache, R.R., Barbe, A., Goldman, A., Gillis, J.R., Brown, L.K., Toth, A., Flaud, J.M. and Camy-Peret, C. (1983), Applied Optics 22, 2247-2236.

Rothman, L.S., Goldman, A., Gillis, J.R., Picket, H.M., Pynnter, R.L., Husson, N. and Chedlin, A. (1983), Applied Optics 22, 1616, 1627.

Shin, J.B. and Nicholls, R.W. (1978), Proc. 11th International Symposium on Shock Tubes and Waves, pp. 140-147 (eds. B. Ahlborn and R. Russell), University of Washington Press.

Tatum, J.B. (1967), Astrophys. J. Supp. 14, 21-55.

Whiting, E.E. (1968), Journ. Quant. Spect. Rad. Transf. 8, 1379-1384.

Whiting, E.E. and Nicholls, R.W. (1974), Astrophys. J. Supp. 235, 1-20.

Whiting, E.E., Patterson, J.A., Nicholls, R.W. and Kovacs, I. (1973), Journ. Mol. Spect. 47, 84-98.

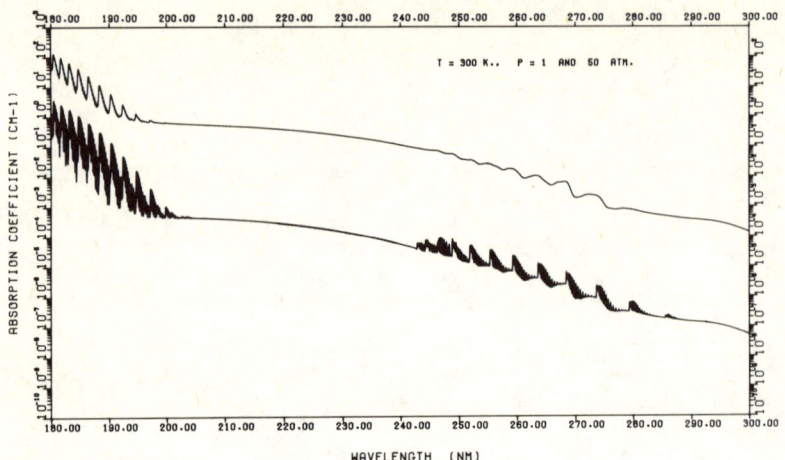

Figure 1: Synthetic 300°K O_2 Absorption Spectra at 1 (lower curve) and 50 (upper curve) Atmospheres.

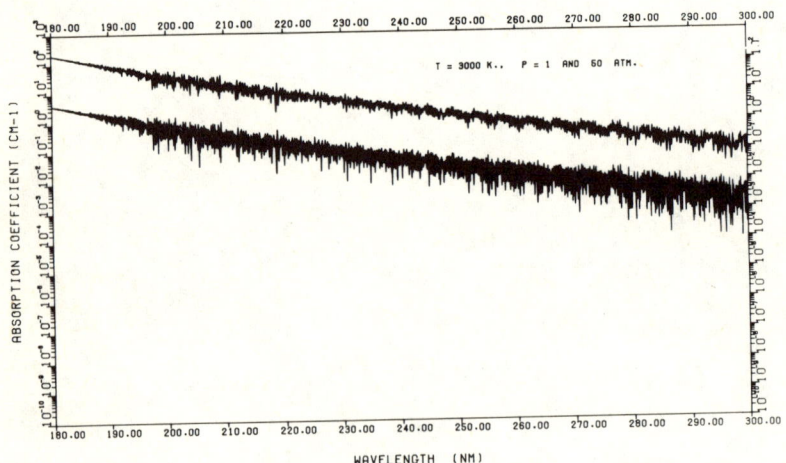

Figure 2: Synthetic 3000°K O_2 Absorption Spectra at 1 (lower curve) and 50 (upper curve) Atmospheres.

COMMENTS ON THE MEASUREMENT OF TRANSITION PROBABILITIES

Martin C E Huber and Ursula Pauls
Swiss Federal Institute of Technology Zurich (ETHZ)
Institute of Astronomy and Physical Chemistry Laboratory
ETH-Zentrum
CH-8092 Zurich
Switzerland

Developments in the measurements of oscillator strengths, that are of current and future interest to astrophysics, have taken place in the past few years:

- there is increased emphasis on methods which are independent of level populations (i.e., of number densities and thus equilibrium in the vapour or plasma under investigation),

- clear progress has been made towards higher precision (and accuracy) and

- the dynamic range of astrophysically relevant oscillator strengths was further extended towards weaker lines.

Efforts have expanded to determine branching fractions*, which can be used - together with the numerous, already measured lifetimes - to obtain absolute transition probabilities (see e.g. Whaling 1976, Wiese 1979, Hannaford et al. 1982). In addition, several methods - summarised under the name "Ladenburg methods" (Humber and Sandeman 1986) - have been developed, which combine data from emission and absorption or dispersion measurements to form a set of oscillator strengths that is independent of number densities (see e.g., Cardon, Smith and Whaling 1979, Cardon et al. 1982, Kock, Kroll and Schnehage 1984).

Techniques for evaluating spectral interferograms, as they are produced for dispersion measurements - particularly with the hook method (cf. Huber and Sandeman 1986) - have been perfected: the hook vernier (Sandeman 1979), for example, increases the sensitivity (and thus also the precision) of hook measurements by a factor of 25 over that of the conventional hook evaluation. High precision is also obtained by fitting calculated patterns to the observed spectral interferograms (Hill 1986).

Precision measurements based on low-noise spectroscopy in absorption have been reported for several 3d-elements by the Oxford Group [see, e.g., for Ti I and II Blackwell, Menon and Petford (1983 and 1982, respectively), for Cr I Blackwell, Booth, Menon and Petford (1986), for Mn I Booth et al. (1984) and for Fe I Blackwell, Petford and Simmons (1982)]. An important astrophysical result based on such precision laboratory-data was the unequivocal demonstration of an enery-dependence of the solar photospheric iron-abundance, if determined by use of local thermal equilibrium (LTE) from Fe-I lines (Blackwell, Booth and Petford 1984). Accordingly, non-LTE effects must now be taken into consideration in modelling the solar photosphere. By use of precision data, Simmons and Blackwell (1982) have also obtained evidence on the complex nature of the microturbulence parameter.

---

* The designation "branching fraction" is used here - instead of the more common "branching ratio" - to indicate the measures of relative photon flux for a complete set of decay channels from a common upper level: the sum of all branching fractions belonging to a given upper level adds up to one. "Branching ratio", accordingly, can be used for relative photon fluxes without this constraint.

Use of a Fourier-transform spectrometer (FTS) can even further reduce the noise in spectra. The signal-to-noise (S/N) ratio that can be achieved by a FTS is in principle equivalent to that of a scanning spectrometer with the same spectral resolution and optical aperture ratio; but given the entrance apertures, namely a sizeable hole versus a narrow slit, the FTS can work with a considerably higher photon flux than a conventional spectrometer: it has a throughput advantage (Brault 1985). This advantage is further enhanced in the case of an emission spectrum, and especially, if a narrow wavelength band is isolated by use of a suitable filter (Brault 1985). An example of the dramatic gain in S/N obtained by restricting the wavelength window seen by the FTS detectors is shown in figure 1. From such a spectrum, the photon flux of extremely weak Fe-II lines (solar equivalent widths $W_\lambda/\lambda \simeq 3 \cdot 10^{-6}$ and transition probabilities $A \approx 10^4$ s^{-1}) originating in a hollow-cathode discharge can be measured and subsequently branching fractions can be determined. (Note that the majority of the weak lines observed in figure 1 is unidentified; the single lines appearing in the low-noise spectrum can be assigned to Fe or the Ne carrier gas, based on their Doppler widths, but most of these lines represent hitherto unknown transitions).

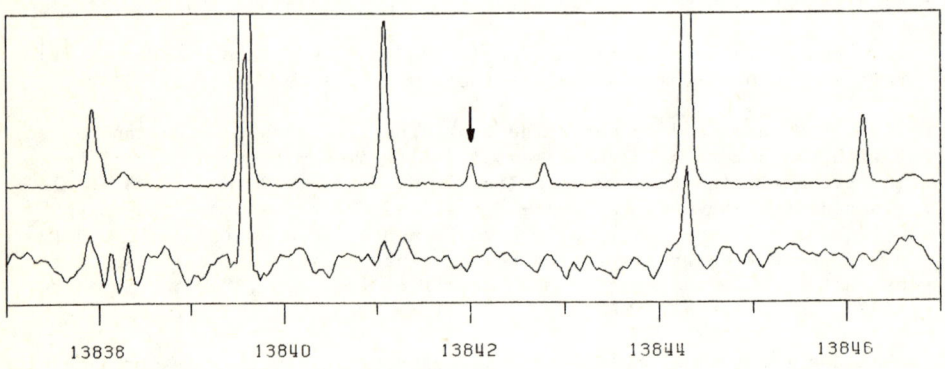

Figure 1
Comparison of two spectra (13838-13846 cm^{-1}) of the same hollow-cathode source recorded by us on the Fourier-transform spectrometer (FTS) at the U.S. National Solar Observatory on Kitt Peak (Brault 1979). The bandwidth seen by the FTS detectors is 2 nm in the upper and 300 nm - 1μm in the lower spectrum: a dramatic lowering of the noise level results. Note that the intensity scale is linear, and is expanded by a factor of ca. ten for the low-noise spectrum. The Fe-II line at 722.2394 nm (W_λ = 20 mÅ, $A \simeq 10^4$ s^{-1}) is marked by an arrow.

One of the problems of precision measurements is that the stated precision - in the case of the Oxford measures this reaches 0.5 percent - cannot easily be assessed by independent measurements. To date, independent emission measures have shown agreement with the Oxford absorption data for Cr I (Blackwell, Menon and Petford 1984) to the three-percent level (Tozzi, Brunner and Huber 1985).

A further extension of the dynamic range of the Oxford Fe-I measures by a factor of ten has just been reported (Blackwell, Booth, Haddock and Petford 1986). Such measurements are of great importance for stars cooler than the Sun. There, Fe-I lines that are suitable for solar studies become too strong for the stellar-atmosphere work that can be performed with the great photon collectors of the future. However, assessing the precision is again very difficult, because even a test by use of low-noise solar spectra is unreliable, given the extremely small equivalent widths of the solar Fe-I lines in question.

In conclusion, we mention that several methods for measuring oscillator strengths, which are based on non-linear processes, have been demonstrated in the past years. [Note that for linear as well as for non-linear optical processes, the coupling of the electronmagnetic field with atoms or molecules takes place through electric-dipole (and higher-order) moments and thus can be expressed in terms of oscillator strengths.] Non-linear methods have, of course, not yet reached the maturity and experimental perfection of the classical dispersion, absorption and emission methods, yet they have been proven in concept.

In view of their potential it would be worthwhile exploring the suitability of non-linear processes for the measurements of oscillator strengths of astrophysical interest. A review of the relevant processes and applications is given by Huber and Sandeman (1986).

References

Blackwell,D.E.,Booth,A.J.,Haddock,D.J.and Petford,A.D.1986,Mon.Not.R.astr.Soc. (in press).
Blackwell,D.E.,Booth,A.J.,Menon,S.L.R.and Petford,A.D.1986,Mon.Not.R.astr.Soc. (in press).
Blackwell,D.E.,Booth,A.J., and Petford, A.D. 1984, Astron. Astrophys. **132**, 236.
Blackwell,D.E.,Menon,S.L.R. and Petford,A.D. 1982, Mon.Not.R.astr.Soc. **201**, 603.
Blackwell,D.E.,Menon,S.L.R. and Petford, A.D. 1983, Mon.Not.R.astr.Soc. **204**, 883.
Blackwell,D.E.,Menon,S.L.R. and Petford, A.D. 1984, Mon.Not.R.astr.Soc. **207**, 533.
Blackwell,D.E.,Petford,A.D.,Shallis,M.J. and Simmons,G.J.1982,Mon.Not.R.astr.Soc. **199**, 43.
Booth,A.J.,Blackwell,D.E.,Petford,A.D., and Shallis,M.J.1984,Mon.Not.R.astr.Soc. **208**, 147.
Brault,J.W.1985, in High Resolution in Astronomy,Adv.Course of the Swiss Soc. of Astron. and Astrophys.,Saas-Fee 1985 (Geneva: Observatoire Cantonal) chap.3.2 and 5.1.1.
Brault,J.W.1979, Osservazioni e Memorie dell' Osservatorio Astrofisico di Arcetri **106**, 33.
Cardon,B.L., Smith,P.L., Scalo,J.M., Testerman, L. and Whaling, W. 1982, Astrophys. J.**260**, 395.
Cardon,B.L., Smith,P.L. and Whaling, W. 1979, Phys.Rev.A **20**, 2411.
Hannaford,P.,Lowe,R.M., Grevesse,N., Biemont,E. and Whaling,W. 1982, Astrophys.J. **261**, 736.
Hill,W.T.III 1986, Appl. Optics (submitted).
Huber,M.C.E. and Sandeman, R.J. 1986, Rep.Prog.Phys. (in press).
Sandeman,R.F. 1979, Appl. Optics **18**, 3873.
Simmons,G.J. and Blackwell, D.E. 1982, Astron. Astrophys. **112**,209.
Tozzi, G.P., Brunner, A.J. and Huber, M.C.E. 1985, Mon.Not.R.ast.Soc. **217**, 423-434.
Whaling,W. 1976, in Beam-Foil Spectroscopy, vol.2, eds. I.A.Sellin and D.J. Peggs (New York: Plenum Press) p. 705.
Wiese, W.L. 1979, in Progress in Atomic spectroscopy, eds. W. Hanle and H. Kleinpoppen (New York: Plenum Press) Part B, p. 1101.

MEASUREMENTS OF TRANSITION PROBABILITIES FOR INTERSYSTEMS LINES OF ATOMIC IONS USED IN DIAGNOSIS OF ASTROPHYSICAL PLASMAS.

Peter L Smith
Harvard-Smithsonian Center for Astrophysics
60 Garden Street
CAMBRIDGE
MA 02138
U S A

1. ASTROPHYSICAL MOTIVATION

In many low-charge ions of astrophysically abundant light elements, the spins of the ground and first-excited terms are different. Because of spin-orbit interactions, the states of these terms are mixtures of LS basis states, and as a consequence, transitions with $\Delta S = 1$, i.e., 'intersystem transitions', can occur. The transition probabilities (A-values) for such lines in low-Z ions are about 10^4 sec^{-1} to 10^2 sec^{-1}. These values are of the same order of magnitude as the collisional de-excitation rates in many low-density astronomical objects. Consequently, intersystem lines, which are not readily seen in the laboratory, are significant features in many astronomical spectra. IUE spectra that show such lines in novae, hot stars, cool stars, symbiotic stars, binary stars, variable stars, Herbig-Haro objects, H II regions, planetary nebulae, quasars, and galaxies are discussed in Kondo et al. (1982). Moreover, because of the commensurate radiative and collisional de-excitation rates for the upper levels, the ratios of intersystem-line intensities to allowed-line intensities are density sensitive in many objects.

Line ratio techniques for determining electron densities (and temperatures) in astrophysical plasmas have been reviewed most recently by Doschek (1985), who gives many references to other work. All such procedures are only as accurate as the atomic data used. The work reported here describes the first measurements of the necessary radiative transition probabilities.

2. EXPERIMENTAL METHOD

The apparatus, method, and procedure have been reviewed by Smith et al. (1984) and will be discussed only briefly here. Ions were created by electron bombardment on gases (Si^{++}, O^{++}, N^+, C^{++}) or in a laser-produced plasma (Al^+), and stored in a radio-frequency ion trap. In such a trap, electric fields are used to create a potential well that can hold ions in an approximately collision-free environment (see Wineland et al. 1983). A delay, of the order of tens of microseconds, followed the creation and storage of the ions. During this period, allowed transitions occured, thus eliminating cascade repopulation of the metastable levels during the measurement period that followed. The radiative lifetimes were measured by studying the time dependence of the intensity of the intersystem line. Then the ions were ejected from the trap. The create-store-delay-measure-eject cycle was repeated at a rate of 10 to 100 Hz. Smith et al. (1984) show the timing sequence in a schematic manner, discuss the collisional loss rates for the metastable ions, which are small relative to the measured A-values, and present details of the data analysis.

3. RESULTS

Our results and references to details of the measurements are presented in Table 1. For ions such as O^{++} and N^+, which have two intersystem decay transitions from the metastiable level, our lifetime measurement technique gives only the sum of the transition probabilities for the decays. Several of our measurements have uncertainties that are less than ±10 percent, and are, therefore, more accurate than any of the calculated values (which are given in the papers referenced in Table 1). Our results have shown that the more sophisticated theoretical techniques can produce accurate transition probabilities for the intersystem lines used in astrophysical plasma diagnosis, but in some cases, inaccurate values have been used in electron density determinations for some astrophysical plasmas.

TABLE 1

Measured Transition Probabilities for Intersystem Transitions

ION	TRANSITION				λ(nm)	$A(10^3 \text{ sec}^{-1})$	NOTE
Si III	$3s^2$	$^1S_0$	$- 3s3p$	$^3P_1^0$	189.203	16.7 (± 1.0)	a
O III	$2s^3 2p^3$	$^4P_{2,3}$	$- 2s3p^4$	$^6S_3^5$	166.080 166.615	0.82($\neq$.05)	b
Al II	$3s^3$	$^2S_0$	$- 3s3p$	$^4P_2^5$	266.916	3.33($\neq$.23)	c
N II	$2s^3 2p^3$	$^4P_{2,3}$	$- 2s2p^4$	$^4S_3^5$	213.968 214.355	0.24($\neq$.03)	d
C III	$2s^3$	$^2S_0$	$- 2s2p$	$^4P_2^5$	190.873	0.07($\neq$.02)	e

a Kwong et al. (1983a)
b Johnson et al. (1984)
c Johnson et al. (1986)
d Knight (1982)
e Kwong et al. (1983b)

ACKNOWLEDGEMENTS

This work was performed in collaboration with H S Kwong, R D Knight, B C Johnson, and W H Parkinson. It was supported in part by NASA grants NGL-22-007-006 and NSG-7034 to Harvard College and by the Smithsonian Institution. Presentation of this paper was made possible in part by grants from the Smithsonian Institution Foreign Currency Fund and from the U S National Committee of the IAU.

REFERENCES

Doschek, G.A., 1985, in Autoinization, ed. A. Temkin (Plenum: New York) p.171.
Johnson, B.C., P.L. Smith, and R.D. Knight 1984, Astrophys. J. **281**, 477.
Johnson, B.C., P.L. Smith, and W.H. Parkinson 1986, in preparation; also Measurements of Radiative Lifetimes of Spin-Changing Transitions in Ions of Astrophysical Interest, B.C. Johnson, Ph.D. Thesis, Harvard University (1985).
Knight, R.D. 1982, Phys. Rev. Lett. **48**, 792.
Kondo, Y., J.M. Mead, and R.D. Chapman 1982, Advances in Ultraviolet Astronomy: Fours Years of IUE Research, NASA CP-2238 (Greenbelt, MD: NASA).
Kwong, H.S., B.C. Johnson, P.L. Smith, and W.H. Parkinson 1983a, Phys. Rev. A **27**, 39.
Kwong, H.S., P.L. Smith, W.H. Parkinson, and R.D. Knight 1983b, Bull. Am. Phys. Soc. **28**, 781; also, P.L. Smith, H.S.Kwong, B.C. Johnson, and W.H. Parkinson, Bull. Am. Astron. Soc. **15**, 703 (1983).
Smith, P.L., B.C. Johnson, H.S. Kwong, and W.H. Parkinson 1984, Physica Scripta **T8**, 88.
Wineland, D.J., W.M. Itano, R.S. Van Dyck 1983, in Advances in Atomic and Molecular Physics **19**, eds. B. Bederson and D.R. Bates (New York: Acedemic Press) p. 135.

SPECTRA DUE TO DIELECTRONIC RECOMBINATION

J Dubau
Observatoire de Paris
92195 Meudon Principal Cedex
and
F Bely-Dubau
Observatoire de Nice
BP 139
06003 Nice Cedex

Dielectronic recombination (DR) is an electron-ion process particularly effective in high temperature plasmas such as those observed in the Solar Corona, Supernovae remnants in fusion plasmas (Tokamak and laser produced plasmas). This process is a resonant capture process of projectile electrons by a target ion as one of the target electron is excited, thereby forming an intermediate autionising state which can decay radiatively to a singly excited state. DR plays an important role on the establishment of ionisation equilibrium in the plasma and is also responsible for spectral lines appearing as satellites of the resonance lines of the target ion. The analysis and interpretation of such satellite lines in terms of plasma diagnostics has been widely used in soft X-ray spectroscopy during the last decade, and has given reliable estimates of the physical parameters of the plasma (electron and ion temperatures and densities). In the case of H-like and He-like resonance lines, high resolution spectra have been obtained in Tokamak for Z = 14 - 28 and most of the satellites have been clearly identified. To help the reader to go further we give some references of solar studies [2,3,4,0], Tokamak [6,7], laser plasma [8,9].

REFERENCES

1. Gabriel, A.H., Jordan, C. (1972) in "Case studies in Atomic Collisions Physics II", (eds. E.W. McDaniel and M.R.C. Mcdowell) North Holland, Amsterdam P210.
2. Dubau, J., and Volonte, S., (1980), Ref. Prog. Phys. **43**, 199.
3. Bely-Dubau, F., Gabriel, A.H., and Volonte, S. (1979), Mon. Not. R. Astron. Soc., **187**, 801.
4. Antonucci, E., Gabriel, A.H., Doyle, J.G., Dubau, J., Faucher, P., Jordan, C., and Veck, N., (1984), Astron. Astrophys. **133**, 239.
5. Bitter, Von Goder, S., Cohen, S., Hill, K.W., Sesnic, S., Tenny, F., Timberland, J., Safronova, V.I., Vainshtein, L.A., Dubau, J., Loulergue, M., Bely-Dubau, F., Steenman-clark, L., (1984), A29, 661.
6. T F R group, Cornille, M., Dubau, J., Loulergue, M., 1985, **A32**, 3000.
7. Boiks, V.A., Lisina, T.G., Pikuz, S.A., Skoliliv, I., Yu, and Faenov, A., Ya. (1982). Opt. Spectrosc. (VSSR) **52**, 226.
8. Audebert, P., Geindre, J.P., Gautier, J.C., Chenais-Popovis, C., Cornille, M., Dubau, J., and Faucher, P., (1985). Phys. Rev. **A32**, 3120.

COMMISSION 15: PHYSICAL STUDY OF COMETS, MINOR PLANETS AND METEORITES (L'ETUDE PHYSIQUE DES COMETES, DES PETITES PLANETES ET DES METEORITES)

Report of Meetings 26 and 27 November 1985

PRESIDENT: C. R. Chapman   VICE PRESIDENT: L. Kresak

ACTING PRESIDENT: J. Rahe

I. BUSINESS SESSION

Commission 15 had one business and three scientific sessions. During the business session new officers and members of the Organizing Committee were elected:

President: L. Kresak

Vice President: J. Rahe

Organizing Committee: M. F. A'Hearn, C. Arpigny, C. R. Chapman, O. V. Dobrovolsky, H. Fechtig, A. W. Harris, H. F. Haupt, L. M. Shul'man, J. T. Wasson, S. Wyckoff, V. Zappala

Tribute was paid to those members of Commission 15 who had passed away since the last General Assembly: Drs. Bappu, Beyer, Demenko, Krinov, Opik, Simonenko, Swings, Vseksvyatskij.

Three working groups were established and it was suggested that additional members be added to the ones already proposed:

Working Group on Comets: C. Arpigny (chairman), J. C. Brandt, O. V. Dobrovolsky, A. C. Levasseur-Regourd

Working Group on Minor Planets: A. W. Harris (chairman), H. F. Haupt, V. Zappala

Working Group on Meteorites: J. T. Wasson (chairman), M. Fulchignoni, Y. Yavnel.

It is expected that these Working Groups coordinate the corresponding Commission 15 related activities, and assist the Commission 15 President in submitting his report to the IAU General Secretary before the next General Assembly.

A "Catalogue of Cometary Emissions" covering the spectral range from the ultraviolet to the near infrared is being prepared in Liege. A preliminary version of this catalogue will be available in early 1986 in form of a computer printout. Interested persons may obtain further information from C. Arpigny, Institut d'Astrophysique, Universite de Liege B-4200 OUGREE-LIEGE (Belgium).

II. SCIENTIFIC SESSIONS

One scientific session of Commission 15 was dedicated to "Asteroid Research". It was chaired by P. Farinella. Several short presentations reviewed a number of activities related to physical studies of asteroids, which mostly imply or require some coordination and cooperative effort at an international level.

P. Farinella reported on an international workshop held in Pisa, Italy, on July 30 - August 2, 1985, on "Catastrophic Disruption of Asteroids and Satellites". He stressed the usefulness of discussions and cooperation among planetary astronomers, solid-state physicists, and researchers carrying out laboratory experiments on related phenomena, which appear very important for an understanding of the evolution of small solar system bodies.

V. Zappala reviewed the present status of the determination of asteroid rotational poles, discussing both the different methods currently used and their relationships, and the preliminary statistics of the results for a few tens of objects. A larger sample of pole data appears to be needed to reach firm conclusions on the existence of anisotropical pole distributions at least for large asteroids. These conclusions will constrain future models of asteroid collisional evolution.

T. Gehrels discussed the possibility of observing small asteroids during space missions like Galileo, Venera and CRAF, which will carry out flyby studies of large asteroids. Hundreds of asteroids with a diameter of a few km will be within range of the cameras of these spacecraft. Earth-based telescopes could identify suitable candidates, refine their orbits and provide pointing data for spacecraft CCD cameras.

H. Schober presented a report on the planned flyby of 476 Hedwig by the CRAF mission, and described an accepted proposal for asteroid observations by the Hubble Space Telescope.

A. H. Harris gave a report on the activity of the Commission 15 working group formed in Patras to prepare an updated and comprehensive file of asteroid rotational properties. The file which will be ready within a short time, will include a standard list of data to be used for statistical studies, a list of all relevant references and a more detailed set of data to be used for planning the observational work. Results from the statistics of asteroidal rotational periods were discussed.

T. Bowell provided information on a new system to predict asteroid magnitudes as a function of phase angles, which was also discussed by Commission 20. It allows a direct comparison between asteroid and satellite magnitudes.

Z. Knezevic described the joint work on asteroid families in progress by groups in Belgrade, Nice, Turin, and Pisa. An essential condition for future physical studies of families appears to be a better procedure to define proper orbital elements, since the available analytical theories appear to be unable to provide stable elements over timescales comparable with the ages of families.

L. K. Kristensen and P. Gammelgaard presented their recent work on the analysis of photometric lightcurves of 51 Nemausa. Using many observations it was shown that the lightcurves have four maxima, while constant B-V values probably exclude albedo variations. The rotation periods in B and V give an external test of mean errors. Inconsistencies were found in the proposed new slope parameter G for the magnitude vs. phase angle relationship.

D. Olsson-Steel summarized the results of a study on the impact rates of asteroids against terrestrial planets. Using Kessler's (1981, Icarus $\underline{48}$, 39) general method, lifetimes of planet-crossing asteroid populations were calculated for Mercury, Venus, the Earth, and Mars. For asteroids larger than 1 km, the derived impact rates are one per 5 million, 300,000, 160,000 and 300,000 years, respectively.

Finally C. Keay reported briefly on the current status of the theory of anomalous sounds produced by large meteor fireballs, soliciting attention to this topic by the scientific community represented in Commission 15.

Two sessions focussed on "Comet Research" and "Comet Halley". They were chaired by L. Kresak and J. C. Brandt, respectively.

J. C. Brandt described the International Cometary Explorer (ICE) encounter with the plasma tail of comet Giacobini-Zinner (G/Z) on September 11, 1985 at a tailward distance of 8,000 km. The spacecraft was in the comet for roughly 3-1/2 hours with about 20 minutes of this in the central tail region.

The initial scientific results indicate: (1) that the magnetic field capture and draping model, originated by Alfven, is correct; (2) that comets contain energetic ions, probably produced by the "pick up" process; (3) the presence of intense plasma wave activity; (4) that the central plasma tail is dense and cold; (5) that the principal ions are in the $H_2O^+ - H_3O^+$ group; (6) that the bow wave, as seen on the flanks, is not a shock but an extended interaction region; (7) that impacts of micron-sized dust particles with the spacecraft were detected.

The ICE spacecraft apparently survived the encounter with comet G/Z unscathed and is on the way to become an upstream monitor of Halley's Comet. It may detect comet Halley in early April 1986.

W. F. Huebner presented an update on modeling coma chemistry and solar wind interaction. The following processes are now taken into account: (1) energy balance at the nuclear surface - coma interface, (2) chemical kinetics including dissociation, ionization, and dissociative ionization, (3) coma energy balance and optical depth effects, (4) multifluid flow for the rapidly escaping atomic and molecular hydrogen, the heavier bulk fluid, and the plasma with separate temperatures for electrons and the remainder of the gas, (5) transition from a collision dominated inner region to free molecular flow of neutrals in the outer region, (6) pickup of cometary ions by the solar wind, (7) counter and cross streaming of neutrals with respect to the plasma which, outside of the contact surface, also contains solar wind ions, and (8) magnetic fields carried by the solar wind.

The electron temperature and density and the outstream velocity that his model predicts for comet Giacobini-Zinner are in good agreement with the measured values from the ICE intercept at 8000km from the nucleus, just at the onset of the plasma tail: $T_e=2\times10^4 K$, $n_e=700 cm^{-3}$, $v=1 km/s$.

J. Rahe summarized recent observations of comet Halley and discussed the physical properties which could be derived from these measurements. K. R. Sivaraman presented observations of comet Halley obtained in India during the last months.

P. D. Feldman described observations of comet Halley obtained with the International Ultraviolet Explorer satellite between Sept. 12 and Nov. 4, 1985. The OH (0,0) band brightness (in a 10" x 15" aperture) is 150 R on Sept. 12.1 and 340 R on Sept. 21.9. The derived OH production rate (using the vectorial model) is $2.0 \times 10^{28} s^{-1}$ for Sept. 12.1 and $4.1 \times 10^{28} s^{-1}$ for Sept. 21.9. In addition, 700 R of HI Lyman-a (in a similar aperture) was detected on Sept. 22.1, which is consistent with a primarily water ice source for the H and OH. Further observations on Oct. 19 and Nov. 4 showed the water production rate to be increasing.

A. C. Levasseur-Regourd described the Halley Optical Probe Experiment (HOPE) on board the Giotto spacecraft; it is designated to provide in situ photopolarimetric data in both the dust cloud and the gaseous atmosphere in Halley's coma. The changes in number density and grain size distribution, the spatial distribution of various emissions, and the gas to dust mass production ratio, should be obtained as a function of the probe position in Halley's coma.

V. G. Shkodrov described the observations of P/Halley and P/Giacobini-Zinner, carried out at the Rozhen Astronomical Observatory with the 2-m and the Schmidt telescopes between November 1984 and November 1985. He pointed out that observations made during this time seem to indicate that comet Halley's nucleus is not single.

COMMISSION 16: PHYSICAL STUDY OF PLANETS AND SATELLITES
ETUDE PHYSIQUE DES PLANETES ET DES SATELLITES

Report of Meetings, 20, 22, and 25 November 1985

PRESIDENT: V.G. Tejfel. SECRETARY: R.A. West.

Three regular half-day sessions of the Commission were held with topics as listed below. In addition, on 27 November 1985, the Commission (together with Commission 4) heard a report (given by M. E. Davies) of the IAU/IAG/COSPAR Working Group on Cartographic Coordinates and Rotational Elements of the Planets and Satellites.

20 November 1985

ADMINISTRATIVE MATTERS

During the meeting the following administrative matters were discussed: 13 new members of the commission were approved. V.G. Tejfel read his proposals for resolutions to form a working group on variable phenomena in the solar system and for a reference book on the physical characteristics of planets. These were discussed by the Commission.

22 November 1985

CHAIRPERSON: A. Dollfus. TITLE: Variable Phenomena in the Solar System. I.

SCIENTIFIC PRESENTATIONS

The following communications were presented during the meeting of the Commission, in order of their presentation:

- M.E. Davies: Voyager Imaging Science and Prospective Uranus Encounter.
- D. Morrison: Recent Observations of Volcanic Activity on Io.
- D. Bonneau and R. Foy: First Direct Measurements of the Diameter of the Large Satellites of Uranus and of Neptune.
- R.A. West: Vertical Cloud Structure in the Jovian Atmosphere.
- V.G. Tejfel: Polarization of Polar Regions on Jupiter (short communication).

25 November 1985

CHAIRPERSON: D. Morrison. TITLE: Variable Phenomena in the Solar System. II.

SCIENTIFIC PRESENTATIONS:

The following communications were presented during the meeting of the Commission, in order of their presentation:

- G. Pettengill: Surface Composition of Venus from Radio/Radar Observations.
- O.N. Rzhiga: Detailed Radar Mapping of the Venus Surface by the Venera-15 and Venera-16 Spacecraft.
- S. Kumar: The Escape of Hydrogen and Deuterium from Venus and Implications on Planetary Evolution.

GENERAL BUSINESS SESSION

I. ADMINISTRATIVE MATTERS

During the meeting the following administrative matters were discussed: As new officers of the Commission were proposed and unanimously nominated by the Commission for approval by the Executive-Committee of the Union: President, G.E. Hunt; Vice Presidents, A. Brahic and D. Morrison. As members of the Organizing Committee were elected by the Commission: J.L. Bertaux, J. Burns, Chen Daohan, D. Cruikshank, M. Davies, T. Encrenaz, D. Gautier, M. Marov, H. Masursky, A. Morozhenko, T. Owen, V. Shevchenko, B. Smith, and V. Tejfel.

II. RESOLUTIONS

The following resolutions were discussed and approved by the commission. They were submitted for approval by the XIX General Assembly of the IAU. The first of these was adopted by the General Assembly as Resolution C8:

Commission 16 proposes the formation of a special working group on "Planetary Surveys" to coordinate ground-based and space-based observations of variable phenomena on the surfaces and in the atmospheres of the planets and satellites. These observations must be regular and are meant to serve in planning future space missions and to complement spacecraft encounter data. They also will contribute to the understanding of possible correlations of solar activity and planetary phenomena.

The second resolution was assigned to category D by the executive committee. Commission 16 recommends to continue work on the preparation of the reference book, Physical Properties of Planets and Satellites and to appoint Drs. T. Gehrels, G. Hunt, and V. Tejfel as joint organizers who will propose the chapter titles of this book, chapter editors, and probable authors.

Commission 19: Rotation of the Earth (Rotation de la Terre)

Report of Meetings: 20-27 November 1985

President: Ya. A. Yatskiv   Vice-President: W. J. Klepczynski

20 November (Wednesday) 1330:   Commission 19 - Organizing Committee, I

The Organizing Committee met to discuss the schedule of meetings and appointments to be made during the general assembly.

21 November (Thursday) 1100:   Commission 19 - Organizing Committee, II

The organizing committee met to discuss proposed resolutions.

21 November (Thursday) 1400:   Commission 19 - Business meeting, I

President: Ya. A. Yatskiv   Secretary: W. J. Klepczynski

The agenda for the business session was approved by the membership. The results of the election of officers for the coming triennium were announced: Klepczynski was elected president and B. Kolaczck and M. Feissel were elected Vice-Presidents.

There were 12 appointments to the Organizing Committee: F. E. Barlier, P. Brosche, W. E. Carter, D. M. Djurovic, I. I. Mueller, M. G. Rochester, B. F. Schutz, J. Vondrak, G. S. Wilkins, Ya. Y. Yatskiv, Ye Shu-Hua, K. Yokoyama. Twenty eight new members were brought into the commission, bringing the total membership of the commission to 162. Eight consultants to the commission were approved: R. Anderle, E. A. Flinn, Kawajiri, M. Schuh, V. I. Sergienk, D. E. Smith, P. Wilson, and X.-y Zhu.

The President thanked everyone for their contribution to the triennial commission report. Professor I. Mueller then gave a short report on the Merit/Cotes project, summarizing its recommendations. Dr. Teleki gave a very short report in the work of Djurovic and Stajic concerning the possibility of solar activity as the ultimate cause of the 55 day cyclic variation noted by M. Feissel and D. Gambis.

The IPMS report was given by Dr. K. Yokoyama in two parts. The first part summarized the activities of the IPMS during the last three years. New software was developed to derive Earth Rotation Parameters (ERP's) from the new techniques on a daily basis. The R.M.S. of the differences between the data set as derived by the IMPS and that of the IRIS campaign is about 1.5 ms. Optical Astrometry still seems to be good in UT1 but not in PM. There is an agreement of about 0.2ms between the optical data set when compared with data sets corrected for the variation in the Earth's Atmospheric Angular Momentum. The second part of the report was concerned with ILOM. Presently, a review of the International Latitude Observatory at Mizusawa is being made. ILOM is planning to establish 2 VLBI stations within Japan in 5 years. ILOM plans to deal with optical data only until 1988.

Dr. Feissel's report in the activities of the BIH covered:

1.) Improved methods for reducing optical observations as developed at BIH by Li Zheng-xin of Shanghai Observatory;

2.) The introduction of Space Geodesy Techniques into the BIH Terrestrial System; and

3.) The BIH combined solutions for ERP's.

While the analysis of the improved optical observations is not complete, sufficient data exists to allow a comparison of a somewhat long series of optical observations with observations from the new techniques. Since 1962, no detectable drift of the BIH system is found. Dr. Feissel also reports that, starting with 1985, the BIH Terrestrial System is accessable through 34 points located throughout the world.

Dr. McCarthy of USNO reported on some of the details of Symposium #128, Earth Rotation and Reference Frames for Geodesy and Geodynamics to be held 20-24 October 1986 in the Washington, D.C. area.

Dr. Ye Shu-hua was confirmed as Commission 19 representative to the IPMS.

22 November (Friday) 1400: Merit/Cotes Project

CHAIRMAN: Session 1:   D. D. MaCarthy   SECRETARY: G. A. Wilkins
          Session 2:   Y. A. Yatskiv

ABSTRACT. A joint Meeting of Commissions 19 and 31 was held during the IAU General Assembly at Delhi to consider the recommendations for a new international Earth-rotation service put forward by the IAU/IUGG Joint Working Groups on the rotation of the Earth and the conventional terrestrial reference system. Wilkins gave summaries of the MERIT programme of activities to monitor Earth rotation and intercompare the techniques of observation and analysis and of the COTES programme to establish the basis of a new conventional terrestrial reference system. He reviewed the recommendations of the groups, and then introduced a draft resolution of the commissions. An amendment on the continuation of the use of the technique of optical astrometry was accepted and the resolution was then adopted without objection. Four papers on work related to the MERIT/COTES programmes were then presented. Paquet discussed the agreement in the results by different techniques for polar motion and universal time. Preuss presented a paper by Campbell and Schuh on short-period variations in Earth-rotation determined by VLBI. Dickey discussed the intercomparisons between the Earth-orientation parameters obtained by different techniques and then reviewed the close correlation between the length of the day and the angular momentum of the atmosphere. Finally, Vicente and Verbeiren presented new techniques for processing time and polar motion series.

1. REVIEW OF THE MERIT/COTES REPORT

The Chairman of the first session, D. D. McCarthy, Acting President of Commission 31 on Time, opened the first session of the Joint Meeting of Commissions 19 and 31 at 2 pm on Friday, 1985 November 22, during the XIXth General Assembly of the International Astronomical Union. The meeting was attended by about 60 persons.

G. A. Wilkins, the chairman of the IAU/IUGG Joint Working Group on the Rotation of the Earth, drew attention to the summary report that he had prepared with I. I. Mueller, the Chairman of the IAU/IUGG Joint Working Group on the Establishment and Maintenance of the Conventional Terrestrial Reference System. (This summary report has been reproduced immediately before this report on the meeting in Delhi.) He then summarised the objectives of the two working groups and the programme of activities that had been carried out during the past seven years. The MERIT programme (to Monitor Earth Rotation and Intercompare the Techniques of observation and analysis) involved six different techniques of observation, namely: Optical Astrometry, the Doppler Tracking of Satellites, Laser Ranging to Geodetic Satellites (SLR) and to the Moon (LLR), and radio interferometric observations of quasars using connected-elements and Very-Long-Baseline (VLBI) systems. The stimulus provided by the MERIT Short Campaign (in 1980) and the MERIT Main Campaign (in 1983/4) did much to foster the development of the new techniques of observation based on laser ranging and radio interferometry. The Earth-rotation parameters (universal time, length of day, and the coordinates of the pole) are now determined with much higher precision and better time-resolution. The analyses of the observational data clearly demonstrated the close correlation between the rotation of the crust of the Earth (as indicated by the observed length of the day) and the angular momentum of the atmosphere. Special observations and analyses were also made to determine the coordinates of the stations and the differences between the reference systems implicit in each technique; this work represents a major contribution to the COTES programme to establish and maintain a new conventional terrestrial reference system. It is now established that the tidal motions and relative drifts of the stations must be taken into account in the determination of Earth-rotation parameters and of geodetic coordinates of high precision.

The MERIT and COTES Working Groups met at the Third MERIT Workshop, which was held at Columbus, Ohio, on 1985 July 29-30, and on August 3; after reviewing the results of the campaigns they adopted three recommendations concerning the future international services for monitoring the rotation of the Earth and the adoption of new conventional terrestrial and celestial reference systems. The background to these recommendations on the reference systems is described in the paper by Wilkins in the report of the proceedings of the Joint Discussion on Reference Systems. It was recommended that the new service should be based initially on three techniques, namely VLBI, SLR and LLR, and that the general organization of the service should be similar to that adopted during the MERIT Campaigns. In the meantime the MERIT/COTES activities should be continued to ensure the continuing availability of high-quality data on Earth-rotation and to provide further data for use in defining the new reference systems.

2. RESOLUTION ON THE EARTH-ROTATION SERVICE

The Chairman (McCarthy) asked Wilkins to introduce the draft resolution on the implementation of the MERIT/COTES recommendations that had been previously circulated to all members of Commissions 19 and 31. The preamble to the resolution recognises the success of the MERIT/COTES activities, thanks all those concerned, and endorses the report and the recommendations. The main purpose of the resolution was to obtain the authority of the Union to proceed with the preparations for the setting up of a new service which would replace the International Polar Motion Service and the Earth-Rotation Service of the Bureau International de l'Heure. The Provisional Directing Board for the new service would be expected to put forward specific proposals for consideration at the general assembly of the International Union of Geodesy and Geophysics in Vancouver in 1987. It would also act as the steering committee for the extension of the MERIT/COTES programme until the new service is in operations.

The Chairman then drew attention to the amendments put forward by W. Markowitz and W. J. Klepczynski. B. Guinot and I. I. Mueller pointed that there was a separate resolution dealing with the responsibility for leap-seconds in UTC and that the provisional Directing Board would consider this matter in its review of the functions of the new service; Klepczynski withdrew the amendment.

Paquet then introduced an amendment to insert a new clause to the effect that an optical astrometric network should be maintained for the determination of UT1. He considered that the report gave the impression that such observations were no longer of value. His view was supported by Yatskiv and others who suggested minor changes to clarify the amendment. Yokoyama considered that the amendment was not necessary since the main resolution implied that the technique would continue to be used until 1988 by which time the new techniques should be able to meet the requirements for the rapid determination of UT1. After further discussion an amended version of the original amendment was adopted without objection. The amended resolution was then put to the meeting and was adopted without objection. The text of the resolution was edited by the IAU Resolutions Committee and was adopted without objection by the General Assembly on November 28. The full text of the resolution is given as annex 3 to the Joint Summary Report.

In adjourning the meeting for tea, McCarthy drew attention to the availability of the MERIT Standards document (US Naval Observatory Circular No. 167) and to the fact that some additions were in preparation. Mueller stated that the Proceedings of the Columbus Conference would be published shortly, and that the Proceedings of the Workshop and the Catalogue of MERIT/COTES data would be published in 1986; bibliographic details are given at the end of the Joint Summary Report.

3. REPORT ON SESSION 2: THE ROTATION OF THE EARTH

The Chairman of the second session, Y. A. Yatskiv, President of Commission 19 on the Rotation of the Earth, introduced the speakers who presented papers on various aspects of the methods of analysis, the intercomparison of the series obtained by different techniques, and the interpretation of the results on the variations in the rate of rotation of the Earth and on the motion of the pole of rotation.

3.1 P. Paquet presented his paper on the Agreement in Polar Motion and UT measurements during the MERIT Campaign. He had compared five series of values of the coordinates of the pole obtained during the MERIT Main Campaign (1983/4): two series were based on optical astrometric data provided by the BIH and IMPS analysis centres, and other series were based on Doppler observations of the NOVA satellite, satellite laser ranging and VLBI. He also compared two series of UT1 obtained by optical astrometry with that obtained by VLBI.

For the polar-motion data he determined and removed the annual and Chandler components. He then fitted the residuals by a smooth curve corresponding to a Vondrak filter of approximately 30 days, formed residuals with respect to this curve, and then carried out various correlation tests between the data for the different techniques. He found that for periods greater than 30 days there is a high correlation between the residuals for the various techniques, but that for periods less than one month the series are not correlated except between the SLR and Doppler series for the x-component. He considered that the Doppler results could be further improved by the use of better models for the analysis of data from NOVA satellites.

For the UT1-data (optical astrometry and VLBI only) he removed a linear drift and found residuals from a smooth curve by the Vondrak method. He then found that the three series of residuals showed an irregular variation with a period of about 50 days, and he claimed that the IPMS results for periods over 30 days are of high quality. He confirmed the very good performance of the VLBI technique and concluded that the activity in optical astrometry should be continued.

In the ensuing discussion Feissel displayed tables of operational time series on polar motion: a considerable improvement in the SLR results as a consequence of the adoption of better models was apparent. She supported Paquet's view that the Doppler results could also be improved.

3.2   E. Preuss presented a paper by Campbell and Schuh on Short-period variations of Earth-rotation determined by VLBI. The VLBI observations had been made during the MERIT Campaigns, in sessions lasting only one hour on the 6000-km baseline between Westford (USA) and Wettzell (GFR); data were obtained each day over a 64-day period and for 18 days at an interval of 15 days. The data were analysed at the IRIS analysis centre at the National Geodetic Survey (USA) to determine UT1, and a spectral analysis was then carried out. Terms with periods 9.1, 13.6 and 29.2 days were found and were claimed to be in good agreement with the Yoder model of tidal effects. McCarthy commented that USNO had obtained different periods at different times of the year from analyses of daily data from the connected-elements interferometer at Greenbank (USA). Guinot stated that N. Capitaine had also found that the 13-day term varied over an interval of 4 years.

3.3   J. O. Dickey first presented a paper by Dickey, Eubanks, Newhall, Spieth, Steppe, Sovers and Williams on work carried out at the Jet Propulsion Laboratory on Earth-orientation: analysis, intercomparisons and implications. She drew attention to the extent of the MERIT-related work at JPL: the use of the Deep Space Network (DSN) as a VLBI system to obtain regular estimates of Earth-orientation; and the operation of an LLR analysis centre, including the regular production of UT0 data. She also highlighted the recent advances in LLR at CERGA (Grasse, France), McDonald Observatory (Texas, USA) and Haleakala (Hawaii, USA) in both the quality and the quantity of the observational data.

An intercomparison of Earth-rotation and polar-motion results from a variety of services has made it possible to evaluate the accuracy of the various measurement techniques. The period considered was from 1983.5 to 1985.0. The UT1 data from LLR and VLBI (both IRIS and DSN) agree to within their formal errors; the LLR formal errors are too large, probably reflecting an overconservative analysis. On the other hand, the formal errors given by BIH from optical astrometry are significantly too small, attributable, at least in part, to seasonal errors. Differencing the different data with respect to a smoothed IRIS multi-baseline determination, the RMS differences are 0.2, 0.4, 0.5 ms for the Westford/Wettzell VLBI, the LLR, and the DSN results, respectively. In contrast, differencing the optical data with respect to a combined space-based smoothed series gives an RMS difference of 1.2 ms. Polar motion determinations from SLR by the University of Texas and from intercontinental VLBI appear to be at least as accurate as is expected from the measurement formal errors, while the errors in the results obtained by BIH from optical astrometry and in the results obtained by the Defence Mapping Agency from the Doppler measurements of the NOVA satellite are substantially larger than can be explained by their formal errors. There are no apparent periodic systematic errors in the SLR and VLBI results, but the optical astrometric and the Doppler series show appproximately annual systematic errors. The RMS difference between SLR and IRIS is about 2 milliarcseconds (mas); while the RMS differences of SLR with respect to Doppler and optical astrometry are 13 mas and 19 mas respectively for the x-component and 11 mas and 14 mas for the y-component. These results are indicative of the greatly improved performance of the new techniques.

Nutation estimates from long-duration VLBI experiments conducted by the DSN and reduced at JPL were intercompared with similar estimates from the IRIS/POLARIS data as reduced at Harvard University. The two series have and RMS difference of 1.6 mas or less. After removal of the Wahr (or IAU 1980) theory of nutation, there exist large (2 mas) annual and smaller (1 mas) semi-annual oscillations, as well as linear trends (about 1 mas/year), in both obliquity and longitude. The period of the free core nutation is about 430 d, rather than 460 d, and the damping time is about a decade; these results are consistent with interactions between core and mantle due to bumps, on the scale of 1 to 2 km, at the boundary.

In the discussion Yatskiv said that the period of 430 d has also been derived in the USSR and Dickey confirmed that the nutation corrections are consistent with those obtained by Herring.

3.4 J. O. Dickey then presented a paper by Eubanks, Dickey and Steppe on Atmospheric Angular Momentum and Earth Rotation. She began by drawing attention to the dramatic impact of the new technologies on the study of polar motion and of the variations in the rate of rotation of the Earth. The development of space geodesy has greatly increased the accuracy and precision of Earth-orientation measurements, while the analysis of global weather data for operational weather forecasting now routinely provides high quality estimates of the atmospheric excitation of the Earth's rotation. The combination of these data types has also increased the understanding of the Earth's angular momentum balance in general and, in particular, of the atmospheric as well as the non-atmospheric excitations of Earth-orientation changes.

Particular emphasis was placed on the recent data from the MERIT Main Campaign. Geodetic estimates of changes in the length of day (LOD) were compared with the corresponding meteorological excitation estimates for the period from 1983 September 1, through 1984 October 1. The geodetic excitation estimates were obtained from a Kalman smoothing of data from VLBI and LLR, while meteorological values were provided by the U. S. National Meteorological Center (NMC) and from calculations by the U. K. Meteorological office based on the results of the European Centre for Medium Range Weather Forecasting (EC). There were significant seasonal discrepancies between the EC and the NMC wind-term estimates, but seasonal errors in the pressure terms seem to be small. Changes in the EC weather-analysis software on 1984 February 1 caused a large step function change in the EC pressure term but had no observable effect on the EC wind data. The sudden jump in the pressure term was followed by a slow rebound which restored about 10% of the change over a period of 10 to 14 days. There is in general excellent agreement between the EC and NMC data (RMS difference is 0.062 ms for the wind term; 0.022 ms for the pressure term) and the geodetic LOD estimates (RMS difference is 0.072 ms for EC pressure plus wind; 0.087 ms for the NMC pressure plus wind). Anomalously high values of atmospheric angular momentum and length of day were observed in late January 1983. This signal in the time series of these two coupled quantities appears to have been a consequence of the El-Nino equatorial-Pacific warming event of 1982-83. The atmospheric estimates from both the EC and NMC results were compared with the LOD estimates; no appreciable time delay could be detected. A combined LOD series beginning in 1962 was formed by including the modern space techniques as well as the classical optical results. Studies using the longer data sets suggest that there may have been similar LOD changes during previous El-Ninos and that some of the interannual changes in the LOD are related to the Southern Oscillation. These studies reveal a correlation of -0.5 between the interannual fluctuations in the Southern Oscillation Index and in the length of day. Studies of the relation between the length of the day and the angular momentum of the atmosphere are being coordinated by IAG Special Study 5.98 (of which Dickey is the active chairman).

The discussion turned to the contribution of the ocean to the changes in the angular momentum of the Earth. It was recognised that the short-term changes due to the ocean are much less than those due to the atmosphere and that they are more difficult to monitor, but nevertheless it was agreed that it would be worthwhile attempting to obtain data on the oceanic contribution.

3.5   The session concluded with two short papers on new techniques for processing time and polar-motion series. R. O. Vicente had treated polar motion as a complex time series and had analysed data from the BIH Annual Reports for 1981-83 and also more recent data. He showed various plots and stressed the need to "be careful" in interpreting the results of the comparisons between different techniques. R. Verbieren presented a paper on the Computation of pole coordinates from the MERIT Campaign with least-squares collocations. He pointed out that the correct computation of a final set of coordinates of the pole at prespecified equidistant epochs from all available observational series is a severe statistical problem. He considered that least-squares collocation (using a technique introduced by Moritz) offers the possibility of combining in one computational step the determinations of both the coordinates of the pole and biases in the reduction constants and reference systems. The different series are introduced with appropriate weight through the use of their covariance matrices. The application of this method to the MERIT Main Campaign shows its power by giving x, y values for every day with an accuracy of 0.004; it shows the high-level of accuracy of the SLR and VLBI results and of the determination by DMA from Doppler observation of the NOVA satellite. He considered that it also shows that the methods of classical astrometry provide good results in spite of their much higher noise level. In answer to a question be stated that he had not yet used the method for prediction purposes as the series are too short.

3.6   The Chairman (Yatskiv) closed the session at 5:40 pm by thanking all the speakers and by congratulating the coordinators and other participants on the undoubted success of the MERIT/COTES Campaigns.

November 26, 1986 (Tuesday) 0900:   Commission 19 - Scientific I (Statistical Properties and Prediction of Earth Rotation Parameters)

Chairman: Y. Yatskiv   Secretary: J. O. Dickey

This joint meeting of Commissions 7, 8, 19 and 31 provided a very valuable forum for the presentation and discussion of recent work on the prediction of Earth rotation. Predictions require the most accurate observational data as a basis and a mathematical model which accurately represents the Earth's motion. The three papers given here each represented a different technique for investigating the statistical properties of Earth rotation. Morgan and Xing utilized a multi-channel Wiener filter. Eubanks and co-workers used a multi-dimensional Kalman filter. McCarthy considered the amplitude spectrum as a function of frequency with additional variations at selected periods. Discussion by both the audience and speakers addressed the motivations and requirements for Earth rotation prediction.

P. Morgan (Canberra College of Advanced Education, Australia) presented the highlights of a collaborative effort with C. Xing studying the prediction of the Earth rotation vector using a multi-channel Wiener filter. This filter was developed for the enhancement (smoothing) and prediction by means of a realizable linear operator. The Wiener filter is a linear, least squares, time-invariant process that assumes equally spaced data and uses the statistical information (regularity) of the known or historical data to predict the future behavior of the time series. The formulation of the Wiener filter is best accomplished in the frequency domain; however, the implementation is best done in the time domain using the recursive algorithms of Robinson (1976). The wobble components (x and y) and the rotation component are first detrended with polynominals models up to degree 2 and then have a number of forced periods removed from them. These periods include the Chandler and annual periods for the rotational component. The detrended series are then multiplexed into the linear Wiener filter which uses the measured series, advanced by the desired prediction interval, as the desired output series. The output from the Wiener filter is then reconstituted according to the pre-processor models. Error correcting post-processing is then performed over the last few data points to take into consideration conditions prevailing at the boundary between the known data and the prediction span. The post-processing modelling includes harmonic and polynominal terms. The output of the filter yields residuals whose magnitudes vary linearly with the distance into the prediction span.

D. McCarthy of the Naval Observatory (USNO) - Washington, D. C. addressed the activities of the USNO Earth Orientation Parameters Service. The goals are to provide high accuracy Earth orientation data with the shortest possible delay between observation and dissemination and to predict Earth orientation. Earth orientation observations, parameters and predictions are routinely distributed in the USNO - Time Service Publication - Series 7. Requirements for the various data sources were evaluated and include timeliness (rapid turnaround), precision and consistency. For polar motion prediction, the power law behavior of the amplitude spectrum is considered as a function of frequency (the resultant power is -0.84 ± 0.06) with the additional variations included at periods of 180 and 98 days. For Earth rotation (UT1) predictions, the amplitude is parameterized as a function of frequency (with the resultant power of -1.10 ± 0.07). Additional variations with the periods of 180, 111, 150, 7 and 122 days as well as Earth tides and seasonal variations are included for UT1. To evaluate the "predictability", variations were treated as discrete Fourier series. He concluded that UT1 - UTC may vary by ± 0.2 miliseconds in one day from prediction while x and y may vary by ± 1 milliarcseconds in two days from prediction. High frequency sampling of Earth orientation parameters may be a requirement. Better predictions would result from improvements in data accuracy, density and processing time and from modeling advancements.

J. Dickey (Jet Propulsion Laboratory/California Institute of Technology, USA) presented the paper entitled "Predictions and Smoothing of Earth Orientation Changes Using a Kalman Filter", by T. M. Eubanks, D. D. Morabito, and J. A. Steppe (all at JPL). A Kalman filter is being developed to provide prediction and smoothing of Earth rotation and polar motion for spacecraft navigation by the Deep Space Network. This filter uses stochastic models to account for the rapid changes in the Earth orientation primarily driven by the atmosphere by using the known physics for the relation between orientation and excitation (X_1, X_2 and X_3) (Barnes et al., 1983). The derivations of these models from atmospheric angular momentum and the development of the filter were outlined. The implementation of this filter and its performance under a variety of ideal measurement strategies as well as with actual data were discussed. The UT1 excitation (X_3 or length of day) is an integrated random walk with white noise forcing; while the polar motion excitation (X_1 and X_2) is an isotropic random walk driven by equal white noise forcing on both components. The model for the X_2 component includes a second order autoregressive (AR) oscillator with a resonance at 1 year and a damping time of 3 years. The white noise component of the X's account for the high frequency spectrum of the atmospheric excitation; the AR term accounts for the seasonal X_2 oscillation, and the random walk allows the filter to follow any secular drift in the pole position. Its advantages include: 1) data from any source or multiple sources may be used; 2) data may be evenly spaced and be of varying quality; 3) one, two or three dimensional measurements of any sets of components may be utilized; and 4) excitation estimates are provided automatically.

November 26, 1985 (Tuesday) 1100: Commission 19 - Scientific Session II

Chairman: D. McCarthy   Secretary: J. O. Dickey

This session included the discussion of Earth rotation observations, analysis, and implications on a broad spectrum of time scales. Morrison began the session presenting results from ancient and medieval eclipse data as well as occultations together spanning close to three millennia. For these long time periods, variations in Earth rotation (ΔT = the difference between ephemeris time and Universal Time) amount to hours. On the other end of the spectrum, Kolaczek discussed polar motion oscillations with periods ranging from 10 to 100 days. Kakuta addressed collocations of geophysical observations at the Earth Orientation Parameter observing sites and considered the analysis of measurements using running yearly means.

L. V. Morrison (Royal Greenwich Observatory, United Kingdom) presented the highlights of a collaborative effort with F. R. Stephenson studying observations of secular, non-tidal changes in Earth rotation. Occultations of stars by the moon, and solar and lunar eclipses were analysed for variations in the Earth's rotation over the past 2700 years. Lunar and solar tidal braking is shown to be the dominant long-term mechanism reducing the Earth's rate of rotation. Adopting the value of -26"/century square for the lunar tidal acceleration, the rate of change in length of day is ±2.40 ms/century. A parabolic formulation is used to parametrize ΔT; two separate fits yield the best representation of the data. A parabolic fit for data from A.D. 948 indicates a rate of lengthening of the day by +1.4 ms/century; while data prior to this date implies a lengthening of +2.4 ms/century. There are also non-tidal changes that vary on timescales ranging from decades to millennia. The magnitude and temporal behavior of these non-tidal variations were evaluated. (F. R. Stephenson and L. V. Morrison, Phil. Trans. R. Soc. Lond. A313 47-70, 1984)

B. Kolaczek (Space Research Center, Poland) discussed the results of recent studies by Kolaczek and Kosek on short periodical oscillations of polar motion determined by different techniques during the MERIT campaign. Several short period terms with periods ranging from 10-100 days have been detected by the Maximum Entropy Analysis and the Ormsby filter. The amplitudes of these terms range from 12 mas in the case of the BIH-Astrometric results as published in the MERIT circular to 2 mas for the CSR-LAGEOS(84 L 01) and IRIS-VLBI(85 FEB 01) solutions. Similiar short period terms were detected in the equatorial components of the atmospheric excitation functions X_1 and X_2 of the European Centre for Medium Range Weather Forecasting.

C. Kakuta (International Latitude Observatory, Japan) reported on an intercomparison study of Earth rotation (UT1-UTC) and latitude as determined by optical and radio interferometric techniques and discussed their geophysical implications. Non-tidal deformations of the Earth affect the Earth rotation parameters and may also change the terrestrial reference system. Geophysical measurements of the Earth's deformations may provide information for monitoring the terrestrial reference system. Determinations of Earth orientation parameters using data from several stations will be useful for evaluting variations of the terrestrial reference system as well as for studying excitations of the Earth's rotational motion.

November 27, 1985 (Wednesday) 1400: Commission 19 - Scientific Session III

CHAIRMAN: W. J. Klepczynski   SECRETARY: M. Feissel

G. Wilkins reported on the work at the Royal Greenwich Observatory on the Rotation of the Earth. Regular observations of Satellite Laser Ranging (SLR) started on 1983 October 2; the operation of the PZT ended on 1984 June 30. Analyses of SLR data from the RGO station as well as from the global world network were conducted and determination of a Set of Station Coordinates and of the Earth Rotation Parameters during MERIT Campaign (1983-84) were made. This included a study of time series of measurements of individual stations, and evaluation of the Earth rotation information from a single station, in view of Rapid Service determinations. Studies of the short-term oscillations in UT1 and determinations of UT1 from occultations of stars by the Moon have been realized. The possibility of tracking geostationary satellites by photography, for the improvement of the geopotential model has been investigated.

J. Popelar reported on Earth Rotation monitoring in Canada. The two PZT's in Calgary and Ottawa continued regular observations; a global reprocessing of the total series of observations is in project. The Doppler stations in the two sites have been upgraded (TRANET 2 stations); continuous geodetic monitoring of the sites with GPS receivers is planned starting with 1986; studies of the geopotential model have been performed. The implementation of the Canadian Geodetic Long Base Interferometry (CGLBI) is progressing; it will consist primarily of a master station (25-32m diameter antenna), a reference sub-station (antenna over 9m in diameter), and a mobile station (antenna under 7m in diameter), the permanent sites will be tied to their surroundings by the monitoring of a local network (40 - 100 km) of GPS receivers and gravity stations; the data processing facility, located in Ottawa, will operate in real time; the different parts of the equipment and software are currently under development and testing; the network should be operational by 1990.

Z. Li (Shanghai Observatory) reported on a new determination of the Earth Rotation Parameters from optical astrometry that he performed during his stay at the Bureau International de l'Heure (Observatoire de Paris). The series covers 20 years (1962-1982); it is based on a series of observations obtained at 136 stations, including some (about 50%) series revised since their use in the operational work of BIH, and several new series made available more recently. The computation involved 500 000 group results of latitude or universal time, referred to the IAU 1976-1980 system of constants; it is based on a new approach in which group unknown (function of the local sideral time of observations) and their time derivatives are adjusted simultaneously with the Earth Rotation Parameters and the classical auxiliary unknowns z and w. The five time series obtained for x, y (pole coordinates), UT1-UTC, and the auxiliary unknowns z and w have a sampling time of 5 days from 1962 through the end of 1981. The individual determinations have an uncertainly smaller than those of the operational BIH series; moreover, they provide the first high resolution series prior to 1967.

S. Debarbat reported on work made at Paris Observatory in connection with the introduction of new constants and new references. The observation program of the astrolabe for the Earth's rotation has been continued; guidelines for the computation of apparent positions of stars in the new IAU 1976 system for the case of astrolabe have been published by F. Chollet. The detectability of diurnal nutation by SLR has been studied by D. Gambis, who has also determined Earth rotation parameters and station coordinates for the MERIT campaign. N. Capitaine has clarified the concept of the Celestial Ephemeris Pole as defined by IAU in 1981 and studied its realization by the different techniques of observation, either geometrical or physical.

Resolution concerning classical astrometric observations approved at final meeting of General Assembly:

COMMISSION 19

"NOTING that the new International Earth Rotation Service, to become operational in 1988, already depends considerably on radio and laser-ranging techniques and eventually may be based on these methods; and

RECOGNIZING that classical astrometric determinations of latitude and Universal Time might be valuable for studies of long-period variations in Earth rotation, for improvement of star catalogues and for studying geophysical phenomena such as changes of the local vertical, variations in refraction and the possible prediction of earthquake activity;

RECOMMENDS that a working group be established to study the future role of classical astrometric observations and to report on this study to the IAU at its next General Assembly."

COMMISSION 20: POSITIONS AND MOTIONS OF MINOR PLANETS, COMETS, AND SATELLITES
(POSITIONS ET MOUVEMENTS DES PETITES PLANETES, DES COMETES ET DES SATELLITES)

Report of Meetings, 21, 23, and 27 November 1985

PRESIDENT: E Roemer   SECRETARIES: M P Candy, J Hers

21 November 1985

ADMINISTRATION I
   The President noted with pleasure the number of members of the Commission present. M P Candy and J Hers were confirmed as secretaries without objection. All present stood for a moment in remembrance of members, or former members, of the Commission who had died during the triennium: Gerald E Merton (1893-1983), S K Vsekhsvyatskij (1905-1984), Harley W Wood (1911-1984).

   The President then reported that C J Van Houten, currently the Vice President, felt not in a position to accept nomination as President, and that Y Kozai, in addition to serving as chairman of the Working Group on Satellites for a number of years, had served earlier as President of Commission 7. Approval was given to nominations of Kozai to be President, and of Yu V Batrakov to be Vice President for the triennium 1985-88. Kozai had proposed J-E Arlot as new chairman of the Working Group on Satellites, and the President proposed that R L Millis succeed Gordon Taylor, who was retiring as chairman of the Working Group on the Prediction of Occultations by Minor Planets and Satellites. Both recommendations were made after considerable correspondence with members of the Working Groups, and both were then approved.

   B G Marsden would continue as Director of the Minor Planet Center, but Ľ Kresák, who was to be President of Commission 15, wished to be replaced as chairman of the Working Group on Comets. It was agreed that Roemer would replace him. It was also agreed that K Aksnes, who would continue as chairman of the Satellite Nomenclature Liaison Committee (SNLC) and representative of Commission 20 on the Working Group on Planetary System Nomenclature (WGPSN), would become a member of the Organizing Committee. The membership of the Organizing Committee was then confirmed: Aksnes, Arlot, Candy, Kresák, L K Kristensen, Marsden, Millis, Roemer, D K Yeomans, and Y-Z Zhang.

   A number of new members of the Commission were approved. Since some additional members were also approved later, the complete list, and that of consultants for 1985-88, is given below in the report of the administrative session on 27 November.

   The President then reported on the status of resolutions. A request had been made for a subvention of SwF5000 per year for the triennium 1986-88 to help support the work of the Minor Planet Center. [Only SwF3100 was included in the proposed IAU budget. Roemer and Marsden met with the Finance Committee and the General Secretary to discuss alternative forms of IAU support for the Center.] Approval of permanent designations for several satellites had been recommended by SNLC; these would be presented and voted upon during the session on satellites. The new protocol for magnitude ephemerides of minor planets recommended by the *ad hoc* committee, a proposal concerning names for minor planets, and a recommendation about cometary nomenclature would be presented and discussed during the sessions on minor planets and comets, respectively. Votes would be deferred until the second administrative session.

SCIENTIFIC SESSION: MINOR PLANETS
   Marsden took the chair and reported first on progress during the triennium.

Among the points touched upon were the desirability of sending positions to the Minor Planet Center in machine-readable form, the availability of a new edition of the observation tape, and the extreme competitiveness of the work on identifications. Minor planets have been numbered up to 3330, but six of the numbered planets are still lost. There are problems with the publication of observations. Some journals won't accept them, and the volume in the *MPC*s is very large. Could publication in printed form be discontinued? Elements and ephemerides are available through the phone-in service, but overseas users encounter difficulties in using the system.

The matter of the transition from the reference frame B1950.0 to J2000.0 was discussed next. Though the system J2000.0 is much more accurate than B1950.0, star catalogs on the new system do not yet exist. P K Seidelmann noted that the best instructions for the transformation are those in the *Astronomical Almanac* and by Aoki et al. in *Astr Astrophys* 128, 263 (1983). But attention has to be given to the reference frame used for the observations, and some uncertainties will remain until the FK5 catalog is available. The consensus was that positions determined on the basis of star catalogs in the B1950.0 system should be reported in that system, but that there would be a gradual introduction of the J2000.0 system for the most accurate new observations.

E Bowell presented the report of the *ad hoc* Committee on Magnitude Ephemerides for Minor Planets. A two-parameter system for calculating minor planet magnitude ephemerides had been developed. One of the parameters is the reduced magnitude H at zero phase angle. The second, termed the slope parameter G, is a measure of the gradient of a minor planet's phase curve. A wide spectrum of observational data, from the steep phase curves of low-albedo minor planets to the flatter phase curves of icy satellites, is well fitted. The system reproduces the observed opposition surge and the non-linear drop-off in brightness at large phase angles, and is valid for phase angles $0° < \alpha \lesssim 120°$.

A Harris then reported on behalf of E F Tedesco on the photometric data base. Absolute magnitudes H and slope parameters G have been derived for all numbered minor planets through 3318 using the Lumme-Bowell-Harris phase function. The data base includes 10 234 magnitude records from 147 different references. Updated lists of H and G will be provided annually for publication in a suitable place, probably *Efemeridy Malykh Planet* (*EMP*). The observation data set, detailed tabulations of H and G, including error bars and the phase angle statistics, and a complete description of the methodology employed, are a part of the IRAS data base and will be published soon by Tedesco.

L K Kristensen discussed extensive photometric work on 51 Nemausa done by himself and P Gammelgaard, particularly as it pertained to the proposed new magnitude formulae and parameters. He expressed the opinion that the proposed system is based on too few observations, especially at zero phase, to allow the general extrapolation to zero phase required to obtain H. Such extrapolation would not preserve the true accuracy of observations. The zero point and scale of the slope parameter G are not defined. To obtain a stable system of magnitudes, with fundamental parameters defined with an accuracy comparable to the mean errors of photoelectric observations, he proposed that H and G be transformed to quantities that would specify the straight line that best approximates the phase curve in the interval of observation. Outside the range of the opposition effect, but in the range of most observations, this line approximates phase curves within ± 0.02 mag. Only for unusual planets would the error in magnitudes predicted this way occasionally exceed 0.1 mag.

A straw vote taken following extensive discussion showed that a substantial number of people were undecided on the relative merits of the two points of view. A special study session was arranged to consider the matter further.

B Morando reported on HIPPARCOS and minor planets. Sixty-three minor planets have been proposed for observations by HIPPARCOS, but some of them might prove too faint. The conditions of observation are such that the minor planets will be observed 43° from quadrature, the phase angle lying roughly between 15° and 21°. The problem, if one wants to reach the few milliarcseconds nominal accuracy of HIPPARCOS, is to determine the position of the photocenter to that accuracy. Data are being gathered covering semiaxes of figure, the scattering law, position of the pole, and rotation. It is hoped that the HIPPARCOS observations will lead to improved knowledge of these quantities.

L V Morrison told of observations of minor planets with the Carlsberg Automatic Meridian Circle on La Palma. Some 664 observations of position and magnitude of 14 minor planets were obtained in the interval May-December 1984. The list is to be increased to include most of the minor planets on the HIPPARCOS program. The instrument is fully automated, with a photoelectric scanning slit micrometer. The magnitude limit is m_v = 13.0, and some 850 objects can be observed each night. Comparison of positions of minor planets with numerical integrations supplied by Bec-Borsenberger show considerable differences, especially in the case of Flora. Residuals of 2" for that object would be too large to permit observation by HIPPARCOS.

Marsden remarked that with the implementation of computer-controlled typesetting, the Institute for Theoretical Astronomy had invited suggestions for revisions in the form and contents of the *EMP*. Several suggestions were made, including listing of Δ, r, and β for each date of the opposition ephemerides, as is now done in the extended ephemerides for bright planets. It was also suggested that the positions of the antisun and Moon be put in a header line along with the dates of opposition included on each page. L E Doggett invited suggestions for revisions in the form of the minor planet data published in the *Astronomical Almanac*.

Names suggested for minor planets are currently subject to review by a small committee. Some recent proposals have led to a recommendation to prohibit names glorifying individuals or events known principally for recent political or military activities or implications. Gehrels cited the recent appearance in the *MPC*s of a citation for a minor planet named for a highly respected and recently deceased member of the staff of the Kitt Peak Observatory, immediately followed by one for a cat. He thought that minor planets should not be named for animals. And who is to judge what is politics? He proposed abolishing the committee and leaving the matter to the discretion of the Director of the Minor Planet Center. But there should be the possibility of appeal in case of an adverse decision on a proposed name. Since this topic, too, generated animated discussion, the President suggested that decisions be deferred until the second administrative session, to give time for exchange of views.

Marsden then explained the background of a question that had arisen concerning the proper spelling of the name of the planet 1148. Erroneous transliteration (from French to Russian to German) caused the name to be spelled incorrectly when it was introduced in the *A.N.* as well as in recent editions of the *EMP* and other standard references. The explanation of the name in *The Names of the Minor Planets* (Cincinnati 1955, 1968) is correct, and it was agreed that henceforth the original French spelling, Rarahu (rather than Raraju), should be used.

23 November 1985

The President presented a resolution that arose from Joint Discussion I, Reference Frames, calling for Presidents of interested IAU Commissions to form a Working Group to study a number of problems related to reference frames and make recommendations to the XXth General Assembly in 1988. The resolution had been acted upon by Commissions 4 and 7 and was being considered by other Commissions that co-sponsored the Joint Discussion. After some discussion, the resolution was approved except for a point concerning possible restructuring of IAU Commissions, which was regarded as outside the purview of Commission 20. A slightly revised version appears as Resolution C2: Reference System in the report of the XIXth IAU General Assembly.

SCIENTIFIC SESSION: COMETS

Ľ Kresák remarked, in taking the chair, about the effectiveness of the international cooperation that had been organized for observations of P/Halley. He hoped that it would continue after the Halley campaign was over. Marsden confirmed that much interest in astrometry had been aroused by the campaigns on P/Giacobini-Zinner and P/Halley, but as many as half of the observations are of too poor quality to be useful, apparently as a result of lack of experience. There are also many gross errors in times. The *MPC*s have been overcrowded with observations of these two comets. On the other hand, very few observations were made of two rather bright comets that were discovered in the field with P/Halley. There is need to encourage observations of *all* comets, but especially the bright ones, for which accurate ephemerides are needed for radio observations. Astrophysicists don't understand how difficult it is to provide reliable ephemerides with insufficient data. Marsden noted that S Nakano had given particularly valuable help in reporting promptly the positions measured by Japanese observers.

Marsden then presented the report of the Comet Nomenclature Committee. A recommendation was made concerning provisional designations if more than 26 comet discoveries or recoveries are made in a calendar year, and for names of comets discovered from spacecraft. Comets discovered with the Infrared Astronomy Satellite had received the name IRAS, but those discovered with the UK Schmidt had been named for the people involved. The question arose as to when a discoverer should be credited and when a corporation. The times of perihelion of the several sungrazing comets discovered from the SOLWIND spacecraft were well defined even though the observations extended over only a few hours and were reduced some considerable time after they were made. These, and some other observations, as of P/Gunn in 1954, P/Schwassmann-Wachmann 1 in 1908, and an observation of P/Smirnova-Chernykh in 1967, one of several observations of comets recognized by Nakano, can be interpolated into the Roman numeral sequence. In some other cases, in which there are only isolated observations, there is a problem in distinguishing which objects are real, let alone in calculating an orbit. Some images of possible comets are found months after the plates were taken. Those reports that seem to refer to real objects are mentioned on the *IAU Circulars* and in the annual reports in the *QJRAS*. It was generally agreed that nothing further should be done about unconfirmed reports when there were too few observations for an orbit. However, observers should be encouraged to make enough observations to secure their discoveries.

T Gehrels then reported on his astrometric work with a charge coupled device (CCD) in the scanning mode. The 91-cm Newtonian f/5 reflector, modified to f/3.8 with a relay lens, of the Steward Observatory on Kitt Peak, is now in operation as a dedicated telescope for comets and minor planets. With drive off and the CCD charges transferred at sidereal rate, an average of six minor planets and six SAO stars are observed per set of three scans covering 1.2 square degrees, which takes 1½ hours to complete. Follow-up astrometry is done for objects with high inclination and other interesting features. Since the drive is off, the method has the advantages of a transit instrument. The precision of positions published in the *MPC*s is about $\pm 0''\!.8$. Additional information will appear in a paper by Gehrels, Marsden and Scotti in the *Astron J* in 1986.

Marsden congratulated Gehrels on recoveries of periodic comets, that of P/Shajn-Schaldach not having been expected at all. In response to a question about the magnitude limit, Gehrels stated that all known minor planets in the fields scanned were found; most discoveries are of new objects. For comets there is not a direct correlation between magnitudes from CCD scanning and the "nuclear" magnitudes from long-focus photography, as by Roemer. Marsden noted that $m_2(TG)\ 17 \approx m_2(ER)\ 19$.

Cl. Froeschlé described work he had done with H Scholl on the dynamical evolution of meteor streams in resonance with Jupiter. They had investigated numerically the dynamical evolution of particle ring systems orbiting the Sun in a 2/1 mean motion resonance with Jupiter. Only the main forces, the gravitational forces exerted by the Sun and Jupiter, were taken into account. Highly inclined rings located at resonance centers may give rise to the formation of arcs of rings. These arcs evolve separately due to the different possible modes for the motions of the ring particles' nodal lines, regression or progression. The mechanism may also explain the recently discovered arcs of a ring around Neptune. The paper will appear soon in *Astron Astrophys*.

SCIENTIFIC SESSION: SATELLITES

Kozai opened the session by summarizing and updating the report of the Working Group on Satellites, which had been included in the Report of Commission 20 in *Trans IAU XIXA*. He noted in particular a paper by Ferraz-Mello in *Celestial Mechanics* in which updated elements of many of the satellites were presented. The Bureau des Longitudes has introduced a new series of books containing ephemerides for the satellites of Jupiter, Saturn, and Uranus, and of phenomena and configurations of the brighter satellites of Jupiter and Saturn. V Abalakin provided information about some of the work in progress on satellite theories in the USSR and offered assistance in obtaining copies of publications.

On behalf of Aksnes, Seidelmann then presented the report of SNLC. The purpose of the committee, which was created during the IAU General Assembly in Patras in 1982, is to investigate and provide documentation on the orbits of newly discovered satellites and to give advice on when the orbits are sufficiently well known to warrant assignment of permanent designations (Roman numerals) and names to the satellites by Commission 20 and WGPSN, respectively. Documentation should include the measured satellite positions and residuals, and a complete set of orbital elements.

During the triennium SNLC investigated the orbits of 12 satellites: 1979 J1, 1979 J2, 1979 J3; 1980 S1, 1980 S3, 1980 S6, 1980 S13, 1980 S25, 1980 S26, 1980 S27, 1980 S28; and 1978 P1. Nomenclature for 1979 J2, 1980 S1, 1980 S3, 1980 S6, 1980 S13 and 1980 S25 was approved at the Patras meeting and later endorsed by the IAU Executive Committee, even though orbital parameters for some of these satellites are still incomplete. Following receipt of further data from S P Synnott, SNLC in July 1983 recommended approval of permanent designations, as well as names recommended by WGPSN, for 1979 J1, 1979 J3 and 1980 S28. These recommendations received IAU endorsement in September 1983. More recently SNLC has reviewed the orbital information on 1980 S26, 1980 S27 and 1978 P1. Additional data on the first two satellites were provided by Synnott in 1983 and 1984, and the breakthrough on Pluto's satellite came with the successful observation in January and February 1985 of the long awaited occultations and transits between Pluto and its satellite. Further, a new, quite satisfactory, orbit has been derived by D J Tholen (*Astron J* 90, 2353, 1985) from 19 speckle interferometric observations. The committee therefore recommends adoption of permanent designations as follows:

```
          Saturn XVI    =  1980 S27
          Saturn XVII   =  1980 S26
          Pluto I       =  1978 P1
```

This recommendation was then put to a vote of the Commission and received unanimous approval.

It was noted further that Aksnes had reviewed the orbital status of the recently discovered faint Jovian and Saturnian satellites in a paper published in *Stability of the Solar System and Its Minor Natural and Artificial Bodies*, ed V G Szebehely (Dordrecht: Reidel) 1985. WGPSN has decided to publish an annual gazetteer of planetary and satellite nomenclature. A soft-cover edition is to be produced under the direction of H Masursky at the U S Geological Survey in Flagstaff in the fall of

1985. P Millman is working on a more elaborate hard-cover version, to be published in 1986 and at intervals of 6-12 years thereafter.

Seidelmann then described techniques for observations of faint satellites with a CCD. Both short (1^s) and long (30^s-1^m) exposures are made to extend the dynamic range. Positions of faint satellites can then be measured with respect to brighter ones. Orientation is calibrated from star trails and the scale from star fields. Objects observed include several of the Jovian satellites, Nereid, and the Uranian system, including Miranda, in preparation for the Voyager encounter. Reductions made by fitting to a 3-dimensional surface yield an external precision, determined by comparison with JPL orbits, of $\pm 0\rlap{.}''10$ in x and $\pm 0\rlap{.}''16$ in y. Plans for observations with the Space Telescope include both positions and colors of Pluto's satellite, a search for additional faint satellites of the major planets and for possible satellites of minor planets as well as observations of the rings of Jupiter, Saturn, Uranus, and Neptune in search for shepherding satellites. In response to questions, Seidelmann stated that CCDs may be better suited to observation of the Jovian and Saturnian satellites than to the satellites of Uranus, but that they had not succeeded in observing the co-orbital satellites in spite of repeated attempts. Observation of satellites librating around the Lagrangian points was recognized as of great importance.

Seidelmann called attention, on behalf of Tedesco, to the opportunity for observation of mutual events in the Pluto-Charon system. Predictions for 1986 events and observing suggestions have been given by Tholen (*Astron J* 90, 2639, 1985).

Arlot spoke next, on behalf also of Morando and W Thuillot, to report on the campaign PHEMU 85 to observe mutual phenomena of the Galilean satellites. More than 300 events were observable around the world in a series that began in July 1985 and extends into early 1986. Jupiter was near opposition during the early part of the series, with events best observable from the Southern Hemisphere. Photoelectric timing of events gives satellite positions of significantly higher accuracy than can be obtained from direct photographic observations. Complementary campaigns were also organized by the Hydrographic Office and the Tokyo Astronomical Observatory in Japan and at the Perth Observatory. In addition, Aksnes used telescopes at Cerro Tololo and on La Palma.

Chr. Veillet reported on his researches concerning masses of the Uranian satellites. Various authors have pointed out that it is not possible to derive the mass of these satellites from their pericenter motions as determined using constant eccentricities. But the Laplacian quasi-commensurability between the mean motions of the three inner satellites and the high inclination (4°) of Miranda's orbit, permits the evaluation of the masses of Ariel and Umbriel in a way only slightly dependent on the mass of the outer satellites. Determination of Miranda's nodal precession motion and of the resonance effects on the longitudes by Veillet (1983) and Jacobson (1985, JPL pre-Voyager ephemeris) yield densities of 1.03 ± 0.36 (0.94) for Ariel and 1.33 ± 0.45 (1.46) for Umbriel, assuming density 1 (respectively 2) for Titania and Oberon. The mass of Miranda is found as $(0.24\pm0.06)\ 10^{-5}$ Uranus mass, yielding a radius 346 ± 85 km. Only this latter mass could be improved from the Voyager flyby. Further improvements for the other satellites need a complete redetermination of the orbital elements, including varying eccentricities, combined with Voyager radius measurements.

27 November 1985

SCIENTIFIC SESSION: OCCULTATIONS

The President presented a report on behalf of G E Taylor, retiring chairman of the Working Group. Taylor had chaired the Group since its formation at Grenoble in 1976. His retirement as chairman was occasioned by his retirement from the Royal Greenwich Observatory. He thought it might be time for a new leader, and he supported the proposal for extension of the activities of the Working Group to include coordination of observations in addition to identification of possible events and improvement of predictions. He endorsed the nomination of Millis as the new chairman.

Considerable progress has been made in the prediction and observation of occultations by minor planets and satellites in the nine years of activity. Among the results have been the discovery of the rings of Uranus, the confirmation of the satellite of Pluto, and enormous improvement in our knowledge of the sizes of some of the minor planets.

The major obstacle to the improvement of the observational success rate lies in obtaining accurate last-minute astrometry, followed by rapid reduction and updating of predictions. There are not enough observatories doing this work. It will be some time before we can hope to get digitized data from an astrometric telescope in space, but what a step forward that would be, with no need to worry about cloud cover and no need for taking, developing, and measuring plates. Although the astrographic telescope at RGO will no longer be used for last minute astrometry, Taylor hoped that other observers would get involved and that the new chairman would receive support in that aspect of the work. Taylor retains his personal interest, and hopes to see himself an occultation of a star by a minor planet. Those present gave enthusiastic support to the President's suggestion that she convey greetings and appreciation to Taylor for his dedicated service.

Bowell then gave an overview on behalf of Millis of plans for the Working Group. It is proposed that the Working Group identify those occultations that warrant a serious prediction refinement effort, orchestrate that effort, and serve as a conduit through which prediction updates can be quickly disseminated to observers around the world. Selection of events would be guided by knowledge of which objects are intrinsically more interesting on physical grounds. Many members of the Working Group would serve as regional coordinators in various parts of the world, both for dissemination of information and for coordination of arrangements for observations. The group at Lowell Observatory has been involved for a number of years in the identification of events, and they expect to continue that work. With the new 18-inch astrograph, they are now in a good position to contribute also to the refinement of predictions.

A list of potential members of the Working Group was compiled by so selecting among individuals who had been active previously in identifying, refining predictions for, or observing events as to obtain the best possible geographical coverage. Several new contacts developed at the General Assembly are expected to lead to extension of the network to additional areas. The Commission agreed to the President's proposal that Millis be allowed flexibility in defining the formal membership of the Working Group and in appointing regional coordinators.

ADMINISTRATION II

Several additional new members of the Commission were approved. The complete list was then: J-E Arlot, M E Bailey, Z C Chol, J Churms, G De Sanctis, A Dollfus, G Forti (reinstatement), J B Gibson, H F Haupt, J Henrard, J-L Heudier, V Ivanova, T P Kisseleva, L E da Silva Machado, H S Mahra, A Mrkos, C D Murray, S Nakano, H J Reitsema, H J Schober, V G Shkodrov, I Stellmacher, J Svoreň, S P Synnott, D B Taylor, G B Valsecchi, C Veillet, R Vieira Martins, H Wroblewski, and S A Yabushita. J A Bruwer and J Kovalevsky have resigned from the Commission.

The following were approved as Consultants to the Commission for 1985-88: C M Bardwell, K I Churyumov, R W Farquhar, W Ferreri, E I Kazimirchak-Polonskaya, Z M Pereyra, V Protitch-Benishek, N Samojlova-Yakhontova, T Seki, and A L Whipple.

Composition of Working Groups was then confirmed as follows:

Comets: N A Belyaev, M P Candy, A Gilmore, Ľ Kresák, B G Marsden, S Nakano, H Rickman, E Roemer (chm), G Sitarski, R M West, P Wild, D K Yeomans.

Satellites: K Aksnes, J-E Arlot (chm), S Ferraz-Mello, P Ianna, J Lieske, T Nakamura, D Pascu, M Rapaport, P K Seidelmann, V Shor, D B Taylor.

Marsden reported that near-unanimous agreement on a new protocol for magnitude ephemerides of minor planets had eventually been reached at the study session, which had been attended by 10-15 of those most seriously interested. There had also been further discussion of possible changes in the contents and format of the *EMP*, in particular the section that contains the standard opposition ephemerides.

Bowell then presented a slightly revised version of the resolution for formal consideration by the Commission. In the course of discussion, the idea grew that it might be advisable to switch from the B to the V photometric system along with the introduction of the new protocol for magnitude ephemerides. Though photographic magnitudes probably would continue to dominate for another ten years, V magnitudes are more appropriate for CCD observations. Much of the primary photoelectric data is in the V system. Gehrels and Harris agreed that it should be possible for Tedesco to furnish V absolute magnitudes. The proposed resolution was then put to a vote and approved nearly unanimously as follows:

(1) Commission 20 recommends that the minor planet magnitude system put forward by the *ad hoc* Committee on Magnitude Ephemerides be adopted for use in publications that conform with the policies of the Commission. A formula for the prediction of the apparent magnitude of a minor planet is

$$V = 5 \log r \Delta + H - 2.5 \log \left[(1 - G)\Phi_1 + G \Phi_2 \right],$$

where r and Δ are, respectively, the heliocentric and geocentric distances (in AU), H is the absolute magnitude (in the V band unless otherwise specified) at solar phase angle $\alpha = 0°$, G is termed the slope parameter, and Φ_1 and Φ_2 are two phase functions approximated by

$$\Phi_i = \exp \left[-A_i (\tan \tfrac{1}{2} \alpha)^{B_i} \right]; \quad i = 1, 2$$

$$A_1 = 3.33 \qquad A_2 = 1.87 \qquad B_1 = 0.63 \qquad B_2 = 1.22$$

(2) It is recommended that, for numbered minor planets, values of H and G be published annually in the *Efemeridy Malykh Planet*, that files of photometric data be maintained and frequently updated, and that the files be overseen and approved for publication by a standing committee.

(3) If G cannot be satisfactorily determined, and in the absence of albedo or taxonomic class, it is sufficient to adopt the value G = 0.25. If further sophistication is desired, it is appropriate to adopt instead G = 0.15 if the minor planet appears (even in the absence of available proper elements) to belong to the Nysa family or to have semi-major axis > 2.50 AU (unless it is an Apollo object), or G = 0.40 if it appears to belong to Williams family 190.

It was noted that the combined effect of the new definition of absolute magnitude and the conversion from B to V is that $H \simeq B(1,0) - 1.0$. Further, it was suggested that it would be useful to flag objects that are known to have large light variations due to rotation or aspect effects, and that a summary table and references to more detailed information might usefully be included in the *EMP*. It was also agreed that the membership of the standing committee to oversee publication of photometric data for minor planets should consist of the President of the Commission, the Director of the Minor Planet Center, and a liaison to the keepers of the photometric files. Membership for 1985-88 would then be Kozai, Marsden, and Bowell.

Discussion then turned to further suggestions that had arisen in the study session concerning the contents and format of the *EMP*. A list was made of the ideas that had substantial support, and the President was instructed to communicate them to the ITA as a response to their invitation for suggestions.

Some refinements were made in the wording of a resolution concerning names for minor planets, and the resolution was then approved as follows:

> Names proposed for minor planets will not be accepted if, in the opinion of the Minor Planet Names Committee, they are too nearly similar to those of other minor or major planets or natural satellites, or are in questionable taste. Names should be pronounceable, preferably expressible as a single word, and no more than sixteen characters long. Names glorifying individuals or events principally known for their political or military activities or implications are considered unsuitable unless at least one hundred years have elapsed since the individuals died or the events concerned took place. Objects involved with the Jovian triangular libration points should be named in accordance with the tradition of honoring heroes of the Trojan War. In a disputed case, the proposer may appeal the committee's decision at a general meeting of Commission 20, provided that due written notice is given to the President of the Commission.

It was agreed that edited citations should be referred back to the discoverer before they appear in the *MPC*s. The Minor Planet Names Committee is composed of the President and Vice President of the Commission and the Director of the Minor Planet Center. Members for 1985-88 thus are Kozai, Batrakov and Marsden.

The following resolution on comet nomenclature, incorporating points made earlier, was then approved:

> Commission 20 supports the report of the Comet Nomenclature Committee, specifically with regard to the following points:
>
> (1) The sequence of provisional designations should be a, b, ... z, a_1, b_1, ... z_1, a_2, b_2, ... It is acceptable to replace the subscripts with full-size Arabic numerals.
>
> (2) The use of appellations *IRAS*, *SOLWIND*, etc., is encouraged in cases of corporate discoveries. The inclusion of Arabic numerals, as in the *SOLWIND* case, is recognized as being inconsistent with the normal practice of supplying such numerals only for short-period comets, but is considered unavoidable.
>
> (3) The interpolation into the tabulation of Roman numeral designations of past comets for which adequate orbital data are available is encouraged.
>
> (4) The use of new provisional designations for old comets or possible comets for which orbital data are not available is not recommended. Discoverers are urged to ensure that enough accurate observations are secured for the computation of satisfactory orbits.

The Comet Nomenclature Committee is composed of the President of the Commission, the chairman of the Working Group on Comets, and the Director of the Telegram Bureau. Members for 1985-88 are then Kozai, Roemer, and Marsden.

It had been agreed after consultations that Roemer would represent Commission 20 in the Working Group to be established under the resolution that arose from Joint Discussion I, Reference Frames.

Membership of the Satellite Nomenclature Liaison Committee consists of the President and Vice President of the Commission, the Chairman of the Working Group on Satellites, and two others selected by the President from among the members of the Working Group on Satellites. The membership for 1985-88 is: Kozai, Batrakov, Arlot, Aksnes (chm and representative to WGPSN) and Seidelmann (vice chm and alternate representative to WGPSN).

Bowell then proposed a vote of thanks to the retiring President, noting that she had served the Commission well for much of two terms. Applause followed and the meeting was closed.

Addendum:

The Fifth Edition of the *Catalogue of Cometary Orbits*, published jointly by the Central Bureau for Astronomical Telegrams and the Minor Planet Center, is now available. It contains 1187 sets of orbital elements and is complete for comets observed through 1985 December. Inquiries may be addressed to:

> Minor Planet Center
> Smithsonian Astrophysical Observatory
> 60 Garden Street
> Cambridge, MA 02138
> U.S.A.

COMMISSION 21 : LIGHT OF THE NIGHT SKY (LUMIERE DU CIEL NOCTURNE)

Report of the Meetings November 20, 22, 23, 27

PRESIDENT: R.H. Giese   SECRETARIES: A.C. Levasseur-Regourd
                                                      K. Mattila

ADMINISTRATIVE SESSION

20 November 1985

The President welcomed the few members attending the General Assembly at New Delhi and reported the sad news about the dead of Dr. F. Link, presenting a short necrology taking into account the very fruitful and versatile work of Dr. Link in many fields of Astronomy and Geophysics including the field of Commission 21. Members stood up in silence in memoriam of this outstanding scientist and friendly colleague.

The President reported about activities in the passed triennium: A circular plus Newsletters No 7 and No 8 were sent out forward latest information in preparation of IAU Colloquium No 85 and of the General Assembly and to ask for inputs and suggestions for these events, membership of the new Organizing Committee and concerning "Reports on Astronomy". The President acknowledged with thanks the valuable response and stated that the contribution of Commission 21 to Reports on Astronomy was based on drafts kindly prepared by R. Dumont, K. Mattila and H. Tanabe and on very helpful inputs from other members of the Organizing Committee. It was further reported about preparations and the final success of IAU Coll. No 85 "Properties and Interactions of Interplanetary Dust" at Marseille, which was organized by our commission, supported by Commissions 15, 22 and 49 and by COSPAR. 91 papers were presented by 78 participants from 14 countries. The proceedings are now available at Reidel Publ. Comp., Dordrecht, 1985 (R.H. Giese and P. Lamy, eds).

Concerning membership and officers it was agreed to propose to the IAU Executice Committee following list for approval:
  President: K. Mattila
  Vice-President: A.C. Levasseur-Regourd
  Organizing Committee: R. Dumont, Yu.I. Galperin, R.H. Giese, M.S. Hanner,
                        P. Lamy, T. Mukai, H. Tanabe, J.L. Weinberg
  New Members: In addition to E. Grün, who was already included in the latest IAU Commission 21 member list the following names were added
  a) V.V. Agashe, O.I. Belkovich, S. Bowyer, S. Hong, J. Houck, F. Paresce,
      M. Woolfson.
  b) J.J. Lopez-Moreno, M. Lopez-Puertas, A. Molina, R. Rodrigo,
      M.L. Sanchez-Saavedra, N.N. Shefov.
  c) F. Giovane.

List a) refers to experts, who were invited by the Commission and who agreed to become members in order to strengthen or to extend the expertise of Commission 21 concerning night sky radiation outside the visual range. List b) and c) refer to new IAU members proposed by the national adhering bodies or by presidents, respectively.

It was further agreed to invite M.G. Hauser (IRAS) to become a member or consultant. One proposal of the national bodies was not yet adopted by Commission 21 since it seemed to be out of the field of the Commission. The following List of Concultants was adopted to be given to the IAU General Secretary:

J. Buitrago	D.J. Kessler	J. Michael	G.H. Schwehm
C. Classen	W. Kokott	A. Mujica	G.N. Toller
A. Frey	G. Lopez	A.W. Peterson	M.R. Torr
B.A. Gustafson	J. Maucherat	H. Radoski	K. Weiß-Wrana
M.G. Hauser	J.A.M. McDonnell	R. Schaefer	H. Zook
	R.D. Mercer	A. Schulz	

It was noted, that about 1/3 of the commission members did never respond to any circular or questionaire. It should be checked on an individual base if this is due to mailing problems or to definite change of the field of interest.

Future activities were presented for discussion by K. Mattila (Colloquium on galactic and extragalactic backgrounds) and by H. Tanabe (Dust Colloquium 1989 Japan).

The President thanked the members for their cooperation and those present for participation. The incoming President expressed the members' thanks to the retiring president for his work on behalf of the Commission during the past triennium.

SCIENTIFIC SESSIONS

22 November 1985
(Common Session with Commission No 22)

The review R.H. Giese: 3D-Models of the Zodiacal Cloud was included in this session. Complete program: See report of Commission 22.

23 November 1985
(Chair persons: A.C. Levasseur-Regourd, H. Tanabe)

P.V. Kulkarni: Temperature Measurements in the Earth's Atmosphere and on the Stellar Objects with Simple Optical Techniques. - Ground based observations related to optical measurements from natural and artificial atmospheric glows were briefly summarised. It was shown that with natural airglow photoelectric filter photometric and spectroscopic techniques can be used to estimate the temperature in the 85-90 km region of the earths atmosphere from measurement of OH rotational vibrational band intensities.
In the 250-300 km altitude region, line width of 6300 A emission by a Fabry-Perot high resolution spectrometer gives satisfactory results. F.P. also is sussessfully used to estimate the temperature profile in the altitude range 120-270 km region by monitoring an artificially released Na vapour trail. With Ba-Sr release of blobs in the upper atmosphere temperature of those regions can be estimated from experimentally determined diffusion coefficients.
In the stellar situation, only one application of F.P.I mapping coronal temperatures at the time of two total Solar eclipses (1980, India; 1983 Indonesia) was given (literature: Proc. Ind.Acad.Sci. 89, 109 (1980), JATP 32. 1235 (1970), Plan. Space Sci. 23, 273 (1984)).

H. Tanabe, A. Miyashita and A. Takechi: A Star-eliminating Photometer. - A new photoelectric photometer was designed and constructed, which measures the brightness of only the extended light source with automatic elimination of the starlight in the field of view by utilizing the difference between extended and point light sources.

The photometer has a square field stop with a moving wire in the focal plane of a telescope. During the motion of the wire from one end to the other of the field stop, inverse pulses are produced when the wire covers star images. The photometer with a micro-computer measures the total light flux and depth of each inverse pulse, and calculates instantly the surface brightness of the extended light source.

The photometer can be applied, in future, for a star count by inverse use of the star-elimination by attaching to a large Schmidt telescope. It could be, also, developed to an automatic star-eliminating airglow-separating photometer with utilizing the difference between line and continuous spectra.

J. Houck and M.G. Hauser: IRAS Observations of the Zodiacal Light - A Progress Report. - The IRAS sky survey yielded a very extensive picture of bright diffuse thermal emission from interplanetary dust grains at wavelengths from 12 to 100 micrometers. These data provide insights into the character, spatial distribution, and perhaps origin of these particles which both confirm and expand upon those gaied from previous zodiacal light studies. The zodiacal emission is a major component of the large-scale infrared brightness of the sky at all IRAS wavelengths. It must therefore be carefully characterized both to facilitate study of large-angular-scale sources in the Galaxy and beyond using IRAS data, and to permit optimum design of future space infrared instruments.

A model has been developed which identifies the narrow emission bands discovered by IRAS as arising from debris in the orbits of the Eos, Themis and Koronos asteroid families.

The Infrared Astronomical Satellite (IRAS) was developed and operated by the Netherlands Agency for Aerospace Programs (NIVR), the United States National Aeronautics and Space Administration (NASA), and the United Kingdom Science and Engineering Research Council (SERC).

For additional details see: M.G. Hauser and J.R. Houck, "The Zodiacal Background in the IRAS Data", Proceed. of the Noordwijk meeting, Light on Dark Matter, June 1985. - S.F. Dermott, P.D. Nicholson and B. Wolven, Preliminary Analysis of the IRAS Solar System Dust Data, "Asteroids, Comets, Meteors II", ed. Lagerkvist and Rickman, Uppsala (1985).

A.C. Levasseur-Regourd and R. Dumont: Temperatures and Albedos of Zodiacal Dust, as Inferred from IRAS and ZIP Measurements. - Localizing the information in some points of the line-of sight in optical studies of the zodiacal light has been performed with the method of the "nodes of lesser uncertainty" (see Planet. Space Sci., 31, 1381, 1983 and 33, 1, 1985). An extension to the thermal case is being made on the data of IRAS and of the zodiacal infrared project ZIP (CRAS 300, II, 109, 1985 and IAU Coll. 85, 207, 1985).

There still exist two nodes where the local elemental contribution to the integrated infrared brightness I_ν can be retrieved with lower uncertainty than elsewhere. The knowledge, als well from IRAS as from ZIP, of I_ν at two frequencies allows to determine the temperature of the dust, assumed to radiate like a grey body. Then, the emissivities allow to reach the global emissivity for all wavelengths, which in comparison to the optical scattering coefficient allows to reach the albedo.

The temperature of the dust is found to be (250 ± 25) K at 1 AU. The heliocentric decrease from 1 to 1.4 AU seems to be smaller than would result from a uniformity of composition of the dust. The albedo exhibits a negative heliocentric gradient, from ~ 0.08 at 1 AU to ~ 0.06 at 1.4 AU. These results definitely invalidate the old assumption of a homogeneous zodiacal cloud.

The papers V.V. Agashe: Determination of Mesopheric Temperatures from Ground Based Intensity Measurements - and S.Hong: The Connection between IR and Visual

Zodiacal Light - were only read by title since authors had to cancel attendance due to unexpected obligations.

<u>27 November 1985</u>
(Chairman K. Mattila)

P.C. van der Kruit: Pioneer 10: Surface Photometry of the Galaxy. - The background starlight experiment on Pioneer 10 has provided a distribution of surface brightness across the sky comparable to results of surface photometry of edge-on external galaxies. After reasonable correction for diffuse Galactic light these data can be used at higher Galactic latitude so set constraints on two parameters of the old disk population of stars that have not been determined up till now with great precision. These are the radial scalelength (or e-folding) of the disk population (5.5 ± 1.0 kpc) and the colour of the Galactic disk as seen by an outside observer (B-V) = 0.84 ± 0.15). These parameters are fundamental for an understanding of Galactic structure and evolution, and comparison of our Galaxy to others.

K. Mattila: Some Recent Results on the Galactic and Extragalactic Components of the Light of the Night Sky. - In first part of the talk a review was given on the progress in this field during the past three years. An extensive photographic UBVR surface photometry of the whole Milky Way has been carried out by Schmidt-Kaler et al. Further analysis of the unique Pioneer 10/11 Milky Way photometry (spacecraft outside the zodiacal cloud) has been performed by Toller who has derived information on: (1) the large scale structure of the Galaxy; (2) the scattering properties of the interstellar grains which give rise to the diffuse galactic light; and (3) the extragalactic component of the night sky, for which an upper limit of 3.9 $S_{10}(V)_{G2V}$ at the 2σ level at 4400 A was obtained. The long-standing question of polarization of the Milky Way light has been investigated by Leinert and Richter using observations from the Helios space probes. A marginal detection of polarization was reported with a polarization direction perpendicular to the galactic plane. References to these and some further studies are given in Reports on Astronomy (1981-84), pp. 232-234.

In the second part of the talk a report was given on a study by Mattila, Schnur and Laureijs of the optical and infrared surface brightness in the area of the high galactic latitude dark nebula L1642 (l = 211°, b = -37°). A good correlation was observed between the p.e. optical surface brightness (scattered light) and the 100 μm IRAS flux (thermal emission) as well as the 100 μm optical depth of the dust. The data will be used to derive the albedo and scattering asymmetry parameter of interstellar dust grains. Also, the IR optical depths will facilitate the determination of the extragalactic background light using the method as presented by the author in Astron. Astrophys. <u>47</u>, 77 (1976).

S. Bowyer: The Diffuse Far Ultraviolet Background from 1200 to 2000 A. - Over the past 10 years, substantial progress has been made in characterizing the diffuse Far Ultraviolet backgrond. Ten years ago the <u>intensity</u> of the background was virtually unknown; published data on the parameter varied by almost three orders of magnitude. It was generally reported that the flux was isotropic, and hence was probably cosmological in origin. The spectrum of the flux was unknown.

A very significant step was made with data obtained with an instrument flown as part of the Apollo-Soyuz mission (Paresce et al, Ap.J. <u>240</u>, 387, 1980). These data showed the flux was non-isotropic, and was loosely correlated with the total neutral galactic hydrogen column as derived from 21 cm radio data. This was direct proof that the majority of the flux was galactic in origin. These results were immediately confirmed by Maucherat-Joubert et al, Astron. & Astroph., <u>88</u>, 323 (1980). At the time we have very little evidence as to the source mechanism or mechanisms producing the galactic flux. A substantial number of candidate processes have been advanced including the scattering of starlight by interstellar dust, molecular hydrogen fluorescence, two photon emission, emission from an intermediate temperature ($\sim 10^5$ K) interstellar medium, and radiation from old supernovea remnants.

The existing data, when extrapolated to zero hydrogen columm, indicate the existence of an isotropic, diffuse flux of $\sim$ 500 photons/cm^{-2}sec^{-1}st^{-1}A^{-1} which must emanate from the interstellar medium, the summed radiation of galaxies, or some more exotic extragalactic sources.

The task ahead is to identify what processes, galactic and extragalactic are producing the observed flux.

J. Holberg: Voyager Data on the EUV and Far UV Diffuse Background. - New limits are presented on the extreme and far UV diffuse backgrounds, for both line and continuum emission, in the 500 to 1200 A region. These limits are derived from a single spectrum of extremely long integration time (1.5 x 10 s) obtained with the Voyager 2 ultraviolet spectrometer in the direction of the north galactic pole. This spectrum can be explained solely on the basis of resonant scattering from interplanetary HI and HeII. Limits on any additional sources of sky background radiation corresponding to less than 100 to 200 photons cm^{-2}s^{-1}sr^{-1}A^{-1}, for continuum emission, and 10^4 photons cm^{-2}s^{-1}sr^{-1} for line emission are demonstrated. Comparisons of these upper limits with existing measurements and upper limits is discussed. A first detection of resonantly scattered interplanetary HI Lyman γ (973 A) emission is reported. Additional Voyager observations obtained at lower galactic latitudes exist and indicate the presence of a stellar-like continuum. Both the spectral shape of this continuum and its intensity point to an origin associated with the scattering of starlight from interstellar dust.

A. Davidsen and F. Paresce: The Spectrum of the Diffuse Far UV Background. - The spectrum of the diffuse far UV background has gone through several peculiar gyrations since its measurement was first attempted by means of a 12° x 12° scanning spectrometer on Apollo 17 in trans-earth coast. From the single wide feature having profound cosmological implications of intensity $\simeq$ 300 photons cm^{-2}s^{-1}sr^{-1}A^{-1} at 1400 A reported by Henry et al., 1978 (Ap.J., 223, 437) through the featureless flat continuum with a possible rise beyond 1650 A observed by Anderson et al., 1979 (Ap.J., 234, 415) by means of a 1.4 x 5.8°, 60 A resolution rocket-borne spectrometer, it has progressed to the possibly complex set of weak emission lines superimposed on a very weak continuum advocated by Feldman, Brune and Henry, 1981 (Ap.J. 249, L51). This last result stems from a reanalysis of the Anderson et al, 1979 data consisting essentially of adding together the individual spectra obtained in different directions towards the north galactic pole in the original flight. This technique allows the reasonably confident identification of four prominent features in the 1200 to 1650 A band at 1400, 1490, 1550, 1660 A that cannot be attributed to atmospheric emission.

The reanalysis carried out by Feldman et al., 1981 was prompted by the suggestion made earlier by Jakobson and Paresce, 1981 (Ap.J., 96, 23) that the hot galactic corona could be the source of very weak line emission in the far UV (at most $\simeq$ 200 photons cm^{-2}s^{-1}sr^{-1} in the CIV, 1549 A line). Using a simple two component constant temperature model of the disc and coronal gas and the relation between the expected line intensity I_ℓ and the collisional excitation rate $\gamma_\ell(T)$: $I_\ell = (n_e n_i \beta \gamma_\ell(T))/8\pi$ Jakobson and Paresce, 1981 predicted the appearance of a number of lines i the UV background. In this equation n_e and n_i are the electron and ion densities and ß is the scale height of the gas. The most prominent in the 1200 to 1650 A band were predicted to be the SiIV, 1398, OIV], 1406, NIV], 1487, CIV, 1549 and OIII], 1663 A lines. The interesting fact that they corresponded almost exactly to the observed features was not lost on the authors involved although the almost two factors of ten discrepancy in intensities tended to cloud the issue somewhat.

Paresce et al., 1983 (Ap.J. 266, L107) suggested a simple explanation for the large discrepancy, however. Rather than fixing the gas temperature in the manner adopted by Jakobsen and Paresce, we allowed this parameter to vary freely to investigate possible observationally allowed regions of the emision measure (EM) - temperature plane. Surprisingly, such an internally consistent region was found at EM $\simeq 10^{-1} cm^{-6} pc$ and $T = 2.5 \cdot 10^5 K$. In other words, a single emitting plasma of ≥ 160 parsec extent at this temperature could explain the Feldman et al., 1981 data. Many other lines are expected to be prominent both in the far UV and in the EUV below $\simeq 200$ A where interstellar absorption is less severe and a measurement of any of these would clearly add significantly to our understanding of the hot phase of the ISM.

The important point to be made in this context is simply that since we have no information at present on either the extent or the distribution of this gas, if it exists, it is impossible to predict accurately the intensity of these lines as a function of view direction. If our calculations are correct and the line emission is due to a disc gas, the expected line intensities at intermediate or low galactic latitudes could become substantial contributors to the far UV background even in broad and observations. Thus, this component cannot be safely ignored when considering possible sources of galactic emission on which the extragalactic component has to be determined.

COMMISSION 22 - METEORS AND INTERPLANETARY DUST
(METEORES ET LA POISSIERE INTERPLANETAIRE)

Report of Meetings, 20 and 22 December 1985

PRESIDENT: O.I. Belkovich.   SECRETARY: C.S.L. Keay.

20 November 1985

I. OFFICERS AND MEMBERSHIP
   The President reported that he had received no communication for a very considerable time from the current Vice-President, D.E. Brownlee. As it has been customary for the Vice-President to proceed to Presidency of the Commission, the President sought the guidance of the Meeting. Following discussion of the situation it was agreed that P.B. Babadzhanov, who had been nominated for incoming Vice-President by the Organising Committee, should be nominated as President for the coming term. This recommendation was endorsed without dissent.

   Nominations for the other Officers of the Commission were proposed and endorsed as follows: Vice-President, C.S.L. Keay; Organising Committee: W.J. Baggaley, O.I. Belkovich, W.G. Elford, H. Fechtig, M.S. Hanner, I. Hasegawa, J.A.M. McDonnell, J. Stohl, K. Tomita.

   The President announced the names of 11 new members: F. Akira, G. Cevolani, I. Kapisinsky, C. Koeberl, U. Marvin, D. Meisel, K. Nakazawa, J.A. Nuth, P. Pecina, K. Yamakoshu, J. Zvolankova. Consultants appointed for the next term are: G.V. Andreev, K.B. Hindley, F. Herz, B. Lokanadham, V.G. Kruchinenko, J.W. Mason, Y. Obrumov, D. Ohlsson-Steel, M.S. Rao, G. Schwehm, Y. Yabu.

   The President noted with regret the deaths of seven members of the Commission since the last meeting: C.L. Hemenway, L.A. Katasev, A. Kizilirmak, E.L. Krinov, D.W.R. McKinley, E. Opik, A.N. Simonenko.

II REPORT OF THE COMMISSION
   The President tabled the Report and thanked those members who prepared the specialist sections.

III PROJECTS AND WORKING GROUPS

1. International Halley Watch

   P.B. Babadzhanov reported that the International Halley Watch Committee is anxious that as many radar observations and and meteor spectra as possible should be obtained for the Eta Aquarid and Orionid meteor streams during 1986. It is particularly important to have observations of meteor rates that are continuous in longitude.

   The meeting was advised that three further meteor observatories wish to participate in the I.H.W. program. They are at Hyderabad and Waltair in India, and Adelaide in Australia.

2. GLOBMET Project

C.S.L. Keay reported that the First GLOBMET Symposium held in Dushanbe, U.S.S.R., in August 1985, had been very successful, attracting about 130 participants. Some 80 scientific papers were presented covering all of the the objectives of the GLOBMET Project, which is concerned with improving measurements and modelling of meteor distributions, influx and interaction with the atmosphere.

3. Canadian Fireball Network

The meeting noted with dismay the recent closure of the MORP network in Canada despite the Resolution recommending its continuation passed by the previous meeting of Commission 22 and endorsed by the I.A.U.

IV SCIENTIFIC PRESENTATION

K. Tomita: Japanese Observations of the Giacobinids.

In the early evening of 1985 October 8 meteors of the Giacobinid stream were observed in the north-eastern part of Japan. At the Dodaira station of the Tokyo Astronomical Observatory, eight meteors were detected during the period from 1040h to 1130h UT by a SIT TV camera. One of them, recorded on the 11th frame (one frame corresponding to 1/30th second), was a stationary meteor and its position was at RA = 17h 57.5m, DEC = $+55°$ 10' (1950.0).

Many amateur observers visually recorded the meteors, among them K. Gomi at Nagano, who commenced observing at 0930h UT and counted 128 meteors during ten minutes. According to his observations, the maximum rate is estimated to have occurred before 0930h UT, and is therefore 0.15 of a day earlier than had been predicted.

22 November 1985

I MEMBERSHIP

In addition to the 11 new members announced at the previous meeting, two addtional names were recorded as new members: A. Carusi and A.C. Levassuer-Regourd. The membership was now 93, plus 11 consultants.

II PROJECTS

1. I.A.U. Meteor Data Center, Lund, Sweden

The I.A.U. draft budget for 1986-1988 did not include funding for the Meteor Data Center. The President and Secretary were invited by the General Assembly Finance Committee to argue the case for the continuation of the annual maintenance grant toward the operation of the Data Center. They reported that members of Commission 22 were strongly of the opinion that the Data Center should continue in operation and drew attention to the Report of the Director (Dr B.A. Lindblad) in the I.A.U. Information Bulletin 54 which stated that requests for data from Czechoslovakia, Great Britain, U.S.A. and the U.S.S.R. had been processed during 1984. (The Finance Committee subsequently approved continuation of I.A.U. support at the rate of 1000 Swiss francs per annum until 1988).

2. *Symposium*

Plans were announced for a Symposium at Tucson, Arizona, U.S.A., during May 19 - 22, 1987, on the Subject of Interplanetary Matter and Meteoroids.

III GENERAL BUSINESS

The retiring President thanked all those members of the Commission who had assisted him in his work and extended his best wishes for success to the incoming Officers.

IV SCIENTIFIC PRESENTATIONS

P.B. Babadzhanov and Yu. V. Obrubov: Dynamics of Meteor Streams.

The results of investigations of the evolution of the Geminid and Quadrantid meteor streams show that under the influence of planetary perturbations the streams may be flat, but they may thicken as well depending on the range of variation of orbital inclinations. Eventually, due to planetary perturbations, a meteor stream may take such a shape that it gives rise to several active meteor showers at different solar longitudes. Thus the Geminid stream is also associated with the Canis Minorids and the daytime Sextantids as well as a Delta Leonid stream which has not yet been observed. Likewise the Quadrantids have evolved to produce the Carinids and the North and South Delta Aquarids as well as the daytime Arietids and Alpha Cetids. The calculated and theoretical values of geocentric radiants and orbital elements are in good accord with observations.

I. Hasegawa: Draconid (Giacobinid) Meteors in 1985.

Just after evening twilight on 1985 October 8, a strong appearance of the Draconid meteor shower was observed in Japan. During the ten minutes from 1000h UT, the corrected hourly rate of visual meteors probably exceeded 180-200. The rates clearly decreased during the period of observation, so the maximum seems to have occurred before 1000h UT (mean solar longitude $194.57°$, equinox of 1950.0). No Draconid meteors were observed after 1300h UT. The magnitude distribution exhibited a maximum at nearly 3.

D. Ohlsson-Steel: Meteor Research at Adelaide, Australia.

Meteor-related research at Adelaide is proceeding on a number of fronts:
1. Radar meteors - experimental
    a. The height distribution to radio magnitude +7 has been determined at a frequency of 2 MHz where the echo ceiling is inconsequential. As expected, the results contradict measurements at shorter wavelengths: the peak is found at ~104 km, with many meteors seen to at least 130 km. This suggests that missing mass is present as a faint, dense, high-velocity, high-altitude component.
    b. To confirm the above result it is planned to make observations in 1986 at 6 MHz, and simultaneously at 6 and 54 MHz.
    c. A further possibility is to attempt to use a 54 MHz tropospheric scatter radar to measure radiants (narrow beam) and velocities (high Doppler shifts) using the head echoes from meteors moving radially towards the station: decelerations and good orbits are hoped for.
    d. The Jindalee over-the-horizon radar in central Australia is also being used for meteor observations via forward scatter, back scatter, and multiple bounces

from the ground and F-region.

2. Meteors – Theory

a. The lifetimes of meteoroids against catastrophic collisions with zodiacal dust particles have been calculated for characteristic orbits.

b. Planetary close encounters have been shown to account for the production of sporadic meteors from streams.

c. The theory of the separation of electrons and ions during the diffusion of underdense trains may explain some peculiarities of radar meteors.

3. Asteroids

The impact rates of asteroids upon the terrestrial planets have been reported to Commission 15.

V.V. Andreev and O.I. Belkovich: The Distribution of Meteoroidal matter at the distance of 1 au from the Sun.

In order to minimise the influence of observational selectivity a new combined method was proposed for deriving a mathematical model of the sporadic meteor distribution in the vicinity of the Earth's orbit. The method combines measurements of the Fresnel velocity distribution vs. elongation angle and determinations of the radiant distribution over the celestial sphere by the rotating antenna technique. The method is based on transformation of the three-dimensional distribution of radiant coordinates and velocities into the distribution of any three orbital elements that are related to the meteor orbits intersecting the heliocentric sphere of radius 1 au. The best first approximation of the inclination distribution of meteor orbits was found to be:

$$p(i) = \begin{cases} 6.33 \exp(-i/23) \sin i & 0° < i < 90° \\ 0.059 \exp(i/118) \sin i & 90° < i < 180° \end{cases}$$

The concentrations of perihelion distances near to the orbits of the three innermost planets have been established for meteor orbits having small inclinations.

R.H. Giese: Three-dimensional Models of the Zodiacal Cloud.

Present models of the three-dimensional distribution of interplanetary dust derived from the zodiacal light are based on the assumption of a flattened zodiacal cloud having homogeneous physical properties. Number density n can be expressed by a decrease according to a power law (-1.3) with solar distance r, and concentration towards the plane of symmetry by a factor f which is a function of ecliptic latitude only. Models assuming a bimodal function of f suggesting a second population of dust particles concentrated in high inclination orbits can be definitely excluded by obvious discrepancies with respect to observations of brightness along circles of low (15) elongation about the Sun. The other models - although different in detail - show a monotonous decrease of n above the ecliptic (symmetry) plane, typically to one half of the density within less than 0.3 au near the Earth's orbit. If the questionable assumption homogeniety is abandoned, the isodensity surfaces of n are modified. Taking the hypothetical example that the decrease of particle albedo with r would follow a power law (-0.5), as claimed by some recent publications, then n would decrease with r according to a power law of 0.8 only. Therefore the isodensity surfaces should be squeezed towards the Sun inside 1 au and stretched out outside. Such modifications modifications are not negligible, however more information on spatial changes f of dust properties is needed to justify advanced modelling.

COMMISSION 24: PHOTOGRAPHIC ASTROMETRY (ASTROMETRIE PHOTOGRAPHIQUE)

Report of Meetings on 22, 25, and 26 November 1985

PRESIDENT: W. Gliese   SECRETARY: J.L. Russell

Activities of Commission 24 at the XIX General Assembly consisted of three sessions concerned with matters specifically connected with Commission 24 itself, one Joint Meeting with Commission 8, participation in a Joint Meeting on "HIPPARCOS" (Commissions 7, 8, 24, 25, 33, 37) and in Joint Discussion I "Reference Frames" (Commissions 7, 8, 19, 20, 24, 31, 33, 40).

Business Meeting: 22 November

The President welcomed the members present at the meeting. He especially expressed his joy at seeing again Willem Luyten, who has been an active astronomer since before the founding of the IAU; with great pleasure the President welcomed also Kaj Strand, a successful member of the Commission on Parallaxes and Proper Motions (now Commission 24) for many decades.

The President called for a moment of silence to honour the memory of the members lost by death since the meeting at Patras: W. Dieckvoss, N.V. Fatchikhin, I.V. Gavrilov, D.J.K. O'Connell, N.E. Wagman.

The President announced that, as a result of the mailed ballots from the members of the Commission, the following names were proposed for the new officers for 1985-1988:
President: Arthur R. Upgren
Vice-President: William van Altena
Organising Committee: A.N. Argue, T.E. Corbin, Ch. de Vegt, W. Gliese, I.I. Kanaev, T.E. Lutz, J.D. Stock.

V.V. Lavdovskij resigned from IAU membership, and E. de Graeve has resigned as consultant.

The following were confirmed as new members of the commission: U. Bastian, N.M. Bronnikova, G.G. Douglass, D. Harwood, V.S. Kislyuk, J. Kovalevsky, L.A. Marschall, D.G. Monet, R.-Sh. Pan, M.A.C. Perryman, D. Quin, S. Roeser, H. Ruder, G.-Ch. Shi, J.-J. Wang, G.L. White, H. Wroblewski, S.M. Younis, X.-H. Zhou. In the following days Y. Requième, new President of Commission 8, applied for membership in Commission 24 and was announced to the General Secretary.

K.-G. Steinert and C.E. Lopez were proposed and approved as new Consultants of Commission 24.

The report on the Commission work 1981 to 1984 was approved with one addition - that after the termination of research at Sydney Observatory, not only the telescope and astrometric equipment was transferred to Macquarie University, but also the astrographic plates, which are in a fair state of preservation.

The Commission approved the following reports of its two Working Groups:

COMMISSION 24

WORKING GROUP ON PARALLAX STANDARD STARS

A.R. Upgren, Chairman

The Working Group on Parallax Standard Stars reported no activity since the completion of the first part of its mission, the selection of the standard stars. The commission approved the continuation of the working group, giving it the responsibility to monitor all parallax programs and their observations of standard stars and regions. The working group will first determine the observations which each parallax program has made since the standard list was published. Upgren wished to be replaced as chairman but remain a member of the group, and he was replaced by Lutz.

WORKING GROUP ON OPTICAL-RADIO REFERENCE FRAME

Chr. de Vegt, Chairman

1. Introduction

Following the last IAU General Assembly 1982 in Patras, two meetings of the WG were held in 1985 at Washington (USNO 23.6.85) and Aussois/France (informal, during Hipparcos INCA conference 3.-6.6.85). An interim report of work status was given during IAU Symposium 109 (Astrometric Techniques, Gainesville/USA, January 1984) by the chairman (in press).

2. Report on WG activities

2.1 In fulfilment of the main task of the WG, to provide a candidate list of suitable objects for the construction of a future extragalactic reference frame, a catalog of 233 sources has been compiled and published in Astronomy and Astrophysics (Vol. 130, 191, 1984). Extensive further optical and radio observations however are necessary to improve the present data on source structure, photometry and astrometry in both domains. In addition more objects south of -40° declination are needed for homogeneous sky coverage.

The question of any revision and extension of the present list was discussed. In particular the problem of a selection of additional quasars near the galaxy candidates was addressed by the chairman. It was pointed out by the radio astronomers that any inclusion of additional (weaker than 1 Jy) sources could not be expected within the next few years, mainly because of the large amount of radio observations necessary for a detailed investigation of source structure and background confusion problems. It is estimated that about 900 sources down to 0.1 Jy may be available in principle.

The WG agreed that extensive work has to be concentrated now on detailed optical and radio investigations of the present list for many years to get full insight in the astrophysical and astrometric properties of the sources.

2.2 Radio Stars. During the last years great progress has been made to obtain precise astrometric results for radio stars both in the radio (VLA, VLBI) and optical domain. The WG agreed that radio stars will provide the most suitable link candidates to connect the present galactic and future extragalactic reference frames. Therefore the problem of providing a similar list for radio stars was discussed. It was agreed that no ad hoc list of optimally suitable candidates can be produced due to the very heterogeneous physical properties of radio star emission and related optical structure. More information is required. However, as a first step, a provisional list of about 50 radio stars has been compiled, based on the data provided mainly by the radio astronomers of the WG. The list is suggested to

PHOTOGRAPHIC ASTROMETRY

First provisional List of Optical-Radio Star Candidates

Object	h m s (1950.0)	° ' "	Remarks
UU PSC	00 12 24.115	08 32 36.36	5)
Zeta And	00 44 40.968	23 59 43.95	
39 CET	01 14 03.932	-02 45 46.70	
Algol	03 04 54.360	40 45 52.46	3) 4) 5)
CC CAS	03 10 07.408	59 22 38.48	5)
UX ARI	03 23 33.150	28 32 29.0	1) 3) 5)
HR 1099	03 34 13.130	00 25 28.00	1) 3) 5)
b PER	04 14 28.585	50 10 27.09	2) 5)
RZ ERI	04 41 24.006	-10 46 29.31	
12 CAM	05 01 50.604	58 57 15.46	
HR 1890	05 32 53.307	-04 31 30.89	4)
SIG.ORI E	05 36 16.396	-02 37 17.58	4)
CHI 1 ORI	05 51 25.196	20 16 07.36	
ALPHA ORI	05 52 27.780	07 23 57.70	4) 5)
CQ AUR	06 00 39.261	31 19 51.10	
HD 50896	06 52 08.115	-23 51 51.74	2)
RY GEM	07 24 32.914	15 45 42.87	
KQ PUP	07 31 30.082	-14 24 51.94	2)
SIGMA GEM	07 40 11.382	29 00 22.55	5)
54 CAM	07 58 31.922	57 24 49.18	2) 5)
TY PYX	08 57 34.039	-27 37 10.53	5)
SAO 81134	09 57 13.283	24 47 36.76	4)
DM UMA	10 52 36.414	60 44 11.25	
RW UMA	11 38 04.982	52 16 31.31	
SS BOO	12 02 04.827	38 56 34.42	
DK DRA	12 13 21.347	72 49 45.09	4)
RS CVN	13 08 17.734	36 12 01.33	2) 5)
FK COM	13 28 24.641	24 29 24.48	2)
HR 5110	13 32 34.145	37 26 16.19	2) 3) 5)
RV LIB	14 33 01.226	-17 49 07.46	
RW CRB	15 37 12.525	29 47 01.01	
SIG CRB A	16 12 47.632	33 59 00.57	2) 3) 5)
ALP SCO A	16 26 20.201	-26 19 22.42	2) 5)
WW DRA	16 38 21.839	60 47 49.66	
EPS UMI	16 51 00.905	82 07 21.54	
Z HER	17 55 51.297	15 08 33.74	2)
9 SGR	18 00 48.402	-24 21 48.72	2) 5)
FR SCT	18 20 34.026	-12 42 27.74	2)
BETA LYR	18 48 13.932	33 18 12.32	2) 5)
U SGE	19 16 37.057	19 31 04.09	4)
CYG X-1	19 56 28.870	35 03 55.00	3) 5)
V444 CYG	20 17 42.591	38 34 24.25	4)
BD+433571	20 18 46.686	43 41 42.93	2) 5)
CYG OB2 5	20 30 34.838	41 08 03.91	2) 5)
ER VUL	21 00 16.429	27 36 33.36	
AD CAP	21 37 03.793	-16 13 58.52	
VV CEP	21 55 14.413	63 23 13.41	2)
RT LAC	21 59 28.742	43 38 56.48	2) 5)
AR LAC	22 06 39.256	45 29 47.29	2) 3) 5)
SZ PSC	23 10 50.602	02 24 10.83	2) 3) 5)
LAMB. AND	23 35 06.520	46 11 13.82	3) 5)
II PEG	23 52 30.500	28 21 19.00	2) 5)

199

References

1) Johnston, Wade, Florkowski, de Vegt (1985), AJ 90, 1343
2) Florkowski, Johnston, Wade, de Vegt: Stellar Radio Astrometry I,
   AJ 90, 2381, 1985
   de Vegt, Florkowski, Johnston, Wade: Stellar Radio Astrometry II,
   AJ 90, 2387, 1985
   Johnston, de Vegt, Florkowski, Wade: Stellar Radio Astrometry III,
   AJ 90, 2390, 1985
3) Lestrade, Preston, Mutel, Niell, Phillips (1985) HIPPARCOS Second FAST Thinkshop p.87
4) Johnston et al., to be published
5) de Vegt, (1982) Abh.HS X,3, p.119

undergo changes. Especially south of −40° declination detailed information is lacking; special instrumental and institutional support is urgently needed.

2.3 WG-activities with Hipparcos. Most of the WG-members are involved now in the Hipparcos project, contributing especially to the problem of the ERL (Hipparcos subgroup 2130, Argue, task leader). Extragalactic candidate sources for the ER have been selected almost entirely from the WG candidate catalog; a similar list of radio stars is under investigation. Extensive detailed studies of optical and radio source structure have been started and in particular, the determination of precise radio positions of stars relative to the extragalactic VLBI-based reference frame is in progress, using the VLA and different VLBI-configurations.

3. Future activities

The WG agreed that a substantial improvement of the present candidate list or similar work on a list of radio stars can be achieved only on a longer time scale, due to expected slow accumulation of high quality observational data both in the optical and radio domain. In addition to the general problem of the extragalactic reference frame needs more evaluation.

Especially radio work on the southern hemisphere needs immediate support to avoid serious unbalance on the global distribution of reference frame sources. It is expected that our Australian colleagues will be heavily engaged in this difficult task. Furthermore it is obvious that the successful construction of a new extragalactic reference frame is only possible if a detailed knowledge of the physical nature of the source candidate is available.

To improve this situation the WG strongly suggests an IAU Colloquium or Symposium in spring 1988 which would bring together astrometrists and experts from astrophysics to discuss both problems of the reference frame and the physical nature of the reference "point" sources.

Finally, for the coming three year period the WG suggests Dr. D.L. Jauncey, CSIRO, for chairmanship.

For the coming triennial period two IAU meetings are proposed which will be sponsored by Commission 24:

1. A symposium to be held 1987 at Paris as Centenary of the Carte du Ciel/Astrographic Catalogue organised by the Observatoire de Paris (S. Débarbat) and Commissions 24 and 41 (History of Astronomy). After long discussion this meeting was approved as Symposium No. 133 "Mapping the Sky - Past Heritage and Future Directions", proposed date in the first week of June 1987 at Paris.

2. J. Dommanget proposed a Colloquium "Wide Components in Double and Multiple Systems" to be held at Brussels in June 1987. It was been approved also by the IAU as Colloquium No.97.

The question of the name of Commission 24 was discussed again, without any decision for a change.

A recommendation was made and accepted by the commission that authors of future publications containing trigonometric parallax results make every effort to quote all references containing previous trigonometric parallaxes published since the closing date of the Fourth Edition of the General Catalogue of Trigonometric Stellar Parallaxes, currently being completed at Yale. This practice will facilitate the work of data acquisition and compilation in the future.

Scientific Meeting: Miscellaneous: 22 November

The first scientific meeting was planned with the title "Astrographic Catalogue Work". However, as the proposed Chairman was not able to attend the General Assembly, no program was available.

The President reported on the questionnaire by C. Jaschek, concerning the Astrographic Catalogue (AC) and the Carte du Ciel (CdC) in CDS Bull. 27, 197 (1984), CDS Bull. 28, 169 (1985) and CDS Bull. 29, 61 (1985) and he read the conclusions drawn by Jaschek from the answers he had received. Westerhout supplemented Jaschek's statement, that the only large effort to get the published measures onto magnetic tape was the one by Lacroute for all of the French zones, by announcing that the USNO has been working to get all of the other zones onto tape.

As several observatories have already started programs which can be regarded as a revival of AC work, the Symposium No.133 at Paris 1987 will certainly become a fruitful meeting.

The following papers were presented during the meeting:

J. Russell: Guide Star Catalog for the Space Telescope, currently being compiled at STScI, is based on Palomar and UK Schmidt survey plates. Each plate is digitized into a 14 000 by 14 000 grid of 25 micrometer pixels. Star positions are centroided and reduced with a 10 plate constant model in each coordinate. The catalog will be complete to at least 14th mag, with relative accuracy of 0.25 arcsec and 0.15 mag and absolute accuracy of 2 arcsec and 0.7 mag. This catalog of about 20 million positions, magnitudes and stellar/non-stellar classifications will be published in 1987-88. At the time of the General Assembly nearly half of the 1500 plates had been scanned and 10% had been completely processed and cataloged.

F. Ghigo, R.M. Humphreys, and R. Landau: Description and First Results of the Redeveloped Automatic Plate Scanner at the University of Minnesota. A redevelopment of the Automatic Plate Scanner allows measurement of a pair of the sky survey sized (35 cm square) plates in 2 1/2 hours. The repeatability of position measurements is 1-2 micrometers. Stellar magnitudes can be calibrated to an accuracy of $0^m.1$ or better. The machine is available for use by the astronomical community.

H.G. Walter: A Data Base of Radio Stars for the Extragalactic Link. Radio emission of stars provides the basis for relating directly the optical and radio reference frames. An inventory of radio star data is presented and its suitability for frame linkage is examined, taking account the HIPPARCOS astrometric space mission.

G.L. White and D.L. Jauncey: Astrometric work on the Southern Optical-Radio Tie Frame. A catalogue of 101 quasars south of declination -30° has been prepared for possible inclusion in the inertial reference frame. Accurate radio positions (at the 0.1 arcsec level) and optical positions (~ 0.2 arcsec) have been measured for a small number of these sources. This project is continuing.

N.V. Kharchenko, *A.B. Onegina*, S.P. Rybka, and A.I. Yatsenko: At Kiev-Goloseevo proper motions of about 20 000 stars in 100 sky areas were obtained with respect to galaxies. The s.e. of a proper motion is $\pm 0\rlap{.}''008$. Sources of errors were studied and the magnitude equation was determined.

Joint Meeting of Commissions 8 and 24: Common Interests 25:November

The members of both commissions met together. Two resolutions were discussed. However, since the resolutions had been introduced for Commission 8 only, the discussion and voting on them is thus included in the Commission 8 report.

The proposed meetings in Paris, Brussels, and Beograd were discussed. The final meeting endorsements are summarized elsewhere in this and the Comm. 8 report.

The matter of renaming both commissions was discussed. There was general agreement that both commissions have names which do not reflect completely their current interests and that it was a good idea to discuss the names at the same time because both are "astrometric" commissions. However, the session concluded with no consensus on new names.

Scientific Meeting: Parallaxes and Proper Motions

27 November

Chairman Strand opened the session remembering that thirty years ago the current president, W. Gliese, attended his first meeting of the Commission "Parallaxes and Proper Motions" at which time Strand became its president. He pointed to the great progress made in these past decades; the following papers will demonstrate the situation and status in 1985.

W.F. van Altena: The new General Catalogue of Trigonometric Stellar Parallaxes. The new edition of the Yale Parallax Catalogue contains the relative parallaxes, average reference star magnitudes, proper motions and source of publication for 13 253 parallax determinations of 7 562 systems through early 1985. In addition weighted mean absolute parallaxes are given and UBV photoelectric photometry, MK spectral types and numerous cross identifications are listed when available. Printed versions of the Catalogue will be available through the Yale University Observatory in 1986. Preliminary results of a study at Yale of the determination of absolute parallaxes with respect to faint galaxies indicate a peak accuracy of $\pm 0\rlap{.}''002$ (s.e.) for stars as faint as $m \sim 19$.

G. Westerhout: CCD Parallax Work at Flagstaff. A CCD array is being used with the 1.55 m USNO Flagstaff Astrometric Reflector. There are 160 stars on the program. Initial results on about 20 stars show that the reference frame (5-12 stars) is better than 1 mas; stars with good reference frames and good corrections for differential colour refraction have parallax accuracies of 1 mas. Future work includes installation of a Tectronics 2048 × 2048 CCD and on-line data reduction equipment. It is expected that 1 mas for faint and bright stars will be routine products of the USNO at a rate of 100+ parallaxes per year.

W.J. Luyten: Results of the Proper Motion Surveys. Luyten summarized his proper motion surveys, the Bruce Survey on Harvard plates and the Palomar Survey. The published proper motions from both surveys amount to about 200 000, while data for another 250 000 are ready on tape to be sent to the NASA Goddard Space Center at Greenbelt/Maryland. Luyten has begun handblinking some ESO Schmidt pairs of plates in the South.

A.R. Upgren: New Results on Nearby Dwarf Stars. Parallaxes and proper motions obtained by the Van Vleck Observatory for several hundred lower main-sequence stars are limited to the brighter dwarfs of the McCormick lists because only these have radial velocities which allow space motions to be determined. It is now shown that this selection does not introduce a bias in the transverse velocities of the Van Vleck program stars. Thus they are representative of the McCormick stars in general, as well as those of the Catalogue of Nearby Stars. They are likely to be representative of the solar neighbourhood as well.

W. *Gliese* and H. Jahreiss: A Third Catalogue of Nearby Stars will include all stars known to be nearer than 25 parsecs. A 1984 preliminary version of the new Yale Catalogue of Trigonometric Stellar Parallaxes contained nearly 2 000 objects with parallaxes exceeding $0\!''\!0394$. From luminosity determinations based on spectral type and/or colour, about 700 more stars can be added. Probably the new Yale Catalogue will not change significantly the currently used spectral type-luminosity and colour-luminosity relations.

C.A. Murray: Proper Motions and Parallaxes with the UK Schmidt Telescope. A catalogue of positions, proper motions and parallaxes for 6125 stars brighter than B = 17.5 in an area of 20 square degrees near the South Galactic Pole, has been derived from plates taken between 1975 and 1981 with the UK Schmidt-Telescope and measured on the GALAXY machine at RGO. The zero point of proper motions has been deduced from the observed parallactic motion and this has also been used to calibrate the trigonometric parallaxes for magnitude dependent errors. The average external error of a proper motion component and parallax is about $\pm 0\!''\!015$.

Finally the President read to the Commission members an abstract about a future project by de Vegt: An All Sky High Precision Astrometric Catalog Project down to the 17th Magnitude Using an Integrated High Speed Telescope-Measuring System.

COMMISSION 25: STELLAR PHOTOMETRY AND POLARIMETRY
(PHOTOMETRIE ET POLARIMETRIE STELLAIRES)

Report of Meetings November 19,20,23

PRESIDENT: J.Tinbergen   SECRETARIES: I.S.McLean
                                             J.Tinbergen
                                             R.R.Shobbrook

The Commission 25 programme at the General Assembly in New Delhi consisted of 3 joint meetings, reported mainly elsewhere, and 3 of Commission 25 by itself, reported here. The subjects of the joint meetings were (chairmen/organisers in brackets):

Nov. 20   Synthetic Photometry (Buser)
Nov. 21   Precision Photometry in Clusters (Richer)
Nov. 23   Hipparcos (Kovalevski)

Of these, the first originated within our commission, the other two we co-sponsored.

The subjects of our own meetings were (chairmen/organisers in brackets):

Nov. 19   Polarimetry with the Space Telescope and with Ground-based Facilities (McLean)
Nov. 20   Business and short scientific reports (Tinbergen)
Nov. 23   The Zero Point of the Beta Index for B Stars (Shobbrook)

This programme reflected our primary task, viz. promoting good techniques in the fields of photometry and polarimetry while leaving applications to other commissions.

Synthetic Photometry: a Working Group?

The joint meeting on Synthetic Photometry was initiated by Commission 25, sponsored jointly by Commissions 25,29,36 and 45 and was organised by R.Buser. The main purpose of the meeting was to review the techniques and potentials of synthetic photometry. The papers presented will be published in Highlights of Astronomy.

At the end of the meeting, a general discussion about the possible formation of a Working Group showed that there is sufficient active support to proceed further with this idea. During the discussion, concern was expressed that such a Working Group might interfere with the work of the Hubble Space Telescope calibration team or with that of the Working Group on Standard Stars. It was stressed that this must not happen, but rather that these two groups should be supported and complemented. It was also agreed that spectrometry should be included within the scope of any such Working Group.

With these constraints, Buser (who is on our Organising Committee) will sound the community for further active support and, in consultation with the presidents of the commissions involved, will take further action as appropriate.

Polarimetry with the Space Telescope and with Ground-based Facilities (I.S.McLean)

The meeting was quite well attended considering its unusual time-slot prior to the Inaugural Ceremony and was effective in drawing together a remarkably wide range of very interesting topics involving polarimetry. It was noted that, while all the talks referred to observational techniques and instrumentation, each speaker also demonstrated the worth of polarimetry by discussing real astrophysical results. Seven papers were presented and these are summarised below.

POLARIMETRY WITH THE SPACE TELESCOPE: O.L.Lupie and H.S.Stockman

The paper was given by Dr Olivia Lupie and embodied a description of the HST itself and each of the instruments having a polarimeter capability. Dr Lupie briefly reviewed the proposed calibration schemes and so-called "standard" targets, and appealed for as much ground-based support as possible for the improvement of the lists of standard sources. A written report of polarimetry with the Space Telescope is available from Dr Lupie at the Space Telescope Science Institute at Baltimore.

POLARIMETRY WITH VERY LARGE TELESCOPES: J.Tinbergen

In this short talk Dr Tinbergen emphasised that regrettably, at least some of the New Technology very large telescope designs do not preserve the polarization of the incident radiation. He indicated that with the support of the Commission he hoped to lodge a resolution at the XIXth General Assembly, which would be distributed to institutions responsible for large telescope projects, and which would underline the seriousness of the matter (see Business meeting - J.T.).

EXPERIENCE WITH A FIBER-LINKED GRATING SPECTROPOLARIMETER: R.Ostreicher

A unique spectropolarimeter system employing a grating spectrometer linked by an optical fiber to the telescope and a linear array detector was described in this presentation. Actual observational results were reported, in particular the polarization of R Aquarii.

INFRARED POLARIMETRY FROM 1-10 MICRONS: D.Aitken

In this presentation Dr Aitken gave a fine review of the physics and the techniques of infrared polarimetry. He described some recent and quite remarkable polarization observations of Orion and of the Galactic Centre and outlined for us some of the astrophysical implications. It was very clear to all that infrared polarimetry had undergone a significant transformation in recent years.

SIMULTANEOUS FIVE-COLOUR (UBVRI) POLARIMETRY OF STRONGLY INTERACTING CLOSE BINARIES: V.Piirola

Dr Piirola described his highly successful multi-channel photo-polarimeter system and reported UBVRI polarization observations of SS433. From these measurements he derives both the interstellar polarization and the intrinsic polarization. He finds that the average intrinsic polarization, probably due to electron scattering, is approximately parallel to the direction of the jets. A possible model was discussed. Dr Piirola also reported simultaneous observations of circular and linear polarization in five colours of AM Her binaries, but deferred discussion to the following review by Dr Bailey.

POLARIMETRY OF AM HERCULIS BINARIES: J.Bailey

In this talk the speaker carefully set out the physics of the cyclotron emission models for these compact magnetic objects and demonstrated the importance of obtaining simultaneous photometry and polarimetry over a wide wavelength baseline (0.4-2.2 micron). The circular polarization of these sources is very large. He described the dual optical/IR polarimeter used to make his measurements and reported detailed observations of several sources.

OPTICAL AND RADIO POLARIZATION OBSERVATIONS OF NEARBY GALAXIES: R.Wielebinski
   The final paper was mainly an observational comparison of optical and radio polarization maps of galaxies. Dr Wielebinski showed several sources, discussed the implications and appealed for a greater awareness of the possibilities of combining optical and radio polarization data.

ACKNOWLEDGEMENT
   I.S.McLean would like to take this opportunity to thank all the participants in this session, speakers and audience, for a stimulating meeting.

Business and short scientific reports (J.Tinbergen)

   At the business meeting of Commission 25, the proposal, by the outgoing organising committee, for new Commission Officers was adopted unanimously. The very useful practice of teaming up a photometric president with a polarimetric vice-president (or v.v.) has been turned into a "tradition" with the election of Dr McLean as vice-president. Our officers for the 1985/88 triennium are:

President: F.G.Rufener   Vice-president: I.S.McLean

Organising Committee: R.Buser, J.Dachs, P.J.Edwards, I.S.Glass, D.Kilkenny, J.S.Miller, E.F.Milone, A.J.Penny, V.Piirola, N.M.Shakhovskoj, J.Tinbergen, F.J.Vrba, R.Wielebinski, A.T.Young

   The coming triennium will see widespread adoption of multi-element detectors with their own observational procedures and reduction algorithms. This shift in emphasis is beginning to be reflected in the composition of the organising committee: in addition to mainstream fields, its expertise now extends over very precise and/or differential photometry, IDS and CCD techniques, synthetic photometry and radio polarimetry. This development may eventually lead to the commission merging into a much larger technical commission, or to a general reallocation of scopes of the IAU technical commissions. In our commission's field of interest, wavelength limits are vague, "stellar" is becoming an unnecessary and illogical limitation and the dividing line between multi-colour photometry and spectrometry is fast disappearing. The next two triennia may well see a gradual move towards a commission structure more tailored to modern astronomical practice. If so, our new officers face a complex task.

   The other main item of business was the resolution on polarization fidelity of large telescopes. This resolution, as published with background considerations in IAU Bulletin no 54, had been redrafted at P.Notni's suggestion to remove the limitation to "internal properties" and "point sources". The meeting suggested a further clarification, viz. that degradation of high-accuracy photometric data by telescope polarization effects could be irreversible. The text printed below resulted from this. Unfortunately, in cleaning it up, the Resolutions Committee reduced its scope to an apparent complaint by polarimetric specialists rather than the wide-ranging warning it was intended to be. This is why the less elegant but more relevant original resolution, passed by Commissions 25 and 9, is printed here:

Commissions 25 and 9 of the IAU,

<u>considering</u>   that certain properties of astronomical objects express themselves only in the polarization of the observed radiation,
<u>recognising</u>   that, astronomical polarizations being generally small at most wavelengths, high signal-to-noise and therefore relatively large telescopes are needed for polarization observations, and
<u>moreover,</u>   that high-accuracy photometric observations in general, while

	increasingly becoming both necessary and feasible, could be irrevocably degraded by polarization effects in the telescope,
affirm	that high-accuracy observations of relatively bright objects are as legitimate a use of large telescopes as is the detection of very faint objects, and therefore
recommend	that, in reaching the compromises involved in the design of the very largest telescopes, due weight is given to the need to use particularly those telescopes for the most accurate photometric observations, such as polarimetry, in the decades to come.

The remaining business items were two:

-- p.261 of Reports on Astronomy contains a collating error by Reidel; the correct text is available from J.Tinbergen
-- the list of new members and consultants was displayed and approved; see elsewhere in this volume for an up-to-date commission listing.

In the science part of the meeting, mention was made of J.S.Miller's new CCD polarimeter (his and I.S.McLean's being the only 2 known to be in operation) and of the LEST Technical Report by Stenflo and Povel on optical demodulation (this might be the best technique for pressing the CCD into precision polarimetric service). Short papers were then given on Calgary's rapid chopping differential photometry system (E.F.Milone), on UKIRT infrared array progress (I.S.McLean) and on annual and volcanic-origin extinction variations for a 10-year period at La Silla (F.G.Rufener). Details may be obtained from the authors.

The Zero Point of the Beta Index for B Stars (R.R.Shobbrook)

At this session R.Shobbrook and E.Schmidt demonstrated the existence of some problems with the zero point of the Beta index for B stars, in some regions of the sky.

Shobbrook first presented evidence from a 1980 paper (MNRAS 192, 821, fig. 2) which shows differences of about 0.015 mag for the southern sky between the Crawford, Barnes and Golson (Astr. J., 76, 621, 1971: CBG) and the Gronbech and Olsen (Astr. Astrophys. Suppl. 27, 442, 1976: G&O) Beta indices for Bright Star Catalogue (BSC) B stars. Such errors are significant, especially for B supergiants, since the slope $\Delta M_V / \Delta \beta > 50$ for such stars and the distance moduli may be in error by about 0.7 mag.

The new Balona and Shobbrook (MNRAS 211, 375, 1984) calibration of M_V in terms of β and c_o, determined from 13 clusters with uvbyβ photometry down to the late-B ZAMS, is believed to be the best that can be accomplished with the present data. However, figure 6 of that paper clearly indicates that $(V_o - M_V)$ does vary with M_V for some clusters. The simplest (though admittedly not necessarily the correct) explanation for these apparent 'evolutionary' effects is that for some clusters there are errors in the Beta index zero points of up to 0.01 to 0.02 mag. The variation of $(V_o - M_V)$ with V_o occurs because such an error in β corresponds to 0.5 - 1.0 mag for the most luminous members but only 0.1 to 0.3 mag for the late B dwarfs.

There is some evidence that the G&O scale is correct for the southern sky. Observations of members of NGC 2547 (Shobbrook: MNRAS, in press), whose Beta indices were determined using nearby G&O stars as secondary standards, are in excellent agreement (to 0.001 mag) with the β scales of Eggen (Ap.J. 238, 627, 1984) and of Lynga & Wramdemark (Astr.Astrophys. 132, 58, 1984), who both referred their measurements directly to the primary Beta standards. NGC 2547 is in a region where the mean difference in β between the G&O and the CBG BSC stars is a maximum. This result therefore indicates that the G&O stars in this region are accurately on the scale defined by the primary standards.

Finally, Shobbrook compared 13 clusters which have distance moduli determined both from the Beta index (using the Balona & Shobbrook calibration) and by Mermilliod (Astr. Astrophys. Suppl. 44, 467, 1981) from UBV ZAMS fitting. For 9 of the clusters, the mean difference in distance modulus (in the sense BS - Mer) was only +0.03 (s.d. 0.21) mag. For the other 4 clusters, the differences are -0.7 for Chi Per, +1.1 for NGC 3114 and +0.6 for NGC 2362 and NGC 6231. It is not suggested at present that these problems are solely due to an error in the Beta indices, since ZAMS fitting for the youngest clusters is notoriously risky in the UBV system. However, it would be prudent to remeasure a few stars in each cluster, preferably against the primary Beta standards, to check the zero points.

Schmidt's contribution mainly addressed the problem of the calibration of the zero point of the Cepheid period/luminosity relation. Hβ photometry of B stars in clusters containing Cepheids results in a luminosity scale which is fainter by 0.1 to 0.4 mag than that determined from UBV studies of the same clusters. On the other hand, ultraviolet studies of the companions of Cepheids results in a scale which is a further 0.6 mag fainter than that from Hβ photometry and theoretical studies indicate a scale brighter by 0.2 to 0.4 mag. A calibration based on Walraven photometry and that using the recent Balona and Shobbrook β calibration are in excellent agreement if the Pleiades modulus is given the same value in both calibrations. Unacceptable discrepancies do remain, but there is no reason to assume that the Hβ scale of M_V is in error by more than a very few tenths of a magnitude.

To produce convergence of the various estimates of the Cepheid luminosities, each method needs to be investigated in detail. Schmidt suggested several possible difficulties with the Hβ scale, as follows:

a) Internal consistency of the photometric systems: Balona & Shobbrook showed that zero point errors between clusters are a plausible source of inconsistencies in the luminosity calibration for some clusters. One possible cause for such an error is the failure to apply a colour term to the β transformation which may be necessary for some filter sets. This may be important for some of the Cepheid clusters which tend to suffer more reddening than do most of the calibrating clusters. This can be checked by reobserving a few B stars in each cluster containing Cepheids and taking care to determine colour effects accurately. Schmidt has made a preliminary investigation by comparing a few stars in each of three Cepheid clusters and three calibrating clusters. Although the existence of a colour term was suspected, there was no evidence suggesting that a revision of the scale was necessary. More detailed checks are still required, however.

b) Emission line stars: because emission at Hβ causes the inferred luminosity and distance to be too large, Be stars must be excluded from the determination of the average cluster moduli. This would most effectively be accomplished by measuring a photometric Hα index in conjunction with the Beta index. As has been shown by several authors, the Be stars may thus be easily identified.

c) Cluster membership: membership studies based on motions are needed for all clusters with Cepheids.

d) Extension to fainter stars: the Hβ studies of clusters with Cepheids have been restricted to the B stars. The extension to fainter A and F stars would provide a valuable check on the moduli. Such an extension is now becoming feasible with the advent of new panoramic detectors.

COMMISSION 26 : DOUBLE AND MULTIPLE STARS (ETOILES DOUBLES ET MULTIPLES)

Report of Meetings.

Business Section, Wednesday, November 20, at 4:00 P.M.

The President opened the meeting and welcomed those in attendance : Fracastoro (chairman), Batten, Dommanget, Fredrick*, Heintz, Hidayat, Rakos, J. Russell, Strand, van Dessel.

I   Commission President 1985-1988

Commission members unanimously propose to the IAU Executive Committee that prof. K. D. Rakos serve as President of Commission 26 for the 1985-1988 triennium.

II   Election of the Vice President

The chairman announces that he has received 13 votes by mail. It is unanimously decided to accept them as valid. Then, votes are expressed by secret ballot from the 10 members attending the session. The final result (13+10) is: McAlister 11 votes,   Heintz 9 votes.
At this point Heintz leaves.

III   Organizing Committee

It is unanimously decided to maintain the O.C. at the present number. For the 1985-1988 triennium, Fracastoro, as past president , becomes an ex officio member of the O.C. and Dommanget and Heintz remain for a second term. The terms of Franz, Poveda and Scarfe have just expired; therefore, 3 members must be elected. The chairman announces that 13 members voted by mail. Adding the votes of the 9 members now attending the session, the result is :   Harrington 13 votes, Couteau 12 votes Kiseliov 9 votes and Abt   7 votes.

IV   New members

It is decided that the following persons can be accepted as members of Commission 26, in view of their scientific activity: Bacchus, Bernacca, Eichhorn, Fletcher, Luyten, Morel, Zukevic, and also Radiman and Zinnecker if they are already IAU members.

V   Report for IAU Transactions XIX A

At the proposal of Strand, the report concerning the activity of the commission during the 1982-1984 triennium is unanimously approved.

At 5:30 P.M. the session ends.

Addendum to the Business Meeting.

At 1:05 P.M. on Thursday, November 21, Fracastoro is informed by another Italian delegate (Felli), who shares the same mailbox, that he has some correspondence in the mailbox. Immediately, Fracastoro collects two envelopes, containing votes expressed by two commission members from the USSR. At the IAU Secretariat's office, Fracastoro is advised to inform the commission as soon as possible. This is done before the scientific session of Friday, November 22 begins.
Fracastoro asks first whether or not these votes should be taken into account and added to the previous score. He is strongly in favor of accepting these votes. However, when his proposal is voted on, the majority is against (4 to 1). Consequently, the case comes to an end.

* *Frederick declares to be a member of Commission 26, although he is not listed in the official list received from the IAU Secretariat.*

2nd and 3rd Session (scientific), Friday November 22.   Chairman K. Aa. Strand

K. D. Rakos :

Mass ratio of unresolved third companions associated with the visual binary components.

Three groups of visual binaries have been used to investigate the cosmic scatter in the main sequence. To the first group belong 147 main sequence binaries, the observations being published by Rakos et al. and by Hurly and Warner. The second group is formed by 84 binary systems on the main sequence, selected from the photometry published by Eggen. The third group consists of common proper motion pairs of the same spectral types (stars from the solar neighborhood later than F0). In general the standard deviation of typical B-V and ΔV values for the first and second group does not exceed 0.05 magnitudes. A histogram can be constructed for each group placing all secondaries on the main sequence and considering the difference Δm between the magnitude of primaries and the main sequence.

All three histograms show a local maximum in the distribution of Δm for Δm = 0.2 magnitudes. This maximum can be introduced by the unresolved companions. If we assume that one of the primary components has a close unresolved companion, and the triple structure (AB-C) is responsible for the Δm deviation from the main sequence, we can estimate the luminosity - or mass - distribution of the unseen companion B. As long as the mass ratio $q = M_B/M_A$ is not smaller than 0.5, the distribution is undistinguishable to the van Rhijn luminosity function. For smaller tertiary masses the number of unseen companions decreases rapidly.

From the histograms it can also be seen that the number of hypothetical third companions is larger for the earlier spectral types and twice more frequent for primaries than for secondaries, independent of the spetral type.

The basic formation mechanism for close binaries seems to be the fission or the fragmentation of a gaseous disk. The fragmentation of a rotating protostellar cloud is probably responsible for the formation of binary stars with periods exceeding 100 years. The mass distribution of close companions may be different from van Rhijn's function. According to Lucy, the instability of the 3rd harmonic of an elongated triaxial rotating ellipsoid yields a mass ratio of the components of about 0.7.

In future work, an improvement can be achieved by measuring R-I colors and the precise radial velocities to check the hypothesis of the unresolved companion. Also very extensive photometry of about 400 common proper motion stars is in preparation.

M. Shara, H. McAlister, D. Hutter and O. Franz :

Estimating the percentage of binaries among potential guide stars for Space Telescope   (read by J. Russell).

The fine guidance sensors for the Space Telescope cannot lock onto guide stars with Δm < 2.0 mag. and separation from 0.02 to 5 arcsec, so knowing the percentage of undiscovered close binaries is essential for planning the operation of the spacecraft. This study involved a list of 683 stars from the Yale Bright Star Catalog, chosen from entries with magnitude between 5.0 and 6.5, declination $-20°$ to $+60°$, and without regard to previously known duplicity. The stars were observed with speckle interferometry on the CFH telescope at Mauna Kea, Hawaii. The use of bright stars made possible a complete search with Δm < 2.0 mag. and a separation range of 0.15 to 1.2 arcsec. The results include: 62 previously known binaries were resolved (mean separation 0.6 arcsec); 52 new binary systems were discovered (mean separation 0.17 arcsec); 52 of the known and 40 of the new binaries are luminosity class V; 10 of the known and 12 of the new binaries are class III and IV; extrapolating from this discovery rate, there are about 300 class V and about 350 class III doubles still to be discovered in the Bright Star Catalog. Correcting for this percentage of giants among the potential guide stars, for random orientations and for the binaries with separations less than 0.15 arcsec, about 20% of possible guide stars will be binaries and unusable for Space Telescope operation.

J. Dommanget :

The future of double and multiple star research.

L'astronomie des étoiles doubles et multiples semble atteindre aujourd'hui un point de stagnation particulièrement néfaste au sein de la commission, alors que les étoiles doubles (tout autant celles aux composantes écartées que les couples serrés) apparaissent de plus comme etant les objets les plus fréquents du milieu stellaire.

Les raisons de cette situation se situent sans doute d'abord dans l'intérêt croissant pour l'astrophysique, la radioastronomie, la cosmologie, etc, mais aussi et surtout dans l'attitude des membres de la commission 26. Notre conception de l'astronomie des étoiles doubles est désuète: elle continue à admettre pour piliers fondamentaux des recherches l'observation, le calcul d'orbites, la détermination des masses stellaires, le raffinement de la relation masse-luminosité, parfois aussi l'étude des perturbations des compagnons obscurs et de masses faibles. Bien que ces recherches soient irremplaçables, elles n'en restent pas moins trop restreintes et peu attrayantes pour les jeunes chercheurs, alors que l'astronomie des étoiles doubles et multiples recèle les bases d'immenses domaines à explorer touchant à la photométrie, la spectroscopie, l'évolution stellaire, l'évolution galactique etc.

Curieusement pourtant, il existe chez les collègues des autres commissions un courant sous-jacent favorable aux étoiles doubles, mais qu'ils ne semblent pas pouvoir exprimer par manque de stimulant au sein de la notre. Deux faits importants peuvent être avancés à ce sujet. La proposition d'organiser un colloque sur les composantes écartées d'étoiles doubles et de systèmes multiples a trouvé plus d'échos en dehors de notre commission qu'en dedans. En réponse à l'appel lancé par l'Agence Spatiale Européenne (ESA) pour constituer le catalogue des étoiles à observer par le satellite astrométrique Hipparcos, plus de 200 propositions furent reçues, dont seulement 23 concernant les étoiles doubles de toutes natures. Mais quelques unes seulement sont dues à des membres de la commission 26. On peut se demander ce qu'il en sera des programmes d'observation proposés au Space Telescope.

Aussi, le seul remède à la situation actuelle consiste dans un accroissement sensible de l'éventail des voies de recherche au sein de la commission 26 et dans un recrutement de nouveaux membres intéressés ou déjà impliqués dans leurs développements. Il devrait également comporter une propagande systématique pour ces aspects négligés de l'astronomie des étoiles doubles et multiples.

J. Dommanget :

Wide components in double and multiple systems.

Astrometric observations as well as astrophysical observations of components of medium wide and wide pairs do not favor a particularly great enthusiasm. But we have to decide whether we want to know more about binaries, or if we just want to conduct our activities in small areas, ignoring the characteristics of the whole domain which we are interested in. Actually, wide components and even medium wide ones appear more and more as subjects for discussion concerning the reality of their physical existence, because of their interest for binary and galactic evolutions (see: Lembang-Bamberg IAU Colloquium No. 80 and Bamberg International Conference). Common proper motion stars (mainly discovered by Luyten), moving groups (studied by Schutte and Eggen) and very wide pairs with separations of several parsecs (considered by Upgren and Vandervoort) for instance are different classes of such objects that should lead to increased attention.

The frequency of such objects appears to be much higher than expected: systematic exploration of photographic plates, for instance, reveals that generally half of the pairs with medium wide separations are ignored in the catalogs. Such a situation will lead to further problems as to what kind of systems should be retained for introducing in double star catalogs (index for instance). Other catalog problems have to do with accurate positions and identifications.

The recognition on the sky and the final census of all binary and multiple systems are of the highest interest for correct statistics; for instance, in the case of establishment of the distribution diagrams of various physical parameters (e.g. magnitudes, spectra and also true separation of their components). All these diagrams are of fundamental importance for a better understanding of the formation and the evolution of the binary and multiple systems, and indirectly for stellar and galactic origin and evolution, as succintly reported by Dommanget in a paper given at the Lembang-Bamberg IAU Colloquium No. 80. An IAU colloquium on "Wide components in double and multiple systems" has been proposed and officially supported by Commission 26 during the XIX General Assembly in New Delhi.

The main aim of this colloquium would be restricted more to practical problems than to theoretical ones. It should make possible a better evaluation of the subject at a time where it is urgently needed.

A. H. Batten :

Parallax and masses for 70 Ophiuchi.

Batten reports about his continuing high-dispersion spectroscopic observations of both components of 70 Ophiuchi. It was recently possible to publish values for the parallax of the system and the masses of the components, derived from these spectroscopic observations, and they agree well with the corresponding values derived from the astrometric observations, thus strengthening the credibility of very small paralaxes obtained spectroscopically for similar systems. The masses of the components of 70 Oph must now be among the best determined. In 20 years, no residuals from the velocity curve as large as 1 km/s have been found and most are much smaller. Thus it has been possible to eliminate the 3 rd body postulated for this system, just as Strand had shown earlier, that supposed perturbations in the plane of the sky were not real. The radial velocity of the primary component is now increasing again. Similar observations of other systems show that the components of many visual binaries are not main-sequence stars, and they are very valuable for testing theories of stellar evolution.

R. Pannunzio, G. Massone, V. Zappalà and R. Morbidelli :

Statistical analysis of the errors of visual double stars observations (read by V. Zappalà).

A selected sample of well-known binary orbits and 3141 observations of visual binaries, obtained by 12 observers of the old generation, have been analyzed, comparing the average residuals in separation ρ and position angle θ to the actual and the apparent ρ, allowing to evidence systematic and accidental errors. A systematic positive increase of the residuals in ρ for decreasing ρ's has been detected inside the resolving power of the used telescopes. Moreover, a negative and a positive systematic displacement appear for large separations in the $(O-C)_\rho$ and $(O-C)_\theta$ plots respectively, for which no physical explanation seems to be plausible. The analysis of the standard deviations allows to show a different behavior for the accidental errors inside and outside the resolving power. In the inner zone they slightly increase with decreasing separations, and depend on the apparent separation between the components, as seen in the eye-piece. In the outer zone, the accidental errors increase with increasing separations and strongly depend on the actual separation. In general, the single plots relative to each observer of our sample do not differ sensibly from the general plot obtained by averaging all the 3141 observations. The contribution of the "resolving power effect" on the orbit determination has been qualitatively investigated, pointing out some possible tests for its quantification. Similar studies on modern observers and on observations obtained by interferometric and photographic techniques are in progress. Furthermore, taking into account the results obtained by the present statistical study, a systems of weights, depending on the separations of the binaries and on the telescopes size, is given.

M. G. Fracastoro recalls a paper published in "Atti della Accademia Nazionale dei Lincei (vol. LXXIII, p. 226, December 1982), in which the Yale Catalog of Bright Stars, by D. Hoffleit, and the Catalog of Nearby Stars, by Gliese, have been analysed in order to determine the frequency of couples, in view of the mission of the astrometric satellite Hipparcos. In particular, the observed number of these pairs N_{obs} is higher than that theoretically foreseen by the Poisson formula N_{exp}, up to separations of 10 arcsec, namely well beyond the resolving power of any telescope, as shown by the following table, where the ratio $\log N_{obs}/N_{exp}$ is given, in terms of magnitude of the secondary and of separation.

m_2 \ s"	4:5	5:6	6:7	7:8	8:9	9:10	10:11	11:12	12:13
0.1-0.9	5.7	5.7	5.2	4.7	3.7	2.9	2.2	1.4	–
1.0-1.9	4.6	4.6	4.3	3.9	3.3	2.8	2.2	1.7	1.2
2.0-3.9	3.8	3.8	3.6	3.2	2.8	2.2	1.8	1.3	0.8
4.0-7.9	3.0	3.1	2.8	2.6	2.2	1.7	1.2	0.7	0.4
8.0-16	2.7	2.6	2.3	2.0	1.5	1.1	0.7	0.3	-0.2
16 - 32	2.1	1.7	1.5	1.3	0.9	0.5	0.1	-0.2	-0.5
32 - 64	1.0	1.2	0.8	0.7	0.4	0.1	-0.4	-0.9	-1.4
64 -128	0.3	0.1	0.1	-0.2	-0.4	-0.9	-1.3	-1.8	-2.4
128-256	-0.4	-0.4	-0.7	-1.1	-1.5	-1.7	-2.2	-2.6	-3.7

COMMISSION 27: VARIABLE STARS (ETOILES VARIABLES)

Report of Meetings: November 20, 21 and 26, 1985

PRESIDENT: Norman H. Baker   SECRETARY: John R. Percy

November 20, 1985

SESSION ON FLARE STARS

C.J. Butler: "Coordinated EXOSAT and Optical Observations of Flare Stars and Coronal Heating". By considering the relation between X-ray, U-band and total luminosity, and by studying the time variability of each, the author reached the conclusion that the coronae of flaring stars may represent the superposition of large numbers of microflares.

R. Stern: "Thermal X-ray Emission in Solar and Stellar Flares". Observations were described which were relevant to the question of the sequence of events which occur in solar and stellar flares, starting with the still-poorly-understood initial flare event. The main difference between solar and stellar flares is the large volumes occupied by the emitting material in the latter case. Hyades dwarfs have thermal X-ray temperatures similar to that of the flaring sun.

D. Gibson: "EXOSAT Observations of YY Gem". The advantage of observing the flare stars in this system is that they form an eclipsing system; the disadvantage is that the emission cannot be separated from any from the A-type components. An eclipse of the flaring region shows that this is very small. Slow, low-level variations were also observed.

L.N. Mirzoyan and E.S. Parsamyan: "Flare Stars in Star Clusters and Associations". The following aspects were discussed: mean flare frequency in clusters, occurrence of classical flare star activity in T Tauri stars, change of average luminosity of flare stars as a function of age of cluster, and the classification of flare types.

M. Tsvetkov, H. Duerbeck and W. Seitter: "A Search for Flare Stars with the GPO Astrograph at La Silla - ESO". Multiple exposures on 2°×2° plates taken with this astrograph have been used to discover flare stars. The results were compared with those using other instruments.

L.N. Mavridis and S. Avgoloupis: "The Activity Cycle of EV Lac" (read by N. Baker). From a long and homogeneous data set, the authors have discovered an interesting cycle in the mean quiescent luminosity and in the flare rate in this star. The cycle length was about five years.

P. Feldman: "Radio Emission from FK Comae Stars". FK Comae stars are single stars of G-K III type which show high rotation and high levels of activity. High-luminosity radio flares have been detected in two of these stars, but optical observations suggest that both are binary stars, and therefore not FK Comae stars.

This session was organized by L. Mavridis. About 50 people attended.

BUSINESS MEETING

The Commission president, N.H. Baker, introduced members of the Organizing Committee, and outlined the agenda for the meeting.

1. J.R. Percy reported that, thanks to a resolution of the 1982 IAU General Assembly, and to a financial contribution by the IAU, the American Association of Variable Star Observers (AAVSO) had been able to obtain additional funds, and to begin publication of their archival data in the form of monographs on individual stars.

2. M. Breger reported on the IAU Archives of Unpublished Observations of Variable Stars. The archives serve as a depository for files of observations, thus ensuring that the observations will be available at any future time. Contributors should send three copies of the observations, with a descriptive cover sheet, to the coordinator (M. Breger), who assigns a file number. The copies are then deposited at the Royal Astronomical Society (England), Centre de Données Stellaires (France) and the Odessa Astronomical Observatory (USSR). Copies of the files may be obtained by writing to one of these institutions. A list of new files is published regularly in the IBVS and the Publications of the Astronomical Society of the Pacific. Files were received at an average rate of 16 per year from 1980-1985, with 25 being received in the most recent year. A total of 156 files has been received to date. Approximately 20 requests for copies are received each year.

The coordinator recommended that each file should contain observations of only one star, unless several stars have been described in a single publication. In these cases, observations of all stars can be included in a single file. Reference to the publication should be included on the cover sheet. This recommendation was accepted.

There was some discussion of the possibility of depositing observations on magnetic tape or diskette, but it was decided that a paper copy would be more permanent, and less likely to become technologically obsolescent!

3. N.H. Baker reported on a meeting on November 19 on the topic of the designation of astronomical objects. The situation in the field of variable stars is much better than in most other fields! Commission 5 is working on the problem, and it was recommended that Commission 27 should maintain an interest in this work. In variable star work, authors are urged to use multiple designations where possible.

4. B. Szeidl (editor) reported on the Information Bulletin on Variable Stars. A total of 2814 issues have been published since the birth of the IBVS 24 years ago; 600 have been published since 1982; the number per year continues to increase. Papers are not formally refereed, though about 1/3 are refused for one reason or another. This results in variable scientific quality, but quick "turn-around" time (3-5 weeks). The IBVS is not intended for detailed discussions, or for papers not requiring urgent publication (the latter guideline is not always adhered to). The IBVS does not accept papers based on visual observations. The IBVS is currently sent to 350 institutions and 200 individuals.

5. N. Baker reported on the current status of the General Catalogue of Variable Stars; see also IAU IB #54. The second volume of the current edition (Cyg-Ori) has just been published. Members of Commission 27 receive copies free of charge.

6. N. Baker reported on membership in Commission 27. The IAU Secretariat, in order to "clean" their membership files, has written to members to confirm their commission membership. There are presently 250 members of Commission 27. A list

of possible new members was read and approved.

7. The following symposia and colloquia are planned:

"Advances in Helio- and Astro-seismology", July 7-11, 1986 in Aarhus, Denmark, approved as IAU Symposium 123 by the IAU Executive Committee.

"Circumstellar Material in Close Binaries", summer 1987 in Victoria, Canada. Cosponsored by Commission 42. Commission 27 agreed to cosponsor this meeting.*

"Atmospheric Phenomena as Manifestation of Internal Evolution of Stars", August 1987 in Tokyo, Japan. Cosponsored by several other commissions. Commission 27 agreed to cosponsor this meeting.

"Solar and Stellar Flares", in 1988 in Palo Alto, USA, and "Flare Stars in Stellar Clusters and Associations and in the Solar Neighbourhood" in 1988 at the Byurakan Observatory, USSR (in honour of the 80th birthday of V. Ambartsumyan). It was noted that, although both of these meetings are scientifically useful, there is overlap of content and conflict of schedule. It was agreed that the president and vice-president of Commission 27 should discuss these concerns with the organizers. Some problems were encountered in obtaining visas for a previous USSR meeting; this matter was referred to the IAU Executive Committee.

8. N. Baker proposed that the 1985-1988 Organizing Committee be the same as in 1982-1985, except that M.A. Smith be added. M. Breger becomes vice-president and B. Szeidl becomes president. This proposal was approved. The Organizing Committee therefore consists of A.N. Cox, R.E. Gershberg, M. Jerzykiewicz, L.N. Mavridis, L.N. Mirzoyan, J.R. Percy, M.A. Smith, A.M. van Genderen and B. Warner.

9. There was some discussion about the possible revival of the Working Group on Flare Stars. There is a need for cooperation and coordination of observations, especially between satellites and ground-based facilities. This matter was referred to the following session on Coordinated Multisite Observations.

SCIENTIFIC SESSION

The following two papers were presented:

A.N. Cox: "The Puzzling B Star Pulsations". The author reviewed the persistent problem of what is the pulsation mechanism in B stars, and proposed a new theory based on nuclear driving. In the presence of a molecular weight gradient (caused by the shrinking convective core), this mechanism may be capable of driving pulsations in a low-order ($\ell=1$) g mode.

H. Deasy: "Mass Loss from Cepheids". The author has used the IRAS catalogue to search for IR excesses in Cepheids and nonvariable yellow supergiants. Long-period Cepheids have such excesses, which are attributed to mass loss. The mass loss does not seem to be sufficient to explain all of the Cepheid mass discrepancy.

*Not yet approved by IAU Executive Committee.

SESSION ON COORDINATED MULTISITE OBSERVATIONS

The following report was prepared by the Secretary for publication in the XIX IAU GA Newspaper Mandakini:

"The potential value of observations from two or more sites, using different techniques or frequencies, is well known. It was noted at Joint Discussion II, for example, that thanks to a coordinated "campaign", the recent eclipse of Epsilon Aurigae was studied in unprecedented detail. Furthermore, the availability of measurements from the radio to the X-ray region has made it possible to construct greatly improved models of this and related objects. At Joint Discussion III, several speakers noted that coordinated observations from different longitudes are absolutely essential in determining accurate and reliable pulsation periods in the Sun and other stars -- particularly Delta Scuti and Be stars. The International Halley Watch is perhaps the most ambitious example of coordinated multi-site observations.

C. Sterken and J. Christensen-Dalsgaard have recently proposed that a new IAU Commission (or more likely, a working group) should be established to assist in organizing such coordinated observations. On 21 November, Commissions 12 and 27 cosponsored an informal discussion of this proposal. Several advantages were pointed out. A working group might publish a newsletter containing information on the success (or failure) of coordinated campaigns. It could circulate a list of astronomers and observatories which would be interested in and available for campaigns (many "local" observatories have far more flexibility in scheduling than do the "national" ones). It might even be possible to set up a quasi-permanent network of observatories, somewhat analogous to the *ad-hoc* VLBI network in the USA. It might devise ways to simplify the simultaneous submission of complex proposals to several ground-based or space-based observatories. At the very least, the working group could keep the astronomical community aware of the problems (and solutions) in making multi-site observations, and could encourage and assist observers to do better planning. It was realized that the success of a working group depends on having some enthusiastic and active organizers, and that a formal administrative structure can sometimes do more harm than good.

In the end, no consensus was reached about what to do, and no decisions were taken. Nevertheless, the discussion was useful in that it demonstrated the great interest in and variety of opinions about the topic. Sterken and Christensen-Dalsgaard plan to organize a workshop on 'Coordinated Multi-site Observations' in 1987, somewhere in Europe, and would be happy to hear from prospective participants."

November 26, 1985

SESSION ON T TAURI STARS

I. Appenzeller: "High Resolution Spectroscopy of T Tauri Stars". Spectra of S CrA and VV CrA, obtained at a resolution of 20,000 and a S/N of 25 to 50, were described. They showed interesting profile shapes and variability in emission lines; absorption lines of several elements (e.g. Li) were detected. This work is in press (AA Suppl.)

V. Pirronello: "Ice Stability during Eruptive Phases in T Tauri Stars". Ice grains may be expected in T Tauri envelopes, and have been observed in HL Tau (by Cohen). This paper described calculations and experiments relevant to the destruction of grains by sublimation and sputtering during FU Ori-type outbursts. The surfaces of comet nuclei may show evidence for radiation chemistry reactions which occurred during these outbursts.

A.L. Gyulbudaghian: "Trapezium-Like Tight Systems Containing Red Dwarf Stars" (read by N. Baker). Observations of Trapezium-like systems of typical dimensions 0.1 pc were described. Since these are dynamically unstable, they must be young. Both systems containing OB stars and systems containing red dwarf stars were investigated.

This session was organized by M. Cohen. About 35 people attended.

COMMISSION 28: GALAXIES (GALAXIES)

Report of meetings 20, 21, 22, 23, 25, 26 and 27 November 1985

PRESIDENT: V.C. Rubin   SECRETARY: P.C. van der Kruit

Commission 28 was involved in three all-day Joint Discussions, of which the proceedings will be published in Highlights of Astronomy, volume 7:
Radio Astronomy and Cosmology (Commission 28, 40, 47), 21 November,
Evolution in Young Populations in Galaxies (Commissions 25, 28, 33,
   34, 37, 45), 26 November,
Supernovae (Commissions 27, 28, 34, 35, 40, 44, 47, 48), 27 November.
Two sessions were held jointly with other commissions: Stellar Orbits and Structure and Evolution of Galaxies (with Commission 33) on 22 November and Galaxy Radial Velocities (with Commission 30) on 20 November. Ten sessions were held by commission 28 alone. Programs of these and abstracts of some of the papers follow below. The session on Galaxy Radial Velocities is covered in the report of Commission 30.

20 November: Business and Working Group on Internal Motions

CHAIR: V.C. Rubin

The following items were covered in the business session:

New officers: the commission unanimously elected P.C. van der Kruit as the new president and G.A. Tammann as the new vice-president. Outgoing members of the Organizing Committee are J.A. Graham, P.W. Hodge, M.-H. Ulrich-Demoulin and B.E. Westerlund, who were thanked for their contribution. Elected for a second term were S.d'Odorico, J. Einasto, I.D. Karachentsev, D. Lynden-Bell, A. Toomre and K.-I. Wakamatsu. The outgoing president V.C. Rubin was also elected for an additional term. Newly elected members are E. Khachikian, Li Qi-Bin, J. Lequeux and H. Quintana.
New members: A total of 117 names of new members were proposed before and during the General Assembly and one of a consultant. These were admitted. Commission 28 now has 473 members and 1 consultant. Some attendants urged the new president to ask non-active members to resign and so reduce the membership.
Draft report: A member of participants indicated the usefulness of the report. It was subsequently adopted.
Working groups: The WG on Internal Motions needs a new chairperson. Names were proposed during the General Assembly upon the president's request. After the General Assembly C.J. Peterson was appointed.
Resolutions: None
Other business: The members were informed on the other meetings of Commission 28 during the General Assembly.

In the scientific session of the Working Group on Internal Motions the following papers were presented:
J.S. Gallagher (USA): Echelle measurements of emission line kinematics in galaxies
R. Giovanelli (USA): UGC 12591 - the most rapidly rotating disk galaxy.
A. Sandquist (Sweden): ^{12}CO in NGC 1365 and M31.
M.A. Kazaryan (USSR): Spectra of active galaxies including Seyfert, narrow-line galaxies and absorption-emission line systems observed with the 2.6m Byurakan and USSR 6m telescopes.

Gallagher:
The echelle spectograph on the Kitt Peak Mayall 4-m telescope has enabled observations to be obtained for emission lines with FWHM velocity resolutions of 12-30 km/s. Results are briefly reviewed from programs to study Sc spiral nuclei, Seyfert galaxies with extended radio emission, and diffuse ionized gas in blue irregular galaxies. These show that the echelle data provide improved precision in determining velocity field from line centroids and furthermore reveal a variety of local kinematic anomalies through the line profiles. For example, in the M51 nuclear region disturbed gas with "nuclear" emission characteristics extends over the inner ~1 kpc and in blue galaxies disturbed velocity fields are associate with ionized filaments and may be due to energy input from young stars.

Giovanelli:
Work done in collaboration with M.P. Haynes and V.Rubin. Arecibo 21cm observations of the Soa galaxy UGC 12591 exhibit a line profile with a width of about 1000 km/s, the broadest ever observed. The radio observations are corroborated by an optical rotation curve (Hα + [N II]) obtained at the 5m telescope on Mt. Palomar. The total mass within R(25) inferred from the rotation curve is 2×10^{12} M$_\odot$. Implications on the mass-rotational velocity relation, luminosity-linewidth relation and the gas content of early-type galaxies were discussed.

20 November: Working Group on Space Schmidt Telescopes

CHAIR: F. Bertola

During the business section Dr. Lequeux of Marseille was elected chairman and Dr. K. Henise vice-chairman.
During the scientific section Dr. Lequeux reported on a proposal submitted to ESA in answer to a call for ideas of a small low cost 50cm Cassegrain telescope to be placed on the Eureka platform. Dr. Tovmassian remarked that a similar instrument, especially concerned for studies of galaxies and quasars, called GLAZAR, will be put in orbit soon by the Soviet Union. Dr. R. Barban of Padova, Italy, reported on the work done in Italy for the frame of one meter class Schmidt telescopes. Dr. H. Smith, University of Texas described the situation, which at the present is not promising, in the United States concerning the Space Schmidt. Dr. Heatherthorn of US Naval Observatory described the present status of development of the detectors for Space Schmidt. Dr. Davidson of Johns Hopkins University provides some information on the UIT project.

20 November: Working Group on Supernovae

CHAIR: V. Trimble

Participating members of Commission 28 and other interested IAU Members and Invited Participants agreed (a) that the Working Group should exist; (b) that its membership should consist of the current mailing list (about 120) plus others who desire to join, and (c) that nominations and an election for a vice-chairman should take place during 1986, the vice-chairman to become chairman in the course of the XXth General Assembly in 1988.
After some discussion, participants agreed upon a system of assigning temporary and permanent names to supernovae as they are discovered. This recommendation was endorsed by the General Assembly as Resolution C11 and appears elsewhere in this volume.
In addition to its involvement in the organization of Joint Discussion VII (Supernovae), the Working Group had one session of its own, at which 10 contributed papers were presented. A list of these, with exceedingly brief summaries, follows.

"Some Possible Identification between Chinese Stars and SNR's" Wang Z.-R. (PRC). Seven tentative new associations were suggested, on the basis of approximate positional coincidence and congruence of distance and age. Events date from 832 BCE to 1523 CE, and remnants include some wellknown ones (Kes 25; RCW 103). Confirmation of even one or two of these would enormously increase our understanding of young SNR's.

"Supernova Rates in Flocculent and Grand Design Spirals" M.L. McCall (Canada). Spiral galaxies with and without grand designs attributable to spiral density waves do not differ significantly in their supernova rate or in the proportion of Type II events. The implication is that massive star formation is not much inhibited by the absence of a density wave.

"Gravitational Radiation by a Rotating Stellar Core during Collapse" Yu Y.Q. (PRC). Numerical simulation of the collapse of rotating configurations indicates that gravitational radiation is not very efficient at carrying off angular momentum. This leads to a sort of barrier at $a/M = 1$, and the possibility of formation of naked singularities.

"Supernova and Cosmic Rays" M.M. Shapiro (USA), Production of a substantial fraction of cosmic rays by supernova of Type I may account for many features of their composition and spectra.

"Non-LTE Effects in SNe Photospheres" R. Wehrse (GFR). The color temperature of Type II Supernova at visible wavelengths is typically somewhat higher than the effective temperature, owing to scattering effects. This must be taken into account when measuring SN distances by the Baade-Wesselink method.

"The Jodrell Bank Pulsar Search" F.G. Smith (UK). A search over a wide range of periods and dispersion measures had identified 32 new pulsars with large dispersion measure, indicating distances in the range 5-10 kpc. Most are within 60° of $l = 0°$. Two new young objects with periods near 0.1 sec have slowing-down ages of 22,000 and 17,000 years, filling in the gap between Vela and longer-period objects.

"Contributions of SNR's to the Diffuse X-Ray Background on the Galactic Ridge" K. Koyama (Japan). SNR's are responsible for a significant portion, but probably not all, of this relative smooth background.

"VLBI Observations of the Compact Sourses in M82" N. Bartel (USA). About 20 bright sources, unresolved at 0".15, have been mapped with the VLA. At least a couple have declined measurably in brightness over
1-2 yr. Many (not the variable ones) have been resolved with VLBI, indicating sizes near 1pc. The shape is never symmetric. They are thus too extended for recent supernovae and too bright (many times Cas A) for intermediate-age remnants and so apparently represent a new class of source.

"Phase Transitions of High Density Matter and Supernova Explosions" K. Sato (Japan). A phase transition to pion condensate, strange quark matter, etc. during gravitational collapse of a stellar core can either weaken or strengthen the outgoing shock needed to produce a supernova explosion. In particular, a strong first-order transition would greatly increase the chances of the shock getting out with sufficient energy.

"Supernova-Induced Star Formation (IRAS Data)" T.N. Rengarajan (USA). A search of the IRAS data base for point sources in the vicinity of known supernova rements uncovered a significant excess (555 where 389 would be expected), 2/3 associated with six remnants, and most around the edges rather than the centers. Although the remnants (Pup A, W28, IC 443, etc) are relatively old, near 10^4 yr, their ages are still much less than the free-fall collapse time of a protostellar cloud. Thus the excess must reflect the general association of supernova with regions of active star formation, rather than direct causality.

22 November: New Results (2 sessions)

CHAIR: P.C. van der Kruit

The following papers were presented of which some abstracts are included below:

S. van den Bergh (Canada): RR Lyrae Variables in M31.
J. Hutchings (Canada): IUE spectra of OB stars in M31 and M33.
I.A. Issa (Egypt): Distance Estimates of the LMC, SMC and M33 from the apparent radius distribution of HII-regions.
R. Wielebinski (GFR): Radio continuum - far infrared - blue light correlations in galaxies.
D.A. Hunter, W. Rice, F. Gillet, J. Gallagher (USA): IRAS observations of irregular galaxies.
S. Jörsater (ESO): On the nature of spiral arms in spiral galaxies.
P.O. Lindblad (Sweden): Peculiar gas motions in the nuclear region of NGC 1365.
G. Rydbeck, A. Hjarmarson, L.E.B. Johansson, O.E.H. Rydbeck (Sweden): Density wave related molecular cloud spiral arms in M51.
E. Skillman, H. van Woerden (Netherlands): HI observation of dwarf irregulars in the Local Group.
F. Sakhibov, M.A. Smiznov (USSR): The corotation radius in spiral galaxies with application to M33.
J. Gallagher, O.A. Hunter, A. Sandage (USA): UBV colors of Virgo cluster irregular galaxies.
S.D. Mathur (India): Small oscillations of collisionless gravitating systems.
T. Padmanabhan, M.M. Vasanthi (India): Cosmological scenario with unstable dark matter.
R. Cowsik, S. Ghosh (India): Effect of dark matter on the structure and dynamics of galaxies and galactic clusters.
M. McAdam (Australia): A radio Head-Tail Source in IC 2082.
K. Taylor (UK): TAURUS observations of ionized gas in Centaurus A.

Van den Bergh:
Eighteen RR Lyrae Suspects have been found in a halo field 40' from the nucleus of M31. Photometry of 16 CCD frames of this field yields a period of 0.599±0.08 days for the first of these objects that has so far been studied in detail.

Issa:
The size distribution of HII regions in the LMC, SMC and M33 is used to estimate their distances. The values deduced are 42.66±2.51, 89.73±5.03 and 7.55±47.14 kpc respectively.

Hunter:
Normal, non-interacting irregular galaxies appear to be unusual among actively star-forming galaxies in lacking numerous optically dark nebulae. IRAS data are being used to explore the properties of thermal emission from dust in a sample of irregulars with a range in star formation rates. In general the irregulars are a smooth extension of the spiral galaxies to hotter dust color-temperatures and higher L_{IR}/L_B ratios. No correlation is seen with metallicity. The $L_{IR}/L_{H\alpha}$ ratios are consistent with the bulk of the star formation being optically visible.

Jörsater:
Mosaic Multicolour CCD photometry has been done on a number of large southern barred spirals. The best studied case, NGC 1365, shows a clear difference of colours between the bar and the arms. The arms also contain a red stellar component, but this may be stars rather recently formed in the spiral shock, rather than an actual density wave in the old population. The relatively little shear in the star formation region of the arms is suggested to imply that the intensity of star formation depends on the distance from corotation.

Hjalmarson:
Data from a continuing CO (J=1-0) mapping of the dense (molecular) cloud distribution in M51 where presented. Although CO is observed in arm as well as interarm regions there is a pronounced signal enhancement (observed arm-interarm contrast ratio ~2-3). Streaming in the sense predicted by density wave theory is observed across the arms.

Van Woerden:
Neutral hydrogen distribution and motions in dwarf irregulars in the Local Group have been mapped with the WSRT with ~100 pc resolution. Derivation of rotation curves and mass distributions is hampered by noncircular motions in some objects. In Sextans A and Pegasus the velocity dispersions are ~8 km/s. Sextans A shows strong HI maxima bordering on young star complexes. Skillman finds a critical surface density of 10×10^{20} atoms/cm^2 in 4 dwarfs, over 500 pc scales.

Sakhibov:
The observational detection method of the corotation radius based on the analysis of the age gradient in spiral arm starformation complexes is proposed. Direction and velocity of the starformation wave in the complexes depend on the distance from the nucleus of the M33. The corotation radius is R = 5 kpc in M33.

Gallagher:
UBV photoelectric photometry has been obtained for 60 irregular galaxies chosen from the Virgo cluster galaxy morphological classifications by Binggeli, Sandage, and Tamman published in the Astronomical Journal. Virgo Irrs have a larger spread in color than is found in field Irr samples, which is due to the presence of abnormally red Irr systems and very blue BCDs in Virgo. Red Irr galaxies are concentrated in the galaxy density enhancement associated with NGC 4472 and have an overall spatial distribution like that of the dEs. Irrs with normal colours seen to roughly follow the more extended surface density distribution defined by spirals. The BCDs which are blue (not all are!) are perhaps noteworthy for their spread in radial velocity, as pointed out by Karachentev and colleagues in the USSR.
Complex relationships between galaxy properties and cluster environment parameters thus extend to Irr members of Virgo, but we do not find evidence for hypothesized evolutionary connections between Irr and dE galaxies in clusters.

Mathur:
We investigate the evolution of small perturbations on a stable, equilibrium configuration of a self gravitating collisionless system. Astrophysical appplications are for galactic halos and elliptical galaxies. The system is described by the (linearized) collisionless Boltzmann equation. We find that the spectrum of oscillations is absolutely continuous - and all small perturbations mix away (asymptotically) to leave the system in equilibrium. We achieve our results by applying techniques from the theory of linear operators to the evolution operator of the perturbations.

Padmanabhan:
Cosmological scenarios with one unstable heavy neutrino ν_H of mass m_H which decays into a stable light neutrino ν_L of mass m_L, are considered. In the early phase of the universe ν_H dominates the expansion, and density fluctuations in ν_H grow. When ν_L becomes non-relativistic it condenses on to the ν_H potential wells and these perturbations go non-linear. The ν_H decays after this epoch and the decay product ν_L provides the large scale structure in the universe. Consistent scenarios are possible if m_H is in the range of (150-200)eV and m_L is about 10eV and the lifetime of ν_H is about 10^{11} sec.

McAdam:
Radio emission from the source 0427-53 surrounds the $13^{m}.9$ IC 2082, a giant cD galaxy with a double nucleus in a cluster of more than 80 giant cD galaxy with a double nucleus in a cluster of more than 80 members. Radio maps at 843 MHz (Molonglo, 43

arcsec beam) and 1415 MHz (Fleurs, 20 arcsec) show a wide angle tail structure and a double head. The midpoint is identified with the fainter optical nucleus which has a velocity of 240 km s^{-1} with respect to the rest of the galaxy and the cluster centre. This nucleus is thus in orbit and not yet captured.

The radio tail has a sharp bend 5 arcmin (150 kpc) from the nucleus, suggesting tangential drag in a dark cluster medium with ordered flow.

22 November: Stellar Orbits and Structure and Evolution in Galaxies
(jointly with Commission 33)

CHAIR: L. Martinet

The programme consisted of two parts, in each of which three papers were presented:
a. Information inferred from orbits for studies of structure and evolution in dynamics
L. Martinet (Switzerland): Recent developments in the field-an overview.
T. de Zeeuw (USA): Construction of triaxial galaxies.
A. Wilkinson (UK): Integrals of motion of the stellar orbits in a tri-axial N-body model of an E-galaxy.
b. Other topics in dynamics:
H. de Jonghe (Belgium/USA): On the non-uniqueness of anisotropic distribution functions for spherical systems.
T.R. Bontekoe (Netherlands): Dynamical friction in a binary galaxy model.
Yong Yi (China): The general solutions of Poisson's equation in 3-D for disk-like galaxies and their applications.

The abstracts of these papers are given below.

Martinet:
We review recent results on the orbital structure of perturbed integrable systems, the constraints for equilibrium models of SB galaxies, the sensitivity of irregular orbits to various populations, the advantages of action-angle variables, the importance of the complex instability of periodic orbit in 3-D problems.

De Zeeuw:
The orbits in triaxial elliptical galaxies resemble closely those in the separable potentials first classified by Stackel, for which three exact integrals of motion exist that are explicitly known. It is shown that the existence of more that one basic orbit family ensures that for a given triaxial mass model many phase space distribution functions exist that are consistent with it. This means that the dynamical structure of elliptical galaxies is very rich.

Wilkinson:
The potential of a relaxed n-body model of a triaxial elliptical galaxy can be closely approximated by a Stackel potential. This enables us to determine the orbital content of the model accurately for the first time. The distribution function of the model is also completely known. This enables us to test the equilibrium state of the model, the nature of the distribution function, and yields the phase space density of each crucial type. Fitting a Stackel potential therefore finishes a quick en efficient way of investigating the dynamics of suitable body models.

De Jonghe:
A given spherical mass density allows an infinity of distribution functions depending on energy and total angular momentum. Only one of them is isotropic, but in order to determine a unique distribution function when full anisotropy is allowed, we need all its moments. Consequently, a given mass density together with radial and tangential velocity dispersions do not suffice. This fact is illustrated with a family of distribution functions that yield the well-known Plummer model, for which many of the dynamical variables can be analytically calculated.

Bontekoe:
Numerical experiments simulating the merger between two unequal galaxies yield measurements of the coefficient of dynamical friction. In these simulations the main galaxy consists of 5000 selfgravitating pointmasses, while the satellite galaxy is represented by a constant Plummer potential of 1/10 of the mass and radius of the main galaxy.
In case that the main galaxy is an index 3 polytrope the satellite starts at the outer boundary on a circular orbit. The decay of the satellite orbit is characterized by high values of ln Λ in the outer regions, dropping to a nearly constant value between the 90 and 10 percent mass radius of the main galaxy. This constant value of ln Λ indicates that the dynamical friction process is a local process in the spirit of Chandrasekhar (1943) and thus global responses of the main galaxy, e.g. tides, are dynamically unimportent.
However, in a similar experiment using a King-model as main galaxy we find almost linearly decreasing values of ln Λ with separation of the two galaxies. (In this model the satellite started at the 90 percent mass radius on a circular orbit; the King model has index 1.5). This means that the purely local description, as for the polytrope, is not valid anymore, but we can not explain the fundamental difference. A strictly linear relation between ln Λ and galaxy separation predicts a linear decrease of the galaxy separation with time (see Lin & Tremaine 1983). The thus defined merger time agrees exactly with the time found by our simulation.
In a third experiment the satellite starts at the outer boundary of the King-model with one tenth of the circular velocity. The resulting orbit has an excentricity 0.99 with the first pericentre at the core radius. The merging process yields no circularisation of the satelite orbit due to dynamical friction. The semi-major axis of the orbital parameters decreases almost linearly with time, whilst the excentricity remains constant.

Yong Yi:
It is necessary to consider the spiral galaxies in three-dimensional space because they are not infinitely thin disk. Taking the solution of the Poisson equation of infintely thin disk to be the Green function we will be able to obtain two solutions of the Poisson equation for disk galaxies of finite thickness. One is a rigorous solution from which we can estimate the thickness of spiral galaxies, the other one is a general solution from which we can get the accurate patterns of spiral arms. With the help of the two solutions, we can get the density distribution within galaxies, the gravitational potential of the disk for spiral galaxies and the velocity dispersion distribution of stars within the galaxies, etc.

23 November : Working Group on the Magellanic Clouds (two sessions).

CHAIR: M.W. Feast

Feast will continue as chairman of the WG, but it was agreed that a small (4 member) committee would be useful. The chairman would arrange for a list of names to be proposed at the next General Assembly.
The following papers were read:
J. Lequeux (France): The ultraviolet extinction in the Magellanic Clouds.
J. Lequeux (France): The kinematics of the Magellanic Clouds.
J. Lequeux (France): C-stars and emission line objects in the SMC.
M.W. Feast (South-Africa): The distances of the Magellanic Clouds.
R. Kraft (USA): The stellar population near NGC 121.
R.D. Cannon (UK): The stellar population in the SMC.
R. Foy (France): The chemical composition of an SMC supergiant.
A. Bianchi (Italy): The ultraviolet extinction in the Magellanic Clouds.
D.J. Helfound (USA): Einstein X-ray results in the LMC.
I. Appenzeller (GFR): B[e] supergiants in the Magellanic Clouds.
A. Preite-Martinez (Italy): SNR's in the LMC.

R. Wielebinski (GFR): Radio continuum from the Magellanic Clouds.
A. Florsch (France): Catalogues of Magellanic Cloud objects. A communication from M.C. Lartet (France) on nomenclature in the Magellanic Clouds was noted.

25 November: Working Group on photometry and spectrophotometry of Galaxies

CHAIR: J.-L. Nieto

i) Report on the activity of the WG during the triennum 1982-1985. About 320 papers were quoted in the WG report published in the IAU Transactions, but a small percentage of them comes from the WG members. A specialized meeting in connection with WG interests, entitled 'New aspects of Galaxy Photometry', was organised in Toulouse (France) in September 1984, five years after the Austin meeting.
ii) Continuity of the WG as such and membership. It was suggested to try to open the WG to colleagues very active in the field of IR and UV Photometry and Spectrophotometry.
iii) Dr. J.-L. Nieto was reelected as chairman for the triennum 1985-1988.
In addition, a meeting of interest to WG members 'Structure and Dynamics of Elliptical Galaxies' to be held in Princeton in May 1986 (IAU Symposium, 127) was presented as well as the content of another WG group meeting entitled 'Structures and Substructures in Galaxies' that will be proposed as IAU Colloquium possibly in Japan at the turn of 1988.
At Dr. de Vaucouleurs' request, recent publications from the University of Texas Monographs in Astronomy were also presented to the audience.
Six communications were given:
P. van der Kruit (Netherlands): The central surface brightness of spiral disks.
K. Kodaira, M. Watanabe, S. Okamura (Japan): Bulge-disk models based on digital photometry.
V. Afanasjev, M. Capaccioli, H. Lorenz (Italy): NGC 1023/1023a: an interacting system?.
J. Gallagher, D.A. Hunter, M. Peck, H. Bushouse (USA): Spectral analysis of blue galaxies.
F. Bertola, G. Galletta, W.W. Zeilinger (Italy/Austria): Photometry of dust-lane ellipticals.
J.-L. Nieto, F. Machetto, M. Perryman, S. di Serego Alighieri, G. Lelievre (France, USA): UV observations of the nucleus of M31 with the ESA Photon Counting Detector.

Van der Kruit:
In collaboration with M. de Vries all background disk galaxies on the J-plates of the Palomar-Westerbork survey with an estimated angular size of ≥ 2 arcmin were digitized for surface photometry. All 51 systems were in the UGC catalogue and the diameter given in the UGC are at about $\mu_B \sim 26.5$ mag arcsec^{-2}. The face-on extrapolated central surface brightness μ_B of all systems is in the range 21.78 ± 0.70 mag arcsec^{-2} (r.m.s.) and evidence is presented that this is not a result of observational selection.

Kodaira:
Systematic properties of spheroid and disk of 167 selected galaxies in the Virgo Cluster and the Ursa Major Clouds are investigated on the basis of compisite models profiles consisting of $r^{\frac{1}{4}}$-law spheroid and an exponential-law disk. To each galaxy a model is assigned which shows the smallest r.m.s. of residuals in profile among the ten models nearest to the actual galaxy in the three-dimensional parameter space of diameter, mean surface brightness and mean concentration index. (log D_{26}, SB, X1(P)), introduced in the previous papers of the present authors. The main results are: (1) the structural parameters of disks are confined in the narrow ranges (~4 mag in brightness-scale parameter and 0.7 dex in log (length-scale parameter) while those of spheroids occupy very wide ranges (9 mag and 1.9 dex, respectively). (2)

With increasing morphological type index, the spheroid becomes larger and fainter. (3) There is a very tight correlation between the spheroid-to-disk ratio of the length-scale parameter and that of the brightness-scale parameter, indicating an interplay between spheroids and disks at the time of their formation. (4) Elliptical galaxies seem to have a disk, if any, either more compact or looser and larger than those of spiral galaxies. (4) Lenticular galaxies behave as intermediate between spiral and elliptical galaxies.

Kodaira:
Parameters repesenting the velocity-scales within galaxies are introduced based on the length-scale parameters and the surface-brightness parameters derived from digital surface photometric data in the previous study on the bulge-disk models, under the assumption of the constant mass-to-luminosity ratio. One velocity scale, which corresponds to the rotational velocity in the disk, shows a high correlation to the observed width of the HI 21 cm line and also to the absolute visual magnitude. This correlation can be explained as the results of the empirical laws found for the bulge-disk models in the previous paper, namely the narrow range of the length-scale of the disk, and the tight interplay of the ratios of the length- and the surface-brightness scales between the disk and the bulge.

Lorenz:
Kinematical information on the NGC 1023 system was obtained from medium dispersion spectra taken with the Photon Counting Multichannel Spectrophotometer at the 6 m telescope of the USSR Academy of Sciences. The velocity difference between the galaxy and the cloud is 127 ± 30 km s^{-1} On the basis of a detailed HI study by Sancisi et al. (1984) and photometric investigation it was concluded that NGC 1023A is a Magellanic-type galaxy at M_B = 17.5 mag which lost 90% if its gas content in a tidal encounter of the system with NGC 1023. The timescale of the evolution of the system is on the order of 10^8 years. The gas surrounding NGC 1023 has probably an external nature and is not driven out by internal sources of the S0 galaxy.

Gallagher:
Stellar populations are potentially important probes of galaxy evolutionary histories. A number of groups have approached this problem by using theoretical models to synthesize spectra which can then be compared with observations. For several years we have been utilizing more empirical mathematical programming methods developed by M. Peck to analyze spectrophotometric measurements of blue galaxies. These models are very effective in reproducing spectra, but provide only limited information on evolutionary histories. We are therefore exploring methods to continue the theoretical and mathematical spectral synthesis methods with other measurements of galaxies in order to improve this situation, e.g. by extending the diagnostic techniques described by Gallagher, Hunter, and Tutukov (1984, Ap. J. 284, 544).

Galletta:
In the frame of a more extended study on the structure of a number of dust lane ellipticals, we present the photometric results concerning 8 of these galaxies. They are: Anon 0151-4S, NGC 1947, NGC 2534, IC 4320, NGC 5266, NGC 5363 and NGC 5485, all showing a dust lane along their apparent minor axis, and Anon 1029-459 in which the dust lane lies on the apparent major axis and is strongly warped outward. The plates come from the 4m AAT of siding spring and from the 1.22 m Asiago Telescope. The structure of the dust lane is well defined in NGC 5266 as a warped ring of dust with a small HII region in the inner part, while in NGC 1947 it consists of a number of rings which late to the spiral arms. All these objects follow the $R^{\frac{1}{4}}$ law for their luminosity profile.

Nieto:
The nucleus of M31 was observed at the CFH in a near-UV band (3750 Å, $\Delta\lambda$ = 30 Å) for 4000 seconds. The FWHM is 0.″75. The resulting image shows:
1) numerous dust features, including a thin nuclear dust lane, oriented along the

major axis of the galaxy and cutting the nucleus at no more than 0".3. from the center,
(ii) strong ellipticity and orientation changes as well as an off-centering of the very central nuclear isophotes with respect to the outer ones.
The dust appears to be correlated with the light distribution and with the stellar potential since the dust lies in the equatorial plane of the nucleus.

25 November: Binary, multiple and merging galaxies (two sessions)

CHAIR: F. Bertola

K. Wakanamatsu (Japan): Structure of the polar ring galaxy NGC 4650A.
F. Bertola, G. Galletta (Italy): Dynamics of dust lane ellipticals.
W. Sparks, J.V. Wall (UK): Dust in elliptical galaxies.
S. van den Bergh (Canada): Galaxy mergers and globular clusters.
C. Dupraz, F. Combes (France): Shells around elliptical galaxies.
J.-L. Nieto, Ph. Prugniel (France): Tidally truncated elliptical galaxies.
Y. Terzian (USA): HI observations of binary galaxies.
H. Bushouse, J. Gallagher (USA): Star formation in violently interacting spiral galaxies.
J. Mahoney, B. Burke, J.M. van de Hulst (USA/Netherlands): Quantitative study of the interaction in NGC 4038/39.
S.M. Alladin (India): The merger time of binary galaxies.
T.K. Chatterjee (India): Impulsive approximation studies of ring formation.

Wakamatsu:
From the surface photometry of NGC 4650A, we found that the gaseous component is distributed in a ring and the stellar component in a disk, making a polar stellar disk. After forming the innner stellar disk, the gas in the innner region of the polar disk would have fallen into the nuclear region due to shock waves that was caused by interaction of the gas and the deep potential well of the S0 disk. This mechanism is similar to that of galactic shocks in spiral and barred galaxies.

Galletta:
The kinematics of a sample of dust lane ellipticals are studied. The result is that when the dust lane lies along the minor axis, the rotation axis of the gas is perpendicular to that of the stars, indicating an external origin for the gas.
Three out of four warped dust lane ellipticals exhibit prograde motions in the warps, suggesting that the warps are transient phenomena, if stellar counterstreaming is negligable.
While ellipticals with the dust lane perpendicular to the major axis are both fast than slow rotators, those with the dust lane parallel to the major axis are only fast rotators.
In addition, the first of the two above type greatly outnumber those with the dust lane parallel to the major axis, if any of this latter type indeed exist.

Sparks:
To search for discrete dust features, B-(reconstructed I) colour maps were derived from CCD images of 30 nearby elliptical galaxies. Both chaotic and smooth, symmetric dust-lanes were found, the former in radio ellipticals, the latter in radio-quiet ellipticals. Red nuclei are common in active ellipticals. The data indicate that although on occasion radio emission my be triggered by external gas accretion, the dominant mechanisms are regular internal processes.

Van den Bergh:
1. Spirals. Van den Bergh and Morbey have shown that clusters of all ages in the Magellanic Clouds are significantly more flattened than their galactic counterparts. This shows that our own Milky Way system did not form by the merger of ancestral

Magellanic-type irregulars. Furthermore the rarity of carbon stars in the galactic halo indicates that no Magellanic-type irregular has merged with the Milky Way during the last few billion years. An additional argument against the idea that mergers were a major factor in galaxy building is provided by the radial metallicity gradient of galactic globular clusters and the even stronger correlation between cluster diameter and galactocentric distance that is observed in the Galaxy, M31 and the peculiar elliptical NGC 5128. Not unexpectedly the specific globular cluster frequency, i.e. the number of globulars per $M_V = -15$ of galaxy luminosity, depends strongly on Hubble-types and ranges from S = 3 in the galaxy NGC 4594 to 0.3 in the Sb galaxy NGC 4565.

2. Ellipticals. Perhaps the strongest argument against the notion that most ellipticals form from merging spirals is provided by the fact that elliptical galaxies typically have specific globular cluster frequencies $S \sim 3$ in the field and $S \sim 6$ in Virgo, These values are an order of magnitude higher than those which are observed in spirals of type Sb. This discrepancy is reduced by only a factor of 2 by taking into account the fact that spiral galaxies will fade in brightness as star formation dies out following the removal of gas during mergers. It is therefore concluded that field ellipticals contain ~5 times and Virgo ellipticals ~10 times as many globular clusters as would be expected if they had been formed by merger of typical spirals.

Combes:
With test particle simulations, we show that shells can help to discriminate between oblate and prolate galaxies. Shells form much more easily during a merger with a prolate elliptical. Dwarf elliptical companions produce shells as well as spiral companions. In numerous shells galaxies, the shell system provides evidence for an extended mass component around the elliptical.

Nieto:
We have approached the problem of the evolution of tidally truncated elliptical galaxies from both observational (Photometry with FWHM 0".5-0".7 and spectroscopy with the CFH Telescope) and theoretical points of view. We have found a fourth tidally truncated elliptical: NGC 4478 and a fourth characteristics common to the four cases: the isophotes are more circular outwards than inwards. This can be explained by the orbital time of the compact about the nearby massive companion being much larger than the dynamical time of its external parts. A simple model for decay with a variable mass for the compact shows that the decay time is much larger than the Hubble time: we conclude that <u>either</u> dynamical friction formulae do not apply <u>or</u> compactness is intrinsic to the formation of this type of ellipticals <u>or</u> there must have been a strongly inelastic encounter after which the decay has occured as described by our model (e.g. slowly).

Terzian:
Observations of HI in binary galaxies, and in small groups, performed with the Arecibo 100-ft radio telescope were described (Schneider, Helou, Salpeter and Terzian 1985). M/L ratios were derived both from the HI rotation curves, and from pair dynamics, and these are of order 5 to 20, with a maximum dynamical M/L~80.

Gallagher:
A morphologically selected sample of 100 violently interacting spiral systems has been chosen from the <u>UGC</u> on the basis of the presence of tidal tails or other pronounced structural disturbances. Optical spectrophotometry reveals a wide range in stellar content and emission line strengths. On average, the interacting systems have significantly enhanced levels of star deformation as compared with field spirals, but are deficient in detectable active nuclei (Seyferts, Liners). Many galaxies have only old stellar populations in their muclei, indicating that nuclear star formation or activity are not necessary consequences of interactions. Spatially extended star formation is also common in these systems, and young stars may play an important role in the visibility of tidal tails and other large scale features. A goal of this project will be to combine these data as a means to estimate the level of evolutionary change produced by strong interactions.

Burke:
A study of the hydrogen distribution and dynamics in the interacting system NGC 4038/39 has been made, using the VLA of the National Radio Astronomy Observatory. A simple Toomre-type calculation indicates that the galaxies are interacting for the first time. The best fit to the combined radio and optical data is given by introducing a small amount of dynamical friction (in our admittedly simplified model), but no massive, extended halo seems to be needed. On the contrary, any halo must contain a relatively small fraction of the total mass if it is greatly extended beyond the luminous matter.

Alladin:
The merger times of binary galaxies obtained from the analytic formulae derived under the impulsive and adiabatic approximations are compared with the numerical results by Borne (1984 Ap. J.). The agreement with the numerical results is found to be quite good with the impulsive approximation formulae.

Chatterjee:
A study of the formation of ring galaxies, in the light of disk-sphere collisions, (under the impulsive approximation), indicate that the formation and properties of the rings are closely related to the fractional change in binding energy of the disk galaxy. A relationship exists between the size of the ring and the collision parameters, which enables us to interpret some prominent ring galaxies.

COMMISSION 29: STELLAR SPECTRA (SPECTRES STELLAIRES)

Report of the Business and Scientific Meetings, 20-27 November 1985

PRESIDENT: J. Jugaku   SECRETARY: R.F. Garrison

Business Meeting

The President welcomed those present at the meeting. He first listed the proposed agenda on a viewgraph which was adopted by those present. The Commission then elected R.F. Garrison as Secretary for the meeting.

The President asked for a moment of silence in memory of Ch. Bertaud, T. Dunham Jr., O.A. Mel'nikov, A. Przybylski, and J. Tech who passed away during the last triennium.

The President then gave a report on the activities of the Commission since the last meeting in Patras. The main activities were the following.

1. Symposia and Colloquia: The Commission sponsored or cosponsored 8 IAU meetings in 1982-1985. For details, see Transactions IAU, XIXA, p. 353 (1985). The Organizing Committee has also approved sponsorship of five more IAU meetings in 1986-1987; they are Symposium No. 122, "Circumstellar Matter," Heidelberg, FRG, June 23-27, 1986; Colloquium No. 92, "Physics of Be Stars," Boulder, Colorado, USA, August 18-22, 1986; Colloquium No. 94, "Physics of Formation of Fe II Lines outside LTE," Capri, Italy, July 1986; "Circumstellar Matter in Close Binaries," Victoria, Canada, June 1987; and Symposium No. 132, "The Impact of Very High S/N Spectroscopy on Stellar Physics," Paris, France, June 1987.

2. Working Groups: The Commission sponsored two Working Groups together with Commission 45: Be Stars and Ap Stars. It also sponsored a third Working Group on Standard Stars together with Commissions 30 and 45. The report on the Working Group on Be stars was given by its Chairperson, A. Slettebak, while those of the other two were read by the President. Chairperson of the third Working Group, L.E. Pasinetti, suggested that the name of the Working Group be changed to "Working Group on Standard Stars and Calibration Methods." This suggestion was discussed and it was deferred to the meeting of the Working Group (see the Report of Commission 45). C.O.R. Jaschek and P.C. Keenan proposed the creation of a new Working Group on Peculiar Red Giants. This proposal was endorsed and afterwards it was also endorsed by Commission 45.

3. Sponsorship of the International Register of Stellar Spectroscopists. The IRSS is a list of stellar spectroscopists giving the names, addresses, and reseach fields irrespective of their IAU memberships. It is edited by W. Bonsack and A. Slettebak. A. Slettebak reported that its fourth edition would be distributed shortly among those interested.

The President presented a draft of By-laws of Commission 29 which were prepared by the Organizing Committee in the past three years. P.C. Keenan suggested to remove the second sentence of item 3(a) in the by-laws and his suggestion was unanimously accepted.

The election of new officers and of the Organizing Committee was carried out. Those elected were: President, G. Cayrel de Strobel; Vice President, P.S. Conti; Organizing Committee, M.S. Bessell, A.M. Boesgaard, J. Jugaku, D.L. Lambert, O.H.

Levato, M.A. Smith, J. Smolinski, M. Spite, and S.C. Wolff.

R.F. Griffin described the present status of Mount Wilson Observatory and proposed to submit a resolution to the General Assembly concerning the closing of the Observatory. The motion was carried and R.F. Griffin accepted to discuss this matter with other relevant commissions. This proposal became Resolution C12 of the General Assembly and is printed in the resolution section of these proceedings.

Scientific Meeting

A meeting on "Nucleosynthesis in the Galaxy from Studies of Low Mass Stars" was held on 27 November. The following papers were presented:

- J. Audouze: Nucleosynthesis in the Galaxy from studies of low mass stars
- J. Audouze: Chemical evolution of galaxies and abundances in low mass stars
- R. Kraft: Recent work on CNO in dwarfs and subdwarfs
- C. Pilachowski, C. Sneden, and D. Vandenberg: The carbon isotope ratio in metal poor giants
- D. Lambert: Abundances of light elements (Na-Ti) in disk and halo stars
- J. Audouze: Abundances of the very light elements D, He, Li and their cosmological implications
- M. Bessell and J. Norris: Non-solar abundance ratios in extreme Population II stars
- D. Duncan: Li in Population II stars
- F. Spite and M. Spite: Nitrogen enhancement in metal-poor dwarfs: from inside or outside?
- G. Cayrel de Strobel: On a possible metallicity gradient in SMR stars
- M. Grenon: The metallicity gradient in the disk population
- C. Pilachowski: On the existence of a metallicity gradient in the halo of the Milky Way
- R. Cayrel: A new interpretation of the metallicity histogram of globular clusters
- B. Gustafsson: Final remarks

The Proceedings are being edited by G. Cayrel de Strobel and M. Parthasarathy and will be published by the Indian Academy of Science, Bangalore.

Working Groups

WG on Ap Stars: The Working Group has elected the following Organizing Committee: C.R. Cowley (chairperson), B. Hauck, M. Jaschek, J. Jugaku, V. Khokhlova, D.W. Kurtz, and G.J. Michaud.

WG on Peculiar Red Giants: The Working Group has elected the following Organizing Committee: A. Alksnis, R.F. Garrison, R.F. Griffin, C.O.R. Jaschek (co-chairperson), M. Jaschek, H.R. Johnson, P.C. Keenan (co-chairperson), T. Lloyd-Evans, E.E. Mendoza V., F.R. Querci, H.B. Richer, V. Straizys, T. Tsuji, P.R. Wood, and S.B. Yorka.

The following papers were presented at the meeting:
- R. Foy: Spectral-features-selected variations in the diameter of Mira stars
- M.S. Vardya: A few comments on S and M Mira variables
- H. Richer: Search for carbon and M-type stars in external galaxies
- R. Wing: Temperature of carbon stars
- V. Straizys: Simultaneous photometric determination of the metallicity and the O/C ratio for late-type stars

Joint Discussion

Commission 29 participated in Joint Discussion V "Stellar Activity: Rotaion and Magnetic Fields."

COMMISSION 30: RADIAL VELOCITIES (VITESSES RADIALES)

Report of Meetings in New Delhi, November, 1985

President: A. G. Davis Philip

A.   Business Meeting, November 20, 1985

1. New Members of the Commission

The names of the new members proposed for Commission 30 were listed on the overhead projector screen. They are;

M. Breger	University of Vienna
G. Burki	Geneva Observatory
R. Burnage	Haute Provence Observatory
B. Campbell	Dominion Astrophysical Obs.
B. W. Carney	Univ. of North Carolina
J. -M. Carquillat	OPMT, Toulouse
W. D. Cochran	University of Texas
B. J. Hrivnak	Valpariso University
T. H. Kadouri	Space & Astron. Res. Cntr., Iraq
R. D. McClure	Dominion Astrophysical Obs.
V. Popov	Pulkova Observatory
Y. Romanov	Odessa Observatory
Myron Smith	National Solar Observatory
K. Yoss	University of Illinois

These new members were confirmed by a unanimous vote of those present. J. Sahade has resigned from Commission 30.

2. Officers of the Commission

Concerning the Organizing Committee, M. Duflot, D. Hube and J. Sahade leave the committee. After some discussion A. Florsch and R. McClure were appointed as new members. For the period 1985 - 88 the Organizing Committee will consist of J. Andersen, A. Florsch, D. Latham, E. Maurice, M. Mayor, R. McClure and D. Philip.

J. Andersen becomes President and D. W. Latham was elected Vice President for the term 1985 - 88.

3. Commission Report

The commission report, as printed in Volume XIXa of the transactions, was not the correct, final version. Copies of the correct version were passed out to all members of Commission 30 that were present at the meeting and other copies were given to J. Andersen for mail distribution after the General Assembly. Griffin had noted that many of his papers were missed in the Bibliography which was constructed from the Astronomy and Astrophysics Abstracts and which appeared at the end of the commission report. Philip pointed out that this problem arose because many of the Griffin papers were listed under spectroscopic binaries by Astronomy & Astrophysics Abstracts and therefore did not appear as entries under radial velocities.

4. New Committee on Radial Velocity Standards

J. Andersen proposed that a committee be formed of those people who are working actively on the problem of establishing, new, accurate radial velocity standards. Andersen was appointed as chairman with members B. Campbell, D. W. Latham, M. Mayor and R. D. McClure. The main aim is to help in setting up a list of radial velocity standards with the required accuracy for the new techniques of measuring radial velocity. The goal is to achieve an accuracy of ± 0.1 km/s for the mean velocities of each standard star and for the absolute zero point of the whole system. It was agreed that the list of 34 stars selected by Mayor and Maurice (1985 in IAU Colloquium No. 88, Stellar Radial Velocities, A. G. D. Philip and D. W. Latham, eds., L. Davis Press, Schenectady, p. 229) should be observed intensively over the next few years, and that absolute calibrations of the zero point should be carried out compared to solar system objects. A provisional list will be presented to the General Assembly in Baltimore in 1988.

5. Commission Support of Proposed Meetings

The commission voted to co-sponsor the following meetings:

The Second Conference on Faint Blue Stars (Tucson, Arizona, May 30 - June 3, 1987, D. Philip is Chairman of SOC).
Very High Signal/Noise Spectroscopy: A New Era for Stellar Physics (Paris, France - June, 1986, G. Cayrel is Chairman of SOC).
Circumstellar Matter in Close Binaries (Victoria, BC, Canada, June, 1987, A. Batten is Chairman of SOC).

6. Report of Working Groups

Bibliographic Stellar Radial Velocities Catalogue. Barbier reported that this catalogue is complete up to 1980 and is to be published in an Astron. Astrophys. Suppl. It contains 11,350 references. The magnetic tape is available at the Marseille Observatory. The data have been sent in to the Strasbourg Data Center and are included in the SIMBAD data base. The data up to 1985 will be included by the end of 1986.

Mean Stellar Radial Velocity Catalogue. Barbier reported that the rules governing the inclusion of a velocity in the catalogue were defined at the XVII General Assembly in Patras. Last year at IAU Colloquium No. 88 in Schenectady she presented the mean radial velocities of about 100 stars, as an example, to show that the program is working well. At the present time the mean radial velocity catalogue is ready up to 1975, including stars at 21, 22 and 23 hours which were not included in Evan's Catalogue.

The commission 30 members, present at the meeting, expressed their thanks to Dr. Barbier for her great efforts in preparing these two catalogues.

B. Joint Meeting with Commissions 28, 30 and 40

Galaxy Redshift Surveys - November 20, 1985

The program for this meeting was as follows:

D. W. Latham	Redshift Surveys and the Large-Scale Structure of the Universe
R. Giovanelli	Radio Surveys
M. J. Geller J. T. Huchra D. W. Latham V. DeLapparent	Optical Surveys
L. A. N. DaCosta	Southern Surveys: An Overview
J. W. Menzies I. M. Coulson	Southern Surveys: The South African Astronomical Observatory Contributions
A. Fairall	Red Shifts from South Africa

The main thrust of the session was to review the latest efforts to extend redshift surveys of galaxies and clusters of galaxies and also to review the plans for the continuation of this work. The individual papers are being published in the Highlights of Astronomy, under the editorship of D. W. Latham. Latham chaired the joint meeting.

C.   First Scientific Meeting

Progress in Radial Velocity - November 22, 1985]

The program was as follows:

D. W. Latham	Review of IAU Colloquium No. 88, Stellar Radial Velocities
J. Andersen G. Hill	Cross Correlation Velocities for Early-Type Stars
M. Mayor	CORAVEL Results
D. W. Latham	Digital Speedometry
A. Florsch	Radial Velocity Projects in France
B. Carney	Halo Survey
R. Griffin	Recent Results

D.   Second Scientific Meeting

Standard Radial Velocity Stars - November 22, 1985

The program was as follows:

    A. H. Batten   The Present Situation Concerning Radial Velocity Standards

    M. Mayor   CORAVEL Measures of Radial Velocity Standards: The Absolute Zero Point of Radial Velocities

    D. W. Latham   Digital Speedometry of Radial Velocity Standards

E.   <u>Joint Meeting of Commissions 29 and 30</u>

    <u>Working Group on Standard Stars</u> - November 26, 1985

The program was as follows:

    D. Philip   The Microfiche of Standard Stars

    H. Neckel   The Absolute Energy Distributions of the Sun, of the Solar Analogs 16 CygB, 16 CygA, VB 64 and of the Standard Stars Alpha Lyrae and 29 Psc

    G. Cayrel   Remarks Concerning Solar Analogs

    M. L. Malagnini
    C. Morossi
    L. Rossi   Does a Solar Twin Really Exist?

    I. N. Glushneva
    E. A. Makarova
    A. V. Kharitonov   Energy Distribution, Photometry and Physical Characteristics of the Sun and Star-Solar Analogs

    L. Rossi
    A. Altamore
    C. Rossi   A Study of the Solar Analogs in the Ultraviolet

    M. Fracassini
    L. E. Pasinetti   Photometry of Comet Halley: Solar Analogs Selected Along the Path (November 1986 - May 1986)

F.   <u>Report by R. P. Stefanik, D. W. Latham and R. E. McCrosky concerning Measurements of Selected IAU Radial Velocity Standard Stars</u>

    The results for the 9 standard stars observed at CfA are presented in Table I. The Henry Draper Catalog number, 1950 right ascension and declination, apparent visual magnitude, mean velocity and RMS, number of exposures, difference between the maximum and minimum velocities for the star, and the interval in days between the first and most recent observation are shown. The mean RMS residual is about 0.5 km/s, and this is the precision that is expected for a single measurement. Since there are roughly 20 observations per star, the mean velocities for these stars should be good to about 0.1 or 0.2 km/s, assuming that there are no serious systematic errors, an assumption which may not be valid.

TABLE I.

Summary of CfA Velocity Measurements

HD	RA (1950)	Dec	m_V	<CfA>	RMS	N	Delta	Days
8779	01 23 53.6	-00 39 29	6.4	-3.57 ± 0.54		22	2.42	1508
26162	04 06 14.9	+19 28 43	5.5	24.93 ± 0.55		23	2.07	1513
66141	07 59 39.9	+02 28 24	4.4	72.19 ± 0.98		13	4.06	1417
89449	10 17 01.0	+19 43 31	4.8	6.45 ± 0.50		16	1.39	1635
92588	10 38 51.5	-01 28 42	6.3	42.41 ± 0.22		5	0.59	1407
114762	13 09 54.5	+17 46 55	7.3	49.59 ± 0.55		23	2.07	1520
136202	15 16 45.4	+01 57 12	5.1	54.77 ± 0.37		17	1.47	1222
182572	19 22 35.1	+11 50 10	5.2	-100.18 ± 0.46		40	1.76	1522
213014	22 25 45.8	+17 00 28	7.3	-39.57 ± 0.39		35	1.59	1513

In Table II the mean velocities for each of the nine stars are compared with the mean velocities reported for the CORAVELs (Mayor and Maurice 1985 in IAU Colloquium No. 88, Stellar Radial Velocities, A. G. D. Philip and D. W. Latham, eds., L. Davis Press, Schenectady, p. 299) and for the Victoria spectrometer (Fletcher, J., Harris, H., McClure, R. and Scarfe, C. 1982 Publ. Astron. Soc. Pacific 94, 1017).

TABLE II

Comparison of CfA with CORAVEL and Victoria

HD	CfA-COR	CfA-Vic	COR-Vic	Comment
8779	1.11	1.23	0.12	Suspected Variable
26162	0.60	0.11	-0.49	Suspected Variable
66141	1.00	0.45	-0.55	
89449	0.50	0.22	-0.28	Suspected Variable
92588	0.13	-0.38	-0.51	Few CfA observations
114762	0.43	-0.01	-0.44	
136202	0.39	0.26	-0.13	
182572	0.34	0.11	-0.23	
213014	0.80	0.15	-0.65	

In the comparison of the zero points of these three velocity systems, HD

8779 was eliminated because the differences from one system to the next are large. The differences between the velocity systems are found

V(CfA) - V(COR) = 0.52 ± 0.27 km/s
V(CfA) - V(Vic) = 0.11 ± 0.24 km/s

where the error is the RMS of the differences for individual stars. These errors are encouragingly small. For example, if we assume that all three systems have about the same error for the determination of the mean velocity of an individual IAU standard, then this error must be less than about 0.2 km/s for the eight stars included in the above comparison. If the eight stars are not affected by systematic effects such as slow velocity variations with time, then the error one might expect for the comparison of the mean zero points should be root eight smaller, or something like 0.1 km/s. These errors suggest that it should be possible to determine the velocity zero point of a selected subset of IAU standards to an accuracy of 0.1 km/s, but only if the systematic errors can be controlled.

G.   Continuation of the Bibliography of Radial Velocity Papers from A & A Abstracts, Volumes 38 and 39

In the appendix of IAU Colloquium No. 88, Stellar Radial Velocities, there is a bibliography of radial velocity papers taken from volumes 30 -34 and 37 of Astronomy and Astrophysics Abstracts. Since the proceedings of this meeting were published, two additional volumes of the abstracts have appeared, Volumes 38 and 39. The abstract numbers of the additional papers are listed below, segregated by topic.

Topic	Abstract Numbers		
Associations, OB	39.152007	39.157008	
Bibliography	39.002100		
CCD Detectors	39.036188		
CLusters, Globular	38.154029	38.154049	38.154078
	39.154045	39.154051	39.154076
	39.154077	39.154088	
CLusters, Hyades	38.153002	39.153051	
Clusters, Open	38.153043	39.152007	39.153034
	39.153051		
Data Processing	39.111008		
Galaxies	38.157048	38.157150	38.157205
	39.002095	39.036072	39.036073
	39.157130	39.157213	
Galaxies, Clusters of	38.160016	39.160064	39.160141
	39.160144		
Galaxies, Elliptical	39.157159		
Galaxies, Magellanic Clouds	39.156026	39.156034	
Galaxies, Markarian	38.158050		
Galaxies, Spiral	39.157058		
H II Regions	39.132017		
Methods of Observation	38.036117	39.034123	39.034125
	39.034126	39.034127	39.036072
	39.036118	39.036184	39.036185
	39.036187	39.036188	39.036189

RADIAL VELOCITIES

Topic	Abstract Numbers		
	39.036191	39.111010	
Nebulae, Planetary	39.134012	39.134040	
Solar Atmosphere	38.036141		
Solar Corona	38.074001		
Solar Radio Bursts	39.077018		
Spectra, Stellar	39.036117	39.064088	
Spectrometers	39.034084		
Stars	38.111020	38.111021	39.002077
	39.111099		
Stars, A	39.111004		
Stars, AM Her	39.117385		
Stars, AM	39.120034		
Stars, Barium	39.111030	39.111036	
Stars, Be	38.112028	39.112100	
Stars, Beta Cephei	39.122123		
Stars, Binaries, Cataclysmic	39.117331	39.117386	
Stars, Binaries, Close	38.117080	38.120017	39.117251
Stars, Binaries, Contact	39.117060	39.117142	
Stars, Binaries, Eclipsing	38.119082	38.119087	38.120018
	39.119007	39.119042	39.119098
	39.119106	39.119110	
Stars, Binaries, Semi-Detached	38.117206	38.120001	38.120002
Stars, Binaries, Spectroscipic	38.120019	38.120022	38.120023
	39.013060	39.036186	39.111002
	39.120002	39.120006	39.120011
	39.120024	39.120025	39.120032
	39.120044		
Stars, Binaries, X-Ray	38.119095	39.117200	39.117384
Stars, Blue Stragglers	38.153029		
Stars, Bright	38.111024	39.111002	
Stars, Carbon	39.111012		
Stars, Catalogues	38.002019	38.002035	39.002098
	39.111023		
Stars, Cepheids	38.122093	39.122039	39.122187
Stars, Cepheids, Dwarf	39.122184		
Stars, Cool	39.111026		
Stars, Delta Scuti	38.123107		
Stars, F Dwarfs	39.155153		
Stars, F Supergiants	38.114058		
Stars, Fundamental	39.041022		
Stars, Giants, Late-Type	39.111039	39.111041	
Stars, Giants, Pop II	39.155100		
Stars, Halo	39.111004	39.111046	39.120035
	39.120036		
Stars, Hg-Mn	38.120001		
Stars, High-Velocity	38.111009		
Stars, Horizontal-Branch	39.111001		
Stars, K Giants	38.155048	38.155049	
Stars, Late Type	38.111024	38.114057	39.111002
	39.114008	39.114099	
Stars, Magnetic	38.118029		
Stars, Mira	38.122079		
Stars, Nearby	39.111024	39.120018	

Topic	Abstract Numbers			
Stars, Novae	39.124228			
Stars, Novae, Dwarf	39.117202			
Stars, OB	39.111022			
Stars, Peculiar	39.111008			
Stars, Pop II	39.111032			
Stars, Pre Main-Sequence	39.121004			
Stars, Proper Motion	39.111036			
Stars, RR Lyrae	39.122188			
Stars, RS CVn	39.117141			
Stars, Red Giants	39.155154			
Stars, SU UMa	38.117176			
Stars, Standard	38.111003	39.111040	39.111042	
	39.111043	39.111044		
Stars, Subdwarfs	38.111004	38.126062	39.117157	
Stars, Sun	39.036208			
Stars, Supergiants, Late-Type	39.111038			
Stars, Symbiotic	38.117248	39.117279		
Stars, Variable, Late-Type	38.122090			
Stars, Variable, Nova-Like	39.117202			
Stars, Variable, Semiregular	38.111013			
Stars, W UMa	38.117228	39.117280	39.119009	
Stars, White Dwarfs	39.117330			
Stars, Wolf-Rayet	38.111026	38.114052		
Surveys	39.111034	39.111035	39.111037	
	39.111045			
X-Ray Sources	38.119095	38.142047		

COMMISSION 31: TIME (L'HEURE)

Report of Meetings: 20, 21, 22, 26, and 27 November 1985

Acting PRESIDENT: D. D. McCarthy   SECRETARY: H.F. Fliegel

20 November 1985

A Joint Discussion on Reference Frames was held including Commissions 4, 7, 8, 19, 20, 24, 31, 33, and 40, chaired by J. Hughes.

21 November 1985

REPORTS, AND ADMINISTRATIVE MATTERS:

Following a meeting of the Organizing Committee, an administrative meeting of Commission 31 was held. D. McCarthy announced that he was acting in place of Commission President G. Hemmleb, who has been ill.

The Report of the Commission President has been prepared by G. Hemmleb and was summarized by D. McCarthy; and has been published in "Highlights of Astronomy". The success of the MERIT campaign was noted, and members were reminded that the MERIT/COTES resolutions have been presented for possible approval to Commission 31. The IAU Symposium 109, sponsored by Commission 31, has been held, in which the need for the exact definition of entities which define astronomical time was identified as a special concern of Commission 31. The transfer of responsibility for International Atomic Time (TAI) from the Bureau International de l'Heure (BIH) to the Bureau International des Poids et Mesures (BIPM) was noted. There is a continuing need for astronomical observations of Universal time (UT) and Ephemeris Time (ET) from cooperating observatories. Thirty three laboratories including 140 clocks now contribute to the formation of TAI, and seasonal variation of commercial cesium clocks from primary cesium standards has been measured. Time synchronization experiments have been carried out using TV, LORAN-C, portable clocks, Very Long Baseline Interferometry (VLBI) and Global Positioning System (GPS) and other satellites. Accuracies have been reported in the 1 to 10 nanosecond range. GPS receivers have been deployed at the US National ureau of Standards (NBS), US Naval Observatory (USNO), BIH (at Paris Observatory), Physikalisch-Technische Bundesanstalt (PTB), Technische Universitat Graz (TUG), Tokyo Astronomical Observatory (TAO), and Australia.

B. Guinot presented the Report of the BIH. He noted that the number of clocks now used to form TAI is rapidly increasing, thanks to better methods of time transfer from remote laboratories (for example, GPS). That number is now 150, up from 140 at the time the Report of the Commission President was prepared (see above). As the first step in forming TAI, an intermediate time scale, Echelle Atomique Libre (EAL), is computed from the data of all available clocks. The stability of EAL has been 0.5 parts in 10 to the 14th for averaging times of 4 to 6 months. Then the duration of the scale unit of EAL is determined using the primary standards of the National Research Council of Canada (NRC), PTB, NBS, and the US Naval Research Laboratory (NRL). No steering has been necessary during 1984-1985. However, an annual variation has been discovered, traced to the clocks and not to the transmission methods or media, of approximately 1 microsecond amplitude over a year. The appropriate statistical treatment of the timing data is therefore still to be determined. Thirty nine laboratories now contribute to TAI, and 10 organizations form local independent scales of time, which are useful for intercomparison and statistical studies. More than half the contributing laboratories contribute to the BIH via GPS time transfers. The ΓIH has confirmed reports that 10 nanosecond accuracies are possible using GPS time transfer; but, because of imperfectly calibrated receiver delays, station location errors, software errors, and occasional inadequacies in local laboratory timing signals, the accuracy actually attained until now has been about 50 nanoseconds. The BIH now imposes a maximum value on the weight assigned to clocks, but is considering the possibility that very good clocks may be underrepresented by this procedure.

The Report to Commission 31 on the 10th Meeting of the Consultative Committee for the Definition of the Second (CCDS) had been prepared by W. Markowitz, and was read by D. McCarthy. The topic of chief interest to the IAU was the proposed transfer of responsibility for TAI from BIH to the BIPM. Other topics include atomic standards, time and frequency transfer, and relativity.

J. Pilkington's Report on the activities of CCIR Study Group 7 was read by H. Fliegel. The principal work consisted of updating previous reports dealing with dissemination of time and frequency, the assessment of frequency standards, and the formation of time scales. CCIR Report 439-3 is the working standard on time and relativity. The Recommendations Reports of the CCIR, Volume VII (1982), and the Conclusions of the Interim Meeting of Study Group 7 (CCIR Doc 7/77, 1984) are recommended for study to everyone in the field of time and frequency.

Draft Resolution 1 concerning the transfer of responsibility for TAI from the BIH to the BIPM was discussed, slightly amended, and passed unanimously (see below).

D. McCarthy reported that approval has been received for the proposed IAU Symposium 128, "Earth Rotation and Reference Frames", to be held in Washington DC, USA, 20 - 24 October 1986. The Academia Sinica proposes a Symposium on Time Frequency in Shanghai in 1987. It was suggested that 1987 might no longer be open, but members recommended that the IAU sponsor the Symposium for the earliest mutually agreeable date.

The members approved the following nominations for 1982-1985:

President: D. McCarthy   Vice President: P. Paquet

Organizing Committee: S. Aoki, J. Benavente, N. Blinov, G. Hemmleb, J. Kovalevsky, Y. Miao, I. Mueller, J. Pilkington, E. Proverbio, B. Guinot, S. Ye.

Representatives: to CCDS, J. Benavente; to FAGS, J. Kovalevsky; to CCIR Study Group 7, J. Pilkington; to the CCDS Working Group on TAI, G. Winkler; to the BIH Directing Board, P. Paquet and G. Winkler.

The members approved without objection the proposed list of new members and consultants to the Commission.

22 November

ADMINISTRATIVE MATTERS:

D. McCarthy reported that the Draft Resolution approved in the Joint Discussion on Reference Frames included the recommendation to commit the task of the exact definition of TAI to a special working group. He therefore requested that Commission 31 Draft Resolution 2, concerning a definition of TAI, not be brought to a vote, but invited discussion on TAI. B. Guinot stated that there is a need to define the frame of reference of TAI. Although the reference frame of TAI has been defined as geocentric, in practice the unit of TAI has been the second as realized by clocks at mean sea level. Guinot believed that the definition of the TAI second given by the CCDS should be acceptable, but reported that some colleagues apparently object to the term "proper time" when applied to atomic clocks. In the ensuing discussion, it was reported that some astronomers object to the application of the term "coordinate time" to TAI. Commission members agreed to refer the issue to the forthcoming working group.

D. McCarthy requested that Draft Resolution 3, concerning the definition of the work of Commission 31, be withdrawn in favor of the Draft Resolution from the Joint Discussion on Reference Frames (Resolution B2); and he invited discussion. There was general agreement that the members of Commission 31 would not object to an expansion of its role to include the definition and establishment of reference systems, nor to a change in the name of the Commission, if a consensus could be reached with other Commissions and the IAU leadership.

A joint meeting was held in the afternoon between Commissions 19 and 31 dealing with the MERIT/COTES Project (see Commission 19 Report).

26 November

There was a joint meeting uniting Commission 7, 8, 19, and 31.

SCIENTIFIC PRESENTATIONS:

B. Kolaczek, in a joint paper with W. Kosek, reviewed the search for periodic oscillations in polar coordinates with periods of 10 - 100 days. Quick look data taken by different techniques and reported in the MERIT Circulars were examined using Maximum Entropy Analysis and the Ormsby filter. Amplitudes ranged from several milliarcseconds for CSR - LAGEOS laser ranging, IRIS - VLBI and DMA - Doppler to 12 mas for BIH - Astronometry. Improved data sets showed only 2 mas amplitude for CSR - LAGEOS (84 L 01) and IRIS - VLBI (85 Feb. 01). Similar short period terms were detected in the equatorial components of the atmospheric excitation function chi 1, chi 2 of the European Centre for Medium Range Weather Forecasting (ECMRWF).

C. Kakuta presented results of collocations of geophysical observations at certain Earth Orientation Parameters (EOP) observing sites. Differences were formed between determinations of UT1 by two different observing techniques -- between the Connected Element Radio Interferometer (CERI, at Green Bank, West Virginia) and the USNO PZT at Washington; between VLBI Goldstone-Orroral and VLBI Goldstone-Madrid; between the International Latitude Station (ILS) at Mizusawa and the USNO PZT at Washington. The Mizusawa - Washington UT1 differences appear to correlate with mean sea level at Truk Island. An apparent displacement of Orroral westward appears to coincide with the 1982-1983 El Nino. Absolute gravity measurements made at the Esashi Earth Tides Station have been compared to Mizusawa UT1 residuals.

L. Morrison summarized the astronomical observations of secular non-tidal variations. For the period from 700 BCE to 500 CE, about 40 Babylonian records of the time of lunar eclipses are available, and about 12 Chinese observations of total or near solar eclipses, of which 6 are valuable, plus one Babylonian record. For the period from 500 to 1600 CE, a number of untimed solar eclipses are available, and about 30 timed solar and lunar eclipses. From 1620 CE forward, records of 37000 lunar occultations are available, plus eclipse data. When all the data are assembled, no one parobolic curve can be fitted to all the data, as one might expect if the variation in UT were entirely from lunar tidal deceleration of Earth rotation. A two acceleration model has been used, with values of 2.4 msec/day/century prior to 1000 CE, and 1.4 msec/day/century thereafter.

27 November

SCIENTIFIC PRESENTATIONS:

S. Leschiutta made a brief introduction to the subject of time transfer and synchronization, reviewing the history of increases in accuracy which followed the introduction of each new method: HF, VLF, LORAN-C, and space techniques.

S. Starker's abstract was read in absentia by D. McCarthy. An experiment was performed on NASA Space Shuttle mission D-1, in which a cesium and rubidium clock on the shuttle were compared to several cesium techniques. After a few initial difficulties, 74 hours of useful data were obtained. The predicted Einstein relativistic effect agreed with the data to better than 1%.

W. Klepczynski surveyed the latest developments in timekeeping at the US Naval Observatory. It was noted that the accuracy of timekeeping has, on the average, improved by a factor of 10 every 10 years, until it has attained 1 nanosecond. Beginning in 1983, the USNO introduced H-masers into its master clock system. Two active Smithsonian Astrophysical Observatory (SAO) VLG 11B active masers are used at the USNO at present. One maser is employed as master, the other as slave; and the two can be switched. Frequency residuals between the two masers for one day averaging times are typically 0.5 parts in 10 to the 15th. The output from the two masers is fed into a microstepper, and the microstepper output is adjusted to track USNO UTC. At present, the draft of the master maser with respect to UTC(USNO) is about 2 parts in 10 to the 15th. Over a four week period, the master hydrogen maser showed a drift of about 1 nanosecond relative to UTC. An experiment using GPS single channel receivers (STI 502) between USNO and TAO provided good statistics on the accuracy of the common view technique. The r.m.s. residual for two hours of observation, 6 second data points, was 22 nanoseconds.

H. Fliegel reviewed the current status of the Global Positioning System (GPS) satellite constellation. Navstars 3,4,6,8,9,10, and 11 are currently operational. The ground tracks of Navstars 4,8, and 9 are more closely spaced than they will be in the final GPS configuration, in order to provide good coverage over the test facility at Yuma, Arizona; therefore they are also well placed for experiments over India, central Asia, and Japan. There are currently no US restrictions on export of GPS receivers designed for Clear Access (C/A) use. Satellite ephemerides are to be made generally available by the US National Geodetic Survey (NGS). H. Fliegel has agreed to act as coordinator for non-US users who need information or contacts using GPS.

R. Kaarls' contribution was presented in absentia by S. Leschiutta. A number of experiments were made by five European laboratories between 1980 and 1983 using direct TV satellites. The precision was about 10 nanoseconds after a few seconds operation; after the fact, biases of about 200 nanoseconds were measured between this TV method and GPS.

S. Leschiutta presented the results of several time transfer experiments between Italy and China using the Italian SIRIO-1 satellite. Using 3-meter antennas and MITREX equipment, two way ranging employing pseudo-random code and a spread spectrum technique, 200 picosecond resolution was obtained; and the uncertainty was estimated to be about 150 nanoseconds.

H. Fliegel reminded all users of LORAN-C that the US Coast Guard is scheduled to phase out all its non-US chains in favor of GPS by 1992, unless special arrangements are made to transfer operations to other, interested nations. S. Leschiutta commented that new LORAN-C chains have recently been put into operation in Scandinavia and in Saudi Arabia; the LORAN-C technique will no doubt continue to be used in Europe and in Asia.

Jean Gaignebet reported that the LASSO package has been integrated into another metrology satellite, but will probably not be launched before early 1987, and will not be visible from the US. However, San Fernando in Spain, Grasse in France, and Wettzell in Germany are prepared to transfer time using LASSO.

There was a joint meeting with Commission 19 and 31 described in the report of Commission 19.

RESOLUTION C1: Astronomical Constants
COMMISSIONS 4, 7, 8, 19, & 31

RECOGNIZING the importance of ensuring that the IAU system of astronomical constants is rigorously defined and is well suited to current applications,
INVITES the President of IAU Commission 4,7,8,19, and 31 to form a working group to serve in collaboration with the appropriate special study group of the International Association of Geodesy which will

1. review current determinations of astronomical and geodetic constants,
2. provide for informational purposes the current best estimates of the values, accuracies and sources of these constants,
3. propose appropriate changes in the relevant definitions and values of the constants of the IAU system.
4. urge all authors to specify completely the values and accuracies, as well as the sources, of the constants used in their work, and
5. submit a preliminary report in 1987.

RESOLUTION C2: Reference Systems
COMMISSION 4, 7, 8, 19, 20, 24, 31, 33, & 40

RECOGNIZING
1. the existence of inconsistent reference systems based upon different theories and modes of observations,
2. the significant improvement in the accuracy of observations using new techniques, and
3. the importance of a space-fixed reference system, independent of the mode of observation, for use in astronomy and geodesy and satisfying the requirements of relativistic theories,

INVITES the Presidents of interested IAU Commissions (for example, 4, 7, 8, 19, 20, 24, 31, 33 and 40) to form an IAU Working Group, with appropriate sub-groups devoted to specialized topics, under the overall chairmanship of the Chairman of the Joint Discussion on Reference Frames, which will report to the XXth General Assembly in 1988 with recommendations for
1. the definition of the Conventional Terrestrial and Conventional Celestial Reference Systems,
2. ways of specifying practical realizations of these systems,
3. methods of determining the relationships between these realizations, and
4. a revision of the definitions of dynamical and atomic time to ensure their consistency with appropriate relativistic theories, and

INVITES the President of the International Association of Geodesy to appoint a representative to the Working Group for appropriate coordination on matters relevant to Geodesy.

THE INTERNATIONAL ASTRONOMICAL UNION

RECOGNIZING the highly significant improvements in the determination of the orientation of the Earth in space as a consequence of the MERIT/COTES* programmes of observation and analysis, and

RECOGNIZING the importance for scientific research and operational purposes of regular Earth-orientation monitoring and of the establishment and maintenance of a new Conventional Terrestrial Reference Frame,

THANKS all the organizations and individuals who have contributed to the developments and implementation of the MERIT and COTES programmes and the the operations of the International Polar Motion Service and the Bureau International de l' Heure, and

ENDORSES the final report and recommendations of the MERIT and COTES Joint Working Groups, and

DECIDES
1. to establish in consultation with IUGG a new International Earth Rotation Service within the Federation of Astronomical and Geophysical Services (FAGS) for monitoring Earth-rotation and for the maintenance of the Conventional Terrestrial Reference System; the new service is to replace both the IPMS and the BIH as from 1988 January 1;
2. to extend the MERIT/COTES programmes of observation, analysis, intercomparison and distribution of results until the new service is in operation;
3. to recommend that an optical astrometric network be maintained for the rapid determination of UT1 for so long as this is recognized to be useful;
4. to set up a Provisional Directing Board to submit recommendations on the terms of reference, structure and composition of the new service, and to serve as the steering committee for the extended MERIT/COTES programmes; and

INVITES National Committees for Astronomy and for Geodesy and Geophysics to submit proposals for the hosting of individual components of the new service by national organizations and observatories; and

URGES the participants in Project MERIT to continue to determine high-precision data on Earth rotation and reference systems and to make the results available to the BIH until the new service is in operation.

*MERIT = A programme of international collaboration to monitor Earth-rotation and intercompare the techniques of observation and analysis.

*COTES = A programme of international collaboration to establish and maintain a new conventional terrestrial reference system.

COMMISSION 33: STRUCTURE AND DYNAMICS OF THE GALACTIC SYSTEM
(STRUCTURE ET DYNAMIQUE DU SYSTEME GALACTIQUE)

Report of Meetings on 21, 22, 23, 25 and 27 November 1985

PRESIDENT: R. Wielen   SECRETARY: W.B. Burton

21 November 1985

Business Meeting of Commission 33

The President pointed out that Commission 33 had its 60th anniversary in 1985. He especially welcomed Professor Willem J. Luyten, who is one of the original members of the Commission since 1925. Professor Jan H. Oort and Professor Luyten are active members of the Commission 33 for now sixty years !

The Commission unanimously approved the election of the new President, W.B. Burton, and Vice-President, M. Mayor. The Commission unanimously accepted the list of names proposed for the new Organizing Committee: continuing until 1988: L. Blitz, G. Lynga, R. Wielen as outgoing president; for the period 1985-1991: J.N. Bahcall, L. Balazs, J.J. Binney, J. Einasto, M. Tosa. The list of IAU members who wish to join Commission 33 was read and adopted: K. Aizu, A.K. Ambastha, J.N. Bahcall, B. Barberis, B. Basu, J.A.R. Caldwell, T. Ciurla, R.S. Cohen, J. Colin, E. Costa, A. Dekel, J.R. Ducati, D. Egret, S.M. Fall, B. Fuchs, D. Galletto, A. Habe, M.R.S. Hawkins, F.P. Israel, D.R. Jiang, S.G. Lee, F.J. Lockman, P.J. McGregor, S. Mikkola, D.G. Monet, M. Morris, S. Ninkovic, M. Nishida, C.A. Olano, J. Palous, Z.Y. Qian, N. Reid, J.X. Rong, M.L. Sanchez-Saavedra, G.X. Song, L. Sparke, Y. Tong, E.I. Vega, S.M. Younis, B. Zhang, J.L. Zhao, C. Zhen.

The scientific scope of Commission 33 was discussed. It was agreed that Commission 33 should continue to act as a joint forum for all astronomers interested in stellar dynamics in general, even if their work is not directly concerned with the Galaxy. Furthermore, basic astronomical data of relevance for galactic astronomy should also remain within the scope of the Commission.

The President reported on the Joint Discussion I on Reference Frames. During this Joint Discussion, a resolution on reference systems was adopted. Commission 33 endorsed this resolution (given elsewhere in this volume) and proposed R. Wielen for membership in the planned Working Group.

The Working Group on Galactic Constants (Chairman: F.J. Kerr, Vice-Chairman: D. Lynden-Bell) has prepared a proposal for the recommendation of new values of galactic constants. A critical review on galactic constants has been written by Kerr and Lynden-Bell. It is planned to publish this review in the IAU "Highlights of Astronomy", Volume 7. The President announced that a decision on the proposal of the Working Group will be taken during the scientific session on galactic constants.

Galactic Constants

Chairman: F.J. Kerr. The following papers were presented:

F.J. Kerr: Report of the working group
M.W. Feast: Recent work on R_o
M.J. Reid, M.H. Schneps, J.M. Moran, C.R. Gwinn, R. Genzel, D. Downes,
   B. Ronnang (presented by F.J. Kerr): First results on the parallax of the galactic center
S. Ninkovic: Orbital eccentricity study for the spherical component of our Galaxy
V.C. Rubin: Recent work on rotation in other galaxies
A.A. Stark, L. Blitz, M. Fich: The rotation curve in the outer Galaxy
J.P. Ostriker: The mass of the galactic halo
M. Mayor: The solar motion

After a lengthy discussion, the Commission adopted with overwhelming majority (approximately 57 in favour, none against, and 3 abstentions) a resolution on galactic constants which is given at the end of this report. The Working Group on Galactic Constants shall continue its work, especially by investigating the problems connected with the solar motion.

22 November 1985

Information Inferred from Orbits for Studies of the Structure and Evolution of Galaxies

Joint Meeting of Commission 28 and 33. Chairman: L. Martinet. The following papers were presented:

L. Martinet: Recent developments in the field (Overview)
T. de Zeeuw: Construction of triaxial galaxies
A. Wilkinson: Integrals of motion of the stellar orbits in a triaxial N-body model of an elliptical galaxy
H. de Jonghe: On the non-uniqueness of anisotropic distribution functions for spherical systems
T. Bontekoe: Dynamical friction in a binary galaxy model
Y. Tong: The general solution of the Poisson equation in 3-D for disk-like galaxies.

23 November 1985

HIPPARCOS

Joint Meeting of Commissions 8, 24, 25, 33 and 37. Chairman: J. Kovalevsky. The proceedings of this meeting will be published in the IAU "Highlights of Astronomy".

25 November 1985

The Galactic Center

Chairman: M. Morris. The following papers were presented:

M. Morris: Review of recent VLA and other radio observations of the galactic center
A. Sandqvist: HCO^+ in the inner 5 pc of the Galaxy

C. Salter: A new 3-mm map of the galactic center at 70" resolution
R. Sramek: Parallax measurements of the compact radio source at the galactic center
N. Kaifu: Molecular and continuum observations of the galactic center
J.H. van Gorkom, U.J. Schwarz, J.D. Bregman: H76α recombination line observations of the galactic center
R. Wielebinski: Radio continuum and polarization results from Bonn
A.A. Stark: ^{13}CO mapping of the galactic center

27 November 1985

The Large-Scale Distribution of
Interstellar Matter in the Galaxy

Chairman: W.B. Burton. The following papers were presented:

B. Pagel: Metallicity gradients
P. Wanner: Isotope-ratio gradients
A.A. Stark: Galactic molecular cloud distribution
W. Hermsen: Distribution of molecular hydrogen from γ-radiation
H. Walker: Galactic dust distribution
F.J. Lockmann: The thickness of the HI layer

The Large-Scale Distribution of
Stars in the Galaxy

Chairman: K.C. Freeman. The following papers were presented:

K.C. Freeman: Overview of galactic components
P. van der Kruit: Pioneer X results - surface photometry of the Galaxy
A. Murray: Probing the South Galactic Cap
A.R. Upgren: Properties of nearby dwarf stars
J. MacConnell: Near-IR photographic survey for southern red supergiants

Joint Discussions

A full account of Joint Discussion I on "Reference Frames" and of Joint Discussion VI on the "Evolution in Young Populations in Galaxies", both co-sponsored by Commission 33, is given in the IAU "Highlights of Astronomy".

Resolution on Galactic Constants
adopted by IAU Commission 33 on 21 November 1985

Whereas (1) many recent determinations of R_o (the distance of the Sun from the center of the Galaxy) and Θ_o (the circular rotation velocity at the Sun) have departed from the commonly used values of 10 kpc and 250 km s^{-1} respectively,

and (2) although these quantities are still not known with great precision, determinations reported in the last decade have led to mean values of R_o, Θ_o, and the Oort constants A and B, which form a physically reasonable set,

and (3) there are practical advantages in agreeing on conventional values which can provide a good basis for comparisons between the work of different astronomers, while being consistent with the best determinations available, to within their uncertainties,

Commission 33 recommends, on the basis of the conclusions of its Working Group on Galactic Constants, use of the values

$$R_o = 8.5 \text{ kpc}$$
$$\Theta_o = 220 \text{ km s}^{-1}$$

in cases where standardization on a common set of galactic parameters is desirable.

The estimated uncertainty in these values, based on the dispersion among recent determinations, is $\pm$ 1 kpc and $\pm$ 20 km s^{-1}.

The Commission makes no recommendation concerning A and B, but notes that the above values of R_o and Θ_o imply that the quantity $(A-B) = 25.9$ km s^{-1} kpc^{-1}.

R. Wielen
President, IAU Commission 33

COMMISSION 34: INTERSTELLAR MATTER
(MATIERE INTERESTELLAIRE)

Report of Meetings 21-25 November 1985

PRESIDENT: M. Peimbert
VICE-PRESIDENT: J. Lequeux   SECRETARY: D.C.V. Mallik

Business session, 21 November

The President summarized the Commission report for the triennium 1982-85 and thanked members of the Organizing Committee for their cooperation and help in preparing the report. The President said that a few important fields had been left out in the report for lack of space. He mentioned in particular Intergalactic Medium and hoped that in the next report it would be possible to include this field. The report had already been mailed to all members of the Commission. Two circulars had also been sent out to all members of the Commission during this period which summarized the Commission activities. The members expressed their appreciation for being kept well informed.

Membership and Structure

The President remarked that Commission 34 is one of the largest of the Union with 470 existing members. During this triennium another 90 have expressed their desire to join the Commission and are going to be accepted at the end of this General Assembly; on the other hand, there have been only four resignations.

The President felt that considering the large size of the Commission and the diversity of subjects under its scope, one or more Working Groups could be formed in addition to the WG on nomenclature. A preliminary discussion on the structure of the Commission took place and as a consequence of it, a committee under the chairmanship of J. Lequeux and consisting of J.S. Mathis, P.A. Shaver, Y. Terzian and P.G. Wannier was formed to study this problem. It was also suggested that the Commission should take an active interest in the field of intergalactic medium.

The Committee on the structure of the Commission met after the business meeting and decided as a first step to form a Working Group on Planetary Nebulae and PN precursors, the chairman will be Y. Terzian. The Committee will report on his work at the next General Assembly.

Organizing Committee

The President then drew up the slate of officers and members of the Organizing Committee for 1985-1988. Based on recommendations that he had received from members and past officers of the Commission, he proposed J.S. Mathis as the new Vice-President and K. de Boer, D.R. Flower, H.J. Habing, B.M. Shustov and D.G. York as new members of the OC. As has been the convention, the present Vice-President J. Lequeux will take over as President at the end of the XIX General Assembly. The continuing members of the OC are: S. D'Odorico, B.G. Elmegreen, P.A. Shaver, P.G. Wannier and M. Peimbert. The proposed slate of officers and OC members was approved by the Assembly. E.B. Kostyakova, U. Mebold, Y. Terzian and V. Radhakrishnan will retire from the OC at the end of the current General Assembly.

Symposia, Colloquia and Scientifc Sessions

Commission 34 sponsored two IAU symposia and cosponsored one colloquium just prior and after the XIX General Assembly. The President briefly commented upon the highly successful symposium on Star Forming Regions held in Tokyo earlier in the month, as well as on the colloquium on Hydrogen Deficient Stars and Related Objects held in Mysore, during the same time. In addition, the Commission has cosponsored two Joint Discussions (Evolution in Young Populations in Galaxies and Supernovae) and four Commission meetings on important recent results during the XIX General Assembly. Brief reports on the Commission meetings are given below while the proceedings of the Joint Discussions will be published in Highlights of Astronomy Volume 7.

The future symposia sponsored and cosponsored by the Commission and already approved by the Executive Committee are Symposium 122 on Circumstellar Matter, to be held in Heidelberg in June 23-27 and Symposium 131 on Planetary Nebulae, to be held in Mexico City in late 1987. The President mentioned two other meetings which may be of interest to members of the Commission: the meeting on Interstellar Processes in Wyoming, USA in July 2-8 1986, and the one on Interaction of Supernova Remnants with the Interstellar Medium in Canada in June 1987.

Working Group on ISM Nomenclature

The activities of the Working Group were reported by its chairperson, Dr. Hélène Dickel. A preliminary manuscript describing a scheme for the "Designation and Nomenclature for Diffuse Radiating Sources" resulted from the deliberations of the WG and was circulated at the meeting. A list of resolutions concerning designations was presented and briefly explained. The publication of the report and the continuation of the WG was approved and the members of the WG were thanked for their efforts to clarify the nomenclature situation.

The resolutions of the WG on Nomenclature of Commission 34 and similar ones from Commission 40 (radio astronomy) endorsed by Commission 48 (high energy astrophysics) were coordinated by Commission 5 (documentation and astronomical data). The resulting recommendations for source designations from Commission 5 will be published in the 1986 revision of the IAU Style Book. Editors of astronomical Journals will be urged to include written requests to both authors and referees to insure that designation of astronomical sources are well documented and conform to the IAU rules.

Other Matters

The President then asked Y. Terzian to report on the plans for a new catalog on planetary nebulae. Terzian mentioned the plans of Perek and Kohoutek to bring out an updated version of their Catalog on Galactic Planetary Nebulae. Furthermore, Mme. A. Acker of Strasbourgh Stellar Data Centre, has also been compiling a Catalog of Planetary Nebulae. Terzian described his efforts in convincing the authors to collaborate rather than duplicate the work; the Commission endorsed Dr. Terzian's endeavours on this matter. It is hoped that by the time of the next IAU Symposium on Planetary Nebulae in Mexico, the new catalog would be ready.

COMMISSION 34 MEETINGS

Dust

Chairperson: J.S. Mathis   21 November 1985

Dr. D.C.B. Whittet (UK) summarized observations. Interstellar depletions of heavy elements (e.g., Ti) from the gas phase are well correlated with the average density of the interstellar medium, showing that they accrete onto grains. The only spectral feature in the entire 0.1 - 0.3 μm wavelength range is the 0.22 μm "bump"; this places some constraints on candidate materials for grains because many substances have strong absorptions which are not observed. The diffuse interstellar bands and the optical continuous extinction are not correlated with any ultraviolet extinction properties. The 9.7 μm and 18 μm (amorphous silicates) are seen in both emission and absorption. The 3.3 and 3.4 μm absorption features are due to the C-H stretch, but it is not yet possible to pin down the specific molecule. For cold clouds, the 3.07 μm feature is fitted well by a mixture of water and ammonia ices, the 4.67 μm band is CO, and the 4.62 μm is CN.

Dr. Peter Martin (Canada) summarized several current theories of grains in general terms. The only unanimous conclusion is that amorphous silicates are present and that there is a large range in particle sizes. Several suggestions have been made for the origin of the 0.22 μm bump: OH^- ions on the surfaces of silicate grains (Duley); very small graphite (Greenberg); a range of graphites (Mathis "MRN"); hydrogenated amorphous carbon. The visual extinction is usually attributed to more than one substance, such as silicates and amorphous carbon.

Dr. Bruce Draine (USA) discussed formation and destruction of grains. Grains form in the atmospheres of both carbon and oxygen-rich stars, but it seems impossible to understand their formation in the best-studied star (α Ori) on the basis of a steady state. If formation is episodic, present theory cannot make realistic predictions. Destruction caused by both sputtering and grain-grain collisions in supernova shocks should deplete grains in a time scale of 3×10^8 yrs if there is no formation within the interstellar medium itself. If there are many small grains, their total area is large enough to allow appreciable coagulation and accretion on this time scale. Thus, the grain size distribution in the interstellar medium must be heavily modified through growth processes as well as by shattering.

New Results on Planetary Nebulae

Chairperson: Y. Terzian   25 November 1985

R. Gathier reported on individual distances of several planetary nebulae. He showed that the positions of the PN central stars in the HR diagram indicate discrepancies with theory, concerning evolutionary timescales, and range of central star luminosities. D. Schönberner discussed a correlation between the absolute magnitude of central stars and the excitation of their nebulae. P. Wannier and R. Sahai presented observations of oxygen isotopic abundances from circumstellar envelopes of giant stars. ^{17}O is found to be overabundant in all the cases investigated. B. Zuckerman reported that he used the UKIRT telescope together with I. Gatley to observe H_2 molecules from bi-polar PN, including the Dumbbell and NGC 2346. Y. Terzian presented λ6 cm VLA maps of several PN performed in collaboration with C. Bignell and J. van Gorkom. Several of these objects were observed with an angular resolution of 0.4". These observations will be used with future ones to determine nebular expansions and distances. H.C. Bhatt and D.C.V. Mallik reported on the two dense PN Cn1-1 and M1-2. The high densities indicate young objects, but the central stars show stellar spectra of F and G types and from IRAS observations low luminosities have been found indicating discrepancies with stellar evolutionary tracks. S. Kwok presented a VLA survey of stellar PN at λ6 cm with high resolution, and indicated that most have been spatially resolved. Y. Terzian, J. Phillips and H. Payne described a survey of OH in PN and indicated the detection of 1612 MHz OH line in Vg2-2 and NGC 6302. They strongly emphasifed that at these high sensitivities OH confusion from the galactic plane is extremely important. H.C. Bhatt and S.R. Pottasch reported on dust temperatures in PN from IRAS data, and suggested that the dust in PN may mainly be carbon dominated. Two papers were read in absentia. One by E. Kostyakova on the variability of IC 4997; and one by C.T. Hua on monochromatic images of small compact nebulae.

Important Recent Results I

Chairperson: J. Lequeux   25 November 1985

The following contributed papers were presented:

Observations of Nova RS Oph – R. Davis.

Optical Spectrum of RS Oph – T.P. Prabhu.

Theoretical Model of RS Oph – F.D. Kahn.

2D Spectroscopy of Planetary Nebulae with a CCD Echelle – C.D. Mc Keith.

Molecular Astronomy in China – Xing Jun.

Models of Molecular Clouds – W. Boland.

Diffuse Bands and Polarization – W.B. Somerville.

Chemical Composition of NGC 2363 – S. Torres-Peimbert, M. Peimbert and M. Peña.

Important Recent Results II

Chairperson: F.D. Kahn   25 November 1985

The following contributed papers were presented:

On the Nebular N/O abundance ratio – R. Rubin.

Abundances in Liners – L. Binette.

Ionization in the Galactic Halo – J.N. Bregman and P.J. Harrington.

ISM work at Nançay – I. Kazes.

Very Low Frequency Recombination Lines – K.R. Anantaramaiah.

Dust in H II Regions – M.A. Gordon, P.R. Jewell, M. Kaftan-Kassim, and C.J. Salter.

BG 2107+49, a Peculiar H II Region – L. Higgs.

COMMISSION 35: STELLAR CONSTITUTION (CONSTITUTION DES ETOILES)

Report of Meetings 20 November 1985

PRESIDENT: A. N. Cox   VICE PRESIDENT: D. Sugimoto

20 November 1985

I. BUSINESS MEETING

The business meeting was started by a moment of silence with the members standing in respect for three members that had passed away since the last meeting in 1982. These names were read by the President: John P. Cox, Sven Rosseland, and Stefan Temesvary.

New officers were voted for in an election held between March and June 1985, and the results were reported by the President in a letter dated June 30 to all the members of the Commission. The President, Vice President and the Organizing Committee members were endorsed by the Executive Committee at the General Assembly after the Commission 35 business meeting. The new President is Daiichiro Sugimoto (Japan), who was elected Vice President before the Patras 1982 General Assembly and was expected to succeed to be President in 1985. The election for recommendations to the Executive Committee resulted in Andre Maeder (Switzerland) as the new Vice President.

The election by the rules agreed upon at the Montreal General Assembly resulted in a tie for second place among four candidates for the Organizing Committee. It was decided by the President to offer all five new names, and this was accepted at the business meeting. The current Organizing Committee then consists of the five carryover people and the five new ones: P. Bodenheimer (USA), C. Chiosi (Italy) A. N. Cox (USA), D. O. Gough (UK), R. Kippenhahn (FRG), Y. Osaki (Japan), J.-L. Tassoul (Canada), J. W. Truran (USA), V. Weidemann (FRG), and J. C. Wheeler (USA). The incoming President asked A. Tutukov to be a special consultant to him and the Organizing Committee to secure close contact with the USSR and its neighborhood.

Due to the new computer now in the Secretariat in Paris, it has been possible to computerize the lists of members of the IAU and the various Commissions. Letters were sent to all IAU members by the Secretariat to see if they still want to remain members and to secure current addresses, etc. Numerous changes in Commission memberships resulted, and the actual Commission 35 membership is still being sorted out by the new President at the time of this report after the General Assembly. There are new members who have been IAU members for some time, and there are new members who have been just elected to the IAU. There are also resignations from some who have been listed by mistake. Professor Ludwig Biermann has asked to be deleted from membership of Commission 35 so that he can pursue other interests in other Commissions. The total membership now is approximately 291.

The President brought to the Commission a proposal to form a Working Group on Theoretical Solar Models and Solar Oscillations. The problem is that the several groups calculating solar models and their predicted oscillation frequencies differ more among themselves than do the many observations. The causes of these differences can be many, and they need to be discovered and eliminated to take advantage of the very accurate observed frequencies. Professors Christensen-Dalsgaard and Ulrich are already working with the several theoretical groups on this problem, and it was decided by the Commission that formation of a Working Group was not necessary from any point of view. The Working Group will not be formed, but the activity of the solar oscillation predictors was strongly endorsed.

Three proposed IAU sponsored meetings were discussed by the Commission. Endorsement by many of the Organizing Committee members had already been secured at the time of the business meeting. A Symposium on planetary Nebulae to be held in Mexico in October 1987 was made by J. B. Kaler (USA). Another Symposium on Atmospheric Phenomena as Manifestation of Internal Evolution of Stars was proposed for Tokyo in August 1987 by K. Nomoto (Japan). This now has been changed to be a Colloquium. A Colloquium on Faint Blue Stars to be held in Tucson in the Spring of 1987 was proposed by A. G. D. Philip (USA). All these conferences were enthusiastically supported by the Commission with some possible better definition for the Atmospheric Phenomena proposal. The Executive Committee has accepted the first and last of these three conferences at the time of this report.

Da Run Xiong from Nanjing, China gave a brief report on the work being done in his country, because the research from this newly adhering country is not so well known to the members. He also discussed briefly his work on nonlocal convection theory.

The usual triennial report of Commission 35 was published along with the other Commission reports in volume 19a Reports on Astronomy in mid-1985. Scientific reports were Massive Stars (R. M. Humphreys), Rotation in Late Type Stars (W. Benz), Helioseismology (Christiansen-Dalsgaard), Planetary Nebula Central Stars (E. M. Sion), Pulsations in Hot Degenerate Dwarf Stars (A. N. Cox and S. D. Kawaler), and White Dwarfs (V. Weidemann).

II. SCIENTIFIC MEETINGS

There were five scientific meetings of Commission 35, two concerned with our main business of stellar structure and evolution, two as a Joint Commission meeting with Commissions 25, 37, and 45 on Precision Photometry of Clusters, and a brainstorming session mostly with Commission 27 and 35 members on the theoretical problems in the B star pulsations. Only the first two of these sessions are reported here. The Joint Commission meeting will be reported as a Commission 37 meeting, and the brainstorming session was organized informally by M. Aizenman with no program or conclusions.

Commission 35 also jointly sponsored three Joint Discussions: Solar and Stellar Nonradial Oscillations, Stellar Activity: Rotation and Magnetic Fields, and Supernovae. Proceedings of these three will be found in Highlights of Astronomy.

20 November 1985

Stellar Surface Abundances

A. Tutukov	Introduction
A. Maeder	Chemistry of Massive Stars
D. Schoenberner	OBN Stars
D. Lambert	Chemistry of AGB Stars
R. Kraft	Carbon and Nitrogen Abundances in the Atmospheres of Giants and Subgiants of Metal Poor Galactic Clusters
G. Michaud	Abundance Anomalies and Stellar Hydrodynamics
K. Hunger	On the IAU Colloquium 87 "Hydrogen Deficient Stars and Related Objects"

In a brief introduction, Tutukov pointed out that 97 per cent of the stars display abundances that are normal while only 3 per cent are peculiar. Three processes contribute to the peculiar abundances - mass loss, internal mixing in the star, and accretion. In the first case, the mass loss can be because of a binary companion or because of a stellar wind. The mixing in stars has many causes, among them being convection, semi-convection, circulation, diffusion, and shear flows. Accretion can occur from a close binary or from interstellar matter.

Maeder discussed the mass loss, convective dredgeup, overshooting, and turbulent diffusion as causes for the abundance peculiarities seen in massive stars. Nitrogen 14, a common isotope in normal stellar material is enhanced in hydrogen burning by the CNO cycle and then later is converted to Neon 22 in the helium burning stages in the core of massive stars. Two alpha particle captures make this element disappear as evolution proceeds. When this processed material reaches the surface, continuously for the large mass loss and deep dredgeup case for 100 solar mass, the C/N ratio drops at first from its normal value of 4.1 to 0.02 and then increases to over 1000 as the nitrogen disappears. The same drop and rise is also seen for the O/N ratio. The various Wolf-Rayet stages are then: WNL-hydrogen and nitrogen seen, WNE-hydrogen exhausted but nitrogen not yet depleted, WC-all H and N gone but considerable C from helium burning, WO-both C and O present as a result of the very hot helium burning in layers now exposed by mixing and dredgeup.

Schoenberner presented some observational data for 4 OBN stars (HD 89137, HD 14633 θ Car, and HD 48279) and 3 ones (τ Sco 10 Lac, and 15 Mon) with normal compositions. Non-LTE atmospheres were constructed to analyze the CNO abundances. Incomplete CN cycle H burning is verified for masses in the range of 20 to 30 solar mass. Overshooting of the convection and large mass loss are both needed to explain large He overabundances in the surface exposed material. Mixing and processing of envelope matter occurs right at the early main sequence stages.

Lambert discussed the AGB stars with their preflash helium shell flash compositions and their compositions after the s-processing flashes. Spectral classes M, MS, and S were studied using synthetic spectra. Data on the s-process abundances correlate with the carbon abundance as expected. Ratios of the carbon 12 to carbon 13 and carbon 12 to oxygen 16 were also presented.

The old population CNO abundance question was discussed by Kraft, who has measured these abundances in metal poor galactic clusters. The giants and subgiants, which have dredged up some processed material, show the expected decrease in carbon and increase in nitrogen. For a given C/Fe ratio, however, the clusters show more nitrogen enhancement than field stars. The issue of whether this difference could be due to a primordial variation as seen in metal poor globular clusters like M92 and M15 was discussed. The actual rate of nitrogen enhancement and carbon depletion among the subgiants was presented.

Michaud discussed the lithium abundance question for the dwarf stars. The relative importance of lithium settling and dredge up of the lithium is important in this case. As evolution brings these stars to cooler surface effective temperatures, the convection zone depth increases rapidly. Then the lithium that has settled away is retrieved to bring the surface abundance back to the value for the hotter stars. This happens because the settling time grows longer for the cooler temperatures, and the lithium cannot get away. It is also important to consider at the same time the helium settling, because the depth of the convection zone depends on the helium abundance. Mass loss rates cannot be larger than about 10^{-15} solar mass per year, or the settling velocity will be reversed by the mass loss flow.

Finally, Hunger gave a very brief report on the IAU Colloquium held just before the General Assembly in Mysore. He presented the whole array of hydrogen deficient star types and concluded that most, if not all, were the result of normal stellar evolution.

20 November 1985

General Scientific Session

D. Sugimoto	Eddington Luminosity of X-Ray Bursts and the Distance to the Galactic Center
T. Lee	The Astrophysical Source of the Extinct Ration-Nuclides in the Early Solar System, New Clues from Recent Cr Data
C.-K. Chou	Angular Momentum Transport in a Protostar Disk System
C. Mohan and M. Saxena	Effects of Rotation and Tidal Distortion on the Periods of Small Adiabatic Radial and Nonradial Modes of Oscillations of Stellar Models
H. A. Hill	The Sun's Gravitational Quadrupole Moment Inferred From the Fine Structure of the Acoustic and Gravity Normal Spectra of the Sun
C. A. Rouse	Periods of g-Modes in Solar Models with a High Z Iron-like Core

The aim of this paper is to draw the attention to the possible importance of x-ray bursts as a standard candle to determine for instance distances to the galactic center.

Recently, the structure of x-ray bursting neutron stars has been studied by solving the stellar structure with a steadily mass-losing envelope. The theory is being elaborated by T. Ebisuzaki and K. Nomoto taking into account the transport on non-equilibrium radiation through the extended envelope of the neutron star. Since the time scale of the x-ray bursts is much shorter than the dynamical time scale of the neutron star, the luminosity cannot exceed the Eddington limit by more than several percent.

There are also, from the study of the observations during the burst, good physical reasons to assume that the strongest burst of an individual source is equal to the Eddington luminosity. Therefore, the distance can be determined. With the distances obtained for different x-ray bursts, their spatial distribution was plotted and the center of their distribution was found to be situated at the distance of 7 kpc in the direction of the galactic center.

The author emphasizes the astrophysical information provided by the study of isotopic ratios in rocks and meteorites and discusses the interest of chromium isotopes. The chromium isotopic anomalies of $^{53}Cr/^{52}/Cr$ and $^{54}/Cr/^{52}/Cr$ in Allende refractory inclusions have been analyzed; the author discusses the results of Birck and Allègre, who found isotopic anomalies in the range of 5×10^{-4}, with an excess of ^{54}Cr and a deficiency in ^{53}Cr. The author examines the various astrophysical sites where the above enhancements can be produced and concludes that reactions associated with ^{28}Si-burning in supernovae of Type II could explain the above anomalies.

The author uses properties of the model of density waves in the framework of the models of collapsing clouds with an emphasis on the transport of angular momentum in rotating protostellar disks. The application of stability criteria to such disks leads to the results that the growth of the stellar angular momentum in the protostar is limited by the transfer in the disk. The author considers that the effect of the viscous damping is likely to cause the excess of angular momentum to be deposited within a small region close to the protostar.

The authors present the explicit mathematical formulations of the eigenvalue problems of radial and nonradial modes of oscillations of stellar models incorporating the effects of both rotation as well as tidal distortion forces. The mathematical formulations of these eigenvalue problems have been developed making use of the averaging technique of Kippenhahn and Thomas (1970), "Stellar Rotation," ed. A. Stettbak, p. 20) for incorporating the effects of rotation and tidal distortions on the equilibrium structure of a star in conjunction with Kopal's method (1972. Adv. Astron. Astrophys. Vol. 9, p. 1) for evaluating various equilibrium structure parameters on the Roche equipotential surfaces. The effectiveness of the approach developed is illustrated by the authors, who apply their method to 10, 5, and 2 $M_\odot$ main-sequence models.

The fine structure of 30 low-order, low-degree acoustic mode multiplets, the average fine structure of $\ell = 1$ and 2 intermediate-order acoustic model multiplets, and 31 low-degree gravity mode multiplets have been analyzed to infer the internal rotation of the Sun and to place upper limits on the internal magnetic fields. The multiplet classifications are based on differential velocity observations of Kotov et al. and differential radius observations. No evidence at the 7 standard deviation level was obtained to support the Duvall and Harvey findings. However, this test did yield evidence of intermediate degree f-modes with a multiplet fine structure consistent with that found in the results of previous works by H. A. Hill; this represents the first observational evidence of those f-modes located just above the asymptotic g-mode limit. The inferred angular velocity distribution based explicitly on the properties of the 63 multiplets is quite similar to the results of Hill, Rabaey and Rosenwald and yields a gravitational quadrupole moment J_2 of $= 7.7 \times 10^{-6}$, which has important implications for planetary tests of theories of gravitation.

The variable period spacings of the $\ell = 1,2,3$, and 4 degree g-mode of oscillation for the high Z core (HZC) solar model are discussed and found to be consistent with the high μ-gradient and central density of the HZC model. The properties of the HZC model were compared with the observed frequencies of oscillation in the five minute band and agreements were found by the author. The effect of point spacing on the predicted frequencies is also emphasized, and it is shown that it is necessary to calculate the solar model and the oscillation frequency eigenvalues with space steps smaller than 10^{-3} to as small as 10^{-5}. From his results, the author suggests that the high Z core model is favored.

COMMISSION 36: THE THEORY OF STELLAR ATMOSPHERES (LA THEORIE DES ATMOSPHERES STELLAIRES)

Report of the Meetings on November 20th, 21st and 27th

BUSSINESS SESSIONS

November 20th, 9.00 - 9.20 and November 27th, 11.00 - 11.25

CHAIRMAN: B. Gustafsson   SECRETARY: R. Wehrse

I.   COMMISSION ACTIVITIES
     The President outlined the activities of the Commission in the period following the 18th General Assembly.

II. NEW MEMBERS
     The membership of the following 28 new members was approved:
     S. Baird, D.N. Brown, K.L. Chan, P. Chen, S. Drake, D. Dravins, J.M. Fontenla, W.R. Hamann, U. Heber, H. Herold, T.H. Kaduri, A.P. Linnell, C. Liu, J. Madej, S. Massaglia, Å. Nordlund, R. Pallavicini, M.G. Rovira, G. Scharmer, W. Schmutz, K.P. Simon, L. Snezhko, H. Spruit, P. Ulmschneider, T. Watanabe, R. White, L.A. Willson, H. Wöhl.

III. NOMINATION OF PRESIDENT AND VICE PRESIDENT
     The nomination of K. Kodaira as President and of D.F. Gray as Vice President was approved unanimously.

IV. ORGANIZING COMMITTEE
     The following names for the new Organizing Committee were approved:
     J. Cassinelli, L. Cram, B. Gustafsson, A.G. Hearn, I. Hubeny, W. Kalkofen, R. Kudritzki, D. Mihalas, M. Seaton, A.B. Underhill.

V.   ENDORSEMENT OF RESOLUTION
     The Commission endorsed a resolution in favour of the continuation of solar and stellar research activities at Mt Wilson Observatory.

VI. FUTURE ACTIVITIES
     The President outlined the planned activities, which were subsequently discussed.

SCIENTIFIC SESSIONS

1)   MEETING: November 20th, 9.20, continued on
               November 27th, 11.25 - 11.45

     Radiative Transfer   Chairman: W. Kalkofen

     Papers on the following subjects were given:

          G.B. Rybicki: The physical principles underlying escape probability methods in radiative transfer.

R. Wehrse: The generalization of the formal solution of the scalar transfer equation to the set of coupled equations occuring in radiative transfer, subject to integral constraints.

A. Peraiah: The discrete space theory for the solution of the partial differential equation, describing the radiative transfer in an atmosphere with spherical symmetry.

W. Kalkofen: An operator perturbation method employing differential equations for the solution of the radiative transfer equation in conservation form of a two-level atomic model with complete redistribution.

D. Mohan Rao: Time-dependent, multi-dimensional radiative transfer in stellar atmospheres. (Work for a thesis under the supervision of A. Peraiah at Bangalore University).

K.E. Rangarajan: Spectral line formation in stellar atmospheres. (Work for a thesis under the supervision of A. Peraiah at Bangalore University).

Most of the papers presented will appear in a book under the title "Numerical Radiative Transfer", to be published in 1986 by Cambridge University Press. Reviews of numerical methods can be found in "Progress in Stellar Spectral Line Formation Theory", from a conference held in Trieste in 1984, in "Methods of Radiative Transfer", from a session of Commission 36 at the IAU meeting in Patras, Greece, in 1982, and in a review paper by Auer (1986).

The main topics treated at the session in Delhi and in "Numerical Radiative Transfer" are operator perturbation methods and problems encountered in their efficient use, and the transfer of polarized radiation. Additional discussions concern escape probability methods and line blanketing.

References

Methods in Radiative Transfer, 1984, W. Kalkofen ed., Cambridge University Press, Cambridge, UK.

Progress in Stellar Spectral Line Formation Theory, 1985, J.E. Beckman & L. Crivellari eds., D. Reidel Publ. Co., Dordrecht, Holland.

Numerical Radiative Transfer, 1986, W. Kalkofen ed., Cambridge University Press, Cambridge, UK.

Auer, L.H.: 1986, Methods in Comp. Phys., in press.

2) JOINT SESSION (with Commissions 25, 29 and 45): November 20th, 14.00

Synthetic Photometry   Chairman: R. Buser

The minutes of this session are published in Highlights of Astronomy, Vol 7.

3) JOINT SESSION (with Commission 14): November 21st, 16.00

Atomic and Molecular Data for Studies of
Stellar Atmospheres   Chairman: A.H. Gabriel

ATOMIC DATA FOR ANALYSING EARLY-TYPE STELLAR SPECTRA

A.E. Lynas-Gray, Dept. of Physics and Astronomy, University College London.

Among the atomic data needed, for analysing early-type stellar spectra, are oscillator strengths and photoionization cross-sections for the prediction of line strengths together with continuous and line opacities. Electron collision excitation and ionisation rates are also needed if an attempt to solve the radiative transfer equation, simultaneously with the statistical equilibrium equations, is to be made. Some indication of the atomic data that has become available since 1982 is presented below; no attempt has been made to provide a complete bibliography.

A useful review of atomic data, with assessment, available before 1982 has been prepared by Mendoza (1983). Precision oscillator strengths can now be used to obtain large numbers of moderately accurate oscillator strengths, from line intensities (Cowley 1983, Cowley & Corliss 1983) and Kurucz & Peytremann's (1975) tabulation (Blackwell et al. 1983). Variational wavefunctions have been used by Kono & Hattori (1984) to calculate accurate oscillator strengths for neutral helium.

Saraph (1985) has reviewed recent work on photoionization cross-sections. It should be noted that detailed resonance structures have not been incorporated into model atmosphere calculations so far. Bell et al. (1983) have provided recommendations for ground-state electron collision ionisation cross-sections, though the Montagne et al. (1984) recommendation should be adopted for neutral helium. Aggarwal (1983), Aggarwal et al. (1984) and Cochrane & McWhirter (1983) have recommended electron collision excitation cross-sections and rates for HI, He I and Li-like ions, respectively.

References

Aggarwal, K.M.: 1983, Mon. Not. R. astr. Soc. 202, 15P.
Aggarwal, K.M., Kingston, A.E. and McDowell, M.R.C.: 1984, Astrophys. J. 278, 874.
Bell, K.L., Gilbody, H.B., Hughes, J.G., Kingston, A.E. and Smith, F.J.: 1983, J. Phys. Chem. Ref. Data 12, 891.
Blackwell et al.: 1983, Mon. Not. R. astr. Soc. 204, 141.
Cochrane, D.M. and McWhirter, R.W.P.: 1983, Physica Scripta 28, 25.
Cowley, C.R.: 1983, Mon. Not. R. astr. Soc. 202, 417.
Cowley, C.R. and Corliss, C.H.: 1983, Mon. Not. R. astr. Soc. 203, 651.
Kono, A. and Hattori, S.: 1984, Phys. Rev. A. 29, 2981.
Kurucz, R.L. and Peytremann, E.: 1975, SAO Special Report 362.
Mendoza, C.: 1983, IAU Symposium 103, 143.
Montagne, R.G., Harrison, M.F.A. and Smith, A.C.H.: 1984, J. Phys. B: At. Mol. Phys. 17, 3295.
Saraph, H.E.: 1985, Proceedings of CCP2/CEC Atomic Data Workshop (in press)

COMMENT BY W.L. WIESE:

In addition to the recent data sources on transition probabilities of Fe I cited in the preceding paper, the 1981 data tables by J.R. Fuhr et al. (J. Phys. Chem. Ref. Data 10, 305-565) are worth noting. The significance of these transition probability tables is that all data are critically evaluated and the roughly 2000 selected transitions are estimated to have typically accuracies within the 25 to 50% range and sometimes much better.

COMMENT BY C. COWLEY:

First of all let me say that astronomical spectroscopists owe a great dept of gratitude to Dr. Wiese and his group for their compilations of critically evaluated oscillator strengths for a large number of lines of astrophysical interest. These values provide a world-wide standard to be used in the analysis of cosmic plasmas. I do not see how we could do our research without them.

I judge from his comment that we still have not clarified the point that we have massive needs that extend beyond these critical evaluations. Our light sources are so powerful, and so contaminated (relative to a good laboratory experiment) that we need data even for lines that have not yet been seen in the laboratory.

This is the reason for our efforts to make use of the Meggers, Corliss, and Scribner TABLES OF SPECTRAL LINE INTENSITIES, as well as the Corliss-Tech work on Fe I (NBS Monograph 108). An immense effort was put into this same basic Fe I data set by Boyarchuk and Savanov (Izv. Crimean. Astrophys. Obs. Vol 70). The new volume of Critical Evaluations by the NBS will contain some 1950 Fe I lines, and will certainly ease the need for reliance on the older systems of measurement. Modern measurements have still not exceeded in number the 3288 Fe I lines of Corliss and Tech, and in the domain of the rare earth lines, there are only a few hundred modern measurements, so our need of Monograph 145 remains strong.

THE EFFECTS OF ATOMIC-LINE ABSORPTION ON STELLAR ATMOSPHERES
R.L. Kurucz, Harvard-Smithsonian Center for Astrophysics

In 1983, working with Lucio Rossi from Frascati and with John Dragon and Rod Whitaker at Los Alamos, I finally completed line lists for all diatomic molecules that produce important opacity in G and K stars. Once the line data were ready, I computed new distribution function opacity tables. The calculations involved 17,000,000 atomic and molecular lines, 3,500,000 wavelength points, 50 temperatures, and 20 pressures, and took a large amount of computer time.

As a test the opacities were used to compute a theoretical solar model, to predict solar fluxes and intensities from empirical models, and (with fudging) to produce improved empirical models that are able to match the Ca II H and K line profiles and both the UV and IR intensities formed near the temperature minimum. The work on empirical models is in collaboration with Avrett and Loeser.

There are several regions between 200 and 350 nm where the predicted solar intensities are several times higher than observed, say 85% blocking instead of the 95% observed. The integrated flux error of these regions is several per cent of the total. In a flux constant model this error is balanced by a flux error in the red. The model predicts the wrong colors. After many experiments with convection and opacities, and after synthesizing the spectrum in detail, I have determined that this discrepancy is caused by missing iron group atomic lines that go to excited configurations that have not been observed in the laboratory. Most laboratory work has been done with emission sources that cannot strongly populate these configurations. Stars, however, show lines in absorption without difficulty.

I have used Bob Cowan's Hartree-Fock programs at Los Alamos to compute Slater single- and configuration interaction integrals for the lowest 50 configurations of the first 10 stages of ionization for elements up through Zn and for the first 5 for heavier elements. These calculations allow me to determine eigenvectors by combining least squares fits for levels that have been observed with computed integrals (scaled) for higher configurations. Each least squares iteration takes a significant amount of time on a Cray and many iterations are required. Thus far I have completed new line lists only for Fe I and II, but they produce the strongest effect on the spectrum. Radiative, Stark, and van der Waals damping constants and Lande g values are automatically produced for each line. The complexity of these calculations is illustrated by this table,

	Fe I		Fe II	
	even	odd	even	odd
number of configurations	26	20	22	16
number of levels	5401	5464	5723	5198
largest Hamiltonian matrix	1069	1094	1102	1049
number of least squares parameters	963	746	729	541
(many fixed at scaled HF)				
total number of lines saved	583,814		1,112,322	

I have computed the blocking in the solar ultraviolet spectrum produced by Fe
lines, once with the old line data mentioned above, and again with the newly
calculated data. The increase in opacity is dramatic. Many moderate strength lines
appear to fill in the gaps. I expect there to be similar effects in hotter stars.
I am currently trying to get computer time from the NSF to complete the iron group
elements including higher stages of ionization. Once the line data are complete, I
will recompute the opacity tables for a range of abundances, then compute new grids
of models, and synthesize spectra for comparison to observed spectra.

ATOMIC DATA SUMMARY
W.L. Wiese, NBS, Washington DC

Astronomers may obtain compilations of evaluated atomic data, as well as
bibliographies of relevant literature references, from two principal sources
(information and data material is usually freely supplied):
- (a) The traditional source are the data centers of the National Bureau of
  Standards (NBS, USA, Address: Gaithersburg, MD. 20899), specifically the
  Data Center on Atomic Energy levels and Wavelengths (Director: W. Martin)
  and the Data Center on Atomic Transition Probabilities (Director: W. Wiese).
  Also, NBS operates a Molecular Spectroscopy Data Center (Director: F. Lovas).
  At the Joint Institute of Laboratory Astrophysics (JILA) Boulder, CO.,
  an information center on Low Energy Atomic and Molecular Collision Data is
  operated (Director J. Gallagher).

- (b) Due to worldwide data needs by the magnetic fusion research community,
  atomic spectroscopy data and especially data on atomic collisions
  (excitation, ionisation, charge transfer and recombination cross-sections
  and rates) are collected by the Center on Atomic Data for Controlled
  Fusion, Oak Ridge National Laboratories, USA; the Information Center at
  the Institute for Plasma Physics, University of Nagoya, Japan; the Japan
  Atomic Energy Research Institute, Tokai; and the Atomic Data Center,
  Queen's University, Belfast, Ireland. In addition, the International
  Atomic Energy Agency, Vienna, Austria publishes a quarterly Information
  Bulletin containing a comprehensive listing of current literature
  references on atomic data.

EFFECTS OF MOLECULAR LINE ABSORPTION ON STELLAR ATMOSPHERES
B. Gustafsson, Stockholm Observatory, Sweden

The general effects of molecular blanketing have been discussed in several
reviews, see, e.g., Gustafsson and Olander (1979), Carbon (1979 and 1984) and Johnson
(1986). Very schematically one finds that the line absorption generally causes a
heating of the deeper layers (backwarming), while the upper layers may be cooled or
heated. Cooling of the upper layers is in particular produced by absorption at
wavelengths on the red side of the Planck maximum or, at shorter wavelengths, by
absorption distributed through the atmosphere. Heating of the surface layers may
result from absorption on the blue side of the Planck maximum, in particular if the
absorption is concentrated towards the surface. An example of cooling molecular
absorption is the absorption of CO VR lines in K stars, while the TiO line absorption
heats the upper layers of M stars.

A number of facts should be noted:
- (i) The surface effects are severely overestimated in LTE models if the lines in
  reality are formed in scattering instead of in absorption processes.
- (ii) The collective effects of numerous individually very weak lines may well be of
  very great importance for the atmospheric structure - an example are the
  effects of "hot" combination bands of HCN on carbon-star atmospheres
  (Eriksson et al. 1984, Jørgensen et al. 1985).
- (iii) There is an interesting coupling between the effects of a spherical extension

of the atmosphere and the effects of molecular absorption (Watanabe and
Kodaira 1978, 1979, Schmid-Burgk et al. 1981).
(iv) The picture outlined above changes qualitatively as a result of convective
overshoot – in a realistic upper photosphere convection cools the gas
(through adiabatic expansion) while radiation heats it (Nordlund 1985).

An interesting example of the complexity of the effects of molecular and
metal-line absorption on the upper photospheric layers is the effects of line
blanketing in the solar temperature-minimum region. Ayres (1981) and collaborators
(Ayres and Testerman 1981, Ayres et al. 1985) have traced two components in the solar
atmosphere, one hotter component visible in the Ca II H and K lines, and one cooler
which shows up in the analysis of CO VR lines. Kneer (1983) has argued that two
different equilibrium states might exist in classical model atmospheres; Nordlund
(1985), however, finds that this is not the case if the metal lines are assumed to
be formed in absorption. However, if the lines in the ultraviolet are assumed to be
formed in scattering there are two different solutions to the model-atmosphere
problem with very different surface temperature.

Two numerical methods for taking line blanketing into account are in current
use - the Opacity Sampling (OS) method and the Opacity Distribution Function (ODF)
method. The first-mentioned method is more flexible for individual models and easier
to generalize to non-LTE cases and complex velocity fields, while the second method
is more economical for the calculation of extensive grids of models. The two methods
tend to give very similar results – for cool carbon stars, however, the ODF models
deviate since the polyatomic opacity in the surface layers is not well correlated in
wavelength with the absorption from diatomic molecules at greater depths (Ekberg et
al. 1985). Similar problems are expected to occur for cool M stars where water
vapour is an efficient absorber in the surface layers.

References

Ayres, T.R.: 1981, Astrophys. J. 244, 1064.
Ayres, T.R. and Testerman, L.: 1981, Astrophys. J. 245, 1124.
Ayres, T.R., Testerman, L. and Brault, J.: 1985, preprint.
Carbon, D.F.: 1979, Ann. Rev. Astron. Astrophys. 17, 513.
Carbon, D.F.: 1984, in Methods in Radiative Transfer, ed. W. Kalkofen,
    Cambridge University Press, p. 394.
Ekberg, U., Eriksson, K. and Gustafsson, B.: 1985, Astron. Astrophys. subm.
Eriksson, K., Gustafsson, B., Jørgensen, U.G. and Nordlund, Å.: 1984, Astron.
    Astrophys. 132, 137.
Gustafsson, B. and Olander, N.: 1979, Phys. Scripta 20, 570.
Johnson, H.R.: 1986, review for Monograph Ser. on Non-Thermal Phenomena in Stellar
    Atmospheres, CNRS/NASA, in press.
Jørgensen, U.G., Almlöf, J., Gustafsson, B., Larsson, M. and Siegbahn, P.: 1985,
    J. Chem. Phys. 83, 3034.
Kneer, F.: 1983, Astron. Astrophys. 128, 311.
Nordlund, Å.: 1985, in Theoretical Problems in High Resolution Solar Physics, ed.
    H.U. Schmidt, MPA 212, p. 1.
Schmid-Burgk, J., Scholz, M. and Wehrse, R.: 1981, Monthly Notices Roy. Astron. Soc.
    194, 383.
Watanabe, T. and Kodaira, K.: 1978, Publ. Astron. Soc. Japan 30, 21.
Watanabe, T. and Kodaira, K.: 1979, Publ. Astron. Soc, Japan 31, 6.

IMPORTANT MOLECULAR DATA FOR THE ANALYSIS OF LATE-TYPE STELLAR SPECTRA
D.L. Lambert, Department of Astronomy, University of Texas

I have taken the point of view that a 15 minutes talk is not the forum for
presenting lists of what we need from the community of physical chemists and chemical
physicists, which anyhow is grossly underrepresented here so that pleas for more
data, even specific pleas, are likely to be unheard.

Therefore, I shall just provide a few illustrations of what we need, and what is available, using as my basis two recent studies of CNO abundances of M, MS and S stars (Smith and Lambert 1985a and b) and of carbon stars (Lambert et al. 1986). The goal in these studies was to investigate how surface CNO-abundances change as asymptotic-giant-branch stars evolve and dredge up carbon from the helium-burning shell.

The molecular lines that are used in these studies are the CO V-R lines (for obtaining the carbon abundance in oxygen-rich stars and the oxygen abundance in carbon stars), OH V-R lines (for O-C in oxygen-rich stars), the C_2 Phillips and Ballik-Ramsay lines (for the C-O in carbon stars) and the Red CN system lines (for obtaining the nitrogen abundances). As extra checks we used the NH V-R lines for the nitrogen abundances and the CH V-R lines for C-O in carbon stars.

The basic need for physical data includes dissociation energies for observed molecules, and other molecules that are important in the molecular equilibria, and oscillator strengths - if LTE were not assumed one would also require a vast amount of collision cross-sections.

From this body of required data I select the following topics for discussion: the NH V-R frequencies, the f values for the CN Red system and the C_2 Phillips system, the f values for the OH V-R lines and the dissociation energy of CN.

The NH V-R lines were first detected in stellar spectra. The frequencies were obtained from rotational constants, derived from analyses of electronic systems, principally $A^3\Pi - X^3\Sigma^-$. Now, a few low V-R lines have been measured by tunable diode laser spectroscopy (Bernath and Amano, 1982). Fourier-transform spectra of the A-X system have also been obtained (Bernath, private communication) which has led to more accurate energy levels and reliable predictions of the frequencies of the V-R lines. This is important since the crowded spectra of late-type stars make precise frequencies necessary for a safe identification. Moreover, each transition is a triplet and the spacing is critical to the analysis of the strengths of saturated lines.

Recent experimental determinations of the CN $A^2\Pi$ lifetimes and studies based on solar observations tend to converge, although the reason for the departure of the Duric et al. (1978) results for low v' values is still unknown. However, the POL-CI and the SCF+CI calculations of Cartwright and Hay (1982) and Lavendy et al. (1984), respectively, give significantly longer lifetimes, although the CASSCF calculations of Larson et al. (1983) agree much better with the experiments. The reason for the discrepancy between the calculations needs further investigation.

For the C_2 A'Π state the experimental lifetimes are significantly longer that the calculated ones. The optimum stellar lines come from the $\Delta v=-2$ sequence and experiments for determining the relevant branching ratios would be of great help (cf. Lambert et al. 1986).

The OH V-R lines is a real success story, involving ab initio calculations, experimental measurements and the use of the Sun as a laboratory furnace. Grevesse et al. (1984) and Sauval et al. (1984) have shown that very consistent results with other oxygen abundance indicators are obtained with available physical data for the V-R and the pure rotational lines, provided that a realistic solar model is used.

The dissociation energy of CN is still unknown - recent determinations range from 7.5 to 8.0 eV. At determinations of nitrogen abundances in cool carbon stars an uncertainty of this magnitude is fatal since it leads to an uncertainty in the CN abundance of about 0.5 dex or more. Moreover, the N abundance as derived from the CN lines is proportional to the square of the CN abundance, due to the fact that most of the nitrogen is bound in N_2 molecules. Therefore, the uncertainty in $D^o_0(CN)$ leads to an uncertainty of one order of magnitude in the nitrogen abundance.

In concluding, I would like to stress the importance of a continuous flow of information between stellar spectroscopists and physicists/chemists. We - the stellar spectroscopists - must make a strong effort to understand the methods of experimental molecular physics and theoretical quantum chemistry. How do we achieve this? We need a continuous education, through books, short courses, and meetings like the present one.

References

Bernath, P.S., and Amano, T.: 1982, J. Mol. Spectrosc. 95, 359.
Cartwright, D.C., and Hay, P.J.: 1982, Astrophys. J. 257, 383.
Duric, N., Erman, P., and Larsson, M.: 1978, Phys. Scripta 18, 39.
Grevesse, N., Sauval, A.J., and van Dishoeck, E.F.: 1984, Astron. Astrophys. 141, 10.
Lambert, D.L., Gustafsson, B., Eriksson, K., and Hinkle, K.H.: 1986, Astrophys. J. Suppl. Ser., in press.
Larsson, M,. Siegbahn, P.E.M., and Ågren, H.: 1983, Astrophys. J. 272, 369.
Lavendy, H., Gandara, G., Robbie, J.M.: 1984, J. Mol. Spectrosc. 106, 395.
Sauval, A.J., Grevesse, N., Brault, J.W., Stokes, G.M. and Zander, R.: 1984, Astrophys. J. 282, 330.
Smith, V.V., and Lambert, D.L.: 1985a, Astrophys. J. 294, 326.
Smith, V.V., and Lambert, D.L.: 1985b, in preparation.

4) MEETING: November 27th, 9.00

Non-Classical Problems, Non-Classical Atmospheres
                                    Chairman: K. Kodaira

Topics were selected from an observational point of view to complement the theoretical sessions.

1. D. Gray presented a review on the progress in measuring macroturbulence in stellar atmospheres and its systematic dependence on the spectral type and the luminosity class of stars. He interpreted the systematic trend in terms of the convective activity in the outer envelope. Furthermore he reported on observations of curving of the bisector of absorption line profiles and of its variability, which required a new theoretical interpretation. Some basic questions were asked about the Fourier method used for the derivations of macroturbulence fields.

2. U. Heber discussed the results of NLTE fine analyses of spectra of 25 helium-poor hot subdwarfs. Their effective temperatures range from 25000 to 40000K. Helium is strongly underabundant, in some cases by more than a factor of 100. This is accompanied by large deficiencies of carbon and silicon in some stars. These peculiarities were explained by gravitational settling. As regards the evolutionary status of these stars, he concluded that they are extended horizontal branch stars with a very low envelope mass (less than 0.01 solar masses) and behave like helium main sequence stars. According to his view, the blue horizontal-branch stars observed in the globular cluster NGC6752 are the natural population-II counterpart to the field subdwarfs. During the discussion a question was raised about the different behavior of nitrogen relative to carbon. Heber answered that an extended low-rate mass-loss may play an important role in this concern.

3. D.M. Gibson was invited to discuss "Radio Aspects of Stellar Atmospheres". He pointed out the importance of radio data for determining macro-physical parameters such as mass-loss rates, structure of stellar coronae, and stellar activities. His HR diagram, showing the distribution of radio-loud stars, called a debate how far this is affected by selection effects.

4. K. Koyama demonstrated how we can understand the X-ray bursts under the assumption that the peak luminosity reaches the Eddington limit in a consistent manner with the current picture of neutron stars. The decay phase of the X-ray bursts can well be interpreted with a scattering atmosphere dominated by Compton processes. The high red-shift of the Fe absorption line detected by the Temma satellite was accounted for by the high surface gravity of a neutron star. This self-consistent picture led to a claim that the distance to the galactic center should be 6-7kpc. This last point, however, was questioned by some attendants.

5)   MEETING: November 27th, 11.45 - 12.30

Miscellaneous scientific papers

Chairman: B. Gustafsson

1. R.K. Prinja reported on a study on narrow absorption components and variability in UV P Cygni profiles of early-type stars (together with I.D. Howarth). An extensive set of velocity and column density measurements of both narrow absorption components and 'underlying' P Cygni profiles have been obtained for a sample of 21 main sequence, giant, and supergiant stars with spectral types B1 to O4. To study variability characteristics every unsaturated resonance line doublet profile was modelled in each of 322 uniformly extracted high resolution IUE spectra. Mass-loss rates are given for 19 stars. Variability in $\dot{M}$ is usually at the 10% level, on timescales of a day or longer; changes of a factor of 2 or greater are not observed. Narrow components were observed at least some of the time in every star with unsaturated P Cygni profiles. In many cases Copernicus observations taken a decade earlier show features at the same velocities. Multiple components (up to 3) are not uncommon. Central velocities and velocity dispersions average 0.8 and 0.06 of the terminal velocity respectively, while column densities are typically ~ 14.5 dex cm^{-2}, and show substantial (>2X) variability in most stars. This variability appears to be unrelated to changes in the underlying profiles. A search for correlations between different aspects of our data reveal few systematic trends, but there is a suggestion that rapid rotators may be exceptional in some respects.

2. Underhill reported on the ongoing analysis of Wolf-Rayet spectra which she and A.K. Bhatia (Goddard Space Flight Center) are carrying out. The results indicate that Wolf-Rayet stars are like B1 to O9 stars except that they have conspicuous, hot mantles. In WC stars the electron temperature in the mantle is between 5×10^4 and 10^5 K; in WN stars it is greater than 10^5 K. The composition of Wolf-Rayet atmospheres is normal (solar), just as for OB stars.

3. R. Wehrse summarized recent work on white dwarf atmospheres, as a complement to his contribution in the Com. 36 Reports on Astronomy, IAU Trans. XIXA.

The meeting ended in a general discussion of the need, and the adequate form for the Reports on Astronomy.

COMMISSION 37: STAR CLUSTERS AND ASSOCIATIONS
(AMAS STELLAIRES ET ASSOCIATIONS)

Report of Meetings, 20, 21 and 25 November, 1985

PRESIDENT: K.C. Freeman   SECRETARY: C. Pilachowski

Commission 37 participated in Joint Discussion VI: Evolution in Young Populations in Galaxies (Chairman: G. Lynga). The report will be found in Highlights of Astronomy. In addition to the Commission sessions reported below, Commission 37 also participated in the joint session HIPPARCOS, with Commissions 7, 8, 24, 25 and 33.

20 November, 1985

SCIENTIFIC SESSION
    This session was on Cluster Dynamics and Formation (Chair: R. Mathieu). The following papers were presented.

    J. Ostriker: "Fokker-Planck Studies of Core Collapse and the Effects of Binaries".

    D. Heggie: "Recent Progress in the Study of Core Evolution".

    D. Sugimoto: "Gravothermal Oscillations in N-body Systems".

    E. Bettwieser: "Observational Consequences of Gravothermal Oscillations".

    M. Mayor: "Radial Velocity Studies of ω Centauri".

    J. Grindlay: "X-ray Binary Clues to Globular Cluster Evolution".

21 November, 1985

BUSINESS SESSION
    The President summarised the changes in membership of the Commission. Two members have resigned, three have died, and 35 new members have been proposed. Following a postal ballot for one new member, the membership of the Organising Committee for 1986-1988 was accepted:

    D.C. Heggie   President
    G.L.H. Harris   Vice-President
    J. Hesser
    P. Nissen
    C. Pilachowski
    G. Salukvadze
    K. Freeman   Past President

    In the period 1982-1985, Commission 37 supported ten proposals for IAU Symposia and Colloquia.

SCIENTIFIC SESSION
    This was a poster session, followed by discussion of the posters (Chair: C. Pilachowski). The following poster papers were shown.

J.C. Mermilliod: Present State of Data Compilation for Stars in Open Clusters (Fall, 85).

P. Bottinelli, R. Capuzzo Dolcetta: "Synthetic Evolutionary Properties of Magellanic Cloud Globular Clusters". Synthetic models were computed in order to study the evolution with time of integrated luminosities and colors for stellar systems having total masses and chemical compositions similar to those characteristic of the Magellanic Cloud globular clusters.

A. Aiad: "Proper Motion of the Open Star Cluster NGC 2301". Proper motions of 190 stars in the field of NGC 2301 were determined from photographic plates taken with the Alger astrograph in the years 1898 and 1973.

A. Aiad: "Proper Motion of the Open Star Cluster NGC 2324". Proper motions of 98 stars in the field of NGC 2324 were determined from photographic plates taken with the Alger astrograph in the years 1905 and 1973.

J. Andersen, A. Blecha, J. Storm, M. Walker: "High Resolution Deep Photometry of Star Clusters in the Magellanic Clouds". The main observational difficulties in obtaining deep color-magnitude diagrams of star clusters in the Magellanic Clouds are crowding and field star contamination, The 9-cm McMullan electronographic camera provides the high spatial resolution and large field to overcome these difficulties. (MNRAS 211, 695; Astron. Astrophys. 150, L12; paper in press).

R.G. Noble, R.D. Cannon, W.K. Griffiths: "A New CM Diagram for Faint Stars in Omega Centauri". CCD photometry has been obtained for several hundred faint stars in ω Cen. Despite the much higher precision of these data, the upper main sequence is only slightly better defined than in previous photographic work. This indicates that the exveptional metallicity spread seen among the red giants also exists on the main sequence, and is probably primordial. The relatively faint location of the main sequence turnoff found previously is also confirmed; the horizontal branch appears to be anomalously bright.

R. Sagar, R.D. Cannon, M. Hawkins: "NGC 5824 and the Mass of the Galaxy". We have obtained a new CMD for NGC 5824 using CCD data. The cluster has a strong BHB starting at V = 18.5 and extending to out limit at V = 20.5; it is thus more distant by about 0.6 mag than was previously believed. The color of the giant branch indicates [Fe/H] = -1.7. NGC 5824 lies on the line corresponding to a galactic mass of 2.3×10^{11} $M_\odot$ in Lynden-Bell's velocity-distance diagram, and thus does not require the existence of an unseen massive halo to remain bound to the Galaxy.

R.D. Cannon, R. Sagar: "New Electronographic CM Diagrams for NGC 1851 and the Fornax Dwarf". BV data to V = 21 have enabled us to locate the main sequence turnoff in NGC 1851 and the horizontal branch in Fornax. The turnoff occurs 3.4 magnitudes below the HB in NGC 1851, which therefore has an age of 15-18 billion years. The Fornax dwarf spheroidal galaxy has a predominantly red HB and a distance modulus of about 20.7. The exceptional width of the Fornax giant branch is confirmed (rms spread in B-V is 0.2).

Jean Brodie, David Hanes: "Metallicity Determinations for Globular Clusters through Spectrophotometry of their Integrated Light". Using an appropriately weighted combination of 16 indices of absorption line strengths measured in low dispersion spectra of the integrated light of globular clusters, we determine metallicities for 36 clusters in the Galaxy. We confirm that Zinn's 1980 scale suffers a systematic error in the region of intermediate metallicity and show that our estimates are insensitive to HB morphology. We

STAR CLUSTERS AND ASSOCIATIONS 277

apply a similar method with modified calibration to determine metallicities for the nuclei of six galaxies.

Jean Brodie, David Hanes: "Metal Abundances in the M87 Globular Cluster System". We have derived metallicities from low dispersion spectra of six globular clusters associated with M87. Their mean metallicity is [Fe/H] = -0.5, considerably more metal rich than the average (-1.2) for Milky Way clusters. A significant fraction of M87 clusters appear to have higher metallicity than any known in the Galaxy, although the metal poorest cluster in our sample has metallicity comparable to the low abundance Galactic globulars. We see no correlation of cluster metallicity with galactocentric distance, but there is some evidence that less luminous clusters are metal poorer, which may reflect some form of self enrichment in the most massive clusters.

R.D. Mathieu: "The Dynamics of Open Star Clusters".

S. Wramdemark, G. Lynga, L. Johansson: "CO Observations in the Eastern Part of W5". We have studied the small nebula S201 and the bright rim IC 1848A using the Onsala 20-m telescope. Results are interpreted as different centers of star formation, also related to known infrared sources.

Zh. Anosova, V. Orlov: "Dynamical Evolution of Triple Systems". This is a review of investigations on the dynamical evolution of triple systems by numerical experiments. The full text is in Transactions of the Astronomical Observatory of Leningrad State University, vol XL, 1985, pp 66-144.

R.D. Mathieu: "The Stellar Kinematics of Star-Forming Regions".

SCIENTIFIC SESSION
This joint session with Commissions 25, 35 and 45 was on Precision Photometry of Clusters (Chair: H. Richer). The following papers were presented.

D. Vandenbergh: "A Theoretical Overview: What Can be Learned from Precision Cluster Photometry".

P. Stetson: "Techniques of Data Reduction in Crowded Fields".

K. Janes: "Precision Photometry in Open Clusters".

R. Buonanno: "Precision Photometry in Globular Clusters I".

A. Penny: "Precision Photometry in Globular Clusters II".

B. Carney: "The Main Sequence of the Draco System".

H. Richer: "White Dwarfs in M71".

25 November, 1985

SCIENTIFIC SESSION
This session was on Star Clusters and Space Telescope (Chair: R. Cannon). The following papers were presented.

F. Fusi-Pecci and A. Renzini: "Star Cluster Research with the HST: Programs and Simulations.

K. Freeman: "Globular Cluster Kinematics (Internal and Systemic)".

H. Richer: "Ground-based Searches for White Dwarfs in Clusters".

R. Sagar: "Use of ST to Study Local and Extragalactic Clusters".

M. Bessell: "Spectroscopic Programs".

A. Penny: "Some Considerations for Large-scale Programs".

M. Kontizas and E. Kontizas: "Magellanic Cloud Clusters".

H. Zinnecker: "The M87 Cluster System".

COMMISSION 38: EXCHANGE OF ASTRONOMERS - (ECHANGE DES ASTRONOMES)

Committee of the Executive Committee

Report of Meeting on November 21, 1985

PRESIDENT: F.B. WOOD   SECRETARY: F.G. SMITH

Officers and Organizing Committee

The Commission agreed that the Vice-President, Prof. E.A. Müller should become President for the next term of office, and proposed Prof. F.G. Smith as new Vice-President. It was also agreed that retiring Presidents would normally serve ex officio for three years on the Organizing Committee, and other members of the OC would normally serve for six years. Consequently, the following members retire now from the OC: J. Delhaye, D.A. MacRae, P.M. Routly, all former Presidents of the Commission. Furthermore K. Sahade, being proposed for President of the Union, retires for the time being from the OC of the Commission. To replace them the following persons were proposed for membership in the OC: A. Florsch (France), Y. Kozai (Japan), K. Ch. Leung (USA), and G. Swarup (India).

Some changes in the regular Commission membership were also proposed.

Consequently, the following slate of officers and members for 1985-1988 was proposed by the Commission for ratification at the second session of the XIX General Assembly:

President: E.A. Müller   Rennweg 15
                                      CH-4052 Basel
                                      Switzerland

Vice-President: F.G. Smith   NRAL
                                      Jodrell Bank
                                      Macclesfield
                                      Cheshire SK11 9DL
                                      U.K.

Organizing Committee: A.A. Boyarchuk, A. Florsch, Y. Kozai, K. Ch. Leung,
                          G. Swarup, C.R. Tolbert, H.H. Voigt, F.B. Wood (ex
                          officio), Ye Shu-Hua.

Members: A.A. Al-Sabti, J. Delhaye, G. Godoli, H. Haupt,
                          E. van den Heuvel, D. MacRae, M. Marik, S. Nin-kovic,
                          Il-Seong Nha, S. Okoye, A. Opolski, A. Reiz, P.M. Routly,
                          G. Ruben, J. Sahade, E.V.P. Smith, J.P. Wild.

Guidelines for the Travel Grant Scheme

The guidelines were discussed and it was suggested that more emphasis should be placed on the need to use the lowest possible air fares. Thus, in item (5) of the guidelines the first sentence was changed to read as follows:

"The amount of the grant will be governed by the cost of a single
return economy air fare, and limited to the least expensive fare
such as PEX, APEX, etc.) between the home and host institutions
and normally is to be used by the applicant for such travel."

No other changes in the guidelines were proposed. The complete text of the guidelines and the application procedure will be published in the IAU information Bulletin No. 55, 1986.

Joint Meeting of Commissions 5, 38 and 46

The President mention d that Commission 5 wished to have a joint meeting with Commissions 38 and 46 on "Documentation for Developing Countries", the meeting to take place on November 25, 1985. Commission 38 will attend as an observer only, since it deals exclusively with travel grants and these are not restricted to developing countries only. Commission 5 will report on the joint meeting.

COMMISSION 40: RADIO ASTRONOMY (RADIO ASTRONOMIE)

Report of Meetings, 20, 21, 22, 23, 25, 26, 27 November 1985

PRESIDENT: K. Kellermann.   SECRETARY: M. Gordon.

Business Meetings

I. NEW OFFICERS

John Baldwin (UK) was elected President and Peter Mezger (FRG) Vice-President for the period 1986-1988. Continuing members of the Organizing Committee are: G. Setti (Italy), R. Strom (Netherlands), S. Wang (China), T. Wilson (FRG), and K. Kellermann (USA). The following were selected as new members of the SOC: A. Barrett (USA), D. Jauncey (Australia), V. Kapahi (India), A. Baudry (France), N. Kaifu (Japan), L. Matveyenko (USSR), R. Booth (Sweden), G. Nicolson (South Africa), and E. Seaquist (Canada).

II. NEW MEMBERS

Approximately 130 new members were approved by the Commission bringing the total membership to about 690. Commission 40 thus continues as the largest Commission in the Union creating significant communication problems. The Commission felt, however, that the membership should not be restricted.

III. IUCAF/CCIR REPRESENTATION

G. Swarup (India) and R. Schilizzi (Netherlands) were unanamously approved to continue as IAU representatives to IUCAF. John Whiteoak (Australia) continues as representative to CCIR while A. R. Thompson (USA) will replace L. Doherty (Canada) who asked to step down. The President thanked all of these individuals for their continued efforts on the behalf of radio astronomy.

IV. IUCAF/CCIR ACTIVITIES

John Findlay, Chairman of IUCAF, reported on recent IUCAF activities. A major concern of IUCAF has been the 1667 MHz down link associated with the VEGA mission. IUCAF has publicized the problem so that radio astronomers may be aware of potential interference. So far no difficulties have been reported, but a potential problem exists with OH observations of Halley's Comet at the time of the VEGA encounter.

There is also increasing pressure for more geostationary satellites operating in the 1300-1700 MHz band where high technology receivers are not needed. Spurious radiation may be a problem and radio astronomy may effectively lose a ten degree band centered on the geostationary arc to these devices. Also, mobile services may be growing in this band.

Findlay discussed the future of IUCAF in light of the work being carried out by CCIR. There was general agreement that the continued activity of a non-government body such as IUCAF is important and that IUCAF should remain active.

Vern Pankonin reviewed CCIR activities as neither of the two IAU representatives to CCIR attended the General Assembly. More and more satellites are using spread spectrum and frequency flexible techniques which blanket L-band from 1400 to 1700 MHz. CCIR is recommending extra protection from out of band emissions from these satellites. The 1982 IAU list of important frequencies has also been inserted into the CCIR system.

The concerns of radio astronomy appear to be adequately represented in CCIR activities.

V. IAU REPRESENTATIVE TO URSI
John Baldwin was appointed as the IAU representative to URSI.

VI. RESOLUTIONS
Resolutions concerning the growing threat to radio astronomy from spacecraft emissions in the 1.4 to 1.7 GHz band, the importance of CCIR's Study Group II Report, and the need to coordinate planned spacecraft transmissions through IUCAF, were introduced by K. Kellermann on behalf of Commission 40, and by John Findlay on behalf of IUCAF. After discussion and redrafting by an ad hoc committee of Findlay, Pankonin, Schilizzi, and Smith, these resolutions were unanimously passed by the Commission, and later approved by the General Assembly as Resolutions B3 and B4.

R. Schilizzi introduced a resolution concerning the coordinating of space and ground VLBI activities throughout the world. The resolution was unanamously approved by the Commission and endorsed by the General Assembly as Resolution B5.

VII. SYMPOSIA
Commission 40 has participated in the planning of six symposia (No's 107, 109, 110, 112, 115, and 119) during the period 1983-1985. The Commission agreed to support the planned symposia on Activity in Galaxies (No. 121), Observational Cosmology (No. 124), Neutron Stars (No. 125), VLBI (No. 129), and the Large Scale Structure of the Universe (No. 130).

The proposals put forth by Richard Wielebinski to hold a symposium on Magnetic Fields in Astrophysical Plasmas in 1987 in Germany, and by Frank Drake for a SETI symposium in Hungary in 1987, were enthusiastically endorsed by the Commission.

VIII. REPORT OF WORKING GROUPS

(a) Nomenclature

Burke (Chairman), Baldwin, Dickel, Felli, Heinz, Kapahi, Johnston, and Strom.

There was considerable discussion of the use of coordinate names together with J2000 coordinates. The Commission endorsed a resolution jointly sponsored by Commissions 5, 34, 40, and 48 recommending appropriate nomenclature.

(b) Protection of Spectral Lines

Robinson (Chairman), Swarup, Baudry, Doherty, Morimoto, Pankonin, Slish, Turner, Wilson, Kaufmann, Webster, Tiuri, Schilizzi, and Hjalmarson.

In the absence of the Chairman, G. Swarup reported on the activities of the Working Group and reported on a poll made by the Chairman. The following lines were found to be used relatively infrequently and poorly supported with receivers: Formaldehyde (14.488 GHz), DCO^+ (72.039 GHz), water (183.310 GHz), and CS (195.962 GHz). Usage of these lines should be monitored between now and 1988 to judge whether they should continue to be classified as "astrophysically important." All other lines are observed frequently and their protection should be maintained. No changes were recommended to the 0-275 GHz list which will keep the IAU list consistent with the CCIR SG2 list confirmed at the September 1985 SG2

"Final" meeting in Geneva. The possible addition of C_3H_2 at 18.343 GHz was deferred to 1988.

The President thanked the Chairman, B. Robinson, for his work on behalf of the radio astronomy community, and the Commission unanamously endorsed the proposal that Robinson continue as WG Chairman until 1988.

(c) VLBI

Johnston, (Chairman), Ananthakrishnan, Backer, Bajaja, Booth, Legg, Cohen, Grueff, Manchester, Kaufman, Kellermann, Kus, Matveyenko, Morimoto, Nicolson, Pauliny-Toth, Schilizzi, Wilkinson, Yeh, Aobenz, and Biraud.

The VLBI Working Group is compiling a list of observatories active in VLBI and the available instrumentation at each facility. The WG aims to improve the flow of information between the observatories and scientists needing these facilities, to help in the coordination of observations requiring the simultaneous use of facilities throughout the world, and the planning of new space and ground facilities. Individual scientists are encouraged to contact a member of the WG for more information.

IX. MISCELLANEOUS BUSINESS

The future of Commission 40 and the preparation of the Triennial Report was vigorously discussed. Radio astronomy is impacting essentially all areas of astronomy and it is no longer feasible for the Commission to be involved in all activities of the Union concerning radio astronomy. The Commission felt that the Triennial Report forms a valuable service and is part of the historical record of radio astronomy. It should continue in roughly its present form.

The President thanked the members and particularly the Scientific Organizing Committee for their help and support during the past three years. The Commission joined in thanking the local hosts in India for their splendid hospitality and efficient organizing of the activities in New Delhi.

Scientific Sessions

I. COMMISSION MEETINGS

Instrumentation, November 20, 1985   Chairman: M. Gordon

3-Station Solar Wind Observatory in India	Hari Om Vats
Solar 3 arc min Fan Beam Array at 35 MHz	Ch.V. Sastry
RRI 1.5-meter Millimeter Telescope	V. Radhakrishnan
New Technology 15-meter Dishes	P. de Jonge
Millimeter Receivers at IRAM	D. Emerson
Instrumentation on the NRAO 12-meter Millimeter Telescope	M. Gordon
Shanghai-Kashima VLBI Experiment	T. S. Wan

Solar Systems and Galactic Research, November 26, 1985   Chairman: N. Kaifu

Observations of Star Forming Regions	N. Kaifu
Recent Developments at DRAO A New Galactic Plane Survey	L. Higgs
Reports from MPIfR	R. Wielebinski
Interstellar Scattering at Low Galactic Latitudes	R. Rao
Low Frequency Pulsar Observations	A. Deshpande
Southern Hemisphere Recombination Line Survey	M. J. Gaylard
MM and IR Observations of Bipolar Outflow Source G35.2N	G. H. McDonald
Reports from ARO	P. Feldman
Radio Bursts at cm-wavelength	V. N. Ikhsanova

Properties of Lunar Surface Obtained from Observations N. S. Soboleva and
  with RATAN 600   M. N. Nougolna
Observations of the Moon at 1.2 and 3 mm with the 30-m   D. T. Emerson
  Telescope

Extragalactic Research, November 26, 1985   Chairman: R. Strom

Report from the MPIfR   R. Wielebinski
Edge-on Galaxies   S. Sukumar
The Giant Radio Galaxy MSH 05-22   C. V. Subrahmanya
New Results from the VLA: 3C 75, Cygnus A, M87   F. Owen
Entrainment and Evolution of Radio Jets   D. De Young
327 MHz WSRT Observations of the Perseus Cluster   A. G. de Bruyn
The Cambridge-VLA Rotation Measure Survey   J. P. Leahy
The MG Survey   B. F. Burke
1144-379: A Rapidly Variable BL Lac Source   D. Bramwell
VLA Observations of Rapid Variability in OJ 287   D. Roberts
Report on VLBI at the MPIfR   E. Preuss

Reports from Observatories, November 26, 1985   Chairman: S. G. Wang

New Developments at the VLA   R. Sramek
Report from the Hat Creek Observatory   J. Welch
Report from Nancay   I. Kazes
Report from China   S. G. Wang
Report from Hartebeesthoek   M. J. Gaylard
Report from Nobeyama   N. Kaifu
Report from Ooty   M. Joshi
Report from Bologna   G. Tofani
Report from OVRO   M. Cohen
Report from Westerbork   A. G. de Bruyn
Report from the Soviet VLB Network   L. Matveyenko

New Radio Telescopes, November 22, 1985

   This all day meeting is summarized in "Highlights of Astronomy." The
meeting consisted of three sessions.

       Recently Completed Radio Telescopes   Chairman: K. Kellermann
       Telescopes Now Under Construction   Chairman: V. Radhakrishnan
       Radio Telescopes of the Future   Chairman: R. Wielebinski

II. JOINT MEETINGS
       Galaxy Radical Velocities (28,30,40)   November 20
       Twenty-five Years of Radio SETI (40,52) November 23

III. JOINT DISCUSSIONS
       Reference Frames   November 20
       Radio Astronomy and Cosmology   November 21
       Stellar Activity   November 25
       Supernovae   November 27

COMMISSION 41: HISTORY OF ASTRONOMY (HISTOIRE DE L'ASTRONOMIE)

Report of Meetings, 20 and 21 November 1985

PRESIDENT: O. Pedersen   SECRETARY: J.A. Eddy

I. BUSINESS SECTION

The members paid a silent tribute to the memory of colleagues who had died since the last General Assembly, viz. Jean Filliozat, Eric Forbes, François Link, Derek J. de Solla Price, and Edward Rosen, whose names were read by the President.

The following were elected members of the Organising Committee for the period 1985-1988:

Eddy, Dr. John A. (Boulder): President
North, Prof. John D. (Groningen): Vice President
Débarbat, Dr. Suzanne (Paris)
Eelsalu, Prof. Heino (Tartu)
Xi Zezong, Prof. (Beijing)
Pedersen, Prof. Olaf (Aarhus): Past President

Commission 41 has now 102 ordinary members who are listed elsewhere in this volume. During the meeting the following 44 historians of astronomy were elected consultant members for 1985-1988: A. Aaboe, W.B. Ashworth, A. Aveni, J.A. Bennett, V. Bialas, R. Billard, P. Bulgakov, L.E. Doggett, A.I. Eremeeva, B.R. Goldstein, A.A. Gorstein, P. Galluzzi, N.S. Hetherington, E.S. Kennedy, D. King, J.A. Klimishin, H.G. Körber, H. Labat, J.E. Lankford, N.B. Lavrova, F.R. Maddison, Y. Maeyama, R. Mercier, N.I. Nevskaja, A. Rodrigues, B.A. Rosenfeld, G. Rosinska, G. Saliba, J. Samsó, A. Segonds, A.R. Serban, R.W. Smith, Z. Sokolovskaya, G.J. Toomer, G.L'E. Turner, B.L. van der Waerden, A. Van Helden, J. Vernet, D.W. Waters, S.R. Weart, S. Weng, R.S. Westman, D.T. Whiteside, M. Yano.

The following resolution was proposed by the Swedish National Committee of the IAU:

> Commission 41 notes with great satisfaction that an action has been initiated in Sweden and Denmark, aiming at an improvement of the state of the remains of Tycho Brahe's observatories on the island of Hven. This site is of unique importance in the history of astronomy: here modern practical astronomy was born, and the long series of observations made here by Tycho and his assistants provided the material from which the laws of planetary motions were derived by Kepler and Newton. The Commission urges the relevant authorities to make every possible effort to ascertain that the ruins are being preserved and the site as a whole kept in a state worthy of its glorious past.

It was unanimously adopted and passed on to the final session of the General Assembly where it was carried by acclamation.

Reports on recent activity in the history of astronomy in Canada and France had been received from Prof. Kennedy and Dr. Débarbat respectively.

At the meeting Dr. Débarbat reported on a proposal for an international symposium to be held in Paris in 1887 to commemorate the centenary of the Carte du Ciel. Both Commission 41 and the International Union of the History and Philosophy of Science will be involved in the planning of this meeting together with the Observatoire de Paris.

The President drew attention to the many important publications which have appeared during 1982-85 as a testimony to the increasing interest in the history of astronomy on a scholarly basis. Only a few major works can be mentioned here, such as Vol. 4 A of the General History of Astronomy, edited by O. Gingerich (Cambridge 1985), the new English translation of the Almagest by G.J. Toomer (London 1984), and N. Swerdlow and O. Neugebauer: Mathematical Astronomy in Copernicus's De Revolutionibus 1-2 (New York 1984), as well as two volumes of collected articles, Astronomy and History by O. Neugebauer (New York 1983) and Studies in the Islamic Exact Sciences by E.S. Kennedy (Beirut 1983). A Source Book of Indian Astronomy, edited by B.V. Subbarayappa and K.V. Sarma (Bombay 1985) had been published on the occasion of the General Assembly.

II. SCIENTIFIC SESSIONS

The meetings of the Commission comprised three scientific sessions which were attended by about 20 members and a similar number of other astronomers or historians of astronomy. Since several members had chosen to present their recent work to the IAU Colloquium 91 on "The History of Oriental Astronomy" which was held in New Delhi on 13-16 November just before the General Assembly, the number of papers read to the Commission was not as large as usual. Nevertheless, they covered a great variety of subject matters. A major paper on "Modern Astronomy in Medieval India - Persian Sources" by S.M.R. Ansari described the penetration of "Western" astronomy into India in the 18th century and later; in the absence of the author it was read by one of his colleagues and is already available as a separate publication. P. Kunitzsch spoke of "Star Tables in Medieval Astronomy" and R. Mercier about "Mean Motions in the Toledan Tables". In both these papers the interconnections between the astronomy of different cultural areas were exposed by means of new source material. More recent topics were dealt with by George Wilkins who painted a vivid portrait of "Simon Newcomb - born 1835", and by K. Lang who spoke of the "Highlights of Solar System Astronomy in the 20th Century". Among a number of brief communications was a paper by Maria Firneis on "Newly Discovered Instruments of the Viennese School" and a survey of "Astronomy at Strassbourg" by K. Fischer, followed by a presentation of "Recent Evidence Pertaining to Compasses Employed in 19th-Century Surveying in Canada" by J.E. Kennedy and an artistic exposition of "Sundrawing Ancient and Modern" by J. Saad-Cook.

COMMISSION 42:   CLOSE BINARY STARS (ETOILES DOUBLES SERREES)

Report of Meetings, 20 and 26 November 1985

PRESIDENT: A.H. Batten   SECRETARY: A. Sanyal

   The business meetings of the Commission were held in Room D, Vigyan Bhavan, from 11.00-12.30 on November 20 and from 9.00-10.30 on November 26.

   The President welcomed all members of the Commission, and others present, and especially three former presidents. He announced that he had appointed Dr. A. Sanyal to act as secretary during the Commission's meetings. He pointed out that a provisional agenda had been mailed to all members, and suggested some additional items. There being no objections, the revised agenda was adopted.

   The President asked members to stand in silence while the names of eleven members who had died in the previous three years were read. He drew attention in particular to M.K.V. Bappu, former President of the Union and D.J.K. O'Connell S.J., former President of the Commission.

   A letter from J.A. Mattei was read expressing thanks for the Commission's support of the publication of the AAVSO's archival observations of cataclysmic variables. Funds (including an IAU grant) had been found, and extracts from the first monograph of the series (on SS Cyg) were available for inspection at the meeting.

   The President announced that, after considerable discussion, the Organizing Committee had recommended to the IAU Executive that the new President should be J. Smak and the new Vice-President R.H. Koch. These recommendations were endorsed by the Commission. The composition of the new Organizing Committee was discussed and the following was agreed to: K.D. Abhyankar, J. Andersen, E. Budding, A.M. Cherepashchuk, D.M. Gibson, M. Kitamura, Y. Kondo, K.-C. Leung, J. Rahe, M. Rodonó and G. Shaviv, with A.H. Batten as past-president, remaining for one more term. Nearly sixty people had indicated an interest in becoming members of the Commission. This large number was partly caused by the revision of the IAU files, which revealed several persons with obvious claims to membership whose formal election had been overlooked. The President suggested that detailed consideration of these names should be delegated the Organizing Committee which would meet in Delhi. This course was adopted and, at the second session, forty-one names were recommended to the Commission, which endorsed them.

   The President noted some comments he had received on the Report published in IAU Transactions XIXA (pp. 583-606). D.M. Popper requested that the first sentence of the paragraph OB Binaries (p. 597) be replaced by the following:

> Popper (ApJ **262**, p. 641) pointed out that the luminosity ratios in O-type double-lined binaries, all probably contact systems, are much closer to unity than expected from their mass-ratios. These systems, superficially at least, are high-mass counterparts of the W-UMa systems.

J. Sahade points out that the lines and transitions named in the paragraph on β Lyrae (p. 597) are not forbidden, but intercombination. P. Harmanec draws attention to his work on Be stars (p. 597) in BAC **34**, 324, **35**, 164, 193; Hvar Obs. Bull. **7** (1), 55. S. Rucinski suggests references to the X-ray survey of

W UMa stars (Cruddace and Dupree ApJ 277, 263) and to work on uv emissions (Vilhu and Rucinski AAp 127, 5; MN 202, 1221) should be added (pp. 600-1). He emphasizes the relevance of work discussed on pp. 603 and 605 to W UMa systems and draws attention to Duerbeck's determination (IAU Coll. 80, 363) of the galactic density and scale height of these systems.

The President also invited discussion of the form of the report, particularly questioning the value of the two lengthy tables (nos. 2 and 5 in the present report). Several members found these tables useful and the consensus was that they should be retained. Some reorganization of the report might be necessary and should be left to the incoming Organizing Committee.

T. Herczeg discussed the form and content of the Bibliography and Program Notes. Members of the Commission felt that it should continue in substantially its present form, and that it does not duplicate the work of CDS Strasbourg. Nevertheless close cooperation with CDS seems very desirable. The existence of alternative designations for many objects creates problems. The Commission felt that extensive cross-indexing is desirable. The Commission approved a resolution seeking a subsidy of $500 from the IAU for the publication of the Bibliography and expressed its appreciation to Prof. Herczeg for his work as Editor.

The President reported on correspondence with Dr. Paul Schmidtke about the on-line data centre that the IAPPP wishes to set up in conjunction with automatic photoelectric telescopes both operating and planned. Several members of the Commission reported that they were impressed by this group, and a motion of support for their efforts was passed unanimously.

There was some discussion of the Commission's proposal for an IAU symposium on "Circumstellar Matter in Close Binary Systems". The meeting had been proposed for Victoria, B.C. in June 1987, (or possibly Europe later the same summer) but had not been accepted in its original form by the Executive. The committee set up to formulate the proposal will continue its work and the Commission will be informed as soon as definite decisions have been made. The Executive proposal to combine Commissions 42 and 26 was also discussed; the majority of those present opposed the plan.

The preparation of catalogues of both photometric and spectroscopic orbital elements was discussed. A motion was passed supporting the preparation by A.H. Batten of an Eighth Catalogue of Orbital Elements of Spectroscopic Binary Systems.

The President reported that he had had some correspondence about the inclusion of a representative list of eclipsing binaries (with current coordinates) in the Astronomical Ephemeris. He invited suggestions from Commission members. M. Rodonó reported on the meeting that considered the setting up of a working group on coordinated multisite observations. He felt that personal contact between interested parties would be more useful than formal Commission involvement, but agreed to keep the Commission informed. K.-C. Leung reported briefly on the meeting just concluded in Beijing on close binaries. A letter from A.J. Wesselink proposing coordinated observations of SZ Cam and AL Vel was read.

At the end of the second business session, short scientific contributions were made by T.J. Herczeg, A. Gimenez, L.P.R. Vaz, F. Vakili and Z. Kviz.

Report of Scientific Session "New Techniques in the Observation and Analysis of Close Binary Systems" held on 23 November, 1985.

PRESIDENT: A.H. Batten   SECRETARY: A. Sanyal

Abstracts of Papers

Observations of Close Binaries in Planetary Nebulae
E. Budding

Study of those central stars of planetary nebulae that happen to be binaries is useful for three main reasons: (i) determination of parameters of the nebulae, (ii) possible detection of certain "exhibitionist" phases in binary evolution, (iii) study of the production and properties of hot subdwarfs. The observations described provide an example of the application of two-dimensional CCD arrays to the study of close binaries and were made with the 1.9-m Mt. Stromlo telescope. Only preliminary results are available from the survey, but the spectrum of the known binary in the centre of NGC 2346 has been studied.

W Ursae Majoris Stars: Observations and Analysis
Albert P. Linnell

High speed UBVRI photometry provides improved discrimination among proposed models for W Ursae Majoris. From among proposed models for W-type light curves, only the Rucinski hot secondary model and a model which combines a warm secondary with a fixed backside spot on the larger component acceptably simulate the observational data. Both models are consistent with the Webbink mass-circulation scenario. They are inconsistent with a barotropic photosphere. The best-fitting UBVRI model does not represent the ANS data by Eaton, Wu, and Rucinski. An unreasonable change in T_{eff} does provide a representation of the ANS data, but that resolution undoubtedly is spurious. If the far-uv characteristics of other W-type systems mimic W UMa, and if the model difficulty persists after introduction of synthetic spectra, the observational indication is that there is an excess in the uv surface brightnesses of the primary components.

The Use of the Cross-Correlation Technique to Obtain Radial Velocities of Contact Binaries
B.J. Hrivnak and E.F. Milone

Radial velocity studies are seriously lacking for contact binaries, and most of the older studies are of poor quality. This is due to the faintness of the systems for time-resolved spectroscopy and the broad-lined, blended nature of the spectra. The cross-correlation technique permits one to overcome this latter difficulty, as was initially demonstrated by McLean (1981). In this paper we present a detailed discussion of the procedure for reducing data obtained with photographic plates or a Reticon detector. Several cross-correlation function profiles are illustrated for light ratios of 1 and 6, and the importance of excluding the broad lines and bands is demonstrated. Examples of the good velocity curves presently being obtained are presented. Tests indicate the internal consistency of the technique. In contrast to some earlier reported discrepancies, good agreement is now found between these new spectroscopic mass-ratios and the photometric mass-ratios for totally eclipsing systems. At least 22 systems have now been studied spectroscopically by this technique, and a promising data base is emerging with which to compare models of the structure and evolution of the contact binaries.

Accretion-Disk Structure in Algols and Cataclysmics from Time-Resolved Spectrophotometry
R.K. Honeycutt, R. Kaitchuk and E. Schlegel

The techniques of data acquisition, data reduction, and data display for time-resolved spectrophotometry are described, with particular attention to the use of image-analysis tools. We give the results of a survey with these techniques of 104 eclipses in 52 short-period Algol systems. It is found that when the relative radius of the primary star is plotted versus mass ratio, the systems with permanent disks, transient disks and without disks segregate into three distinct regions of the plot, indicating that the location of the stream impact is responsible for these three kinds of behavior. We also report evidence that several Algols with transient disk systems have a bifurcation on the leading side of the disk. For the cataclysmics we show characteristic time-resolved spectrophotometry of nova-like and of U Gem-type systems, with emphasis on the S-wave phenomenon. We derive the distribution of stream impact positions based on S-wave phasing, and point out the systematic changes in the S-wave in WZ Sge. Finally it is shown that many of the puzzling kinds of behavior in the time-resolved spectrophotometry of nova-like systems can be understood by assuming that the emission-line profiles arise not only from a rotating disk but also from an accretion-disk wind.

Applications of the Reticon to Observations of Binaries
M. Parthasarathy

Recent work on various binary systems undertaken with Reticon detectors was described. Particular emphasis was laid on the possibility of discovering weak secondary spectra in (e.g.) Algol-type systems. The value of this work for the determination of accurate masses and dimensions was pointed out. Also discussed was the potential for making accurate abundance analyses and its value in the study of binary evolution.

Cross-Correlation Radial Velocities of Early-type Binaries
J. Andersen and G. Hill

Cross-correlation techniques for determining radial-velocity curves of early-type binaries have been explored in a well-studied system. Deep-lined synthetic template spectra and noise filtering are found to be essential.

Voyager Observations of Close Binaries
R.S. Polidan (presented by J. Holberg)

The value of spacecraft for the observation of close binaries - especially in the uv - was illustrated by a discussion of the systems observed with Voyager. Particular reference was made to observations of ε Aur during the recent eclipse.

Doppler-Imaging Techniques
D. Gibson

The recently developed Doppler-imaging technique was described. It has proved particularly useful in the study of RSCVn systems, confirming the spot interpretation of their out-of-eclipse variability. The study of plages on the surface of the components of AR Lac was described in some detail.

COMMISSION 44: ASTRONOMY FROM SPACE (L'ASTRONOMIE A PARTIR DE L'ESPACE)

Report of Meeting

PRESIDENT: M. Oda

VICE PRESIDENT: Y. Kondo

I. BUSINESS MEETING

The following officers and members of the Organizing Committee were elected: President: Y. Kondo; Vice President: K. Pounds; Organizing Committee: A. A. Boyarchuk, G. W. Clark, G. Courtes, M. Grewing, E. B. Jenkins, F. Macchetto, M. Oda, J. Rahe, G. B. Sholomitzcy, Y. Tanaka, J. Truemper, K. A. van der Hucht, A. J. Willis.

The formation of an international "Working Group on Far and Extreme UV Spectroscopy" was proposed, with J. Linskey as chairman.

II. SCIENCE MEETINGS

Commission 44 sponsored eight science sessions which were devoted to "Future Space Programmes"; they were chaired by R. Bonnet, B. Burke, Y. Kondo, M. Oda, J. Rahe, and A. Willis. Brief summaries of the meetings, some of which are received from the speakers, are given in the following. Commission 44 co-sponsored several Joint Sessions which are not included in this report.

1. Topics Related to Space Station (SS)

B. Burke reviewed discussions at the US Space Science Board meetings "Major Directions 1995-2015" which focussed on future strategies of space science in general and of each discipline, including astronomy and astrophysics. Naturally the US plan for the SS has deep impacts on the strategy. Discussions of a possible lunar base and specifically of large scale space interferometers were reviewed.

It was reported that the TFSUSS (so-called Banks Committee) consisting of US scientists with international observers as an advisory group to OSSA of NASA discusses various possible modes of scientific uses of the SS. In the committee's sub-meeting held in October at Napa Valley, California the necessity of establishing an international group of scientists (i.e., a forum, arena, or even an institution) to reflect demands of the scientific community to the SS was emphasized and a report will be issued shortly.

R. D. Chapman commented on SS and discussed four points. First, he described the organization of the SS programme: a level A programme at NASA Headquarters with broad overview of the activity, a level B programme at the Johnson Space Center with responsibility for integrating the entire programme, and level C projects at other NASA centers with responsibilities for parts of the programme. Then he outlined the objectives of the programme, and pointed out that the ability to construct, deploy, service, and maintain scientific payloads will change the way space science will be done in the future. He then described the SS infrastructure which includes the core station, co-orbiting and polar platforms. In effect, free-fliers and other platforms (SPARTAN, Leasecraft) will become part of the infrastructure through servicing and maintenance. Finally, he discussed several issues such as instrument component compatibility, data system transparency and telescience.

R. M. Bonnet reviewed the European views on the SS. Being invited by NASA in 1983 to participate in the SS project, ESA signed in June 1985 a MOU which covers activities during Phase B. In January, the Council of ESA approved all the components of ESA's Long Term Plan. Among this Plan, the Columbus project represents what Europe intends to develop as their participation to the SS. Columbus consists of the Pressurized Module, Service Module, a co-orbiting and a polar Platform, and a Service Vehicle.

The Europeans foresee that the SS will be mostly used for assembling larger structures in space, servicing and repairing of larger facilities. The station could be used also as a transportation mode for planetary or interplanetary missions.

ESA's Space Science Directorate, in its long term plan "Space Science, Horizon 2000" which was reviewed in a separate session, has identified two missions which would be suited to and would take advantage of the SS. These are a high-throughput Heterodyne Spectroscopy Mission (FIRST) to be placed on the co-orbiting platform, and a Cometary Mission which uses the station as a transportation mode and as quarantine facility.

At this stage, there are no concrete plans, but studies are proceeding to better define the SS capabilities.

A more economic approach may be to base the design of these elements on the re-utilization of modules or of devices which have been developed by ESA for the Spacelab programme, such as the EUREKA platform, a shuttle launched and retrievable platform. As soon as the SS is operational, could EUREKA serve as one of the co-orbiting platforms and be used as an element to assemble larger platforms. The first flight of EUREKA is scheduled for 1988; 80% will be devoted to microgravity and 20% to astronomy.

2. Agencies/Report (1)

NASA: C. Pellerin discussed NASA's space plans. The present status of facility-type space observations with advanced technology was reviewed including Hubble ST (launch scheduled in '86), GRO ('88), AXAF ('92?), and SIRTF ('94?). As examples of moderate missions EUVE, COBE, STE, Cosmic Ray Explorer, Gravity Probe with the precision gyro-scope which has been developed in the past 20 years were discussed.

SMM will be taken back around 1988 and repeatedly be used as a spacecraft bus for moderate missions: EUVE may be the first of these and XTE may follow.

As a research base, a concept of SPARTAN has emerged from the rocket program and the Get-Away-Specials aboard the shuttle. Future SPARTAN and SPARTAN from SS are discussed.

FUSE (LYMAN), QUASAT, ST follow-on instruments, life science, how to proceed in infrared astronomy were reviewed as well as shuttle, SS, advanced SS, LDR etc.

USSR: The original plan for this session included an updated progress report of the USSR programmes for the 1980s, similar to the one presented at the Patras General Assembly. While the overall report was not presented, for the France-Soviet gamma ray experiment, SIGMA, a written report was contributed, and the radio interferometer was introduced as RADIOASTRON.

H. M. Tovmassian reported the progress of the GLASAR experiment to search for galaxies and quasars with excess UV-emission. This new survey in the far UV has been planned in Byurakan where the concept of the activity of the nuclei of galaxies was originally developed and where a survey to find galaxies with excess UV-emission had been undertaken in the mid-sixties.

The telescope has been manufactured and will be launched in the next year. With two star trackers the expected area to be photographed is about 9000 square degrees. The Geneva observatory is taking part in the programme.

NASA/EUVE: S. Bowyer presented the progress report on the Extreme UV Explorer which is to provide the first all-sky map of extreme ultraviolet sources in the 80 to 800 A band and to provide spectra of some of the sources.

3. Agencies Report (2)

The Near Term Activities at the ESA: R. M. Bonnet reported on ESA's activities for the period extending from the present time to 1992, including projects already approved and in the construction phase.

EXOSAT, an X-ray observatory operating in the medium and low energy range, which was launched in May 1983 carries Imaging Telescopes operating at low energies and a set of medium energy Gas Scintillation Proportional Counters. A long period of uninterrupted observations allowed by the high eccentric orbit of EXOSAT has made it possible to obtain the best light curves ever obtained on binaries. During its 2.5 years of observation many new objects have been observed. The end of the mission is foreseen for October 1986.

ESA's mission to Comet Halley, GIOTTO, was launched on the second of July 1985 with the Ariane-1 launcher. Its interplanetary trajectory will be corrected a few days prior to encounter in order to achieve a flyby of Halley at a distance of 500 km. GIOTTO carries experiments in particular to study the environment of the nucleus of the comet and to take pictures of it with a resolution of 30 meters. The camera proved already its capability by taking pictures of Vega, Jupiter and of the earth.

ULYSSES, the new name for the International Solar Polar Mission (ISPM), prepared jointly by ESA and NASA, will observe the sun from above the poles and make measurements of the solar wind and of the interplanetary medium at high ecliptic latitude ($>70°$). The satellite is scheduled for launch in May 1986. It has been agreed that ULYSSES will fly over the South Pole first in mid 1989.

ESA is also participating in the Hubble Space Telescope (HST) by providing the solar arrays of the satellite, building one of the focal plane device, the Faint Object Camera (FOC), and participating in the operation at the ST Science Institute in Baltimore.

HIPPARCOS which is to be launched after HST by an Ariane launcher in 1988 is an astrometric mission whose primary objectives are to measure the positions, parallaxes of more than 100,000 stars with an accuracy of 0.002 arcseconds, and the proper motion of these stars with anaccuracy of 0.002 arcseconds per year. In addition astrometric and photometric measurements with a lower accuracy but for a much larger number of stars will be performed. The payload is now being tested.

ESA is now preparing ISO, the Infrared Space Observatory, to be launched at the end of 1992 (which will be presented in more detail in another session).

The SOHO, CLUSTER and other Cornerstones of ESA's Long Term Programme were included in the presentation by J. Bleeker on "ESA's Report Horizon 2000" (ESA-SP-1070, 1984). (see page 299)

Japan's Space Programme/ISAS: M. Oda reviewed Japan's presently ongoing and future programmes which have been approved or are under planning. First, the organizational structure of space development in Japan was reviewed; it was emphasized that science and application are promoted under separate organizations sponsored by different governmental agencies and that scientific programmes are conducted within the academic circle including rocketry and spacecraft engineering. With this policy technical capability and scientific demands are well matched and academic autonomy and flexibility are assured but the scale of the programme is limited. The ISAS as one of the Inter-University Research Institutes plays the role of executing space science programmes, coordinating the activities of the universities and producing the long-term strategies for space science. The strategy recommended to the government in 1975 was that the backbone are small and moderate missions at a constant pace fitting to the scientific demands, with big projects and international programmes scattered at intervals of several years.

Currently operating astronomical missions are TENMA, the X-ray astronomy satellite specialized by the cm^2 Gas Scintillation Proportional Counters, and two Halley's Comet missions, SAKIGAKE and SUISEI. SUISEI had started to take Lyman alpha pictures of Halley's hydrogen coma. The X-ray astronomy satellite ASTRO-C to be launched in February 1987 will carry 5000 cm^2 proportional counters, an all sky monitor and a gamma ray burst monitor. The mission is under collaboration with UK and US scientists.

Apart from missions for space plasmas, a programme for the next solar maximum, HESP (SOLAR-A), is proposed for 1991 launch. An X-ray astronomy satellite, ASTRO-D, with a large throughput X-ray telescope with medium spatial and spectral resolution is considered for proposal for 1992-3.

A space VLBI with highly eccentric orbit with an apogee of 50,000 km is under study for mid 1990's hopefully with coherent relationships to QUASAT.

Japan's approach to the SS was discussed. An MOU was signed for Japan's participation in the Phase B study: Japan is to contribute a pressurized experimental module with an attached platform. Whether the module is an autonomous body for operation and utilization and whether the use of Japanese scientists are limited to this module is uncertain to ISAS and also to NASA: i.e. are international contribution recognized as those to the whole SS system or parts with clean interfaces? Concensus among scientists is that, if the use is limited to the module, few modes of the use are expected for scientific observation except for microgravity experiments: scientists find free-fliers much more useful. The small free-flying platform system somewhat similar to EUREKA appeals to Japan scientists for basic technological studies.

4. Hubble Space Telescope and Related Topics

A status report on the preparations for Hubble Space Telescope operations were presented by E. J. Schreier, as a guide for potential proposers. The status of the science instruments was discussed, and possible changes in their nominal characteristics indicated, based on current testing. Potential limitations during early operations were pointed out, such as observing minor planets with small aperture instruments.

In the area of HST proposals, scientists are reminded that proposal material (forms, guidelines, policies, descriptive material and instructions) were sent out to the community at the end of October; the deadline for proposals for the first cycle is 28 February 1986. If changes in the launch schedule become known in time, attempts will be made to extend the deadline. The recommendations of the working groups in the various disciplines were reviewed. Three possible key projects were recommended: Ho and the distance scale; QSO absorption lines; and a medium-deep survey. Also discussed was the likelihood of adding a remote entry capability for proposers. The STScI is developing a system which may allow users to submit proposals electronically from their home or institutional computers, via a commercial packet-switched network; this would also give proposers an opportunity to validate their proposals (for snytax and internal consistency), using STScI developed software.

Finally, the status of scientific data processing for HST users was reviewed. In particular, one has to deal with the three key elements of a distributed data analysis system: portable software; a state-of-the-art archive; and a communications network linking users and resources. In the first area are mentioned the status and anticipated capabilities of the Science Data Analysis System (SDAS -- the basic applications software designed for HST), the Image Reduction and Analysis Facility (IRAF -- a system being developed by the National Optical Astronomy Observatories with STScI collaboration, which will be used as a command language and environment for SDAS, and which in addition will have its own set of applications software), and the Calibration Data Base System (CDBS -- a set of applications and systems software and a relational data base to handle HST instrument calibration and engineering data.) IRAF has been demonstrated to be portable on a variety of systems (VAX/VMS, VAX/UNIX, SUN/UNIX, MICROVAX, MV10000/AOS-VS). SDAS has been integrated with IRAF but, as the current time, is limited to VMS systems. In the second area, the status of a joint effort by the STScI and the European Coordinating Facility is discussed to develop a prototype optical disk archive; it is currently anticipated that this facility is available by launch. Lastly, it was pointed out that NASA is sponsoring a pilot project which will link several astronomical institutions and will also provide external user access. The network is presently being set up and may be set up early enough to allow remote user access to HST data.

J. M. Mead reported on the data coverage across the spectrum. Approximately 500 machine-readable catalogs of non-solar system objects are now available at both the Astronomical Data Center of the NASA/Goddard Space Flight Center and at the Centre de Donnees Stellaires (CDS) of the Strasbourg Observatory. Using this data base, we have developed several tools which are currently providing identification and analysis support for archival data from the Infrared Astronomy Satellite (IRAS) and the International Ultraviolet Explorer (IUE) satellite:

1) the <u>Catalog of Infrared Observations</u> (CIO) (Gezari, Schmitz and Mead, 1982), which contains data on all objects observed in the infrared since 1965 as obtained from journal articles;
2) the <u>Combined List of Astronomical Sources</u> (CLAS) (Mead and Hill, 1983), which merges 25 catalogs containing potential candidates for optical identifications by IRAS;
3) the <u>Bibliographical Index of Objects Observed by IUE</u> (Mead, Kondo and Boggess, 1983), which lists the objects observed by IUE and reported in the astronomical literature;
4) the <u>Data Inventory of Space-based Celestial Observations</u> (Version 1.0) (DISCO) (Brotzman, Hill and Mead, 1985), which provides a directory of non-solar system objects observed by 16 space experiments.

All of these data tools should be useful not only for planning future space observations, but also for correlative analyses across the spectrum.

5. Missions Under Preparation

<u>ROSAT (German X-ray Observatory)</u>: B. Aschenbach described the ROSAT

mission. ROSAT (Röntgensatellit) is the next X-ray astronomy mission to be launched. The scientific payload consists of two independent instruments: a large X-ray telescope (6 - 100 Å) and a smaller XUV telescope - the Wide Field Camera - (60 - 300 Å) which are looking parallel. The primary scientific objective is to perform the first all-sky X-ray and XUV surveys with imaging telescopes leading to an improvement in sensitivity by several orders of magnitude compared with previous surveys. A large number of new sources is expected to be discovered ($\geq 10^5$ sources in the X-ray band) and located with an accuracy of ≤ 1 arcmin in the X-ray band and ≤ 2 arcmin in the XUV band, respectively.

After completion of the sky survey which will take half a year, the instruments will be used in a pointing mode for detailed investigations of selected sources with respect to spatial structures, spectra and time variability. In this mode ROSAT will be available for guest observers. For the X-ray telescope three different detectors can be commanded into the focal plane: either one of two redundant position sensitive proportional counters (PSPC) providing coarse spatial (30") as well as simultaneous spectral resolution (E/ E 2) or a channel plate detector (HRI) with very good spatial resolution (7") but no spectral information at all. The XUV Wide Field Camera is equipped with two redundant channel plate detectors and thin film absorption filters are used for spectral discrimination. More details about the instrumentation and the scientific objectives can be found in the papers by Trumper, Phys. Scr. T7, 209 (1984) and Pye et al. in 'X-Ray Emission from Active Galactic Nuclei", eds. W. Brinkmann and J. Trumper, p. 261 (1984).

The project as a whole is now well in the hardware phase with the engineering models of the spacecraft, the telescope, the X-ray focal plane instrumentation and the Wide Field Camera being completed. The nominal environmental tests, i.e. vibration of the telescope, static load tests of the spacecraft structure, model survey tests of the complete spacecraft and the first integrated system tests have successfully been run. Quite a success was also obtained in the development areas: the PSPC could be built with an energy resolution of 40% at 1 keV maintaining the required spatial resolution and the X-ray mirror prototype system shows an angular resolution of 4 arcsec half energy width. Flight hardware is now being built and the project is on schedule which encourages us to see ROSAT being launched by the Space Shuttle at the end of 1987.

ASTRO-C: K. Koyama presented a progress report of ASTRO-C which will be launched in February 1987. Main objective is the study of time variability of galactic and extragalactic X-ray sources. The prime instrument is a set of large area proportional counters of total effective area about 5000 cm^2 which are being produced in collaboration with Leicester University and Rutherford-Appleton Laboratory. The spacecraft also carries an all sky monitor and a gamma ray burst monitor which is under collaboration with Los Alamos National Laboratory.

The ASTRO Observatory: The progress of this new shuttle-borne observatory for ultraviolet astronomy was reviewed by A. Davidsen. The ASTRO Observatory consists of three distinct instruments. The Hopkins Ultraviolet Telescope (HUT) is for probing the far and extreme ultraviolet designed and built by members of the Johns Hopkins University. HUT consists of a 90 cm f/2.0 mirror. The Ultraviolet Imaging Telescope developed at NASA's GSFC is for the deep, wide-field ultraviolet photography. UIT has a 38 cm diameter mirror of f/9 and a 40 arcmin field of view which covers the spectrum from 1200 to 3200 A. The Wisconsin Ultraviolet Photopolarimetry Experiment (WUPPE) is designed to measure the polarization of ultraviolet light from celestial objects. The instrument is capable of detecting both linear and circular polarization simultaneously, and thus well suited for the examination of intense magnetic fields in compact objects. The Wide Field Camera developed by NASA's MSFC is designed to photograph comet Halley.

6. Planned Missions (1)

The SAX Mission (Italian X-ray Astronomy Satellite): The SAX Mission proposed to the Italian National Space Plan was described by G. Spada. The original proposal was submitted by a consortium of Italian institutes, with participation of Holland ESA. SAX has the objective of carrying out systematic and comprehensive observations of X-ray sources in the 0.1-200 keV energy region. The mission is for spectral and timing measurements on a variety of selected sources and for monitoring the sky in the 2-30 keV range for the investigation of long term variability and the study of the transients. Coaligned narrow field of view instruments are 4 Concentrator/Spectrometer, a High Pressure Gas Scintillation Proportional Counter and a Crystal Scintillator (Phoswich). The wide field of view instrument consists of three Wide Field Cameras. The nominal launch date is the end of 1989.

The Infrared Space Observatory (ISO): R. Bonnet described the ISO mission. It is an approved mission of the ESA Science Programme. It is a cryogenically cooled 60 cm Ritchey Cheretien Telescope equipped with a set of four focal plane instruments; a camera, a photometer, and a short and a long wavelength spectrometer. ISO will have a sensitivity 10^3 to 10^4 times larger than IRAS and will therefore be able to detect very weak sources. The main scientific aims of the mission are to study extragalactic astronomy, cosmology and star formation processes. The overall length of ISO is 5.2 m and it weighs 1800 kg; it will be on a 12 hour ecliptical orbit with an apogee of 39400 km.

The Solar Optical Telescope (SOT): This shuttle borne solar telescope was discussed by S. Jordan; it will be built under the management of the GSFC. It was reported that the telescope will have a 1.3 m diameter primary mirror with an on-axis Gregorian optical system which provides diffraction-limited viewing in the visible of 0.1 arcsecond or a resolution of 73 km on the sun. The coordinated instrument package consists of the photographic filtergraph, the tunable filtergraph and the spectrograph system. SOT will become operative in early 1990's.

SIGMA:
A written comment was contributed by Cezarsky on this French gamma ray imaging experiment on the board Soviet satellite GRANAT. The instrument consists of the NaI detector with the coded mask covering the energy range of 30 keV - 2 Mev. The instrument with a diameter of 1.2 m and a length of 3.5 m weighs 1000 kg. In the field of view of 7.3° x 7.3° the source positional accuracy is 1.5 arcmin. Expected launch date is December 1987.

7. and 8. Planned Missions (2)

XMM (European X-ray Telescope Mission)*: J. Bleeker. HESP: Japan's next solar maximum mission was presented by Y. Uchida. The mission is planned to be performed in collaboration of solar physicists in US and Japan. It will follow the collaborative efforts between SMM and HINOTORI during the last Solar maximum. The scientific instruments consist of a hard X-ray imager, soft X-ray mirror telescope, X-ray spectrometer, Bragg spectrometer, gamma-ray spectrometer, and a solar irradiance-meter. The hard X-ray imager covering 10-80 keV as a Fourier-synthesis type modulation collimator provides a spatial resolution of 8". The soft X-ray mirror telescope for 0.1-1 keV provides a 2.5" resolution. The launch date is expected to be mid 1991.

QUASAT (NASA-ESA Space VLBI): R. T. Schilizzi described that the idea of very-long-baseline radio interferometers with separations of elements larger than the size of the Earth has led to the QUASAT concept of space VLBI. The early concepts for this joint ESA/NASA mission were discussed at a Workshop on QUASAT held in Vienna in June 1984 (ESA Document SP-213) and reviewed again at a Workshop held in Charlottesville in May 1985. The mission concept involves a free-flying satellite carrying a radio telescope in an elliptical orbit around the Earth in conjunction with the major ground-based VLBI networks in Europe, USA, USSR, and Australia. The perigee is 5700 km, the apogee 12500 km, and the inclination 63°. The angular resolution expected for QUASAT at its wavelengths of 1.35, 6 and 18 cm will be 90, 400 and 1200 micro arcsec respectively. The main goal of the QUASAT mission is to probe the nuclei of radiogalaxies and quasars more deeply and with greater detail than is possible with ground-based networks alone.

As study has shown that the overall mission, which is based on current VLBI and spacecraft engineering practices, is technically feasible. As far as the spacecraft is concerned the major new aspect is the deployable antenna. Antenna configurations have been studied in ESA and in NASA.

Discussion of the requirements for a potential multi-nation space VLBI mission involving QUASAT, the RADIOASTRON satellite of the USSR and a Japanese satellite is going on.

Indian X-ray and Gamma Ray Satellite Experiments: T. M. K. Marar reported on two scientific payloads which are planned and are under preparation to be flown on the Stretched Rohini Series of Satellites (SROSS). The first of these, SROSS-1, will carry a gamma ray burst payload to monitor gamma ray bursts in the energy range of 20-3000 keV. It will measure temporal variations with high time resolution. It will also measure temporal evolution of burst energy spectra searching for cyclotron lines and other features. The instrument consists of a main and a redundant scintillation detector.

An Indian X-ray astronomy satellite on the fourth SROSS mission during 1989-1990 is proposed for the study of temporal variability and spectral characteristics of X-ray sources jointly by scientists from TIFR and ISRO. The instrument consists of X-ray telescopes for pointed mode and scan mode observations. A secondary payload on the observatory is for the study of transient X-ray novae.

Chinese Space Programmes were presented in a poster by Ma Yugian.

LYMAN/FUSE (Far UV Spectroscopic Explorer), report by A. Boggess and M. Grewing.

The scientific importance of the 900-1200 Angstrom region follows from the concentration of atomic and ionic resonance lines in this region. These as well as the lines from molecular hydrogen are powerful diagnostic tools for studying the temperatures, densities, and abundances in hot, thin plasmas that exist in interstellar and intergalactic space, in the photospheres, chromospheres, and coronae of stars, in galactic nuclei, and within our own solar system.

A telescope with large collecting area, an efficient spectrograph (either of Rowland or of échelle type), and high quantum-yield detectors have to be combined : a grazing incidence telescope similar to ones used in x-ray astronomy is probably the best choice to achieve a large effective collecting area over the 900-1200 Angstrom range, and to maintain good efficiency throughout the entire EUV band, down to 100 Angstrom.

Studies of this project, which have so far implied scientists from ESA, NASA and Australia, are being presently actively pursued.

HORIZON 2000 : J.A.M. Bleeker.

The Horizon 2000 long term plan for European space science is based on four major elements, identified as cornerstones, with a time line to about 2007. These cornerstones represent the top priority in the domains of solar-system sciences, astronomy and astrophysics in Europe. The cornerstones include :
- A Solar Terrestrial Physics (STP) programme comprising a solar and heliospheric observatory and a four-probe space plasma mission.
- A mission to Primordial Bodies, this will most probably entail a comet-nucleus sample return.
- A High Throughput X- ray spectroscopy mission between 0.2-10 keV.
- A High Throughput heterodyne spectroscopy mission between 0.1-1 mm.

Apart from these major elements a number of smaller scale missions are realized in the same time frame like ULYSSES, the ESA part in HST, HIPPARCOS, ISO and about four more projects in this class. In addition space science payloads based on the EURECA-platform are envisaged.
Reference document ESA SP-1070.

*A EUROPEAN MISSION ON HIGH THROUGHPUT X-RAY SPECTROSCOPY : J.A.M. Bleeker

This mission is a cornerstone programme in the Horizon 2000 long term plan. The scientific objectives require a powerful imaging instrument with the highest possible collecting area to arrive at the photon statistics required for high quality spectral measurements and a large dynamic range in energy (wide band spectroscopy). The prime design drivers for this mission can be stated as :
- Energy band : 0.2-10 keV (bulk of the X-ray photons).
- Throughput : Optimized for the 2-8 keV band : $A_{eff} \geq 10^4$ cm^2 (2 keV), 5.10^3 cm^2 (8 keV).
  Angular resolution : $\leq$ 30 arcsec half power width at 7 keV.
- Spectral resolution : a wide range of spectral resolving power $\lambda/\Delta\lambda$ = 10 $\rightarrow$ few 10^3 is called for.
  Considering the emphasis of the mission objectives low resolution spectroscopy ($\lambda/\Delta\lambda$ = 100) should be ensured at full throughput.
- Time resolution : one millisecond for spectral variability studies.
- Time scale : technological development 1986-1992, hardware development 1992-1997, launch 1997/1998, operations 10 years minimum.

COMMISSION 45: STELLAR CLASSIFICATION
(CLASSIFICATION STELLAIRE)

Report of Meetings 20, 21, 22, 23, 25, 26, and 27 November 1985

PRESIDENT: V. Straižys   SECRETARY: N. Houk

I. JOINT DISCUSSION

   Evolution in Young Populations in Galaxies (26 November, with Commissions 25, 28, 33, 34, and 37).

II. JOINT MEETINGS

   1. Synthetic Spectroscopy and Photometry (20 November, with Commissions 25, 29, and 36).

   This meeting was organized by R. Buser. The following papers were presented:
   R. Buser, Principles and scope of synthetic photometry.
   B. Gustafsson, Present state of the art (Review of the situation with model stellar atmospheres, synthetic spectra, and the calibration of photometric indices).
   F. Rufener, Passbands and photometric systems.
   D.S. Hayes, Observed stellar energy distributions.
   R.L. Kurucz, Theoretical stellar energy distributions.
   R. Buser, Calibration of the Hubble Space Telescope.

   2. Precision Photometry of Star Clusters (21 November, with Commissions 25, 35, and 37).

   This meeting is outlined in the Report of Commission 37.

III. MEETINGS OF WORKING GROUPS

   1. WG on Be Stars (21 November, with Commission 29). This meeting was organized by A. Slettebak. The following papers were presented:
   Y. Andrillat (with M. and C. Jaschek), The Hα line in Ae and A-shell stars.
   K. Apparao, V/R variability in Be stars.
   D. Baade, Mass loss and nonradial oscillations of Omega Orionis.
   P. Barker, Hα observations and coordinated IUE spectroscopy of Be stars.
   F. Vakili, The angular diameter of the envelope around Gamma Cas.
   P. Harmanec and J. Percy, Report on the Be-star photometric campaign.

A. Slettebak, Be stars in open clusters.

The new Organizing Committee of the WG has been elected: D. Baade (Chairman), P. Barker, V. Doazan, J.M. Marlborough, G. Peters, A. Slettebak, and T.P. Snow.

2. <u>WG on Standard Stars</u> (26 November, with Commissions 29 and 30). This meeting was organized by L.E. Pasinetti and chaired by A. Batten. The following papers were presented:
A.G.D. Philip, The microfiche on standard stars.
H. Neckel, The absolute energy distributions of the Sun and of the solar analogs.
M.L. Malagnini, C. Morossi, and L. Rossi, Does a solar twin really exists?
L. Rossi, A. Altamore, and C. Rossi, A study of the solar analogs in the ultraviolet.
M. Fracassini and L.E. Pasinetti, Photometry of the comet Halley: solar analogs selected along the path.
I.N. Glushneva, E.A. Makarova, and A.V. Kharitonov, Energy distribution, photometry and physical characteristics of the Sun and stars – solar analogs.
After the IAU General Assembly Patras meeting the WG publishes "Standard Star Newsletter" (ed. L.E. Pasinetti, Milano). It is distributed to the members of the WG, to the members of the WG on Spectroscopic and Photometric Data and, on request, to all astronomers and institutions interested in the subject. The Newsletter is published twice a year, so far six issues have appeared. The WG participated in organizing the IAU Symposium No. 111 "Calibration of Fundamental Stellar Quantities", held in Como, Italy, in May 1984.
The new Organizing Committee of the WG has been elected: A. Batten (Chairman), S.J. Adelman, R.F. Garrison, I.N. Glushneva, R.F. Griffin, D.S. Hayes, C. Jaschek, P.C. Keenan, H. Neckel, L.E. Pasinetti.

3. <u>WG on Ap Stars</u> (23 November, with Commission 29). The report of this meeting may be found in the Commission 29 Report.

4. <u>WG on Peculiar Red Giants</u> (27 November, with Commission 29). This meeting is outlined in the Commission 29 Report.

5. <u>WG on Spectroscopic and Photometric Data.</u> No meeting of this WG took place during the General Assembly. Its Chairman A.G.D. Philip presented the following report:
1. Twice a year the WG was pulled to obtain a list of newly prepared catalogues and catalogues in preparation. This information has been sent to C. Jaschek for publication in the Bulletin of the Strasbourg Data Center.
2. A microfiche of standard stars was prepared under the editorship of Philip and Egret, containing all the available data concerning standard stars in various systems as well as other stars of particular interest. The first version of the microfiche has been published in IAU Symposium No. 111 "Calibration of Fundamental Stellar Quantities" volume. A revised version is now under preparation.

IV. SCIENTIFIC MEETINGS OF THE COMMISSION

1. <u>Spectral and Photometric Classification of Population II Stars</u> (25 November). The meeting was organized and chaired by A.G. Davis Philip. The following papers were presented:
    C. Pilachowski, High resolution spectroscopy of population II stars.
    R. Foy, Determination of the atmospheric parameters of late-type stars from low resolution spectra.
    D.J. MacConnell, Recognition of low metallicity stars from moderate dispersion plates.
    N. Houk, The statistics of weak-lined stars in the Michigan 10° Catalog.
    F. Spite, The relation between photometry, spectral classification and high dispersion spectroscopy.
    B. Gustafsson, Model atmosphere analysis of population II stars.
    D. Hayes, Field horizontal branch stars in the T_{eff}, log g plane.
    B. Hauck, Population II stars in the Geneva system.
    A.G.D. Philip, Field horizontal branch stars.
    B. Carney, A new survey of local halo stars.
    A. Ardeberg.
    F. Fusi-Pecci.

2. <u>Difficulties of Extrapolation of the MK System to Other Spectral Ranges</u> (22 November). The meeting was organized and chaired by A. Heck. The following papers were presented:
    R.F. Garrison, The MK process.
    A. Heck, The IUE ultraviolet classification system.
    M.F. McCarthy, Extrapolation of the MK system to the near infrared.
    K. Nandy, Luminosity effects on the major ultraviolet stellar lines.
    J.C. Rountree, Spectral classification of B-stars in the ultraviolet.

V. ADMINISTRATIVE MEETING (22 November).

1. Report of the President

Since the XVIII General Assembly in Patras, Commission 45 co-sponsored the following meetings: (1) The Workshop "The MK Process and Stellar Classification", held in Toronto, Canada, June 6-10, 1983, (2) IAU Symposium No.105 "Observational Tests of the Stellar Evolution Theory", held in Geneva, Switzerland, September 12-16, 1983, (3) IAU Colloquium No. 78 "Astronomy with Schmidt-Type Telescopes", held in Asiago, Italy, August 30 - September 2, 1983, (4) IAU Symposium No. 108 "Structure and Evolution of the Magellanic Clouds", held in Tübingen, FRG, September 5-8, 1983, (5) IAU Symposium No. 111, Calibration of Fundamental Stellar Quantities, held in Como, Italy, May 24-29, 1984, (6) The Colloquium "Cool Stars with Excess of Heavy Elements", held in Strasbourg, France, July 3-6, 1984, (7) IAU Colloquium No. 90 "Upper Main Sequence Stars With Anomalous Abundances", held in Crimea, USSR, May 14-17, 1985.
    Five circular letters were sent by the President to the Commission members during 1982-1985 containing information on the Commission matters. The Commission 45 Report for 1982-1984 was published in the Transactions of the IAU, vol. 19 A. The sections of the report were written by R.F. Garrison, D.J. MacConnell, A. Heck, V. Straižys, and C.Jaschek.

2. Membership

The following Commission 45 members resigned: P. Boyce, A.A. Hoag, J. Landi-Dessy, A. Maeder, S.D. Sinvhal. The following astronomers asking membership in Commission 45 were accepted: L.Celis, Ch. Corbally, D. Egret, I. Fukuda, L. Labhardt, S.G. Lee, C. Marossi, B. Nicolet, P. North, B.S. Rautela. This brings the total number of the Commission members to 119.

3. Election of Officers and Organizing Committee

The new officers and Organizing Committee members were elected as follows: President R.F. Garrison (Canada), Vice-President M. Golay (Switzerland), OC members J.J. Claria (Argentina), A. Heck (France), N. Houk (USA), T. Lloyd-Evans (South Africa), D.J. MacConnell (USA), E.H. Olsen (Danmark), and V. Straižys (USSR).

4. Reports of Working Groups of the Commission

The Commission was informed about the activity of the following working groups: WG on Be stars, WG on Ap stars, WG on Standard Stars, and WG on Spectroscopic and Photometric Data. It was decided to extend the time of existence of these working groups. Additionally, a new working group, called "Peculiar Red Giants" involving Commissions 29 and 45 has been proposed by P.C. Keenan and C. Jaschek in a letter to the Commission Presidents. The Administrative Meeting expressed the support to the creation of this WG.

5. Future Symposia and Colloquia

The meeting was informed about the future scientific meetings which are in the scope of interests of the Commission members: (1) IAU Colloquium No. 92 "Physics of Be Stars", Boulder, Colorado, August 18-22, 1986; (2) IAU Symposium No. 126 "The Harlow Shapley Symposium on Globular Cluster Systems in Galaxies", to be held in Cambridge, USA, August 25-29, 1986; (3) IAU Colloquium "The Second Conference on Faint Blue Stars", to be held in Tucson, Arizona, Spring of 1987. Dr. A. Slettebak informed that W.K. Bonsack have prepared the Third edition of the "International Register of Stellar Spectroscopists".

6. Resolution

The Commission supported the Resolution proposed by R.F. Garrison and addressed to the Time Allocating Committees for large telescopes. The Resolution recommends that a small fraction of observing time on large telescopes should be allocated for the program establishing a set of faint MK standards. The program proposed by a group of experts is planning to establish MK standards of 9-11 and 14-16 magnitudes in the equatorial belt, within $\pm 10°$ of declination. Astronomers are requested to propose candidate stars suitable as faint MK standards of different spectral types and luminosities.

COMMISSION 46 : TEACHING OF ASTRONOMY (ENSEIGNEMENT DE L'ASTRONOMIE)

Report of Meetings : 19, 21, 25, 26, 27 November 1985

ACTING PRESIDENT : C. Iwaniszewska   SECRETARY : J.R. Percy, D. McNally

Session 1, 21 November 1985

BUSINESS SESSION

1. Organizing Committee 1985-1988

President : C. Iwaniszewska
Vice-President : A. Sandqvist
Organizing Committee :
    L. Gouguenheim (ICSU CTS representative)
    J. Kleczek (Secretary of ISYA)
    J.R. Percy (Editor of Newsletter)
    L. Houziaux (Past President, publisher of Newsletter)
    M. Gerbaldi, E.V. Kononovich, R.R. Robbins (Astronomy Education Materials)
    D.G. Wentzel (Chairman of Visiting Lecturers Programme)
    S. Ferraz-Mello, B. Hidajat, Y.K. Miao (Visiting Lecturers Programme)

2. National Representatives

    New National Representatives from Malaysia, Paraguaj, Uruguaj, Peru, Thailand have been proposed; since some of them are not IAU members, they will remain consultants. The National Committees for some countries have still not appointed new representatives.

3. Regular Members

    There was a discussion on membership criteria for Commission 46. Some feel that Commission 46 should consist of national representatives only. Some feel that a few additional members with special qualifications should be admitted, and some feel that all those IAU members who requested to be members of Commission 46 should be admitted. In the end, it was decided to admit a few extra members who had been present at the meeting: H.U. Keller (G.F.R.), J. Kreiner (Poland), P.P. Saxena (India), R.M. West (Denmark), and to defer any further decision until the matter has been discussed with the IAU Executive Committee.

4. International Schools for Young Astronomers (ISYA)

    J.R. Percy read the report from J. Kleczek regarding the ISYA. The report outlined the history of ISYA, the method by which they operated, the reasons for their success, and the persons and institutions who should be thanked for making them possible. Future schools have been proposed as follows: Patiala, India (postponed from 1985), China, Portugal (postponed from earlier), Columbia, Iraq, the Francophone countries of Africa, and Sri Lanka. Which of these schools are held depend primarily on external circumstances. The first three are the ones most likely to be held in the near future.

5. Newsletter

    D. McNally reported on the Newsletter of Commission 46, which he has edited for many years. He asked whether the Newsletter should evolve into a more formal Journal, or be otherwise developed of expanded. It was decided that -especially as the Newsletter had exceeded its current budget- it should be retained in its present form.

6. National Reports and AEM

The response to the National Reports was favorable. There was some discussion, however, on the Astronomy Education Material (AEM). It was noted that, although the AEM is useful in many countries, it is perhaps not necessary to compile the material in all languages, or to distribute all material to all addresses on the mailing list. Perhaps the National Representatives could distribute the AEM more widely in their own countries. It was noted that the authors of the English-language AEM copyrighted it, and that this was not desirable.

D. McNally was thanked for his years of service to the Newsletter. The new editor, J.R. Percy, appealed for contributions to future Newsletters.

7. Colloquium

J.R. Percy discussed a proposal for an IAU Colloquium on the Teaching of Astronomy, to be held probably in 1988 in the area of Washington, USA. The colloquium would be co-sponsored by one or more North American astronomical societies. The members of Commission 46 expressed general support for the proposal.

8. Visiting Lecturers Programme

There was a brief discussion on the Visiting Lecturers Programme. Such a programme is planned for Peru in 1986, and is proceeding favorably. A programme planned for Nigeria is having difficulties.

D. McNally discussed a proposal by himself and R.M. West for a "travelling telescope" which would be sent on loan to developing countries, perhaps in connecttion with a Visiting Lecturers Programme or ISYA. It was hoped that the telescope - of approximately 20 cm aperture and equipped with camera, photometer and spectrograph- could be donated by the manufacturer, and transported from place to place by national airlines, free of charge. There was a general support for this proposal.

Session 2, 21 November 1985

CONTRIBUTED PAPERS

1. Aage Sandqvist "A New Concept in Secondary Education"

After describing the elementary and secondary education system in Sweden, and the place of astronomy in it, the author described a proposal "Astronomy High School". Students in the school receive about 5 hours/week of astronomy, in addition to their other courses, and carry out practical work at the Stockholms Observatory.

2. John R. Percy "A Teacher's Guide to Astronomy"

The author described a self-contained guide to astronomy, intended for senior elementary and secondary school teachers. A preliminary version was reviewed by several teachers and a final version is about submitted for publication.

3. Jing-Kui Qian "An International School for Young Astronomers in China"

The author described such a school, which is being considered for sponsorship by the IAU.

4. John L. Safko and R.F. Whitesell "Radio Astronomy for Undergraduates"

The authors described ways in which a 5.5m steerable radio telescope (operating at 11-16 GHz) and a 3m transit radiotelescope (operating at 1.42 GHz - 21cm) have been incorporated into introductory, intermediate and advanced astronomy courses. As the authors have not been present, their report has been briefly summarized by the President.

Session 3, 25 November 1985
(jointly with Comm. 5 and 38)

DEVELOPPING COUNTRIES

Jean-Claude Pecker presided a special session organized by 3 commissions devoted to documentation problems for developing countries. A report on the present situation has been prepared by Pierre Lantos. There is a large need of research books, astronomical journals, of abstracting services for poorer countries. Help could come from direct actions : from person to person, from group to group, from laboratory to laboratory, country to country, from international societies.

A long discussion which followed has been devoted to many problems : the importance of exchange of astronomers (M.L. Aguilar), the possibility of obtaining basic text-books at lower price (P. Venugopal, J.C. Pecker, S. Gurm), on how to use best the AEM (E.A. Müller, F.G. Wood), and the IAU Information Bulletin or Commission 46 Newsletter (J.R. Percy, E.A. Müller), on the possibility of getting more quickly preprints of useful publications, as for instance Almanacs (W.D. Heintz, M.K. Aly), on the equipment needs of developing countries (M.K. Aly).

Session 4, 25 November 1985

POPULARIZATION OF ASTRONOMY

An additional session of Commission 46 has been devoted to an informal discussion on popularization of astronomy, on methods "how to deal with the general public". The main speakers have been James Cornell from Smithsonian Astrophysical Observatory, President of the International Science Writers Association, and Leif Robinson of "Sky and Telescope". The diffusion of astronomical knowledge to the public may be done either indirectly, by releasing special bulletins to the press, press conferences, etc., or directly -by arranging special evenings at the Observatory for different groups of people, etc. The public expects to receive information on what is done at the Observatory. In order to have accurate knowledge diffused, the scientists ought to work with the press. But of course not everybody is a good communicator.

Jean-Claude Pecker stressed the point that the contact with the press might be much more difficult in poorer countries, where they may have problems, for instance with translation of informations. Astronomers want to pay their duty to the general public, but then the newspapermen ought not put together astronomers and astrologers ! Jean-Louis Heudier told about the French experience -public shows at metro stations, attended during 10 days by some 6 millions people. It seems a very good idea to use interesting astronomical events to popularize astronomy.

Local astronomical societies ought to keep an amount of activity directed towards the general public (J.R. Percy). "Sky and Telescope" editorial office has a telephone service on instant astronomical discoveries, but there are no such possibilities in Europe. The first informations usually reach astronomers through "Sky and Telescope" issues, but they ought to be published and propagated quickly (I. Almar). When dealing with the general public one must bear in mind those who probably will never attend any school, therefore one cannot use a very specialized language (M.L. Aguilar).

Session 5, 26 November 1985

ROLE OF PLANETARIA IN SCIENCE TEACHING

Organized by Mrs. Nirupama Raghavan of Nehru Planetarium, New Delhi, and presided by Hans-Ulrich Keller of Stuttgart Planetarium, this has been one of the longest, most interesting and useful sessions of Commission 46. An extensive report on India has been prepared by N. Raghavan. There are now about twenty planetaria in India and neighbouring countries, the oldest of more than 20 years in Calcutta with an attendance of 600 000 visitors per year, and the youngest, of 1.5 year in New Delhi with 180 000 visitors. Following problems ought to be faced when looking for the educatory role of planetaria: how long will it take to have pupils from all schools of a given area pay a visit at the planetarium, how much money is needed for the maintenance of planetaria and how large their staff ought to be, how can planetaria help in educating teachers, in publishing teaching materials. Nearly all planetaria have telescopes used everyday; Bombay has a travelling mini-planetarium in a tent going through the region with some general programmes. Each of them must have an extensive explanation of astronomical events because of popular astrological beliefs!

H.-U. Keller told about the International Planetarium Directors Conference, as well as on the First European Planetarium meeting at Strasbourg in 1984. Some possibility of collaboration, of exchange of programmes, of training planetarium personnel has been envisaged for European countries. H. Köhler introduced the latest news in ZEISS Planetaria automatization, and G. Reed of West Chester State College gave an invited lecture on planetaria in astronomy education. While the planetarium might be considered as the greatest single teaching instrument ever invented, it must not be used all the time when teaching astronomical concepts. One must bear in mind that astronomy is a science most closely associated with our culture, astronomy is very attractive to the laymen.

Next the planetarium directors spoke about their specific situations: in Pakistan they have school children coming everyday, and teachers' courses free of cost; in Allahabad the audience is composed of pilgrims coming from all over India; in Calcutta the programmes are prepared in 3 languages: Bengali, Hindu, English, with music as a kind of cultural show; in New Delhi secondary school pupils work at easy observing projects, as observing sunspots and solar rotation, as drawing the path of the Moon on a star chart, etc; in Delhi Physics Laboratory Planetarium they arrange science weeks to attract the teachers from science centers.

The session concluded with the final question -how can we help the planetaria- answered: through bringing together people from different parts of the world, through an exchange of experiments and informations, perhaps through more personal contacts or founding of a committee for planetaria of a given part of the world.

TEACHING OF ASTRONOMY

Sessions 6 and 7, 19 November 1985

TEACHERS' MEETING

The traditional meeting with local teachers chaired by the Acting President C. Iwaniszewska has been attended by about 50 teachers from the local Delhi area and about 50 Indian and foreign astronomers. The aims of this session, outlined by K.B. Bhatnagar of Delhi, have been to emphasize the role of astronomy in school curriculum and to establish closer contacts between school teachers and astronomers. K.D. Abhyankar of Hyderabad was of the opinion, that teachers when introducing astronomical notions in the 11-class, should be more interested in recent developments of modern astronomy, for instance they should be able to show that the laws of physics are universal everywhere.

D. McNally of London pointed out that astronomy should have its place in the teaching of physics and mathematics, since there are 3 important areas of impact of astronomy at school:
- when introducing the concept of time in physics,
- when getting at applications of physics to modern astrophysics,
- when looking for simple examples in teaching mathematical computations.

J.C. Bhattacharya of Bangalore spoke on the introduction of simple astronomical experiments at school, beginning with the construction of sundials, ending with observations made by means of self-made simple telescopes. A. Bhatnagar of Udaipur and S.M.N. Ansari of Delhi told about old Indian astronomical traditions, India being still a mine of old documents: manuscripts of astronomical tables translated from arabic and islam into sanscrit.

A paper on "How to introduce the Universe in physics teaching" has been presented by T.P. Padmanabhan from Bombay, while M.P. Chhaya of Delhi presented the point of view of teachers. It is not so easy to popularize astronomy at school level, one ought to have some adequate astronomical literature, but there is nearly none at present. In order to keep the interest of children it would have been highly desirable to have astronomical clubs at school or to organize other out-of-school activities to keep the habit of sky watching. Visits at observatories would of course be highly educational, but a training programme or regular teachers courses would be best. Teachers must be brought to understand themselves all astronomical phenomena. It would of course be very helpful if daily newspapers could bring a monthly star-chart and an astronomical column.

The chairman spoke about popularization of astronomy and teachers training in her country -Poland-, and the session concluded with a show of astronomical slides presented by N. Raghavan of Delhi. It is hoped that as a result of this meeting, a committee of astronomers and teachers will begin its work, helping the everyday work of teachers, so that Henri Poincaré's words will become true: "The stars not only give us that light which strikes our physical eyes, but they also give us the light which illuminates our minds". These have been the concluding words quoted by the organizer of this session, S.M. Alladin of Hyderabad.

On November 27, Commission 46 arranged a special show of American astronomical movies about the Universe and old Egyptian astronomy brought by Aage Sandqvist from Sweden, through the courtesy of Curt Roslund from Göteborg. An audience of a few hundred astronomers and school teachers attended.

COMMISSION 47. COSMOLOGY (COSMOLOGIE)

Report of Meetings on 20, 21, 22, 25 and 27 November 1985

PRESIDENT :   J. Audouze
VICE PRESIDENT : G. Setti   SECRETARY : V. Trimble

I. BUSINESS MEETING (20 November 1985)

Participating members of the Commission endorsed recommandations of the organizing committee that G. Setti (Italy) and K. Sato (Japan) serve as president and vice-president for the period 1985-1988 and that J. Audouze (France, retiring president) and Fang L.-Z. (Chine) be added to the organizing committee, bringing the total membership to nine.

Sixty-one new applications for commission membership were also approved, bringing the total to 283. It was suspected that some of the continuing members may no longer be active in cosmology, and the incoming president plans early in his term to circulate a letter to the membership requesting a verification of current scientific interests, in the expectation that this will somewhat reduce the total. We are not aware of any deaths or resignations from the commission membership during the previous three years.

The president announced that Commission 47 will cosponsor two symposia, 124 on Observational Cosmology (Shangai, China; August 25-29 1986) and 130 on Evolution of Large Scale Structure in the Universe (Balatonfured, Hungary; June 1987) in the coming few years. In addition, at least five other meetings of interest to cosmologists will occur between January and July 1986. We look forward to seeing each other often !

The business session concluded with a presentation on "Cosmological Research in China" by Zou Z.-L., incorporating material that had not been received in time for inclusion in the commission's report earlier this year. In the triennum 1982-85, our colleagues in Chine published nearly two dozen papers on the topics of the very early universe, anisotropy of the background radiation, cosmological parameters, large scale structure of the universe, non-standard cosmological models. Some of the highlights include :

Two investigations by Guo H.Y. et al. (1982 Ch.Astron.Astrophys. 6,243) and Xu D.Y. et al. (Scien. Sinica 27, 81, 1984) on the possible prevention of cosmological singularities when quantum and symmetry-breaking effects are included.

A derivation by Xu C.X. (1984, Ch.Astron.Astrophys. 8,340) of the solar velocity relative to the diffuse X-ray background as measured by HEAO 1 A2. The result, 387±159 km/s, is in good accord with the solar velocity relative to the microwave background.

Several investigations by Chen J.S. et al. (1984, Astron. Astrophys. 134, 306; 1984 Publ. Astron. Soc. Pac. 95, 335; and references cited therein) of the Lyman alpha absorption line forest in large-redshift quasars. They find evidence for moderate evolution in number density of the absorbing clouds and no evidence for power-law clustering like galaxies.

Two investigations by Zhou Y.Y. et al. (Ch.Astron.Astrophys. 9, 20, 1985) and Fang L.Z. et al. (1984 Ch.Astron.Astrophys. 8, 148) attributing the structure in plots of quasars numbers vs. redshift either to observational selection effects or to the universe being multiply connected with the present scale of compactified space being about 600 Mpc.

II. JOINT DISCUSSIONS (21 and 27 November 1985)

The commission cosponsored joint discussions IV (Radio Astronomy and Cosmology) and VII (Supernovae). Vol.7 of Highlights will include many of the talks presented at these. The most relevant to the interests of Commission 47 were (from Supernovae) a presentation by N. Bartel on the combining of optically-measured supernova expansion velocities with VLBI measurements of proper motion expansion, thus providing a dynamical parallax and a distance scale independent of the Hyades, Cepheids, etc; and (from Radio Astronomy and Cosmology) a suggestion by J. Baldwin that pancakes in the process of becoming protogalaxies will introduce a1000km/s wide, 21cm emission feature, now redshifted to near 150 MHz, and having a brightness temperature near 1K and angular size scales of a few minutes of arc. Modification of an existing NRAO installation to search for these emission features is underway.

III. SCIENTIFIC SESSIONS (20, 22 and 25 November 1985)
In the course of the six sessions, 19 papers were presented, as follows:

Large Structures and Radio Sources

1   C.R. Subrahmanya (Australia): The cosmological evolution of radio sources
2   D.N. Schramm (USA): Large scale structures and galaxy formation- fractals and strings

Quasars

3   X.T. He (PRC): Quasars in the Virgo cluster region
4   V. Petrosian (USA): Evolution of quasars
5   P.A. Shaver (ESO): Quasar pairs

Anisotropy of the Blackbody Radiation

6   G.F. Smoot (USA): On the cosmic background radiation
7   K. Shivanadan (USA): Measurements of the cosmic infrared background radiation
8   R.B. Partridge (USA): Aperture synthesis observations of the microwave background

Clusters of galaxies

9   W. Saslaw (USA): Gravity and clustering of galaxies
10   M. Mosconi (Argentina): Clusters of galaxies
11   S.P. Bhavsar (India): Are the filaments real ?
12   R. Cowsik (India) : Self-consistent distribution of galaxies in clusters
13   I.E. Segal (USA): Recent extragalactic observations and rational comparative cosmology

Primordial Nucleosynthesis

14   G. Steigman (USA) : Primordial nucleosynthesis: a status report
15   J. Audouze (France): Primordial nucleosynthesis and strange particles
16   G. Burbidge (USA): How much of the helium is really primordial?
17   G. Steigman (USA): Dark matter and exotic particles: in search of the perfect WIMPS
18   K. Sato (Japan): Baryogenesis in the inflationary universe
19   J.P. Luminet (France): Nucleosynthesis induced by stellar pancakes.

IV. ABSTRACTS OF SOME OF THESE PRESENTATIONS

3. QSO research in China (X.T. He - Dept of Astronomy, Peking Normal Univ., Beijing, China)
There are about 30 astronomers or physicists interested in QSO research in

China. About 150 papers have been published concerned with QSOs. Some observing projects co-operatively with other countries or to apply big telescope time in the world start since 1980's. We are trying to use chinese telescope for QSO observation.

The data have been used from the UK Schmidt and Palomar Schmidt for QSO survey in large area, for example, in Field 297 ($1^h 44^m$, $40°00'$) 1092 QSO candidates have been found in ≃ 40 square degrees. 36 of the candidates were observed spectroscopically 21 of the objects are confirmed as QSOs. 35 new QSOs have been discovered in the Virgo Cluster Region. The bright background QSOs provide a network of probes of absorbing material both in the cluster and surrounding individual galaxies.

Other works include clustering test, qo determination, identification of redshift systems, mechanism of line formation, X-ray studies and so on.

We are doing emission line object and UVX object survey and to observe some BL Lac and active galaxies using the Chinese facilities.

4. Evolution of quasars (V. Petrosian - Stanford Univ., Stanford, California USA)

The relation between the evolution of the luminosity function, derived from statistical analyses of complete samples of quasars (optically selected), in an assumed cosmological model, and the physical evolution of the luminosity and the formation rate of the object were described. It was reiterated that the observed data on optically selected sample disagrees with the requirements of either a pure density or pure luminosity evolution of the luminosity function.
Furthermore, it was stressed that only very special physical evolutions could give rise to such pure evolutionary forms.

In general, at low and moderate redshifts ($z \leq 2$) the low luminosity objects show weak or very little evolution while the high luminosity objects show the strong evolution commonly attributed to all quasars. Such changes in the luminosity function can come about if, for example, the lifetime of the object at each luminosity decreases rapidly with increasing luminosity.

At larger redshift, where only the brightest objects can be studied, the deduced evolution of the luminosity function and, in particular, the rate of the decrease of the evolution or the existence of a redshift cutoff depends critically on the cosmological model.

5. Quasar clustering (P.A. Shaver - ESO, Garching, RFA)

The physical clustering of quasars has been studied by applying a novel technique to a catalogue of nearly 3000 quasars with known redshifts. Clustering is detected on scales less than 10 Mpc, but on larger scales the distribution appears to be random. The same technique has been applied to two large samples of galaxies, permitting a direct comparison. From this it appears that quasars at high redshifts are clustered similarly to galaxies today, suggesting that structure on these scales has not evolved significantly since $z \sim 2$.

6. Large angular scale anisotropy (G. Smoot - Univ. of California, Berkeley, USA)

A review of recent large-angular-scale anisotropy experiments, Berkeley, Princeton and Russian Prognos 9.

The old and new experiments agree roughly about the magnitude and direction of the first order (dipole) anisotropy. The three maps are are shown with and without the host fitted dipole anisotropy. Expected future experiments by Berkeley, Princeton, Cosmic Background Explorer (COBE) are mentioned.

8. Aperture synthesis observations of the Microwave Background (R.B. Partridge, Haverford College, Pennsylvania, USA)

Colleagues from the National Radio Astronomy Observatory, TESRE Bologna, and the Center for Astrophysics and I have used the Very Large Array (VLA) to wake two different sorts of observations of the cosmic microwave background.

1) We searched for fluctuations in the microwave background on scales smaller than usually observed, down to 4". The most recent and sensitive results are at scales 18" - 40". Our technique permits us to subtract instrumental noise to high occuracy. We can also remove evident discrete sources. Even after these corrections are made, a non-zero signal remains. We assume some of this sky noise may be ascribed to weak sources ; we are currently attempting to model and to subtract such effects. Until this process is complete, we prefer to give our preliminary results as upper limits on fluctuations in the cosmic microwave background (at the 95% confidence level) :

On an angular scale of 18", $\Delta T/T \leq 3 + 10^{-4}$
On an angular scale of 40", $\Delta T/T \leq 2 + 10^{-4}$

Earlier work at higher resolution (Knoke, Partridge and Rotner, *Ap.J.* 1984) allows us to fix a limit at 4" : $\Delta T/T \leq 3.2 \times 10^{-3}$.

These results are in good agreement with the results of a similar experiment reported at this assembly by Dr K. Kellerman. They are also close to the predictions discussed here by Dr J. Ostriker (work with E. Vishniac). These predictions are of particular interest since they suggest values of $\Delta T/T = \theta$ (10^{-5}) at smaller angular scales than emerge naturally from inhomogeneities at the epoch of recombination. [work with Dr H. Martin and M. Rotner].

2) The VLA was used at $\lambda=4$ cm to look for the Sunyaev-Zel'dovich effect (the "cooling" of microwave photons by scattering from hot electrons in plasma in clusters of galaxies in one well-studied cluster, Abell 2218. Instrumental effects and the presence of sources limited the accuracy of this preliminary study to ~0.5mK. No convincing evidence for the Sunyaev-Zel'dovich effect was seen at either the nominal cluster center given by Abell, or the center used in the observations of Birkinshaw, Gull and Hardebeck (*Nature*, 1984), but the errors are such that the value reported for the effect by Birkinshaw et al., $\Delta T_{central}$ = - 0.8mK cannot be excluded by our data in its present form. [work with Dr N. Mandolesi and R. Perley].

9. **Gravity and galaxy clustering** (W.C. Saslaw - N.R.A.O, Univ. of Virginia, USA and Inst. of Astronomy, Cambridge, GB)

The asymptotic relaxed state of a statistically homogeneous gravitating distribution of point galaxies in the expanding universe can be described by a distribution function

$$\nabla^2 \phi_g = 4\pi G p_g(o) \, exp\{-\frac{m_g}{kT_g}(\phi_g + \phi_\nu)\}$$

$$\nabla^2 \phi_\nu = 4\pi G p_\nu(o) \, exp\{-\frac{m_\nu}{kT_\nu}(\phi_g + \phi_\nu)\}$$

for the probability of finding any number, N, of galaxies in a volume whos size is related to the average number density of galaxies by $N = nV$ and where $b \equiv -W/2K$ is a ratio of the gravitation correlation energy W to the kinetic energy K of peculiar galaxy motions. This distribution is in good agreement with both N-body computer experiments and with the observed clustering of galaxies in the Zwicky catalog. It includes the distribution of voids and contains much more information than just the low order correlation functions.

Using the cosmic energy equation and the linear solution of the BBGKY hierarchy, one can derive a simple evolution equation for b(t). It applies most accurately to the case of an initially unclustered distribution. The asymptotic value of b for non-linear clustering is $b \to (2\alpha + 1)/(2\alpha + 2)$ where the cosmic scale length $R(t) \propto t^\alpha$. The observed value of b is 0.1 ± 0.05 for the distribution of galaxies in our Universe and this agrees with the asymptotic distribution. The asymptotic value of b for non-linear clustering is $b \to (2\alpha + 1)/(2\alpha + 2)$ where the cosmic scale length $R(t) \propto t^\alpha$. The observed value of b is 0.1 ± 0.05 for the distribution of galaxies in our Universe and this agrees with the asymptotic limit for α near the Einstein-de Sitter value of 2/3.

The high agreement between the predicted f(N) distribution of galaxies will set significant constraints on the amount and distribution of dark matter in our Universe.

10a. On the tidal origin of angular momentum in galaxies (D.G. Lambas[1], M.B. Mosconi[2], J.L. Sersic[3] - [1]CONICOR, [2]Observ. Astronomico, [3]CONICET, Argentina)

We present the results of numerical simulations of relaxating protogalaxies under the tidal action of other similar systems and also clusters of galaxies. It is found that the bimodal behaviour of the observed angular momentum of galaxies can be explained under the assumption of different initial dynamical conditions induced by the evolving structure of the Universe expected in the Adiabatic Picture.

10b. Dynamical effects of dark matter in systems of galaxies (J.F. Navarro[1], D.G. Garcia Lambas[1,2], J.L. Sersic[1,3] - [1]Observ. Astronomico, [2]CONICOR, [3]CONICET, Argentina)

Several N-body experiments were performed in order to simulate the dynamical behaviour of systems of galaxies gravitationally dominated by a massive dark background.

We discuss mass estimates from the dynamics of the luminous component (M_{VT}) under the influence of such a background assuming a constant dark/luminous mass ratio (M_D/M_L) and plausible physical conditions. We extend in this way previous studies (Smith, 1980, 1984) about the dependence of M_{VT} on the relative distributions of dark and luminous matter (Limber, 1959). We found that the observed ratio of the virial theorem mass to luminosity (M_{VT}/L) in systems of galaxies of different sizes could be the result of different stages of their post-virialization evolution as was previously suggested by White and Rees (1978) and Barnes (1983). This evolution is mainly the result of dynamical friction that dark matter exerts on the luminous component. Thus our results give support to the idea that compact groups of galaxies are dynamically more evolved than large clusters, which is expected from the "hierarchical clustering" picture for the formation of such structures.

11. Are the filaments real ? (S.P. Bhavsar - Raman Research Institute, Bangalore, India)

We suggest to treat with caution when interpreting and comparing the visual impressions of filaments in the large scale distribution of luminous matter, both in two and three dimensional data sets. Numerical experiments show how sensitive global impressions obtained by our eye are to minor perturbations which effect nearest neighbour distances. An objective and quantitative method for finding and assessing the significance of filaments is necessary. A graph theoretical construction, the Minimal Spanning Tree (MST) has proven very successful. The construction and preliminary results of applying this method to the Zwicky and CfA Catalogues are detailed in a 1985 paper by Barrow, Bhavsar and Sonada (Mon. Not. R. astr. Soc. 216, 17-35).

12. Distribution of visible and dark matter in clusters and galaxies (R. Cowsik, P. Ghosh - Tata Institute of Fundamental Research, Bombay, India)

The question as to how the galaxies and the dark matter are distributed in clusters important in understanding their dynamics and in estimating Ω of the universe. Nothing that both the galaxies and the constituents of dark matter (referred to here as neutrinos for bravity) satisfy the collisionless Boltzmann equation an attempt is made to derive their distributions self consistently. To this end one assumes that their distributions are Maxwellian, consistent with Jean's theorem, and obtains the equations (with obvious notation) :

$$f(N) = \frac{\bar{N}(1-b)}{N1}[\bar{N}(1-b) + Nb]^{N-1}e^{-\bar{N}(1-b)-Nb}$$

These equations are coupled through the potential term in the exponential and yet their solution is effected in a manner similar to the Lane-Emden equation. The densities calculated thus are projected to get surface densities which reproduce well the observed profiles of galaxies in clusters. The distribution of dark matter, also displayed, is broarder and averages to $\Omega \approx 1$ over distances of ~ 20 Mpc corresponding to typical seperation between clusters.

13. Recent extragalactic observations and rational comparative cosmology (I.E. Segal - Dept of Mathematics, MIT, Cambridge, USA)

An equitable, non-parametric, and statistically efficient method is described for comparative testing of alternative cosmologies, on the basis of complete samples of designated discrete sources. In particular, objective determinations are possible via Monte Carlo statistics for the probabilities of deviations from a non-evolutionary Friedmann model observed in the magnitude-redshift relations of discrete sources. These significantly low probabilities can be considered to establish evolution, which is however incapable of direct observation. Moreover, the Friedmann deviations are just as predicted by the non-parametric, non-evolutionary chronometric theory, which itself fits the observations closely, for which the Friedmanntheory has no explanation. In addition the isotropy and other basic features of the microwave and X-ray cosmic backgrounds are direct non-parametric consequences of the chronometric theory, in which galaxy formation and nucleosynthesis result from large random deviations in a temporally homogeneous stochastic process modelling the Universe. For these and other reasons, the verifiably scientific character of the Friedmann model appears questionable.

14. Primordial Nucleosynthesis : A Status Report (G. Steigman - Bartol Res. Foundation, Univ. of Delaware, USA)

Primordial nucleosynthesis provides a unique probe for testing the standard, hot big bang model of cosmology. A detailed comparison of the theoretical predictions with the primordial abundances of the light elements inferred from observational data permits a test of the consistency of the standard model. As summarized by Boesgaard and Steigman (Ann. Rev. Astron. Astrophys. $\underline{23}$, 319 (1985)), the hot big bang model passes this test with flying colors. Provided that the nucleon to photon ratio exceeds $3-4 \times 10^{-10}$, the observed abundances of D, ^{3}He and ^{7}Li can be accounted foir by primordial nucleosynthesis. The final key test is to compare the predicted a abundance of ^{4}He with the observational data. Recent work of Gallagher, Schramm and Steigman suggests that the SMC should have a nearly primordial abundance of ^{4}He and finds $Y_P \approx 0.235$. This is consistent (within the uncertainties) with the standard model provided that there are at most four flavors of light neutrinos, a result consistent with the limit to neutrino flavors from collider experiments. Accounting for the uncertainties in the Hubble parameter ($45 \leq H_0 \leq 100$) and the microwave temperature ($2.7 \leq T_0 \leq 2.8K$), the above range for the nucleon to photon ratio corresponds to a present nucleon density between 1 and 20 per cent of the critical density ; nucleons may account for all the mass -including the dark mass- inferred from dynamical studies of galactic halos, groups, clusters and superclusters of galaxies.

15. Primordial nucleosynthesis and strange particles (J. Audouze - Institut d'Astrophysique de Paris and Laboratoire R. Bernas, Orsay, France)

Several attempts are currently made to reconcile the predictions of the Big Bang nucleosynthesis regarding the production of D, ^{3}He, ^{4}He and ^{7}Li with the existence of dark matter able to close the Universe or such that $\Omega = 1$ (as favoured by the inflation models). Several possibilities have been investigated by our group : massive neutrinos or gravitinos (Audouze, Lindley and Silk 1985), quark nuggets (Shaeffer, Delbourgo-Salvador and Audouze 1985) and photinos (Salati, Delbourgo-Salvador and Audouze 1986).

16. The very high luminosity infrared galaxies ; Are they primordial ?
(G. Burbidge - Univ. of California, San Diego, USA)

Observations made by IRAS have given us a great deal of new information concerning the far infrared (25μ-100μ) radiation from galaxies. From those data it is now clear that, depending on their far infrared luminosity and bolometric luminosity they can be divided into three groups :
a) Normal spiral and irregular galaxies in which on the average about 30% of the total flux is in the far infrared.
b) Spiral and irregular galaxies in which the far infrared flux dominates, but whose bolometric luminosities in the normal range of 10^{10}-10^{11} L_0.
c) Galaxies in which the far infrared flux dominates, and the bolometric luminosities are very much higher than those of normal galaxies. They lie in the range of 10^{12}-5×10^{12} L_0, and are the most luminous stellar systems known.

It is the galaxies in category (c) that we are speculating are at very early stages in their evolutionary history and may be primordial.

It is very reasonable to suppose that the first stars that form are very massive. In this case they evolve very rapidly and make many heavier elements which are ejected and condense into dust. Thus it is proposed that these massive stars are responsible both for the high luminosity, and the dust which transforms the stellar ultraviolet radiation into far infrared flux from dust clouds with temperatures of about 60-100 degrees Kelvin.

About eight galaxies of this type are known including IC 4553 (Arp 220) and NGC 6240.

Calculations show that the minimum mass in stars with masses in the range 20-120M_0 required to produce a luminosity of $\sim 10^{12} L_0$. Several generations of massive stars are required to produce the necessary conditions so that "ages" for the systems are about 10^8 years are indicated. We suppose that the initial star formation process is triggered by tidal interactions with companion galaxies.

Observational taks of this hypothesis are possible. If the idea is correct, these galaxies should have anomalous abundances of the heavy elements which should be detected both in optical spectra, and by studying the composition of the dust.

I would like to thank the National Science Foundation for research support.

17. Dark matter and exotic particles : In search of the perfect WIMP (G. Steigman - Bartol Res. Foundation, Univ. of Delaware, Newark, USA)

Various exotic particles have been proposed as candidates for the dark matter in the Universe but the nature and amount of dark matter remains a mystery. Dynamical studies of clusters and superclusters suggest a universal mass density only a fraction (0.2 ± 0.1) of the critical density. Such a mass density could be dominated by ordinary (nucleonic) matter. Inflation and the smoothness of the cosmic background radiation argue for a higher density Universe which would require non-nucleonic matter. Hot, warm, cold and decaying dark matter scenarios are reviewed and it is noted that, under the standard assumptions of adiabatic density perturbations with a Harrison-Zeldovich spectrum, none are successful in accounting for the large scale structure of the Universe.

19. Stellar Pancake nucleosynthesis (J.P. Luminet - Groupe d'Astrophysique Relativiste, Observatoire de Paris, Meudon, France)

Currently discussed sites of explosive nucleosynthesis are the primordial universe, supernovae and the surfaces of accreting compact stars. Here I recall that quite ordinary stars, neither massive or compact, can suffer by accident explosive nucleosynthesis when a strong <u>external</u> gravitational field acts as a detonator.

Giant black holes such as those expected in many galactic nuclei provide quite naturally such strong fields. As shown recently by Carter and Luminet, any star penetrating deeply inside the tidal radius of a massive black hole, or any couple of stars colliding at high velocity in the vicinity of a supermassive black hole, will undergo a short-lived phase of high compression and heating in flattened, "pancake" configuration.

The main astrophysical consequences are the ejection of a significant fraction of the stellar debris and the enrichment of the neighbouring interstellar medium in proton-rich isotopes.

COMMISSION 48: HIGH-ENERGY ASTROPHYSICS (ASTROPHYSIQUE DES HAUTES ENERGIE)

Report of Meetings, 19, 20, and 21 November 1985

PRESIDENT: Riccardo Giacconi

INTRODUCTION

The Scientific sessions of Commission 48 at the XIX General Assembly of the International Astronomical Union in Delhi, India, were focussed on "High Energy Galactic Phenomena". The three sessions, which were held on November 19, 20, and 21, dealt with "Cosmic Rays", "Very High and Ultra High Energy Gamma-rays from Compact Objects", and "X-rays and High Energy Gamma-rays". Each session consisted of invited discourses and contributed papers. Below a list of the invited discoursers is given:

COSMIC RAYS:

    Contemporary Problems and Perspectives in the Origin of Cosmic Rays. Dr. Ramanath Cowsik, TATA Institute of Fundamental Research
    Cosmic Rays Above 10 to the Power of 15 EV: Their Origin and Propagation. Dr. Arnold Wolfendale, University of Durham
    Cosmic Ray Acceleration by Shock Waves. Dr. Catherine Cesarsky, Service D'Astrophysique Institute de Recherche Fondamentale

VERY HIGH AND ULTRA HIGH ENERGY GAMMA RAYS FROM COMPACT OBJECTS:

    Model of Theories of High Energy Sources. Dr. Kenneth Brecher, Goddard Space Flight Center
    Experimental Observations of Gamma Rays. Dr. P.V. Ramana-Murthy, TATA Institute of Fundamental Research
    Production of Neutrinos and Other Particles. Dr. Thomas Gaisser, Bartol Research Foundation

X-RAY AND HIGH ENERGY GAMMA RAYS:

    Nonthermal Processes in X-Ray Binaries and AGNs. Dr. Jonathan E. Grindlay, Harvard Observatory Center for Astrophysics
    EXOSAT Observations of the Structure of Low Mass X-Ray Binaries. Dr. Nicholas White, European Space Agency (ESA)
    High-Energy Gamma Rays from Non-Compact Active Sources and Compact Objects. Dr. Wim Hermsen,

Laboratory for Space Research Leiden of the
National Institute for Space Research

In what follows, an abstracted version of the invited discourses is given.

Session 1 - Cosmic Rays.

a) Dr. Ramanath Cowsik reviewed contemporary problems in the study of cosmic rays with particular emphasis on their origin. According to the author:

> The problem of cosmic ray origin centers around finding an explanation for the relativistic corpuscular radiation which permeates the galactic environment with an energy density of $\sim 10^{-12}$ erg cm^{-3} comparable to other forms of energy and as those associated with turbulence and thermal motions of the interstellar gas.
>
> The weight of the observational evidence (Proc. ICRC, La Jolla, 1985, eds. Jones, Adams and Mason) and the theoretical studies suggest that bulk of the cosmic rays up to $\sim 10^4$ GeV/nucleon, originate in numerous compact sources distributed in a thick disc.
>
> The current discussion and debate is around the basic set of questions: a) where are the particles accelerated, in the general interstellar space or in compact sources? b) what is the region of storage, a thick disc or an extended halo? c) are the decreases in the ratio of primaries to secondaries with energy to be interpreted as indicative of sources shrouded with matter (nested leaky-box) through which the particles of higher energy pass with increasing facility before entering the general interstellar medium, or in terms of energy dependent transport out of the galactic volume (simple leaky-box)? d) what is the source material for the cosmic rays? e) how are the chosen particles injected efficiently into the accelerators? and f) what fraction is extragalactic?

Dr. Cowsik presented a summary of the current status of observational evidence and theory regarding the first two questions.

b) Dr. Arnold Wolfendale devoted his contribution mainly to the general problem of galactic vs extragalactic cosmic ray origin. The author pointed out that although a strong case exists for supernova remnants being responsible at energies below about 10^{10} eV, conventional models indicate a near impossibility of SNR being efficient above about 10^{14} eV and at these higher energies objects epitomised by Cygnus X-3 (which produces γ-rays to some 10^{16} eV, and thus protons to 10^{17} eV or so) are likely sources. At still higher energies no good candidates exist, as yet.

The author considers that the main problem above 10^{15} eV is the determination of the transition energy at which extragalactic particles start to predominate. Data from extensive air showers experiments show an increasing concentration of arrival directions from the general direction of the Galactic Plane with increasing energy. This work has been updated recently by Szabelski and collaborators with the inclusion of the latest results from Sydney and Haverah Park. The analysis appears to show a termination of the increased concentration at a little above 10^{19} eV, strongly pointing towards an extragalactic origin above this energy. It is likely that M87, with its dramatic jet is an important source of these extragalactic particles.

c) Dr. Catherine Cesarsky discussed the latest theoretical developments in the study of galactic cosmic ray acceleration by shock wave.

This attractive acceleration mechanism was introduced, a few years ago, and simultaneously, by several groups from all over the world (Krymsky, 1977; Axford et al., 1977; Bell, 1978; Blandford and Ostriker, 1978). The basic ideas are:

1) every time a relativistic particle of energy E crosses a shock of velocity V, it suffers an energy increase proportional to EV/c;

2) if particles can be retained for a long time in the shock vicinity by a scattering mechanism, they can cross the shock a large number of times, and their energy can be boosted by a large factor. Under some general conditions and with some first-order approximations (e.g. test particle approximation for the cosmic rays), the processes predicted a power law spectrum of index (-2) similar enough to that observed.

It seemed that, at last, the acceleration of galactic cosmic rays had been understood: it was the result of the interactions of cosmic rays and supernova shocks in the interstellar medium (Blandford and Ostriker, 1980; Axford, 1980).

It was also immediately realized that, for s.n. shocks in the interstellar medium, the test particle approximation is not valid, and the work on the non-linear case started immediately.

The most recent work by Eichler and Ellison (1985) shows that taking into account more realistic conditions prevailing in the interstellar medium a similar spectrum can be obtained with an efficiency of acceleration of 25%. Unfortunately however, the author points out, the great drawback of the shock acceleration mechanism is that it is slow. Consequently, when applied to realistic shocks, which have a finite lifetime, the theory predicts a high energy cut off. Since non linear effects slow down the acceleration, an upper limit to the maximum energy can be derived with the linear theory. In the case of the acceleration of cosmic rays by a supernova shock, the most optimistic assumptions on the diffusion coefficient lead to a maximum proton energy of about 10^5 GeV. Using a self-consistent theory, where the scattering is due to cosmic ray-generated waves, the maximum proton energy decreases to values as low as 2000 GeV. This is the major problem encountered by this theory in the context of galactic cosmic ray acceleration. It cannot be solved by invoking acceleration of supernova shocks in the galactic halo (Lagage and Cesarky, 1985).

Session 2 - Very High and Ultra High Energy Gamma-Rays from Compact Objects

a) Dr. Ramana-Murthy reviewed the observations of very high and ultra high energy gamma-rays from compact objects. His summary is given in Table I.

b) Dr. Kenneth Brecher reported on the cosmic-ray acceleration in UHE and VHE gamma-ray sources.

VHE ($>10^{12}$ eV) and UHE ($>10^{15}$ eV) γ-ray emission has been reported from the X-ray source Cygnus X-3, as well as several known X-ray binaries, including Her X-1, LMC X-4, Cen X-3 and 4U 0115 + 63. These observations, if subsequently found to be statistically significant, imply the acceleration of high energy cosmic rays in these sources. Any model of the acceleration of these particles would have to satisfy several criteria:

(1) Gamma-rays with energies of up to 10^{16} eV have been detected, implying initial particle energies (hadrons) of 10^{17} eV. assuming that the gamma-rays arise from pion decay.

(2) The total cosmic-ray luminosity of these sources must be at least 10^{38} erg/sec, perhaps as high as 10^{39} erg/sec in the case of Cygnus X-3.

(3) UHE particle production may be the major energy loss mechanism for these sources (though non-relativistic bulk gas ejection may also be important.)

(4) The spectrum of accelerated particles could be mono-energetic, even though the gamma-ray spectrum has an E^{-2} power law photon distribution.

TABLE I. VERY HIGH AND ULTRA HIGH ENERGY GAMMA RAYS FROM COMPACT OBJECTS: OBSERVATIONS

Ser. No.	OBJECT	COORDINATES R.A. h-m	DECL. DEG.	APPROX. DIST. kpc	APPROX. PERIOD	V.H.E.G.R. TIME AVG. INTEGRAL FLUX (≥ 1 TeV) $cm^{-2} s^{-1}$	V.H.E.G.R. TIME AVG. LUMINOSITY ($\geq$ TeV) ergs s^{-1}	U.H.E.G.R. TIME AVG. INTEGRAL FLUX (≥ 1 PeV) $cm^{-2} s^{-1}$	U.H.E.G.R. TIME AVG. LUMINOSITY (≥ 1 PeV) ergs s^{-1}
1.	CRAB PULSAR	05-32	+22	2.0	33.3 ms	4.10^{-12}	6.10^{33}	?	?
2.	VELA PULSAR	08-34	-45	0.5	89.2 ms	3.10^{-12}	3.10^{32}	?	?
3.	PSR 1937+21	19-37	+21	5	1.56 ms	2.10^{-11}	2.10^{35}	?	?
4.	PSR 1953+29	19-53	+29	3.5	6.13 ms	$1.2.10^{-12}$	6.10^{35}	?	?
5.	CYG X-3	20-31	+41	$\gtrsim 11.4$	4.8 hr 12.6 ms	5.10^{-11} ?	$\gtrsim 3.10^{36}$?	2.10^{-14} ?	$\gtrsim 10^{36}$
6.	HER X-1	16-56	+35	5	1.24 s	3.10^{-11}	3.10^{35}	$3.3.10^{-12}$ *	$1.6.10^{37}$ *
7.	4U0115+63	01-15	+63	5	3.61 s	7.10^{-11}	6.10^{35}	?	?
8.	CEN-A	13-22	-43	4400	Steady flux	4.10^{-12}	3.10^{40}	?	?
9.	CRAB NEBULA/PULSAR	05-32	+22	2.0	DITTO	10^{-11}	$1.6.10^{34}$	1.10^{-13}	$1.5.10^{35}$
10.	M 31	00-40	+41	670	DITTO	$2.2.10^{-10}$	4.10^{40}	?	?
11.	VELA X-1	09-00	-40	1.4	8.96 d	?	?	$9.3.10^{-15}$ **	$2.3.10^{34}$ **
12.	LMC X-4	05-32	-66	50	1.41 d	?	?	$4.6.10^{-15}$ ***	10^{38} ***

* Flux and Luminosity about 5.10^{14} eV.
** Flux and Luminosity above 3.10^{15} eV.
*** Flux and Luminosity above 10^{16} eV.

(5) Particle acceleration must be fast (seconds or less), in order for the particles to escape the acceleration region without major energy loss before hitting the presumed gas target giving rise to the observed VHE and UHE gamma-rays.

The author discussed the three different kinds of models proposed to date to account for the cosmic ray flux from these sources.

Pulsar Acceleration. Pulsars are known to accelerate particles to high energies, in the case of the Crab nebula, electrons with energies of at least 10^{11} eV. A similar mechanism may apply to Cygnus X-3, a source with a 4.8 hour gamma-ray periodicity, but with a possible shorter underlying (pulsar?) periodicity in the 1 - 100 ms range. While such a model could be made to fit the properties of Cygnus X-3, it cannot fit the observed luminosity of the four other reported VHE and UHE gamma-ray sources because of their longer observed pulse periods. At least for these sources an alternative energy source, derived from accretion rather than rotational energy, is required.

Shock Acceleration. Since we are dealing with accreting binary systems, the ultimate energy for the accelerated particles comes from accretion, rather than rotation, thus allowing for a long lived source. A standing shock near the polar cap can accelerate protons to high energies. However, these particles suffer severe energy losses by synchrotron radiation and other processes. The maximum accelerated particle energy is achieved by equating the acceleration to loss times and, for reasonable parameters, gives $\gamma_m \lesssim$ few $\times 10^7$.

Unipolar Induction. This model in a sense combines some of the features of pulsar acceleration models with the shock acceleration model, in that the particles are accelerated by a parallel electric field, but the ultimate energy source is from accretion. For weak enough magnetic fields (B $\lesssim 10^9$ gauss) and strong enough accretion rates (L $\gtrsim 10^{39}$ erg/sec), a potential drop of 10^{17} volts can develop. If the Alfven surface lies just above the neutron star surface, the accelerated particle luminosity can equal the accretion luminosity. Since particles, not photons, carry off the energy, the total accretion luminosity can excede the normal (photon) Eddington luminosity by a factor of 10 - 100, thus allowing for the high observed non-thermal luminosity of Cygnus X-3.

The author concludes that any of these models is at least plausible.

c) Dr. Thomas Gaisser reported on the production of neutrinos in close binaries.

If a compact object in a binary system accelerates protons to high energies, it is natural to expect production of high energy secondaries when the accelerated particles collide with target nucleons in the system. The secondaries will include high energy photons from decay of neutral pions produced in the collisions.

Charged pions will also be produced. These will either interact and contribute further to the cascade or decay to muons and neutrinos. Therefore, if this model correctly explains very high energy gamma rays from binaries, then these systems should also be potent neutrino sources. The author made two points:

1) The expected neutrino flux is large, but (because of the small cross section for neutrino interaction) the neutrino-induced signal is too small to explain the reported underground signals from the direction of Cygnus X-3. Neutrino signals <u>may</u>, however, be large enough to show up in underground detectors of large area (≥ 1000 m^2), and they <u>would likely</u> be detectable in an area as large as proposed for a deep underwater muon and neutrino detector ($\sim 10^5$ m^2).

2) Absorption of neutrinos with energies ≥ 2 TeV deep in the companion star is likely to be very significant and may give rise to upper limits on the total cosmic ray luminosity of the system and to mechanism for quenching the high energy signals periodically.

Session 3 - X-Rays and High Energy Gamma-Rays

a) Dr. Jonathan Grindlay discussed nonthermal phenomena in X-ray binaries and AGNs. He pointed out that:

Whereas it is customary to think of the primary processes occurring in and around compact X-ray binaries as being thermal ones, many of these same systems show striking non-thermal behavior in the form of variable radio emission, jets and high energy spectra. In this sense, the physics of these objects may closely resemble that in active galactic nuclei (AGN). Although the possible similarities have been pointed out many times before, particularly for sources such as SS433, work carried out in collaboration with L. Molnar and D. Band points out more directly the possible links between Cygnus X-3, SS433 and AGN.

The peculiar X-ray binary Cygnus X-3 produces relatively intense radio flares with an apparent period of 4.95 hours, which is significantly displaced from the 4.8 hour X-ray period. These flares constitute the normal "quiescent" radio emission from the source and are probably the low end of a size spectrum which extends up to the giant radio outbursts which the source seems to produce each September-October. The radio flares show a striking frequency versus time dependence which strongly suggests an expanding synchrontron source and indeed our recent VLBI observations at 1.3 cm wavelength directly measure the expansion of the source to be 0.4 mas/hr with an axial ratio of about 2:1 in the north-south direction (refs. 2,3). This shows that the binary is ejecting relativistic electrons into jets roughly once each binary orbit, although during the giant flare events the injection is strong enough to be directly resolved in the jets even a month after the outburst. The total energy in relativistic electrons in each "quiescent" radio flare is probably in excess of 10^{41} ergs, so that the total luminosity in non-thermal particles is at least 10^{37} erg/ sec, or comparable to the total observed accretion luminosity of the system. An underlying hard X-ray spectrum (non-thermal) with spectral index comparable to the radio index (0.6) should therefore accompany the flares; Compton-synchrotron models for the source are being calculated.

SS433 shows radio flares which also inject relativistic electrons into the jets and which do in fact accompany apparently non-thermal X-ray flares of the system. The more luminous central-most X-ray source is apparently buried so that only the surrounding non-thermal emission region is seen during flares and the additional thermal emission (producing Doppler-shifted iron line emission) from hot plasma entrained in the jets. Recent calculations for the origin of the non-thermal X-ray emission show that it may be produced by inverse Compton scattering of electrons in a central synchrotron source on the thermal IR and optical photons produced in the accretion disk or companion star and surrounding dust. The detailed calculations show that the non-thermal X-ray flares cannot arise from the same extended region in the jets as the observed radio emission without gross violations of equipartition; instead the X-ray flares arise in a central optically thick synchrotron source from which the plasmons in the jets are probably ejected.

A very similar Compton-synchrotron model for the X-ray emission (as well as far-IR through optical spectrum) for radio-quiet QSOs and AGN (e.g. Seyfert 1s) has been developed.

b) Dr. Nicholas White described EXOSAT Observations of the Structure of Low Mass X-ray Binaries.

The study of the X-ray properties of Low Mass X-ray Binaries (LMXRB) has been revolutionized by the capability of EXOSAT to make long uninterrupted observations lasting up to 80 hours and by its fast (up to 5000 Hz) temporal resolution. Five new orbital periods have been established with periods ranging between 2.9 and 4.4 hours and one at 21. hours. These periods manifest themselves as irratic dips in the X-ray flux that recur periodically. They are thought to be caused by a splash of material at the edge of an accretion disk at the point where the gas stream from the companion collides with the disk.

The orbital period distribution of the LMXRB is centered between 2.9 and 7. hours. Orbital modulations have only been convincingly detected in the lower luminosity system (~10^{37} erg/s). The bright galactic bulge sources with luminosities of ~10^{38} erg/s, do not show dips or eclipses indicating their orbital periods are longer than a few days. This supports the view that there is a dichotomy in the properties of the LMXRB with the high luminosity systems driven by the evolution of a giant, whereas, the lower luminosity systems are driven by gravitational radiation/magnetic braking.

Quasi-periodic Oscillations (QPO) have been discovered by EXOSAT, first from GX5-1, and then subsequently from the following six sources: Cyg X-2, Sco X-1, GX17+2, GX349+2, the rapid burster and 4U1820-30. The QPO frequency ranges from 20-35 Hz in GX5-1 to 2-5 Hz from the rapid burster, with the remainder in between.

The frequency-intensity dependence of the QPO comes in two flavors. First, the highest frequency QPO (>10 Hz) show a strong dependence between frequency and intensity (typically to the power 2 or 3). When the QPO are at a lower frequency they are not strongly correlated with intensity and if anything show a slight anti-correlation with intensity. The typical RMS amplitude of the QPO in these seven sources ranges from 1% up to 10%. In the rapid burster the behavior of the QPO is quite different from the other sources. Here QPO appear both during bursts and in persistent emission found between bursts. The frequency seen during a burst is inversely correlated with the peak flux during the burst. In between bursts, the QPO execute an S shaped pattern in frequency from 2 to 4 Hz and then appear in the following burst at the final frequency seen in the S pattern. The most puzzling aspect of this source is that the QPO are not seen all the time, but only during ~50% of the bursts and only very infrequently in between bursts.

The models for QPO are as varied as the number of different properties being found by EXOSAT. The most promising seems to be the beat frequency model where the rapidly rotating magnetosphere of a millisecond pulsar interacts with the inner accretion disk. This model, if correct, fits in with evolutionary models for the millisecond radio pulsars where the pulsar is re-cycled back to a rapid rotation period in a LMXRB. On the other hand non-magnetic models for QPO are also being proposed where the QPO are generated in an inner accretion disk corona or in the boundary layer between an accretion disk and the neutron star.

c) Dr. Wim Hermsen discussed the observation of high energy gamma-rays from non-compact active sources and compact objects.

The detection of localised sources of high energy gamma radiation, first by SAS-2 and later more comprehensively by COS-B, has led to much discussion regarding the physical nature of the objects that number 25 in the 2CG catalog. Only the Crab and Vela pulsars have been unambiguously identified; the ρ-Oph cloud has subsequently been resolved; and 3C273 and the X-ray source 1E0630+178 have also been proposed as counterparts. The status of the remaining sources is much less clear. An exhaustive review has been given by Bignami and Hermsen and additional papers discussing the observations and possible models can be found in the proceedings of the meeting on 'Galactic Astrophysics and Gamma-Ray Astronomy' (Morfill and Buccheri, 1983) organized during the General Assembly of the IAU in Patras in 1982.

Of the gamma radiation observed above 100 MeV only a few percent is due to the cataloged sources which are viewed against intense background emission from the galactic plane. Detailed modelling of the galactic plane emission as being due to the interactions of cosmic rays with atomic and molecular interstellar gas demonstrates that the large angular scale features of the gamma-ray intensity distributions are well reproduced in this way. The analysis has been extended to small angular scales showing which of the 2CG sources might be due to conventional levels of cosmic rays within clumps of gas and which cannot be so explained. A possible scenario for some of the sources is the interaction of SNR's with interstellar clouds. For such a SNR-cloud coincidence Pollock argues for identification of 2CG006+00 and 2CG078+00 with the synchrotron-emitting region behind the shock front. Alternative explanations have been proposed, e.g. contributions of stellar winds or close binary systems.

COMMISSION 49: THE INTERPLANETARY PLASMA AND THE HELIOSPHERE
PLASMA INTERPLANÉTAIRE ET HELIOSPHÈRE

Report of Meetings 21, 22, 26, 27 November 1985

PRESIDENT: I. W. Roxburgh

27 November 1985

BUSINESS MEETING

The President reported on activity since the last General Assembly in 1982, and that Dr. Burlaga had been nominated for Vice President.

The Commission discussed the suggestion of reconstituting Commission 43 on Astrophysical Plasmas; this suggestion received full support - members emphasising the importance of contacts between those working in heliospheric physics and people working in other areas of plasma astrophysics - and indeed laboratory plasma physics.

22 November 1985

SCIENTIFIC MEETING

The following communications were presented:

B. Buti: Some non linear processes in the Interplanetary Plasma
S. Grzedzielski: Long Magnetospheric Tails in the Heliosphere
S. Grzedzielski: Sources of High Energy Neutral Atoms in the Solar System

21 and 26 November 1985

JOINT MEETINGS WITH COMMISSIONS 10 AND 12.

21st November: Reconnection in Astrophysical Plasmas

26th November: Coronal Activity and Interplanetary Disturbances.

A report on these meetings is given in the report of Commission 12.

COMMISSION 50 : PROTECTION OF EXISTING AND POTENTIAL OBSERVATORY SITES
(PROTECTION DES SITES D'OBSERVATOIRES EXISTANTS ET POTENTIELS)

Report of Meetings, 20 and 21 November 1985

PRESIDENT : A. Hoag

20 November 1985

SCIENTIFIC SESSION I

This meeting, organized and chaired by Vernon Pankonin, National Science Foundation, U.S.A., was begun at 11:00 a.m., and approximately 40 persons attended. Presentations were made as follows:

V. Pankonin, National Science Foundation, U.S.A., started the session with an **Introduction and Overview of Radio Spectrum Management.** There is a wide base of interest in radio frequency interference (RFI) at both the local and the international levels. The noise environment in the higher telecommunication frequencies has been augmented by satellites and at lower frequencies by increased traffic. In spite of care in frequency allocation, technical imperfections such as side band spillover contribute to RFI. International organizations concerned with radio frequency allocations and control include ITU, the International Telecommunications Union; WARC, the World Administrative Radio Conference; CCIR, the International Radio Consultative Committee; and the IFRB, the International Frequency Registration Board. International policies, regulations, and allocations are made in the following way: Scientific requirements are formulated and communicated through the IAU, URSI (Union Radio Scientifique Internationale), or COSPAR (Committee on Space Research) to the Inter-Union Commission on Frequency Allocation for Radio Astronomy and Space Science, IUCAF, whose recommendations then go to the WARC and CCIR and thence to the ITU for regulatory administration. Major international RFI problems remain. Among them are geostationary and navigational satellite transmissions.

G. Swarup, Tata Institute of Fundamental Research, Bangalore, spoke about **Radio Noise Surveys for India's Giant Meter-Wavelength Radio Telescope.** He described plans for a 14-km sized Y-array radio telescope designed to work in the 21-cm to meter wavelength region. Swarup, who represents the IAU in the work of the Inter-Union Commission on Frequency Allocations for Radio Astronomy and Space Science, has discussed these plans at World Administrative Radio Conferences. Sites in India have the advantage of being near the Earth's magnetic equator, but a place must be found having sufficiently low levels of interference from radio, power-line, and vehicle ignition emissions. RFI measures made over extensive ranges of time for a few sites were presented as power versus frequency plots. Additional site factors and future prospects were discussed.

V. Pankonin, NSF, U.S.A., gave a brief description of the administration of **Protection of Radio Astronomy in the U.S.A.** All U.S. governmental radio transmission is coordinated by the National Telecommunications and Information Administration (NTIA). Non-governmental radio transmission is regulated by the Federal Communication Commission (FCC), and overall coordination is achieved by means of an Interdepartmental Radio Advisory Committee. Specific site protection has been provided at Greenbank by the estabishment of a National Radio Quiet Zone and at the VLA by designation of a Coordination Zone where emission in certain frequency bands is limited in a specific geographical area. Protection for components of the Very Long Baseline Array, now being developed in the United States, will be provided by individual Coordination Zones.

A paper by J. Carneiro and A. Magalhâes, Universidade do Porto, Portugal, **Radio Frequency Interference to an Optical Telescope,** was read by J. Osório. The 76-cm telescope of the Astronomical Observatory of the University of Porto has been subject to radio interference problems caused by three transmitters 100 meters distant, each having 10 kw input power. Two RFI problems have been encountered. Until four years ago, when use of an AM transmitter terminated, serious problems were encountered with the servo tracking and pointing system that included frequent destruction of integrated circuit components. Sometimes the broadcast could actually be heard from the servo tracking motor. No fully effective shielding and grounding methods could be found. The two remaining FM transmitters have caused severe interference problems when photon-counting photometric equipment has been used. In this case, empirical changes in shielding, filtering, and grounding have made photometric measures possible.

Weather RADAR - A Possible Threat was the subject of a report by A. Hoag, Lowell Observatory, U.S.A. The U.S. National Weather Service is planning a countrywide system of doppler RADARs for detecting and mapping storms and for weather surveillance at airports. Each of approximately 150 units in the system is to consist of an 8.5-meter parabolic antenna that will transmit a one-degree beam that scans in azimuth at elevations from -1° to 20°. The properties of emission are to be as follows: Frequency = 2700-3000 MHz; Peak Power = 750 kw, Duty Cycle = 0.002; Average Power = 1.5 kw; Bandwidth = 8 MH ; Pulse Width = 1.5-4 μs; Pulse Repetition = 25-1250 sec^{-1}; Rise-Fall times = 0.4/0.4 μs. An experiment has been carried out by the Weather Service and a consortium of Arizona observatories to determine the susceptibility of astronomical equipment to this radiation. Preliminary results indicate that optical through near-infrared detectors, when shielded and grounded according to current practices, show little susceptibility to RADAR output at anticipated power levels but that some submillimeter to radio detection systems may be subject to hazard.

<u>21 November 1985</u>

SCIENTIFIC SESSION II

This was the first of two meetings on the subject of identification and protection of observatory sites chaired by Sidney van den Bergh, Dominion Astrophysical Observatory, Canada. The meeting took place between 9:00 and 10:30, and 26 persons attended.

Results of an **Optical Site Evaluation in Saudi Arabia** were discussed by E. Brosterhus, Dominion Astrophysical Observatory, Canada. He described a May 1982 to October 1985 survey in the Kingdom of Saudi Arabia that was a cooperative venture between the Saudi Arabian National Center for Science and Technology and the National Research Council of Canada. Of the four sites surveyed, three were used to explore the Asir range that borders the Red Sea. Two of these were used to evaluate the highest available sites, which are near the coast, while site number 3 was 35 km inland. The fourth site was on a relatively low escarpment in the central part of the country. Computer-based telescopes were used to collect image-quality and photometric-quality data of unprecedented precision and completeness. These data and meteorological observations indicate that all four sites are of high quality but that the inland site in the Asir range offers more clear hours than the sites near the coast as well as excellent seeing. It appears that there may be many sites having the characteristics of site 3. A supplemental aereal survey has been made to identify additional possibilities that could be considered for final selection and testing.

J. Davis, University of Sydney, Australia, described **An Interferometric Seeing Monitor – Measuring r_0 Directly.** The Chatterton Astronomy Department of the University of Sydney is developing a new high angular resolution stellar interferometer. As part of this program, J. O'Byrne has built an instrument to measure the atmospheric coherence diameter, r_0. A folded-shearing interferometer is used to image a 35-cm telescope aperture, via a three-stage image intensifier, onto a cooled 50×50 Reticon array detector. The aperture image contains interference fringes whose time-averaged profiles correspond to the atmospheres' long exposure MTF. At the end of each several-second exposure, the Reticon output is digitized and analyzed to yield a value of r_0. The instrument has a limiting sensitivity of $B = 2$, and it measures r_0 to ±5%. An observational program is under way to correlate r_0 values with meteorological parameters and with observational data from the 11.4-meter baseline prototype stellar interferometer. The instrument will also be used to compare astronomical sites and, in particular, to evaluate the proposed site for a new 640-meter baseline stellar interferometer.

BUSINESS SESSION I

H. Ables announced that an international conference on **Identification, Optimization, and Protection of Optical Telescope Sites** will be held in Flagstaff, Arizona, during 22-23 May 1986, under the joint sponsorship of the Lowell and U.S. Naval Observatories. R. L. Millis, Lowell Observatory, is the chairman of the organizing committee.

A. Hoag distributed a summary of past IAU Resolutions having to do with protection of observatory sites and of experimental techniques as a basis for consideration of further resolutions concerning contamination of space and of the electromagnetic spectrum.

SCIENTIFIC SESSION III

The second meeting on the subject of identification and protection of observatory sites chaired by S. van den Bergh; 11:00-12:30, with 36 in attendance.

La Palma as an Observing Site and the Protection of Its Future, a report by P. Murdin, Royal Greenwich Observatory, United Kingdom, and F. Sanchez, Instituto de Astrofisica de Canarias (IAC), Tenerife, was presented by Murdin. An extensive review of the quality of the Tenerife and La Palma sites and the astronomical facilities there is the subject of a special issue of **Vistas in Astronomy** (Vol. 28, Pt. 3, 1985). Murdin emphasized the role of the Carlsberg Automatic Meridian Circle at La Palma in providing systematic records of image quality and atmospheric extinction at this major observing site. The La Palma site (2400 meter altitude) has proven to be of such good quality that a major effort has been made to protect its future. As part of the Royal Inauguration of the IAC Observatories in 1985, the Autonomous Government of the Canaries passed a law designed to protect the observatories on La Palma. Outdoor lights emitting 20,000 lumens or more are required to be monochromatic (low-pressure) sodium and beamed downward. Radio transmitters emitting more than 250 watts will be approved by the IAC only if the power at the observatories is less than the Commission 50-recommended upper limit. Any industrial activity above 1500 meters in elevation (the height of the inversion layer) is subject to approval by the IAC. These regulations are retroactive and are expected to be ratified and financed by the national government of Spain in 1986.

In a report of **Site Testing for an Infrared Telescope Near Leh, Ladhak,** A. Bhatnagar of the Udaipur Solar Observatory, India, described an interinstitutional search for a site for infrared and solar coronographic observations in the Ladhak region of the Himalayas. The Ladhak plateau is a high and dry desert region with an annual rainfall of only 120 mm and is relatively free of clouds during the monsoon months when the rest of India is cloudy. Preliminary data from the base (3700 meters) and summit (4100 meters) of Mount Nimmu, near Leh, show promising results as far as precipitable water vapor, sky brightness, wind, microthermals, and cloud cover are concerned. Regular observations of nighttime "seeing," microthermal fluctuations, extinction, and sky brightness are being supplemented by hourly data from an automatic meteorological station. Leh, a town of 5000 persons, and having minimal outdoor lighting, is accessible all year by air and by highway.

A. Ardeberg, Lund Observatory, Sweden, and the European Southern Observatory, summarized an **ESO Site Survey in Northern Chile.** In connection with plans for a Very Large Telescope, sites in the high (2300-6100 meters), dry area of Chile north of -25° latitude are being compared to La Silla, which is well known for good observing conditions. A reference site at Paranal, a 2700-meter coastal mountain, has been occupied since September 1983. Cloud cover, precipitable water vapor, wind, and thermal properties have been monitored continuously. Approximately 80% of the nights at Paranal are photometric, compared to about 60% for La Silla; and it is drier, precipitable water being frequently less than 1 mm. A dozen sites have been selected for further investigation, and several of them are being studied regularly. For the higher inland sites there is some increase in cloud cover and wind, but conditions can be even drier than at Paranal. Systematic image-quality measures will now be made at a number of sites. Test data so far obtained clearly indicate that northern Chile is an excellent place for optical, infrared, and submillimeter astronomy.

G. Teleki, Beograd, Yugoslavia, spoke about **Astrometric Site Selection,** stressing special needs for ground stability, a homogeneous temperature regime, and an even distribution of clear weather day and night for continuity in observations. Global evaluation shows that there are only a very few locations where these conditions may be met. Harlan Smith pointed out in discussion that modern space techniques are providing new information about ground stability.

SCIENTIFIC SESSION IV

The first of two meetings on the subject of atmospheric extinction and volcanism; 14:00-15:30, and 24 in attendance.

G. W. Lockwood, Lowell Observatory, U.S.A., started the first of two sessions on atmospheric extinction that he subsequently chaired by reporting **The Effects of Volcanic Eruption on Atmospheric Transparency.** He illustrated a variety of methods now used to estimate mass loading of the atmosphere by volcanic aerosols. Photometry of sunrise and sunset observed from orbiting SAM and SAGE satellites provides a sensitive means of detecting suspended material and stratospheric clouds. Alternative methods include LIDAR backscatter measures that have been routinely made from Hampton, Virginia, and from Mauna Loa, Hawaii, since 1976. In addition, sun photometers operating at Mauna Loa and at the Atmospheric Physics Department of the University of Arizona in Tucson provide systematic atmospheric transmission data. Nighttime measurements of atmospheric transparency derived from stellar photoelectric photometry have been made frequently with one telescope at Flagstaff (elevation 2200 meters) since 1955. Removal of the normal seasonal variation, which has a range of 0.05 in optical depth, reveals a record of changes strongly associated with volcanic activity. Dips in transparency are seen following eruptions of Agung, Hekla, de Fuego, St. Helens, and El Chichón. The increase in optical depth has been typically on the order of 0.05, except for El Chichón, which was much larger. The spectrophotometric properties of the El Chichón cloud transmission were investigated with a spectrum scanner during and following May 1982 during the course of a detailed Sun-Vega comparison. The Flagstaff data are highly correlated with published transmission values obtained from Mauna Loa (latitude 20°), much less so with values obtained from Davos, Switzerland (latitude 46°).

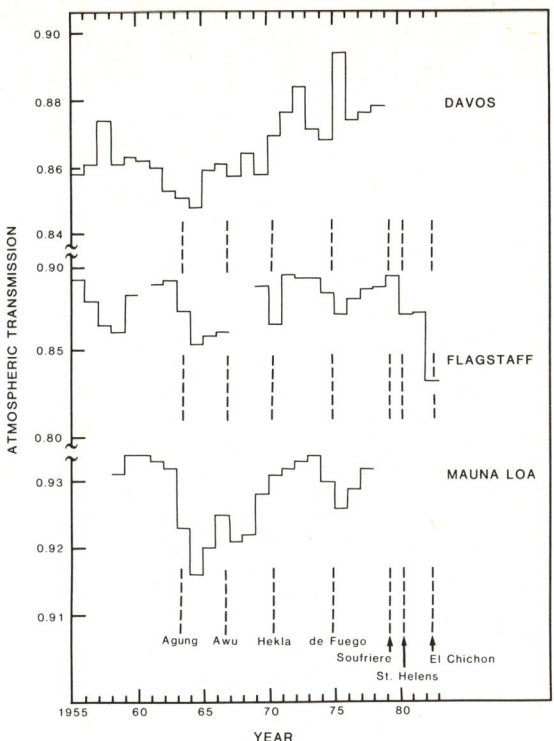

Figure: Annual mean transparency trends at Davos, Switzerland (latitude 46°, elevation 1600 meters) [Hoyt and Fröhlich, 1982 preprint], Flagstaff (latitude 35°, elevation 2200 meters) [Lockwood, unpublished], and Mauna Loa (latitude 20°, elevation 3500 meters) [Mendonca, Hanson, and DeLuisi, 1978 **Science** 202, 513].

While some authors dealt with volcanism as a photometric hazard, a paper by C. Blanco, Osservatorio Astrofisico di Catania, Italy, **Volcanic Activity and Astronomical Observations,** which was read by M. Fracastoro, was used to describe successful observatory operation on the slope of an active volcano. If one avoids those volcanoes on active continental margins, good sites can be found on active volcanoes having suitable orographic properties. Mount Etna is an example. The Serra la Nave station of the Catania Astrophysical Observatore is located on a reverse slope on Mount Etna at an elevation of 1725 meters and 90° from the direction of the prevailing wind. The reverse slope provides protection from lava and from thermal mixing in the boundary layer between subsiding air and the free air of the prevailing wind. Records show that only 20 nights have been rendered nonphotometric by volcanic activity since 1965. Further, systematically collected extinction coefficients show no correlation with volcanic activity or with season, and values are typical of those from other good sites at similar elevations.

B. Hidayat and H. L. Malasan, Bosscha Observatory, Java, described dust phenomena and **Coefficients of Extinction at Lembang** in connection with 1982 eruptions of Mount Galunggung. This volcano, some 60 km southeast of the Bosscha Observatory, persistently emitted dust during March to October 1982 and covered some 10^5 km² of West Java with dust and ash. During the peak of the eruption in July 1982, the daytime sky became as dark as at normal civil twilight, and dust fallout was detected 700 km away in Sumatra. However, a change in wind direction and a clearing rain permitted standard extinction measures to be made on 18 July, when values of k_v near 1 mag. were obtained. Measures resumed in April 1983 and continued through 1984 showed a steady decline toward normal values of $k_v < 0.2$.

BUSINESS SESSION II

Ratification of membership, organizing committee, and officer lists prior to submission to the IAU-UAI Secretariat.

SCIENTIFIC SESSION V

The second meeting on the subject of atmospheric extinction chaired by G. W. Lockwood; 16:00-17:30 and 24 in attendance.

Atmospheric Extinction at Kavalur Observatory During 1980-85, B. S. Shylaja and J. C. Bhattacharyya, Indian Institute of Astrophysics, Bangalore, was discussed by Shylaja. Observations made with an automated spectrum scanner (M. K. V. Bappu, 1977, **Kodaikanal Obs. Bull.** Series A2, 64) on 40 nights have been used to monitor extinction at the Kavalur Observatory in the range $\lambda\lambda 4000$-7700 Å. Components due to Rayleigh scattering, aerosol scattering, and ozone and water vapor absorption have been separately analyzed and correlated with meteorological information. A small systematic increase in aerosol scattering has been observed during the interval 1980-1985. Ozone absorption was significantly larger in 1980-1981 than during the rest of the interval, a change confirmed by the meteorological measures. The spectrophotometric water vapor measures are correlated with meteorological determinations made at Bangalore.

A study of **Extinction Coefficients at the Xinglong Station of the Beijing Observatory** by Huang Lin, Guo Zi-he, Jiang Shi-yang, Zhang Rong-xian, and Zhang Ji-tong was reported by Huang Lin. Visual extinction values collected from site-testing measures made in 1964 and from five photometric programs carried out from then until June 1985 show appreciable scatter and a slight systematic increase with time. Color extinction terms, however, show little change, indicating that variations in k_v may be caused by dust particles which produce neutral extinction. Seasonal analysis shows that a minimum number of large values of k_v occur during June to November, which is the calm rainy season when dust is minimized. From seven to twelve months after the El Chichón eruption there is a group of k_v values having small scatter but larger values than the mean of the 1980s, which may represent a detection of volcanic dust.

F. Rufener, Observatoire de Genève, Switzerland, described **The Evolution of Atmospheric Extinction at La Silla.** He based his discussion on precise values of extinction coefficients determined by the M and D method (Rufener 1964, **Publ. Obs. Genève**, A, 66, 413) on 452 nights uniformly distributed over the period November 1975 to March 1985. The method makes use of repetitive observations of mounting and descending (M and D) stars at about the same air mass and allows for gradual isotropic changes in extinction that may occur during the night. Extinction coefficients in [U], [B], and [V] show monotonic decreases in values during the interval 1975 to April 1982 with color behavior suggesting scattering from small particles distributed around 0.1 nm. This systematic decrease in extinction is attributed to the gradual fallout of particles injected into the atmosphere by the eruptions of Agung (Bali, 1963) and El Fuego (Guatemala, 1974). The gradual decrease in extinction is uniform because volcanic activity in the southern hemisphere remained low during 1975 to April 1982. The extinction coefficients show a small annual variation with minima in July. There is a marked discontinuity in extinction values following the El Chichón eruption with a maximum effect observed seven months after the event, the delay being a result of the latitude difference between El Chichón (17°) and La Silla (-29°).

IAU COMMISSION 51

BIOASTRONOMY: SEARCH FOR EXTRATERRESTRIAL LIFE
BIOASTRONOMIE: RECHERCHE DE LA VIE DANS L'UNIVERS

XIX GENERAL ASSEMBLY

NEW DELHI, NOVEMBER 1985

F. D. Drake, Editor

COMMISSION 51: BIOASTRONOMY: SEARCH FOR EXTRATERRESTRIAL LIFE

Report of Meeting of November 23, 1985

President: M. D. Papagiannis

BUSINESS MEETING

I. NEW OFFICERS

Frank D. Drake (USA) was elected President and George Marx (Hungary) Vice-President for the period 1986-1988. The members of the Scientific Organizing Committee for this period are: J. Billingham (USA), D. Devincenzi (USA), F. Drake (USA), J. Jugaku (Japan), N. Kardashev (USSR), G. Marx (Hungary), M. Papagiannis (USA), R. Pesek (Czechoslovakia), C. Sagan (USA), and V. Troitsky (USSR).

II. NEW MEMBERS

Approximately 37 new members were added to the Commission, bringing the total membership to about 240. The Commission also has 39 consultants, bringing the total number of scientists active in the Commission to about 277. This represents a remarkable growth of the Commission since its inception at the XVIII General Assembly at Patras.

III. SYMPOSIA

The Commission held its first Symposium, IAU Symposium No. 112, at Boston in June, 1984. This successful symposium was attended by several hundred scientists. The Proceedings of the Symposium, now published, is a comprehensive work of nearly 600 pages with more than 50 contributions.

The Commission received an invitation from the Hungarian Academy of Sciences to hold its next Symposium in Hungary. The Commission accepted this invitation, tendered by George Marx, with pleasure. It is proposed that this symposium be held at Lake Balaton, Hungary, during the period June 22-26, 1987. The Chairman of the Local Organizing Committee will be George Marx. It was proposed that the Scientific Organizing Committee for the Symposium include F. D. Drake (USA), P. A. Feldman (Canada), N. S. Kardashev (USSR), A. Martin (UK), R. Pesek (Czechoslovakia), and J. Tarter (USA). The necessary approvals of IAU officials will be sought.

IV. MISCELLANEOUS BUSINESS

The Commission was informed that the Executive Committee of the IAU had chosen to change the name of the Commission to "Bioastronomy: Search for Extraterrestrial Life."

The President of the Commission, M. Papagiannis, presented F.D. Drake with an engraved plaque commemorating the 25th anniversary of the first modern radio search of extraterrestrial life, which Dr. Drake carried out in 1960.

The Commission as a whole expressed its extreme gratitude to Dr. Papagiannis for the very extensive and tireless work he has performed on behalf of the Commission since its inception. It was recognized that he had carried out the preliminary work necessary to bring the Commission into existence, and had then gone to great effort to organize the work of the Commission, as well as Symposium 112. He has gotten this young Commission off to a magnificent start.

SCIENTIFIC SESSION

A brief scientific session was held in New Delhi. The invited speakers were:

- J. Tarter: The Search for Extrasolar Planetary Systems.
- F. Drake: The Beginning and Early Years of SETI.
- B. Zuckerman: Radio vs. Infrared for SETI.
- M. Papagiannis: Recent Progress in SETI and Exciting Plans for the Future.

A panel discussion was then held.

WORKING GROUP FOR PLANETARY SYSTEM NOMENCLATURE (NOMENCLATURE DU SYSTEM PLANETAIRE)
(Committee of the Executive Committee)

Report of meetings on 6 April, 1984, 6 August, 1984, 8 October, 1984, and 24 August, 1985

President, Harold Masursky

A brief report of meetings held in 1983-84, including current membership lists of the Working and Task Groups, is contained in Trans. IAU vol. XIXA, page 725-726, 1985.

Five new Latin terms were adopted for use on any planet or satellite:

Corona (Coronae)	Ovoid-shaped feature
Facula (Faculae)	Bright spot
Fluctus (Fluctus)	Flow terrain
Labes (Labes)	Landslide terrain
Tessera (Tesserae)	Tile-like, polygonal terrain

In specific cases, three additional terms are used:

Landing site name	Lunar features at or near Apollo landing sites
Eruptive center	Active volcanic centers on Io
Large ringed feature	Cryptic ringed features on Callisto

A method of communication faster than standard airmail will be sought so that proposed names and other questions can be considered more promptly.

Working Group Meetings will be held at least once a year in the Soviet Union and in USA or Europe, when a representative group of Working Group members cannot be convened in either location.

A new Task Group for naming surface features on asteroids and comets has been formed. Dr. David Morrison, chairman of the Mercury Task Group, is chairman, pro tempore; a permanent chairman from Europe will be chosen at the first meeting, to be held in Toulouse, France in June, 1986, in connection with the COSPAR meeting. Other members of the new committee are:

Dr. Joseph Veverka, USA	Dr. Leonid Ksanfomaliti, USSR
Dr. Andre Brahic, France	Dr. Yaroslav Yatskiv, USSR
Dr. Marcello Fulchignoni, Italy	Dr. Syuzo Isobe, Japan
Dr. Tomas Gombosi, Hungary	Dr. Y.C. Chang, China

The following 593 names were adopted for features on the Moon, Mercury, Venus, Mars, three satellites of Jupiter, and newly discovered satellites of Saturn and Pluto.

LUNAR NOMENCLATURE

Table 1. 56 names approved as assigned to features on the lunar surface. Positions are given in degrees of latitude and longitude.

CRATER

Name	Lat	Lon	Description
ABETTI, Georgio	19.9N	27.7E	(1882–1982) Italian astronomer
ANDERSSON, Leif E.	49.7S	95.3W	(1943–1979) American astronomer
BOURNOUILLI, Jean	35.0N	60.7E	(1667–1748) Swiss mathematician
CARPENTER, Edwin F.	69.4N	50.9W	(1898–1963) American astronomer
CHADWICK, James	52.7S	101.3W	(1891–1974) British physicist
CHALONGE, Daniel	21.2S	117.3W	(1895–1977) French astronomer
COUDER, Andre	4.8S	92.4W	(1897–1978) French astronomer
DOERFEL, Georg Samuel	69.1S	107.9W	(1643–1688) German astronomer
FLORENSKY, Cyril P.	25.3N	131.5E	(1915–1982) Soviet geochemist
FRYXELL, R.H.	21.3S	101.4W	(1934–1974) American Geologist
GAVRILOV, Igor B.	17.4N	130.9E	(1928–1982) Soviet astronomer
HADLEY, John	25.4N	2.7E	(1682–1793) British instrument maker
HEYROVSKY, Jaroslav	39.6S	95.3W	(1890–1967) Czech chemist
IL'IN, N. Ja.	17.8S	97.5W	(1901–1937) Soviet rocket scientist
LALLEMAND, Andre	14.3S	84.1W	(1904–1978) French astronomer
MCDONALD, Thomas L.	30.4N	20.9W	(unk.–1973) Scottish selenographer
POPOV, C.	17.2N	99.7E	(1880–1966) Bulgarian astronomer
SHULEJKIN, M. V.	27.1S	92.5W	(1884–1939) Soviet radio engineer
UREY, Harold	27.9N	87.4E	(1893–1981) American chemist
YAKOVKIN, A.A.	54.5S	78.8W	(1887–1974) Soviet astronomer

CATENA

Name	Lat	Lon	Description
LITTROW, CATENA	22.2N	29.5E	Named from nearby crater
TARUNTIUS, CATENA	3N	48E	Named from nearby crater
TIMOCHARIS, CATENA	29N	13W	Named from nearby crater

MONS

Name	Lat	Lon	Description
DELISLE, MONS	29.5N	35.8W	Named from nearby crater

RIMA

Name	Lat	Lon	Description
AGRICOLA, RIMA	29N	53W	Named from nearby crater
ARTSIMOVICH, RIMA	27N	39W	Named from nearby crater
BRAYLEY, RIMA	23N	36W	Named from nearby crater
DAWES, RIMA	17.5N	26.6E	Named from nearby crater
DELISLE, RIMA	31N	32W	Named from nearby crater
DIOPHANTUS, RIMA	29N	33W	Named from nearby crater
DRAPER, RIMA	18N	25W	Named from nearby crater
EULER, RIMA	21N	31W	Named from nearby drater
FOCAS, RIMAE	28S	98W	Named from nearby crater
GALILAEI, RIMA	13N	59W	Named from nearby crater
GERARD, RIMAE	46N	84W	Named from nearby crater
HASE, RIMA	29.4S	62.5E	Named from nearby crater
KOPFF, RIMAE	17.4S	89.6W	Named from nearby crater
KRIEGER, RIMA	29.0N	45.6W	Named from nearby crater
MAESTLIN, RIMAE	2N	40W	Named from nearby crater
MAIRAN, RIMA	38N	47W	Named from nearby crater
MAUPERTUIS, RIMAE	52N	23W	Named from nearby crater
MAYER, T., RIMA	13N	31W	Named from nearby crater

MERSENIUS, RIMAE	20.0S	45W	Within crater
MILICHIUS, RIMA	8N	33W	Named from nearby crater
PETTIT, RIMAE	23S	92W	Named from nearby crater
REPSOLD, RIMAE	51N	80W	Named from nearby crater
RICCIOLI, RIMAE	02N	74W	Named from nearby crater
SECCHI, RIMAE	1N	44E	Named from nearby crater
SHEEPSHANKS, RIMA	58N	24E	Named from nearby crater
SUESS, RIMA	6N	47W	Named from nearby crater
SUNG-MEI,	24.6N	11.3E	Chinese female name
TARUNTIUS, RIMAE	5.5N	46.5E	Within crater
VASCO DA GAMA, RIMAE	10N	82W	Named from nearby crater

RUPES

BORIS	30.6N	33.5W	Named from nearby crater
MERCATOR	30S	23W	Named from nearby crater
TOSCANELLI	27.4N	47.5W	Named from nearby crater

MERCURY NOMENCLATURE

Table 2. 32 names approved as assigned to features on Mercury. Positions are given in degrees of latitude and longitude.

CRATER

AL-AKHTAL	59N	97W	(C. 640-710) Arabic poet
BELINSKIJ, V.G.	76S	104W	(1811-1848) Russian literary critic
BJORNSON, BJORNSTJERNE	73N	110W	(1832-1910) Norwegian poet
BRUEGHEL, Pieter	50N	108W	(1525-1569) Flemish painter
BURNS, Robert	54N	116W	(1759-1796) Scottish poet
CÉZANNE, Paul	08S	124W	(1839-1906) French painter
ECHEGARAY, Jose	43N	19W	(1832-1916) Spanish dramatist
FET, Afanasy A.	05S	180W	(1820-1892) Russian poet
FLAUBERT, Gustave	14S	72W	(1821-1880) French novelist
GAINSBOROUGH, Thomas	36S	183W	(1727-1788) English painter
GOGOL, Nikolay	28S	147W	(1809-1852) Russian dramatist
GRIEG, Edvard	51N	14W	(1843-1907) Norwegian composer
HALS, Frans	55S	115W	(1581-1666) Dutch painter
HAN KAN	71.5S	145W	(720-780) Chinese painter
HAUPTMANN, Gerhart	23S	180W	(1862-1946) German novelist
JANÁČEK, Leos	56N	154W	(1854-1928) Czech composer
KŌSHŌ	60N	138W	(13th century) Japanese sculptor
LESSING, Gotthold E.	29S	90W	(1729-1781) German dramatist
LISZT, Franz	16S	168W	(1811-1886) Hungarian composer
ŌKYO, Maruyama	69S	76W	(1733-1795) Japanese painter
RAVEL, Maurice	12S	38W	(1875-1937) French composer
RIMBAUD, Arthur	63S	148W	(1854-1891) French poet
RUDE, Francois	33S	80W	(1784-1855) French sculptor
RŪMĪ	24S	105W	(1207-1273) Persian poet
SIBELIUS, Jean	49S	145W	(1865-1957) Finnish composer
SIMONIDES	29S	45W	(556-468 B.C.) Greek lyric poet
SMETANA, Bedrich	48S	71W	(1824-1884) Czech composer
SOSEKI, Natsume	39N	38W	(1867-1916) Japanese novelist
TAKANOBU, Fujiwara	31N	108W	(1142-1205) Japanese poet artist
THOREAU, Henry David	06N	133W	(1817-1862) American philosopher
VLAMINCK, Maurice de	28N	13W	(1876-1958) French painter
WHITMAN, Walter	41N	111W	(1819-1892) American poet

VENUS NOMENCLATURE

Table 3. 337 names approved as assigned to features on the surface of Venus. Positions are given in degrees of latitude and longitude.

CRATER, LARGE

Name	Lat	Lon	Dates	Description
AKHMATOVA, Anna	61N	308E	(1889-1966)	Soviet poet
ARIADNE	44N	0		Greek goddess; helped Theseus
BARSOVA, Valeria	61N	223E	(1892-1967)	Soviet singer
BARTO, Agniya	45N	146E	(1906-1981)	Soviet poet
BERNHARDT, Sarah	31N	84.5E	(1844-1923)	French actress
BROOKE, Frances	48N	296E	(1724-1789)	Canadian novelist
BROWNING, Elizabeth	28N	5E	(1806-1861)	British poet
CATHER, Willa	47N	107E	(1876-1947)	American novelist
CHRISTIE, Agatha	28N	72.5E	(1891-1976)	British novelist
COCHRAN, Jacqueline	52N	143E	(1906-1980)	French aviator
COTTON, Egenni	71N	300E	(1881-1967)	French physicist
DASHKOVA, Yekaterina	78N	306E	(1743-1810)	Russian philologist
DEKEN, Agatha	47N	288.5E	(1741-1804)	Dutch novelist
DELEDDA, Grazia	76N	126E	(1871-1936)	Italian novelist
DICKINSON, Emily	74N	176E	(1830-1886)	American poet
DUNCAN, Isadora	68N	291.5E	(1878-1927)	American dancer
EDGEWORTH, Maria	32N	23E	(1767-1849)	Irish writer
EFIMOVA, Nina	80N	225E	(1877-1948)	Soviet painter
ERMOLOVA, Mariya	60N	154E	(1853-1928)	Soviet actress
FEDORETS, Valentina	59N	65E	(1923-1976)	Soviet astronomer
FEDOSOVA, Irina	44.5N	172E	(1831-1899)	Russian poet
FERNANDEZ, M.A	76.5N	17.5E	(18th cent.)	Spanish actress
GOLUBKINA, Anna	60.5N	286E	(1864-1927)	Soviet sculptor
HAYASI, Fumiko	54N	244E	(1903-1951)	Japanese writer
KAUFFMAN, Angelica	49.5N	27E	(1741-1807)	German painter
KEMBLE, Fanny	48N	15E	(1809-1893)	English actress
KHATUN, Mihri	40.5N	87.5E	(1456-1514)	Turkish poet
KLENVOA, Mariya	78N	104E	(c1910-c1978)	Soviet geologist
KOIDULA, Lydia	64N	138E	(1843-1886)	Estonian poet
LA FAYETTE, Marie	70N	107E	(1634-1693)	French novelist
LAGERLÖF, Selma	81N	285E	(1858-1940)	Swedish novelist
LANDOWSKA, Wanda	84.5N	83E	(1877-1959)	Polish pianist
LIND, Jenny	50N	355E	(1820-1887)	Swedish singer
MAGNANI, Anna	58	337E	(1908-1973)	Italian actress
MONTESSORI, Maria	59.5N	280E	(1870-1952)	Italian educator
MOSES, A. "Grandma"	34.5N	120E	(20TH cent.)	American painter
NADIRA	43.5N	202E	(1791-1842)	Uzbek poet
MUKHINA, Vera	29N	0.5E	(1889-1953)	Soviet sculptor
OBUKHOVA, Nadezhda	71N	289E	(1886-1961)	Soviet singer
ORLOVA, Lyubov	56N	235E	(1902-1975)	Soviet actress
OSIPENKO, Polina	71N	321E	(1907-1939)	Soviet aviator
PATTI, Adelina	35N	302E	(1843-1919)	Italian singer
POTANINA, Aleksandra	31.5N	53E	(1843-1893)	Russian explorer
PRICHARD, Catharina	44N	11.5E	(1884-1969)	Australian writer
ROSSETTI, Christina	57N	07E	(1830-1894)	English poet
RUDNEVA, Varvara	78N	176E	(1844-1899)	Russian doctor
RUSLANOVA, Lidiya	84N	16E	(1900-1973)	Soviet singer
SEVIGNE, Mari	52.5N	326.5E	(1626-1696)	French writer
SIDDONS, Sarah	62N	341E	(1755-1831)	English actress
TAGLIONI, Maria	42N	122.5E	(1804-1884)	Italian dancer
TSERASKAYA, Lidiya	28N	79E	(1855-1931)	Soviet astronomer
TSVETAYEVA, Marina	64N	147E	(1892-1941)	Soviet poet

UNDSET, Sigrid	52N	59.5E	(1882–1949) Norwegian author
VOLKOVA, Anna	74.5N	243E	(1800–1876) Russian chemist
VOYNICH, Lilian	35.5N	56E	(1864–1960) English writer
WHARTON, Edith	56N	61.5E	(1862–1937) American writer
YABLOCHKINA, Aleksandra	49N	195E	(1866–1964) Soviet actress
ZHILOVA, Maria	66.5N	125.5E	(1870–1934) Russian astronomer
ZVEREVA, Lidiya	45N	283E	(1890–1916) Russian aviator

CRATER, SMALL

ALMEIDA	46N	125E	Portuguese first name
ANTONINA	28N	107E	Russian first name
ANYA	39.5N	297.5	Russian first name
BERTA	62N	322E	Finnish first name
DOLORES	51.5N	200.5E	Spanish first name
ERIKA	72N	176E	German first name
FRIDA	68N	56E	Swedish first name
GALINA	47N	307E	Bulgarian first name
GERDA	46N	92.5E	Danish first name
GINA	78N	76E	Italian first name
GLORIA	68.5N	94.5E	Portuguese first name
INDIRA	64N	289.5E	Hindu first name
IRINA	35N	91E	Russian first name
IVKA	68N	304E	Serbo-Croatian first name
JADWIGA	68.5N	91E	Polish first name
JEANNE	40N	332E	French first name
JOSEFINA	45N	32.5E	Portuguese first name
JULIE	51N	242.5E	Czech, German first name
LAURA	49N	141E	Spanish first name
LENA	39.5N	23E	Russian first name
LILIYA	30N	31E	Russian first name
LOTTA	51N	336E	Swedish first name
LYUDMILA	62N	330E	Russian first name
MAGDA	67N	329.5E	Danish first name
MARGIT	60N	273E	Hungarian first name
MASHA	63N	88E	Russian first name
MONIKA	72.5N	122E	German first name
NANA	50N	75E	Serbo-Croatian first name
NATALIA	67N	273E	Romanian first name
ODILIA	80.5N	206E	Portuguese first name
OLYA	51N	292E	Russian first name
POLINA	42.5N	148E	Russian first name
RADKA	76N	95E	Bulgarian first name
RANI	64.5N	160E	Hindu first name
REGINA	30N	147E	Polish first name
RITA	71N	335E	Hungarian first name
RUTH	43.5N	20E	English first name
SELMA	68.5N	157E	Swedish first name
SIGRID	63.5N	314.5E	Norwegian first name
SIMONE	59.5N	82E	French first name
STEFANIA	51N	333E	Romanian first name
STINA	37.5N	23E	Swedish first name
TAMARA	61.5N	317.5E	Georgian first name
TATYANA	85.5N	217E	Russian first name
TÜNDE	76N	197E	Hungarian first name
ULRIQUE	76N	55.5E	French first name
VALBORG	75.5N	272E	Danish first name
VALENTINA	46.5N	143E	Latin first name
WANDA	71.5N	323E	Polish first name
ZDRAVKA	65N	299E	Bulgarian first name

ZINA	41.5N	320E	Romanian first name
ZLATA	64.5N	334E	Serbo-Croatian first name
ZOYA	68N	237E	Russian first name

CHASMA

ARANYANI CHASMA	73/68N	73/74E	Indian forest goddess
BABA-JAGA CHASMA	55/50N	40/55E	Slavic forest witch
DAURA CHASMA	74/72N	50/56E	Hausa (W. Sudan) great huntress
KAYGUS CHASMATA	49/46N	50/56E	Ketian (Siberia) ruler of forest animals
KOTTRAVEY CHASMA	32/28N	77/78E	Dravidian (India) hunting goddess
KOZHLA-AVA CHASMA	57/55N	47/54E	Marian (Volga Finn) forest goddess
LASDONA CHASMA	72/66N	30/40E	Lithuanian main forest goddess
MEDEINA CHASMA	49/43N	86/93E	Lithuanian forest goddess
MEŽAS-MATE CHASMA	51/49N	47/55E	Latvian forest goddess
MISNE CHASMA	78/78.5N	310/327E	Mansi forest maiden
MORANA CHASMA	71/67N	25/24E	Czech moon goddess
MOTS CHASMA	53/50N	54/57E	Avarian (Caucasus) moon goddess
VARZ CHASMA	72/70N	26/29E	Lezghin (Caucasus) moon goddess
VIRES-AKKA CHASMA	77/71N	335/355E	Saami-Lapp forest goddess

COLLES

AKKRUVA COLLES	40/52N	110/130E	Saami-Lapp fishing goddess
JURATE COLLES	56/63N	150/160E	Lithuanian sea goddess

CORONA

ANAHIT CORONA	76/79N	270/286E	Armenian goddess of fertility
BACHUE CORONA	72/75N	250/270E	Chibcha earth goddess
COATLICUE CORONA	62/65N	271/276E	Aztec earth goddess
DEMETER CORONA	52/57N	290/300E	Greek goddess of fertility
FAKAHOTU CORONA	57/61N	105/110E	Tuamotu mother earth
FERONIA CORONA	66/70N	275/285E	Italian goddess of spring
OTAU CORONA	67/69N	296/300E	Bini goddess of fertility
POMONA CORONA	78/81N	290/308E	Roman goddess of fruits
RANANEIDA CORONA	61/67N	255/270E	Saami-Lapp goddess of spring
RAUNI CORONA	39/42N	270/273E	Finnish goddess of harvest, earth
TUSHOLI CORONA	69/71N	98/103E	Chechen, Inguish (Caucasus) goddess of fertility
VACUNA CORONA	58/63N	90/100E	Sabinian (Anc. Italy) harvest goddess

DORSUM

AHSONNUTLI DORSA	35N/62N	195/200E	Navajo (N. Amer.) light and sky goddess
ALLAT DORSA	60/65N	65/80E	Arabian sky goddess
AUŠKA DORSUM	59/62N	356/2E	Lithuanian goddess of sun rays
AUŠRA DORSA	45/55N	20/28E	Lithaunian dawn goddess
BEN DORSA	71/67N	282/293E	Vietnamese sky goddess
BEZLEA DORSA	24/40N	34/43E	Lithuanian evening light goddess
BREKSTA DORSA	33/36N	306/304E	Lithuanian night darkness goddess
DENNITSA DORSA	66/86N	203/70E	Slavic goddess of day light
DODOLA DORSA	41/51N	277/268E	South Slavic rain goddess
DYAN-MU DORSA	75/78N	44/25E	Chinese goddess of lightning
FRIGG DORSA	46/53N	152/149E	Norse, wife of supreme god odin
HEMERA DORSA	47/57N	245/243	Greek goddess of day

HERA DORSA	27/44N	15/30E	Greek sky goddess, wife of Zeus
IRIS DORSA	45/65N	218/223	Greek goddess of rainbow
IYELE DORSA	42/51N	280/278E	Moldavian wind witch goddess
KAMARI DORSA	50/62N	60/45E	Georgian weather goddess
LAUMA DORSA	55/74N	187/190E	Latvian witch, flies in sky
LUKELONG DORSA	61/78N	185/150E	Polynesian creator of heavens
MARDEZH-AVA DORSA	28/40N	55/68E	Marian (Volga Finn) wind goddess
NEPHELE DORSA	32/50N	130/140E	Greek cloud goddess
OKIPETA DORSA	60/73N	245/225	Greek goddess of whirlwind
PANDROSOS DORSA	50/70N	205/215	Greek dew goddess
SEL-ANYA DORSA	75/84N	80/75	Hungarian wind goddess
SEMUNI DORSA	70/79N	03/10E	Orochian (Siberia) sky goddess
TEZAN DORSA	82/86N	63/0E	Etruscan dawn goddess
TOMEM DORSA	26/39N	02/10E	Ketian (Siberia) "hot" goddess who lives in sky near sun
UNI DORSA	30/38N	120/110E	Etruscan goddess, same as Hera
VARMA-AVA DORSA	58/68N	262/272E	Mordovian wind goddess
VEDMA DORSA	27/59N	152/181E	East Slavic witch
YUMYN-UDYR DORSA	75/84N	90/180E	Marian daughter of main god
ZORILE DORSA	35/44N	340/336E	Moldavian dawn goddess

FOSSA

ARIONROD FOSSAE	31/40N	237/242	Warrior queen in Celtic mythology
BELLONA FOSSAE	33/41N	218/224E	Italian war goddess, wife of Mars
FRIAGABI FOSSAE	50/51N	110/112E	Old English war goddess
HILDR FOSSA	38/45N	44/45E	Norse myth warrior
ILBIS FOSSAE	74/69.5N	240/267	Yukutian (Siberia) goddess of bloodshed
MINERVA FOSSAE	62/67N	250/255E	Roman goddess of war
MIST FOSSAE	38/40N	249/247E	Norse valkyrie
RANGRID FOSSAE	62/63N	350/0E	Norse valkyrie
SIGRUN FOSSAE	48/53N	17/19E	Norse valkyrie

MONS

API MONS	39N	55E	Scythian goddess of earth
ATIRA MONS	52N	267E	Pawnee (N.Amer.) wife of Tirawa
COTIS MONS	44N	234E	Frakian fertility goddess
DANU MONTES	62/58N	317/340E	Celtic mother of gods
FURKI MONS	36N	236E	Chechen and Ingush (Caucasus) goddess, wife of thunder god
MAKOSHA MONS	56/59N	254/257E	East Slavic main goddess
MELIA MONS	61/63N	118/122E	Greek nymph
MENTHA MONS	43N	238E	Roman goddess, personified humankind
RENPET MONS	75/77N	235/238	Egyptian goddess of spring, youth
SEKMET MONS	43/46N	239/242	Ancient Egyptian goddess of war
TEPEV MONS	29N	45E	Kiche (Cent. Amer.) creator goddess
THALLO MONS	74N	232E	Greek goddess, flowering vegetation
VENILIA MONS	31/34N	238/242	Ancient Italian sea goddess

PATERA

ASPASIA PATERA	56N	189E	(c 429 B.C.) famous Greek woman
HATSHEPSUT PATERA	28N	64.5E	(1479 B.C.) Egyptian pharaoh

HIEI CHU PATERA	48N	97E	(2698 B.C.) Chinese woman who started silk industry
HROSWITHA PATERA	36N	34.5E	(1000 A.D.) German historian
IZUMI PATERA	49.5N	194E	(974-1036) Japanese writer
KOTTAUER PATERA	37N	39E	(1410-1471) Austrian
RAZIA PATERA	45N	198E	(13TH cen.) Indian queen
TIPPORAH PATERA	39N	43E	(1500 B.C.) Hebrew medical scholar
TROTULA PATERA	41.5N	19E	(1097 A.D.) Italian physician
YAROSLAVNA PATERA	39N	21E	(12 cen.) Russian, wife of Igor

PLANITIA

BEREGHINYA PLANITIA	25/45N	10/45E	Slavic water spirit
GANIKI PLANITIA	30/50N	180/200E	Orochian water spirit
KAWELU PLANITIA	20/50N	230/260E	Hawaiian heroine
LOUHI PLANITIA	78/90N	360E	Karelo mother of the North
SNEGUROCHKA PLANITIA	75/90N	215/75E	Russian snowgirl
VELLAMO PLANITIA	40/53N	150/165E	Karelo-Finn mermaid
VINMARA PLANITIA	50/65N	190/205E	New Hebrides swan maiden

RUPES

FORNAX RUPES	29/32	200/202E	Roman hearth goddess
GABIE RUPES	68/68.5N	107/120E	Lithuanian hearth goddess
UORSAR RUPES	78/76N	330/345E	Adygan (Caucasus) hearth goddess

TESSERA

ANANKE TESSERA	46/54N	125/140E	Greek goddess of necessity
ATROPOS TESSERA	66/75N	285/314E	Greek, one of three fates
CLOTHO TESSERA	54/60N	329/344E	Greek, one of three fates
DEKLA TESSERA	58/55N	66/83E	Latvian goddess of fate
FORTUNA TESSERA	54/76N	4/95E	Roman goddess of chance
ITZPAPALOTL TESSERA	71/79N	305/0	Aztec goddess of fate
KUTUE TESSERA	37/42N	105/110E	Ulchian toad, brings happiness
LACHESIS TESSERA	41/45N	296/301E	Greek, one of three fates
LAIMA TESSERA	45/58N	31/59E	Latvian goddess of fate
MANZAN-GURME TESSERAE	33/46N	356/003E	Mongol fate woman
MENI TESSERA	45/50N	75/80E	Semitic goddess of fate
MOIRA TESSERA	58/60N	307/314E	Greek goddess of fate
NEMESIS TESSERA	44/46N	193/196E	Greek goddess of fate
MESKHENT TESSERA	65/68N	95/128E	Egyptian goddess of fortune
SHIMTI TESSERA	28/35N	94/102E	Babylonian goddess of fate
VIRILIS TESSERA	52/61N	230/243E	Another name for Roman goddess of chance

THOLUS

ALE THOLUS	68N	246E	Igbo (Nigeria) vegetation goddess
ASHTART THOLUS	49N	247E	Phoenician goddess of love
BAST THOLUS	58N	131E	Egyptian goddess of joy
BRIGIT THOLUS	49N	246E	Celtic goddess of wisdom
NERTUS THOLUS	61.5N	248E	German, Norse vegetation goddess
SEMELE THOLI	64N	203/204	Frygian earth goddess
UPUNUSA THOLUS	67.5N	247E	E. Indonesian earth goddess
WURUNSEMU THOLUS	39.5N	210E	Hatti (Hittite) sun goddess

MARS NOMENCLATURE

Table 4. 146 names approved as assigned to features on the surface of Mars. Positions given in degrees latitude and longitude.

CRATER, LARGE			
BABAKIN	36.5S	71.5W	(1914-1970); Soviet builder of unmanned space stations
BERNARD, P.	23.5S	154.5W	(unknown) French atmospheric scientist
DEJNEV, Semen I.	25.5S	164.5W	(1605-1673) Russian explorer
KOVAL'SKY, M.A.	30.0S	142W	(1821-1884) Russian astronomer
MUTCH, Thomas A.	0.5N	55.0W	(1931-1980) American geologist
TIKHONRAVOV, M.K.	14.0N	324.0W	(1851-1916) Russian rocket scientist
CRATER, SMALL			
AVEIRO	21.5N	80.0W	Town in Portugal
CHIA	1.6N	59.5W	Town in Spain
COBRES	11.8S	153.6W	Town in Argentina
MARCA	10S	1580W	Town in Peru
STEGE	3.8N	59.5W	Town in Denmark
SULAK	18.3N	78.6W	Town in USSR
CATENA			
ALBA CATENA	34.1/36.3N	114.5/114.6W	Classical albedo feature
ARTYNIA CATENA	45.3/49.5N	119.4/119.7W	Classical albedo feature
CERAUNIUS CATENA	36.9/37.8N	107.9/108.3W	Classical albedo feature
CYANE CATENA	35.8/38.2N	118.2/118.8W	Classical albedo feature
ELYSIUM CATENA	17.6/18.1N	209.9/210.6W	Classical albedo feature
HYBLAEUS CATENA	21.7/21.9N	219.2/219.8W	Classical albedo feature
STYGIS CATENA	22.5/24N	209/210W	Classical albedo feature
CAVUS			
AVERNUS CAVI	2.7/4.8S	186.7/188.2W	Classical albedo feature
CHAOS			
ATLANTIS CHAOS	33.5/36S	176/179W	Classical albedo feature
CANDOR CHAOS	6.6/7.7S	72.2/73.2W	Classical albedo feature
ECHUS CHAOS	7/12N	74/76.5W	Classical albedo feature
GALAXIAS CHAOS	33.6/34.5N	211.2/213.6W	Classical albedo feature
GORGONUM CHAOS	37.0/39.0S	19.5/172.0W	Classical albedo feature
ISTER CHAOS	12.0/13.8N	56.1/57.0W	Classical albedo feature
NILUS CHAOS	24.1/27.1N	74.6/80.8W	Classical albedo feature
CHASMA			
BAETIS CHASMA	3.7/5.0S	64.3/65.3W	Classical albedo feature
CETI CHASMA	5/5.3S	67.8/68.7W	Classical albedo feature
ELYSIUM CHASMA	21.7/23.5N	217.7/219.1W	Classical albedo feature
HYBLAEUS CHASMA	21.9/22.5N	218.5/219.2W	Classical albedo feature
HYDRAE CHASMA	6.4/7.3S	61.4/62.3W	Classical albedo feature

COLLES

ALPHEUS COLLES	32/45S	297/304W	Classical albedo feature
ARENA COLLES	29.0N	277W	Classical albedo feature
ASTAPUS COLLES	32/38.5N	268/280W	Classical albedo feature
AVERNUS COLLES	0.0/3.9S	187/190W	Classical albedo feature
GALAXIAS COLLES	34/42N	208/212W	Classical albedo feature
OXIA COLLES	17.5/23N	24/30.5W	Classical albedo feature
SCANDIA COLLES	57/71N	105/165W	Classical albedo feature
TARTARUS COLLES	8/15S	182/194.5W	Classical albedo feature

DORSUM

AESACUS DORSUM	35/38N	205.5/207.8W	Classical albedo feature
AVERNUS DORSA	3.6/10.1N	188.2/190W	Classical albedo feature
JUVENTAE DORSA	3.5N/3.5S	67.8/75W	Classical albedo feature
NILUS DORSA	18.5/23.3N	78.5/79.6W	Classical albedo feature
SACRA DORSA	23.5/4.5S	58/78W	Classical albedo feature
STYX DORSUM	30.4/32.4N	208.0/208.6W	Classical albedo feature
URANIUS DORSUM	21.3/25.1N	73.9/78.7W	Classical albedo feature

FLUCTUS

GALAXIAS FLUCTUS	31.5/33.1N	216.3/218.8W	Classical albedo feature

FOSSA

ALBOR FOSSAE	17.9/18.8N	208.1/210W	Classical albedo feature
CORACIS FOSSAE	32/41S	74/84W	Classical albedo feature
CYANE FOSSAE	27.0/39.0N	116/126W	Classical albedo feature
FORTUNA FOSSAE	2/7N	91.5/94W	Classical albedo feature
GALAXIAS FOSSAE	32.8/39.5N	214.6/222.0W	Classical albedo feature
HALEX FOSSAE	26.5/28.5N	125.5/127W	Classical albedo feature
HYBLAEUS FOSSAE	19.5/23.2N	219.1/226.1W	Classical albedo feature
ICARIA FOSSAE	33/63S	111/155W	Classical albedo feature
JOVIS FOSSAE	16.5/22.5N	114.0/118.0	Classical albedo feature
LABEATIS FOSSAE	25/36N	69/85.5W	Classical albedo feature
MELAS FOSSAE	22/31S	70/77W	Classical albedo feature
NECTARIS FOSSAE	18/29S	56/59W	Classical albedo feature
NOCTIS FOSSAE	0.5N/6S	97/102.5W	Classical albedo feature
OLYMPICA FOSSAE	23/27N	110.5/117W	Classical albedo feature
PYRAMUS FOSSAE	48/53.5N	291/296W	Classical albedo feature
STYGIS FOSSAE	23.0/28.9N	209.0/213.7W	Classical albedo feature
ULYSSES FOSSAE	3.0/18.0N	120.0/126.5W	Classical albedo feature
URANIUS FOSSAE	23/27N	88/91.5W	Classical albedo feature
ZEPHYRUS FOSSAE	25.7/26.7N	214.8/217.2W	Classical albedo feature

LABES

COPRATES LABES	11.4/12.3S	67.2/68.1W	Classical albedo feature
MELAS LABES	8/9.6S	71/72.2W	Classical albedo feature
OPHIR LABES	10.7/11.4S	67.5/69W	Classical albedo feature

MENSA

BAETIS MENSA	5/6.7S	2.4/72.8W	Classical albedo feature
GALAXIAS MENSAE	34/38N	211/217W	Classical albedo feature
NILUS MENSAE	21.8/22.6N	70.2/74.1W	Classical albedo feature
ZEPHYRIA MENSAE	5.7/14.5S	186/191W	Classical albedo feature

MONS

ECHUS MONTES	6/11N	76/78W	Classical albedo feature
HIBES MONTES	2/4.5N	187.5W	Classical albedo feature
OCEANIDUM MONS	55.0S	41.0W	Classical albedo feature
SISYPHI MONTES	68.5/72S	344/349W	Classical albedo feature

TARTARUS MONTES	17/28N	181/193	Classical albedo feature

PATERA
DIACRIA PATERA	35N	132.5W	Classical albedo feature
PENEUS PATERA	57/59S	306/309W	Classical albedo feature

PLANITIA
OLYMPIA PLANITIA	80/83N	120/240W	Classical albedo feature

PLANUM
CHRONIUM, PLANUM	56/63S	212/226W	Classical albedo feature
MALEA PLANUM	59/70S	281/325W	Classical albedo feature

RUPES
ARIMANES RUPES	8.8/11.5S	147.1/147.9W	Classical albedo feature
AVERNUS RUPES	7.6/10.3S	185.7/187.2W	Classical albedo feature
CHALCOPOROS RUPES	42/59.5S	336/342W	Classical albedo feature
CYDNUS RUPES	56.5/61.5N	251/259W	Classical albedo feature
ELYSIUM RUPES	24.4/26.4	211.9/212.8W	Classical albedo feature
PITYUSA RUPES	61/69S	326/335W	Classical albedo feature
TARTARUS RUPES	5.5/7.6S	183.6/184.8W	Classical albedo feature
THYLES RUPES	67.0/77.0S	185/228W	Classical albedo feature

SCOPULUS
CHARYBDIS SCOPULUS	20.0/30.0S	238.5/342.0W	Classical albedo feature
CORONAE SCOPULUS	0.0/37.0S	292/297W	Classical albedo feature
SCYLLA SCOPULUS	21.5/29.5S	340.5/342.5W	Classical albedo feature
TARTARUS SCOPULUS	2.8/12.5S	182/183.8W	Classical albedo feature

SULCUS
AMAZONIS SULCI	3.5/4.2S	144.5/147.1W	Classical albedo feature
APOLLINARIS SULCI	10.0/12.5S	182.9/183.8W	Classical albedo feature
MEDUSAE SULCI	3.5/6.3S	159.6/160.0W	Classical albedo feature
MEMNONIA SULCI	4.0/10.0S	173/180W	Classical albedo feature
SACRA SULCI	22.5/27.5N	68/75W	Classical albedo feature

VALLIS
ABUS VALLIS	4.9/6.0S	147.1/147.4W	Classical river, England
APSUS VALLIS	34.3/36.1N	224.6/225.9W	Classical river, Macedonia
ASOPUS VALLIS	4.1/4.7S	149.5/149.6W	Classical river, Greece
BUVINDA VALLIS	32.5/34.0N	207.8/208.6W	Classical river, Ireland
CUSUS VALLES	12/16.5N	307/311W	Classical river, modern Czechoslovakia
DEVA VALLIS	7.9/8.3S	156.6/156.8W	Classical river, Scotland
DITTAINO VALLES	0.5N/4S	66.2/66.7W	Modern river, Italy
DOANUS VALLIS	60.5/63.0S	26.0/31.0W	Classical river on Ptolemy's map
DRILON VALLIS	6.5/7.9N	51.9/52.8W	Classical river, Macedonia
DUBIS VALLIS	4.9/5.6S	148.0/148.2W	Classical river, France
FRENTO VALLIS	50.0/52.0S	11.0/16.5W	Classical river, Italy
HER DESHER VALLIS	25/26S	47/48.5W	Egyptian name for Mars
HERMUS VALLIS	5.0/5.8S	147.5W	Classical river, Lydia
HRAD VALLIS	33.5/44N	218/231W	Armenian word for Mars
HYPANIS VALLES	9.2/11.1N	45.1/48.1W	Classical river, Scythia
IBERUS VALLIS	21.0/21.9N	207.6/208.6W	Classical river, Spain

ISARA VALLES	5.3/5.5S	146.3/146.5W	Classical river, anc. Belgium; mod. Oise River, France
ITUXI VALLIS	24.8/25.3N	206.8/207.5W	Modern river, Brazil
LABOU VALLIS	6.5/9.5S	153.5/156W	French word for Mars
LIRIS VALLES	9.0/15.0S	299.0/308.0W	Classical river, Italy
MATRONA VALLES	7.6/8.2S	183.6/183.8W	Classical river, France
MINIO VALLIS	3.9/5.3S	151.3/153.4W	Classical river, Italy
MOSA VALLIS	14.0/16.5S	337.5/338.5W	Modern river, England
MUNDA VALLIS	5.3/5.5S	146.1W	Classical river, Lusitania
NARO VALLIS	0.5/5.0S	299.0/300.0W	Classical river, Yugoslavia
NESTUS VALLIS	6.9/7.6S	157.8/158.5W	Classical river, Macedonia
NICER VALLES	6.8/7.6S	157.9/158.6W	Classical river, Magna Germanica
OCHUS VALLES	6.1/7.8N	44.9/45.4W	Classical river, Asia Minor
PADUS VALLIS	4.1/5.0S	150.0W	Classical river, Italy
PATAPSCO VALLIS	23.0/25.5N	206/209W	Modern river, Maryland, USA
RAVIUS VALLES	44.2/47.5N	105.5/114.2W	Classical river, Ireland
SABIS VALLIS	3.5/6.1S	152.1/152.8W	Classical river, Gaul
SABRINA VALLIS	10.4/11.8N	46.9/50.8W	Classical river, England
SENUS VALLIS	5.1/5.5S	146.7/147.1W	Classical river, Ireland
SEPIK VALLIS	0.8/1.3S	65.6/66.1W	Modern river, New Guinea
STURA VALLIS	22.6/23.2N	217.0/218.2W	Classical river, Italy
SUBUR VALLIS	11.4/11.8N	52.9/53.3W	Classical river, anc. Mauretania
TADER VALLES	48/52S	150/155W	Classical river, Spain
TAUS VALLIS	4.8/5.0S	148.3W	Classical river, Caledonia
TINIA VALLES	4.5-4.8S	148.8/149.0W	Classical river, Italy
TINJAR VALLES	36/40N	234/239.5W	Modern river, Sarawak, Malaysia
TISIA VALLES	10.0/12.5S	310.5/315.0W	Classical river, Ukraine
TREBIA VALLES	30.9/33.9N	208.4/211.5W	Classical river, Liguria
TYRAS VALLIS	7.8/8.7N	49.7/50.4W	Classical river, Scythia
VISTULA VALLES	12.7/15.6N	50.5/53.7W	Classical river, ancient Germania

OUTER SOLAR SYSTEM NOMENCLATURE

The names and designations of the following 4 satellites were adopted:

Saturn XII	1980 S6	Helene
Saturn XVI	1980 S27	Prometheus
Saturn XVII	1980 S26	Pandora
Pluto I	1978 P1	Charon

JUPITER NOMENCLATURE

Table 5. 118 names approved as assigned to features on the surfaces of 3 Jovian satellites. Positions given in degrees of latitude and longitude.

EUROPA NOMENCLATURE

CRATER

Name	Lat	Lon	Description
CILIX	1N	182W	Europa's brother
MORVRAN	6S	15W	Celtic son of Tegid
RHIANNON	82S	20W	Celtic heroine
TALIESIN	23S	22W	Celtic magician
TEGID	0.5S	164W	Celtic hero

FLEXUS

Name	Lat	Lon	Description
DELPHI FLEXUS	60/68S	135/234W	Where the cow lead Cadmus
PHOCIS FLEXUS	48S	188/205W	Where the cow lead Cadmus

LINEA

Name	Lat	Lon	Description
ALPHESIBOEA LINEA	21/33S	157/194W	Europa's nephew
ECHION LINEA	12.5/17S	175/202W	A founder of Thebes
PHOENIX LINEA	6/15N	161/202W	Europa's brother
RHADAMANTHYS LINEA	2/23N	175/206W	Europa's son by Zeus
TECTAMUS LINEA	12/25N	172/195W	Father of Asterius
TELEPHASSA LINEA	6/9S	167/191W	Europa's mother
THYNIA LINEA	53/82S	139/160W	Phineus sought Europa here

GANYMEDE NOMENCLATURE

CRATER

Name	Lat	Lon	Description
AGREUS	15N	235W	Hunter god in Tyre
AGROTES	62N	199W	Greatest god of Gebal
AMON	33N	223W	Theban king of gods
ANAT	3S	128W	Babylonian dew goddess
ANDJETI	52/4.5S	15.9W	Egyptian god at Busirus
ANTUM	4N	220W	Babylonian; wife of Anu
ASHIMA	3.7/7S	122	Arabian god of fate
BES	25.7S	180.3	Egyptian god of marriage
GAD	12.5S	137.3	Semitic god of fate
GEB	58.5N	187.5W	Heliopolis earth god
GEINOS	18N	221W	Tyre god of brickmaking
GIR	35.5N	146.5W	Sumerian god of summer
HALIEUS	35.5N	168.5W	Tyre fisherman god
ILAH	23N	161W	First Sumerian sky god
ILUS	11.5S	111W	Ganymede's brother
IRKALLA	31S	114.7	Sumerian goddess of underworld
ISHKUR	0.5N	11.5	Sumerian god of rain
ISIMU	8.5N	2.0	Sumerian god of spring
KADI	48.5	181W	Babylonian goddess of justice
KITTU	0.5S	336.5W	Babylonian god of justice
KULLA	35N	115W	Sumerian god of bricklaying
LUMHA	37.3N	155.4W	Patron of singers
MEHIT	29.7	165.4W	Egyptian lion-headed goddess
MIR	04S	232W	West Semitic god of wind
MISHARU	5S	338.5W	Babylonian god of law
MUSH	13S	115W	Sumerian male deity
NANNA	18.5S	244W	Sumerian god of wisdom
NEHEH	71N	58	Egyptian God of eternity
NIDABA	19N	123.5W	Sumerian grain goddess
NINKI	07S	21.5W	Sumerian; Ea's companion

NINLIL	08N	119W	Chief Assyrian god
NINSUM	13S	140W	Babylonian god of wisdom
SEIMA	16.5N	217W	Aramean mother goddess
SELKET	17N	107.3	Egyptian goddess
TAMMUZ	13N	232W	Sumerian youthful god
THOTH	43S	146W	Egyptian god of science

FACULA

ABYDOS FACULA	34N	154W	Egyptian town
BUSIRIS FACULA	15N	215W	Town on Nile delta
BUTO FACULA	13N	204W	Egyptian swamp
COPTOS FACULA	9N	209W	Egyptian caravan center
DENDERA FACULA	0	256W	Egyptian town
EDFU FACULA	27N	148W	Egyptian town
MEMPHIS FACULA	16N	132W	Egyptian capital of lower kingdom
OMBOS FACULA	4N	338W	Egyptian town
PUNT FACULA	25S	247W	Land east of Egypt
SIWAH FACULA	7.5N	144W	Egyptian oasis
TETTU FACULA	39N	162.5W	Egyptian town
THEBES FACULA	6N	202W	Egyptian ancient capital of upper kingdom

FOSSA

LAKHAMU FOSSA	12/15S	220/230W	Assyro-Babylonian dragon
LAKHMU FOSSAE	2/30N	115/160W	Assyro-Babylonian dragon
ZU FOSSAE	21/60N	118/180W	Akkadian dragon of chaos

SULCUS

ARBELA SULCUS	08/27S	5/335W	Assyrian town
DUKUG SULCUS	78/85N	10/310W	Sumerian chamber of gods
ELAM SULCI	40/60N	188/223W	Babylonian town
ERECH SULCUS	2N/16S	175/182W	Akkadian town
HURSAG SULCUS	8/17S	226/242W	Sumerian mountain
LAGASH SULCUS	5/20S	153/167W	Early Babylonian town
NIPPUR SULCUS	20/46N	170/203W	Sumerian city
SIPPAR SULCUS	6/10S	160/202W	Babylonian town
UR SULCUS	31/65N	175/187W	Babylonian seat of moon worship

IO NOMENCLATURE

CATENA

| RESHET CATENA | 2.5N/1.5S | 304/308W | Aramean sun god |

FLUCTUS

EUBOEA FLUCTUS	43.5/45.5S	351/354.5W	Io passed through
TUNG YO FLUCTUS	11.5/21S	351/03W	Chinese fire god
UTA FLUCTUS	27/37S	15/23W	Assyro-Babylonian hero

MENSA

ECHO MENSA	78/81.5S	17.5/342W	Mother of Iynx
EPAPHUS MENSA	51.5/54.5S	238/243.5W	Son of Io and Zeus
IYNX MENSA	60.5/63S	301/308.5W	Cast a spell on Zeus
PAN MENSA	46/55S	30/42W	Father of Iynx

MONS

| BOÖSAULE MONTES | 0.5/13S | 262/275W | Cave where Epaphus born |
| CRIMEA MONS | 74/78.5S | 238/250 | Io passed through |

EUBOEA MONTES	44/50S	332/343W	Io passed through
PATERA			
AGNI PATERA	40.5S	334.5W	Hindu god of fire
ANGPETU PATERA	21S	10.5W	Dakota name meaning sun
ARAMAZD PATERA	73.5S	339W	Armenian thunder god
CARANCHO PATERA	1.5N	318W	Bolivian legendary hero; received fire from an owl
CATAQUIL PATERA	24S	18.5W	Inca thunder god
DINGIR PATERA	04S	342W	Sumerian sun god
HATCHAWA PATERA	58.5S	36W	Yaroro fire god
HEISEB PATERA	30N	245W	Bushman fire devil
ILMARINEN PATERA	14S	3W	Finnish blacksmith
KAVA PATERA	17S	342W	Persian blacksmith
KHALLA PATERA	6N	304W	Bushman sun
LU HUO PATERA	38S	355W	Chinese hearth god
MAMA PATERA	1S	357W	Chagaba word for sun
MENAHKA PATERA	31S	346W	Mandan name for sun
MITHRA PATERA	59S	267W	Persian god of light
NINURTA PATERA	17S	316W	Babylonian god of sun
PÄIVE PATERA	46S	0	Saami-Lapp sun god
PAUTIWA PATERA	32.5S	347.5W	Hopi name for sun
PODJA PATERA	18S	305W	Tungu spirit, keeps fire
PURGINE PATERA	3S	298W	Mordvinian thunder god
PYERUN PATERA	56S	252W	Slavonic god of thunder
SED PATERA	03S	304W	Carthage & Tyre; chariot rider of sun
SIUN PATERA	49.5S	2.5W	Nanai (Siberia) sun god
SUI JEN PATERA	19.5S	4.5W	Chinese hero; discovered fire
TARANIS PATERA	71S	30W	Celtic thunder god
TOL-AVA PATERA	2N	322.5W	Mordvinian fire goddess
TUNG YO PATERA	19S	2.5W	Chinese fire god
PLANUM			
ARGOS PLANUM	45.5/49.5S	315/323W	Where Zeus captured Io
DANUBE PLANUM	19/23S	256/262W	Io passed through
ETHIOPIA PLANUM	41/46S	16/27W	Io passed through
HYBRISTES PLANUM	52/55.5S	13/27.5W	Io passed through
IOPOLIS PLANUM	30/38S	332/336	Town where Io was worshipped
LYRCEA PLANUM	37/48S	253/275W	Plain where Io was born

CHAPTER V

FUNCTIONS AND STATUTES OF THE UNION

As indicated in the Preface, a new edition of the "Astronomers' Handbook" is being prepared. In particular, a revised "Style Book" is presently being finalized and will be available soon, either as an independent booklet, or in some other printed form.

This chapter is therefore kept very short : it actually reminds the reader of a few specific functions of the Union, and, in addition reproduces the present IAU Statutes and By-laws (only by-law 17 was changed during the XIXth General Assembly, because of a modification in the structure of ICSU).

1) **Useful addresses** : The addresses of the members of the Executive Committee as well as that of the Secretariat can be found elsewhere in these transactions (see page 57).

2) **History of the Union** : The "IAU Over the Years" (1922-1985), i.e. a table containing the names of former Presidents and General Secretaries, the number of Commissions and IAU Members, the places where General Assemblies were held, with the approximative number of participants, has been published recently in IAU Information Bulletin No. 55, pp. 6-7.

3) **IAU Representatives to other Organizations**

a) ICSU : International Council of Scientific Unions

   General Committee : R.M. West* (until September 1986)
   J.-P. Swings (after September 1986)

b)

Organization		Representative 1985-1988
BIH	Bureau International de l'Heure	P. Pâquet
		G. Winkler
CCDS	Consultative Comm. for the Definition of the Second	J. Benavente
	WG on TAI	G. Winkler
CCIR	International Radio Consultative Committee	
	Study Group 2	J. Whiteoak
		A.R. Thompson
	Study Group 7	J. Pilkington
CODATA	Committee for Data for Science and Technology	G. Westerhout
COSPAR	Committee on Space Research	J.-P. Swings
COSTED	Comm. on Sc. & Technology in Developing Countries	S. Ramadurai
CTS	Committee on the Teaching of Science	L. Gouguenheim
EPS	European Physical Society Conference Committee	D. McNally

---

* Also member of the Executive Board of ICSU

FAGS	Federation of Astronomical & Geophysical Services	J. Kovalevsky E. Tandberg-Hanssen
IAF	International Astronautical Federation	
ICSTI	International Council for Scientific & Technical Information	G. Wilkins
IPMS	International Polar Motion Service	Ye Shu-hua
IUCAF	Inter-Union Commission on Frequency Allocation for Radio Astronomy & Space Science	R. Schilizzi G. Swarup
URSI	Union Radio Scientifique Internationale	J. Baldwin
IUWDS	Int. Ursigram & World Day Service	H. Coffey
QBSA	Quarterly Bulletin on Solar Activity	E. Hiei
SCOPE	Scientific Committee on Problems of Environment	R. Cayrel
SCOSTEP	Scientific Committee on Solar-Terrestrial Physics	S.T. Wu
IUPAP	Int. Union of Pure & Applied Physics	V. Trimble
CIE	Compagnie Internationale de l'Eclairage	D. Crawford

4) Services of the IAU

The reader is referred to IAU Transactions XVIII B, pp. 387-394 and references therein for all matters related, e.g. to the Minor Planet Center and the Central Bureau for Astronomical Telegrams. One should note that the telex number of these services remains unchanged (710-320-6842 ASTROGRAM CAM), but that the telephone number has now become (617)-495- 7244.

The Astronomical Telegram Code, as revised in 1980, is also to be found in Transactions XVIIIB (pp. 388-392).

As far as the assignment of designations and names to newly discovered objects in the solar system is concerned, the reader is referred to Transactions XVIIIB, pp. 392-394.

5) Rules for Scientific Meetings

Although essentially unchanged, the rules relevant to the organization of scientific meetings will be displayed shortly in a forthcoming issue of the IAU Information Bulletin.

6) Statutes, by-Laws and Working Rules

These are given on pages 358 through 379 of the present Transactions.

UNION ASTRONOMIQUE INTERNATIONALE

STATUTS

I. Dénomination, Buts et Domicile

1. L'Union Astronomique Internationale (ci-après dénommée l'Union) est une organisation non-gouvernementale, qui a pour buts de :

    (a) développer l'astronomie par la coopération internationale,
    (b) encourager l'étude et le développement de l'astronomie sous tous ses aspects,
    (c) servir et sauvegarder les intérêts de l'astronomie.

2. L'Union a son siège légal à Bruxelles.

II. Affiliation de l'Union

3. L'Union adhère au Conseil International des Unions Scientifiques.

III. Membres de l'Union

4. L'Union a pour membres :

    (a) des personnes morales (Pays adhérents)
    (b) des membres individuels (Membres).

IV. Organisations Affiliées

5. L'Union peut accepter l'affiliation d'organisations internationales non-gouvernementales qui contribuent au développement de l'astronomie.

V. Pays Adhérents

6. Les pays adhèrent à l'Union

    soit :

    (a) par l'intermédiaire de l'organisation par laquelle ils adhèrent au Conseil International des Unions Scientifiques, ou par l'intermédiaire d'un Comité National d'Astronomie approuvé par cette organisation,

    soit :

    (b) s'ils n'adhèrent pas au Conseil International des Unions Scientifiques, par l'intermédiaire d'un Comité National d'Astronomie reconnu par le Comité Exécutif de l'Union.

    (c) Les Organisations ou Comités mentionnés à l'article 6(a) et les Comités Nationaux d'Astronomie mentionnés à l'article 6(b) sont dénommés ci-après organismes adhérents.

7. L'adhésion d'un pays à l'Union est proposée par le Comité Exécutif et approuvée par l'Assemblée Générale ; elle prend fin si le pays se retire de l'Union.

8. Les Pays Adhérents sont répartis en catégories. Le nombre des catégories est fixé par le Règlement. Un pays qui sollicite son adhésion indique la catégorie dans laquelle il désire

INTERNATIONAL ASTRONOMICAL UNION

STATUTES

I. Denomination, Objects and Domicile

1. The International Astronomical Union (referred to as the Union) is a non-governmental organization, whose objects are :
   - (a) to develop astronomy through international co-operation,
   - (b) to promote the study and development of astronomy in all aspects,
   - (c) to further and safeguard the interests of astronomy.

2. The legal domicile of the Union is Brussels.

II. Adherence to the Union

3. The Union adheres to the International Council of Scientific Unions.

III. Composition of the Union

4. The Union is composed of :
   - (a) corporate members (Adhering Countries)
   - (b) individual members (Members).

IV. Affiliated Organizations

5. The Union may admit the affiliation of international non-governmental organizations which contribute to the development of astronomy.

V. Adhering Countries

6. Countries adhere to the Union

   either :
   - (a) through the organization by which they adhere to the International Council of Scientific Unions, or through a National Committee of Astronomy approved by that organization,

   or
   - (b) if they do not adhere to the International Council of Scientific Unions, through a National Committee of Astronomy recognized by the Executive Committee of the Union.
   - (c) The Adhering Organizations and National Committees of Astronomy are referred to as adhering bodies.

7. Adherence of a country to the Union is approved, on the proposal of the Executive Committee, by the General Assembly ; it terminates if the country withdraws from the Union.

8. Adhering Countries are classified in categories. The number of categories shall be specified in the By-laws. A country requesting adherence shall specify the category in

être classé. La proposition peut être refusée par le Comité Exécutif si la catégorie est manifestement inadéquate.

VI. Membres

9. Les Membres sons admis dans l'Union par le Comité Exécutif, sur proposition de l'un des organismes adhérents mentionnés à l'article 6, en considération de leur activité dans une branche de l'astronomie.

VII. Assemblée Générale

10. (a) L'activité de l'Union est dirigée par l'Assemblée Générale des représentants des Pays Adhérents et des Membres. Chaque Pays Adhérent nomme un représentant autorisé à voter en son nom.

    (b) L'Assemblée Générale rédige un Règlement qui précise les modalités d'application des Statuts.

    (c) Elle nomme un Comité Exécutif chargé d'exécuter les décisions de l'Assemblée Générale, et d'administrer l'Union pendant la periode séparant les réunions de deux Assemblées Générales ordinaires successives. Le Comité Exécutif rend compte de sa gestion à l'Assemblée Générale. L'Assemblée Générale, en acceptant le rapport du Comité Exécutif, le décharge de sa responsabilité.

11. (a) Sur les questions concernant l'administration de l'Union, sans implication budgétaire, le vote à l'Assemblée Générale a lieu par Pays Adhérent, chaque pays disposant d'une voix. Les Pays Adhérents qui ne sont pas à jour de leurs cotisations annuelles au 31 décembre de l'année précédant l'Assemblée Générale ne peuvent pas participer aux votes.

    (b) Sur les questions engageant le budget de l'Union, le vote a lieu de même par Pays Adhérent, dans les conditions et avec les réserves prévues à l'article 11(a), le nombre de voix de chaque Pays Adhérent étant égal à l'indice de sa catégorie, définie conformément à l'article 8, augmenté d'une unité.

    (c) Les Pays Adhérents peuvent voter par correspondance sur les questions figurant à l'ordre du jour de l'Assemblée Générale.

    (d) Un scrutin n'est valable que si au moins deux tiers des Pays Adhérents disposant du droit de vote en vertu de l'article 11(a) y prennent part.

12. Sur les questions scientifiques n'engageant pas le budget de l'Union, les Membres de l'Union disposent chacun d'une voix.

13. Sur toutes les questions prévues aux articles 11 et 12, les décisions sont prises à la majorité absolue des suffrages. Cependant, une décision de modification des Statuts n'est valable que si elle a été prise à la majorité des deux tiers des voix des Pays Adhérents qui disposent du droit de vote en vertu de l'article 11(a).

14. Une proposition de modification des Statuts ne peut être discutée que si elle figure, en tant que telle, à l'ordre du jour de l'Assemblée Générale.

VIII. Comité Exécutif

15. Le Comité Exécutif se compose du Président de l'Union, de six Vice-Présidents, du Secrétaire Général et du Secrétaire Général Adjoint, élus par l'Assemblée Générale sur la proposition du Comité Spécial des Nominations.

IX. Commissions de l'Union

16. L'Assemblée Générale crée des Commissions en vue d'assurer la réalisation des buts qu'elle se propose.

which it desires to be classed. The specification may be declined by the Executive Committee if the category proposed is manifestly inadequate.

VI. Members

9. Members are admitted to the Union by the Executive Committee, on the proposal of an adhering body referred to in article 6, with regard to their achievements in some branch of astronomy.

VII. General Assembly

10. (a) The work of the Union is directed by the General Assembly of representatives of Adhering Countries and of Members. Each Adhering Country appoints a representative authorized to vote in its name.

    (b) The General Assembly draws up By-laws governing the application of the Statutes.

    (c) It appoints an Executive Committee to implement the decisions of the General Assembly, and to direct the affairs of the Union in the interval between meetings of two successive ordinary General Assemblies. The Executive Committee reports to the General Assembly. The General Assembly, in accepting the report of the Executive Committee, discharges it of liability.

11. (a) On questions concerning the administration of the Union, not involving its budget, voting at the General Assembly is by Adhering Country, each country having one vote. Adhering Countries which have not paid their annual contributions up to 31 December of the year preceding the General Assembly may not participate in the voting.

    (b) On questions involving the budget of the Union, voting is similarly by Adhering Country, under the same conditions and with the same reservations as in article 11(a), the number of votes for each Adhering Country being one greater than the number of its category, as defined in article 8.

    (c) Adhering Countries may vote by correspondence on questions on the agenda for the General Assembly.

    (d) A vote is valid only if at least two thirds of the Adhering Countries having the right to vote by virtue of article 11(a) participate in it.

12. On scientific questions not involving the budget of the Union the Members of the Union each have one vote.

13. On all questions in articles 11 and 12, decisions are taken by an absolute majority of the votes cast. However, a decision to change the Statutes is only valid if taken with the approval of at least two thirds of the votes of the Adhering Countries having the right to vote by virtue of article 11(a).

14. A motion to change the Statutes can only be discussed if it appears, in specific terms, on the agenda for the General Assembly.

VIII. Executive Committee

15. The Executive Committee consists of the President of the Union, six Vice-Presidents, the General Secretary and the Assistant General Secretary elected by the General Assembly on the proposal of a Special Nominating Committee.

IX. Commissions of the Union

16. The General Assembly forms Commissions for such purposes as it may decide.

X. Représentation Légale de l'Union

17. Le Secrétaire Général est le représentant légal de l'Union.

XI. Budget et Cotisations

18. (a) Pour chaque Assemblée Générale ordinaire, le Comité Exécutif prépare un projet de budget pour la période à courir jusqu'à l'Assemblée Générale ordinaire suivante, ainsi que les comptes de l'Union pour la période précédente. Il les soumet au Comité des Finances pour examen ; ce Comité des Finances est composé de membres nommés par les organismes adhérents, à raison d'un membre par organisme, et il est approuvé par l'Assemblée Générale. Lors de sa première séance pendant l'Assemblée Générale, le Comité des Finances élit un Président parmi ses membres.

(b) Le Comité des Finances examine les comptes de l'Union pour voir si les dépenses engagées ont été conformes aux vœux émis lors de la précédente réunion de l'Assemblée Générale et il s'assure que le budget proposé vise à la poursuite de la politique de l'Assemblée Générale, telle qu'elle est interprétée par le Comité Exécutif. Il présente des rapports sur ces questions qu'il soumet à l'Assemblée Générale pour approbation des comptes, et pour décision sur le budget.

(c) Chaque Pays Adhérent verse annuellement à l'Union un nombre d'unités de cotisation qui est fonction de sa catégorie. Le nombre d'unités de cotisation pour chaque catégorie est fixé par le Règlement.

(d) Le montant de l'unité de cotisation est fixé par l'Assemblée Générale, sur la proposition du Comité Exécutif et avec l'avis du Comité des Finances.

(e) Le paiement des cotisations est à la charge des organismes adhérents. La responsabilité de chaque Pays Adhérent envers l'Union est limitée au montant des cotisations dues par ce pays à l'Union.

(f) Un Pays Adhérent qui cesse d'adhérer à l'Union renonce de ce fait à ses droits sur l'actif de l'Union.

XII. Dissolution de l'Union

19. La décision de dissoudre l'Union n'est valable que si elle est prise à la majorité des trois quarts des voix des Pays Adhérents qui disposent du droit de vote en vertu de l'article 11(a).

XIII. Dévolution de l'Autorité en Cas de Force Majeure.

20. Si, par suite d'événements indépendants de la volonté de l'Union, des circonstances apparaissent qui rendent impossible le respect des clauses de ces Statuts et du Règlement établi par l'Assemblée Générale, les organes et membres du Comité Exécutif de l'Union, dans l'ordre fixé ci-dessous, prendront toutes dispositions qu'ils jugeront nécessaires pour la continuation du fonctionnement de l'Union. Ces dispositions devront être soumises à une autorité supérieure dès que cela deviendra possible, jusqu'à ce qu'une Assemblée Générale extraordinaire puisse être réunie. L'autorité est dévolue dans l'ordre ci-dessous :

l'Assemblée Générale ; une Assemblée Générale extraordinaire ; le Comité Exécutif, réuni ou par correspondance ; Le Président de l'Union ; Le Secrétaire Général ; ou, à défaut de la possibilité de recourir à l'une de ces autorités ou de leur disponibilité, un des Vice-Présidents.

XIV. Clauses Finales

21. Ces Statuts entrent en vigueur le 1er Septembre 1970.

22. Les présents Statuts sont publiés en versions française et anglaise. En cas d'incertitude, la version française fait seule autorité.

X. Legal Representation of the Union

17. The General Secretary is the legal representative of the Union.

XI. Budget and Dues

18. (a) For each ordinary General Assembly the Executive Committee prepares a budget proposal covering the period to the next ordinary General Assembly, together with the accounts of the Union for the preceding period. It submits these to the Finance Committee for consideration ; this Finance Committee consists of one member nominated by each adhering body and approved by the General Assembly. At its first meeting during the General Assembly, the Finance Committee elects a Chairman from among its members.

    (b) The Finance Committee examines the accounts of the Union from the point of view of responsible expenditure within the intent of the previous General Assembly, and it considers whether the proposed budget is adequate to implement the policy of the General Assembly, as interpreted by the Executive Committee. It submits reports on these matters to the General Assembly for approval of the account and decision on the budget.

    (c) Each Adhering Country pays annually to the Union a number of units of contribution according to its category. The number of units of contribution for each category shall be specified in the By-laws.

    (d) The amount of the unit of contribution is determined by the General Assembly, on the proposal of the Executive Committee and with the advice of the Finance Committee.

    (e) The payment of contributions is the responsibility of the adhering bodies. The liability of each Adhering Country in respect of the Union is limited to the amount of that country's dues to the Union.

    (f) An Adhering Country that ceases to adhere to the Union resigns at the same time its rights to a share in the assets of the Union.

XII. Dissolution of the Union

19. The decision to dissolve the Union is only valid if taken with the approval of three quarters of the votes of the Adhering Countries having the right to vote by virtue of article 11(a).

XIII. Emergency Powers

20. If, through events outside the control of the Union, circumstances arise in which it is impracticable to comply with the provisions of these Statutes and of the By-laws drawn up by the General Assembly, the organs and officers of the Union, in the order specified below, shall take such actions as they deem necessary for the continued operation of the Union. Such action shall be reported to a higher authority immediately this becomes practicable until such time as an extraordinary General Assembly can be convened. The following is the order of authority :

    The General Assembly ; an extraordinary General Assembly ; the Executive Committee in meeting or by correspondence ; the President of the Union ; the General Secretary ; or failing the practicability or availability of any of the above, one of the Vice-Presidents.

XIV. Final Clauses

21. These Statutes enter into force on 1 September 1970.

22. The present Statutes are being published in French and English version. In case of doubt, the French version is the only authority.

REGLEMENT

I. Les Membres de l'Union

1. Les demandes d'adhésion des pays à l'Union Astronomique Internationale (ci-après dénommée l'Union) sont examinées par le Comité Exécutif et soumises à l'approbation de l'Assemblée Générale.

2. Les propositions de modifications de la liste des Membres sont, après examen attentif des suggestions des Présidents de Commissions, soumises pour avis au Comité des Nominations, composé d'un représentant de chaque Pays Adhérent désigné par l'organisme adhérent habilité, avant la décision du Comité Exécutif.

3. Les Commissions peuvent, avec l'approbation du Comité Exécutif, coopter des consultants qu'elles jugent en mesure d'apporter une contribution utile à leur travail. L'adhésion des consultants a pour terme le dernier jour de la première Assemblée Générale ordinaire qui suit leur admission, à moins qu'elle ne soit renouvelée.

4. Une organisation affiliée peut participer au travail de l'Union dans les conditions fixées par accord entre l'organisation et le Comité Exécutif.

II. L'Assemblée Générale

5. L'Union se réunit en Assemblée Générale ordinaire régulièrement une fois tous les trois ans. Si le lieu et la date de l'Assemblée Générale ordinaire n'ont pas été décidés lors de la précédente Assemblée Générale, ils sont fixés par le Comité Exécutif et communiqués aux organismes adhérents au moins six mois à l'avance.

6. Le Président peut convoquer, avec l'accord du Comité Exécutif, une Assemblée Générale extraordinaire. Il est tenu de le faire à la demande du tiers des Pays Adhérents.

7. L'Ordre du Jour de chaque Assemblée Générale ordinaire est arrêté par le Comité Exécutif et communiqué aux Organismes Adhérents au moins quatre mois avant le premier jour de la réunion. Il devra inclure la proposition du Comité Exécutif concernant le montant de l'unité de cotisation qui permet l'application de l'article 24.

8. (a) L'Ordre du Jour doit inclure toute motion ou proposition reçue par le Secrétaire Général au moins cinq mois avant le premier jour d'une Assemblée Générale ordinaire, qu'elle émane d'un organisme adhérent, d'une Commission de l'Union, ou d'une Commission mixte dans laquelle l'Union est représentée.

    (b) Une motion ou proposition concernant l'administration ou le budget de l'Union qui ne figure pas à l'Ordre du Jour, préparé par le Comité Exécutif, ou tout amendement à une motion qui figure à l'Ordre du Jour, ne peut être discuté qu'avec l'accord préalable des deux tiers au moins des voix des Pays Adhérents représentés à l'Assemblée Générale et disposant du droit de vote en vertu de l'article 11(a) des Statuts.

9. S'il y a doute sur le caractère administratif ou scientifique d'une question donnant lieu à un vote, l'avis du Président est prépondérant.

10. En cas de partage égal des voix, le Président a voix prépondérante.

11. Le Président peut inviter des représentants d'autres organisations, des scientifiques et de jeunes astronomes à participer à l'Assemblée Générale. Avec l'accord du Comité Exécutif, il peut déléguer ce privilège au Secrétaire Général en ce qui concerne les représentants d'autres organisations, aux organismes adhérents en ce qui concerne les scientifiques et les jeunes astronomes.

BY - LAWS

I. Membership

1. Applications of countries for adherence to the International Astronomical Union (referred to as the Union) are examined by the Executive Committee and submitted to the General Assembly for approval.

2. Proposed changes in the list of Members are, with due regard to the suggestions of the Presidents of Commissions, submitted for advice to the Nominating Committee, consisting of one representative of each Adhering Country designated by the appropriate adhering body, before decision by the Executive Committee.

3. Commissions may, with the approval of the Executive Committee, co-opt consultants whom they consider may contribute to their work. The adherence of consultants expires on the last day of the ordinary General Assembly next following their admission, unless renewed.

4. An affiliated organization may participate in the work of the Union as mutually agreed between the organization and the Executive Committee.

II. General Assembly

5. The Union meets in ordinary General Assembly, as a rule, once every three years. The place and date of the ordinary General Assembly unless determined by the General Assembly at its previous meeting, shall be fixed by the Executive Committee and communicated to the adhering bodies at least six months beforehand.

6. The President, with the consent of the Executive Committee, may summon an extraordinary General Assembly. He must do so at the request of one third of the Adhering Countries.

7. The agenda of business for each ordinary General Assembly is determined by the Executive Committee and is communicated to the adhering bodies at least four months before the first day of the meeting. It shall include the proposal of the Executive Committee in regard to the unit of contribution as called for in article 24.

8. (a) Any motion or proposal received by the General Secretary at least five months before the first day of an ordinary General Assembly, whether from an adhering body, from a Commission of the Union, or from an Inter-Union Commission on which the Union is represented, must be placed on the agenda.

    (b) A motion or proposal concerning the administration or budget of the Union which does not appear on the agenda prepared by the Executive Committee, or any amendment to a motion that appears on the agenda, shall only be discussed with the prior approval of at least two thirds of the votes of Adhering Countries represented at the General Assembly and having the right to vote by virtue of Statute 11(a).

9. If there is doubt as to the administrative or scientific character of a question giving rise to a vote, the President determines the issue.

10. Where there is an equal division of votes, the President determines the issue.

11. The President may invite representatives of other organizations, scientists and young astronomers to participate in the General Assembly. Subject to the agreement of the

III. Le Comité Spécial des Nominations

12. (a) Les propositions pour les élections du Président de l'Union, des six Vice-Présidents, du Secrétaire Général et du Secrétaire Général Adjoint sont soumises à l'Assemblée Générale par le Comité Spécial des Nominations. Ce Comité se compose du Président en fonction et du Président sortant, d'un membre proposé par le Comité Exécutif sortant et n'appartenant ni au Comité Exécutif actuel ni au Comité Exécutif précédent, et de quatre membres élus par le Comité des Nominations parmi douze membres proposés par les Présidents de Commissions. A l'exception du Président en fonction et du Président sortant, les membres actuels et les anciens membres du Comité Exécutif ne doivent pas faire partie du Comité Spécial des Nominations. Les membres du Comité Spécial des Nominations doivent tous appartenir à des pays différents.

    (b) Le Secrétaire Général et le Secrétaire Général Adjoint participent au travail du Comité Spécial des Nominations à titre consultatif.

    (c) Le Comité Spécial des Nominations est nommé par l'Assemblée Générale et est responsable directement devant elle. Il reste en fonction jusqu'à la fin de l'Assemblée Générale ordinaire qui suit immédiatement sa nomination, et il peut combler toute vacance survenant parmi ses membres.

IV. Le Comité Exécutif et ses Membres

13. (a) Le Président de l'Union reste en fonction jusqu'à la fin de l'assemblée Générale ordinaire qui suit immédiatement celle de son élection ; les Vice-Présidents restent en fonction jusqu'à la fin de la deuxième Assemblée Générale ordinaire qui suit celle de leur élection. Ils ne sont pas rééligibles immédiatement pour les mêmes fonctions.

    (b) Le Secrétaire Général et le Secrétaire Général Adjoint restent en fonction jusqu'à la fin de l'Assemblée Générale ordinaire qui suit immédiatement celle de leur élection. Normalement, le Secrétaire Général Adjoint succède au Secrétaire Général, mais l'un et l'autre peuvent être réélus aux mêmes fonctions pour une seconde période consécutive.

    (c) Les élections ont lieu au cours de la dernière réunion de l'Assemblée Générale, les noms des candidats proposés ayant été annoncés au cours d'une réunion antérieure.

14. Le Président sortant et le Secrétaire Général sortant deviennent conseillers du Comité Exécutif jusqu'à la fin de l'Assemblée Générale ordinaire qui suit immédiatement celle de la fin de leur mandat. Ils participent au travail du Comité Exécutif et assistent à ses réunions sans droit de vote.

15. Le Comité Exécutif peut combler toute vacance survenant en son sein. Toute personne ainsi nommée reste en fonction jusqu'à l'Assemblée Générale ordinaire suivante.

16. Le Comité Exécutif peut rédiger et publier des Directives pour expliciter les Statuts et le Règlement.

17. Le Comité Exécutif nomme le représentant de l'Union qui doit siéger au sein du Conseil International des Unions Scientifiques ; si ce représentant n'est pas déjà un membre élu du Comité Exécutif, il devient conseiller.

18. (a) Le Secrétaire Général est responsable auprès du Comité Exécutif des dépenses qu'il engage, qui ne doivent pas dépasser le montant des fonds mis à sa disposition.

    (b) Un bureau administratif, sous la direction du Secrétaire Général, est chargé de la correspondance, de la gestion des fonds de l'Union, et de la conservation des archives.

Executive Committee he may delegate this privilege concerning representatives of other organizations to the General Secretary, and concerning scientists and young astronomers to the adhering bodies.

III. Special Nominating Committee

12. (a) Proposals for elections to the President of the Union, six Vice-Presidents, the General Secretary and the Assistant General Secretary are submitted to the General Assembly by the Special Nominating Committee. This consists of the President and past President of the Union, a member proposed by the retiring Executive Committee, and four members elected by the Nominating Committee from among twelve Members proposed by Presidents of Commissions. Other than the President and immediate past President, present and former members of the Executive Committee shall not serve on the Special Nominating Committee. No two members of the Special Nominating Committee shall belong to the same country.

    (b) The General Secretary and the Assistant General Secretary participate in the work of the Special Nominating Committee in an advisory capacity.

    (c) The Special Nominating Committee is appointed by the General Assembly to which it reports direct. I remains in office until the end to the ordinary General Assembly next following that of its appointment, and it may fill any vacancy occurring among its members.

IV. Officers and Executive Committee

13. (a) The President of the Union remains in office until the end of the ordinary General Assembly next following that of his election ; the Vice-Presidents remain in office until the end of the second ordinary General Assembly following that of their election. They may not be re-elected immediately to the same offices.

    (b) The General Secretary and the Assistant General Secretary remain in office until the end of the ordinary General Assembly next following that of their election. Normally the Assistant General Secretary succeeds the General Secretary though both officers may be re-elected for another term.

    (c) The election takes place at the last session of the General Assembly, the names of the candidates proposed having been announced at a previous session.

14. The retiring President and the retiring General Secretary become advisers to the Executive Committee until the end of the ordinary General Assembly next following that of their retirement. They participate in the work of the Executive Committee and attend its meetings without voting right.

15. The Executive Committee may fill any vacancy occurring among its members. Any person so appointed remains in office until the next ordinary General Assembly.

16. The Executive Committee may draw up and publish Working Rules to implement the Statutes and By-laws.

17. The Executive Committee appoints the Union's representative to the International Council of Scientific Unions ; if not already an elected member of the Executive Committee, this representative will become its adviser.

18. (a) The General Secretary is responsible to the Executive Committee for not incurring expenditure in excess of the funds at his disposal.

    (b) An administrative office, under the direction of the General Secretary, conducts the correspondence, administers the funds, and preserves the archives of the Union.

V. Commissions

19. (a) Les Commissions de l'Union poursuivent les buts scientifiques de l'Union par des moyens tels que l'étude de domaines particuliers de l'Astronomie, l'encouragement de recherches collectives et la discussion de questions relatives aux accords internationaux et à la standardisation.

    (b) Les Commissions de l'Union établissent des rapports sur les sujets qui leur ont été confiés.

20. Chaque Commission se compose de :

    (a) un Président et au moins un Vice-Président élus par l'Assemblée Générale sur la proposition du Comité Exécutif. Ils demeurent en fonction jusqu'à la fin de l'Assemblée Générale ordinaire qui suit immédiatement celle de leur élection. Ils ne sont pas normalement rééligibles,

    (b) un Comité d'Organisation, dont les membres sont désignés par la Commission sous réserve de l'approbation du Comité Exécutif. Le Comité d'Organisation assiste le Président et le(s) Vice-Président(s) dans leur tâche. Une Commission peut décider qu'elle n'a pas besoin de Comité d'Organisation,

    (c) des membres de l'Union, nommés par les Présidents, Vice-Président(s) et Comité d'Organisation, en considération de leurs spécialités ; leur désignation est soumise à confirmation par le Comité Exécutif.

21. Entre deux Assemblées ordinaires, les Présidents de Commissions peuvent coopter, parmi les Membres de l'Union, de nouveaux membres des Comités d'Organisation et des Commissions elles-mêmes.

22. Les Commissions rédigent leur propre règlement. Les décisions sont prises, à l'intérieur des Commissions, par un vote de leurs membres.

VI. Organismes Adhérents

23. Le rôle des organismes adhérents est d'encourager et de coordonner, sur leurs territoires respectifs, l'étude des diverses branches de l'astronomie, particulièrement en ce qui concerne leurs besoins sur le plan international. Ils ont le droit de soumettre au Comité Exécutif des propositions pour discussion par l'Assemblée Générale.

VII. Finances

24. Chaque Pays Adhérent verse à l'Union une cotisation annuelle, qui est un multiple de l'unité de cotisation en fonction de sa catégorie, comme suit :

Catégories définies conformément à l'article 8 des Statuts	:	1	2	3	4	5	6	7	8
Nombre respectif d'unités de cotisations	:	1	2	4	6	10	14	20	30

25. Les ressources de l'Union sont consacrées à la poursuite de ses buts, y compris :

    (a) les frais de publication et les dépenses administratives ;

    (b) l'encouragement des activités astronomiques qui nécessitent la coopération internationale ;

    (c) la cotisation due par l'Union au Conseil International des Unions Scientifiques.

26. Les ressources provenant de dons sont utilisées par l'Union en tenant compte des vœux exprimés par les donateurs.

V. Commissions

19. (a) The Commissions of the Union shall pursue the scientific objects of the Union by activities such as the study of special branches of astronomy, the encouragement of collective investigations, and the discussion of questions relating to international agreements or to standardization.

    (b) The Commissions of the Union shall prepare reports on the work with which they are concerned.

20. Each Commission consists of :

    (a) a President and at least one Vice-President elected by the General Assembly on the proposal of the Executive Committee. They remain in office until the end of the ordinary General Assembly next following that of their election. They are not normally re-eligible,

    (b) an Organizing Committee, whose members are appointed by the Commission subject to the approval by the Executive Committee. The Organizing Committee assists the President and Vice-President(s) in their duties. A Commission may decide that it needs no Organizing Committee,

    (c) Members of the Union, appointed by the President, Vice-President(s) and the Organizing Committee, in consideration of their special interests ; their appointment is subject to the confirmation by the Executive Committee.

21. Between two ordinary General Assemblies, Presidents of Commissions may co-opt, from among Members of the Union, new members to the Organizing Committees and to the Commissions themselves.

22. Commissions draw up their own rules. Decisions within Commissions are taken according to the vote of their members.

VI. Adhering Bodies

23. The functions of the Adhering Bodies are to promote and co-ordinate, in their respective territories, the study of the various branches of astronomy, more especially in relation to their international requirements. They are entitled to submit to the Executive Committee motions for discussions by the General Assembly.

VII. Finances

24. Each Adhering Country pays annually to the Union a number of units of contribution according to its category as follows :

Category as defined in Statute 8 :	1	2	3	4	5	6	7	8
Number of units of contribution :	1	2	4	6	10	14	20	30

25. The income of the Union is to be devoted to its objects, included

    (a) costs of publication and expenses of administration ;

    (b) the promotion of astronomical enterprises requiring international co-operation ;

    (c) the contribution due from the Union to the International Council of Scientific Unions.

26. Funds derived from donations are used by the Union in accordance with the wishes expressed by the donors.

VIII. Publications

27. L'Union a la propriété littéraire de tous les textes imprimés dans ses publications, sauf accord différent.

28. Les Membres de l'Union ont le droit de recevoir les publications de l'Union gratuitement ou à prix réduit, à la discrétion du Comité Exécutif qui décide en fonction de la situation financière de l'Union.

IX. Clauses Finales

29. Ce règlement entre en vigueur le 1er Septembre 1970. Il peut être modifié avec l'approbation de la majorité absolue des voix des Pays Adhérents qui disposent du droit de vote en vertu de l'article 11(a) des Statuts.

30. Le présent règlement est publié en versions française et anglaise. En cas d'incertitude, la version française fait seule autorité.

VIII. Publications

27. The Union has the copyright to all materials printed in its publications, unless otherwise arranged.

28. Members of the Union are entitled to receive the publications of the Union free of charge or at reduced prices at the discretion of the Executive Committee taking due regard of the financial situation of the Union.

IX. Final Clauses

29. These By-laws enter into force on 1 September 1970. They can be changed with the approval of an absolute majority of the votes of the Adhering Countries having the right to vote by virtue of Statute 11(a).

30. The present By-laws are being published in French and English versions. In case of doubt, the French version is the only authority.

UNION ASTRONOMIQUE INTERNATIONALE

DIRECTIVES

I. Publications

1. Les publications de l'Union Astronomique Internationale, approuvées dans le budget par l'Assemblée Générale, sont préparées par le Bureau Administratif de l'Union.

2. Les Commissions de l'Union peuvent, avec l'approbation du Comité Exécutif, avoir leurs propres publications.

3. Le Comité Exécutif décide, sur la proposition du Secrétaire Général, des modalités de distribution des publications de l'Union.

4. Les Membres de l'Union peuvent acquérir les publications de l'Union à un prix réduit.

II. Appartenance à l'Union

A. Pays Adhérents

5. Les demandes d'adhésion à l'Union formulées par les pays sont examinées par le Comité Exécutif compte tenu des points suivants :
   (a) justesse du choix de la catégorie dans laquelle le pays souhaite être classé ;
   (b) situation actuelle de l'Astronomie dans le pays formulant la demande, et ses possibilités de développement ;
   (c) mesure dans laquelle le futur organisme adhérent est représentatif de l'activité astronomique de son pays.

6. Les demandes proposant une contribution annuelle appropriée seront soumises pour décision à l'Assemblée Générale, avec la recommandation du Comité Exécutif.

B. Membres

7. Les personnes proposées pour devenir Membres de l'Union doivent en principe être choisies parmi des astronomes et des chercheurs dont les activités sont liées à l'astronomie, compte tenu de :
   (a) la qualité de leur œuvre scientifique ;
   (b) la mesure dans laquelle leur activité scientifique implique des recherches astronomiques ;
   (c) leur désir de contribuer à la poursuite des buts de l'Union.

8. Les jeunes astronomes doivent être considérés comme pouvant devenir Membres de l'Union dès qu'ils ont fait la preuve de leur capacité (en principe par une thèse de doctorat ou son équivalent) et de leur aptitude (quelques années d'activité fructueuse) à mener une recherche personnelle.

9. Pour les astronomes professionnels, leur contribution à l'astronomie peut consister soit en des recherches personnelles, soit en une collaboration assidue à des programmes importants d'observations.

10. Les autres personnes ne peuvent devenir Membres de l'Union que si certains de leurs travaux originaux concernent étroitement la recherche astronomique.

INTERNATIONAL ASTRONOMICAL UNION

WORKING RULES

I. Publications

1. The publications of the International Astronomical Union, approved in the budget by the General Assembly, are prepared by the Administrative Office of the Union.

2. Commissions of the Union may, with the approval of the Executive Committee, issue their publications independently.

3. The distribution of publications of the Union is decided, on the proposal of the General Secretary, by the Executive Committee.

4. Members may purchase the publications of the Union at reduced prices.

II. Membership

A. Adhering Countries

5. Applications of countries for adherence to the Union are examined by the Executive Committee for
   (a) the adequacy of the category in which the country wishes to be classified ;
   (b) the present state and expected development of astronomy in the applying country ;
   (c) the degree to which the prospective adhering body is representative of its country's astronomical activity.

6. Applications proposing an adequate annual contribution to the Union shall, with the recommendation of the Executive Committee, be submitted to the General Assembly for decision.

B. Members

7. Individuals proposed for Union membership should, as a rule, be chosen from among astronomers and scientists, whose activity is closely linked with astronomy taking into account
   (a) the standard of their scientific achievement
   (b) the extent to which their scientific activity involves research in astronomy
   (c) their desire to assist in the fulfilment of the aims of the Union.

8. Young astronomers should be considered eligible for membership after they have shown their capability (as a rule Ph.D. or equivalent) of and experience (some years of successful activity) in conducting original research.

9. For full time professional astronomers the achievement in astronomy may consist either of original research or of substantial contributions to major observational programmes.

10. Others are eligible for membership only if they are making original contributions closely linked with astronomical research.

11. Huit mois avant une Assemblée Générale ordinaire, il sera demandé aux organismes adhérents de proposer de nouveaux Membres. Les propositions devront parvenir au Secrétaire Général au moins cinq mois avant la première session de l'Assemblée Générale. Les propositions reçues après cette date limite ne seront prises en considération que si des circonstances exceptionnelles justifient le retard.

12. Chaque proposition du nouveau Membre doit être présentée séparément et indiquer le nom, les prénoms et l'adresse postale du candidat (de préférence celle de son Institut ou Observatoire), ses date et lieu de naissance, l'Université devant laquelle il a soutenu sa thèse ou le diplôme équivalent, la date de soutenance, la situation actuelle du candidat, les titres et renseignements bibliographiques de deux ou trois de ses articles ou publications les plus significatifs et, s'il y a lieu, tous les renseignements susceptibles d'être pris en considération par le Comité des Nominations.

13. (a) Les Présidents de Commissions qui désirent suggérer de nouveaux membres doivent adresser leurs suggestions au Secrétaire Général au moins cinq mois avant la première session d'une Assemblée Générale ordinaire. Les propositions devront fournir les mêmes renseignements que ceux mentionnés à l'article 13.

    (b) Le Secrétaire Général fait part de ces suggestions aux organismes adhérents intéressés.

14. Le Secrétaire Général préparera deux listes pour le Comité des Nominations

    (a) l'une contenant les noms des candidats proposés par les organismes adhérents,

    (b) l'autre contenant les noms des candidats proposés par les Présidents de Commissions, mais qui ne sont pas déjà inclus dans les propositions des organismes adhérents.

15. A partir des deux listes mentionnées à l'article 15, le Comité des Nominations prépare les propositions définitives de nouveaux membres de l'Union.

16. Les organismes adhérents peuvent proposer la radiation de Membres ayant abandonné le domaine de l'astronomie pour d'autres activités, à moins qu'ils ne continuent à apporter une contribution à l'astronomie. Ces propositions doivent être portées à la connaissance du Secrétaire Général et du Membre concerné.

17. Le Secrétaire Général publiera la liste alphabétique des Membres de l'Union dans les Transactions de chaque Assemblée Générale ordinaire.

III. Membres des Commissions

18. Les membres des Commissions de l'Union sont cooptés par les Commissions. Cette procédure est régie par des règles établies par les Commissions elle-mêmes.

19. Les Commissions devraient choisir, ou approuver, la liste des membres de leurs commissions compte tenu de la spécialité de ces personnes, en particulier de leur activité scientifique dans le domaine de recherche de la Commission, et de leur contribution au travail de la Commission. Elles peuvent

    (a) inviter les Membres de l'Union à devenir membres de la Commission,

    (b) radier les membres de la Commission qui n'ont pas contribué à son activité,

    (c) accepter ou refuser les demandes présentées par des Membres de l'Union, ou par des personnes proposées comme tels, en vue d'appartenir à la Commission,

    (d) suggérer l'élection comme Membres de l'Union de personnes n'y appartenant pas, ce qui leur permettrait alors de devenir membres de la Commission.

11. Eight months before an ordinary General Assembly, adhering bodies will be asked to propose new Members. The proposals should reach the General Secretary not later than five months before the first session of the General Assembly. Proposals received after the closing date will only be taken into consideration if the delay is justified by exceptional circumstances.

12. Each proposal shall be written separately. It should include the name, first names and postal address of the candidate, preferably that of his/her Institute or Observatory, his/her place and date of birth, the University and the year of his/her Ph.D. or equivalent title, his/her present occupation, titles and bibliographic data of two or three of his/her more important papers or publications, and details, if any, worthy to be considered by the Nominating Committee.

13. (a) Presidents of Union Commissions wishing to suggest new Members for admittance should address their suggestions to the General Secretary five months before the first session of an ordinary General Assembly. The proposals should contain particulars as in article 13.

 (b) The General Secretary notifies the adhering bodies in questions of such suggestions.

14. The General Secretary shall prepare two lists for the Nominating Committee.

 (a) One containing the candidates proposed by the adhering bodies,

 (b) the other containing those suggested by Presidents of Commissions, but not included among the proposals of adhering bodies.

15. The Nominating Committee prepares the final proposals for Union membership from the two lists as mentioned in article 15.

16. Adhering bodies should propose cancellation of Members who have left the field of astronomy for other interests, unless they continue to contribute to astronomy. Such proposals should be announced to the Member concerned and to the General Secretary.

17. The alphabetical list of Union Members will be published by the General Secretary in the Transactions of each ordinary General Assembly.

III. Commission Membership

18. Members of Union Commissions are co-opted by Commissions. The rules governing the procedure of such co-option are drawn up by the Commissions themselves.

19. Commissions should choose, or approve of, Commission members taking into account their special interests, in particular their scientific activity in the appropriate fields of research and their contribution to the work of the Commission. They may,

 (a) invite Members to become members of their Commission,

 (b) remove members who have not contributed to the work of the Commission,

 (c) accept or reject applications for membership from existing or proposed Members,

 (d) suggest non-Members for election as Members, thus enabling them to become members of the Commission.

20. Members may not, as a rule, be members of more than three Commissions.

20. Les Membres de l'Union ne peuvent pas, en règle générale, appartenir à plus de trois Commissions.

21. Les membres de l'Union peuvent demander à être admis dans une Commission en écrivant au Président de cette Commission. Ils ne devraient faire cette demande que si leur propre activité rentre dans le cadre des recherches de la Commission et s'ils sont décidés à contribuer au travail de la Commission.

22. Les membres des Commissions peuvent se retirer d'une Commission en écrivant à son Président.

23. En envoyant leurs propositions de nouveaux Membres, les organismes adhérents peuvent également suggérer le choix d'une Commission pour chaque candidat.

24. Le Secrétaire Général enregistrera et analysera la liste des membres des Commissions ; si cela est nécessaire, il tentera de trouver une solution aux anomalies évidentes.

25. Le Secrétaire Général publiera la liste des membres des Commissions dans les Transactions de chaque Assemblée Générale ordinaire.

IV. Consultants

26. Peuvent être élus Consultants des personnes qui ne sont pas astronomes, mais qui sont susceptibles de servir les intérêts de l'astronomie.

27. Les Commissions doivent en principe envoyer, pour approbation, leurs propositions de consultants au Secrétaire Général au moins cinq mois avant la première session d'une Assemblée Générale ordinaire.

28. Le Secrétaire Général préparera une liste des personnes proposées comme consultants et la soumettra pour approbation au Comité Exécutif.

29. Le Bureau Administratif établira une liste alphabétique des consultants.

30. Les consultants peuvent participer aux réunions de l'Union. Ils peuvent avoir droit de vote dans leurs Commissions respectives. Ils reçoivent gratuitement le Bulletin d'Information de l'Union.

V. Réunions Scientifiques

31. Le Secrétaire Général publiera un règlement pour les réunions scientifiques organisées ou parrainées par l'Union.

VI. Contacts Extérieurs

32. Aucune relation avec des tiers, imputable à l'Union, ne sera entreprise par quiconque membre de l'Union, si ce n'est sous l'autorité du Secrétaire Général.

33. Les représentants de l'Union dans d'autres organisations, en particulier les Comités de l'ICSU et les Commissions Inter-Unions, seront désignés par le Comité Exécutif. Les noms sont proposés par les Présidents des Commissions concernées.

34. Les dépenses encourues par les représentants de l'Union dans d'autres organisations seront remboursées à la discrétion du Secrétaire Général, dans les limites du Budget adopté par l'Assemblée Générale. Les représentants sont priés d'obtenir l'accord préalable du Secrétaire Général avant d'engager ces dépenses.

21. Members may apply for Commission membership by writing to the President of the Commission concerned. Such applications should only be made if the Member is actively engaged in the appropriate field of research and is prepared to contribute to the work of the Commission.

22. Members of Commissions may resign from a Commission by writing to its President.

23. Adhering bodies, in sending in their proposals for new Members, may also suggest one Commission for each candidate.

24. The General Secretary will record and analyse the list of members of Commissions ; if necessary he will try to resolve any outstanding anomalies.

25. The list of Commission members will be published by the General Secretary in the Transactions of each ordinary General Assembly.

IV. Consultants

26. Eligible as Consultants are non-astronomers in a position to further the interest in astronomy.

27. Proposals of Commissons for the approval of consultants should, as a rule, reach the General Secretary not later than five months before the first session of an ordinary General Assembly.

28. The General Secretary shall prepare a list of those proposed for admission as consultants and submit it to the Executive Committee for approval.

29. The Administrative Office will maintain an alphabetical list of consultants.

30. Consultants may participate in the meetings of the Union. They may have voting right in the respective Commission. They receive, free of charge, the Information Bulletin of the Union.

V. Scientific Meetings

31. The General Secretary shall publish rules for scientific meetings organized or sponsored by the Union.

VI. External Contacts

32. No dealings with third parties, attributable to the Union, shall be undertaken by any Member of the Union except on the authority of the General Secretary.

33. Representatives of the Union in other bodies, especially ICSU Committees and ICSU Inter-Union Committees, shall be appointed by the Executive Committee. Nominations are sought from Presidents of appropriate Commissions.

34. Expenses incurred by Representatives of the Union in other bodies will be reimbursed at the discretion of the General Secretary, within the provisions of the Budget Estimate adopted by the General Assembly. Representatives are required to obtain prior approval of the General Secretary before incurring such expenses.

VII. Assemblées Générales

35. Huit mois avant l'Assemblée Générale, le Secrétaire Général envoie aux Comités Nationaux d'Astronomie et aux Organisations Adhérentes le budget préparé par le Comité Exécutif, pour commentaires.

VII. General Assemblies

35. The General Secretary distributes the budget prepared by the Executive Committee to National Committes of Astronomy and/or Adhering Organizations for comments eight months before the General Assembly.

CHAPTER VI
MEMBERSHIP

1) LIST OF ADHERING COUNTRIES

The year of adherence and approximate number of IAU members in the geographical area are indicated (as of May 1986).

Country	Adhering Organization	Year	Members
Arab Republic of Egypt	Academy of Scientific Research & Technology Dept. of Scientific Societies and International Unions 101 Kasr El-Einy Street Cairo	1925	37
Argentina	Consejo Nacional de Investigaciones Cientificas y Técnicas Rivadavia 1917 1033 Buenos Aires	1927	57
Australia	Australian Academy of Sciences PO Box 783 Canberra City, ACT 2601	1939	125
Austria	Osterreichische Akademie der Wissenschaften Dr. Ignaz-Seipel-Platz 2 A-1010 Wien	1955	32
Belgium	Académie Royale de Belgique Palais des Académies Rue Ducale 1 B-1000 Bruxelles	1920	87
Brazil	Conselho Nacional de Desenvolvimento Cientifico e Tecnologico - CNPq Av. W3 Norte, Quadra 507-B Caixa Postal 11-1142 70740 Brasilia DF	1961	39
Bulgaria	Bulgarian Academy of Sciences 7th November Street 1 1000 Sofia	1957	37
Canada	National Research Council of Canada International Relations Ottawa, Ontario K1A OR6	1957	189
Chile	Universidad de Chile Observatorio Astronomico Nacional Casilla 36-D Santiago	1947	23

China	Chinese Astronomical Society Purple Mountain Observatory Academia Sinica Nanking	1935	280
	Astronomy Union Located at Taipei Taipei 115 Taiwan	1959	14
Colombia	Universidad Nacional de Colombia Apartado Aereo No. 5997 Bogota	1967	3
Cuba	Academia de Ciencias de Cuba Capitolio Nacional La Habana 2	1970	0
Czechoslovakia	Czechoslovak Academy of Sciences Narodni 3 Praha 1	1922	79
Denmark	Kongelige Danske Videnskabernes Selskab H.C. Andersen Boulevard 35 DK-1553 København V	1922	48
Finland	Académie des Sciences et Lettres Snellmaninkatu 9-11 Helsinki 17	1948	27
France	Académie des Sciences COFUSI 23, quai Conti F-75006 Paris	1920	414
Germany, DR	Akademie der Wissenschaften der DDR DDR-108 Berlin Otto-Nuschke-Strasse 22/23	1951	46
Germany, FR	Rat Westdeutscher Sternwarte MPI für Physik & Astrophysik Karl-Schwarzschild-str. 1 D-8046 Garching bei München	1951	278
Greece	Academy of Athens Panepistimiou Str. Athens	1920	67
Hungary	Hungarian Academy of Sciences Roosevelt Tér 9 Budapest V	1947	33
India	Indian National Science Academy Bahadur Shah Zafar Marg New Delhi 110002	1964	172

Indonesia	Indonesian Institute of Sciences (LIPI) Gedung Widya Graha Jl. Jend. Gatot Subroto 10 Jakarta Selatan	1979	4
Iran	Geophysics Institute of Tehran University North Amir-Abad Avenue Tehran	1969	10
Iraq	Council for Scientific Research Astronomy and Space Research Center PO Box 255 Baghdad	1976	10
Ireland	The Royal Irish Academy Academy House 19 Dawson Street Dublin 2	1947	27
Israel	The Israel Academy of Sciences and Humanities Albert Einstein Square, Talbieh Jerusalem 91040	1954	37
Italy	Consiglio Nazionale delle Ricerche Piazzale Aldo Moro 7 I-00100 Roma	1920	302
Japan	Science Council of Japan 22-34 Roppongi 7 chome Minato-Ku Tokyo 106	1920	282
Korea, DPR	Academy of Sciences of DPR of Korea Pyongyang	1961	21
Korea, Republic of	Korean National Astronomical Observatory Yoksam-Dong Kangnam-ku Seoul, 134-03	1973	14
Mexico	Instituto de Astronomia, UNAM Apartado Postal 70-264 Ciudad Universitaria 04510 DF Mexico	1921	43
Netherlands	Koninklijke Nederlandse Akademie van Wetenschappen Kloveniersburgwal 29 NL-1011 JV Amsterdam	1922	121
New Zealand	The Royal Society of New Zealand PO Box 12249 Wellington	1964	22

Nigeria	Nigerian Academy of Sciences Faculty of Science, University of Lagos P.M.B. 1004 University of Lagos Post Office Lagos	1985	1
Norway	Det Norske Videnskaps-Akademi i Oslo Drammensveien 78 Oslo 2	1922	21
Poland	Polskiej Akademii Nauk Palac Kultury i Nauki Skrytka pocztowa 24 00-901 Warszawa	1922	82
Portugal	Secçao Portuguesa das Unioes Internacionais Astronomica e Geodesica e Geofisica Praça de Estrela Lisboa 1200	1924	16
Roumania	Roumanian National Committee of Astronomy Astronomical Observatory Cutitul de Argint 5 PO Box 28 75212 Bucarest	1928	16
South Africa	Council for Scientific and Industrial Research International Relations Division PO Box 395 Pretoria 0001	1938	22
Spain	Comision Nacional de Astronomia Instituto Geografico y Cadastral General Ibanez 3 Madrid (3)	1922	73
Sweden	Kungl. Vetenskapsakademien PO Box 50005 S-184 05 Stockholm 50	1925	73
Switzerland	Schweizerische Naturforschende Gesellschaft Zentralsekretariat Postfach 2535 Hirschengraben 11 CH-3001 Bern	1923	53
Turkey	Astronomi Denergi Baskani Universitet Rasthanesi Universite Istanbul	1961	47
United Kingdom	The Royal Society 6, Carlton House Terrace London SW1Y 5AG	1920	501

LIST OF ADHERING COUNTRIES

Uruguay	Ministerio de Relaciones Exteriores Avenida 18 de Julio 1205 Montevideo	1970	1
USA	National Academy of Sciences Office of International Affairs 2101 Constitution Avenue N.W. Washington, DC 20418	1920	1615
USSR	Academy of Sciences of the USSR Foreign Relations Department Leninskij Prospekt 14 Moscow 71	1935	420
Vatican City State	Pontificia Academia delle Scienze I-00120 Citta del Vaticano	1932	7
Venezuela	Comité Nacional de Astronomia CIDA Apartado 264 Mérida 5101-A	1953	8
Yugoslavia	Savez Drustava Matematicara, Fizicara i Astronoma Jugosloslavije Institut za Matematiku i Fiziku Cetinjski put bb 81000 Titograd	1935	31

2. MEMBERSHIP OF COMMISSIONS

COMMISSION No. 4

EPHEMERIDES (EPHEMERIDES)

President : MORANDO BRUNO L DR

Vice-President(s) : SEIDELMANN P KENNETH DR

Organizing Committee: ABALAKIN VICTOR K DR
AOKI SHINKO PROF
CHAPRONT JEAN DR
DUNCOMBE RAYNOR L DR
LEDERLE TRUDPERT DR
LIESKE JAY H DR
YALLOP BERNARD D DR
YAMAZAKI AKIRA DR

Members:
ARIAS DE GREIFF J PROF
BEC-BORSENBERGER ANNICK
CAPITAINE NICOLE
DE CASTRO ANGEL DR
DICKEY JEAN O'BRIEN
FIALA ALAN D DR
FURSENKO M A DR
HAUPT RALPH F
JANICZEK PAUL M DR
KINOSHITA HIROSHI DR
KRASINSKY GEORGE A DR
LI GI MAN
LIU BAO-LIN
O'HANDLEY DOUGLAS A DR
ROMERO PEREZ M PILAR
SHAPIRO IRWIN I PROF
SOCHILINA ALLA S DR
TING YEOU-TSWEN
WILKINS GEORGE A DR
XIAN DING-ZHANG

ARLOT JEAN-EUDES
BRETAGNON PIERRE DR
CATALAN MANUEL DR
DEPRIT ANDRE PROF
DUNHAM DAVID W
FOMINOV ALEXANDR M DR
GLEBOVA NINA I DR
HENRARD JACQUES PROF
JOHNSTON KENNETH J
KLEPCZYNSKI WILLIAM J DR
KUBO YOSHIO
LI HYOK HO
MORRISON LESLIE V
OESTERWINTER CLAUS
ROSSELLO GASPAR
SIMON JEAN-LOUIS MR
STANDISH E MYLES DR
VAN FLANDERN THOMAS DR
WILLIAMS JAMES G DR
ZHAO XIAN-ZI

BANDYOPADHYAY A DR
BRUMBERG VICTOR A DR
CHOLLET FERNAND DR
DI XIAO-HUA
EMERSON BRIAN MR
FRICKE WALTER PROF DR
GONDOLATSCH FRIEDRICH PRF
ILYAS MOHAMMAD DR
KING ROBERT WILSON JR DR
KOLACZEK BARBARA DR
LAHIRI N C
LI NENG-YAO
MUELLER IVAN I PROF
REASENBERG ROBERT D DR
SCHWAN HEINER DR
SINZI AKIRA M DR
TAYLOR GORDON E
WACKERNAGEL H BEAT DR
WINKLER GERNOT M R DR

MEMBERSHIP OF COMMISSIONS

COMMISSION No. 5

DOCUMENTATION AND ASTRONOMICAL DATA

(DOCUMENTATION ET DONNEES ASTRONOMIQUES)

President   : WILKINS GEORGE A DR

Vice-President(s)   : HAUCK BERNARD PROF

Organizing Committee: DLUZHNEVSKAYA O B DR
                      HEINTZ WULFF D DR
                      JASCHEK CARLOS O R PROF
                      LANTOS PIERRE DR
                      MITTON SIMON DR
                      SPITE FRANCOIS M DR
                      WAYMAN PATRICK A PROF
                      WESTERHOUT GART DR
                      WORLEY CHARLES E

Members:
ABALAKIN VICTOR K DR   ABT HELMUT A DR   BAKER NORMAN H PROF
BENACCHIO LEOPOLDO   BESSELL MICHAEL S DR   BIDELMAN WILLIAM P PROF
BOUSKA JIRI DR   DAVIS MORRIS S PROF   DAVIS ROBERT J DR
DEWHIRST DAVID W DR   DICKEL HELENE R DR   DIXON ROBERT S DR
DUCATI JORGE RICARDO DR   DUNCOMBE RAYNOR L DR   EGRET DANIEL DR
FRICKE WALTER PROF DR   GARSTANG ROY H PROF   GRIFFIN ROGER F DR
GROSBOL PREBEN JOHNSON DR   HARTLEY KENNETH F DR   HECK ANDRE DR
HEFELE HERBERT PH D   HEINRICH INGE   JENKNER HELMUT DR
KADLA ZDENKA I DR   KALBERLA PETER   KLECZEK JOSIP DR
LEDERLE TRUDPERT DR   LEQUEUX JAMES DR   LORTET MARIE CLAIRE
LYNGA GOSTA DR   MARTYNOV D YA PROF DR   MCNAMARA DELBERT H DR
MEAD JAYLEE MONTAGUE DR   MEADOWS A JACK PROF   MEIN PIERRE
MERMILLIOD JEAN-CLAUDE DR   NISHIMURA SHIRO DR   OCHSENBEIN FRANCOIS DR
OGORODNIKOV KYRILL P PROF   PAMYATNIKH A A DR   PASINETTI LAURA E PROF
PECKER JEAN-CLAUDE PROF   PHILIP A G DAVIS   RADLOVA L N DR
REMY BATTIAU LILIANE G A   RENSON P F M DR   SCHMADEL LUTZ D DR
SCHMIDT K H DR   SEDMAK GIORGIO PROF   SHAKESHAFT JOHN R DR
SHCHERBINA-SAMOJLOVA I DR   TERASHITA YOICHI PROF   TRITTON SUSAN BARBARA
UESUGI AKIRA DR   VELGHE ALBERT G PROF DR   VOIGT HANS H PROF
WALLACE PATRICK T MR   WARREN WAYNE H JR DR   WEIDEMANN VOLKER PROF
WELLS DONALD C III DR

COMMISSION No. 6

ASTRONOMICAL TELEGRAMS (TELEGRAMMES ASTRONOMIQUES)

President   : MRKOS ANTONIN DR

Vice-President(s) : ROEMER ELIZABETH PROF

Organizing Committee: MARSDEN BRIAN G DR

Members:
  BIRAUD FRANCOIS DR   CANDY MICHAEL P MR   CESCO CARLOS DR
  CUNNINGHAM LELAND E PROF   EVERHART EDGAR DR   GRINDLAY JONATHAN E DR
  HERS JAN MR   KOZAI YOSHIHIDE PROF   MARTYNOV D YA PROF DR
  NAKANO SYUICHI   POUNDS KENNETH A PROF   ROSINO LEONIDA PROF
  XIE GUANG-ZHONG

COMMISSION No. 7
CELESTIAL MECHANICS (MECANIQUE CELESTE)

President   : BRUMBERG VICTOR A DR

Vice-President(s)   : HENRARD JACQUES PROF

Organizing Committee: BATRAKOV YU V DR
                      BHATNAGAR K B DR
                      CANDY MICHAEL P MR
                      CHAPRONT JEAN DR
                      DEPRIT ANDRE PROF
                      FERRAZ-MELLO S PROF DR
                      HADJIDEMETRIOU JOHN D
                      KINOSHITA HIROSHI DR
                      KOVALEVSKY JEAN DR
                      SCHOLL HANS DR
                      SEIDELMANN P KENNETH DR
                      SINCLAIR ANDREW T DR
                      YI ZHAO-HUA

Members:
ABALAKIN VICTOR K DR   AHMED MOSTAFA   AKSENOV E P PROF DR
AKSNES KAARE DR   ALTAVISTA CARLOS A DR   ANTONACOPOULOS GREG PROF
AOKI SHINKO PROF   BAGHOS BALEGH B DR   BALMINO GEORGES G DR
BARBERIS BRUNO   BEC-BORSENBERGER ANNICK   BENEST DANIEL DR
BETTIS DALE G PROF   BOIGEY FRANCOISE   BOZIS GEORGE PROF
BRETAGNON PIERRE DR   BRIEVA EDUARDO PROF   BROOKES CLIVE J DR
BROUCKE ROGER DR   CALAME ODILE DR   CARANICOLAS NICHOLAS DR.
CEFOLA PAUL J DR   CHAPRONT-TOUZE MICHELLE   CHEN ZHEN
CHOI KYU-HONG   CID PALACIOS RAFAEL PROF   CONTOPOULOS GEORGE PROF
COOK ALAN H PROF   COUNSELMAN CHARLES C PROF CUI CHUNFANG
CUI DOU-XING   CUNNINGHAM LELAND E PROF   DANBY J M ANTHONY DR
DAVIS MORRIS S PROF   DEMIN V G PROF DR   DIN HUA
DORMAND JOHN RICHARD DR   DROZYNER ANDRZEJ   DUBOSHIN G N PROF DR
DUNCOMBE RAYNOR L DR   DVORAK RUDOLF DR   EICHHORN HEINRICH K DR
ELIPE SANCHEZ ANTONIO   ELYASBERG P E PROF DR   EMELIANOV NIKOLAJ V DR
ERDI B DR   EVERHART EDGAR DR   FABRE HERVE DR
FARINELLA PAOLO DR   FERNANDEZ SILVIA M. DR   FERRER MARTINEZ SEBASTIAN
FIALA ALAN D DR   FONG CHU-GANG   FROESCHLE CLAUDE DR
GALIBINA I V DR   GALLETTO DIONIGI   GAPOSCHKIN EDWARD M DR
GARFINKEL BORIS DR   GASKA STANISLAW DR   GIACAGLIA GIORGIO E PROF
GOLDREICH P DR   GONZALEZ CAMACHO ANTONIO   GOUDAS CONSTANTINE L PROF
GREBENIKOV E A PROF DR   GREENBERG RICHARD DR   GROUSHINSKY N P PROF DR
HAMID S EL DIN DR   HANSLMEIER ARNOLD   HE MIAO-FU
HEGGIE DOUGLAS C DR   HELALI YHYA E DR   HENON MICHEL C DR
HORI GENICHIRO PROF   HUANG TIANYI   IVANOVA VIOLETA DR
IZVEKOV V A DR   JANICZEK PAUL M DR   JEFFERYS WILLIAM H DR
JOURNET ALAIN   JOVANOVIC BOZIDAR   JUPP ALAN H DR
KATSIS DEMETRIUS DR   KAULA WILLIAM M PROF   KHOLSHEVNIKOV K V DR
KING-HELE DESMOND G DR   KOZAI YOSHIHIDE PROF   KRASINSKY GEORGE A DR
KUSTAANHEIMO PAUL E PROF LALA PETR DR   LAZOVIC JOVAN P PROF
LEMAITRE ANNE DR   LIESKE JAY H DR   LU BEN-KUI
LUNDQUIST CHARLES A DR   MAGNARADZE N G DR   MARCHAL CHRISTIAN DR

MARKELLOS VASSILIS V DR	MARSDEN BRIAN G DR	MATAS VLADIMIR R DR
MAVRAGANIS A G PROF	MEIRE RAPHAEL	MELBOURNE WILLIAM G DR
MERMAN G A DR	MESSAGE PHILIP J DR	MIGNARD FRANCOIS DR
MIKKOLA SEPPO DR	MILANI ANDREA	MOHLER ORREN C PROF
MOONS MICHELE B M M	MORANDO BRUNO L DR	MULHOLLAND J DERRAL DR
MURRAY CARL D	MUSEN PETER DR	MYACHIN V F DR
NACOZY PAUL E DR	NAHON FERNAND PROF	NOBILI ANNA M
NOSKOV BORIS N DR	NOVOSELOV V S PROF DR	O'HANDLEY DOUGLAS A DR
OESTERWINTER CLAUS	OMAROV TUKEN B PROF	ORUS JUAN J PROF
OSORIO JOSE J S P PROF	PAL ARPAD PROF DR	PEALE STANTON J PROF
PETROVSKAYA M S DR	PIERCE A KEITH DR	POPOVIC BOZIDAR PROF DR
ROBINSON WILLIAM J DR	ROY ARCHIE E PROF	RYABOV YU A PROF DR
SAGNIER JEAN-LOUIS DR	SCHUBART JOACHIM DR	SCONZO PASQUALE DR
SEHNAL LADISLAV DR	SESSIN WAGNER DR	SHAPIRO IRWIN I PROF
SHARAF SH G DR	SHTEINS K A DR	SIDLICHOVSKY MILOS DR
SIMA ZDISLAV DR	SIMON JEAN-LOUIS MR	SIRY JOSEPH W
STANDISH E MYLES DR	STELLMACHER IRENE DR	SULTANOV G F ACAD
SUN YI-SUI	SZEBEHELY VICTOR G PROF	TATEVYAN S K DR
TAWADROS MAHET JACOUB DR	TAYLOR DONALD BOGGIA DR	THIRY YVES R PROF
TONG FU	VALTONEN MAURI J PROF	VARVOGLIS H DR
VASHKOV'YAK SOF'YA N DR	VEILLET CHRISTIAN	VILHENA DE MORAES R DR
VINTI JOHN P DR	WALCH JEAN-JACQUES	WALKER IAN WALTER
WILLIAMS CAROL A	WNUK EDWIN	WU LIAN-DA
XU PINXIN	YAROV-YAROVOJ M S DR	YOSHIDA HARUO
YOSHIDA JUNZO PROF	YUASA MANABU DR	ZARE KHALIL DR
ZHANG SHENG-PAN	ZHAO XIAN-ZI	ZHENG JIA-QING
ZHENG XUE-TANG	ZHOU HONG-NAN	ZHU WEN-YAO

COMMISSION No. 8

POSITIONAL ASTRONOMY (ASTRONOMIE DE POSITION)

```
President            : REQUIEME YVES DR

Vice-President(s)    : MIYAMOTO MASANORI DR

Organizing Committee: BENEVIDES SOARES P DR
                      DUMA DMITRIJ P DR
                      HELMER LEIF
                      HUGHES JAMES A DR
                      LINDEGREN LENNART DR
                      LUO DING-JIANG
                      NOEL FERNANDO
                      PINIGIN GENNADIJ I DR
                      QUIJANO LUIS
                      SADZAKOV SOFIJA DR
                      SCHWAN HEINER DR
                      SMITH CLAYTON A JR DR
                      YOSHIZAWA MASANORI DR
```

Members:

ANGUITA CLAUDIO A DR	ARGYRAKOS JEAN PROF DR	BACCHUS PIERRE PROF
BACKER DONALD CH DR	BAGILDINSKIJ BRONISLAV K	BEM JERZY DR
BIEN REINHOLD DR	BILLAUD GERARD J	BRANHAM RICHARD L JR
BROUW W N DR	BYKOV M F DR	CARESTIA REINALDO A DR
CARRASCO GUILLERMO DR	CHA DU JIN	CHAMBERLAIN JOSEPH M DR
CHERNEGA N A A DR	CHIUMIENTO GIUSEPPE	CHLISTOVSKY FRANCA DR
CHOLLET FERNAND DR	CLAUZET LUIZ B FERREIRA	CORBIN THOMAS ELBERT DR
COSTA EDGARDO DR	COUNSELMAN CHARLES C PROF	CRIFO FRANCOISE DR
DAMBARA TAKESHI PROF	DE VEGT CH PROF DR	DEBARBAT SUZANNE V DR
DEJAIFFE RENE J DR	DICK STEVEN J	DJUROVIC DRAGUTIN M DR
DRAVSKIKH A F DR	DUNCOMBE RAYNOR L DR	EICHHORN HEINRICH K DR
FABRICIUS CLAUS V	FEDOROVA RIMMA T DR	FEISSEL MARTINE DR
FOGH OLSEN H J	FOMIN VALERY A DR	FRICKE WALTER PROF DR
GAUSS F STEPHEN	GLIESE WILHELM DR	GRUDLER PIERRE
GUBANOV VADIM S DR	GULYAEV A P DR	HARWOOD DENNIS MR
HEINTZ WULFF D DR	HEMENWAY PAUL D	HOEG ERIK DR
HU NING-SHENG	HUA YING-MIN	HURUKAWA KIICHIRO DR
JACKSON PAUL DR	JIANG CHONG-GUO	JOHNSTON KENNETH J
JOURNET ALAIN	KHARIN A S DR	KLOCK B L DR
KOKURIN YURIJ L DR	KONIN V V DR	KOSIN GENNADIJ S DR
KOVALEVSKY JEAN DR	LACLARE F MR	LACROUTE PIERRE A PROF
LAUSTSEN SVEND DR	LEDERLE TRUDPERT DR	LI DONG-MING
LI NENG-YAO	LI ZHI-FANG	LI ZHIGANG
LOPEZ JOSE A ING	LOYOLA PATRICIO DR	LU CHUN-LIN
MANRIQUE WALTER T PROF	MAO WEI	MAVRIDIS L N PROF
MELCHIOR PAUL J PROF DIR	MIAO YONG-KUAN	MITIC LJUBISA A DR
MORRISON LESLIE V	MURRAY C ANDREW	NEFEDEVA ANTONINA I PROF
NEMIRO ANDREJ A DR PROF	NIKOLOFF IVAN DR	OSORIO JOSE J S P PROF
PERRYMAN MICHAEL A C	PETROV G M DR	PHAM-VAN JACQUELINE MME
PILOWSKI K PROF DR	PODOBED V V DR	POLNITZKY GERHARD DR
POLOZHENTSEV DIMITRIJ DR	POMA ANGELO DR	PROVERBIO EDOARDO PROF

QIAN ZHI-HAN DR	RAIMOND ERNST DR	REIZ ANDERS PROF
ROBERTSON WILLIAM H	ROESER SIEGFRIED DR	ROUSSEAU JEAN-MICHEL MR
RUSSELL JANE L DR	RUSU I DR	SALETIC DUSAN
SANCHEZ MANUEL	SATO KOICHI DR	SCHMEIDLER F PROF DR
SEVARLIC BRANISLAV M PROF	SHEN KAIXIAN	SHI GUANG-CHEN
SIMS KENNETH P DR	SOEDERHJELM STAFFAN DR	SOLARIC NIKOLA
STANGE LOTHAR	STONE RONALD CECIL	TELEKI GEORGE DR
THOBURN CHRISTINE	THOMAS DAVID V DR	WALLACE PATRICK T MR
WALTER HANS G DR	WIELEN ROLAND PROF DR	XIA YI-FEI
XIE LIANGYUN	XU BANG-XIN	XU TONG-QI
YAMAZAKI AKIRA DR	YASUDA HARUO PROF DR	YATSKIV YA S DR
YE SHU-HUA	YU KYUNG-LOH PROF	ZHANG HUI
ZVEREV MITROFAN S PROF DR		

COMMISSION No. 9

INSTRUMENTS AND TECHNIQUES (INSTRUMENTS ET TECHNIQUES)

President   : HUMPHRIES COLIN M DR

Vice-President(s)   : DAVIS JOHN DR

Organizing Committee: BHATTACHARYYA J C PROF
                      ENGVOLD ODDBJOERN DR
                      FORT BERNARD P DR
                      HU NING-SHENG
                      LIVINGSTON WILLIAM C
                      MALIN DAVID F MR
                      STESHENKO N V DR
                      TULL ROBERT G
                      WALKER GORDON A H PROF
                      WILCOCK WILLIAM L PROF
                      WLERICK GERARD DR

Members:

ABLES HAROLD D DR	AIME C DR	ALBRECHT RUDOLF DR
APARICI JUAN DR	ARNAUD JEAN	ATHERTON PAUL DAVID
BABCOCK HORACE W DR	BAO KEREN	BARANNE A DR
BARROSO JR JAIR	BARWIG HEINZ	BAUM WILLIAM A DR
BEER REINHARD DR	BENSAMMAR SLIMANE DR	BINGHAM RICHARD G DR
BLITZSTEIN WILLIAM DR	BONNEAU DANIEL	BORGNINO JULIEN DR
BOYCE PETER B DR	BRAULT JAMES W DR	BRECKINRIDGE JAMES B DR
BREJDO IZABELLA I DR	BURTON W BUTLER DR	CAMPBELL BRUCE DR
CAO CHANGXIN	CHARVIN PIERRE PR	CHELLI ALAIN
CHRISTY JAMES WALTER DR	CLARKE DAVID DR	COHEN RICHARD S
COOKE JOHN ALAN	CORNEJO ALEJANDRO A DR	CRAWFORD DAVID L DR
CURRIE DOUGLAS G DR	DAN XHI-XIANG	DESAI JYOTINDRA N
DOBRONRAVIN PETER DR	DOKUCHAEVA OLGA D DR	DRAVINS DAINIS PROF
DREHER JOHN W	DUCHESNE MAURICE DR	DUNKELMAN LAWRENCE
EDWIN ROGER P	FABRICANT DANIEL G	FEHRENBACH CHARLES PROF
FELLGETT PETER PROF	FLETCHER J MURRAY	FOMENKO ALEXANDR F DR
FORD W KENT JR DR	FOY RENAUD DR	FU DELIAN
GALAN MAXIMINO J	GAO BILIE	GAUSS F STEPHEN
GAY JEAN DR	GILLINGHAM PETER MR	GLASS IAN STEWART DR
GONG SHOU-SHEN	GRAY PETER MURRAY	GRIFFITHS RICHARD E DR
GRIGORJEV VICTOR M DR	GRUNDMANN WALTER	GUTCKE DIETRICH
HALLAM KENNETH L DR	HAMMERSCHLAG ROBERT H DR	HAO YUN-XIANG
HARMER CHARLES F W MR	HARMER DIANNE L MRS	HECKATHORN HARRY M
HEUDIER JEAN-LOUIS DR	HEWITT ANTHONY V DR	HILLIARD R DR
HONEYCUTT R KENT PROF	HOOGHOUDT B G IR	HU JING-YAO
HUANG TIE-QIN	HYSOM EDMUND J	ILYAS MOHAMMAD DR
IOANNISIANI B K DR	ISMAILOV TOFIK K	JAYARAJAN A P MR
JEFFERS STANLEY DR	JELLEY JOHN V PHD	JENKNER HELMUT DR
JIANG SHI-YANG	JONES BARBARA	KARACHENTSEV I D DR
KARPINSKIJ VADIM N DR	KIPPER TONU DR	KISSELL KENNETH E DR
KLOCK B L DR	KOEHLER H PROF DR	KOEHLER PETER
KOPYLOV I M DR	KOROVYAKOVSKIJ YURIJ P DR	KOVACHEV B J DR
KREIDL TOBIAS J N	KUEHNE CHRISTOPH F	LABEYRIE ANTOINE DR

LAQUES PIERRE DR
LEMAITRE GERARD R DR
LOCHMAN JAN
MAHRA H S DR
MCGEE JAMES D PROF
MENG XINMIN
MILLIKAN ALLAN G MR
MORTON DONALD C DR
NELSON JERRY
NISHIMURA SHIRO DR
ODGERS GRAHAM J DR
PERRYMAN MICHAEL A C
PRITCHET CHRISTOPHER J DR
RAKOS KARL D PROF
RICHARDSON E HARVEY DR
RODDIER CLAUDE DR
RUDER HANNS
SANCHEZ MAGRO C DR
SCHULTZ G V DR
SEDMAK GIORGIO PROF
SHCHEGLOV P V DR
SHIVANANDAN KANDIAH DR
SMYTH MICHAEL J DR
SU DING-QIANG
TRAUB WESLEY ARTHUR
VALNICEK BORIS DR
VLADIMIROV SIMEON
WALKER MERLE F PROF
WANG LAN-JUAN
WATSON FREDERICK GARNETT
WOEHL HUBERTUS DR
WU LIN-XIANG
YANG SHI JIE
ZACHAROV IGOR DR
ZHANG YOUYI

LASKER BARRY M DR
LI DEPEI
LYNCH DAVID K
MAILLARD JEAN-PIERRE DR
MCMULLAN DENNIS DR
MERTZ LAWRENCE N DR
MINAROVJECH MILAN
MURRAY STEPHEN S DR
NIEMI AIMO
NUNES ROGERIO S DE SOUSA
PASIAN FABIO
PETFORD A DAVID DR
PROKOF'EVA VALENTINA V DR
RAMSEY LAWRENCE W DR
RING JAMES PROF
RODDIER FRANCOIS PROF
RUSCONI LUIGIA DR
SAXENA A K DR
SCHUMANN JOERG DIETER DR
SERVAN BERNARD
SHEN CHANGJUN
SIM MARY E MISS
SNEZHKO LEONID I
SWINGS JEAN-PIERRE DR
TUEG HELMUT DR
VAN CITTERS GORDON W DR
VRBA FREDERICK J DR
WALLACE PATRICK T MR
WANG YANAN
WEST RICHARD M DR
WORDEN SIMON P DR
WYLLER ARNE A PROF
YAO ZHENG-QIU
ZAMBON GIULIO
ZHU NENGHONG

LELIEVRE GERARD DR
LI TING
MACK PETER
MARTINS DONALD HENRY DR
MEINEL ADEN B PROF
MIKHELSON NIKOLAJ N DR
MORGAN BRIAN LEALAN
NAKAI YOSHIHIRO
NIKONOV V B DR
O'DELL CHARLES R DR
PENNY ALAN JOHN DR
PICAT JEAN-PIERRE DR
RACINE RENE DR
REAY NEWRICK K DR
ROBINSON LLOYD B DR
ROSCH JEAN PROF
RYLOV VALERIJ S DR
SCHROEDER DANIEL J PROF
SCHUSTER HANS-EMIL
SHAKHBAZYAN YURIJ L DR
SHEN PARN-AN
SMITH CHARLES DITTO
STOREY JOHN W V DR
TANGO WILLIAM J. DR
ULICH BOBBY LEE
VELKOV KIRIL
WALKER ALISTAIR ROBIN DR
WAMPLER E JOSEPH PROF
WANG YIMING
WESTPHAL JAMES A PROF
WORSWICK SUSAN
WYNNE CHARLES G PROF
YE BINXUN
ZEALEY WILLIAM J DR

MEMBERSHIP OF COMMISSIONS

COMMISSION No. 10

SOLAR ACTIVITY (ACTIVITE SOLAIRE)

President   : PICK MONIQUE DR

Vice-President(s)   : PRIEST ERIC R PROF

Organizing Committee: ALISSANDRAKIS C PH D
                     FALCIANI ROBERTO DR
                     HIRAYAMA TADASHI PROF
                     KANE SHARAD R DR
                     KAUFMANN PIERRE PROF
                     KUKLIN G V DR
                     VALNICEK BORIS DR
                     VINOD S KRISHAN MRS DR
                     ZHANG BAI-RONG

Members:

ABBASOV ALIK R DR	ABRAMI ALBERTO PROF	AHLUWALIA HARJIT SINGH DR
AI GUOXIANG	ALTROCK RICHARD C DR	ALTSCHULER MARTIN D PROF
ALY M KHAIRY PROF	AMBROZ PAVEL DR	ANDERSON KINSEY A PROF
ANTALOVA ANNA	ANTIOCHOS SPIRO KOSTA	ATHAY R GRANT DR
AVIGNON YVETTE DR	BAGARE S P DR	BALLI EDIBE PROF
BANIN V G DR	BARROW COLIN H DR	BATES J RAPHAEL
BECKERS JACQUES M DR	BEEBE HERBERT A	BELL BARBARA DR
BELVEDERE GAETANO DR	BENZ ARNOLD DR	BHATNAGAR ARVIND DR
BOCCHIA ROMEO DR	BOHN HORST-ULRICH	BONOV ANGEL DR
BOUGERET J L DR	BOYER RENE	BRANDT PETER N
BRAY ROBERT J DR	BROWN JOHN C PROF	BRUECKNER GUENTER E DR
BRUNER MARILYN E DR	BRUZEK ANTON DR	BUMBA VACLAV DR
CADEZ VLADIMIR	CANE HILARY VIVIEN	CARLQVIST PER A DR
CHAMBE GILBERT	CHAPMAN GARY A DR	CHEN BIAO
CHEN ZHENCHENG	CHENG CHUNG-CHIEH DR	CHERTOPRUD V E DR
CHIUDERI-DRAGO FRANCA PR	CHUPP EDWARD L DR	CIMINO MASSIMO A PROF
CLIVER EDWARD W	COFFEY HELEN E MS	COOK JOHN W
COUTREZ RAYMOND A J PROF	COVINGTON ARTHUR E	CRANNELL CAROL JO DR
CSADA IMRE K DR	CULHANE LEONARD PROF	DATLOWE DAYTON DR
DE GROOT T DR	DE JAGER CORNELIS PROF	DENNIS BRIAN ROY DR
DERE KENNETH PAUL	DEUBNER FRANZ-LUDWIG DR	DEZSO LORANT PROF
DING YOU-JI	DIZER MUAMMER PROF	DOLLFUS AUDOUIN PROF
DRYER MURRAY DR	DUBOIS MARC A	DUBOV EMIL E PROF
DULK GEORGE A PROF	DUNCAN ROBERT A PROF	DUNN RICHARD B DR
DWIVEDI BHOLA NATH DR	EDDY JOHN A DR	ELSTE GUNTHER H DR
ELWERT GERHARD PROF	EMSLIE A. GORDON	ENGVOLD ODDBJOERN DR
ENOME SHINZO PROF	ERICKSON WILLIAM C DR	FALCHI AMBRETTA
FANG CHENG	FARNIK FRANTISEK	FEIBELMAN WALTER A DR
FENG KE-JIA	FOKKER AAD D DR	FORTINI TERESA DR
FOSSAT ERIC DR	FRIEDMAN HERBERT DR	FRITZOVA-SVESTKA L DR
GABRIEL ALAN H	GAIZAUSKAS VICTOR DR	GALLOWAY DAVID DR
GARCIA DE LA ROSA JOSE I	GELFREIKH GEORGIJ B DR	GERGELY TOMAS ESTEBAN DR
GILLILAND RONALD LYNN	GILMAN PETER A DR	GLATZMAIER GARY A
GLEISSBERG WOLFGANG PROF	GNEVYSHEVA RAISA S DR	GODOLI GIOVANNI PROF
GOKHALE MORESHWAR HARI PR	GOPASYUK S I DR	GU XIAO-MA

GURTOVENKO E A DR	HAGEN JOHN P	HAGYARD MONA JUNE
HAMMER REINER	HANASZ JAN DR	HANSEN RICHARD T MR
HANSLMEIER ARNOLD	HARVEY JOHN W DR	HAUG EBERHARD DR
HAYWARD JOHN	HEDEMAN E RUTH	HENOUX JEAN-CLAUDE DR
HEYDEN FRANCIS J SJ DR	HIEI EIJIRO DR	HILDNER ERNEST DR
HOLZER THOMAS EDWARD DR	HONG HYON IK	HOOD ALAN
HOWARD ROBERT F DR	HOYNG PETER DR	HUANG YOU-RAN
HUDSON HUGH S DR	HURFORD GORDON JAMES	HYDER C L DR
IOSHPA B A DR	IVANCHUK VICTOR I DR	JAKIMIEC JERZY PROF
JENSEN EBERHART PROF	JOCKERS KLAUS DR	JONES HARRISON PRICE DR
JORDAN STUART D DR	KABURAKI OSAMU DR	KAHLER STEPHEN W DR
KAI KEIZO DR	KALMAN BELA DR	KANG JIN SOK
KANNO MITSUO PROF	KARLICKY MARIAN	KARPEN JUDITH T
KJELDSETH-MOE OLAV DR	KLECZEK JOSIP DR	KLVANA MIROSLAV
KNOSKA STEFAN	KOECKELENBERGH ANDRE DR	KOPECKY MIROSLAV DR
KOSTIK ROMAN I	KOTRC PAVEL	KOUTCHMY SERGE DR
KOVACS AGNES DR	KRAUSE F DR	KRIVSKY LADISLAV DR
KRUEGER ALBRECHT DR	KUBOTA JUN DR	KUENZEL HORST
KUNDU MUKUL R DR	KUPERUS MAX PROF DR	KUROCHKA L N DR
KUROKAWA HIROKI DR	LANDECKER PETER BRUCE DR	LANDMAN DONALD ALAN DR
LANG KENNETH R ASST PROF	LEIBACHER JOHN DR	LEROY JEAN-LOUIS
LI CHUN-SHENG	LI SON JAE	LI WEI BAO
LIN YUANZHANG	LIVSHITS M A DR	LOUGHHEAD RALPH E DR
LOW BOON CHYE	LUO BAO-RONG	LUO XIANHAN
MACKINNON ALEXANDER L	MACQUEEN ROBERT M DR	MACRIS CONSTANTIN J PROF
MAKAROV VALENTINE I	MAKITA MITSUGU DR	MALITSON HARRIET H MS
MALTBY PER PROF	MALVILLE J MCKIM PROF	MANDELSTAM S L PROF
MARTRES MARIE-JOSEPHE	MASON GLENN M	MATSUURA OSCAR T DR
MATTIG W PROF DR	MAXWELL ALAN DR	MCCABE MARIE K MS
MCKENNA LAWLOR SUSAN	MCLEAN DONALD J DR	MEIN PIERRE
MELROSE DONALD B PROF	MERGENTALER JAN PROF	MICHALITSIANOS ANDREW
MICHARD RAYMOND DR	MOGILEVSKIJ EH I DR	MOHLER ORREN C PROF
MOISEEV I G DR	MORENO-INSERTIS FERNANDO	MORETON G E
MORIYAMA FUMIO PROF	MOROZHENKO N N DR	MOTTA SANTO DR
MULLER RICHARD DR	MUSTEL E R PROF DR	NAGASAWA SHINGO PROF
NAKAGAWA YOSHINARI DR	NAKAJIMA HIROSHI	NAMBA OSAMU DR
NELSON GRAHAM JOHN DR	NEUPERT WERNER M DR	NISHI KEIZO DR
NOYES ROBERT W PROF	NUSSBAUMER HARRY PROF	OBRIDKO VLADIMIR N DR
OHKI KENICHIRO DR	ORRALL FRANK Q PROF	PALLAVICINI ROBERTO DR
PAN LIANDE	PARKINSON JOHN H DR	PARKINSON WILLIAM H
PATERNO LUCIO PROF	PEDERSEN BENT M DR	PETROSIAN VAHE PROF
PFLUG KLAUS DR	PHILLIPS KENNETH J H	PIDDINGTON JACK H RES FEL
PNEUMAN GERALD W	POLAND ARTHUR I DR	POLETTO GIANNINA PROF
POLUPAN P N DR	POQUERUSSE MICHEL	PROKAKIS THEODORE J DR
QIAN JING-KUI	RAADU MICHAEL A DR	RABIN DOUGLAS MARK
RAO A PRAMESH DR	RAYROLE JEAN R DR	REES DAVID ELWYN DR
REEVES EDMOND M DR	REEVES HUBERT PROF	ROBERTS BERNARD DR
ROBINSON JR RICHARD D DR	ROCA CORTES TEODORO	ROEMER MAX PROF
ROMANCHUK P R DR	ROMPOLT BOGDAN DR	ROSCH JEAN PROF
ROXBURGH IAN W PROF	ROZELOT JEAN P	RUBASHEV BORIS M DR
RUSIN VOJTECH	RUST DAVID M DR	RUZDJAK VLADIMIR DR
RUZICKOVA-TOPOLOVA B DR	RYBANSKY MILAN	SAEMUNDSON THORSTEINN
SAITO KUNIJI PROF	SAKURAI KUNITOMO PROF	SAKURAI TAKASHI DR
SAWYER CONSTANCE B DR	SCHATTEN KENNETH H DR	SCHLUETER A PROF DR
SCHMAHL EDWARD J DR	SCHMIDT H U DR	SCHMIEDER BRIGITTE DR
SCHOBER HANS J DR	SCHROETER EGON H PROF	SEMEL MEIR DR
SEVERNYJ A B PROF DR	SHAPLEY ALAN H	SHEA MARGARET A DR
SHEELEY NEIL R DR	SHI ZHONG-XIAN	SHIBASAKI KIYOTO
SHINE RICHARD A DR	SILBERBERG REIN DR	SIMNETT GEORGE M

SIMON GUY
SMALDONE LUIGI ANTONIO
SOTIROVSKI PASCAL DR
STELLMACHER GOETZ
STEPANOV V E PROF
STIX MICHAEL DR
SUEMOTO ZENZABURO PROF DR
SYKORA JULIUS DR
TALON RAOUL DR
TANDBERG-HANSSEN EINAR A
TESKE RICHARD G PROF
TIFREA EMILIA DR
TREUMANN RUDOLF A. DR
TSUBAKI TOKIO PROF
URPO SEPPO I
VAN'T VEER FRANS DR
VENKATESAN DORASWAMY DR
VITINSKIJ YURIJ I DR
WANG JIA-LONG
WIEHR EBERHARD DR
WITTMANN AXEL D. PH D
WU FEI
XU AO-AO
YANG HAI SHOU
YOU JIAN-QI
ZELENKA ANTOINE DR
ZHOU DAOQI
ZLOBEC PAOLO DR

SITNIK G F PROF
SMITH DEAN F DR
SPICER DANIEL SHIELDS DR
STENFLO JAN O DR
STESHENKO N V DR
STURROCK PETER A PROF
SUN KAI
SYLWESTER JANUSZ
TAMENAGA TATSUO DR
TANDON JAGDISH NARAIN DR
THOMAS JOHN H PROF
TLAMICHA ANTONIN DR
TRITAKIS BASIL P DR
TUOMINEN JAAKKO V PROF
VAN ALLEN JAMES A PROF
VAUGHAN ARTHUR H DR
VIAL JEAN-CLAUDE
VYALSHIN GENNADIJ F DR
WANG JING-XIU
WILD JOHN PAUL DR
WOEHL HUBERTUS DR
WU SHI TSAN DR
XU ZHENTAO
YE SHI-HUI
YUN HONG-SIK PROF
ZHANG HE-QI
ZHUGZHDA YUZEF D DR
ZOU YI-XIN

SLONIM E M DR
SMOL'KOV GENNADIJ YA DR
SPRUIT HENK C DR
STEPANIAN N N DR
STEWART RONALD T MR
SUDA JAN
SVESTKA ZDENEK DR
TAKAKURA TATSUO PROF EMER
TANAKA KATSUO DR
TANG YU-HUA
THOMAS ROGER J DR
TRELLIS MICHEL DR
TROTTET GERARD DR
UNDERWOOD JAMES H DR
VAN HOVEN GERARD DR
VELKOV KIRIL
VINLUAN RENATO
WALDMEIER MAX PROF DR
WENTZEL DONAT G DR
WILSON PETER R PROF
WOLTJER LODEWIJK PROF
XANTHAKIS JOHN N PROF
YAKOVKIN N A DR
YOSHIMURA HIROKAZU DR
ZAPPALA ROSARIO ALDO DR
ZHANG ZHEN-DA
ZIRIN HAROLD DR
ZWAAN CORNELIS PROF DR

COMMISSION No. 12

RADIATION AND STRUCTURE OF THE SOLAR ATMOSPHERE

(RADIATION ET STRUCTURE DE L'ATMOSPHERE SOLAIRE)

President   : KUPERUS MAX PROF DR

Vice-President(s) : HARVEY JOHN W DR

Organizing Committee: BRUECKNER GUENTER E DR
CHEN BIAO
CHRISTENSEN-DALSGAARD J
CRAM LAWRENCE EDWARD DR
GURTOVENKO E A DR
HIEI EIJIRO DR
KOTOV VALERY DR
NOYES ROBERT W PROF
SIVARAMAN K R DR
STENFLO JAN O DR

Members:
ACTON LOREN W DR
ALISSANDRAKIS C PH D
ALTSCHULER MARTIN D PROF
ARNAUD JEAN
BALIUNAS SALLIE L
BECKMAN JOHN E PROF
BENFORD GREGORY DR
BILLINGS DONALD E PROF
BOCCHIA ROMEO DR
BOHN HORST-ULRICH
BOUGERET J L DR
BRAY ROBERT J DR
BRUZEK ANTON DR
CAVALLINI FABIO
CHAN KWING LAM
CHISTYAKOV VLADIMIR E DR
COOK JOHN W
CUI LIAN-SHU
DELBOUILLE LUC PROF
DOGAN NADIR PROF
DUMONT SIMONE DR
DUVALL THOMAS L JR
ELLIOTT IAN DR
EVANS J V DR
FELDMAN URI
FOSSAT ERIC DR
FRIEDMAN HERBERT DR
GAIZAUSKAS VICTOR DR
GNEVYSHEV MSTISLAV N DR
GOLDBERG LEO PROF
GOPALASWAMY N DR
GREVESSE N DR

ADAM MADGE G DR
ALLEN C W PROF
ANDO HIROYASU DR
ATHAY R GRANT DR
BEARD DAVID B DR
BEEBE HERBERT A
BHATNAGAR ARVIND DR
BLACKWELL DONALD E PROF
BOEHM KARL-HEINZ PROF
BONNET ROGER M DR
BRANDT PETER N
BRECKINRIDGE JAMES B DR
BUMBA VACLAV DR
CEPPATELLI GUIDO DR
CHAPMAN GARY A DR
CHVOJKOVA WOYK E DR
COX ARTHUR N DR
DE JAGER CORNELIS PROF
DEUBNER FRANZ-LUDWIG DR
DRAVINS DAINIS PROF
DUNKELMAN LAWRENCE
EDMONDS FRANK N JR DR
ELSTE GUNTHER H DR
FALCIANI ROBERTO DR
FIALA ALAN D DR
FOUKAL PETER V DR
FROEHLICH CLAUS
GAUR V P
GODOLI GIOVANNI PROF
GOLDMAN MARTIN V
GOPASYUK S I DR
GU XIAO-MA

AIME C DR
ALTROCK RICHARD C DR
ANSARI S M RAZAULLAH PROF
AYRES THOMAS R
BECKERS JACQUES M DR
BEL NICOLE J DR
BHATTACHARYYA J C PROF
BLAMONT JACQUES E PROF
BOEHM-VITENSE ERIKA PROF
BOOK DAVID L
BRAULT JAMES W DR
BRUNER MARILYN E DR
CADEZ VLADIMIR
CHAMBE GILBERT
CHENG CHUNG-CHIEH DR
CLARK THOMAS ALAN DR
CRAIG IAN JONATHAN D DR
DELACHE PHILIPPE J DR
DEZSO LORANT PROF
DUBOV EMIL E PROF
DUNN RICHARD B DR
EINAUDI GIORGIO
EPSTEIN GABRIEL LEO DR
FANG CHENG
FONTENLA JUAN MANUEL DR
FRAZIER EDWARD N DR
GABRIEL ALAN H
GLATZMAIER GARY A
GOKDOGAN NUZHET PROF
GOMEZ MARIA THERESA DR
GORDON CHARLOTTE PROF
HAGYARD MONA JUNE

HAMMER REINER	HASAN SAIYID STRAJUL	HAUPT RALPH F
HILDNER ERNEST DR	HIRAYAMA TADASHI PROF	HOLWEGER HARTMUT PROF
HORTON BRIAN H DR	HOTINLI METIN DR	HOUSE LEWIS L DR
HOWARD ROBERT F DR	HOYNG PETER DR	ILLING RAINER M E
JABBAR SABEH RHAMAN	JEFFERIES JOHN T DR	JONES HARRISON PRICE DR
JORDAN CAROLE DR	JORDAN STUART D DR	JOSHI G C DR
KALKOFEN WOLFGANG DR	KALMAN BELA DR	KANNO MITSUO PROF
KARPEN JUDITH T	KARPINSKIJ VADIM N DR	KATO SHOJI PROF
KAUFMANN PIERRE PROF	KAWAGUCHI ICHIRO PROF	KEIL STEPHEN L
KHETSURIANI T S DR	KNEER FRANZ DR	KONONOVICH EDWARD V DR
KOPECKY MILOSLAV DR	KOSTIK ROMAN I	KOTRC PAVEL
KOUTCHMY SERGE DR	KOYAMA SHIN PROF DR	KRAEMER GERHARD DR
KUBICELA ALEKSANDAR DR	KUKLIN G V DR	KUNDU MUKUL R DR
KUROCHKA L N DR	LA BONTE BARRY JAMES	LABS DIETRICH PROF
LANDI DEGL'INNOCENTI E PR	LANDI DEGL'INNOCENTI M	LANDMAN DONALD ALAN DR
LANDOLFI MARCO	LEIBACHER JOHN DR	LEIGHTON R B PROF
LEROY JEAN-LOUIS	LIN YUANZHANG	LINSKY JEFFREY L DR
LIVINGSTON WILLIAM C	LOCKE JACK L DR	LOPEZ-ARROYO M
LOUGHHEAD RALPH E DR	LUEST REIMAR PROF	MAKAROV VALENTINE I
MAKAROVA ELENA A DR	MAKITA MITSUGU DR	MARIK MIKLOS DR.
MARILLI ETTORE DR	MARISKA JOHN THOMAS	MARMOLINO CIRO
MATSUSHIMA SATOSHI DR	MATTIG W PROF DR	MCKENNA LAWLOR SUSAN
MEIN PIERRE	MERGENTALER JAN PROF	MEWE R DR
MEYER FRIEDRICH DR	MICHARD RAYMOND DR	MIGEOTTE MARCEL V PROF
MIHALAS DIMITRI DR	MILKEY ROBERT W DR	MORENO-INSERTIS FERNANDO
MORIYAMA FUMIO PROF	MOURADIAN ZADIG M DR	MUELLER EDITH A PROF
MULLER RICHARD DR	MUNRO RICHARD H DR	NAMBA OSAMU DR
NECKEL HEINZ DR	NEVEN LUC DR	NICOLAS KENNETH ROBERT
NICOLET MARCEL PROF	NISHI KEIZO DR	NORDLUND AKE DR
ORRALL FRANK Q PROF	OSTER LUDWIG F PROF DR	PANDE MAHESH CHANDRA DR
PAPATHANASOGLOU D DR	PARKINSON WILLIAM H	PASACHOFF JAY M PROF
PECKER JEAN-CLAUDE PROF	PEYTURAUX ROGER H PROF	PFLUG KLAUS DR
PHILLIPS KENNETH J H	PIERCE A KEITH DR	POQUERUSSE MICHEL
PRIEST ERIC R PROF	PROKAKIS THEODORE J DR	RABIN DOUGLAS MARK
REES DAVID ELWYN DR	REEVES EDMOND M DR	RIGHINI-COHEN GIOVANNA DR
RIGUTTI MARIO PROF	ROBERTI GIUSEPPE DR	ROBERTS BERNARD DR
ROCA CORTES TEODORO	RODDIER FRANCOIS PROF	ROLAND GINETTE DR
ROVIRA MARTA GRACIELA	RUSIN VOJTECH	RUTTEN ROBERT J. DR
RYBANSKY MILAN	SAKAI JUN-ICHI	SAKURAI TAKASHI DR
SAMAIN DENYS DR	SCHMAHL EDWARD J DR	SCHOBER HANS J DR
SCHUESSLER MANFRED DR	SCHWARTZ STEVEN JAY	SEATON MICHAEL J PROF
SEMEL MEIR DR	SEVERINO GIUSEPPE	SEVERNYJ A B PROF DR
SHALLIS MICHAEL J DR	SHEELEY NEIL R DR	SHEN LONG-XIANG
SHINE RICHARD A DR	SIMON GEORGE W DR	SIMON GUY
SINGH JAGDEV DR	SINHA K DR	SITNIK G F PROF
SITTERLY CHARLOTTE M DR	SKUMANICH ANDRE PROF	SOBOLEV VLADISLAV M DR
SONG MU-TAO	SOTIROVSKI PASCAL DR	SOUFFRIN PIERRE B DR
SPICER DANIEL SHIELDS DR	STAUDE JUERGEN DR	STEBBINS ROBIN
STEPANOV V E PROF	STIX MICHAEL DR	SUEMOTO ZENZABURO PROF DR
SVESTKA ZDENEK DR	SWENSSON JOHN W DR	TANAKA KATSUO DR
TANDBERG-HANSSEN EINAR A	TEPLITSKAYA R B DR	THOMAS JOHN H PROF
THOMAS RICHARD N DR	TORELLI M DR	TOUSEY RICHARD DR
TRIPATHI B M DR	TSAP T T DR	TSUBAKI TOKIO PROF
UCHIDA YUTAKA PROF	UNNO WASABURO PROF	UUS UNDO DR
VAN HOVEN GERARD DR	VASILEVA GALINA J DR	VAUGHAN ARTHUR H DR
VENKATAKRISHNAN P DR	VIAL JEAN-CLAUDE	VITINSKIJ YURIJ I DR
VOLONTE SERGE DR	VUKICEVIC K M PROF DR	WALDMEIER MAX PROF DR
WANG JING-XIU	WANG ZHEN-YI	WARWICK JAMES W DR

WEISS NIGEL O DR	WENTZEL DONAT G DR	WILSON PETER R PROF
WITTMANN AXEL D. PH D	WOEHL HUBERTUS DR	WORDEN SIMON P DR
WU FEI	WU LIN-XIANG	WYLLER ARNE A PROF
YOSHIMURA HIROKAZU DR	YOU JIAN-QI	YOUSSEF NAHED H DR
ZELENKA ANTOINE DR	ZHOU DAOQI	ZHUGZHDA YUZEF D DR
ZIRIN HAROLD DR	ZIRKER JACK B DR	ZWAAN CORNELIS PROF DR

COMMISSION No. 14

ATOMIC AND MOLECULAR DATA

(DONNEES ATOMIQUES ET MOLECULAIRES)

President   : NICHOLLS RALPH W PROF

Vice-President(s)   : SAHAL-BRECHOT SYLVIE DR

Organizing Committee: GABRIEL ALAN H
                     KATO TAKAKO DR
                     LOVAS FRANCIS JOHN DR
                     MANDELSTAM S L PROF
                     NUSSBAUMER HARRY PROF
                     PARKINSON WILLIAM H
                     RUDZIKAS ZENONAS B
                     WIESE WOLFGANG L DR

Members:

ADELMAN SAUL J DR	ALLEN C W PROF	ANDREW KENNETH L PROF
ARDUINI-MALINOVSKY M. DR	ARTRU MARIE-CHRISTINE DR	BAIRD KENNETH M DR
BARNARD HANNES A J DR	BARROW RICHARD F DR	BATES DAVID R PROF
BELY OLEG DR	BELY-DUBAU FRANCOISE	BERRINGTON KEITH ADRIAN
BIEMONT EMILE DR	BLACK JOHN HARRY DR	BLAHA MILAN DR
BRANSCOMB L M DR	BRAULT JAMES W DR	BROMAGE GORDON E DR
BURGESS ALAN DR	CARBON DUANE F DR	CARROLL P KEVIN PROF
CARVER JOHN H PROF	COOK ALAN H PROF	CORLISS C H DR
CZYZAK STANLEY J DR	DALGARNO ALEXANDER PROF	DELSEMME ARMAND H PROF DR
DESESQUELLES JEAN DR	DIERCKSEN GEERD H F PH D	DRESSLER KURT PROF
DUBAU JACQUES DR	DUFAY MAURICE PROF	EDLEN BENGT PROF
ENGELHARD E J G PROF DR	EPSTEIN GABRIEL LEO DR	FAUCHER PAUL DR
FEAUTRIER NICOLE DR	FELENBOK PAUL DR	FINK UWE DR
FLOWER DAVID R DR	GARSTANG ROY H PROF	GARTON W R S PROF
GLAGOLEVSKIJ JU V DR	GOLDBACH CLAUDINE MME	GRANT IAN P DR
GREEN LOUIS C PROF	HEDDLE DOUGLAS W O PROF	HEFFERLIN RAY A PROF
HERMAN RENEE DR	HEROLD HEINZ	HERZBERG GERHARD DR
HESSER JAMES E DR	HOUSE LEWIS L DR	HUBER MARTIN C E DR
HUEBNER WALTER F DR	HUMPHREYS CURTIS JUDSON	ILIEV ILIAN
IRWIN ALAN W DR	JACQUINOT PIERRE DR	JOHNSON DONALD R DR
JOHNSON FRED M PROF DR	JOLY FRANCOIS DR	JORDAN CAROLE DR
JORDAN H L DR DIREKTOR	KENNEDY EUGENE T	KESSLER KARL G DR
KIM ZONG DOK	KING R B DR	KINGSTON ARTHUR E PROF
KIPPER TONU DR	KOHL JOHN L DR	LAGERQVIST ALBIN PROF
LANDMAN DONALD ALAN DR	LANG JAMES DR	LAUNAY JEAN-MICHEL DR
LAWRENCE G M DR	LAYZER DAVID PROF	LE DOURNEUF MARYVONNE
LOCHTE-HOLTGREVEN W PROF	LOULERGUE MICHELLE DR	LUTZ BARRY L DR
MAILLARD JEAN-PIERRE DR	MARTIN WILLIAM C DR	MASON HELEN E DR
MCWHIRTER R W PETER DR	MEWE R DR	MIGEOTTE MARCEL V PROF
MOHLER ORREN C PROF	MONFILS ANDRE G PROF	MUMMA MICHAEL JON
NEVIN THOMAS E PROF	NEWSOM GERALD H PROF	NOLLEZ GERARD DR
OBI SHINYA PROF	OETKEN L DR	OKA TAKESHI DR
OMONT ALAIN PROF	ORTON GLENN S DR	PEACH GILLIAN DR
PETRINI DANIEL DR	PETROPOULOS BASIL CH DR	PETTINI MARCO

PFENNIG HANS H DR	PHILLIPS JOHN G PROF	PROKOF'EV VLADIMIR K PROF
QUERCI FRANCOIS R DR	RAO K NARAHARI	RICHTER JOHANNES PROF
ROSS JOHN E R DR	ROUEFF EVELYNE M A DR	RUDER HANNS
SCHADEE AERT DR	SCHRIJVER JOHANNES DR	SEATON MICHAEL J PROF
SHORE BRUCE W	SITTERLY CHARLOTTE M DR	SMITH GEOFFREY DR
SMITH PETER L DR	SMITH WM HAYDEN PROF	SOMERVILLE WILLIAM B DR
SORENSEN GUNNAR DR	STEENMAN-CLARK LOIS DR	STREL'NITSKIJ VLADIMIR DR
SUMMERS HUGH P DR	SWINGS JEAN-PIERRE DR	TAKAYANAGI KAZUO PROF
TATUM JEREMY B DR	TOUSEY RICHARD DR	TOZZI GIAN PAOLO
TREFFTZ ELEONORE E DR	VAN REGEMORTER HENRI DR	VAN RENSBERGEN WALTER DR
VARSAVSKY C M DR	VARSHALOVICH DIMITRIJ PR	VOELK HEINRICH J PROF
VOLONTE SERGE DR	VUJNOVIC VLADIS DR	WARES GORDON W DR
WENIGER SCHAME DR	WILSON ROBERT PROF	WINNEWISSER GISBERT DR
WUNNER GUENTER	YOUNG LOUISE GRAY DR	ZEIPPEN CLAUDE DR
ZIRIN HAROLD DR		

COMMISSION No. 15

PHYSICAL STUDY OF COMETS, MINOR PLANETS AND METEORITES

(ETUDE PHYSIQUE DES COMETES, DES PETITES PLANETES ET DES METEORITES)

President   : KRESAK LUBOR DR

Vice-President(s) : RAHE JURGEN PROF

Organizing Committee: A'HEARN MICHAEL F DR
ARPIGNY CLAUDE PROF
CHAPMAN CLARK R DR
DOBROVOLSKY OLEG V PROF
FECHTIG HUGO DR
HARRIS ALAN WILLIAM DR
HAUPT HERMANN F PROF
SHUL'MAN L M DR
WASSON JOHN T
WYCKOFF SUSAN DR
ZAPPALA VINCENZO PROF

Members:
ALLEGRE CLAUDE PROF
ANDERS EDWARD PROF
ANDRIENKO DMITRY A DR
BABADZHANOV PULAT B DR
BARKER EDWIN S DR
BEARD DAVID B DR
BIRCH PETER MR
BLAMONT JACQUES E PROF
BOBROVNIKOFF NICHOLAS DR
BOUSKA JIRI DR
BOWELL EDWARD L G DR
BRANDT JOHN C DR
BRECHER AVIVA DR PROF
BROWN ROBERT HAMILTON
BRUNK WILLIAM E DR
BURLAGA LEONARD F DR
BURNS JOSEPH A PROF
CANDY MICHAEL P MR
CEPLECHA ZDENEK DR
CHAPMAN ROBERT D DR
CHEN DAO-HAN
CHEREDNICHENKO V I DR
COCHRAN WILLIAM DAVID DR
COSMOVICI BATALLI C DR
CRISTESCU CORNELIA G DR
CROVISIER JACQUES
CRUIKSHANK DALE P DR
DANKS ANTHONY C DR
DE SANCTIS GIOVANNI
DEGEWIJ JOHAN DR
DELSEMME ARMAND H PROF DR
DERMOTT STANLEY F
DEUTSCHMAN WILLIAM A DR
DI MARTINO MARIO
DONN BERTRAM D
DOSSIN F DR
DOUGLAS A VIBERT DR
DRYER MURRAY DR
ERSHKOVICH ALEXANDER PROF
EVERHART EDGAR DR
EVIATAR AHARON PROF
FARINELLA PAOLO DR
FELDMAN PAUL DONALD DR
FERNANDEZ JEAN-CLAUDE DR
FESTOU MICHEL C DR
FULCHIGNONI MARCELLO PROF
GEHRELS TOM PROF
GERARD ERIC DR
GIBSON DAVID MICHAEL DR
GIESE RICHARD H PROF
GIOVANE FRANK
GRADIE JONATHAN CAREY
GREENBERG J MAYO DR
GROSSMAN LAWRENCE PROF
GRUDZINSKA STEFANIA DR
GRUEN EBERHARD DR
GUSTAFSON BO A S
HALLIDAY IAN DR
HANNER MARTHA S DR
HAPKE BRUCE W DR
HARTMANN WILLIAM K
HARWIT MARTIN PROF
HASER LEO N K DR
HERZBERG GERHARD DR
HU ZHONG-WEI
HUEBNER WALTER F DR
HUGHES DAVID W DR
IBADINOV KHURSANDKUL DR
IP WING-HUEN
IRVINE WILLIAM M PROF
ISOBE SYUZO DR
IVANOVA VIOLETA DR
JACKSON WILLIAM M DR
JOCKERS KLAUS DR
JOHNSON TORRENCE V DR
KELLER HANS ULRICH DR
KNACKE ROGER F DR
KOEBERL CHRISTIAN DR
KOHOUTEK LUBOS DR
KONOPLEVA VARVARA P DR
KOWAL CHARLES THOMAS
KRISHNA SWAMY K S DR
LAMY PHILIPPE DR
LANCASTER BROWN PETER
LARSON HAROLD P DR
LARSON STEPHEN M
LEBOFSKY LARRY ALLEN
LEVASSEUR-REGOURD A.C. PR
LEVIN BORIS J DR
LILLER WILLIAM DR
LILLIE CHARLES F DR
LINDSEY CHARLES ALLAN
LIU LIN-ZHONG

LUEST RHEA DR
MALAISE DANIEL J DR
MARTEL MARIE-THERESE DR
MCCORD THOMAS B DR
MEISEL DAVID D DR
MILLER FREEMAN D PROF
MOORE ELLIOTT P PROF
MUMMA MICHAEL JON
NEWBURN RAY L JR
PAOLICCHI PAOLO DR
PILCHER CARL BERNARD DR
REMY BATTIAU LILIANE G A
ROEMER ELIZABETH PROF
SCHMIDT MAARTEN PROF
SIVARAMAN K R DR
SOLC MARTIN
TAKEDA HIDENORI DR
TOMITA KOICHIRO MR
VANYSEK VLADIMIR PROF
WALLIS MAX K DR
WEHINGER PETER A DR
WHIPPLE FRED L DR
WOOD JOHN A DR
YAVNEL ALEXANDER A DR
ZHOU XING-HAI

LUPISHKO DMITRIJ F
MARAN STEPHEN P DR
MATSON DENNIS L DR
MCCROSKY RICHARD E DR
MENDIS DEVAMITTA ASOKA DR
MILLMAN PETER M DR
MORRISON DAVID PROF
NAKAMURA TSUKO DR
O'DELL CHARLES R DR
PELLAS PAUL DR
PITTICH EDUARD M DR
REVELLE DOUGLAS ORSON DR
SCALTRITI FRANCO DR
SCHOBER HANS J DR
SMOLUCHOWSKI ROMAN PROF
SPINRAD HYRON PROF
TANABE HIROYOSHI DR
TYPHOON LEE
VEEDER GLENN J DR
WANG SI-CHAO
WEISSMAN PAUL ROBERT
WILKENING LAUREL L DR
WOOLFSON MICHAEL M PROF
YEOMANS DONALD K DR

LYTTLETON RAYMOND A PROF
MARSDEN BRIAN G DR
MATSUURA OSCAR T DR
MCKENNA LAWLOR SUSAN
MILET BERNARD L DR
MOEHLMANN DIEDRICH
MRKOS ANTONIN DR
NEFF JOHN S
O'KEEFE JOHN A DR
PEREZ-DE-TEJADA H A DR
PROISY PAUL E DR
RICKMAN HANS DR
SCHMIDT H U DR
SEKANINA ZDENEK DR
SNYDER LEWIS E
SVOREN JAN
TEDESCO EDWARD F
VALSECCHI GIOVANNI B
VEVERKA JOSEPH DR
WATERFIELD REGINALD L DR
WETHERILL GEORGE W
WISNIEWSKI WIESLAW Z
WOSZCZYK ANDRZEJ PROF
ZELLNER BENJAMIN H DR

COMMISSION No. 16

PHYSICAL STUDY OF PLANETS AND SATELLITES

(ETUDE PHYSIQUE DES PLANETES ET DES SATELLITES)

President   : HUNT GARRY E DR

Vice-President(s) : BRAHIC ANDRE DR
                    MORRISON DAVID PROF

Organizing Committee: BERTAUX J L DR
                      BURNS JOSEPH A PROF
                      CHEN DAO-HAN
                      CRUIKSHANK DALE P DR
                      DAVIES MERTON E MR
                      ENCRENAZ THERESE DR
                      GAUTIER DANIEL
                      MAROV MIKHAIL YA PROF
                      MASURSKY HAROLD DR
                      MOROZHENKO A V DR
                      OWEN TOBIAS C PROF
                      SHEVCHENKO VLADISLAV V DR
                      SMITH BRADFORD A PROF
                      TEJFEL VIKTOR G DR

Members:
AKABANE TOKUHIDE DR   ANDERS EDWARD PROF   APPLEBY JOHN F
ARTHUR DAVID W G   ATREYA SUSHIL K   BARROW COLIN H DR
BATTANER EDUARDO DR   BAUM WILLIAM A DR   BAZILEVSKY ALEXANDR T
BEEBE RETA FAYE DR   BEER REINHARD DR   BELTON MICHAEL J S DR
BENDER PETER L DR   BERGE GLENN L DR   BERGSTRALH JAY T DR
BHATIA R K DR   BLAMONT JACQUES E PROF   BOBROV M S DR
BONDARENKO L N DR   BOSMA PIETER B DR   BOYCE PETER B DR
BOYER CHARLES   BRECHER AVIVA DR PROF   BROADFOOT A LYLE DR
BROWN ROBERT HAMILTON   BRUNK WILLIAM E DR   BYSTROV NIKOLAI F DR
CALAME ODILE DR   CALDWELL JOHN JAMES   CAMERON WINIFRED S MRS
CAMICHEL HENRI DR   CAMPBELL DONALD B   CATALANO SANTO DR
CHAMBERLAIN JOSEPH W PROF   CHAPMAN CLARK R DR   COLLINSON EDWARD H
COLOMBO G PROF DR   CONNES JANINE DR   COOK ALLAN F DR
COUNSELMAN CHARLES C PROF   DE MOTTONI Y PALACIOS DR   DE PATER IMKE
DEGEWIJ JOHAN DR   DERMOTT STANLEY F   DICKEL JOHN R
DICKEY JEAN O'BRIEN   DOLLFUS AUDOUIN PROF   DRAKE FRANK D PROF
DZHAPIASHVILI VICTOR P DR   EL-BAZ FAROUK DR   ELLIOT JAMES L DR
ELSTON WOLFGANG E PROF   ESHLEMAN VON R PROF   ESPOSITO LARRY W
FARINELLA PAOLO DR   FIELDER GILBERT DR   FINK UWE DR
FOX KENNETH DR   FOX W E MR   FUJIWARA AKIRA DR
GALKIN LEONID S A DR   GEAKE JOHN E DR   GEHRELS TOM PROF
GEISS JOHANNES PROF   GICLAS HENRY L MR   GIERASCH PETER J DR
GOLD THOMAS PROF   GOLDREICH P DR   GOLDSTEIN RICHARD M DR
GOODY R M   GORENSTEIN PAUL DR   GOUDAS CONSTANTINE L PROF
GREEN JACK PROF   GROSSMAN LAWRENCE PROF   GUERIN PIERRE DR
GUEST JOHN E DR   GULKIS SAMUEL DR   GURSHTEIN A A DR
HABIBULLIN SH T PROF DR   HAGFORS T DR   HALL JOHN S DR

HALLIDAY IAN DR
HOLBERG JAY B
HU ZHONG-WEI
INGRAO HECTOR C
JEFFREYS HAROLD PROF SIR
KARANDIKAR R V PROF
KISLYUK VITALIJ S DR
KSANFOMALITI L V DR
KUZMIN ARKADII D PROF DR
LARSON STEPHEN M
LEWIS J S
LOPEZ-PUERTAS MANUEL
MAHRA H S DR
MAYER CORNELL H
MEADOWS A JACK PROF
MILLIS ROBERT L DR
MOEHLMANN DIEDRICH
MOROZ V I PROF DR
MUMMA MICHAEL JON
NESS NORMAN F DR
OTTELET I J DR
PETROPOULOS BASIL CH DR
RAO M N DR
RUNCORN S K PROF
SAGAN CARL DR
SCHLOERB F. PETER
SHIMIZU TSUTOMU PROF EMER
SJOGREN WILLIAM L MR
SODERBLOM LARRY DR
STROBEL DARRELL F
SYNNOTT STEPHEN P
TOMBAUGH CLYDE W PROF
TYLER JR G LEONARD DR
VEVERKA JOSEPH DR
WASSERMAN LAWRENCE H DR
WEST ROBERT ALAN
WILDEY ROBERT L PROF DR
WOOD JOHN A DR
YODER CHARLES F
ZHANG MING-CHANG

HERZBERG GERHARD DR
HOREDT GEORG PAUL DR
HUBBARD WILLIAM B PROF
IRVINE WILLIAM M PROF
JOHNSON TORRENCE V DR
KAULA WILLIAM M PROF
KOPAL ZDENEK PROF
KUMAR SHIV S PROF
LANE ARTHUR LONNE DR
LEIKIN G A DR
LOCKWOOD G WESLEY DR
LUMME KARI A DR
MARTYNOV D YA PROF DR
MCCORD THOMAS B DR
MIDDLEHURST BARBARA M MS
MILLMAN PETER M DR
MOLINA ANTONIO
MOUTSOULAS MICHAEL PROF
MURPHY ROBERT E DR
NEUKUM G DR
PANG KEVIN
PETTENGILL GORDON H PROF
RODRIGO RAFAEL
RUSKOL EUGENIA L DR
SAGITOV M U DR
SHAPIRO IRWIN I PROF
SHOEMAKER EUGENE M
SMITH HARLAN J PROF
SONETT CHARLES P PROF
STROM ROBERT G PROF
TERRILE RICHARD JOHN
TRAFTON LAURENCE M DR
VAN ALLEN JAMES A PROF
WALKER ROBERT M A PROF
WASSON JOHN T
WETHERILL GEORGE W
WILLIAMS IWAN P DR
WOOLFSON MICHAEL M PROF
YOUNG ANDREW T DR

HIDE RAYMOND PROF
HOVENIER J W DR
HUNTEN DONALD M PROF
IWASAKI KYOSUKE DR
JURGENS RAYMOND F
KILADZE R I DR
KOWAL CHARLES THOMAS
KURT V G DR
LARSON HAROLD P DR
LEVIN BORIS J DR
LOPEZ-MORENO JOSE JUAN
LUTZ BARRY L DR
MATSON DENNIS L DR
MCELROY M B DR
MIKHAIL JOSEPH SIDKY DR
MIYAMOTO SIGENORI PROF
MOORE PATRICK DR
MULHOLLAND J DERRAL DR
NAKAGAWA YOSHITSUGU DR
O'KEEFE JOHN A DR
PAOLICCHI PAOLO DR
POLLACK JAMES B DR
ROSCH JEAN PROF
SAFRONOV VICTOR S DR
SAISSAC JOSEPH DR
SHIMIZU MIKIO PROF
SINTON WILLIAM M
SMOLUCHOWSKI ROMAN PROF
STONE EDWARD C DR
STRONG JOHN D PROF
THOMPSON THOMAS WILLIAM
TROITSKY V S PROF DR
VAN FLANDERN THOMAS DR
WALLACE LLOYD V DR
WEIMER THEOPHILE P F DR
WHITAKER EWEN A
WILLIAMS JAMES G DR
WOSZCZYK ANDRZEJ PROF
YOUNG LOUISE GRAY DR

COMMISSION No. 19

ROTATION OF THE EARTH (ROTATION DE LA TERRE)

President : KLEPCZYNSKI WILLIAM J DR

Vice-President(s) : FEISSEL MARTINE DR
KOLACZEK BARBARA DR

Organizing Committee: BARLIER FRANCOIS E DR
BROSCHE PETER PROF
CARTER WILLIAM EUGENE
DJUROVIC DRAGUTIN M DR
MUELLER IVAN I PROF
ROCHESTER MICHAEL G PROF
SCHUTZ BOB EWALD
VONDRAK JAN DR
WILKINS GEORGE A DR
YATSKIV YA S DR
YE SHU-HUA
YOKOYAMA KOICHI DR

Members:
ABRAHAM HENRY J M
BARRETO LUIZ MUNIZ PROF
BLINOV N S DR
CANNON WAYNE H DR
CHEN XING
DAVIES JOHN G DR
DICKEY JEAN O'BRIEN
ELSMORE BRUCE DR
FEDOROV E P PROF
GAIGNEBET JEAN DR
GROTEN ERWIN PROF
HAN TIANQI
HIDE RAYMOND PROF
HURUKAWA KIICHIRO DR
JEFFREYS HAROLD PROF SIR
KALMYKOV A M DR
KOKURIN YURIJ L DR
LEDERLE TRUDPERT DR
LIESKE JAY H DR
MANABE SEIJI DR
MCCARTHY DENNIS D DR
MELCHIOR PAUL J PROF DIR
MILOVANOVIC VLADETA DR
MOCZKO JANUSZ DR
NAUMOV VITALIJ A DR
O'HORA NATHY P J
OOE MASATSUGU DR
PAN XIAO-PEI
PERDOMO RAUL
POPELAR JOSEF DR
RANDIC LEO PROF DR
ARABELOS DIMITRIOS DR
BENDER PETER L DR
BONANOMI J DR
CAPITAINE NICOLE
CHIUMIENTO GIUSEPPE
DEBARBAT SUZANNE V DR
DICKMAN STEVEN R
ENSLIN HEINZ DR
FLIEGEL HENRY F
GAO BUXI
GUINOT BERNARD R PROF
HELLWIG HELMUT WILHELM DR
HOSOYAMA KENNOSHUKE DR
IIJIMA SHIGETAKA PROF
JI HONG-QING
KING ROBERT WILSON JR DR
KOSTINA LIDIJA D DR
LEFEBVRE MICHEL DR
LUO DING-JIANG
MARKOWITZ WILLIAM DR
MEINIG MANFRED DR
MERRIAM JAMES B
MIRONOV NIKOLAY T
MORGAN PETER DR
NIEMI AIMO
OKAMOTO ISAO DR
ORTE ALBERTO
PAQUET PAUL EG DR
PILKINGTON JOHN D H DR
PRODAN Y I DR
REN JIANG-PING
BANG YONG GOL
BILLAUD GERARD J
CALAME ODILE DR
CATALAN MANUEL DR
CURRIE DOUGLAS G DR
DEJAIFFE RENE J DR
DRAMBA C PROF
FANSELOW JOHN LYMAN
FONG CHU-GANG
GAPOSCHKIN EDWARD M DR
HALL R GLENN DR
HEMMLEB GERHARD DR
HUA YING-MIN
JAKS WALDEMAR DR
KAKUTA CHUICHI DR
KNOWLES STEPHEN H DR
LAMBECK KURT PROF
LI ZHENG-XIN
LUO SHI-FANG
MATSAKIS DEMETRIOS N
MELBOURNE WILLIAM G DR
MIETELSKI JAN S DR
MIYADI MASASI DR
MORRISON LESLIE V
NOBILI ANNA M
OKAZAKI SEICHI DR
OTERMA LIISI PROF
PARIJSKIJ N N PROF
POMA ANGELO DR
PROVERBIO EDOARDO PROF
ROBERTSON DOUGLAS S

RUDER HANNS	RUNCORN S K PROF	RUSU I DR
RYKHLOVA LIDIJA V DR	SADZAKOV SOFIJA DR	SAKHAROV VLADIMIR I DR
SANCHEZ MANUEL	SASAO TETSUO DR	SATO KOICHI DR
SEKIGUCHI NAOSUKE PROF	SEVARLIC BRANISLAV M PROF	SEVILLA MIGUEL J DR
SHAPIRO IRWIN I PROF	SHI GUANG-CHEN	SIDORENKOV NIKOLAY S
SILVERBERG ERIC C DR	SLADE MARTIN A III DR	SMITH F GRAHAM PROF
SMITH HUMPHRY M	SMYLIE DOUGLAS E DR	STANILA GEORGE DR
STEPHENSON F RICHARD DR	STOYKO ANNA	SUGAWA CHIKARA DR
SUN YONGXIANG	TAKAGI SHIGETSUGU DR	TAPLEY BYRON D DR
TARADY VLADIMIR K DR	TELEKI GEORGE DR	THOMAS DAVID V DR
TORAO MASAHISA	TSAO MO PROF	TSUBOKAWA IETSUNE DR
VEILLET CHRISTIAN	VEIS GEORGE PH D	VICENTE RAIMUNDO O PROF
WAKO KOJIRO DR	WAN TONG-SHAN	WANG ZHENG MING
WARD WILLIAM R DR	WILLIAMS JAMES G DR	WILSON P DR
WINKLER GERNOT M R DR	WU SHOU-XIAN	XIA JIONGYU
XIAO NAI-YUAN	XU TONG-QI	YANG FUMIN
YUMI S PROF DR	ZHANG GUO-DONG	ZHAO MING
ZHENG DA-WEI	ZHU YONG-HE	

COMMISSION No. 20

POSITIONS AND MOTIONS OF MINOR PLANETS, COMETS AND SATELLITES

(POSITIONS ET MOUVEMENTS DES PETITES PLANETES, DES COMETES ET DES SATELLITES)

President : KOZAI YOSHIHIDE PROF

Vice-President(s) : BATRAKOV YU V DR

Organizing Committee: AKSNES KAARE DR
ARLOT JEAN-EUDES
KRESAK LUBOR DR
KRISTENSEN LEIF KAHL DR
MARSDEN BRIAN G DR
MILLIS ROBERT L DR
ROEMER ELIZABETH PROF
YEOMANS DONALD K DR
ZHANG YU-ZHE

Members:
A'HEARN MICHAEL F DR
BABADZHANOV PULAT B DR
BELYAEV NIKOLAJ A DR
BIELICKI MACIEJ DR
BOWELL EDWARD L G DR
CALAME ODILE DR
CHERNYKH N S DR
CRISTESCU CORNELIA G DR
DE SANCTIS GIOVANNI
DIRIKIS M A DR
DVORAK RUDOLF DR
EVDOKIMOV YU V DR
FORTI GIUSEPPE DR
FROESCHLE CLAUDE DR
GEHRELS TOM PROF
GILMORE ALAN C MR
HARRIS ALAN WILLIAM DR
HELIN ELEANOR FRANCIS
HERS JAN MR
HURUKAWA KIICHIRO DR
IZVEKOV V A DR
KISSELEVA TAMARA P
KOWAL CHARLES THOMAS
LINDBLAD BERTIL A DR
MACHADO LUIZ E. DA SILVA
MILET BERNARD L DR
MRKOS ANTONIN DR
NACOZY PAUL E DR
NOBILI ANNA M
PIERCE DAVID ALLEN
PROTICH MILORAD B
REITSEMA HAROLD J

ABALAKIN VICTOR K DR
BAILEY MARK EDWARD
BENEST DANIEL DR
BIEN REINHOLD DR
BRANHAM RICHARD L JR
CANDY MICHAEL P MR
CHIO CHOL ZONG
CUNNINGHAM LELAND E PROF
DEBEHOGNE HENRI DR SC
DOLLFUS AUDOUIN PROF
EDMONDSON FRANK K PROF
EVERHART EDGAR DR
FRANKLIN FRED A DR
GALIBINA I V DR
GIBSON JAMES
GREENBERG RICHARD DR
HASEGAWA ICHIRO DR
HEMENWAY PAUL D
HEUDIER JEAN-LOUIS DR
IANNA PHILIP A
KHATISASHVILI ALFEZ SH DR
KLEMOLA ARNOLD R DR
LAGERKVIST CLAES-INGVAR
LOMB NICHOLAS RALPH DR
MAHRA H S DR
MINTZ BLANCO BETTY MRS
MULHOLLAND J DERRAL DR
NAKAMURA TSUKO DR
OTERMA LIISI PROF
PITTICH EDUARD M DR
QUIJANO LUIS
RICKMAN HANS DR

AREND S DR
BEC-BORSENBERGER ANNICK
BENNETT JOHN CAISTER MR
BOERNGEN FREIMUT DR PH
BURNS JOSEPH A PROF
CARUSI ANDREA
CHURMS JOSEPH
DE PASCUAL MARTINEZ M DR
DELSEMME ARMAND H PROF DR
DUNHAM DAVID W
ELLIOT JAMES L DR
FERRAZ-MELLO S PROF DR
FREITAS MOURAO R R DR
GARFINKEL BORIS DR
GICLAS HENRY L MR
HARRINGTON ROBERT S DR
HAUPT HERMANN F PROF
HENRARD JACQUES PROF
HURNIK HIERONIM PROF
IVANOVA VIOLETA DR
KIANG TAO PROF
KOHOUTEK LUBOS DR
LIESKE JAY H DR
LOVAS MIKLOS
MCCROSKY RICHARD E DR
MORANDO BRUNO L DR
MURRAY CARL D
NAKANO SYUICHI
PASCU DAN DR
POPOVIC BOZIDAR PROF DR
RAPAPORT MICHEL DR
ROBERTSON WILLIAM H

SAGNIER JEAN-LOUIS DR	SCHMADEL LUTZ D DR	SCHOBER HANS J DR
SCHOLL HANS DR	SCHRUTKA-RECHTENSTAMM PR.	SCHUBART JOACHIM DR
SCHUSTER HANS-EMIL	SEIDELMANN P KENNETH DR	SEKANINA ZDENEK DR
SHELUS PETER J DR	SHKODROV V G DR	SHOR VIKTOR A DR
SHTEINS K A DR	SINCLAIR ANDREW T DR	SITARSKI GRZEGORZ PROF
STELLMACHER IRENE DR	STROBEL WILLI DR	SULTANOV G F ACAD
SVOREN JAN	SYNNOTT STEPHEN P	TAYLOR DONALD BOGGIA DR
TAYLOR GORDON E	TOMITA KOICHIRO MR	TORRES CARLOS DR
VAGHI SERGIO DR	VALSECCHI GIOVANNI B	VAN FLANDERN THOMAS DR
VAN HOUTEN C J DR	VAN HOUTEN-GROENEVELD I	VEILLET CHRISTIAN
VIEIRA MARTINS ROBERTO DR	VU DUONG TUYEN DR	WASSERMAN LAWRENCE H DR
WEISSMAN PAUL ROBERT	WEST RICHARD M DR	WHIPPLE FRED L DR
WILD PAUL PROF	WILLIAMS IWAN P DR	WILLIAMS JAMES G DR
WROBLEWSKI HERBERT DR	YABUSHITA SHIN A PROF	ZADUNAISKY PEDRO E PROF
ZAPPALA VINCENZO PROF	ZHANG JIA-XIANG	ZIOLKOWSKI KRZYSZTOF DR

COMMISSION No. 21

LIGHT OF THE NIGHT SKY (LUMIERE DU CIEL NOCTURNE)

President   : MATTILA KALEVI DR

Vice-President(s)   : LEVASSEUR-REGOURD A.C. PR

Organizing Committee: DUMONT RENE DR
                      GALPERIN YU I PROF
                      GIESE RICHARD H PROF
                      HANNER MARTHA S DR
                      LAMY PHILIPPE DR
                      MUKAI TADASHI DR
                      TANABE HIROYOSHI DR
                      WEINBERG J L DR

Members:

ALVAREZ P	ANDERSON KINSEY A PROF	ANGIONE RONALD J DR
BAGGALEY WILLIAM J PROF	BANOS COSMAS J DR	BATES DAVID R PROF
BELKOVICH O I DR	BLACKWELL DONALD E PROF	BLAMONT JACQUES E PROF
BOWYER C STUART PROF	BROADFOOT A LYLE DR	CHAMBERLAIN JOSEPH W PROF
COOK ALLAN F DR	DACHS JOACHIM PROF DR	DIVARI N B DR
DUFAY MAURICE PROF	DUNKELMAN LAWRENCE	ELSAESSER HANS PROF
FECHTIG HUGO DR	FELDMAN PAUL DONALD DR	FISHKOVA LUISA M PROF
FRACASSINI MASSIMO PROF	GIOVANE FRANK	GREENBERG J MAYO DR
GRUEN EBERHARD DR	HALLIDAY IAN DR	HARWIT MARTIN PROF
HAUG ULRICH PROF	HENRY RICHARD C. PROF.	HOFMANN WILFRIED DR
HONG SEUNG SOO DR	HOUCK JAMES R	HURUHATA MASAAKI PROF
IVANOV-KHOLODNY G S DR	JAMES JOHN F MR	JARRETT ALAN H PROF
JOUBERT MARTINE	KAPLAN J DR	KARANDIKAR R V PROF
KARYGINA ZOYA V DR	KOUTCHMY SERGE DR	KULKARNI PRABHAKAR V PROF
LEINERT CHRISTOPH DR	LILLIE CHARLES F DR	LOPEZ-MORENO JOSE JUAN
LOPEZ-PUERTAS MANUEL	MATSUMOTO TOSHIO DR	MEGRELISHVILI T G PROF
MISCONI NEBIL YOUSIF DR	MOLINA ANTONIO	MORGAN DAVID H DR
MUKAI SONOYO DR	NAWAR SAMIR DR	NEUZIL LUDEK DR
NEY EDWARD P PROF	NICOLET MARCEL PROF	NISHIMURA TETSUO DR
PARESCE FRANCESCO DR	PEARSE REGINALD W B DR	PFLEIDERER JORG PROF
PITZ ECKHART DR	RAPAPORT MICHEL DR	RIPKEN HARTMUT W DR
ROACH FRANKLIN E	ROBLEY R DR	RODRIGO RAFAEL
ROOSEN ROBERT G DR	ROZHKOVSKIJ DIMITRIJ A	SANCHEZ FRANCISCO PROF
SANCHEZ MAGRO C DR	SANCHEZ-SAAVEDRA M LUISA	SAXENA P P DR
SHAROV A S DR	SHEFOV NICOLAI N	SOBERMAN ROBERT K DR
SPARROW JAMES G DR	STAUDE HANS JAKOB PH D	TOROSHLIDZE TEIMURAZ I DR
TRUTSE YU L DR	TYSON JOHN A DR	VAN ALLEN JAMES A PROF
VAN DE HULST H C PROF DR	WALLACE LLOYD V DR	WEILL GILBERT M DR
WENIGER SCHAME DR	WITT ADOLF N DR	WOLSTENCROFT RAMON D DR
WOOLFSON MICHAEL M PROF	ZERULL REINER H DR	

MEMBERSHIP OF COMMISSIONS

COMMISSION No. 22

METEORS AND INTERPLANETARY DUST

(METEORES ET LA POUSSIERE INTERPLANETAIRE)

President   : BABADZHANOV PULAT B DR

Vice-President(s)   : KEAY COLIN S L PROF

Organizing Committee: BAGGALEY WILLIAM J PROF
                      BELKOVICH O I DR
                      ELFORD WILLIAM GRAHAM DR
                      FECHTIG HUGO DR
                      HANNER MARTHA S DR
                      HASEGAWA ICHIRO DR
                      MCDONNELL J A M PROF
                      STOHL JAN DR
                      TOMITA KOICHIRO MR

Members:

ABBOTT WILLIAM N DR	ANDERS EDWARD PROF	BEARD DAVID B DR
BHANDARI N DR	BIBARSOV RAVIL'SH DR	BLACKWELL ALAN TREVOR
BROWNLEE DONALD E PROF	CARUSI ANDREA	CEPLECHA ZDENEK DR
CEVOLANI GIORDANO	CLIFTON KENNETH ST	COOK ALLAN F DR
DAVIES JOHN G DR	FIREMAN EDWARD L	FORTI GIUSEPPE DR
GIESE RICHARD H PROF	GLASS BILLY PRICE DR	GOSWAMI J N DR
GRUEN EBERHARD DR	HAJDUK ANTON DR	HAJDUKOVA MARIA
HALLIDAY IAN DR	HARVEY GALE A DR	HASEGAWA HIROICHI DR
HAWKINS GERALD S DR	HEY JAMES STANLEY DR	HODGE PAUL W PROF
HONG SEUNG SOO DR	HOPPE J A PROF DR	HUGHES DAVID W DR
JACCHIA LUIGI G DR	JENNISON ROGER C PROF	JONES JAMES DR
KAISER THOMAS R PROF	KAPISINSKY IGOR	KASHCHEEV B L PROF DR
KOEBERL CHRISTIAN DR	KOSTYLEV K V DR	KRAMER KH N DR
KRESAK LUBOR DR	KRESAKOVA MARGITA DR	KVIZ ZDENEK DR
LAMY PHILIPPE DR	LEBEDINETS V N DR	LEVASSEUR-REGOURD A.C. PR
LEVIN BORIS J DR	LINDBLAD BERTIL A DR	LOVELL SIR BERNARD PROF
MARVIN URSULA B DR	MCCROSKY RICHARD E DR	MCINTOSH BRUCE A DR
MEISEL DAVID D DR	MILES HOWARD G MR	MILLMAN PETER M DR
MISCONI NEBIL YOUSIF DR	NAKAZAWA KIYOSHI DR	NEWBURN RAY L JR
NUTH JOSEPH A III	O'KEEFE JOHN A DR	PADEVET VLADIMIR DR
PECINA PETR	PLAVEC ZDENKA DR	POLNITZKY GERHARD DR
PORUBCAN VLADIMIR DR	RAJCHL JAROSLAV DR	REVELLE DOUGLAS ORSON DR
RIPKEN HARTMUT W DR	ROOSEN ROBERT G DR	RUSSELL JOHN A PROF
SEKANINA ZDENEK DR	SHAO CHENG-YUAN	SIMEK MILOS DR
SOBERMAN ROBERT K DR	SOUTHWORTH R B DR	TEDESCO EDWARD F
TERENTJEVA ALEXANDRA K DR	VERNIANI FRANCO PROF	WEINBERG J L DR
WETHERILL GEORGE W	WHIPPLE FRED L DR	WILLIAMS IWAN P DR
WOOD JOHN A DR	WOOLFSON MICHAEL M PROF	YAMAKOSHI KAZUO
YAVNEL ALEXANDER A DR	YEOMANS DONALD K DR	ZVOLANKOVA JUDITA

COMMISSION No. 24

PHOTOGRAPHIC ASTROMETRY (ASTROMETRIE PHOTOGRAPHIQUE)

President   : UPGREN ARTHUR R DR

Vice-President(s) : VAN ALTENA WILLIAM F PROF

Organizing Committee: ARGUE A NOEL MR
                      CORBIN THOMAS ELBERT DR
                      DE VEGT CH PROF DR
                      GLIESE WILHELM DR
                      KANAEV IVAN I DR
                      LUTZ THOMAS E DR
                      STOCK JURGEN D

Members:

ABHYANKAR KRISHNA D PROF	BASTIAN ULRICH	BENEDICT GEORGE F DR
BLAAUW ADRIAAN PROF DR	BOUIGUE ROGER PROF	BRANHAM RICHARD L JR
BRONNIKOVA NINA M	BROSCHE PETER PROF	CHIU LIANG-TAI GEORGE
CHRISTY JAMES WALTER DR	CHURMS JOSEPH	CLUBE S V M DR
CONNES PIERRE DR	CREZE MICHEL DR	CUDWORTH KYLE MCCABE DR
DAHN CONARD CURTIS DR	DELHAYE JEAN PROF	DEUTSCH ALEKSANDR N PROF
DOMMANGET J DR	DOUGLASS GEOFFREY G	EICHHORN HEINRICH K DR
ELSMORE BRUCE DR	FALLON FREDERICK W DR	FANSELOW JOHN LYMAN
FIRNEIS FRIEDRICH J DR	FIRNEIS MARIA G DR	FRACASTORO MARIO G PROF
FRANZ OTTO G DR	FREDRICK LAURENCE W PROF	FRESNEAU ALAIN DR
GALLOUET LOUIS DR	GATEWOOD GEORGE DIRECTOR	GICLAS HENRY L MR
GOYAL A N DR	HANSON ROBERT B DR	HARRINGTON ROBERT S DR
HARWOOD DENNIS MR	HEINTZ WULFF D DR	HEMENWAY PAUL D
HERSHEY JOHN L DR	HEUDIER JEAN-LOUIS DR	HILL GRAHAM DR
HOFFLEIT E DORRIT DR	HUGHES JAMES A DR	IANNA PHILIP A
JAHREISS HARTMUT DR	JEFFERYS WILLIAM H DR	JOHNSTON KENNETH J
JONES BURTON DR	JONES DEREK H P DR	KISLYUK VITALIJ S DR
KLEMOLA ARNOLD R DR	KLOCK B L DR	KOLCHINSKIJ I G DR
KOVALEVSKY JEAN DR	LACROUTE PIERRE A PROF	LAPUSHKA K K DR
LATYPOV A A DR	LE POOLE RUDOLF S DR	LIPPINCOTT SARAH LEE DR
LOZINSKIJ A M DR	LU PHILLIP K DR	LUYTEN WILLEM J PROF
MARSCHALL LAURENCE A	MATIAGIN VALERY S DR	MCALISTER HAROLD A DR
MENNESSIER MARIE-ODILE DR	MEURERS JOSEPH PROF DR	MONET DAVID G
MURRAY C ANDREW	NICHOLSON WILLIAM	OJA TARMO PROF
ONEGINA A B DR	PAN RONG-SHI	PERRYMAN MICHAEL A C
PODOBED V V DR	POLOZHENTSEV DIMITRIJ DR	POTTER HEINO I DR
PROCHAZKA FRANZ V DR	QIN DAO	QUIJANO LUIS
REQUIEME YVES DR	RIZVANOV NAUFAL G DR	ROBERTSON WILLIAM H
ROEMER ELIZABETH PROF	ROESER SIEGFRIED DR	RUDER HANNS
RUSSELL JANE L DR	SANDERS W L PROF	SHI GUANG-CHEN
SIMS KENNETH P DR	SMITH CLAYTON A JR DR	STANGE LOTHAR
STEIN JOHN WILLIAM	STONE RONALD CECIL	STRAND KAJ AA DR
THOMAS DAVID V DR	TURON-LACARRIEU C DR	VALBOUSQUET ARMAND DR
VAN DE KAMP PETER	VASILEVSKIS STANISLAUS	WALTER HANS G DR
WAN LAI	WANG JIA-JI	WASSERMAN LAWRENCE H DR
WESSELINK ADRIAAN J DR	WESTERHOUT GART DR	WHITE GRAEME LINDSAY DR
WROBLEWSKI HERBERT DR	YOUNIS SAAD M	ZHOU XING-HAI

COMMISSION No. 25

STELLAR PHOTOGRAPHY AND POLARIMETRY
(PHOTOMETRIE ET POLARIMETRIE STELLAIRE)

President   : RUFENER FREDY G PROF

Vice-President(s) : MCLEAN IAN S DR

Organizing Committee: BUSER ROLAND DR
DACHS JOACHIM PROF DR
EDWARDS PAUL J DR
GLASS IAN STEWART DR
KILKENNY DAVID DR
MILLER JOSEPH S PROF
MILONE EUGENE F PROF
PENNY ALAN JOHN DR
PIIROLA VILPPU E DR
SHAKHOVSKOJ N M DR
TINBERGEN JAAP DR
VRBA FREDERICK J DR
WIELEBINSKI RICHARD PROF
YOUNG ANDREW T DR

Members:
ABLES HAROLD D DR
ANGEL J ROGER P PROF
ARNAUD JEAN
AXON DAVID
BECK RAINER
BESSELL MICHAEL S DR
BORGMAN JAN DR PROF
BROWN DOUGLAS NASON
CELIS LEOPOLDO DR
COYNE GEORGE V DR
DENOYELLE JOZEF KIC
DUCATI JORGE RICARDO DR
FERNIE J DONALD PROF
GALLOUET LOUIS DR
GHOSH S K DR
GRAHAM JOHN A DR
GREWING MICHAEL PROF
HARDIE R PROF
HAYES DONALD S DR
HILDITCH RONALD W DR
HOLMBERG ERIK B PROF
HYLAND A R HARRY DR
JERZYKIEWICZ MIKOLAJ DR
KEMP JAMES C
KOCH ROBERT H DR
KVIZ ZDENEK DR
LANDSTREET JOHN D PROF
LENZEN RAINER DR

ADELMAN SAUL J DR
ANTHONY-TWAROG BARBARA J
ARSENIJEVIC JELISAVETA
BAHNG JOHN D R PROF
BECKER WILHELM PROF
BLANCO VICTOR M DR
BORRA ERMANNO F DR
BRUCK HERMANN A PROF
CHUGAJNOV P F DR
CRAWFORD DAVID L DR
DESHPANDE M R DR
EELSALU HEINO DR
FITCH WALTER S DR
GEHRELS TOM PROF
GOLAY MARCEL PROF
GRAUER ALBERT D
GUTIERREZ-MORENO A DR MRS
HARWOOD DENNIS MR
HECK ANDRE DR
HILL PHILIP W DR
HU JING-YAO
IRWIN ALAN W DR
JOSHI SURESH CHANDRA DR
KEPLER S O
KRON GERALD E DR
LABHARDT LUKAS
LASKARIDES PAUL G ASSPROF
LI SIN HYONG

ALBRECHT RUDOLF DR
ARGUE A NOEL MR
ASHOK N M DR
BALDINELLI LUIGI DR
BEHR ALFRED PROF EMERITUS
BOOKMYER BEVERLY B DR
BREGER MICHEL DR
CARNEY BRUCE WILLIAM
COUSINS A W J DR
DAHN CONARD CURTIS DR
DUBOUT RENEE
FEINSTEIN ALEJANDRO DR
FORTE JUAN CARLOS DR
GEHRZ ROBERT DOUGLAS DR
GOY GERALD PROF
GRENON MICHEL DR
HALL DOUGLAS S DR
HAUCK BERNARD PROF
HENSBERGE HERMAN
HILTNER W ALBERT PROF
HUANG LIN
IYENGAR K V K PROF
KAWARA KIMIAKI
KNUDE JENS KIRKESKOV DR
KUNKEL WILLIAM E DR
LANDOLT ARLO U PROF
LASKER BARRY M DR
LOCKWOOD G WESLEY DR

LUB JAN DR	LUNA HOMERO G. DR	MARKKANEN TAPIO DR
MARRACO HUGO G DR	MASANI A PROF	MAYER PAVEL DR
MCCARTHY MARTIN F DR	MENDOZA V EUGENIO E DR	MENZIES JOHN W DR
MIANES PIERRE DR	MINTZ BLANCO BETTY MRS	MITCHELL RICHARD MR
MOFFETT THOMAS J PROF	MORENO HUGO PROF	MOROSSI CARLO
MORRIS STEPHEN C DR	MULLER A B DR	MUMFORD GEORGE S PROF
NICOLET BERNARD	NIKONOV V B DR	NOTNI P DR
OESTREICHER ROLAND	OSAWA KIYOTERU DR	OSKANYAN V S DR
PAGE ARTHUR MR	PEDREROS MARIO DR	PEL JAN WILLEM DR
PERRY CHARLES L DR	PFAU WERNER	PFEIFFER RAYMOND J
PHILIP A G DAVIS	RAO P VIVEKANANDA DR	ROBINSON EDWARD LEWIS DR
ROSLUND CURT DR	RYBKA EUGENIUSZ PROF DR	RYDGREN ALFRED ERIC JR DR
SARMA M B K PROF	SCHMIDT EDWARD G	SCHMIDT HANS PROF
SCHOENEICH W DR	SHAWL STEPHEN J DR	SMITH CHARLES DITTO
SMYTH MICHAEL J DR	STEINLIN ULI PROF	STOCK JURGEN D
STOCKMAN HERVEY S JR DR	STONE REMINGTON P S DR	STRAIZYS V PROF DR
STROHMEIER WOLFGANG PROF	SULLIVAN DENIS JOHN DR	SZKODY PAULA DR
TANDON S N PROF	TAPIA-PEREZ SANTIAGO	THOMAS JOHN A PROF
TODORAN IOAN DR	TOLBERT CHARLES R DR	TRODAHL HARRY JOSEPH DR
ULRICH BRUCE T PROF	URECHE VASILE DR	VARDANIAN R A DR
VAUGHAN ARTHUR H DR	VELGHE ALBERT G PROF DR	VERMA R P DR
VISVANATHAN NATARAJAN DR	WALKER ALISTAIR ROBIN DR	WALLENQUIST AAKE A E PROF
WALRAVEN TH DR	WANG CHUAN-JIN	WARREN WAYNE H JR DR
WEISTROP DONNA DR	WESSELINK ADRIAAN J DR	WESSELIUS PAUL R DR
WHITE NATHANIEL M DR	WILLSTROP RODERICK V DR	WINIARSKI MACIEJ
WISNIEWSKI WIESLAW Z	WOLSTENCROFT RAMON D DR	WOO JONG OK
WRAMDEMARK STIG S O DR	YAMASHITA YASUMASA PROF	YIN JI-SHENG

MEMBERSHIP OF COMMISSIONS

COMMISSION No. 26

DOUBLE AND MULTIPLE STARS (ETOILES DOUBLES ET MULTIPLES)

President : RAKOS KARL D PROF

Vice-President(s) : MCALISTER HAROLD A DR

Organizing Committee: ABT HELMUT A DR
COUTEAU PAUL PROF
DOMMANGET J DR
FRACASTORO MARIO G PROF
HARRINGTON ROBERT S DR
KISELYOV ALEXEJ A DR

Members:
ALLEN CHRISTINE
BAIZE PAUL DR
CABRITA EZEQUIEL DR
CULVER ROGER BRUCE DR
DOCOBO DURANTEZ JOSE A
FEKEL FRANCIS C
FRANZ OTTO G DR
GEYER EDWARD H PROF DR
HIDAJAT BAMBANG PROF DR
KULIKOVSKIJ P G DR
LUYTEN WILLEM J PROF
MIKKOLA SEPPO DR
MULLER PAUL
POVEDA ARCADIO DR
SCARDIA MARCO
STRAND KAJ AA DR
VAN DE KAMP PETER
WALKER RICHARD L
WILSON RAYMOND H DR
AREND S DR
BATTEN ALAN H DR
CESTER BRUNO PROF
DADAEV ALEKSANDR N DR
DUNHAM DAVID W
FERRER OSVALDO EDUARDO DR
FREITAS MOURAO R R DR
HERNANDEZ CARLOS ALBERTO
HOLDEN FRANK
KUMSISHVILI J I DR
MAGALASHVILI N L DR
MORBEY CHRISTOPHER L
PANNUNZIO RENATO
RUSSELL JANE L DR
SCARFE COLIN D DR
TAPIA MAURICIO DR
VAN DER HUCHT KAREL A DR
WEIS EDWARD W DR
WORLEY CHARLES E
BACCHUS PIERRE PROF
BERNACCA P L PROF
CHEN ZHEN
DEUTSCH ALEKSANDR N PROF
EICHHORN HEINRICH K DR
FLETCHER J MURRAY
GATEWOOD GEORGE DIRECTOR
HERSHEY JOHN L DR
ISHIDA GORO DR
LIPPINCOTT SARAH LEE DR
MEYER CLAUDE DR
MOREL PIERRE JACQUES DR
POPOVIC GEORGIJE DR
SALUKVADZE G N DR
SHUL'BERG A M DR
VALBOUSQUET ARMAND DR
VAN DESSEL EDWIN LUDO DR
WIETH-KNUDSEN NIELS P DR
YAN LIN-SHAN

COMMISSION No. 27

VARIABLE STARS (ETOILES VARIABLES)

President   : SZEIDL BELA DR

Vice-President(s) : BREGER MICHEL DR

Organizing Committee: COX ARTHUR N DR
GERSHBERG R E DR
JERZYKIEWICZ MIKOLAJ DR
MAVRIDIS L N PROF
MIRZOYAN L V DR PROF
PERCY JOHN R PROF
SMITH MYRON A ASST PROF
VAN GENDEREN A M DR
WARNER BRIAN PROF

Members:

AIAD A DR	AIZENMAN MORRIS L DR	ALANIA I F DR
ANDO HIROYASU DR	ANTONELLO ELIO	ARELLANO FERRO ARMANDO
ARKHIPOVA V P DR	ARSENIJEVIC JELISAVETA	ASTERIADIS GEORGIOS DR
BAADE DIETRICH DR	BAGLIN ANNIE DR	BAKER NORMAN H PROF
BAKOS GUSTAV A PROF	BALONA LUIS ANTERO DR	BARNES III THOMAS G DR
BARTOLINI CORRADO	BARWIG HEINZ	BASTIEN PIERRE DR
BATESON FRANK M OBE DR	BATH GEOFFREY T DR	BELSERENE EMILIA P
BELVEDERE GAETANO DR	BERTHOMIEU GABRIELLE DR	BESSELL MICHAEL S DR
BIANCHINI ANTONIO DR	BOCHONKO D RICHARD DR	BOLTON C THOMAS PROF
BOND HOWARD E DR	BOPP BERNARD W DR	BOULON JACQUES J DR
BOYARCHUK A A DR	BOYARCHUK MARGARITA E DR	BURKI GILBERT DR
BUSKO IVO C DR	BUTLER C JOHN DR	BUTLER DENNIS DR
BYRNE PATRICK B DR	CATCHPOLE ROBIN MICHAEL	CHAVIRA ENRIQUE SR
CHEREPASHCHUK A M PROF	CHRISTENSEN-DALSGAARD J	CHRISTY ROBERT F DR
CHUGAJNOV P F DR	COGAN BRUCE C DR	COHEN MARTIN DR
CONNOLLY LEO PAUL	CONTADAKIS MICHAEL E DR	COULSON IAIN
COUTTS-CLEMENT CHRISTINE	DE GROOT MART DR	DELGADO ANTONIO JESUS
DEMERS SERGE DR	DEUPREE ROBERT G DR	DICKENS ROBERT J DR
DUPUY DAVID L DR	DZIEMBOWSKI WOJCIECH PROF	EDWARDS PAUL J DR
EFREMOV YURY N DR	EL-BASSUNY ALAWY A A	ESKIOGLU A NIHAT
EVANS ANEURIN	EVANS NANCY REMAGE DR	FADEYEV YURI A
FEAST MICHAEL W DR	FEIBELMAN WALTER A DR	FERLAND GARY JOSEPH
FERNIE J DONALD PROF	FITCH WALTER S DR	FRIEDJUNG MICHAEL DR
FROLOV M S DR	GAHM GOESTA F DR	GALLAGHER III JOHN S DR
GARRIDO RAFAEL	GASCOIGNE S C B DR	GEYER EDWARD H PROF DR
GIBSON DAVID MICHAEL DR	GIEREN WOLFGANG P DR	GLAGOLEVSKIJ JU V DR
GODOLI GIOVANNI PROF	GORBATSKY VITALIJ G PROF	GOUGH DOUGLAS O DR
GRAHAM JOHN A DR	GRASDALEN GARY L DR	GRYGAR JIRI DR
GUERRERO GIANANTONIO DR	GUINAN EDWARD FRANCIS DR	GURM HARDEV S PROF
GURSKY HERBERT DR	HACKWELL JOHN A DR	HAEFNER REINHOLD DR
HAISCH BERNHARD MICHAEL	HALL DOUGLAS S DR	HANSEN CARL J PROF
HARMANEC PETR DR	HARO GUILLERMO DR	HEISER ARNOLD M DR
HERBIG GEORGE H DR	HERR RICHARD B DR	HERS JAN MR
HESSER JAMES E DR	HEYDEN FRANCIS J SJ DR	HILL HENRY ALLEN DR

HILL PHILIP W DR
HURUHATA MASAAKI PROF
JARZEBOWSKI TADEUSZ DR
KADOURI TALIB HADI
KEPLER S O
KIPPENHAHN RUDOLF PROF
KRAUTTER JOACHIM DR
KUBIAK MARCIN A DR
KUNKEL WILLIAM E DR
LAGO MARIA TERESA V T MS
LASKARIDES PAUL G ASSPROF
LEUNG KAM CHING PROF
LUUD LAURI DR
MAFFEI PAOLO PROF
MAKARENKO EKATERINA N DR
MARGRAVE THOMAS EWING JR
MATTEI JANET AKYUZ DR
MENNESSIER MARIE-ODILE DR
MILONE LUIS A DR
MORRISON NANCY DUNLAP DR
NATHER R EDWARD
NIKOLOV ANDREJ DR
ODGERS GRAHAM J DR
OSAWA KIYOTERU DR
PAPOUSEK JIRI
PATERNO LUCIO PROF
PIIROLA VILPPU E DR
PROVOST JANINE DR
RAO N KAMESWARA
ROBINSON EDWARD LEWIS DR
ROMANO GIULIANO PROF
ROUNTREE JANET DR
SAMUS NIKOLAI N DR
SAREYAN JEAN-PIERRE DR
SAWYER-HOGG HELEN B DR
SCHWARTZ PHILIP R DR
SHARA MICHAEL DR
SINVHAL SHAMBHU DAYAL DR.
SMITH HARLAN J PROF
STEPIEN KAZIMIERZ DR
STOBIE ROBERT S DR
SZECSENYI-NAGY GABOR DR
TAMMANN G ANDREAS PROF DR
TJIN-A-DJIE HERMAN R E DR
TSESEVICH V P PROF DR
TUTUKOV A V DR
VALTIER JEAN-CLAUDE DR
VIOTTI ROBERTO DR
WAELKENS CHRISTOFFEL
WALLERSTEIN GEORGE PROF
WEHLAU AMELIA DR
WESSELINK ADRIAAN J DR
WILLSON LEE ANNE DR
WISNIEWSKI WIESLAW Z
XIONG DA-RUN

HOFFLEIT E DORRIT DR
HUTCHINGS JOHN B DR
JIANG SHI-YANG
KANYO SANDOR DR
KHOLOPOV P N DR
KOPYLOV I M DR
KREINER JERZY MAREK DR
KUHI LEONARD V PROF
KURTZ DONALD WAYNE DR
LANDOLT ARLO U PROF
LEDOUX P J PROF
LOCKWOOD G WESLEY DR
MADORE BARRY FRANCIS DR
MAHMOUD FAROUK M A B DR
MANNINO GIUSEPPE PROF
MARTIN WILLIAM L DR
MAYALL MARGARET W
METZ KLAUS DR
MOFFETT THOMAS J PROF
MUMFORD GEORGE S PROF
NEFF JOHN S
NUGIS TIIT
OLAH KATALIN DR
OSKANYAN V S DR
PARSAMYAN ELMA S DR
PELTIER LESLIE C
POPOVA MALINA D PROF DR
PSKOVSKIJ JU P DR
RENSON P F M DR
RODGERS ALEX W DR
ROMANOV YURI S DR
RUSSEV RUSCHO DR
SANDMANN WILLIAM HENRY
SARMA M B K PROF
SCHMIDT EDWARD G
SCHWARZENBERG-CZERNY A
SHERWOOD WILLIAM A DR
SMAK JOSEPH I PROF
STARRFIELD SUMNER PROF
STERKEN CHRISTIAAN LEO DR
STROHMEIER WOLFGANG PROF
SZKODY PAULA DR
TEMPESTI PIERO PROF
TORRES CARLOS ALBERTO DR
TSIOUMIS ALEXANDROS DR
TYLENDA ROMUALD DR
VAN AGT S L TH J DR
VOGT NIKOLAUS DR
WALKER EDWARD N MR
WALRAVEN TH DR
WEHLAU WILLIAM H PROF
WHITELOCK PATRICIA ANN DR
WILSON LIONEL DR
WOOD PETER R DR
YAO BAO-AN

HOUK NANCY DR
IBEN ICKO JR PROF
JONES ALBERT F MR
KARP ALAN HERSH DR
KIM TU HWAN
KRAFT ROBERT P PROF
KRZEMINSKI WOJCIECH DR
KUMSISHVILI J I DR
KWEE K K DR
LANEY CLIFTON
LEITE SCHEID PAULO DR
LUB JAN DR
MAEDER ANDRE PROF
MAHRA H S DR
MANTEGAZZA LUCIANO
MASANI A PROF
MCNAMARA DELBERT H DR
MILONE EUGENE F PROF
MORGULEFF NINA ING
MURDIN PAUL G DR
NIARCHOS PANAYIOTIS PH D
O'DONOGHUE DARRAGH
OPOLSKI ANTONI PROF
PAPALOIZOU JOHN C B DR
PARTHASARATHY M DR
PETERSEN J O DR
PRINGLE JAMES E DR
RAKOS KARL D PROF
RICHTER G A DR
RODONO MARCELLO DR
ROSINO LEONIDA PROF
SADIK AZIZ R DR
SANYAL ASHIT DR
SATO NAONOBU PROF
SCHOEMBS ROLF DR
SCUFLAIRE RICHARD DR
SHOBBROOK ROBERT R DR
SMEYERS PAUL PROF
STELLINGWERF ROBERT F DR
STIFT MARTIN JOHANNES DR
SZABADOS LASZLO PH D
TAKEUTI MINE DR
TERZAN AGOP DR
TREMKO JOZEF DR
TSVETKOV MILCHO K DR
USHER PETER D DR
VAN HOOF A PROF EM
WACHMANN A A PROF DR
WALKER MERLE F PROF
WEBBINK RONALD F DR
WENZEL W DR
WILLIAMON RICHARD M
WING ROBERT F PROF
WRIGHT FRANCES W DR
ZUCKERMAN BEN M DR

COMMISSION No. 28

GALAXIES (GALAXIES)

President   : VAN DER KRUIT PIETER C DR

Vice-President(s)   : TAMMANN G ANDREAS PROF DR

Organizing Committee: D'ODORICO SANDRO DR
                      EINASTO JAAN DR
                      KARACHENTSEV I D DR
                      KHACHIKIAN E YE PROF
                      LEQUEUX JAMES DR
                      LI QI-BIN
                      LYNDEN-BELL DONALD PROF
                      QUINTANA HERNAN DR
                      RUBIN VERA C DR
                      TOOMRE ALAR DR
                      WAKAMATSU KEN-ICHI DR

Members:

AARONSON MARC	ABLES HAROLD D DR	ADELMAN SAUL J DR
AGUERO ESTELA L DR	AHMAD FAROOQ DR	ALCAINO GONZALO DR
ALFVEN HANNES PROF	ALLADIN SALEH MOHAMED DR	ALLEN RONALD J DR
ALLOIN DANIELLE DR	AMBARTSUMIAN V A PROF DR	ANDRILLAT YVETTE DR
ARDEBERG ARNE L PROF	ARP HALTON DR	ATHANASSOULA EVANGELIE DR
AZZOPARDI MARC MR	BABADZHANIANC MICHAIL DR	BAHCALL JOHN N PROF
BAILEY MARK EDWARD	BALKOWSKI-MAUGER CH DR	BALLABH G M DR
BARBIERI CESARE PROF	BARBON ROBERTO PROF	BASU BAIDYANATH PROF
BATTANER EDUARDO DR	BAUM WILLIAM A DR	BECK RAINER
BENEDICT GEORGE F DR	BERGERON JACQUELINE A DR	BERGVALL NILS AKE SIGVARD
BERKHUIJSEN ELLY M DR	BERMAN ROBERT HIRAM DR	BERTOLA FRANCESCO PROF
BETTONI DANIELA DR	BHATTACHARYYA TARA DR	BIAN YU-LIN
BIERMANN PETER L DR	BINETTE LUC	BINGGELI BRUNO
BINNEY JAMES J DR	BIRKINSHAW MARK	BLACKMAN CLINTON PAUL DR
BOERNGEN FREIMUT DR PH	BOESHAAR GREGORY ORTH DR	BOKSENBERG ALEC PROF
BONDI HERMANN PROF SIR	BORCHKHADZE TENGIZ M DR	BOSMA ALBERT DR
BOTTINELLI LUCETTE DR	BRACCESI ALESSANDRO PROF	BRECHER KENNETH PROF
BRINKMANN WOLFGANG	BRODIE JEAN P	BROSCHE PETER PROF
BRUCK MARY T DR	BRUSTON PAUL DR	BURBIDGE E MARGARET PROF
BURBIDGE GEOFFREY R PROF	BURNS JACK O'NEAL JR	BURSTEIN DAVID
BUTCHER HARVEY R PROF DR	BYRD GENE G DR	CAMPUSANO LUIS E
CANNON RUSSELL D DR	CAPACCIOLI MASSIMO DR	CARRANZA G J DR
CARRASCO LUIS DR	CARSWELL ROBERT F DR	CARTER DAVID DR
CASINI CATERINA DR	CHAMARAUX PIERRE DR	CHEN JIAN-SHENG
CHEN ZHENCHENG	CHINCARINI GUIDO L DR	CHU YAOQUAN
CHUVAEV K K DR	CLAVEL JEAN	COLIN JACQUES DR
CONTOPOULOS GEORGE PROF	CORWIN HAROLD G JR	COUCH WARRICK DR
COURTES GEORGES PROF	COWSIK RAMANATH	DANKS ANTHONY C DR
DAVIDSEN ARTHUR FALNES DR	DAVIES RODNEY D PROF	DE LA NOE JEROME DR
DE SILVA L.N.K. DR	DE VAUCOULEURS GERARD PR	DEJONGHE HERWIG BERT DR
DEKEL AVISHAI	DENISYUK EDVARD K DR	DI FAZIO ALBERTO
DI SEREGO ALIGHIERI S DR	DI TULLIO GRAZIELLA DR	DIBAY E A DR
DICKENS ROBERT J DR	DICKEY JOHN M	DONNER KARL JOHAN DR

DOTTORI HORACIO A DR	DRESSEL LINDA L	DRESSLER ALAN
DREW JANET	DUBOIS PASCAL DR	DUBOUT RENEE
DULTZIN DEBORAH DR	DUVAL MARIE-FRANCE	EDMUNDS MICHAEL GEOFFREY
EFSTATHIOU GEORGE	EKERS RONALD D DR	EL-BASSUNY ALAWY A A
ELMEGREEN DEBRA MELOY	ELVIUS AINA M PROF	EMERSON DAVID
ESIPOV V F DR	FABBIANO GIUSEPPINA	FABER SANDRA M PROF
FABRICANT DANIEL G	FAIRALL ANTHONY P PROF	FALL S MICHAEL DR
FEAST MICHAEL W DR	FEHRENBACH CHARLES PROF	FEITZINGER JOHANNES PROF
FIELD GEORGE B PROF	FLIN PIOTR	FLORSCH ALPHONSE DR
FOLTZ CRAIG B.	FOMENKO ALEXANDR F DR	FORD HOLLAND C RES PROF
FORD W KENT JR DR	FRASER C W DR	FREEMAN KENNETH C PROF
FRIED JOSEF WILHELM DR	FROGEL JAY ALBERT DR	FTACLAS CHRIST
FUCHS BURKHARD DR	GALLAGHER III JOHN S DR	GALLETTA GIUSEPPE PROF
GAMALELDIN ABDULLA I DR	GASCOIGNE S C B DR	GERHARD ORTWIN
GHIGO FRANCIS D DR	GIOVANARDI CARLO	GIOVANELLI RICCARDO DR
GLASS IAN STEWART DR	GOPALA RAO U V MR	GOSS W MILLER PROF
GOTTESMAN STEPHEN T DR	GOUGUENHEIM LUCIENNE	GRAHAM JOHN A DR
GRANDI STEVEN ALDRIDGE DR	GRASDALEN GARY L DR	GRIFFITHS RICHARD E DR
GUNN JAMES E PROF	GURZADIAN G A PROF DR	HAGEN-THORN V A DR
HAMABE MASARU DR	HARA TETSUYA DR	HARDY EDUARDO
HARMS RICHARD JAMES DR	HARO GUILLERMO DR	HE XIANG-TAO
HEESCHEN DAVID S DR	HEIDMANN JEAN DR	HICKSON PAUL DR
HINTZEN PAUL MICHAEL N DR	HJALMARSON AKE G DR	HODGE PAUL W PROF
HOLMBERG ERIK B PROF	HOYLE FRED SIR	HSIANG YAN-YU
HUANG JIE-HAO	HUANG KE-LIANG	HUCHRA JOHN PETER DR
HUCHTMEIER WALTER K DR	HUMPHREYS ROBERTA M PROF	HUNSTEAD RICHARD W DR
HUNTER JAMES H PROF	ILLINGWORTH GARTH D DR	ISSA ALI DR
JOHNSON HUGH M DR	JOLY MONIQUE	JONES THOMAS WALTER DR
JOSHI MOHAN N PROF	JUGAKU JUN DR	KAFATOS MINAS DR
KALAFI MANOUCHER	KALINKOV MARIN P DR	KALLOGLIAN ARSEN T DR
KANEKO NOBORU DR	KAUFMAN MICHELE DR	KEEL WILLIAM C
KELLERMANN KENNETH I DR	KENNICUTT ROBERT C JR	KIANG TAO PROF
KIM YUL	KING IVAN R PROF	KINMAN THOMAS D DR
KIRSHNER ROBERT PAUL DR	KLEIN ULRICH	KNAPP GILLIAN R DR
KOCHHAR R K DR	KODAIRA KEIICHI PROF	KOGOSHVILI NATELA G
KOJOIAN GABRIEL DR	KOPYLOV I M DR	KORMENDY JOHN DR
KOROVYAKOVSKIJ YURIJ P DR	KRISHNA GOPAL	KRON RICHARD G
KRUMM NATHAN ALLYN	KUEHR HELMUT	KUMAR C KRISHNA DR
KUNDU MUKUL R DR	KUSTAANHEIMO PAUL E PROF	LAFON JEAN-PIERRE J DR
LARSON RICHARD B PROF	LASKER BARRY M DR	LAUBERTS ANDRIS DR
LAUSBERG ANDRE DR	LAYZER DAVID PROF	LEACOCK ROBERT JAY
LELIEVRE GERARD DR	LI JING	LI XIAO-QING
LIN CHIA C PROF	LINDBLAD PER OLOF PROF	LIPOVETSKY V A
LIU RU-LIANG	LIU YONG-ZHEN	LO KWOK-YUNG DR
LORENZ HILMAR	LOVELACE RICHARD V E DR	LOW FRANK J DR
LUGGER PHYLLIS M	LUMINET JEAN-PIERRE	LYNDS BEVERLY T DR
LYNDS ROGER C DR	LYUTY VICTOR M DR	MA ER
MACCHETTO FERDINANDO DR	MACGILLIVRAY HARVEY T DR	MACKAY CRAIG D DR
MADORE BARRY FRANCIS DR	MALAGNINI MARIA LUCIA	MARCELIN MICHEL
MARK JAMES WAI-KEE DR	MARQUES DOS SANTOS P PROF	MARTINET LOUIS PROF
MATERNE JUERGEN DR	MATHEWSON DONALD S PROF	MAURICE ERIC N
MAVRIDES STAMATIA DR	MAY ANDREW	MAYALL NICHOLAS U ASTRON
MCBREEN BRIAN PHILIP DR	MCCLURE ROBERT D PROF	MCCREA WILLIAM SIR
MCGIMSEY BEN Q JR DR	MCVITTIE GEORGE C PROF	MEIER DAVID L
MEIKLE WILLIAM P S	MENON T K PROF	MILLER HUGH R PROF
MILLER JOSEPH S PROF	MILLER RICHARD H DR	MILLS BERNARD Y PROF
MIRABEL IGOR FELIX DR	MOLES MARIANO J DR	MOORWOOD ALAN F M
MORBEY CHRISTOPHER L	MORGAN WILLIAM W PROF	MOSS CHRISTOPHER DR
MOULD JEREMY R	MURDOCH HUGH S DR	MURRAY STEPHEN S DR
MUZZIO JUAN C PROF	MacALPINE GORDON M	NARLIKAR JAYANT V PROF
NEUGEBAUER GERRY DR	NIETO JEAN-LUC	NISHIMURA MASAKI

NOONAN THOMAS W PROF	O'CONNELL ROBERT WEST DR	OEMLER AUGUSTUS JR DR
OKAMURA SADANORI DR	OKE J BEVERLEY PROF	OLEAK H DR
OMER GUY C JR PROF	OORT JAN H PROF	OSMAN ANAS MOHAMED DR
OSTERBROCK DONALD E PROF	OWEN FRAZER NELSON DR	PACHNER JAROSLAV PROF
PACHOLCZYK ANDRZEJ G PROF	PAGE THORNTON L DR	PAN NING-BAO
PAN RONG-SHI	PASTORIZA MIRIANI G DR	PATUREL GEORGES
PAYNE DAVID G	PEIMBERT MANUEL DR	PENG QIU-HE
PETERS WILLIAM L III DR	PETERSON BRADLEY MICHAEL	PETERSON CHARLES JOHN DR
PETROU MARIA DR	PHILLIPS MARK M DR	PISMIS DE RECILLAS PARIS
PIZZICHINI GRAZIELLA	POPOV VASIL NIKOLOV	POVEDA ARCADIO DR
PRABHU TUSHAR P	PRENDERGAST KEVIN H PROF	PRESS WILLIAM H DR
PREVOT-BURNICHON M.L. DR	PRITCHET CHRISTOPHER J DR	PRONIK I I DR
PRONIK V I DR	PROUST DOMINIQUE	REAVES GIBSON PROF
RICHER HARVEY B DR	RICKARD LEE J DR	RINDLER WOLFGANG PROF
ROBERTS MORTON S DR	ROBERTS WILLIAM W JR PROF	ROBERTSON JAMES GORDON DR
ROBINSON I PROF	ROESER HERMANN-JOSEF DR	ROOD HERBERT J
ROSA MICHAEL RICHARD DR	ROSE JAMES ANTHONY	ROSINO LEONIDA PROF
ROTS ARNOLD H DR	RUDNICKI KONRAD PROF	RYDBECK GUSTAF H B DR
SADLER ELAINE MARGARET	SALVADOR-SOLE EDUARDO	SANCISI RENZO DR
SANDERS ROBERT DR	SANTIN PAOLO DR	SARAZIN CRAIG L DR
SARGENT WALLACE L W DR	SASLAW WILLIAM C PROF	SASTRY SHANKARA K
SAVAGE ANN DR	SCHMIDT MAARTEN PROF	SCHUECKING E L DR
SCHULTZ G V DR	SCHULZ HARTMUT DR	SCHUSTER HANS-EMIL
SCHWARZ ULRICH J DR	SCHWEIZER FRANCOIS DR	SCIAMA DENNIS W DR
SCOTT ELIZABETH L PROF	SCOVILLE NICHOLAS Z	SEARLE LEONARD DR
SEGAL IRVING E DR	SEIDEN PHILIP E	SELLWOOD JEREMY ARTHUR
SERRANO ALFONSO DR	SERSIC J L DR	SETTI GIANCARLO PROF
SHAHBAZIAN ROMELIA K DR	SHAKESHAFT JOHN R DR	SHAVER PETER A DR
SHEN BENJAMIN S P PROF	SHERWOOD WILLIAM A DR	SHIELDS GREGORY A DR
SHORE STEVEN N	SHOSTAK G SETH DR	SIMIEN FRANCOIS DR
SIMKIN SUSAN M DR	SITKO MICHAEL L	SMITH BRUCE F DR
SMITH HARDING E JR DR	SMITH HAYWOOD C DR	SMITH MALCOLM G DR
SOBOUTI YOUSEF PROF	SOLTAN ANDRZEJ MARIA DR	SONG GUO-XUAN
SPARKS WILLIAM BRIAN	SPINRAD HYRON PROF	STEIN WAYNE A PROF
STIBBS DOUGLAS W N PROF	STONE REMINGTON P S DR	STROM ROBERT G PROF
SU HONG-JUN	SUBRAHMANYAM P V DR	SULENTIC JACK W DR
SULLIVAN WOODRUFF T III	TAKARADA KATSUO DR	TAKASE BUNSHIRO PROF
TALBOT RAYMOND J JR DR	TEREBIZH VALERY YU DR	TERLEVICH ROBERTO JUAN
TERZIAN YERVANT PROF	THOMPSON LAIRD A DR	THONNARD NORBERT DR
THUAN TRINH XUAN DR	TIFFT WILLIAM G PROF	TOMITA KENJI PROF
TONG YI	TOVMASSIAN H M DR	TREDER H J PROF DR
TREMAINE SCOTT DUNCAN	TRIMBLE VIRGINIA L DR	TRINCHIERI GINEVRA
TULLY RICHARD BRENT DR	TURNER EDWIN L DR	TYSON JOHN A DR
ULRICH MARIE-HELENE D DR	URBANIK MAREK DR	VAN ALBADA TJEERD S DR
VAN DEN BERGH SIDNEY PROF	VAN DER HULST JAN M DR	VAN DER LAAN H PROF DR
VAN WOERDEN HUGO PROF DR	VARMA RAM KUMAR PROF	VERON MARIE-PAULE DR
VERON PHILIPPE DR	VISVANATHAN NATARAJAN DR	VORONTSOV-VEL'YAMINOV B A
WARD MARTIN JOHN	WARNER JOHN W DR	WEEDMAN DANIEL W PROF
WEHINGER PETER A DR	WELCH GARY A DR	WESTERLUND BENGT E PROF
WHITE SIMON DAVID MANION	WHITFORD ALBERT E PROF	WHITMORE BRADLEY C
WIELEBINSKI RICHARD PROF	WIELEN ROLAND PROF DR	WIITA PAUL JOSEPH
WILD PAUL PROF	WILKINSON ALTHEA	WILLIAMS BARBARA A
WILLIAMS ROBERT E DR	WILLIAMS THEODORE B DR	WILLS BEVERLEY J DR
WILLS DEREK DR	WILSON ALBERT G DR	WINKLER KARL-HEINZ A DR
WLERICK GERARD DR	WOOD ROGER DR	WOOSLEY S E PROF
WORRALL DIANA MARY	WRAY JAMES D DR	WYNN-WILLIAMS C G DR
YOUNG JUDITH SHARN	YOUNIS SAAD M	ZASOV ANATOLE V DR
ZEL'DOVICH YA B ACAD	ZHOU YOU-YUAN	ZINN ROBERT J DR
ZOU ZHEN-LONG		

MEMBERSHIP OF COMMISSIONS

COMMISSION No. 29

STELLAR SPECTRA (SPECTRES STELLAIRES)

President : CAYREL DE STROBEL GIUSA

Vice-President(s) : CONTI PETER S DR

Organizing Committee: BESSELL MICHAEL S DR
BOESGAARD ANN M PROF
JUGAKU JUN DR
LAMBERT DAVID L PROF
LEVATO ORLANDO HUGO DR
SMITH MYRON A ASST PROF
SMOLINSKI JAN DR
SPITE MONIQUE DR
WOLFF SIDNEY C DR

Members:
ABHYANKAR KRISHNA D PROF
ALLER LAWRENCE HUGH
APPENZELLER IMMO PROF
BARATTA GIOVANNI BATTISTA
BIDELMAN WILLIAM P PROF
BOUIGUE ROGER PROF
BRANDI ELISANDE ESTELA DR
BUES IRMELA D DR
CAMPBELL BRUCE DR
CASTELLI FIORELLA DR
CAYREL ROGER DR
COWLEY ANNE P DR
DIVAN LUCIENNE DR
DOLIDZE MADONA V DR
DWORETSKY MICHAEL M DR
FEAST MICHAEL W DR
FOY RENAUD DR
FRINGANT ANNE-MARIE DR
GEHREN THOMAS PH D
GILRA DAYA P DR
GOEBEL JOHN H DR
GRAY DAVID F PROF
GRIFFIN ROGER F DR
GUTHRIE BRUCE N G DR
HARMANEC PETR DR
HARO GUILLERMO DR
HENIZE KARL G ASTRONAUT
HIRAI MASANORI DR
HUANG CHANG-CHUN
HYLAND A R HARRY DR
JOHNSON HOLLIS R PROF
KHOKHLOVA V L DR
KITCHIN CHRISTOPHER R DR
KOMAROV N S DR
KOVACHEV B J DR
ABT HELMUT A DR
ANDRILLAT HENRI L PROF
ASLANOV I A DR
BARRY DON C DR
BOGGESS ALBERT DR
BOULON JACQUES J DR
BROWN DOUGLAS NASON
BURKHART CLAUDE DR
CARNEY BRUCE WILLIAM
CATALANO SANTO DR
CLIMENHAGA JOHN L PROF
COWLEY CHARLES R PROF
DOAZAN VERA DR
DRAVINS DAINIS PROF
EDMONDS FRANK N JR DR
FERNANDEZ-FIGUEROA M J DR
FREIRE FERRERO RUBENS G
FUJITA YOSHIO PROF
GERBALDI MICHELE DR
GLAGOLEVSKIJ JU V DR
GOLDBERG LEO PROF
GREENSTEIN J L PROF
GROTH HANS G PROF DR
HACK MARGHERITA PROF
HARMER CHARLES F W MR
HEARNSHAW JOHN B DR
HERBIG GEORGE H DR
HIRATA RYUKO
HUBENY IVAN
JASCHEK CARLOS O R PROF
KEENAN PHILIP C PROF EMER
KING R B DR
KODAIRA KEIICHI PROF
KOPYLOV I M DR
KRAFT ROBERT P PROF
AIKMAN G CHRIS L
ANDRILLAT YVETTE DR
BAADE DIETRICH DR
BERGER JACQUES G DR
BONSACK WALTER K PROF
BOYARCHUK A A DR
BRUHWEILER FRED C JR
BUSCOMBE WILLIAM PROF
CASSATELLA ANGELO DR
CATCHPOLE ROBIN MICHAEL
CODE ARTHUR D
DE GROOT MART DR
DOBRONRAVIN PETER DR
DUNCAN DOUGLAS KEVIN DR
FARAGGIANA ROSANNA PROF
FLOQUET MICHELE DR
FRIEDJUNG MICHAEL DR
GARRISON ROBERT F PROF
GERSHBERG R E DR
GLUSHNEVA I N DR
GRATTON LIVIO PROF
GRIFFIN RITA E M DR
GUSTAFSSON BENGT DR
HAGEN WENDY ANNE
HARMER DIANNE L MRS
HEINTZE J R W DR
HERMAN RENEE DR
HOUZIAUX L PROF
HUBERT-DELPLACE A.-M. DR
JASCHEK MERCEDES DR
KHARITONOV ANDREJ V DR
KIPPER TONU DR
KOGURE TOMOKAZU DR
KOUBSKY PAVEL
KREMPEC-KRYGIER JANINA DR

KUMAJGORODSKAYA RAISA DR	LABS DIETRICH PROF	LAMLA ERICH E DR
LANGER GEORGE EDWARD DR	LARSSON-LEANDER G PROF	LECKRONE DAVID S DR
LESTER JOHN B DR	LOCANTHI DOROTHY DAVIS DR	LUUD LAURI DR
LYNAS-GRAY ANTHONY E	MAITZEN HANS M DR	MALARODA STELLA M DR
MATHYS GAUTIER DR	MCNAMARA DELBERT H DR	MEGESSIER CLAUDE DR
MILLIGAN J E	MOFFAT ANTHONY F J DR	MOOS HENRY WARREN DR
MORGULEFF NINA ING	MOROSSI CARLO	MORRISON NANCY DUNLAP DR
MORTON DONALD C DR	MUSTEL E R PROF DR	NICHOLLS RALPH W PROF
NIKITIN A A DR	NISHIMURA SHIRO DR	OETKEN L DR
OKE J BEVERLEY PROF	ORLOV MIKHAIL DR	OSAWA KIYOTERU DR
PAGEL BERNARD E J PROF	PARSONS SIDNEY B DR	PARTHASARATHY M DR
PASINETTI LAURA E PROF	PATERSON-BEECKMANS F	PEAT D W DR
PEDOUSSAUT ANDRE	PEERY BENJAMIN F PROF	PERRIN MARIE-NOEL DR
PETERS GERALDINE JOAN DR	PETERSON RUTH CAROL DR	PILACHOWSKI CATHERINE DR
PLAVEC MIREK J PROF	POECKERT ROLAND H DR	PRESTON GEORGE W DR
QUERCI FRANCOIS R DR	QUERCI MONIQUE DR	RAMELLA MASSIMO
RAO N KAMESWARA	REGO FERNANDEZ M DR	REIMERS DIETER PROF
RINGUELET ADELA E DR	RODGERS ALEX W DR	ROSSI LUCIO
SADAKANE KOZO DR	SAHADE JORGE PROF	SCHILD RUDOLPH E DR
SCHOLZ GERHARD DR	SEGGEWISS WILHELM PROF	SHCHEGOLEV DIMITRIJ E DR
SHORE STEVEN N	SINNERSTAD ULF E PROF	SLETTEBAK ARNE PROF
SMAK JOSEPH I PROF	SNOW THEODORE P PROF	SPITE FRANCOIS M DR
STALIO ROBERTO DR	STAWIKOWSKI ANTONI DR	STECHER THEODORE P
STENCEL ROBERT EDWARD	SUNTZEFF NICHOLAS B	SVOLOPOULOS SOTIRIOS PROF
SWENSSON JOHN W DR	SWINGS JEAN-PIERRE DR	TAFFARA SALVATORE PROF
TAKADA-HIDAI MASAHIDE DR	THOMPSON G I DR	TSUJI TAKASHI
TUOMINEN ILKKA V DR	UNDERHILL ANNE B DR	UTSUMI KAZUHIKO DR
VAN DER HUCHT KAREL A DR	VAN'T VEER-MENNERET CL DR	VILHU OSMI DR
VIOTTI ROBERTO DR	VOGT STEVEN SCOTT	VOIGT HANS H PROF
VREUX JEAN MARIE DR	WALLERSTEIN GEORGE PROF	WATERWORTH MICHAEL DR
WEGNER GARY ALAN	WEHINGER PETER A DR	WEHLAU AMELIA DR
WEHLAU WILLIAM H PROF	WEISS WERNER W DR	WELLMANN PETER PROF DR
WENIGER SCHAME DR	WILLIAMS PEREDUR M DR	WILSON ROBERT PROF
WING ROBERT F PROF	WOLF BERNHARD PH D	WOOD III H J DR
WRIGHT KENNETH O DR	WYCKOFF SUSAN DR	WYLLER ARNE A PROF
YAMASHITA YASUMASA PROF		

COMMISSION No. 30

RADIAL VELOCITIES (VITESSES RADIALES)

President   : ANDERSEN JOHANNES

Vice-President(s) : LATHAM DAVID W DR

Organizing Committee: FLORSCH ALPHONSE DR
                      MAURICE ERIC N
                      MAYOR MICHEL DR
                      MCCLURE ROBERT D PROF
                      PHILIP A G DAVIS

Members:
ABT HELMUT A DR	AZZOPARDI MARC MR	BALONA LUIS ANTERO DR
BARBIER-BROSSAT M DR	BATTEN ALAN H DR	BEARDSLEY WALLACE R DR
BEAVERS WILLET I DR	BERTIAU FLOR C PROF	BOLTON C THOMAS PROF
BOUIGUE ROGER PROF	BOULON JACQUES J DR	BREGER MICHEL DR
BURKI GILBERT DR	BURNAGE ROBERT	CAMPBELL BRUCE DR
CARNEY BRUCE WILLIAM	CARQUILLAT JEAN-MICHEL	COCHRAN WILLIAM DAVID DR
CRAMPTON DAVID DR	DE JONGE J K DR	DE VAUCOULEURS GERARD PR
DUFLOT MARCELLE DR	EDMONDSON FRANK K PROF	EELSALU HEINO DR
FEHRENBACH CHARLES PROF	FLETCHER J MURRAY	GEORGELIN YVON P DR
GIESEKING FRANK DR	GRIFFIN ROGER F DR	HEINTZE J R W DR
HILL GRAHAM DR	HRIVNAK BRUCE J	HUANG CHANG-CHUN
HUBE DOUGLAS P DR	IMBERT MAURICE DR	KADOURI TALIB HADI
KARACHENTSEV I D DR	KRAFT ROBERT P PROF	MARTIN NICOLE DR
MORBEY CHRISTOPHER L	NORDSTROM BIRGITTA DR	OETKEN L DR
PEDOUSSAUT ANDRE	PERRY CHARLES L DR	POPOV VICTOR S DR
PRESTON GEORGE W DR	PREVOT LOUIS DR	REBEIROT EDITH DR
ROMANOV YURI S DR	RUBIN VERA C DR	SANWAL N B DR
SCARFE COLIN D DR	SMITH MYRON A ASST PROF	STOCK JURGEN D
VAN DESSEL EDWIN LUDO DR	WILLSTROP RODERICK V DR	YOSS KENNETH M DR

MEMBERSHIP OF COMMISSIONS

COMMISSION No. 31

TIME (L'HEURE)

President   : MCCARTHY DENNIS D DR

Vice-President(s)   : PAQUET PAUL EG DR

Organizing Committee: AOKI SHINKO PROF
                      BENAVENTE JOSE
                      BLINOV N S DR
                      GUINOT BERNARD R PROF
                      HEMMLEB GERHARD DR
                      KOVALEVSKY JEAN DR
                      MIAO YONG-RUI
                      MUELLER IVAN I PROF
                      PILKINGTON JOHN D H DR
                      PROVERBIO EDOARDO PROF
                      YE SHU-HUA

Members:
ABELE M K DR   AFANASJEVA PRASKOVYA M DR   ALLAN DAVID W MR
BELOTSERKOVSKIJ DAVID J   BENDER PETER L DR   BILLAUD GERARD J
BONANOMI J DR   CAPRIOLI GIUSEPPE PROF   CARTER WILLIAM EUGENE
CHA GI UNG   CHAMBERLAIN JOSEPH M DR   COSTAIN CECIL C DR
DICKEY JEAN O'BRIEN   DOMINSKI IRENEUSZ DR   DORENWENDT KLAUS DR
DRAMBA C PROF   ENSLIN HEINZ DR   FALLON FREDERICK W DR
FLIEGEL HENRY F   FUJIMOTO MASA-KATSU DR   GAIGNEBET JEAN DR
GOKMEN TARIK ASSOC PROF   GONG HUI-REN   GRUDLER PIERRE
HALL R GLENN DR   HAN TIANQI   HELLWIG HELMUT WILHELM DR
HERS JAN MR   IIJIMA SHIGETAKA PROF   JIN WEN-JING
KAKUTA CHUICHI DR   KESSLER KARL G DR   KLEPCZYNSKI WILLIAM J DR
KOBAYASHI YUKISAYU   LEDERLE TRUDPERT DR   LIANG ZHONG-HUAN
LIESKE JAY H DR   LOZINSKIJ A M DR   LUO DINGCHANG
LUO SHI-FANG   MARKOWITZ WILLIAM DR   MATHUR B S DR
MATSAKIS DEMETRIOS N   MELBOURNE WILLIAM G DR   MELCHIOR PAUL J PROF DIR
MORGAN PETER DR   NAUMOV VITALIJ A DR   NIIMI YUKIO
NOEL FERNANDO   ORTE ALBERTO   PARCELIER PIERRE DR
POPELAR JOSEF DR   PUSHKIN SERGEY B DR   RANDIC LEO PROF DR
ROBERTSON DOUGLAS S   SCHULER WALTER DR   SINZI AKIRA M DR
SMITH HUMPHRY M   SMYLIE DOUGLAS E DR   SONG JIN-AN
STANILA GEORGE DR   STOYKO ANNA   TAKAGI SHIGETSUGU DR
TSAO MO PROF   TSUCHIYA ATSUSHI DR PROF   VICENTE RAIMUNDO O PROF
WACKERNAGEL H BEAT DR   WEBROVA LUDMILA DR   WIETH-KNUDSEN NIELS P DR
WILKINS GEORGE A DR   WINKLER GERNOT M R DR   WU SHOU-XIAN
YANG KE-JUN   YUMI S PROF DR   ZHAI ZAOCHENG
ZHANG JINTONG   ZHAO GANG   ZHENG YING
ZHUANG QIXIANG

MEMBERSHIP OF COMMISSIONS

COMMISSION No. 33

STRUCTURE AND DYNAMICS OF THE GALACTIC SYSTEM

(STRUCTURE ET DYNAMIQUE DU SYSTEME GALACTIQUE)

President   : BURTON W BUTLER DR

Vice-President(s) : MAYOR MICHEL DR

Organizing Committee: BAHCALL JOHN N PROF
BALAZS LAJOS G DR
BINNEY JAMES J DR
BLITZ LEO
EINASTO JAAN DR
LYNGA GOSTA DR
TOSA MAKOTO DR
WIELEN ROLAND PROF DR

Members:

AARSETH SVERRE J DR	AGEKJAN TATEOS A PROF	AIZU KO PROF
ALTENHOFF WILHELM J DR	AMBARTSUMIAN V A PROF DR	AMBASTHA A K DR
ANDRLE PAVEL DR	ANTONOV VADIM A DR	AOKI SHINKO PROF
ARDEBERG ARNE L PROF	ASTERIADIS GEORGIOS DR	ATHANASSOULA EVANGELIE DR
BALDWIN JOHN E DR	BARBANIS BASIL PROF	BARBERIS BRUNO
BASH FRANK N PROF	BASU BAIDYANATH PROF	BAUD BOUDEWIJN DR
BECKER WILHELM PROF	BERKHUIJSEN ELLY M DR	BLAAUW ADRIAAN PROF DR
BLANCO VICTOR M DR	BOULON JACQUES J DR	BURKE BERNARD F DR
CALDWELL JOHN A R	CANE HILARY VIVIEN	CARRASCO LUIS DR
CASWELL JAMES L DR	CHEN ZHEN	CHURCHWELL EDWARD B DR
CIURLA TADEUSZ	CLUBE S V M DR	COHEN RICHARD S
COLIN JACQUES DR	COMINS NEIL FRANCIS	CONTOPOULOS GEORGE PROF
COSTA EDGARDO DR	COURTES GEORGES PROF	CRAMPTON DAVID DR
CRAWFORD DAVID L DR	CREZE MICHEL DR	CUDWORTH KYLE MCCABE DR
CUPERMAN SAMI PROF	DAVIES RODNEY D PROF	DE JONG TEIJE DR
DEKEL AVISHAI	DELHAYE JEAN PROF	DENOYELLE JOZEF KIC
DICKEL HELENE R DR	DICKEL JOHN R	DIETER NANNIELOU H DR
DOWNES DENNIS DR	DRILLING JOHN S	DUCATI JORGE RICARDO DR
DZIGVASHVILI R M DR	EDMONDSON FRANK K PROF	EFREMOV YURY N DR
EGRET DANIEL DR	ELMEGREEN DEBRA MELOY	ELSAESSER HANS PROF
ELVIUS TORD PROF EMERITUS	EVANGELIDIS E DR	FABER SANDRA M PROF
FALL S MICHAEL DR	FEAST MICHAEL W DR	FEHRENBACH CHARLES PROF
FEITZINGER JOHANNES PROF	FENKART ROLF P PROF DR	FITZGERALD M PIM PROF
FREEMAN KENNETH C PROF	FRICKE WALTER PROF DR	FUCHS BURKHARD DR
FUJIMOTO MASA-KATSU DR	GALLETTO DIONIGI	GENKIN IGOR L PROF DR
GEORGELIN YVON P DR	GEORGELIN YVONNE M DR	GILMORE GERARD FRANCIS
GLIESE WILHELM DR	GOLDREICH P DR	GORDON MARK A DR
GRAYZECK EDWIN J DR	GYLDENKERNE KJELD DR	HABE ASAO
HABING H J DR	HAMAJIMA KIYOTOSHI DR	HAUG ULRICH PROF
HAWKINS MICHAEL R S	HAYLI AVRAM PROF	HEILES CARL PROF
HENON MICHEL C DR	HERBST WILLIAM DR	HOBBS ROBERT W DR
HORI GENICHIRO PROF	HUGHES VICTOR A PROF	HULSBOSCH A N M DR
HUMPHREYS ROBERTA M PROF	HUNTER CHRISTOPHER PROF	IKEUCHI SATORU DR
INAGAKI SHOGO DR	INNANEN KIMMO A PROF	IRWIN JOHN B PROF
ISOBE SYUZO DR	ISRAEL FRANK P DR	IWANISZEWSKA CECYLIA DR

IWANOWSKA WILHELMINA PROF   IYE MASANORI DR   JACKSON PETER DOUGLAS DR
JAHREISS HARTMUT DR   JASCHEK CARLOS O R PROF   JIANG DONG-RONG
JOHNSON HUGH M DR   JONES DEREK H P DR   KABURAKI MASAKI PROF
KALANDADZE N B DR   KALNAJS AGRIS J DR   KATO SHOJI PROF
KERR FRANK J DR   KHARADZE E K PROF   KHOLOPOV P N DR
KING IVAN R PROF   KINMAN THOMAS D DR   KLARE GERHARD DR
KNAPP GILLIAN R DR   KOLESNIK IGOR G DR   KOLESNIK L N DR
KORMENDY JOHN DR   KULSRUD RUSSELL M DR   KUTUZOV S A DR
KUZMIN GRIGORI G PROF   LAFON JEAN-PIERRE J DR   LARSON RICHARD B PROF
LECAR MYRON DR   LI JING   LIN CHIA C PROF
LINDBLAD PER OLOF PROF   LOCKMAN FELIX J   LODEN KERSTIN R DR
LODEN LARS OLOF PROF   LUNEL MADELEINE DR   LUYTEN WILLEM J PROF
LYNDEN-BELL DONALD PROF   MACCONNELL DARRELL J DR   MACRAE DONALD A PROF
MANCHESTER RICHARD N DR   MARK JAMES WAI-KEE DR   MAROCHNIK L S PROF DR
MARTINET LOUIS PROF   MATHEWSON DONALD S PROF   MAVRIDIS L N PROF
MCCARTHY MARTIN F DR   MCGREGOR PETER JOHN DR   MENNESSIER MARIE-ODILE DR
MEZGER PETER G PROF   MIKKOLA SEPPO DR   MILLER RICHARD H DR
MIRABEL IGOR FELIX DR   MIRZOYAN L V DR PROF   MIYAMOTO MASANORI DR
MOFFAT ANTHONY F J DR   MONET DAVID G   MONNET GUY J DR
MORRIS MARK ROOT DR   MUENCH GUIDO PROF   MURRAY C ANDREW
MUZZIO JUAN C PROF   NAHON FERNAND PROF   NECKEL TH DR
NELSON ALISTAIR H DR   NIIMI HIDEYUKI DR   NINKOVIC SLOBODAN
NISHIDA MINORU PROF   NISHIDA MITSUGU   OGORODNIKOV KYRILL P PROF
OJA TARMO PROF   OKUDA HARUYUKI DR PROF   OLANO CARLOS ALBERTO DR
OLLONGREN A PROF DR   OORT JAN H PROF   OSTRIKER JEREMIAH P PROF
OVENDEN MICHAEL W PROF   PALMER PATRICK E PROF   PALOUS JAN DR
PAPAYANNOPOULOS TH DR   PAULS THOMAS ALBERT DR   PAVLOVSKAYA E D DR
PEIMBERT MANUEL DR   PEREK LUBOS DR   PERRY CHARLES L DR
PESCH PETER DR   PHILIP A G DAVIS   PILOWSKI K PROF DR
PISMIS DE RECILLAS PARIS   PRICE R MARCUS DR   PRIESTER WOLFGANG PROF
QIAN ZHONG-YU   RAMBERG JOERAN M PROF   REID NEILL
RIEGEL KURT W DR   ROBERTS MORTON S DR   ROBERTS WILLIAM W JR PROF
ROBINSON BRIAN J DR   ROHLFS K PROF DR   RONG JIAN-XIANG
RUBIN VERA C DR   RUIZ MARIA TERESA DR   RYBICKI GEORGE B DR
SAAR ENN DR   SANCHEZ-SAAVEDRA M LUISA   SANDQVIST AAGE DR
SANDULEAK NICHOLAS DR   SANG GAK LEE   SCHMIDT HANS PROF
SCHMIDT K H DR   SCHMIDT MAARTEN PROF   SCHMIDT-KALER TH PROF
SCHWERDTFEGER HANS-M. DR   SEGGEWISS WILHELM PROF   SELLWOOD JEREMY ARTHUR
SHANE WILLIAM W DR   SHAROV A S DR   SHER DAVID DR
SHIMIZU TSUTOMU PROF EMER   SHU FRANK H PROF   SHUTER WILLIAM L H DR
SIMONSON S CHRISTIAN DR   SLETTEBAK ARNE PROF   SOLOMON PHILIP M DR
SONG GUO-XUAN   SPARKE LINDA   SPIEGEL E DR
STECKER FLOYD W DR   STEFANOVITCH-GOMEZ A E DR STEINLIN ULI PROF
STEPHENSON C BRUCE PROF   STIBBS DOUGLAS W N PROF   STROBEL ANDRZEJ DR
STURCH CONRAD R DR   SVOLOPOULOS SOTIRIOS PROF SZEBEHELY VICTOR G PROF
TAMMANN G ANDREAS PROF DR   TERZIDES CHARALAMBOS DR   THE PIK-SIN PROF
THIELHEIM KLAUS O DR   TOBIN WILLIAM   TONG YI
TOOMRE ALAR DR   TOOMRE JURI   TREFZGER CHARLES F DR
TSIOUMIS ALEXANDROS DR   TURON-LACARRIEU C DR   UPGREN ARTHUR R DR
VAN DER KRUIT PIETER C DR   VAN HOOF A PROF EM   VAN WOERDEN HUGO PROF DR
VANDERVOORT PETER O DR   VARSAVSKY C M DR   VEGA E. IRENE DR
VELGHE ALBERT G PROF DR   VENUGOPAL V R DR   VERSCHUUR GERRIT L PROF
VETESNIK MIROSLAV DR   VOROSHILOV V I DR   WAYMAN PATRICK A PROF
WEAVER HAROLD F PROF   WEISTROP DONNA DR   WESTERHOUT GART DR
WESTERLUND BENGT E PROF   WHITE RAYMOND E DR   WHITEOAK J B DR
WHITTET DOUGLAS C B DR   WIELEBINSKI RICHARD PROF   WILSON THOMAS L DR
WOLTJER LODEWIJK PROF   WOODWARD PAUL R DR   WRAMDEMARK STIG S O DR
YOSHII YUZURU DR   YOUNIS SAAD M   YUAN CHI PROF
ZHANG BIN   ZHAO JUN-LIANG

COMMISSION No. 34

INTERSTELLAR MATTER (MATIERE INTERSTELLAIRE)

President   : LEQUEUX JAMES DR

Vice-President(s) : MATHIS JOHN S PROF

Organizing Committee: D'ODORICO SANDRO DR
DE BOER KLAAS SJOERDS DR
ELMEGREEN BRUCE GORDON DR
FLOWER DAVID R DR
HABING H J DR
PEIMBERT MANUEL DR
SHAVER PETER A DR
SHUSTOV BORIS M DR
WANNIER PETER GREGORY DR
YORK DONALD G DR

Members:
AANNESTAD PER ARNE DR   ACKER AGNES PROF DR   AITKEN DAVID K DR
AKABANE KENJI A PROF   ALDROVANDI SUELI M V DR   ALLER LAWRENCE HUGH
ALTENHOFF WILHELM J DR   ANDREW BRYAN H DR   ANDRIESSE CORNELIS D DR
ANDRILLAT HENRI L PROF   ANDRILLAT YVETTE DR   ARKHIPOVA V P DR
ARNY THOMAS T DR   AVERY LORNE W DR   AXFORD W IAN PROF
BAARS JACOB W M DR   BAART EDWARD E PROF   BALDWIN JOHN E DR
BALUTEAU JEAN-PAUL   BANIA THOMAS MICHAEL   BARLOW MICHAEL J DR
BARNES AARON DR   BARRETT ALAN H PROF   BASH FRANK N PROF
BAUDRY ALAIN DR   BECKLIN ERIC E DR   BECKMAN JOHN E PROF
BECKWITH STEVEN V W   BEL NICOLE J DR   BERGERON JACQUELINE A DR
BERKHUIJSEN ELLY M DR   BERNAT ANDREW PLOUS DR   BHATT H C DR
BIANCHI LUCIANA   BIEGING JOHN HAROLD DR   BIGNELL R CARL DR
BINETTE LUC   BIRKLE KURT PH D   BLACK JOHN HARRY DR
BLADES JOHN CHRIS DR   BLAIR GUY NORMAN DR   BLESS ROBERT C PROF
BOCHKAREV NIKOLAY G DR   BODE MICHAEL F   BOESHAAR GREGORY ORTH DR
BOGGESS ALBERT DR   BOHLIN RALPH C DR   BOLAND WILFRIED
BORGMAN JAN DR PROF   BRAND PETER W J L DR   BRAUNSFURTH EDWARD PH D
BRINKMANN WOLFGANG   BROMAGE GORDON E DR   BROWN RONALD D PROF
BRUCK MARY T DR   BRUHWEILER FRED C JR   BURGESS ALAN DR
BURKE BERNARD F DR   BURTON W BUTLER DR   BYSTROVA NATALIJA V DR
CANTO JORGE DR   CAPPA DE NICOLAU CRISTINA   CAPRIOTTI EUGENE R DR
CAPUZZO DOLCETTA ROBERTO   CARRUTHERS GEORGE R DR   CASWELL JAMES L DR
CERRUTI-SOLA MONICA   CERSOSIMO JUAN CARLOS DR   CESARSKY CATHERINE J DR
CESARSKY DIEGO A DR   CHEVALIER ROGER A DR   CHINI ROLF
CHOPINET MARGUERITE DR   CHURCHWELL EDWARD B DR   CLARK FRANK OLIVER DR
CLEGG ROBIN E S DR   CODE ARTHUR D   COHEN MARSHALL H PROF
COLLIN-SOUFFRIN SUZY DR   COSTERO RAFAEL   COURTES GEORGES PROF
COWIE LENNOX LAUCHLAN DR   COX DONALD P PROF   COYNE GEORGE V DR
CROVISIER JACQUES   CRUVELLIER PAUL E DR   CUDABACK DAVID D DR
CUGNON PIERRE DR   CZYZAK STANLEY J DR   DAHN CONARD CURTIS DR
DALGARNO ALEXANDER PROF   DANKS ANTHONY C DR   DAVIES RODNEY D PROF
DE JONG TEIJE DR   DE LA NOE JEROME DR   DEGUCHI SHUJI DR
DEWDNEY PETER E F DR   DIBAY E A DR   DICKEL HELENE R DR

DICKEY JOHN M	DINERSTEIN HARRIET L	DISNEY MICHAEL J PROF
DOKUCHAEVA OLGA D DR	DONN BERTRAM D	DOPITA MICHAEL ANDREW DR
DORSCHNER JOHANN DR	DOTTORI HORACIO A DR	DOWNES DENNIS DR
DRAINE BRUCE T	DRAPATZ SIEGFRIED W DR	DREHER JOHN W
DUBOUT RENEE	DUFOUR REGINALD JAMES	DUPREE ANDREA K DR
DWEK ELI	DYSON JOHN E DR	EL SHALABY MOHAMED
ELITZUR MOSHE	ELLIOTT KENNETH H DR	ELMEGREEN DEBRA MELOY
ELVIUS AINA M PROF	EMERSON JAMES P	ENCRENAZ PIERRE J DR
ESIPOV V F DR	EVANS NEAL J II ASS PROF	FALK SYDNEY W JR DR
FALLE SAMUEL A DR	FAN YING	FAULKNER DONALD J DR
FEDERMAN STEVEN ROBERT	FEIBELMAN WALTER A DR	FEITZINGER JOHANNES PROF
FELLI MARCELLO DR	FELTEN JAMES E DR	FERRINI FEDERICO
FIELD DAVID	FIELD GEORGE B PROF	FIERRO JULIETA
FLANNERY BRIAN PAUL DR	FORD HOLLAND C RES PROF	FORSTER JAMES RICHARD DR
FRIEDEMANN CHRISTIAN DR	FRISCH PRISCILLA	FURNISS IAN
GARDNER FRANCIS F DR	GAUSTAD JOHN E PROF	GAY JEAN DR
GEHRELS TOM PROF	GEORGELIN YVON P DR	GERARD ERIC DR
GEROLA HUMBERTO DR	GEZARI DANIEL YSA DR	GILRA DAYA P DR
GIOVANELLI RICCARDO DR	GODFREY PETER DOUGLAS DR	GOEBEL JOHN H DR
GOLDREICH P DR	GOLDSMITH DONALD W. DR.	GOLDSMITH PAUL F DR
GOLDSTEIN SAMUEL J PROF	GOLDSWORTHY FREDERICK A	GORDON COURTNEY P PROF
GORDON MARK A DR	GOSACHINSKIJ I V DR	GOSS W MILLER PROF
GRAHAM DAVID A	GRASDALEN GARY L DR	GREENBERG J MAYO DR
GREWING MICHAEL PROF	GUELIN MICHEL DR	GUESTEN ROLF
GULL THEODORE R DR	GURZADIAN G A PROF DR	GUSEINOV O H PROF
HACKWELL JOHN A DR	HALL JOHN S DR	HARDEBECK ELLEN G DR
HARRINGTON J PATRICK DR	HARRIS ALAN WILLIAM	HARRIS STELLA
HARTEN RONALD H DR	HARTQUIST THOMAS WILBUR	HARVEY PAUL MICHAEL DR
HAYNES RAYMOND F PROF	HEFELE HERBERT PH D	HEILES CARL PROF
HELFER H LAWRENCE PROF	HENIZE KARL G ASTRONAUT	HENKEL CHRISTIAN
HERZBERG GERHARD DR	HIDAJAT BAMBANG PROF DR	HIGGS LLOYD A DR
HILDEBRAND ROGER H	HILTNER W ALBERT PROF	HIPPELEIN HANS H DR
HJALMARSON AKE G DR	HJELLMING ROBERT M DR	HOBBS LEWIS M DR
HOEGLUND BERTIL PROF	HOLLENBACH DAVID JOHN DR	HOLLIS JAN MICHAEL DR
HONG SEUNG SOO DR	HOUZIAUX L PROF	HUA CHON TRUNG DR
HUGHES VICTOR A PROF	HULSBOSCH A N M DR	HUMMER DAVID G DR
HUTCHINGS JOHN B DR	IRVINE WILLIAM M PROF	ISOBE SYUZO DR
ISSA ALI DR	ITOH HIROSHI DR	JABIR NIAMA LAFTA
JACOBY GEORGE H	JENKINS L F MS	JENNINGS R E PROF
JOHNSON FRED M PROF DR	JOHNSON HUGH M DR	JOHNSTON KENNETH J
JONES FRANK CULVER DR	JURA MICHAEL DR	KAFATOS MINAS DR
KAFTAN MAY A DR	KAHN FRANZ D PROF	KAIFU NORIO DR
KALER JAMES B PROF	KAMIJO FUMIO PROF DR	KAZES ILYA DR
KEGEL WILHELM H PROF	KERR FRANK J DR	KHARADZE E K PROF
KHROMOV G S DR	KIMURA HIROSHI DR	KIRKPATRICK RONALD C DR
KIRSHNER ROBERT PAUL DR	KNACKE ROGER F DR	KNAPP GILLIAN R DR
KNUDE JENS KIRKESKOV DR	KOEPPEN JOACHIM DR	KOHOUTEK LUBOS DR
KOLESNIK IGOR G DR	KOMESAROFF MAX M	KONDO YOJI DR
KOORNNEEF JAN DR	KOSTYAKOVA ELENA B DR	KRAUTTER JOACHIM DR
KREYSA ERNST	KRISHNA SWAMY K S DR	KUIPER THOMAS B H DR
KUMAR C KRISHNA DR	KUNDU MUKUL R DR	KUTNER MARC LESLIE DR
KWITTER KAREN BETH DR	KWOK SUN DR	KYLAFIS NIKOLAOS D DR
LADA CHARLES JOSEPH DR	LANGER WILLIAM DAVID DR	LASKER BARRY M DR
LAURENT CLAUDINE DR	LE SERGEANT D'HENDECOURT	LE SQUEREN ANNE-MARIE DR
LEE TERENCE J DR	LEUNG CHUN MING DR	LILLER WILLIAM DR
LIN CHIA C PROF	LINKE RICHARD ALAN DR	LISZT HARVEY STEVEN
LO KWOK-YUNG DR	LOCKMAN FELIX J	LOREN ROBERT BRUCE DR
LORTET MARIE CLAIRE	LOUISE RAYMOND PROF	LOVAS FRANCIS JOHN DR
LOW FRANK J DR	LOZINSKAYA TAT'YANA A DR	LUCAS ROBERT DR

LUTZ BARRY L DR	LYNDS BEVERLY T DR	MACIEL WALTER J DR
MACLEOD JOHN M DR	MAIHARA TOSHINORI DR	MALLIK D C V DR
MANCHESTER RICHARD N DR	MANFROID JEAN DR	MARTEL MARIE-THERESE DR
MARTIN ROBERT N DR	MASSON COLIN R	MATHEWS WILLIAM G PROF
MATHEWSON DONALD S PROF	MATTILA KALEVI DR	MCCALL MARSHALL LESTER DR
MCCRAY RICHARD DR	MCCREA J DERMOTT	MCGEE RICHARD X DR
MCKEE CHRISTOPHER F PROF	MCKEITH CONAL D DR	MCNALLY DEREK DR
MEABURN J DR	MEBOLD ULRICH DR PROF	MEIER ROBERT R
MELNICK GARY J	MENDEZ ROBERTO H DR	MENON T K PROF
MESZAROS PETER DR	MEZGER PETER G PROF	MICHALITSIANOS ANDREW
MILLAR THOMAS J DR	MILLER JOSEPH S PROF	MILNE DOUGLAS K DR
MININ I N PROF	MINN YOUNG KEY DR	MIYAMA SYOKEN
MIZUNO SHUN	MORGAN DAVID H DR	MORIMOTO MASAKI DR
MORRIS MARK ROOT DR	MORTON DONALD C DR	MOUSCHOVIAS TELEMACHOS CH
MUENCH GUIDO PROF	MUFSON STUART LEE DR	MYERS PHILIP C
NAKADA YOSHIKAZU DR	NANDY KASHINATH DR	NEUGEBAUER GERRY DR
NGUYEN-QUANG RIEU DR	NITTMAN JOHANN	NORDH H LENNART DR
NUSSBAUMER HARRY PROF	NUTH JOSEPH A III	O'DELL CHARLES R DR
O'DELL STEPHEN L	OHTANI HIROSHI DR	OKUDA HARUYUKI DR PROF
OLOFSSON HANS	ONAKA TAKASHI	OORT JAN H PROF
OSAKI TORU DR	OSBORNE JOHN L DR	OSTERBROCK DONALD E PROF
OZERNOY LEONID M PROF	PALLA FRANCESCO	PALMER PATRICK E PROF
PANAGIA NINO DR	PANKONIN VERNON LEE DR	PARKER EUGENE N
PAULS THOMAS ALBERT DR	PECKER JEAN-CLAUDE PROF	PENSTON MICHAEL V DR
PENZIAS ARNO A DR	PERINOTTO MARIO PROF	PERSI PAOLO
PETERS WILLIAM L III DR	PETROSIAN VAHE PROF	PHILLIPS ANTHONY PAUL
PHILLIPS THOMAS GOULD DR	PISMIS DE RECILLAS PARIS	POEPPEL WOLFGANG G L DR
POTTASCH STUART R PROF	PRASAD SHEO S	PREITE-MARTINEZ ANDREA DR
PRICE R MARCUS DR	PRONIK I I DR	PSKOVSKIJ JU P DR
PUGET JEAN-LOUP DR	QIN ZHI-HAI	RADHAKRISHNAN V PROF
RAIMOND ERNST DR	RAYMOND JOHN CHARLES	RENGARAJAN T N DR
REYNOLDS RONALD J DR	RICKARD LEE J DR	RIGHINI-COHEN GIOVANNA DR
ROBBINS R ROBERT PROF	ROBERTS WILLIAM W JR PROF	ROBINSON BRIAN J DR
RODRIGUEZ LUIS F	ROESER HANS-PETER	ROGER ROBERT S DR
ROGERS ALAN E E DR	ROHLFS K PROF DR	ROSA MICHAEL RICHARD DR
ROSE WILLIAM K DR	ROSINO LEONIDA PROF	ROXBURGH IAN W PROF
ROZHKOVSKIJ DIMITRIJ A	ROZYCZKA MICHAL	RUBIN ROBERT HOWARD
RUBIN VERA C DR	SABANO YUTAKA DR	SABBADIN FRANCO DR
SALINARI PIERO	SALPETER EDWIN E PROF	SANAHUJA BLAS
SANCHEZ-SAAVEDRA M LUISA	SANCISI RENZO DR	SANDELL GORAN HANS L DR
SANDQVIST AAGE DR	SARAZIN CRAIG L DR	SATO FUMIO DR
SATO SHUJI DR	SAVAGE BLAIR D DR	SAVEDOFF MALCOLM P PROF
SCALO JOHN MICHAEL	SCHALEN CARL PROF	SCHATZMAN EVRY PROF
SCHERB FRANK PROF	SCHEUER PETER A G DR	SCHMID-BURGK J DR PROF
SCHMIDT THOMAS DR	SCHMIDT-KALER TH PROF	SCHULTZ G V DR
SCHULZ ROLF ANDREAS	SCHWARTZ PHILIP R DR	SCHWARTZ RICHARD D
SCHWARZ ULRICH J DR	SCOTT EUGENE HOWARD	SEATON MICHAEL J PROF
SEKI MUNEZO DR	SHAH GHANSHYAM A DR	SHANE WILLIAM W DR
SHAO CHENG-YUAN	SHAPIRO STUART L	SHARPLESS STEWART PROF
SHAWL STEPHEN J DR	SHCHEGLOV P V DR	SHERWOOD WILLIAM A DR
SHIELDS GREGORY A DR	SHU FRANK H PROF	SHULL JOHN MICHAEL
SHUTER WILLIAM L H DR	SILBERBERG REIN DR	SILK JOSEPH I PROF
SILVESTRO GIOVANNI	SIMONS STUART DR	SINGH PATAN DEEN DR
SITKO MICHAEL L	SIVAN JEAN-PIERRE DR	SKILLING JOHN DR
SMITH BARHAM W DR	SMITH HOWARD ALAN	SMITH PETER L DR
SNELL RONALD L	SNOW THEODORE P PROF	SOBOLEV V V DR
SOFIA SABATINO PROF	SOLC MARTIN	SOLOMON PHILIP M DR
SOMERVILLE WILLIAM B DR	SPITZER LYMAN JR DR	STANGA RUGGERO
STECHER THEODORE P	STROM RICHARD G DR	SU BUMEI

SUN JIN
TARAFDAR SHANKAR P DR
TERZIAN YERVANT PROF
THOMPSON A RICHARD DR
TORRES-PEIMBERT SILVIA DR
TREFFERS RICHARD R
ULRICH MARIE-HELENE D DR
VAN DE HULST H C PROF DR
VANDEN BOUT PAUL A
VERSCHUUR GERRIT L PROF
VINER MELVYN R DR
VRBA FREDERICK J DR
WEAVER HAROLD F PROF
WEISHEIT JON C DR
WEYMANN RAY J PROF
WHITELOCK PATRICIA ANN DR
WHITWORTH ANTHONY PETER
WILLIAMS DAVID A PROF
WILLIS ALLAN J DR
WINNBERG ANDERS DR
WOLTJER LODEWIJK PROF
WOOTTEN HENRY ALWYN
WYNN-WILLIAMS C G DR

YABUSHITA SHIN A PROF
ZEALEY WILLIAM J DR
ZHOU ZHEN-PU

TAKAKUBO KEIYA PROF
TAYLOR KENNETH N R PROF
THADDEUS PATRICK PROF
THONNARD NORBERT DR
TOSI MONICA
TURNER BARRY E DR
UNNO WASABURO PROF
VAN DER LAAN H PROF DR
VANYSEK VLADIMIR PROF
VIDAL JEAN-LOUIS DR
VISVANATHAN NATARAJAN DR
WALKER GORDON A H PROF
WEBSTER B LOUISE DR
WENDKER HEINRICH J PROF
WHITE GLENN J
WHITEOAK J B DR
WICKRAMASINGHE N C PROF
WILLIAMS IWAN P DR
WILLNER STEVEN PAUL DR
WITT ADOLF N DR
WOODWARD PAUL R DR
WRIGHT EDWARD L DR
XIANG DELIN

YORKE HAROLD W DR
ZEILIK MICHAEL II DR
ZIMMERMANN HELMUT DR

TAMURA SHINICHI DR
TENORIO-TAGLE G DR
THE PIK-SIN PROF
THRONSON HARLEY ANDREW JR
TOWNES CHARLES HARD DR
TURNER KENNETH C DR
URASIN LIRIK A DR
VAN WOERDEN HUGO PROF DR
VARSHALOVICH DIMITRIJ PR
VIDAL-MADJAR ALFRED DR
VORONTSOV-VEL'YAMINOV B A
WATT GRAEME DAVID
WEILER KURT W DR
WESSELIUS PAUL R DR
WHITE RICHARD L
WHITTET DOUGLAS C B DR
WILLIAMS D DR
WILLIAMS ROBERT E DR
WILSON ROBERT W DR
WOLSTENCROFT RAMON D DR
WOOLF NEVILLE J
WU CHI CHAO DR
XING JUN

YOUNIS SAAD M
ZHANG CHENG-YUE
ZUCKERMAN BEN M DR

COMMISSION No. 35
STELLAR CONSTITUTION (CONSTITUTION DES ETOILES)

President   : SUGIMOTO DAIICHIRO PROF

Vice-President(s) : MAEDER ANDRE PROF

Organizing Committee: BODENHEIMER PETER PROF
                      CHIOSI CESARE S DR
                      COX ARTHUR N DR
                      GOUGH DOUGLAS O DR
                      KIPPENHAHN RUDOLF PROF
                      OSAKI YOJI DR
                      TASSOUL JEAN-LOUIS PROF
                      TRURAN JAMES W JR
                      WEIDEMANN VOLKER PROF
                      WHEELER J CRAIG PROF

Members:
AIAD A DR   AIZENMAN MORRIS L DR   ANAND S P S DR
ANGELOV TRAJKO   APPENZELLER IMMO PROF   ARAI KENZO DR
ARIMOTO NOBUO DR   ARNETT W DAVID PROF   ARNOULD MARCEL L DR
AUDOUZE JEAN PROF   BAGLIN ANNIE DR   BAKER NORMAN H PROF
BAYM GORDON ALAN DR   BEAUDET GILLES DR   BECKER STEPHEN A
BENZ WILLY   BERTHOMIEU GABRIELLE DR   BISNOVATYI-KOGAN G S DR
BLUDMAN SIDNEY A PROF   BOCCHIA ROMEO DR   BOEHM KARL-HEINZ PROF
BONDI HERMANN PROF SIR   BROWNLEE ROBERT R DR   BUCHLER J ROBERT PROF
BURBIDGE GEOFFREY R PROF   CALLEBAUT DIRK K DR   CALOI VITTORIA DR
CAMERON ALASTAIR G W PROF   CANAL RAMON M DR   CAPUTO FILIPPINA DR
CARSON T R DR   CASTELLANI VITTORIO PROF   CASTOR JOHN I DR
CAUGHLAN GEORGEANNE R   CHAN KWING LAM   CHANDRASEKHAR S PROF
CHEVALIER CLAUDE DR   CHITRE SHASHIKUMAR M DR   CHIU HONG-YEE DR
CHKHIKVADZE IAKOB N   CHRISTENSEN-DALSGAARD J   CHRISTY ROBERT F DR
COHEN JEFFREY M DR   CONNOLLY LEO PAUL   COWAN JOHN J DR
COWLING THOMAS G PROF   D'ANTONA FRANCESCA DR   DAS MRINAL KANTI
DAVIS CECIL G JR   DE GREVE JEAN-PIERRE DR   DE JAGER CORNELIS PROF
DE LOORE CAMIEL PROF   DEARBORN DAVID PAUL S DR   DEINZER W PROF DR
DEMARQUE P PROF   DESPAIN KEITH HOWARD DR   DEUPREE ROBERT G DR
DINGENS P PROF DR   DLUZHNEVSKAYA O B DR   DURISEN RICHARD H DR
DZIEMBOWSKI WOJCIECH PROF   EDWARDS ALAN CH DR   EDWARDS TERRY W
EGGLETON PETER P DR   EMINZADE T A DR   ENDAL ANDREW S DR
EPSTEIN ISADORE PROF   ERGMA E V DR   ERIGUCHI YOSHIHARU DR
EZER-ERYURT DILHAN PROF   FADEYEV YURI A   FAULKNER DONALD J DR
FAULKNER JOHN PROF   FLANNERY BRIAN PAUL DR   FONTAINE GILLES DR
FORBES J E DR   FOSSAT ERIC DR   FOWLER WILLIAM A PROF
FRANTSMAN YU L DR   FRICKE KLAUS DR   FUJIMOTO MASAYUKI DR
GABRIEL MAURICE R DR   GALLINO ROBERTO   GEROYANNIS VASSILIS S DR
GIANNONE PIETRO PROF   GINGOLD ROBERT ARTHUR DR   GIRIDHAR SUNETRA DR
GLATZMAIER GARY A   GONG SHU-MO   GOOSSENS MARCEL DR
GRAHAM ERIC DR   GURM HARDEV S PROF   HAMADA TETSUO PROF
HAYASHI CHUSHIRO PROF   HEILESEN POUL MARTIN DR   HILF EBERHARD R H PH D
HITOTSUYANAGI JUICHI PROF   HOSHI REIUN DR   HOYLE FRED SIR
HUANG RUN-QIAN   HUMPHREYS ROBERTA M PROF   IBEN ICKO JR PROF

ILIEV ILIAN
ISHIZUKA TOSHIHISA DR
KAEHLER HELMUTH DR
KHOZOV GENNADIJ V
KOCHHAR R K DR
KOVETZ ATTAY PROF
KUMAR SHIV S PROF
LAMB DONALD QUINCY JR DR
LASKARIDES PAUL G ASSPROF
LEBOVITZ NORMAN R PROF
LI HEN
LITTLETON JOHN E

IMSHENNIK V S DR
ITOH NAOKI DR
KAMINISHI KEISUKE PROF
KIGUCHI MASAYOSHI DR
KOESTER DETLEV DR
KOZLOWSKI MACIEJ DR
KUSHWAHA R S PROF
LAMB SUSAN ANN DR
LASOTA JEAN-PIERRE DR
LEDOUX P J PROF
LI ZONG-WEI
LIVIO MARIO

ISAAK GEORGE R PROF
JAMES RICHARD A DR
KATO MARIKO
KING DAVID S PROF
KOTHARI D S DR
KROOK M DR
LABAY JAVIER
LARSON RICHARD B PROF
LATOUR JEAN J
LEPINE JACQUES R D DR
LINNELL ALBERT P PROF
MAHESWARAN MURUGESAPILLAI

MALLIK D C V DR
MASSEVICH ALLA G DR
MCCREA J DERMOTT
MEYER-HOFMEISTER E DR
MIYAJI SHIGEKI DR
MOORE DANIEL R DR
MOSS DAVID L DR
NAKAMURA TAKASHI DR
NARIAI HIDEKAZU
NISHIDA MINORU PROF
ODELL ANDREW P
OSTRIKER JEREMIAH P PROF
PANDE GIRISH CHANDRA PROF
PINES DAVID PROF
POPOVA MALINA D PROF DR
PRENTICE ANDREW J R DR
RAEDLER K H DR
REIZ ANDERS PROF
ROUSE CARL A DR
SACKMANN I JULIANA DR
SALPETER EDWIN E PROF
SAVONIJE GERRIT JAN DR
SCHATZMAN EVRY PROF
SCHUTZ BERNARD F PROF
SEARS RICHARD LANGLEY DR
SHAVIV GIORA DR
SHUSTOV BORIS M DR
SILVESTRO GIOVANNI
SMITH ROBERT CONNON DR
SOUFFRIN PIERRE B DR
SREENIVASAN S RANGA PROF
STIBBS DOUGLAS W N PROF
SUDA KAZUO PROF
TAKAHARA MARIKO
THIELEMANN FRIEDRICH-KARL
TOHLINE JOEL EDWARD
TUOMINEN ILKKA V DR
TYPHOON LEE
UNNO WASABURO PROF
VAN DER BORGHT RENE PROF
VAN RIPER KENNETH A DR
VAUCLAIR GERARD P DR
WARD RICHARD A DR
WEIGERT ALFRED PROF
WILSON ROBERT E PROF
WRIGHT GEOFFREY A E DR
YUNGELSON LEV R
ZIOLKOWSKI JANUSZ DR

MARX GYORGY PROF.
MAZUREK THADDEUS JOHN DR
MELIK-ALAVERDIAN YU DR
MICHAUD GEORGES J DR
MOELLENHOFF CLAUS DR
MORGAN JOHN ADRIAN
MUELLER EWALD
NAKANO TAKENORI DR
NARITA SHINJI DR
NOELS ARLETTE DR
OHYAMA NOBORU PROF
PACZYNSKI BOHDAN PROF
PAPALOIZOU JOHN C B DR
PINOTSIS ANTONIS D DR
PORFIR'EV V V DR
PROVOST JANINE DR
RAMADURAI SOURIRAJA DR
RENZINI ALVIO PROF
ROXBURGH IAN W PROF
SAIO HIDEYUKI DR
SATO KATSUHIKO PROF
SCALO JOHN MICHAEL
SCHILD HANSRUEDI
SCHWARZSCHILD MARTIN PROF
SEIDOV ZAKIR F DR
SHIBAHASHI HIROMOTO DR
SIENKIEWICZ RYSZARD DR
SION EDWARD MICHAEL
SOBOUTI YOUSEF PROF
SPARKS WARREN M DR
STARRFIELD SUMNER PROF
STRITTMATTER PETER A PROF
SWEET PETER A PROF
TASSOUL MONIQUE DR
THOMAS HANS-CHRISTOPH DR
TOOMRE JURI
TUOMINEN JAAKKO V PROF
UCHIDA JUICHI DR
UUS UNDO DR
VAN DER RAAY HERMAN B
VANDENBERG DON DR
VILA SAMUEL C PROF
WEAVER THOMAS A DR
WEISS NIGEL O DR
WINKLER KARL-HEINZ A DR
XIONG DA-RUN
ZAHN JEAN-PAUL DR

MASANI A PROF
MAZZITELLI ITALO DR
MESTEL LEON PROF
MITALAS ROMAS ASSOC PROF
MONAGHAN JOSEPH J DR
MORRIS STEPHEN C DR
NADYOZHIN D K DR
NAKAZAWA KIYOSHI DR
NEWMAN MICHAEL JOHN DR
NOMOTO KEN'ICHI DR
OKAMOTO ISAO DR
PAMYATNIKH A A DR
PHILLIPS MARK M DR
PLAVEC MIREK J PROF
POVEDA ARCADIO DR
QU QIN-YUE
REEVES HUBERT PROF
ROOD ROBERT T DR
RUBEN G PROF DR
SAKASHITA SHIRO PROF
SAVEDOFF MALCOLM P PROF
SCHATTEN KENNETH H DR
SCHRAMM DAVID N PROF
SCUFLAIRE RICHARD DR
SENGBUSCH KURT V DR
SHIBATA YUKIO DR
SIGNORE MONIQUE DR
SMEYERS PAUL PROF
SOFIA SABATINO PROF
SPIEGEL E DR
STELLINGWERF ROBERT F DR
STROMGREN BENGT PROF
SWEIGART ALLEN V DR
TAYLER ROGER J PROF
TJIN-A-DJIE HERMAN R E DR
TRIMBLE VIRGINIA L DR
TUTUKOV A V DR
ULRICH ROGER K PROF
VAN DEN HEUVEL EDWARD P J
VAN HORN HUGH M PROF
VARDYA M S DR
VILHU OSMI DR
WEBBINK RONALD F DR
WILLSON LEE ANNE DR
WOOD PETER R DR
YORKE HAROLD W DR
ZHEVAKIN S A PROF DR

MEMBERSHIP OF COMMISSIONS

COMMISSION No. 36

THEORY OF STELLAR ATMOSPHERES

(THEORIE DES ATMOSPHERES STELLAIRES)

President   : KODAIRA KEIICHI PROF

Vice-President(s) : GRAY DAVID F PROF

Organizing Committee: CASSINELLI JOSEPH P DR
                      CRAM LAWRENCE EDWARD DR
                      GUSTAFSSON BENGT DR
                      HEARN ANTHONY G DR
                      HUBENY IVAN
                      KALKOFEN WOLFGANG DR
                      KUDRITZKI ROLF-PETER PH D
                      MIHALAS DIMITRI DR
                      SEATON MICHAEL J PROF
                      UNDERHILL ANNE B DR

Members:

ABHYANKAR KRISHNA D PROF	ALLER LAWRENCE HUGH	ALTROCK RICHARD C DR
ARPIGNY CLAUDE PROF	ATHAY R GRANT DR	AUER LAWRENCE H DR
AUMAN JASON R PROF	AVRETT EUGENE H DR	BAIRD SCOTT R
BASCHEK BODO PROF	BELL ROGER A DR	BERNAT ANDREW PLOUS DR
BERTOUT CLAUDE	BLANCO CARLO DR	BLESS ROBERT C PROF
BOEHM KARL-HEINZ PROF	BOEHM-VITENSE ERIKA PROF	BOEHME SIEGFRIED DR
BOESGAARD ANN M PROF	BROWN DOUGLAS NASON	BUES IRMELA D DR
CARBON DUANE F DR	CARSON T R DR	CASTOR JOHN I DR
CAYREL DE STROBEL GIUSA	CAYREL ROGER DR	CHAN KWING LAM
CHEN PEISHENG	CONTI PETER S DR	COWLEY CHARLES R PROF
CRIVELLARI LUCIO	CUNY YVETTE J DR	DAVIS CECIL G JR
DELACHE PHILIPPE J DR	DOMKE HELMUT PH D	DRAKE STEPHEN A
DRAVINS DAINIS PROF	DUFTON PHILIP L DR	DUMONT SIMONE DR
DUPREE ANDREA K DR	EDMONDS FRANK N JR DR	ELSTE GUNTHER H DR
ERIKSSON KJELL DR	EVANGELIDIS E DR	FARAGGIANA ROSANNA PROF
FINN G D DR	FONTENLA JUAN MANUEL DR	FOY RENAUD DR
FRISCH HELENE DR	FRISCH URIEL DR	FROESCHLE CHRISTIANE D DR
GAIL HANS-PETER DR	GEBBIE KATHARINE B DR	GOKDOGAN NUZHET PROF
GORDON CHARLOTTE PROF	GOUGH DOUGLAS O DR	GRANT IAN P DR
GREENSTEIN J L PROF	GREVESSE N DR	GRININ VLADIMIR P DR
GROTH HANS G PROF DR	GUSSMANN E A DR	HACK MARGHERITA PROF
HAISCH BERNHARD MICHAEL	HAMANN WOLF-RAINER	HARDORP JOHANNES PROF
HARTMANN LEE WILLIAM	HEASLEY JAMES NORTON	HEBER ULRICH
HEKELA JAN DR	HEROLD HEINZ	HITOTSUYANAGI JUICHI PROF
HOLWEGER HARTMUT PROF	HOLZER THOMAS EDWARD DR	HOTINLI METIN DR
HOUSE LEWIS L DR	HUMMER DAVID G DR	HUNGER KURT PROF
HUTCHINGS JOHN B DR	IVANOV VSEVOLOD V DR PROF	JEFFERIES JOHN T DR
JOHNSON HOLLIS R PROF	KADOURI TALIB HADI	KAMP LUCAS WILLEM DR
KANDEL ROBERT S DR	KARP ALAN HERSH DR	KHOKHLOVA V L DR
KLEIN RICHARD I DR	KOESTER DETLEV DR	KOLESOV A K DR
KONTIZAS EVANGELOS DR	KRISHNA SWAMY K S DR	KUHI LEONARD V PROF

KUMAR SHIV S PROF
LAMBERT DAVID L PROF
LINSKY JEFFREY L DR
MARLBOROUGH J M PROF
MATSUSHIMA SATOSHI DR
MNATSAKANIAN MAMIKON A DR
MUKAI SONOYO DR
NAGIRNER DMITRIJ I DR
NEVEN LUC DR
ORRALL FRANK Q PROF
PALLAVICINI ROBERTO DR
PECKER JEAN-CLAUDE PROF
PHILLIPS JOHN G PROF
QUERCI FRANCOIS R DR
RAMSEY LAWRENCE W DR
ROVIRA MARTA GRACIELA
SAITO KUNIJI PROF
SCHARMER GOERAN BJARNE
SCHMUTZ WERNER
SEDLMAYER ERWIN DR
SIMON KLAUS PETER
SITNIK G F PROF
SNIJDERS MATTHEUS A J
SOUFFRIN PIERRE B DR
SPITE MONIQUE DR
STEIN ROBERT F ASSOC PROF
STROM STEPHEN E
THOMAS RICHARD N DR
TSUJI TAKASHI
ULMSCHNEIDER PETER PROF
VAN'T VEER FRANS DR
VARDYA M S DR
WEBER STEPHEN VANCE
WELLMANN PETER PROF DR
WILLSON LEE ANNE DR
WOEHL HUBERTUS DR
YANOVITSKIJ EDGARD G DR
ZWAAN CORNELIS PROF DR

KURUCZ ROBERT L DR
LEIBACHER JOHN DR
LIU CAIPIN
MASSAGLIA SILVANO
MICHAUD GEORGES J DR
MUELLER EDITH A PROF
MUSTEL E R PROF DR
NARIAI KYOJI DR
NORDLUND AKE DR
OXENIUS JOACHIM DR
PANEK ROBERT J DR
PERAIAH ANNAMANENI DR
POTTASCH STUART R PROF
QUERCI MONIQUE DR
REIMERS DIETER PROF
RUTTEN ROBERT J. DR
SAKHIBULLIN NAIL A DR
SCHMALBERGER DONALD C DR
SCHOENBERNER DETLEF PROF
SHINE RICHARD A DR
SIMON THEODORE
SKUMANICH ANDRE PROF
SOBOLEV V V DR
SPIEGEL E DR
SPRUIT HENK C DR
STEPIEN KAZIMIERZ DR
SWIHART THOMAS L DR
TOOMRE JURI
UENO SUEO PROF
UNNO WASABURO PROF
VAN'T VEER-MENNERET CL DR
VIIK TONU DR
WEHRSE RAINER DR
WHITE RICHARD L
WILSON PETER R PROF
WRIGHT KENNETH O DR
YORKE HAROLD W DR

KUSHWAHA R S PROF
LINNELL ALBERT P PROF
MADEJ JERZY
MATSUMOTO MASAMICHI PROF
MIYAMOTO SIGENORI PROF
MUENCH GUIDO PROF
MUTSCHLECNER J PAUL DR
NEFF JOHN S
O'MARA BERNARD J PROF
PAGEL BERNARD E J PROF
PASINETTI LAURA E PROF
PETERS GERALDINE JOAN DR
PRADERIE FRANCOISE DR
RACHKOVSKY D N DR
ROSS JOHN E R DR
RYBICKI GEORGE B DR
SAPAR ARVED DR
SCHMID-BURGK J DR PROF
SCHOLZ M PROF
SHIPMAN HENRY L DR
SIMONNEAU EDUARDO DR
SNEZHKO LEONID I
SOBOUTI YOUSEF PROF
SPITE FRANCOIS M DR
STALIO ROBERTO DR
STIBBS DOUGLAS W N PROF
TARAFDAR SHANKAR P DR
TRAVING GERHARD PROF
UESUGI AKIRA DR
VAN REGEMORTER HENRI DR
VARDAVAS ILIAS MIHAIL
WATANABE TETSUYA
WEIDEMANN VOLKER PROF
WICKRAMASINGHE N C PROF
WILSON S J
WYLLER ARNE A PROF
ZAHN JEAN-PAUL DR

MEMBERSHIP OF COMMISSIONS

COMMISSION No. 37

STAR CLUSTERS AND ASSOCIATIONS

(AMAS STELLAIRES ET ASSOCIATIONS)

President   : HEGGIE DOUGLAS C DR

Vice-President(s)   : HARRIS GRETCHEN L H DR

Organizing Committee: FREEMAN KENNETH C PROF
                      HESSER JAMES E DR
                      NISSEN POUL E PROF
                      PILACHOWSKI CATHERINE DR
                      SALUKVADZE G N DR

Members:

AARSETH SVERRE J DR	ABOU-EL-ELLA MOHAMED S DR	AGEKJAN TATEOS A PROF
AIAD A DR	ALCAINO GONZALO DR	ALFARO EMILIO JAVIER
ALKSNIS ANDREJS DR	ALLEN CHRISTINE	AURIERE MICHEL
BALAZS BELA A DR	BARKHATOVA KLAUDIA PROF	BECKER WILHELM PROF
BIJAOUI ALBERT DR	BLAAUW ADRIAAN PROF DR	BOUVIER PIERRE PROF
BURKHEAD MARTIN S	BUTLER DENNIS DR	BYRD GENE G DR
CALLEBAUT DIRK K DR	CANNON RUSSELL D DR	CAPUZZO DOLCETTA ROBERTO
CARNEY BRUCE WILLIAM	CHAVARRIA-K. CARLOS	CHRISTIAN CAROL ANN
CHUN MUN-SUK DR	CLARIA JUAN DR	COLIN JACQUES DR
CUDWORTH KYLE MCCABE DR	CUFFEY J MR	DA COSTA GARY STEWART DR
DAUBE-KURZEMNIECE I A DR	DEMARQUE P PROF	DEMERS SERGE DR
DICKENS ROBERT J DR	DLUZHNEVSKAYA O B DR	EFREMOV YURY N DR
EINASTO JAAN DR	EL-BASSUNY ALAWY A A	ELMEGREEN BRUCE GORDON DR
FALL S MICHAEL DR	FEAST MICHAEL W DR	FEINSTEIN ALEJANDRO DR
FITZGERALD M PIM PROF	FORTE JUAN CARLOS DR	GASCOIGNE S C B DR
GOLAY MARCEL PROF	GREEN ELIZABETH M. DR	GRIFFITHS WILLIAM K
GRINDLAY JONATHAN E DR	GRUBISSICH C PROF DR	HANES DAVID A DR
HARRIS HUGH C	HARRIS WILLIAM E DR	HARVEL CHRISTOPHER ALVIN
HASSAN S M DR	HAWARDEN TIMOTHY G DR	HAZEN MARTHA L DR
HENON MICHEL C DR	HERBST WILLIAM DR	HILLS JACK G DR
IBEN ICKO JR PROF	ILLINGWORTH GARTH D DR	INAGAKI SHOGO DR
ISHIDA KEIICHI PROF	JANES KENNETH A DR	JONES DEREK H P DR
JOSHI U C DR	KADLA ZDENKA I DR	KAMP LUCAS WILLEM DR
KHOLOPOV P N DR	KILAMBI G C DR	KING IVAN R PROF
KONTIZAS EVANGELOS DR	KONTIZAS MARY DR	KRON GERALD E DR
LADA CHARLES JOSEPH DR	LAPASSET EMILIO DR	LARSSON-LEANDER G PROF
LAVAL ANNIE DR	LLOYD-EVANS THOMAS DR	LODEN LARS OLOF PROF
LU PHILLIP K DR	LYNDEN-BELL DONALD PROF	LYNGA GOSTA DR
MAEDER ANDRE PROF	MARKKANEN TAPIO DR	MARRACO HUGO G DR
MARTINS DONALD HENRY DR	MAYOR MICHEL DR	MENON T K PROF
MENZIES JOHN W DR	MERMILLIOD JEAN-CLAUDE DR	MEURERS JOSEPH PROF DR
MIKKOLA SEPPO DR	MOFFAT ANTHONY F J DR	MOULD JEREMY R
MURRAY C ANDREW	NEMEC JAMES	NESCI ROBERTO
NEWELL EDWARD B DR	OGURA KATSUO DR	OSBORN WAYNE DR
OSMAN ANAS MOHAMED DR	PARSAMYAN ELMA S DR	PEDREROS MARIO DR

PENNY ALAN JOHN DR   PETERSON CHARLES JOHN DR   PETROVSKAYA M S DR
PHILIP A G DAVIS   PISKUNOV ANATOLY E   POPOVA MALINA D PROF DR
POVEDA ARCADIO DR   PRITCHET CHRISTOPHER J DR QIAN BO-CHEN
RAJAMOHAN R DR   RAM SAGAR DR   RICHER HARVEY B DR
ROSINO LEONIDA PROF   ROTH MARIA LUISE PH D   ROUNTREE JANET DR
RUPRECHT JAROSLAV DR   RUSSEVA TATJANA   SAMUS NIKOLAI N DR
SANDERS W L PROF   SAWYER-HOGG HELEN B DR   SCARIA K K DR
SCHILD HANSRUEDI   SCHUSTER HANS-EMIL   SHAROV A S DR
SHAWL STEPHEN J DR   SHER DAVID DR   SHOBBROOK ROBERT R DR
SIMODA MAHIRO PROF   STETSON PETER B. DR   SUGIMOTO DAIICHIRO PROF
SZECSENYI-NAGY GABOR DR   TERZAN AGOP DR   THE PIK-SIN PROF
TSVETKOVA KATIA   TURNER DAVID G DR   UPGREN ARTHUR R DR
VAN ALTENA WILLIAM F PROF VAN DEN BERGH SIDNEY PROF VANDENBERG DON DR
VOGT NIKOLAUS DR   WALKER GORDON A H PROF   WALKER MERLE F PROF
WALLENQUIST AAKE A E PROF WAN LAI   WARREN WAYNE H JR DR
WEAVER HAROLD F PROF   WEHLAU AMELIA DR   WHITE RAYMOND E DR
WIELEN ROLAND PROF DR   WRAMDEMARK STIG S O DR   WU HSIN-HENG DR
ZHAO JUN-LIANG   ZINN ROBERT J DR

MEMBERSHIP OF COMMISSIONS

COMMISSION No. 38

EXCHANGE OF ASTRONOMERS (ECHANGE DES ASTRONOMES)

```
President          : MUELLER EDITH A PROF

Vice-President(s)  : SMITH F GRAHAM PROF

Organizing Committee: BOYARCHUK A A DR
                      FLORSCH ALPHONSE DR
                      KOZAI YOSHIHIDE PROF
                      LEUNG KAM CHING PROF
                      SWARUP GOVIND PROF
                      TOLBERT CHARLES R DR
                      VOIGT HANS H PROF
                      WOOD F BRADSHAW PROF
                      YE SHU-HUA
```

Members:
AL-SABTI ABDUL ADIM DR	DELHAYE JEAN PROF	GODOLI GIOVANNI PROF
HAUPT HERMANN F PROF	MACRAE DONALD A PROF	MARIK MIKLOS DR.
NHA IL-SEONG DR	NINKOVIC SLOBODAN	OKOYE SAMUEL E PROF
OPOLSKI ANTONI PROF	REIZ ANDERS PROF	ROUTLY PAUL M DR
RUBEN G PROF DR	SAHADE JORGE PROF	SMITH ELSKE V P DR
VAN DEN HEUVEL EDWARD P J	WILD JOHN PAUL DR	

MEMBERSHIP OF COMMISSIONS
COMMISSION No. 40
RADIO ASTRONOMY (RADIOASTRONOMIE)

President   : BALDWIN JOHN E DR

Vice-President(s)   : MEZGER PETER G PROF

Organizing Committee: BARRETT ALAN H PROF
                         BAUDRY ALAIN DR
                         BOOTH ROY S PROF
                         JAUNCEY DAVID L DR
                         KAIFU NORIO DR
                         KAPAHI V K DR
                         KELLERMANN KENNETH I DR
                         MATVEYENKO L I DR
                         NICOLSON GEORGE D DR
                         SEAQUIST ERNEST R PROF
                         SETTI GIANCARLO PROF
                         STROM RICHARD G DR
                         WANG SHOU-GUAN
                         WILSON THOMAS L DR

Members:

ABDULLA SHAKER ABDUL AZIZ	ABLES J G DR	ABRAMI ALBERTO PROF
ADE PETER A R DR	AIZU KO PROF	AKABANE KENJI A PROF
ALEXANDER JOSEPH K	ALLEN RONALD J DR	ALLER HUGH D DR
ALLER MARGO F DR	ALTENHOFF WILHELM J DR	ANANTHARAMAIAH K R DR
ANDREW BRYAN H DR	APARICI JUAN DR	ARGYLE P E DR
ARNAL MARCELO EDMUNDO DR	ASSOUSA GEORGE ELIAS DR	AVERY LORNE W DR
AVIGNON YVETTE DR	AXON DAVID	BAARS JACOB W M DR
BAART EDWARD E PROF	BAATH LARS B DR	BACKER DONALD CH DR
BAGRI DURGADAS S	BALKLAVS A E DR	BARROW COLIN H DR
BARTEL NORBERT HARALD DR	BASH FRANK N PROF	BASU DIPAK DR
BATES RICHARD HEATON T DR	BATTY MICHAEL DR	BECK RAINER
BENZ ARNOLD DR	BERGE GLENN L DR	BERKHUIJSEN ELLY M DR
BHANDARI RAJENDRA DR	BHONSLE RAJARAM V PROF	BIEGING JOHN HAROLD DR
BIERMANN PETER L DR	BIGNELL R CARL DR	BIRAUD FRANCOIS DR
BIRKINSHAW MARK	BLAIR DAVID GERALD	BLANDFORD ROGER DAVID DR
BLUM EMILE-JACQUES DR	BOEHME A DR	BOEHME SIEGFRIED DR
BOISCHOT ANDRE DR	BOLTON JOHN G	BORIAKOFF VALENTIN
BOWERS PHILLIP F	BRACEWELL RONALD N PROF	BRAUDE SEMION YA PROF AG
BREGMAN JACOB D IR	BRIDLE ALAN H PROF	BRODERICK JOHN DR
BROTEN NORMAN W	BROUW W N DR	BROWNE IAN W A DR
BURBIDGE GEOFFREY R PROF	BURKE BERNARD F DR	CAROUBALOS C A PROF
CARR THOMAS D PROF	CASWELL JAMES L DR	CHAN KWING LAM
CHEN HONGSHENG	CHIKADA YOSHIHIRO DR	CHINI ROLF
CHRISTIANSEN WAYNE A	CHRISTIANSEN WILBUR PROF	CLARK BARRY G DR
CLARK DAVID H DR	CLARK FRANK OLIVER DR	COHEN MARSHALL H PROF
COHEN RAYMOND J DR	COHEN RICHARD S	COLE TREVOR WILLIAM PROF
CONDON JAMES J DR	CONKLIN EDWARD K	CONWAY ROBIN G DR
CORDES JAMES M	COSTAIN CARMAN H DR	COTTON WILLIAM D Jr
COUTREZ RAYMOND A J PROF	COVINGTON ARTHUR E	CRANE PATRICK C

CROOM DAVID L DR
CUI ZHENXING
DAVIES JOHN G DR
DAVIS ROBERT J DR
DE LA NOE JEROME DR
DELANNOY JEAN DR
DENT WILLIAM A PROF
DICKEL JOHN R
DIXON ROBERT S DR
DOUGLAS JAMES N PROF
DRAKE FRANK D PROF
DREHER JOHN W
DYSON F J DR

CROVISIER JACQUES
DAINTREE EDWARD J DR
DAVIES RODNEY D PROF
DE GROOT T DR
DE YOUNG DAVID S DR
DENISSE JEAN-FRANCOIS DR
DEWDNEY PETER E F DR
DICKEY JOHN M
DOHERTY LORNE H DR
DOWNES DENNIS DR
DRAKE STEPHEN A
DUFFETT-SMITH PETER JAMES
EKERS RONALD D DR

CUDABACK DAVID D DR
DAISHIDO TSUNEAKI PROF
DAVIS MICHAEL M DR
DE JAGER CORNELIS PROF
DEGAONKAR S S DR
DENNISON BRIAN KENNETH
DICKEL HELENE R DR
DIETER NANNIELOU H DR
DOUBINSKIJ B A DR
DOWNS GEORGE S DR
DRAVSKIKH A F DR
DULK GEORGE A PROF
ELGAROY OYSTEIN PROF

ELLIS G R A PROF
ENOME SHINZO PROF
ERIKSEN GUNNAR PROF
EWING MARTIN S
FELDMAN PAUL A DR
FERETTI LUIGINA
FLEISCHER ROBERT DR
FOMALONT EDWARD B DR
FRIBERG PER
FUERST ERNST DR
GARAY GUIDO DR
GEBLER KARL-HEINZ DR
GENT HUBERT MR
GHIGO FRANCIS D DR
GIOIA ISABELLA M DR
GOLDSMITH PAUL F DR
GONZE ROGER F J IR
GOSACHINSKIJ I V DR
GRAHAM DAVID A
GUELIN MICHEL DR
GULKIS SAMUEL DR
HADDOCK FRED T DR
HAN FU
HANBURY BROWN ROBERT PROF
HARRIS DANIEL E DR
HASCHICK AUBREY
HAYNES RAYMOND F PROF
HEIDMANN JEAN DR
HEWISH ANTONY PROF
HILF EBERHARD R H PH D
HJALMARSON AKE G DR
HOANG BINH DY DR
HOEGLUND BERTIL PROF
HOOGHOUDT B G IR
HUGHES PHILIP
HUNSTEAD RICHARD W DR
INATANI JUNJI
JANSSEN MICHAEL ALLEN
JIN SHEN-ZENG
JOHNSTON KENNETH J
JOSHI MOHAN N PROF
KAI KEIZO DR
KANG GON IK
KAUFMANN PIERRE PROF
KENDERDINE SIDNEY DR

ELSMORE BRUCE DR
EPSTEIN EUGENE E DR
ESHLEMAN VON R PROF
FANTI ROBERTO
FELLI MARCELLO DR
FIELD GEORGE B PROF
FLETT ALISTAIR M
FORT DAVID NORMAN DR
FRIEDMAN HERBERT DR
FUKUI YASUO DR
GARDNER FRANCIS F DR
GELDZAHLER BERNARD J
GENZEL REINHARD DR
GIBSON DAVID MICHAEL DR
GIOVANNINI GABRIELE
GOLDSTEIN SAMUEL J PROF
GORDON MARK A DR
GOSS W MILLER PROF
GREGORY PHILIP C DR
GUESTEN ROLF
GULL STEPHEN F DR
HAGEN JOHN P
HAN WENJUN
HANKINS TIMOTHY HAMILTON
HARTEN RONALD H DR
HASLAM C GLYN T DR
HAZARD CYRIL DR
HEILES CARL PROF
HEY JAMES STANLEY DR
HILLS RICHARD E DR
HJELLMING ROBERT M DR
HOBBS ROBERT W DR
HOGG DAVID E DR
HOWARD WILLIAM E III DR
HUGHES VICTOR A PROF
IKHSANOV ROBERT N DR
INOUE MAKOTO DR
JENKINS CHARLES R
JOHANSSON LARS ERIK B DR
JOLY FRANCOIS DR
KAFTAN MAY A DR
KAKINUMA TAKAKIYO T PROF
KARDASHEV N S DR
KAWABATA KINAKI PROF
KERR FRANK J DR

ELWERT GERHARD PROF
ERICKSON WILLIAM C DR
EVANS KENTON DOWER DR
FEIX GERHARD DR
FELTEN JAMES E DR
FINDLAY JOHN W DR
FOKKER AAD D DR
FRATER ROBERT H DR
FU QI JUN
GALT JOHN A DR
GAYLARD MICHAEL JOHN
GELFREIKH GEORGIJ B DR
GERGELY TOMAS ESTEBAN DR
GINZBURG VITALY L PROF
GOLD THOMAS PROF
GOLDWIRE HENRY C JR
GORGOLEWSKI STANISLAW PR
GOWER J F R DR
GREWING MICHAEL PROF
GUIDICE DONALD A DR
HACHENBERG OTTO PROF DR
HAMILTON P A DR
HANASZ JAN DR
HARDEE PHILIP
HARTZ THEODORE R DR
HAYNES MARTHA P
HEESCHEN DAVID S DR
HENKEL CHRISTIAN
HIGGS LLOYD A DR
HIRABAYASHI HISASHI DR
HO PAUL T P
HOEGBOM JAN A DR
HOLLIS JAN MICHAEL DR
HUCHTMEIER WALTER K DR
HULSBOSCH A N M DR
IKHSANOVA VERA N DR
JAFFE WALTER JOSEPH DR
JENNISON ROGER C PROF
JOHNSON DONALD R DR
JONES DAYTON L
KAHN FRANZ D PROF
KALBERLA PETER
KASUGA TAKASHI
KAZES ILYA DR
KESTEVEN MICHAEL J L DR

KISLYAKOV ALBERT G DR
KOJOIAN GABRIEL DR
KRAUS JOHN D PROF
KRISHNAMOHAN S DR
KRUEGEL ENDRIK DR
KULKARNI PRABHAKAR V PROF
KUNDU MUKUL R DR
KUTNER MARC LESLIE DR
LADA CHARLES JOSEPH DR
LANG KENNETH R ASST PROF
LARGE MICHAEL I DR
LEBLANC YOLANDE DR
LEQUEUX JAMES DR
LI GYONG WON
LILLEY EDWARD A PROF
LO KWOK-YUNG DR
LONGAIR M S PROF
LOVELL SIR BERNARD PROF
LUO XIANHAN
MACDONALD GEOFFREY H DR
MACLEOD JOHN M DR
MANDOLESI NAZZARENO
MARSCHER ALAN PATRICK
MASSON COLIN R
MATTILA KALEVI DR
MCADAM W BRUCE DR
MCKENNA LAWLOR SUSAN
MEEKS M LITTLETON DR
MICHALEC ADAM
MILNE DOUGLAS K DR
MOFFET ALAN T PROF
MORAN JAMES M DR
MORITA KAZUHIKO
MORRIS DAVID DR
MURDOCH HUGH S DR
MYERS PHILIP C
NAN REN-DONG
OKOYE SAMUEL E PROF
OWEN FRAZER NELSON DR
PALMER PATRICK E PROF
PARIJSKIJ YU N DR
PASACHOFF JAY M PROF
PAYNE DAVID G
PENG YUN-LOU
PETERS WILLIAM L III DR
PICK MONIQUE DR
PORCAS RICHARD DR
PRICE R MARCUS DR
QIAN SHAN-JIE
RAIMOND ERNST DR
RAY THOMAS P
REBER GROTE DR
REYES FRANCISCO DR
RICKETT BARNABY JAMES DR
ROBERTS DAVID HALL DR
ROBERTSON DOUGLAS S
ROBINSON JR RICHARD D DR
ROENNAENG BERNT O DR

KLEIN ULRICH
KOMESAROFF MAX M
KREYSA ERNST
KRISHNAN THIRUVENKATA MR
KUIJPERS H. JAN M.E. DR
KULKARNI V K DR
KURIL-CHIK V N DR
KUZMIN ARKADII D PROF DR
LAFFINEUR MARIUS MR
LANGER WILLIAM DAVID DR
LASENBY ANTHONY
LEGG THOMAS H DR
LEUNG CHUN MING DR
LI HONG-WEI
LINKE RICHARD ALAN DR
LOCKE JACK L DR
LOREN ROBERT BRUCE DR
LOZINSKAYA TAT'YANA A DR
LYNE ANDREW G DR
MACDONALD JAMES
MACRAE DONALD A PROF
MARAN STEPHEN P DR
MARTIN ROBERT N DR
MATHESON DAVID NICHOLAS
MAXWELL ALAN DR
MCCULLOCH PETER M DR
MCLEAN DONALD J DR
MEIER DAVID L
MILEY G K DR
MILOGRADOV-TURIN JELENA
MOISEEV I G DR
MORIMOTO MASAKI DR
MORIYAMA FUMIO PROF
MORRIS MARK ROOT DR
MUTEL ROBERT LUCIEN
NADEAU DANIEL DR
NGUYEN-QUANG RIEU DR
OORT JAN H PROF
PACHOLCZYK ANDRZEJ G PROF
PANKONIN VERNON LEE DR
PARKER EDWARD A DR
PAULINY TOTH IVAN K K DR
PEARSON TIMOTHY J
PENZIAS ARNO A DR
PETTENGILL GORDON H PROF
PONSONBY JOHN E B DR
PRESTON ROBERT ARTHUR
PRIESTER WOLFGANG PROF
QIU YU-HAI
RAMATY REUVEN DR
RAZIN V A DR
REICH WOLFGANG
RIBES JEAN-CLAUDE DR
RIIHIMAA JORMA J DR
ROBERTS JAMES A DR
ROBERTSON JAMES GORDON DR
RODRIGUEZ LUIS F
ROESER HANS-PETER

KO HSIEN C PROF
KOTELNIKOV V A ACAD
KRISHNA GOPAL
KRONBERG PHILIPP DR
KUIPER THOMAS B H DR
KUNDT WOLFGANG PROF DR
KUS ANDRZEJ JAN DR
KWOK SUN DR
LAING ROBERT
LANTOS PIERRE DR
LE SQUEREN ANNE-MARIE DR
LEPINE JACQUES R D DR
LI CHUN-SHENG
LIANG SHI-GUANG
LITTLE LESLIE T DR
LOCKMAN FELIX J
LORENZ HILMAR
LU YANG
MACCHETTO FERDINANDO DR
MACHALSKI JERZY DR
MANCHESTER RICHARD N DR
MARQUES DOS SANTOS P PROF
MASLOWSKI JOZEF DR
MATSAKIS DEMETRIOS N
MAYER CORNELL H
MCGEE JAMES D PROF
MEBOLD ULRICH DR PROF
MENON T K PROF
MILLS BERNARD Y PROF
MIRABEL IGOR FELIX DR
MOLCHANOV A P PROF
MORISON IAN MR
MORRAS RICARDO DR.
MULLER C A PROF JR
MUXLOW THOMAS
NAGNIBEDA VALERY G DR
O'SULLIVAN JOHN DAVID DR
OSTERBROCK DONALD E PROF
PADMAN RACHAEL
PAPAGIANNIS MICHAEL D PRO
PARRISH ALLAN DR.
PAULS THOMAS ALBERT DR
PEDLAR ALAN DR
PERLEY RICHARD ALAN
PHILLIPS THOMAS GOULD DR
POOLEY GUY DR
PREUSS EUGEN DR
PUSCHELL JEFFERY JOHN
RADHAKRISHNAN V PROF
RAO A PRAMESH DR
READHEAD ANTHONY C S DR
REID MARK JONATHAN DR
RICKARD LEE J DR
RILEY JULIA M DR
ROBERTS MORTON S DR
ROBINSON BRIAN J DR
ROEDER ROBERT C PROF
ROGER ROBERT S DR

ROGERS ALAN E E DR
ROMNEY JONATHAN D DR
RUBIO MONICA DR
RYDBECK OLOF E H PROF
SALOMONOVICH A E DR
SANDELL GORAN HANS L DR
SATO FUMIO DR
SCALISE JR EUGENIO DR
SCHLICKEISER REINHARD DR
SCHULZ ROLF ANDREAS
SCOTT JOHN S DR
SEIRADAKIS JOHN HUGH DR
SHAVER PETER A DR
SHOLOMITSKY G B DR
SIMON PAUL A DR
SLEE O B DR
SMITH DEAN F DR
SOBOLEVA N S DR
SPENCER JOHN HOWARD
STAHR-CARPENTER M DR
STEINBERG JEAN-LOUIS DR
STONE R G DR
SWARUP GOVIND PROF
TAKAGI KOJIRO PROF
TANAKA RIICHIRO PROF
THOMASSON PETER DR
TLAMICHA ANTONIN DR
TOMASI PAOLO DR
TRITTON KEITH P DR
TSURUTA SACHIKO DR
TURNER KENNETH C DR
UKITA NOBUHARU
UNWIN STEPHEN C
VALTAOJA ESKO
VAN DER KRUIT PIETER C DR
VAN NIEUWKOOP J DR IR
VARSAVSKY C M DR
VERON PHILIPPE DR
VIVEKANAND M DR
WALMSLEY C MALCOLM DR
WANG JING-SHENG
WARNER PETER J DR
WEI MINGZHI
WELLINGTON KELVIN DR
WESTFOLD KEVIN C PROF
WIELEBINSKI RICHARD PROF
WILLIAMS D DR
WILLS DEREK DR
WILSON ROBERT W DR
WINNBERG ANDERS DR
WOLTJER LODEWIJK PROF
WRIGHT ALAN E DR
WU XINJI
YANG JIAN
ZAITSEV VALERII V DR
ZHENG YI-JIA
ZLOBEC PAOLO DR

ROGSTAD DAVID H DR
ROWSON BARRIE DR
RUDNICK LAWRENCE DR
RYZHKOV N F DR
SALPETER EDWIN E PROF
SARMA N V G PROF
SAUNDERS RICHARD D.E.
SCHEUER PETER A G DR
SCHMIDT MAARTEN PROF
SCHWARTZ PHILIP R DR
SCOTT PAUL F DR
SHAFFER DAVID B DR
SHERIDAN K V DR
SHUTER WILLIAM L H DR
SINHA RAMESHWAR P
SLYSH V I DR
SMITH F GRAHAM PROF
SOFUE YOSHIAKI PROF
SPENCER RALPH E DR
STANLEY G J
STEWART PAUL DR
SULLIVAN WOODRUFF T III
SWENSON GEORGE W JR PROF
TAKAKUBO KEIYA PROF
TARTER JILL C DR
THOMPSON A RICHARD DR
TOFANI GIANNI PROF
TOVMASSIAN H M DR
TROITSKY V S PROF DR
TURLO ZYGMUNT DR
TURTLE A J DR
ULRICH BRUCE T PROF
URPO SEPPO I
VAN DE HULST H C PROF DR
VAN DER LAAN H PROF DR
VAN WOERDEN HUGO PROF DR
VELUSAMY T DR
VERSCHUUR GERRIT L PROF
WADE CAMPBELL M DR
WALSH DENNIS DR
WANNIER PETER GREGORY DR
WARWICK JAMES W DR
WELCH WILLIAM J PROF
WENDKER HEINRICH J PROF
WHITEOAK J B DR
WILD JOHN PAUL DR
WILLIS ANTHONY GORDON DR
WILLSON ROBERT FREDERICK
WILSON WILLIAM J DR
WINNEWISSER GISBERT DR
WOODSWORTH ANDREW W.DR
WU HUAI-WEI
XU PEI-YUAN
YIN QI-FENG
ZHAO REN-YANG
ZHOU TI-JIAN
ZUCKERMAN BEN M DR

ROHLFS K PROF DR
RUBIN ROBERT HOWARD
RYDBECK GUSTAF H B DR
SAIKIA D J DR
SANAMIAN V A DR
SASTRY CH V
SAVAGE ANN DR
SCHILIZZI RICHARD T DR
SCHULTZ G V DR
SCHWARZ ULRICH J DR
SEIELSTAD GEORGE A
SHAKESHAFT JOHN R DR
SHIMMINS ALBERT JOHN
SIEBER WOLFGANG PH D
SLADE MARTIN A III DR
SMITH ALEX G PROF
SMOL'KOV GENNADIJ YA DR
SOROCHENKO R L DR
SRAMEK RICHARD A DR
STANNARD DAVID DR
STEWART RONALD T MR
SUZUKI HIROKO
TABARA HIROTO DR
TAKAKURA TATSUO PROF EMER
TERZIAN YERVANT PROF
THUM CLEMENS DR
TOLBERT CHARLES R DR
TOWNES CHARLES HARD DR
TROLAND THOMAS HUGH
TURNER BARRY E DR
UDAL'TSOV V A DR
ULRICH MARIE-HELENE D DR
VALLEE JACQUES P DR
VAN DER HULST JAN M DR
VAN GORKOM JACQUELINE H
VANDEN BOUT PAUL A
VENUGOPAL V R DR
VINER MELVYN R DR
WALL JASPER V DR
WAN TONG-SHAN
WARDLE JOHN F C PROF
WEAVER HAROLD F PROF
WELIACHEW LEONID DR
WESTERHOUT GART DR
WICKRAMASINGHE N C PROF
WILKINSON PETER N DR
WILLS BEVERLEY J DR
WILSON ANDREW S DR
WINK JOERN ERHARD DR
WITZEL ARNO DR
WOOTTEN HENRY ALWYN
WU SHENGYIN
XU ZHI-CAI
YOUNIS SAAD M
ZHELEZNIAKOV VLADIMIR V
ZIEBA STANISLAW DR

COMMISSION No. 41

HISTORY OF ASTRONOMY (HISTOIRE DE L'ASTRONOMIE)

```
President            : EDDY JOHN A DR

Vice-President(s)    : NORTH JOHN DAVID PROF

Organizing Committee: DEBARBAT SUZANNE V DR
                      EELSALU HEINO DR
                      PEDERSEN OLAF PROF
                      XI ZE-ZONG
```

Members:
ANSARI S M RAZAULLAH PROF / ARGYRAKOS JEAN PROF DR / BADOLATI ENNIO
BERENDZEN RICHARD DR / BISHOP ROY L DR / BO SHU-REN
BRUNET JEAN-PIERRE DR / CARLSON JOHN B / CHEN ZUN-GUI
CIMINO MASSIMO A PROF / CORNEJO ALEJANDRO A DR / DADIC ZARKO DR
DARIUS JON DR / DEEMING T J DR / DEKKER E DR
DEWHIRST DAVID W DR / DICK STEVEN J / DOBRZYCKI JERZY PROF
DOUGLAS A VIBERT DR / DeVORKIN DAVID H / EDMONDSON FRANK K PROF
ERPYLEV N P DR / EVANS DAVID S PROF / FERNIE J DONALD PROF
FERRARI D'OCCHIEPPO K DR / FIRNEIS MARIA G DR / FLORIDES PETROS S PROF
FREIESLEBEN H C DR / FREITAS MOURAO R R DR / GINGERICH OWEN PROF
HAWKINS GERALD S DR / HAYLI AVRAM PROF / HEGGIE DOUGLAS C DR
HERRMANN DIETER DR / HORSKY ZDENEK DR / HOSKIN MICHAEL A DR
HOWSE H DEREK / HYSOM EDMUND J / IDLIS G M DR
JACKISCH GERHARD DR / KENNEDY JOHN E PROF / KHROMOV G S DR
KIANG TAO PROF / KING HENRY C DR / KOTSAKIS DEMETRIUS PROF
KULIKOVSKIJ P G DR / KUNITZSCH PAUL PROF. / LANG KENNETH R ASST PROF
LEVY JACQUES R DR / LI ZHI-SEN / LIU JINYI
MALIN STUART / MCKENNA LAWLOR SUSAN / MEADOWS A JACK PROF
MERLEAU-PONTY J PROF / MOESGAARD KRISTIAN P / NAKAYAMA SHIGERU DR
OMER GUY C JR PROF / ORCHISTON WAYNE DR / OSTERBROCK DONALD E PROF
PELSENEER JEAN KON PROF / PETERSON CHARLES JOHN DR / PETRI WINFRIED PROF DR
PINGREE DAVID PROF / POGO ALEXANDER DR / PORTER NEIL A PROF
POULLE EMMANUEL PROF / PROKAKIS THEODORE J DR / QUAN HEJUN
RONAN COLIN A / RYBKA EUGENIUSZ PROF DR / RYBKA PRZEMYSLAW DR
SATO NAONOBU PROF / SHUKLA K / SIGNORE MONIQUE DR
SOLC MARTIN / STEPHENSON F RICHARD DR / SULLIVAN WOODRUFF T III
SUNDMAN ANITA DR / SVOLOPOULOS SOTIRIOS PROF / SWERDLOW NOEL PROF
TATON RENE PROF / THOREN VICTOR E PROF / VERDET JEAN-PIERRE DR
WATTENBERG D PROF / WHITAKER EWEN A / WHITE GRAEME LINDSAY DR
WHITROW GERALD JAMES PROF / WILSON CURTIS A / WRIGHT HELEN
XU ZHENTAO / YABUUTI KIYOSHI PROF / YEOMANS DONALD K DR
ZHANG PEIYU / ZHUANG WEIFENG / ZOSIMOVICH IRINA D

MEMBERSHIP OF COMMISSIONS

COMMISSION No. 42

CLOSE BINARY STARS (ETOILES DOUBLES SERREES)

President   : SMAK JOSEPH I PROF

Vice-President(s)   : KOCH ROBERT H DR

Organizing Committee: ABHYANKAR KRISHNA D PROF
                     ANDERSEN JOHANNES
                     BATTEN ALAN H DR
                     BUDDING EDWIN DR
                     GIBSON DAVID MICHAEL DR
                     KITAMURA M PROF
                     KONDO YOJI DR
                     LEUNG KAM CHING PROF
                     RAHE JURGEN PROF
                     RODONO MARCELLO DR
                     SHAVIV GIORA DR

Members:

AL-NAIMY HAMID M K DR	ANTONOPOULOU E DR	AWADALLA NABIL SHOUKRY DR
BARKER BRUCE MICHAEL	BARTOLINI CORRADO	BATH GEOFFREY T DR
BLITZSTEIN WILLIAM DR	BOLTON C THOMAS PROF	BONAZZOLA SILVANO DR
BOOKMYER BEVERLY B DR	BOPP BERNARD W DR	BREINHORST ROBERT A DR
BROGLIA PIETRO DR	BROWNLEE ROBERT R DR	BRUHWEILER FRED C JR
BUNNER ALAN N DR	BUSSO MAURIZIO	CATALANO SANTO DR
CESTER BRUNO PROF	CHAMBLISS CARLSON R DR	CHANMUGAM GANESAR PROF
CHAPMAN ROBERT D DR	CHEN KWAN-YU PROF	CHEREPASHCHUK A M PROF
CHOCHOL DRAHOMIR	CHOI KYU-HONG	CILLIE G G PROF
CLAUSEN JENS VIGGO LEKTOR	COLLINS GEORGE W II PROF	COWLEY ANNE P DR
CRISTALDI SALVATORE DR	DADAEV ALEKSANDR N DR	DE GREVE JEAN-PIERRE DR
DE GROOT MART DR	DE KORT JULES J DR	DE LOORE CAMIEL PROF
DELGADO ANTONIO JESUS	DEMIRCAN OSMAN DR	DOUGHTY NOEL A DR
DRECHSEL HORST DR	DUERBECK HILMAR W DR	EATON JOEL A DR
EGGLETON PETER P DR	FAULKNER JOHN PROF	FEKEL FRANCIS C
FERRARI D'OCCHIEPPO K DR	FIRMANI CLAUDIO A PROF	FLANNERY BRIAN PAUL DR
FRACASSINI MASSIMO PROF	FRACASTORO MARIO G PROF	FRANK JUHAN
FRANTSMAN YU L DR	FREDRICK LAURENCE W PROF	FRIEDJUNG MICHAEL DR
GARMANY CATHERINE D DR	GEYER EDWARD H PROF DR	GIANNONE PIETRO PROF
GIMENEZ ALVARO	GIOVANNELLI FRANCO DR	GIURICIN GIULIANO
GRYGAR JIRI DR	GUINAN EDWARD FRANCIS DR	GULLIVER AUSTIN FRASER DR
GURSKY HERBERT DR	GUSEINOV O H PROF	GYLDENKERNE KJELD DR
HALL DOUGLAS S DR	HAMMERSCHLAG-HENSBERGE G	HARMANEC PETR DR
HAZLEHURST JOHN DR	HEINTZ WULFF D DR	HENSLER GERHARD
HERCZEG TIBOR J PROF DR	HILDITCH RONALD W DR	HILL GRAHAM DR
HJELLMING ROBERT M DR	HOFFMANN MARTIN DR	HOLT STEPHEN S
HONEYCUTT R KENT PROF	HORAK TOMAS B DR	HRIVNAK BRUCE J
HUBE DOUGLAS P DR	HUTCHINGS JOHN B DR	IBANOGLU C DR
IMBERT MAURICE DR	IRWIN JOHN B PROF	JABBAR SABEH RHAMAN
JASCHEK CARLOS O R PROF	JOSS PAUL CHRISTOPHER DR	JURKEVICH IGOR DR
KADOURI TALIB HADI	KANDPAL CHANDRA D	KARETNIKOV VALENTIN G R
KAWABATA SHUSAKU PROF	KEMP JAMES C	KOPAL ZDENEK PROF

KOUBSKY PAVEL
KRAUTTER JOACHIM DR
KRON KATHERINE GORDON
KUMSIASHVILY MZIA I DR
KWEE K K DR
LANDOLT ARLO U PROF
LAVROV M I PROF
LUCY LEON B PROF
MACERONI CARLA
MARDIROSSIAN FABIO
MARTYNOV D YA PROF DR
MAYER PAVEL DR
MEYER-HOFMEISTER E DR
MIYAJI SHIGEKI DR
MUMFORD GEORGE S PROF
NATHER R EDWARD
NORDSTROM BIRGITTA DR
OLSON EDWARD C PROF
PACZYNSKI BOHDAN PROF
PATKOS LASZLO DR
PIIROLA VILPPU E DR
POPPER DANIEL M PROF
RAFERT JAMES BRUCE
REFSDAL S PROF DR
ROBERTSON JOHN ALISTAIR
ROVITHIS-LIVANIOU HELEN
SADIK AZIZ R DR
SANWAL N B DR
SCALTRITI FRANCO DR
SCHOBER HANS J DR
SEMENIUK IRENA DR
SHU FRANK H PROF
SIMMONS JOHN FRANCIS L
SMITH ROBERT CONNON DR
SOLHEIM JAN ERIK
STARRFIELD SUMNER PROF
SUGIMOTO DAIICHIRO PROF
SZAFRANIEC ROZALIA DR
THOMPSON KEITH DR
TRIMBLE VIRGINIA L DR
URECHE VASILE DR
VAN HOOF A PROF EM
VAZ LUIZ PAULO RIBEIRO
WALKER RICHARD L
WARD MARTIN JOHN
WEBBINK RONALD F DR
WEILER EDWARD J DR
WILLIAMON RICHARD M
WOOD DAVID B DR
YAMASAKI ATSUMA DR
ZIOLKOWSKI JANUSZ DR

KRAFT ROBERT P PROF
KREINER JERZY MAREK DR
KRUSZEWSKI ANDRZEJ PROF
KURPINSKA-WINIARSKA M DR
LACY CLAUD H DR
LAPASSET EMILIO DR
LINNELL ALBERT P PROF
LYUTY VICTOR M DR
MAGALASHVILI N L DR
MARILLI ETTORE DR
MATTEI JANET AKYUZ DR
MCCLUSKEY GEORGE E JR DR
MEZZETTI MARINO
MOCHNACKI STEPHAN W DR
NAKAMURA YASUHISA
NELSON BURT DR
OKAZAKI AKIRA DR
OSAKI YOJI DR
PADALIA T D DR
PETERS GERALDINE JOAN DR
PLAVEC MIREK J PROF
PRINGLE JAMES E DR
RAHUNEN TIMO
REUNING ERNEST G DR
ROBINSON EDWARD LEWIS DR
ROXBURGH IAN W PROF
SAHADE JORGE PROF
SANYAL ASHIT DR
SCARFE COLIN D DR
SCHOEFFEL EBERHARD F DR
SHAKURA NICHOLAJ I DR
SHUL'BERG A M DR
SINVHAL SHAMBHU DAYAL DR.
SOBIESKI STANLEY DR
SPARKS WARREN M DR
STENCEL ROBERT EDWARD
SUNDMAN ANITA DR
SZKODY PAULA DR
TODORAN IOAN DR
TSESEVICH V P PROF DR
VAN DEN HEUVEL EDWARD P J
VAN PARADIJS JOHANNES DR
VETESNIK MIROSLAV DR
WALKER WILLIAM S G
WARGAU WALTER FRIEDRICH
WEHLAU WILLIAM H PROF
WELLMANN PETER PROF DR
WILLIAMS ROBERT E DR
WOOD F BRADSHAW PROF
ZHAI DI-SHENG
ZUIDERWIJK EDWARDUS J

KRAICHEVA ZDRAVSKA DR
KRIZ SVATOPLUK DR
KRZEMINSKI WOJCIECH DR
KVIZ ZDENEK DR
LAMB DONALD QUINCY JR DR
LARSSON-LEANDER G PROF
LIU XUEFU
MACDONALD JAMES
MAMMANO AUGUSTO DR
MARINO BRIAN F ENG
MAUDER HORST PROF DR
MERRILL JOHN E DR
MILONE EUGENE F PROF
MORGAN THOMAS H DR
NARIAI KYOJI DR
NHA IL-SEONG DR
OLIVER JOHN PARKER DR
OVENDEN MICHAEL W PROF
PARTHASARATHY M DR
PICCIONI ADALBERTO
POLIDAN RONALD S
QIAO GUOJUN
RAKOS KARL D PROF
RITTER HANS DR
ROVITHIS PETER DR
RUCINSKI SLAWOMIR M DR
SAIJO KEIICHI
SAVONIJE GERRIT JAN DR
SCHMIDT HANS PROF
SEGGEWISS WILHELM PROF
SHEN LIANG-ZHAO
SIMA ZDISLAV DR
SISTERO ROBERTO F DR
SOEDERHJELM STAFFAN DR
SRIVASTAVA J B DR
STROHMEIER WOLFGANG PROF
SVECHNIKOV M A DR
TAN HUISONG
TREMKO JOZEF DR
TUTUKOV A V DR
VAN HAMME WALTER
VAN'T VEER FRANS DR
VILHU OSMI DR
WALTER KURT PROF DR
WARNER BRIAN PROF
WEIGERT ALFRED PROF
WESSELINK ADRIAAN J DR
WILSON ROBERT E PROF
WRIGHT KENNETH O DR
ZHU CI-SHENG

COMMISSION No. 44

ASTRONOMY FROM SPACE (L'ASTRONOMIE A PARTIR DE L'ESPACE)

President   : KONDO YOJI DR

Vice-President(s)   : POUNDS KENNETH A PROF

Organizing Committee: BOYARCHUK A A DR
                               CLARK GEORGE W PROF
                               COURTES GEORGES PROF
                               GREWING MICHAEL PROF
                               JENKINS EDWARD B DR
                               MACCHETTO FERDINANDO DR
                               ODA MINORU PROF
                               RAHE JURGEN PROF
                               SHOLOMITSKY G B DR
                               TANAKA YASUO PROF
                               TRUEMPER JOACHIM PROF
                               VAN DER HUCHT KAREL A DR
                               WILLIS ALLAN J DR

Members:

ACTON LOREN W DR	AGRAWAL P C DR	ALEXANDER JOSEPH K
ASCHENBACH BERND PH D	AYRES THOMAS R	BALIUNAS SALLIE L
BENEDICT GEORGE F DR	BERGERON JACQUELINE A DR	BERNACCA P L PROF
BIANCHI LUCIANA	BLAMONT JACQUES E PROF	BLEEKER JOHAN A M DR IR
BLESS ROBERT C PROF	BOGGESS ALBERT DR	BOGGESS NANCY W DR
BOHLIN RALPH C DR	BOKSENBERG ALEC PROF	BONNET ROGER M DR
BOUGERET J L DR	BOWYER C STUART PROF	BOYD ROBERT L F PROF
BRANDT JOHN C DR	BRINKMAN BERT C DR	BRUECKNER GUENTER E DR
BRUHWEILER FRED C JR	BRUNER MARILYN E DR	BUMBA VACLAV DR
BUNNER ALAN N DR	BURGER MARIJKE DR	BURTON WILLIAM M
CAMPBELL MURRAY F	CARROLL P KEVIN PROF	CARVER JOHN H PROF
CATURA RICHARD C DR	CHAPMAN ROBERT D DR	CHARLES PHILIP ALLAN
CHOCHOL DRAHOMIR	CHUBB TALBOT A DR	CLARK THOMAS ALAN DR
CODE ARTHUR D	COURVOISIER THIERRY J.-L.	CRANNELL CAROL JO DR
CULHANE LEONARD PROF	DAVIDSEN ARTHUR FALNES DR	DAVIS ROBERT J DR
DE JAGER CORNELIS PROF	DENNIS BRIAN ROY DR	DI COCCO GUIDO
DOLAN JOSEPH F DR	DUNKELMAN LAWRENCE	DUPREE ANDREA K DR
EL-RAEY MOHAMED E DR	FABRICANT DANIEL G	FARAGGIANA ROSANNA PROF
FAZIO GIOVANNI G DR	FELDMAN PAUL DONALD DR	FERRARI TONIOLO MARCO
FICHTEL CARL E DR	FISHER PHILIP C	FISHMAN GERALD J
FITTON BRIAN DR	FREDGA KERSTIN PROF	FRIEDMAN HERBERT DR
FU CHENG-QI	FURNISS IAN	GABRIEL ALAN H
GEZARI DANIEL YSA DR	GIACCONI RICCARDO PROF	GILRA DAYA P DR
GLASER HAROLD DR	GOLD THOMAS PROF	GOLDBERG LEO PROF
GONDHALEKAR PRABHAKAR DR	GREYBER HOWARD D DR	GRIFFITHS RICHARD E DR
GULL THEODORE R DR	GURSKY HERBERT DR	GUSEINOV O H PROF
HACK MARGHERITA PROF	HADDOCK FRED T DR	HALLAM KENNETH L DR
HAN ZHENG-ZHONG	HANG HENG-RONG	HARMS RICHARD JAMES DR
HARTZ THEODORE R DR	HARVEY CHRISTOPHER C DR	HARVEY PAUL MICHAEL DR

HAYAKAWA SATIO PROF HEARN ANTHONY G DR HECKATHORN HARRY M
HEISE JOHN DR HELMKEN HENRY F DR HENIZE KARL G ASTRONAUT
HENOUX JEAN-CLAUDE DR HENSBERGE HERMAN HINTEREGGER HANS E DR
HOFFMAN JEFFREY ALAN DR HOLBERG JAY B HOLT STEPHEN S
HOUZIAUX L PROF HOWARTH IAN DONALD HOYNG PETER DR
HU WEN-RUI HUBER MARTIN C E DR INOUE HAJIME DR
JAMAR CLAUDE A J DR JORDAN CAROLE DR JORDAN STUART D DR
KAFATOS MINAS DR KARPINSKIJ VADIM N DR KASTURIRANGAN K DR
KOCH-MIRAMOND LYDIE DR KRAEMER GERHARD DR KRAUSHAAR WILLIAM L PROF
KURT V G DR LAMERS H J G L M DR LECKRONE DAVID S DR
LEMAIRE PHILIPPE DR LEWIN WALTER H G PROF LI TIPEI
LI ZHONGYUAN LINDBLAD BERTIL A DR LINSKY JEFFREY L DR
LINSLEY JOHN LOVELL SIR BERNARD PROF LUEST REIMAR PROF
MA YU-QIAN MALAISE DANIEL J DR MALITSON HARRIET H MS
MANARA ALESSANDRO A DR MANDELSTAM S L PROF MANDOLESI NAZZARENO
MARAN STEPHEN P DR MARAR T M K MAROV MIKHAIL YA PROF
MATSUOKA MASARU DR MCCLUSKEY GEORGE E JR DR MCCRACKEN KENNETH G DR
MCWHIRTER R W PETER DR MELNICK GARY J MEWE R DR
MICHALITSIANOS ANDREW MIYAMOTO SIGENORI PROF MODISETTE JERRY L PROF
MONET DAVID G MONFILS ANDRE G PROF MOOS HENRY WARREN DR
MORGAN THOMAS H DR MORTON DONALD C DR MUELLER EDITH A PROF
MURDOCK THOMAS LEE NESS NORMAN F DR NEUPERT WERNER M DR
NORDH H LENNART DR NOVICK ROBERT NOVOTNY VACLAV
NOYES ROBERT W PROF O'MONGAIN EON ODA NAOKI
OGAWARA YOSHIAKI OKUDA TORU OLTHOF HINDERICUS DR
OWEN TOBIAS C PROF PACINI FRANCO PROF PAPAGIANNIS MICHAEL D PRO
PARKINSON JOHN H DR PARKINSON WILLIAM H PETERS GERALDINE JOAN DR
PETERSON LAURENCE E PROF PHILLIPS KENNETH J H PINKAU K PROF
POLIDAN RONALD S PRICE STEPHAN DONALD PROKOF'EV VLADIMIR K PROF
PROSZYNSKI MIECZYSLAW RAO RAMACHANDRA V PROF REES MARTIN J PROF
REEVES EDMOND M DR RENSE WILLIAM A DR RIGHINI-COHEN GIOVANNA DR
ROMAN NANCY G DR ROSENDHAL JEFFREY D DR RUBEN G PROF DR
RUDER HANNS SALOMONOVICH A E DR SATO KATSUHIKO PROF
SAVAGE BLAIR D DR SCHOENEICH W DR SCHULTZ G V DR
SCHWARTZ DANIEL A DR SCHWARTZ STEVEN JAY SELVELLI PIERLUIGI DR
SEVERNYJ A B PROF DR SHIVANANDAN KANDIAH DR SILVESTRO GIOVANNI
SIMON PAUL A DR SIMON PAUL C DR SMITH BRADFORD A PROF
SMITH HARLAN J PROF SMITH HOWARD ALAN SMITH LINDA J
SNOW THEODORE P PROF SOFIA SABATINO PROF SPADA GIANFRANCO DR
SPEER R J DR SPITZER LYMAN JR DR STACHNIK ROBERT V
STAUBERT RUDIGER PROF DR STECHER THEODORE P STEINBERG JEAN-LOUIS DR
STENCEL ROBERT EDWARD STERN ROBERT ALLAN STIER MARK T
STOCKMAN HERVEY S JR DR STONE R G DR SU WAN-ZHEN
TAKAKURA TATSUO PROF EMER THOMAS ROGER J DR TOVMASSIAN H M DR
TRAUB WESLEY ARTHUR UNDERHILL ANNE B DR UNDERWOOD JAMES H DR
UPSON WALTER L II DR VALNICEK BORIS DR VAN BEEK FRANK DR
VAN DE HULST H C PROF DR VAN DUINEN R J DR VAN SPEYBROECK LEON P DR
VIAL JEAN-CLAUDE VIDAL-MADJAR ALFRED DR VIOTTI ROBERTO DR
WALSH DENNIS DR WAMSTEKER WILLEM DR WANG SHUI
WARNER JOHN W DR WEILER EDWARD J DR WEINBERG J L DR
WESSELIUS PAUL R DR WESTPHAL JAMES A PROF WILLNER STEVEN PAUL DR
WILSON ROBERT PROF WRAY JAMES D DR WU CHI CHAO DR
WUNNER GUENTER YAMASHITA KOJUN DR ZARNECKI JAN CHARLES DR
ZOMBECK MARTIN V DR ZOU HUI-CHENG

COMMISSION No. 45

STELLAR CLASSIFICATION (CLASSIFICATION STELLAIRE)

President   : GARRISON ROBERT F PROF

Vice-President(s) : GOLAY MARCEL PROF

Organizing Committee: CLARIA JUAN DR
HECK ANDRE DR
HOUK NANCY DR
LLOYD-EVANS THOMAS DR
MACCONNELL DARRELL J DR
OLSEN ERIK H
STRAIZYS V PROF DR

Members:
ALBERS HENRY PROF
BARBIER-BROSSAT M DR
BELL ROGER A DR
BUSCOMBE WILLIAM PROF
CESTER BRUNO PROF
CORBALLY CHRISTOPHER
CRAWFORD DAVID L DR
EGRET DANIEL DR
FEHRENBACH CHARLES PROF
GERBALDI MICHELE DR
GUETTER HARRY HENDRIK
HALLAM KENNETH L DR
HENIZE KARL G ASTRONAUT
HILTNER W ALBERT PROF
JASCHEK CARLOS O R PROF
KHARADZE E K PROF
LEVATO ORLANDO HUGO DR
LUTZ JULIE H DR
MALARODA STELLA M DR
MCNAMARA DELBERT H DR
MORGAN WILLIAM W PROF
NANDY KASHINATH DR
NORTH PIERRE
OSAWA KIYOTERU DR
PASINETTI LAURA E PROF
PRESTON GEORGE W DR
ROUNTREE JANET DR
SANG GAK LEE
SCHMIDT-KALER TH PROF
SINNERSTAD ULF E PROF
STEPHENSON C BRUCE PROF
STROMGREN BENGT PROF
WALKER GORDON A H PROF
WESSELIUS PAUL R DR
WILLIAMS JOHN A DR
WYCKOFF SUSAN DR

ARDEBERG ARNE L PROF
BARRY DON C DR
BIDELMAN WILLIAM P PROF
BUSER ROLAND DR
CHEREPASHCHUK A M PROF
COWLEY ANNE P DR
DIVAN LUCIENNE DR
ELVIUS TORD PROF EMERITUS
FRACASSINI MASSIMO PROF
GEYER EDWARD H PROF DR
GURZADIAN G A PROF DR
HAUCK BERNARD PROF
HERMAN RENEE DR
HUANG LIN
JASCHEK MERCEDES DR
KRON GERALD E DR
LODEN KERSTIN R DR
LYNGA GOSTA DR
MCCARTHY MARTIN F DR
MEAD JAYLEE MONTAGUE DR
MORGULEFF NINA ING
NICOLET BERNARD
NOTNI P DR
OSBORN WAYNE DR
PERRY CHARLES L DR
RAUTELA B S DR
RUDKJOBING MOGENS PROF
SANWAL N B DR
SEITTER WALTRAUT C PROF
SLETTEBAK ARNE PROF
STOCK JURGEN D
UPGREN ARTHUR R DR
WARNER BRIAN PROF
WEST RICHARD M DR
WING ROBERT F PROF
YOSS KENNETH M DR

BAHNG JOHN D R PROF
BARTAYA R A DR
BLANCO VICTOR M DR
CELIS LEOPOLDO DR
CHRISTY JAMES WALTER DR
CRAMPTON DAVID DR
DUFLOT MARCELLE DR
FEAST MICHAEL W DR
FUKUDA ICHIRO
GLAGOLEVSKIJ JU V DR
HACK MARGHERITA PROF
HAYES DONALD S DR
HILL PHILIP W DR
HUMPHREYS ROBERTA M PROF
KEENAN PHILIP C PROF EMER
LABHARDT LUKAS
LOW FRANK J DR
MAEHARA HIDEO DR
MCCLURE ROBERT D PROF
MENDOZA V EUGENIO E DR
MOROSSI CARLO
NIKONOV V B DR
OJA TARMO PROF
PARSONS SIDNEY B DR
PHILIP A G DAVIS
ROMAN NANCY G DR
SANDULEAK NICHOLAS DR
SCHILD RUDOLPH E DR
SHARPLESS STEWART PROF
STEINLIN ULI PROF
STROBEL ANDRZEJ DR
WALBORN NOLAN R DR
WARREN WAYNE H JR DR
WESTERLUND BENGT E PROF
WU HSIN-HENG DR
ZDANAVICIUS KAZIMERAS DR

COMMISSION No. 46
TEACHING OF ASTRONOMY
(L'ENSEIGNEMENT DE L'ASTRONOMIE)

President   : IWANISZEWSKA CECYLIA DR

Vice-President(s)   : SANDQVIST AAGE DR

Organizing Committee: FERRAZ-MELLO S PROF DR
                      GERBALDI MICHELE DR
                      GOUGUENHEIM LUCIENNE
                      HIDAJAT BAMBANG PROF DR
                      HOUZIAUX L PROF
                      KLECZEK JOSIP DR
                      KONONOVICH EDWARD V DR
                      MIAO YONG-KUAN
                      PERCY JOHN R PROF
                      ROBBINS R ROBERT PROF
                      WENTZEL DONAT G DR

Members:
ACKER AGNES PROF DR   AIAD A DR   ANDRILLAT HENRI L PROF
ANSARI S M RAZAULLAH PROF   BARKHATOVA KLAUDIA PROF   BAROCAS VINICIO PROF
BATTANER EDUARDO DR   BOTEZ ELVIRA DR   BOTTINELLI LUCETTE DR
BRAES L L E DR   BRIEVA EDUARDO PROF   BUSCOMBE WILLIAM PROF
BYRNE PATRICK B DR   CATALA MARIA ASUNCION   DR   CHAMBERLAIN JOSEPH M DR
CLARKE DAVID DR   CLIMENHAGA JOHN L PROF   CODINA LANDABERRY SAYD J
DARIUS JON DR   DOMINKO FRAN PROF DR   DOUGHTY NOEL A DR
DUPUY DAVID L DR   ELGAROY OYSTEIN PROF   FAWELL DEREK R DR
FENG KE-JIA   FERNANDEZ-FIGUEROA M J DR   FUENMAYOR FRANCISCO J DR
GURM HARDEV S PROF   HAUCK BERNARD PROF   HAUPT HERMANN F PROF
HERNANDEZ CARLOS ALBERTO   HOFF DARREL BARTON   ISOBE SYUZO DR
JARRETT ALAN H PROF   JORGENSEN HENNING E PROF   KELLER HANS ULRICH DR
KENNEDY JOHN E PROF   KITCHIN CHRISTOPHER R DR   KOURGANOFF VLADIMIR PROF
KREINER JERZY MAREK DR   LAGO MARIA TERESA V T MS   LOMB NICHOLAS RALPH DR
MADDISON RONALD CH DR   MARIK MIKLOS DR.   MARSH JULIAN C D
MAVRIDIS L N PROF   MCCARTHY MARTIN F DR   MCNALLY DEREK DR
MILES HOWARD G MR   MUELLER EDITH A PROF   NICOLOV NIKOLAI S DR
OJA HEIKKI DR   OKOYE SAMUEL E PROF   OSBORN WAYNE DR
OSORIO JOSE J S P PROF   OTHMAN MAZLAN   OWAKI NAOAKI DR
PASACHOFF JAY M PROF   PEERY BENJAMIN F PROF   PROVERBIO EDOARDO PROF
RAMADURAI SOURIRAJA DR   REGO FERNANDEZ M DR   RIGUTTI MARIO PROF
ROBINSON LEIF J   RODGERS ALEX W DR   RODRIGUEZ LUIS F
ROSLUND CURT DR   ROY ARCHIE E PROF   SAFKO JOHN L
SANWAL N B DR   SAXENA P P DR   SCHLOSSER WOLFHARD PROF
SCHMIDT THOMAS DR   SEYMOUR P A H   SHEN CHUN-SHAN
SHIPMAN HENRY L DR   TAYLOR KENNETH N R PROF   TRITTON KEITH P DR
TROCHE-BOGGINO A E DR   VUJNOVIC VLADIS DR   WEST RICHARD M DR
WOO JONG OK   YI ZHAO-HUA   ZEALEY WILLIAM J DR
ZEILIK MICHAEL II DR   ZIMMERMANN HELMUT DR

MEMBERSHIP OF COMMISSIONS

COMMISSION No. 47
COSMOLOGY (COSMOLOGIE)

President   : SETTI GIANCARLO PROF

Vice-President(s) : SATO HUMITAKA PROF

Organizing Committee: AUDOUZE JEAN PROF
DE VAUCOULEURS GERARD PR
GUNN JAMES E PROF
HAYAKAWA SATIO PROF
LI ZHI-FANG
LONGAIR M S PROF
NOVIKOV I D DR
TAMMANN G ANDREAS PROF DR
TRIMBLE VIRGINIA L DR

Members:
AIZU KO PROF
ANDRILLAT HENRI L PROF
BARBERIS BRUNO
BARROW JOHN DAVID
BEL NICOLE J DR
BHAVSAR SUKETU P
BOKSENBERG ALEC PROF
BONNOR W B PROF
CALVANI MASSIMO DR
CHENG FU-ZHEN
CHU YAOQUAN
DADHICH NARESH DR
DATTA BHASKAR DR
DAVIS MICHAEL M DR
DEMARET JACQUES DR
DOROSHKEVICH A G DR
EHLERS JURGEN PROF
FABER SANDRA M PROF
FANG LI-ZHI
FLORIDES PETROS S PROF
FORMAN WILLIAM RICHARD DR
FUKUI TAKAO DR
GIURICIN GIULIANO
GOLDSMITH DONALD W. DR.
GREGORY STEPHEN ALBERT DR
HACYAN SHAHEN DR.
HARMS RICHARD JAMES DR
HAYASHI CHUSHIRO PROF
HELLER MICHAEL PROF
HUCHRA JOHN PETER DR
IYER B R DR
JIANG SHUDING
JUSZKIEWICZ ROMAN
KASPER U DR
ALFVEN HANNES PROF
AULUCK FAQIR CHAND PROF
BARDEEN JAMES M PROF
BASU DIPAK DR
BERGERON JACQUELINE A DR
BICKNELL GEOFFREY V DR
BOND JOHN RICHARD
BRECHER KENNETH PROF
CAVALIERE ALFONSO G PROF
CHINCARINI GUIDO L DR
COHEN JEFFREY M DR
DANESE LUIGI DR
DAVIDSON WILLIAM PROF
DE ZOTTI GIANFRANCO DR
DICKE ROBERT H PROF
DULTZIN DEBORAH DR
EINASTO JAAN DR
FALK SYDNEY W JR DR
FELTEN JAMES E DR
FONG RICHARD
FRENK CARLOS S
GALLETTO DIONIGI
GODART ODON PROF
GORET PHILIPPE DR
GREYBER HOWARD D DR
HARA KEN NOSUKE DR
HARRISON EDWARD R PROF
HE XIANG-TAO
HEWETT PAUL
ICKE VINCENT DR
JAROSZYNSKI MICHAL
JONES BERNARD J T DR
KAPOOR RAMESH CHANDER
KATO SHOJI PROF
ALLAN PETER M
BALDWIN JOHN E DR
BARNOTHY JENO DR PROF
BECKMAN JOHN E PROF
BERTOLA FRANCESCO PROF
BLUDMAN SIDNEY A PROF
BONDI HERMANN PROF SIR
BURBIDGE GEOFFREY R PROF
CHENG FU-HUA
CHITRE DATTAKUMAR M DR
CONDON JAMES J DR
DAS P K DR
DAVIES PAUL CHARLES W
DEKEL AVISHAI
DIONYSIOU DEMETRIOS D DR
DYER CHARLES CHESTER DR
ELLIS GEORGE F R PROF
FALL S MICHAEL DR
FIELD GEORGE B PROF
FORD HOLLAND C RES PROF
FUJIMOTO MITSUAKI DR
GELLER MARGARET JOAN
GOLD THOMAS PROF
GRATTON LIVIO PROF
GRISHCHUK L P DR
HARDY EDUARDO
HAWKING STEPHEN W PROF
HEIDMANN JEAN DR
HOYLE FRED SIR
IKEUCHI SATORU DR
JAUNCEY DAVID L DR
JOSHI MOHAN N PROF
KARACHENTSEV I D DR
KAWABATA KINAKI PROF

KELLERMANN KENNETH I DR
KEMBHAVI AJIT K
KIM JIK SU
KODAMA HIDEO
KORMENDY JOHN DR
KOVETZ ATTAY PROF
KOZLOVSKY B Z DR
KUSTAANHEIMO PAUL E PROF
LACHIEZE-REY MARC
LAKE KAYLL WILLIAM DR
LASOTA JEAN-PIERRE DR
LAUSBERG ANDRE DR
LAYZER DAVID PROF
LEQUEUX JAMES DR
LIEBSCHER DIERCK-E DR
LIU LIAO
LIU YONG-ZHEN
LU TAN
LUMINET JEAN-PIERRE
MACCALLUM MALCOLM A H
MANDOLESI NAZZARENO
MARANO BRUNO
MARDIROSSIAN FABIO
MAREK JOHN
MATERNE JUERGEN DR
MATHER JOHN CROMWELL
MATSUMOTO TOSHIO DR
MAVRIDES STAMATIA DR
MCCREA J DERMOTT
MCCREA WILLIAM SIR
MCVITTIE GEORGE C PROF
MERAT PARVIZ
MESZAROS PETER DR
MEZZETTI MARINO
MISNER CHARLES W PROF
MORRISON PHILIP PROF
MULLER RICHARD A
NARIAI HIDEKAZU
NARLIKAR JAYANT V PROF
NEEMAN YUVAL PROF
NISHIDA MINORU PROF
NOERDLINGER PETER D PROF
NOONAN THOMAS W PROF
NOTTALE LAURENT
O'CONNELL ROBERT WEST DR
OEMLER AUGUSTUS JR DR
OKOYE SAMUEL E PROF
OMER GUY C JR PROF
OMNES ROLAND PROF
OORT JAN H PROF
OZERNOY LEONID M PROF
OZSVATH I PROF
PAAL GYORGY DR
PACHNER JAROSLAV PROF
PADMANABHAN T DR
PAGE DON NELSON
PAN RONG-SHI
PEACOCK JOHN ANDREW
PECKER JEAN-CLAUDE PROF
PEEBLES P JAMES E
PENZIAS ARNO A DR
PERRYMAN MICHAEL A C
PERSIDES SOTIRIOS C
PETERSON BRUCE A DR
PETROSIAN VAHE PROF
PRESS WILLIAM H DR
QU QIN-YUE
RAMELLA MASSIMO
REES MARTIN J PROF
REEVES HUBERT PROF
RINDLER WOLFGANG PROF
RIVOLO ARTHUR REX
ROBERTS DAVID HALL DR
ROBINSON I PROF
ROEDER ROBERT C PROF
ROXBURGH IAN W PROF
RUBIN VERA C DR
RUDDY VINCENT P DR
RUDNICK LAWRENCE DR
SAAR ENN DR
SALVADOR-SOLE EDUARDO
SAPAR ARVED DR
SARGENT WALLACE L W DR
SASAKI MISAO
SATO KATSUHIKO PROF
SAVAGE ANN DR
SCHATZMAN EVRY PROF
SCHEUER PETER A G DR
SCHMIDT MAARTEN PROF
SCHNEIDER JEAN
SCHRAMM DAVID N PROF
SCHUECKING E L DR
SCHULTZ G V DR
SCIAMA DENNIS W DR
SCOTT ELIZABETH L PROF
SEGAL IRVING E DR
SEIDEN PHILIP E
SEIELSTAD GEORGE A
SERSIC J L DR
SHAVER PETER A DR
SHAVIV GIORA DR
SHIVANANDAN KANDIAH DR
SIGNORE MONIQUE DR
SIMON RENE L E PROF
SISTERO ROBERTO F DR
SIVARAM C DR
SMITH HARDING E JR DR
SMOOT III GEORGE F.
SOKOLOWSKI LECH
SPYROU NICOLAOS PROF
STECKER FLOYD W DR
STEIGMAN GARY PROF
STEWART JOHN MALCOLM DR
STOEGER WILLIAM R DR
STRUBLE MITCHELL F
SUBRAHMANYA C R
SUNYAEV R A DR
SURDEJ JEAN M G
TARTER JILL C DR
TAUBER GERALD E PROF
TAYLER ROGER J PROF
THOMPSON LAIRD A DR
THUAN TRINH XUAN DR
TIFFT WILLIAM G PROF
TOMIMATSU AKIRA DR
TOMITA KENJI PROF
TREDER H J PROF DR
TREMAINE SCOTT DUNCAN
TREVESE DARIO
TULLY RICHARD BRENT DR
TURNER EDWIN L DR
TURNER MICHAEL S
VAIDYA P C PROF
VAN DER LAAN H PROF DR
VANYSEK VLADIMIR PROF
VISHNIAC ETHAN T
VISHVESHWARA C V PROF
VON BORZESZKOWSKI H H DR
WAGONER ROBERT V PROF
WANG RENCHUAN
WEBSTER ADRIAN S DR
WEINBERG STEVEN DR
WESSON PAUL S DR
WHEELER JOHN A DR
WHITE SIMON DAVID MANION
WHITROW GERALD JAMES PROF
WILKINSON DAVID T
WILL CLIFFORD M DR
WILSON ALBERT G DR
WOLTJER LODEWIJK PROF
WRIGHT EDWARD L DR
XANTHOPOULOS B C DR
XIANG SHOUPING
XIAO XING HUA
YANG LAN-TIAN
ZEL'DOVICH YA B ACAD
ZEL'MANOV A L DR
ZHANG JIA-LU
ZHANG ZHEN-JIU
ZHOU YOU-YUAN
ZHU SHI-CHANG
ZHU XINGFENG
ZIEBA ANDRZEJ PROF
ZIEBA STANISLAW DR
ZOU ZHEN-LONG
ZUIDERWIJK EDWARDUS J

MEMBERSHIP OF COMMISSIONS

COMMISSION No. 48

HIGH-ENERGY ASTROPHYSICS

(ASTROPHYSIQUE DES HAUTES ENERGIES)

President   : CESARSKY CATHERINE J DR

Vice-President(s)   : SUNYAEV R A DR

Organizing Committee: CLARK GEORGE W PROF
                      GIACCONI RICCARDO PROF
                      PACINI FRANCO PROF
                      QU QIN-YUE
                      SALPETER EDWIN E PROF
                      SCHEUER PETER A G DR
                      SCHRAMM DAVID N PROF
                      TRIMBLE VIRGINIA L DR
                      TRUEMPER JOACHIM PROF
                      WOLFENDALE ARNOLD W PROF
                      WOLTJER LODEWIJK PROF

Members:
ABRAMOWICZ MAREK DR   ADAMS DAVID J DR   AHLUWALIA HARJIT SINGH DR
AIZU KO PROF   ALFVEN HANNES PROF   ALVAREZ LUIS DR
APPARAO K M V DR   ARNOULD MARCEL L DR   ARONS JONATHAN
ASCHENBACH BERND PH D   ASSEO ESTELLE DR   AUDOUZE JEAN PROF
AVNI YORAM DR   AXFORD W IAN PROF   BAAN WILLEM A
BARNOTHY JENO DR PROF   BASU DIPAK DR   BAYM GORDON ALAN DR
BECKER ROBERT HOWARD   BEGELMAN MITCHELL CRAIG   BENFORD GREGORY DR
BERGERON JACQUELINE A DR   BICKNELL GEOFFREY V DR   BIERMANN PETER L DR
BISWAS SUKUMAR DR   BLANDFORD ROGER DAVID DR   BLEEKER JOHAN A M DR IR
BLUDMAN SIDNEY A PROF   BONAZZOLA SILVANO DR   BONOMETTO SILVIO A DR
BOYD ROBERT L F PROF   BRECHER KENNETH PROF   BUNNER ALAN N DR
BURBIDGE GEOFFREY R PROF   BURROWS ADAM SETH   CAMERON ALASTAIR G W PROF
CASH WEBSTER C JR   CASSE MICHEL DR   CATURA RICHARD C DR
CAUGHLAN GEORGEANNE R   CAVALIERE ALFONSO G PROF   CHANDRASEKHAR S PROF
CHITRE SHASHIKUMAR M DR   CHUBB TALBOT A DR   CHUPP EDWARD L DR
COHEN JEFFREY M DR   COLLIN-SOUFFRIN SUZY DR   CONDON JAMES J DR
COWIE LENNOX LAUCHLAN DR   COWSIK RAMANATH   CRUISE ADRIAN MICHAEL DR
CULHANE LEONARD PROF   CURIR ANNA   DA COSTA JOSE MARQUES DR
DAUTCOURT G DR   DAVIDSEN ARTHUR FALNES DR   DAVIDSON WILLIAM PROF
DAVIS LEVERETT JR PROF   DAVIS MICHAEL M DR   DE FELICE FERNANDO DR
DE GRAAF T DR   DE YOUNG DAVID S DR   DEBRUNNER HERMANN DR
DENNIS BRIAN ROY DR   DEWITT BRYCE S DR   DICKE ROBERT H PROF
DISNEY MICHAEL J PROF   DOLAN JOSEPH F DR   DRAKE FRANK D PROF
DUTHIE JOSEPH G PROF   EDWARDS PAUL J DR   EICHLER DAVID DR
EILEK JEAN   EVANS W DOYLE   FABIAN ANDREW C DR
FANG LI-ZHI   FAZIO GIOVANNI G DR   FELTEN JAMES E DR
FENTON K B DR   FERRARI ATTILIO DR   FICHTEL CARL E DR
FIELD GEORGE B PROF   FISHER PHILIP C   FORMAN WILLIAM RICHARD DR
FOWLER WILLIAM A PROF   FRIEDMAN HERBERT DR   GAISSER THOMAS K

GALEOTTI PIERO PROF   GARMIRE GORDON P PROF   GINZBURG VITALY L PROF
GOLD THOMAS PROF   GOLDSMITH DONALD W. DR.   GONZALES-A WALTER D DR
GREISEN KENNETH I PROF   GREWING MICHAEL PROF   GREYBER HOWARD D DR
GRIFFITHS RICHARD E DR   GRINDLAY JONATHAN E DR   GUNN JAMES E PROF
GURSKY HERBERT DR   GUSEINOV O H PROF   HALL ANDREW NORMAN
HANG HENG-RONG   HARWIT MARTIN PROF   HAWKING STEPHEN W PROF
HAYAKAWA SATIO PROF   HEISE JOHN DR   HELFAND DAVID JOHN
HENRIKSEN RICHARD N DR   HENRY RICHARD C. PROF.   HOANG BINH DY DR
HOFFMAN JEFFREY ALAN DR   HOLLOWAY NIGEL J DR   HOYLE FRED SIR
HUANG KE-LIANG   ICHIMARU SETSUO DR   INOUE HAJIME DR
IPSER JAMES R PROF   ISRAEL WERNER PROF   ITO KENSAI A PROF
JACKSON J C DR   JAFFE WALTER JOSEPH DR   JELLEY JOHN V PHD
JOKIPII J R PROF   JONES FRANK CULVER DR   JONES THOMAS WALTER DR
JOSS PAUL CHRISTOPHER DR   KAFKA PETER   KAHN FRANZ D PROF
KATZ JONATHAN I   KELLERMANN KENNETH I DR   KELLOGG EDWIN M DR
KEMBHAVI AJIT K   KIRK JOHN DR   KOCH-MIRAMOND LYDIE DR
KOCHAROV GRANT E PROF   KONDO MASAAKI DR   KOZLOWSKI MACIEJ DR
KREISEL E PROF   KRISTIANSSON KRISTER PROF   KULSRUD RUSSELL M DR
KUNDT WOLFGANG PROF DR   KURT V G DR   LAMB DONALD QUINCY JR DR
LAMB FREDERICK K PROF   LAMB SUSAN ANN DR   LAMPTON MICHAEL
LASHER GORDON JEWETT DR   LATTIMER JAMES M DR   LEA SUSAN MAUREEN DR
LI QI-BIN   LI TIPEI   LI ZONG-WEI
LINSLEY JOHN   LIU JINYI   LIU RU-LIANG
LONGAIR M S PROF   LOVELACE RICHARD V E DR   LU TAN
LUEST REIMAR PROF   LUMINET JEAN-PIERRE   LYNDEN-BELL DONALD PROF
MA YU-QIAN   MACCACARO TOMMASO DR   MACCAGNI DARIO
MACCHETTO FERDINANDO DR   MARTIN INACIO MALMONGE DR   MASON KEITH OWEN
MATSUOKA MASARU DR   MAZUREK THADDEUS JOHN DR   MCBREEN BRIAN PHILIP DR
MCCRAY RICHARD DR   MELROSE DONALD B PROF   MESTEL LEON PROF
MESZAROS PETER DR   MEYER FRIEDRICH DR   MEYER JEAN-PAUL DR
MICHEL F CURTIS PROF   MILLER JOHN C DR   MIYAJI SHIGEKI DR
MIYAMOTO SIGENORI PROF   MORRISON PHILIP PROF   NEEMAN YUVAL PROF
NOVICK ROBERT   O'CONNELL ROBERT F PROF   O'SULLIVAN DENIS F
ODA MINORU PROF   OKOYE SAMUEL E PROF   OSTRIKER JEREMIAH P PROF
OZERNOY LEONID M PROF   PACHOLCZYK ANDRZEJ G PROF   PAGE CLIVE G DR
PALUMBO GIORGIO G C DR   PARKER EUGENE N   PARKINSON JOHN H DR
PAULINY TOTH IVAN K K DR   PENG QIU-HE   PEROLA G C DR
PETERSON BRUCE A DR   PETERSON LAURENCE E PROF   PETROSIAN VAHE PROF
PIDDINGTON JACK H RES FEL   PINKAU K PROF   PORTER NEIL A PROF
POUNDS KENNETH A PROF   PREUSS EUGEN DR   QUINTANA HERNAN DR
RADHAKRISHNAN V PROF   RAMADURAI SOURIRAJA DR   RAUBENHEIMER BAREND C
RAZDAN HIRALAL   REES MARTIN J PROF   REEVES HUBERT PROF
RENGARAJAN T N DR   ROSNER ROBERT   ROSSI BRUNO B PROF
SALVATI MARCO   SANDERS WILTON TURNER III   SARTORI LEO PROF
SASLAW WILLIAM C PROF   SAVEDOFF MALCOLM P PROF   SCARGLE JEFFREY D DR
SCHATTEN KENNETH H DR   SCHATZMAN EVRY PROF   SCHILIZZI RICHARD T DR
SCHWARTZ DANIEL A DR   SCIAMA DENNIS W DR   SCOTT JOHN S DR
SEIELSTAD GEORGE A   SETTI GIANCARLO PROF   SHAHAM JACOB PROF
SHAKURA NICHOLAJ I DR   SHAPIRO MAURICE M PROF   SHAVER PETER A DR
SHAVIV GIORA DR   SHIELDS GREGORY A DR   SHUKRE C S DR
SIGNORE MONIQUE DR   SILBERBERG REIN DR   SKILLING JOHN DR
SMITH BARHAM W DR   SMITH F GRAHAM PROF   SOFIA SABATINO PROF
SPADA GIANFRANCO DR   STAUBERT RUDIGER PROF DR   STEIGMAN GARY PROF
STEPANIAN A A DR   STOCKMAN HERVEY S JR DR   STRONG IAN B DR
STURROCK PETER A PROF   SWANK JEAN HEBB   TAKAHARA FUMIO DR
TAYLER ROGER J PROF   TERRELL NELSON JAMES JR   THORNE KIP S PROF
TOMIMATSU AKIRA DR   TRURAN JAMES W JR   TSURUTA SACHIKO DR

VAN RIPER KENNETH A DR	VIDAL NISSIM V DR	VOELK HEINRICH J PROF
WANG SHOU-GUAN	WANG ZHEN-RU	WEAVER THOMAS A DR
WEBSTER ADRIAN S DR	WEISHEIT JON C DR	WEISSKOPF MARTIN CH DR
WENTZEL DONAT G DR	WESTFOLD KEVIN C PROF	WHEELER JOHN A DR
WILL CLIFFORD M DR	WILSON JAMES R DR	WOLSTENCROFT RAMON D DR
WORRALL DIANA MARY	YANG HAI SHOU	YANG LAN-TIAN
YOU JUNHAN	ZEL'DOVICH YA B ACAD	ZHANG HE-QI
ZHANG JIA-LU	ZHANG ZHEN-JIU	ZOMBECK MARTIN V DR

COMMISSION No. 49

THE INTERPLANETARY PLASMA AND THE HELIOSPHERE

(LE PLASMA INTERPLANETAIRE ET L'HELIOSPHERE)

President   : GRZEDZIELSKI STANISLAW PR

Vice-President(s)   : BURLAGA LEONARD F DR

Organizing Committee: BUTI BIMLA PROF
　　　　　　　　　　　CUPERMAN SAMI PROF
　　　　　　　　　　　DOLGINOV ARKADY Z PROF DR
　　　　　　　　　　　FAHR HANS JOERG PROF DR
　　　　　　　　　　　KELLER HORST UWE DR
　　　　　　　　　　　PARESCE FRANCESCO DR
　　　　　　　　　　　ROXBURGH IAN W PROF
　　　　　　　　　　　SUESS STEVEN T DR

Members:
AHLUWALIA HARJIT SINGH DR	ANANTHAKRISHNAN S	ANDERSON KINSEY A PROF
ANTONUCCI ESTER DR	BARNES AARON DR	BARROW COLIN H DR
BARTH CHARLES A PROF	BERTAUX J L DR	BLACKWELL DONALD E PROF
BLUM PETER PROF	BOCHSLER PETER	BONNET ROGER M DR
BRANDT JOHN C DR	CHAMBERLAIN JOSEPH W PROF	CHEN BIAO
DE JAGER CORNELIS PROF	DELACHE PHILIPPE J DR	DRYER MURRAY DR
DURNEY BERNARD DR	DYSON JOHN E DR	ERGMA E V DR
ESHLEMAN VON R PROF	EVIATAR AHARON PROF	FEYNMAN JOAN DR
FIELD GEORGE B PROF	GOSLING JOHN T DR	HABBAL SHADIA RIFAI
HARVEY CHRISTOPHER C DR	HEYVAERTS JEAN DR	HOLLWEG JOSEPH V
HOLZER THOMAS EDWARD DR	IONSON JAMES ALBERT	JOKIPII J R PROF
KAKINUMA TAKAKIYO T PROF	LAFON JEAN-PIERRE J DR	LEVY EUGENE H DR
LI XIAO-QING	LOTOVA N A DR	LUEST REIMAR PROF
MACQUEEN ROBERT M DR	MANGENEY ANDRE DR	MASON GLENN M
MATSUURA OSCAR T DR	MAVROMICHALAKI HELEN DR	MENDIS DEVAMITTA ASOKA DR
MESTEL LEON PROF	MICHEL F CURTIS PROF	MOUSSAS XENOPHON PH D
NAKAGAWA YOSHINARI DR	PARKER EUGENE N	PERKINS FRANCIS W DR
PFLUG KLAUS DR	PNEUMAN GERALD W	RAADU MICHAEL A DR
READHEAD ANTHONY C S DR	REAY NEWRICK K DR	RICKETT BARNABY JAMES DR
RIDDLE ANTHONY C DR	RIPKEN HARTMUT W DR	ROACH FRANKLIN E
ROSNER ROBERT	RUSSELL CHRISTOPHER T	SARRIS EMMANUEL T PH D
SASTRI HANUMATH J DR	SAWYER CONSTANCE B DR	SCHATZMAN EVRY PROF
SCHERB FRANK PROF	SCHINDLER KARL PROF DR	SCHMIDT H U DR
SCHREIBER ROMAN	SCHWARTZ STEVEN JAY	SETTI GIANCARLO PROF
SEVERNYJ A B PROF DR	SHAWHAN STANLEY D DR	SHEA MARGARET A DR
SMITH DEAN F DR	SONETT CHARLES P PROF	STOKER PIETER H
STONE R G DR	STURROCK PETER A PROF	TRITAKIS BASIL P DR
VAINSTEIN L A DR	VAN ALLEN JAMES A PROF	VERHEEST FRANK ASSOC PROF
VINOD S KRISHAN MRS DR	WALLIS MAX K DR	WATANABE TAKASHI DR
WELLER CHARLES S DR	WILD JOHN PAUL DR	WU SHI TSAN DR

COMMISSION No. 50

PROTECTION OF EXISTING AND POTENTIAL OBSERVATORY SITES

(PROTECTION DES SITES D'OBSERVATOIRES EXISTANTS ET POTENTIELS)

President   : VAN DEN BERGH SIDNEY PROF

Vice-President(s)   : CRAWFORD DAVID L DR

Organizing Committee: BLANCO CARLO DR
                      BLANCO VICTOR M DR
                      COYNE GEORGE V DR
                      JIANG SHI-YANG
                      KOZAI YOSHIHIDE PROF
                      MURDIN PAUL G DR
                      PANKONIN VERNON LEE DR
                      SHCHEGLOV P V DR
                      TORRES CARLOS ALBERTO DR
                      TREMKO JOZEF DR
                      WALKER MERLE F PROF

Members:

ALY M KHAIRY PROF	ARDEBERG ARNE L PROF	ARIAS DE GREIFF J PROF
BARRETO LUIZ MUNIZ PROF	BENSAMMAR SLIMANE DR	BHATTACHARYYA J C PROF
BURSTEIN DAVID	CAYREL ROGER DR	DAVIS JOHN DR
DOMMANGET J DR	DUNKELMAN LAWRENCE	EDWARDS PAUL J DR
GALAN MAXIMINO J	GOEBEL ERNST DR	HELMER LEIF
HIDAJAT BAMBANG PROF DR	HOAG ARTHUR A DR	HUANG YINLIANG
KUBICELA ALEKSANDAR DR	LEIBOWITZ ELIA M DR	MAHRA H S DR
MARKKANEN TAPIO DR	MARX SIEGFRIED DR	MATTIG W PROF DR
MAVRIDIS L N PROF	MCCARTHY MARTIN F DR	MENZIES JOHN W DR
NELSON BURT DR	OSORIO JOSE J S P PROF	SANCHEZ FRANCISCO PROF
SCHILIZZI RICHARD T DR	SMITH F GRAHAM PROF	TORRES CARLOS DR
WAYMAN PATRICK A PROF	WOOLF NEVILLE J	WOSZCZYK ANDRZEJ PROF
WU MING-CHAN	ZHANG BAI-RONG	

COMMISSION No. 51

BIOASTRONOMY : SEARCH FOR EXTRATERRESTRIAL LIFE

(BIOASTRONOMIE : RECHERCHE DE LA VIE DANS L'UNIVERS)

```
President          :  DRAKE FRANK D PROF

Vice-President(s)  :  KARDASHEV N S DR
                      MARX GYORGY PROF.

Organizing Committee: BROWN RONALD D PROF
                      CONNES PIERRE DR
                      GATEWOOD GEORGE DIRECTOR
                      GOLDBERG LEO PROF
                      JUGAKU JUN DR
                      PACINI FRANCO PROF
                      REES MARTIN J PROF
                      TROITSKY V S PROF DR
```

Members:

AL-NAIMY HAMID M K DR	AL-SABTI ABDUL ADIM DR	ALMAR IVAN PROF
AMBARTSUMIAN V A PROF DR	ANDO HIROYASU DR	BAKOS GUSTAV A PROF
BALAZS BELA A DR	BALL JOHN A DR	BANIA THOMAS MICHAEL
BARBIERI CESARE PROF	BASU BAIDYANATH PROF	BASU DIPAK DR
BAUM WILLIAM A DR	BEAUDET GILLES DR	BECKMAN JOHN E PROF
BECKWITH STEVEN V W	BEEBE RETA FAYE DR	BENEST DANIEL DR
BERENDZEN RICHARD DR	BERNACCA P L PROF	BILLINGHAM JOHN
BIRAUD FRANCOIS DR	BLESS ROBERT C PROF	BOWYER C STUART PROF
BOYCE PETER B DR	BRACEWELL RONALD N PROF	BRODERICK JOHN DR
BURKE BERNARD F DR	CAMPBELL BRUCE DR	CAMPUSANO LUIS E
CARLSON JOHN B	CARR THOMAS D PROF	CHAISSON ERIC J PROF
CHOU KYONG CHOL PROF	CLARK THOMAS A DR	CURRIE DOUGLAS G DR
DAIGNE GERARD G	DARIUS JON DR	DAVIS MICHAEL M DR
DAWE JOHN ALAN DR	DE GRAAFF W DR	DE JAGER CORNELIS PROF
DE JONGE J K DR	DE LOORE CAMIEL PROF	DE VINCENZI DONALD DR
DELSEMME ARMAND H PROF DR	DICK STEVEN J	DIXON ROBERT S DR
DORSCHNER JOHANN DR	DOWNS GEORGE S DR	DYSON F J DR
ECCLES MICHAEL J DR	ELLIS GEORGE F R PROF	EPSTEIN EUGENE E DR
EVANS NEAL J II ASS PROF	FAZIO GIOVANNI G DR	FEJES ISTVAN DR
FELDMAN PAUL A DR	FIELD GEORGE B PROF	FIRNEIS FRIEDRICH J DR
FIRNEIS MARIA G DR	FISHER PHILIP C	FRACASSINI MASSIMO PROF
FREDRICK LAURENCE W PROF	FUJIMOTO MASA-KATSU DR	FUJIMOTO MITSUAKI DR
GEHRELS TOM PROF	GHIGO FRANCIS D DR	GINZBURG VITALY L PROF
GIOVANNELLI FRANCO DR	GODOLI GIOVANNI PROF	GOLDSMITH DONALD W. DR.
GOTT III J RICHARD	GOUDIS CHRISTOS D PROF	GREENBERG J MAYO DR
GREENSTEIN J L PROF	GREGORY PHILIP C DR	GULKIS SAMUEL DR
GUNN JAMES E PROF	GURM HARDEV S PROF	HADDOCK FRED T DR
HAISCH BERNHARD MICHAEL	HAJDUK ANTON DR	HARRINGTON ROBERT S DR
HARRISON EDWARD R PROF	HART MICHAEL H DR	HECK ANDRE DR
HEESCHEN DAVID S DR	HEIDMANN JEAN DR	HERCZEG TIBOR J PROF DR
HERSHEY JOHN L DR	HEUDIER JEAN-LOUIS DR	HIRABAYASHI HISASHI DR

HOANG BINH DY DR
HUNTEN DONALD M PROF
IDLIS G M DR
JASTROW ROBERT
JONES ERIC M
KAUFMANN PIERRE PROF
KLEIN MICHAEL J DR
KOCH ROBERT H DR
KRAUS JOHN D PROF
KUZMIN ARKADII D PROF DR
LIPPINCOTT SARAH LEE DR
MAFFEI PAOLO PROF
MARTIN ANTHONY R DR
MAVRIDIS L N PROF
MENDOZA V EUGENIO E DR
MIRABEL IGOR FELIX DR
MORRIS MARK ROOT DR
MULLER RICHARD A
ODA MINORU PROF
OSTRIKER JEREMIAH P PROF
PAPAGIANNIS MICHAEL D PRO
PESEK RUDOLPH PROF
PONSONBY JOHN E B DR
PURCELL EDWARD M PROF
RAJAMOHAN R DR
ROBINSON LEIF J
RUBIN ROBERT HOWARD
SAKURAI KUNITOMO PROF
SCARGLE JEFFREY D DR
SCHOBER HANS J DR
SEIRADAKIS JOHN HUGH DR
SHIMIZU MIKIO PROF
SINGH H P
SMITH F GRAHAM PROF
SNYDER LEWIS E
STEIN JOHN WILLIAM
SULLIVAN WOODRUFF T III
TAVAKOL REZA
TERZIAN YERVANT PROF
TOVMASSIAN H M DR
TURNER EDWIN L DR
VAN DE KAMP PETER
VAZQUEZ MANUEL DR
VOGT NIKOLAUS DR
WATSON FREDERICK GARNETT
WIELEBINSKI RICHARD PROF
WOLSTENCROFT RAMON D DR

HOLLIS JAN MICHAEL DR
HUNTER JAMES H PROF
IRVINE WILLIAM M PROF
JEFFERS STANLEY DR
KAFATOS MINAS DR
KELLER HANS ULRICH DR
KNOWLES STEPHEN H DR
KOEBERL CHRISTIAN DR
KSANFOMALITI L V DR
LAQUES PIERRE DR
LODEN LARS OLOF PROF
MARGRAVE THOMAS EWING JR
MATSAKIS DEMETRIOS N
MCALISTER HAROLD A DR
MILET BERNARD L DR
MORIMOTO MASAKI DR
MORRISON PHILIP PROF
NAKAGAWA YOSHINARI DR
OLIVER BERNARD M PROF
OWEN TOBIAS C PROF
PARIJSKIJ YU N DR
POLLACK JAMES B DR
PROCHAZKA FRANZ V DR
QUINTANA HERNAN DR
REAY NEWRICK K DR
ROOD ROBERT T DR
RUSSELL JANE L DR
SANCISI RENZO DR
SCHATZMAN EVRY PROF
SEEGER CHARLES LOUIS III
SHAPIRO MAURICE M PROF
SHOSTAK G SETH DR
SIVARAM C DR
SMITH HARLAN J PROF
SOFUE YOSHIAKI PROF
STRAIZYS V PROF DR
TAKADA-HIDAI MASAHIDE DR
TEDESCO EDWARD F
THADDEUS PATRICK PROF
TOWNES CHARLES HARD DR
VALBOUSQUET ARMAND DR
VAN FLANDERN THOMAS DR
VENUGOPAL V R DR
VON HOERNER SEBASTIAN DR
WELCH WILLIAM J PROF
WILLIAMS IWAN P DR
ZUCKERMAN BEN M DR

HOROWITZ PAUL PROF
HYSOM EDMUND J
ISRAEL FRANK P DR
JENNISON ROGER C PROF
KAFKA PETER
KELLERMANN KENNETH I DR
KOCER D DR
KOTSAKIS DEMETRIUS PROF
KUIPER THOMAS B H DR
LILLEY EDWARD A PROF
LOVELL SIR BERNARD PROF
MAROV MIKHAIL YA PROF
MATSUDA TAKUYA PROF
MCDONOUGH THOMAS R DR
MINN YOUNG KEY DR
MOROZ V I PROF DR
MOUTSOULAS MICHAEL PROF
NIARCHOS PANAYIOTIS PH D
OLLONGREN A PROF DR
PAGE THORNTON L DR
PEREK LUBOS DR
PONNAMPERUMA CYRIL PROF
PROVERBIO EDOARDO PROF
QUINTANA JOSE M DR
RIIHIMAA JORMA J DR
ROWAN-ROBINSON MICHAEL DR
SAGAN CARL DR
SANG GAK LEE
SCHILD RUDOLPH E DR
SEIELSTAD GEORGE A
SHEN CHUN-SHAN
SHUTER WILLIAM L H DR
SLYSH V I DR
SMITH HOWARD ALAN
STALIO ROBERTO DR
STURROCK PETER A PROF
TARTER JILL C DR
TEJFEL VIKTOR G DR
TOLBERT CHARLES R DR
TRIMBLE VIRGINIA L DR
VALLEE JACQUES P DR
VARSHALOVICH DIMITRIJ PR
VERSCHUUR GERRIT L PROF
WALLIS MAX K DR
WETHERILL GEORGE W
WILLSON ROBERT FREDERICK

WORKING GROUP FOR PLANETARY SYSTEM NOMENCLATURE
(GROUPE DE TRAVAIL POUR LA NOMENCLATURE DU SYSTEME PLANETAIRE)

President : H. Masursky

Members :

K. Aksnes	G.E. Hunt	M. Ya. Marov	P.M. Millman
D. Morrison	T.C. Owen	V.V. Shevchenko	B.A. Smith
V.G. Tejfel			

Task Groups :

<u>1) Lunar Nomenclature :</u>

V.V. Shevchenko (Chairman)

A. Dollfus	F. El-Baz	H. Masursky	P.M. Millman
S.K. Runcorn	E.A. Whitaker		

<u>2) Mercury Nomenclature :</u>

D. Morrison (Chairman)

D.P. Campbell	M.E. Davies	A. Dollfus	N.P. Erpylev
J.E. Guest			

<u>3) Venus Nomenclature :</u>

M. Ya. Marov (Chairman)

A.T. Basilevsky	D.B. Campbell	R.M. Goldstein	R.F. Jurgens
H. Masursky	G.H. Pettengill	Y.F. Tjuflin	

<u>4) Mars Nomenclature :</u>

B.A. Smith (Chairman)

A. Dollfus	M. Ya. Marov	Ya. Martynov	H. Masursky
S. Miyamoto	C. Sagan		

<u>5) Outer Solar System Momenclature</u>

T.C. Owen (Chairman)

K. Aksnes	A.T. Basilevsky	R. Beebe	M.S. Bobrov
A. Brahic	M.E. Davies	N.P. Erpylev	H. Masursky
B.A. Smith	V.G. Tejfel		

<u>6) Surface on Asteroid and Comets</u>

D. Morrison (Chairman pro tempore)

J. Veverka	A. Brahic	M. Fulchignoni	T. Gombosi
L. Ksantfomaliti	Y. Yatskiv	Y.C. Chang	S. Isobe

3. ALPHABETICAL LIST OF MEMBERS

Note : In Commission Membership :

    the letter "P" means President,
    the letter "V" means Vice-President,
    the letter "C" means Member of the Organizing Committee.

Reprints of this list are available from the IAU Publisher:

    D. Reidel Publishing Company
    P.O. Box 17
    3000 AA Dordrecht
    Holland

LIST OF MEMBERS

A'HEARN MICHAEL F DR
ASTRONOMY PROGRAM
UNIVERSITY OF MARYLAND
COLLEGE PARK MD 20742
U.S.A.
TEL: 301-454-6076
TLX: 710-826-0352
COM: 15C,20.

AANNESTAD PER ARNE DR
PHYSICS DEPT
ARIZONA STATE UNIVERSITY
TEMPE AZ 85287
U.S.A.
TEL: 602-965-3644
TLX:
COM: 34.

AARONSON MARC
STEWARD OBSERVATORY
UNIVERSITY OF ARIZONA
TUCSON AZ 85721
U.S.A.
TEL: 602-621-4514
TLX:
COM: 28

AARSETH SVERRE J DR
INSTITUTE OF ASTRONOMY
MADINGLEY ROAD
CAMBRIDGE CB3 0HA
U.K.
TEL: 62204
TLX: 817297 ASTRON G
COM: 33,37

ABALAKIN VICTOR K DR
USSR ACADEMY OF SCIENCES
CENTR. ASTR. OBSERVATORY
PULKOVO
196140 LENINGRAD M-140
U.S.S.R.
TEL: 298-2242
TLX: 12261 FENIKS
COM: 04C,05,07,20

ABBASOV ALIK R DR
SCIENT. & INDUSTRIAL ASS.
OF COSMIC RESEARCH
159 LENIN PROSPECT
370106 BAKU
U.S.S.R.
TEL:
TLX:
COM: 10

ABBOTT WILLIAM N DR
UNIVERSITY OF ATHENS
MICHALACOPOULOU 42
GR-11528 ATHENS
GREECE
TEL: 7213352
TLX:
COM: 22.

ABDALA JOSE DR
AVENIDA SOLANO P.B.No. 1
EDIFICIO ARAGUANEY
CHACAITO CARACAS 1050
VENEZUELA
TEL:
TLX:
COM:

ABDULLA SHAKER ABDUL AZIZ
ASTRONOMY & SPACE RES CTR
COUNCIL FOR SCI RESEARCH
P O BOX 2441
JADIRIYAH, BAGHDAD
IRAQ
TEL: 7765127
TLX: 2187 BATHILMI IK
COM: 40

ABELE M K DR
LATVIAN STATE UNIVERSITY
ASTRONOMICAL OBSERVATORY
226098 RIGA
U.S.S.R.
TEL:
TLX:
COM: 31

ABHYANKAR KRISHNA D PROF
DEPT OF ASTRONOMY
OSMANIA UNIVERSITY
HYDERABAD 500 007
INDIA
TEL: 851672
TLX:
COM: 24,29,36,42C

ABLES HAROLD D DR
FLAGSTAFF STATION
US NAVAL OBSERVATORY
P O BOX 1149
FLAGSTAFF AZ 86002
U.S.A.
TEL: 602-779-5132
TLX: 26230 ASTRO
COM: 09,25,28

ABLES J G DR
CSIRO
DIVISION OF RADIOPHYSICS
P.O.BOX 76
EPPING NSW 2121
AUSTRALIA
TEL:
TLX:
COM: 40

ABOU-EL-ELLA MOHAMED S DR
HELWAN OBSERVATORY (HIAG)
HELWAN-CAIRO
EGYPT
TEL: 780645
TLX: 93070 HIAG UN
COM: 37

ABRAHAM HENRY J M
12 MORTLOCK CIRCUIT
KALEEN ACT 2617
AUSTRALIA
TEL: 062-413685
TLX:
COM: 19.

ABRAHAM ZULEMA DR
INST.PESQUISAS ESPACIAIS
CRAAM RADIO OBSERVATORIO
CAIXA POSTAL 515
12200 S. JOSE DOS CAMPOS
BRAZIL
TEL: 011-826-6588
TLX: 01134061
COM:

ABRAMI ALBERTO PROF
OSSERVATORIO ASTRONOMICO
DI TRIESTE
I-34131 TRIESTE
ITALY
TEL:
TLX:
COM: 10,40

ABRAMOWICZ MAREK DR
INTERNATIONAL SCHOOL FOR
ADVANCED STUDIES
I-34014 TRIESTE
ITALY
TEL: 040-224281
TLX: 460392 ICTP
COM: 48.

ABT HELMUT A DR
KITT PEAK NATIONAL OBS
BOX 26732
TUCSON AZ 85726
U.S.A.
TEL: 602-325-9215
TLX: 0666-484 AURA NOAC
COM: 05,26C,29,30

ABU EL ATA NABIL DR
BUREAU DES LONGITUDES
77 AVE DENFERT-ROCHEREAU
F-75014 PARIS
FRANCE
TEL: 1-43-20-13-30
TLX: 270776 F
COM:

ACKER AGNES PROF DR
OBSERVATOIRE
UNIVERSITE DE STRASBOURG
11 RUE DE L UNIVERSITE
F-67000 STRASBOURG
FRANCE
TEL: 88)35-43-00
TLX: 890506 STAROBS
COM: 34,46

ACTON LOREN W DR
LOCKHEED PALO ALTO
RESEARCH LAB
3251 HANOVER ST
PALO ALTO CA 94306
U.S.A.
TEL: 415-858-4067
TLX: 346409
COM: 12,44

ADAM MADGE G DR
DEPT OF ASTROPHYSICS
SOUTH PARKS ROAD
OXFORD OX1 3RQ
U.K.
TEL:
TLX:
COM: 12

ADAMS A N MR
6549 N. 35TH ROAD
ARLINGTON VA 22213
U.S.A.
TEL: 703-532-8246
TLX:
COM:

ADAMS DAVID J DR
ASTRONOMY DEPT
THE UNIVERSITY
LEICESTER LE1 7RH
U.K.
TEL: 0533-554455
TLX: 341198
COM: 48

ADAMS THOMAS F DR
LOS ALAMOS SCIENTIFIC LAB
G-7 MS329
P O BOX 1663
LOS ALAMOS NM 87545
U.S.A.
TEL: 505-667-6384
TLX:
COM:

ADE PETER A R DR
PHYSICS DEPT
QUEEN MARY COLLEGE
MILE END ROAD
LONDON E1 4NS
U.K.
TEL: 01-980-4811
TLX: 893753\
COM: 40

ADEL ARTHUR F PROF EMER
N. ARIZONA UNIVERSITY
BOX 5679
FLAGSTAFF AZ 86001
U.S.A.
TEL: 602-774-6597
TLX:
COM:

ADELMAN SAUL J DR
LAB ASTRON & SOLAR PHYS
NASA/GSFC. CODE 681
GREENBELT MD 20771
U.S.A.
TEL: 301-344-7445
TLX: 710-828-9716
COM: 14,25,28

ADJABSHIRIZADEH ALI
CTR FOR ASTRON RESEARCH
KHADJEH NASSIR ALDDIN OBS
UNIVERSITY OF TABRIZ
TABRIZ 51664
IRAN
TEL: 0098-041-32564
TLX:
COM:

ADOLFSSON TORD DR
KRAGEHOLMSGATAN 12
S-216 19 MALMOE
SWEDEN
TEL: 040-157586
TLX:
COM:

AFANASJEVA PRASKOVYA M DR
PULKOVO OBSERVATORY
196140 LENINGRAD
U.S.S.R.
TEL: 298-22-42
TLX:
COM: 31

AGEKJAN TATEOS A PROF
UNIVERSITY OF LENINGRAD
OBSERVATORY
199178 LENINGRAD
U.S.S.R.
TEL:
TLX:
COM: 33,37

AHNERT P DR
ZNTRLINST. F. ASTROPHYSIK
STERNWARTEN-STR. 25 53/17
DDR-6400 SONNEBERG
GERMANY. D.R.
TEL: SONNEBERG 2287
TLX:
COM:

AIZU KO PROF
PHYSICS DEPT
RIKKYO UNIVERSITY
NISHI-IKEBUKURO 3
TOSHIMAKU TOKYO 171
JAPAN
TEL: 03 9851 2414
TLX:
COM: 33,40,47,48

AL-SABTI ABDUL ADIM DR
PHYSICS DEPT
SCIENCE COLLEGE
BAGHDAD UNIVERSITY
JADIRIYAH. BAGHDAD
IRAQ
TEL: 5552340
TLX:
COM: 38,51.

AGRAWAL P C DR
TATA INSTITUTE OF
FUNDAMENTAL RESEARCH
HOMI BHABHA RD. COLABA
BOMBAY 400 005
INDIA
TEL: 219-111 x 336
TLX: 113009 TIFR IN
COM: 44

AI GUOXIANG
BEIJING ASTRONOMICAL OBS
BEIJING
CHINA. PEOPLE'S REP.
TEL:
TLX:
COM: 10

AKABANE KENJI A PROF
TOKYO ASTRONOMICAL
OBSERVATORY
OSAWA MITAKA
TOKYO 181
JAPAN
TEL: 0267-98-2831
TLX: 3329005 TAO NRO J
COM: 34,40,

ALANIA I F DR
ABASTUMANI ASTROPHYSICAL
OBSERVATORY
383762 ABASTUMANI.GEORGIA
U.S.S.R.
TEL: 225/244
TLX:
COM: 27,

AGRINIER BERNARD L MR
C E N SACLAY
BP NO 2
F-91190 GIF-YVETTE
FRANCE
TEL:
TLX:
COM:

AIAD A DR
DEPT OF ASTRONOMY
FACULTY OF SCIENCE
CAIRO UNIVERSITY
GEZA ORMAN
EGYPT
TEL:
TLX:
COM: 27,35,37,46

AKABANE TOKUHIDE DR
HIDA OBSERVATORY
KAMITAKARA
GIFU 506-13
JAPAN
TEL: 0578-6-2311
TLX:
COM: 16

ALBERS HENRY PROF
VASSAR COLLEGE OBS
POUGHKEEPSIE NY 12601
U.S.A.
TEL: 914-452-7000
TLX:
COM: 45,

AGUERO ESTELA L DR
OBSERVATORIO ASTRONOMICO
LAPRIDA NO 854
5000 CORDOBA
ARGENTINA
TEL: 36876 40613
TLX: 51822 BUCOR
COM: 28

AIKAWA TOSHIKI DR
ASTRONOMICAL INSTITUTE
TOHOKU UNIVERSITY
AOBAYAMA SENDAI
JAPAN
TEL:
TLX:
COM:

AKCAYLI MELEK M A DR
EGE UNIVERSITY
FEN FAKULTESI
GOK BILIMLERI ENSTITUSU
BORNOVA-IZMIR
TURKEY
TEL:
TLX:
COM:

ALBRECHT RUDOLF DR
SPACE TELESCOPE EUROPEAN
COORDINATING FACILITY
KARL-SCHWARZSCHILD-STR 2
D-8046 GARCHING B MUNCHEN
GERMANY, F.R.
TEL: 89-320-06-287
TLX: 528 282 22 EO D
COM: 09,25.

AHLUWALIA HARJIT SINGH DR
DEPT PHYSICS & ASTRONOMY
UNIVERSITY OF NEW MEXICO
800 YALE BLVD N E
ALBUQUERQUE NM 87131
U.S.A.
TEL: 505-277-2941
TLX: 660461
COM: 10,48,49

AIKMAN G CHRIS L
DOMINION ASTROPHYS OBS
5071 W SAANICH ROAD
RR 5
VICTORIA BC V8X 4M6
CANADA
TEL: 604-388-3975
TLX: 049-7295
COM: 29,

AKSENOV E P PROF DR
STERNBERG STATE
ASTRONOMICAL INSTITUTE
UNIVERSITETSKIJ PROSP 13
119899 MOSCOW
U.S.S.R.
TEL: 139-28-58
TLX:
COM: 07

ALCAINO GONZALO DR
INSTITUTO ISAAC NEWTON
CASILLA 8-9
CORREO 9
SANTIAGO
CHILE
TEL: 472-013
TLX: c/o ESO 240853 ESOGO
COM: 28,37

AHMAD FAROOQ DR
DEPT OF PHYSICS
UNIVERSITY OF KASHMIR
SRINAGAR 190 006 KASHMIR
INDIA
TEL: 71559
TLX:
COM: 28

AIME C DR
DEPT D'ASTROPHYSIQUE
UNIVERSITE DE NICE
PARC VALROSE
F-06034 NICE CEDEX
FRANCE
TEL: 93-51-91-00
TLX:
COM: 09,12,

AKSNES KAARE DR
NORWEGIAN DEFENCE
RESEARCH ESTABLISHMENT
P O BOX 25
N-2007 KJELLER
NORWAY
TEL: 2-737650
TLX: 76528
COM: 07,20C,W

ALDROVANDI RUBEN DR
INST DI FISICA TEORICA
RUA PAMPLONA 145
01405 SAO PAULO SP
BRAZIL
TEL: 288-5643
TLX:
COM:

AHMED IMAM IBRAHIM PROF
DEPT OF ASTRONOMY
FACULTY OF SCIENCE
CAIRO UNIVERSITY
GIZA CAIRO
EGYPT
TEL:
TLX:
COM:

AITKEN DAVID K DR
PHYSICS DEPT (RAAF)
UNIV OF MELBOURNE
PARKVILLE VIC 3052
AUSTRALIA
TEL: 03-341-6818
TLX: 35185 UNIMEL
COM: 34

AKYOL MUSTAFA U PROF
EGE UNIVERSITY
SCIENCE FACULTY
DEPT OF ASTRONOMY
BORNOVA-IZMIR
TURKEY
TEL: 51-180110
TLX:
COM:

ALDROVANDI SUELI M V DR
INST. ASTRON E GEOFISICO
AV MIGUEL STEFANO 4200
04301 SAO PAULO SP
BRAZIL
TEL: 011-275-3720
TLX: 1136221 IAGM BR
COM: 34

AHMED MOSTAFA
ASTRONOMY & METEOROL DEPT
FACULTY OF SCIENCE
CAIRO UNIVERSITY
CAIRO
EGYPT
TEL:
TLX:
COM: 07

AIZENMAN MORRIS L DR
DIV ASTRONOMICAL SC
NATL SCI FOUNDATION
RM 615, 1800 G ST NW
WASHINGTON DC 20550
U.S.A.
TEL: 202-357-7643
TLX:
COM: 27,35,

AL-NAIMY HAMID M K DR
ASTRONOMY & SPACE RES CTR
COUNCIL FOR SCI RESEARCH
P O BOX 2441
JADIRIYAH. BAGHDAD
IRAQ
TEL: 7765127
TLX: 212187
COM: 42,51,

ALECIAN GEORGES DR
OBSERVATOIRE DE PARIS
SECTION DE MEUDON
DAF
F-92195 MEUDON PL CEDEX
FRANCE
TEL: 1-45-34-75-70
TLX:
COM:

ALEXANDER JOHN B
ROYAL GREENWICH OBS
HERSTMONCEUX CASTLE
HAILSHAM BN27 1RP
U.K.
TEL: 0323-833171
TLX: 87451
COM:

ALLAMANDOLA LOUIS JOHN DR
NASA-AMES RESEARCH CENTER
MAIL STOP N-245-6
MOFFET FIELD CA 94035
U.S.A.
TEL: 415-694-6890
TLX: 348408 NASA AMES
COM:

ALLER HUGH D DR
DEPT OF ASTRONOMY
UNIVERSITY OF MICHIGAN
810 DENNISON BUILDING
ANN ARBOR MI 48109-1090
U.S.A.
TEL: 313-764-3466
TLX:
COM: 40

ALTROCK RICHARD C DR
AIR FORCE GEOPHYSICS LAB
NATIONAL SOLAR OBS
SUNSPOT NM 88349
U.S.A.
TEL: 505-434-1390
TLX: 066-484
COM: 10,12,36

ALEXANDER JOSEPH K
GODDARD SPACE FLIGHT CTR
CODE 690
GREENBELT MD 20771
U.S.A.
TEL: 301-344-8112
TLX:
COM: 40,44.

ALLAN DAVID W MR
NATL BUREAU OF STANDARDS
BOULDER CO 80303
U.S.A.
TEL: 303-497-5637
TLX: 910-940-5906
COM: 31

ALLER LAWRENCE HUGH
ASTRONOMY DEPT
UNIVERSITY OF CALIFORNIA
MATH-SCIENCES BLDG
LOS ANGELES CA 90024
U.S.A.
TEL: 213-825-3515
TLX: 910-342-7597
COM: 29,34,36

ALTSCHULER MARTIN D PROF
DEPT RAD THERAPY,BOX 522
HOSP UNIV OF PENNSYLVANIA
3400 SPRUCE ST
PHILADELPHIA PA 19104
U.S.A.
TEL: 215-662-6472
TLX:
COM: 10,12

ALFARO EMILIO JAVIER
INST DE ASTROFISICA DE
ANDALUCIA, APDO 2144
PROFESOR ALBAREDA 1
18080 GRANADA
SPAIN
TEL: 958-12-13-00
TLX: 78573 IAAG E
COM: 37

ALLAN PETER M
ASTRONOMY DEPT
UNIVERSITY OF MANCHESTER
MANCHESTER M13 9PL
U.K.
TEL: 061-273-7121
TLX:
COM: 47

ALLER MARGO F DR
ASTRONOMY DEPT
UNIVERSITY MICHIGAN
ANN ARBOR MI 48109-1090
U.S.A.
TEL: 313-764-3465
TLX: 810-223-6056
COM: 40

ALURKAR S K DR
PHYSICAL RESEARCH LAB
NAVRANGPURA
AHMEDABAD 380 009
INDIA
TEL: 462129
TLX: 121-397 PRL IN
COM:

ALFVEN HANNES PROF
DEPT OF PLASMA PHYSICS
ROYAL INST OF TECHNOLOGY
S-100 44 STOCKHOLM
SWEDEN
TEL: 46-08-787 70 00
TLX: 10389 KTHB
COM: 28,47,48.

ALLEGRE CLAUDE PROF
INST DE PHYSIQUE DU GLOBE
4 PLACE JUSSIEU
F-75005 PARIS
FRANCE
TEL:
TLX:
COM: 15

ALLOIN DANIELLE DR
OBSERVATOIRE DE PARIS
SECTION DE MEUDON
F-92195 MEUDON PL CEDEX
FRANCE
TEL: 1-45-34-75-70
TLX: 201571 F
COM: 28

ALVAREZ HECTOR DR
DEPTO DE ASTRONOMIA
UNIVERSIDAD DE CHILE
CASILLA 36-D
SANTIAGO
CHILE
TEL: 2294101
TLX:
COM:

ALISSANDRAKIS C PH D
LAB OF ASTROPHYSICS
UNIVERSITY OF ATHENS
PANEPISTIMIOPOLIS
GR-15771 ATHENS
GREECE
TEL: 1-735122
TLX:
COM: 10C,12

ALLEN C W PROF
MT STROMLO OBSERVATORY
WODEN P.O. ACT 2606
AUSTRALIA
TEL:
TLX:
COM: 12,14

ALMAR IVAN PROF
KONKOLY OBSERVATORY
BOX 67
H-1525 BUDAPEST
HUNGARY
TEL: 366-621
TLX: 227460
COM: 51

ALVAREZ LUIS DR
LAWRENCE BERKELEY LAB.
BLDG 50B, RM 5239
1 CYCLOTRON ROAD
BERKELEY CA 94720
U.S.A.
TEL: 415-486-4400
TLX:
COM: 48

ALKSNE ZENTA DR
RADIOASTROPHYSICAL
OBSERVATORY
TURGENEVA 19
226524 RIGA LATVIA
U.S.S.R.
TEL: 226796 RIGA
TLX:
COM:

ALLEN CHRISTINE
INSTITUTO DE ASTRONOMIA
UNAM
APDO POSTAL 70-264
04510 MEXICO DF
MEXICO
TEL:
TLX:
COM: 26,37.

ALTAMORE ALDO
ISTITUTO ASTRONOMICO
UNIVERSITA DI ROMA
VIA G.M. LANCISI 29
I-00161 ROMA
ITALY
TEL: 06-8442977
TLX: 613255 INFRO
COM:

ALVAREZ P
INST DE ASTROFISICA DE
CANARIAS
LA LAGUNA
38071 TENERIFE
SPAIN
TEL:
TLX: 92640 IACE E
COM: 21

ALKSNIS ANDREJS DR
RADIOASTROPHYSICAL
OBSERVATORY
226524 RIGA LATVIA
U.S.S.R.
TEL: 226796 RIGQ
TLX:
COM: 37

ALLEN DAVID A DR
ANGLO AUSTRALIAN OBS
PO BOX 296
EPPING NSW 2121
AUSTRALIA
TEL: 02-868-1666
TLX: 23999 OSYD AA
COM:

ALTAVISTA CARLOS A DR
OBSERVATORIO ASTRONOMICO
PASEO DEL BOSQUE
1900 LA PLATA
ARGENTINA
TEL: 21-7308
TLX: 31151 BULAP
COM: 07

ALY M KHAIRY PROF
HELWAN OBSERVATORY
24 MONTAZAH STR.
HELIOPOLIS, CAIRO
EGYPT
TEL: 448832
TLX:
COM: 10,50

ALLADIN SALEH MOHAMED DR
DEPT OF ASTRONOMY
OSMANIA UNIVERSITY
HYDERABAD 500 007
INDIA
TEL: 71116
TLX:
COM: 28

ALLEN RONALD J DR
ASTRONOMY DEPARTMENT
UNIVERSITY OF ILLINOIS
1011 W. SPRINGFIELD AVE
URBANA IL 61801
U.S.A.
TEL: 217-333-3090
TLX: 510-011-969
COM: 28,40

ALTENHOFF WILHELM J DR
MPI FUER RADIOASTRONOMIE
AUF DEM HUEGEL 69
D-5300 BONN
GERMANY, F.R.
TEL: 0228-525293
TLX: 0886440 MPIFR D
COM: 33,34,40

AMBARTSUMIAN V A PROF DR
BYURAKAN ASTROPHYSICAL
OBSERVATORY
378433 ARMENIA
U.S.S.R.
TEL: 52-45-80
TLX:
COM: 28,33,51

AMBASTHA A K DR
UDAIPUR SOLAR OBSERVATORY
11 VIDYA MARG
UDAIPUR 313 001
INDIA
TEL: 25626
TLX:
COM: 33

ANDERSON BRYAN DR
UNIVERSITY OF MANCHESTER
JODRELL BANK
MACCLESFIELD SK11 9DL
U.K.
TEL:
TLX:
COM:

ANDREWS PETER J DR
ROYAL GREENWICH OBS
HERSTMONCEUX CASTLE
HAILSHAM BN27 1RP
U.K.
TEL: 0323-83-3171
TLX: 87451
COM:

ANGELOV TRAJKO
INSTITUTE OF ASTRONOMY
UNIVERSITY OF BELGRADE
STUDENTSKI TRG 16
YU-11000 BELGRADE
YUGOSLAVIA
TEL: 011-638-715
TLX:
COM: 35

AMBROZ PAVEL DR
ASTRONOMICAL INSTITUTE
CZECH. ACAD. OF SCIENCES
251 65 ONDREJOV
CZECHOSLOVAKIA
TEL: 72-46-25
TLX: 121579 ASTR C
COM: 10

ANDERSON CHRISTOPHER M DR
WASHBURN OBSERVATORY
UNIVERSITY OF WISCONSIN
MADISON WI 53706
U.S.A.
TEL: 608-262-0492
TLX:
COM:

ANDRIENKO DMITRY A DR
VLADIMIRSKAYA 51 53
FLAT 13
252003 KIEV
U.S.S.R.
TEL: 25-07-75
TLX: 132201
COM: 15

ANGIONE RONALD J DR
ASTRONOMY DEPT
SAN DIEGO STATE UNIV
SAN DIEGO CA 92182
U.S.A.
TEL: 619-265-6183
TLX:
COM: 21

ANAND S P S DR
APPLIED RESEARCH CORP.
8201 CORPORATE DRIVE
SUITE 920
LANDOVER MD 20785
U.S.A.
TEL: 301-459-8442
TLX:
COM: 35

ANDERSON KINSEY A PROF
SPACE SCIENCES LAB
UNIVERSITY OF CALIFORNIA
BERKELEY CA 94720
U.S.A.
TEL: 415-642-1313
TLX: 910-3667945 UC SPACE
COM: 10,21,49

ANDRIESSE CORNELIS D DR
VALERIUSLAAN 15
NL-6865 JA DOORWERTH
NETHERLANDS
TEL: 085-332794
TLX:
COM: 34

ANGUITA CLAUDIO A DR
OBS ASTRONOMICO NACIONAL
UNIVERSIDAD DE CHILE
CASILLA 36-D
SANTIAGO
CHILE
TEL: 229-4101
TLX:
COM: 08

ANANTHAKRISHNAN S
RADIOASTRONOMY CENTER
TATA INSTITUTE OF
FUNDAMENTAL RESEARCH
OOTACAMUND 643 001
INDIA
TEL: 2032
TLX: 853241
COM: 49

ANDERSON KURT S
NEW MEXICO STATE UNIV
DEPT OF ASTRONOMY
LAS CRUCES NM 88003
U.S.A.
TEL: 505-646-1032
TLX: 210-983-0549 NMSUCI
COM:

ANDRILLAT HENRI L PROF
LABORATOIRE D'ASTRONOMIE
UNIV DES SCIENCES & TECH-
NIQUES DU LANGUEDOC
F-34060 MONTPELLIER
FRANCE
TEL: 67-63-25-37
TLX: 490944 USTMONT F
COM: 29,34,46,47

ANILE ANGELO M
DIPARTIMENTO MATEMATICA
CITTA UNIVERSITARIA
I-95123 CATANIA
ITALY
TEL: 095-330533
TLX:
COM:

ANANTHARAMAIAH K R DR
NTL RADIO ASTRONOMY OBS
VLA SITE
PO BOX 6
SOCORRO NM 87801
U.S.A.
TEL: 505-772-4306
TLX: 9109881710
COM: 40

ANDO HIROYASU DR
TOKYO ASTRONOMICAL
OBSERVATORY
OSAWA 2-21-1, MITAKA
TOKYO 181
JAPAN
TEL: 0422-32-5111
TLX: 2822307 TAOMK J
COM: 12,27,51

ANDRILLAT YVETTE DR
LABORATOIRE D'ASTRONOMIE
UNIV DES SCIENCES & TECH-
NIQUES DU LANGUEDOC
F-34060 MONTPELLIER
FRANCE
TEL:
TLX:
COM: 28,29,34

ANSARI S M RAZAULLAH PROF
PHYSICS DEPT
ALIGARH MUSLIM UNIVERSITY
ALIGARH UP 202 001
INDIA
TEL: 4568
TLX:
COM: 12,41,46

ANDERS EDWARD PROF
ENRICO FERMI INSTITUTE
UNIVERSITY OF CHICAGO
5640 S. ELLIS AVE
CHICAGO IL 60637
U.S.A.
TEL: 312-962-7108
TLX: 6871133
COM: 15,16,22

ANDREW BRYAN H DR
CHIEF, PROGRAM SERVICES
NATL RESEARCH COUNCIL
OTTAWA ONT K1A 0R6
CANADA
TEL: 613-993-3731
TLX: 0533145
COM: 34,40

ANDRLE PAVEL DR
ASTRONOMICAL INSTITUTE
CZECH. ACAD. OF SCIENCES
BUDECSKA 6
120 23 PRAHA 2
CZECHOSLOVAKIA
TEL: 25-87-57
TLX: 122486
COM: 33

ANTALOVA ANNA
ASTRONOMICAL INSTITUTE
CZECH. ACAD. OF SCIENCES
05 960 TATRANSKA LOMNICA
CZECHOSLOVAKIA
TEL:
TLX:
COM: 10

ANDERSEN JOHANNES
COPENHAGEN UNIVERSITY OBS
BRORFELDEVEJ 23
DK-4340 TOLLOSE
DENMARK
TEL: 45-3-488195
TLX: 44155 DANAST DK
COM: 30P,42C,

ANDREW KENNETH L PROF
DEPT OF PHYSICS
PURDUE UNIVERSITY
W LAFAYETTE IN 47907
U.S.A.
TEL: 317-494-5540
TLX:
COM: 14

ANGEL J ROGER P PROF
STEWARD OBSERVATORY
UNIVERSITY OF ARIZONA
TUCSON AZ 85721
U.S.A.
TEL: 602-621-6541
TLX: 467175
COM: 25

ANTHONY-TWAROG BARBARA J
DEPT PHYSICS & ASTRONOMY
UNIVERSITY OF KANSAS
LAWRENCE KS 66045
U.S.A.
TEL: 913-864-4933
TLX:
COM: 25

ANDERSEN TORBEN BRENDER
OPTICAL DESIGN GR. B/254E
LOCKHEED RESEARCH LABS
3251 HANOVER STREET
PALO ALTO CA 94304
U.S.A.
TEL:
TLX:
COM:

ANDREWS DAVID A DR
ARMAGH OBSERVATORY
ARMAGH BT61 9DG
U.K.
TEL:
TLX:
COM:

ANGELETTI LUCIO DR
OSSERVATORIO ASTRONOMICO
VIA DEL PARCO MELLINI 84
I-00136 ROMA
ITALY
TEL: 34-70-56
TLX:
COM:

ANTIOCHOS SPIRO KOSTA
NAVAL RESEARCH LABORATORY
CODE 4170 SA
4555 OVERLOOK AVE. S.W.
WASHINGTON DC 20375
U.S.A.
TEL: 202-767-6199
TLX:
COM: 10

LIST OF MEMBERS

ANTONACOPOULOS GREG PROF
DEPT OF ASTRONOMY
UNIVERSITY OF PATRAS
GR-26110 PATRAS
GREECE
TEL: 991-145
TLX:
COM: 07

APPARAO K M V DR
TATA INSTITUTE OF
FUNDAMENTAL RESEARCH
HOMI BHABHA RD
BOMBAY 400 005
INDIA
TEL: 219111x341
TLX: 011-3009 TIFR IN
COM: 48

ARELLANO FERRO ARMANDO
INSTITUTO DE ASTRONOMIA
APDO POSTAL 70-264
CIUDAD UNIVERSITARIA
04510 MEXICO DF
MEXICO
TEL: 905-548-5305
TLX: 01760155 CICME
COM: 27

ARLOT JEAN-EUDES
BUREAU DES LONGITUDES
77 AVE DENFERT-ROCHEREAU
F-75014 PARIS
FRANCE
TEL: 1-43-20-13-30
TLX:
COM: 04,20C

ANTONELLO ELIO
OSSERVATORIC ASTRONOMICO
VIA E. BIANCHI 46
I-22055 MERATE
ITALY
TEL: 039-592035
TLX:
COM: 27

APPENZELLER IMMO PROF
LANDESSTERNWARTE
KOENIGSTUHL
D-6900 HEIDELBERG 1
GERMANY, F.R.
TEL: 06221-10036
TLX: 461789 MPIA D
COM: 29,35

AREND S DR
AVENUE DE SATURNE 11
B-1180 BRUXELLES
BELGIUM
TEL:
TLX:
COM: 20,26

ARNAL MARCELO EDMUNDO DR
INSTITUTO ARGENTINO
DE RADIOASTRONOMIA
CASILLA DE CORREO 5
1894 VILLA ELISA Bs.As.
ARGENTINA
TEL: 21-43793
TLX:
COM: 40

ANTONOPOULOU E DR
DEPT OF ASTRONOMY
UNIVERSITY OF ATHENS
PANEPISTIMIOPOLIS
GR-15771 ATHENS
GREECE
TEL:
TLX:
COM: 42

APPLEBY JOHN F
JET PROPULSION LAB.
CALTECH MS 183-301
4800 OAK GROVE DIRVE
PASADENA CA 91109
U.S.A.
TEL: 818-354-3943
TLX:
COM: 16

ARGUE A NOEL MR
INSTITUTE OF ASTRONOMY
THE OBSERVATORIES
MADINGLEY ROAD
CAMBRIDGE CB3 0HA
U.K.
TEL: 0223-62204-63
TLX: 817297 ASTRON G
COM: 24C,25

ARNAUD JEAN
OBSERVATOIRE PIC-DU-MIDI
ET TOULOUSE
14 AVENUE EDOUARD BELIN
F-31400 TOULOUSE
FRANCE
TEL: 61-25-21-01
TLX: 521880
COM: 09,12,25

ANTONOV VADIM A DR
LENINGRAD STATE UNIV
ASTRONOMICAL OBSERVATORY
199178 LENINGRAD
U.S.S.R.
TEL:
TLX:
COM: 33

ARABELOS DIMITRIOS DR
DEPT GEODESY & SURVEYING
UNIVERSITY THESSALONIKI
GR-54006 THESSALONIKI
GREECE
TEL: 003031-99-2693
TLX: 412181 AUTH GR
COM: 19

ARGYLE P E DR
DOMINION RADIO ASTROPHY-
SICAL OBSERVATORY
BOX 248
PENTICTON BC V2A 6K3
CANADA
TEL:
TLX:
COM: 40

ARNETT W DAVID PROF
ENRICO FERMI INSTITUTE
933 E. 56TH AVE
CHICAGO IL 60637
U.S.A.
TEL: 312-962-8208
TLX: 910-221-5617
COM: 35

ANTONUCCI ESTER DR
ISTITUTO DI FISICA
UNIVERSITA DI TORINO
CORSO D'AZEGLIO 46
I-10125 TORINO
ITALY
TEL: 011-657694
TLX: 211041 INFNTO I
COM: 49

ARAI KENZO DR
DEPT OF PHYSICS
KUMAMOTO UNIVERSITY
KUMAMOTO 860
JAPAN
TEL: 096-344-2111
TLX:
COM: 35

ARGYRAKOS JEAN PROF DR
193 PATISSION ST
GR-11253 ATHENS
GREECE
TEL: 8677000
TLX:
COM: 08,41

ARNOULD MARCEL L DR
INSTITUT D'ASTRONOMIE
UNIV LIBRE DE BRUXELLES
B-1050 BRUXELLES
BELGIUM
TEL: 2-649-00-30
TLX: 23069 UNILIB
COM: 35,48

ANZER ULRICH DR
MPI FUER PHYSIK UND
ASTROPHYSIK
KARL-SCHWARZSCHILD-STR 1
D-8046 GARCHING B MUNCHEN
GERMANY, F.R.
TEL: 089-32990
TLX: 524629 ASTRO D
COM:

ARDAVAN HOUSHANG DR
INSTITUTE OF ASTRONOMY
MADINGLEY ROAD
CAMBRIDGE CB3 0HA
U.K.
TEL: 0223-62204
TLX: 81292 CAULAB G
COM:

ARIAS DE GREIFF J PROF
OBSERVATORIO NACIONAL
APARTADO 2584
BOGOTA 1, D.E.
COLOMBIA
TEL:
TLX:
COM: 04,50

ARNQUIST WARREN N DR
8127 DELGANY AVE
PLAYA DEL REY CA 90291
U.S.A.
TEL: 213-821-2724
TLX:
COM:

AOKI SHINKO PROF
TOKYO ASTRONOMICAL
OBSERVATORY
MITAKA
TOKYO 181
JAPAN
TEL: 0422-32-5111
TLX: 2822307 TAOMTK J
COM: 04C,07,31C,33

ARDEBERG ARNE L PROF
LUND OBSERVATORY
BOX 43
S-221 00 LUND
SWEDEN
TEL: 046-10-72-90
TLX: 33199 OBSNOT S
COM: 28,33,45,50

ARIMOTO NOBUO DR
OBSERVATOIRE DE PARIS
SECTION DE MEUDON
LAM
F-92195 MEUDON PL CEDEX
FRANCE
TEL: 1-45-34-75-70
TLX: 207571 F
COM: 35

ARNY THOMAS T DR
DEPT PHYSICS & ASTRONOMY
UNIV OF MASSACHUSETTS
GRC TOWER B
AMHERST MA 01003
U.S.A.
TEL: 413-545-2194
TLX:
COM: 34

APARICI JUAN DR
DEPTO DE ASTRONOMIA
UNIVERSIDAD DE CHILE
CASILLA 36-D
SANTIAGO
CHILE
TEL: 2294101
TLX: 440005 ATTN OBSERNAL
COM: 09,40

ARDUINI-MALINOVSKY M. DR
CNES
2 PLACE MAURICE QUENTIN
F-75039 PARIS CEDEX 01
FRANCE
TEL:
TLX:
COM: 14

ARKHIPOVA V P DR
STERNBERG STATE ASTRO-
NOMICAL INSTITUTE
119899 MOSCOW V-234
U.S.S.R.
TEL: 139-26-57
TLX:
COM: 27,34

ARONS JONATHAN
DEPT OF ASTRONOMY
UNIVERSITY OF CALIFORNIA
601 CAMPBELL HALL
BERKELEY CA 94720
U.S.A.
TEL: 415-642-4730
TLX: 820181 UCB AST RAL
COM: 48

ARP HALTON DR
MT WILSON & LAS CAMPANAS
OBSERVATORIES
813 SANTA BARBARA ST
PASADENA CA 91101
U.S.A.
TEL: 213-577-1122
TLX:
COM: 28

ARPIGNY CLAUDE PROF
INSTITUT D'ASTROPHYSIQUE
UNIVERSITE DE LIEGE
AVENUE DE COINTE 5
B-4200 COINTE-OUGREE
BELGIUM
TEL: 041-529980-263
TLX:
COM: 15C.36

ARSENIJEVIC JELISAVETA
ASTRONOMICAL OBSERVATORY
VOLGINA 7
YU-11050 BEOGRAD
YUGOSLAVIA
TEL:
TLX:
COM: 25,27

ARTHUR DAVID W G
US GEOLOGICAL SURVEY
FLAGSTAFF AZ 86001
U.S.A.
TEL:
TLX:
COM: 16

ARTRU MARIE-CHRISTINE DR
OBSERVATOIRE DE PARIS
SECTION DE MEUDON
ASTROPHYS FONDAMENTALE
F-92195 MEUDON PL CEDEX
FRANCE
TEL: 1-45-34-75-70
TLX: 201571
COM: 14

ARTZNER GUY
LPSP
BP 10
F-91370 VERRIERES-LE-B.
FRANCE
TEL: 1-69-20-10-60
TLX: 600252
COM:

ASCHENBACH BERND PH D
MPI F. PHYSIK & ASTROPHYS
INST F. EXTRATERR PHYSIK
KARL-SCHWARZSCHILD-STR 1
D-8046 GARCHING B MUNCHEN
GERMANY. F.R.
TEL:
TLX:
COM: 44,48

ASHOK N M DR
PHYS. RESEARCH LABORATORY
NAVRANGPURA
AHMEDABAD 380 009
INDIA
TEL: 462129
TLX: 121397
COM: 25

ASLAN ZEKI DR
ASTRONOMY DEPT
UNIVERSITY OF ANKARA
FEN FAKULTESI
ANKARA
TURKEY
TEL: 41-232105
TLX:
COM:

ASLANOV I A DR
SHEMAKHA ASTROPHYSICAL
OBSERVATORY
373243 AZERBAIDZAN
U.S.S.R.
TEL:
TLX:
COM: 29

ASSEO ESTELLE DR
CENTRE PHYSIQUE THEORIQUE
ECOLE POLYTECHNIQUE
F-91128 PALAISEAU
FRANCE
TEL: 1-69-41-82-00
TLX: 691596
COM: 48

ASSOUSA GEORGE ELIAS DR
DEPT TERRESTR. MAGNETISM
CARNEGIE INST. WASHINGTON
5241 BROAD BRANCH RD N.W.
WASHINGTON DC 20015
U.S.A.
TEL:
TLX: 440427 MAGN UI
COM: 40

ASTERIADIS GEORGIOS DR
DEPT GEODESY & SURVEYING
UNIVERSITY THESSALONIKI
GR-54006 THESSALONIKI
GREECE
TEL: 003-31-992693
TLX: 412181 AUTH GR
COM: 27,33

ATAC TAMER
KANDILLI OBSERVATORY
BOSPHORUS UNIVERSITY
CENGELKOY
ISTANBUL
TURKEY
TEL: 3320240
TLX: 26411 BOUN TR
COM:

ATANASIJEVIC IVAN DR
FACULTY OF SCIENCES
NL-6500 GL NIJMEGEN
NETHERLANDS
TEL:
TLX:
COM:

ATHANASSOULA EVANGELIE DR
OBSERVATOIRE DE MARSEILLE
2 PLACE LE VERRIER
F-13248 MARSEILLE CEDEX 4
FRANCE
TEL: 91-95-90-88
TLX: 420241 F
COM: 28,33

ATHAY R GRANT DR
HIGH ALTITUDE OBSERVATORY
P O BOX 3000
BOULDER CO 80307
U.S.A.
TEL: 303-497-1556
TLX:
COM: 10,12,36

ATHERTON PAUL DAVID
ASTRONOMY GROUP
BLACKETT LABORATORY
IMPERIAL COLLEGE
LONDON SW7
U.K.
TEL:
TLX:
COM: 09

ATKINSON ROBERT D'E DR
SWAIN HALL WEST 319
ASTRONOMY DEPT
INDIANA UNIVERSITY
BLOOMINGTON IN 47401
U.S.A.
TEL:
TLX:
COM:

ATREYA SUSHIL K
UNIVERSITY OF MICHIGAN
DEPT ATM. & OCEANIC SCI.
SPACE RESEARCH BLDG
ANN ARBOR MI 48109-2143
U.S.A.
TEL: 313-764-3335
TLX: 8102236056
COM: 16

AUBIER MONIQUE G DR
OBSERVATOIRE DE PARIS
SECTION DE MEUDON
F-92195 MEUDON PL CEDEX
FRANCE
TEL: 1-45-34-75-30
TLX: 270912
COM:

AUDOUZE JEAN PROF
INSTITUT D'ASTROPHYSIQUE
98 BIS BOULEVARD ARAGO
F-75014 PARIS
FRANCE
TEL: 1-43-20-14-25
TLX:
COM: 35,47C,48

AUER LAWRENCE H DR
HIGH ALTITUDE OBSERVATORY
P O BOX 1470
BOULDER CO 80302
U.S.A.
TEL:
TLX:
COM: 36

AUGARDE RENEE DR
OBSERVATOIRE DE MARSEILLE
2 PLACE LE VERRIER
F-13248 MARSEILLE CEDEX 4
FRANCE
TEL: 91-95-90-88
TLX: 420241
COM:

AULUCK FAQIR CHAND PROF
DEPT OF PHYSICS AND
ASTROPHYSICS
UNIVERSITY OF DELHI
DELHI 110 007
INDIA
TEL: 2918993
TLX:
COM: 47

AUMAN JASON R PROF
DEPT GEOPHYS & ASTRONOMY
UNIV OF BRITISH COLOMBIA
VANCOUVER BC V6T 1WS
CANADA
TEL: 604-228-2892
TLX:
COM: 36

AUNER GERHARD DR
INSTITUT FUER ASTRONOMIE
TUERKENSCHANZSTR 17
A-1180 WIEN
AUSTRIA
TEL:
TLX:
COM:

AURIEMMA GIULIO DR
DIPARTIMENTO DI FISICA
UNIVERSITA DI ROMA
PIAZZALE A. MORO 2
I-00187 ROMA
ITALY
TEL: 396-4976336
TLX: 613255 INFNRO
COM:

AURIERE MICHEL
OBSERVATOIRE PIC-DU-MIDI
ET TOULOUSE
F-65200 BAGNERES-DE-B.
FRANCE
TEL: 62-95-19-69
TLX:
COM: 37

AUVERGNE MICHEL
OBSERVATOIRE DE NICE
BP 139
F-06003 NICE CEDEX
FRANCE
TEL: 93-89-04-20
TLX:
COM:

AVCIOGLU KAMURAN PROF DR
UNIVERSITY OBSERVATORY
UNIVERSITE
ISTANBUL
TURKEY
TEL: 90-1-522-35-97
TLX:
COM:

AVERY LORNE W DR
HERZBERG INST ASTROPHYS
NATIONAL RESEARCH COUNCIL
OTTAWA ONT K1A 0R6
CANADA
TEL: 613-993-6060
TLX:
COM: 34,40

AVIGNON YVETTE DR
OBSERVATOIRE DE PARIS
SECTION DE MEUDON
F-92195 MEUDON PL CEDEX
FRANCE
TEL: 1-45-34-75-30
TLX: 200590
COM: 10,40

AYRES THOMAS R
UNIVERSITY OF COLORADO
CTR/ASTROPHYS & SPACE AST
CAMPUS BOX 391
BOULDER CO 80309
U.S.A.
TEL: 303-492-5320
TLX: 755842 JILA
COM: 12,44

BABADZHANOV PULAT B DR
ASTROPHYSICAL INSTITUTE
TADJIK ACAD OF SCIENCES
734670 DUSHANBE
U.S.S.R.
TEL:
TLX:
COM: 15,20,22P

BAGARE S P DR
INDIAN INSTITUTE OF
ASTROPHYSICS
BANGALORE 560 034
INDIA
TEL: 566585/566497
TLX: 845763 IIAB IN
COM: 10

AVNI YORAM DR
WEIZMANN INSTITUTE
REHOVOT 76100
ISRAEL
TEL: 8-482331
TLX: 361900 WIX
COM: 48

AZZOPARDI MARC MR
ESO
KARL-SCHWARZSCHILD-STR 2
D-8046 GARCHING B MUNCHEN
GERMANY, F.R.
TEL: 49-8932006-0
TLX: 52828222 EO D
COM: 28,30

BABCOCK HORACE W DR
MT WILSON & LAS CAMPANAS
OBSERVATORIES
813 SANTA BARBARA ST
PASADENA CA 91101
U.S.A.
TEL: 818-577-1122
TLX:
COM: 09

BAGGALEY WILLIAM J PROF
PHYSICS DEPT
UNIVERSITY OF CANTERBURY
CHRISTCHURCH
NEW ZEALAND
TEL: 482-009 x767
TLX: 4144 UNICANT NZ
COM: 21,22C

AVRETT EUGENE H DR
CENTER FOR ASTROPHYSICS
60 GARDEN STREET
CAMBRIDGE MA 02138
U.S.A.
TEL: 617-495-7423
TLX: 921428 SATELLITE CAM
COM: 36

BAADE DIETRICH DR
ST/ECF
C/O ESO
KARL-SCHWARZSCHILD-STR 2
D-8046 GARCHING B MUNCHEN
GERMANY, F.R.
TEL: 49-89-32006388
TLX: 05 282820 EO D
COM: 27,29

BACCHUS PIERRE PROF
LABORATOIRE D'ASTRONOMIE
UNIVERSITE DE LILLE I
1 IMP. DE L'OBSERVATOIRE
F-59000 LILLE
FRANCE
TEL: 20-52-44-24
TLX:
COM: 08,26

BAGHOS BALEGH B DR
HELWAN OBSERVATORY (HIAG)
HELWAN-CAIRO
EGYPT
TEL: 780645 - 934948
TLX: 93070 HIAG
COM: 07

AWAD MERVAT EL-SAID DR
DEPT OF ASTRONOMY
FACULTY OF SCIENCE
CAIRO UNIVERSITY
CAIRO
EGYPT
TEL:
TLX:
COM:

BAAN WILLEM A
ARECIBO OBSERVATORY
PO BOX 995
ARECIBO PR 00613
U.S.A.
TEL: 809-878-2612
TLX: 385638
COM: 48

BACKER DONALD CH DR
RADIO ASTRONOMY LAB
UNIVERSITY OF CALIFORNIA
601 CAMPBELL HALL
BERKELEY CA 94720
U.S.A.
TEL: 415-NGC-5118
TLX: 820181 UCB AST RAL
COM: 08,40

BAGILDINSKIJ BRONISLAV K
PULKOVO OBSERVATORY
196140 LENINGRAD
U.S.S.R.
TEL:
TLX:
COM: 08

AWADALLA NABIL SHOUKRY DR
HELWAN OBSERVATORY (HIAG)
HELWAN-CAIRO
EGYPT
TEL: 780645
TLX: 93070 HIAG UN
COM: 42

BAARS JACOB W M DR
MPI FUER RADIOASTRONOMIE
AUF DEM HUEGEL 69
D-5300 BONN
GERMANY, F.R.
TEL: 0228-525310
TLX: 886440
COM: 34,40

BADOLATI ENNIO
VIA GIUSEPPE COTRONEI 11
I-80129 NAPOLI
ITALY
TEL: 081-243245
TLX:
COM: 41

BAGLIN ANNIE DR
OBSERVATOIRE DE NICE
BP 139
F-06003 NICE CEDEX
FRANCE
TEL: 93-89-04-20
TLX: 460004
COM: 27,35

AXFORD W IAN PROF
MPI FUER AERONOMIE
POSTFACH 20
D-3411 KATLENBURG-LINDAU
GERMANY, F.R.
TEL: 05556-41-414
TLX: 965527
COM: 34,48

BAART EDWARD E PROF
DEPT OF PHYSICS
RHODES UNIVERSITY
P O BOX 94
GRAHAMSTOWN 6140
SOUTH AFRICA
TEL: 0461-7128
TLX: 244226
COM: 34,40

BAECK NICOLE A L DR
191 E-3 PLEIN
B-9218 GENT
BELGIUM
TEL:
TLX:
COM:

BAGRI DURGADAS S
RADIOASTRONOMY CENTER
TATA INSTITUTE OF
FUNDAMENTAL RESEARCH
OOTACAMUND 643 001
INDIA
TEL:
TLX: 8458488 TIFR IN
COM: 40

AXON DAVID
UNIVERSITY OF MANCHESTER
NUFFIELD RADIO ASTRON LAB
JODRELL BANK
MACCLESFIELD SK11 9DL
U.K.
TEL: 0477-71321
TLX: 36149
COM: 25,40

BAATH LARS B DR
PL 8481
S-439 00 ONSALA
SWEDEN
TEL:
TLX: 2400 ONSPACE S
COM: 40

BAEK CHANG RYONG
PHYSICS DEPT
KIM IL SONG UNIVERSITY
TAESONG DISTRICT
PYONGYANG
KOREA DPR
TEL:
TLX:
COM:

BAHCALL JOHN N PROF
INST FOR ADVANCED STUDY
OLDEN LN., BLDG E
PRINCETON NJ 08540
U.S.A.
TEL: 609-734-8054
TLX: 837680
COM: 28,33C

AYDIN CEMAL PROF DR
FEN FAKULTESI
ASTRONOMI BOLUMU
BESEVLER, ANKARA
TURKEY
TEL: 232105-94
TLX:
COM:

BABADZHANIANC MICHAIL DR
ASTRONOMICAL OBSERVATORY
LENINGRAD UNIVERSITY
LENINGRAD
U.S.S.R.
TEL:
TLX:
COM: 28

BAERENTZEN JORN
ELMEHOJVEJ 66
DK-8270 HOJBJERG
DENMARK
TEL: 45-6-272428
TLX:
COM:

BAHNER KLAUS DR
LANDESSSTERNWARTE &
MPI FUER ASTRONOMIE
KOENIGSTUHL
D-6900 HEIDELBERG 1
GERMANY, F.R.
TEL: 06221-528223
TLX: 461789 MPIA D
COM:

BAHNG JOHN D R PROF
DEARBORN OBSERVATORY
NORTHWESTERN UNIVERSITY
EVANSTON IL 60201
U.S.A.
TEL: 312-491-8645
TLX:
COM: 25,45

BAKER JAMES GILBERT DR
14 FRENCH DRIVE
BEDFORD NH 03102
U.S.A.
TEL: 603-472-5860
TLX:
COM:

BALDWIN RALPH B
6190 GATEHOUSE DR S.E.
GRAND RAPIDS MI49506
U.S.A.
TEL: 619-949-6190
TLX:
COM:

BALLI EDIBE PROF
UNIVERSITE RASATHANESI
ISTANBUL
TURKEY
TEL:
TLX:
COM: 10

BAILEY JEREMY A
ANGLO-AUSTRALIAN OBS
P.O. BOX 296
EPPING NSW 2121
AUSTRALIA
TEL: 02-868-1666
TLX: 23999 AAOSYD AA
COM:

BAKER NORMAN H PROF
MPI FUER ASTROPHYSIK
KARL-SCHWARZSCHILD-STR 1
D-8046 GARCHING B MUNCHEN
GERMANY, F.R.
TEL: 089-32990
TLX:
COM: 05,27,35

BALICK BRUCE PROF
UNIVERSITY OF WASHINGTON
ASTRONOMY DEPT FM-20
SEATTLE WA 98195
U.S.A.
TEL: 206-543-7683
TLX: 4740096 UW UI
COM:

BALMINO GEORGES G DR
CNES/GRGS/BGI
18 AVE EDOUARD BELIN
F-31055 TOULOUSE CEDEX
FRANCE
TEL: 61-27-44-27
TLX: 531081 CNEST B F
COM: 07

BAILEY MARK EDWARD
DEPT OF ASTRONOMY
THE UNIVERSITY
MANCHESTER M13 9PL
U.K.
TEL: 061-273-7121
TLX:
COM: 20,28

BAKOS GUSTAV A PROF
DEPT OF PHYSICS
UNIVERSITY WATERLOO
WATERLOO ONT N2L 3G1
CANADA
TEL: 519-885-1211
TLX:
COM: 27,51

BALIUNAS SALLIE L
CENTER FOR ASTROPHYSICS
60 GARDEN STREET
CAMBRIDGE MA 02138
U.S.A.
TEL: 617-495-7415
TLX:
COM: 12,44

BALONA LUIS ANTERO DR
S A A C
P O BOX 9
OBSERVATORY 7935
SOUTH AFRICA
TEL: 47-0025
TLX: 20309
COM: 27,30

BAIRD GEORGE A DR
PHYSICS DEPT
UNIVERSITY COLLEGE
BELFIELD
DUBLIN 4
IRELAND
TEL:
TLX:
COM:

BALAZS BELA A DR
INSTITUT FUER ASTRONOMIE
TUERKENSCHANZSTR 17
A-1180 WIEN
AUSTRIA
TEL: 141019
TLX:
COM: 37,51

BALKLAVS A E DR
RADIOPHYSICAL OBSERVATORY
LATVIAN ACADEMY OF SCI.
TURGENEVA 19
226524 RIGA
U.S.S.R.
TEL:
TLX:
COM: 40

BALUTEAU JEAN-PAUL
OBS DE HAUTE-PROVENCE
F-04870 ST-MICHEL-L'OBS.
FRANCE
TEL: 92-76-63-68
TLX: 410690 F
COM: 34

BAIRD KENNETH M DR
DIVISION OF PHYSICS
NATL RES COUNCIL CANADA
MONTREAL ROAD
OTTAWA ONT K1A OR6
CANADA
TEL:
TLX:
COM: 14

BALAZS LAJOS G DR
KONKOLY OBSERVATORY
BOX 67
H-1525 BUDAPEST
HUNGARY
TEL: 166-506,166-426
TLX: 227460 KONOB H
COM: 33C

BALKOWSKI-MAUGER CH DR
OBSERVATOIRE DE PARIS
SECTION DE MEUDON
DAF
F-92195 MEUDON PL CEDEX
FRANCE
TEL: 1-45-34-75-70
TLX:
COM: 28

BANDERMANN L W DR
21131 GRENOLA DRIVE
CUPERTINO CA 95014
U.S.A.
TEL:
TLX:
COM:

BAIRD SCOTT R
DEPT PHYSICS & ASTRONOMY
BENEDICTINE COLLEGE
N-14
ATCHISON KS 66002-1499
U.S.A.
TEL: 913-367-5340
TLX:
COM: 36

BALDINELLI LUIGI DR
C P 1630
I-40100 BOLOGNA
ITALY
TEL: 051-227002
TLX:
COM: 25

BALL JOHN A DR
NEROC HAYSTACK OBS
OFF ROUTE 40
WESTFORD MA 01886
U.S.A.
TEL: 617-692-4764
TLX:
COM: 51

BANDYOPADHYAY A DR
POSITIONAL ASTRONOMY CTR
P-546, BLOCK N
NEW ALIPORE
CALCUTTA 700 053
INDIA
TEL: 450321
TLX:
COM: 04

BAIZE PAUL DR
6 RUE DAUBIGNY
F-75017 PARIS
FRANCE
TEL:
TLX:
COM: 26

BALDWIN JACK A DR
CERRO TOLOLO
INTERAMERICAN OBSERVATORY
CASILLA 603
LA SERENA
CHILE
TEL: 213352
TLX:
COM:

BALLABH G M DR
DEPT OF ASTRONOMY
OSMANIA UNIVERSITY
HYDERABAD 500 007
INDIA
TEL: 71951 x 247
TLX:
COM: 28

BANG YONG GOL
PYONGYANG ASTRON OBS
ACADEMY OF SCIENCES DPRK
TAESONG DISTRICT
PYONGYANG
KOREA DPR
TEL:
TLX:
COM: 19

BAJAJA E DR
INSTITUTO ARGENTINO
DE RADIOASTRONOMIA
CASILLA NO 5
1894 VILLA ELISA (Bs.As.)
ARGENTINA
TEL: 021-4-3793
TLX:
COM:

BALDWIN JOHN E DR
CAVENDISH LABORATORY
MADINGLEY ROAD
CAMBRIDGE CB3 0HE
U.K.
TEL: 223-66477
TLX: 81292 CAVLAB G
COM: 33,34,40P,47

BALLARIO M C PROF
OSSERVATORIO ASTRONOMICO
DI ARCETRI
VIA S LEONARDO
I-50100 FIRENZE
ITALY
TEL:
TLX:
COM:

BANIA THOMAS MICHAEL
DEPT OF ASTRONOMY
BOSTON UNIVERSITY
725 COMMONWEALTH AVE
BOSTON MA 02215
U.S.A.
TEL: 617-353-3652
TLX: 95129 BOS UNIV BSN
COM: 34,51

LIST OF MEMBERS

BANIN V G DR
SIBIZMIR
P B 4
664697 IRKUTSK 33
U.S.S.R.
TEL: 6-02-65
TLX:
COM: 10

BANOS COSMAS J DR
ASTRONOMICAL INSTITUTE
NATIONAL OBSERVATORY
P O 20048
GR-11810 ATHENS
GREECE
TEL: 3461-191
TLX: 215530 OBSA GR
COM: 21

BANOS GEORGE J PROF
UNIVERSITY OF IOANNINA
PHYSICS DEPT
DIV OF ASTRO-GEOPHYSICS
GR-45332 IOANNINA
GREECE
TEL: 0651-91697
TLX: 322160
COM:

BAO KEREN
NANJING ASTRONOMICAL
INSTRUMENT FACTORY
NANJING
CHINA. PEOPLE'S REP.
TEL: 46191
TLX: 34136 GLYNJ c/c NAIF
COM: 09

BARANNE A DR
OBSERVATOIRE DE MARSEILLE
2 PLACE LE VERRIER
F-13248 MARSEILLE CEDEX 4
FRANCE
TEL: 91-95-90-88
TLX:
COM: 09

BARATTA GIOVANNI BATTISTA
OSSERVATORIO ASTRONOMICO
DI ROMA
V. DEL PARCO MELLINI 84
I-00136 ROMA
ITALY
TEL: 06-347-056
TLX:
COM: 29

BARBANIS BASIL PROF
DEPT OF ASTRONOMY
UNIVERSITY THESSALONIKI
GR-54006 THESSALONIKI
GREECE
TEL: 0030-31-991357
TLX: 412181
COM: 33

BARBARO G DR
OSSERVATORIO ASTRONOMICO
VICOLO DELL'OSSERVATORIO
I-35100 PADOVA
ITALY
TEL: 49-66-14-99
TLX: 430176 UNPADU I
COM:

BARBERIS BRUNO
IST. DI FISICA MATEMATICA
UNIVERSITA DI TORINO
VIA CARLO ALBERTO 10
I-10123 TORINO
ITALY
TEL: 011-539214
TLX:
COM: 07,33,47

BARBIER-BROSSAT M DR
OBSERVATOIRE DE MARSEILLE
2 PLACE LE VERRIER
F-13248 MARSEILLE CX 04
FRANCE
TEL: 91-95-90-88
TLX: 420241 F
COM: 30,45

BARBIERI CESARE PROF
INSTITUTO DI ASTRONOMIA
UNIVERSITA DI PADOVA
VIC. DELL'OSSERVATORIO 5
I-35100 PADOVA
ITALY
TEL: 049-661499
TLX: 430176 UNPADU I
COM: 28,51

BARBON ROBERTO PROF
OSSERVATORIO ASTROFISICO
I-36012 ASIAGO VI
ITALY
TEL: 0424-62665
TLX: 430110 SETURIST
COM: 28

BARCZA SZABOLCS DR
KONKOLY OBSERVATORY
THEGE UT 13/17
BOX 67
H-1525 BUDAPEST
HUNGARY
TEL: 1-166-426
TLX: 227460 KONOB H
COM:

BARDEEN JAMES M PROF
PHYSICS DEPT FM-15
UNIVERSITY OF WASHINGTON
SEATTLE WA 98195
U.S.A.
TEL: 206-545-2394
TLX: 4740096 UW UI
COM: 47

BARKAT ZALMAN PROF
RACAH INST. OF PHYSICS
HEBREW UNIV. OF JERUSALEM
JERUSALEM 91904
ISRAEL
TEL: 02-584490
TLX: 25391
COM:

BARKER BRUCE MICHAEL
DEPT PHYSICS & ASTRONOMY
UNIVERSITY OF ALABAMA
UNIVERSITY AL 35486
U.S.A.
TEL: 205-348-5050
TLX:
COM: 42

BARKER EDWIN S DR
ASTRONOMY DEPT
UNIVERSITY OF TEXAS
RLM 15.308
AUSTIN TX 78712
U.S.A.
TEL: 512-471-4461
TLX: 910-874-1351
COM: 15

BARKER PAUL K DR
DEPT OF ASTRONOMY
UNIV OF WESTERN ONTARIO
LONDON ONT N6A 3K7
CANADA
TEL:
TLX:
COM:

BARKHATOVA KLAUDIA PROF
STATE UNIVERSITY
LENIN ST 51
620083 SVERDLOVSK
U.S.S.R.
TEL:
TLX:
COM: 37,46

BARLAI KATALIN DR
KONKOLY OBSERVATORY
THEGE UT 13/17
BOX 67
H-1525 BUDAPEST
HUNGARY
TEL: 36-1-366-621
TLX: 227460 KONOB H
COM:

BARLETTI RAFFAELE ENG
OSSERVATORIO ARCETRI
LARGO E. FERMI 5
I-50125 FIRENZE
ITALY
TEL: 055-220034
TLX: 572268
COM:

BARLIER FRANCOIS E DR
CERGA
AVENUE COPERNIC
F-06130 GRASSE
FRANCE
TEL: 93-36-58-49
TLX: 470865
COM: 19C

BARLOW MICHAEL J DR
DEPT PHYSICS & ASTRONOMY
UNIVERSITY COLLEGE LONDON
GOWER STREET
LONDON WC1E 6BT
U.K.
TEL: 01-387-7050
TLX: 28722 UCPHYS G
COM: 34

BARNARD HANNES A J DR
PHYSICS DEPT
UNIV OF BRITISH COLUMBIA
6224 AGRICULTURE ROAD
VANCOUVER BC V6T 2A6
CANADA
TEL: 604-228-2894
TLX: 04-508576
COM: 14

BARNES AARON DR
NASA AMES RESEARCH CENTER
CODE 245-3
MOFFETT FIELD CA 94035
U.S.A.
TEL: 415-694-5506
TLX:
COM: 34,49

BARNES III THOMAS G DR
DEPT OF ASTRONOMY
UNIVERSITY OF TEXAS
MC DONALD OBSERVATORY
AUSTIN TX 78712
U.S.A.
TEL: 512-471-4461
TLX: 910-874-1351
COM: 27

BARNOTHY JENO DR PROF
833 LINCOLN STREET
EVANSTON IL 60201
U.S.A.
TEL: 312-328-5729
TLX:
COM: 47,48

BAROCAS VINICIO PROF
11 YEWLANDS AVENUE
FULWOOD
PRESTON PR2 4QR
U.K.
TEL: 0772-719249
TLX:
COM: 46

BARRETO LUIZ MUNIZ PROF
OBSERVATORIO NACIONAL
RUA GENERAL BRUCE 586
20921 RIO DE JANEIRO
BRAZIL
TEL: 021-580-6087
TLX: 2121288 CNPq BR
COM: 19,50

BARRETT ALAN H PROF
DEPT OF PHYSICS
MIT
ROOM 26-331
CAMBRIDGE MA 02139
U.S.A.
TEL: 617-253-5283
TLX:
COM: 34,40C

BARROSO JR JAIR
CNPé - OBSERVATORIO NACL
LAB. NACL. ASTROFISICA
CAIXA POSTAL 21
37500 ITAJUBA MG
BRAZIL
TEL: 035-6220788
TLX: 031-2603
COM: 09

BARROW COLIN H DR
NATIONAL SPACE DATA CTR
NASA/GSFC, CODE 601
GREENBELT MD 20771
U.S.A.
TEL:
TLX:
COM: 10,16,40,49

BARROW JOHN DAVID
ASTRONOMY CENTRE
UNIVERSITY OF SUSSEX
FALMER
BRIGHTON BN1 9QH
U.K.
TEL: 0273-606755
TLX: 877159 UNISEX G
COM: 47

BARROW RICHARD F DR
PHYSICAL CHEMISTRY
LABORATORY
SOUTH PARKS RD
OXFORD OX1 3QZ
U.K.
TEL: 0865-53322
TLX:
COM: 14

BARRY DON C DR
DEPT OF ASTRONOMY
UNIV. OF SOUTHERN CALIF.
LOS ANGELES CA 90089
U.S.A.
TEL: 213-743-2764
TLX:
COM: 29,45.

BARTAYA R A DR
ABASTUMANI ASTROPHYSICAL
OBSERVATORY
383762 ABASTUMANI,GEORGIA
U.S.S.R.
TEL: 237 ABASTUMANI
TLX: 327409
COM: 45

BARTEL NORBERT HARALD DR
CENTER FOR ASTROPHYSICS
60 GARDEN STREET
CAMBRIDGE MA 02138
U.S.A.
TEL: 617-495-9278
TLX: 921428
COM: 40

BARTH CHARLES A PROF
LASP
UNIVERSITY OF COLORADO
BOX 392
BOULDER CO 80309
U.S.A.
TEL: 303-492-7502
TLX:
COM: 49

BARTHOLDI PAUL DR
OBSERVATOIRE DE GENEVE
CH-1290 SAUVERNY
SWITZERLAND
TEL: 22-55-26-11
TLX: 27720 OBSG CH
COM:

BARTOLINI CORRADO
DIPARTIMENTO ASTRONOMIA
UNIVERSITA DI BOLOGNA
VIA ZAMBONI 33
I-40126 BOLOGNA
ITALY
TEL: 051-226677
TLX: 211664
COM: 27,42

BARWIG HEINZ
INST F ASTRON & ASTROPHYS
UNIVERSITAET MUNCHEN
SCHEINERSTR. 1
D-8000 MUNCHEN 80
GERMANY, F.R.
TEL: 089-98-90-21
TLX: 529815 UNIVM D
COM: 09,27

BASART JOHN P
COOVER HALL
IOWA STATE UNIVERSITY
AMES IA 50011
U.S.A.
TEL: 515-294-2663
TLX:
COM:

BASCHEK BODO PROF
INSTITUT F THEORETISCHE
ASTROPHYSIK
IM NEUENHEIMER F 561
D-6900 HEIDELBERG
GERMANY, F.R.
TEL: 49-6221-562837
TLX: 461515 UNIHD D
COM: 36

BASH FRANK N PROF
ASTRONOMY DEPT
UNIVERSITY OF TEXAS
R.L. MOORE BLDG
AUSTIN TX 78712
U.S.A.
TEL: 512-471-4461
TLX: 910-874-1351
COM: 33,34,40

BASRI GIBOR B
ASTRONOMY DEPARTMENT
UNIVERSITY OF CALIFORNIA
BERKELEY CA 94720
U.S.A.
TEL: 415-642-8198
TLX: 820181 UCB ASTRAL UD
COM:

BASTIAN ULRICH
ASTRONOMISCHES-RECHEN
INSTITUT
MOENCHHOFSTR. 12-14
D-6900 HEIDELBERG
GERMANY, F.R.
TEL: 06221-49026
TLX: 461336 ARI HD D
COM: 24

BASTIEN PIERRE DR
DEPT DE PHYSIQUE
UNIVERSITE DE MONTREAL
C P 6128 SUCC A
MONTREAL PQ H3C 3J7
CANADA
TEL: 514-343-7355
TLX: 055-62425 UDEMPHYSAS
COM: 27

BASTIN J A PROF
PHYSICS DEPT
QUEEN MARY COLLEGE
MILE END ROAD
LONDON E1 4NS
U.K.
TEL:
TLX:
COM:

BASU BAIDYANATH PROF
APPLIED MATHEMATICS DEPT
CALCUTTA UNIVERSITY
92 A P C ROAD
CALCUTTA 700 009
INDIA
TEL:
TLX:
COM: 28,33,51

BASU DIPAK DR
DEPT OF PHYSICS
UNIV OF THE WEST INDIES
ST AUGUSTINE
TRINIDAD AND TOBAGO
TEL: 663-1369-3123
TLX:
COM: 40,47,48,51

BATES BRIAN DR
DEPT OF PURE & APPL PHYS
QUEEN'S UNIVERSITY
BELFAST BT7 1NN
U.K.
TEL: 245133
TLX: 74487
COM:

BATES DAVID R PROF
DEPT OF APPLIED MATHS
QUEEN'S UNIVERSITY
BELFAST BT7 1NN
U.K.
TEL:
TLX:
COM: 14,21

BATES J RAPHAEL
METEOROLOGICAL SERVICE
GLASNEVIN HILL
DUBLIN 9
IRELAND
TEL: 424411
TLX: 25239
COM: 10

BATES RICHARD HEATON T DR
ELECTRICAL & ELECTRONIC
ENG. DEPT
UNIVERSITY OF CANTERBURY
CHRISTCHURCH
NEW ZEALAND
TEL: 03-482-009 x336
TLX:
COM: 40

BATESON FRANK M OBE DR
ASTRONOMICAL RESEARCH LTD
P O BOX 3093
GREERTON TAURANGA
NEW ZEALAND
TEL: 64-075-410-216
TLX: 2880 CPO TG NZ
COM: 27

BATH GEOFFREY T DR
DEPT OF ASTROPHYSICS
UNIVERSITY OF OXFORD
OXFORD OX1 3RQ
U.K.
TEL: 511336
TLX: 83295
COM: 27,42

BATRAKOV YU V DR
INSTITUTE OF THEORETICAL
ASTRONOMY
NABEREZHNAYA KUTUZOVA 10
191187 LENINGRAD
U.S.S.R.
TEL: 272-40-23
TLX: 121578 ITA SU
COM: 07C,20V

BATTANER EDUARDO DR
DPTO FIS TIERRA & COSMOS
FACULTAD DE CIENCIAS
AVDA FUENTENUEVA
GRANADA
SPAIN
TEL: 958-202212-306
TLX:
COM: 16,28,46

BATTEN ALAN H DR
HERZBERG INST ASTROPHYS
DOMINION ASTROPHYS OBS
5071 W SAANICH RD
VICTORIA BC V8X 4M6
CANADA
TEL: 604-388-0009
TLX: 0497295
COM: 26,30,42C

BATTISTINI PIERLUIGI DR
OSSERVATORIO ASTRONOMICO
VIA ZAMBONI 33
I-40126 BOLOGNA
ITALY
TEL:
TLX:
COM:

BATTY MICHAEL DR
CSIRO
DIVISION OF RADIOPHYSICS
P.O. BOX 76
EPPING NSW 2121
AUSTRALIA
TEL: 02-868-0222
TLX: 26230 ASTRO AA
COM: 40

BAUD BOUDEWIJN DR
FOKKER B.V.
SPACE DIVISION
POSTBUS 7600
NL-1117 ZJ SHIPHOL
NETHERLANDS
TEL: 020-5449111
TLX:
COM: 33

BAUDRY ALAIN DR
OBSERVATOIRE DE BORDEAUX
F-33270 FLOIRAC
FRANCE
TEL: 56-86-43-30
TLX:
COM: 34,40C

BAUER CARL A DR
PENNA STATE UNIVERSITY
506 DAVEY
UNIVERSITY PARK PA 16802
U.S.A.
TEL:
TLX:
COM:

BAUM WILLIAM A DR
LOWELL OBSERVATORY
1400 W. MARS HILL RD
FLAGSTAFF AZ 86001
U.S.A.
TEL: 602-774-3358
TLX:
COM: 09,16,28,51

BAUSTIAN W W MR
KITT PEAK NAT OBSERVATORY
PO BOX 26732
950 N. CHERRY AVE
TUCSON AZ 85726
U.S.A.
TEL:
TLX:
COM:

BAUTZ LAURA P DR
DIV ASTRONOMICAL SCIENCES
NATL SCIENCE FOUNDATION
WASHINGTON DC 20550
U.S.A.
TEL: 202-357-9488
TLX:
COM:

BAYM GORDON ALAN DR
DEPT OF PHYSICS
UNIVERSITY OF ILLINOIS
URBANA IL 61801
U.S.A.
TEL: 217-333-4363
TLX: 910-830-6599 PHYSICS
COM: 35,48

BAZILEVSKY ALEXANDR T
VERNADSKY INST GEOCHEM &
ANALYTICAL CHEMISTRY
KOSYGIN STR 19
117334 MOSCOW
U.S.S.R.
TEL:
TLX:
COM: 16

BEALE JOHN S DR
231 MARLBOROUGH ROAD
SWINDON SN3 1NN
U.K.
TEL: 0793-34725
TLX:
COM:

BEARD DAVID B DR
DEPT PHYSICS & ASTRONOMY
UNIVERSITY OF KANSAS
LAWRENCE KS 66045
U.S.A.
TEL: 913-864-3752
TLX:
COM: 12,15,22

BEARDSLEY WALLACE R DR
PO BOX 531
NEWARK CA 94560
U.S.A.
TEL: 415-792-2786
TLX:
COM: 30

BEAUDET GILLES DR
DEPT DE PHYSIQUE
UNIVERSITE DE MONTREAL
CP 6128
MONTREAL PQ H3C 3J7
CANADA
TEL: 514-343-6669
TLX: 055-62425
COM: 35,51

BEAVERS WILLET I DR
ERWIN W FICK OBSERVATORY
IOWA STATE UNIVERSITY
AMES IA 50011
U.S.A.
TEL: 515-294-3667
TLX:
COM: 30

BEC-BORSENBERGER ANNICK
BUREAU DES LONGITUDES
77 AVE DENFERT-ROCHEREAU
F-75014 PARIS
FRANCE
TEL: 1-43-20-12-10
TLX:
COM: 04,07,20

BECK H G
VEB CARL ZEISS
FORSCHUNGSZENTRUM
CARL-ZEISS STR 1
DDR-6900 JENA
GERMANY, D.R.
TEL:
TLX:
COM:

BECK RAINER
MPI FUER RADIOASTRONOMIE
AUF DEM HUEGEL 69
D-5300 BONN 1
GERMANY, F.R.
TEL: 0228-525-320
TLX: 886440 MPIFR D
COM: 25,28,40

BECKER FRIEDRICH DR PROF
WEITLSTR 66-2109
D-8000 MUENCHEN 45
GERMANY, F.R.
TEL: 089-38582109
TLX:
COM:

BECKER ROBERT A DR
PO BOX 4609
CARMEL CA 93921
U.S.A.
TEL:
TLX:
COM:

BECKER ROBERT HOWARD
PHYSICS DEPT
UNIVERSITY OF CALIFORNIA
DAVIS CA 95616
U.S.A.
TEL: 916-752-6921
TLX: 910-531-0785 UC DAVS
COM: 48

BECKER STEPHEN A
LOS ALAMOS NATIONAL LAB.
APPL. THEORET. PHYS. DIV.
PO BOX 1663, MS B220
LOS ALAMOS NM 87545
U.S.A.
TEL: 505-667-8931
TLX: 660495
COM: 35

BECKER WILHELM PROF
ASTRONOMISCHES INSTITUT
UNIVERSITAET BASEL
VENUSSTRASSE 7
CH-4102 BINNINGEN
SWITZERLAND
TEL: 061-22-77-11
TLX:
COM: 25,33,37

BECKERS JACQUES M DR
ADV DEVELOPMENT PROGRAM
NAT OPTICAL ASTR OBS
950 N. CHERRY AVENUE
TUCSON AZ 85726
U.S.A.
TEL: 602-325-9226
TLX: 0666484
COM: 10,12

BECKLIN ERIC E DR
INSTITUTE FOR ASTRONOMY
2680 WOODLAWN DRIVE
HONOLULU HI 96822
U.S.A.
TEL: 808-948-6666
TLX: 723 8459 UHAST HR
COM: 34

BECKMAN JOHN E PROF
INSTITUTO DE FISICA
DE CANARIAS
LA LAGUNA
38071 TENERIFE
SPAIN
TEL:
TLX:
COM: 12,34,47,51

BECKWITH STEVEN V W
DEPT OF ASTRONOMY
SPACE SCIENCES BLDG
CORNELL UNIVERSITY
ITHACA NY 14853
U.S.A.
TEL: 607-256-4805
TLX:
COM: 34,51

BEDOGNI ROBERTO
OSERVATORIO DI ASTRONOMIA
UNIVERSITA DEGLI STUDI
C P 596
I-40100 BOLOGNA
ITALY
TEL: 051-222956
TLX: 211664 INFNBO I
COM:

BEEBE HERBERT A
NEW MEXICO STATE UNIV.
DEPT OF ASTRONOMY
LAS CRUCES NM 88003
U.S.A.
TEL: 505-646-4438
TLX: 910-983-0549 NMSUC
COM: 10,12

BEEBE RETA FAYE DR
NEW MEXICO STATE UNIV
DEPT OF ASTRONOMY
BOX 4500
LAS CRUCES NM 88003
U.S.A.
TEL: 505-646-1938
TLX:
COM: 16,51

BEER REINHARD DR
JET PROPULSION LAB
CALIF INST OF TECHNOLOGY
4800 OAK GROVE DR
PASADENA CA 91109
U.S.A.
TEL: 818-354-4748
TLX:
COM: 09,16

BEGELMAN MITCHELL CRAIG
JILA
UNIVERSITY OF COLORADO
CAMPUS BOX 440
BOULDER CO 80309
U.S.A.
TEL: 303-492-7856
TLX: 755842 JILA
COM: 48

BEGGS DENIS W MR
INSTITUTE OF ASTRONOMY
MADINGLEY ROAD
CAMBRIDGE CB3 0HA
U.K.
TEL: 0223-62204
TLX: 817297 ASTRON G
COM:

BEHR ALFRED PROF EMERITUS
ESCHENWEG 3
D-3406 BOVENDEN
GERMANY, F.R.
TEL: 0551-8897
TLX:
COM: 25

BEINTEMA DOUWE A DR
SPACE RESEARCH DEPT
UNIVERSITY OF GRONINGEN
P O BOX 800
NL-9700 AV GRONINGEN
NETHERLANDS
TEL: 050-116631
TLX: 53572
COM:

BEKENSTEIN JACOB D DR
PHYSICS DEPT
BEN GURION UNIVERSITY
P O B 653
BEERSHEVA 84105
ISRAEL
TEL: 057-664271
TLX: 5253 UNASI IL
COM:

BEL NICOLE J DR
OBSERVATOIRE DE PARIS
SECTION DE MEUDON
F-92195 MEUDON PL CEDEX
FRANCE
TEL: 1-45-34-75-70
TLX:
COM: 12,34,47

BELKOVICH O I DR
ENGELHARDT ASTRONOMICAL
OBSERVATORY
OBSERVATORY STATION
422526 KAZAN
U.S.S.R.
TEL: 324827
TLX:
COM: 21,22C

BELTON MICHAEL J S DR
SOLAR SYSTEM PROGRAM
NATL OPTICAL ASTRON OBS
950 N. CHERRY AVE
TUCSON AZ 85726
U.S.A.
TEL: 602-327-5511
TLX: 666-484 AURA KNPO TU
COM: 16

BENDER PETER L DR
JILA
UNIVERSITY OF COLORADO
BOULDER CO 80309
U.S.A.
TEL: 303-492-6793
TLX: 755842 JILA
COM: 16,19,31

BENVENUTI PIERO DR
ST/ECF
C/O ESO
KARL-SCHWARZSCHILD-STR 2
D-8046 GARCHING B MUNCHE
GERMANY, F.R.
TEL: 49-89-32006291
TLX: 52828222 EO D
COM:

BELL BARBARA DR
CENTER FOR ASTROPHYSICS
60 GARDEN STREET
CAMBRIDGE MA 02138
U.S.A.
TEL: 617-495-2688
TLX:
COM: 10

BELVEDERE GAETANO DR
ISTITUTO DI ASTRONOMIA
CITTA UNIVERSITARIA
I-95125 CATANIA
ITALY
TEL: 39-95-330533
TLX: 970359 ASTRCT I
COM: 10,27

BENDINELLI ORAZIO
DIPT. DI ASTRONOMIA
VIA ZAMBONI 33
I-40126 BOLOGNA
ITALY
TEL: 051-226677/956
TLX: 211664 INFNBO I
COM:

BENZ ARNOLD DR
GRUPPE F RADIOASTRONOMIE
INSTITUT FUR ASTRONOMIE
ETH-ZENTRUM
CH-8092 ZUERICH
SWITZERLAND
TEL: 1-256-42-23
TLX: 53178 ETHBI CH
COM: 10,40

BELL BURNELL S JOCELYN DR
ROYAL OBSERVATORY
BLACKFORD HILL
EDINBURGH EH9 3HJ
U.K.
TEL: 031-667-3321
TLX: 72383 ROEDIN G
COM:

BELY OLEG DR
OBSERVATOIRE DE NICE
BP 139
F-06003 NICE CEDEX
FRANCE
TEL: 93-89-04-20
TLX:
COM: 14

BENEDICT GEORGE F DR
DEPT OF ASTRONOMY
UNIVERSITY OF TEXAS
AUSTIN TX 78712
U.S.A.
TEL: 512-471-4461
TLX:
COM: 24,28,44

BENZ WILLY
LOS ALAMOS NAT LABORATORY
T-6, MS B-288
LOS ALAMOS NM 87545
U.S.A.
TEL: 505-667-7645
TLX:
COM: 35

BELL KENNETH LLOYD DR
DEPT OF APPLIED MATH
QUEEN'S UNIVERSITY
BELFAST BT7 1NN
U.K.
TEL: 245133
TLX: 74487 QUBADM
COM:

BELY-DUBAU FRANCOISE
OBSERVATOIRE DE NICE
BP 139
F-06003 NICE CEDEX
FRANCE
TEL: 93-89-04-20
TLX:
COM: 14

BENEST DANIEL DR
OBSERVATOIRE DE NICE
BP 139
F-06003 NICE CEDEX
FRANCE
TEL: 93-89-04-20
TLX: 460004
COM: 07,20,51

BERENDZEN RICHARD DR
PRESIDENT'S OFFICE
THE AMERICAN UNIVERSITY
WASHINGTON DC 20016
U.S.A.
TEL: 202-885-2121
TLX:
COM: 41,51

BELL MORLEY B
HERZBERG INST ASTROPHYS
NATL RESEARCH COUNCIL
OTTAWA ONT K1A 0R6
CANADA
TEL: 613-993-6060
TLX: 0533715
COM:

BELYAEV NIKOLAJ A DR
INST FOR THEORETICAL
ASTRONOMY
10 KUTUZOV QUAY
191187 LENINGRAD
U.S.S.R.
TEL: 279-06-67
TLX: 121578 ITA SU
COM: 20

BENEVIDES SOARES P DR
INST ASTRON. E GEOFISICO
CAIXA POSTAL 30627
01051 SAO PAULO SP
BRAZIL
TEL: 11-275-3720
TLX: 1136221 IAGM BR
COM: 08C

BERG RICHARD A DR
DEPT PHYSICS & ASTRONOMY
UNIVERSITY OF ROCHESTER
419 SPACE SCIENCE CTR
ROCHESTER NY 14627
U.S.A.
TEL:
TLX:
COM:

BELL ROGER A DR
ASTRONOMY PROGRAM
UNIVERSITY OF MARYLAND
COLLEGE PARK MD 20742
U.S.A.
TEL: 301-454-6282
TLX: 887294
COM: 36,45

BEM JERZY DR
ASTRONOMICAL OBSERVATORY
WROCLAW UNIVERSITY
UL. KOPERNIKA 11
51-622 WROCLAW
POLAND
TEL:
TLX:
COM: 08

BENFORD GREGORY DR
PHYSICS DEPT
UNIVERSITY OF CALIFORNIA
IRVINE CA 92717
U.S.A.
TEL: 714-856-5147
TLX:
COM: 12,48

BERGE GLENN L DR
OWENS VALLEY RADIO OBS
CALTECH 170-25
PASADENA CA 91125
U.S.A.
TEL: 818-356-6969
TLX: 675425
COM: 16,40

BELOTSERKOVSKIJ DAVID J
VNIIFTRI
GOSSTANDART USSR
LENINSKY PROSPECT 9
117049 MOSCOW
U.S.S.R.
TEL: 236-40-44
TLX: 411378 GOST
COM: 31

BENACCHIO LEOPOLDO
OSSERVATORIO ASTRONOMICO
VICOLO DELL'OSSERVATORIO
I-35122 PADOVA
ITALY
TEL: 49-66-14-99
TLX: 430176 UNPADU I
COM: 05

BENNETT JOHN CAISTER MR
90 MALAN STREET
RIVIERA
PRETORIA 0084
SOUTH AFRICA
TEL: 012-704895
TLX:
COM: 20

BERGEAT JACQUES G DR
OBSERVATOIRE DE LYON
F-69230 ST-GENIS-LAVAL
FRANCE
TEL: 78-56-07-05
TLX:
COM:

BELSERENE EMILIA P
MARIA MITCHELL OBS
3 VESTAL STREET
NANTUCKET MA 02554
U.S.A.
TEL: 617-228-9273
TLX:
COM: 27

BENAVENTE JOSE
INSTITUTO Y OBSERVATORIO
DE MARINA
SAN FERNANDO (CADIZ)
SPAIN
TEL: 956-883548
TLX: 76108 IOM E
COM: 31C

BENSAMMAR SLIMANE DR
OBSERVATOIRE DE PARIS
SECTION DE MEUDON
F-92195 MEUDON PL CEDEX
FRANCE
TEL: 1-45-34-75-30
TLX: 270912 OBSASTR F
COM: 09,50

BERGER CHRISTIANE DR
CERGA
AVENUE COPERNIC
F-06130 GRASSE
FRANCE
TEL: 93-36-58-49
TLX: 470865 F
COM:

LIST OF MEMBERS

BERGER JACQUES G DR
OBSERVATOIRE DE PARIS
61 AVE DE L OBSERVATOIRE
F-75014 PARIS
FRANCE
TEL: 1-43-20-12-10
TLX:
COM: 29

BERGER XAVIER DR
LAB ECOTHERMIQUE SOLAIRE
SOPHIA ANTIPOLIS
BP 21
F-06562 VALBONNE CEDEX
FRANCE
TEL: 93-65-34-00
TLX: 970134 F
COM:

BERGERON JACQUELINE A DR
INSTITUT D'ASTROPHYSIQUE
98 BIS BOULEVARD ARAGO
F-75014 PARIS
FRANCE
TEL: 1-43-20-14-25
TLX: 270070 c/o IAP
COM: 28,34,44,47,48

BERGSTRALH JAY T DR
JPL
M/X 183-301
4800 OAK GROVE DRIVE
PASADENA CA 91109
U.S.A.
TEL: 818-354-2296
TLX:
COM: 16

BERGVALL NILS AKE SIGVARD
ASTRONOMISKA OBSERVATORIE
BOX 515
S-751 20 UPPSALA
SWEDEN
TEL:
TLX:
COM: 28

BERKHUIJSEN ELLY M DR
MPI FUER RADIOASTRONOMIE
AUF DEM HUEGEL 69
D-5300 BONN 1
GERMANY, F.R.
TEL:
TLX: 886440 MPIFR D
COM: 28,33,34,40

BERMAN ROBERT HIRAM DR
MIT
RM 36-227
77 MASSACHUSETTS AVE
CAMBRIDGE MA 02139
U.S.A.
TEL: 617-253-1000
TLX:
COM: 28

BERNACCA P L PROF
OSSERVATORIO ASTROFISICO
DELL'UNIVERSITA'
I-36012 ASIAGO (VICENZA)
ITALY
TEL: 424-62505
TLX: 430110 SETOUR
COM: 26,44,51

BERNAT ANDREW PLOUS DR
COMPUTER SCIENCE DEPT
UNIVERSITY OF TEXAS
AT EL PASO
EL PASO TX 79968
U.S.A.
TEL: 915-747-5494
TLX:
COM: 34,36

BERRINGTON KEITH ADRIAN
DEPT OF APPLIED MATHS
QUEEN'S UNIVERSITY
BELFAST BT7 1NN
U.K.
TEL:
TLX:
COM: 14

BERTAUX J L DR
SERVICE D'AERONOMIE
BP NO 3
F-91370 VERRIERES-LE-B.
FRANCE
TEL: 1-69-20-31-16
TLX: 692400 F
COM: 16C,49

BERTHOMIEU GABRIELLE DR
OBSERVATOIRE DE NICE
BP 139
F-06003 NICE CEDEX
FRANCE
TEL: 93-89-04-20
TLX: 46004 F
COM: 27,35

BERTIAU FLOR C PROF
ASTRONOMISCH INSTITUT
KATHOLIEKE UNIV LEUVEN
CELESTIJNENLAAN 200B
B-3030 HEVERLEE
BELGIUM
TEL: 016-227401
TLX:
COM: 30

BERTIN GIUSEPPE PROF
SCUOLA NORMALE SUPERIORE
PIAZZA DEI CAVALIERI
I-56100 PISA
ITALY
TEL: 50-597265
TLX: 590548 SNSPI I
COM:

BERTOLA FRANCESCO PROF
OSSERVATORIO ASTRONOMICO
VICOLO DELL'OSSERVATORIO
I-35100 PADOVA
ITALY
TEL: 49-66-14-99
TLX: 430176 UNPADU I
COM: 28,47

BERTOUT CLAUDE
INSTITUT D'ASTROPHYSIQUE
98 BIS BOULEVARD ARAGO
F-75014 PARIS
FRANCE
TEL: 1-43-20-14-25
TLX:
COM: 36

BESSELL MICHAEL S DR
MT STROMLO OBSERVATORY
WODEN P.O. ACT 2606
AUSTRALIA
TEL: 062-881111
TLX: 62270 CANOPUS AA
COM: 05,25,27,29C

BETTIS DALE G PROF
TICOM
UNIVERSITY OF TEXAS
AT AUSTIN
AUSTIN TX 78712
U.S.A.
TEL:
TLX:
COM: 07

BETTONI DANIELA DR
OSSERVATORIO ASTRONOMICO
VICOLO DELL'OSSERVATORIO
I-35122 PADOVA
ITALY
TEL: 49-66-14-99
TLX: 430176 UNPADU I
COM: 28

BEUERMANN KLAUS P PROF
INSTITUT FUR ASTRONOMIE &
ASTROPHYSIK ZU BERLIN
ERNST-REUTER-PL 7
D-1000 BERLIN
GERMANY, F.R.
TEL:
TLX:
COM:

BHANDARI N DR
PHYSICAL RESEARCH LAB
NAVRANGPURA
AHMEDABAD 380 009
INDIA
TEL: 462129
TLX: 0121397
COM: 22

BHANDARI RAJENDRA DR
RAMAN RESEARCH INSTITUTE
BANGALORE 560 080
INDIA
TEL: 812-360122
TLX: 845671 RRI IN
COM: 40

BHATIA PREM K DR
DEPT OF MATHEMATICS
UNIVERSITY OF JODHPUR
JODHPUR 342 001
INDIA
TEL:
TLX:
COM:

BHATIA R K DR
DEPT OF ASTRONOMY
OSMANIA UNIVERSITY
HYDERABAD 500 007
INDIA
TEL:
TLX:
COM: 16

BHATIA V B DR
DEPT PHYSICS & ASTROPHYS
DELHI UNIVERSITY
DELHI-110 007
INDIA
TEL: 2918993
TLX:
COM:

BHATNAGAR ARVIND DR
UDAIPUR SOLAR OBSERVATORY
11 VIDYA MARG
UDAIPUR RAJAS 313 001
INDIA
TEL: 25626-23861
TLX:
COM: 10,12

BHATNAGAR K B DR
ZAKIR HUSSAIN COLLEGE
UNIVERSITY OF DELHI
AJMERI GATE
NEW DELHI 110 006
INDIA
TEL: 522802
TLX:
COM: 07C

BHATT H C DR
INDIAN INSTITUTE OF
ASTROPHYSICS
SARJAPUR ROAD
BANGALORE 560 034
INDIA
TEL: 566585
TLX: 845763 IIAB IN
COM: 34

BHATTACHARYYA J C PROF
INDIAN INSTITUTE OF
ASTROPHYSICS
BANGALORE 560 034
INDIA
TEL: 566583/566585
TLX: 0845763 IIAB IN
COM: 09C,12,50

BHATTACHARYYA TARA DR
JOGAMAYA DEVI COLLEGE
92 SYAMAPRADAD MUKERJEE
CALCUTTA 700 026
INDIA
TEL:
TLX:
COM: 28

BHAVSAR SUKETU P
RAMAN RESEARCH INSTITUTE
BANGALORE 560 080
INDIA
TEL: 0812-30122
TLX: 845671
COM: 47

BHONSLE RAJARAM V PROF
PHYSICAL RESEARCH LAB
NAVRANGPURA
AHMEDABAD 380 009
INDIA
TEL: 462129
TLX: 121397 PRL IN
COM: 40

BIAN YU-LIN
BEIJING ASTRONOMICAL OBS
BEIJING 100080
CHINA. PEOPLE'S REP.
TEL:
TLX: 22040 BAOAS CN
COM: 28

BIELICKI MACIEJ DR
ASTRONOMICAL OBSERVATORY
AL. UJAZDOWSKIE 4
00-478 WARSZAWA
POLAND
TEL:
TLX:
COM: 20

BILLINGS DONALD E PROF
UNIVERSITY OF COLORADO
DEPT OF ASTROGEOPHYSICS
BOULDER CO 80309
U.S.A.
TEL:
TLX:
COM: 12

BIRKLE KURT PH D
MPI FUER ASTRONOMIE
KOENIGSTUHL
D-6900 HEIDELBERG 1
GERMANY, F.R.
TEL:
TLX:
COM: 34

BIANCHI LUCIANA
OSSERVATORIO ASTRONOMICO
DI TORINO
I-10025 PINO TORINESE
ITALY
TEL: 11-842040
TLX: 213236 TO ASTR I
COM: 34,44

BIEMONT EMILE DR
INSTITUT D'ASTROPHYSIQUE
UNIVERSITE DE LIEGE
AVENUE DE COINTE 5
B-4200 COINTE-OUGREE
BELGIUM
TEL: 41-52-99-80
TLX:
COM: 14

BINETTE LUC
ESO
KARL-SCHWARZSCHILD-STR 2
D-8046 GARCHING B MUNCHEN
GERMANY, F.R.
TEL: 089-320-06-0
TLX: 5282820 EO D
COM: 28,34

BISHOP ROY L DR
DEPT OF PHYSICS
ACADIA UNIVERSITY
WOLFVILLE NS B0P 1X0
CANADA
TEL: 902-542-2201
TLX:
COM: 41

BIANCHINI ANTONIO DR
OSSERVATORIO ASTROFISICO
I-36012 ASIAGO
ITALY
TEL: 424-62665
TLX:
COM: 27

BIEN REINHOLD DR
ASTRONOMISCHES RECHEN-
INSTITUT
MOENCHHOFSTR 12-14
D-6900 HEIDELBERG
GERMANY, F.R.
TEL: 06221/49026
TLX: 461336 ARIHD D
COM: 08,20

BINGGELI BRUNO
ASTRONOMISCHES INSTITUT
UNIVERSITAET BASEL
VENUSSTRASSE 7
CH-4102 BINNINGEN
SWITZERLAND
TEL:
TLX:
COM: 28

BISIACCHI GIANFRANCO DR
INSTITUTO DE ASTRONOMIA
UNAM
APDO POSTAL 70-264
04510 MEXICO DF
MEXICO
TEL: 548-4537
TLX: 1760155 CICME
COM:

BIBARSOV RAVIL'SH DR
ASTROPHYSICAL INSTITUTE
TADJIK ACAD OF SCIENCES
734670 DUSHANBE
U.S.S.R.
TEL:
TLX:
COM: 22

BIERMANN PETER L DR
MPI FUER RADIOASTRONOMIE
AUF DEM HUEGEL 69
D-5300 BONN 1
GERMANY, F.R.
TEL: 228-525279
TLX: 886440 MPIFR D
COM: 28,40,48

BINGHAM RICHARD G DR
ROYAL GREENWICH OBS
HERSTMONCEUX CASTLE
HAILSHAM BN27 1RP
U.K.
TEL: 0323-833171
TLX: 87451
COM: 09

BISNOVATYI-KOGAN G S DR
SPACE RESEARCH INSTITUTE
USSR ACADEMY OF SCIENCES
PROFSOYUZNAYA 84/32
117810 MOSCOW
U.S.S.R.
TEL: 333-31-22
TLX: 411498 STARSU
COM: 35

BICAK JIRI DR
DEPT OF MATH PHYSICS
CHARLES UNIVERSITY
HOLESOVICKACH 2
180 00 PRAHA 8
CZECHOSLOVAKIA
TEL: 849951
TLX:
COM:

BIGNELL R CARL DR
NRAO-VLA
PO BOX 0
SOCORRO NM 87801
U.S.A.
TEL: 505-772-4242
TLX: 910-988-1710
COM: 34,40

BINNEY JAMES J DR
DEPT THEORETICAL PHYSICS
1 KEBLE ROAD
OXFORD OX1 3NP
U.K.
TEL: 865-53281
TLX: 83295 NUCLOX G
COM: 28,33C

BISWAS SUKUMAR DR
COSMIC RAY GROUP
TATA INST FUND RESEARCH
HOMI BHABHA RD
BOMBAY 400 005
INDIA
TEL: 91-22-219111
TLX: 113009 TIFR IN
COM: 48

BICKNELL GEOFFREY V DR
MOUNT STROMLO & SIDING
SPRING OBSERVATORIES
PRIVATE BAG
WODEN P.O. ACT 2606
AUSTRALIA
TEL: 61-62-88-1111
TLX: 62270 AA
COM: 47,48

BIJAOUI ALBERT DR
OBSERVATOIRE DE NICE
BP 139
F-06003 NICE CEDEX
FRANCE
TEL: 93-89-04-20
TLX: 460004
COM: 37

BIRAUD FRANCOIS DR
OBSERVATOIRE DE PARIS
SECTION DE MEUDON
F-92195 MEUDON PL CEDEX
FRANCE
TEL: 1-45-34-75-30
TLX: 270912
COM: 06,40,51

BJORNSSON CLAES-INGVAR
STOCKHOLM OBSERVATORY
S-133 00 SALTSJOEBADEN
SWEDEN
TEL: 08-7170195
TLX: 12972 SOBSERV S
COM:

BIDELMAN WILLIAM P PROF
WARNER & SWASEY OBS
CASE WESTERN RESERVE UNIV
CLEVELAND OH 44106
U.S.A.
TEL: 216-368-6699
TLX:
COM: 05,29,45

BILLAUD GERARD J
CERGA
AVENUE COPERNIC
F-06130 GRASSE
FRANCE
TEL: 93-36-58-49
TLX: 470865
COM: 08,19,31

BIRCH PETER MR
PERTH OBSERVATORY
BICKLEY, W. AUSTR. 6076
AUSTRALIA
TEL: 09-2938-255
TLX:
COM: 15

BLAAUW ADRIAAN PROF DR
KAPTEYN LABORATORY
P O BOX 800
NL-9700 AV GRONINGEN
NETHERLANDS
TEL: 050-634084
TLX: 53572 STARS NL
COM: 24,33,37

BIEGING JOHN HAROLD DR
RADIO ASTRONOMY LAB
CAMPBELL HALL
UNIVERSITY OF CALIFORNIA
BERKELEY CA 94720
U.S.A.
TEL: 415-642-6931
TLX:
COM: 34,40

BILLINGHAM JOHN
LIFE SCIENCE DIVISION
NASA AMES RESEARCH CTR
MOFFETT FIELD CA 94035
U.S.A.
TEL: 415-694-5181
TLX: 348408 NASA AMES MOF
COM: 51

BIRKINSHAW MARK
HARVARD UNIVERSITY
DEPT OF ASTRONOMY
60 GARDEN STREET
CAMBRIDGE MA 02138
U.S.A.
TEL: 617-495-9092
TLX: 921428
COM: 28,40

BLACK JOHN HARRY DR
STEWARD OBSERVATORY
UNIVERSITY OF ARIZONA
TUCSON AZ 85721
U.S.A.
TEL: 602-621-6531
TLX: 467175
COM: 14,34

LIST OF MEMBERS

BLACKMAN CLINTON PAUL DR
CARLSTON LODGE
CAMPSIE RD, TORRANCE
GLASGOW G64 4HD
U.K.
TEL:
TLX:
COM: 28

BLACKWELL ALAN TREVOR
METEORITE PROJECT
BOX 464
SUB POST OFFICE 6
SASKATOON S7N 0W0
CANADA
TEL:
TLX:
COM: 22

BLACKWELL DONALD E PROF
DEPT OF ASTROPHYSICS
SOUTH PARKS ROAD
OXFORD OX1 3RQ
U.K.
TEL: 0865-511336
TLX:
COM: 12,21,49

BLADES JOHN CHRIS DR
SPACE TELESCOPE SCI INST
HOMEWOOD CAMPUS
3700 SAN MARTIN DR
BALTIMORE MD 21218
U.S.A.
TEL: 301-338-4805
TLX: 6849101 STSCI UW
COM: 34

BLAHA MILAN DR
LAB FOR PLASMA & FUSION
UNIVERSITY OF MARYLAND
COLLEGE PARK MD 20742
U.S.A.
TEL: 301-454-7089
TLX:
COM: 14

BLAIR DAVID GERALD
PHYSICS DEPARTMENT
UNIVERSITY OF W.AUSTRALIA
NEDLANDS WA 6009
AUSTRALIA
TEL:
TLX: 92992 AA
COM: 40

BLAIR GUY NORMAN DR
ASTRONOMY PROGRAM
DEPT EARTH & SPACE SCS
SUNY
STONY BROOK NY 11794
U.S.A.
TEL:
TLX:
COM: 34

BLAMONT JACQUES E PROF
C N E S
2 PLACE MAURICE QUENTIN
F-75039 PARIS CEDEX 01
FRANCE
TEL: 1-45-08-76-12
TLX: 214674
COM: 12,15,16,21,44

BLANCO CARLO DR
OSSERVATORIO ASTROFISICO
CITTA UNIVERSITARIA
I-95125 CATANIA
ITALY
TEL: 095-330533
TLX: 970359 ASTRCT I
COM: 36,50C

BLANCO VICTOR M DR
CERRO TOLOLO
INTERAMERICAN OBSERVATORY
CASILLA 603
LA SERENA
CHILE
TEL: 213352
TLX: 34-645227 AURA CT
COM: 25,33,45,50C

BLANDFORD ROGER DAVID DR
THEORETICAL ASTROPHYSICS
CALTECH 130-33
PASADENA CA 91125
U.S.A.
TEL: 213-356-4200
TLX: 675429
COM: 40,48

BLASIUS KARL RICHARD DR
1125 RUBIO
ALTADENA CA 91001
U.S.A.
TEL:
TLX:
COM:

BLECHA ANDRE BORIS G DR
16 RUE ET. DUMONT
CH-1204 GENEVE
SWITZERLAND
TEL:
TLX:
COM:

BLEEKER JOHAN A M DR IR
SPACE RESEARCH LABORATORY
BENELUXLAAN 21
NL-3527 HS UTRECHT
NETHERLANDS
TEL: 030-937145
TLX: 47224 ASTRO NL
COM: 44,48

BLESS ROBERT C PROF
ASTRONOMY DEPT
UNIVERSITY OF WISCONSIN
475 N. CHARTER ST
MADISON WI 53706
U.S.A.
TEL: 608-262-1715
TLX:
COM: 34,36,44,51

BLINOV N S DR
STERNBERG STATE
ASTRONOMICAL INSTITUTE
117234 MOSCOW
U.S.S.R.
TEL: 139-10-49
TLX:
COM: 19,31C

BLITZ LEO
ASTRONOMY PROGRAM
UNIVERSITY OF MARYLAND
COLLEGE PARK MD 20742
U.S.A.
TEL: 301-454-3001
TLX: 710-826-0352
COM: 33C

BLITZSTEIN WILLIAM DR
DEPT ASTRON & ASTROPHYS
UNIV OF PENNSYLVANIA
DAVID RITTENHOUSE LAB E1
PHILADELPHIA PA 19104
U.S.A.
TEL: 215-898-7899
TLX: 834621
COM: 09,42

BLUDMAN SIDNEY A PROF
DEPT OF PHYSICS
UNIV OF PENNSYLVANIA
PHILADELPHIA PA 19104
U.S.A.
TEL: 215-898-8151
TLX: 831908
COM: 35,47,48

BLUM EMILE-JACQUES DR
5 RUE MIZON
F-75015 PARIS
FRANCE
TEL: 43-20-99-72
TLX:
COM: 40

BLUM PETER PROF
INSTITUT F. ASTROPHYSIK
UNIVERSITAET BONN
AUF DEM HUEGEL 71
D-5300 BONN
GERMANY, F.R.
TEL: 0228-73-36-65
TLX: 0886440 MPIFR
COM: 49

BO SHU-REN
INST F.HISTORY OF NAT SCI
1 GONG YUAN WEST ROAD
BEIJING
CHINA, PEOPLE'S REP.
TEL:
TLX:
COM: 41

BOBROV M S DR
ASTRONOMICAL COUNCIL
USSR ACADEMY OF SCIENCES
PYATNITSKAYA UL 48
109017 MOSCOW
U.S.S.R.
TEL: 231-39-80
TLX: 412623 SCSTP SU
COM: 16

BOBROVNIKOFF NICHOLAS DR
1623 VISALIA AVE
BERKELEY CA 94707
U.S.A.
TEL: 415-524-9244
TLX:
COM: 15

BOCCHIA ROMEO DR
OBSERVATOIRE DE BORDEAUX
AVENUE P SEMIROT
F-33270 FLOIRAC
FRANCE
TEL: 56-86-43-30
TLX:
COM: 10,12,35

BOCHKAREV NIKOLAY G DR
STERNBERG STATE ASTR INST
1179899 MOSCOW
U.S.S.R.
TEL:
TLX:
COM: 34

BOCHONKO D RICHARD DR
DEPT MATH & ASTRONOMY
UNIVERSITY OF MANITOBA
WINNIPEG MB R3T 2M8
CANADA
TEL: 204-474-9501
TLX:
COM: 27

BOCHSLER PETER
PHYSIKALISCHES INSTITUT
UNIVERSITAET BERN
SIDLERSTRASSE 5
CH-3012 BERN
SWITZERLAND
TEL: 0041-31-65-4419
TLX: 32320 PHYBE CH
COM: 49

BODE MICHAEL F
SCHOOL OF PHYSICS & ASTR
LANCASHIRE POLYTECHNIC
PRESTON PR1 2TQ
U.K.
TEL:
TLX:
COM: 34

BODENHEIMER PETER PROF
LICK OBSERVATORY
UNIVERSITY OF CALIFORNIA
SANTA CRUZ CA 95064
U.S.A.
TEL: 408-429-2064
TLX:
COM: 35C

BOEHM KARL-HEINZ PROF
ASTRONOMY DEPT
UNIVERSITY OF WASHINGTON
SEATTLE WA 98195
U.S.A.
TEL: 206-543-2888
TLX: 4740096
COM: 12,35,36

BOEHM-VITENSE ERIKA PROF
ASTRONOMY DEPT
UNIVERSITY OF WASHINGTON
FM 20
SEATTLE WA 98195
U.S.A.
TEL: 206-543-4858
TLX:
COM: 12,36

BOEHME A DR
HEINRICH HERTZ INSTITUTE
SOLAR TERRESTR PHYSICS
TELEGRAFENBERG
DDR-1500 POTSDAM
GERMANY, D.R.
TEL:
TLX:
COM: 40

BOEHME SIEGFRIED DR
ASTRONOMISCHES
RECHEN INSTITUT
MOENCHHOFSTR 12-14
D-6900 HEIDELBERG
GERMANY, F.R.
TEL: 06221-49026
TLX:
COM: 36,40

BOERNER GERHARD DR
MPI F PHYSIK & ASTROPHYS
FOEHRINGER RING 6
D-8000 MUENCHEN
GERMANY, F.R.
TEL:
TLX:
COM:

BOERNGEN FREIMUT DR PH
ZNTRLINST. F. ASTROPHYSIK
KARL-SCHWARZSCHILD-OBS
DDR-6901 TAUTENBURG
GERMANY, D.R.
TEL: JENA 23530
TLX:
COM: 20,28

BOESGAARD ANN M PROF
INSTITUTE FOR ASTRONOMY
2680 WOODLAWN DR
HONOLULU HI 96822
U.S.A.
TEL: 808-948-8756
TLX: 723-8459
COM: 29C,36

BOESHAAR GREGORY ORTH DR
SPACE TELESCOPE INSTITUTE
HOMEWOOD CAMPUS
3700 SAN MARTIN DRIVE
BALTIMORE MD 21218
U.S.A.
TEL:
TLX:
COM: 28,34

BOGGESS ALBERT DR
GODDARD SPACE FLIGHT CTR
CODE 689
GREENBELT MD 20771
U.S.A.
TEL: 301-344-5975
TLX:
COM: 29,34,44

BOGGESS NANCY W DR
NASA HEADQUARTERS
CODE EZ
WASHINGTON DC 20546
U.S.A.
TEL: 202-453-1469
TLX:
COM: 44

BOHANNAN BRUCE EDWARD
SOMMERS-BAUSCH OBS
UNIVERSITY OF COLORADO
BOX 391
BOULDER CO 80309
U.S.A.
TEL: 303-492-8782
TLX:
COM:

BOHLIN J DAVID DR
NASA HEADQUARTERS
CODE EZ
WASHINGTON DC 20546
U.S.A.
TEL: 202-453-1466
TLX: 89530
COM:

BOHLIN RALPH C DR
SPACE TELESCOPE SCI INST
HOMEWOOD CAMPUS
3700 SAN MARTIN DRIVE
BALTIMORE MD 21218
U.S.A.
TEL: 301-338-4804
TLX: 6849101 STSCI UWI
COM: 34,44

BOHN HORST-ULRICH
INST F ASTRON & ASTROPHYS
AM HUBLAND
D-8700 WURZBURG
GERMANY, F.R.
TEL: 0931-888-5032
TLX:
COM: 10,12

BOHRMANN ALFRED PROF
SCHAERSTR 23
D-2050 HAMBURG 80
GERMANY, F.R.
TEL: 7399800
TLX:
COM:

BOIGEY FRANCOISE
IMTA, LAB MECAN. CELESTE
UNIVERSITE P & M CURIE
4 PLACE JUSSIEU, TOUR 66
F-75230 PARIS CEDEX 05
FRANCE
TEL:
TLX:
COM: 07

BOISCHOT ANDRE DR
OBSERVATOIRE DE PARIS
SECTION DE MEUDON
F-92195 MEUDON PL CEDEX
FRANCE
TEL: 1-45-34-75-30
TLX: 200590 CNET
COM: 40

BOKSENBERG ALEC PROF
ROYAL GREENWICH OBS
HERSTMONCEUX CASTLE
HAILSHAM BN27 1RP
U.K.
TEL: 323-833171
TLX: 87451 RGOBSY G
COM: 28,44,47

BOLAND WILFRIED
ASTRON/ZWO
KONINGIN SOPHIESTRAAT 124
NL-2595 TM DEN HAAG
NETHERLANDS
TEL: 070-824231
TLX: 31660 TNOGV
COM: 34

BOLCAL CETIN DR
BOGAZICI UNIVERSITY
KANDILLI OBS & EARTHQUAKE
INST., CENGELKOI
ISTANBUL
TURKEY
TEL: 13320240
TLX: 26401 BOUNTR
COM:

BOLDT ELIHU DR
GODDARD SPACE FLIGHT CTR
CODE 661
GREENBELT MD 20771
U.S.A.
TEL: 301-344-5853
TLX: 89675 NASCOM-GBLT
COM:

BOLEY FORREST I
WILDER LABORATORY
DARTMOUTH COLLEGE
HANOVER NH 03755
U.S.A.
TEL: 603-646-2966
TLX:
COM:

BOLTON C THOMAS PROF
DAVID DUNLAP OBSERVATORY
P O BOX 360
RICHMOND HILL ONT L4C 4Y6
CANADA
TEL: 416-884-9652
TLX:
COM: 27,30,42

BOLTON JOHN G
39 PANORAMA CRESCENT
BUDERIM QLD 4556
AUSTRALIA
TEL: 071-453374
TLX:
COM: 40

BONANOMI J DR
OBSERVATOIRE DE NEUCHATEL
CH-2000 NEUCHATEL
SWITZERLAND
TEL:
TLX:
COM: 19,31

BONAZZOLA SILVANO DR
OBSERVATOIRE DE PARIS
SECTION DE MEUDON
F-92195 MEUDON PL CEDEX
FRANCE
TEL: 1-45-34-75-70
TLX:
COM: 42,48

BOND HOWARD E DR
SPACE TELESCOPE SCI INST
3700 SAN MARTIN DRIVE
BALTIMORE MD 21218
U.S.A.
TEL: 301-338-4718
TLX: 6849101
COM: 27

BOND JOHN RICHARD
DEPT OF PHYSICS
STANFORD UNIVERSITY
STANFORD CA 94305
U.S.A.
TEL: 415-497-1775
TLX:
COM: 47

BONDARENKO L N DR
STERNBERG STATE ASTR INST
PROSPECT 13
UNIVERSITETSKY PROSP. 13
119899 MOSCOW V-234
U.S.S.R.
TEL: 139-3721
TLX:
COM: 16

BONDI HERMANN PROF SIR
DEPT OF ENERGY
THAMES HOUSE SOUTH
MILLBANK
LONDON SW1P 4QJ
U.K.
TEL:
TLX:
COM: 28,35,47

BONET JOSE A
INSTITUTO DE ASTROFISICA
DE CANARIAS
38071 TENERIFE
SPAIN
TEL:
TLX:
COM:

BONEV BONU K MR
PEOPLE'S ASTR.OBSERVATORY
ST. AVGUSTA TRAIANA 29/8
6000 STARA ZAGORA
BULGARIA
TEL:
TLX:
COM:

BONIFAZI ANGELO DR
OSSERVATORIO ASTRONOMICO
I-40100 BOLOGNA
ITALY
TEL:
TLX:
COM:

BONNEAU DANIEL
CERGA
OBSERVATOIRE DU CALERN
F-06460 ST VALLIER DE T.
FRANCE
TEL: 93-42-62-70
TLX: 461402
COM: 09

BONNET ROGER M DR
ESA
8-10 RUE MARIO NIKIS
F-75738 PARIS CEDEX 15
FRANCE
TEL: 1-42-73-71-07
TLX: ESA 202746
COM: 12,44,49

BONNOR W B PROF
1 SOUTH BANK TERRACE
SURBITON, SURREY KT6 6DG
U.K.
TEL: 1-399-1103
TLX:
COM: 47

BONOLI FABRIZIO
OSSERVATORIO ASTRONOMICO
UNIVERSITARIO
C P 596
I-40100 BOLOGNA
ITALY
TEL: 051-222956
TLX: 211664 INFNBO I
COM:

BONOMETTO SILVIO A DR
ISTITUTO DI FISICA
G GALILEI 8
VIA MARZOLO
I-35100 PADOVA
ITALY
TEL:
TLX:
COM: 48

BONOV ANGEL DR
ASTRONOMICAL OBSERVATORY
P O BOX 36
1504 SOFIA 4
BULGARIA
TEL:
TLX:
COM: 10

BONSACK WALTER K PROF
INSTITUTE FOR ASTRONOMY
2680 WOODLAWN DRIVE
HONOLULU HI 96822
U.S.A.
TEL: 805-948-8138
TLX: 723-8459 UHAST HR
COM: 29

BOOK DAVID L
USE NAVAL RESEARCH LAB
CODE 4040
WASHINGTON DC 20375
U.S.A.
TEL:
TLX:
COM: 12

BOOKMYER BEVERLY B DR
DEPT OF PHYS & ASTRONOMY
CLEMSON UNIVERSITY
CLEMSON SC 29631
U.S.A.
TEL: 803-656-3417
TLX:
COM: 25,42

BOOTH ANDREW J
DEPT OF ASTROPHYSICS
SOUTH PARKS ROAD
OXFORD OX1 3RJ
U.K.
TEL: 511336 OXFORD
TLX:
COM:

BOOTH ROY S PROF
ONSALA SPACE OBSERVATORY
S-439 00 ONSALA
SWEDEN
TEL: 46-300-62590
TLX: 2400 ONSPACE S
COM: 40C

BOPP BERNARD W DR
DEPT PHYSICS & ASTRONOMY
UNIVERSITY OF TOLEDO
TOLEDO OH 43606
U.S.A.
TEL: 419-537-2274
TLX:
COM: 27,42

BORCHKHADZE TENGIZ M DR
ABASTUMANI ASTROPHYSICAL
OBSERVATORY
383762 ABASTUMANI,GEORGIA
U.S.S.R.
TEL:
TLX:
COM: 28

BORD DONALD JOHN
DEPT OF NATURAL SCIENCES
UNIVERSITY OF MICHIGAN
DEARBORN
DEARBORN MI 48128
U.S.A.
TEL: 313-593-5483
TLX:
COM:

BORDERIES NICOLE
OBSERVATOIRE PIC-DU-MIDI
ET TOULOUSE
14 AVENUE EDOUARD BELIN
F-31400 TOULOUSE
FRANCE
TEL: 61-25-21-01
TLX: 530776
COM:

BORGMAN JAN DR PROF
KAPTEYN OBSERVATORY
MENSINGHEWEG 20
NL-9301 KA RODEN DR
NETHERLANDS
TEL:
TLX:
COM: 25,34

BORGNINO JULIEN DR
DEPT D'ASTROPHYSIQUE
UNIVERSITE DE NICE
PARC VALROSE
F-06034 NICE CEDEX
FRANCE
TEL:
TLX:
COM: 09

BORIAKOFF VALENTIN
NAIC
420 SPACE SCIENCES BLDG
CORNELL UNIVERSITY
ITHACA NY 14853
U.S.A.
TEL: 607-256-3734
TLX: 932454
COM: 40

BORRA ERMANNO F DR
DEPT DE PHYSIQUE
UNIVERSITE LAVAL
STE FOY PQ G1K 7P4
CANADA
TEL: 418-656-7405
TLX: 051-31621
COM: 25

BOSMA ALBERT DR
OBSERVATOIRE DE MARSEILLE
2 PLACE LE VERRIER
F-13248 MARSEILLE CEDEX 4
FRANCE
TEL: 91-95-90-88
TLX: 420241
COM: 28

BOSMA PIETER B DR
DEPT PHYSICS & ASTRONOMY
FREE UNIVERSITY
DE BOELELAAN 1081
NL-1081 HV AMSTERDAM
NETHERLANDS
TEL: 020-5485338
TLX:
COM: 16

BOSMAN-CRESPIN DENISE
BLVD D AVROY 68
BTE 093
B-4000 LIEGE
BELGIUM
TEL: 0032-41-237486
TLX:
COM:

BOTEZ ELVIRA DR
INSTITUT D'ENSEIGNEMENT
SUPERIEUR
13 RUE EM. BODNARAS
5800 SUCEAVA
RUMANIA
TEL: 98716147
TLX:
COM: 46

BOTTINELLI LUCETTE DR
OBSERVATOIRE DE PARIS
SECTION DE MEUDON
RADIOASTRONOMIE
F-92195 MEUDON PL CEDEX
FRANCE
TEL: 1-45-34-75-30
TLX: 270912 OBSASTR F
COM: 28,46

BOUGERET J L DR
OBSERVATOIRE DE PARIS
SECTION DE MEUDON
DESPA
F-92195 MEUDON PL CEDEX
FRANCE
TEL: 1-45-34-75-30
TLX: 204464
COM: 10,12,44

BOUIGUE ROGER PROF
LABORATOIRE D'ASTRONOMIE
UNIVERSITE PAUL SABATIER
F-31062 TOULOUSE CEDEX
FRANCE
TEL: 61-55-68-25
TLX:
COM: 24,29,30

BOULESTEIX JACQUES
OBSERVATOIRE DE MARSEILLE
2 PLACE LE VERRIER
F-13248 MARSEILLE
FRANCE
TEL: 91-95-90-88
TLX: 420241F
COM:

BOULON JACQUES J DR
OBSERVATOIRE PARIS
61 AVE DE L'OBSERVATOIRE
F-75014 PARIS
FRANCE
TEL: 1-43-20-12-10
TLX:
COM: 27,29,30,33

BOUSKA JIRI DR
DEPT OF ASTRONOMY
CHARLES UNIVERSITY
SVEDSKA 8
150 00 PRAHA
CZECHOSLOVAKIA
TEL: 42-2-540395
TLX: 121673 MFF
COM: 05,15

BOUVIER PIERRE PROF
OBSERVATOIRE DE GENEVE
CH-1290 SAUVERNY
SWITZERLAND
TEL:
TLX:
COM: 37

BOWELL EDWARD L G DR
LOWELL OBSERVATORY
1400 W. MARS HILL RD
FLAGSTAFF AZ 86001
U.S.A.
TEL: 602-774-3358
TLX:
COM: 15,20

BOWEN EDWARD G DR
5010 MAXWELL AVE
WEST RIVER MD 20881
U.S.A.
TEL:
TLX:
COM:

BOWERS PHILLIP F
NAVAL RESEARCH LABORATORY
CODE 4134
WASHINGTON DC 20375
U.S.A.
TEL: 202-767-2495
TLX:
COM: 40

BOWYER C STUART PROF
UNIVERSITY OF CALIFORNIA
ASTRONOMY DEPT
BERKELEY CA 94720
U.S.A.
TEL: 415-642-1648
TLX: 910-366 7945
COM: 21,44,51

BOYNTON PAUL EDWARD DR
ASTRONOMY DEPT
UNIVERSITY OF WASHINGTON
SEATTLE WA 98195
U.S.A.
TEL:
TLX:
COM:

BRANCH DAVID R DR
DEPT PHYSICS & ASTRONOMY
UNIVERSITY OF OKLAHOMA
NORMAN OK 73019
U.S.A.
TEL: 405-325-3961
TLX: 9108306521
COM:

BRANSCOMB L M DR
NAT BUREAU OF STANDARDS
WASHINGTON DC 20025
U.S.A.
TEL:
TLX:
COM: 14

BOYARCHUK A A DR
CRIMEAN ASTROPHYSICAL OBS
USSR ACADEMY OF SCIENCES
NAUCHNY
334413 CRIMEA
U.S.S.R.
TEL:
TLX:
COM: 27,29,38C,44C

BOZIS GEORGE PROF
DEPT THEORET MECHANICS
UNIVERSITY THESSALONIKI
GR-54006 THESSALONIKI
GREECE
TEL: 031-992845
TLX:
COM: 07

BRAND PETER W J L DR
DEPT OF ASTRONOMY
UNIVERSITY OF EDINBURGH
ROYAL OBSERVATORY
EDINBURGH EH9 3HJ
U.K.
TEL: 031-667-3321
TLX: 72383 ROE EDIN G
COM: 34

BRANSON NICHOLAS J B A DR
BOARD OF GRADUATE STUDIES
4 MILL LANE
CAMBRIDGE CB2 1RZ
U.K.
TEL:
TLX:
COM:

BOYARCHUK MARGARITA E DR
CRIMEAN ASTROPHYSICAL
OBSERVATORY
P/O NAUCHNYJ
334413 CRIMEA
U.S.S.R.
TEL:
TLX:
COM: 27

BOZKURT SUKRU DR
EGE UNIVERSITY
OBSERVATORY
P K 21
BORNOVA-IZMIR
TURKEY
TEL: 180 306 (HOME)
TLX:
COM:

BRANDI ELISANDE ESTELA DR
OBSERVATORIO ASTRONOMICO
UNIVERSIDAD NACIONAL
PASEO DEL BOSQUE
1900 LA PLATA
ARGENTINA
TEL: 21-1761
TLX:
COM: 29

BRATIJCHUK MATRONA V
UZHGOROD STATE UNIVERSITY
294000 UZHGOROD
U.S.S.R.
TEL:
TLX:
COM:

BOYCE PETER B DR
AMERICAN ASTRON SOCIETY
2000 FLORIDA AVE N.W.
SUITE 300
WASHINGTON DC 20009
U.S.A.
TEL: 202-328-2010
TLX: 257588 AASW UR
COM: 09,16,51

BRACCESI ALESSANDRO PROF
DIPTO DI ASTRONOMIA
VIA ZAMBONI 33
I-40126 BOLOGNA
ITALY
TEL: 222956
TLX: 211664 INFNBOI
COM: 28

BRANDIE GEORGE W DR
ENVIRONMENTAL ENGINEERING
DUPUIS HALL
QUEEN'S UNIVERSITY
KINGSTON ONT K7L 3N6
CANADA
TEL:
TLX:
COM:

BRAUDE SEMION YA PROF AG
INST RADIOPHY ELECTR
UKRAINIAN ACADEMY OF SCI
310085 KHARKOV
U.S.S.R.
TEL: 441092
TLX:
COM: 40

BOYD ROBERT L F PROF
MULLARD SPACE SCIENCE LAB
UCL
HOLMBURY ST MARY
DORKING, SURREY RH5 6NS
U.K.
TEL:
TLX:
COM: 44,48

BRACEWELL RONALD N PROF
STANFORD UNIVERSITY
DURAND 329 A
STANFORD CA 94305
U.S.A.
TEL: 415-497-3545
TLX:
COM: 40,51

BRANDT JOHN C DR
GODDARD SPACE FLIGHT CTR
CODE 680
GREENBELT MD 20771
U.S.A.
TEL: 301-344-8701
TLX:
COM: 15,44,49

BRAULT JAMES W DR
NATL SOLAR OBSERVATORY
PO BOX 26732
950 N. CHERRY AVE
TUCSON AZ 85726
U.S.A.
TEL: 325-9363
TLX: 666484 AURA NOAO TUC
COM: 09,12,14

BOYDAG-YILDIZDOGDU F S
ACADEMY OF ISTANBUL
ENG & ARCHITECTURE
DEPT OF PHYSICS
ISTANBUL
TURKEY
TEL:
TLX:
COM:

BRAES L L E DR
STERREWACHT
POSTBUS 9513
NL-2300 RA LEIDEN
NETHERLANDS
TEL:
TLX:
COM: 46

BRANDT PETER N
KIEPENHEUER INSTITUT FUER
SONNENPHYSIK
SCHONECKSTR. 6
D-7800 FREIBURG BR.
GERMANY, F.R.
TEL: 0761-32864
TLX: 7721552
COM: 10,12

BRAUN ARIE
RACAH INST. OF PHYSICS
HEBREW UNIV. OF JERUSALEM
JERUSALEM 91904
ISRAEL
TEL: 02-584521
TLX: 25391 HU IL
COM:

BOYER CHARLES
64 BIS RUE ALFRED DUMERIL
F-31400 TOULOUSE
FRANCE
TEL: 61-52-90-34
TLX:
COM: 16

BRAHDE ROLF
INST THEORET ASTROPHYSICS
UNIVERSITY OF OSLO
N-0315 BLINDERN, OSLO 3
NORWAY
TEL: 2-456508
TLX:
COM:

BRANDUARDI-RAYMONT G
MULLARD SPACE SCIENCE LAB
HOLMBURY ST MARY
DORKING, SURREY RH5 6NT
U.K.
TEL: 030-670-292
TLX: 859185
COM:

BRAUNINGER HEINRICH DR
MPI F PHYSIK & ASTROPHYS
INST F. EXTRATERR. PHYSIK
D-8046 GARCHING B MUNCHEN
GERMANY, F.R.
TEL: 089-3299-566
TLX:
COM:

BOYER RENE
OBSERVATOIRE DE PARIS
SECTION DE MEUDON
DASOP
F-92195 MEUDON PL CEDEX
FRANCE
TEL: 1-45-34-75-30
TLX:
COM: 10

BRAHIC ANDRE DR
OBSERVATOIRE DE PARIS
SECTION DE MEUDON
F-92195 MEUDON PL CEDEX
FRANCE
TEL: 1-45-34-75-70
TLX:
COM: 16V

BRANHAM RICHARD L JR
CENTRO REGIONAL DE INVEST
CIENTIFICAS Y TECNOL.
CASILLA DE CORREO 131
5500 MENDOZA
ARGENTINA
TEL: 061-2411794
TLX: 55438 CYTME AR
COM: 08,20,24

BRAUNSFURTH EDWARD PH D
IM HAARMANNSBOCH 99A
D-4630 BOCHUM
GERMANY, F.R.
TEL:
TLX:
COM: 34

LIST OF MEMBERS

BRAY ROBERT J DR
CSIRO
DIV OF APPLIED PHYSICS
P.O. BOX 218
LINDFIELD NSW 2070
AUSTRALIA
TEL: 467-6354
TLX: 26296
COM: 10,12

BRAZ MARIA ALCINA DR
INST ASTRON. E GEOFISICO
UNIVERSIDADE DE SAO PAULO
CAIXA POSTAL 30627
01051 SAO PAULO
BRAZIL
TEL:
TLX: 011-36211
COM:

BRECHER AVIVA DR PROF
BOSTON UNIVERSITY
145 BAY STATE ROAD
BOSTON MA 02215
U.S.A.
TEL: 617-353-2314
TLX:
COM: 15,16

BRECHER KENNETH PROF
DEPT OF ASTRONOMY
BOSTON UNIVERSITY
725 COMMONWEALTH AVE
BOSTON MA 02215
U.S.A.
TEL: 617-353-3423
TLX: 95-1289 BIS UNIV BSN
COM: 28,47,48

BRECKINRIDGE JAMES B DR
JPL/CALTECH
MS 183-301
4800 OAK GROVE DR
PASADENA CA 91103
U.S.A.
TEL: 213-354-6785
TLX: 675429
COM: 09,12

BREGER MICHEL DR
INSTITUT FUER ASTRONOMIE
TUERKENSCHANZSTR 17
A-1180 WIEN
AUSTRIA
TEL: 222-34-53-605
TLX: 133099 VIAST A
COM: 25,27V,30

BREGMAN JACOB D IR
NETHERLANDS FOUNDATION
RADIOASTRONOMY
POSTBUS 2
NL-7990 AA DWINGELOO
NETHERLANDS
TEL: 05219-7244
TLX: 42043
COM: 40

BREGMAN JOEL N
NRAO
EDGEMONT ROAD
CHARLOTTESVILLE VA 22903
U.S.A.
TEL: 804-296-0235
TLX:
COM:

BREINHORST ROBERT A DR
ASTRONOMISCHES INSTITUT
STERNWARTE
AUF DEM HUEGEL 71
D-5300 BONN 1
GERMANY, F.R.
TEL: 0228-733660
TLX:
COM: 42

BREJDO IZABELLA I DR
PULKOVO OBSERVATORY
196140 LENINGRAD
U.S.S.R.
TEL: 297-94-59
TLX:
COM: 09

BRETAGNON PIERRE DR
BUREAU DES LONGITUDES
77 AVE DENFERT-ROCHEREAU
F-75014 PARIS
FRANCE
TEL: 1-43-20-12-10
TLX:
COM: 04,07

BRIDLE ALAN H PROF
NRAO
EDGEMONT ROAD
CHARLOTTESVILLE VA 22903
U.S.A.
TEL: 804-296-0375
TLX: 910-997-0174
COM: 40

BRIEVA EDUARDO PROF
OBSERVATORIO NACIONAL
APARTADO 2584
BOGOTA 1, D.E.
COLOMBIA
TEL: 423786
TLX:
COM: 07,46

BRIHAYE CHARLES C A DR
UNIV LIBRE DE BRUXELLES
50 AVE F.D. ROOSEVELT
B-1050 BRUXELLES
BELGIUM
TEL: 02-6876928
TLX:
COM:

BRINI D PROF
LABORATORIO TESRE
VIA CASTAGNOLI 1
I-40100 BOLOGNA
ITALY
TEL:
TLX:
COM:

BRINKMAN BERT C DR
SPACE RESEARCH LABORATORY
BENELUXLAAN 21
NL-3527 HS UTRECHT
NETHERLANDS
TEL:
TLX:
COM: 44

BRINKMANN WOLFGANG
MPI F PHYS & ASTROPHYSIK
INST F EXTRATERR PHYSIK
KARL-SCHWARZSCHILD-STR 1
D-8046 GARCHING B MUNCHEN
GERMANY, F.R.
TEL: 893299877
TLX: 05215845 XTER D
COM: 28,34

BRIOT DANIELLE DR
OBSERVATOIRE DE PARIS
61 AVE DE L'OBSERVATOIRE
F-75014 PARIS
FRANCE
TEL: 1-43-20-12-10
TLX: 270776
COM:

BROADFOOT A LYLE DR
UNIVERSITY OF ARIZONA
LUNAR & PLANETARY LAB.
3625 EAST AJO WAY
TUCSON AZ 85713
U.S.A.
TEL: 602-621-4303
TLX: 910-952-1143
COM: 16,21

BRODERICK JOHN DR
PHYSICS DEPT
VPI & SU
BLACKSBURG VA 24061
U.S.A.
TEL: 703-961-5321
TLX: 910-3331861 VPIBKS
COM: 40,51

BRODIE JEAN P
SPACE SCIENCES LABORATORY
UNIVERSITY OF CALIFORNIA
BERKELEY CA 94720
U.S.A.
TEL: 415-642-1579
TLX: 910-366-7945
COM: 28

BROGLIA PIETRO DR
OSSERVATORIO ASTRONOMICO
VIA E. BIANCHI 46
I-22055 MERATE COMO
ITALY
TEL: 039-592035
TLX:
COM: 42

BROMAGE GORDON E DR
ASTROPHYSICS GROUP
RUTHERFORD APPLETON LAB
CHILTON DIDCOT OX11 0QX
U.K.
TEL: 235-21900
TLX: 83159
COM: 14,34

BRONNIKOVA NINA N
PULKOVO OBSERVATORY
196140 LENINGRAD
U.S.S.R.
TEL:
TLX:
COM: 24

BROOKES CLIVE J DR
EARTH, SATELLITE RES UNIT
DEPT OF MATHEMATICS
ASTON UNIVERSITY
BIRMINGHAM B4 7ET
U.K.
TEL: 21-359-3611
TLX: 335787
COM: 07

BROSCHE PETER PROF
OBSERVATORIUM HOHER LIST
UNIV STERNWARTE BONN
D-5568 DAUN
GERMANY, F.R.
TEL: 06592-2150
TLX:
COM: 19C,24,28

BROSTERHUS E B F DR
C/O LOCKHEED CITY
PO BOX 6308
JEDDAH
SAUDI ARABIA
TEL: 02-656-2501x355
TLX:
COM:

BROTEN NORMAN W
HERZBERG INST ASTROPHYS
NATL RESEARCH COUNCIL
OTTAWA ONT K1A 0R6
CANADA
TEL: 613-593-6060
TLX:
COM: 40

BROUCKE ROGER DR
7203 RUNNING ROPE CIRCLE
AUSTIN TX 78731
U.S.A.
TEL: 512-345-6435
TLX:
COM: 07

BROUW W N DR
RADIOSTERRENWACHT
POSTBUS 2
NL-7990 AA DWINGELOO
NETHERLANDS
TEL: 31-5219-7244
TLX: 42043 SRZM NL
COM: 08,40

BROWN ALEXANDER
JILA
UNIVERSITY OF COLORADO
BOULDER CO 80309
U.S.A.
TEL: 303-492-8962
TLX: 755842 JILA
COM:

BROWN DOUGLAS NASON
UNIVERSITY OF WASHINGTON
DEPT OF ASTRONOMY, FM-20
SEATTLE WA 98195
U.S.A.
TEL: 2065434313/2888
TLX:
COM: 25,29,36

BROWN HARRISON DR
3005 LA MANCHA DRIVE
ALBUQUERQUE NM 87104
U.S.A.
TEL:
TLX:
COM:

BROWN JOHN C PROF
DEPT OF ASTRONOMY
GLASGOW UNIVERSITY
GLASGOW G12 8QQ
U.K.
TEL: 041-339-8855
TLX: 778421 GLASUL
COM: 10

BROWN ROBERT HAMILTON
JPL/CALTECH
MS 183-501
4800 OAK GROVE DRIVE
PASADENA CA 91109
U.S.A.
TEL: 818-354-2517
TLX:
COM: 15,16

BROWN ROBERT L DR
NRAO
EDGEMONT ROAD
CHARLOTTESVILLE VA 22901
U.S.A.
TEL: 804-296-0232
TLX: 910-997-0174
COM:

BROWN RONALD D PROF
CHEMISTRY DEPT
MONASH UNIVERSITY
WELLINGTON ROAD
CLAYTON VIC 3168
AUSTRALIA
TEL:
TLX:
COM: 34,51C

BROWNE IAN W A DR
NUFFIELD RADIO ASTR LABS
JODRELL BANK
MACCLESFIELD SK119DL
U.K.
TEL: 0477-71321
TLX: 36149
COM: 40

BROWNLEE DONALD E PROF
DEPT OF ASTRONOMY
UNIVERSITY OF WASHINGTON
SEATTLE WA 98195
U.S.A.
TEL: 206-543-2888
TLX:
COM: 22

BROWNLEE ROBERT R DR
MS F670
LOS ALAMOS SCIENTIFIC LAB
LOS ALAMOS NM 87544
U.S.A.
TEL: 505-662-6427
TLX:
COM: 35,42

BRUCH ALBERT
ASTRONOMISCHES INSTITUT
DER UNIVERSITAET MUENSTER
DOMAGKSTR 75
D-4400 MUENSTER
GERMANY, F.R.
TEL:
TLX:
COM:

BRUCK HERMANN A PROF
CRAIGOWER
PENICUIK EH26 9LA
U.K.
TEL: 968-75918
TLX:
COM: 25

BRUCK MARY T DR
ROYAL OBSERVATORY
EDINBURGH EH9 3HJ
U.K.
TEL: 31-667-3321
TLX: 72383
COM: 28,34

BRUECKNER GUENTER E DR
CODE 4160
NAVAL RESEARCH LABORATORY
WASHINGTON DC 20375-5000
U.S.A.
TEL: 202-767-3287
TLX:
COM: 10,12C,44

BRUHWEILER FRED C JR
NASA/GSFC
CODE 685/CSC
GREENBELT MD 20771
U.S.A.
TEL:
TLX:
COM: 29,34,42,44

BRUMBERG VICTOR A DR
INST OF THEORET ASTRONOMY
10 KUTUZOV QUAY
192187 LENINGRAD
U.S.S.R.
TEL: 27 8-88-92
TLX: 121578 ITA SU
COM: 04,07P

BRUNER MARILYN E DR
LOCKHEED PALO ALTO RES
LAB DEPT 91-20 BLDG 255
3251 HANOVER ST
PALO ALTO CA 94304
U.S.A.
TEL: 415-858-4023
TLX: 346409 LMSC
COM: 10,12,44

BRUNET JEAN-PIERRE DR
OBS PIC-DU-MIDI TOULOUSE
14 AVENUE EDOUARD BELIN
F-31400 TOULOUSE
FRANCE
TEL: 61-25-21-01
TLX: 503776
COM: 41

BRUNK WILLIAM E DR
OFF SPACE SC CODE SL
NASA HEADQUARTERS
400 MARYLAND AVE S W
WASHINGTON DC 20546
U.S.A.
TEL: 202-453-1596
TLX:
COM: 15,16

BRUSTON PAUL DR
L P S P
BP NO 10
F-91371 VERRIERES-LE-B.
FRANCE
TEL: 1-69-20-10-60
TLX:
COM: 28

BRUZEK ANTON DR
SCHWAIGHOFSTR 7
D-7800 FREIBURG
GERMANY, F.R.
TEL: 0761-78522
TLX:
COM: 10,12

BRUZUAL GUSTAVO
C I D A
APARTADO POSTAL 264
MERIDA 5101-A
VENEZUELA
TEL: 58-74-639930
TLX: 74174 CIDA VC
COM:

BUCHLER J ROBERT PROF
DEPT OF PHYSICS
UNIVERSITY OF FLORIDA
GAINESVILLE FL 32611
U.S.A.
TEL: 904-373-9942
TLX:
COM: 35

BUDDING EDWIN DR
CARTER OBSERVATORY
P O BOX 2909
WELLINGTON 1
NEW ZEALAND
TEL: 04-728-167
TLX: 30172 NATOBS NZ
COM: 42C

BUES IRMELA D DR
REMEIS STERNWARTE
STERNWARTSTR 7
D-8600 BAMBERG
GERMANY, F.R.
TEL: 0951-57708
TLX:
COM: 29,36

BUFF JAMES S DR
DEPT PHYSICS & ASTRONOMY
DARTMOUTH COLLEGE
HANOVER NH 03755
U.S.A.
TEL:
TLX:
COM:

BUHL DAVID DR
INFRARED & RADIO ASTR BR.
NASA/GSFC, CODE 693
GREENBELT MD 20771
U.S.A.
TEL: 301-344-8810
TLX:
COM:

BUITRAGO JESUS
INST DE ASTROFISICA DE
CANARIAS
LA LAGUNA
38071 TENERIFE
SPAIN
TEL: 922-26-22-11
TLX: 92640
COM:

BUJARRABAL VALENTIN
CENTRO ASTRON DE YEBES
O A N
APARTADO CORREOS 148
19080 GUADALAJARA
SPAIN
TEL: 11-223358
TLX:
COM:

BUMBA VACLAV DR
ASTRONOMICAL INSTITUTE
CZECH. ACAD. OF SCIENCES
251 65 ONDREJOV
CZECHOSLOVAKIA
TEL: 72-45-25
TLX: 121579 ASTR C
COM: 10,12,44

BUNNER ALAN N DR
PERKIN-ELMER CORP
M S 897
100 WOOSTER HEIGHTS RD
DANBURY CT 06810
U.S.A.
TEL: 203-797-6339
TLX:
COM: 42,44,48

BUONANNO ROBERTO
OSS ASTRON SU MONTE MARIO
VIA DEL PARCO MELLINI 84
I-00136 ROMA
ITALY
TEL: 06-347056
TLX:
COM:

BURBIDGE E MARGARET PROF
CTR ASTROPHYS & SPACE SCI
UNIVERSITY OF CALIFORNIA
SAN DIEGO MC C-011
LA JOLLA CA 92093
U.S.A.
TEL: 619-452-4477
TLX:
COM: 28

BURBIDGE GEOFFREY R PROF
CTR ASTROPHYS & SPACE SCI
UNIVERSITY OF CALIFORNIA
SAN DIEGO MC C-011
LA JOLLA CA 92093
U.S.A.
TEL: 619-452-6626
TLX:
COM: 28,35,40,47,48

LIST OF MEMBERS

BURGER J J DR IR
ESTEC
POSTBUS 299
NL-2200 AG NOORDWIJK
NETHERLANDS
TEL: 31-1719-84404
TLX: 39098
COM:

BURGER MARIJKE DR
ASTROPHYSICAL INSTITUTE
VRIJE UNIV BRUSSEL
PLEINLAAN 2
B-1050 BRUSSEL
BELGIUM
TEL: 02-6413525
TLX: 81051 VUBCO B
COM: 44

BURGESS ALAN DR
DEPT OF APPLIED MATHS
SILVER STREET
CAMBRIDGE CB3 9EW
U.K.
TEL:
TLX:
COM: 14,34

BURGESS DAVID D PROF
BLACKETT LABORATORY
IMPERIAL COLLEGE OF
SCIENCE & TECHNOLOGY
LONDON SW7 2BZ
U.K.
TEL: 1-589-5111x6931
TLX:
COM:

BURKE BERNARD F DR
DEPT OF PHYSICS
M I T RM 26-335
CAMBRIDGE MA 02139
U.S.A.
TEL: 617-253-2572
TLX: 92-1473
COM: 33,34,40,51

BURKE J ANTHONY DR
DEPT OF PHYSICS
UNIVERSITY OF VICTORIA
VICTORIA BC V8W 2Y2
CANADA
TEL: 604-721-7743
TLX:
COM:

BURKHART CLAUDE DR
OBSERVATOIRE DE LYON
F-69230 ST-GENIS-LAVAL
FRANCE
TEL: 78-56-07-05
TLX: 310-926
COM: 29

BURKHEAD MARTIN S
ASTRONOMY DEPT
INDIANA UNIVERSITY
SWAIN HALL WEST
BLOOMINGTON IN 47405
U.S.A.
TEL: 812-335-6917
TLX:
COM: 37

BURKI GILBERT DR
OBSERVATOIRE DE GENEVE
CH-1290 SAUVERNY
SWITZERLAND
TEL: 22-55-26-11
TLX: 27720
COM: 27,30

BURLAGA LEONARD F DR
NASA/GSFC
CODE 692
GREENBELT MD 20771
U.S.A.
TEL:
TLX:
COM: 15,49V

BURNAGE ROBERT
OBS DE HAUTE-PROVENCE
F-04870 ST-MICHEL-L'OBS.
FRANCE
TEL: 92-76-63-68
TLX:
COM: 30

BURNS JACK O'NEAL JR
PHYSICS & ASTRONOMY DEPT
UNIVERSITY OF NEW MEXICO
800 YALE BLVD N.E.
ALBUQUERQUE NM 87131
U.S.A.
TEL: 505-277-2705
TLX:
COM: 28

BURNS JOSEPH A PROF
CORNELL UNIVERSITY
THURSTON HALL
ITHACA NY 14850
U.S.A.
TEL: 607-256-4875
TLX: 937478
COM: 15,16C,20

BURROWS ADAM SETH
DEPT OF PHYSICS
SUNY
STONY BROOK NY 11794
U.S.A.
TEL: 516-246-6810
TLX:
COM: 48

BURSA MILAN DR
ASTRONOMICAL INSTITUTE
CZECH. ACAD. OF SCIENCES
BUDECSKA 6
120 23 PRAHA 2
CZECHOSLOVAKIA
TEL: 25-05-51
TLX: 122486
COM:

BURSTEIN DAVID
ARIZONA STATE UNIVERSITY
DEPT OF PHYSICS
TEMPE AZ 85281
U.S.A.
TEL:
TLX:
COM: 28,50

BURTON W BUTLER DR
STERREWACHT
POSTBUS 9513
NL-2300 RA LEIDEN
NETHERLANDS
TEL: 71-148-333
TLX: 39058 ASTRONL
COM: 09,33P,34

BURTON WILLIAM M
SPACE & ASTROPHYS DIV
RUTHERFORD APPLETON LAB
CHILTON DIDCOT OX11 0QX
U.K.
TEL: 235-21900
TLX: 83159
COM: 44

BUSCOMBE WILLIAM PROF
ASTRONOMY DEPT
NORTHWESTERN UNIVERSITY
EVANSTON IL 60201
U.S.A.
TEL: 312-491-7527
TLX:
COM: 29,45,46

BUSER ROLAND DR
ASTRONOMISCHES INSTITUT
UNIVERSITAET BASEL
VENUSSTRASSE 7
CH-4102 BINNINGEN
SWITZERLAND
TEL: 061-227711
TLX:
COM: 25C,45

BUSKO IVO C DR
MCT/INST PESQUISAS ESPEC.
AVE DOS ASTRONAUTAS 1758
JARDIM DA GRANJA
12200 SAO JOSE DOS CAMPOS
BRAZIL
TEL: 22-9977 x 392
TLX: 011-33530 INPE BR
COM: 27

BUSSO MAURIZIO
OSSERVATORIO ASTRONOMICO
DI TORINO
I-10025 PINO TORINESE
ITALY
TEL: 011-84-1067
TLX: 213239 TO ASTR I
COM: 42

BUTCHER HARVEY R PROF DR
KAPTEYN ASTRONOM INSTITUT
POSTBUS 800
NL-9700 AV GRONINGEN
NETHERLANDS
TEL: 05908-19631
TLX: 53567 KSWRO NL
COM: 28

BUTI BIMLA PROF
PHYSICAL RESEARCH LAB
NAVRANGPURA
AHMEDABAD 380 009
INDIA
TEL: 462129
TLX: 121-397 IN
COM: 49C

BUTLER C JOHN DR
ARMAGH OBSERVATORY
COLLEGE HILL
ARMAGH BT61 9DG N IRELAND
U.K.
TEL: 0861-522-928
TLX: 747937 ARMOBS G
COM: 27

BUTLER DENNIS DR
ASTRONOMY DEPT
YALE UNIVERSITY
BOX 2023 YALE STATION
NEW HAVEN CT 06520
U.S.A.
TEL:
TLX:
COM: 27,37

BUTTERWORTH PAUL
DEPT OF CHEMISTRY
HOWARD UNIVERSITY
WASHINGTON DC 20059
U.S.A.
TEL:
TLX:
COM:

BYARD PAUL L DR
DEPT OF ASTRONOMY
OHIO STATE UNIVERSITY
174 W. 18TH AVE
COLUMBUS OH 43210
U.S.A.
TEL: 614-422-1773
TLX:
COM:

BYKOV M F DR
ASTRONOMICAL INSTITUTE
UZBEKIAN ACADEMY OF SCI
700000 TASHKENT
U.S.S.R.
TEL:
TLX:
COM: 08

BYRD GENE G DR
DEPT PHYSICS & ASTRONOMY
UNIVERSITY OF ALABAMA
BOX 1921
UNIVERSITY AL 35486
U.S.A.
TEL: 205-348-5050
TLX:
COM: 28,37

BYRNE PATRICK B DR
ARMAGH OBSERVATORY
ARMAGH BT61 9DG
U.K.
TEL: 44-861-522928
TLX: 747937 ARMOBS G
COM: 27,46

BYSTROV NIKOLAI F DR
PULKOVO OBSERVATORY
196140 LENINGRAD
U.S.S.R.
TEL: 297-94-81
TLX:
COM: 16

BYSTROVA NATALIJA V DR
SPECIAL ASTROPHYSICAL
OBSERVATORY
LENINGRAD BRANCH
196140 LENINGRAD
U.S.S.R.
TEL: 297-9452
TLX:
COM: 34

CALAMAI G PROF
OSSERVATORIO ASTROFISICO
DI ARCETRI
I-50125 FIRENZE
ITALY
TEL:
TLX:
COM:

CALVO MANUEL
DPTO DE ASTRONOMIA
UNIVERSIDAD DE ZARAGOZA
50009 ZARAGOZA
SPAIN
TEL: 357011
TLX:
COM:

CAMPBELL MURRAY F
DEPT PHYSICS & ASTRONOMY
COLBY COLLEGE
WATERVILLE ME 04901
U.S.A.
TEL: 207-872-3251
TLX:
COM: 44

CABRITA EZEQUIEL DR
OBS ASTRONOMICO DE LISBOA
TAPADA DA LISBOA
1300 LISBOA
PORTUGAL
TEL: 637351-634669
TLX:
COM: 26

CALAME ODILE DR
CERGA
AVENUE COPERNIC
F-06130 GRASSE
FRANCE
TEL: 93-36-58-49
TLX: 470865
COM: 07,16,19,20

CAMARENA BADIA VICENTE PR
DPTO MATEMATICA APLICADA
ETS INGENIEROS INDUSTR
UNIVERSIDAD DE ZARAGOZA
50009 ZARAGOZA
SPAIN
TEL:
TLX:
COM:

CAMPOS L M BRAGA DA COSTA
INST SUPERIOR TECNICO
AVE ROVISCO PAIS
1096 LISBOA CODEX
PORTUGAL
TEL: 800525
TLX: 63423 ISTUTL P
COM:

CACCIANI ALESSANDRO PROF
DIPARTIMENTO DI FISICA
UNIVERSITA LA SAPIENZA
PIAZZALE ALDO MORO 2
I-00185 ROMA
ITALY
TEL: 6-4976-265
TLX: 613255 INFNRO
COM:

CALDWELL JOHN A R
MT WILSON & LAS CAMPANAS
OBSERVATORIES
813 SANTA BARBARA STREET
PASADENA CA 91101
U.S.A.
TEL: 818-577-1122
TLX:
COM: 33

CAMERON ALASTAIR G W PROF
HARVARD COLLEGE OBS
60 GARDEN STREET
CAMBRIDGE MA 02138
U.S.A.
TEL: 617-495-5374
TLX:
COM: 35,48

CAMPUSANO LUIS E
DEPTO DE ASTRONOMIA
UNIVERSIDAD DE CHILE
CASILLA 36-D
SANTIAGO
CHILE
TEL: 02-2294101
TLX: 44001 PBCZ
COM: 28,51

CACCIARI CARLA DR
SPACE TELESCOPE SC INST
HOMEWOOD CAMPUS
3700 SAN MARTIN DR
BALTIMORE MD 21218
U.S.A.
TEL: 301-338-4912
TLX: 6849101
COM:

CALDWELL JOHN JAMES
E.S.S. DEPT
SUNY AT STONY BROOK
STONY BROOK NY 11794
U.S.A.
TEL: 516-246-7148
TLX: 5102287767 SUNY ADMI
COM: 16

CAMERON WINIFRED S MRS
5087 IRON SPRINGS RD
PRESCOTT AZ 86301
U.S.A.
TEL:
TLX:
COM: 16

CANAL RAMON M DR
DPTO FIS TIERRA & COSMOS
UNIVERSIDAD DE BARCELONA
08014 BARCELONA
SPAIN
TEL:
TLX:
COM: 35

CACCIN BRUNO
DIPARTIMENTO DI FISICA
MOSTRA D'OLTREMARE
PAD. 20
I-80125 NAPOLI
ITALY
TEL: 7253428
TLX: 7200359 INFNNA
COM:

CALLEBAUT DIRK K DR
DEPT OF PHYSICS UIA
UIA
UNIVERSITEITSPLEIN 1
B-2610 WILRIJK-ANTWERPEN
BELGIUM
TEL: 3-828-2528
TLX: 33646
COM: 35,37

CAMICHEL HENRI DR
24, AVE C. FLAMMARION
F-31500 TOULOUSE
FRANCE
TEL: 61-48-96-91
TLX:
COM: 16

CANAVAGGIA RENEE DR
OBSERVATOIRE DE PARIS
61 AVE DE L'OBSERVATOIRE
F-75014 PARIS
FRANCE
TEL: 1-43-20-12-10
TLX: 270776
COM:

CADEZ ANDREJ DR
DEPT OF PHYSICS
UNIVERSITY OF LJUBLJANA
JADRANSKA 19
61000 LJUBLJANA
YUGOSLAVIA
TEL: 265-061
TLX:
COM:

CALOI VITTORIA DR
CNR
ASTROFISICA SPAZIALE
C P 67
I-00044 FRASCATI
ITALY
TEL: 39-6-942-5654
TLX: 610261 CNR FRA
COM: 35

CAMPBELL BRUCE DR
DOMINION ASTROPHYS OBS
5071 W SAANICH ROAD
VICTORIA BC V8X 4M6
CANADA
TEL: 604-388-3753
TLX:
COM: 09,29,30,51

CANDY MICHAEL P MR
PERTH OBSERVATORY
WALNUT ROAD
BICKLEY 6076 WEST AUSTR
AUSTRALIA
TEL:
TLX:
COM: 06,07C,15,20

CADEZ VLADIMIR
INSTITUTE OF PHYSICS
P.O. BOX 57
YU-11001 BEOGRAD
YUGOSLAVIA
TEL: 011-212-219
TLX: 11002 INFIZ YU
COM: 10,12

CALVANI MASSIMO DR
ISTITUTO DI ASTRONOMIA
VIC. DELL'OSSERVATORIO 5
I-35122 PADOVA
ITALY
TEL: 049-661499
TLX: 430176 UNPADU I
COM: 47

CAMPBELL DONALD B
ARECIBO OBSERVATORY
BOX 995
ARECIBO PR 00613
U.S.A.
TEL: 809-878-2612
TLX: 385638
COM: 16

CANE HILARY VIVIEN
NASA/GSFC
CODE 661
GREENBELT MD 20771
U.S.A.
TEL: 301-344-7794
TLX:
COM: 10,33

CAHN JULIUS H PROF
UNIVERSITY OF ILLINOIS
DEPARTMENT OF ASTRONOMY
1011 WEST SPRINFIELD AVE
URBANA IL 61801
U.S.A.
TEL: 217-333-3090
TLX:
COM:

CALVET NURIA DR
C I D A
APARTADO POSTAL 264
MERIDA 5101-A
VENEZUELA
TEL: 58-74-639930
TLX: 74174 CIDA VC
COM:

CAMPBELL JAMES W
ROYAL OBSERVATORY
BLACKFORD HILL
EDINBURGH EH9 3HJ
U.K.
TEL:
TLX:
COM:

CANFIELD RICHARD C DR
2680 WOODLAWN DRIVE
HONOLULU HI 96822
U.S.A.
TEL:
TLX:
COM:

LIST OF MEMBERS

CANIZARES CLAUDE R PROF
M I T
RM 37-501
CAMBRIDGE MA 02139
U.S.A.
TEL: 617-253-7500
TLX: 921473 MITCAM
COM:

CANNON RUSSELL D DR
ROYAL OBSERVATORY
BLACKFORD HILL
EDINBURGH EH9 3HJ
U.K.
TEL: 31-667-3321
TLX: 72283 ROEDIN G
COM: 28,37

CANNON WAYNE H DR
DEPTS/PHYS/EARTH & ATM SC
YORK UNIVERSITY
4700 KEELE STREET
DOWNSVIEW ONT M3J1P3
CANADA
TEL: 416-667-6410
TLX: 06524736
COM: 19

CANTO JORGE DR
INSTITUTO DE ASTRONOMIA
UNAM
APDO POSTAL 70-264
04510 MEXICO DF
MEXICO
TEL: 5485305
TLX: 1760155 CIC ME
COM: 34

CANTU ALBERTO M DR
IST CIBERNETICA/BIOFISICA
CNR
I-16032 CAMOGLI (GE)
ITALY
TEL: 0185-770646
TLX:
COM:

CAO CHANGXIN
NANJING ASTRONOMICAL
INSTRUMENT FACTORY
NANJING
CHINA, PEOPLE'S REP.
TEL: 46191
TLX: 34136 GLYNJ CN
COM: 09

CAPACCIOLI MASSIMO DR
OSSERVATORIO ASTRONOMICO
UNIVERSITY OF PADOVA
VIC. DELL'OSSERVATORIO 5
I-35122 PADOVA
ITALY
TEL: 049-66-14-99
TLX: 430176 UNDAPU I
COM: 28

CAPEN CHARLES F
SOLIS LACUS OBSERVATORY
RT.2, BOX 262 E
CUBA MO 65453
U.S.A.
TEL:
TLX:
COM:

CAPITAINE NICOLE
OBSERVATOIRE DE PARIS
61 AVE DE L'OBSERVATOIRE
F-75014 PARIS
FRANCE
TEL: 1-43-20-12-10
TLX: 270776
COM: 04,19

CAPLAN JAMES
OBSERVATOIRE DE MARSEILLE
2 PLACE LE VERRIER
F-13248 MARSEILLE CEDEX 4
FRANCE
TEL: 91-95-90-88
TLX:
COM:

CAPPA DE NICOLAU CRISTINA
INSTITUTO ARGENTINO DE
RADIOASTRONOMIA
CASILLA DE CORREO No. 5
1894 VILLA ELISA (Bs.As.)
ARGENTINA
TEL:
TLX:
COM: 34

CAPRIOLI GIUSEPPE PROF
OSSERVATORIO ASTRONOMICO
DI ROMA
VIA TRIONFALE 204
I-00136 ROMA
ITALY
TEL:
TLX:
COM: 31

CAPRIOTTI EUGENE R DR
DEPT OF ASTRONOMY
OHIO STATE UNIVERSITY
5058 ALPHEUS SMITH LAB
COLUMBUS OH 43210
U.S.A.
TEL: 614-422-1773
TLX:
COM: 34

CAPUTO FILIPPINA DR
CNR ASTROFISICA SPAZIALE
C P 67
I-00044 FRASCATI, ROMA
ITALY
TEL: 06-942-5651
TLX: 610261 CNR FRA
COM: 35

CAPUZZO DOLCETTA ROBERTO
ISTITUTO ASTRONOMICO
UNIVERSITA LA SAPIENZA
VIA G.M. LANCISI 29
I-00161 ROMA
ITALY
TEL: 06-867525
TLX:
COM: 34,37

CARANICOLAS NICHOLAS DR.
DEPT OF ASTRONOMY
UNIVERSITY THESSALONIKI
GR-54006 THESSALONIKI
GREECE
TEL: 031-991357/59
TLX:
COM: 07

CARBON DUANE F DR
LICK OBSERVATORY
UCSC
SANTA CRUZ CA 95064
U.S.A.
TEL: 408-429-2149
TLX:
COM: 14,36

CARDONA OCTAVIO DR
INST NAC DE ASTROFISICA
OPTICA Y ELECTRONICA
APDO POSTAL 216
72000 PUEBLA, PUE.
MEXICO
TEL: 22-470500
TLX:
COM:

CARDUS ALMEDA J O MR
OBSERVATORIO DEL EBRO
ROQUETES (TARRAGONA)
SPAIN
TEL: 977-500511
TLX:
COM:

CARESTIA REINALDO A DR
OBS ASTRONOMICO F AGUILAR
AV BENAVIDEZ 8175 OESTE
RIVADAVIA
5400 SAN JUAN
ARGENTINA
TEL: 0054-064-231615
TLX:
COM: 08

CARLBERG RAYMOND GARY DR
PHYSICS DEPARTMENT
YORK UNIVERSITY
TORONTO ONT M3J IP3
CANADA
TEL: 416-667-3851
TLX:
COM:

CARLETON NATHANIEL P DR
SMITHSONIAN ASTROPHYSICAL
OBSERVATORY
60 GARDEN ST
CAMBRIDGE MA 02138
U.S.A.
TEL: 617-495-7405
TLX: 921428 SATELLITE CAM
COM:

CARLQVIST PER A DR
DEPT OF PLASMA PHYSICS
ROYAL INST OF TECHNOLOGY
S-100 44 STOCKHOLM 70
SWEDEN
TEL: 46-8-787-7697
TLX:
COM: 10

CARLSON JOHN B
ASTRONOMY PROGRAM
UNIVERSITY OF MARYLAND
COLLEGE PARK MD 20742
U.S.A.
TEL: 301-454-4460
TLX:
COM: 41,51

CARNEY BRUCE WILLIAM
DEPT OF PHYS & ASTRONOMY
PHILLIPS HALL 039A
UNIV OF NORTH CAROLINA
CHAPEL HILL NC 27514
U.S.A.
TEL: 919-962-3023
TLX:
COM: 25,29,30,37

CAROFF LAWRENCE J
SPACE SCIENCE DIVISION
NASA AMES RESEARCH CTR
MS 245-6
MOFFETT FIELD CA 94035
U.S.A.
TEL: 415-694-5523
TLX:
COM:

CAROUBALOS C A PROF
LAB ELECTRONIC PHYSICS
UNIVERSITY OF ATHENS
KTHRIA TYPA-ILISSIA
ATHENS 144
GREECE
TEL: 7244096/11119
TLX: 215530 OBSA GR
COM: 40

CARPENTER LLOYD DR
EG/G WASH ANALYT SERV CTR
5000 PHILADELPHIA WAY
LANHAM MD 20706
U.S.A.
TEL:
TLX:
COM:

CARQUILLAT JEAN-MICHEL
OBSERVATOIRE PIC-DU-MIDI
ET TOULOUSE
14 AVENUE EDOUARD BELIN
F-31400 TOULOUSE
FRANCE
TEL: 61-25-21-01
TLX:
COM: 30

CARR BERNARD JOHN
SCHOOL OF MATHEM. SCI.
QUEEN MARY COLLEGE
MILE END ROAD
LONDON E1 4NS
U.K.
TEL: 01-980-4811
TLX:
COM:

CARR THOMAS D PROF
DEPT OF ASTRONOMY
UNIVERSITY OF FLORIDA
GAINESVILLE FL 32611
U.S.A.
TEL: 904-392-2066
TLX:
COM: 40,51

CARRANZA G J DR
LAPRIDA 880
5000 CORDOBA
ARGENTINA
TEL:
TLX:
COM: 28

CARRASCO GUILLERMO DR
OBS ASTRONOMICO NACIONAL
UNIVERSIDAD DE CHILE
CASILLA 36-D
SANTIAGO
CHILE
TEL: 229-40-02
TLX:
COM: 08

CARRASCO LUIS DR
INSTITUTO DE ASTRONOMIA
UNAM
APDO POSTAL 70-264
04510 MEXICO DF
MEXICO
TEL: 905-548-5305
TLX:
COM: 28,33

CARROLL P KEVIN PROF
PHYSICS DEPT
UNIVERSITY COLLEGE
BELFIELD
DUBLIN 4
IRELAND
TEL:
TLX:
COM: 14,44

CARRUTHERS GEORGE R DR
SPACE SCIENCE DIVISION
US NAVAL RESEARCH LAB
CODE 7123
WASHINGTON DC 20375
U.S.A.
TEL: 202-767-2764
TLX:
COM: 34

CARSON T R DR
UNIVERSITY OBSERVATORY
BUCHANAN GARDENS
ST ANDREWS, FIFE KY16 9SS
U.K.
TEL:
TLX:
COM: 35,36

CARSWELL ROBERT F DR
INSTITUTE OF ASTRONOMY
MADINGLEY ROAD
CAMBRIDGE CB3 0HA
U.K.
TEL: 223-62204
TLX: 817297 ASTRON G
COM: 28

CARTER DAVID DR
MT STROMLO OBSERVATORY
PRIVATE BAG
WODEN P.O. ACT 2606
AUSTRALIA
TEL: 062-88-1111
TLX: 62270 AA
COM: 28

CARTER WILLIAM EUGENE
NATL GEODETIC SURVEY
N/CG114
ADV. TECHNOL. SECT., GRDL
ROCKVILLE MD 20852
U.S.A.
TEL: 301-443-8423
TLX:
COM: 19C,31

CARUSI ANDREA
IST ASTROFISICA SPAZIALE
REP PLANETOLOGIA
VIALE DELL'UNIVERSITA' 11
I-00185 ROMA
ITALY
TEL: 39-6-4956951
TLX: 610261 CNRFRA
COM: 20,22

CARVER JOHN H PROF
RESEARCH SCHOOL OF
PHYSICAL SCIENCES
AUSTRALIAN NAT UNIVERSITY
CANBERRA ACT 2601
AUSTRALIA
TEL: 062-492476
TLX: 62615 RPHYS
COM: 14,44

CASANOVAS JUAN DR
VATICAN OBSERVATORY
I-00120 VATICAN CITY
VATICAN CITY STATE
TEL: 698-3411/5266
TLX: 2020 VATOBS VA
COM:

CASH WEBSTER C JR
LASP
UNIVERSITY OF COLORADO
BOX 392
BOULDER CO 80309
U.S.A.
TEL: 303-492-8208
TLX:
COM: 48

CASINI CATERINA DR
VIA P. NERI 2
I-20146 MILANO
ITALY
TEL:
TLX:
COM: 28

CASSATELLA ANGELO DR
LAB ASTROFISICA SPAZIALE
C P 67
I-00044 FRASCATI, ROMA
ITALY
TEL:
TLX:
COM: 29

CASSE NICHEL DR
SECTION D'ASTROPHYSIQUE
CEN SACLAY
BP NO 2
F-91190 GIF-S-YVETTE
FRANCE
TEL:
TLX:
COM: 48

CASSINELLI JOSEPH P DR
ASTRONOMY DEPT
UNIVERSITY OF WISCONSIN
475 N. CHARTER ST.
MADISON WI 53706
U.S.A.
TEL: 608-262-1752
TLX: 265452 UOFWISC MDS
COM: 36C

CASTELLANI VITTORIO PROF
ISTITUTO ASTRONOMICO
UNIVERSITA DI ROMA
VIA LANCISI 29
I-00161 ROMA
ITALY
TEL: 6-86-7525
TLX:
COM: 35

CASTELLI FIORELLA DR
OSSERVATORIO ASTRONOMICO
VIA TIEPOLO 11
I-34131 TRIESTE
ITALY
TEL: 40-793921
TLX: 461137 OAT I
COM: 29

CASTELLI JOHN P
AFGL-PHP
HANSCOM AFB
BEDFORD MA 01731
U.S.A.
TEL:
TLX:
COM:

CASTOR JOHN I DR
LAWRENCE LIVERMORE
NATIONAL LAB L-23
PO BOX 808
LIVERMORE CA 94550
U.S.A.
TEL: 415-422-4664
TLX: 910-3868339 LLNL
COM: 35,36

CASWELL JAMES L DR
CSIRO
DIVISION OF RADIOPHYSICS
P.O.BOX 76
EPPING NSW 2121
AUSTRALIA
TEL: 02-868-0222
TLX: 26230 ASTRO AA
COM: 33,34,40

CATALA MARIA ASUNCION DR
DPTO FIS TIERRA & COSMOS
UNIVERSIDAD DE BARCELONA
08028 BARCELONA
SPAIN
TEL: 330-73-11/244
TLX:
COM: 46

CATALAN MANUEL DR
INSTITUTO Y OBSERVATORIO
DE MARINA
SAN FERNANDO (CADIZ)
SPAIN
TEL: 883548
TLX: 76108
COM: 04,19

CATALANO FRANCESCO A DR
ISTITUTO DI ASTRONOMIA
CITTA UNIVERSITARIA
I-95125 CATANIA
ITALY
TEL: 095-330533
TLX: 970359 ASTRCT I
COM:

CATALANO SANTO DR
ISTITUTO DI ASTRONOMIA
CITTA UNIVERSITARIA
I-95125 CATANIA
ITALY
TEL: 095-330533
TLX: 970359 ASTRCT I
COM: 16,29,42

CATCHPOLE ROBIN MICHAEL
S A A O
P O BOX 9
OBSERVATORY 7935
SOUTH AFRICA
TEL: 470025
TLX: 5720309
COM: 27,29

CATO B TORGNY DR
NORDISK TELESATELLITSTAT.
BOX 107
S-457 00 TANUMSHEDE
SWEDEN
TEL: 46-525-291-55
TLX: 20164 NORDSAT S
COM:

CATURA RICHARD C DR
LOCKHEED RESEARCH LAB
DEPT 91-20 BLDG 255
3251 HANOVER ST
PALO ALTO CA 94304
U.S.A.
TEL: 415-858-4066
TLX: 346409 LMSC SUVL
COM: 44,48

CAUGHLAN GEORGEANNE R
PHYSICS DEPT
MONTANA STATE UNIVERSITY
BOZEMAN MT 59717
U.S.A.
TEL: 406-994-6170
TLX:
COM: 35,48

CAVALIERE ALFONSO G PROF
ASTROFISICA, DIP FISICA
II UNIVERSITA DI ROMA
VIA O. RAIMONDO
I-00173 ROMA
ITALY
TEL:
TLX:
COM: 47,48

CAVALLINI FABIO
OSSERVATORIO ASTROFISICO
DI ARCETRI
LARGO E. FERMI 5
I-50125 FIRENZE
ITALY
TEL: 055-220034
TLX: 572268
COM: 12

CAYREL DE STROBEL GIUSA
OBSERVATOIRE DE PARIS
SECTION DE MEUDON
F-92195 MEUDON PL CEDEX
FRANCE
TEL: 1-45-34-75-70
TLX:
COM: 29P,36

CAYREL ROGER DR
OBSERVATOIRE DE PARIS
61 AVE DE L'OBSERVATOIRE
F-75014 PARIS
FRANCE
TEL: 1-43-20-12-10
TLX: 270776
COM: 29,36,50

CAZENAVE ANNY DR
CNES-GRGS
18 AVENUE EDOUARD BELIN
F-31400 TOULOUSE
FRANCE
TEL: 61-27-40-11
TLX: 531081
COM:

CAZZOLA P DR
OSSERVATORIO ASTRONOMICO
VICOLO DELL'OSSERVATORIO
I-35100 PADOVA
ITALY
TEL:
TLX:
COM:

CEFOLA PAUL J DR
MAIL STATION 64
C S DRAPER LAB
555 TECHNOLOGY SQ
CAMBRIDGE MA 02139
U.S.A.
TEL: 617-258-1787
TLX:
COM: 07

CELIS LEOPOLDO DR
PONTIF. UNIV. CATOLICA
DEPTO DE ASTRONOMIA
CASILLA 6014
SANTIAGO
CHILE
TEL:
TLX:
COM: 25,45

CELNIKIER LUDWIK DR
OBSERVATOIRE DE PARIS
SECTION DE MEUDON
F-92195 MEUDON PL CEDEX
FRANCE
TEL: 1-45-34-75-70
TLX:
COM:

CEPLECHA ZDENEK DR
ASTRONOMICAL INSTITUTE
CZECH. ACAD. OF SCIENCES
OBSERVATORY
251 65 ONDREJOV
CZECHOSLOVAKIA
TEL: 724525
TLX: 121579
COM: 15,22

CEPPATELLI GUIDO DR
OSSERVATORIO ASTROFISICO
DI ARCETRI
LARGO E. FERMI 5
I-50125 FIRENZE
ITALY
TEL: 055-22-00-34
TLX: 572268
COM: 12

CERRUTI-SOLA MONICA
OSSERVATORIO ASTROFISICO
DI ARCETRI
LARGO E. FERMI 5
I-50125 FIRENZE
ITALY
TEL: 055-220034
TLX: 572268 ARCETR I
COM: 34

CERSOSIMO JUAN CARLOS DR
INSTITUTO ARGENTINO DE
RADIOASTRONOMIA
CASILLA DE CORREO No. 5
1894 VILLA ELISA (Bs.As.)
ARGENTINA
TEL:
TLX:
COM: 34

CESARSKY CATHERINE J DR
INST RECHERCHE FONDAMENT.
CEN SACLAY
DPHG/SAP-BAT 28
F-91191 GIF/YVETTE CEDEX
FRANCE
TEL: 1-69-08-39-12
TLX: 690860
COM: 34,48P

CESARSKY DIEGO A DR
INSTITUT D'ASTROPHYSIQUE
98 BIS BD ARAGO
F-75014 PARIS
FRANCE
TEL: 1-43-20-14-25
TLX:
COM: 34

CESCO CARLOS DR
OBS ASTRONOMICO F AGUILAR
BENAVIDEZ 8175 OESTE
5407 MARQUESADO (S.J.)
ARGENTINA
TEL:
TLX:
COM: 06,

CESTER BRUNO PROF
OSSERVATORIO ASTRONOMICO
VIA TIEPOLO 11
I-34131 TRIESTE
ITALY
TEL: 40-793921/221
TLX: 461137 OAT I
COM: 26,42,45

CEVOLANI GIORDANO
FISBAT-CNR
VIA DE' CASTAGNOLI 1
I-40126 BOLOGNA
ITALY
TEL: 519593/94
TLX: 511350
COM: 22

CHA DU JIN
PYONGYANG ASTRON OBS
ACADEMY OF SCIENCES DPRK
TAESONG DISTRICT
PYONGYANG
KOREA DPR
TEL:
TLX:
COM: 08

CHA GI UNG
PYONGYANG ASTRON OBS
ACADEMY OF SCIENCES DPRK
TAESONG DISTRICT
PYONGYANG
KOREA DPR
TEL:
TLX:
COM: 31

CHAFFEE FREDERIC H DR
MULTIPLE MIRROR TELES OBS
UNIVERSITY OF ARIZONA
TUCSON AZ 85721
U.S.A.
TEL:
TLX:
COM:

CHAISSON ERIC J PROF
DEPT OF ASTRONOMY
HAVERFORD COLLEGE
HAVERFORD PA 19041
U.S.A.
TEL: 215-896-1146
TLX:
COM: 51

CHAMARAUX PIERRE DR
OBSERVATOIRE DE PARIS
SECTION DE MEUDON
F-92195 MEUDON
FRANCE
TEL: 1-45-34-75-30
TLX: 270912
COM: 28

CHAMBE GILBERT
OBSERVATOIRE DE PARIS
SECTION DE MEUDON
DASOP
F-92195 MEUDON PL CEDEX
FRANCE
TEL: 1-45-34-75-30
TLX:
COM: 10,12

CHAMBERLAIN JOSEPH M DR
ADLER PLANETARIUM
1300 S. LAKE SHORE DR
CHICAGO IL 60605
U.S.A.
TEL: 312-322-0325
TLX:
COM: 08,31,46

CHAMBERLAIN JOSEPH W PROF
SPACE PHYS & ASTRON DEPT
RICE UNIVERSITY
HOUSTON TX 77001
U.S.A.
TEL: 713-527-8101
TLX: 556457
COM: 16,21,49

CHAMBLISS CARLSON R DR
DEPT OF PHYS SCIENCES
KUTZTOWN UNIVERSITY
KUTZTOWN PA 19530
U.S.A.
TEL: 215-683-4439
TLX:
COM: 42

CHAN KWING LAM
APPLIED RESEARCH CORP.
8201 CORPORATE DRIVE
LANDOVER MD 20785
U.S.A.
TEL: 301-459-8442
TLX:
COM: 12,35,36,40

CHANDRA SUBHASH
MIS PHILIPS LABS
345 SCARBOROUGH RD
BRIAR CLIFF NY 10510
U.S.A.
TEL:
TLX:
COM:

CHANDRASEKHAR S PROF
LAB ASTRON & SPACE RES
933 E 56TH STRET
CHICAGO IL 60637
U.S.A.
TEL: 312-962-7860
TLX:
COM: 35,48

CHANMUGAM GANESAR PROF
PHYS & ASTRONOMY DEPT
LOUISIANA STATE UNIV
BATON ROUGE LA 70803-4001
U.S.A.
TEL: 504-388-6894
TLX:
COM: 42

CHAPMAN CLARK R DR
PLANETARY SCI INSTITUTE
2030 E SPEEDWAY
SUITE 201
TUCSON AZ 85719
U.S.A.
TEL: 602-881-0332
TLX:
COM: 15C,16

CHAPMAN GARY A DR
SAN FERNANDO OBSERVATORY
DEPT PHYSICS & ASTRONOMY
CALIFORNIA STATE UNIV.
NORTHRIDGE CA 91330
U.S.A.
TEL: 818-885-2775
TLX:
COM: 10,12

CHAPMAN ROBERT D DR
CODE PD311
LYNDON JOHNSON SPACE CTR
HOUSTON TX 77058
U.S.A.
TEL:
TLX:
COM: 15,42,44

CHAPRONT JEAN DR
BUREAU DES LONGITUDES
77 AVE DENFERT-ROCHEREAU
F-75014 PARIS
FRANCE
TEL: 1-43-20-12-10
TLX:
COM: 04C,07C

CHAPRONT-TOUZE MICHELLE
BUREAU DES LONGITUDES
77 AVE DENFERT-ROCHEREAU
F-75014 PARIS
FRANCE
TEL: 1-43-20-12-10
TLX:
COM: 07

CHARLES PHILIP ALLAN
DEPT OF ASTROPHYSICS
UNIVERSITY OF OXFORD
SOUTH PARKS ROAD
OXFORD OX1 3RQ
U.K.
TEL: 0865-511336x506
TLX: 83295 NUCLOX
COM: 44

CHARVIN PIERRE PR
OBSERVATOIRE DE PARIS
61 AVE DE L'OBSERVATOIRE
F-75014 PARIS
FRANCE
TEL: 1-43-20-12-10
TLX: OBS PARIS 270 776
COM: 09

CHAU WAI Y PROF
PHYSICS DEPT
QUEEN'S UNIVERSITY
KINGSTON ONT K7L 3N6
CANADA
TEL: 613-547-3526
TLX:
COM:

CHAVARRIA-K. CARLOS
INSTITUTO DE ASTRONOMIA
UNAM
APARTADO POSTAL 70-264
04510 MEXICO DF
MEXICO
TEL:
TLX:
COM: 37

CHAVIRA ENRIQUE SR
INAOE
AP POSTALES 216 y 51
72000 PUEBLA, PUE.
MEXICO
TEL: 47-05-00
TLX:
COM: 27

CHE-BOHNENSTENGEL ANNE
SUELZBRACKRING 39A
D-2050 HAMBURG 80
GERMANY, F.R.
TEL: 040-7238550
TLX:
COM:

CHELLI ALAIN
OBSERVATOIRE DE LYON
F-69230 ST-GENIS-LAVAL
FRANCE
TEL: 78-56-07-05
TLX:
COM: 09

CHEN BIAO
PURPLE MOUNTAIN
OBSERVATORY
NANJING
CHINA, PEOPLE'S REP.
TEL: 46700
TLX: 34144 PMONJ CN
COM: 10,12C,49

CHEN CHUAN-LE
BEIJING ASTRONOMICAL OBS
ACADEMIA SINICA
BEIJING
CHINA, PEOPLE'S REP.
TEL: 281698 BEIJING
TLX: 22040 BAOAS CN
COM:

CHEN DAO-HAN
PURPLE MOUNTAIN
OBSERVATORY
NANJING
CHINA, PEOPLE'S REP.
TEL: 31096
TLX: 34144 PMONJ CN
COM: 15,16C

CHEN HONGSHENG
BEIJING ASTRONOMICAL OBS
ACADEMIA SINICA
BEIJING
CHINA, PEOPLE'S REP.
TEL:
TLX: 9053
COM: 40

CHEN JIAN-SHENG
BEIJING ASTRONOMICAL OBS
ACADEMIA SINICA
PEKING
CHINA, PEOPLE'S REP.
TEL:
TLX: 22040 BAOAS CN
COM: 28

CHEN KWAN-YU PROF
DEPT PHYSICS & ASTRONOMY
UNIVERSITY OF FLORIDA
GAINESVILLE FL 32611
U.S.A.
TEL: 904-392-2055
TLX:
COM: 42

CHEN PEISHENG
YUNNAN OBSERVATORY
P.O. BOX 110
KUNMING, YUNNAN PROVINCE
CHINA, PEOPLE'S REP.
TEL: 72946
TLX:
COM: 36

CHEN XIAO-ZHONG
BEIJING PLANETARIUM
BEIJING
CHINA, PEOPLE'S REP.
TEL:
TLX:
COM:

CHEN XING
SHANGHAI OBSERVATORY
ACADEMIA SINICA
SHANGHAI
CHINA, PEOPLE'S REP.
TEL: 380696
TLX: 33164 SHAO CN
COM: 19

CHEN ZHEN
PURPLE MOUNTAIN
OBSERVATORY
NANJING
CHINA, PEOPLE'S REP.
TEL: 46700
TLX: 34144 PMONJ CN
COM: 07,26,33

CHEN ZHENCHENG
BEIJING ASTRONOMICAL OBS
ACADEMIA SINICA
BEIJING
CHINA, PEOPLE'S REP.
TEL: 281698
TLX: 9053
COM: 10,28

CHEN ZUN-GUI
BEIJING PLANETARIUM
BEIJING
CHINA, PEOPLE'S REP.
TEL:
TLX:
COM: 41

CHENG CHUNG-CHIEH DR
NAVAL RESEARCH LABORATORY
CODE 4175CC
WASHINGTON DC 20375
U.S.A.
TEL: 202-767-2350
TLX:
COM: 10,12

CHENG FU-HUA
CENTER FOR ASTROPHYSICS
UNIV SCIENCE & TECHNOLOGY
HEFEI, ANHUI
CHINA, PEOPLE'S REP.
TEL: 63300-526 HEFEI
TLX: 90028 USTC CN
COM: 47

CHENG FU-ZHEN
CENTER FOR ASTROPHYSICS
UNIV SCIENCE & TECHNOLOGY
HEFEI, ANHUI PROVINCE
CHINA, PEOPLE'S REP.
TEL: 63300 HEFEIx527
TLX: 90028 USTC CN
COM: 47

CHEREDNICHENKO V I DR
KIEV POLYTECHNICAL INST
252056 KIEV
U.S.S.R.
TEL:
TLX:
COM: 15

CHEREPASHCHUK A M PROF
STERNBERG STATE
ASTRONOMICAL INSTITUTE
119899 MOSCOW
U.S.S.R.
TEL: 139-38-38
TLX:
COM: 27,42,45

CHERNEGA N A A DR
OBSERVATORY OF THE
KIEV UNIVERSITY
OBSERVATORNAYA 3
252053 KIEV
U.S.S.R.
TEL: 26 23-91
TLX:
COM: 08

CHERNYKH N S DR
CRIMEAN ASTROPHYSICAL
OBSERVATORY
NAUCHNY
334413 CRIMEA
U.S.S.R.
TEL:
TLX:
COM: 20

CHERTOPRUD V E DR
HYDROMETEOROLOGICAL CTR
OF THE USSR
123376 MOSCOW
U.S.S.R.
TEL:
TLX:
COM: 10

CHEVALIER CLAUDE DR
OBS DE HAUTE-PROVENCE
F-04870 ST-MICHEL-L'OBS.
FRANCE
TEL: 92-76-63-68
TLX: 410690
COM: 35

CHEVALIER ROGER A DR
DEPT OF ASTRONOMY
UNIVERSITY OF VIRGINIA
PO BOX 3818
CHARLOTTESVILLE VA 22903
U.S.A.
TEL: 804-924-4889
TLX:
COM: 34

CHIAN ABRAHAM CHIAN-LONG
INST PESQUISAS ESPACIAIS
CAIXA POSTAL 515
12200 S. JOSE DOS CAMPOS
BRAZIL
TEL: 0123-22-9977
TLX: 011-33530
COM:

CHIKADA YOSHIHIRO DR
NOBEYAMA RADIO OBS
TOKYO ASTRONOMICAL OBS
NOBEYAMA
NAGANO PREF 384-13
JAPAN
TEL: 267-98-2831
TLX: 3329005 TAONRO
COM: 40

LIST OF MEMBERS

CHINCARINI GUIDO L DR
OSSERVATORIO ASTRONOMICO
VIA E BIANCHI 46
I-22055 MERATE
ITALY
TEL: 0039-39-596412
TLX:
COM: 28,47

CHINI ROLF
MPI FUER RADIOASTRONOMIE
AUF DEM HUEGEL 69
D-5300 BONN
GERMANY, F.R.
TEL:
TLX: 886440
COM: 34,40

CHIO CHOL ZONG
PYONGYANG ASTRON OBS
ACADEMY OF SCIENCES DPRK
TAESONG DISTRICT
PYONGYANG
KOREA DPR
TEL:
TLX:
COM: 20

CHIOSI CESARE S DR
ISTITUTO DI ASTRONOMIA
UNIVERSITA DI PADOVA
I-35100 PADOVA
ITALY
TEL: 49-66-1499
TLX: 430176 UNDAPU I
COM: 35C

CHISTYAKOV VLADIMIR E DR
USSURIISK SOLAR STATION
PRIMORSKY KRAY
692533 GORNOTAEZHNOE
U.S.S.R.
TEL: 91121 USSURIISK
TLX: 213954 SOLNZE
COM: 12

CHITRE DATTAKUMAR M DR
COMPUTER SCIENCES CORP.
8728 COLESVILLE RD
SILVER SPRING MD 20910
U.S.A.
TEL:
TLX:
COM: 47

CHITRE SHASHIKUMAR M DR
TATA INSTITUTE OF
FUNDAMENTAL RESEARCH
HOMI BHABHA RD
BOMBAY 400 005
INDIA
TEL: 219111
TLX: 011-3009 TIFR IN
COM: 35,48

CHIU HONG-YEE DR
NUDD BLDG RM 828
COLUMBIA UNIVERSITY
NEW YORK NY 10027
U.S.A.
TEL:
TLX:
COM: 35

CHIU LIANG-TAI GEORGE
T J WATSON RESEARCH CTR
BOX 218
YORKTOWN HEIGHTS NY 10598
U.S.A.
TEL: 914-945-2436
TLX:
COM: 24

CHIUDERI CLAUDIO PROF
OSSERVATORIO ASTROFISICO
DI ARCETRI
LARGO E. FERMI 5
I-50125 FIRENZE
ITALY
TEL:
TLX:
COM:

CHIUDERI-DRAGO FRANCA PR
OSSERVATORIO ASTROFISICO
DI ARCETRI
LARGO E. FERMI 5
I-50125 FIRENZE
ITALY
TEL:
TLX:
COM: 10

CHIUMIENTO GIUSEPPE
OSS ASTRONOMICO DI TORINO
STRADA OSSERVATORIO 20
I-10025 PINO TORINESE
ITALY
TEL: 011-841067
TLX: 213236 TOASTR I
COM: 08,19

CHKHIKVADZE IAKOB N
ABASTUMANI ASTROPHYSICAL
OBSERVATORY
383762 ABASTUMANI,GEORGIA
U.S.S.R.
TEL: 2-78
TLX: 327409
COM: 35

CHLISTOVSKY FRANCA DR
OSSERVATORIO ASTRONOMICO
DI BRERA
VIA BRERA 28
I-20121 MILANO
ITALY
TEL:
TLX:
COM: 08

CHMIELEWSKI YVES DR
OBSERVATOIRE DE GENEVE
CHEMIN DES MAILLETTES 51
CH-1290 SAUVERNY
SWITZERLAND
TEL: 22-55 26 11
TLX: 2 77 20 OBSG CH
COM:

CHOCHOL DRAHOMIR
ASTRONOMICAL INSTITUTE
CZECH. ACAD. OF SCIENCES
059 60 TATRANSKA LOMNICA
CZECHOSLOVAKIA
TEL: 969-967866
TLX: 78377
COM: 42,44

CHOI KYU-HONG
DEPT ASTRON & METEOROLOGY
YONSEI UNIVERSITY
134 SHINCHON, SUDAEMUN
SEOUL 120
KOREA, REPUBLIC
TEL: 02-392-0131
TLX:
COM: 07,42

CHOI WON CHOL
PYONGYANG ASTRON OBS
ACADEMY OF SCIENCES DPRK
TEASONG DISTRICT
PYONGYANG
KOREA DPR
TEL: 5-3134, 5-3239
TLX:
COM:

CHOLLET FERNAND DR
OBSERVATOIRE DE PARIS
61 AVE DE L'OBSERVATOIRE
F-75014 PARIS
FRANCE
TEL: 1-43-20-12-10
TLX: 270776 OBS PARIS
COM: 04,08

CHOPINET MARGUERITE DR
57, RUE THIERS
F-92100 BOULOGNE
FRANCE
TEL: 1-47-61-11-44
TLX:
COM: 34

CHOU KYONG CHOL PROF
KYUNG HEE UNIVERSITY
DEPT ASTRONOMY & SPACE
SCIENCE
SEOUL 131
KOREA, REPUBLIC
TEL: 966-0061/5
TLX:
COM: 51

CHRISTENSEN-DALSGAARD J
INSTITUTE OF ASTRONOMY
UNIVERSITY OF AARHUS
DK-8000 AARHUS C
DENMARK
TEL: 06-12-88-99
TLX: 64767 AAUSCI DK
COM: 12C,27,35

CHRISTIAN CAROL ANN
CANADA-FRANCE-HAWAII
TELESCOPE CORPORATION
BOX 1597
KAMUELA HI 96743
U.S.A.
TEL: 808-885-7944
TLX: 633147 CFHT
COM: 37

CHRISTIANSEN WAYNE A
DEPT PHYSICS & ASTRONOMY
UNIVERSITY OF N. CAROLINA
CHAPEL HILL NC 27514
U.S.A.
TEL: 919-962-3011
TLX:
COM: 40

CHRISTIANSEN WILBUR PROF
MOUNT STROMLO OBSERVATORY
PRIVATE BAG
WODEN P.O. ACT 2606
AUSTRALIA
TEL: 062-881111
TLX: 62270
COM: 40

CHRISTOPHE-GLAUME J DR
SERVICE D'AERONOMIE
BP NO 3
F-91370 VERRIERES-LE-B.
FRANCE
TEL: 1-69-20-10-60
TLX:
COM:

CHRISTY JAMES WALTER DR
EXPLORATORY DEVLPT STAFF
US NAVAL OBSERVATORY
WASHINGTON DC 20390
U.S.A.
TEL:
TLX:
COM: 09,24,45

CHRISTY ROBERT F DR
CALTECH
PASADENA CA 91125
U.S.A.
TEL: 213-795-6811
TLX:
COM: 27,35

CHU YAOQUAN
CENTER FOR ASTROPHYSICS
UNIV SCIENCE & TECHNOLOGY
HEFEI, ANHUI
CHINA, PEOPLE'S REP.
TEL:
TLX: 90028 USTC CN
COM: 28,47

CHU YOU-HUA
UNIVERSITY OF ILLINOIS
ASTRONOMY DEPT
1011 W. SPRINGFIELD AVE.
URBANA IL 61801
U.S.A.
TEL: 217-333-5535
TLX:
COM:

CHUBB TALBOT A DR
5023 N. 38TH STREET
ARLINGTON VA 22207
U.S.A.
TEL:
TLX:
COM: 44,48

CHUGAJNOV P F DR
CRIMEAN ASTROPHYSICAL OBS
USSR ACADEMY OF SCIENCES
NAUCHNIIY
334413 CRIMEA
U.S.S.R.
TEL:
TLX:
COM: 25,27

CHUN MUN-SUK DR
DEPT ASTR & METEOROLOGY
COLLEGE OF SCIENCE
YONSEI UNIVERSITY
SEOUL
KOREA, REPUBLIC
TEL: 2-392-0131
TLX:
COM: 37

CHUPP EDWARD L DR
PHYSICS DEPT
UNIV OF NEW HAMPSHIRE
DEMERITT HALL
DURHAM NH 03824
U.S.A.
TEL: 603-862-2750
TLX: 950030
COM: 10,48

CHURCHWELL EDWARD B DR
WASHBURN OBSERVATORY
UNIVERSITY OF WISCONSIN
475 N. CHARTER ST
MADISON WI 53706
U.S.A.
TEL: 608-262-7857
TLX: 265452 UOFWISC MDS
COM: 33,34

CHURMS JOSEPH
S A A O
P O BOX 9
OBSERVATORY 7935
SOUTH AFRICA
TEL: 021-47-0025
TLX:
COM: 20,24

CHUVAEV K K DR
CRIMEAN ASTROPHYSICAL OBS
USSR ACADEMY OF SCIENCES
NAUCHNY
334413 CRIMEA
U.S.S.R.
TEL:
TLX:
COM: 28

CHVOJKOVA WOYK E DR
ASTRONOMICAL INSTITUTE
CZECH. ACAD. OF SCIENCES
BUDECSKA 6
120 23 PRAHA 2
CZECHOSLOVAKIA
TEL:
TLX:
COM: 12

CIATTI FRANCO DR
OSSERVATORIO ASTROFISICO
I-36012 ASIAGO VI
ITALY
TEL: 0424-62665
TLX: 430110 SETURIST
COM:

CID PALACIOS RAFAEL PROF
FACULTAD DE CIENCIAS
DEPT DE ASTRONOMIA
CIUDAD UNIVERSITARIA
50009 ZARAGOZA
SPAIN
TEL: 357011
TLX:
COM: 07

CILLIE G G PROF
4 MINSERIE STREET
STELLENBOSCH 7600
SOUTH AFRICA
TEL: 02231-3515
TLX:
COM: 42

CIMINO MASSIMO A PROF
VIA A. CADLOLO 19
I-00136 ROMA
ITALY
TEL: 06-34-92-598
TLX:
COM: 10,41

CIURLA TADEUSZ
ASTRONOMICAL INSTITUTE
WROCLAW UNIVERSITY
UL. KOPERNIKA 11
51-622 WROCLAW
POLAND
TEL:
TLX:
COM: 33

CLARIA JUAN DR
OBSERVATORIO ASTRONOMICO
LAPRIDA 854
5000 CORDOBA
ARGENTINA
TEL: 26876 OR 40613
TLX: 51822 BUCOR
COM: 37,45C

CLARK ALFRED JR PROF
UNIVERSITY OF ROCHESTER
DEPT OF MECHANICAL
ROCHESTER NY 14627
U.S.A.
TEL:
TLX:
COM:

CLARK BARRY G DR
NRAO
VLA PROJECT
PO BOX 0
SOCORRO NM 87801
U.S.A.
TEL: 505-772-4011
TLX: 9109881710
COM: 40

CLARK DAVID H DR
SIENCE DIVISION. SCIENCE
NORTH STAR AVE.
SWINDON SN2 1ET
U.K.
TEL: 0793-26222
TLX: 449466
COM: 40

CLARK FRANK OLIVER DR
DEPT PHYSICS & ASTRONOMY
UNIVERSITY OF KENTUCKY
LEXINGTON KY 40506
U.S.A.
TEL: 606-257-3376
TLX:
COM: 34,40

CLARK GEORGE W PROF
MIT
ROOM 37-611
CAMBRIDGE MA 02139
U.S.A.
TEL: 617-253-5842
TLX:
COM: 44C,48C

CLARK THOMAS A DR
NASA/GSFC
CODE 974
GREENBELT MD 20771
U.S.A.
TEL: 301-344-5957
TLX:
COM: 51

CLARK THOMAS ALAN DR
UNIVERSITY OF CALGARY
PHYSICS DEPARTMENT
2500 UNIVERSITY DRIVE NW
CALGARY ATL T2N 1N4
CANADA
TEL: 403-284-5392
TLX:
COM: 12,44

CLARKE DAVID DR
THE UNIVERSITY
DEPARTMENT OF ASTRONOMY
GLASGOW G12 8QQ
U.K.
TEL: 41-339-8855
TLX: 778421
COM: 09,46

CLARKE JOHN T
NASA/GSFC
HUBBLE SPACE TELESCOPE
CODE 681
GREENBELT MD 20771
U.S.A.
TEL: 301-344-5781
TLX: 710-828-9716
COM:

CLARKE THOMAS R DR
MCLAUGHLIN PLANETARIUM
ROYAL ONTARIO MUSEUM
100 QUEENS PARK CRESCENT
TORONTO ONT M5S 2C6
CANADA
TEL: 416-998-8551
TLX:
COM:

CLAUSEN JENS VIGGO LEKTOR
COPENHAGEN UNIVERSITY OBS
BRORFELDEVEJ 23
DK-4340 TOLLOSE
DENMARK
TEL: 45-3-488195
TLX: 44155 DANAST
COM: 42

CLAUZET LUIZ B FERREIRA
INST ASTRON. E GEOFISICO
UNIVERSIDADE DE SAO PAULO
CAIXA POSTAL 30627
01051 SAO PAULO
BRAZIL
TEL:
TLX:
COM: 08

CLAVEL JEAN
IUE OBSERVATORY
ESA
APARTADO 54065
28080 MADRID
SPAIN
TEL: 34-1-401-9661
TLX: 42444 VILSE
COM: 28

CLAYTON DONALD D PROF
DEPT SPACE PHYS & ASTRON
RICE UNIVERSITY
HOUSTON TX 77001
U.S.A.
TEL: 713-527-8101
TLX:
COM:

CLEGG PETER E DR
QUEEN MARY COLLEGE
MILE END ROAD
LONDON E1 4NS
U.K.
TEL: 01-980-4811
TLX: 893-750 GMCUOL
COM:

CLEGG ROBIN E S DR
UNIVERSITY COLLEGE LONDON
DEPT PHYSICS & ASTRONOMY
GOWER STREET
LONDON WC1E 6BT
U.K.
TEL: 01-387-7050x382
TLX: 28722 UCPHYS G
COM: 34

CLEMENT MAURICE J PROF
UNIVERSITY OF TORONTO
DEPARTMENT OF ASTRONOMY
TORONTO ONT M5S 1A7
CANADA
TEL: 416-978-4833
TLX: 06-986766
COM:

CLIFTON KENNETH ST
NASA MARSHALL SPACE
FLIGHT CENTER, ES 63
HUNTSVILLE AL 35812
U.S.A.
TEL: 205-453-2305
TLX: 594416
COM: 22

CLIMENHAGA JOHN L PROF
UNIVERSITY OF VICTORIA
DEPARTMENT OF PHYSICS
VICTORIA BC V8W 2Y2
CANADA
TEL: 604-721-7741
TLX: 0497222
COM: 29,46

CLIVER EDWARD W
US AIR FORCE GEOPHYS LAB
SPACE PHYSICS DIVISION
HANSCOM AIR FORCE BASE
BEDFORD MA 01731
U.S.A.
TEL: 617-861-3975
TLX: 928123 AFGL HANSCOM
COM: 10

CLUBE S V M DR
UNIVERSITY OF OXFORD
DEPT OF ASTROPHYSICS
SOUTH PARKS ROAD
OXFORD OX1 3RQ
U.K.
TEL: 0865-511336
TLX:
COM: 24,33

COFFEEN DAVID L DR
BOX 151
HASTINGS/HUDSON NY 10706
U.S.A.
TEL: 914-478-2594
TLX:
COM:

COHEN RAYMOND J DR
NUFFIELD RADIO ASTR LABS
JODRELL BANK
MACCLESFIELD SK11 9DL
U.K.
TEL: 0477-71321
TLX: 36149
COM: 40

COLLINS GEORGE W II PROF
THE OHIO STATE UNIVERSITY
174 W. 18TH AVENUE
COLUMBUS OH 43210
U.S.A.
TEL: 614-422-5467
TLX:
COM: 42

CLUTTON-BROCK MARTIN DR
UNIVERSITY OF MANITOBA
DEPARTMENT OF MATHEMATICS
WINNIPEG MB R3T 2N2
CANADA
TEL: 204-261-9255
TLX:
COM:

COFFEY HELEN E MS
NOAA
NGDC E/GC2
325 BROADWAY
BOULDER CO 80303
U.S.A.
TEL: 303-497-6223
TLX: 592811 NOAA MASC BDR
COM: 10

COHEN RICHARD S
COLUMBIA UNIVERSITY
INST FOR SPACE STUDIES
2880 BROADWAY
NEW YORK NW 10025
U.S.A.
TEL: 212-678-5611
TLX:
COM: 09,33,40

COLLINSON EDWARD H
THE COPSE
CHURCH LANE
PLAYFORD
IPSWICH, SUFFOLK IP6 9DR
U.K.
TEL: 62-22-57 IPSWIC
TLX:
COM: 16

COCHRAN WILLIAM DAVID DR
UNIVERSITY OF TEXAS
DEPARTMENT OF ASTRONOMY
AUSTIN TX 78712
U.S.A.
TEL: 512-471-4461
TLX:
COM: 15,30

COGAN BRUCE C DR
MOUNT STROMLO OBSERVATORY
PRIVATE BAG
WODEN P.O. ACT 2606
AUSTRALIA
TEL: 062-88-1111
TLX: 62270
COM: 27

COHN HALDAN N
INDIANA UNIVERSITY
DEPT OF ASTRONOMY
SWAIN WEST 319
BLOOMINGTON IN 47405
U.S.A.
TEL: 812-335-4174
TLX:
COM:

COLOMB FERNANDO R DR
INSTITUTO ARGENTINO DE
RADIOASTRONOMIA
CASILLA DE CORREO NO 5
1894 VILLA ELISA (Bs.As.)
ARGENTINA
TEL: 021-43793
TLX: 18052 CICYT AR
COM:

COCKE WILLIAM JOHN PROF
STEWARD OBSERVATORY
UNIVERSITY OF ARIZONA
TUCSON AZ 85721
U.S.A.
TEL: 602-621-6540
TLX:
COM:

COHEN JEFFREY M DR
UNIV OF PENNSYLVANIA
PHYSICS DEPARTMENT
PHILADELPHIA PA 19174
U.S.A.
TEL:
TLX:
COM: 35,47,48

COLBURN DAVID S DR
NASA AMES RESEARCH CENTER
245-3
MOFFETT FIELD CA 94035
U.S.A.
TEL: 415-965-5000
TLX: 34-8408
COM:

COLOMBO G PROF DR
ISTITUTO MECCANICA APPL
UNIVERSITA DI PADOVA
VIA F. MARZOLO 9
I-35100 PADOVA
ITALY
TEL:
TLX:
COM: 16

CODE ARTHUR D
WASHBURN OBSERVATORY
UNIVERSITY OF WISCONSIN
475 N. CHARTER ST
MADISON WI 53706
U.S.A.
TEL: 608-262-9594
TLX:
COM: 29,34,44

COHEN JUDITH DR
KITT PEAK NATIONAL OBS
P O BOX 26732
TUCSON AZ 85726
U.S.A.
TEL:
TLX:
COM:

COLE TREVOR WILLIAM PROF
SCHOOL OF ELECTRICAL ENG
UNIVERSITY OF SYDNEY
SYDNEY NSW 2006
AUSTRALIA
TEL: 02-692-2682
TLX:
COM: 40

COMA JUAN CARLOS
INSTITUTO Y OBSERVATORIO
DE MARINA
SAN FERNANDO (CADIZ)
SPAIN
TEL:
TLX:
COM:

CODINA LANDABERRY SAYD J
OBSERVATORIO NACIONAL
RUA GENERAL BRUCE 586
20921 RIO DE JANEIRO RJ
BRAZIL
TEL: 580-7313 x 267
TLX: 21288
COM: 46

COHEN LEON PROF
HUNTER COLLEGE
DEPARTMENT OF PHYSICS
695 PARK AVE
NEW YORK NY 10021
U.S.A.
TEL: 212-570-5696
TLX:
COM:

COLGATE STIRLING A DR
THEORETICAL DIVISION
LOS ALAMOS SCIENTIFIC LAB
MS 275 B
LOS ALAMOS NM 87545
U.S.A.
TEL: 505-667-2897
TLX:
COM:

COMBES FRANCOISE DR
OBSERVATOIRE DE PARIS
SECTION DE MEUDON
DEMIRM
F-92195 MEUDON PL CEDEX
FRANCE
TEL: 1-45-34-75-30
TLX: 270912 OBSASTR
COM:

CODINA VIDAL J M DR
FABRA OBSERVATORY
GRAN VIA DE LOS CORTES
CATALANES 679
08013 BARCELONA
SPAIN
TEL: 34-3-2454766
TLX:
COM:

COHEN MARSHALL H PROF
CALTECH
105-24
PASADENA CA 91125
U.S.A.
TEL: 213-356-4000
TLX: 675425
COM: 34,40

COLIN JACQUES DR
OBSERVATOIRE DE BESANCON
41B AVE DE L'OBSERVATOIRE
F-25000 BESANCON
FRANCE
TEL: 81-80-22-66
TLX:
COM: 28,37,33

COMBES MICHEL
OBSERVATOIRE DE PARIS
SECTION DE MEUDON
F-92195 MEUDON PL CEDEX
FRANCE
TEL: 1-45-34-75-30
TLX:
COM:

COELHO BALSA MARIO C DR
RUA TRINDADE COELHO 21
2o DTO
3000 COIMBRA
PORTUGAL
TEL:
TLX:
COM:

COHEN MARTIN DR
UNIVERSITY OF CALIFORNIA
RADIO ASTRONOMY LAB
601 CAMPBELL HALL
BERKELEY CA 94720
U.S.A.
TEL: 415-642-2833
TLX: 820181 UCB AST RALUD
COM: 27

COLLIN-SOUFFRIN SUZY DR
OBSERVATOIRE DE PARIS
SECTION DE MEUDON
F-92195 MEUDON PL CEDEX
FRANCE
TEL: 1-45-34-75-70
TLX: 201571 F
COM: 34,48

COMINS NEIL FRANCIS
UNIVERSITY OF MAINE
DEPT PHYSICS & ASTRONOMY
BENNETT HALL
ORONO ME 04469
U.S.A.
TEL: 207-581-1037
TLX:
COM: 33

CONCONI PAOLO DR
OSSERVATORIO ASTRONOMICO
DI BRERA
I-22055 MILANO
ITALY
TEL:
TLX:
COM:

CONTOPOULOS GEORGE PROF
UNIVERSITY OF ATHENS
ASTRONOMY DEPARTMENT
PANEPISTIMIOPOLIS
GR-15771 ATHENS
GREECE
TEL: 01-7243-211
TLX:
COM: 07,28,33

CORBALLY CHRISTOPHER
VATICAN OBS. RES. GROUP
STEWARD OBSERVATORY
UNIVERSITY OF ARIZONA
TUCSON AZ 85721
U.S.A.
TEL: 602-621-3225
TLX: 467175
COM: 45

COSMOVICI BATALLI C DR
IST FISICA SPAZIO INTERPL
CNR
I-00044 FRACASTI (ROMA)
ITALY
TEL: 39-6-9423801
TLX: 610261 I
COM: 15

CONDON JAMES J DR
NRAO
EDGEMONT ROAD
CHARLOTTESVILLE VA 22903
U.S.A.
TEL: 804-296-0211
TLX: 910-997-0174
COM: 40,47,48

CONWAY ROBIN G DR
NUFFIELD RADIO ASTR LABS
JODRELL BANK
MACCLESFIELD SK11 9DL
U.K.
TEL: 0477-71321
TLX: 36149
COM: 40

CORBIN THOMAS ELBERT DR
US NAVAL OBSERVATORY
ASTROMETRY DEPT
WASHINGTON DC 20390
U.S.A.
TEL: 202-653-1557
TLX: 710-8221970
COM: 08,24C

COSTA EDGARDO DR
DEPTO DE ASTRONOMIA
UNIVERSIDAD DE CHILE
CASILLA 36-D
SANTIAGO
CHILE
TEL:
TLX:
COM: 08,33

CONKLIN EDWARD K
FORTH INC
111 N. SEPULVEDA BLVD 300
MANHATTAN BEACH CA 90266
U.S.A.
TEL:
TLX: 275182 FORT UR
COM: 40

COOK ALAN H PROF
DEPT PHYS/UNIV CAMRIDGE
THE MASTER'S LODGE
SELWYN COLLEGE
CAMBRIDGE CB3 9DQ
U.K.
TEL: 223-62381 Ex 29
TLX: 81292 CAVLAB
COM: 07,14

CORDES JAMES M
CORNELL UNIVERSITY
SPACE SCIENCE BUILDING
ITHACA NY 14853
U.S.A.
TEL: 607-256-3734
TLX: 932 458
COM: 40

COSTA ENRICO
IST ASTROFISICA SPAZIALE
C P 67
I-00044 FRASCATI
ITALY
TEL: 6-942-5655
TLX: 610261 I
COM:

CONNES JANINE DR
CIRCE
BP 63
F-91406 ORSAY CEDEX
FRANCE
TEL: 1-69-28-76-75
TLX: FACORS 692166 F
COM: 16

COOK ALLAN F DR
CENTER FOR ASTROPHYSICS
HCO-SAO
60 GARDEN STREET
CAMBRIDGE MA 02138
U.S.A.
TEL: 617-495-7229
TLX: 921428
COM: 16,21,22

CORDOVA FRANCE A D
LOS ALAMOS NATIONAL LAB
MS D436
LOS ALAMOS NM 87545
U.S.A.
TEL: 505-667-3904
TLX:
COM:

COSTA VICTOR DR
INSTITUTO DE ASTROFISICA
DE ANDALUCIA
APARTADO 2144
18080 GRANADA
SPAIN
TEL: 58-121311
TLX: 78573 IAGG E
COM:

CONNES PIERRE DR
SERVICE D'AERONOMIE
B P 3
F-91370 VERRIERES-LE-B.
FRANCE
TEL: 1-69-20-10-60
TLX: 692400
COM: 24,51C

COOK JOHN W
8032 SLEEPY VIEW LN
SPRINGFIELD VA 22153
U.S.A.
TEL: 202-767-2161
TLX:
COM: 10,12

CORLISS C H DR
FOREST HILLS LABORATORY
2955 ALBEMARLE STREET NW
2955 ALBEMARLE STR N.W.
WASHINGTON DC 20008
U.S.A.
TEL: 202-362-6085
TLX:
COM: 14

COSTAIN CARMAN H DR
DOMINION RADIO ASTROPHY-
SICAL OBSERVATORY
BOX 248
PENTICTON BC V2A 6K3
CANADA
TEL: 604-497-5321
TLX: 048-88127
COM: 40

CONNOLLY LEO PAUL
SOUTHEAST MISSOURI STATE
UNIVERSITY
DEPT OF PHYSICS
CAPE GIRARDEAU MO 63701
U.S.A.
TEL: 314-651-2167
TLX:
COM: 27,35

COOKE B A DR
LEICESTER UNIVERSITY
X-RAY ASTRONOMY GROUP
PHYSICS DEPARTMENT
LEICESTER LE1 7RH
U.K.
TEL: 533-554455 E188
TLX: 341664
COM:

CORNEJO ALEJANDRO A DR
INAOE
AP POSTALES 216 y 51
72000 PUEBLA, PUE.
MEXICO
TEL: 47-05-00
TLX:
COM: 09,41

COSTAIN CECIL C DR
DIVISION OF PHYSICS
NAT RES COUNCIL OF CANADA
OTTAWA K1A 0R6
CANADA
TEL:
TLX:
COM: 31

CONTADAKIS MICHAEL E DR
DEPT GEODESY & SURVEYING
UNIVERSITY THESSALONIKI
UNIV. BOX 503
GR-54006 THESSALONIKI
GREECE
TEL: 003-031-99-2693
TLX: 412181 AUTH GR
COM: 27

COOKE JOHN ALAN
UNIVERSITY OF EDINBURGH
DEPT OF ASTRONOMY
ROYAL OBSERVATORY
EDINBURGH EH9 3HJ
U.K.
TEL: 031-667-3221
TLX: 72383 ROEDIN G
COM: 09

CORNIDE MANUEL
DEPTO DE ASTROFISICA
FACULTAD DE FISICA
UNIVERSIDAD COMPLUTENSE
28040 MADRID
SPAIN
TEL: 449-53-16
TLX: 47273 FFUC
COM:

COSTERO RAFAEL
INSTITUTO DE ASTRONOMIA
UNAM
APDO POSTAL 70-264
04510 MEXICO DF
MEXICO
TEL: 548-5305
TLX: 1760155 CICME
COM: 34

CONTI PETER S DR
JILA
UNIVERSITY OF COLORADO
BOX 440
BOULDER CO 80309
U.S.A.
TEL: 303-492-8913
TLX: 755842 JILA
COM: 29V,36

CORADINI ANGIOLETTA
IST ASTROFISICA SPAZIALE
REP PLANETOLOGIA
VIALE UNIVERSITA 11
I-00185 ROMA
ITALY
TEL: 39-6-495-6951
TLX: 680489 CNR FRA
COM:

CORWIN HAROLD G JR
UNIVERSITY OF TEXAS
DEPARTMENT OF ASTRONOMY
RLM 15.308
AUSTIN TX 78712-1083
U.S.A.
TEL: 512-471-4461
TLX: 9108741351
COM: 28

COTTON WILLIAM D Jr
NRAO
EDGEMONT ROAD
CHARLOTTESVILLE VA 22901
U.S.A.
TEL: 804-296-0319
TLX: 5105875482
COM: 40

COTTRELL PETER LEDSAM
DEPT OF PHYSICS
UNIVERSITY OF CANTERBURY
CHRISTCHURCH 1
NEW ZEALAND
TEL: 03-482-009
TLX: 4144 NZ
COM:

COUCH WARRICK DR
ANGLO-AUSTRALIAN OBS
P.O. BOX 296
EPPING NSW 2121
AUSTRALIA
TEL: 02-868-1666
TLX: 23999 AAOSYD AA
COM: 28

COULSON IAIN
S A A O
P O BOX 9
OBSERVATORY 7935
SOUTH AFRICA
TEL: 021-47-00-25
TLX: 5720309
COM: 27

COUNSELMAN CHARLES C PROF
MIT
DEPT EARTH & PLANET SCI
ROOM 54-620
CAMBRIDGE MA 02139
U.S.A.
TEL: 617-253-7902
TLX: 921473 MIT CAM
COM: 07,08,16

COUPINOT GERARD DR
OBS PIC-DU-MIDI/TOULOUSE
9 PONT DE LA MOULETTE
F-65200 BAGNERES
FRANCE
TEL: 61-95-19-69
TLX:
COM:

COURTES GEORGES PROF
OBSERVATOIRE DE MARSEILLE
2 PLACE LE VERRIER
F-13248 MARSEILLE CEDEX 4
FRANCE
TEL: 91-95-90-88
TLX: 410584
COM: 28,33,34,44C

COURVOISIER THIERRY J.-L.
ST-ECF
ESO
KARL-SCHWARZSCHILD-STR 2
D-8046 GARCHING B MUNCHEN
GERMANY, F.R.
TEL: 89-320-06-280
TLX: 528222 EOD
COM: 44

COUSINS A W J DR
S A A O
P O BOX 9
OBSERVATORY 7935
SOUTH AFRICA
TEL: 021-47-0025
TLX: 5720309
COM: 25

COUTEAU PAUL PROF
OBSERVATOIRE DE NICE
BP 139
F-06003 NICE CEDEX
FRANCE
TEL: 93-89-04-20
TLX:
COM: 26C

COUTREZ RAYMOND A J PROF
6 RUE EGIDE BOUVIER
B-1160 BRUXELLES
BELGIUM
TEL:
TLX:
COM: 10,40

COUTTS-CLEMENT CHRISTINE
UNIVERSITY OF TORONTO
DEPARTMENT OF ASTRONOMY
TORONTO ONT M5S 1A7
CANADA
TEL: 1-416-978-5186
TLX:
COM: 27

COVINGTON ARTHUR E
131 COLLEGE STREET
KINGSTON ON K7L 4L7
CANADA
TEL:
TLX:
COM: 10,40

COWAN JOHN J DR
UNIVERSITY OF OKLAHOMA
DEPT PHYSICS & ASTRONOMY
NORMAN OK 73019
U.S.A.
TEL: 405-325-3961
TLX:
COM: 35

COWIE LENNOX LAUCHLAN DR
SPACE TELESCOPE SCI INST
HOMEWOOD CAMPUS
3700 SAN MARTIN DR
BALTIMORE MD 21218
U.S.A.
TEL:
TLX:
COM: 34,48

COWLEY ANNE P DR
ARIZONA STATE UNIVERSITY
PHYSICS DEPARTMENT
TEMPE AZ 85287
U.S.A.
TEL: 602-965-2919
TLX:
COM: 29,42,45

COWLEY CHARLES R PROF
UNIVERSITY OF MICHIGAN
ASTRONOMY DEPARTMENT
ANN ARBOR MI 48109-1090
U.S.A.
TEL: 313-764-3437
TLX: 810-2236056
COM: 29,36

COWLING THOMAS G PROF
19 HOLLIN GARDENS
HEADINGLEY
LEEDS LS16 5NL
U.K.
TEL: 785-342 LEEDS
TLX:
COM: 35

COWSIK RAMANATH
TATA INSTITUTE OF
FUNDAMENTAL RESEARCH
HOMI BHABHA RD
BOMBAY 400 005
INDIA
TEL:
TLX:
COM: 28,48

COX ARTHUR N DR
LOS ALAMOS NATIONAL LAB
P O BOX 1663
LOS ALAMOS NM 87545
U.S.A.
TEL: 505-667-7648
TLX: 910-988-1773
COM: 12,27C,35C

COX DONALD P PROF
UNIVERSITY OF WISCONSIN
DEPT OF ASTRONOMY
1150 UNIVERSITY AVENUE
MADISON WI 53706
U.S.A.
TEL: 608-262-5916
TLX:
COM: 34

COYNE GEORGE V DR
SPECOLA VATICANA
I-00120 CITTA D. VATICANO
VATICAN CITY STATE
TEL: 06-698-3411
TLX: 2020 VAT OBS
COM: 25,34,50C

CRAIG IAN JONATHAN D DR
DEPT APPLIED MATHEMATICS
UNIVERSITY OF WAIKATO
HAMILTON
NEW ZEALAND
TEL: 62889
TLX:
COM: 12

CRAINE ERIC RICHARD DR
WESTERN RESEARCH CO
5061 W CAMINO DE GIRASOL
TUCSON AZ 85745
U.S.A.
TEL: 602-743-7377
TLX:
COM:

CRAM LAWRENCE EDWARD DR
CSIRO
DIV OF APPLIED PHYSICS
P.O.BOX 218
LINDFIELD NSW 2070
AUSTRALIA
TEL:
TLX:
COM: 12C,36C

CRAMPTON DAVID DR
DOMINION ASTROPHYS OBS
5071 W SAANICH ROAD
RR 5
VICTORIA BC V8X 4M6
CANADA
TEL: 604-388-3900
TLX: 0497295
COM: 30,33,45

CRANE PATRICK C
NRAO
PO BOX 0
SOCORRO NM 87801
U.S.A.
TEL: 505-772-4011
TLX: 910-988-1710
COM: 40

CRANE PHILIPPE
ESO
KARL-SCHWARZSCHILD-STR 2
D-8046 GARCHING B MUNCHEN
GERMANY, F.R.
TEL: 49-89-79-20-98
TLX: 528-28222 EO D
COM:

CRANNELL CAROL JO DR
NASA-GSFC
CODE 682
GREENBELT MD 20771
U.S.A.
TEL: 301-344-5007
TLX: 89675
COM: 10,44

CRAWFORD DAVID L DR
KITT PEAK NATIONAL OBS
BOX 26732
950 N. CHERRY AVENUE
TUCSON AZ 85726
U.S.A.
TEL: 602-325-9346
TLX: 666-484 AURA NOAO
COM: 09,25,33,45,50V

CREZE MICHEL DR
OBSERVATOIRE DE BESANCON
41 AVE DE L'OBSERVATOIRE
F-25000 BESANCON
FRANCE
TEL: 81-80-22-66
TLX: OBSBES 361144 F
COM: 24,33

CRIFO FRANCOISE DR
OBSERVATOIRE DE PARIS
SECTION DE MEUDON
DEPEG
F-92195 MEUDON PL CEDEX
FRANCE
TEL: 1-45-34-75-30
TLX: 201571 F
COM: 08

CRISTALDI SALVATORE DR
OSSERVATORIO ASTROFISICO
CITTA UNIVERSITARIA
VIALE ARTALE ALAGONA 75
I-95126 CATANIA
ITALY
TEL: 33-07-34
TLX: 970359 ASTRCT I
COM: 42

CRISTESCU CORNELIA G DR
ASTRONOMICAL OBSERVATORY
CUTITUL DE ARGINT 5
75212 BUCAREST
RUMANIA
TEL: 23-68-92
TLX:
COM: 15,20

CRIVELLARI LUCIO
OSSERVATORIO ASTRONOMICO
DI TRIESTE
VIA G.B. TIEPOLO 11
I-34131 TRIESTE
ITALY
TEL: 39-40-793221
TLX: 461137 OAT I
COM: 36

CROOM DAVID L DR
RUTHERFORD APPLETON LAB
CHILTON DIDCOT OX11 0QX
U.K.
TEL: 0235-21900
TLX: 83159
COM: 40

CROVISIER JACQUES
OBSERVATOIRE DE PARIS
SECTION DE MEUDON
F-92195 MEUDON PL CEDEX
FRANCE
TEL: 1-45-34-75-30
TLX: 270912
COM: 15,34,40

CRUIKSHANK DALE P DR
INSTITUTE FOR ASTRONOMY
2680 WOODLAWN DRIVE
HONOLULU HI 96822
U.S.A.
TEL: 808-948-6664
TLX: 723-8459
COM: 15,16C

CRUISE ADRIAN MICHAEL DR
MULLARD SPACE SCIENCE LAB
HOLMBURY ST MARY
DORKING RH5 6NS
U.K.
TEL: 30-670-292
TLX: 859-185
COM: 48

CRUTCHER RICHARD M DR
UNIV ILLINOIS ASTRON DEPT
341 ASTRON BLDG
1011 W. SPRINGFIELD AVE
URBANA IL 61801
U.S.A.
TEL: 217-333-9581
TLX:
COM:

CRUVELLIER PAUL E DR
LAB ASTRONOMIE SPATIALE
TRAVERSE DU SIPHON
LES TROIS LUCS
F-13012 MARSEILLE
FRANCE
TEL: 91-66-08-32
TLX: 420584
COM: 34

CRUZ-GONZALEZ IRENE
INSTITUTO DE ASTRONOMIA
UNAM
APDO POSTAL 70-264
04510 MEXICO DF
MEXICO
TEL: 905-548-5306
TLX:
COM:

CSADA IMRE K DR
KONKOLY OBSERVATORY
THEGE UT 13/17
BOX 67
H-1121 BUDAPEST
HUNGARY
TEL: 166-426
TLX: 227460
COM: 10

CUDABACK DAVID D DR
UNIVERSITY OF CALIFORNIA
RADIO ASTRONOMY LAB
BERKELEY CA 94720
U.S.A.
TEL: 415-642-5724
TLX: 820181 UCB AST
COM: 34,40

CUDWORTH KYLE MCCABE DR
YERKES OBSERVATORY
UNIVERSITY OF CHICAGO
BOX 258
WILLIAMS BAY WI 53191
U.S.A.
TEL: 414-245-5555
TLX:
COM: 24,33,37

CUFFEY J MR
NEW MEXICO STATE UNIV
DEPT EARTH SCI & ASTRON
UNIVERSITY PARK NM 88001
U.S.A.
TEL:
TLX:
COM: 37

CUGNON PIERRE DR
OBS ROYAL DE BELGIQUE
AVENUE CIRCULAIRE 3
B-1180 BRUXELLES
BELGIUM
TEL: 2-375-24-84
TLX: 21565 OBSBEL
COM: 34

CUGUSI LEONINO DR
DIPT DI SCIENZE FISICHE
UNIVERSITA DI CAGLIARI
VIA OSPEDALE 72
I-09100 CAGLIARI
ITALY
TEL: 70-664770
TLX:
COM:

CUI CHUNFANG
WUHAN TECHNICAL UNIV. OF
SURVEYING AND MAPPING
WUHAN
CHINA, PEOPLE'S REP.
TEL:
TLX:
COM: 07

CUI DOU-XING
CHANGCHUN ARTIFICIAL
SATELLITE OBSERVATORY
P.O. BOX 1067
CHANGCHUN
CHINA, PEOPLE'S REP.
TEL: 42859
TLX: 2421 CHANGCHUN
COM: 07

CUI LIAN-SHU
DEPT OF ASTRONOMY
NANJING UNIVERSITY
NANJING
CHINA, PEOPLE'S REP.
TEL: 37551
TLX: 0909
COM: 12

CUI ZHENXING
BEIJING ASTRONOMICAL OBS
BEIJING 100080
CHINA, PEOPLE'S REP.
TEL: 282100 BEIJING
TLX: 22040 BAOAS CN
COM: 40

CULHANE LEONARD PROF
MULLARD SPACE SCIENCE LAB
UNIVERSITY COLLEGE LONDON
HOLMBURY ST MARY
DORKING RH5 6NS
U.K.
TEL: 030670-292
TLX: 859185 UCMSSL G
COM: 10,44,48

CULVER ROGER BRUCE DR
COLORADO STATE UNIVERSITY
DEPARTMENT OF PHYSICS
FORT COLLINS CO 80523
U.S.A.
TEL: 303-491-6206
TLX: 9109309000 ENGRCSUFT
COM: 26

CUNNINGHAM LELAND E PROF
UNIVERSITY OF CALIFORNIA
DEPARTMENT OF ASTRONOMY
BERKELEY CA 94720
U.S.A.
TEL:
TLX:
COM: 06,07,20

CUNY YVETTE J DR
OBSERVATOIRE DE PARIS
SECTION DE MEUDON
F-92195 MEUDON PL CEDEX
FRANCE
TEL: 1-45-34-75-30
TLX:
COM: 36

CUPERMAN SAMI PROF
TEL-AVIV UNIVERSITY
DEPT PHYSICS & ASTRONOMY
RAMAT-AVIV 69978
ISRAEL
TEL: 03-42021/425697
TLX: 342171 VERSY IL
COM: 33,49C

CURIR ANNA
OSS ASTRONOMICO DI TORINO
STRADA OSSERVATORIO 20
I-10025 PINO TORINESE
ITALY
TEL: 011-84-10-67
TLX: 213236 TO ASTR I
COM: 48

CURRIE DOUGLAS G DR
UNIVERSITY OF MARYLAND
DEPT PHYSICS & ASTRONOMY
COLLEGE PARK MD 20742
U.S.A.
TEL: 301-454-3405
TLX:
COM: 09,19,51

CZYZAK STANLEY J DR
OHIO STATE UNIVERSITY
DEPT PHYSICS & ASTRONOMY
174 W. 18TH ST
COLUMBUS OH 43210
U.S.A.
TEL: 614-422-6543
TLX:
COM: 14,34

D'ANTONA FRANCESCA DR
OSSERVATORIO ASTRONOMICO
I-00100 ROMA
ITALY
TEL:
TLX:
COM: 35

D'ODORICO SANDRO DR
ESO
KARL-SCHWARZSCHILD-STR 2
D-8046 GARCHING B MUNCHEN
GERMANY, F.R.
TEL: 89-320-06-00
TLX: 528-28-222
COM: 28C,34C

DA COSTA GARY STEWART DR
YALE UNIVERSITY
DEPARTMENT OF ASTRONOMY
BOX 6666
NEW HAVEN CT 06511
U.S.A.
TEL: 203-436-3460
TLX:
COM: 37

DA COSTA JOSE MARQUES DR
INST PESQUISAS ESPACIAIS
INPE
CAIXA POSTAL 515
12200 S. JOSE DOS CAMPOS
BRAZIL
TEL: 0123-229977
TLX: 1133530 INPEBR
COM: 48

DA ROCHA VIEIRA E DR
INSTITUTO DE FISICA
UNIVISERSIDADE FEDERAL
DO RIO GRANDE DO SUL
90000 PORTO ALEGRE RS
BRAZIL
TEL: 0512-21-7666
TLX: 0511055 UFRSBR
COM:

DA SILVA A V C S
OBSERVATORIO ASTRONOMICO
UNIVERSIDADE SANTA CLARA
3000 COIMBRA
PORTUGAL
TEL:
TLX:
COM:

DA SILVA LICIO DR
OBSERVATORIO NACIONAL
RUA GENERAL BRUCE 586
20921 RIO DE JANEIRO RJ
BRAZIL
TEL: 580-7313
TLX: 21288
COM:

DACHS JOACHIM PROF DR
ASTRONOMISCHES INSTITUT
RUHR-UNIVERSITAET
POSTFACH 102148
D-4630 BOCHUM 1
GERMANY, F.R.
TEL: 0234-7003454
TLX: 0825860
COM: 21,25C

DADAEV ALEKSANDR N DR
PULKOVO OBSERVATORY
196140 LENINGRAD
U.S.S.R.
TEL:
TLX:
COM: 26,42

DADHICH NARESH DR
DEPT OF MATHEMATICS
UNIVERSITY OF POONA
PUNE 411 007
INDIA
TEL: 56061/91
TLX:
COM: 47

DADIC ZARKO DR
ZAVOD ZA POVIJEST
ZNANOSTI JAZU
ANTE KOVACICA 5
41000 ZAGREB
YUGOSLAVIA
TEL: 3841-440124
TLX:
COM: 41

DAHN CONARD CURTIS DR
US NAVAL OBSERVATORY
PO BOX 1149
FLAGSTAFF AZ 86002
U.S.A.
TEL: 602-779-5132
TLX:
COM: 24,25,34

DAIGNE GERARD G
OBSERVATOIRE DE PARIS
SECTION DE MEUDON
DERAD
F-92195 MEUDON PL CEDEX
FRANCE
TEL: 1-45-34-75-30
TLX:
COM: 51

DAINTREE EDWARD J DR
NUFFIELD RADIO ASTRO LABS
JODRELL BANK
MACCLESFIELD SK11 9DL
U.K.
TEL: 0477-71321
TLX: 36149
COM: 40

DAISHIDO TSUNEAKI PROF
DEPARTMENT OF SCIENCE
SCHOOL OF EDUCATION
WASEDA UNIVERSITY
SHINJUKU-KU TOKYO 160
JAPAN
TEL: 2-203-4141
TLX: 2323280 WASEDA J
COM: 40

DALGARNO ALEXANDER PROF
CENTER FOR ASTROPHYSICS
60 GARDEN STREET
CAMBRIDGE MA 02138
U.S.A.
TEL: 617-495-4403
TLX: 921428
COM: 14,34

DALLAPORTA N PROF
ISTITUTO DI ASTRONOMIA
UNIVERSITA
VIC. DELL'OSSERVATORIO 5
I-35100 PADOVA
ITALY
TEL: 49-66-14-99
TLX:
COM:

DALTABUIT ENRIQUE DR
INSTITUTO DE ASTRONOMIA
UNAM
APDO POSTAL 70-264
04510 MEXICO DF
MEXICO
TEL:
TLX:
COM:

DAMBARA TAKESHI PROF
SHIZUOKA UNIVERSITY
OTANI SHIZUOKA 422
JAPAN
TEL:
TLX:
COM: 08

DAMLE S V DR
TATA INSTITUTE OF
FUNDAMENTAL RESEARCH
BOMBAY 400 005
INDIA
TEL: 219111
TLX: 0113009 TIFR IN
COM:

DAN XHI-XIANG
SHANGHAI OBSERVATORY
ACADEMIA SINICA
SHANGHAI
CHINA, PEOPLE'S REP.
TEL: 386191
TLX: 33164 SHAO CN
COM: 09

DANBY J M ANTHONY DR
DEPARTMENT OF MATHEMATICS
N.CAROLINA STATE UNIV
RALEIGH NC 27695-8205
U.S.A.
TEL: 919-737-3210
TLX:
COM: 07

DANESE LUIGI DR
OSSERVATORIO ASTRONOMICO
VIC. DELL'OSSERVATORIO 5
I-35100 PADOVA
ITALY
TEL: 49-66-1499
TLX: 430176 UNPADU-I
COM: 47

DANKS ANTHONY C DR
DEPT PHYSICS & ASTRONOMY
MICHIGAN STATE UNIVERSITY
EAST LANSING MI 48224
U.S.A.
TEL: 517-353-2986
TLX: 810-251-0737 MSU INT
COM: 15,28,34

DANZIGER I JOHN DR
ESO
KARL-SCHWARZSCHILD-STR 2
D-8046 GARCHING B MUNCHEN
GERMANY, F.R.
TEL:
TLX:
COM:

DAPPEN WERNER
HIGH ALTITUDE OBSERVATORY
NCAR
PO BOX 3000
BOULDER CO 80307A
U.S.A.
TEL: 303-497-1512
TLX: 45694
COM:

DARIUS JON DR
SCIENCE MUSEUM
LONDON SW7 2DD
U.K.
TEL: 01-589-3456x643
TLX: 21200 SCMLIB G
COM: 41,46,51

DAS MRINAL KANTI
DEPT PHYSICS/DELHI UNIV
SRI VENKATESWARA COLLEGE
DHAULA KUAN
NEW DELHI 110 021
INDIA
TEL:
TLX:
COM: 35

DAS P K DR
INDIAN INSTITUTE OF
ASTROPHYSICS
BANGALORE 560 034
INDIA
TEL: 566585
TLX: 845763 IIAB IN
COM: 47

DATLOWE DAYTON DR
LOCKHEED PALO A. RES LAB
DEPT 91-20 BLDG 255
3251 HANOVER STREET
PALO ALTO CA 94304
U.S.A.
TEL: 415-858-4074
TLX:
COM: 10

DATTA BHASKAR DR
INDIAN INSTITUTE OF
ASTROPHYSICS
BANGALORE 560 034
INDIA
TEL: 566585 / 566497
TLX: 845763 IIAB IN
COM: 47

DAUBE-KURZEMNIECE I A DR
RADIOASTROPHYSICAL OBS
LATVIAN ACAD OF SCIENCES
TURGENEVA 19
226524 RIGA LATVIA
U.S.S.R.
TEL: 226796
TLX:
COM: 37

DAUTCOURT G DR
ZNTRLINST. F. ASTROPHYSIK
STERNWARTE BABELSBERG
ROSA-LUXEMBURG-STR 17A
DDR-1502 POTSDAM
GERMANY, D.R.
TEL:
TLX:
COM: 48

DAVIDSEN ARTHUR FALNES DR
DEPT PHYSICS & ASTRONOMY
JOHNS HOPKINS UNIVERSITY
CHARLES & 34TH STREETS
BALTIMORE MD 21218
U.S.A.
TEL: 301-338-7370
TLX:
COM: 28,44,48

DAVIDSON KRIS DR
SCHOOL PHYS & ASTRONOMY
UNIVERSITY OF MINNESOTA
116 CHURCH ST S.E.
MINNEAPOLIS MN 55455
U.S.A.
TEL: 612-373-7795
TLX:
COM:

DAVIDSON WILLIAM PROF
25 PADDOCK CLOSE
MANSFIELD NOTTS NG21 9PL
U.K.
TEL:
TLX:
COM: 47,48

DAVIES JOHN G DR
NUFFIELD RADIO ASTRO LABS
JODRELL BANK
MACCLESFIELD SK11 9DL
U.K.
TEL: 0477-71321
TLX: 36174
COM: 19,22,40

DAVIES MERTON E MR
THE RAND CORPORATION
1700 MAIN STREET
SANTA MONICA CA 90406
U.S.A.
TEL: 213-393-0411
TLX:
COM: 16C

DAVIES PAUL CHARLES W
SCHOOL OF PHYSICS
THE UNIVERSITY
NEWCASTLE/TYNE NE1 7RU
U.K.
TEL:
TLX:
COM: 47

DAVIES RODNEY D PROF
NUFFIELD RADIO ASTR LABS
JODRELL BANK
MACCLESFIELD SK11 9DL
U.K.
TEL: 0477-71321
TLX: 36149
COM: 28,33,34,40

DAVIS CECIL G JR
UNIVERSITY OF CALIFORNIA
LOS ALAMOS NATIONAL LAB
GROUP P-15 / MS D 406
LOS ALAMOS NM 87545
U.S.A.
TEL: 505-667-5908
TLX:
COM: 35,36

DAVIS JOHN DR
SCHOOL OF PHYSICS
UNIVERSITY OF SYDNEY
SYDNEY NSW 2006
AUSTRALIA
TEL: 02-692-3604
TLX: 26169 UNISYD AA
COM: 09V,50

DAVIS LEVERETT JR PROF
CALTECH 405-47
PASADENA CA 91125
U.S.A.
TEL: 818-356-4243
TLX:
COM: 48

DAVIS MARC DR
DEPARTMENT OF ASTRONOMY
UNIVERSITY OF CALIFORNIA
601 CAMPBELL HALL
BERKELEY CA 94720
U.S.A.
TEL: 415-642-5156
TLX: 820181 UCB AST
COM:

DAVIS MICHAEL M DR
ARECIBO OBSERVATORY
PO BOX 995
ARECIBO PR 00613
U.S.A.
TEL: 809-878-2612
TLX: 385638
COM: 40,47,48,51

DAVIS MORRIS S PROF
DEPT PHYSICS & ASTRONOMY
UNIVERSITY OF N. CAROLINA
284 PHILLIPS HALL 039A
CHAPEL HILL NC 27514
U.S.A.
TEL: 919-962-3011
TLX:
COM: 05.07

DAVIS RICHARD J DR
NUFFIELD RADIO ASTR LABS
JODRELL BANK
MACCLESFIELD SK11 9DL
U.K.
TEL:
TLX:
COM:

DAVIS ROBERT J DR
OPTICAL/INFRARED ASTR DIV
SMITHSONIAN ASTROPHYS OBS
60 GARDEN ST, MS 20
CAMBRIDGE MA 02138
U.S.A.
TEL: 616-495-7335
TLX: 921428 SATELLITE CAM
COM: 05,40,44

DAVOUST EMMANUEL
OBSERVATOIRE DE BESANCON
F-25044 BESANCON CEDEX
FRANCE
TEL: 81-80-22-66
TLX: 361144 OBS BES
COM:

DAWE JOHN ALAN DR
A.N.U.
SIDING SPRING OBSERVATORY
PRIVATE BAG
COONABARABRAN 2857
AUSTRALIA
TEL: 068-426-221
TLX: 63945 AA CANOPUS
COM: 51

DE BERGH CATHERINE DR
OBSERVATOIRE DE PARIS
SECTION DE MEUDON
LAB ASTRO INFRAROUGE
F-92195 MEUDON PL CEDEX
FRANCE
TEL:
TLX:
COM:

DE BIASE G A DR
OSSERVATORIO ASTRONOMICO
VIA DEL PARCO MELLINI 84
I-00136 ROMA
ITALY
TEL:
TLX:
COM:

DE BOER KLAAS SJOERDS DR
ASTRONOMISCHES INSTITUT
UNIVERSITAET BONN
AUF DEM HUEGEL 71
D-5300 BONN
GERMANY, FED. REP.
TEL: 48-228-733656
TLX: 886440
COM: 34C

DE BRITO e ABREU J C DR
RUA DO OLIVAL 142
1200 LISBOA
PORTUGAL
TEL: 60-72-00
TLX:
COM:

DE BRUYN A. GER DR
RADIOSTERREWACHT
POSTBUS 2
NL-7990 AA DWINGELOO
NETHERLANDS
TEL: 5219-7244
TLX: 42043 SRZM NL
COM:

DE CASTRO ANGEL DR
NATL ASTRONOMICAL OBS
ALFONSO XII-3
APARTADO 12354
28014 MADRID
SPAIN
TEL: 91-2271935
TLX: 234651GCE
COM: 04

DE CASTRO ELISA
DEPTO DE ASTROFISICA
FACULTAD DE FISICA
UNIVERSIDAD COMPLUTENSE
28040 MADRID
SPAIN
TEL: 449-53-16
TLX: 47273 FFUC
COM:

DE FELICE FERNANDO DR
DIP. FISICA G. GALILEI
CITTA UNIVERSITATIA
VIA MARZOLO 8
I-35100 PADOVA
ITALY
TEL: 049-844-278
TLX: 430308 DFGGPDI
COM: 48

DE FREITAS PACHECO J A DR
OBSERVATORIO DE SAO PAULO
UNIVERSIDADE DE SAO PAULO
CAIXA POSTAL 30627
01051 SAO PAULO
BRAZIL
TEL: 021-717-3518
TLX:
COM:

DE GRAAF T DR
INSTITUUT VOOR FONETISCHE
WETENSCHAPPEN
GROTE ROZENSTRAAT 31
NL-9712 TG GRONINGEN
NETHERLANDS
TEL:
TLX:
COM: 48

DE GRAAFF W DR
APPELGAARDE 117
NL-3992 JD HOUTEN
NETHERLANDS
TEL:
TLX:
COM: 51

DE GRAAUW TH DR
ESA
SPACE SCIENCE DEPARTMENT
POSTBUS 299
NL-2200 AG NOORDWIJK
NETHERLANDS
TEL:
TLX:
COM:

DE GREVE JEAN-PIERRE DR
ASTROPHYSICAL INSTITUTE
VRIJE UNIV BRUSSEL
PLEINLAAN 2
B-1050 BRUSSELS
BELGIUM
TEL: 32-2-6413498
TLX: 61051 VUBCO
COM: 35,42

DE GROOT MART DR
ARMAGH OBSERVATORY
COLLEGE HILL
ARMAGH BT61 9DG
U.K.
TEL: 0861-522928
TLX: 747937 ARMOBS G
COM: 27,29,42

DE GROOT T DR
STERREKUNDIG INSTITUUT
ZONNENBURG 2
NL-3512 NL UTRECHT
NETHERLANDS
TEL:
TLX: 47224 ASTRO
COM: 10,40

DE JAGER CORNELIS PROF
ASTRONOMICAL OBSERVATORY
LAB FOR SPACE RESEARCH
BENELUXLAAN 21
NL-3527 HS UTRECHT
NETHERLANDS
TEL: 31-30-937145
TLX: 47224
COM: 10,12,35,40,44,49,51

DE JAGER GERHARD PROF
DEPT PHYSICS/ELECTRONICS
RHODES UNIVERSITY
P O BOX 94
GRAHAMSTOWN 6140
SOUTH AFRICA
TEL: 0461-7128
TLX: 244226
COM:

DE JONG TEIJE DR
ASTRONOMICAL INSTITUTE
ROETERSTRAAT 15
NL-1018 WB AMSTERDAM
NETHERLANDS
TEL: 20-5223004
TLX: 16460 FACWN NL
COM: 33,34

DE JONGE J K DR
DEPARTMENT OF ASTRONOMY
UNIVERSITY OF PITTSBURGH
RIVERVIEW PARK
PITTSBURGH PA 15214
U.S.A.
TEL:
TLX:
COM: 30,51

LIST OF MEMBERS

DE KORT JULES J DR
HOUTLAAN 4
NL-6500 GV NIJMEGEN
NETHERLANDS
TEL:
TLX:
COM: 42

DE PATER IMKE
UNIVERSITY OF CALIFORNIA
ASTRONOMY DEPT
601 CAMPBELL HALL
BERKELEY CA 94720
U.S.A.
TEL: 415-642-1947
TLX:
COM: 16

DE YOUNG DAVID S DR
KITT PEAK NAT OBSERVATORY
PO BOX 26732
TUCSON AZ 85726
U.S.A.
TEL: 602-327-5511
TLX:
COM: 40,48

DEGAONKAR S S DR
PHYS. RESEARCH LABORATORY
NAVRANGPURA
AHMEDABAD 380 009
INDIA
TEL: 462129
TLX: 121397 PRL IN
COM: 40

DE KORTE PIETER A J DR
LAB SPACE RESEARCH LEIDEN
WASSENAARSEWEG 78
PO BOX 9504
NL-2300 RA LEIDEN
NETHERLANDS
TEL: 071-148333
TLX: 39058 ASTRO NL
COM:

DE RUITER HANS RUDOLF
IST DI RADIOASTRONOMIA
VIA IRNERIO 46
I-40126 BOLOGNA
ITALY
TEL: 51-23-28-56
TLX: 211664 INFN BO I
COM:

DE ZOTTI GIANFRANCO DR
ISTITUTO DI ASTRONOMIA
VIC. DELL'OSSERVATORIO 5
I-35122 PADOVA
ITALY
TEL: 49-66-1499
TLX: 430176 UNPADU I
COM: 47

DEGEWIJ JOHAN DR
MODDERMANSTRAAT 66
NL-2313 GS LEIDEN
NETHERLANDS
TEL:
TLX:
COM: 15,16

DE LA HERRAN V JOSE ENG
INSTITUTO DE ASTRONOMIA
MEXICO
APDO POSTAL 971
MEXICO 1 DF
MEXICO
TEL:
TLX:
COM:

DE SABBATA V PROF DR
ISTITUTO DI FISICA
UNIVERSITA DI BOLOGNA
VIA IRNERIO 46
I-40100 BOLOGNA
ITALY
TEL: 260991/051
TLX:
COM:

DEARBORN DAVID PAUL S DR
LAWRENCE LIVERMORE LAB
L-23
PO BOX 808
LIVERMORE CA 94550
U.S.A.
TEL:
TLX:
COM: 35

DEGUCHI SHUJI DR
UNIV ILLINOIS AT URBANA
CHAMPAIGN, DEPT/PHYSICS
1110 W. GREEN STREET
URBANA IL 61801
U.S.A.
TEL: 217-333-0864
TLX: 9103808375 PHYSICSDO
COM: 34

DE LA NOE JEROME DR
OBSERVATOIRE DE BORDEAUX
AVENUE PIERRE SEMIROT
BP 21
F-33270 FLOIRAC
FRANCE
TEL: 56-86-43-30
TLX:
COM: 28,34,40

DE SANCTIS GIOVANNI
OSSERVATORIO ASTRONOMICO
DI TORINO
STRADA OSSERVATORIO 20
I-10025 PINO TORINESE
ITALY
TEL: 011-841067
TLX: 213236 TOASTR I
COM: 15,20

DEBARBAT SUZANNE V DR
OBSERVATOIRE DE PARIS
61 AVE DE L'OBSERVATOIRE
F-75014 PARIS
FRANCE
TEL: 1-43-20-12-10
TLX: 270776
COM: 08,19,41C

DEHARVENG JEAN-MICHEL DR
LAS
TRAVERSE DU SIPHON
LES TROIS LUCS
F-13012 MARSEILLE
FRANCE
TEL: 91-66-08-32
TLX: 420 584 F
COM:

DE LA REZA RAMIRO DR
OBSERVATORIO NACIONAL
RUA GENERAL BRUCE 586
20921 RIO DE JANEIRO RJ
BRAZIL
TEL: 580-7313
TLX: 21288
COM:

DE SILVA L.N.K. DR
INSTITUTE OF FUND.STUDIES
380/72 BAUDDHALOKA MAW.
COLOMBO 7
SRI LANKA
TEL: 597538
TLX: 21700 IFS CE
COM: 28

DEBEHOGNE HENRI DR SC
OBS ROYAL DE BELGIQUE
AVENUE CIRCULAIRE 3
B-1180 BRUXELLES
BELGIUM
TEL: 02-3743801
TLX: 21565 B
COM: 20

DEHARVENG LISE DR
OBSERVATOIRE DE MARSEILLE
2 PLACE LE VERRIER
F-13248 MARSEILLE CEDEX 4
FRANCE
TEL: 91-95-90-88
TLX:
COM:

DE LOORE CAMIEL PROF
ASTROPHYSICAL INSTITUTE
VRIJE UNIVERSITEIT
PLEINLAAN 2   76
B-1050 BRUSSEL
BELGIUM
TEL: 32-2-6413496
TLX: 61051 VUBCO B
COM: 35,42,51

DE VAUCOULEURS GERARD PR
DEPARTMENT OF ASTRONOMY
UNIVERSITY OF TEXAS
RLM 15.212
AUSTIN TX 78712
U.S.A.
TEL: 512-471-4461
TLX:
COM: 28,30,47C

DEBRUNNER HERMANN DR
PHYSIKALISCHES INSTITUTE
UNIVERSITAET BERN
SIDLERSTRASSE 5
CH-3000 BERN
SWITZERLAND
TEL: 31-65-40-51
TLX: 32320 CH
COM: 48

DEINZER W PROF DR
UNIVERSITAETS-STERNWARTE
GEISMARLANDSTR 11
D-3400 GOETTINGEN
GERMANY, F.R.
TEL: 0551-395044
TLX: 96753 USTERN D
COM: 35

DE MOTTONI Y PALACIOS DR
PRIVATE OBSERVATORY
VIA ROSSELLI 15-23
I-16145 GENOVA
ITALY
TEL: 30-19-32
TLX:
COM: 16

DE VEGT CH PROF DR
HAMBURGER STERNWARTE
GOJENBERGSWEG 112
D-2050 HAMBURG 80
GERMANY, F.R.
TEL: 40-7252-4128
TLX: 21788 HAMST
COM: 08,24C

DEEMING T J DR
DIGICON GEOPHYSICAL CORP
3701 KIRBY DRIVE
HOUSTON TX 77098
U.S.A.
TEL: 713-526-5611
TLX: 762577
COM: 41

DEJAIFFE RENE J DR
OBS ROYAL DE BELGIQUE
AVENUE CIRCULAIRE 3
B-1180 BRUXELLES
BELGIUM
TEL: 3752484
TLX: 21565 OBSBEL B
COM: 08,19

DE PASCUAL MARTINEZ M DR
OBSERVATORIO ASTRONOMICO
ALFONSO XII 3 & 5
28014 MADRID
SPAIN
TEL: 2270107
TLX: 23475 IGC
COM: 20

DE VINCENZI DONALD DR
LIFE SCIENCES DIVISION
NASA HEADQUARTERS
CODE EBR
WASHINGTON DC 20546
U.S.A.
TEL: 202-453-1525
TLX:
COM: 51

DEERENBERG A.J.M. DR
SPACE RESEARCH LABORATORY
HUYGENS LABORATORY
WASSENAARSEWEG 78
NL-2300 RA LEIDEN
NETHERLANDS
TEL: 71-148333
TLX:
COM:

DEJONGHE HERWIG BERT DR
INST FOR ADVANCED STUDY
OLDEN LANE
PRINCETON NJ 08540
U.S.A.
TEL: 609-734-8084
TLX: 229734 IAS UR
COM: 28

DEKEL AVISHAI
YALE UNIVERSITY
DEPT OF ASTRONOMY
NEW HAVEN CT 06511
U.S.A.
TEL: 203-436-3460
TLX:
COM: 28,33,47

DELGADO ANTONIO JESUS
INST DE ASTROFISICA DE
ANDULACIA, APDO 2144
PROFESOR ALBAREDA 1
18080 GRANADA
SPAIN
TEL: 58-12-13-00
TLX: 78573 IAAG E
COM: 27,42

DEMIN V G PROF DR
STERNBERG STATE
ASTRONOMICAL INSTITUTE
UNIVERSITETSKIJ PROSP 13
119889 MOSCOW
U.S.S.R.
TEL: 139-36-81
TLX:
COM: 07

DENNISON P A DR
TRINITY HALL
UNIVERSITY OF CAMBRIDGE
CAMBRIDGE CB3
U.K.
TEL:
TLX:
COM:

DEKKER E DR
MUSEUM BOERHAAVE
STEENSTRAAT 1A
NL-2312 BS LEIDEN
NETHERLANDS
TEL: 071-123084
TLX:
COM: 41

DELHAYE JEAN PROF
OBSERVATOIRE DE PARIS
61, AVE DE L'OBSERVATOIRE
F-75014 PARIS
FRANCE
TEL: 1-43-20-12-10
TLX: 270776
COM: 24,33,38

DEMIRCAN OSMAN DR
PHYSICS DEPARTMENT
MIDDLE EAST & TECHN UNIV
ANKARA
TURKEY
TEL: 237100/3253 ANK
TLX: 42761 OOTK TR
COM: 42

DENOYELLE JOZEF KIC
KON STERRENWACHT V.BELGIE
RINGLAAN 3
B-1180 BRUSSEL
BELGIUM
TEL: 02-375-2484
TLX: 21565 OBSBEL
COM: 25,33

DEL RIO GERARDO DR
NATL ASTRONOMICAL OBS
ALFONSO XII-3
28014 MADRID
SPAIN
TEL: 91-2270107/1935
TLX: 23465 IGCE
COM:

DELLI SANTI SAVERIO
OSSERVATORIO ASTRONOMICO
UNIVERSITARIO
C P 596
I-40100 BOLOGNA
ITALY
TEL: 051-222956
TLX: 211664 INFNBO I
COM:

DENIS CARLO DR
INSTITUT D'ASTROPHYSIQUE
UNIVERSITE DE LIEGE
AVENUE DE COINTE 5
B-4200 COINTE-OUGREE
BELGIUM
TEL: 41-52-99-80
TLX: 41264
COM:

DENT WILLIAM A PROF
DEPT PHYSICS & ASTRONOMY
TOWER B
UNIV OF MASSACHUSETTS
AMHERST MA 01003
U.S.A.
TEL: 413-545-3665
TLX:
COM: 40

DELABOUDINIERE J.-P.
LPSP
BP 10
F-91370 VERRIERES-LE-B.
FRANCE
TEL:
TLX:
COM:

DELSEMME ARMAND H PROF DR
DEPT OF PHYS & ASTRONOMY
UNIVERSITY OF TOLEDO
2801 W BANCROFT STREET
TOLEDO OH 43606
U.S.A.
TEL: 419-537-2654
TLX:
COM: 14,15,20,51

DENISSE JEAN-FRANCOIS DR
48 RUE MR. LE PRINCE
F-75006 PARIS
FRANCE
TEL: 1-43-29-48-74
TLX:
COM: 40

DEPRIT ANDRE PROF
CTR FOR APPLIED MATHS
NATL BUREAU OF STANDARDS
GAITHERSBURG MD 20899
U.S.A.
TEL: 301-921-2631
TLX:
COM: 04,07C

DELACHE PHILIPPE J DR
OBSERVATOIRE DE NICE
BP 139
F-06003 NICE CEDEX
FRANCE
TEL: 93-89-04-20
TLX: 460004
COM: 12,36,49

DEMARCQ JEAN ING
OBSERVATOIRE DE NICE
BP 139
F-06003 NICE CEDEX
FRANCE
TEL: 93-55-89-65
TLX:
COM:

DENISYUK EDVARD K DR
ASTROPHYSICAL INSTITUTE
480068 ALMA ATA
U.S.S.R.
TEL:
TLX:
COM: 28

DERE KENNETH PAUL
NAVAL RESEARCH LABORATORY
CODE 4163
WASHINGTON DC 20375
U.S.A.
TEL: 202-767-2517
TLX:
COM: 10

DELANNOY JEAN DR
IRAM
DOMAINE UNIVERSITAIRE
VOIE 10
F-38406 ST MARTIN D'HERES
FRANCE
TEL: 76-42-33-83
TLX: 980753
COM: 40

DEMARET JACQUES DR
158/033 AV L'OBSERVATOIRE
B-4000 LIEGE
BELGIUM
TEL: 041-52-72-61
TLX: 41264 ASTROLIEGE
COM: 47

DENNIS BRIAN ROY DR
NASA-GSFC
CODE 682
GREENBELT MD 20771
U.S.A.
TEL: 301-344-6604
TLX: 89675
COM: 10,44,48

DERMAN I ETHEM DR
ANKARA UNIVERSITY
FACULTY OF SCIENCE
ANKARA
TURKEY
TEL: 41-236550/0109
TLX:
COM:

DELBOUILLE LUC PROF
INSTITUT D'ASTROPHYSIQUE
UNIVERSITE DE LIEGE
AVENUE DE COINTE 5
B-4200 COINTE-OUGREE
BELGIUM
TEL: 041 52 99 80
TLX: 41264 ASTRLG B
COM: 12

DEMARQUE P PROF
YALE UNIV OBSERVATORY
260 WHITNEY AVENUE
PO BOX 666
NEW HAVEN CT 06511
U.S.A.
TEL: 203-436-8246
TLX:
COM: 35,37

DENNISON BRIAN KENNETH
HULBERT CTR FOR SPACE RES
CODE 4130
NAVAL RESEARCH LABORATORY
WASHINGTON DC 20375-5000
U.S.A.
TEL:
TLX:
COM: 40

DERMENDJIEV VLADIMIR DR
DEPT OF ASTRONOMY
BULGARIAN ACAD SCIENCES
72 LENIN BLVD
1784 SOFIA
BULGARIA
TEL: 73-41-269
TLX: 23561 ECF BAN BG
COM:

DELCROIX ANDRE J S DR
19A RUE E VANDERVELDE
B-7230 FRAMERIES
BELGIUM
TEL:
TLX:
COM:

DEMERS SERGE DR
DEPARTEMENT DE PHYSIQUE
UNIVERSITE DE MONTREAL
CP 6128 SUCC A
MONTREAL PQ H3C 3J7
CANADA
TEL: 514-343-6718
TLX: 05562425
COM: 27,37

DENNISON EDWIN W DR
ELECTRONIC VISION CO
11526 SORRENTO
VALLEY ROAD
SAN DIEGO CA 92121
U.S.A.
TEL:
TLX:
COM:

DERMOTT STANLEY F
CORNELL UNIVERSITY
CTR RADIOPHYS & SPACE RES
SPACE SICENCES BLDG
ITHACA NY 14853
U.S.A.
TEL: 607-256-3727
TLX:
COM: 15,16

LIST OF MEMBERS

DESAI JYOTINDRA N
PHYSICAL RESEARCH LAB
ROOM No. 763
AHMEDABAD 380 009
INDIA
TEL: 462-129
TLX: 121397
COM: 09

DESESQUELLES JEAN DR
UNIVERSITE LYON I
CAMPUS LA DOUA
F-69622 VILLEURBANNE CDX
FRANCE
TEL: 78-89-81-24
TLX: 380273 IPN F
COM: 14

DESHPANDE M R DR
PHYSICAL RESEARCH LAB
NAVRANGPURA
AHMEDABAD 380 009
INDIA
TEL: 462129
TLX: 121397 PRL IN
COM: 25

DESPAIN KEITH HOWARD DR
LOS ALAMOS NATIONAL LAB
X-2, MS B220
PO BOX 1663
LOS ALAMOS NM 87545
U.S.A.
TEL: 505-667-2388
TLX:
COM: 35

DEUBNER FRANZ-LUDWIG DR
INST F ASTRONOMIE &
ASTROPHYSIK
AM HUBLAND
D-8700 WUERZBURG
GERMANY, F.R.
TEL: 0931-8885030
TLX: 68671 UNIWBG D
COM: 10,12

DEUPREE ROBERT G DR
ESS-5 MS F665
LOS ALAMOS NATIONAL LAB
PO BOX 1663
LOS ALAMOS NM 87545
U.S.A.
TEL: 505-667-8215
TLX:
COM: 27,35

DEUTSCH ALEKSANDR N PROF
PULKOVO OBSERVATORY
196140 LENINGRAD
U.S.S.R.
TEL:
TLX:
COM: 24,26

DEUTSCHMAN WILLIAM A DR
MATH. & PHYSICS DEPT
SEMON HALL
OREGON INST OF TECHNOLOGY
KLAMATH FALLS OR 97601
U.S.A.
TEL: 503-882-6321
TLX:
COM: 15

DEVINNEY EDWARD J DR
100 UNION AVE
DELANCO NJ 08075
U.S.A.
TEL: 609-764-1250
TLX:
COM:

DEWDNEY PETER E F DR
DOMINION RADIO ASTROPHYS
OBSERVATORY
P O BOX 248
PENTICTON BC V2A 6K3
CANADA
TEL: 604-497-5321
TLX: 048-88127
COM: 34,40

DEWHIRST DAVID W DR
INSTITUTE OF ASTRONOMY
THE OBSERVATORIES
MADINGLEY ROAD
CAMBRIDGE CB3 0HA
U.K.
TEL: 0233-62204
TLX: 817297 ASTRON G
COM: 05,41

DEWITT BRYCE S DR
DEPT OF PHYSICS
UNIVERSITY OF TEXAS
AUSTIN TX 78712
U.S.A.
TEL: 512-471-5055
TLX: 9108741305
COM: 48

DEWITT JOHN H JR
3602 HOODS HILL RD
NASHVILLE TN 37215
U.S.A.
TEL: 615-383-8272
TLX:
COM:

DEWITT-MORETTE CECILE PR
DEPT OF PHYSICS
UNIVERSITY OF TEXAS
9.220 R.L. MOORE HALL
AUSTIN TX 78712
U.S.A.
TEL: 512-471-1052
TLX: 910-8741305
COM:

DEZSO LORANT PROF
HELIOPHYSICAL OBSERVATORY
H-4010 DEBRECEN
HUNGARY
TEL: 52-11-015
TLX: 72517 DEOBS L
COM: 10,12

DI COCCO GUIDO
ISTITUTO TE.S.R.E.-C.N.R.
VIA DE CASTAGNOLI 1
I-40126 BOLOGNA
ITALY
TEL: 051-519593
TLX: 511350 CNR BO I
COM: 44

DI FAZIO ALBERTO
OSSERVATORIO ASTRONOMICO
DI ROMA
VIA DEL PARCO MELLINI 84
I-00136 ROMA
ITALY
TEL: 06-34-70-56
TLX: 613103 PPRMT I
COM: 28

DI MARTINO MARIO
OSSERVATORIO ASTRONOMICO
DI TORINO
STRADA OSSERVATORIO 20
I-10025 PINO TORINESE
ITALY
TEL: 011-841067
TLX: 213236 TO ASTR I
COM: 15

DI SEREGO ALIGHIERI S DR
SPACE TELESCOPE EUROPEAN
COORDINATING FACILITY
KARL-SCHWARZSCHILD-STR 2
D-8046 GARCHING B MUNCHEN
GERMANY, F.R.
TEL:
TLX:
COM: 28

DI TULLIO GRAZIELLA DR
OSSERVATORIO ASTRONOMICO
I-35100 PADOVA
ITALY
TEL:
TLX:
COM: 28

DI XIAO-HUA
PURPLE MOUNTAIN
OBSERVATORY
NANJING
CHINA, PEOPLE'S REP.
TEL: 37609
TLX: 34144 PMONJ CN
COM: 04

DIBAY E A DR
STERNBERG ASTRONOMICAL
INSTITUTE
117234 MOSCOW
U.S.S.R.
TEL:
TLX:
COM: 28,34

DICK STEVEN J
BLACK BIRCH ASTRON. OBS.
NSF ANTARCTICA
DETACHMENT CHRISTCHURCH
FPO S.FRANCISCO CA 96690
U.S.A.
TEL: 640-578-7164
TLX:
COM: 08,41,51

DICKE ROBERT H PROF
JOSEPH HENRY LABS
PHYSICS DEPT
PRINCETON UNIVERSITY
PRINCETON NJ 08540
U.S.A.
TEL: 609-452-4317
TLX:
COM: 47,48

DICKEL HELENE R DR
ASTRONOMY DEPT
UNIVERSITY OF ILLINOIS
1011 W. SPRINFIELD AVE
URBANA IL 61801-3000
U.S.A.
TEL: 217-333-5602
TLX: 910-245-2434 AST
COM: 05,33,34,40

DICKEL JOHN R
341 ASTRONOMY BLDG
UNIVERSITY OF ILLINOIS
1011 W. SPRINGFIELD AVE
URBANA IL 61801
U.S.A.
TEL: 217-333-5532
TLX: 910-245-2434 PURCH
COM: 16,33,40

DICKENS ROBERT J DR
RUTHERFORD APPLETON LAB
SPACE & ASTROPHYS DIV
CHILTON DIDCOT OX11 0QX
U.K.
TEL: 0235-21900
TLX: 83159
COM: 27,28,37

DICKEY JEAN O'BRIEN
JPL - CALTECH
MS 138-208
4800 OAK GROVE DRIVE
PASADENA CA 91109
U.S.A.
TEL: 213-354-8235
TLX: 675429
COM: 04,16,19,31

DICKEY JOHN M
UNIVERSITY OF MINNESOTA
DEPT OF ASTRONOMY
116 CHURCH STREET S.E.
MINNEAPOLIS MN 55455
U.S.A.
TEL: 612-373-3308
TLX:
COM: 28,34,40

DICKINSON DALE F DR
LOCKHEED RESEARCH LAB
92-20 205
3251 HANOVER ST
PALO ALTO CA 94304
U.S.A.
TEL: 415-424-2701
TLX:
COM:

DICKMAN STEVEN R
DEPT GEOLOGICAL SCIENCES
STATE UNIV OF NEW YORK
BINGHAMTON NY 13901
U.S.A.
TEL: 607-777-4378
TLX:
COM: 19

DIERCKSEN GEERD H F PH D
MPI F. PHYSIK & ASTROPHYS
KARL-SCHWARZSCHILD-STR 1
D-8046 GARCHING B MUNCHEN
GERMANY, F.R.
TEL: 89-32-990
TLX: 524629 ASTRO D
COM: 14

DIETER NANNIELOU H DR
CLAY ROAD
N. THERFORD VT 05054
U.S.A.
TEL: 802-333-4079
TLX:
COM: 33,40

DIONYSIOU DEMETRIOS D DR
DEPT OF ASTRONOMY
UNIVERSITY OF ATHENS
GR-15771 ATHENS
GREECE
TEL:
TLX:
COM: 47

DLUZHNEVSKAYA O B DR
ASTRONOMICAL COUNCIL
USSR ACADEMY OF SCIENCES
PYATNITSKAYA UL 48
109017 MOSCOW
U.S.S.R.
TEL: 231-54-61
TLX: 412623 SCSTP SU
COM: 05C,35,37

DOGAN NADIR PROF
ANKARA UNIVERSITY
FEN FAKULTESI
ANKARA
TURKEY
TEL:
TLX:
COM: 12

DIMITRIJEVIC MILAN
ASTRONOMSKA OPSERVATORIJA
VOLGINA 7
YU-11050 BEOGRAD
YUGOSLAVIA
TEL: 419357
TLX:
COM:

DIRIKIS M A DR
LATVIAN STATE UNIVERSITY
ASTRONOMICAL OBSERVATORY
226098 RIGA
U.S.S.R.
TEL:
TLX:
COM: 20

DOAZAN VERA DR
OBSERVATOIRE DE PARIS
61 AVE DE L'OBSERVATOIRE
F-75014 PARIS
FRANCE
TEL: 1-43-20-12-10
TLX: OBS PARIS 270776
COM: 29

DOHERTY LORNE H DR
HERZBERG INST ASTROPHYS
NATIONAL RESEARCH COUNCIL
OTTAWA ONT K1A 0R6
CANADA
TEL: 613-993-6060
TLX: 053-367-15
COM: 40

DIN HUA
DEPT OF ASTRONOMY
NANJING UNIVERSITY
NANJING
CHINA, PEOPLE'S REP.
TEL:
TLX: 34151 PRCNU CN
COM: 07

DISNEY MICHAEL J PROF
UNIV COLLEGE CARDIFF
DEPT APPLIED MATHS/ASTR
PO BOX 78
CARDIFF CF1 1XL
U.K.
TEL: 222-44211 x2692
TLX: 49365 ULIBCF
COM: 34,48

DOBRITSCHEV V M MR
DEPT OF ASTRONOMY
BULGARIAN ACAD SCIENCES
7TH NOVEMBER STR 1
1000 SOFIA
BULGARIA
TEL: 7341
TLX: 23561 ECFBAN BG
COM:

DOHERTY LOWELL R PROF
ASTRONOMY DEPT
UNIVERSITY OF WISCONSIN
475 N. CHARTER ST
MADISON WI 53706
U.S.A.
TEL: 608-262-1249
TLX:
COM:

DINERSTEIN HARRIET L
UNIV OF TEXAS AT AUSTIN
ASTRONOMY DEPT
RLM 15.308
AUSTIN TX 78712
U.S.A.
TEL: 512-471-3449
TLX: 910-874-1351
COM: 34

DIVAN LUCIENNE DR
INSTITUT D'ASTROPHYSIQUE
98 BIS BOULEVARD ARAGO
F-75014 PARIS
FRANCE
TEL: 1-43-20-14-25
TLX: INAG 270-0-70
COM: 29,45

DOBRONRAVIN PETER DR
CRIMEAN ASTROPHYSICAL
OBSERVATORY
NAUCHNYJ 2-2
334413 CRIMEA
U.S.S.R.
TEL:
TLX:
COM: 09,29

DOKUCHAEVA OLGA D DR
STERNBERG ASTRONOMICAL
INSTITUTE
UNIVERSITY PROSPECT 13
119899 MOSCOW
U.S.S.R.
TEL:
TLX:
COM: 09,34

DINESCU A DR
INSTITUT DE GEODESIE
PHOTOGRAMM CARTOGRAPHIE
1A BLVD DE L'EXPOSITION
78334 BUCAREST
RUMANIA
TEL:
TLX:
COM:

DIVARI N B DR
ODESSA POLYTECHNICAL INST
270004 ODESSA
U.S.S.R.
TEL:
TLX:
COM: 21

DOBROVOLSKY OLEG V PROF
INST OF ASTROPHYSICS
SVIRIDENKO ST 22
734670 DUSHANBE
U.S.S.R.
TEL:
TLX:
COM: 15C

DOLAN JOSEPH F DR
NASA/GSFC
CODE 681
GREENBELT MD 20771
U.S.A.
TEL: 301-344-5920
TLX: 89675
COM: 44,48

DING YOU-JI
YUNNAN OBSERVATORY
P O BOX 110
KUNMING, YUNNAN PROVINCE
CHINA, PEOPLE'S REP.
TEL: 22034
TLX: 64040 YUOBS CN
COM: 10

DIXON ROBERT S DR
OHIO STATE UNIVERSITY
RADIO OBSERVATORY
2015 NEIL AVE
COLUMBUS OH 43210
U.S.A.
TEL: 614-422-6789
TLX:
COM: 05,40,51

DOBRZYCKI JERZY PROF
HISTORY OF SCIENCE
POLISH ACAD OF SCIENCES
GWIAZDZISTA 27/169
01-814 WARSZAWA
POLAND
TEL: 33-22-03
TLX:
COM: 41

DOLGINOV ARKADY Z PROF DR
IOFFE PHYSICAL TECH INST
194021 LENINGRAD
U.S.S.R.
TEL:
TLX:
COM: 49C

DINGENS P PROF DR
KORTRIJKSE STEENWEG 763
B-9000 GENT
BELGIUM
TEL: 091-221966
TLX:
COM: 35

DIZER MUAMMER PROF
KANDILI OBSERVATORY
BUGAZICI UNIVERSITY
CENGELKOV
ISTANBUL
TURKEY
TEL: 3320277
TLX:
COM: 10

DOCOBO DURANTEZ JOSE A
OBSERVATORIO ASTRONOMICO
RAMON MARIA ALLER
P O BOX 197
SANTIAGO DE COMPOSTELA
SPAIN
TEL:
TLX:
COM: 26

DOLIDZE MADONA V DR
ABASTUMANI ASTROPHYSICAL
OBSERVATORY
383762 ABASTUMANI,GEORGIA
U.S.S.R.
TEL:
TLX:
COM: 29

DINULESCU NICOLAE I PROF
SOSEANA KISELEFF 13
SECTOR 1
72168 BUCAREST
RUMANIA
TEL:
TLX:
COM:

DJUROVIC DRAGUTIN M DR
DEPT OF ASTRONOMY
FACULTY OF SCIENCES
STUDENTSKI TRG 16
YU-11000 BEOGRAD
YUGOSLAVIA
TEL: 011-420-221
TLX:
COM: 08,19C

DODD RICHARD J DR
CARTER OBSERVATORY
P O BOX 2909
WELLINGTON 1
NEW ZEALAND
TEL: 728-167
TLX: 30172 NATOBS NZ
COM:

DOLLFUS AUDOUIN PROF
OBSERVATOIRE DE PARIS
SECTION DE MEUDON
F-92195 MEUDON PL CEDEX
FRANCE
TEL: 1-45-34-75-30
TLX:
COM: 10,16,20

DOMINKO FRAN PROF DR
SARANOVICEVA 11
YU-61000 LJUBLJANA
YUGOSLAVIA
TEL: 061-322-210
TLX:
COM: 46

DOMINSKI IRENEUSZ DR
ASTRONOMICAL LATITUDE OBS
BOROWIEC
62-035 KORNIK
POLAND
TEL:
TLX:
COM: 31

DOMKE HELMUT PH D
ZNTRLINST. F. ASTROPHYSIK
ROSA-LUXEMBURG-STR 17A
DDR-1502 POTSDAM
GERMANY, D.R.
TEL:
TLX:
COM: 36

DOMMANGET J DR
OBS ROYAL DE BELGIQUE
AVENUE CIRCULAIRE 3
B-1180 BRUXELLES
BELGIUM
TEL: 2-375-24-84
TLX: 21565
COM: 24,26C,50

DONG IL ZUN
PYONGYANG ASTRON OBS
ACADEMY OF SCIENCES DPRK
TAESONG DISTRICT
PYONGYANG
KOREA DPR
TEL:
TLX:
COM:

DONN BERTRAM D
NASA/GSFC
CODE 691
GREENBELT MD 20771
U.S.A.
TEL: 301-344-6859
TLX: 89675
COM: 15,34

DONNER KARL JOHAN DR
NORDITA
BLEGDAMSVEJ 17
DK-2100 COPENHAGEN
DENMARK
TEL:
TLX:
COM: 28

DOPITA MICHAEL ANDREW DR
MOUNT STROMLO OBSERVATORY
PRIVATE BAG
WODEN P.O. ACT 2606
AUSTRALIA
TEL: 062-88-1111
TLX: 62270 CANOPUS AA
COM: 34

DORENWENDT KLAUS DR
PHYSIKALISCH-TECHNISCHES
BUNDESANSTALT
BUNDESALLEE 100
D-3300 BRAUNSCHWEIG
GERMANY, F.R.
TEL: 531-592-12-10
TLX: 9-52-822 PTB D
COM: 31

DORMAND JOHN RICHARD DR
MATHEMATICS DEPARTMENT
TEESSIDE POLYTECHNIC
MIDDLESBROUGH
CLEVELAND TS1 3BA
U.K.
TEL: 642-218121x4365
TLX:
COM: 07

DOROSHKEVICH A G DR
INST OF APPLIED MATHS
USSR ACADEMY OF SCIENCES
125047 MOSCOW
U.S.S.R.
TEL: 972-37-14
TLX:
COM: 47

DORSCHNER JOHANN DR
UNIV STERNWARTE JENA
SCHILLERGAESSCHEN 2
DDR-6900 JENA
GERMANY, D.R.
TEL: 8222637
TLX: 5886134
COM: 34,51

DOS REIS M PROF
OBSERVATORIO ASTRONOMICO
3000 COIMBRA
PORTUGAL
TEL:
TLX:
COM:

DOSCHEK GEORGE A DR
NAVAL RESEARCH LABORATORY
CODE 4170
WASHINGTON DC 20375
U.S.A.
TEL: 202-767-6473
TLX:
COM:

DOSSIN F DR
INSTITUT D'ASTROPHYSIAUE
UNIVERSITE DE LIEGE
AVENUE DE COINTE 5
B-4200 COINTE-OUGREE
BELGIUM
TEL: 041-52-99-80
TLX: 41264 ASTRLG
COM: 15

DOTTORI HORACIO A DR
INSTITUTO DE FISICA
UNIV RIO GRANDE DO SUL
AV. BENTO GONCALVES 9500
90049 PORTO ALEGRE - RS
BRAZIL
TEL: 0055-512-364677
TLX: 511055 UFRS BR
COM: 28,34

DOUBINSKIJ B A DR
INST OF RADIO & ELECTRON
USSR ACADEMY OF SCIENCES
103907 MOSCOW
U.S.S.R.
TEL:
TLX:
COM: 40

DOUGHTY NOEL A DR
PHYSICS DEPARTMENT
UNIVERSITY OF CANTERBURY
CHRISTCHURCH 1
NEW ZEALAND
TEL:
TLX:
COM: 42,46

DOUGLAS A VIBERT DR
402 67 SYDENHAM STR.
KINGSTON ONT K7L 3H2
CANADA
TEL: 613-542-7007
TLX:
COM: 15,41

DOUGLAS JAMES N PROF
DEPARTMENT OF ASTRONOMY
UNIVERSITY OF TEXAS
R.L. MOORE HALL
AUSTIN TX 78712-1083
U.S.A.
TEL: 512-471-4461
TLX: 910874-1351
COM: 40

DOUGLASS GEOFFREY G
US NAVAL OBSERVATORY
34TH & MASSACHUSETTS AVE
WASHINGTON DC 20390
U.S.A.
TEL: 202-653-1457
TLX:
COM: 24

DOWNES ANN JULIET B
MULLARD RADIO AST OBS
CAVENDISH LABORATORY
MADINGLEY ROAD
CAMBRIDGE CB3 0HE
U.K.
TEL: 0233-66477
TLX: 81292
COM:

DOWNES DENNIS DR
IRAM
VOIE 10
DOMAINE UNIVERSITAIRE
F-38406 ST-MARTIN-D'HERES
FRANCE
TEL: 76-42-33-83
TLX: 980753
COM: 33,34,40

DOWNS GEORGE S DR
MIT LINCOLN LABORATORY
ROOM B285
PO BOX 73
LEXINGTON MA 02173
U.S.A.
TEL:
TLX:
COM: 40,51

DOYLE JOHN GERARD
ARMAGH OBSERVATORY
ARMAGH BT61 9DG
U.K.
TEL: 861-522-928
TLX: 747937
COM: STELLAR ATMOSPHERES

DRAINE BRUCE T
PRINCETON UNIVERSITY OBS
PRINCETON UNIVERSITY
PEYTON HALL
PRINCETON NJ 08544
U.S.A.
TEL: 609-452-3574
TLX:
COM: 34

DRAKE FRANK D PROF
DIV OF NATURAL SCIENCES
BOARD OF STUDIES ASTR &
ASTROPHYS UNIV CALIFORNIA
SANTA CRUZ CA 95064
U.S.A.
TEL: 408-429-2931
TLX:
COM: 16,40,48,51P

DRAKE STEPHEN A
GODDARD SPACE FLIGHT CTR
CODE 602.6, UVSP
GREENBELT MD 20771
U.S.A.
TEL: 301-344-7985
TLX:
COM: 36,40

DRAMBA C PROF
ASTRONOMICAL OBSERVATORY
P O BOX 28
CUTITUL DE ARGINT 5
75212 BUCAREST
RUMANIA
TEL: 753998: 193407
TLX:
COM: 19,31

DRAPATZ SIEGFRIED W DR
MPI F EXTRATERRESTRISCHE
PHYSIK
KARL-SCHWARZSCHILD-STR 1
D-8046 GARCHING B MUNCHEN
GERMANY, F.R.
TEL: 089-3299880
TLX: 05215845
COM: 34

DRAVINS DAINIS PROF
LUND OBSERVATORY
BOX 43
S-221 00 LUND
SWEDEN
TEL: 46-10 70 00
TLX: 33199 OBSNOT S
COM: 09,12,29,36

DRAVSKIKH A F DR
SPECIAL ASTROPHYSICAL OBS
USSR ACADEMY OF SCIENCES
LENINGRAD BRANCH
196140 LENINGRAD
U.S.S.R.
TEL: 297-94-52
TLX:
COM: 08,40

LIST OF MEMBERS

DRECHSEL HORST DR
DR-REMEIS-STERNWARTE
ASTR INST UNIV ERLANGEN-N
STERNWARTSTR 7
D-8600 BAMBERG
GERMANY, F.R.
TEL: 0951-57708
TLX: 629830 UNIER D
COM: 42

DREHER JOHN W
M I T
DEPT OF PHYSICS
ROOM 26-315
CAMBRIDGE MA 02139
U.S.A.
TEL: 617-253-8519
TLX: 921473
COM: 09,34,40

DRESSEL LINDA L
DEPT SPACE PHYS & ASTRON
RICE UNIVERSITY
PO BOX 1892
HOUSTON TX 77251
U.S.A.
TEL: 713-527-8101
TLX:
COM: 28

DRESSLER ALAN
MT WILSON & LAS CAMPANAS
OBSERVATORIES
813 SANTA BARBARA STREET
PASADENA CA 91101-1292
U.S.A.
TEL: 818-304-0245
TLX: 675425 CALTECH PSD
COM: 28

DRESSLER KURT PROF
LAB PHYSIK CHEMIE
ETH-ZENTRUM
CH-8092 ZUERICH
SWITZERLAND
TEL: 256-4441
TLX: 53178 ETHBI CH
COM: 14

DREVER RONALD W P DR
DEPT NATURAL PHILOSOPHY
GLASGOW UNIVERSITY
GLASGOW G12 8QQ
U.K.
TEL:
TLX:
COM:

DREW JANET
DEPT OF ASTROPHYSICS
SOUTH PARKS ROAD
OXFORD OX1 3RQ
U.K.
TEL: 0865-511336
TLX: 851-83295
COM: 28

DRILLING JOHN S
DEPT PHYSICS & ASTRONOMY
LOUISIANA STATE UNIV
BATON ROUGE LA 70803
U.S.A.
TEL: 504-388-6795
TLX:
COM: 33

DROFA VASILIY K DR
DEPARTMENT OF ASTRONOMY
KIEV UNIVERSITY
252127 KIEV
U.S.S.R.
TEL:
TLX:
COM:

DROZYNER ANDRZEJ
INSTITUTE OF ASTRONOMY
N. COPERNICUS UNIVERSITY
UL. CHOPINA 12/18
87-100 TORUN
POLAND
TEL: 26017 x 53
TLX: 0552234 ASTR PL
COM: 07

DRYER MURRAY DR
SPACE ENVIRONMENT LAB
NOAA ERL (R/E/SE)
325 BROADWAY
BOULDER CO 80303
U.S.A.
TEL: 303-497-3978
TLX: 592811 NOAA MASC BDR
COM: 10,15,49

DUBAU JACQUES DR
OBSERVATOIRE DE PARIS
SECTION DE MEUDON
F-92195 MEUDON PL CEDEX
FRANCE
TEL: 1-45-34-75-30
TLX: 270912 OBSASTR
COM: 14

DUBNER GLORIA DR
INSTITUTO ARGENTINO DE
RADIOASTRONOMIA
CASILLA DE CORREO No. 5
1894 VILLA ELISA (Bs.As.)
ARGENTINA
TEL: 21-4-3793
TLX: 22414 CEDOC AR
COM:

DUBOIS MARC A
CEA
D R F C
BP 6
F-92260 FONTENAY-AUX-ROS.
FRANCE
TEL: 1-46-54-78-81
TLX:
COM: 10

DUBOIS PASCAL DR
OBS DE STRASBOURG
11 RUE DE L'UNIVERSITE
F-67000 STRASBOURG
FRANCE
TEL: 88-35-43-00
TLX:
COM: 28

DUBOSHIN G N PROF DR
STERNBERG STATE
ASTRONOMICAL INSTITUTE
UNIVERSITETSKIJ PROSP. 13
119899 MOSCOW
U.S.S.R.
TEL: 139-23-82 HOME
TLX:
COM: 07

DUBOUT RENEE
OBSERVATOIRE DE LYON
AVENUE CHARLES ANDRE
F-69230 ST-GENIS-LAVAL
FRANCE
TEL: 78-56-07-05
TLX: 310926
COM: 25,28,34

DUBOV EMIL E PROF
WDC-B2
MOLODEZHNAYA 3
117296 MOSCOW
U.S.S.R.
TEL: 923-55-71
TLX: 411478 SGC SU
COM: 10,12

DUCATI JORGE RICARDO DR
INSTITUTO DE FISICA
UFRGS
AV. BENTO GONCALVES 9500
90000 PORTO ALEGRE - RS
BRAZIL
TEL: 512-364677
TLX: 051-1055 UFRS BR
COM: 05,25,33

DUCHESNE MAURICE DR
OBSERVATOIRE DE PARIS
61 AVE DE L'OBSERVATOIRE
F-75014 PARIS
FRANCE
TEL: 1-43-20-12-10
TLX:
COM: 09

DUERBECK HILMAR W DR
ASTRONOMISCHES INSTITUT
UNIVERSITAT MUENSTER
DOMAGKSTR. 75
D-4400 MUENSTER
GERMANY, F.R.
TEL: 251-833561
TLX:
COM: 42

DUERST JOHANNES DR
LANGWIES
CH-8821 SCHONENBERG
SWITZERLAND
TEL: 01-788-1785
TLX:
COM:

DUFAY MAURICE PROF
UNIVERSITE CLAUDE-BERNARD
LYON I
43 BD DU 11 NOVEMBRE
F-69621 VILLEURBANNE
FRANCE
TEL:
TLX:
COM: 14,21

DUFFETT-SMITH PETER JAMES
MULLARD RADIO ASTRON OBS
CAVENDISH LABORATORY
MADINGLEY ROAD
CAMBRIDGE CB3 OHE
U.K.
TEL: 0223-66477
TLX: 81292 CAVLAB G
COM: 40

DUFLOT MARCELLE DR
OBSERVATOIRE DE MARSEILLE
2 PLACE LE VERRIER
F-13248 MARSEILLE CEDEX 4
FRANCE
TEL: 91-95-90-88
TLX: 420241
COM: 30,45

DUFOUR REGINALD JAMES
SPACE PHYSICS DEPARTMENT
RICE UNIVERSITY
204K SPACE SCIENCE BLDG.
HOUSTON TX 77001
U.S.A.
TEL: 713-527-8101
TLX:
COM: 34

DUFTON PHILIP L DR
DEPT PURE & APPLIED PHYS
QUEEN'S UNIVERSITY
BELFAST BT7 1NN
U.K.
TEL: 245133
TLX: 74487
COM: 36

DULEY WALTER W PROF
PHYSICS DEPARTMENT
YORK UNIVERSITY
4700 KEELE STREET
DOWNSVIEW ONT M3J1P3
CANADA
TEL: 416-667-3040
TLX:
COM:

DULK GEORGE A PROF
DEPT ASTROPHYS SCIENCES
UNIVERSITY OF COLORADO
CB 391
BOULDER CO 80309
U.S.A.
TEL: 303-492-8788
TLX:
COM: 10,40

DULTZIN DEBORAH DR
INSTITUTO DE ASTRONOMIA
UNAM
APDO POSTAL 70-264
04510 MEXICO DF
MEXICO
TEL: 548-53-05/6
TLX: 1760155 CICME
COM: 28,47

DUMA DMITRIJ P DR
MAIN ASTRONOMICAL OBS
UKRAINIAN ACADEMY OF SCI
252127 KIEV
U.S.S.R.
TEL: 66-31-10
TLX: 131406 SKY SU
COM: 08C

DUMONT RENE DR
OBSERVATOIRE DE BORDEAUX
B.P. 21
F-33270 FLOIRAC
FRANCE
TEL: 56-86-43-30
TLX:
COM: 21C

DUMONT SIMONE DR
INSTITUT D'ASTROPHYSIQUE
98 BIS BOULEVARD ARAGO
F-75014 PARIS
FRANCE
TEL: 1-43-20-14-25
TLX:
COM: 12,36

DUPUY DAVID L DR
DEPARTMENT OF PHYSICS
VIRGINIA MILITARY INST
LEXINGTON VA 24450
U.S.A.
TEL: 703-463-6225
TLX:
COM: 27,46

DUVALL THOMAS L JR
NATL SOLAR OBSERVATORY
PO BOX 26732
950 N. CHERRY AVE
TUCSON AZ 85726
U.S.A.
TEL: 602-325-9338
TLX: 666-484 AURA-KPNO-
COM: 12

DYSON F J DR
INST FOR ADVANCED STUDY
PRINCETON NJ 08540
U.S.A.
TEL: 609-734-8055
TLX:
COM: 40,51

DUNCAN DOUGLAS KEVIN DR
SPACE TELESCOPE SC INST
HOMEWOOD CAMPUS
3700 SAN MARTIN DR
BALTIMORE MD 21218
U.S.A.
TEL: 301-338-4935
TLX:
COM: 29

DURGAPRASAD N DR
TATA INSTITUTE OF
FUNDAMENTAL RESEARCH
BOMBAY 400 005
INDIA
TEL: 219111 x 342
TLX: 0113009
COM:

DVORAK RUDOLF DR
INSTITUT FUER ASTRONOMIE
UNIVERSITAETSSTERNWARTE
TUERKENSCHANZSTRASSE 17
A-1180 VIENNA
AUSTRIA
TEL: 222-34-53-600
TLX:
COM: 07,20

DYSON JOHN E DR
ASTRONOMY DEPARTMENT
UNIVERSITY OF MANCHESTER
MANCHESTER M13 9PL
U.K.
TEL: 061-273-7121
TLX: 668932
COM: 34,49

DUNCAN ROBERT A PROF
CSIRO
DIVISION OF RADIOPHYSICS
P.O.BOX 76
EPPING NSW 2121
AUSTRALIA
TEL:
TLX:
COM: 10

DURISEN RICHARD H DR
DEPARTMENT OF ASTRONOMY
INDIANA UNIVERSITY
SWAIN WEST 319
BLOOMINGTON IN 47405
U.S.A.
TEL: 812-335-6921
TLX:
COM: 35

DWEK ELI
NASA/GSFC CODE 697
LAB EXTRATERRESTR.PHYSICS
GREENBELT MD 20771
U.S.A.
TEL: 301-344-6209
TLX:
COM: 34

DZHAPIASHVILI VICTOR P DR
ABASTUMANI ASTROPHYSICAL
OBSERVATORY
383762 ABASTUMANI,GEORGIA
U.S.S.R.
TEL:
TLX:
COM: 16

DUNCOMBE RAYNOR L DR
DEPT OF AEROSPACE ENG
UNIVERSITY OF TEXAS
AUSTIN TX 78712
U.S.A.
TEL: 512-471-4239
TLX: 704265 CSRUTX UD
COM: 04C,05,07,08

DURNEY BERNARD DR
NATL SOLAR OBSERVATORY
SACRAMENTO PEAK, NOAO
SUNSPOT NM 88349
U.S.A.
TEL: 505-434-1390
TLX:
COM: 49

DWIVEDI BHOLA NATH DR
DEPT APPLIED PHYSICS,I.T.
BANARAS HINDU UNIVERSITY
VARANASI 221 005
INDIA
TEL:
TLX: 0545-208 TECH IN
COM: 10

DZIEMBOWSKI WOJCIECH PROF
ASTRONOMICAL CENTER
UL. BARTYCKA 18
00-716 WARSAW
POLAND
TEL:
TLX:
COM: 27,35

DUNHAM DAVID W
SYSTEM SCIENCES DIVISION
COMPUTER SCIENCES CORP
8728 COLESVILLE ROAD
SILVER SPRING MD 20910
U.S.A.
TEL: 301-589-1545
TLX: 7108259636 CSCSSD
COM: 04,20,26

DUROUCHOUX PHILIPPE
CEA CEN/SACLAY
DPHG/SAP
F-91191 GIF/YVETTE CEDEX
FRANCE
TEL: 69-08-33-76
TLX: 690860 PHYSPAC F
COM:

DWORETSKY MICHAEL M DR
DEPT PHYSICS & ASTRONOMY
UNIVERSITY COLLEGE LONDON
GOWER STREET
LONDON WC1E 6BT
U.K.
TEL: 01-387-7050
TLX: 28722
COM: 29

DZIGVASHVILI R M DR
ABATSUMANI ASTROPHYSICAL
OBSERVATORY
383762 ABASTUMANI
U.S.S.R.
TEL:
TLX:
COM: 33

DUNKELMAN LAWRENCE
LUNAR & PLANETARY LAB
UNIVERSITY OF ARIZONA
PO BOX 36241
TUCSON AZ 85740
U.S.A.
TEL: 602-621-6963
TLX:
COM: 09,12,21,44,50

DURRANT CHRISTOPHER J DR
DEPT APPLIED MATHEMATICS
UNIVERSITY OF SYDNEY
SYDNEY NSW 2006
AUSTRALIA
TEL: 02-692-3373
TLX: 20056 FISHLIB AA
COM:

DYCK M DR
KITT PEAK NATIONAL OBS
PO BOX 26732
TUCSON AZ 85726
U.S.A.
TEL:
TLX:
COM:

DeVORKIN DAVID H
NATL AIR & SPACE MUSEUM
SMITHSONIAN INSTITUTION
WASHINGTON DC 20560
U.S.A.
TEL: 202-357-2828
TLX:
COM: 41

DUNN RICHARD B DR
NATL SOLAR OBSERVATORY
SUNSPOT NM 88349
U.S.A.
TEL: 505-434-1390
TLX:
COM: 10,12

DUTHIE JOSEPH G PROF
UNIVERSITY OF ROCHESTER
DEPT PHYSICS & ASTRONOMY
ROCHESTER NY 14627
U.S.A.
TEL:
TLX:
COM: 48

DYER CHARLES CHESTER DR
PHYS SCS GR RM S-650
SCARBOROUGH COLLEGE
UNIVERSITY OF TORONTO
TORONTO ONT M1C 1A4
CANADA
TEL: 416-284-3318
TLX:
COM: 47

EATON JOEL A DR
INDIANA UNIVERSITY
ASTRONOMY DEPT
SWAIN HALL WEST 319
BLOOMINGTON IN 47405
U.S.A.
TEL: 812-335-4176
TLX:
COM: 42

DUPREE ANDREA K DR
SOLAR & STELLAR DIVISION
CENTER FOR ASTROPHYSICS
60 GARDEN STREET
CAMBRIDGE MA 02138
U.S.A.
TEL: 617-495-7489
TLX: 921428 SATELLITE CAM
COM: 34,36,44

DUVAL MARIE-FRANCE
OBSERVATOIRE DE MARSEILLE
2 PLACE LE VERRIER
F-13248 MARSEILLE CEDEX 4
FRANCE
TEL: 91-95-90-88
TLX:
COM: 28

DYER EDWARD R DR
3226 DAVIS STREET N.W.
WASHINGTON DC 20007
U.S.A.
TEL:
TLX:
COM:

ECCLES MICHAEL J DR
SUNNYSIDE
BALLENCRIEFF TOLL
BATHGATE EH48 4LD
U.K.
TEL: 0506-53989
TLX: 727484
COM: 51

ECHEVERRIA ROMAN JUAN M.
APDO POSTAL 877
22860 ENSENADA, B. CALIF.
MEXICO
TEL:
TLX:
COM:

EDWARDS TERRY W
DEPT PHYSICS & ASTRONOMY
UNIVERSITY OF MISSOURI
COLUMBIA MO 65211
U.S.A.
TEL: 314-882-3036
TLX:
COM: 35

EHLERS JURGEN PROF
MPI FUER PHYSIK UND
ASTROPHYSIK
KARL-SCHWARZSCHILD-STR 1
D-8046 GARCHING B MUNCHEN
GERMANY, F.R.
TEL: 089-3299-9444
TLX: 524629 ASTRO D
COM: 47

EL EID MOUNIB DR
UNIVERSITAETSSTERNWARTE
GEISMARLANDSTR 11
D-3400 GOETTINGEN
GERMANY, F.R.
TEL:
TLX:
COM:

EDDY JOHN A DR
UNIV. CORP FOR ATMOS. RES
PO BOX 3000
BOULDER CO 80307
U.S.A.
TEL: 303-497-1150
TLX: 45694
COM: 10,41P

EDWIN ROGER P
UNIVERSITY OBSERVATORY
BUCHANAN GARDENS
ST ANDREWS, FIFE KY16 9LZ
U.K.
TEL:
TLX:
COM: 09

EICHHORN HEINRICH K DR
DEPARTMENT OF ASTRONOMY
UNIVERSITY OF FLORIDA
231 SPACE SC RES BLDG
GAINESVILLE FL 32611
U.S.A.
TEL: 904-392-2052
TLX:
COM: 07,08,24,26

EL SHALABY MOHAMED
ASTRONOMY & METEOROL DEPT
CAIRO UNIVERSITY
CAIRO
EGYPT
TEL:
TLX:
COM: 34

EDLEN BENGT PROF
DEPARTMENT OF PHYSICS
UNIVERSITY OF LUND
SOELVEGATAN 14
S-223 62 LUND
SWEDEN
TEL: 046-107730
TLX:
COM: 14

EELSALU HEINO DR
TARTU OBSERVATORY
202444 TORAVERE, ESTONIA
U.S.S.R.
TEL: 41469 TARTU
TLX:
COM: 25,30,41C

EICHLER DAVID DR
ASTRONOMY PROGRAM
UNIVERSITY OF MARYLAND
COLLEGE PK MD 20742
U.S.A.
TEL: 301-454-6448
TLX: 710-8260352
COM: 48

EL-BASSUNY ALAWY A A
HIAG
HELWAN-CAIRO
EGYPT
TEL: 782683
TLX: 9703 HIAG UN
COM: 27,28,37

EDMONDS FRANK N JR DR
DEPARTMENT OF ASTRONOMY
UNIVERSITY OF TEXAS
RLM 15.212
AUSTIN TX 78712
U.S.A.
TEL: 512-471-4461
TLX:
COM: 12,29,36

EFREMOV YU I DR
INST FOR APPLIED MATHS
USSR ACADEMY OF SCIENCES
125047 MOSCOW
U.S.S.R.
TEL:
TLX:
COM:

EILEK JEAN
PHYSICS DEPARTMENT
NEW MEXICO TECH
SOCORRO NM 87801
U.S.A.
TEL: 505-835-5433
TLX:
COM: 48

EL-BAZ FAROUK DR
ITEK OPTICAL SYSTEMS
10 MAGUIRE ROAD
LEXINGTON MA 02173
U.S.A.
TEL: 617-276-2532
TLX: 923456
COM: 16

EDMONDSON FRANK K PROF
GOETHE LINK OBSERVATORY
INDIANA UNIVERSITY
319 A SWAIN HALL WEST
BLOOMINGTON IN 47405
U.S.A.
TEL: 812-335-6918
TLX:
COM: 20,30,33,41

EFREMOV YURY N DR
STERNBERG ASTRON INST
UNIVERSITETSKY PROSP. 13
119899 MOSCOW
U.S.S.R.
TEL: 139 26 57
TLX:
COM: 27,33,37

EINASTO JAAN DR
TARTU ASTROPHYSICAL OBS
ESTONIAN ACAD OF SCIENCES
202444 TORAVERE, ESTONIA
U.S.S.R.
TEL:
TLX:
COM: 28C,33C,37,47

EL-RAEY MOHAMED E DR
DEPT ENVIRONMENT. STUDIES
INST GRADUATE STUD.& RES.
UNIVERSITY OF ALEXANDRIA
ALEXANDRIA
EGYPT
TEL:
TLX:
COM: 44

EDMUNDS MICHAEL GEOFFREY
DEPT APPLIED MATH & AST
UNIVERSITY COLLEGE
PO BOX 78
CARDIFF CF1 1XL
U.K.
TEL: 0222-44211
TLX: 498635 ULIBCF
COM: 28

EFSTATHIOU GEORGE
INSTITUTE OF ASTRONOMY
MADINGLEY ROAD
CAMBRIDGE CB3 0HA
U.K.
TEL: 0223-62204
TLX: 817297 ASTRON G
COM: 28

EINAUDI GIORGIO
SCUOLA NORMALE SUPERIORE
DI PISA
PIAZZA DEI CAVALIERI
I-56100 PISA
ITALY
TEL: 050-597-325
TLX: 590548 SNSPI
COM: 12

EL-SHAARAWY M B DR
HELWAN OBSERVATORY
HELWAN-CAIRO
EGYPT
TEL:
TLX:
COM:

EDWARDS ALAN CH DR
DEPT OF ASTROPHYSICS
SOUTH PARKS ROAD
OXFORD OX1 3RQ
U.K.
TEL:
TLX:
COM: 35

EGGLETON PETER P DR
INSTITUTE OF ASTRONOMY
MADINGLEY ROAD
CAMBRIDGE CB3 0HA
U.K.
TEL: 223-62204
TLX: 817297 ASTRON G
COM: 35,42

EINICKE O H LEKTOR
UNIVERSITY OBSERVATORY
OESTER VOLDGADE 3
DK-1350 COPENHAGEN K
DENMARK
TEL: 1-14-17-90
TLX:
COM:

ELFORD WILLIAM GRAHAM DR
DEPARTMENT OF PHYSICS
UNIVERSITY OF ADELAIDE
GPO BOX 498
ADELAIDE 5001
AUSTRALIA
TEL: 02-228-5321
TLX: 89141 UNIVQD AA
COM: 22C

EDWARDS PAUL J DR
MT STROMLO OBSERVATORY
PRIVATE BAG
WODEN P.O. ACT 2606
AUSTRALIA
TEL: 062-88-1111
TLX: 68270 AA
COM: 25C,27,48,50

EGRET DANIEL DR
OBS DE STRASBOURG
11 RUE DE L'UNIVERSITE
F-67000 STRASBOURG
FRANCE
TEL: 88-35-43-00
TLX: 890506 STAROBS
COM: 05,33,45

EKERS RONALD D DR
NRAO
PO BOX 0
SOCORRO NM 87801
U.S.A.
TEL: 505-772-4297
TLX: 910-9881710
COM: 28,40

ELGAROY OYSTEIN PROF
INST THEORET ASTROPHYSICS
UNIVERSITY OF OSLO
P O BOX 1029
N-0315 BLINDERN, OSLO 3
NORWAY
TEL: 02-456-504
TLX:
COM: 40,46

ELIPE SANCHEZ ANTONIO
DPTO FIS TIERRA & COSMOS
UNIVERSIDAD DE ZARAGOZA
50009 ZARAGOZA
SPAIN
TEL: 976-357011
TLX: 58198
COM: 07

ELLIS RICHARD S
PHYSICS DEPARTMENT
DURHAM UNIVERSITY
SOUTH ROAD
DURHAM DH1 3LE
U.K.
TEL:
TLX:
COM:

ELVIS MARTIN S DR
HARVARD SMITHSONIAN CTR
FOR ASTROPHYSICS
60 GARDEN STREET
CAMBRIDGE MA 02138
U.S.A.
TEL: 617-495-7442
TLX: 921428
COM:

EMERSON JAMES P
DEPT OF PHYSICS
QUEEN MARY COLLEGE
MILE END ROAD
LONDON E1 4NS
U.K.
TEL: 01-980-4811
TLX: 893750 QMCUOL G
COM: 34

ELITZUR MOSHE
DEPT PHYSICS & ASTRONOMY
UNIVERSITY OF KENTUCKY
LEXINGTON KY 40506-0055
U.S.A.
TEL: 606-257-4720
TLX:
COM: 34

ELMEGREEN BRUCE GORDON DR
IBM
THOMAS J. WATSON RES CTR
PO BOX 218
YORKTOWN HEIGHTS NY 10598
U.S.A.
TEL: 914-945-2448
TLX: 137456
COM: 34C,37

ELVIUS AINA M PROF
STOCKHOLM OBSERVATORY
S-133 00 SALTSJOEBADEN
SWEDEN
TEL: 08-7170195
TLX: 12972 SOBBSERV S
COM: 28,34

EMINZADE T A DR
SHEMAKHA ASTROPHYSICAL
OBSERVATORY
373243 AZERBAIDZAN
U.S.S.R.
TEL:
TLX:
COM: 35

ELLDER JOEL DR
ONSALA SPACE OBSERVATORY
S-430 34 ONSALA
SWEDEN
TEL:
TLX:
COM:

ELMEGREEN DEBRA MELOY
IBM
THOMAS J. WATSON RES CTR
PO BOX 218
YORKTOWN HEIGHTS NY 10598
U.S.A.
TEL: 914-945-2448
TLX: 137456
COM: 28,33,34

ELVIUS TORD PROF EMERITUS
NORRLANDSGATAN 34F
S-752 29 UPPSALA
SWEDEN
TEL: 018-100857
TLX:
COM: 33,45

EMSLIE A. GORDON
UNIVERSITY OF ALABAMA
DEPT OF PHYSICS
HUNTSVILLE AL 35899
U.S.A.
TEL: 205-895-6167
TLX:
COM: 10

ELLIOT JAMES L DR
DEPT EARTH & PLANET SCI
MIT
BLDG 54-422A
CAMBRIDGE MA 02139
U.S.A.
TEL: 617-253-6308
TLX: 921473 MIT CAM
COM: 16,20

ELSAESSER HANS PROF
MPI FUR ASTRONOMIE
KOENIGSTUHL
D-6900 HEIDELBERG
GERMANY, F.R.
TEL: 62-21-528-200
TLX:
COM: 21,33

ELWERT GERHARD PROF
LEHRSTUHL F THEORETISCHE
ASTROPHYSIK
UNIVERSITAET TUEBINGEN
D-7400 TUEBINGEN
GERMANY, F.R.
TEL: 07071/296483
TLX: 7-262714 ALT D
COM: 10,40

ENCRENAZ PIERRE J DR
ECOLE NORMALE SUPERIEURE
24 RUE LHOMOND
F-75005 PARIS
FRANCE
TEL: 1-43-29-12-35
TLX:
COM: 34

ELLIOTT IAN DR
DUNSINK OBSERVATORY
DUBLIN 15
IRELAND
TEL: 1-387-959
TLX: 31687 DIAS EI
COM: 12

ELSMORE BRUCE DR
CAVENDISH LABORATORY
MADINGLEY ROAD
CAMBRIDGE CB3 0HE
U.K.
TEL: 0223-66477
TLX: 81292
COM: 19,24,40

ELYASBERG P E PROF DR
SPACE RESEARCH INSTITUTE
USSR ACADEMY OF SCIENCES
PROFSOYUZNAYA ST. 84/32
117810 MOSCOW
U.S.S.R.
TEL: 333-31-22
TLX: 411498 IKI STAR SU
COM: 07

ENCRENAZ THERESE DR
OBSERVATOIRE DE PARIS
SECTION DE MEUDON
GROUPE PLANETES
F-92195 MEUDON PL CEDEX
FRANCE
TEL: 1-45-34-75-30
TLX: 204464
COM: 16C

ELLIOTT KENNETH H DR
DEPT OF SPACE RESEARCH
UNIVERSITY OF BIRMINGHAM
PO BOX 363
BIRMINGHAM B15 2TT
U.K.
TEL: 021-472-1301
TLX: 338938
COM: 34

ELST ERIC WALTER DR
KONINGLIJKE STERRENWACHT
VAN BELGIE
RINGLAAN 3
B-1180 BRUSSEL
BELGIUM
TEL:
TLX:
COM:

EMELIANOV NIKOLAJ V DR
STERNBERG STATE
ASTRONOMICAL INSTITUTE
UNIVERSITETSKIJ PROSP 13
119899 MOSCOW
U.S.S.R.
TEL: 139-37-64
TLX:
COM: 07

ENDAL ANDREW S DR
APPLIED RESEARCH CO
8201 CORPORATE DRIVE
LANDOVER MD 20785
U.S.A.
TEL: 301-459-8442
TLX:
COM: 35

ELLIS G R A PROF
UNIVERSITY OF TASMANIA
P.O.BOX 252C
HOBART, TASMANIA
AUSTRALIA
TEL:
TLX: 58150
COM: 40

ELSTE GUNTHER H DR
DEPARTMENT OF ASTRONOMY
UNIVERSITY OF MICHIGAN
DENNISON BUILDING
ANN ARBOR MI 48109
U.S.A.
TEL: 313-764-3444
TLX:
COM: 10,12,36

EMERSON BRIAN MR
ROYAL GREENWICH OBS
HERSTMONCEUX CASTLE
HAILSHAM BN27 1RP
U.K.
TEL: 031-667-3321
TLX: 72383 ROEDIN G
COM: 04

ENGELHARD E J G PROF DR
PHYS-TECHN-BUNDESANSTAET
BRAUNSCHWEIG
SACKRING 34
D-3300 BRAUNSCHWEIG
GERMANY, F.R.
TEL: 0531-56365
TLX:
COM: 14

ELLIS GEORGE F R PROF
DEPT APPLIED MATHEMATICS
UNIVERSITY OF CAPE TOWN
RONDEBOSCH 7700
SOUTH AFRICA
TEL: 698531
TLX: 5721439
COM: 47,51

ELSTON WOLFGANG E PROF
DEPARTMENT OF GEOLOGY
UNIVERSITY OF NEW MEXICO
ALBUQUERQUE NM 87131
U.S.A.
TEL: 505-277-5339
TLX: 660461
COM: 16

EMERSON DAVID
DEPT OF ASTRONOMY
UNIVERSITY OF EDINBURGH
ROYAL OBSERVATORY
EDINBURGH EH9 3HJ
U.K.
TEL:
TLX:
COM: 28

ENGIN SEMANUR PROF
DEPARTMENT OF ASTRONOMY
UNIVERSITY OF ANKARA
FEN FAKULTESI
ANKARA
TURKEY
TEL:
TLX:
COM:

ENGVOLD ODDBJOERN DR
INST THEORET ASTROPHYSICS
UNIVERSITY OF OSLO
P O BOX 1029
N-0315 BLINDERN, OSLO 3
NORWAY
TEL:
TLX:
COM: 09C.10

ERCAN E. NIHAL
KANDILLI OBSERVATORY
BOSPHORUS UNIVERSITY
CENGELKOY
ISTANBUL 81220
TURKEY
TEL: 3320240/41
TLX: 26411 BOUN TR
COM:

ERSHKOVICH ALEXANDER PROF
DEPT GEOPHYS & PLANET SCI
TEL-AVIV UNIVERSITY
TEL-AVIV 69978
ISRAEL
TEL: 03-413505
TLX: 342171 VERSY IL
COM: 15

ESTALELLA ROBERT
DPTO FIS TIERRA & COSMOS
UNIVERSIDAD DE BARCELONA
DIAGONAL 645
08028 BARCELONA
SPAIN
TEL: 330-73-11/298
TLX:
COM:

ENOME SHINZO PROF
TOYOKAWA OBSERVATORY
NAGOYA UNIVERSITY
13 HONOHARA 3-CHOME
TOYOKAWA 442
JAPAN
TEL: 5338-6-3154
TLX: 4322-310 TYKW J
COM: 10,40

ERDI B DR
ASTRONOMICAL DEPARTMENT
LORAND EOTVOS UNIVERSITY
KUN BELA TER 2
H-1083 BUDAPEST
HUNGARY
TEL: 141019
TLX:
COM: 07

ERTAN A YENER DR
EGE UNIVERSITY
FEN FAKULTESI
ASTRONOMI BOLUMU
BORNOVA-IZMIR
TURKEY
TEL:
TLX:
COM:

EVANGELIDIS E DR
PLASMA PHYSICS DIVISION
NUCOR / PELINDABA
PRIVATE BAG X256
PRETORIA 0001
SOUTH AFRICA
TEL: 27-12-21-3311
TLX: 30253 SA
COM: 33,36

ENSLIN HEINZ DR
DEUTSCHES HYDROGRAPHISCH.
INSTITUT
POSTFACH 220
D-2000 HAMBURG 4
GERMANY, F.R.
TEL: 040-31-905-194
TLX: 0211138
COM: 19,31

ERGMA E V DR
ASTRONOMICAL COUNCIL
USSR ACADEMY OF SCIENCES
PYATNITSKAYA UL 48
109017 MOSCOW
U.S.S.R.
TEL: 231-54-61
TLX: 412623 SCSTP SU
COM: 35,49

ERUSHEV N N DR
CRIMEAN ASTROPHYS OBS
USSR ACADEMY OF SCIENCES
334413 NAUCHNIY / CRIMEA
U.S.S.R.
TEL:
TLX:
COM:

EVANS ANEURIN
DEPT OF PHYSICS
UNIVERSITY OF KEELE
KEELE ST5 5BG
U.K.
TEL: 0782-621111
TLX: 36113 UNKLIB G
COM: 27

EPPS HARLAND WARREN PROF
DEPARTMENT OF ASTRONOMY
UNIVERSITY OF CALIFORNIA
MATH SCI RM 8983
LOS ANGELES CA 90024
U.S.A.
TEL: 213-825-3025
TLX: 910-3427597
COM:

ERICKSON WILLIAM C DR
ASTRONOMY PROGRAM
UNIVERSITY OF MARYLAND
COLLEGE PARK MD 20742
U.S.A.
TEL: 301-454-6453
TLX: 7108260352
COM: 10,40

ESHLEMAN VON R PROF
DURAND 221
STANFORD UNIVERSITY
STANFORD CA 94305
U.S.A.
TEL: 415-497-3531
TLX:
COM: 16,40,49

EVANS DAVID S PROF
DEPARTMENT OF ASTRONOMY
UNIVERSITY OF TEXAS
AUSTIN TX 78712
U.S.A.
TEL: 512-471-4461
TLX:
COM: 41

EPSTEIN EUGENE E DR
THE AEROSPACE CORPORATION
2118 PATRICIA AVE
LOS ANGELES CA 90025
U.S.A.
TEL: 213-648-6798
TLX: 664460
COM: 40,51

ERIGUCHI YOSHIHARU DR
DEPT EARTH SC & ASTRONOMY
COLL ARTS & SC/UNIV TOKYO
KOMABA MEGURO
TOKYO 153
JAPAN
TEL: 03-467-1171x439
TLX: 25510 UNITOKYO
COM: 35

ESIPOV V F DR
STERNBERG STATE ASTR INST
117234 MOSCOW
U.S.S.R.
TEL:
TLX:
COM: 28,34

EVANS J V DR
COMSAT LABORATORIES
22300 COMSAT DR
CLARKSBURG MD 20871
U.S.A.
TEL: 301-428-4422
TLX: 908753
COM: 12

EPSTEIN GABRIEL LEO DR
NASA-GSFC
CODE 682
GREENBELT MD 20771
U.S.A.
TEL:
TLX:
COM: 12,14

ERIKSEN GUNNAR PROF
INST THEORET ASTROPHYSICS
UNIVERSITY OF OSLO
P O BOX 1029
N-0315 BLINDERN, OSLO 3
NORWAY
TEL: 02-45-65-15
TLX:
COM: 40

ESKIOGLU A NIHAT
DEVLET MUHENDISLIK
MIMARLIK AKADEMISI
ADAPAZARI, SAKARYA
TURKEY
TEL:
TLX:
COM: 27

EVANS JOHN W DR
1 BAYA ROAD
ELDORADO
SANTA FE NM 87503
U.S.A.
TEL:
TLX:
COM:

EPSTEIN ISADORE PROF
ASTRONOMY DEPARTMENT
COLUMBIA UNIVERSITY
PUPIN PHYSICAL LABS
NEW YORK NY 10027
U.S.A.
TEL: 212-280-3280
TLX: 125953 COLUMBIA
COM: 35

ERIKSSON KJELL DR
ASTRONOMISKA
OBSERVATORIET
BOX 515
S-751 20 UPPSALA
SWEDEN
TEL: 18-11-24-88
TLX: 76024 UNIVUPS S
COM: 36

ESPOSITO F PAUL PROF
DEPT OF PHYSICS 11
UNIVERSITY OF CINCINNATI
CINCINNATI OH 45221
U.S.A.
TEL: 513-475-2233
TLX:
COM:

EVANS KENTON DOWER DR
PHYSICS DEPARTMENT
THE UNIVERSITY
LEICESTER LE1 7RN
U.K.
TEL: 0533 554455
TLX: 341664
COM: 40

EPSTEIN RICHARD I DR
LOS ALAMOS NATIONAL LAB
MS 436
LOS ALAMOS NM 87545
U.S.A.
TEL: 505-667-9595
TLX:
COM:

ERPYLEV N P DR
ASTRONOMICAL COUNCIL
USSR ACADEMY OF SCIENCES
PYATNITSKAYA UL 48
109017 MOSCOW
U.S.S.R.
TEL: 231-54-61
TLX: 412623 SCSTP SU
COM: 41

ESPOSITO LARRY W
LASP
UNIVERSITY OF COLORADO
CAMPUS BOX 392
BOULDER CO 80309
U.S.A.
TEL: 303-492-7325
TLX:
COM: 16

EVANS NANCY REMAGE DR
COMPUTER SCIENCES CORP
IUE OBS - CODE 684.9
GODDARD SPACE FLIGHT CTR
GREENBELT MD 20771
U.S.A.
TEL: 301-344-7537
TLX:
COM: 27

EVANS NEAL J II ASS PROF
DEPARTMENT OF ASTRONOMY
UNIVERSITY OF TEXAS
AUSTIN TX 78712
U.S.A.
TEL: 512-471-4461
TLX:
COM: 34,51

EVANS ROGER G DR
RUTHERFORD LABORATORY
DIDCOT 0X11 0QX
U.K.
TEL: 0235-21900
TLX: 83159 RUTHLB G
COM:

EVANS W DOYLE
390 EL CONEJO
LOS ALAMOS NM 87544
U.S.A.
TEL: 505-667-3644
TLX:
COM: 48

EVDOKIMOV YU V DR
ENGELHARDT ASTR OBS
422526 KAZAN
U.S.S.R.
TEL:
TLX:
COM: 20

EVERHART EDGAR DR
985 DICK MOUNTAIN DR.
BAILEY CO 80421
U.S.A.
TEL:
TLX:
COM: 06,07,15,20

EVIATAR AHARON PROF
DEPT GEOPHYS & PLANET SCI
TEL-AVIV UNIVERSITY
TEL-AVIV 69978
ISRAEL
TEL: 03-420620
TLX: 342171 VERSY IL
COM: 15,49

EWEN HAROLD I DR
60 BEAVER ROAD
WESTON MA 02193
U.S.A.
TEL:
TLX:
COM:

EWING MARTIN S
CALTECH 102-24
PASADENA CA 91125
U.S.A.
TEL: 818-356-4970
TLX: 675425
COM: 40

EZER-ERYURT DILHAN PROF
MIDDLE EAST TECHN UNIV
ANKARA
TURKEY
TEL: 23-71-00 x 3255
TLX: 42761 ODTK TR
COM: 35

FABBIANO GIUSEPPINA
HARVARD-SMITHSONIAN CTR
FOR ASTROPHYSICS
60 GARDEN STREET
CAMBRIDGE MA 02138
U.S.A.
TEL: 617-495-7204
TLX: 921428 SATELLITE CAM
COM: 28

FABER SANDRA M PROF
LICK OBSERVATORY
UNIVERSITY OF CALIFORNIA
SANTA CRUZ CA 95064
U.S.A.
TEL: 408-429-2944
TLX:
COM: 28,33,47

FABIAN ANDREW C DR
INSTITUTE OF ASTRONOMY
MADINGLEY ROAD
CAMBRIDGE CB3 0HE
U.K.
TEL:
TLX:
COM: 48

FABRE HERVE DR
2 AVENUE MARECHAL FOCH
F-06310 BEAULIEU/MER
FRANCE
TEL:
TLX:
COM: 07

FABRICANT DANIEL G
HARVARD-SMITHSONIAN CTR
FOR ASTROPHYSICS
60 GARDEN STREET
CAMBRIDGE MA 02138
U.S.A.
TEL: 617-495-7398
TLX: 921428 SATELLITE CAM
COM: 09,28,44

FABRICIUS CLAUS V
COPENHAGEN UNIVERSITY OBS
BRORFELDE
DK-4340 TOLLOSE
DENMARK
TEL: 45-3-488-195
TLX: 44155 DANAST DK
COM: 08

FADEYEV YURI A
ASTRONOMICAL COUNCIL
USSR ACADEMY OF SCIENCES
PYATNITSKAYA STR 48
109017 MOSCOW
U.S.S.R.
TEL: 231-54-61
TLX: 412623 SCSTP SU
COM: 27,35

FAELTHAMMAR CARL GUNNE PR
DEPT OF PLASMA PHYSICS
ROYAL INST OF TECHNOLOGY
S-100 44 STOCKHOLM 70
SWEDEN
TEL: 0-8-687-7685
TLX: 10389 KTHB
COM:

FAHLMAN GREGORY G DR
DEPT GEOPHYS & ASTRONOMY
UNIV OF BRITISH COLUMBIA
2075 WESBROOK PLACE
VANCOUVER BC V6T 1W5
CANADA
TEL: 604-228-4891
TLX: 04542425
COM:

FAHR HANS JOERG PROF DR
INSTITUT FUR ASTROPHYSIK
DER UNIVERSITAET BONN
AUF DEM HUEGEL 71
D-5300 BONN
GERMANY, F.R.
TEL: 0228-733677
TLX: 886440 MPI
COM: 49C

FAHY EDWARD F PROF
PHYSICS DEPT
UNIVERSITY COLLEGE
CORK
IRELAND
TEL: 021-26871
TLX: 26050
COM:

FAIRALL ANTHONY P PROF
DEPT OF ASTRONOMY
UNIVERSITY OF CAPE TOWN
RONDEBOSCH 7700
SOUTH AFRICA
TEL: 21-698531 x 629
TLX: 5721439
COM: 28

FALCHI AMBRETTA
OSSERVATORIO ASTROFISICO
DI ARCETRI
LARGO E. FERMI 5
I-50125 FIRENZE
ITALY
TEL: 55-220-034
TLX: 572268 ARCETR I
COM: 10

FALCIANI ROBERTO DR
OSSERVATORIO ASTROFISICO
DI ARCETRI
LARGO E. FERMI 5
I-50125 FIRENZE
ITALY
TEL: 55-220034
TLX: 572268 ARCETR I
COM: 10C,12

FALK SYDNEY W JR DR
DEPT OF ASTRONOMY
UNIVERSITY OF TEXAS
AUSTIN TX 78712
U.S.A.
TEL:
TLX:
COM: 34,47

FALL S MICHAEL DR
SPACE TELESCOPE SCI INST
HOMEWOOD CAMPUS
3700 SAN MARTIN DRIVE
BALTIMORE MD 21218
U.S.A.
TEL:
TLX:
COM: 28,33,37,47

FALLE SAMUEL A DR
DEPT APPLIED MATHEMATICS
UNIVERSITY OF LEEDS
LEEDS LS2 9JT
U.K.
TEL: 532-431-751
TLX:
COM: 34

FALLER JAMES E PROF
JILA/NBS
UNIVERSITY OF COLORADO
BOULDER CO 80309
U.S.A.
TEL: 303-492-8509
TLX: 755842 JILA
COM:

FALLON FREDERICK W DR
N/CG 114ASTRONOMY
NOAA/NGS
6010 EXECUTIVE BLVD
ROCKVILLE MD 20852
U.S.A.
TEL:
TLX:
COM: 24,31

FAN YING
DEPT OF ASTRONOMY
BEIJING NORMAL UNIVERSITY
BEIJING 100082
CHINA, PEOPLE'S REP.
TEL: 653531-6285
TLX:
COM: 34

FANG CHENG
DEPT OF ASTRONOMY
NANJING UNIVERSITY
NANJING
CHINA, PEOPLE'S REP.
TEL: 34651-2882
TLX: 34151 PRCNU CN
COM: 10,12

FANG LI-ZHI
CENTER FOR ASTROPHYSICS
UNIV SCIENCE & TECHNOLOGY
HEFEI, ANHUI PROVINCE
CHINA, PEOPLE'S REP.
TEL: 63300
TLX:
COM: 47,48

FANSELOW JOHN LYMAN
JET PROPULSION LAB
M/S 264-748
4800 OAK GROVE DR
PASADENA CA 91109
U.S.A.
TEL: 213-354-6323
TLX: 675429
COM: 19,24

512 LIST OF MEMBERS

FANTI CARLA GIOVANNINI
ISTITUTO RADIOASTRONOMIA
VIA IRNERIO 46
I-40126 BOLOGNA
ITALY
TEL: 051-232856/57
TLX: 211664 INFN BO
COM:

FAWELL DEREK R DR
UNIVERSITY OF LONDON OBS
MILL HILL PARK
LONDON NW7 2QS
U.K.
TEL:
TLX:
COM: 46

FEDOROV E P PROF
MAIN ASTRON OBSERVATORY
UKRAINIAN ACAD SCIENCES
252127 KIEV
U.S.S.R.
TEL: 66-31-10
TLX: 131406 SKY SU
COM: 19

FEIX GERHARD DR
RUHR UNIVERSITAET BOCHUM
DEPT XII
POSTFACH 102148
D-4630 BOCHUM
GERMANY, F.R.
TEL: 0234-700-2051
TLX: 0825860
COM: 40

FANTI ROBERTO
ISTITUTO DI FISICA
UNIVERSITA DI BOLOGNA
VIA IRNERIO 46
I-40126 BOLOGNA
ITALY
TEL: 232856/57
TLX: 211664 INFN BO
COM: 40

FAY THEODORE D DR
MAIL STOP 19
TELEDYNE BROWN ENG
CUMMINGS RES PARK
HUNTSVILLE AL 35807
U.S.A.
TEL:
TLX:
COM:

FEDOROVA RIMMA T DR
ASTRONOMICAL OBSERVATORY
NIKOLAEV BRANCH OF THE
MAIN ASTRONOMICAL OBS
327000 NIKOLAEV
U.S.S.R.
TEL: 37-57-14
TLX:
COM: 08

FEJES ISTVAN DR
FOMI SATELLITE
GEODETIC OBSERVATORY
BOX 546
H-1373 BUDAPEST
HUNGARY
TEL:
TLX:
COM: 51

FARAGGIANA ROSANNA PROF
OSSERVATORIO ASTRONOMICO
VIA TIEPOLO 11
I-34131 TRIESTE
ITALY
TEL: 040-793921
TLX: 461137 OAT I
COM: 29,36,44

FAZIO GIOVANNI G DR
CENTER FOR ASTROPHYSICS
HCO/SAO
60 GARDEN ST
CAMBRIDGE MA 02138
U.S.A.
TEL: 617-495-7458
TLX: 921428 SATELLITE CAM
COM: 44,48,51

FEGAN DAVID J DR
PHYSICS DEPT
UNIVERSITY COLLEGE
BELFIELD
DUBLIN 4
IRELAND
TEL: 692244
TLX: 32693 UCD EI
COM:

FEKEL FRANCIS C
VANDERBILT UNIVERSITY
DYER OBSERVATORY
NASHVILLE TN 37235
U.S.A.
TEL: 615-322-2804
TLX:
COM: 26,42

FARINELLA PAOLO DR
ISTITUTO DI MATEMATICA
UNIVERSITA' DI PISA
VIA BUONARROTI 2
I-56100 PISA
ITALY
TEL:
TLX:
COM: 07,15,16

FEAST MICHAEL W DR
S A A O
P O BOX 9
OBSERVATORY 7935
SOUTH AFRICA
TEL: (27)21 47 00 25
TLX: 520309
COM: 27,28,29,33,37,45

FEHRENBACH CHARLES PROF
LES MAGNANARELLES
LOURMARIN
F-84160 CADENET
FRANCE
TEL: 90-68-00-28
TLX:
COM: 09,28,30,33,45

FELDMAN PAUL A DR
HERZBERG INST ASTROPHYS
NATL RESEARCH COUNCIL
100 SUSSEX DR
OTTAWA ONT K1A 0R6
CANADA
TEL: 613-993-6060
TLX: 0533715
COM: 40,51

FARNIK FRANTISEK
ASTRONOMICAL INSTITUTE
CZECH. ACAD. OF SCIENCES
ONDREJOV OBSERVATORY
251 65 ONDREJOV
CZECHOSLOVAKIA
TEL: 204-999-201/202
TLX: 121579
COM: 10

FEAUTRIER NICOLE DR
OBSERVATOIRE DE PARIS
SECTION DE MEUDON
F-92195 MEUDON PL CEDEX
FRANCE
TEL: 1-45-34-75-70
TLX:
COM: 14

FEIBELMAN WALTER A DR
NASA/GSFC
CODE 685
GREENBELT MD 20771
U.S.A.
TEL: 301-344-5272
TLX:
COM: 10,27,34

FELDMAN PAUL DONALD DR
DEPT PHYSICS & ASTRONOMY
JOHNS HOPKINS UNIVERSITY
BALTIMORE MD 21218
U.S.A.
TEL: 301-338-7339
TLX: 710-234-1090
COM: 15,21,44

FAUCHER PAUL DR
OBSERVATOIRE DE NICE
BP 139
F-06003 NICE CEDEX
FRANCE
TEL: 93-89-04-20
TLX: 460004
COM: 14

FECHTIG HUGO DR
SANSERWEG 3
D-6906 LEIMEN
GERMANY, F.R.
TEL:
TLX:
COM: 15C,21,22C

FEINSTEIN ALEJANDRO DR
OBSERVATORIO ASTRONOMICO
1900 LA PLATA
ARGENTINA
TEL: 021-21-7308
TLX: 31216 CESLA AR
COM: 25,37

FELDMAN U DR
DEPT PHYSICS & ASTRONOMY
TEL-AVIV UNIVERSITY
RAMAT-AVIV 69978
ISRAEL
TEL:
TLX:
COM:

FAULKNER DONALD J DR
MT STROMLO & SIDING
SPRING OBSERVATORIES
PRIVATE BAG
WODEN P.O. ACT 2606
AUSTRALIA
TEL: 062-88-1111
TLX: 62270
COM: 34,35

FEDERICI LUCIANA
DIPTO DI ASTRONOMIA
UNIVERSITA DEGLI STUDI
C P 596
I-40100 BOLOGNA
ITALY
TEL:
TLX: 211664 INFN BO I
COM:

FEISSEL MARTINE DR
OBSERVATOIRE DE PARIS
61 AVE DE L'OBSERVATOIRE
F-75014 PARIS
FRANCE
TEL: 1-43-20-12-10
TLX:
COM: 08,19V

FELDMAN URI
HOLBURT CTR FOR SPACE RES
NAVAL RESEARCH LABORATORY
WASHINGTON DC 20375
U.S.A.
TEL: 202-767-3286
TLX:
COM: 12

FAULKNER JOHN PROF
LICK OBSERVATORY
UNIVERSITY OF CALIFORNIA
SANTA CRUZ CA 95064
U.S.A.
TEL: 408-429-2815
TLX:
COM: 35,42

FEDERMAN STEVEN ROBERT
M/S 183-601
JET PROPULSION LAB
4800 OAK GROVE DRIVE
PASADENA CA 91109
U.S.A.
TEL: 818-354-2274
TLX:
COM: 34

FEITZINGER JOHANNES PROF
ASTRONOMISCHES INSTITUT
RUHR UNIVERSITAET BOCHUM
POSTFACH 102148
D-4630 BOCHUM
GERMANY, F.R.
TEL: 0234-700-3450
TLX: 825860-1 RUB D
COM: 28,33,34

FELENBOK PAUL DR
OBSERVATOIRE DE PARIS
SECTION DE MEUDON
F-92195 MEUDON PL CEDEX
FRANCE
TEL: 1-45-34-75-70
TLX:
COM: 14

FELLGETT PETER PROF
DEPT OF CYBERNETICS
3 EARLY GATE
WHITEKNIGHTS
READING RG6 2AL
U.K.
TEL: 0734-65758
TLX: 847813
COM: 09

FELLI MARCELLO DR
OSSERVATORIO ASTROFISICO
DI ARCETRI
LARGO E. FERMI 5
I-50125 FIRENZE
ITALY
TEL: 055-220034
TLX: 572268
COM: 34,40

FELTEN JAMES E DR
8569 GREENBELT RD, NO.204
GREENBELT MD 20770
U.S.A.
TEL: 301-552-1526
TLX:
COM: 34,40,47,48

FENG HESHENG
YUNNAN OBSERVATORY
KUNMING
CHINA, PEOPLE'S REP.
TEL:
TLX:
COM:

FENG KE-JIA
DEPT OF ASTRONOMY
BEIJING NORMAL UNIVERSITY
BEIJING 80
CHINA, PEOPLE'S REP.
TEL: 65-3531 x 6967
TLX:
COM: 10,46

FENKART ROLF P PROF DR
ASTRONOMISCHES INSTITUT
UNIVERSITAET BASEL
VENUSSTRASSE 7
CH-4102 BINNINGEN
SWITZERLAND
TEL: 061-22-77-11
TLX:
COM: 33

FENTON K B DR
DEPT OF PHYSICS
UNIVERSITY OF TASMANIA
P.O.BOX 252C
HOBART, TASMANIA 7001
AUSTRALIA
TEL: 002-202411
TLX: 58150 AA
COM: 48

FERETTI LUIGINA
ISTITUTO RADIOASTRONOMIA
VIA IRNERIO 46
I-40126 BOLOGNA
ITALY
TEL: 51-232856
TLX: 211664 INFN BO I
COM: 40

FERLAND GARY JOSEPH
ASTRONOMY DEPARTMENT
OHIO STATE UNIVERSITY
COLUMBUS OH 43210
U.S.A.
TEL: 614-422-1773
TLX: 8104821715
COM: 27

FERNANDEZ JEAN-CLAUDE DR
OBSERVATOIRE DE NICE
BP 139
F-06003 NICE CEDEX
FRANCE
TEL: 93-89-04-20
TLX:
COM: 15

FERNANDEZ SILVIA M. DR
OBSERVATORIO ASTRONOMICO
LAPRIDA 854
5000 CORDOBA
ARGENTINA
TEL: 51-40613/36876
TLX: 51822 BUCOR
COM: 07

FERNANDEZ-FIGUEROA M J DR
ASTROFIS FAC DE FISICAS
UNIVERSIDAD COMPLUTENSE
CIUDAD UNIVERSITARIA
28040 MADRID
SPAIN
TEL: 4-49-53-16
TLX: 47273 FF UC
COM: 29,46

FERNIE J DONALD PROF
DAVID DUNLAP OBSERVATORY
P O BOX 360
RICHMOND HILL ONT L4C 4Y6
CANADA
TEL: 416-884-9562
TLX: 06-986766 TELEXPERTS
COM: 25,27,41

FERRARI ATTILIO DR
ISTITUTO DI FISICA
GENERALE DELL'UNIVERSITA
CORSO M. D'AZEGLIO 46
I-10125 TORINO
ITALY
TEL: 011-657694
TLX: 211041 INFN TO I
COM: 48

FERRARI D'OCCHIEPPO K DR
OESTERREICHISCHE AKADEMIE
DER WISSENSCHAFTEN
DR-IGNAZ-SEIPEL-PLATZ 2
A-1010 WIEN
AUSTRIA
TEL: 052-22-81991
TLX: 01-12628
COM: 41,42

FERRARI TONIOLO MARCO
IST ASTROFISICA SPAZIALE
C P 67
I-00044 FRASCATI
ITALY
TEL: 06-9425651
TLX: 610261 CNR-FRA I
COM: 44

FERRAZ-MELLO S PROF DR
UNIVERSIDADE DE SAO PAULO
DEPT ASTRONOMIA
CAIXA POSTAL 30627
01051 SAO PAULO SP
BRAZIL
TEL: 11-549-6709
TLX: 1136221 IAGM BR
COM: 07C,20,46C

FERRER MARTINEZ SEBASTIAN
DPTO FIS TIERRA & COSMOS
UNIVERSIDAD DE ZARAGOZA
50009 ZARAGOZA
SPAIN
TEL: 976-357011
TLX: 58198
COM: 07

FERRER OSVALDO EDUARDO DR
UNIV NACIONAL DE LA PLATA
FACULDAD DE CIENCIAS
ASTRON Y GEOFISICAS
1900 LA PLATA
ARGENTINA
TEL:
TLX:
COM: 26

FERRIN IGNACIO
UNIVERSIDAD DE LOS ANDES
FACULTAD DE CIENCIAS
DEPTO DE FISICA
MERIDA 5101
VENEZUELA
TEL:
TLX:
COM:

FERRINI FEDERICO
ISTITUTO DI ASTRONOMIA
UNIVERSITA DI PISA
PIAZZA TORRICELLI 2
I-56100 PISA
ITALY
TEL: 050-43343
TLX:
COM: 34

FESTOU MICHEL C DR
OBSERVATOIRE DE BESANCON
41B AVE DE L'OBSERVATOIRE
F-25044 BESANCON
FRANCE
TEL: 81 80 22 66
TLX: 361144
COM: 15

FEYNMAN JOAN DR
JPL
144-218
4800 OAK GROVE DRIVE
PASADENA CA 91109
U.S.A.
TEL: 818-354-3454
TLX: 675429
COM: 49

FIALA ALAN D DR
NAUTICAL ALMANAC OFFICE
US NAVAL OBSERVATORY
348 MASSACHUSETTS AVE NW
WASHINGTON DC 20390
U.S.A.
TEL: 202-653-1274
TLX: 710-822-1970
COM: 04,07,12

FICARRA ANTONINO DR
IST DI RADIOASTRONOMIA
VIA IRNERIO 46
I-40126 BOLOGNA
ITALY
TEL: 51-232856
TLX: 211664 INFN BO I
COM:

FICHTEL CARL E DR
GODDARD SPACE FLIGHT CTR
CODE 660
GREENBELT MD 20771
U.S.A.
TEL: 301-344-6281
TLX: 89675
COM: 44,48

FIELD DAVID
SCHOOL OF CHEMISTRY
CANTOCKS CLOSE
BRISTOL BS8 1TS
U.K.
TEL: 0272-24161 x505
TLX: 444174 BUPHYS
COM: 34

FIELD GEORGE B PROF
CENTER FOR ASTROPHYSICS
60 GARDEN ST
CAMBRIDGE MA 02138
U.S.A.
TEL: 617-495-4721
TLX: 921428 SATELLITE CAM
COM: 28,34,40,47,48,49,51

FIELDER GILBERT DR
LUNAR AND PLANETARY UNIT
E S DEPT
LANCASTER UNIVERSITY
LANCASTER LA1 4YR
U.K.
TEL: 65201
TLX: 65111
COM: 16

FIERRO JULIETA
INSTITUTO DE ASTRONOMIA
UNAM
APDO POSTAL 70-264
04510 MEXICO DF
MEXICO
TEL:
TLX:
COM: 34

FILIPOV LATCHEZAR
CENTR LAB FOR SPACE RES
BULGARIAN ACAD SCIENCES
MOSKOVA STR 6
1000 SOFIA
BULGARIA
TEL: 87-09-78
TLX: 23351 CLSR BG
COM:

FILLOY EMILIO MANUEL E.E.
INSTITUTO ARGENTINO DE
RADIOASTRONOMIA
CASILLE DE CORREO No. 5
1894 VILLA ELISA (Bs.As.)
ARGENTINA
TEL: 4-3793
TLX:
COM:

FINDLAY JOHN W DR
NRAO
EDGEMONT ROAD
CHARLOTTESVILLE VA 22901
U.S.A.
TEL:
TLX: 910-997-0174
COM: 40

FIROR JOHN W DR
NCAR
PO BOX 3000
BOULDER CO 80307
U.S.A.
TEL: 303-497-1600
TLX: 45694
COM:

FISHMAN GERALD J
MSFC/ASTROPHYSICS BRANCH
SPACE SCIENCES LAB ES-62
HUNTSVILLE AL 35812
U.S.A.
TEL: 205-453-0117
TLX:
COM: 44

FLETT ALISTAIR M
UNIVERSITY OF ABERDEEN
DEPTARTMENT OF PHYSICS
ABERDEEN AB9 2UE
U.K.
TEL: 0224-40241
TLX: 73458 UNIABN G
COM: 40

FINK UWE DR
LUNAR & PLANETARY LAB
UNIVERSITY OF ARIZONA
TUCSON AZ 85721
U.S.A.
TEL: 602-621-2736
TLX: 9109521143
COM: 14,16

FISCHEL DAVID DR
SYSTEMS & APPL.SCIENCE CO
5809 ANNAPOLIS RD
HYATTSVILLE MD 20784
U.S.A.
TEL:
TLX:
COM:

FITCH WALTER S DR
STEWARD OBSERVATORY
UNIVERSITY OF ARIZONA
TUCSON AZ 85721
U.S.A.
TEL: 602-621-6522
TLX: 467175
COM: 25,27

FLIEGEL HENRY F
3730 EL MORENO AVENUE
PO BOX 8682
LA CRESCENTA CA 91214
U.S.A.
TEL:
TLX:
COM: 19,31

FINN G D DR
INSTITUTE FOR ASTRONOMY
2525 CORREA ROAD
HONOLULU HI 96822
U.S.A.
TEL:
TLX:
COM: 36

FISCHER JACQUELINE
NAVAL RESEARCH LABORATORY
CODE 4138F
WASHINGTON DC 20375
U.S.A.
TEL: 202-767-3058
TLX:
COM:

FITTON BRIAN DR
ESTEC, ASTRONOMY DIVISION
POSTBUS 299
NL-2200 AG NOORDWIJK
NETHERLANDS
TEL: 31-2524-4635
TLX:
COM: 44

FLIN PIOTR
JAGIELLONIAN UNIVERSITY
OBSERVATORY
UL. ORLA 171
30-244 KRAKOW
POLAND
TEL:
TLX: 0322297 UJ PL
COM: 28

FINZI ARRIGO DR
DEPT OF MATHEMATICS
TECHNION, I.I.T.
HAIFA 32000
ISRAEL
TEL:
TLX: 46406 TECON IT
COM:

FISCHER STANISLAV DR
ASTRONOMICAL INSTITUTE
CZECH. ACAD. OF SCIENCES
BUDECSKA 6
120 23 PRAHA 2
CZECHOSLOVAKIA
TEL: 252438
TLX: 122486 ASTRC
COM:

FITZGERALD M PIM PROF
DEPT OF PHYSICS
UNIVERSITY OF WATERLOO
WATERLOO ONT N2L 3G1
CANADA
TEL: 519-885-1572
TLX:
COM: 33,37

FLOQUET MICHELE DR
OBSERVATOIRE DE PARIS
SECTION DE MEUDON
DEPEG
F-92195 MEUDON PL CEDEX
FRANCE
TEL: 1-45-34-75-30
TLX:
COM: 29

FIREMAN EDWARD L
SMITHSONIAN ASTROPHYSICAL
OBSERVATORY
60 GARDEN ST
CAMBRIDGE MA 02138
U.S.A.
TEL: 617-495-7271
TLX:
COM: 22

FISHER J RICHARD
NRAO
PO BOX 2
GREEN BANK WV 24944
U.S.A.
TEL: 304-456-2011
TLX: 710-938-1530
COM:

FIX JOHN D DR
DEPT PHYSICS & ASTRONOMY
UNIVERSITY OF IOWA
IOWA CITY IA 52240
U.S.A.
TEL: 319-353-7064
TLX: 910-525-1398
COM:

FLORENTIN-NIELSEN RALPH
COPENHAGEN UNIVERSITY OBS
BRORFELDEVEJ 23
DK-4340 TOLLOSE
DENMARK
TEL: 3-488195
TLX: 44155 DANAST
COM:

FIRMANI CLAUDIO A PROF
INSTITUTO DE ASTRONOMIA
UNAM
APDO POSTAL 70-264
04510 MEXICO DF
MEXICO
TEL: 905-548-3712
TLX: 1760155 CICME
COM: 42

FISHER PHILIP C
RUFFNER ASSOCIATES
PO BOX 7070
MENLO PARK CA 94026
U.S.A.
TEL:
TLX:
COM: 44,48,51

FLANNERY BRIAN PAUL DR
EXXON RES & ENGINEERING
ROUTE 22 EAST
ANNANDALE NJ 08801
U.S.A.
TEL: 201-730-2540
TLX: 136140 EXXONRES
COM: 34,35,42

FLORIDES PETROS S PROF
SCHOOL OF MATHEMATICS
TRINITY COLLEGE
DUBLIN 2
IRELAND
TEL: 772941
TLX: 25442 TCD EI
COM: 41,47

FIRNEIS FRIEDRICH J DR
INST INFO PROC/OEAW
SONNENFELSGASSE 19/2
A-1010 WIEN
AUSTRIA
TEL:
TLX:
COM: 24,51

FISHER RICHARD R DR
HIGH ALTITUDE OBSERVATORY
PO BOX 3000
BOULDER CO 80307
U.S.A.
TEL: 303-494-5151
TLX:
COM:

FLEISCHER ROBERT DR
ROUTE 1
BOX 41 A
KEEDYSVILLE MD 21756
U.S.A.
TEL: 301-432-8870
TLX:
COM: 40

FLORSCH ALPHONSE DR
OBS DE STRASBOURG
11 RUE DE L'UNIVERSITE
F-67000 STRASBOURG
FRANCE
TEL: 88-35-43-00
TLX: 890506 STAROBS F
COM: 28,30C,38C

FIRNEIS MARIA G DR
INSTITUT FUER ASTRONOMIE
TUERKENSCHANZSTR 17
A-1180 WIEN
AUSTRIA
TEL: 0222-34-53-60
TLX:
COM: 24,41,51

FISHKOVA LUISA M PROF
ABASTUMANI ASTROPHYSICAL
OBSERVATORY
383762 ABASTUMANI,GEORGIA
U.S.S.R.
TEL:
TLX:
COM: 21

FLETCHER J MURRAY
DOMINION ASTROPHYS OBS
5071 W SAANICH ROAD
RR 5
VICTORIA BC V8X 4M6
CANADA
TEL: 604-388-3905
TLX: 049-7295
COM: 09,26,30

FLOWER DAVID R DR
DEPT OF PHYSICS
UNIVERSITY OF DURHAM
DURHAM DH1 3LE
U.K.
TEL: 0385-64971
TLX: 537351
COM: 14,34C

LIST OF MEMBERS

FOGARTY WILLIAM G DR
2511 PLOVER RD
WISCONSIN RAPIDS WI 54494
U.S.A.
TEL:
TLX:
COM:

FONG CHU-GANG
SHANGHAI OBSERVATORY
ACADEMIA SINICA
SHANGHAI
CHINA, PEOPLE'S REP.
TEL: 386191
TLX: 33164 SHAO CN
COM: 07,19

FORREST WILLIAM JOHN
DEPT PHYSICS & ASTRONOMY
UNIVERSITY OF ROCHESTER
ROCHESTER NY 14627
U.S.A.
TEL: 716-275-4343
TLX:
COM:

FOSSAT ERIC DR
OBSERVATOIRE DE NICE
BP 139
F-06003 NICE CEDEX
FRANCE
TEL: 93-89-04-20
TLX: 460004
COM: 10,12,35

FOGH OLSEN H J
COPENHAGEN UNIVERSITY OBS
BRORFELDEVEJ 23
DK-4340 TOLLOSE
DENMARK
TEL: 03-488195
TLX: 44155 DQNQST
COM: 08

FONG RICHARD
DEPT OF PHYSICS
UNIVERSITY OF DURHAM
SOUTH ROAD
DURHAM DH1 3LE
U.K.
TEL: 64971
TLX: 537351
COM: 47

FORSTER JAMES RICHARD DR
CSIRO
DIVISION OF RADIOPHYSICS
P.O.BOX 76
EPPING NSW 2121
AUSTRALIA
TEL: 02-868-0222
TLX: 26230 ASTRO AA
COM: 34

FOUKAL PETER V DR
ATMOSPH/ENVIRON/RESEA INC
840 MEMORIAL DRIVE
CAMBRIDGE MA 02139
U.S.A.
TEL: 617-547-6207
TLX:
COM: 12

FOKKER AAD D DR
STERREWACHT SONNENBORGH
ZONNENBURG 2
NL-3512 NL UTRECHT
NETHERLANDS
TEL: 030-312-841
TLX:
COM: 10,40

FONTAINE GILLES DR
DEPT OF PHYSICS
UNIVERSITY OF MONTREAL
P O BOX 6128
MONTREAL PQ H3C 3J7
CANADA
TEL: 514-343-6680
TLX: 05562425
COM: 35

FORT BERNARD P DR
OBSERVATOIRE PIC-DU-MIDI
ET TOULOUSE
14 AVENUE EDOUARD BELIN
F-31400 TOULOUSE
FRANCE
TEL: 61-25-21-01
TLX: 530776 F
COM: 09C

FOWLER WILLIAM A PROF
CALTECH 106-38
PASADENA CA 91125
U.S.A.
TEL: 818-356-4272
TLX:
COM: 35,48

FOLTZ CRAIG B.
MULT. MIRROR TEL. OBS.
UNIVERSITY OF ARIZONA
TUCSON AZ 85721
U.S.A.
TEL: 602-621-1269
TLX: 467175
COM: 28

FONTENLA JUAN MANUEL DR
INST. DE ASTRONOMIA y
FISICA DEL ESPACIO
C.C. 67, SUC. 28
1428 CAPITAL
ARGENTINA
TEL:
TLX:
COM: 12,36

FORT DAVID NORMAN DR
733 LONSDALE ROAD
OTTAWA K1K 0J9
CANADA
TEL:
TLX:
COM: 40

FOX KENNETH DR
DEPT PHYSICS & ASTRONOMY
UNIVERSITY OF TENNESSEE
503 PHYSICS
KNOXVILLE TN 37996-1200
U.S.A.
TEL: 615-974-2288
TLX:
COM: 16

FOMALONT EDWARD B DR
NRAO
PO BOX 0
SOCORRO NM 87801
U.S.A.
TEL:
TLX: 910-988-1710
COM: 40

FORBES J E DR
PO BOX 88120
INDIANAPOLIS IN 46208
U.S.A.
TEL:
TLX:
COM: 35

FORTE JUAN CARLOS DR
UNIV NACIONAL DE LA PLATA
FAC DE CIENCIAS ASTRON Y
GEOFISICAS
1900 LA PLATA
ARGENTINA
TEL:
TLX:
COM: 25,37

FOX W E MR
BRITISH ASTRONOMICAL ASS
40 WINDSOR ROAD
NEWARK NOTTINGHAMS
U.K.
TEL: 0636-704-932
TLX:
COM: 16

FOMENKO ALEXANDR F DR
SPECIAL ASTROPHYS OBS
NIZHNIJ ARKHYZ
357140 STAVROPOLSKIJ KRAJ
U.S.S.R.
TEL:
TLX:
COM: 09,28

FORD HOLLAND C RES PROF
SPACE TELESCOPE SCI INST
JOHNS HOPKINS UNIVERSITY
HOMEWOOD CAMPUS
BALTIMORE MD 21218
U.S.A.
TEL: 301-338-4803
TLX:
COM: 28,34,47

FORTI GIUSEPPE DR
OSSERVATORIO ASTROFISICO
DI ARCETRI
LARGO E. FERMI 5
I-50125 FIRENZE
ITALY
TEL: 055-22-00-34
TLX: 572268 ARCETR
COM: 20,22

FOY RENAUD DR
CERGA
OBSERVATOIRE DE CALERN
CAUSSOLS
F-06460 ST VALLIER DE T.
FRANCE
TEL:
TLX: 461402 CERGLOBS
COM: 09,29,36

FOMIN VALERY A DR
PULKOVO OBSERVATORY
196140 LENINGRAD
U.S.S.R.
TEL:
TLX:
COM: 08

FORD W KENT JR DR
DEPT TERRESTR. MAGNETISM
CARNEGIE INST. WASHINGTON
5241 BROAD BRANCH RD N.W.
WASHINGTON DC 20015
U.S.A.
TEL: 202-966-0863
TLX: 440427 MAGN UI
COM: 09,28

FORTINI TERESA DR
OSSERVATORIO ASTRONOMICO
I-00136 ROMA
ITALY
TEL:
TLX:
COM: 10

FRACASSINI MASSIMO PROF
DIPARTIMENTO DI FISICA
UNIVERSITA DI MILANO
VIA CELORIA 16
I-20133 MILANO
ITALY
TEL: 02-2392275
TLX: 334687 INFNMI
COM: 21,42,45,51

FOMINOV ALEXANDR M DR
INST OF THEORET ASTRONOMY
USSR ACADEMY OF SCIENCES
10 KUTUZOV QUAY
191187 LENINGRAD
U.S.S.R.
TEL: 278-88-98
TLX: 121578 ITA SU
COM: 04

FORMAN WILLIAM RICHARD DR
SMITHSONIAN ASTROPHYS OBS
60 GARDEN STREET
CAMBRIDGE MA 02138
U.S.A.
TEL: 617-495-7210
TLX: 92-1428
COM: 47,48

FOSBURY ROBERT A E DR
ST/ECF
C/O ESO
KARL-SCHWARZSCHILD-STR 2
D-8046 GARCHING B MUNCHEN
GERMANY, F.R.
TEL: 49-89-32006235
TLX: 52828222 EO D
COM:

FRACASTORO MARIO G PROF
VIA MONVISO 3
I-10025 PINO TORINESE
ITALY
TEL: 011-840493
TLX:
COM: 24,26C,42

FRANCESCHINI ALBERTO
DIPATIMENTO DI FISICA
UNIVERSITA DI PADOVA
VIA MARZOLO 8
I-35100 PADOVA
ITALY
TEL:
TLX:
COM:

FRATER ROBERT H DR
CHIEF CSIRO DIV RADIOPHYS
P.O.BOX 76
EPPING NSW 2121
AUSTRALIA
TEL: 02-868-0222
TLX: 26230 ASTRO
COM: 40

FRENCH RICHARD G
M I T
DEPT OF EARTH, ATMOSPH.
CAMBRIDGE MA 02139
U.S.A.
TEL: 617-253-3392
TLX:
COM:

FRIEDEMANN CHRISTIAN DR
UNIV STERNWARTE JENA
SCHILLERGAESSCHEN 2
DDR-6900 JENA
GERMANY, D.R.
TEL: 8222637/27122
TLX: 05886134
COM: 34

FRANDSEN SOEREN PROF
INSTITUTE OF ASTRONOMY
UNIVERSITY OF AARHUS
DK-8000 AARHUS C
DENMARK
TEL: 6-128899
TLX: 64767 AAUSCI DK
COM:

FRAZIER EDWARD N DR
TRW
1 SPACE PARK
REDONDO BEACH CA 90278
U.S.A.
TEL: 213-535-4723
TLX:
COM: 12

FRENK CARLOS S
PHYSICS DEPARTMENT
UNIVERSITY OF DURHAM
SOUTH ROAD
DURHAM DH1 3LE
U.K.
TEL: 0385-64971
TLX: 537351 DURLIB G
COM: 47

FRIEDJUNG MICHAEL DR
INSTITUT D'ASTROPHYSIQUE
98 BIS BOULEVARD ARAGO
F-75014 PARIS
FRANCE
TEL: 1-43-20-14-25
TLX: 270776 OBS
COM: 27,29,42

FRANK JUHAN
MPI F PHYS & ASTROPHYSIK
INSTITUT FUR ASTROPHYSIK
KARL-SCHWARZSCHILD-STR 1
D-8046 GARCHING B MUNCHEN
GERMANY, F.R.
TEL: 089-3299-0
TLX: 524629 ASTRO D
COM: 42

FREDGA KERSTIN PROF
SWEDISH BOARD F SPACE ACT
BOX 4006
S-171 54 SOLNA
SWEDEN
TEL: 08-733-6486
TLX: 17128 SPACECO S
COM: 44

FRESNEAU ALAIN DR
SPACE TELESCOPE SCI INST
HOMEWOOD CAMPUS
3700 SAN MARTIN DRIVE
BALTIMORE MD 21218
U.S.A.
TEL: 301-338-4800
TLX: 6849101
COM: 24

FRIEDLANDER MICHAEL PROF
DEPARTMENT OF PHYSICS
WASHINGTON UNIVERSITY
ST LOUIS MO 63130
U.S.A.
TEL: 314-889-6279
TLX:
COM:

FRANKLIN FRED A DR
PLANETARY SCIENCE DIV
CENTER FOR ASTROPHYSICS
60 GARDEN STREET
CAMBRIDGE MA 02138
U.S.A.
TEL: 617-495-7230
TLX:
COM: 20

FREDRICK LAURENCE W PROF
LEANDER MCCORMICK OBS
BOX 3818
UNIVERSITY STATION
CHARLOTTESVILLE VA 22903
U.S.A.
TEL: 804-924-4905
TLX: 510-587-5453 (TWX)
COM: 24,42,51

FRIBERG PER
ONSALA SPACE OBSERVATORY
CHALMERS UNIV TECHNOLOGY
S-439 00 ONSALA
SWEDEN
TEL: 0300-60650
TLX: 8542400 ONSPACE
COM: 40

FRIEDMAN HERBERT DR
US NAVAL RESEARCH LAB
CODE 7100
WASHINGTON DC 20375
U.S.A.
TEL:
TLX:
COM: 10,12,40,44,48

FRANSSON CLAES
STOCKHOLM OBSERVATORY
S-133 00 SALTSJOEBADEN
SWEDEN
TEL: 46-871-70195
TLX: 12972 SOBSERV S
COM:

FREEMAN KENNETH C PROF
MT STROMLO OBSERVATORY
PRIVATE BAG
WODEN PO
CANBERRA ACT 2606
AUSTRALIA
TEL: 062-881111
TLX: 62270 CANOPUS AA
COM: 28,33,37C

FRICKE KLAUS DR
UNIVERSITAETSSTERNWARTE
UNIVERSITAET GOETTINGEN
GEISMARLANDSTR 11
D-3400 GOETTINGEN
GERMANY, F.R.
TEL: 1149551-395051
TLX: 96753 USTERN D
COM: 35

FRINGANT ANNE-MARIE DR
OBSERVATOIRE DE PARIS
61 AVE DE L'OBSERVATOIRE
F-75014 PARIS
FRANCE
TEL: 1-43-20-12-10
TLX:
COM: 29

FRANTSMAN YU L DR
RADIOASTROPHYSICAL OBS
LATVIAN ACAD OF SCIENCES
226524 RIGA
U.S.S.R.
TEL: 226006
TLX:
COM: 35,42

FREIESLEBEN H C DR
APP 6120
FLORENTINER STR 20
D-7000 STUTTGART 75
GERMANY, F.R.
TEL: 0711-4702-6120
TLX:
COM: 41

FRICKE WALTER PROF DR
ASTRONOMISCHES RECHEN
INSTITUT
MOENCHHOFSTR 12-14
D-6900 HEIDELBERG
GERMANY, F.R.
TEL: 06221-412608
TLX: 461336 ARIHD D
COM: 04,05,08,33

FRISCH HELENE DR
OBSERVATOIRE DE NICE
BP 139
F-06003 NICE CEDEX
FRANCE
TEL: 93-89-04-20
TLX: 460004
COM: 36

FRANZ OTTO G DR
LOWELL OBSERVATORY
1400 W. MARS HILL RD
FLAGSTAFF AZ 86001
U.S.A.
TEL: 602-774-3358
TLX:
COM: 24,26

FREIRE FERRERO RUBENS G
OBS DE STRASBOURG
11 RUE DE L'UNIVERSITE
F-67000 STRASBOURG
FRANCE
TEL: 88-35-43-00
TLX:
COM: 29

FRIDMAN ALEKSEY M
ASTRONOMICAL COUNCIL
USSR ACADEMY OF SCIENCES
PYATNITSKAYA STR 48
109017 MOSCOW
U.S.S.R.
TEL: 231-54-61
TLX: 412623 SCSTP SU
COM:

FRISCH PRISCILLA
UNIVERSITY OF CHICAGO
ASTRONOMY & ASTROPHYS CTR
5640 S. ELLIS AVE
CHICAGO IL 60637
U.S.A.
TEL: 312-962-8211
TLX: 910-221-5617
COM: 34

FRASER C W DR
UNIVERSITY OBSERVATORY
BUCHANAN GARDENS
ST ANDREWS FIFE KY16 9LZ
U.K.
TEL:
TLX:
COM: 28

FREITAS MOURAO R R DR
MUSEU ASTR E CIENCIAS
AFINS/CNPQ RUA GEN BRUCE
SAN CRISTOVAO
20921 RIO DE JANEIRO
BRAZIL
TEL: 580-7154/7204
TLX: 22653
COM: 20,26,41

FRIED JOSEF WILHELM DR
MPI FUER ASTRONOMIE
KOENIGSTUHL
D-6900 HEIDELBERG 1
GERMANY, F.R.
TEL: 06221-5281
TLX: 461789 MPIA D
COM: 28

FRISCH URIEL DR
OBSERVATOIRE DE NICE
BP 139
F-06003 NICE CEDEX
FRANCE
TEL: 93-89-04-20
TLX: 460004
COM: 36

FRITZOVA-SVESTKA L DR
DOPPERSTRAAT 147
NL-3752 JC BUNSCHOTEN
NETHERLANDS
TEL: 03499-84403
TLX:
COM: 10

FROEHLICH CLAUS
WORLD RADIATION CENTER
PHYSIKALISCH-METEOROL OBS
POSFACH 173
CH-7260 DAVOS-DORF
SWITZERLAND
TEL: 41-083-521-31
TLX: 74732 PMOD CH
COM: 12

FROESCHLE CHRISTIANE D DR
OBSERVATOIRE DE NICE
BP 139
F-06003 NICE CEDEX
FRANCE
TEL:
TLX:
COM: 36

FROESCHLE CLAUDE DR
OBSERVATOIRE DE NICE
BP 139
F-06003 NICE CEDEX
FRANCE
TEL: 93-89-04-20
TLX:
COM: 07,20

FROESCHLE MICHEL DR
CERGA
AVENUE COPERNIC
F-06130 GRASSE
FRANCE
TEL:
TLX:
COM:

FROGEL JAY ALBERT DR
NAT OPTICAL ASTR OBS
PO BOX 26732
950 N. CHERRY AVENUE
TUCSON AZ 85726
U.S.A.
TEL: 602-327-5511
TLX:
COM: 28

FROLOV M S DR
ASTRONOMICAL COUNCIL
USSR ACADEMY OF SCIENCES
PYATNITSKAYA UL 48
109017 MOSCOW
U.S.S.R.
TEL: 231-54-61
TLX: 412623 SCSTP SU
COM: 27

FROST KENNETH J DR
NASA/GSFC, SPACE STATION
OFFICE, CODE 600.2
GREENBELT RD, BLDG 16
GREENBELT MD 20771
U.S.A.
TEL: 301-344-8824
TLX:
COM:

FRYE GLENN M PROF
PHYSICS DEPARTMENT
CASE WESTERN RESERVE UNIV
ROCK BUILDING
CLEVELAND OH 44106
U.S.A.
TEL: 216-368-2997
TLX:
COM:

FTACLAS CHRIST
SPACE SCIENCE DIVISION
PERKIN-ELMER CORP.
100 WOOSTER HEIGHTS RD
DANBURY CT 06810-7589
U.S.A.
TEL: 203-797-6448
TLX:
COM: 28

FU CHENG-QI
SHANGHAI OBSERVATORY
ACADEMIA SINICA
SHANGHAI
CHINA, PEOPLE'S REP.
TEL: 386191
TLX: 33164 SHAO CN
COM: 44

FU DELIAN
BEIJING ASTRONOMICAL OBS
ACADEMIA SINICA
BEIJING
CHINA, PEOPLE'S REP.
TEL: 282070
TLX: 22040 BAOAS CN
COM: 09

FU QI JUN
ASTRONOMY PROGRAM
UNIVERSITY OF MARYLAND
COLLEGE PARK MD 20742
U.S.A.
TEL: 301-454-3001
TLX: 710-826-0352
COM: 40

FU-SHONG KUO
DEPARTMENT OF PHYSICS
INST PHYSICS & ASTRONOMY
NAT CENTRAL UNIVERSITY
CHUNG LI
CHINA, TAIWAN
TEL:
TLX:
COM:

FUCHS BURKHARD DR
ASTRONOMISCHES-RECHEN
INSTITUT
MOENCHHOFSTR. 12-14
D-6900 HEIDELBERG 1
GERMANY, F.R.
TEL: 06221-49026
TLX: 461336 ARIHD D
COM: 28,33

FUCHS JOSEPH PROF DR
MARIATROSTERSTR.111/9
A-8043 GRAZ
AUSTRIA
TEL:
TLX:
COM:

FUENMAYOR FRANCISCO J DR
UNIVERSIDAD DE LOS ANDES
FACULDAD DE CIENCIAS
DEPARTAMENTO DE FISICA
MERIDA 5101
VENEZUELA
TEL: 074-63-99-30
TLX: 74173 CDCH-ULA
COM: 46

FUERST ERNST DR
MPI FUER RADIOASTRONOMIE
AUF DEM HUEGEL 69
D-5300 BONN
GERMANY, F.R.
TEL:
TLX: 886440 MPIFR D
COM: 40

FUJIMOTO MASA-KATSU DR
TOKYO ASTR OBSERVATORY
TOKYO 181
JAPAN
TEL: 0422-32-5111
TLX: 2822307
COM: 31,33,51

FUJIMOTO MASAYUKI DR
NIIGATA UNIVERSITY
FACULTY OF EDUCATION
8050 IKARASHI-2
NIIGATA 950-21
JAPAN
TEL:
TLX:
COM: 35

FUJIMOTO MITSUAKI DR
DEPT OF PHYSICS
NAGOYA UNIVERSITY
NAGOYA 464
JAPAN
TEL:
TLX:
COM: 47,51

FUJITA YOSHIO PROF
DEPARTMENT OF ASTRONOMY
UNIVERSITY OF TOKYO
BUNKYO KU
TOKYO 113
JAPAN
TEL: 423-74-4186
TLX:
COM: 29

FUJIWARA AKIRA DR
DEPARTMENT OF PHYSICS
KYOTO UNIVERSITY
KITASHIRAKAWA SAKYOKU
KYOTO 606
JAPAN
TEL: 075-751-2111
TLX: 5422693 LIBKYU J
COM: 16

FUKUDA ICHIRO
DEPT OF PHYSICS
KANAZAWA INST OF TECHNOL.
7-1 OGIGAOKA, NONOICHI
ISHIKAWA 921
JAPAN
TEL: 0762-48-1100
TLX: 5122456 KIT LC J
COM: 45

FUKUI TAKAO DR
DEPT OF LIBERAL ARTS
DOKKYO UNIVERSITY
SAKAE-MACHI 600
SOKA SAITAMA
JAPAN
TEL: 0489-42-1111
TLX:
COM: 47

FUKUI YASUO DR
DEPARTMENT OF PHYSICS
NAGOYA UNIVERSITY
FUROCHO CHIKUSAKU
NAGOYA 464
JAPAN
TEL: 052-781-5111
TLX: 4477323 SCUNAG J
COM: 40

FULCHIGNONI MARCELLO PROF
IST. ASTROFISICA SPAZIALE
E.N.R.-C.N.R.
I-00044 FRASCATI
ITALY
TEL: 39-6-4956951
TLX: 680489 CNR FRA
COM: 15

FURENLID INGEMAR K DR
DEPT PHYSICS & ASTRONOMY
GEORGIA STATE UNIVERSITY
ATLANTA GA 30303
U.S.A.
TEL: 404-658-2932
TLX:
COM:

FURNISS IAN
DEPT PHYSICS & ASTRONOMY
UNIVERSITY COLLEGE LONDON
GOWER STREET
LONDON WC1E 6BT
U.K.
TEL: 01-387-7050
TLX: 28722 UCPHYS
COM: 34,44

FURSENKO M A DR
INST OF THEORET ASTRONOMY
USSR ACADEMY OF SCIENCES
10 KUTUZOV QUAY
191187 LENINGRAD
U.S.S.R.
TEL: 278-88-98
TLX: 121578 ITA SU
COM: 04

FUSCO-FEMIANO ROBERTO
IST. ASTROFISICA SPAZIALE
C P 67
I-00044 FRASCATI
ITALY
TEL: 9425655
TLX: 610261
COM:

FUSI PECCI FLAVIO
OSSERVATORIO ASTRONOMICO
UNIVERSITARIO
C P 596
I-40100 BOLOGNA
ITALY
TEL: 51-222956
TLX: 211664 INFNBO-I
COM:

LIST OF MEMBERS

GABRIEL ALAN H
RUTHERFORD APPLETON LAB
SPACE & ASTROPHYS DIV
CHILTON DIDCOT OX11 0QX
U.K.
TEL: 235-21900 E6206
TLX: 83159
COM: 10,12,14C,44

GABRIEL MAURICE R DR
INSTITUT D'ASTROPHYSIQUE
UNIVERSITE DE LIEGE
AVENUE DE COINTE 5
B-4200 COINTE-OUGREE
BELGIUM
TEL: 041-52-99-80
TLX: 41264 ASTRLG
COM: 35

GAHM GOESTA F DR
STOCKHOLM OBSERVATORY
S-133 00 SALTSJOEBADEN
SWEDEN
TEL: 08-717-0637
TLX: 12972
COM: 27

GAIGNEBET JEAN DR
CERGA
AVENUE COPERNIC
F-06130 GRASSE
FRANCE
TEL: 93-36-58-49
TLX: 470865 F
COM: 19,31

GAIL HANS-PETER DR
INST THEORET ASTROPHYSIK
DER UNIVERSITAET
IM NEUENHEIMER FELD 294
D-6900 HEIDELBERG 1
GERMANY, F.R.
TEL:
TLX:
COM: 36

GAISSER THOMAS K
BARTOL RESEARCH FOUND.
UNIVERSITY OF DELAWARE
NEWARK DE 19716
U.S.A.
TEL: 302-451-8111
TLX: 510-666-0805 BARTOL
COM: 48

GAIZAUSKAS VICTOR DR
HERZBERG INST ASTROPHYS
NATL RESEARCH COUNCIL
OTTAWA ONT K1A 0R6
CANADA
TEL: 613-993-7395
TLX: 0533715 NRCOTT
COM: 10,12

GALAL A A DR
HELWAN OBSERVATORY
HELWAN-CAIRO
EGYPT
TEL: 780645, 782683
TLX:
COM:

GALAN MAXIMINO J
OBSERVATORIO ASTRONOMICO
ALFONSO XII-3
28014 MADRID
SPAIN
TEL: 9-227-0107
TLX: 92640
COM: 09,50

GALEOTTI PIERO PROF
IST. DI COSMO-GEOFISICA
CORSO FIUME 4
I-10133 TORINO
ITALY
TEL: 0039-11-658979
TLX: 224379 COSMOT I
COM: 48

GALIBINA I V DR
INSTITUTE OF THEORETICAL
ASTRONOMY
10 KUTUZOV QUAY
191187 LENINGRAD
U.S.S.R.
TEL: 186-19-74
TLX: 121578 ITA SU
COM: 07,20

GALKIN LEONID S A DR
CRIMEAN ASTROPHYSICAL OBS
334413 PO NAUCHNY, CRIMEA
U.S.S.R.
TEL: 32945
TLX:
COM: 16

GALLAGHER III JOHN S DR
NAT OPTICAL ASTR OBS
KITT PEAK NAT OBSRVATORY
BOX 26732
TUCSON AZ 85726
U.S.A.
TEL: 602-329-5511
TLX: 066484 AURA KPUO TUC
COM: 27,28

GALLET ROGER M
964 7TH STREET
BOULDER CO 80302
U.S.A.
TEL:
TLX:
COM:

GALLETTA GIUSEPPE PROF
ISTITUTO DI ASTRONOMIA
UNIVERSITA DI PADOVA
VIC. DELL'OSSERVATORIO 5
I-35122 PADOVA
ITALY
TEL: 49-66-1499
TLX: 430176 UNPADU I
COM: 28

GALLETTO DIONIGI
IST. DI FISICA MATEMATICA
UNIVERSITA DI TORINO
VIA CARLO ALBERTO 10
I-10123 TORINO
ITALY
TEL: 011-539214
TLX:
COM: 07,33,47

GALLINO ROBERTO
IST. DI FISICA GENERALE
DELL'UNIVERSITA'
CORSO M. D'AZEGLIO 46
I-10125 TORINO
ITALY
TEL: 011-655103
TLX: 21104 INFN TO
COM: 35

GALLOUET LOUIS DR
OBSERVATOIRE DE PARIS
61 AVE DE L'OBSERVATOIRE
F-75014 PARIS
FRANCE
TEL: 1-43-20-12-10
TLX:
COM: 24,25

GALLOWAY DAVID R
MPI FUER ASTROPHYSIK
KARL-SCHWARZSCHILD-STR 1
D-8046 GARCHING B MUNCHEN
GERMANY, F.R.
TEL: 89-3299-9428
TLX: 524629 ASTRO D
COM: 10

GALPERIN YU I PROF
SPACE RESEARCH INSTITUTE
USSR ACADEMY OF SCIENCES
117810 MOSCOW GSP-7
U.S.S.R.
TEL: 333-31-22
TLX: 411498 STAR SU
COM: 21C

GALT JOHN A DR
DOMINION RADIO ASTROPHY-
SICAL OBSERVATORY
P O BOX 248
PENTICTON BC V2A 6K3
CANADA
TEL: 604-497-5321
TLX: 048-88127
COM: 40

GAMALELDIN ABDULLA I DR
HELWAN OBSERVATORY
HELWAN-CAIRO
EGYPT
TEL: 780645
TLX:
COM: 28

GAMMELGAARD PETER MAG SCI
INSTITUTE OF ASTRONOMY
UNIVERSITY OF AARHUS
LANGELANDSGADE
DK-8000 AARHUS C
DENMARK
TEL: 06-128899
TLX: 64767 AAUSCI DK
COM:

GAO BILIE
INAOE
AP POSTALES 216 y 51
72000 PUEBLA, PUÉ.
MEXICO
TEL:
TLX:
COM: 09

GAO BUXI
INSTITUTE OF GEODESY &
GEOPHYSICS
XU DONG LU
WUCHAN, HUBEI
CHINA, PEOPLE'S REP.
TEL: 813805
TLX:
COM: 19

GAPOSCHKIN EDWARD M DR
55 FARMCREST AVE
LEXINGTON MA 02173
U.S.A.
TEL: 617-862-2538
TLX:
COM: 07,19

GARAY GUIDO DR
ESO
KARL-SCHWARZSCHILD-STR 2
D-8046 GARCHING B MUNCHEN
GERMANY, F.R.
TEL:
TLX:
COM: 40

GARCIA DE LA ROSA JOSE I
INST DE ASTROFISICA DE
CANARIAS
LA LAGUNA
38071 TENERIFE
SPAIN
TEL: 922-26-22-11
TLX: 92640
COM: 10

GARCIA-BARRETO JOSE A
INST. DE ASTRONOMIA UNAM
OBS ASTRONOMICO NACIONAL
APDO POSTAL 877
22860 ENSENADA, B. CALIF.
MEXICO
TEL: 667-830-93
TLX:
COM:

GARCIA-PELAYO JOSE DR
INST DE ASTROFISICA DE
ANDULACIA, APDO 2144
PROFESOR ALBAREDA 1
18080 GRANADA
SPAIN
TEL: 25-61-03
TLX:
COM:

GARDNER FRANCIS F DR
CSIRO
DIVISION OF RADIOPHYSICS
P.O.76
EPPING NSW 2121
AUSTRALIA
TEL: 02-868-0222
TLX: 26230 AA
COM: 34,40

GARFINKEL BORIS DR
YALE UNIVERSITY OBS
NEW HAVEN CT 06520
U.S.A.
TEL: 203-436-3460
TLX:
COM: 07,20

LIST OF MEMBERS

GARLICK GEORGE F DR
267 SOUTH BELOIT AVE
LOS ANGELES CA 90049
U.S.A.
TEL: 213-472-3512
TLX:
COM:

GARY DALE E
CALTECH
SOLAR ASTRONOMY 264-33
PASADENA CA 91125
U.S.A.
TEL: 818-356-3863
TLX: 675425 CALTECH PSD
COM:

GAUTIER DANIEL
OBSERVATOIRE DE PARIS
SECTION DE MEUDON
F-92195 MEUDON PL CEDEX
FRANCE
TEL: 1-45-34-75-70
TLX: 201571 LAM
COM: 16C

GEHRZ ROBERT DOUGLAS DR
DEPT PHYSICS & ASTRONOMY
UNIVERSITY OF WYOMING
UNIV STATION BOX 3905
LARAMIE WY 82071
U.S.A.
TEL: 307-766-6176
TLX:
COM: 25

GARMANY CATHERINE D DR
JILA
UNIVERSITY OF COLORADO
BOULDER CO 80309
U.S.A.
TEL: 303-492-7836
TLX:
COM: 42

GASCOIGNE S C B DR
MT STROMLO OBSERVATORY
WCDEN P.O. ACT 2606
AUSTRALIA
TEL:
TLX:
COM: 27,28,37

GAY JEAN DR
CERGA
AVENUE COPERNIC
F-06130 GRASSE
FRANCE
TEL: 93-36-58-49
TLX: 470865
COM: 09,34

GEISS JOHANNES PROF
PHYSIK INSTITUT
UNIVERSITAET BERN
SIDLERSTRASSE 5
CH-3012 BERN
SWITZERLAND
TEL: 31-65-44-02
TLX: 32320
COM: 16

GARMIRE GORDON P PROF
PENNA STATE UNIVERSITY
525 DAVEY LAB
UNIVERSITY PARK PA 16802
U.S.A.
TEL: 814-865-0418
TLX: 842510 PENNSTBSTR SC
COM: 48

GASKA STANISLAW DR
INSTITUTE ASTRONOMY
UL CHOPINA 12-18
87-100 TORUN
POLAND
TEL:
TLX:
COM: 07

GAYLARD MICHAEL JOHN
NATL INST FOR TELECOM RES
CSIR
P O BOX 3718
JOHANNESBURG 2000
SOUTH AFRICA
TEL: 011-6424692
TLX: 321006
COM: 40

GELDZAHLER BERNARD J
NAVAL RESEARCH LABORATORY
CODE 4121.6
WASHINGTON DC 20375
U.S.A.
TEL:
TLX:
COM: 40

GARNIER ROBERT ING
OBSERVATOIRE DE LYON
F-69230 ST-GENIS-LAVAL
FRANCE
TEL: 78-56-07-05
TLX:
COM:

GATEWOOD GEORGE DIRECTOR
ALLEGHENY OBSERVATORY
OBSERVATORY STATION
PITTSBURG PA 15214
U.S.A.
TEL: 412-321-2400
TLX:
COM: 24,26,51C

GEAKE JOHN E DR
PHYSICS DEPT
UMIST
MANCHESTER M60 1QD
U.K.
TEL: 61-236-3311
TLX: 666094
COM: 16

GELFREIKH GEORGIJ B DR
PULKOVO OBSERVATORY
196140 LENINGRAD
U.S.S.R.
TEL:
TLX:
COM: 10,40

GARRIDO RAFAEL
INST DE ASTROFISICA DE
ANDULACIA, APDO 2144
PROFESOR ALBAREDA 1
18080 GRANADA
SPAIN
TEL: 58-121311
TLX: 78753
COM: 27

GATLEY IAN
UK INFRARED TELESCOPE
665 KOMOHANA STREET
HILO HI 96720
U.S.A.
TEL: 808-961-3756
TLX: 633135
COM:

GEBBIE KATHARINE B DR
JILA
UNIVERSITY OF COLORADO
BOULDER CO 80309
U.S.A.
TEL: 303-492-7825
TLX: 755842 JILA
COM: 36

GELLER MARGARET JOAN
CENTER FOR ASTROPHYSICS
60 GARDEN STREET
CAMBRIDGE MA 02138
U.S.A.
TEL: 617-495-7409
TLX: 921428 SATELLITE CAM
COM: 47

GARRISON ROBERT F PROF
DAVID DUNLAP OBSERVATORY
P O BOX 360
RICHMOND HILL ONT L4C 4Y6
CANADA
TEL: 416-884-9562
TLX: 06-986766
COM: 29,45P

GAUR V P
UTTAR PRADESH STATE OBS
MANORA PEAK
NAINITAL (UP) 263 129
INDIA
TEL: 2136
TLX:
COM: 12

GEBLER KARL-HEINZ DR
RADIOASTRONOMISCHES INST
DER UNIVERSITAET BONN
AUF DEM HUEGEL 71
D-5300 BONN 1
GERMANY, F.R.
TEL: 0228-733662
TLX:
COM: 40

GENKIN IGOR L PROF DR
PHYSICS FACULTY
KAZAKH STATE UNIVERSITY
KOMSOMOLSKAYA 96
480012 ALMA ATA
U.S.S.R.
TEL: 67-70-18
TLX:
COM: 33

GARSTANG ROY H PROF
JILA
UNIVERSITY OF COLORADO
BOULDER CO 80309
U.S.A.
TEL: 303-492-7795
TLX: 755842 JILA
COM: 05,14

GAUSS F STEPHEN
US NAVAL OBSERVATORY
WASHINGTON DC 20390
U.S.A.
TEL: 202-653-1510
TLX:
COM: 08,09

GEHRELS TOM PROF
LUNAR LABORATORY
UNIVERSITY OF ARIZONA
TUCSON AZ 85721
U.S.A.
TEL: 602-621-6970
TLX:
COM: 15,16,20,25,34,51

GENT HUBERT MR
PROSPECT HOUSE
SHERWOOD LANE
WORCESTER WR2 4NX
U.K.
TEL: 422 186
TLX:
COM: 40

GARTON W R S PROF
BLACKETT LABORATORY
IMPERIAL COLLEGE
LONDON SW7 2BZ
U.K.
TEL: 0233-21657
TLX: 261503
COM: 14

GAUSTAD JOHN E PROF
DEPT OF ASTRONOMY
SWARTHMORE COLLEGE
SWARTHMORE PA 19081
U.S.A.
TEL: 215-447-7271
TLX:
COM: 34

GEHREN THOMAS PH D
INST ASTRON & ASTROPHYSIK
UNIVERSITAETS STERNWARTE
SCHEINERSTRASSE 1
D-8000 MUENCHEN 80
GERMANY, F.R.
TEL: 89-98-90-21
TLX: 529815 UNIVERS D
COM: 29

GENZEL REINHARD DR
DEPT OF PHYSICS
UNIVERSITY OF CALIFORNIA
563 BIRGE HALL
BERKELEY CA 94720
U.S.A.
TEL: 415-642-6577
TLX: 9103667114
COM: 40

LIST OF MEMBERS

GEORGELIN YVON P DR
OBSERVATOIRE DE MARSEILLE
2 PLACE LE VERRIER
F-13248 MARSEILLE CEDEX
FRANCE
TEL: 91-95-90-88
TLX: 420241 F
COM: 30,33,34

GEORGELIN YVONNE M DR
OBSERVATOIRE DE MARSEILLE
2 PLACE LE VERRIER
F-13248 MARSEILLE CEDEX 4
FRANCE
TEL: 91-95-90-88
TLX: 420241 F
COM: 33

GERARD ERIC DR
OBSERVATOIRE DE PARIS
SECTION DE MEUDON
F-92195 MEUDON PL CEDEX
FRANCE
TEL: 1-45-34-75-30
TLX: 270912 OBSASTR.
COM: 15,34

GERBAL DANIEL DR
OBSERVATOIRE DE PARIS
SECTION DE MEUDON
F-92195 MEUDON PL CEDEX
FRANCE
TEL: 1-45-34-75-70
TLX:
COM:

GERBALDI MICHELE DR
INSTITUT D'ASTROPHYSIQUE
98 BIS BD ARAGO
F-75014 PARIS
FRANCE
TEL: 1-43-20-14-25
TLX:
COM: 29,45,46C

GERGELY TOMAS ESTEBAN DR
DIV ASTRONOMICAL SCIENCES
NATL SCIENCE FOUNDATION
1800 G STREET N.W.
WASHINGTON DC 20550
U.S.A.
TEL: 202-357-9696
TLX:
COM: 10,40

GERHARD ORTWIN
MPI F PHYS & ASTROPHYSIK
INSTITUT FUR ASTROPHYSIK
KARL-SCHWARZSCHILD-STR 1
D-8046 GARCHING B MUNCHEN
GERMANY, F.R.
TEL: 089-3299-0
TLX: 524629 ASTRO D
COM: 28

GERLEI OTTO
HELIOPHYSICAL OBSERVATORY
H-4010 DEBRECEN
HUNGARY
TEL:
TLX:
COM:

GEROLA HUMBERTO DR
IBM CORPORATION
DEPT K64/282
5600 COTTLE ROAD
SAN JOSE CA 95193
U.S.A.
TEL:
TLX:
COM: 34

GEROYANNIS VASSILIS S DR
DEPT OF ASTRONOMY
UNIVERSITY OF PATRAS
GR-26110 PATRAS
GREECE
TEL:
TLX:
COM: 35

GERSHBERG R E DR
CRIMEAN ASTROPHYSICAL OBS
NAUCHNY
334413 CRIMEA
U.S.S.R.
TEL:
TLX:
COM: 27C,29

GESZTELYI LIDIA
DEBRECEN HELOPHYSICAL
OBSERVATORY
BOX 30
H-4010 DEBRECEN
HUNGARY
TEL: 52-11-015
TLX: 072517 DEOBS H
COM:

GEYER EDWARD H PROF DR
OBSERVATORIUM HOHER LIST
UNIVERSITAET BONN
D-5568 DAUN/EIFEL
GERMANY, F.R.
TEL: 06592-2150
TLX:
COM: 26,27,42,45

GEZARI DANIEL YSA DR
NASA-GSFC
CODE 693
GREENBELT MD 20771
U.S.A.
TEL: 301-344-1432
TLX:
COM: 34,44

GHIGO FRANCIS D DR
DEPT OF ASTRONOMY
UNIVERSITY OF MINNESOTA
116 CHURCH ST S.E.
MINNEAPOLIS MN 55455
U.S.A.
TEL: 612-376-8644
TLX:
COM: 28,40,51

GHOBROS ROSHDY AZER DR
HELWAN OBSERVATORY
HELWAN-CAIRO
EGYPT
TEL:
TLX:
COM:

GHOSH P DR
TATA INSTITUTE OF
FUNDAMENTAL RESEARCH
BOMBAY 400 005
INDIA
TEL: 21-9111 EXT 260
TLX: 011-3009
COM:

GHOSH S K DR
TATA INSTITUTE OF
FUNDAMENTAL RESEARCH
HOMI BHABHA RD
BOMBAY 400 005
INDIA
TEL: 219111
TLX: 011-3009 TIFR IN
COM: 25

GIACAGLIA GIORGIO E PROF
ESCOLA POLITECNICA
UNIVERSIDADE DE SAO PAULO
CAIXA POSTAL 8174
05508 SAO PAULO
BRAZIL
TEL: 55-11-32237
TLX:
COM: 07

GIACCONI RICCARDO PROF
SPACE TELESCOPE SCI INST
HOMEWOOD CAMPUS
3700 SAN MARTIN DRIVE
BALTIMORE MD 21218
U.S.A.
TEL: 301-338-4711
TLX: 6849101 ST SCI
COM: 44,48C

GIACHETTI RICCARDO PROF
UNIVERSITY OF FLORENCE
DEPARTMENT OF PHYSICS
LARGO E. FERMI 2
I-50100 FIRENZE
ITALY
TEL: 229-8141
TLX: 572570
COM:

GIAMPAPA MARK S
NATL SOLAR OBSERVATORY
PO BOX 26732
950 N. CHERRY AVE
TUCSON AZ 85726-6732
U.S.A.
TEL: 602-327-5511
TLX: 0666484 AURA NOACTUC
COM:

GIANNONE PIETRO PROF
OSSERVATORIO ASTRONOMICO
VIALE DEL PARCO MELLINI 8
I-00136 ROMA
ITALY
TEL: 3452794
TLX:
COM: 35,42

GIANNUZZI MARIA A DR
DIPT. DI MATEMATICA
UNIV DI ROMA LA SAPIENZA
PIAZZA GRAMSCI 5
I-00041 ALBANO/LAZIALE
ITALY
TEL: 06-932-11-01
TLX:
COM:

GIBSON DAVID MICHAEL DR
PHYSICS DEPT
NM INST MINING TECHN
CAMPUS STATION
SOCORRO NM 87801
U.S.A.
TEL: 505-835-5340
TLX:
COM: 15,27,40,42C

GIBSON JAMES
JPL, CALTECH
MAIL STOP 138-307
4800 OAK GROVE DRIVE
PASADENA CA 91103
U.S.A.
TEL: 818-354-3074
TLX: 675429
COM: 20

GICLAS HENRY L MR
120 E. ELM AVE
FLAGSTAFF AZ 86001
U.S.A.
TEL: 602-774-4769
TLX:
COM: 16,20,24

GIERASCH PETER J DR
DEPARTMENT OF ASTRONOMY
CORNELL UNIVERSITY
ITHACA NY 14853
U.S.A.
TEL: 607-256-3507
TLX:
COM: 16

GIEREN WOLFGANG P DR
OBSERVATORIO NACIONAL
APARTADO 2584
BOGOTA 1, D.E.
COLOMBIA
TEL: 2442834/2140772
TLX:
COM: 27

GIESE RICHARD H PROF
RUHR UNIVERSITAET BOCHUM
BEREICH EXTRATERR PHYSIK
NB 7/30
D-4630 BOCHUM
GERMANY, F.R.
TEL: 0234-700-3441
TLX: 0825860 RUB D
COM: 15,21C,22

GIESEKING FRANK DR
OBSERVATORIUM HOHER LIST
UNIVERSITAET BONN
D-5568 DAUN
GERMANY, F.R.
TEL:
TLX:
COM: 30

GIETZEN JOSEPH W
ROYAL GREENWICH OBS
HAILSHAM BN27 1RP
U.K.
TEL: 32-181-3171
TLX:
COM:

LIST OF MEMBERS

GILLILAND RONALD LYNN
HIGH ALTITUDE OBSERVATORY
NCAR
PO BOX 3000
BOULDER CO 80307
U.S.A.
TEL: 303-497-1565
TLX: 45694
COM: 10

GILLINGHAM PETER MR
ANGLO AUSTRALIAN OBS
PRIVATE BAG
COONABARABRAN NSW 2357
AUSTRALIA
TEL: 068-42-1122
TLX: 63945AA
COM: 09

GILMAN PETER A DR
HIGH ALTITUDE OBSERVATORY
NCAR
PO BOX 3000
BOULDER CO 80307
U.S.A.
TEL: 303-497-1560
TLX:
COM: 10

GILMORE ALAN C MR
MT JOHN OBSERVATORY
P O BOX 57
LAKE TEKAPO
NEW ZEALAND
TEL: 64-5-056-813
TLX:
COM: 20

GILMORE GERARD FRANCIS
INST OF ASTRONOMY
MADINGLEY ROAD
CAMBRIDGE CB3 0HA
U.K.
TEL: 0223-62204
TLX: 81797 ASTRON G
COM: 33

GILMOZZI ROBERTO
IST. ASTROFISICA SPAZIALE
C P 67
I-00044 FRASCATI
ITALY
TEL:
TLX:
COM:

GILRA DAYA P DR
SM SYSTEMS & RESEARCH CO.
8401 CORPORATE DRIVE
SUITE 450
LANDOVER MD 20785
U.S.A.
TEL: 301-763-4483
TLX:
COM: 29,34,44

GIMENEZ ALVARO
DEPTO DE ASTROFISICA
FACULTAD DE FISICA
UNIVERSIDAD COMPLUTENSE
28040 MADRID
SPAIN
TEL: 1-449-53-16
TLX: 47273 FFUC
COM: 42

GINGERICH OWEN PROF
CENTER FOR ASTROPHYSICS
60 GARDEN STREET
CAMBRIDGE MA 02138
U.S.A.
TEL: 617-495-7216
TLX: 921428 SATELLITE CAM
COM: 41

GINGOLD ROBERT ARTHUR DR
MOUNT STROMLO & SIDING
SPRING OBSERVATORIES
PRIVATE BAG
WODEN P.O. ACT 2606
AUSTRALIA
TEL: 062-881111
TLX: 62270 AA
COM: 35

GINZBURG VITALY L PROF
P N LEBEDEV PHYS INST
LENINSKY PROSPECT 53
117924 MOSCOW B 333
U.S.S.R.
TEL:
TLX:
COM: 40,48,51

GIOIA ISABELLA M DR
HARVARD-SMITHSONIAN CTR
FOR ASTROPHYSICS
60 GARDEN STREET
CAMBRIDGE MA 02138
U.S.A.
TEL: 617-495-7138
TLX: 921428 SATELLITE CAM
COM: 40

GIOVANARDI CARLO
OSSERVATORIO ASTROFISICO
DI ARCETRI
LARGO E. FERMI 5
I-50125 FIRENZE
ITALY
TEL: 55-220034
TLX: 572268 ARCETR I
COM: 28

GIOVANE FRANK
SPACE ASTRONOMY LAB
1810 NW 6TH STREET
GAINESVILLE FL 32609
U.S.A.
TEL: 904-392-5450
TLX: 810-825-2308
COM: 15,21

GIOVANELLI RICCARDO DR
ARECIBO OBSERVATORY
NAT ASTR & IONOSPHERE CTR
PO BOX 995
ARECIBO PR 00613
U.S.A.
TEL: 809-878-2612
TLX: 385638
COM: 28,34

GIOVANNELLI FRANCO DR
IST. ASTROFISICA SPAZIALE
CP 67
I-00044 FRASCATI
ITALY
TEL: 39-6-942-565155
TLX: 610261 CNRFRAI
COM: 42,51

GIOVANNINI GABRIELE
IST. DI RADIOASTRONOMIA
VIA IRNERIO 46
I-40126 BOLOGNA
ITALY
TEL: 51-232-856
TLX: 211664 INFN BO I
COM: 40

GIRAUD EDMOND
LAB ASTROPHYS. THEORIQUE
COLLEGE DE FRANCE
98 BIS BOULEVARD ARAGO
F-75014 PARIS
FRANCE
TEL: 1-43-20-14-25
TLX:
COM:

GIRIDHAR SUNETRA DR
INDIAN INSTITUTE OF
ASTROPHYSICS
BANGALORE 560 034
INDIA
TEL:
TLX: 845763 IIAB IN
COM: 35

GIURICIN GIULIANO
OSSERVATORIO ASTRONOMICO
VIA G.B. TIEPOLO 11
I-34131 TRIESTE
ITALY
TEL: 40-793221/921
TLX: 461137 OAT I
COM: 42,47

GLAGOLEVSKIJ JU V DR
SPEC ASTROPHYSICAL OBS
USSR ACADEMY OF SCIENCES
NIZHNIJ ARKHYZ
357147 STAVROPOLSKIJ KRAJ
U.S.S.R.
TEL: 93-577
TLX:
COM: 14,27,29,45

GLASER HAROLD DR
1346 BONITA ST
BERKELEY CA 94709
U.S.A.
TEL: 415-527-1860
TLX:
COM: 44

GLASNER SHIMON AMI
RACAH INST. OF PHYSICS
HEBREW UNIV. OF JERUSALEM
JERUSALEM 91904
ISRAEL
TEL: 02-58-4521
TLX: 25391 HUIL
COM:

GLASPEY JOHN W DR
DEPARTEMENT DE PHYSIQUE
UNIVERSITE DE MONTREAL
CP 6128 SUCCURSALE A
MONTREAL QUE H3C 3J7
CANADA
TEL: 514-343-6682
TLX: 05562425
COM:

GLASS BILLY PRICE DR
DEPT OF GEOLOGY
UNIVERSITY OF DELAWARE
NEWARK DE 19716
U.S.A.
TEL: 302-451-8458
TLX:
COM: 22

GLASS IAN STEWART DR
S A A O
P O BOX 9
OBSERVATORY 7935
SOUTH AFRICA
TEL: 021-47-00-25
TLX: 57-20309 SA
COM: 09,25C,28

GLASSGOLD ALFRED E PROF
NEW YORK UNIVERSITY
PHYSICS DEPARTMENT
4 WASHINGTON PLACE
NEW YORK NY 10003
U.S.A.
TEL: 212-598-2020
TLX: 235128 NYU UR
COM:

GLATZMAIER GARY A
LOS ALAMOS NATIONAL LAB
ESS-5 M5F665
LOS ALAMOS NM 87545
U.S.A.
TEL: 505-667-7647
TLX:
COM: 10,12,35

GLEBOCKI ROBERT PROF
INSTITUTE OF THEORETICAL
PHYSICS & ASTROPHYSICS
UL. WITA STWOSZA 57
80-952 GDANSK
POLAND
TEL: 41-87-00
TLX: 0512706 IFAS
COM:

GLEBOVA NINA I DR
INST THEORET ASTRONOMY
USSR ACADEMY OF SCIENCES
10 KUTUZOV QUAY
191187 LENINGRAD
U.S.S.R.
TEL: 278-88-98
TLX: 121578 ITA SU
COM: 04

GLEDHILL JOHN A PROF
DEPT OF PHYSICS
RHODES UNIVERSITY
P O BOX 94
GRAHAMSTOWN 6140
SOUTH AFRICA
TEL: 027-461-7128
TLX: 244226 SA
COM:

GLEISSBERG WOLFGANG PROF
BUCHENWEG 12
D-6374 OBERURSEL
GERMANY, F.R.
TEL:
TLX:
COM: 10

GLENCROSS WILLIAM M DR
DEPT PHYSICS & ASTRONOMY
UNIVERSITY COLLEGE LONDON
GOWER STREET
LONDON WC1E 6BT
U.K.
TEL: 01-387-7050
TLX: 28722 UCPHYS G
COM:

GLIESE WILHELM DR
ASTRONOMISCHES RECHEN
INSTITUT
MOENCHHOFSTR 12-14
D-6900 HEIDELBERG 1
GERMANY, F.R.
TEL: 06221-49026
TLX: 461336 ARIHD D
COM: 08,24C,33

GLUSHNEVA I N DR
STERNBERG STATE ASTR INST
UNIVERSITETSKIJ PROSP. 13
119899 MOSCOW
U.S.S.R.
TEL: 139-20-46
TLX:
COM: 29

GNEDIN YURIJ N DR
LENINGRAD PHYS TECH INST
ACADEMY OF SCIENCES
194021 LENINGRAD
U.S.S.R.
TEL:
TLX:
COM:

GNEVYSHEV MSTISLAV N DR
PULKOVO OBSERVATORY
196140 LENINGRAD
U.S.S.R.
TEL:
TLX:
COM: 12

GNEVYSHEVA RAISA S DR
PULKOVO OBSERVATORY
196140 LENINGRAD
U.S.S.R.
TEL:
TLX:
COM: 10

GODART ODON PROF
RUE DE CHATEAU 96
B-1288 ROUSVAL
BELGIUM
TEL: 010-613-817
TLX:
COM: 47

GODFREY PETER DOUGLAS DR
CHEMISTRY DEPARTMENT
MONASH UNIVERSITY
WELLINGTON ROAD
CLAYTON VIC 3168
AUSTRALIA
TEL: 03-541-0811
TLX: 32691 AA
COM: 34

GODOLI GIOVANNI PROF
ISTITUTO DI ASTRONOMIA
UNIVERSITA
LARGO E. FERMI 5
I-50125 FIRENZE
ITALY
TEL: 220034
TLX: 572268 ARCETR I
COM: 10,12,27,38,51

GODWIN JON GUNNAR DR
UNIVERSITY OBSERVATORY
SOUTH PARKS ROAD
OXFORD OX1 3RQ
U.K.
TEL: 0865-511336/507
TLX: 83295 NUCLOX G
COM:

GOEBEL ERNST DR
INSTITUT FUER ASTRONOMIE
UNIVERSITAET WIEN
TUERKENSCHANZSTR 17
A-1180 WIEN
AUSTRIA
TEL: 0222-345360186
TLX:
COM: 50

GOEBEL JOHN H DR
SPACE SC DIVISION 244/7
SPACE TECHNOL RES BRANCH
NASA-AMES RESEARCH CTR
MOFFETT FIELD CA 94035
U.S.A.
TEL: 415-694-6525
TLX:
COM: 29,34

GOELBASI ORHAN DR
B.U. KANDILLI OBSERVATORY
CENGELKOY
ISTANBUL
TURKEY
TEL: 90-1332-02-41/2
TLX: 26411 BOUNTR
COM:

GOKDOGAN NUZHET PROF
UNIVERSITY OBSERVATORY
UNIVERSITY OF ISTANBUL
ISTANBUL
TURKEY
TEL:
TLX:
COM: 12,36

GOKHALE MORESHWAR HARI PR
INDIAN INSTITUTE OF
ASTROPHYSICS
BANGALORE 560 034
INDIA
TEL: 566585
TLX: 0845763 IIAB IN
COM: 10

GOKMEN TARIK ASSOC PROF
UNIVERSITY OBSERVATORY
ISTANBUL 34
TURKEY
TEL:
TLX:
COM: 31

GOLAY MARCEL PROF
OBSERVATOIRE DE GENEVE
CHEMIN DES MAILLETTES 51
CH-1290 SAUVERNY
SWITZERLAND
TEL: 022-55-26-11
TLX: 27720 OBSG CH
COM: 25,37,45V

GOLD THOMAS PROF
CTR F/RADIOPHYS & SP RES
SPACE SCIENCE BLDG
CORNELL UNIVERSITY
ITHACA NY 14853
U.S.A.
TEL: 607-256-5284
TLX: 937478
COM: 16,40,44,47,48

GOLDBACH CLAUDINE MME
INSTITUT D'ASTROPHYSIQUE
98 BIS BOULEVARD ARAGO
F-75014 PARIS
FRANCE
TEL: 1-43-20-14-25
TLX:
COM: 14

GOLDBERG LEO PROF
KITT PEAK NAT OBSERVATORY
PO BOX 26732
TUCSON AZ 85726
U.S.A.
TEL:
TLX: 666484 AURA KNPO TUC
COM: 12,29,44,51C

GOLDMAN MARTIN V
DEPARTMENT ASTRO-GEOPHYS
UNIVERSITY OF COLORADO
CAMPUS BOX 391
BOULDER CO 80309
U.S.A.
TEL: 303-492-8896
TLX:
COM: 12

GOLDREICH P DR
CALTECH
PASADENA CA 91125
U.S.A.
TEL: 213-356-6193
TLX:
COM: 07,16,33,34

GOLDSMITH DONALD W. DR.
INTERSTELLAR MEDIA
2153 RUSSELL STREET
BERKELEY CA 94705
U.S.A.
TEL: 415-848-1989
TLX:
COM: 34,47,48,51

GOLDSMITH PAUL F DR
DEPT PHYSICS & ASTRONOMY
GRC TOWER B, ROOM 626
UNIV OF MASSACHUSETTS
AMHERST MA 01003
U.S.A.
TEL:
TLX:
COM: 34,40

GOLDSMITH S DR
DEPT PHYSICS & ASTRONOMY
TEL-AVIV UNIVERSITY
TEL-AVIV 69978
ISRAEL
TEL: 03-420-303
TLX: 342171 VERSY
COM:

GOLDSTEIN RICHARD M DR
JPL - CALTECH
MS 183-701
4800 OAK GROVE DRIVE
PASADENA CA 91011
U.S.A.
TEL: 818-354-6999
TLX:
COM: 16

GOLDSTEIN SAMUEL J PROF
UNIVERSITY OF VIRGINIA
PO BOX 3818
CHARLOTTESVILLE VA 22903
U.S.A.
TEL:
TLX:
COM: 34,40

GOLDSWORTHY FREDERICK A
SCHOOL OF MATHEMATICS
UNIVERSITY OF LEEDS
LEEDS LS2 9JT
U.K.
TEL: 0532-431751
TLX:
COM: 34

GOLDWIRE HENRY C JR
UNIVERSITY OF CALIFORNIA
LANL
PO BOX 808 L-451
LIVERMORE CA 94550
U.S.A.
TEL: 415-423-0160
TLX:
COM: 40

GOLLNOW H DR
MOUNT STROMLO OBSERVATORY
CANBERRA ACT
AUSTRALIA
TEL:
TLX:
COM:

GOLUB LEON DR
HARVARD COLLEGE OBS
60 GARDEN ST
CAMBRIDGE MA 02138
U.S.A.
TEL: 617-495-7177
TLX:
COM:

GOMES ALERCIO M DR
R GAVIAO PEIXOTO 13
AP 1401
ICARAI 24000
NITEROJ ERJ
BRAZIL
TEL:
TLX:
COM:

LIST OF MEMBERS

GOMEZ GONZALEZ JESUS DR
PASEO IMPERIAL 29 6H
MADRID 5
SPAIN
TEL:
TLX:
COM:

GOMEZ MARIA THERESA DR
OSSERVATORIO ASTRONOMICO
I-80131 NAPOLI
ITALY
TEL:
TLX:
COM: 12

GOMIDE FERNANDO DE MELLO
DEPARTAMENTO DE FISICA
INST TECN. DE AERONAUTICA
12225 S. JOSE DOS CAMPOS
BRAZIL
TEL: 0123-22-9088
TLX: 0113393 CTAE BR
COM:

GONCZI GEORGES
OBSERVATOIRE DE NICE
BP 139
F-06003 NICE CEDEX
FRANCE
TEL: 93-89-04-20
TLX:
COM:

GONDHALEKAR PRABHAKAR DR
RUTHERFORD & APPLETON LAB
CHILTON DIDCOT X11 0QX
U.K.
TEL: 0235-21900
TLX: 83159
COM: 44

GONDOLATSCH FRIEDRICH PRF
ASTRONOMISCHES RECHEN
INSTITUT
MOENCHHOFSTR 12-14
D-6900 HEIDELBERG
GERMANY, F.R.
TEL: 06221-49026
TLX:
COM: 04

GONG HUI-REN
SHANGHAI OBSERVATORY
ACADEMIA SINICA
SHANGHAI
CHINA, PEOPLE'S REP.
TEL: 386191
TLX: 33164 SHAO CN
COM: 31

GONG SHOU-SHEN
SHANGHAI OBSERVATORY
ACADEMIA SINICA
SHANGHAI
CHINA, PEOPLE'S REP.
TEL: 386191
TLX: 33164 SHAO CN
COM: 09

GONG SHU-MO
PURPLE MOUNTAIN
OBSERVATORY
NANJING
CHINA, PEOPLE'S REP.
TEL: 46700
TLX: 34144 PMO CN
COM: 35

GONZALES-A WALTER D DR
INST PESQUISAS ESPACIAIS
INPE
CAIXA POSTAL 515
12200 S. JOSE DOS CAMPOS
BRAZIL
TEL: 0623-229977
TLX: 011-33530 INPE BR
COM: 48

GONZALEZ CAMACHO ANTONIO
INST DE ASTRON & GEODESIA
FAC DE CIENCIAS MATEM.
UNIVERSIDAD COMPLUTENSE
28040 MADRID
SPAIN
TEL: 91-2442501
TLX:
COM: 07

GONZALEZ G
INAOE
AP POSTALES 216 y 51
72000 PUEBLA, PUE.
MEXICO
TEL:
TLX:
COM:

GONZE ROGER F J IR
OBS ROYAL DE BELGIQUE
AVE CIRCULAIRE 3
B-1180 BRUXELLES
BELGIUM
TEL: 375-24-84
TLX: 21565
COM: 40

GOODE PHILIP R
NJ INST. OF TECHNOLOGY
DEPT OF PHYSICS
323 HIGH STREET
NEWARK NJ 07102
U.S.A.
TEL: 201-596-3562
TLX:
COM:

GOODY R M
CEPP
PIERCE HALL
29 OXFORD STREET
CAMBRIDGE MA 02138
U.S.A.
TEL: 617-495-4517
TLX:
COM: 16

GOOSSENS MARCEL DR
ASTRONOMISCH INSTITUUT
KATHOLIEKE UNIV LEUVEN
CELESTIJNENLAAN 200 B
B-3030 HEVERLEE
BELGIUM
TEL:
TLX:
COM: 35

GOPALA RAO U V MR
SATELLITE METEOROLOGY
INDIAN METEO. DEPT.
LODI ROAD / MAUSAM BHAVAN
NEW DELHI 110 003
INDIA
TEL:
TLX:
COM: 28

GOPALASWAMY N DR
ASTRONOMY PROGRAM
UNIVERSITY OF MARYLAND
COLLEGE PARK MD 20742
U.S.A.
TEL: 301-454-6649
TLX: 62891478
COM: 12

GOPASYUK S I DR
CRIMEAN ASTROPHYS OBS
USSR ACADEMY OF SCIENCES
334413 NAUCHNIY CRIMEA
U.S.S.R.
TEL:
TLX:
COM: 10,12

GORBATSKY VITALIJ G PROF
LENINGRAD UNIVERSITY
ASTRONOMICAL OBSERVATORY
BIBLIOTECHNAJA PL 2
198904 LENINGRAD
U.S.S.R.
TEL: 257-94-91
TLX:
COM: 27

GORDON CHARLOTTE PROF
11 RUE TOURNEFORT
F-75005 PARIS
FRANCE
TEL:
TLX:
COM: 12,36

GORDON COURTNEY P PROF
HAMPSHIRE COLLEGE
AMHERST MA 01002
U.S.A.
TEL:
TLX:
COM: 34

GORDON I M DR
INST OF RADIO PHYS & ELEC
310085 KHARKOV
U.S.S.R.
TEL:
TLX:
COM:

GORDON KURTISS J PROF
FIVE COLL ASTR DEPT
HAMPSHIRE COLLEGE
AMHERST MA 01002
U.S.A.
TEL: 413-549-4600
TLX:
COM:

GORDON MARK A DR
TUCSON OPERATIONS
NRAO, SUITE 100
2010 N FORBES BLVD
TUCSON AZ 85705
U.S.A.
TEL: 602-882-8250
TLX:
COM: 33,34,40

GORENSTEIN MARC V
HARVARD-SMITHSONIAN CTR
FOR ASTROPHYSICS
60 GARDEN STREET / MS-42
CAMBRIDGE MA 02138
U.S.A.
TEL: 617-495-9296
TLX:
COM:

GORENSTEIN PAUL DR
CENTER FOR ASTROPHYSICS
60 GARDEN STREET
CAMBRIDGE MA 02138
U.S.A.
TEL: 617-495-7250
TLX: 921428 SATELLITE CAM
COM: 16

GORET PHILIPPE DR
SECTION D'ASTROPHYSIQUE
CEN SACLAY
F-91190 GIF/YVETTE
FRANCE
TEL: 1-69-08-44-63
TLX: 690860 PHYSPAC F
COM: 47

GORGOLEWSKI STANISLAW PR
KATEDRA RADIOASTR/CHAIR
RADIOASTR/COPERNICUS UNIV
UL. CHOPINA 12/18
87-100 TORUN
POLAND
TEL: 20651 TORUN
TLX: 0552324 TRAO PL
COM: 40

GOSACHINSKIJ I V DR
LENINGRAD BRANCH OF SAO
PULKOVO
196140 LENINGRAD
U.S.S.R.
TEL: 2979452
TLX: 321262
COM: 34,40

GOSLING JOHN T DR
LOS ALAMOS NATIONAL LAB
ESS 8 - MS D 438
LOS ALAMOS NM 87545
U.S.A.
TEL: 505-667-5389
TLX: 660495
COM: 49

GOSS W MILLER PROF
KAPTEYN ASTRONOMICAL INST
POSTBUS 800
NL-9700 AV GRONINGEN
NETHERLANDS
TEL: 31-50-116653/95
TLX: 53572 STARS NL
COM: 28,34,40

GOSWAMI J N DR
PHYS. RESEARCH LABORATORY
NAVRANGPURA
AHMEDABAD 380 009
INDIA
TEL: 462129
TLX: 0121397 PRL IN
COM: 22

GOTT III J RICHARD
DEPT ASTROPHYSICAL SC
PRINCETON UNIVERSITY
PRINCETON NJ 08540
U.S.A.
TEL: 609-452-3813
TLX:
COM: 51

GOTTESMAN STEPHEN T DR
DEPARTMENT OF ASTRONOMY
UNIVERSITY OF FLORIDA
GAINESVILLE FL 32611
U.S.A.
TEL: 904-392-2050/52
TLX: 8108252308
COM: 28

GOTTLIEB CARL A DR
GODDARD INST/SPACE STUD
2880 BROADWAY
NEW YORK NY 10025
U.S.A.
TEL: 212-678-5566
TLX:
COM:

GOTTLIEB KURT
46 JENNINGS STREET
CURTIN ACT 2605
AUSTRALIA
TEL: 062-814166
TLX:
COM:

GOUDAS CONSTANTINE L PROF
DEPT OF MATHEMATICS
UNIVERSITY OF PATRAS
GR-26110 PATRAS
GREECE
TEL: 991-889
TLX: 312239 EFAP GR
COM: 07,16

GOUDIS CHRISTOS D PROF
UNIVERSITY OF PATRAS
DEPT OF ASTRONOMY
GR-26110 PATRAS
GREECE
TEL:
TLX:
COM: 51

GOUGH DOUGLAS O DR
INSTITUTE OF ASTRONOMY
MADINGLEY ROAD
CAMBRIDGE CB3 0HA
U.K.
TEL: 223-62204
TLX: 817297
COM: 27,35C,36

GOUGUENHEIM LUCIENNE
OBSERVATOIRE DE PARIS
SECTION DE MEUDON
RADIOASTRONOMIE
F-92195 MEUDON PL CEDEX
FRANCE
TEL: 1-45-34-75-30
TLX:
COM: 28,46C

GOULD ROBERT J PROF
PHYSICS DEPARTMENT B-019
UNIV/CALIF AT SAN DIEGO
LA JOLLA CA 92093
U.S.A.
TEL: 619-452-3649
TLX:
COM:

GOUTTEBROZE PIERRE DR
LPSP
BP 10
F-91370 VERRIERES-LE B.
FRANCE
TEL: 1-69-20-10-60
TLX:
COM:

GOWER ANN C DR
UNIV OF VICTORIA OBS
VICTORIA BC V8W 2Y2
CANADA
TEL:
TLX:
COM:

GOWER J F R DR
1615 MCTAVISH ROAD
R R 2
SIDNEY BC V8L 3S1
CANADA
TEL: 604-656-5457
TLX:
COM: 40

GOY GERALD PROF
OBSERVATOIRE DE GENEVE
CH-1290 SAUVERNY
SWITZERLAND
TEL: 22-552-611
TLX: 27720 OBSG CH
COM: 25

GOYAL A N DR
DEPT OF MATHEMATICS
UNIVERSITY OF RAJASTHAN
JAIPUR 302 004
INDIA
TEL: 74060 (HOME)
TLX:
COM: 24

GRABOSKE HAROLD C JR
LAWRENCE LIVERMORE LAB
PO BOX 808
LIVERMORE CA 94550
U.S.A.
TEL: 415-422-7262
TLX:
COM:

GRABOWSKI BOLESLAW DR
INSTITUTE OF PHYSICS
UL OLESKA 48
UL. OLESKA 48
45-951 OPOLE
POLAND
TEL: 358-41
TLX: 0732230 WSP PL
COM:

GRADIE JONATHAN CAREY
HAWAII INST OF GEOPHYSICS
DIV PLANETARY GEOSCIENCES
UNIVERSITY OF HAWAII
HONOLULU HI 96822
U.S.A.
TEL: 808-948-6488
TLX:
COM: 15

GRADSZTAJN E DR
DEPT PHYSICS & ASTRONOMY
TEL-AVIV UNIVERSITY
RAMAT-AVIV 69978
ISRAEL
TEL:
TLX:
COM:

GRAHAM DAVID A
MPI FUER RADIOASTRONOMIE
AUF DEM HUEGEL 69
D-5300 BONN 1
GERMANY, F.R.
TEL: 228-525282
TLX: 886440 MPIFR D
COM: 34,40

GRAHAM ERIC DR
PO BOX 579
SANDIA PARK NM 87047
U.S.A.
TEL: 505-281-3184
TLX: 643351
COM: 35

GRAHAM JOHN A DR
DEPT TERRESTR. MAGNETISM
CARNEGIE INST. WASHINGTON
5241 BROAD BRANCH RD N.W.
WASHINGTON DC 20015
U.S.A.
TEL: 202-966-0863
TLX: 440427 MAGN UI
COM: 25,27,28

GRAHL BERND H DR
MPI FUER RADIOASTRONOMIE
AUF DEM HUEGEL 69
D-5300 BONN
GERMANY, F.R.
TEL: 02257-3112
TLX: 8869114 MPIR D
COM:

GRAINGER JOHN F DR
PHYSICS DEPARTMENT
UMIST
MANCHESTER M60 1QD
U.K.
TEL: 061-236-3311
TLX: 666094
COM:

GRANDI STEVEN ALDRIDGE DR
NATL OPTICAL ASTRONOMY
OBSERVATORY
PO BOX 26732
TUCSON AZ 85726
U.S.A.
TEL: 602-327-5511
TLX:
COM: 28

GRANT IAN P DR
PEMBROKE COLLEGE
OXFORD OX1 1DW
U.K.
TEL: 0865-242-271
TLX:
COM: 14,36

GRASDALEN GARY L DR
DEPT PHYSICS & ASTRONOMY
UNIVERSITY OF WYOMING
PO BOX 3905 UN STA
LARAMIE WY 82071
U.S.A.
TEL: 307-766-4385
TLX:
COM: 27,28,34

GRATTON LIVIO PROF
IST. ASTROFISICA SPAZIALE
C P 67
I-00044 FRASCATI
ITALY
TEL:
TLX:
COM: 29,47

GRAUER ALBERT D
DEPT PHYSICS & ASTRONOMY
UALR
33RD & UNIVERSITY
LITTLE ROCK AR 72204
U.S.A.
TEL: 501-569-3275
TLX:
COM: 25

GRAY DAVID F PROF
DEPT OF ASTRONOMY
UNIV OF WESTERN ONTARIO
PHYSICS-ASTRONOMY BLDG
LONDON ONT N6A 3K7
CANADA
TEL: 519-679-3184
TLX:
COM: 29,36V

GRAY PETER MURRAY
ANGLO-AUSTRALIAN OBS.
P.O. BOX 296
EPPING NSW 2121
AUSTRALIA
TEL: 02-868-1666
TLX: 23999 AAOSYD AA
COM: 09

GRAYZECK EDWIN J DR
PHYSICS DEPARTMENT
UNIVERSITY OF NEVADA
4505 S MARYLAND PARKWAY
LAS VEGAS NV 89154
U.S.A.
TEL: 702-739-3507
TLX:
COM: 33

GREBENIKOV E A PROF DR
INST THEOR & EXPER PHYS
117259 MOSCOW
U.S.S.R.
TEL:
TLX:
COM: 07

GREC GERARD
DEPT D'ASTROPHYSIQUE
UNIVERSITE DE NICE
PARC VALROSE
F-06034 NICE
FRANCE
TEL:
TLX:
COM:

GREEN ELIZABETH M. DR
MT STROMLO OBSERVATORY
AUSTRALIAN NAT UNIVERSITY
PRIVATE BAG
WODEN P.O. ACT 2606
AUSTRALIA
TEL: 062-88-1111
TLX: 62270 AA
COM: 37

GREEN JACK PROF
DEPT OF GEOLOGY
CALIF STATE UNIVERSITY
LONG BEACH CA 90840
U.S.A.
TEL: 213-498-4809
TLX:
COM: 16

GREEN LOUIS C PROF
HAVERFORD COLLEGE
7901 COLLEGE AVENUE
HAVERFORD PA 19041
U.S.A.
TEL: 215-649-0265
TLX:
COM: 14

GREEN ROBIN M DR
DEPT OF ASTRONOMY
GLASGOW UNIVERSITY
GLASGOW G12 8QQ
U.K.
TEL: 041-339-8855
TLX: 778421
COM:

GREENBERG J MAYO DR
HUYGENS LABORATORY
UNIVERSITY OF LEIDEN
WASSENAARSEWEG 78
NL-2300 RA LEIDEN
NETHERLANDS
TEL: 31-71-148333
TLX: 39058 ASTRO NL
COM: 15,21,34,51

GREENBERG RICHARD DR
PLANETARY SCIENCE INST
2030 E SPEEDWAY
TUCSON AZ 85719
U.S.A.
TEL: 602-881-0332
TLX:
COM: 07,20

GREENSTEIN GEORGE PROF
ASTRONOMY DEPARTMENT
AMHERST COLLEGE
AMHERST MA 01002
U.S.A.
TEL: 413-542-2075
TLX:
COM:

GREENSTEIN J L PROF
PALOMAR OBSERVATORY
CALIF INST OF TECHNOLOGY
1201 E CALIFORNIA ST
PASADENA CA 91125
U.S.A.
TEL: 818-356-4006
TLX:
COM: 29,36,51

GREGORINI LORETTA
IST. DI RADIOASTRONOMIA
VIA IRNERIO 46
I-40126 BOLOGNA
ITALY
TEL: 51-232-856
TLX: 211664 INFN BOI
COM:

GREGORY PHILIP C DR
PHYSICS DEPT
UNIV OF BRITISH COLUMBIA
6224 AGRICULTURAL RD
VANCOUVER BC V6T 1W5
CANADA
TEL: 604-228-6417
TLX: 04-508576
COM: 40,51

GREGORY STEPHEN ALBERT DR
PHYSICS DEPT
BOWLING GREEN STATE UNIV
BOWLING GREEN OH 43403
U.S.A.
TEL:
TLX:
COM: 47

GREGUL A YA DR
OBSERVATORY OF THE KIEV
UNIVERSITY
OBSERVATORNAYA 3
252053 KIEV
U.S.S.R.
TEL: 262391
TLX:
COM:

GREISEN KENNETH I PROF
336 FOREST HOME DR
ITHACA NY 14850
U.S.A.
TEL: 607-257-1650
TLX:
COM: 48

GRENIER SUZANNE
OBSERVATOIRE DE PARIS
SECTION DE MEUDON
F-92195 MEUDON PL CEDEX
FRANCE
TEL: 1-45-34-75-30
TLX: 201571 LAM
COM:

GRENON MICHEL DR
OBSERVATOIRE DE GENEVE
CH-1290 SAUVERNY
SWITZERLAND
TEL: 22-55-26-11
TLX: 27720 OBSG CH
COM: 25

GREVESSE N DR
INSTITUT D'ASTROPHYSIQUE
UNIVERSITE DE LIEGE
AVENUE DE COINTE 5
B-4200 COINTE-OUGREE
BELGIUM
TEL: 41-52-99-80
TLX: 41264
COM: 12,36

GREWING MICHAEL PROF
ASTRON INST UNIVERSITAET
WALDHAUSERSTR 64
D-7400 TUEBINGEN
GERMANY, F.R.
TEL: 7071-292486
TLX: 07262714 AIT D
COM: 25,34,40,44C,48

GREYBER HOWARD D DR
10123 FALLS ROAD
POTOMAC MD 20854
U.S.A.
TEL:
TLX:
COM: 44,47,48

GRIFFIN RITA E M DR
THE OBSERVATORIES
MADINGLEY ROAD
CAMBRIDGE CB3 0HA
U.K.
TEL: 223-62204
TLX: 817297 ASTRON G
COM: 29

GRIFFIN ROGER F DR
THE OBSERVATORIES
MADINGLEY ROAD
CAMBRIDGE CB3 0HA
U.K.
TEL: 44-223-62204
TLX: 817297 ASTRON G
COM: 05,29,30

GRIFFITH JOHN S PROF
DEPT OF MATH SCIENCE
LAKEHEAD UNIVERSITY
THUNDER BAY ONT P7B 5E1
CANADA
TEL: 807-345-2121
TLX:
COM:

GRIFFITHS RICHARD E DR
SPACE TELESCOPE SCI INST
3700 SAN MARTIN DRIVE
BALTIMORE MD 21218
U.S.A.
TEL: LEEDS 431751
TLX: 556473 UNILDS G
COM: 09,28,44,48

GRIFFITHS WILLIAM K
DEPT OF PHYSICS
THE UNIVERSITY
LEEDS LS2 9JT
U.K.
TEL:
TLX:
COM: 37

GRIGORJEV VICTOR M DR
SIBERIAN INST/TERR MAGN
IONOSPH RADIO WAVE PROP
P BOX 4
664697 IRKUTSK
U.S.S.R.
TEL:
TLX:
COM: 09

GRINDLAY JONATHAN E DR
HARVARD OBSERVATORY
CENTER FOR ASTROPHYSICS
60 GARDEN STREET
CAMBRIDGE MA 02138
U.S.A.
TEL: 617-495-7204
TLX: 921428 SATELLITE CAM
COM: 06,37,48

GRININ VLADIMIR P DR
CRIMEAN ASTROPHYS OBS
USSR ACADEMY OF SCIENCES
334413 NAUCHNY CRIMEA
U.S.S.R.
TEL:
TLX:
COM: 36

GRISHCHUK L P DR
STERNBERG STATE ASTR INST
119899 MOSCOW V-234
U.S.S.R.
TEL: 139-50-06
TLX:
COM: 47

GROOTE DETLEF
HAMBURGER STERNWARTE
GOJENSBERGWEG 112
D-2050 HAMBURG 80
GERMANY, F.R.
TEL: 040-72524112
TLX:
COM:

GROSBOL PREBEN JOHNSON DR
ESO
KARL-SCHWARZSCHILD-STR 2
D-8046 GARCHING B MUNCHEN
GERMANY, F.R.
TEL: 089-320-06-237
TLX: 52828222 EOD
COM: 05

GROSS PETER G PROF
WARNER SWASEY OBSERVATORY
1975 TAYLOR RD
CLEVELAND OH 44112
U.S.A.
TEL:
TLX:
COM:

GROSSMAN ALLEN S PROF
ERWIN FICK OBSERVATORY
IOWA STATE UNIVERSITY
AMES IA 50011
U.S.A.
TEL: 515-294-3666
TLX:
COM:

GRUDLER PIERRE
C E R G A
AVE COPERNIC
F-06130 GRASSE
FRANCE
TEL: 93-36-58-49
TLX:
COM: 08,31

GUARNIERI ADRIANO DR
OSSERVATORIO ASTRONOMICO
VIA ZAMBONI 33
I-40126 BOLOGNA
ITALY
TEL:
TLX:
COM:

GUESTEN ROLF
MPI FUER RADIOASTRONOMIE
AUF DEM HUEGEL 69
D-5300 BONN 1
GERMANY, F.R.
TEL: 49-228-525-379
TLX:
COM: 34,40

GROSSMAN LAWRENCE PROF
DEPT GEOPHYSICAL SCIENCES
UNIVERSITY OF CHICAGO
5734 SOUTH ELLIS AVE
CHICAGO IL 60637
U.S.A.
TEL: 312-962-8153
TLX:
COM: 15,16

GRUDZINSKA STEFANIA DR
ASTRONOMICAL INSTITUTE
UL. CHOPINA 12/18
87-100 TORUN
POLAND
TEL: 20655
TLX: 0552234 ASTR PL
COM: 15

GUBANOV VADIM S DR
PULKOVO OBSERVATORY
196140 LENINGRAD
U.S.S.R.
TEL: 297-94-81
TLX:
COM: 08

GUETTER HARRY HENDRIK
US NAVAL OBSERVATORY
PO BOX 1149
FLAGSTAFF AZ 86002
U.S.A.
TEL: 602-779-5132
TLX:
COM: 45

GROSSMANN-DOERTH U DR
KIEPENHEUER INSTITUT
FUER SONNENPHYSIK
SCHOENECKSTR 6
D-7800 FREIBURG
GERMANY, F.R.
TEL: 0761-32864
TLX: 7721552 KIS D
COM:

GRUEFF GAVRIL DR
LAB. DI RADIOASTRONOMIA
CNR
VIA IRNERIO 46
I-40126 BOLOGNA
ITALY
TEL:
TLX:
COM:

GUDMUNDSSON EINAR H
RAUNVISINDASTOFNUN
HASKOLANS
DUNHAGA 3
IS-107 REYKJAVIK
ICELAND
TEL: 21340
TLX: 2307 ISINFO
COM:

GUIBERT JEAN DR
OBSERVATOIRE DE PARIS
61 AVE DE L'OBSERVATOIRE
F-75014 PARIS
FRANCE
TEL: 1-43-20-12-10
TLX:
COM:

GROTEN ERWIN PROF
INST/PHYSIKALISCHE GEOD
PETERSENSTR 13
D-6100 DARMSTADT
GERMANY, F.R.
TEL: 0-6151-16-3109
TLX: 419579 TH D
COM: 19

GRUEN EBERHARD DR
MPI FUER KERNPHYSIK
POSTFACH 103 980
D-6900 HEIDELBERG
GERMANY, F.R.
TEL: 6621-516478
TLX: 461666 MPIHD D
COM: 15,21,22

GUDUR N DR
EGE UNIVERSITY
OBSERVATORY
P K 21
BORNOVA-IZMIR
TURKEY
TEL: 9051180110/2326
TLX:
COM:

GUIDICE DONALD`A DR
A F GEOPHYSICS LABORATORY
HANSCOM AFB
BEDFORD MA 01731
U.S.A.
TEL: 617-861-3989
TLX:
COM: 40

GROTH EDWARD J III
PHYSICS DEPT
PRINCETON UNIVERSITY
JADWIN HALL
PRINCETON NJ 08544
U.S.A.
TEL: 609-452-4361
TLX:
COM:

GRUNDMANN WALTER
DOMINION ASTROPHYS OBS
5071 W SAANICH ROAD
VICTORIA BC V8X 4M6
CANADA
TEL: 604-388-3157
TLX: 0497295
COM: 09

GUELIN MICHEL DR
INST RADIOASTR MILLIMETR
VOIE 10
DOMAINE UNIV DE GRENOBLE
F-38406 ST MARTIN D'HERES
FRANCE
TEL: 76-42-33-83
TLX: 980753 IRAM
COM: 34,40

GUINAN EDWARD FRANCIS DR
DEPT OF ASTRONOMY
VILLANOVA UNIVERSITY
VILLANOVA PA 19085
U.S.A.
TEL: 215-527-2100
TLX:
COM: 27,42

GROTH HANS G PROF DR
INST ASTRON & ASTROPHYS
UNIVERSITAT MUENCHEN
SCHEINERSTRASSE 1
D-8000 MUENCHEN 80
GERMANY, F.R.
TEL: 089-989021
TLX:
COM: 29,36

GRYGAR JIRI DR
INSTITUTE OF PHYSICS
CZECH. ACAD. OF SCIENCES
250 68 REZ
CZECHOSLOVAKIA
TEL: 84-42-41
TLX: 122626 CS
COM: 27,42

GUERIN PIERRE DR
INSTITUT D'ASTROPHYSIQUE
98 BIS BOULEVARD ARAGO
F-75014 PARIS
FRANCE
TEL: 1-43-20-14-25
TLX: 270070
COM: 16

GUINOT BERNARD R PROF
B.I.P.M.
PAVILLON DE BRETEUIL
F-92310 SEVRES
FRANCE
TEL: 1-45-34-00-51
TLX: 201067 BIPM
COM: 19,31C

GROUSHINSKY N P PROF DR
STERNBERG STATE
ASTRONOMICAL INSTITUTE
UNIVERSITETSKIJ PROSP. 13
119899 MOSCOW
U.S.S.R.
TEL:
TLX:
COM: 07

GRZEDZIELSKI STANISLAW PR
SPACE RESEARCH CENTER
POLISH ACAD OF SCIENCES
UL. ORDONA 21
01-237 WARSZAWA
POLAND
TEL:
TLX:
COM: 49P

GUERRERO GIANANTONIO DR
OSSERVATORIO ASTRONOMICO
VIA E. BIANCHI 46
I-22055 MERATE (COMO)
ITALY
TEL: 039-592035
TLX:
COM: 27

GULKIS SAMUEL DR
JET PROPULSION LABORATORY
4800 OAK GROVE DRIVE
PASADENA CA 91109
U.S.A.
TEL: 213-354-5708
TLX:
COM: 16,40,51

GRUBISSICH C PROF DR
VIA AOSTA 34/5
I-35142 PADOVA
ITALY
TEL: 049-38301
TLX:
COM: 37

GU XIAO-MA
YUNNAN OBSERVATORY
ACADEMIA SINICA
KUNMING
CHINA, PEOPLE'S REP.
TEL: 72946 KUNMING
TLX: 64040 YUOBS CN
COM: 10,12

GUEST JOHN E DR
UNIVERSITY OF LONDON OBS
MILL HILL PARK
LONDON NW7 2QS
U.K.
TEL: 01-959-7367
TLX: 28722 UCPHYS
COM: 16

GULL STEPHEN F DR
CAVENDISH LABORATORY
MADINGLEY ROAD
CAMBRIDGE CB3 0HE
U.K.
TEL: 223-66477
TLX: 81292
COM: 40

GULL THEODORE R DR
LAB. ASTRON & SOLAR PHYS
NASA/GSFC, CODE 683.0
GREENBELT MD 20771
U.S.A.
TEL: 301-344-8060
TLX: 710-8289716
COM: 34,44

GULLIVER AUSTIN FRASER DR
DEPT OF PHYSICS
UNIVERSITY OF ALBERTA
EDMONTON ALB T6G 2J1
CANADA
TEL:
TLX:
COM: 42

GULMEN OMUR DR
EGE UNIVERSITY
OBSERVATORY
P K 21
BORNOVA-IZMIR
TURKEY
TEL: 90-51-180110
TLX:
COM:

GULYAEV A P DR
STERNBERG ASTR INSTITUT
UNIVERSITETSKIJ PROSP. 13
119899 MOSCOW
U.S.S.R.
TEL: 139-19-70
TLX:
COM: 08

GULYAEV RUDOLF A DR
IZMIRAN
AKADEMGORODOK
142092 MOSCOW REGION
U.S.S.R.
TEL:
TLX:
COM:

GUNN JAMES E PROF
DEPT ASTROPHYSICAL SCI
PRINCETON UNIVERSITY
PEYTON HALL
PRINCETON NJ 08544
U.S.A.
TEL: 609-452-3802
TLX:
COM: 28,47C,48,51

GUO QUAN SHI
PURPLE MOUNTAIN
OBSERVATORY
NANJING
CHINA, PEOPLE'S REP.
TEL:
TLX:
COM:

GURM HARDEV S PROF
DEPT/ASTR & SPACE SCI
PANJABI UNIVERSITY
PATIALA 147 002
INDIA
TEL: 73262 x 96
TLX:
COM: 27,35,46,51

GURSHTEIN A A DR
INST HIST OF SCI & TECHN
USSR ACADEMY OF SCIENCES
STAROPANSKY 1/5
103012 MOSCOW
U.S.S.R.
TEL:
TLX:
COM: 16

GURSKY HERBERT DR
NAVAL RESEARCH LABORATORY
CODE 4100
WASHINGTON DC 20375
U.S.A.
TEL: 202-767-6343
TLX:
COM: 27,42,44,48

GURTOVENKO E A DR
MAIN ASTRONOMICAL OBS
UKRAINIAN ACAD OF SCI
252127 KIEV
U.S.S.R.
TEL: 66-10-65
TLX: 131406
COM: 10,12C

GURZADIAN G A PROF DR
BYURAKAN ASTROPHYS OBS
378433 ARMENIA
U.S.S.R.
TEL:
TLX:
COM: 28,34,45

GUSEINOV O H PROF
INSTITUTE OF PHYSICS
NARIMANOV AVENUE 33
370143 BAKU
U.S.S.R.
TEL: 39-39-51
TLX:
COM: 34,42,44,48

GUSEJNOV R EH DR
SHEMAKHA ASTROPHYS OBS
373243 AZERBAIDZAN
U.S.S.R.
TEL:
TLX:
COM:

GUSSMANN E A DR
ZNTRLINST. F. ASTROPHYSIK
ROSA-LUXEMBURG-STR 17A
DDR-1502 POTSDAM
GERMANY, D.R.
TEL:
TLX:
COM: 36

GUSTAFSON BO A S
LUND OBSERVATORY
BOX 43
S-221 00 LUND
SWEDEN
TEL:
TLX:
COM: 15

GUSTAFSSON BENGT DR
ASTRONOMICAL OBSERVATORY
BOX 515
S-751 20 UPPSALA
SWEDEN
TEL:
TLX:
COM: 29,36C

GUTCKE DIETRICH
CARL-ZEISS STR 1
DDR-6900 JENA
GERMANY, D.R.
TEL:
TLX:
COM: 09

GUTHRIE BRUCE N G DR
ROYAL OBSERVATORY
BLACKFORD HILL
EDINBURGH EH9 3HJ
U.K.
TEL: 031-667-3321
TLX: 72383 ROEDIN G
COM: 29

GUTIERREZ-MORENO A DR MRS
DEPTO DE ASTRONOMIA
UNIVERSIDAD DE CHILE
CASILLA 36-D
SANTIAGO
CHILE
TEL: 2294101/2294002
TLX:
COM: 25

GYLDENKERNE KJELD DR
COPENHAGEN UNIVERSITY OBS
BRORFELDEVEJ 23
DK-4340 TOLLOSE
DENMARK
TEL: 3-488-195
TLX: 44155
COM: 33,42

HABBAL SHADIA RIFAI
HARVARD SMITHSONIAN CTR
FOR ASTROPHYSICS
60 GARDEN STREET
CAMBRIDGE MA 02138
U.S.A.
TEL: 617-495-7348
TLX: 921428
COM: 49

HABE ASAO
DEPT OF PHYSICS
HOKKAIDO UNIVERSITY
SAPPORO
HOKKAIDO 060
JAPAN
TEL: 11-711-2111
TLX:
COM: 33

HABIBULLIN SH T PROF DR
KAZAN UNIV OBSERVATORY
LENIN STREET 18
420008 KAZAN
U.S.S.R.
TEL: 323641
TLX:
COM: 16

HABING H J DR
STERREWACHT
POSTBUS 9513
NL-2300 RA LEIDEN
NETHERLANDS
TEL: 071-148333
TLX: 39058
COM: 33,34C

HACHENBERG OTTO PROF DR
RADIOASTRONOMISCHES INST
UNIVERSITAET BONN
AUF DEM HUEGEL 71
D-5300 BONN
GERMANY, F.R.
TEL:
TLX:
COM: 40

HACHISU IZUMI DR
DEPT OF AERONAUTICAL
ENGINEERING
KYOTO UNIVERSITY
KYOTO 606
JAPAN
TEL: 075-751-2111
TLX: 05422693 LIBKYU J
COM:

HACK MARGHERITA PROF
OSSERVATORIO ASTRONOMICO
VIA TIEPOLO 11
I-34131 TRIESTE
ITALY
TEL: 40-793921
TLX: 461137 OAT I
COM: 29,36,44,45

HACKWELL JOHN A DR
DEPT PHYSICS & ASTRONOMY
UNIVERSITY OF WYOMING
LARAMIE WY 82070
U.S.A.
TEL: 307-766-6296
TLX:
COM: 27,34

HACYAN SHAHEN DR.
INSTITUTO DE ASTRONOMIA
UNAM
APDO POSTAL 70-264
04510 MEXICO DF
MEXICO
TEL: 905-548-5305
TLX: 1760155 CICME
COM: 47

HADDOCK FRED T DR
UNIVERSITY OF MICHIGAN
937 PHYSICS-ASTRONOMY BLG
ANN ARBOR MI 48104
U.S.A.
TEL: 313-764-3430
TLX:
COM: 40,44,51

HADJIDEMETRIOU JOHN D
DEPT THEORET MECHANICS
UNIVERSITY THESSALONIKI
GR-54006 THESSALONIKI
GREECE
TEL: 031-99-14-10
TLX:
COM: 07C

LIST OF MEMBERS

HAEFNER REINHOLD DR
UNIVERSITAETS STERNWARTE
SCHEINERSTRASSE 1
D-8000 MUENCHEN 80
GERMANY, F.R.
TEL: 089-98-9021
TLX:
COM: 27

HAEMEEN ANTTILA KAARLE A
ASTRONOMY DEPT
UNIVERSITY OF OULU
SF-90100 OULU 10
FINLAND
TEL:
TLX:
COM:

HAENSEL PAWEL DR
N COPERNICUS ASTRONOMICAL
CENTER
UL. BARTYCKA 18
00-716 WARSAW
POLAND
TEL: 410828
TLX: 81 3978 ZAPLAN PL
COM:

HAERENDEL G DR
MPI F. PHYSIK & ASTROPHYS
INST F. EXTRATERR. PHYSIK
D-8046 GARCHING B MUNCHEN
GERMANY, F.R.
TEL: 089-3299-516
TLX: 05215845 XTERD
COM:

HAGEN JOHN P
613 W. PARK AVE
STATE COLLEGE PA 16803
U.S.A.
TEL: 1-814-237-3031
TLX:
COM: 10,40

HAGEN WENDY ANNE
WHITIN OBSERVATORY
WELLESLEY COLLEGE
WELLESLEY MA 02181
U.S.A.
TEL: 617-235-0320
TLX:
COM: 29

HAGEN-THORN V A DR
ASTRONOMICAL OBSERVATORY
BIBLIOTECHNAJA PL.2
198904 LENINGRAD
U.S.S.R.
TEL: 257-94-91
TLX:
COM: 28

HAGFORS T DR
NATL ASTRON & IONOSPHERE
CTR., SPACE SCIENCES BLDG
CORNELL UNIVERSITY
ITHACA NY 14853
U.S.A.
TEL: 607-256-3734
TLX: 932454
COM: 16

HAGYARD MONA JUNE
NASA MARSHALL SFC
CODE ES52
HUNTSVILLE AL 35812
U.S.A.
TEL: 205-453-5687
TLX: 594416 NASA/MSFC HTV
COM: 10,12

HAIKALA LAURI K
OBS & ASTROPHYSICS LAB
UNIVERSITY OF HELSINKI
TAHTITORNINMAKI
SF-00130 HELSINKI 13
FINLAND
TEL: 1912948 HELSINK
TLX: 124690 UNIH SF
COM:

HAISCH BERNHARD MICHAEL
LOCKHEED PALO ALTO
RESEARCH LABORATORY
DIV 91-20 BLDG 255
PALO ALTO CA 94304
U.S.A.
TEL: 415-858-4073
TLX: 346409
COM: 27,36,51

HAJDUK ANTON DR
ASTRONOMICAL INSTITUTE
SLOVAK ACAD. OF SCIENCES
842 28 BRATISLAVA
CZECHOSLOVAKIA
TEL: 427-375157
TLX: 93373 SEIS
COM: 22,51

HAJDUKOVA MARIA
DEPARTMENT OF ASTRONOMY
COMENIUS UNIVERSITY
MLYNSKA DOLINA
842 15 BRATISLAVA
CZECHOSLOVAKIA
TEL: 427-320-003
TLX:
COM: 22

HALL ANDREW NORMAN
DEPT OF ASTROPHYSICS
SOUTH PARKS ROAD
OXFORD OX1 3RQ
U.K.
TEL:
TLX:
COM: 48

HALL DONALD N DR
INSTITUTE FOR ASTRONOMY
2680 WOODLAWN DRIVE
HONOLULU HI 96822
U.S.A.
TEL: 808-948-8312
TLX: 723-8459
COM:

HALL DOUGLAS S DR
DYER OBSERVATORY
VANDERBILT UNIVERSITY
NASHVILLE TN 37235
U.S.A.
TEL: 615-373-4897
TLX: 554323
COM: 25,27,42

HALL JOHN S DR
110 RED BUTTE DRIVE
SEDONA AZ 86336
U.S.A.
TEL: 602-284-1738
TLX:
COM: 16.34

HALL R GLENN DR
3612 SPRING STREET
CHEVY CHASE MD 20815
U.S.A.
TEL: 301-652-7221
TLX:
COM: 19.31

HALLAM KENNETH L DR
LAB ASTRON & SOLAR PHYS
NASA/GSFC
GREENBELT MD 20771
U.S.A.
TEL: 301-344-6083
TLX:
COM: 09,44,45

HALLIDAY IAN DR
HERZBERG INST ASTROPHYS
NATL RESEARCH COUNCIL
OTTAWA ONT K1A 0R6
CANADA
TEL: 613-990-0704
TLX: 0533715
COM: 15,16,21,22

HAMABE MASARU DR
KISO OBS OF TOKYO ASTRON
OBSERVATORY
MITAKE-MURA, KISO
NAGANO 397-01
JAPAN
TEL: 26452-3360
TLX: 3347577 KSOOBS J
COM: 28

HAMADA TETSUO PROF
DEPT OF PHYSICS
IBARAKI UNIVERSITY
310 MITO
JAPAN
TEL: 0292-252-35-24
TLX:
COM: 35

HAMAJIMA KIYOTOSHI DR
KISO BRANCH OF THE
TOKYO ASTRON OBSERVATORY
MITAKEMURA KISOGUN
NAGANOKEN 397-01
JAPAN
TEL:
TLX:
COM: 33

HAMDY M A M DR
HELWAN OBSERVATORY
HELWAN-CAIRO
EGYPT
TEL: 780645-782683
TLX: 93070 HIAG UN
COM:

HAMID S EL DIN DR
DEPT OF ASTRONOMY
FACULTY OF SCIENCE
UNIVERSITY FOUAD
GIZA CAIRO
EGYPT
TEL:
TLX:
COM: 07

HAMILTON P A DR
UNIVERSITY OF TASMANIA
PHYSICS DEPARTMENT
BOX 252C G.P.O.
HOBART, TASMANIA
AUSTRALIA
TEL: 002-20-2419
TLX: 58150
COM: 40

HAMANN WOLF-RAINER
INST THEOR PHYS & STERNW
UNIVERSITAET KIEL
OLSHAUSENSTR
D-2300 KIEL
GERMANY, F.R.
TEL: 0431-8804101
TLX: 292706 IAPKI D
COM: 36

HAMMER REINER
KIEPENHEUER INSTITUT FUER
SONNENPHYSIK
SCHOENECKSTR 6
D-7800 FREIBURG
GERMANY, F.R.
TEL: 0761-32864
TLX: 7721552 KIS D
COM: 10,12

HAMMERSCHLAG ROBERT H DR
ASTRONOMICAL INSTITUTE
STERREWACHT SONNENBORGH
ZONNENBURG 2
NL-3512 NL UTRECHT
NETHERLANDS
TEL: 030-312-841
TLX:
COM: 09

HAMMERSCHLAG-HENSBERGE G
ASTRONOMICAL INSTITUTE
UNIVERSITY OF AMSTERDAM
ROETERSSTRAAT 15
NL-1018 WB AMSTERDAM
NETHERLANDS
TEL: 0-20-522-3004
TLX: 16460
COM: 42

HAMZAOGLU ESAT E H DR
KING SAUD UNIVERSITY
COLLEGE OF SCIENCE
P.O. BOX 2455
RIYADH 11453
SAUDI ARABIA
TEL:
TLX:
COM:

HAN FU
PURPLE MOUNTAIN
OBSERVATORY
NANJING
CHINA, PEOPLE'S REP.
TEL: 33738
TLX: 34144 PMONJ CN
COM: 40

LIST OF MEMBERS

HAN TIANQI
INST OF GEODESY & GEOPHYS
ACADEMIA SINICA
WUCHANG HUBEI
CHINA, PEOPLE'S REP.
TEL: 813712-570
TLX:
COM: 19,31

HAN WENJUN
BEIJING ASTRONOMICAL OBS
ACADEMIA SINICA
BEIJING
CHINA, PEOPLE'S REP.
TEL: 281698
TLX:
COM: 40

HAN ZHENG-ZHONG
PURPLE MOUNTAIN
OBSERVATORY
NANJING
CHINA, PEOPLE'S REP.
TEL: 33583
TLX: 34144
COM: 44

HANASZ JAN DR
ASTRONOMICAL INSTITUTE
POLISH ACAD OF SCIENCES
UL. CHOPINA 12/18
87-100-TORUN
POLAND
TEL: 260-37
TLX: 0552234 ASTR PL
COM: 10,40

HANBURY BROWN ROBERT PROF
SCHOOL OF PHYSICS
UNIVERSITY OF SYDNEY
SYDNEY NSW 2006
AUSTRALIA
TEL: 02-692-2934
TLX: 26169 UNISYD
COM: 40

HANES DAVID A DR
QUEEN'S UNIVERSITY
PHYSICS DEPT
ASTRONOMY GROUP
KINGSTON ONT K7L 3N6
CANADA
TEL: 613-547-5750
TLX:
COM: 37

HANG HENG-RONG
PURPLE MOUNTAIN
OBSERVATORY
NANJING
CHINA, PEOPLE'S REP.
TEL: 33583
TLX: 34144 PMONJ CN
COM: 44,48

HANKINS TIMOTHY HAMILTON
THAYER SCHOOL OF ENGINRG
DARTHMOUTH COLLEGE
HANOVER NH 03755
U.S.A.
TEL: 603-646-2230
TLX: 7103661828
COM: 40

HANNER MARTHA S DR
JPL
MS T1166
4800 OAK GROVE DR.
PASADENA CA 91109
U.S.A.
TEL: 818-354-4100
TLX: 675429
COM: 15,21C,22C

HANSEN CARL J PROF
JILA
UNIVERSITY OF COLORADO
BOX 440
BOULDER CO 80309
U.S.A.
TEL: 303-492-7811
TLX: 755842 JILA
COM: 27

HANSEN LEIF LECTURER
UNIVERSITY OBSERVATORY
OESTER VOLDGADE 3
DK-1350 COPENHAGEN K
DENMARK
TEL: 1-14 17 90
TLX: 44155 DANAST DK
COM:

HANSEN RICHARD T MR
ENGINEERING 138, VAMC
150 S. HUNTINGTON AVE
BOSTON MA 02130
U.S.A.
TEL: 6177342534-HOME
TLX:
COM: 10

HANSLMEIER ARNOLD
INSTITUT FUER ASTRONOMIE
KARL-FRANZENS-UNIVERSITAT
UNIVERSITAETSPLATZ 5
A-8010 GRAZ
AUSTRIA
TEL: 0316-380-5275
TLX:
COM: 07,10

HANSON ROBERT B DR
LICK OBSERVATORY
UNIVERSITY OF CALIFORNIA
SANTA CRUZ CA 95064
U.S.A.
TEL: 408-429-2755
TLX:
COM: 24

HANSSON NILS DR
LUND OBSERVATORY
BOX 43
S-221 00 LUND
SWEDEN
TEL: 46-107000
TLX: 33533 LUNIVER S
COM:

HAO YUN-XIANG
DEPT OF ASTRONOMY
BEIJING NORMAL UNIVERSITY
BEIJING
CHINA, PEOPLE'S REP.
TEL: 656531-6285
TLX:
COM: 09

HAPKE BRUCE W DR
DEPT GEOL & PLANETARY SCI
UNIVERSITY OF PITTSBURGH
321 OLD ENGINEERING HALL
PITTSBURGH PA 15235
U.S.A.
TEL: 412-624-4719
TLX:
COM: 15

HARA KEN NOSUKE DR
SENDAI-SHIRITSU-JOSHI
SENIOR HIGH SCHOOL
KASHIWAGI 3-3-1
SENDAI 980
JAPAN
TEL:
TLX:
COM: 47

HARA TETSUYA DR
DEPT OF PHYSICS
KYOTO SANGYO UNIVERSITY
KITAKU KAMIGAMO
KYOTO 603
JAPAN
TEL: 075-701-2151
TLX: 5422661 KSU J
COM: 28

HARDEBECK ELLEN G DR
3106 TUMBLEWEED RD
BISHOP CA 93514
U.S.A.
TEL:
TLX:
COM: 34

HARDEE PHILIP
DEPT PHYSICS & ASTRONOMY
UNIVERSITY OF ALABAMA
BOX 1921
UNIVERSITY AL 35486
U.S.A.
TEL: 205-348-5050
TLX:
COM: 40

HARDIE R PROF
DYER OBSERVATORY
VANDERBILT UNIVERSITY
NASHVILLE TN 37235
U.S.A.
TEL:
TLX:
COM: 25

HARDORP JOHANNES PROF
DEPT EARTH SPACE SC
STATE UNIVERSITY
STONY BROOK NY 11794
U.S.A.
TEL: 516-246-4048
TLX:
COM: 36

HARDY EDUARDO
DEPT DE PHYSIQUE
UNIVERSITE LAVAL
FAC DES SCS & DE GENIE
QUEBEC G1K 7P4
CANADA
TEL: 418-656-2960
TLX: 369-5131621
COM: 28,47

HARMANEC PETR DR
ASTRONOMICAL INSTITUTE
CZECH. ACAD. OF SCIENCES
251 65 ONDREJOV
CZECHOSLOVAKIA
TEL: 724525
TLX: 121579
COM: 27,29,42

HARMER CHARLES F W MR
SIRA LTD, OPTC SYST DEPT
RESEARCH & DEVLPT DIV
SOUTH HILL
CHISLEHURST KENT BR27 5EH
U.K.
TEL: 01-467-2636
TLX: 896649
COM: 09,29

HARMER DIANNE L MRS
ROYAL GREENWICH OBS
HERSTMONCEUX CASTLE
HAILSHAM BN27 1RP
U.K.
TEL: 0323-833171
TLX: 87451
COM: 09,29

HARMS RICHARD JAMES DR
APPLIED RESEARCH CORP.
8201 CORPORATE DRIVE
SUITE 920
LANDOVER MD 20785
U.S.A.
TEL: 301-459-8442
TLX:
COM: 28,44,47

HARNDEN FRANK R Jr
HARVARD-SMITHSONIAN CTR
FOR ASTROPHYSICS
60 GARDEN STREET
CAMBRIDGE MA 02138
U.S.A.
TEL: 617-495-7143
TLX: 921428 SATELLITE CAM
COM:

HARO GUILLERMO DR
INAOE
AP POSTALES 216 y 51
72000 PUEBLA, PUE.
MEXICO
TEL: 91-22-47-05-00
TLX:
COM: 27,28,29

HARRINGTON J PATRICK DR
ASTRONOMY PROGRAM
UNIVERSITY OF MARYLAND
COLLEGE PK MD 20742
U.S.A.
TEL: 301-454-5944
TLX: 7108260352
COM: 34

HARRINGTON ROBERT S DR
US NAVAL OBSERVATORY
WASHINGTON DC 20390
U.S.A.
TEL: 202-653-1533
TLX:
COM: 20,24,26C,51

HARRIS ALAN WILLIAM
SERC
RUTHERFORD APPLETON LAB
CHILTON, DIDCOT OX11 0QX
U.K.
TEL: 0235-21900
TLX: 83159 RUTHLB G
COM: 34

HARROWER GEORGE A DR
LAKEHEAD UNIVERSITY
THUNDER BAY ONT P7B 5E1
CANADA
TEL: 807-345-2121
TLX:
COM:

HARTQUIST THOMAS WILBUR
MPI FUR PHYS & ASTROPHYS
KARL-SCHWARZSCHILD-STR 1
D-8046 GARCHING B MUNCHEN
GERMANY, F.R.
TEL: 089-3299-838
TLX: 05215845 XTERR D
COM: 34

HARWIT MARTIN PROF
ASTRONOMY DEPT.
SPACE SCIENCE BLDG
CORNELL UNIVERSITY
ITHACA NY 14853
U.S.A.
TEL: 607-256-4805
TLX:
COM: 15,21,48

HARRIS ALAN WILLIAM DR
JPL
MS 183-501
4800 OAK GROVE DRIVE
PASADENA CA 91109
U.S.A.
TEL: 818-354-6741
TLX: 675429/9105883294/69
COM: 15C,20

HART MICHAEL H DR
PHYSICS DEPT
TRINITY UNIVERSITY
SAN ANTONIO TX 78284
U.S.A.
TEL:
TLX:
COM: 51

HARTWICK F DAVID A DR
UNIVERSITY OF VICTORIA
DEPT OF PHYSICS
VICTORIA BC V8W 2Y2
CANADA
TEL: 604-721-7742
TLX: 049-7222
COM:

HARWOOD DENNIS MR
PERTH OBSERVATORY
BICKLEY, W. AUSTR. 6076
AUSTRALIA
TEL: 09-2938-255
TLX:
COM: 08,24,25

HARRIS DANIEL E DR
CENTER FOR ASTROPHYSICS
60 GARDEN STREET
CAMBRIDGE MA 02138
U.S.A.
TEL: 617-495-7148
TLX: 921428
COM: 40

HARTEN RONALD H DR
RCA ASTRO ELECTRONIC
TB-1
PO BOX 800
PRINCETON NJ 08540
U.S.A.
TEL: 609-426-3551
TLX:
COM: 34,40

HARTZ THEODORE R DR
915 MOUNTAINVIEW AVENUE
OTTAWA ONT K2B 5G3
CANADA
TEL: 613-596-1211
TLX:
COM: 40,44

HASAN SAIYID STRAJUL
INDIAN INSTITUTE OF
ASTROPHYSICS
BANGALORE 560 034
INDIA
TEL:
TLX:
COM: 12

HARRIS GRETCHEN L H DR
DEPT OF PHYSICS
UNIVERSITY OF WATERLOO
WATERLOO ON N2L 3G1
CANADA
TEL: 519-885-1211
TLX: 069-55259
COM: 37V

HARTL HERBERT DR
INSTITUT FUER ASTRONOMIE
UNIVERSITAETSSTR 4
A-6020 INNSBRUCK
AUSTRIA
TEL:
TLX:
COM:

HARVEL CHRISTOPHER ALVIN
6161 STEVEN'S FOREST RD
COLUMBIA MD 21045
U.S.A.
TEL: 301-964-0211
TLX:
COM: 37

HASCHICK AUBREY
HAYSTACK OBSERVATORY
WESTFORD MA 01886
U.S.A.
TEL: 617-692-4764
TLX:
COM: 40

HARRIS HUGH C
US NAVAL OBSERVATORY
PO BOX 1149
FLAGSTAFF AZ 86002
U.S.A.
TEL: 602-779-5132
TLX:
COM: 37

HARTLEY KENNETH F DR
RUTHERFORD APPLETON LAB
CHILTON, DIDCOT OX11 0QX
U.K.
TEL: 0235-21900
TLX: 83159
COM: 05

HARVEY CHRISTOPHER C DR
OBSERVATOIRE DE PARIS
SECTION DE MEUDON
F-92195 MEUDON PL CEDEX
FRANCE
TEL: 1-45-34-75-30
TLX: 204464
COM: 44,49

HASEGAWA HIROICHI DR
DEPT OF PHYSICS
KYOTO UNIVERSITY
SAKYO-KU
KYOTO 606
JAPAN
TEL: 0757512111x3833
TLX:
COM: 22

HARRIS STELLA
DEPT OF PHYSICS
QUEEN MARY COLLEGE
MILE END ROAD
LONDON E1 4NS
U.K.
TEL: 1-980-4811x4050
TLX: 893750
COM: 34

HARTMANN LEE WILLIAM
CENTER FOR ASTROPHYSICS
60 GARDEN STREET
CAMBRIDGE MA
U.S.A.
TEL: 617-495-7487
TLX:
COM: 36

HARVEY GALE A DR
INST F GESCHICHTE D
NATURWISSENSCHAFTEN
GOETHE UNIVERSITAET
D-6000 FRANKFURT
GERMANY, F.R.
TEL:
TLX:
COM: 22

HASEGAWA ICHIRO DR
2-3-11 SAIDAIJINOGAMI
NARA 631
JAPAN
TEL: 81-742-46-2055
TLX:
COM: 20,22C

HARRIS WILLIAM E DR
DEPT OF PHYSICS
MCMASTER UNIVERSITY
HAMILTON ONT L8S 4M1
CANADA
TEL: 416-525-9140
TLX: 369-0618347
COM: 37

HARTMANN WILLIAM K
PLANETARY SCIENCE INST
2030 E SPEEDWAY
SUITE 201
TUCSON AZ 85719
U.S.A.
TEL: 602-881-0332
TLX:
COM: 15

HARVEY JOHN W DR
NATL SOLAR OBSERVATORY
PO BOX 26732
950 N. CHERRY AVE
TUCSON AZ 85726
U.S.A.
TEL: 602-327-5511
TLX:
COM: 10,12V

HASER LEO N K DR
MPI F. EXTRATERR. PHYSIK
D-8046 GARCHING B MUNCHEN
GERMANY, F.R.
TEL: 89-329-98-03
TLX: 5215845 XTER D
COM: 15

HARRISON EDWARD R PROF
DEPT PHYSICS & ASTRONOMY
UNIV OF MASSACHUSETTS
AMHERST MA 01003
U.S.A.
TEL: 413-545-2194
TLX:
COM: 47,51

HARTOOG MARK RICHARD DR
LICK OBSERVATORY
UNIVERSITY OF CALIFORNIA
SANTA CRUZ CA 95064
U.S.A.
TEL:
TLX:
COM:

HARVEY PAUL MICHAEL DR
DEPT OF ASTRONOMY
UNIVERSITY OF TEXAS
AUSTIN TX 78712
U.S.A.
TEL: 512-471-4461
TLX: 910-874-1351
COM: 34,44

HASLAM C GLYN T DR
MPI FUER RADIOASTRONOMIE
AUF DEM HUEGEL 69
D-5300 BONN
GERMANY, F.R.
TEL:
TLX: 886440 MPIFR D
COM: 40

HASSAN S M DR
HIAG
HELWAN-CAIRO
EGYPT
TEL: 780645, 782683
TLX: 93070
COM: 37

HAUPT WOLFGANG DR
GIRONDELLE 105
D-4630 BOCHUM
GERMANY, F.R.
TEL:
TLX:
COM:

HAYASHI CHUSHIRO PROF
MOMOYAMA YOGORO-CH01
FUSHIMI-KU
KYOTO 612
JAPAN
TEL: 075-611-1062
TLX:
COM: 35,47

HAZEN MARTHA L DR
HARVARD COLLEGE OBS
60 GARDEN STREET
CAMBRIDGE MA 02138
U.S.A.
TEL: 617-495-3362
TLX:
COM: 37

HAUBOLD HANS JOACHIN
ZNTRLINST. F. ASTROPHYSIK
AKAD. WISSENSCHAFTEN DDR
ROSA-LUXEMBURG-STR 17A
DDR-1502 POTSDAM-BABELSB.
GERMANY, D.R.
TEL:
TLX:
COM:

HAVLEN ROBERT J DR
NRAO
EDGEMONT ROAD
CHARLOTTESVILLE VA 22901
U.S.A.
TEL: 804-296-0223
TLX: 9109970174
COM:

HAYES DONALD S DR
PO BOX 1907
SCOTTSDALE AZ 85252
U.S.A.
TEL: 602-947-3572
TLX:
COM: 25,45

HAZER S DR
FACULTY OF SCIENCE
DEPT OF ASTRONOMY
P K 21
BORNOVA-IZMIR
TURKEY
TEL:
TLX:
COM:

HAUCK BERNARD PROF
INSTITUT D'ASTRONOMIE
UNIVERSITE DE LAUSANNE
CH-1290 CHAVANNES-DES-B.
SWITZERLAND
TEL: 022-55-26-11
TLX: 27720 OBSG CH
COM: 05V,25,45,46

HAVNES OVE DR
AURORAL OBSERVATORY
UNIVERSITY OF TROMSO
P O 953
N-9001 TROMSO
NORWAY
TEL: 83-86060
TLX: 64124 AUROB N
COM:

HAYLI AVRAM PROF
OBSERVATOIRE DE LYON
F-69230 ST-GENIS-LAVAL
FRANCE
TEL: 78-56-07-05
TLX: 310916
COM: 33,41

HAZLEHURST JOHN DR
HAMBURGER STERNWARTE
GOJENSBERGSWEG 112
D-2050 HAMBURG 80
GERMANY, F.R.
TEL:
TLX:
COM: 42

HAUG EBERHARD DR
MOZARTSTRASSE 20
D-7430 METZINGEN
GERMANY, F.R.
TEL: 07071/296483
TLX:
COM: 10

HAWARDEN TIMOTHY G DR
ROYAL OBSERVATORY
BLACKFORD HILL
EDINBURGH EH9 3HJ
U.K.
TEL: 031-667-3321
TLX: 72383 ROEDIN G
COM: 37

HAYMES ROBERT C PROF
DEPT SPACE PHYS & ASTRON
RICE UNIVERSITY
HOUSTON TX 77001
U.S.A.
TEL: 713-527-4045
TLX: 556457
COM:

HE MIAO-FU
SHANGHAI OBSERVATORY
ACADEMIA SINICA
SHANGHAI
CHINA, PEOPLE'S REP.
TEL: 386191
TLX: 33164 SHAO CN
COM: 07

HAUG ULRICH PROF
HAMBURGER STERNWARTE
GOJENSBERGWEG 112
D-2050 HAMBURG 80
GERMANY, F.R.
TEL: 040-7252-4131
TLX: 217884 HAMST D
COM: 21,33

HAWKING STEPHEN W PROF
DEPT OF APPLIED MATHS
AND THEORETICAL PHYSICS
SILVER STREET
CAMBRIDGE CB3 9EW
U.K.
TEL: 223-351645
TLX: 81240 CAMSPL G
COM: 47,48

HAYNES MARTHA P
ASTRONOMY DEPT
CORNELL UNIVERSITY
SPACE SCIENCES BUILDING
ITHACA NY 14853
U.S.A.
TEL: 607-256-3734
TLX: 932454
COM: 40

HE XIANG-TAO
DEPT OF ASTRONOMY
BEIJING NORMAL UNIVERSITY
BEIJING
CHINA, PEOPLE'S REP.
TEL: 656531-6285
TLX:
COM: 28,47

HAUGE OIVIND DR
INST THEORET ASTROPHYSICS
UNIVERSITY OF OSLO
P O BOX 1029
N-0315 BLINDERN, OSLO 3
NORWAY
TEL: 245-65-06
TLX:
COM:

HAWKINS GERALD S DR
CONSUL 906
2400 VIRGINIA QVE NW
WASHINGTON DC 20037
U.S.A.
TEL: 202-485-2050
TLX:
COM: 22,41

HAYNES RAYMOND F PROF
CSIRO
DIVISION OF RADIOPHYSICS
P.O.BOX 76
EPPING NSW 2121
AUSTRALIA
TEL:
TLX:
COM: 34,40

HEAP SARA R DR
NASA/GSFC
CODE 672
GREENBELT MD 20771
U.S.A.
TEL: 301-344-5359
TLX:
COM:

HAUPT HERMANN F PROF
INSTITUT FUER ASTRONOMIE
DER UNIVERSITAET
UNIVERSTAETSPLATZ 5
A-8010 GRAZ
AUSTRIA
TEL: 0316-380-5271
TLX: 31078A
COM: 15C,20,38,46

HAWKINS MICHAEL R S
ROYAL OBSERVATORY
BLACKFORD HILL
EDINBURGH EH9 3HJ
U.K.
TEL:
TLX:
COM: 33

HAYWARD JOHN
DEPT OF MATHEMATICS
NAPIER COLLEGE
COLINTON RD.
EDINBURGH EH10 5DT
U.K.
TEL:
TLX:
COM: 10

HEARN ANTHONY G DR
STERREKUNDIG INSTITUUT
SERVAAS BOLWERK 13
NL-3512 NK UTRECHT
NETHERLANDS
TEL: 030-312-841
TLX: 427224 ASTRO NL
COM: 36C,44

HAUPT RALPH F
3701 DULWICK DRIVE
SILVER SPRING MD 20906
U.S.A.
TEL: 301-598-7868
TLX:
COM: 04,12

HAYAKAWA SATIO PROF
DEPT PHYSICS/ASTROPHYSICS
NAGOYA UNIVERSITY
FUROKO CHIKUSAKU
NAGOYA 464
JAPAN
TEL: 052-781-5111
TLX: 4477323 SCUNAG J
COM: 44,47C,48

HAZARD CYRIL DR
INSTITUTE OF ASTRONOMY
MADINGLEY ROAD
CAMBRIDGE CB3 0HE
U.K.
TEL:
TLX:
COM: 40

HEARNSHAW JOHN B DR
DEPT OF PHYSICS
UNIVERSITY OF CANTERBURY
PRIVATE BAG
CHRISTCHURCH
NEW ZEALAND
TEL: 03-482009 x 771
TLX: 4144 UNICANT NZ
COM: 29

HEASLEY JAMES NORTON
INSTITUTE FOR ASTRONOMY
2680 WOODLAWN DRIVE
HONOLULU HI 96822
U.S.A.
TEL: 808-948-6826
TLX: 7238459 UHAST HR
COM: 36

HEBER ULRICH
INST THEOR PHYS & STERNW
UNIVERSITAET KIEL
LEIBNIZSTR
D-2300 KIEL 1
GERMANY, F.R.
TEL: 0431-880-4103
TLX: 292706 IAPKI D
COM: 36

HECK ANDRE DR
OBS DE STRASBOURG
11 RUE DE L'UNIVERSITE
F-67000 STRASBOURG
FRANCE
TEL: 88-35-43-00
TLX: 890506 STAROBS F
COM: 05,25,45C,51

HECKATHORN HARRY M
US NAVAL RESEARCH LAB
CODE 4143-2
WASHINGTON DC 20375
U.S.A.
TEL: 202-767-2764
TLX:
COM: 09,44

HECKMAN TIMOTHY M
UNIVERSITY OF MARYLAND
ASTRONOMY PROGRAM
COLLEGE PARK MD 20742
U.S.A.
TEL: 301-454-3001
TLX: 7108260352 ASTR CORP
COM:

HEDDLE DOUGLAS W O PROF
DEPT OF PHYSICS
ROYAL HOLLOWAY & BEDFORD
NEW COLLEGE
EGHAM, SURREY TW20 0EX
U.K.
TEL: 0784-35351
TLX: 935504
COM: 14

HEDEMAN E RUTH
14 N. MAIN ST, NO 201
CLARKSTON MI 48016
U.S.A.
TEL:
TLX:
COM: 10

HEESCHEN DAVID S DR
NRAO
EDGEMONT ROAD
CHARLOTTESVILLE VA 22901
U.S.A.
TEL:
TLX: 910-997-0174
COM: 28,40,51

HEFELE HERBERT PH D
MPI FUER ASTRONOMIE
KONIGSTUHL
D-6900 HEIDELBERG 1
GERMANY, F.R.
TEL:
TLX:
COM: 05,34

HEFFERLIN RAY A PROF
PHYSICS DEPT
SOUTHERN COLLEGE
DRAWER H
COLLEGEDALE TN 37315-0370
U.S.A.
TEL: 615-238-2869
TLX:
COM: 14

HEGGIE DOUGLAS C DR
UNIVERSITY OF EDINBURGH
DEPARTMENT OF MATHEMATICS
KING'S BUILDINGS
EDINBURGH EH9 3JZ
U.K.
TEL: 31-667-1081
TLX: 727442 UNIVEDG
COM: 07,37P,41

HEGYI DENNIS J ASSOC PROF
RANDALL LABORATORY
UNIVERSITY OF MICHIGAN
ANN ARBOR MI 48109
U.S.A.
TEL: 313-764-5448
TLX: 810-2236056
COM:

HEIDMANN JEAN DR
OBSERVATOIRE DE PARIS
SECTION DE MEUDON
F-92195 MEUDON PL CEDEX
FRANCE
TEL: 1-45-34-75-30
TLX: 270912 F
COM: 28,40,47,51

HEILES CARL PROF
ASTRONOMY DEPT
UNIVERSITY OF CALIFORNIA
BERKELEY CA 94720
U.S.A.
TEL: 415-642-4510
TLX: 820181 UCB AST RAL
COM: 33,34,40

HEILESEN POUL MARTIN DR
BOULEVARDEN 14 2 16
DK-2800 LYNGBY
DENMARK
TEL:
TLX:
COM: 35

HEINRICH INGE
ASTRONOMISCHES RECHEN-
INSTITUT
MOENCHHOFSTR 12-14
D-6900 HEIDELBERG 1
GERMANY, F.R.
TEL:
TLX:
COM: 05

HEINTZ WULFF D DR
DEPT OF ASTRONOMY
SWARTHMORE COLLEGE
SWARTHMORE PA 19081
U.S.A.
TEL: 215-447-7265
TLX:
COM: 05C,08,24,42

HEINTZE J R W DR
STERREWACHT SONNENBORGH
SERVAAS BOLWERK 13
NL-3512 NK UTRECHT
NETHERLANDS
TEL:
TLX:
COM: 29,30

HEISE JOHN DR
SPACE RESEARCH LABORATORY
BENELUXLAAN 21
NL-3527 HS UTRECHT
NETHERLANDS
TEL: 31-30937145
TLX: 47224 ASTRO NL
COM: 44,48

HEISER ARNOLD M DR
DYER OBSERVATORY
VANDERBILT UNIVERSITY
BOX 1803-STA B
NASHVILLE TN 37235
U.S.A.
TEL: 615-373-4897
TLX:
COM: 27

HEKELA JAN DR
ASTRONOMICAL INSTITUTE
CZECH. ACAD. OF SCIENCES
251 65 ONDREJOV
CZECHOSLOVAKIA
TEL:
TLX:
COM: 36

HELALI YHYA E DR
HELWAN OBSERVATORY
HELWAN-CAIRO
EGYPT
TEL:
TLX:
COM: 07

HELFAND DAVID JOHN
COLUMBIA ASTROPHYSICS LAB
538 WEST 120TH ST
NEW YORK NY 10027
U.S.A.
TEL: 212-280-2150
TLX:
COM: 48

HELFER H LAWRENCE PROF
DEPT PHYSICS & ASTRONOMY
UNIVERSITY OF ROCHESTER
ROCHESTER NY 14627
U.S.A.
TEL: 716-275-4377
TLX:
COM: 34

HELIN ELEANOR FRANCIS
JET PROPULSION LAB
MS 183-501
4800 OAK GROVE DRIVE
PASADENA CA 91109
U.S.A.
TEL: 818-354-4606
TLX: 67-5429
COM: 20

HELLER MICHAEL PROF
POWSTANCOW WARSAWY 13/94
33-110 TARNOW
POLAND
TEL:
TLX:
COM: 47

HELLWIG HELMUT WILHELM DR
FREQUENCY & TIME SYSTEMS
34 TOZER ROAD
BEVERLY MA 01915
U.S.A.
TEL: 617-927-8220
TLX: 940518
COM: 19,31

HELMER LEIF
COPENHAGEN UNIVERSITY OBS
BRORFELDE
DK-4340 TOLLOSE
DENMARK
TEL: 3-488195
TLX: 44155
COM: 08C,50

HELMKEN HENRY F DR
SMITHSONIAN ASTROPHYS OBS
60 GARDEN ST
CAMBRIDGE MA 02138
U.S.A.
TEL:
TLX:
COM: 44

HELT BODIL E
UNIVERSITY OBSERVATORY
OESTER VOLDGADE 3
DK-1350 COPENHAGEN K
DENMARK
TEL: 1-14 17 90
TLX: 44155 DANAST DK
COM:

HEMENWAY PAUL D
ASTRONOMY DEPT
UNIVERSITY OF TEXAS
R.L. MOORE HALL 15.308
AUSTIN TX 78712
U.S.A.
TEL: 512-471-4461
TLX:
COM: 08,20,24

HEMMLEB GERHARD DR
ZENTRALINSTITUT FUR
PHYSIK DER ERDE
TELEGRAFENBERG A 17
DDR-1500 POTSDAM
GERMANY, D.R.
TEL: 4551
TLX: 15305 VDE PDM DD
COM: 19,31C

LIST OF MEMBERS

HEMPE KLAUS
GRENZWEG 24 B
D-2057 REINBEK
GERMANY, F.R.
TEL: 040-710-56-28
TLX:
COM:

HENIZE KARL G ASTRONAUT
CODE CB
NASA/JOHNSON SPACE CENTER
HOUSTON TX 77058
U.S.A.
TEL: 713-483-2411
TLX:
COM: 29,34,44,45

HENKEL CHRISTIAN
MPI FUER RADIOASTRONOMIE
AUF DEM HUEGEL 69
D-5300 BONN 1
GERMANY, F.R.
TEL:
TLX: 886440 MPIFR D
COM: 34,40

HENON MICHEL C DR
OBSERVATOIRE DE NICE
B P 139
F-06003 NICE CEDEX
FRANCE
TEL: 93-89-04-20
TLX: 460004
COM: 07,33,37

HENOUX JEAN-CLAUDE DR
OBSERVATOIRE DE PARIS
SECTION DE MEUDON
F-92195 MEUDON PL CEDEX
FRANCE
TEL: 1-45-34-75-30
TLX:
COM: 10,44

HENRARD JACQUES PROF
FACULTES UNIV DE NAMUR
RUE DE BRUXELLES 61
B-5000 NAMUR
BELGIUM
TEL: 81-22-90-61
TLX: 59222
COM: 04,07V,20

HENRIKSEN RICHARD N DR
ASTRONOMY GROUP
DEPT OF PHYSICS
QUEEN'S UNIVERSITY
KINGSTON ONT K7L 3N6
CANADA
TEL: 613-547-5536
TLX:
COM: 48

HENRY RICHARD C. PROF.
DEPT PHYSICS & ASTRONOMY
JOHNS HOPKINS UNIVERSITY
BALTIMORE MD 21218
U.S.A.
TEL: 301-338-7350
TLX:
COM: 21,48

HENSBERGE HERMAN
ASTROPHYSICAL INSTITUTE
VRIJE UNIV BRUSSEL
PLEINLAAN 2
B-1050 BRUSSEL
BELGIUM
TEL: 02-641-3468
TLX: 61051 VUBCO
COM: 25,44

HENSLER GERHARD
INST F ASTRON & ASTROPHYS
UNIVERSITAET MUENCHEN
SCHEINERSTR. 1
D-8000 MUNCHEN 80
GERMANY, F.R.
TEL: 089-98-90-21
TLX: 529815 UNIVM D
COM: 42

HERBIG GEORGE H DR
LICK OBSERVATORY
UNIVERSITY OF CALIFORNIA
SANTA CRUZ CA 95064
U.S.A.
TEL: 408-429-2772
TLX:
COM: 27,29

HERBST ERIC DR
DEPT OF PHYSICS
DUKE UNIVERSITY
DURHAM NC 27706
U.S.A.
TEL: 919-684-8180
TLX: 802829 DUKTELECOM-DU
COM:

HERBST WILLIAM DR
ASTRONOMY DEPT
WESLEYAN UNIVERSITY
MIDDLETOWN CT 06457
U.S.A.
TEL: 203-347-9411
TLX:
COM: 33,37

HERCZEG TIBOR J PROF DR
DEPT PHYSICS & ASTRONOMY
UNIVERSITY OF OKLAHOMA
NORMAN OK 73019
U.S.A.
TEL: 405-325-3961
TLX:
COM: 42,51

HERMAN RENEE DR
7 BIS RUE TRUDON
F-92160 ANTONY
FRANCE
TEL:
TLX:
COM: 14,29,45

HERNANDEZ CARLOS ALBERTO
OBSERVATORIO ASTRONOMICO
PASEO DEL BOSQUE
1900 LA PLATA
ARGENTINA
TEL:
TLX:
COM: 26,46

HEROLD HEINZ
THEORETISCHE ASTROPHYSIK
UNIVERSITY TUEBINGEN
AUF DER MORGENSTELLE 12,C
D-7400 TUEBINGEN
GERMANY, F.R.
TEL: 07071/292043
TLX:
COM: 14,36

HERR RICHARD B DR
PHYSICS DEPT
UNIVERSITY OF DELAWARE
NEWARK DE 19716
U.S.A.
TEL: 302-451-2673
TLX:
COM: 27

HERRMANN DIETER DR
ARCHENHOLD OBSERVATORY
AND PUBLIC OBSERVATORY
DDR-1193 BERLIN-TREPTOW
GERMANY, D.R.
TEL: 272-8871 x 494
TLX:
COM: 41

HERS JAN MR
P O BOX 48
SEDGEFIELD 6573
SOUTH AFRICA
TEL: 04455-736
TLX:
COM: 06,20,27,31

HERSHEY JOHN L DR
US NAVAL OBSERVATORY
34TH & MASSACHUSETTS AVE
WASHINGTON DC 20390
U.S.A.
TEL: 202-653-1554
TLX: 710-822-1970
COM: 24,26,51

HERZBERG GERHARD DR
HERZBERG INST ASTROPHYS
NATL RESEARCH COUNCIL
OTTAWA ONT K1A 0R6
CANADA
TEL: 613-990-0917
TLX: 0533715
COM: 14,15,16,34

HESSER JAMES E DR
DOMINION ASTROPHYS OBS
5071 W SAANICH ROAD
RR 5
VICTORIA BC V8X 4M6
CANADA
TEL: 604-388-3974
TLX: 0497295
COM: 14,27,37C

HEUDIER JEAN-LOUIS DR
CERGA
CAUSSOLS
06460 ST VALLIER-DE-THIEY
FRANCE
TEL: 93-42-62-70
TLX: 461402 CERGOBS F
COM: 09,20,24,51

HEWETT PAUL
INSTITUTE OF ASTRONOMY
UNIVERSITY OF CAMBRIDGE
MADINGLEY ROAD
CAMBRIDGE CB3 OHA
U.K.
TEL: 223-62204
TLX: 817297 ASTRON G
COM: 47

HEWISH ANTONY PROF
CAVENDISH LABORATORY
MADINGLEY RD
CAMBRIDGE CB3 OHE
U.K.
TEL: 0223-66477
TLX: 81292
COM: 40

HEWITT ANTHONY V DR
LORAL ELECTRONIC SYSTEMS
YONKERS NY 10710-0800
U.S.A.
TEL: 914-968-2500
TLX:
COM: 09

HEY JAMES STANLEY DR
4 SHORTLANDS CLOSE
EASTBOURNE BN22 0JE
U.K.
TEL:
TLX:
COM: 22,40

HEYDEN FRANCIS J SJ DR
MANILA OBSERVATORY
P O BOX 1231
MANILA
PHILIPPINES
TEL: 999-417
TLX:
COM: 10,27

HEYVAERTS JEAN DR
OBSERVATOIRE DE PARIS
SECTION DE MEUDON
F-92195 MEUDON PL CEDEX
FRANCE
TEL: 1-45-34-75-70
TLX: 201571
COM: 49

HIBBS ALBERT R MGR PLANS
JET PROPULSION LAB
4800 OAK GROVE DRIVE
PASADENA CA 91103
U.S.A.
TEL: 818-354-2430
TLX: 67-5429
COM:

HICKSON PAUL DR
DEPT GEOPHYS & ASTRONOMY
UNIV OF BRITISH COLUMBIA
2219 MAIN MALL
VANCOUVER BC V6T 1W5
CANADA
TEL: 604-228-2267
TLX: 045-4245
COM: 28

HIDAJAT BAMBANG PROF DR
BOSSCHA OBSERVATORY
LEMBANG, JAVA
INDONESIA
TEL: 6001 LEMBANG
TLX: 28234 BD ITB
COM: 26,34,46C,50

HILF EBERHARD R H PH D
PAUL-WAGNER-STR 56
D-6100 DARMSTADT
GERMANY, F.R.
TEL:
TLX:
COM: 35,40

HILTNER W ALBERT PROF
DEPT OF ASTRONOMY
UNIVERSITY OF MICHIGAN
ANN ARBOR MI 48109
U.S.A.
TEL: 313-764-3452
TLX:
COM: 25,34,45

HIRAYAMA TADASHI PROF
TOKYO ASTRONOMICAL OBS
OSAWA MITAKA
TOKYO 181
JAPAN
TEL: 0422-32-5111
TLX: 2822307 TAOMTK-J
COM: 10C,12

HIDALGO MIGUEL A DR
FACULTAD DE CIENCIAS
FISICAS
CIUDAD UNIVERSITARIA
50009 ZARAGOZA
SPAIN
TEL:
TLX:
COM:

HILL GRAHAM DR
DOMINION ASTROPHYS OBS
5071 W SAANICH ROAD
RR 5
VICTORIA BC V8X 4M6
CANADA
TEL: 602-388-3935
TLX:
COM: 24,30,42

HINKLE KENNETH H
KPNO, NOAO
PO BOX 26732
950 N. CHERRY AVE
TUCSON AZ 85726
U.S.A.
TEL: 602-327-5511
TLX: 0666-484 AURA NOAO T
COM:

HIRST WILLIAM P
1 CLIFFORD CRESCENT
BERGVLIET 7800
SOUTH AFRICA
TEL:
TLX:
COM:

HIDE RAYMOND PROF
GEOPHYSICAL FLUID
DYNAMICS LABORATORY
METEOROLOGICAL OFFICE
BRACKNELL, BERKS RG12 2SZ
U.K.
TEL: 0344-420242
TLX: 849801
COM: 16,19

HILL HENRY ALLEN DR
DEPT OF PHYSICS
UNIVERSITY OF ARIZONA
BLDG 81
TUCSON AZ 85721
U.S.A.
TEL: 602-621-6784
TLX: 910-9521143
COM: 27

HINTEREGGER HANS E DR
AIR FORCE GEOPHYSICS LAB
HANSCOM FIELD
BEDFORD MA 01731
U.S.A.
TEL:
TLX:
COM: 44

HIRTH WOLFGANG ERNST PH D
THEODOR-HEUS-STR 18
D-5354 WEILERSWIST
GERMANY, F.R.
TEL:
TLX:
COM:

HIEI EIJIRO DR
TOKYO ASTRONOMICAL OBS
OSAWA MITAKA
TOKYO 181
JAPAN
TEL: 0422-32-5111
TLX: 2822307
COM: 10,12C

HILL PHILIP W DR
UNIVERSITY OBSERVATORY
BUCHANAN GARDENS
ST ANDREWS KY16 9LZ
U.K.
TEL: 0334-76161
TLX: 76213 SAULIB G
COM: 25,27,45

HINTZEN PAUL MICHAEL N DR
NASA/GSFC
CODE 681
GREENBELT MD 20771
U.S.A.
TEL: 301-334-5101
TLX:
COM: 28

HITOTSUYANAGI JUICHI PROF
KATAHIRA 1-CHOME
4-6-402
SENDAI 980
JAPAN
TEL: 0222-27-9351
TLX:
COM: 35,36

HIGGS LLOYD A DR
DOMINION RADIO ASTROPHY-
SICAL OBSERVATORY
NRC, P O BOX 248
PENTICTON BC V2A 6K3
CANADA
TEL: 604-497-5321
TLX: 048-88127
COM: 34,40

HILLEBRANDT WOLFGANG PH D
MPI F. PHYSIK & ASTROPHYS
KARL-SCHWARZSCHILD-STR 1
D-8046 GARCHING B MUNCHEN
GERMANY, F.R.
TEL: 49-89-32999409
TLX:
COM:

HIPPELEIN HANS H DR
MPI FUER ASTRONOMIE
KOENIGSTUHL
D-6900 HEIDELBERG
GERMANY, F.R.
TEL:
TLX:
COM: 34

HJALMARSON AKE G DR
ONSALA SPACE OBSERVATORY
S-439 00 ONSALA
SWEDEN
TEL: 300-60653
TLX: 2400 ONSPACE
COM: 28,34,40

HILDEBRAND ROGER H
ENRICO FERMI INSTITUTE
UNIVERSITY OF CHICAGO
5640 S ELLIS AVE
CHICAGO IL 60637
U.S.A.
TEL: 312-962-7581
TLX:
COM: 34

HILLIARD R DR
OPTOMECHANICS RESEARCH
PO BOX 36522
TUCSON AZ 85740
U.S.A.
TEL: 602-887-4304
TLX:
COM: 09

HIRABAYASHI HISASHI DR
NOBEYAMA RADIO OBS
NOBEYAMA
MINAMIMAKI
NAGANO 384-13
JAPAN
TEL: 2679-8-2831
TLX: 3329005
COM: 40,51

HJELLMING ROBERT M DR
NRAO
PO BOX 0
SOCORRO NM 87801
U.S.A.
TEL: 505-772-4011
TLX: 910-988-1710
COM: 34,40,42

HILDITCH RONALD W DR
UNIVERSITY OBSERVATORY
BUCHANAN GARDENS
ST ANDREWS KY16 9LZ
U.K.
TEL: 0334-76161
TLX: 76213 SAULIB G
COM: 25,42

HILLS JACK G DR
LOS ALAMOS NATL LAB
THEORETICAL DIVISION T6
MS B-288
LOS ALAMOS NM 87545
U.S.A.
TEL: 505-667-9152
TLX:
COM: 37

HIRAI MASANORI DR
FUKUOKA UNIVERSITY
OF EDUCATION
729 MUNAKATA
FUKUOKA 811-41
JAPAN
TEL: 094-032-2381
TLX:
COM: 29

HO PAUL T P
HARVARD UNIVERSITY
60 GARDEN STREET
CAMBRIDGE MA 02138
U.S.A.
TEL: 617-495-3627
TLX: 921428
COM: 40

HILDNER ERNEST DR
NCAR
HIGH ALTITUDE OBSERVATORY
PO BOX 3000
BOULDER CO 80307
U.S.A.
TEL: 303-497-1541
TLX: 45694
COM: 10,12

HILLS RICHARD E DR
CAVENDISH LABORATORY
MADINGLEY ROAD
CAMBRIDGE CB3 0HE
U.K.
TEL: 0223-66477
TLX: 81282
COM: 40

HIRATA RYUKO
DEPT OF ASTRONOMY
KYOTO UNIVERSITY
SAKYO-KU
KYOTO 606
JAPAN
TEL:
TLX:
COM: 29

HOAG ARTHUR A DR
LOWELL OBSERVATORY
1400 W. MARS HILL RD
FLAGSTAFF AZ 86001
U.S.A.
TEL: 602-774-3358
TLX:
COM: 50

HOANG BINH DY DR
OBSERVATOIRE DE PARIS
SECTION DE MEUDON
LAM
F-92195 MEUDON PL CEDEX
FRANCE
TEL: 1-45-34-75-70
TLX: 201571
COM: 40,48,51

HOBBS LEWIS M DR
YERKES OBSERVATORY
UNIVERSITY OF CHICAGO
BOX 258
WILLIAMS BAY WI 53191
U.S.A.
TEL: 414-245-5555
TLX:
COM: 34

HOBBS ROBERT W DR
COMPUTER TECHN ASSOCIATES
1 MARYLAND CORPORATE CTR
7501 FORBES BLVD/S. 201
LANHAM MD 20706
U.S.A.
TEL: 301-464-5300
TLX:
COM: 33,40

HODGE PAUL W PROF
ASTRONOMY FM20
UNIVERSITY OF WASHINGTON
SEATTLE WA 98195
U.S.A.
TEL: 206-543-2888
TLX: 9104740096
COM: 22,28

HOEG ERIK DR
UNIVERSITY OBSERVATORY
OESTER VOLDGADE 3
DK-1350 COPENHAGEN K
DENMARK
TEL: 1-14-17-90
TLX: 44155 DANAST
COM: 08

HOEGBOM JAN A DR
STOCKHOLM OBSERVATORY
S-133 00 SALTSJOEBADEN
SWEDEN
TEL: 08-7170195
TLX: 12972 SOBSERV S
COM: 40

HOEGLUND BERTIL PROF
ONSALA SPACE OBSERVATORY
S-439 00 ONSALA
SWEDEN
TEL: 0300-60652
TLX: 2400
COM: 34,40

HOEKSTRA ROEL DR
SPACE RESEARCH LABORATORY
BENELUXLAAN 21
NL-3527 HS UTRECHT
NETHERLANDS
TEL: 30-937-145
TLX: 47224
COM:

HOESSEL JOHN GREG
WASHBURN OBSERVATORY
UNIV OF WISCONSIN-MADISON
475 N. CHARTER STREET
MADISON WI 53706
U.S.A.
TEL: 608-262-1752
TLX:
COM:

HOEY MICHAEL J DR
PHYSICS DEPT
UNIVERSITY COLLEGE
BELFIELD
DUBLIN 4
IRELAND
TEL:
TLX:
COM:

HOFF DARREL BARTON
DEPT OF EARTH SCIENCES
UNIV OF NORTHERN IOWA
CEDAR FALLS IA 50614
U.S.A.
TEL: 319-273-2389
TLX:
COM: 46

HOFFLEIT E DORRIT DR
DEPT OF ASTRONOMY
YALE UNIVERSITY
BOX 6666
NEW HAVEN CT 06511
U.S.A.
TEL:
TLX:
COM: 24,27

HOFFMAN JEFFREY ALAN DR
NASA-JSC
CODE CB-4
HOUSTON TX 77058
U.S.A.
TEL: 713-483-2411
TLX:
COM: 44,48

HOFFMANN MARTIN DR
OBSERVATORIUM HOHER LIST
STERNWARTE DER
UNIVERSITAET BONN
D-5568 DAUN
GERMANY, F.R.
TEL:
TLX:
COM: 42

HOFMANN WILFRIED DR
ASTRONOMISCHES RECHEN-
INSTITUT
MOENCHHOFSTR 12-14
D-6900 HEIDELBERG 1
GERMANY, F.R.
TEL: 06221-49026
TLX: 461336 ARIHD D
COM: 21

HOGG DAVID E DR
NRAO
EDGEMONT RD
CHARLOTTESVILLE VA 22901
U.S.A.
TEL: 804-296-0220
TLX: 910-997-0174
COM: 40

HOLBERG JAY B
LUNAR & PLANETARY LAB
UNIVERSITY OF ARIZONA
3625 EAST AJO WAY
TUCSON AZ 85713
U.S.A.
TEL: 602-621-4301
TLX:
COM: 16,44

HOLDEN FRANK
2 COLWICH CRESCENT
KINGSTON HILL
STAFFORD ST16 3XP
U.K.
TEL: 0785-53120
TLX:
COM: 26

HOLLENBACH DAVID JOHN DR
NASA/AMES RESEARCH CENTER
MS 245-6
MOFFETT FIELD CA 94035
U.S.A.
TEL: 415-997-6426
TLX:
COM: 34

HOLLIS JAN MICHAEL DR
NASA/GSFC
CODE 685
GREENBELT MD 20771
U.S.A.
TEL: 301-344-7591
TLX:
COM: 34,40,51

HOLLOWAY NIGEL J DR
SAFETY & RELIABILITY DIR.
WIGSHAW LANE
CULCHETH
WARRINGTON WA3 4NE
U.K.
TEL: 095-31244
TLX: 629301
COM: 48

HOLLWEG JOSEPH V
DEPT OF PHYSICS
UNIV OF NEW HAMPSHIRE
DEMERITT HALL
DURHAM NH 03824
U.S.A.
TEL: 603-862-3869
TLX:
COM: 49

HOLMAN GORDON D
NASA/GSFC
CODE 682
GREENBELT MD 20771
U.S.A.
TEL: 301-344-7921
TLX:
COM:

HOLMBERG ERIK B PROF
ENELIDEN 2
S-433 00 PARTILLE
SWEDEN
TEL: 031-265842
TLX:
COM: 25,28

HOLT STEPHEN S
NASA/GSFC
CODE 660
GREENBELT MD 20771
U.S.A.
TEL: 301-344-8801
TLX:
COM: 42,44

HOLWEGER HARTMUT PROF
INST THEOR PHYS & STERNW
UNIVERSITAET KIEL
OLSHAUSENSTR
D-2300 KIEL
GERMANY, F.R.
TEL: 8804107 KIEL
TLX: 292706 IAPKI D
COM: 12,36

HOLZER THOMAS EDWARD DR
HIGH ALTITUDE OBSERVATORY
NCAR
PO BOX 3000
BOULDER CO 80307
U.S.A.
TEL: 303-497-1536
TLX: 45694
COM: 10,36,49

HONEYCUTT R KENT PROF
ASTRONOMY DEPT
INDIANA UNIVERSITY
SWAIN HALL WEST
BLOOMINGTON IN 47405
U.S.A.
TEL: 812-335-6916
TLX:
COM: 09,42

HONG HYON IK
FACULTY OF PHYSICS
KIM IL SONG UNIVERSITY
TAESONG DISTRICT
PYONGYANG
KOREA DPR
TEL:
TLX:
COM: 10

HONG SEUNG SOO DR
DEPT OF ASTRONOMY
COLLEGE NATURAL SCIENCES
SEOUL NATIONAL UNIVERSITY
SEOUL 151
KOREA, REPUBLIC
TEL: 877-2131
TLX: 29664
COM: 21,22,34

HOOD ALAN
APPLIED MATHS DEPT
THE UNIVERSITY
ST ANDREWS, FIFE
U.K.
TEL: 0334-76161
TLX: 76213
COM: 10

HOOGHOUDT B G IR
PRINSENLAAN 10
NL-2341 KT OEGSTGEEST
NETHERLANDS
TEL: 49-71-172524
TLX:
COM: 09,40

HOPPE J A PROF DR
UNIVERSITY OBSERVATORY
SONNENBERGSTR 12, 336-09
DDR-6900 JENA
GERMANY, D.R.
TEL: 24223
TLX:
COM: 22

HORAK TOMAS B DR
GEOFYZIKA NP
JECNA 29A
612 00 BRNO
CZECHOSLOVAKIA
TEL: 772110
TLX: 62512 UGFBO C
COM: 42

HORAK ZDENEK PROF DR
VIETNAMSKA 2
160 00 PRAHA 6
CZECHOSLOVAKIA
TEL:
TLX:
COM:

HOREDT GEORG PAUL DR
DFVLR
D-8031 WESSLING
GERMANY, F.R.
TEL:
TLX:
COM: 16

HORI GENICHIRO PROF
DEPT OF ASTRONOMY
UNIVERSITY OF TOKYO
BUNKYO
TOKYO 113
JAPAN
TEL: 03-8122111x4251
TLX: 33659 UTYOSCI J
COM: 07,33

HOROWITZ PAUL PROF
DEPT OF PHYSICS
HARVARD UNIVERSITY
CAMBRIDGE MA 02138
U.S.A.
TEL: 617-495-3265
TLX: 4992111
COM: 51

HORSKY JAN PROF
DEPT OF THEORETICAL PHYS
PURKYNE UNIVERSITY
KOTLARSKA 2
611 37 BRNO
CZECHOSLOVAKIA
TEL: 51112
TLX:
COM:

HORSKY ZDENEK DR
ASTRONOMICAL INSTITUTE
CZECH. ACAD. OF SCIENCES
BUDECSKA 6
120 23 PRAHA 2
CZECHOSLOVAKIA
TEL:
TLX:
COM: 41

HORTON BRIAN H DR
DEPT OF PHYSICS
UNIVERSITY OF ADELAIDE
GPO BOX 498
ADELAIDE S A 5001
AUSTRALIA
TEL:
TLX:
COM: 12

HORVATH ANDRAS DR
TIT PLANETARIUM &
URANIA OBSERVATORY
BOX 46
H-1476 BUDAPEST
HUNGARY
TEL: 334-525
TLX:
COM:

HORWITZ GERALD PROF
RACAH INST. OF PHYSICS
HEBREW UNIV. OF JERUSALEM
JERUSALEM 91904
ISRAEL
TEL: 584592
TLX:
COM:

HOSHI REIUN DR
DEPT OF PHYSICS
RIKKYO UNIVERSITY
NISHI-IKEBUKURO 3-CH
TOSHIMA-KU TOKYO 171
JAPAN
TEL: 03-985-2414
TLX:
COM: 35

HOSKIN MICHAEL A DR
CHURCHILL COLLEGE
CAMBRIDGE CB3 0DS
U.K.
TEL: 0223-358381
TLX:
COM: 41

HOSKING ROGER J PROF
MATHEMATICS DEPT
UNIVERSITY OF WAIKATO
PRIVATE BAG
HAMILTON
NEW ZEALAND
TEL: 62-889 x 683
TLX:
COM:

HOSOKAWA YOSHIMASA H PROF
SAKIGAOKA 3-4-9
FUNABASHI-CITY
CHIBA PREFECTURE 274
JAPAN
TEL: 0474-48-6679
TLX:
COM:

HOSOYAMA KENNOSHUKE DR
INTL LATITUDE OBSERVATORY
HOSHIGAOKA 2-CHOME
MIZUSAWA IWATE 023
JAPAN
TEL:
TLX:
COM: 19

HOTINLI METIN DR
UNIVERSITE
RASATHANESI
BEYAZIT, ISTANBUL
TURKEY
TEL:
TLX:
COM: 12,36

HOUCK JAMES R
ASTRONOMY DEPT
CORNELL UNIVERSITY
220 SPACE SCIENCE BLDG
ITHACA NY 14853
U.S.A.
TEL: 607-256-4806
TLX: 937478 ITCA
COM: 21

HOUK NANCY DR
DEPT OF ASTRONOMY
UNIVERSITY OF MICHIGAN
1045 PHYS-ASTRO BLDG
ANN ARBOR MI 48109
U.S.A.
TEL: 313-764-3436
TLX:
COM: 27,45C

HOUSE FRANKLIN C DR
HEIDENREICHSTR 42
D-6100 DARMSTADT
GERMANY, F.R.
TEL: 06151-422412
TLX:
COM:

HOUSE LEWIS L DR
HIGH ALTITUDE OBSERVATORY
NCAR
PO BOX 3000
BOULDER CO 80303
U.S.A.
TEL: 303-494-5151
TLX:
COM: 12,14,36

HOUZIAUX L PROF
INSTITUT D'ASTROPHYSIQUE
UNIVERSITE DE LIEGE
AVENUE DE COINTE 5
B-4200 COINTE-OUGREE
BELGIUM
TEL: 31-41520180x494
TLX: 41264 ASTRLG B
COM: 29,34,44,46C

HOVENIER J W DR
FREE UNIVERSITY
DEPT PHYSICS & ASTRONOMY
DE BOELELAAN 1081
NL-1081 HV AMSTERDAM
NETHERLANDS
TEL: 20-540-2414
TLX:
COM: 16

HOWARD ROBERT F DR
NATL SOLAR OBSERVATORY
PO BOX 26732
TUCSON AZ 85726-6732
U.S.A.
TEL: 602-327-5511
TLX: 0666484 AURA NOAOTUC
COM: 10,12

HOWARD W MICHAEL DR
L. LIVERMORE NATIONAL LAB
L-297
LIVERMORE CA 94550
U.S.A.
TEL: 415-422-4138
TLX:
COM:

HOWARD WILLIAM E III DR
US NAVAL SPACE COMMAND
31 WOODLAWN TERRACE
FREDERICKSBURG VA 22405
U.S.A.
TEL: 703-663-7841
TLX:
COM: 40

HOWARTH IAN DONALD
DEPT PHYSICS & ASTRONOMY
UNIVERSITY COLLEGE LONDON
GOWER STREET
LONDON WC1E 6BT
U.K.
TEL: 01-387-7050
TLX: 28722
COM: 44

HOWSE H DEREK
12 BARNFIELD ROAD
RIVERHEAD
SEVENOAKS, KENT TN13 2AY
U.K.
TEL: 0732-454366
TLX:
COM: 41

HOYLE FRED SIR
COCKLEY MOOR
DOCKRAY
PENRITH, CUMBRIA CA11 CLG
U.K.
TEL:
TLX:
COM: 28,35,47,48

HOYNG PETER DR
SPACE RESEARCH LABORATORY
BENELUXLAAN 21
NL-3527 HS UTRECHT
NETHERLANDS
TEL: 030-937-145
TLX: 47224 ASTRO NL
COM: 10,12,44

HRIVNAK BRUCE J
VALPARAISO UNIVERSITY
PHYSICS DEPT
VALPARAISO IN 46383
U.S.A.
TEL: 219-464-5379
TLX:
COM: 30,42

HSIANG YAN-YU
BEIJING ASTRONOMICAL OBS
ACADEMIA SINICA
BEIJING
CHINA, PEOPLE'S REP.
TEL: 281698
TLX: 22040 BAOAS CN
COM: 28

HSIANG-KUANG TSENG
INST PHYSICS & ASTRONOMY
NATL CENTRAL UNIVERSITY
CHUNG LI
CHINA, TAIWAN
TEL:
TLX:
COM:

HUANG JIE-HAO
ASTROPHYSICS INSTITUTE
NANJING UNIVERSITY
NANJING
CHINA, PEOPLE'S REP.
TEL:
TLX: 34151 PRCNU CN
COM: 28

HUANG YONGWEI
BEIJING ASTRONOMICAL OBS.
ACADEMIA SINICA
BEIJING
CHINA, PEOPLE'S REP.
TEL: 281698
TLX: 9053
COM:

HUCHRA JOHN PETER DR
CENTER FOR ASTROPHYSICS
60 GARDEN STREET
CAMBRIDGE MA 02138
U.S.A.
TEL: 617-495-7375
TLX: 921428 SATELLITE CAM
COM: 28,47

HU JING-YAO
BEIJING OBSERVATORY
BEIJING
CHINA, PEOPLE'S REP.
TEL: 281698
TLX: 22040 BAOAS CN
COM: 09,25

HUANG KE-LIANG
ASTRONOMY DEPARTMENT
NANJING UNIVERSITY
NANJING
CHINA, PEOPLE'S REP.
TEL: 34651 EXT 2882
TLX: 34151 PRCNU CN
COM: 28,48

HUANG YOU-RAN
DEPT OF ASTRONOMY
NANJING UNIVERSITY
NANJING
CHINA, PEOPLE'S REP.
TEL:
TLX: 34151
COM: 10

HUCHTMEIER WALTER K DR
MPI FUER RADIOASTRONOMIE
AUF DEM HUEGEL 69
D-5300 BONN 1
GERMANY, F.R.
TEL: 228-525-215
TLX: 886440 MPIFR D
COM: 28,40

HU NING-SHENG
NANJING ASTRONOMICAL
INSTRUMENT FACTORY
NANJING JIANGSU PROVINCE
CHINA, PEOPLE'S REP.
TEL: 46191
TLX: 34136 GLYNJ CN :NAIF
COM: 08,09C

HUANG KUN-YI
PURPLE MOUNTAIN
OBSERVATORY
NANJING
CHINA, PEOPLE'S REP.
TEL: 32893
TLX: 34144 PMONJ CN
COM:

HUBBARD WILLIAM B PROF
PLANETARY SCIENCES DEPT
UNIVERSITY OF ARIZONA
TUCSON AZ 85721
U.S.A.
TEL: 602-621-6942
TLX: 9109521143
COM: 16

HUDSON HUGH S DR
PHYSICS C-011
UCSD
LA JOLLA CA 92093
U.S.A.
TEL: 619-452-4476
TLX:
COM: 10

HU WEN-RUI
INSTITUTE OF MECHANICS
ACADEMIA SINICA
BEIJING
CHINA, PEOPLE'S REP.
TEL: 28-4185
TLX: 22474 ASCHI CN
COM: 44

HUANG LIN
BEIJING ASTR OBSERVATORY
ACADEMIA SINICA
BEIJING
CHINA, PEOPLE'S REP.
TEL: 28-16-98
TLX: 22040 BAOAS CN
COM: 25,45

HUBE DOUGLAS P DR
DEPT OF PHYSICS
UNIVERSITY OF ALBERTA
EDMONTON ALB T6G2J1
CANADA
TEL: 403-432-5410
TLX:
COM: 30,42

HUEBNER WALTER F DR
LOS ALAMOS NATL OBS
T-4, MS B-212
LOS ALAMOS NM 87545
U.S.A.
TEL: 505-667-5751
TLX: 660495/9109881773TWX
COM: 14,15

HU ZHONG-WEI
DEPT OF ASTRONOMY
NANJING UNIVERSITY
NANJING
CHINA, PEOPLE'S REP.
TEL: 37651
TLX: 0909
COM: 15,16

HUANG RUN-QIAN
YUNNAN OBSERVATORY
KUNMING, YUNNAN PROVINCE
CHINA, PEOPLE'S REP.
TEL:
TLX: 64040 YUOBS CN
COM: 35

HUBENET HENRI DR
STERRENWACHT SONNENBORGH
ZONNENBURG 2
NL-3512 NL UTRECHT
NETHERLANDS
TEL: 30-312841
TLX:
COM:

HUGHES DAVID W DR
DEPT OF PHYSICS
THE UNIVERSITY
SHEFFIELD S3 7RH
U.K.
TEL: 0742-78555
TLX: 54348 ULSHEF G
COM: 15,22

HUA CHON TRUNG DR
CANADA-FRANCE-HAWAII
TELESCOPE CORP.
PO BOX 1597
KAMUELA HI 96743
U.S.A.
TEL: 808-885-7944
TLX: 633147 CFHT
COM: 34

HUANG TIANYI
DEPT OF ASTRONOMY
NANJING UNIVERSITY
NANJING
CHINA, PEOPLE'S REP.
TEL:
TLX: 34151 PRCNU CN
COM: 07

HUBENY IVAN
ASTRONOMICAL INSTITUTE
CZECH. ACAD. OF SCIENCES
ONDREJOV OBSERVATORY
251 65 ONDREJOV
CZECHOSLOVAKIA
TEL: 724525 PRAGUE
TLX: 121579 ASTR C
COM: 29,36C

HUGHES JAMES A DR
US NAVAL OBSERVATORY
WASHINGTON DC 20390
U.S.A.
TEL:
TLX:
COM: 08C,24

HUA YING-MIN
P.O. BOX 18
LINTONG
XIAN
CHINA, PEOPLE'S REP.
TEL:
TLX:
COM: 08,19

HUANG TIE-QIN
NANJING ASTRONOMICAL
INSTRUMENT FACTORY
NANJING
CHINA, PEOPLE'S REP.
TEL: 46191
TLX: 34136 GLYNJ CN
COM: 09

HUBER MARTIN C E DR
INSTITUTE OF ASTRONOMY
ETH-ZENTRUM
CH-8092 ZUERICH
SWITZERLAND
TEL: 01-256-3632
TLX: 53178 RTHBI CH
COM: 14,44

HUGHES PHILIP
DEPT OF ASTRONOMY
UNIVERSITY OF MICHIGAN
ANN ARBOR MI 48109-1090
U.S.A.
TEL: 313-764-3430
TLX:
COM: 40

HUANG CHANG-CHUN
PURPLE MOUNTAIN OBS
NANJING
CHINA, PEOPLE'S REP.
TEL: 46700/42817
TLX: 34144 PMONJ CN
COM: 29,30

HUANG YINLIANG
YUNNAN OBSERVATORY
KUNMING, YUNNAN PROVINCE
CHINA, PEOPLE'S REP.
TEL: 72946
TLX: 64040
COM: 50

HUBERT-DELPLACE A.-M. DR
OBSERVATOIRE DE PARIS
SECTION DE MEUDON
F-92195 MEUDON PL CEDEX
FRANCE
TEL: 1-45-34-75-30
TLX: 270912 OBSASTR
COM: 29

HUGHES VICTOR A PROF
DEPT OF PHYSICS
QUEEN'S UNIVERSITY
KINGSTON ONT K7L 3N6
CANADA
TEL: 613-547-6633
TLX:
COM: 33,34,40

HUGUENIN G RICHARD
MULTITECH CORPORATION
PO BOX 109
SOUTH DEERFIELD RES PARK
SOUTH DEERFIELD MA 01373
U.S.A.
TEL: 413-665-8551
TLX: 3719862 TRUB
COM:

HULSBOSCH A N M DR
STERRENKUNDIG INSTITUT
KATHOLIEKE UNIVERSITEIT
TOERNOOIVELD
NL-6525 ED NIJMEGEN
NETHERLANDS
TEL: 080-558833
TLX: 48228
COM: 33,34,40

HUMMER DAVID G DR
JILA
UNIVERSITY OF COLORADO
BOX 440
BOULDER CO 80309
U.S.A.
TEL: 303-492-7837
TLX: 755842 JILA
COM: 34,36

HUMPHREYS CURTIS JUDSON
6065 RIVERSIDE AVE
RIVERSIDE CA 92506
U.S.A.
TEL: 714-684-5652
TLX:
COM: 14

HUMPHREYS ROBERTA M PROF
ASTRONOMY DEPT
UNIVERSITY OF MINNESOTA
116 CHURCH STREET S.E.
MINNEAPOLIS MN 55455
U.S.A.
TEL: 612-373-9747
TLX:
COM: 28,33,35,45

HUMPHRIES COLIN M DR
UK SCHMIDT TELESCOPE
COONABARABRAN NSW 2357
AUSTRALIA
TEL: 068-42-1622
TLX: 63945 CANOPUS AA
COM: 09P

HUNDHAUSEN ARTHUR DR
HIGH ALTITUDE OBSERVATORY
PO BOX 3000
BOULDER CO 80302
U.S.A.
TEL:
TLX:
COM:

HUNGER KURT PROF
INST THEOR PHYS & STERNW
NEUE UNIV PHYSIK ZENTRUM
OLSHAUSENST 40 N61C
D-2300 KIEL 1
GERMANY, F.R.
TEL: 0431-880-4110
TLX: 292706IAOKI
COM: 36

HUNSTEAD RICHARD W DR
SCHOOL OF PHYSICS
UNIVERSITY OF SYDNEY
SYDNEY NSW 2006
AUSTRALIA
TEL: 02-692-3871
TLX: 26169 UNISYD AA
COM: 28,40

HUNT GARRY E DR
ATMOSPHERIC PHYSICS GROUP
BLACKETT LABORATORY
IMPERIAL COLLEGE
LONDON SW7 2BZ
U.K.
TEL: 01-589-5111
TLX: 261503
COM: 16P,W

HUNTEN DONALD M PROF
LUNAR AND PLANETARY LAB
UNIVERSITY OF ARIZONA
TUCSON AZ 85721
U.S.A.
TEL: 602-621-4002
TLX:
COM: 16,51

HUNTER CHRISTOPHER PROF
MATHEMATICS DEPT
FLORIDA STATE UNIVERSITY
TALLAHASSEE FL 32306
U.S.A.
TEL: 904-644-2488
TLX:
COM: 33

HUNTER DEIDRE ANN
DEPT TERRESTR. MAGNETISM
CARNEGIE INST. WASHINGTON
5241 BROAD BRANCH RD N.W.
WASHINGTON DC 20015
U.S.A.
TEL: 202-966-0863
TLX: 440427 MAGN UI
COM:

HUNTER JAMES H PROF
DEPT OF ASTRONOMY
T W BRYANT SPACE SCI BLDG
UNIVERSITY OF FLORIDA
GAINESVILLE FL 32611
U.S.A.
TEL: 904-392-1078
TLX:
COM: 28,51

HURFORD GORDON JAMES
CALTECH 264-33
PASADENA CA 91125
U.S.A.
TEL: 818-356-3866
TLX: 675425
COM: 10

HURNIK HIERONIM PROF
ASTRONOMICAL OBSERVATORY
A MICKIEWICZ UNIVERSITY
SLONECZNA 36
60-286 POZNAN
POLAND
TEL: 679-670
TLX:
COM: 20

HURUHATA MASAAKI PROF
TOKYO ASTRONOMICAL OBS
OSAWA 2-CHOME
MITAKA TOKYO 181
JAPAN
TEL: 0422-32-5111
TLX:
COM: 21,27

HURUKAWA KIICHIRO DR
TOKYO ASTRONOMICAL OBS
OSAWA MITAKA
181 TOKYO
JAPAN
TEL:
TLX:
COM: 08,19,20

HUT PIET
INST. FOR ADVANCED STUDY
PRINCETON NJ 08540
U.S.A.
TEL: 609-734-8075
TLX: 229734 IAS UR
COM:

HUTCHEON RICHARD J DR
X-RAY ASTRONOMY GROUP
PHYSICS DEPT
UNIVERSITY OF LEICESTER
LEICESTER LE1 7RH
U.K.
TEL:
TLX:
COM:

HUTCHINGS JOHN B DR
DOMINION ASTROPHYS OBS
5071 W SAANICH ROAD
VICTORIA BC V8X 4M6
CANADA
TEL: 604-388-3909
TLX: 049-7295
COM: 27,34,36,42

HYDER C L DR
HIGH ALTITUDE OBSERVATORY
PO BOX 3000
BOULDER CO 80307
U.S.A.
TEL:
TLX:
COM: 10

HYLAND A R HARRY DR
MT STROMLO OBSERVATORY
WODEN P.O. ACT 2606
AUSTRALIA
TEL: 062-881111
TLX: 62270 CANOPUS AA
COM: 25,29

HYNEK J ALLEN PROF
DEARBORN OBSERVATORY
NORTHWESTERN UNIVERSITY
EVANSTON IL 60201
U.S.A.
TEL: 312-864-1861
TLX:
COM:

HYSOM EDMUND J
65 DALKEITH ROAD
HARPENDEN, HERTS AL5 5PP
U.K.
TEL:
TLX:
COM: 09,41,51

HYUN JONG-JUNE PROF
SEOUL NATIONAL UNIVERSITY
SINLIM-DONG
KWANAK-KU
SEOUL 151
KOREA, REPUBLIC
TEL: 877-3010/2542
TLX:
COM:

IANNA PHILIP A
UNIVERSITY OF VIRGINIA
PO BOX 3818
CHARLOTTESVILLE VA 22903
U.S.A.
TEL: 804-924-4898
TLX:
COM: 20,24

IANNINI GUALBERTO DR
OBSERVATORIO ASTRONOMICO
LAPRIDA 854
5000 CORDOBA
ARGENTINA
TEL:
TLX:
COM:

IBADINOV KHURSANDKUL DR
INSTITUTE OF ASTROPHYSICS
TADJIK ACAD OF SCIENCES
734670 DUSHANBE
U.S.S.R.
TEL:
TLX:
COM: 15

IBANEZ S. MIGUEL H. DR
UNIVERSIDAD DE LOS ANDES
FACULDAD DE CIENCIAS
DEPTO DE FISICA
MERIDA
VENEZUELA
TEL: 639930/637477
TLX: 74174 CIDA
COM:

IBANOGLU C DR
EGE UNIVERSITY
FACULTY OF SCIENCE
BORNOVA-IZMIR
TURKEY
TEL: 180110-2332
TLX:
COM: 42

IBBETSON PETER AARON DR
WISE OBSERVATORY
TEL-AVIV UNIVERSITY
RAMAT AVIV
TEL-AVIV 69978
ISRAEL
TEL: 972-3-413788
TLX: 342171 VERSY IL
COM:

IBEN ICKO JR PROF
ASTR DEPT/UNIV ILLINOIS
349 ASTRONOMY BLDG
1011 W SPRINGFIELD AVE
URBANA IL 61801
U.S.A.
TEL: 217-333-3090
TLX: 9102452434 AST
COM: 27,35,37

IBRAHIM JORGA
DEPT OF ASTRONOMY
INSTITUTE OF TECHNOLOGY
JALAN TAMANSARI 64
BANDUNG
INDONESIA
TEL:
TLX:
COM:

ICHIMARU SETSUO DR
DEPT OF PHYSICS
UNIVERSITY OF TOKYO
BUNKYO-KU
TOKYO 113
JAPAN
TEL: 03-812-2111
TLX: UTPHYSIC J23472
COM: 48

ICKE VINCENT DR
STERREWACHT LEIDEN
POSTBUS 9513
NL-2300 RA LEIDEN
NETHERLANDS
TEL: 71-148333
TLX: 39058 ASTRO NL
COM: 47

IDLIS G M DR
INSTITUTE FOR HISTORY OF
SCIENCES AND TECHNOLOGY
USSR ACADEMY OF SCIENCES
103012 MOSCOW
U.S.S.R.
TEL: 2281969
TLX:
COM: 41,51

IIJIMA SHIGETAKA PROF
MUSASHI INSTITUTE OF
TECHNOLOGY
TAMAZUTSUMI, SETAGAYA-KU
TOKYO 158
JAPAN
TEL: 03-703-3111
TLX:
COM: 19,31

IKEUCHI SATORU DR
TOKYO ASTRONOMICAL OBS
UNIVERSITY OF TOKYO
MITAKA TOKYO 181
JAPAN
TEL: 0422-32-5111
TLX: 2822307
COM: 33,47

IKHSANOV ROBERT N DR
PULKOVO OBSERVATORY
196140 LENINGRAD
U.S.S.R.
TEL:
TLX:
COM: 40

IKHSANOVA VERA N DR
PULKOVO OBSERVATORY
196140 LENINGRAD
U.S.S.R.
TEL:
TLX:
COM: 40

ILIEV ILIAN
NATL ASTRON OBSERVATORY
P.O. BOX 136
BG-4700 SMOLYAN
BULGARIA
TEL: 73-41-559
TLX: 23561
COM: 14,35

ILL MARTON J DR
KONKOLY OBSERVATORY
TOTH KALMAN U 19
H-6501 BAJA
HUNGARY
TEL: 11-064
TLX: 281303
COM:

ILLES ALMAR ERZSEBET DR
KONKOLY OBSERVATORY
BOX 67
H-1525 BUDAPEST
HUNGARY
TEL: 366-621
TLX: 227460
COM:

ILLING RAINER M E
HIGH ALTITUDE OBSERVATORY
NCAR
PO BOX 3000
BOULDER CO 80307
U.S.A.
TEL: 303-497-1537
TLX:
COM: 12

ILLINGWORTH GARTH D DR
SPACE TELESCOPE SCI INST
HOMEWOOD CAMPUS
3700 SAN MARTIN DRIVE
BALTIMORE MD 21218
U.S.A.
TEL: 301-338-4730
TLX: 6849101
COM: 28,37

ILYAS MOHAMMAD DR
SCHOOL OF PHYSICS
UNIVERSITI SAINS MALAYSIA
11800 USM
PENANG
MALAYSIA
TEL: 883822
TLX: 40254 MA
COM: 04,09

IMBERT MAURICE DR
OBSERVATOIRE DE MARSEILLE
2 PLACE LE VERRIER
F-13248 MARSEILLE CEDEX 4
FRANCE
TEL: 91-95-90-88
TLX:
COM: 30,42

IMSHENNIK V S DR
INSTITUTE OF THEORETICAL
AND EXPERIMENTAL PHYSICS
B. CHEREMUSHKINSKAYA 25
117259 MOSCOW
U.S.S.R.
TEL: 123-02-92
TLX: 411059 CERII SU
COM: 35

INAGAKI SHOGO DR
DEPT OF ASTRONOMY
FACULTY OF SCIENCE
UNIVERSITY OF KYOTO
KYOTO 606
JAPAN
TEL: 075-751-2111
TLX: 5422693 LIBKYU J
COM: 33,37

INATANI JUNJI
NOBEYAMA RADIO OBS
TOKYO ASTRON OBSERVATORY
NOBEYAMA, MINAMISAKU
NAGANO 384-13
JAPAN
TEL: 267-98-2831
TLX: 3329005 TAO NRO J
COM: 40

INGRAO HECTOR C
CAMBRIDGE SYSTEMS CORP.
58 HUNDREDS RD
WELLESLEY HILLS MA 02181
U.S.A.
TEL: 617-235-3711
TLX:
COM: 16

INNANEN KIMMO A PROF
CRESS PHYSICS DEPT
YORK UNIVERSITY
4700 KEELE STREET
NORTH YORK ON M3J1P3
CANADA
TEL: 416-667-3837
TLX: 06524736
COM: 33

INOUE HAJIME DR
INSTITUTE OF SPACE &
ASTRONAUTICAL SCIENCE
KOMABA, MEGURO-KU
TOKYO 153
JAPAN
TEL: 03-467-1111
TLX: 24550 J
COM: 44,48

INOUE MAKOTO DR
NOBEYAMA RADIO OBS
MINAMI-MAKIMURA
MINAMI-SAKU
NAGANO 384-13
JAPAN
TEL: 0267-98-2831
TLX: 3329005 TAONKKRO J
COM: 40

INOUE TAKESHI PROF
KYOTO SANGYO UNIVERSITY
KAMIGAMO
KYOTO 603
JAPAN
TEL: 075-701-2151
TLX: 5422661 KSU J
COM:

IOANNISIANI B K DR
MAIN ASTRONOMICAL OBS
PULKOVO
196140 LENINGRAD
U.S.S.R.
TEL:
TLX:
COM: 09

IONSON JAMES ALBERT
NASA/GSFC
CODE 682
GREENBELT MD 20771
U.S.A.
TEL: 301-344-6184
TLX:
COM: 49

IOSHPA B A DR
INST OF TERR MAGNETITISM
AND IONOSPHERE
142092 TROITSK
U.S.S.R.
TEL: 2321921
TLX: 412623 SCP
COM: 10

IP WING-HUEN
MPI FUER ASTRONOMIE
D-3411 KATLENBURG-LINDAU
GERMANY, F.R.
TEL: 0049-555-6416
TLX: 0965527
COM: 15

IPSER JAMES R PROF
DEPT OF PHYSICS
UNIVERSITY OF FLORIDA
WILLIAMSON HALL
GAINESVILLE FL 32611
U.S.A.
TEL: 904-392-0521
TLX:
COM: 48

IRELAND JOHN G DR
C/O 13 GORDON ROAD
BELVEDERE, KENT DA17 6EA
U.K.
TEL:
TLX:
COM:

IRIARTE B MR
INAOE
AP POSTALES 216 y 51
72000 PUEBLA, PUE.
MEXICO
TEL: 22-47-05-00
TLX:
COM:

IRVINE WILLIAM M PROF
RODIO ASTRONOMY
UNIV OF MASSACHUSETTS
619 GRC TOWER B
AMHERST MA 01003
U.S.A.
TEL: 413-545-0733
TLX: 955491 UNIV MASS AMS
COM: 15,16,34,51

IRWIN ALAN W DR
DEPT OF PHYSICS
UNIVERSITY OF VICTORIA
P O BOX 1700
VICTORIA V8W 2Y2
CANADA
TEL: 604-721-7700
TLX: 049-7222
COM: 14,25

IRWIN JOHN B PROF
2744 N. TYNDALL AVE
TUCSON AZ 85719
U.S.A.
TEL: 602-623-7423
TLX:
COM: 33,42

ISAAK GEORGE R PROF
DEPT OF PHYSICS
UNIVERSITY OF BIRMINGHAM
P O BOX 363
BIRMINGHAM B15 2TT
U.K.
TEL: 021-472-1301
TLX: 338938 SPAPHY G
COM: 35

ISERN JORGE DR
C/SEPULVEDA 83-6-3A
08015 BARCELONA
SPAIN
TEL:
TLX:
COM:

ISHIDA GORO DR
BROADCAST UNIVERSITY
23-11 AKABANE-NISHI
1 CHOME, KITA-KU
TOKYO 115
JAPAN
TEL: 03-909-3871
TLX:
COM: 26

ISHIDA KEIICHI PROF
TOKYO ASTRONOMICAL OBS
2-21-1 OSAWA MITAKA
TOKYO 181
JAPAN
TEL: 04-22-32-5211
TLX: 2822307 TAOMTK J
COM: 37

ISHIZAWA TOSHIAKI A PROF
DEPT OF ASTRONOMY
UNIVERSITY OF KYOTO
KYOTO 606
JAPAN
TEL: 075-751-2111
TLX: 5422693 LIBKYU J
COM:

ISHIZUKA TOSHIHISA DR
DEPT OF PHYSICS
IBARAKI UNIVERSITY
2-1-1 BUNKYO
MITO 310
JAPAN
TEL: 0292-26-1621
TLX:
COM: 35

ISMAILOV TOFIK K
SPACE RES SCI INDUSTR ENT
AZERBAIJAN SSR ACAD SCI
PROSPECT LENINA 159
370106 BAKU
U.S.S.R.
TEL: 62-93-88
TLX: 142407
COM: 09

ISOBE SYUZO DR
TOKYO ASTRONOMICAL OBS
MITAKA
TOKYO 181
JAPAN
TEL: 0422-32-5211
TLX: 02822307 TAOMTK J
COM: 15,33,34,46

ISRAEL FRANK P DR
STERREWACHT
POSTBUS 9513
NL-2300 RA LEIDEN
NETHERLANDS
TEL:
TLX:
COM: 33,51

ISRAEL GUY MARCEL DR
SERVICE D'AERONOMIE CNRS
BP 3
F-91370 VERRIERES-LE-B.
FRANCE
TEL: 1-69-20-10-60
TLX:
COM:

ISRAEL WERNER PROF
PHYSICS DEPT
UNIVERSITY OF ALBERTA
ALBERTA
EDMONTON AL T6G 2J1
CANADA
TEL: 403-432-3552
TLX: 0372979
COM: 48

ISSA ALI DR
HELWAN OBSERVATORY
HELWAN-CAIRO
EGYPT
TEL: 780645, 782683
TLX: 93070
COM: 28,34

ISSERSTEDT JOERG DR
INSTITUT FUER ASTRONOMIE
UND ASTROPHYSIK
AM HUBLAND
D-8700 WUERZBURG
GERMANY, F.R.
TEL:
TLX:
COM:

ITO KENSAI A PROF
RIKKYO UNIVERSITY
DEPT OF PHYSICS
NISHI-IKEBUKURO
TOKYO 171
JAPAN
TEL: 03-985-2384
TLX:
COM: 48

ITOH HIROSHI DR
DEPT OF ASTRONOMY
UNIVERSITY OF KYOTO
KYOTO 606
JAPAN
TEL: 075-751-2111
TLX:
COM: 34

ITOH NAOKI DR
DEPT OF PHYSICS
SOPHIA UNIVERSITY
7-1 KIOI-CHO CHIYODA-KU
TOKYO 102
JAPAN
TEL: 03-238-3431
TLX:
COM: 35

IVANCHUK VICTOR I DR
KIEV UNIVERSITY
OBSERVATORNAYA 3
252053 KIEV
U.S.S.R.
TEL:
TLX:
COM: 10

IVANOV GEORGI R DR
UNIVERSITY OF SOFIA
DEPT OF ASTRONOMY
ANTON IVANOV STR 5
1126 SOFIA
BULGARIA
TEL:
TLX:
COM:

IVANOV VSEVOLOD V DR PROF
ASTRONOMICAL OBSERVATORY
LENINGRAD UNIVERSITY
BIBLIOTECHNAJA PL 2
198904 LENINGRAD
U.S.S.R.
TEL: 257-94-91
TLX:
COM: 36

IVANOV-KHOLODNY G S DR
INST OF APPLIED GEOPHYS
USSR ACADEMY OF SCIENCES
107150 MOSCOW
U.S.S.R.
TEL:
TLX:
COM: 21

IVANOVA VIOLETA DR
DEPT OF ASTRONOMY AND
NATL ASTRON OBSERVATORY
72 LENIN BLVD
1784 SOFIA
BULGARIA
TEL: 7341-559
TLX: 23561 ECF BAN BG
COM: 07,15,20

IVES JOHN CHRISTOPHER MR
ESTEC
POSTBUS 299
NL-2200 AG NOORDWIJK
NETHERLANDS
TEL: 01719-83629
TLX: 39098
COM:

IWANISZEWSKA CECYLIA DR
INSTITUTE OF ASTRONOMY
N COPERNICUS UNIVERSITY
UL. CHOPINA 12/18
87-100 TORUN
POLAND
TEL: 2-60-18
TLX: 0552234 ASTR PL
COM: 33,46P

IWANOWSKA WILHELMINA PROF
INSTITUTE OF ASTRONOMY
UL. CHOPINA 12/18
87-100 TORUN
POLAND
TEL: 260-18
TLX: 86412 PL
COM: 33

IWASAKI KYOSUKE DR
KWASAN OBSERVATORY
YAMASHINA
KYOTO 607
JAPAN
TEL: 075-581-1235
TLX: 5422693 LIBKYUJ
COM: 16

IYE MASANORI DR
TOKYO ASTRONOMICAL OBS
UNIVERSITY OF TOKYO
MITAKA 181
JAPAN
TEL: 0422-32-511x313
TLX: 2822307 TAOMTK J
COM: 33

IYENGAR K V K PROF
TATA INSTITUTE OF
FUNDAMENTAL RESEARCH
BOMBAY 400 005
INDIA
TEL: 219111 x339
TLX: 113009 TIFR IN
COM: 25

IYER B R DR
RAMAN RESEARCH INSTITUTE
BANGALORE 560 080
INDIA
TEL: 360122
TLX: 845671-RRI IN
COM: 47

IZVEKOV V A DR
INSTITUTE OF THEORETICAL
ASTRONOMY
10 KUTUZOV QUAY
191187 LENINGRAD
U.S.S.R.
TEL: 272-40-23
TLX: 121578 ITA SU
COM: 07,20

JAAKKOLA TOIVO S
OBSERVATORY
TAHTITORNINMAKI
SF-00130 HELSINKI 13
FINLAND
TEL: 35-801-912907
TLX: 124690 UNIH SF
COM:

LIST OF MEMBERS

JABBAR SABEH RHAMAN
ASTRONOMY & SPACE RES CTR
COUNCIL FOR SCI RESEARCH
P O BOX 2441
JADIRIYAH, BAGHDAD
IRAQ
TEL: 7765127
TLX: 213976 SRC IK
COM: 12,42

JABIR NIAMA LAFTA
ASTRONOMY & SPACE RES CTR
COUNCIL FOR SCI RESEARCH
P O BOX 2441
JADIRIYAH, BAGHDAD
IRAQ
TEL: 7765127
TLX: 213976 SRC IK
COM: 34

JACCHIA LUIGI G DR
CENTER FOR ASTROPHYSICS
60 GARDEN ST
CAMBRIDGE MA 02138
U.S.A.
TEL: 617-495-7213
TLX:
COM: 22

JACKISCH GERHARD DR
ZNTRLINST. F. ASTROPHYSIK
STERNWARTE BABELSBERG
DDR-6400 SONNEBERG
GERMANY, D.R.
TEL:
TLX:
COM: 41

JACKSON C
HILLTOP OBSERVATORY
P O BOX 33
HAENERTSBURG, N TRANSVAAL
SOUTH AFRICA
TEL:
TLX:
COM:

JACKSON J C DR
16 THE PARK
NEWARK NG24 1SO
U.K.
TEL:
TLX:
COM: 48

JACKSON PAUL DR
INSTITUT FUER ASTRONOMIE
DER UNIVERSITAET WIEN
TUERKENSCHANZSTR 17
A-1180 WIEN
AUSTRIA
TEL:
TLX:
COM: 08

JACKSON PETER DOUGLAS DR
ASTRONOMY PROGRAM
UNIVERSITY OF MARYLAND
COLLEGE PK MD 20742
U.S.A.
TEL: 301-454-6302
TLX:
COM: 33

JACKSON WILLIAM M DR
DEPT OF CHEMISTRY
UNIVERSITY OF CALIFORNIA
ROOM 214
DAVIS CA 95616
U.S.A.
TEL: 916-752-0503
TLX:
COM: 15

JACOBS KENNETH C DR
PHYSICS DEPT
HOLLINS COLLEGE
BOX 9661
ROANOKE VA 24020
U.S.A.
TEL: 703-362-6478
TLX:
COM:

JACOBSEN THEODOR S PROF
6205 17TH AVE N.E.
SEATTLE WA 98115
U.S.A.
TEL: 206-523-5245
TLX:
COM:

JACOBY GEORGE H
KITT PEAK NAT OBSERVATORY
PO BOX 26732
TUCSON AZ 85726
U.S.A.
TEL: 602-325-9292
TLX:
COM: 34

JACQUINOT PIERRE DR
LABORATOIRE AIME COTTON
BAT 505
UNIVERSITE PARIS SUD
F-91405 ORSAY CEDEX
FRANCE
TEL:
TLX:
COM: 14

JAEGER FRIEDRICH W PROF
TELEGRAFENBERG A 33
DDR-1500 POTSDAM
GERMANY, D.R.
TEL: 4551
TLX:
COM:

JAFFE DANIEL T
UNIVERSITY OF CALIFORNIA
SPACE SCIENCES LABORATORY
BERKELEY CA 94720
U.S.A.
TEL: 415-642-1930
TLX:
COM:

JAFFE WALTER JOSEPH DR
SPACE TELESCOPE SCI INST
3700 SAN MARTIN DRIVE
BALTIMORE MD 21218
U.S.A.
TEL: 301-338-4762
TLX: 684 9101 STSCI
COM: 40,48

JAHREISS HARTMUT DR
ASTRONOMISCHES RECHEN-
INSTITUT
MOENCHHOFSTR 12-14
D-6900 HEIDELBERG 1
GERMANY, F.R.
TEL: 06221/49026
TLX: 461 336 ARIHD D
COM: 24,33

JAKIMIEC JERZY PROF
ASTRONOMICAL INSTITUTE
UL. KOPERNIKA 11
51-622 WROCLAW
POLAND
TEL: 482434
TLX: 0712791 UWRPL
COM: 10

JAKOBSEN PETER
ASTROPHYSICS DEPARTMENT
ESA SPACE SCI DEPT/ESTEC
POSTBUS 299
NL-2200 AG NORDWIJK
NETHERLANDS
TEL: 31-171-983-3614
TLX: 39098
COM:

JAKS WALDEMAR DR
ASTRONOMICAL LATITUDE OBS
SPACE RESEARCH CENTRE PAS
BOROWIEC
62-035 KORNIK
POLAND
TEL: POZNAN 170187
TLX: 0412623 AOS PL
COM: 19

JAMAR CLAUDE A J DR
IAL SPACE/UNIV DE LIEGE
AVENUE DU PRE-AILY
B-4900 ANGLEUR-LIEGE
BELGIUM
TEL: 41676760
TLX: 41320 IAL SP
COM: 44

JAMES JOHN F MR
SCHUSTER LABORATORY
THE UNIVERSITY
MANCHESTER M13 9PL
U.K.
TEL: 061-273-7121
TLX:
COM: 21

JAMES RICHARD A DR
DEPARTMENT OF ASTRONOMY
THE UNIVERSITY
MANCHESTER M13 9PL
U.K.
TEL:
TLX:
COM: 35

JAMESON RICHARD F DR
ASTRONOMY DEPARTMENT
THE UNIVERSITY
LEICESTER LE1 7RH
U.K.
TEL: 0533-554455
TLX: 341198
COM:

JANES KENNETH A DR
ASTRONOMY DEPT
BOSTON UNIVERSITY
725 COMMONWEALTH AVE
BOSTON MA 02215
U.S.A.
TEL: 617-353-2627
TLX: 95-1289 BOS UNIV BSN
COM: 37

JANICZEK PAUL M DR
US NAVAL OBSERVATORY
348 MASSACHUSETTS AVE NW
348 MASSACHUSETTS AVE N.W
WASHINGTON DC 20390-5100
U.S.A.
TEL: 202-653-1569
TLX: 710-822-1970
COM: 04,07

JANKOVICS ISTVAN DR
KONKOLY OBSERVATORY
BOX 67
H-1525 BUDAPEST
HUNGARY
TEL: 166-426
TLX: 227460
COM:

JANSSEN MICHAEL ALLEN
JET PROPULSION LAB
MAIL STOP 183-301
4800 OAK GROVE DRIVE
PASADENA CA 91109
U.S.A.
TEL: 213-354-7247
TLX:
COM: 40

JARNEFELT GUSTAF J PROF
LAAJASUONTIE 27
SF-00320 HELSINKI 32
FINLAND
TEL:
TLX:
COM:

JAROSZYNSKI MICHAL
WARSAW UNIVERSITY
OBSERVATORY
AL. UJAZDOWSKIE 4
00-478 WARSAW
POLAND
TEL: 29 40 11
TLX: 813978 ZAPAN PL
COM: 47

JARRETT ALAN H PROF
BOYDEN OBSERVATORY
P O BOX 334
BLOEMFONTEIN 9300
SOUTH AFRICA
TEL: 051-37605
TLX: 267666 SA
COM: 21,46

JARZEBOWSKI TADEUSZ DR
ASTRONOMICAL INSTITUTE
KOPERNIKA 11
51-622 WROCLAW
POLAND
TEL:
TLX:
COM: 27

JASCHEK CARLOS O R PROF
OBSERVATOIRE
11 RUE DE L'UNIVERSITE
F-67000 STRASBOURG
FRANCE
TEL: 88-35-43-00
TLX: 890506 STAROBS
COM: 05C,29,33,42,45

JASCHEK MERCEDES DR
OBSERVATOIRE
11 RUE DE L'UNIVERSITE
F-67000 STRASBOURG
FRANCE
TEL: 88-35-43-00
TLX: 890506 STAROBS F
COM: 29,45

JASTROW ROBERT
INST FOR SPACE STUDIES
2880 BROADWAY
NEW YORK NY 10025
U.S.A.
TEL:
TLX:
COM: 51

JAUNCEY DAVID L DR
CSIRO
DIVISION OF RADIOPHYSICS
P.O.BOX 76
EPPING NSW 2121
AUSTRALIA
TEL: 062-46-5558
TLX: 26230 ASTRO
COM: 40C,47

JAVET PIERRE PROF
AVENUE DE BEAUMONT 36
CH-1012 LAUSANNE
SWITZERLAND
TEL:
TLX:
COM:

JAYARAJAN A P MR
INDIAN INSTITUTE OF
ASTROPHYSICS
KORAMANGALA
BANGALORE 560 034
INDIA
TEL: 566585
TLX: 0845-763 IIAB IN
COM: 09

JEFFERIES JOHN T DR
NATL OPTICAL ASTR OBS
1002 N WARREN AVE
TUCSON AZ 85719
U.S.A.
TEL: 602-881-1960
TLX: 0666484 AURANOAOTUC
COM: 12,36

JEFFERS STANLEY DR
CRESS PHYSICS DEPT
YORK UNIVERSITY
4700 KEELE ST
DOWNSVIEW ONT M3J 1P3
CANADA
TEL: 416-667-3851
TLX:
COM: 09,51

JEFFERYS WILLIAM H DR
ASTRONOMY DEPARTMENT
UNIVERSITY OF TEXAS
AUSTIN TX 78712
U.S.A.
TEL: 512-471-4461
TLX:
COM: 07,24

JEFFREYS HAROLD PROF SIR
160 HUNTINGDON ROAD
CAMBRIDGE CB3 0LB
U.K.
TEL: 356153
TLX:
COM: 16,19

JELLEY JOHN V PHD
29 ABBOTT ROAD
ABINGDON OX14 2DT
U.K.
TEL: 0235-21040
TLX:
COM: 09,48

JENKINS CHARLES R
ROYAL GREENWICH OBS
HERSTMONCEUX CASTLE
HAILSHAM BN27 1RP
U.K.
TEL: 0323-833171
TLX: 87451
COM: 40

JENKINS EDWARD B DR
PRINCETON UNIVERSITY OBS
PRINCETON NJ 08544
U.S.A.
TEL: 609-452-3826
TLX: 322409 ASTRO PRIN
COM: 44C

JENKINS L F MS
YALE UNIV OBSERVATORY
BOX 2023 YALE STATION
NEW HAVEN CT 06520
U.S.A.
TEL:
TLX:
COM: 34

JENKNER HELMUT DR
SPACE TELESOPE SCI INST
3700 SAN MARTIN DRIVE
BALTIMORE MD 21218
U.S.A.
TEL: 301-338-4842
TLX: 6849101 STSCI
COM: 05,09

JENNER DAVID C DR
DEPT OF ASTRONOMY
UNIVERSITY OF WASHINGTON
FM-20
SEATTLE WA 98195
U.S.A.
TEL: 206-543-6182
TLX:
COM:

JENNINGS R E PROF
DEPT PHYSICS & ASTRONOMY
UNIVERSITY COLLEGE LONDON
GOWER STREET
LONDON WC1E 6BT
U.K.
TEL: 01-387-7050
TLX: 28722
COM: 34

JENNISON ROGER C PROF
ELECTRONICS LABORATORY
UNIVERSITY OF KENT
CANTERBURY CT2 7NT
U.K.
TEL:
TLX: 965449
COM: 22,40,51

JENSCH A
PESTALOZZISTR 9
DDR-6900 JENA
GERMANY, D.R.
TEL:
TLX:
COM:

JENSEN EBERHART PROF
INST THEORET ASTROPHYSICS
UNIVERSITY OF OSLO
P O BOX 1029
N-0315 BLINDERN, OSLO 3
NORWAY
TEL: 02-456502
TLX: 72425N UNIOS
COM: 10

JERZYKIEWICZ MIKOLAJ DR
ASTRONOMICAL INSTITUTE
WROCLAW UNIVERSITY
KOPERNIKA 11
51-622 WROCLAW
POLAND
TEL: 48-24-34
TLX: 0712791 UWR PL
COM: 25,27C

JI HONG-QING
INTNATL LATITUTDE STATION
TIANJIN
CHINA, PEOPLE'S REP.
TEL:
TLX:
COM: 19

JIANG CHONG-GUO
YUNNAN OBSERVATORY
P.O. BOX 110
KUNMING
CHINA, PEOPLE'S REP.
TEL: 72946
TLX:
COM: 08

JIANG DONG-RONG
SHANGHAI OBSERVATORY
ACADEMIA SINICA
SHANGHAI
CHINA, PEOPLE'S REP.
TEL: 386191
TLX: 33164 SHAO CN
COM: 33

JIANG SHI-YANG
BEIJING ASTRONOMICAL OBS
BEIJING
CHINA, PEOPLE'S REP.
TEL: 28-1698
TLX: 22040 BADAS CN
COM: 09,27,50C

JIANG SHUDING
GRADUATE SCHOOL
UNIV SCIENCE & TECHNOLOGY
P.O. BOX 3908
BEIJING
CHINA, PEOPLE'S REP.
TEL: 817031-253
TLX:
COM: 47

JIANG ZHAOJI
BEIJING ASTRONOMICAL OBS
BEIJING
CHINA, PEOPLE'S REP.
TEL:
TLX: 22040 BAOAS CN
COM:

JIN SHEN-ZENG
BEIJING ASTRONOMICAL OBS
ACADEMIA SINICA
BEIJING
CHINA, PEOPLE'S REP.
TEL: 28 1698
TLX: 22040 BAOBS CN
COM: 40

JIN WEN-JING
SHANGHAI OBSERVATORY
ACADEMIA SINICA
80 NAN DAN ROAD
SHANGHAI
CHINA, PEOPLE'S REP.
TEL: 386191
TLX: 33164 SHAO CN
COM: 31

JOCKERS KLAUS DR
MPI FUER AERONOMIE
POSTFACH 20
D-3411 KATLENBURG-LINDAU
GERMANY, F.R.
TEL: 05556-411
TLX: 965527 AERLI D
COM: 10,15

JOHANSEN KAREN T LEKTOR
COPENHAGEN UNIVERSITY OBS
BRORFELDEVEJ 23
DK-4340 TOLLOSE
DENMARK
TEL: 03-488195
TLX:
COM:

JOHANSSON LARS ERIK B DR
ARVESGAERDE 18
S-417 44 GOETEBORG
SWEDEN
TEL:
TLX:
COM: 40

LIST OF MEMBERS

JOHNSON DONALD R DR
NATL BUREAU OF STANDARDS
BLDG 221, ROOM A363
GAITHERSBURG MD 20899
U.S.A.
TEL: 301-921-2828
TLX:
COM: 14,40

JOHNSON FRED M PROF DR
DEPT PHYSICS & ASTRONOMY
CALIFORNIA STATE UNIV
FULLERTON CA 92634
U.S.A.
TEL: 714-773-3366
TLX:
COM: 14,34

JOHNSON HOLLIS R PROF
ASTRONOMY DEPT
INDIANA UNIVERSITY
SWAIN WEST 319
BLOOMINGTON IN 47405
U.S.A.
TEL: 812-335-4172
TLX: 272279
COM: 29,36

JOHNSON HUGH M DR
DEPT 91-20 BLDG 255
LOCKHEED MISSILES
3170 PORTER DRIVE
PALO ALTO CA 94304
U.S.A.
TEL: 415-858-4087
TLX: 346409 LMSC-SUVL
COM: 28,33,34

JOHNSON TORRENCE V DR
JET PROPULSION LABORATORY
MAILSTOP 183-301
4800 OAK GROVE DRIVE
PASADENA CA 91109
U.S.A.
TEL: 818-354-2761
TLX: 67-5429
COM: 15,16

JOHNSTON KENNETH J
NAVAL RESEARCH LABORATORY
CODE 7134
WASHINGTON DC 20375
U.S.A.
TEL: 202-767-2351
TLX:
COM: 04,08,24,34,40

JOKIPII J R PROF
DEPT OF PLANET. SCIENCES
UNIVERSITY OF ARIZONA
TUCSON AZ 85721
U.S.A.
TEL: 602-621-4256
TLX:
COM: 48,49

JOLY FRANCOIS DR
UNIVERSITE DE BORDEAUX 1
123 RUE LAMARTINE
F-33400 TALENCE
FRANCE
TEL:
TLX:
COM: 14,40

JOLY MONIQUE
OBSERVATOIRE DE PARIS
SECTION DE MEUDON
F-92195 MEUDON PL CEDEX
FRANCE
TEL: 1-45-34-75-70
TLX: 201571 F
COM: 28

JONES ALBERT F MR
31 RANUI RD, STOKE
NELSON
NEW ZEALAND
TEL: 054-73-905
TLX:
COM: 27

JONES BARBARA
UNIV OF CALIFORNIA AT
SAN DIEGO
CASS/C-011
LA JOLLA CA 92093
U.S.A.
TEL: 714-452-4474
TLX:
COM: 09

JONES BERNARD J T DR
NORDITA
BLEGDAMSVEJ 17
DK-2100 COPENHAGEN
DENMARK
TEL: 01-42-16-16
TLX: 15216 NBI DK
COM: 47

JONES BURTON DR
LICK OBSERVATORY
UNIVERSITY OF CALIFORNIA
SANTA CRUZ CA 95064
U.S.A.
TEL: 408-429-2384
TLX:
COM: 24

JONES DAYTON L
JET PROPULSION LABORATORY
MAIL CODE 138-307
4800 OAK GROVE DRIVE
PASADENA CA 91109
U.S.A.
TEL: 818-354-6734
TLX: 675429
COM: 40

JONES DEREK H P DR
ROYAL GREENWICH OBS
HAILSHAM BN27 1RP
U.K.
TEL: 323-833171
TLX: 87451 RGOBSY G
COM: 24,33,37

JONES ERIC M
LOS ALAMOS NATL LAB
MS F-665
LOS ALAMOS NM 87545
U.S.A.
TEL: 505-667-6386
TLX:
COM: 51

JONES FRANK CULVER DR
NASA/GSFC
CODE 665
GREENBELT MD 20771
U.S.A.
TEL: 301-344-5506
TLX: 710-828-9716
COM: 34,48

JONES HARRISON PRICE DR
KITT PEAK NATL OBS
SOLAR STATION
900 N. CHERRY AVENUE
TUCSON AZ 85726
U.S.A.
TEL: 602-325-9354
TLX:
COM: 10,12

JONES JAMES DR
DEPT OF PHYSICS
UNIV OF WESTERN ONTARIO
LONDON ONT N6A 5B9
CANADA
TEL:
TLX:
COM: 22

JONES JANET E DR
NORDITA
BLEGDAMSVEJ 17
DK-2100 COPENHAGEN
DENMARK
TEL: 01-42-16-16
TLX: 15216 NBI DK
COM:

JONES THOMAS WALTER DR
DEPT OF ASTRONOMY
UNIVERSITY OF MINNESOTA
116 CHURCH ST SE
MINNEAPOLIS MN 55455
U.S.A.
TEL: 612-373-3307
TLX:
COM: 28,48

JORDAN CAROLE DR
DEPT THEORETICAL PHYSICS
OXFORD UNIVERSITY
1 KEBLE ROAD
OXFORD OX1 3NP
U.K.
TEL: 865-53281
TLX: 83295 NUCLOX
COM: 12,14,44

JORDAN H L DR DIREKTOR
INSTITUT F. PLASMAPHYSIK
KERNFORSCHUNGSANLAGE
JUELICH GMBH PF 365
D-5170 JUELICH 1
GERMANY, F.R.
TEL:
TLX:
COM: 14

JORDAN STUART D DR
LAB ASTRO SOLAR PHYSICS
NASA/GSFC, CODE 682
GREENBELT MD 20771
U.S.A.
TEL: 301-344-8811
TLX: 89675
COM: 10,12,44

JORDEN PAUL RICHARD
ROYAL GREENWICH OBS
HERSTMONCEUX CASTLE
HAILSHAM BN27 1RP
U.K.
TEL:
TLX:
COM:

JORGENSEN HENNING E PROF
UNIVERSITY OBSERVATORY
OESTER VOLDGADE 3
DK-1350 COPENHAGEN K
DENMARK
TEL: 1-14 17 90
TLX: 44155
COM: 46

JOSEPH J H DR
DEPT GEOPHYS & PLANET SCI
TEL-AVIV UNIVERSITY
RAMAT-AVIV 69978
ISRAEL
TEL: 3-420-633
TLX: 342171 VERSY IL
COM:

JOSEPH ROBERT D DR
BLACKETT LABORATORY
ASTROPHYSICS GROUP
IMPERIAL COLLEGE
LONDON SW7 2BZ
U.K.
TEL: 1-589-5111x6660
TLX: 261503
COM:

JOSHI G C DR
UTTAR PRADESH STATE OBS
MANORA PEAK
NAINITAL 263 129
INDIA
TEL:
TLX:
COM: 12

JOSHI MOHAN N PROF
RADIO ASTRONOMY CENTER
TIFR
POST BOX 8
UDHAGAMANDALAM 643 001
INDIA
TEL: 2032
TLX: 8458488 TIFR IN
COM: 28,40,47

JOSHI SURESH CHANDRA DR
UP STATE OBSERVATORY
MANORA PEAK
NAINI TAL 263 129
INDIA
TEL: 2136
TLX:
COM: 25

JOSHI U C DR
PHYSICAL RESEARCH LAB
NAVRANGPURA
AHMEDABAD 380 009
INDIA
TEL: 462-129
TLX: 121397
COM: 37

JOSS PAUL CHRISTOPHER DR
MIT
ROOM 6-203
CAMBRIDGE MA 02139
U.S.A.
TEL: 617-243-4845
TLX:
COM: 42,48

JURGENS RAYMOND F
JET PROPULSION LAB
MS 238/420
4800 OAK GROVE DRIVE
PASADENA CA 91109
U.S.A.
TEL: 818-354-4974
TLX: 675429
COM: 16

KAFATOS MINAS DR
PHYSICS DEPT
GEORGE MASON UNIVERSITY
FAIRFAX VA 22030
U.S.A.
TEL: 703-323-2303
TLX:
COM: 28,34,44,51

KAITCHUCK RONALD H
OHIO STATE UNIVERSITY
DEPT OF ASTRONOMY
174 WEST 18TH AVENUE
COLUMBUS OH 43210
U.S.A.
TEL: 614-422-4579
TLX:
COM:

JOUBERT MARTINE
LAB D'ASTRONOMIE SPATIALE
TRAVERSE DU SIPHON
LES TROIS LUCS
F-13012 MARSEILLE
FRANCE
TEL: 91-66-08-32
TLX: 420584
COM: 21

JURKEVICH IGOR DR
3130 PORT WAY
ANNAPOLIS MD 21403
U.S.A.
TEL: 202-767-2003
TLX:
COM: 42

KAFKA PETER
MPI F. PHYSIK & ASTROPHYS
INSTITUT FUR ASTROPHYSIK
KARL-SCHWARZSCHILD-STR 1
D-8046 GARCHING B MUNCHEN
GERMANY, F.R.
TEL: 89-3299-0
TLX: 524629 ASTRO D
COM: 48.51

KAKINUMA TAKAKIYO T PROF
RESEARCH INSTITUTE
OF ATMOSPHERICS
NAGOYA UNIVERSITY
TOYOKAWA AICHI 442
JAPAN
TEL: 05338-6-3154
TLX:
COM: 40,49

JOURNET ALAIN
CERGA
AVENUE COPERNIC
F-06130 GRASSE
FRANCE
TEL: 93-36-58-49
TLX: 470865
COM: 07,08

JUSZKIEWICZ ROMAN
COPERNICUS ASTRON CENTER
UL. BARTYCKA 18
00-716 WARSAW
POLAND
TEL:
TLX:
COM: 47

KAFTAN MAY A DR
NRAO
PO BOX 2
GREEN BANK VA 24944
U.S.A.
TEL:
TLX: 710-938-1530
COM: 34,40

KAKUTA CHUICHI DR
INTL LATITUDE OBSERVATORY
HOSHIGAOKA 2-12
MIZUSAWA IWATE 023
JAPAN
TEL: 0197-24-7111
TLX: 837628 ILSMIZ J
COM: 19,31

JOVANOVIC BOZIDAR
FACULTY OF AGRICULTURE
INST WATERRANGING
VELJKA VLAHOVICA 2
YU-21000 NOVI SAD
YUGOSLAVIA
TEL: 009382158366
TLX:
COM: 07

KABURAKI MASAKI PROF
4-24-9 KICHIJYOJI
MINAMI MUSASHINO
TOKYO 180
JAPAN
TEL:
TLX:
COM: 33

KAHLER STEPHEN W DR
AIR FORCE GEOPHYSICS LAB
SPACE PHYSICS DIV (PHP)
HANSCOM AIR FORCE BASE
BEDFORD MA 01731
U.S.A.
TEL: 617-861-3975
TLX:
COM: 10

KALAFI MANOUCHER
CTR FOR ASTRON RESEARCH
DEPT OF PHYSICS
TABRIZ UNIVERSITY
TABRIZ
IRAN
TEL: 041-32564
TLX:
COM: 28

JUGAKU JUN DR
TOKYO ASTRONOMICAL OBS
OSAWA MITAKA
TOKYO 181
JAPAN
TEL:
TLX: 2822307 TAOMTKJ
COM: 28,29C,51C

KABURAKI OSAMU DR
ASTRONOMICAL INSTITUTE
TOHOKU UNIVERSITY
SENDAI 980
JAPAN
TEL: 222-22-1800
TLX: 852256
COM: 10

KAHN FRANZ D PROF
DEPT OF ASTRONOMY
THE UNIVERSITY
MANCHESTER M13 9PL
U.K.
TEL: 61-273-7121
TLX: 668932 MCHRUL G
COM: 34,40,48

KALANDADZE N B DR
ABASTUMANY ASTROPHYSICAL
OBSERVATORY
383762 ABASTUMANI
U.S.S.R.
TEL: 227
TLX: 327409 TERMIT
COM: 33

JUNG JEAN DR
THOMSON
173 BD HAUSSMANN
F-75379 PARIS CEDEX 08
FRANCE
TEL: 1-45-61-96-00
TLX: 204780 TCSF
COM:

KADLA ZDENKA I DR
PULKOVO OBSERVATORY
196140 LENINGRAD
U.S.S.R.
TEL:
TLX:
COM: 05,37

KAI KEIZO DR
TOKYO ASTRON OBSERVATORY
21-2-1 OSAWA, MITAKA
TOKYO 181
JAPAN
TEL: 0422-32-5111
TLX: 3329005 TAONRO J
COM: 10,40

KALBERLA PETER
RADIOASTRONOMISCHES INST
DER UNIVERSITAET BONN
AUF DEM HUEGEL 71
D-5300 BONN 1
GERMANY, F.R.
TEL: 0228-733645
TLX: 0886440
COM: 05,40

JUPP ALAN H DR
DEPT APPL MATH THEOR PHYS
UNIVERSITY OF LIVERPOOL
PO BOX 147
LIVERPOOL L69 3BX
U.K.
TEL: 051-709-6022
TLX: 627095
COM: 07

KADOURI TALIB HADI
ASTRONOMY & SPACE RES CTR
COUNCIL FOR SCI RESEARCH
P O BOX 2441
JADIRIYAH, BAGHDAD
IRAQ
TEL: 00-96417765127
TLX: 213976 SRC IK
COM: 27,30,36,42

KAIFU NORIO DR
TOKYO ASTRON OBSERVATORY
OSAWA MITAKA
TOKYO 181
JAPAN
TEL: 0422-32-5111
TLX:
COM: 34,40C

KALER JAMES B PROF
UNIVERSITY OF ILLINOIS
349 ASTRONOMY BLDG
1011 W. SPRINGFIELD
URBANA IL 61801
U.S.A.
TEL: 217-333-9382
TLX: 910-2452434 AST
COM: 34

JURA MICHAEL DR
UCLA
DEPT OF ASTRONOMY
MATH SCIENCES BLDG
LOS ANGELES CA 90024
U.S.A.
TEL: 213-825-4302
TLX:
COM: 34

KAEHLER HELMUTH DR
HAMBURGER STERNWARTE
GOJENBERGSWEG 112
D-2050 HAMBURG 80
GERMANY, F.R.
TEL:
TLX:
COM: 35

KAISER THOMAS R PROF
DEPT OF PHYSICS
THE UNIVERSITY
SHEFFIELD S3 7RH
U.K.
TEL: 0742-78555x4277
TLX: 547216 UGSHEF G
COM: 22

KALINKOV MARIN P DR
DEPT OF ASTRONOMY
BULGARIAN ACAD SCIENCES
72 LENIN BLVD
1784 SOFIA
BULGARIA
TEL:
TLX: 22774 CLANP BG
COM: 28

LIST OF MEMBERS

KALKOFEN WOLFGANG DR
SMITHSONIAN ASTROPHYSICAL
OBSERVATORY
60 GARDEN STREET
CAMBRIDGE MA 02138
U.S.A.
TEL: 617-495-7285
TLX:
COM: 12,36C

KALLOGLIAN ARSEN T DR
BYURAKAN ASTROPHYSICAL
OBSERVATORY
378433 BYURAKAN, ARMENIA
U.S.S.R.
TEL:
TLX:
COM: 28

KALMAN BELA DR
HELIOPHYSICAL OBSERVATORY
PO BOX 30
H-4010 DEBRECEN
HUNGARY
TEL: 52-11-015
TLX: 72517 DEOBS H
COM: 10,12

KALMYKOV A M DR
ASTRONOMICAL INSTITUTE
UZBEK ACADEMY OF SCIENCES
700000 TASHKENT
U.S.S.R.
TEL:
TLX:
COM: 19

KALNAJS AGRIS J DR
MT STROMLO OBSERVATORY
WODEN P.O. ACT 2606
AUSTRALIA
TEL: 062-881111-248
TLX: 62270 AA
COM: 33

KAMEL OSMAN M DR
FACULTY OF SCIENCES
ASTRONOMY DEPT
CAIRO UNIVERSITY
GIZA-CAIRO
EGYPT
TEL:
TLX:
COM:

KAMIJO FUMIO PROF DR
DEPT OF ASTRONOMY
UNIVERSITY OF TOKYO
YAYOI BUNKYO KU
TOKYO 113
JAPAN
TEL:
TLX:
COM: 34

KAMINISHI KEISUKE PROF
FACULTY OF SCIENCE
KUMAMOTO UNIVERSITY
KUROKAMI 2 CHOME
860 KUMAMOTO
JAPAN
TEL: 096-344-2111
TLX:
COM: 35

KAMP LUCAS WILLEM DR
DEPT OF ASTRONOMY
BOSTON UNIVERSITY
725 COMMONWEALTH AVE
BOSTON MA 02215
U.S.A.
TEL:
TLX:
COM: 36,37

KAMPER KARL W DR
DAVID DUNLAP OBSERVATORY
RICHMOND HILL, ON L4C 4Y6
CANADA
TEL: 416-884-9562
TLX:
COM:

KANAEV IVAN I DR
PULKOVO OBSERVATORY
196140 LENINGRAD
U.S.S.R.
TEL:
TLX:
COM: 24C

KANDEL ROBERT S DR
LABORATOIRE METEOROLOGIE
DYNAMIQUE
ECOLE POLYTECHNIQUE
F-91128 PALAISEAU CEDEX
FRANCE
TEL: 1-69-41-82-00
TLX: 691596 ECOLEX F
COM: 36

KANDEMIR GUELCIN
ISTANBUL TECHNICAL UNIV
FEN FAKULTESI, FIZIK B
MASLAK
ISTANBUL
TURKEY
TEL: 1609109
TLX:
COM:

KANDPAL CHANDRA D
UP STATE OBSERVATORY
NAINITAL U.P. 263 129
INDIA
TEL: 2136,2325
TLX:
COM: 42

KANE SHARAD R DR
SPACE SCIENCES LAB
UNIVERSITY OF CALIFORNIA
BERKELEY CA 94720
U.S.A.
TEL: 415-642-1719
TLX: 910-366-7945
COM: 10C

KANEKO NOBORU DR
DEPT OF PHYSICS
FACULTY OF SCIENCE
HOKKAIDO UNIVERSITY
060 SAPPORO
JAPAN
TEL: 11-716-2111
TLX: 932510 HOKUSC J
COM: 28

KANG GON IK
PYONGYANG ASTRON OBS
ACADEMY OF SCIENCES DPRK
TAESONG DISTRICT
PYONGYANG
KOREA DPR
TEL:
TLX:
COM: 40

KANG JIN SOK
PYONGYANG ASTRON OBS
ACADEMY OF SCIENCES DPRK
TAESONG DISTRICT
PYONGYANG
KOREA DPR
TEL:
TLX:
COM: 10

KANNO MITSUO PROF
HIDA OBSERVATORY
KAMITAKARA
GIFU-KEN 506-13
JAPAN
TEL: 0578-6-2311
TLX:
COM: 10,12

KANYO SANDOR DR
KONKOLY OBSERVATORY
BOX 67
H-1525 BUDAPEST
HUNGARY
TEL: 166-426
TLX:
COM: 27

KAPAHI V K DR
TIFR CENTRE
INDIAN INST OF SCI CAMPUS
POST BOX NO 1234
BANGALORE 560 012
INDIA
TEL: 362816
TLX: 8458488 TIFR IN
COM: 40C

KAPISINSKY IGOR
ASTRONOMICAL INSTITUTE
SLOVAK ACAD. OF SCIENCES
842 28 BRATISLAVA
CZECHOSLOVAKIA
TEL: 427-375157
TLX: 093355
COM: 22

KAPLAN J DR
DEPT OF PHYSICS
UNIVERSITY OF CALIFORNIA
LOS ANGELES CA 90024
U.S.A.
TEL:
TLX:
COM: 21

KAPLAN LEWIS D DR
ATMOSPH. & ENVIRONMENTAL
RESEARCH, INC.
840 MEMORIAL DRIVE
CAMBRIDGE MA 02139
U.S.A.
TEL: 617-547-6207
TLX: 951417 AERC
COM:

KAPOOR RAMESH CHANDER
INDIAN INSTITUTE OF
ASTROPHYSICS
BANGALORE 560 034
INDIA
TEL: 566585
TLX: 845-763 IIAB IN
COM: 47

KARAALI SALIH DR
ISTANBUL UNIVERSITY
FACULTY OF SCIENCE
DEPT ASTRONOMY & SPACE SC
ISTANBUL 34
TURKEY
TEL: 5224200/610
TLX:
COM:

KARACHENTSEV I D DR
SPEC ASTROPHYS OBS
ACAD OF SC/STAVROPOLSKIJ
ZELENCHUKSKAJA
357147 N ARKHYZ
U.S.S.R.
TEL:
TLX:
COM: 09,28C,30,47

KARANDIKAR R V PROF
DEPT OF ASTRONOMY
OSMANIA UNIVERSITY
HYDERABAD 500 007
INDIA
TEL: 71251
TLX:
COM: 16,21

KARDASHEV N S DR
SPACE RESEARCH INSTITUTE
USSR ACADEMY OF SCIENCES
117810 MOSCOW
U.S.S.R.
TEL:
TLX:
COM: 40,51V

KARETNIKOV VALENTIN G R
ODESSA STATE UNIVERSITY
PARK SHEVCHENKO
ASTRONOMICAL OBSERVATORY
270014 ODESSA
U.S.S.R.
TEL: 25-03-56
TLX:
COM: 42

KARLICKY MARIAN
ASTRONOMICAL INSTITUTE
CZECH. ACAD. OF SCIENCES
OBSERVATORY
251 65 ONDREJOV
CZECHOSLOVAKIA
TEL:
TLX:
COM: 10

KARP ALAN HERSH DR
SCIENTIFIC CENTER IBM
1530 PAGE MILL RD
PALO ALTO CA 94304
U.S.A.
TEL: 415-855-3127
TLX:
COM: 27,36

LIST OF MEMBERS

KARPEN JUDITH T
NAVAL RESEARCH LABORATORY
CODE 4175 K
WASHINGTON DC 20375
U.S.A.
TEL: 202-767-3441
TLX:
COM: 10,12

KARPINSKIJ VADIM N DR
PULKOVO OBSERVATORY
196140 LENINGRAD
U.S.S.R.
TEL:
TLX:
COM: 09,12,44

KARYGINA ZOYA V DR
ASTROPHYSICAL INSTITUTE
480068 ALMA-ATA
U.S.S.R.
TEL:
TLX:
COM: 21

KASHCHEEV B L PROF DR
KHARKOV INSTITUTE FOR
RADIOELECTRONICS
310059 KHARKOV
U.S.S.R.
TEL:
TLX:
COM: 22

KASPER U DR
ZNTRLINST. F. ASTROPHYSIK
STERNWARTE BABELSBERG
ROSA-LUXEMBURG-STR 17A
DDR-1502 POTSDAM
GERMANY, D.R.
TEL:
TLX:
COM: 47

KASTURIRANGAN K DR
ISRO SATELLITE CENTER
AIRPORT ROAD
VIMANAPURA POST
BANGALORE 560 017
INDIA
TEL: 54779
TLX: 0845-325 & 769
COM: 44

KASUGA TAKASHI
NOBEYAMA RADIO OBS
TOKYO ASTRON OBSERVATORY
NOBEYAMA, MINAMISAKU
NAGANO 384-13
JAPAN
TEL: 267-98-2831
TLX: 3329005 TAO NRO J
COM: 40

KATGERT PETER DR
STERREWACHT
POSTBUS 9513
NL-2300 RA LEIDEN
NETHERLANDS
TEL: 071-148333
TLX: 39058
COM:

KATGERT-MERKELIJN J K DR
STERREWACHT
HUYGENS LABORATORIUM
WASSENAARSEWEG 78
NL-2300 RA LEIDEN
NETHERLANDS
TEL:
TLX:
COM:

KATO MARIKO
DEPT OF ASTRONOMY
KEIO UNIVERSITY
4-1-1 HIYOSHI KOULOKU-KU
YOKOHAMA-SHI 223
JAPAN
TEL: 44-63-1111
TLX:
COM: 35

KATO SHOJI PROF
DEPT OF ASTRONOMY
UNIVERSITY OF KYOTO
KITASHIRAKAWA OIWAKE
SAKYOKU KYOTO 606
JAPAN
TEL: 075-751-2111
TLX: 5422693 LIBKYU J
COM: 12,33,47

KATO TAKAKO DR
INST OF PLASMA PHYSICS
NAGOYA UNIVERSITY
RURO-CHO CHIKUSA-KU
NAGOYA 464
JAPAN
TEL: 052-781-5111
TLX: 0447-3691 IPPJNU J
COM: 14C

KATSIS DEMETRIUS DR
12 RUE VARNIS
GR-17124 NEA SMYRNE
GREECE
TEL: 9336014
TLX:
COM: 07

KATZ JONATHAN I
DEPT OF PHYSICS
WASHINGTON UNIVERSITY
ST LOUIS MI 63130
U.S.A.
TEL: 314-889-6202
TLX:
COM: 48

KATZ JOSEPH DR
RACAH INST. OF PHYSICS
HEBREW UNIV. OF JERUSALEM
JERUSALEM 91904
ISRAEL
TEL: 58-46-04
TLX: 25391 HUIL
COM:

KAUFMAN MICHELE DR
PHYSICS DEPT
OHIO STATE UNIVERSITY
174 W. 18TH AVE
COLUMBUS OH 43210
U.S.A.
TEL: 614-422-5713
TLX:
COM: 28

KAUFMANN JENS PETER DR
INSTITUT FUR ASTRONOMIE
TECHNISCHE UNIVERSITAT
HARDENBERGSTR. 36
D-1000 BERLIN 12
GERMANY, F.R.
TEL: 030-3145462
TLX: 184262 TUBLN D
COM:

KAUFMANN PIERRE PROF
INST PESQUISAS ESPACIAIS
CRAAM RADIO OBSERVATORY
CAIXA POSTAL 515
12200 S. JOSE DOS CAMPOS
BRAZIL
TEL: 55-0123-229977
TLX: 11 34061 INPE BR
COM: 10C,12,40,51

KAULA WILLIAM M PROF
1283 BARTONSHIRE WAY
POTOMAC MD 20854
U.S.A.
TEL:
TLX:
COM: 07,16

KAWABATA KINAKI PROF
DEPT OF PHYSICS
NAGOYA UNIVERSITY
FUROCHO CHIKUSAKU
NAGOYA 464
JAPAN
TEL:
TLX:
COM: 40,47

KAWABATA KIYOSHI
DEPT OF PHYSICS, COLL.SCI
SCIENCE UNIV OF TOKYO
1-3 KAGURAZAKA, SHINJUKU
TOKYO
JAPAN
TEL: 3-260-4271
TLX:
COM:

KAWABATA SHUSAKU PROF
KYOTO GAKUEN UNIVERSITY
SOGABE-MACHI KAMEOKA
KYOTO 621
JAPAN
TEL: 07712-2-2001
TLX:
COM: 42

KAWAGUCHI ICHIRO PROF
DEPT OF ASTRONOMY
FACULTY OF SCIENCE
UNIVERSITY OF KYOTO
606 KYOTO
JAPAN
TEL:
TLX:
COM: 12

KAWARA KIMIAKI
CERRO TOLOLO
INTERAMERICAN OBSERVATORY
CASILLA 603
LA SERENA
CHILE
TEL: 56-51-213352
TLX: 34-620301
COM: 25

KAWATA YOSHIYUKI DR
KANAZAWA INSTITUTE
OF TECHNOLOGY
NONOICHO
KANAZAWA MINAMI 921
JAPAN
TEL:
TLX:
COM:

KAZANTZIS PANAYOTIS DR
DEPT OF ASTRONOMY
UNIVERSITY OF GLASGOW
GLASGOW G12 8QQ
U.K.
TEL:
TLX:
COM:

KAZES ILYA DR
OBSERVATOIRE DE PARIS
SECTION D'ASTROPHYSIQUE
F-92195 MEUDON PL CEDEX
FRANCE
TEL: 1-45-34-75-30
TLX:
COM: 34,40

KEAY COLIN S L PROF
PHYSICS DEPT
NEWCASTLE UNIVERSITY
NEWCASTLE NSW 2308
AUSTRALIA
TEL: 049-685-235
TLX: 28194 NEWUN AA
COM: 22V

KEEL WILLIAM C
STERREWACHT LEIDEN
POSTBUS 9513
NL-2300 RA LEIDEN
NETHERLANDS
TEL: 31-71-148333
TLX: 39058 ASTRO NL
COM: 28

KEENAN PHILIP C PROF EMER
PERKINS OBSERVATORY
BOX 449
DELAWARE OH 43015
U.S.A.
TEL: 614-363-1257
TLX: 810-482-1715
COM: 29,45

KEGEL WILHELM H PROF
INST THEORETISCHE PHYSIK
UNIVERSITAT FRANKFURT
ROBERT-MAYER-STR 8-10
D-6000 FRANKFURT/MAIN 1
GERMANY, F.R.
TEL: 069-7982357
TLX: 413932 UNIF D
COM: 34

KEIL STEPHEN L
AIR FORCE GEOPHYSICS LAB
SOLAR RESEARCH BRANCH
SACRAMENTO PEAK OBS
SUNSPOT NM 88349
U.S.A.
TEL: 505-434-1390
TLX:
COM: 12

KELEMEN JANOS
KONKOLY OBSERVATORY
HUNGARIAN ACADEMY OF SCI
BOX 67
H-1525 BUDAPEST
HUNGARY
TEL: 754-122
TLX: 227460 KONOB H
COM:

KELLER CHARLES F
UNIVERSITY OF CALIFORNIA
LOS ALAMOS NATIONAL LAB
BOX 1663 MS F665
LOS ALAMOS NM 87545
U.S.A.
TEL: 505-667-5648
TLX:
COM:

KELLER GEOFFREY
DEPT OF ASTRONOMY
OHIO STATE UNIVERSITY
174 W. 18TH AVENUE
COLUMBUS OH 43210
U.S.A.
TEL: 614-422-6279
TLX:
COM:

KELLER HANS ULRICH DR
OBSERVATORY & PLANETARIUM
POB 161
NECKARST 47
D-7000 STUTTGART
GERMANY, F.R.
TEL: 0711-291004
TLX: 721855 STBST D
COM: 15,46,51

KELLER HORST UWE DR
MPI FUER AERONOMIE
POSTFACH 20
D-3411 KATLENBURG-LINDAU
GERMANY, F.R.
TEL: 05556-41-419
TLX: 965527 AERLI D
COM: 49C

KELLERMANN KENNETH I DR
NRAO
EDGEMONT ROAD
CHARLOTTESVILLE VA 22903
U.S.A.
TEL: 804-296-0240
TLX: 910-997-0174
COM: 28,40C,47,48,51

KELLOGG EDWIN M DR
26 OAKLAND STREET
LEXINGTON MA 02173
U.S.A.
TEL:
TLX:
COM: 48

KEMBHAVI AJIT K
TATA INSTITUTE OF
FUNDAMENTAL RESEARCH
HOMI BHABHA RD
BOMBAY 400 005
INDIA
TEL:
TLX:
COM: 47,48

KEMP JAMES C
PHYSICS DEPT
UNIVERSITY OF OREGON
EUGENE OR 97403
U.S.A.
TEL: 503-687-2952
TLX:
COM: 25,42

KENDERDINE SIDNEY DR
MULLARD RADIO ASTRON LAB
MADINGLEY ROAD
CAMBRIDGE CB3 0HE
U.K.
TEL: 223-66477
TLX: 81292
COM: 40

KENNEDY EUGENE T
SCHOOL OF PHYSICAL SCI
NATL INST F. HIGHER EDUC
GLASNEVIN
DUBLIN 9
IRELAND
TEL: DUBLIN 370071
TLX: 30690 NIHFD
COM: 14

KENNEDY JOHN E PROF
323 LAKE CRESCENT
SASKATOON SASK S7H 3A1
CANADA
TEL: 374-4614
TLX:
COM: 41,46

KENNICUTT ROBERT C JR
DEPT OF ASTRONOMY
UNIVERSITY OF MINNESOTA
116 CHURCH ST S.E.
MINNEAPOLIS MN 55455
U.S.A.
TEL: 612-376-5224
TLX:
COM: 28

KENT STEPHEN M
HARVARD-SMITHSONIAN CTR
FOR ASTROPHYSICS
60 GARDEN STREET
CAMBRIDGE MA 02138
U.S.A.
TEL: 617-495-9681
TLX:
COM:

KEPLER S O
INSTITUTO DE FISICA
UFRGS
AV. BENTO GONCALVES 9500
90049 PORTO ALEGRE - RS
BRAZIL
TEL: 0512-36-4677
TLX: 051-1055 UFRS BR
COM: 25,27

KERES H P PROF DR
INSTITUTE OF PHYSICS
ESTONIAN SSR ACAD SCI
RIIA 142
202400 TARTU
U.S.S.R.
TEL:
TLX:
COM:

KERR FRANK J DR
ASTRONOMY PROGRAM
UNIVERSITY OF MARYLAND
COLLEGE PK MD 20742
U.S.A.
TEL: 301-454-6302
TLX: 710-826-0352
COM: 33,34,40

KERR ROY P PROF
UNIVERSITY OF CANTERBURY
PRIVATE BAG
CHRISTCHURCH
NEW ZEALAND
TEL: 482-009
TLX: NZ 4144
COM:

KESSLER KARL G DR
CTR FOR BASIC STANDARDS
NATL BUREAU OF STANDARDS
B-160 PHYSICS BLDG
GAITHERSBURG MD 20899
U.S.A.
TEL: 301-921-2001
TLX: 197674 TRT
COM: 14,31

KESTEVEN MICHAEL J L DR
DIVISION OF RADIOPHYSICS
CSIRO
P.O.BOX 76
EPPING NSW 2121
AUSTRALIA
TEL: 02-868-0222
TLX: 26230 ASTRO
COM: 40

KHACHIKIAN E YE PROF
BYURAKAN ASTROPHYSICAL
OBSERVATORY
378433 BYURAKAN, ARMENIA
U.S.S.R.
TEL: 56-3453/55-6383
TLX:
COM: 28C

KHARADZE E K PROF
ABASTUMANI ASTROPHYSICAL
OBSERVATORY
383762 GEORGIA
U.S.S.R.
TEL: 998891, 225460
TLX: 327409 TERMIT
COM: 33,34,45

KHARE BISHUN N DR
CTR RADIO PHYS/SPACE RES
CORNELL UNIVERSITY
306 SPACE SCIENCES BLDG
ITHACA NY 14853
U.S.A.
TEL: 607-256-3934
TLX:
COM:

KHARIN A S DR
MAIN ASTRON OBSERVATORY
UKRAINIAN ACAD OF SCIENCE
GOLOSEEVO
252127 KIEV
U.S.S.R.
TEL: 663110
TLX: 131406 SKY
COM: 08

KHARITONOV ANDREJ V DR
ASTROPHYSICAL INSTITUTE
480068 ALMA-ATA
U.S.S.R.
TEL:
TLX:
COM: 29

KHATISASHVILI ALFEZ SH DR
ABASTUMANI ASTROPHYSICAL
OBSERVATORY
383762 ABASTUMANI,GEORGIA
U.S.S.R.
TEL:
TLX:
COM: 20

KHETSURIANI T S DR
ABASTUMANI ASTROPHYSICAL
OBSERVATORY
383762 ABASTUMANI,GEORGIA
U.S.S.R.
TEL:
TLX:
COM: 12

KHOKHLOVA V L DR
ASTRONOMICAL COUNCIL
USSR ACADEMY OF SCIENCES
PYATNITSKAYA UL. 48
109017 MOSCOW
U.S.S.R.
TEL: 231-54-61
TLX: 412623 SCSTP SU
COM: 29,36

KHOLOPOV P N DR
STERNBERG STATE
ASTRONOMICAL INSTITUTE
UNIVERSITETSKIJ. PROSP 13
119899 MOSCOW V-234
U.S.S.R.
TEL:
TLX:
COM: 27,33,37

KHOLSHEVNIKOV K V DR
LENINGRAD STATE UNIV
ASTRONOMICAL OBSERVATORY
BIBLIOTECHNAJA PL. 2
198904 LENINGRAD
U.S.S.R.
TEL: 257-94-88
TLX:
COM: 07

KHOZOV GENNADIJ V
ASTRONOMICAL OBSERVATORY
LENINGRAD STATE UNIV
BIBLIOTECHNAJA PL. 2
198904 LENINGRAD
U.S.S.R.
TEL: 2-57-94-84
TLX: 12168 PHOBOS
COM: 35

KHROMOV G S DR
ASTRONOMICAL COUNCIL
USSR ACADEMY OF SCIENCES
PYATNITSKAYA UL. 48
109017 MOSCOW
U.S.S.R.
TEL:
TLX:
COM: 34,41

LIST OF MEMBERS

KIANG TAO PROF
DUNSINK OBSERVATORY
CASTLEKNOCK
DUBLIN 15
IRELAND
TEL: 387-911
TLX: 31687 DIAS EI
COM: 20,28,41

KILKENNY DAVID DR
S A A O
P O BOX 9
OBSERVATORY 7935
SOUTH AFRICA
TEL: 021-47-0025
TLX: 57-20309 SA
COM: 25C

KING ANDREW R DR
ASTRONOMY DEPT
UNIVERSITY OF LEICESTER
LEICESTER
U.K.
TEL: 0533-554455
TLX: 341198
COM:

KINMAN THOMAS D DR
KITT PEAK NATIONAL OBS
PO BOX 26732
TUCSON AZ 85726
U.S.A.
TEL: 602-327-5511
TLX: 0666-484 AURA NOAO
COM: 28,33

KIASATPOOR AHMAD PROF
PHYSICS DEPT
UNIVERSITY OF ESFAHAN
DANESHGAH E
ESFAHAN
IRAN
TEL: 031-44321
TLX: 31-2295 IRE U
COM:

KIM JIK SU
PYONGYANG ASTRON OBS
ACADEMY OF SCIENCES DPRK
TAESONG DISTRICT
PYONGYANG
KOREA DPR
TEL:
TLX:
COM: 47

KING DAVID S PROF
DEPT PHYSICS & ASTRONOMY
UNIVERSITY OF NEW MEXICO
ALBUQUERQUE NM 87131
U.S.A.
TEL:
TLX:
COM: 35

KINOSHITA HIROSHI DR
TOKYO ASTRONOMICAL
OBSERVATORY
OSAWA MITAKA
181 TOKYO
JAPAN
TEL: 0422-32-5111
TLX:
COM: 04,07C

KIBBLEWHITE EDWARD J DR
INSTITUTE OF ASTRONOMY
MADINGLEY RD
CAMBRIDGE CB3 0HE
U.K.
TEL:
TLX:
COM:

KIM TU HWAN
KOREA ASTR/SPACE SC INST
36-1 WHAAM-DONG JUNG-GU
TAEJEON CHUNGCHUNGNAM-DO
300-31 TAEJEON
KOREA, REPUBLIC
TEL: 042-823-1497
TLX: 45532 K
COM: 27

KING HENRY C DR
TRILLIUM
5 HILBURY CLOSE
CHESHAM BOIS
AMERSHAM, BUCKS HP6 5LB
U.K.
TEL: AMERSHAM 3003
TLX:
COM: 41

KIPPENHAHN RUDOLF PROF
MPI F PHYSIK & ASTROPHYS
KARL-SCHWARZSCHILD-STR 1
D-8046 GARCHING B MUNCHEN
GERMANY, F.R.
TEL: 089-32990
TLX: 524629 ASTRO D
COM: 27,35C

KIGUCHI MASAYOSHI DR
DEPT OF PHYSICS
KYOTO UNIVERSITY
KYOTO 606
JAPAN
TEL:
TLX:
COM: 35

KIM YONG HYOK DR
PYONGYANG ASTRON OBS
ACADEMY OF SCIENCES DPRK
TAESONG DISTRICT
PYONGYANG
KOREA DPR
TEL: 5-3134, 53239
TLX:
COM:

KING IVAN R PROF
ASTRONOMY DEPT
UNIVERSITY OF CALIFORNIA
BERKELEY CA 94720
U.S.A.
TEL: 415-642-2206
TLX: 820181 UCB ASTRAL
COM: 28,33,37

KIPPER TONU DR
TARTU ASTROPHYSICAL OBS
ESTONIAN ACAD OF SCIENCES
TORAVERE
202444 TARTU
U.S.S.R.
TEL:
TLX:
COM: 09,14,29

KIKUCHI SADAEMON PROF
ASTRONOMICAL INSTITUTE
TOHOKU UNIVERSITY
AOBAYAMA
SENDAI 980
JAPAN
TEL:
TLX:
COM:

KIM YONG UK
PYONGYANG ASTRON OBS
ACADEMY OF SCIENCES DPRK
TAESONG DISTRICT
PYONGYANG
KOREA DPR
TEL:
TLX:
COM:

KING R B DR
PO BOX 725
MEDOCINO CA 95460
U.S.A.
TEL:
TLX:
COM: 14,29

KIRAL ADNAN PROF
UNIVERSITY OBSERVATORY
BEYAZIT, ISTANBUL
TURKEY
TEL:
TLX:
COM:

KILADZE R I DR
ABASTUMANI ASTROPHYSICAL
OBSERVATORY
383762 ABASTUMANI,GEORGIA
U.S.S.R.
TEL:
TLX:
COM: 16

KIM YUL
PYONGYANG ASTRON OBS
ACADEMY OF SCIENCES DPRK
TAESONG DISTRICT
PYONGYANG
KOREA DPR
TEL:
TLX:
COM: 28

KING ROBERT WILSON JR DR
DEPT OF EARTH ATMOSPHERIC
MIT 54-620
CAMBRIDGE MA 02139
U.S.A.
TEL: 617-253-7064
TLX: 921473 MIT CAM
COM: 04,19

KIRBIYIK HALIL DR
PHYSICS DEPT
MIDDLE EAST TECHN UNIV
ANKARA 06531
TURKEY
TEL: 237100/3528
TLX: 42761 ODTK TR
COM:

KILAMBI G C DR
DEPT OF ASTRONOMY
OSMANIA UNIVERSITY
HYDERABAD 500 007
INDIA
TEL: 71-251 x 247
TLX:
COM: 37

KIM ZONG DOK
PYONGYANG ASTRON OBS
ACADEMY OF SCIENCES DPRK
TAESONG DISTRICT
PYONGYANG
KOREA DPR
TEL:
TLX:
COM: 14

KING-HELE DESMOND G DR
ROYAL AIRCRAFT ESTABL.
FARNBOROUGH, HANTS
U.K.
TEL: 0252 24461
TLX:
COM: 07

KIRK JOHN DR
MPI F. PHYSIK & ASTROPHYS
KARL-SCHWARZSCHILD-STR 1
D-8046 GARCHING B MUNCHEN
GERMANY, F.R.
TEL: 089-32-99-0
TLX: 524629 ASTRO D
COM: 48

KILAR BOGDAN DR
FACULTY OF GEODESY
LJUBLJANA UNIVERSITY
JAMOVA 2
61000 LJUBLJANA
YUGOSLAVIA
TEL:
TLX:
COM:

KIMURA HIROSHI DR
PURPLE MOUNTAIN
OBSERVATORY
NANJING
CHINA, PEOPLE'S REP.
TEL: 33921
TLX: 34144 PMONJ CN
COM: 34

KINGSTON ARTHUR E PROF
DEPT OF APPLIED MATHS
QUEEN'S UNIVERSITY
BELFAST BT7 1NN
U.K.
TEL: 0232-245133
TLX: 74487 QUB AMD G
COM: 14

KIRKPATRICK RONALD C DR
LOS ALAMOS NATL LAB
MS 220
LOS ALAMOS NM 87545
U.S.A.
TEL: 505-667-4812
TLX:
COM: 34

KIRSHNER ROBERT PAUL DR
DEPT OF ASTRONOMY
HARVARD UNIVERSITY
CAMBRIDGE MA 02138
U.S.A.
TEL: 617-495-7390
TLX:
COM: 28,34

KITCHIN CHRISTOPHER R DR
HATFIELD POLYTECHNIC
OBSERVATORY
BAYFORDBURY
HERTFORD, HERTS SG13 8LD
U.K.
TEL: 0992-558-451
TLX: 262413
COM: 29,46

KLEIN ULRICH
RADIOASTRONOMISCHES INST
DER UNIVERSITAET BONN
AUF DEM HUEGEL 71
D-5300 BONN 1
GERMANY, F.R.
TEL: 0228-73-3644
TLX:
COM: 28,40

KLOCK B L DR
6601 S. HOMESTAKE DRIVE
BOWIE MD 20715
U.S.A.
TEL: 301-262-1506
TLX:
COM: 08,09,24

KISELYOV ALEXEJ A DR
PULKOVO OBSERVATORY
196140 LENINGRAD
U.S.S.R.
TEL:
TLX:
COM: 26C

KJAERGAARD PER DR
UNIVERSITY OBSERVATORY
OESTER VOLDGADE 3
DK-1350 COPENHAGEN K
DENMARK
TEL: 1-14-17-90
TLX: 44155 DANAST DK
COM:

KLEINMANN DOUGLAS E DR
HONEYWELL ELECTRO OPTICS
OPERATION
2 FORBES RD
LEXINGTON MA 02173
U.S.A.
TEL: 617-863-3841
TLX: 92-3477
COM:

KLVANA MIROSLAV
ASTRONOMICAL INSTITUTE
CZECH. ACAD. OF SCIENCES
ONDREJOV OBSERVATORY
251 65 ONDREJOV
CZECHOSLOVAKIA
TEL:
TLX:
COM: 10

KISLYAKOV ALBERT G DR
APPLIED PHYSICS INSTITUTE
ACADEMY OF SCIENCES
ULYANOV STREET 46
603600 GORKY
U.S.S.R.
TEL:
TLX:
COM: 40

KJELDSETH-MOE OLAV DR
INST THEORET ASTROPHYSICS
UNIVERSITY OF OSLO
P O BOX 1029
N-0315 BLINDERN, OSLO 3
NORWAY
TEL: 47-2-456510
TLX: 72425 UNIOS N
COM: 10

KLEMOLA ARNOLD R DR
LICK OBSERVATORY
UCSC
SANTA CRUZ CA 95064
U.S.A.
TEL: 408-429-2907
TLX:
COM: 20,24

KNACKE ROGER F DR
DEPT EARTH & SPACE SCI
SUNY AT STONY BROOK
STONY BROOK NY 11794
U.S.A.
TEL: 516-246-7673
TLX: 5102287767
COM: 15,34

KISLYUK VITALIJ S DR
MAIN ASTRON OBSERVATORY
UKRAINIAN ACAD OF SCIENCE
GOLOSEEVO
252127 KIEV
U.S.S.R.
TEL:
TLX: 131406 SKY US
COM: 16,24

KLARE GERHARD DR
LANDESSTERNWARTE
KOENIGSTUHL
D-6900 HEIDELBERG 1
GERMANY, F.R.
TEL: 06221/10036
TLX:
COM: 33

KLEMPERER W K DR
ELECTROMAG. FIELDS DIV
NATL BUREAU OF STANDARDS
325 BROADWAY
BOULDER CO 80303
U.S.A.
TEL: 303-497-3757
TLX: 592811 NOAA MASC BDR
COM:

KNAPP GILLIAN R DR
DEPT ASTROPHYSICAL SCI
PRINCETON UNIVERSITY
PRINCETON NJ 08544
U.S.A.
TEL: 609-452-3824
TLX:
COM: 28,33,34

KISSELEVA TAMARA P
PULKOVO MAIN ASTRONOMICAL
OBSERVATORY
196140 LENINGRAD
U.S.S.R.
TEL:
TLX:
COM: 20

KLARMANN JOSEPH PROF
WASHINGTON UNIVERSITY
DEPT OF PHYSICS
ST LOUIS MO 63130
U.S.A.
TEL: 314-889-6299
TLX: 650-2557719 MCI
COM:

KLEPCZYNSKI WILLIAM J DR
US NAVAL OBSERVATORY
34 & MASSACHUSETTS AVE NW
WASHINGTON DC 20390
U.S.A.
TEL: 202-653-1521
TLX: 710-822-1970
COM: 04,19P,31

KNEER FRANZ DR
UNIVERSITATS STERNWARTE
GEISMARLANDSTRASSE 11
D-3400 GOETTINGEN
GERMANY, F.R.
TEL: 0551-395-042
TLX: 96753 USTERN D
COM: 12

KISSELL KENNETH E DR
C/O BDM CORPORATION
7915 JONES BRANCH RD
MCLEAN VA 22102
U.S.A.
TEL: 202-453-1484
TLX: 901103
COM: 09

KLECZEK JOSIP DR
ASTRONOMICAL INSTITUTE
251 65 ONDREJOV
CZECHOSLOVAKIA
TEL: 72-45-25
TLX: 121579 ASTR C
COM: 05,10,46C

KLINISHIN I A PROF
PEDAGOGIC INSTITUTE
PUSHKIN STR. 96 APT. 66
284000 IVANOFRANKOV SK
U.S.S.R.
TEL:
TLX:
COM:

KNIFFEN DONALD A DR
NASA/GSFC
CODE 662
GREENBELT MD 20771
U.S.A.
TEL: 301-344-6617
TLX:
COM:

KITAMURA M PROF
TOKYO ASTRONOMICAL
OBSERVATORY
MITAKA TOKYO 181
JAPAN
TEL: 0422-32-5111
TLX: 2822307 TAOMTK J
COM: 42C

KLEIN MICHAEL J DR
SPACE SCIENCES DIVISION
JET PROPULSION LAB
BLDG 264-802
PASADENA CA 91109
U.S.A.
TEL: 818-354-7132
TLX:
COM: 51

KLINGLESMITH DANIEL A DR
NASA/GSFC
CODE 684
GREENBELT MD 20771
U.S.A.
TEL: 301-344-6541
TLX:
COM:

KNOSKA STEFAN
ASTRONOMICAL INSTITUTE
SLOVAK ACAD. OF SCIENCES
059 60 TATRANSKA LOMNICA
CZECHOSLOVAKIA
TEL:
TLX:
COM: 10

KITAMURA SEIICHI DR
SHIGA UNIVERSITY
2 CHOME 5-1
HIRATSU
OTSU SHIGA 520
JAPAN
TEL: 775-37-0081
TLX:
COM:

KLEIN RICHARD I DR
LAWRENCE LIVERMORE LAB
UNIVERSITY OF CALIFORNIA
PO BOX 808:L-23
LIVERMORE CA 94550
U.S.A.
TEL: 415-422-3548
TLX:
COM: 36

KLIORE ARVYDAS JOSEPH DR
JPL
4800 OAK GROVE DRIVE
PASADENA CA 91109
U.S.A.
TEL: 818-354-6164
TLX: 675429
COM:

KNOWLES STEPHEN H DR
NAVAL RESEARCH LABORATORY
CODE 4183
WASHINGTON DC 20375-3000
U.S.A.
TEL: 202-262-2891
TLX:
COM: 19,51

KNUDE JENS KIRKESKOV DR
UNIVERSITY OBSERVATORY
OESTER VOLDGADE 3
DK-1350 COPENHAGEN K
DENMARK
TEL: 1-14-17-90
TLX: 44155 DANAST DK
COM: 25,34

KO HSIEN C PROF
DEPT OF ELECT ENGINEERING
OHIO STATE UNIVERSITY
COLUMBUS OH 43210
U.S.A.
TEL: 614-422-2571
TLX: 24-5334
COM: 40

KOBAYASHI EISUKE DR
SCIENCE INST OF OSAKA
PREFECTURE/13-23 KARITA 4
CHOME/SUMIYOSHI-KU
OSAKA 558
JAPAN
TEL: 06-692-1882
TLX:
COM:

KOBAYASHI YUKISAYU
TOKYO ASTRON OBSERVATORY
2-21-1 OSAWA
MITAKA
TOKYO 181
JAPAN
TEL: 0422-32-5111
TLX: 2822307 TAOMTK
COM: 31

KOCER D DR
BOGAZICI UNIVERSITY
KANDILLI OBSERVATORY
CENGELKOY
ISTANBUL
TURKEY
TEL: 332-02-41
TLX: 26411 BOUN TR
COM: 51

KOCH DAVID G
SMITHSONIAN ASTROPHYSICAL
OBSERVATORY
60 GARDEN STREET
CAMBRIDGE MA 02138
U.S.A.
TEL: 617-495-7479
TLX: 921428 SATELLITE CAM
COM:

KOCH ROBERT H DR
DEPT ASTRON & ASTROPHYS
UNIV OF PENNSYLVANIA
DAVID RITTENHOUSE LAB
PHILADELPHIA PA 19104
U.S.A.
TEL: 215-898-7882
TLX: 834621
COM: 25,42V,51

KOCH-MIRAMOND LYDIE DR
IRF/DPHG/ASTROPHYSIQUE
CEN SACLAY
F-91191 GIF-S/YVETTE CDX
FRANCE
TEL: 1-69-08-43-29
TLX: 690860 PHYSPAC
COM: 44,48

KOCHAROV GRANT E PROF
PHYSICO-TECHNICAL INST
USSR ACADEMY OF SCIENCES
194021 LENINGRAD
U.S.S.R.
TEL: 247-91-67
TLX:
COM: 48

KOCHHAR R K DR
INDIAN INSTITUTE OF
ASTROPHYSICS
BANGALORE 560 034
INDIA
TEL: 566585
TLX: 845763 IIAB IN
COM: 28,35

KODAIRA KEIICHI PROF
TOKYO ASTRONOMICAL OBS
MITAKA
TOKYO 181
JAPAN
TEL: 0422-32-5111
TLX: 2822307 TAOMTK J
COM: 28,29,36P

KODAMA HIDEO
DEPT OF PHYS/FAC OF SCI
UNIVERSITY OF TOKYO
7-3-1 HONGO, BUNKYO-KU
TOKYO 113
JAPAN
TEL: 03-812-2111
TLX:
COM: 47

KOEBERL CHRISTIAN DR
INSTITUTE OF GEOCHEMISTRY
UNIVERSITY OF VIENNA
P.O. BOX 73
A-1094 VIENNA
AUSTRIA
TEL: 222-344630-67
TLX:
COM: 15,22,51

KOECKELENBERGH ANDRE DR
OBS ROYAL DE BELGIQUE
AVENUE CIRCULAIRE 3
B-1180 BRUXELLES
BELGIUM
TEL: 02-375-24-84
TLX: 21565 OBSBEL B
COM: 10

KOEHLER H PROF DR
SAUERBRUCHSTR 6
7920 HEIDENHEIM a.d.BRENZ
GERMANY, F.R.
TEL: 07321-44560
TLX:
COM: 09

KOEHLER JAMES A PROF
PHYSICS DEPT
UNIV OF SASKATCHEWAN
SASKATOON S7N 0W0
CANADA
TEL: 306-966-6442
TLX:
COM:

KOEHLER PETER
CARL-ZEISS STR 1
DDR-6900 JENA
GERMANY, D.R.
TEL:
TLX:
COM: 09

KOENIGSBERGER GLORIA
INSTITUTO DE ASTRONOMIA
UNAM
APDO POSTAL 70-264
04510 MEXICO DF
MEXICO
TEL: 905-548-5305/06
TLX:
COM:

KOEPPEN JOACHIM DR
INST F. THEOR ASTROPHYSIK
DER UNIV HEIDELBERG
IM NEUENHEIMER FELD 561
D-6900 HEIDELBERG
GERMANY, F.R.
TEL: 06221-562988
TLX: 461515 UNIHD D
COM: 34

KOESTER DETLEV DR
INST THEOR PHYS & STERNW
NEUE UNIV PHYS ZENTRUM
OLSHAUSENST GEB N 61C
D-2300 KIEL 1
GERMANY, F.R.
TEL: 0431-8804105
TLX: 292706 IAPKI D
COM: 35,36

KOGOSHVILI NATELA G
ASTROPHYSICAL OBSERVATORY
MOUNT KANOBILI
383762 ABASTUMANI,GEORGIA
U.S.S.R.
TEL: 283
TLX:
COM: 28

KOGURE TOMOKAZU DR
DEPT OF ASTRONOMY
FACULTY OF SCIENCE
UNIVERSITY OF KYOTO
KYOTO 606
JAPAN
TEL: 075-751-2111
TLX: 5422693 LIBKYU J
COM: 29

KOHL JOHN L DR
CENTER FOR ASTROPHYSICS
60 GARDEN ST
CAMBRIDGE MA 02138
U.S.A.
TEL: 617-495-7377
TLX: 921428
COM: 14

KOHOUTEK LUBOS DR
HAMBURGER STERNWARTE
GOJENBERGSWEG 112
D-2050 HAMBURG 80
GERMANY, F.R.
TEL: 40-7252-4112
TLX: 217884 HAMS D
COM: 15,20,34

KOJOIAN GABRIEL DR
DEPT OF PHYSICS
UNIVERSITY OF WISCONSIN
EAU CLAIRE WI 54701
U.S.A.
TEL: 715-836-3148
TLX:
COM: 28,40

KOKURIN YURIJ L DR
LEBEDEV PHYSICAL INST
USSR ACADEMY OF SCIENCES
LENINSKY PROSPEKT 53
117924 MOSCOW
U.S.S.R.
TEL: 135-03-60
TLX: 411479 NEOD SU
COM: 08,19

KOLACZEK BARBARA DR
PLANETARY GEODESY DEPT
POLISH ACAD OF SCIENCES
UL. BARTYCKA 18
00-716 WARSAW
POLAND
TEL: 41-00-41
TLX: 815670 CBK PL
COM: 04,19V

KOLCHINSKIJ I G DR
MAIN ASTRON OBSERVATORY
UKRAINIAN ACADEMY OF SCI
OF SCIENCES
252127 KIEV
U.S.S.R.
TEL:
TLX:
COM: 24

KOLESNIK IGOR G DR
MAIN ASTRON OBSERVATORY
UKRAINIAN ACADEMY OF SCI
GOLOSEEVO
252127 KIEV
U.S.S.R.
TEL: 663110
TLX: 131406 SKY SU
COM: 33,34

KOLESNIK L N DR
MAIN ASTRON OBSERVATORY
UKRAINIAN ACADEMY OF SCI
252127 KIEV
U.S.S.R.
TEL: 66-08-69
TLX: 131406
COM: 33

KOLESOV A K DR
LENINGRAD STATE UNIV
ASTRONOMICAL OBSERVATORY
199178 LENINGRAD
U.S.S.R.
TEL:
TLX:
COM: 36

KOLEV DINITAR ZDRAVKOV
BLOCK D Z-20
SMOLJAN
BULGARIA
TEL:
TLX:
COM:

KOLLBERG ERIK L PROF
ELECTRON PHYSICS 1
CHALMERS UNIV TECHNOLOGY
S-412 96 GOETEBORG
SWEDEN
TEL: 31-810100
TLX: 2400 ONSPACE
COM:

KOMAROV N S DR
ODESSA STATE UNIVERSITY
ASTRONOMICAL OBSERVATORY
SHEVCHENKO PARK
270014 ODESSA
U.S.S.R.
TEL: 220396
TLX:
COM: 29

KOMESAROFF MAX M
CSIRO
DIVISION OF RADIOPHYSICS
P.O. BOX 76
EPPING NSW 2121
AUSTRALIA
TEL: 868-0222
TLX: 26230
COM: 34,40

KONDO MASAAKI DR
COLLEGE GENERAL EDUCATION
UNIVERSITY OF TOKYO
KOMABA 3-8-1 MEGURO-KU
TOKYO 153
JAPAN
TEL:
TLX:
COM: 48

KONDO MASAYUKI DR
TOKYO ASTRONOMICAL OBS
UNIVERSITY OF TOKYO
MITAKA
TOKYO 181
JAPAN
TEL: 0422-32-5111
TLX: 2822307 TAOMTK J
COM:

KONDO YOJI DR
NASA/GSFC
CODE 684
GREENBELT MD 20771
U.S.A.
TEL: 301-344-6247
TLX:
COM: 34,42C,44P

KONIN V V DR
NIKOLAEV BRANCH OF THE
MAIN ASTRONOMICAL OBS
327000 NIKOLAEV
U.S.S.R.
TEL:
TLX:
COM: 08

KONONOVICH EDWARD V DR
STERNBERG ASTRONOMICAL
INSTITUTE
119899 MOSCOW
U.S.S.R.
TEL:
TLX:
COM: 12,46C

KONOPLEVA VARVARA P DR
MAIN ASTRON OBSERVATORY
UKRAINIAN ACADEMY OF SCI
252127 KIEV
U.S.S.R.
TEL: 66 3110
TLX: 131406 SKY SU
COM: 15

KONTIZAS EVANGELOS DR
NATL OBSERVATORY ATHENS
ASTRONOMICAL INSTITUTE
THISSION PO BOX 20048
GR-11810 ATHENS
GREECE
TEL: 01-3461191
TLX: 215530 OBSA GR
COM: 36,37

KONTIZAS MARY DR
DEPT OF ASTRONOMY
UNIVERSITY OF ATHENS
PANEPISTIMIOPOLIS
GR-15771 ZOGRAFOS
GREECE
TEL: 01-7235122
TLX:
COM: 37

KOORNNEEF JAN DR
SPACE TELESCOPE SCI INST
HOMEWOOD CAMPUS
3700 SAN MARTIN DRIVE
BALTIMORE MD 21218
U.S.A.
TEL: 301-338-4802
TLX: 6849101
COM: 34

KOPAL ZDENEK PROF
DEPT OF ASTRONOMY
UNIVERSITY OF MANCHESTER
MANCHESTER M13 9PL
U.K.
TEL: 61-273-7121
TLX: 668932 MCHRUL G
COM: 16,42

KOPECKY MILOSLAV DR
ASTRONOMICAL INSTITUTE
CZECH. ACAD. OF SCIENCES
FRICOVA 1
251 65 ONDREJOV
CZECHOSLOVAKIA
TEL: 724525
TLX: 121579
COM: 10,12

KOPP ROGER A DR
LOS ALAMOS NATIONAL LAB
MS E531
LOS ALAMOS NM 87545
U.S.A.
TEL: 505-667-4398
TLX: 660495 LOS ALAMOS
COM:

KOPYLOV I M DR
SPECIAL ASTROPHYSICAL OBS
USSR ACADEMY OF SCIENCES
STAVROPOL TERRITORY
357140 N ARKHYZ
U.S.S.R.
TEL: 93-159
TLX:
COM: 09,27,28,29

KORCHAK A A DR
INSTITUTE OF TERRESTRIAL
MAGNETISM & IONOSPHERE
142092 TROITSK
U.S.S.R.
TEL:
TLX:
COM:

KORMENDY JOHN DR
DOMINION ASTROPHYS OBS
5071 W SAANICH ROAD
VICTORIA BC V8X 4M6
CANADA
TEL: 604-388-3944
TLX: 0497295 NRC DAO VIC
COM: 28,33,47

KOROVYAKOVSKIJ YURIJ P DR
SPECIAL ASTROPHYS OBS
USSR ACADEMY OF SCIENCES
NIZHNIJ ARKHYZ
357140 STAVROPOLSKIJ KRAJ
U.S.S.R.
TEL:
TLX:
COM: 09,28

KOSIN GENNADIJ S DR
PULKOVO OBSERVATORY
196140 LENINGRAD
U.S.S.R.
TEL:
TLX:
COM: 08

KOSTIK ROMAN I
MAIN ASTRONOMICAL OBS
UKRAINIAN ACADEMY OF SCI
252127 KIEV
U.S.S.R.
TEL: 66-4762
TLX: 131406 SKY SU
COM: 10,12

KOSTINA LIDIJA D DR
PULKOVO OBSERVATORY
196140 LENINGRAD
U.S.S.R.
TEL:
TLX:
COM: 19

KOSTYAKOVA ELENA B DR
STERNBERG STATE
ASTRONOMICAL INSTITUTE
117234 MOSCOW
U.S.S.R.
TEL:
TLX:
COM: 34

KOSTYLEV K V DR
ENGERHARDT ASTRONOMICAL
OBSERVATORY
422526 KAZAN
U.S.S.R.
TEL:
TLX:
COM: 22

KOSUGI TAKEO
NOBEYAMA RADIO OBS OF
TOKYO ASTRON OBSERVATORY
NOBEYAMA, MINAMISAKU
NAGANO 384-13
JAPAN
TEL: 267-98-2034
TLX: 3329005 TAONRO-J
COM:

KOTELNIKOV V A ACAD
INST OF RADIO & ELECTRON
USSR ACADEMY OF SCIENCES
103907 MOSCOW
U.S.S.R.
TEL: 203-60-78
TLX:
COM: 40

KOTHARI D S DR
DEPT OF PHYSICS
UNIVERSITY OF DELHI
NEW DELHI 110 007
INDIA
TEL:
TLX:
COM: 35

KOTOV VALERY DR
CRIMEAN ASTROPHYSICAL OBS
USSR ACADEMY OF SCIENCES
NAUCHNY
334413 CRIMEA
U.S.S.R.
TEL:
TLX:
COM: 12C

KOTRC PAVEL
ASTRONOMICAL INSTITUTE
CZECH. ACAD. OF SCIENCES
251 65 ONDREJOV
CZECHOSLOVAKIA
TEL: 72-45-25
TLX: 121579
COM: 10,12

KOTSAKIS DEMETRIUS PROF
HIPPOCRATES STR 189
GR-11472 ATHENS
GREECE
TEL: 01-642-8331
TLX:
COM: 41,51

KOUBSKY PAVEL
ASTRONOMICAL INSTITUTE
CZECH. ACAD. OF SCIENCES
ONDREJOV OBSERVATORY
251 65 ONDREJOV
CZECHOSLOVAKIA
TEL:
TLX:
COM: 29,42

KOURGANOFF VLADIMIR PROF
20 AVE PAUL APPELL
F-75014 PARIS
FRANCE
TEL: 1-45-40-50-53
TLX:
COM: 46

KOUTCHMY SERGE DR
INSTITUT D'ASTROPHYSIQUE
98 BIS BOULEVARD ARAGO
F-75014 PARIS
FRANCE
TEL: 1-43-20-14-25
TLX: 270070 F
COM: 10,12,21

KOVACHEV B J DR
DEPT ASTRONOMY & NTL OBS
BULGARIAN ACAD SCIENCES
72 LENIN BLVD
1784 SOFIA
BULGARIA
TEL: 758827
TLX: 23561 ECF BAN BG
COM: 09,29

KOVACS AGNES DR
HELIOPHYSICAL OBSERVATORY
HUNGARIAN ACADEMY OF SCI
BOX 30
H-4010 DEBRECEN
HUNGARY
TEL: 52-11-015
TLX: 72517 DEOBS H
COM: 10

KOVACS GEZA DR
KONKOLY OBSERVATORY
HUNGARIAN ACADEMY OF SCI
BOX 67
H-1525 BUDAPEST
HUNGARY
TEL: 1-166-426
TLX: 227460
COM:

KOVAL I K DR
MAIN ASTRONOMICAL OBS
UKRAINIAN ACADEMY OF SCI
GOLOSEEVO
252127 KIEV
U.S.S.R.
TEL: 660869
TLX:
COM:

KOVALEVSKY JEAN DR
CERGA
AVENUE COPERNIC
F-06130 GRASSE
FRANCE
TEL: 93-36-58-49
TLX: 470865
COM: 07C,08,24,31C

KOVAR N S DR
PHYSICS DEPT
UNIVERSITY OF HOUSTON
HOUSTON TX 77004
U.S.A.
TEL:
TLX:
COM:

KOVAR ROBERT P DR
9666 E. ORCHARD DR
ENGLEWOOD CO 80111
U.S.A.
TEL: 303-394-4494
TLX:
COM:

KOVETZ ATTAY PROF
DEPT PHYSICS & ASTRONOMY
TEL-AVIV UNIVERSITY
RAMAT-AVIV 69978
ISRAEL
TEL: 3-420-234
TLX: 342-171 VERSY IL
COM: 35,47

KOWAL CHARLES THOMAS
DEPT OF ASTROPHYSICS
CALTECH
PASADENA CA 91125
U.S.A.
TEL: 818-356-6586
TLX: 675425
COM: 15,16,20

KOYAMA KATSUJI
INST SPACE & ASTRON SCI
4-6-1 KOMABA
MEGURO-KU
TOKYO 153
JAPAN
TEL: 3-467-1111
TLX: 34757 ISASTRO J
COM:

KOYAMA SHIN PROF DR
KAGAWA UNIVERSITY
SAIWAI CHO
TAKAMATSU 760
JAPAN
TEL: 878-61-4141
TLX:
COM: 12

KOZAI YOSHIHIDE PROF
TOKYO ASTRONOMICAL OBS
OSAWA
MITAKA
TOKYO 181
JAPAN
TEL: 0422-32-5111
TLX: 2822307 TAOMTK
COM: 06,07,20P,38C,50C

KOZIEL KAROL PROF DR
ASTRON OBSERVATORY KRAKOW
UL. 22 LIPCA 16
43-460 WISLA
POLAND
TEL: 32-42
TLX:
COM:

KOZLOVSKY B Z DR
DEPT PHYSICS & ASTRONOMY
TEL-AVIV UNIVERSITY
RAMAT-AVIV 69978
ISRAEL
TEL:
TLX:
COM: 47

KOZLOWSKI MACIEJ DR
ASTRONOMICAL OBSERVATORY
WARSAW UNIVERSITY
AL. UJAZDOWSKIE 4
00-478 WARSZAWA
POLAND
TEL:
TLX:
COM: 35,48

KRAEMER GERHARD DR
ASTRONOMISCHES INSTITUT
DER UNIVERSITAET
PHILOSOPHENWEG 37
D-7400 TUEBINGEN
GERMANY, F.R.
TEL:
TLX:
COM: 12,44

KRAFT ROBERT P PROF
LICK OBSERVATORY
UNIVERSITY OF CALIFORNIA
SANTA CRUZ CA 95064
U.S.A.
TEL: 408-429-2991
TLX: 910-5984408
COM: 27,29,30,42

KRAICHEVA ZDRAVSKA DR
DEPT OF ASTRONOMY
BULGARIAN ACAD SCIENCES
7TH NOVEMBER STR 1
1000 SOFIA
BULGARIA
TEL:
TLX:
COM: 42

KRAMER KH N DR
ODESSA STATE UNIVERSITY
ASTRONOMICAL OBSERVATORY
270014 ODESSA
U.S.S.R.
TEL: 22-03-96
TLX:
COM: 22

KRANJC ALDO DR
OSSERVATORIO ASTRONOMICO
DI BRERA
VIA BRERA 28
I-20121 MILANO
ITALY
TEL:
TLX:
COM:

KRASINSKY GEORGE A DR
INST THEORETICAL ASTRON
USSR ACADEMY OF SCIENCES
10 KUTUZOV QUAY
192187 LENINGRAD
U.S.S.R.
TEL: 278-8834
TLX: 121578
COM: 04,07

KRASSOVSKY V I DR
INST PHYSICS OF ATMOSPH
USSR ACADEMY OF SCIENCES
PYSHEVSKY PER. 3
109017 MOSCOW
U.S.S.R.
TEL: 231-88-62
TLX:
COM:

KRAUS JOHN D PROF
RADIO OBSERVATORY
OHIO STATE UNIVERSITY
2015 NEIL AVE
COLUMBUS OH 43210
U.S.A.
TEL: 614-548-7895
TLX:
COM: 40,51

KRAUSE F DR
ZNTRLINST. F. ASTROPHYSIK
ASTROPHYSIKALISCHES OBS
TELEGRAFENBERG
DDR-1500 POTSDAM
GERMANY, D.R.
TEL:
TLX:
COM: 10

KRAUSHAAR WILLIAM L PROF
DEPT OF PHYSICS
UNIVERSITY OF WISCONSIN
1150 UNIVERSITY AVE
MADISON WI 53706
U.S.A.
TEL: 608-262-5916
TLX: 265452
COM: 44

KRAUTTER JOACHIM DR
LANDESSTERNWARTE
KOENIGSTUHL
D-6900 HEIDELBERG
GERMANY, F.R.
TEL: 06221-10036
TLX: 461789 MPIA D
COM: 27,34,42

KREIDL TOBIAS J N
LOWELL OBSERVATORY
1400 W. MARS HILL ROAD
FLAGSTAFF AZ 86001
U.S.A.
TEL: 602-774-3358
TLX:
COM: 09

KREINER JERZY MAREK DR
UL.SENATORSKA 27 M7
30-106 KRAKOW
POLAND
TEL: 12-21-48-37
TLX:
COM: 27,42,46

KREISEL E PROF
EINSTEIN LABORATORIUM
ROSA-LUXEMBURG-STR 17A
DDR-1502 POTSDAM
GERMANY, D.R.
TEL: 762-225
TLX: 15471
COM: 48

KRELOWSKI JACEK DR
INSTITUTE OF ASTRONOMY
N COPERNICUS UNIVERSITY
UL. CHOPINA 12/18
87-100 TORUN
POLAND
TEL: 856-206-55
TLX: 055-2234 ASTR PL
COM:

KREMPEC-KRYGIER JANINA DR
N.COPERNICUS ASTRON CTR
ASTROPHYSICAL LABORATORY
UL. CHOPINA 12/18
87-100 TORUN
POLAND
TEL: 260-18
TLX: 0552234 ASTR PL
COM: 29

KRESAK LUBOR DR
ASTRONOMICAL INSTITUTE
SLOVAK ACAD. OF SCIENCES
DUBRAVSKA CESTA 5
842 28 BRATISLAVA
CZECHOSLOVAKIA
TEL: 427-375157
TLX: 93373 SEIS
COM: 15P,20C,22

KRESAKOVA MARGITA DR
ASTRONOMICAL INSTITUTE
SLOVAK ACAD. OF SCIENCES
DUBRAVSKA CESTA 5
842 28 BRATISLAVA
CZECHOSLOVAKIA
TEL: 427-375157
TLX: 93373 SEIS
COM: 22

KREYSA ERNST
MPI FUER RADIOASTRONOMIE
AUF DEM HUEGEL 69
D-5300 BONN 1
GERMANY, F.R.
TEL: 0228-525269
TLX: 886440 MPIFR D
COM: 34,40

KRIEGER ALLEN S DR
AMERICAN SCIENCE AND
ENGINEERING
FORT WASHINGTON
CAMBRIDGE MA 02139
U.S.A.
TEL: 617-868-1600
TLX:
COM:

KRISHNA GOPAL
RADIOASTRONOMY CENTRE
TIFR
P O BOX 1234
BANGALORE 560 012
INDIA
TEL: 362816
TLX:
COM: 28,40

KRISHNA SWAMY K S DR
ASTROPHYSICS GROUP
TATA INSTITUTE
COLABA
BOMBAY 400 005
INDIA
TEL: 219111
TLX: 113009 TIFR IN
COM: 15,34,36

KRISHNAMOHAN S DR
RADIOASTRONOMY CENTRE
TIFR
P.O.BOX 1234
BANGALORE 560 012
INDIA
TEL: 364062
TLX: 8458488 TIFR IN
COM: 40

KRISHNAN THIRUVENKATA MR
HELIOS ANTENNAS/ELECTRON.
234 AVVAI SHANMUGHAM RD
GOPALAPURAM
MADRAS 600 086
INDIA
TEL: 044-472680
TLX:
COM: 40

KRISTENSEN LEIF KAHL DR
INSTITUTE OF PHYSICS
UNIVERSITY OF AARHUS
NY MUNKEGADE
DK-8000 AARHUS C
DENMARK
TEL:
TLX:
COM: 20C

KRISTENSON HENRIK DR
SLAANBAERSVAEGEN 9
S-381 00 KALMAR
SWEDEN
TEL:
TLX:
COM:

KRISTIAN JEROME DR
MT WILSON & LAS CAMPANAS
OBSERVATORIES
813 SANTA BARBARA STREET
PASADENA CA 91101
U.S.A.
TEL: 818-577-1122
TLX:
COM:

KRISTIANSSON KRISTER PROF
DEPT OF PHYSICS
SOELVEGATAN 14
S-223 62 LUND
SWEDEN
TEL: 046-107726
TLX:
COM: 48

KRIVSKY LADISLAV DR
ASTRONOMICAL INSTITUTE
CZECH. ACAD. OF SCIENCES
251 65 ONDREJOV
CZECHOSLOVAKIA
TEL: 72-45-25 PRAGUE
TLX: 121579 ASTR C
COM: 10

KRIZ SVATOPLUK DR
ASTRONOMICAL INSTITUTE
251 65 ONDREJOV
CZECHOSLOVAKIA
TEL: 204-999201
TLX:
COM: 42

KROGDAHL W S DR
DEPT PHYSICS & ASTRONOMY
UNIVERSITY OF KENTUCKY
LEXINGTON KY 40506
U.S.A.
TEL: 606-272-2659
TLX:
COM:

KROLIK JULIAN H
JOHNS HOPKINS UNIVERSITY
DEPT OF PHYSICS & ASTR
BALTIMORE MD 21218
U.S.A.
TEL: 301-338-7926
TLX:
COM:

KRON GERALD E DR
PINECREST OBSERVATORY
AT QUEEN'S COURT
2929 PONI MOI ROAD
HONOLULU HI 96815
U.S.A.
TEL: 808-922-1514
TLX:
COM: 25,37,45

KRON KATHERINE GORDON
PINECREST OBSERVATORY
AT QUEEN'S COURT
2929 PONI MOI ROAD
HONOLULU HI 96815
U.S.A.
TEL: 808-922-1514
TLX:
COM: 42

KRON RICHARD G
YERKES OBSERVATORY
PO BOX 258
WILLIAMS BAY WI 53191
U.S.A.
TEL: 312-236-5468
TLX:
COM: 28

KRONBERG PHILIPP DR
UNIVERSITY OF TORONTO
DEPT OF ASTRONOMY
60 ST GEORGE STREET
TORONTO ONT M5S 1A7
CANADA
TEL: 416-978-4971
TLX: 06-986766 TOR
COM: 40

KROOK M DR
HARVARD COLLEGE OBS
60 GARDEN STREET
CAMBRIDGE MA 02138
U.S.A.
TEL:
TLX:
COM: 35

KRUEGEL ENDRIK DR
MPI FUER RADIOASTRONOMIE
AUF DEM HUEGEL 69
D-5300 BONN
GERMANY, F.R.
TEL:
TLX: 886440 MPIFR D
COM: 40

KRUEGER ALBRECHT DR
ZNTRLINST. F. ASTROPHYSIK
TELEGRAFENBERG
DDR-1500 POTSDAM
GERMANY, D.R.
TEL: 4551
TLX: 15239 ZIAP DD
COM: 10

KRUEGER E PROF
VIA MAURO MACCHI 65
I-20100 MILANO
ITALY
TEL:
TLX:
COM:

KRUMM NATHAN ALLYN
PHYSICS DEPT
UNIVERSITY OF CINCINNATI
CINCINNATI OH 45221
U.S.A.
TEL: 513-475-2232
TLX:
COM: 28

KRUSZEWSKI ANDRZEJ PROF
ASTRONOMICAL OBSERVATORY
AL. UJAZDOWSKIE 4
00-478 WARSZAWA
POLAND
TEL:
TLX:
COM: 42

KRZEMINSKI WOJCIECH DR
CARNEGIE INST WASHINGTON
LAS CAMPANAS OBSERVATORY
CASILLA 601
LA SERENA
CHILE
TEL: 213032
TLX:
COM: 27,42

KSANFOMALITI L V DR
SPACE RESEARCH INSTITUTE
USSR ACADEMY OF SCIENCES
GSP7 PROFSOYUZNAYA 84/32
117810 MOSCOW
U.S.S.R.
TEL: 333-2322/3122
TLX: 411498 STAR SU
COM: 16,51

KUBIAK MARCIN A DR
WARSAW UNIVERSITY
OBSERVATORY
AL. UJAZDOWSKIE 4
00-478 WARSAW
POLAND
TEL: 295346/294011
TLX: 813978 ZAPAN PL
COM: 27

KUBICELA ALEKSANDAR DR
ASTRONOMSKA OPSERVATORIJA
VOLGINA 7
11050 BEOGRAD
YUGOSLAVIA
TEL: 011-419-357
TLX:
COM: 12,50

KUBO YOSHIO
GEODESY & GEOPHYSICS DIV
HYDROGRAPHIC DEPT
TSUKIJI-5 CHUO-KU
TOKYO 104
JAPAN
TEL: 03-541-3811
TLX: 02522222 JAHYD J
COM: 04

KUBOTA JUN DR
KWASAN OBSERVATORY
YAMASHINA
KYOTO 607
JAPAN
TEL: 75-581-1235
TLX: 5422693 LIBKYU J
COM: 10

KUDRITZKI ROLF-PETER PH D
INST F ASTRON & ASTROPHYS
SCHEINERSTR 1
D-8000 MUNCHEN 80
GERMANY, F.R.
TEL: 089-989021
TLX: 529815 UNIVM D
COM: 36C

KULIKOV K A PROF DR
STERNBERG STATE
ASTRONOMICAL INSTITUTE
117234 MOSCOW
U.S.S.R.
TEL:
TLX:
COM:

KUMAR SHAILENDRA
LUNAR & PLANETARY LAB
UNIVERSITY OF ARIZONA
TUCSON AZ 85721
U.S.A.
TEL:
TLX:
COM:

KUNITZSCH PAUL PROF.
DAVIDSTRASSE 17
D-8000 MUENCHEN 81
GERMANY, F.R.
TEL:
TLX: 916280
COM: 41

KUEHNE CHRISTOPH F
CARL ZEISS COMPANY
POSTFACH 35/36
D-7082 OBERKOCHEN
GERMANY, F.R.
TEL: 07364-202807
TLX: 71375155
COM: 09

KULIKOVSKIJ P G DR
STERNBERG STATE
ASTRONOMICAL INSTITUTE
119899 MOSCOW
U.S.S.R.
TEL:
TLX:
COM: 26,41

KUMAR SHIV S PROF
DEPT OF ASTRONOMY
UNIVERSITY OF VIRGINIA
PO BOX 3818
CHARLOTTESVILLE VA 22903
U.S.A.
TEL: 804-924-4896
TLX:
COM: 16,35,36

KUNKEL WILLIAM E DR
LAS CAMPANAS OBSERVATORY
COLA EL PINO S/N
CASILLA 601
LA SERENA
CHILE
TEL: 51-213032
TLX: 645227 AURA CT
COM: 25,27

KUEHR HELMUT
MPI FUER ASTRONOMIE
KOENIGSTUHL
D-6900 HEIDELBERG
GERMANY, F.R.
TEL: 6221-5281
TLX: 461789 MPIA D
COM: 28

KULKARNI PRABHAKAR V PROF
PHYSICAL RESEARCH LAB
AHMEDABAD 380 009
INDIA
TEL:
TLX: 8458488 TIFR IN
COM: 21,40

KUMSIASHVILY MZIA I DR
ABASTUMANI ASTROPHYSICAL
OBSERVATORY
383762 ABASTUMANI,GEORGIA
U.S.S.R.
TEL: 2-52
TLX: 327409
COM: 42

KUNTH DANIEL
INSTITUT D'ASTROPHYSIQUE
98 BIS BOULEVARD ARAGO
F-75014 PARIS
FRANCE
TEL: 1-43-20-14-25
TLX: 270070 INAG F
COM:

KUENZEL HORST
DIESELSTRASSE 13
DDR-1502 POTSDAM-BABELSB.
GERMANY, D.R.
TEL: 77318
TLX:
COM: 10

KULKARNI V K DR
RADIO ASTRONOMY CENTRE
TATA INSTITUTE OF
FUNDAMENTAL RESEARCH
OOTACAMUND 643 001
INDIA
TEL: 0423-2651
TLX: 853241 RAC IN
COM: 40

KUMSISHVILI J I DR
ABASTUMANI ASTROPHYSICAL
OBSERVATORY
383762 ABASTUMANI,GEORGIA
U.S.S.R.
TEL: 2-79
TLX: 327409
COM: 26,27

KUPERUS MAX PROF DR
ASTRONOMICAL INSTITUTE
ZONNENBURG 2
NL-3512 NL UTRECHT
NETHERLANDS
TEL: 030-312841
TLX: 47224 ASTRO NL
COM: 10,12P

KUHI LEONARD V PROF
ASTRONOMY DEPT
UNIVERSITY OF CALIFORNIA
BERKELEY CA 94720
U.S.A.
TEL: 415-642-3792
TLX:
COM: 27,36

KULSRUD RUSSELL M DR
ASTROPHYSICAL SCIENCES
PRINCETON UNIVERSITY
PRINCETON NJ 08540
U.S.A.
TEL: 609-683-2613
TLX:
COM: 33,48

KUN MARIA DR
KONKOLY OBSERVATORY
HUNGARIAN ACADEMY OF SCI
BOX 67
H-1525 BUDAPEST
HUNGARY
TEL: 166-426/506/590
TLX: 227460 KONOB H
COM:

KUPPERIAN JAMES E DR
GODDARD SPACE FLIGHT CTR
CODE 410
GREENBELT MD 20771
U.S.A.
TEL:
TLX:
COM:

KUIJPERS H. JAN M.E. DR
STERREKUNDIG INSTITUUT
ZONNENBURG 2
NL-3512 NL UTRECHT
NETHERLANDS
TEL: 030-312841
TLX: 47224 ASTRO NL
COM: 40

KULTIMA JOHANNES
GEOPHYSICAL OBSERVATORY
SF-99600 SODANKYLAE
FINLAND
TEL:
TLX:
COM:

KUNCHEV PETER DR
DEPT OF ASTRONOMY
FACULTY OF PHYSICS
ANTON IVANOV STR 5
1126 SOFIA
BULGARIA
TEL:
TLX:
COM:

KURFESS JAMES D
NAVAL RESEARCH LABORATORY
CODE 4150
4555 OVERLOOK AVE S.W.
WASHINGTON DC 20375
U.S.A.
TEL: 202-767-3182
TLX:
COM:

KUIPER THOMAS B H DR
JET PROPULSION LABORATORY
169-5065
PASADENA CA 91109
U.S.A.
TEL: 818-354-5479
TLX: 675429
COM: 34,40,51

KUMAJGORODSKAYA RAISA DR
SPECIAL ASTROPHYSICAL
OBSERVATORY
NIZHNIJ ARKHYZ
357147 STAVROPOLSKIJ KRAJ
U.S.S.R.
TEL: 93-515
TLX:
COM: 29

KUNDT WOLFGANG PROF DR
INSTITUT F ASTROPHYSIK
AUF DEM HUEGEL 71
D-5300 BONN 1
GERMANY, F.R.
TEL: 02226-7400
TLX: 0886440
COM: 40,48

KURIL-CHIK V N DR
STERNBERG STATE
ASTRONOMICAL INSTITUTE
119899 MOSCOW B-234
U.S.S.R.
TEL: 139-10-30
TLX:
COM: 40

KUKLIN G V DR
SIBIZMIR
P B 4
664697 IRKUTSK 33
U.S.S.R.
TEL: 6-02-65
TLX:
COM: 10C,12

KUMAR C KRISHNA DR
DEPT PHYSICS & ASTRONOMY
HOWARD UNIVERSITY
WASHINGTON DC 20059
U.S.A.
TEL: 202-636-6245
TLX:
COM: 28,34

KUNDU MUKUL R DR
ASTRONOMY PROGRAM
UNIVERSITY OF MARYLAND
COLLEGE PARK MD 20742
U.S.A.
TEL: 301-454-3005
TLX: 710-826-0352
COM: 10,12,28,34,40

KUROCHKA L N DR
KIEV STATE UNIVERSITY
ASTRONOMICAL OBSERVATORY
OBSERVATORNAYA STR.3
252053 KIEV
U.S.S.R.
TEL: 26-26-91
TLX:
COM: 10,12

KUROKAWA HIROKI DR
HIDA OBSERVATORY
UNIVERSITY OF KYOTO
KAMITAKARA, YOSHIKI-GUN
GIFU 506-13
JAPAN
TEL: 0578-6-2628
TLX:
COM: 10

KUTNER MARC LESLIE DR
PHYSICS DEPT
RENSSELAER POLYTECHN INST
TROY NY 12180
U.S.A.
TEL: 518-266-6417
TLX:
COM: 34,40

KWITTER KAREN BETH DR
THOMPSON PHYSICS LAB
DEPT PHYSICS & ASTRONOMY
WILLIAMS COLLEGE
WILLIAMSTOWN MA 01267
U.S.A.
TEL: 413-597-2272
TLX:
COM: 34

LABHARDT LUKAS
ASTRONOMISCHES INSTITUT
UNIVERSITAET BASEL
VENUSSTRASSE 7
CH-4102 BINNINGEN
SWITZERLAND
TEL: 0041-61-22-7711
TLX:
COM: 25,45

KURPINSKA-WINIARSKA M DR
ASTRONOMICAL OBSERVATORY
JAGELLONIAN UNIVERSITY
UL. ORLA 171
30-244 KRAKOW
POLAND
TEL:
TLX: 0322297 UJ PL
COM: 42

KUTTER G SIEGFRIED DR
EVERGREEN STATE COLLEGE
OLYMPIA WA 98505
U.S.A.
TEL: 206-866-6000
TLX:
COM:

KWOK SUN DR
DEPT OF PHYSICS
UNIVERSITY OF CALGARY
CALGARY AB T2N 1N4
CANADA
TEL: 403-284-5414
TLX: 038-21545
COM: 34,40

LABRUM NORMAN R MR
CSIRO
DIVISION OF RADIOPHYSICS
P.O.BOX 76
EPPING NSW 2121
AUSTRALIA
TEL:
TLX:
COM:

KURT V G DR
SPACE RESEARCH INSTITUTE
USSR ACADEMY OF SCIENCES
PROFSOYUZNAYA STR 84/32
117810 MOSCOW
U.S.S.R.
TEL: 333-31-22
TLX: 411498 STAR SU
COM: 16,44,48

KUTUZOV S A DR
LENINGRAD STATE UNIV
DEPT OF APPLIED MATHS &
CONTROL PROCESSES
199164 LENINGRAD
U.S.S.R.
TEL:
TLX:
COM: 33

KYLAFIS NIKOLAOS D DR
DEPT OF PHYSICS
UNIVERSITY OF CRETE
PO BOX 470
GR-71110 IRAKLION, CRETE
GREECE
TEL:
TLX:
COM: 34

LABS DIETRICH PROF
LANDESSTERNWARTE
KOENIGSTUHL
D-6900 HEIDELBERG 1
GERMANY, F.R.
TEL: 6221-10036
TLX: 461789 MPIA D
COM: 12,29

KURTZ DONALD WAYNE DR
ASTRONOMY DEPT
UNIVERSITY OF CAPE TOWN
RONDEBOSCH
RONDEBOSCH 7700
SOUTH AFRICA
TEL: 69-8531
TLX: 521439
COM: 27

KUZMANOSKI MIKE
INSTITUTE OF ASTRONOMY
UNIVERSITY OF BELGRADE
STUDENTSKI TRG 16
YU-11000 BELGRADE
YUGOSLAVIA
TEL: 011-638-715
TLX:
COM:

LA BONTE BARRY JAMES
INSTITUTE FOR ASTRONOMY
2680 WOODLAWN DRIVE
HONOLULU HI 96822
U.S.A.
TEL: 808-948-6531
TLX: 723-8459 UHAST HR
COM: 12

LACHIEZE-REY MARC
CEN SACLAY
DPHG/SAP
F-91191 GIF/YVETTE CEDEX
FRANCE
TEL: 1-69=08-62-92
TLX: 690860
COM: 47

KURUCZ ROBERT L DR
SMITHSONIAN ASTROPHYSICAL
OBSERVATORY
60 GARDEN STREET
CAMBRIDGE MA 02138
U.S.A.
TEL: 617-495-7429
TLX: 921428
COM: 36

KUZMIN ARKADII D PROF DR
LEBEDEV PHYSICAL INST
USSR ACADEMY OF SCIENCES
117924 MOSCOW
U.S.S.R.
TEL:
TLX: 411479 NEOD SU
COM: 16,40,51

LA PADULA CESARE
IST. ASTROFISICA SPAZIALE
C P 67
I-00044 FRASCATI
ITALY
TEL:
TLX:
COM:

LACLARE F MR
CERGA
AVENUE COPERNIC
F-06130 GRASSE
FRANCE
TEL:
TLX:
COM: 08

KUS ANDRZEJ JAN DR
RADIOASTRONOMY OBS
COPERNICUS UNIVERSITY
UL. CHOPINA 12/18
87-100 TORUN
POLAND
TEL: 04856-20651
TLX: 0552324 TRAO PL
COM: 40

KUZMIN GRIGORI G PROF
TARTU ASTROPHYSICAL
OBSERVATORY
202444 TORAVERE, ESTONIA
U.S.S.R.
TEL:
TLX:
COM: 33

LABAY JAVIER
DEPTO FIS TIERRA & COSMOS
UNIVERSIDAD DE BARCELONA
DIAGONAL 645
08028 BARCELONA
SPAIN
TEL: 330-7311
TLX:
COM: 35

LACROUTE PIERRE A PROF
2 RUE D'ALISE
F-21000 DIJON
FRANCE
TEL: 80-66-11-54
TLX:
COM: 08,24

KUSHWAHA R S PROF
DEPT OF MATHEMATICS
UNIVERSITY OF JODHPUR
JODHPUR RAJ
INDIA
TEL:
TLX:
COM: 35,36

KVIZ ZDENEK DR
SCHOOL OF PHYSICS
UNIVERSITY OF SOUTH WALES
P.O.BOX 1
KENSINGTON NSW 2033
AUSTRALIA
TEL: 697-45-78
TLX: 26054 AA
COM: 22,25,42

LABEYRIE ANTOINE DR
CERGA
F-06460 ST VALLIER DE T.
FRANCE
TEL: 93-42-62-70
TLX: 461402
COM: 09

LACY CLAUD H DR
DEPT OF PHYSICS
UNIVERSITY OF ARKANSAS
104 PHYSICS BUILDING
FAYETTEVILLE AR 72701
U.S.A.
TEL: 501-575-2506
TLX:
COM: 42

KUSTAANHEIMO PAUL E PROF
DANMARKS TEKN HOJSKOLE
DIA E 451
DK-2800 LYNGBY
DENMARK
TEL: 45-2-883022
TLX:
COM: 07,28,47

KWEE K K DR
STERREWACHT
POSTBUS 9513
NL-2300 RA LEIDEN
NETHERLANDS
TEL: 3171148333
TLX: 39058 ASTRO NL
COM: 27,42

LABEYRIE JACQUES DR
CENTRE DES FAIBLES
RADIOACTIVITES
LAB MIXTE CNRS-CEA
F-91190 GIF-SUR-YVETTE
FRANCE
TEL: 1-69-07-78-28
TLX: 691137 F
COM:

LADA CHARLES JOSEPH DR
STEWARD OBSERVATORY
UNIVERSITY OF ARIZONA
TUCSON AZ 85721
U.S.A.
TEL: 602-621-4878
TLX:
COM: 34,37,40

LAFFINEUR MARIUS MR
21 BLVD BRUNE
F-75014 PARIS
FRANCE
TEL:
TLX:
COM: 40

LAING ROBERT
ROYAL GREENWICH OBS
HERSTMONCEUX CASTLE
HAILSHAM BN27 1RP
U.K.
TEL: 0323-833171
TLX: 87451 RGOBS G
COM: 40

LAMBERT DAVID L PROF
DEPT OF ASTRONOMY
UNIVERSITY OF TEXAS
R L MOORE HALL
AUSTIN TX 78712
U.S.A.
TEL: 512-471-4461
TLX: 9108741351
COM: 29C,36

LANDECKER THOMAS L DR
DOMINION RADIO ASTRO-
PHYSICAL OBSERVATORY
P O BOX 248
PENTICTON BC V2A 6K3
CANADA
TEL: 604-497-5321
TLX: 048-88127
COM:

LAFON JEAN-PIERRE J DR
OBSERVATOIRE DE PARIS
SECTION DE MEUDON
F-92195 MEUDON PL CEDEX
FRANCE
TEL: 1-45-34-75-30
TLX: 204464 F
COM: 28,33,49

LAKE KAYLL WILLIAM DR
DEPT OF PHYSICS
QUEEN'S UNIVERSITY
KINGSTON ONT K7L 3N6
CANADA
TEL: 613-547-3020
TLX:
COM: 47

LAMERS H J G L M DR
LAB FOR SPACE RESEARCH
BENELUXLAAN 21
NL-3527 HS UTRECHT
NETHERLANDS
TEL: 31-30-937145
TLX: 47224 ASTRO NL
COM: 44

LANDI DEGL'INNOCENTI E PR
ISTITUTO DI ASTRONOMIA
UNIVERSITA DI FIRENZE
LARGO E. FERMI 5
I-50125 FIRENZE
ITALY
TEL: 22-00-34
TLX: 572268 ARCETR I
COM: 12

LAGERKVIST CLAES-INGVAR
ASTRONOMICAL OBSERVATORY
BOX 515
S-751 20 UPPSALA
SWEDEN
TEL: 018-113522
TLX: 76024 UNIV UPS S
COM: 20

LAL DEVENDRA
GRD, A-020
SCRIPPS INST OF OCEANGR
LA JOLLA CA 92093
U.S.A.
TEL:
TLX:
COM:

LAMLA ERICH E DR
ASTRONOMISCHE INSTITUTE
STERNWARTE UNIV BONN
AUF DEM HUEGEL 71
D-5300 BONN
GERMANY, F.R.
TEL: 0228-73 36 59
TLX: 088 64 40 MPIFR D
COM: 29

LANDI DEGL'INNOCENTI M
OSSERVATORIO ASTROFISICO
DI ARCETRI
LARGO E. FERMI 5
I-50125 FIRENZE
ITALY
TEL: 39-55-22-00-34
TLX: 572268 ARCETR
COM: 12

LAGERQVIST ALBIN PROF
INSTITUTE OF PHYSICS
VANADISVAEGEN 9
S-113 46 STOCKHOLM
SWEDEN
TEL: 468-16-45-00
TLX: 15433 FYSTO S
COM: 14

LALA PETR DR
ASTRONOMICAL INSTITUTE
CZECH. ACAD. OF SCIENCES
251 65 ONDREJOV
CZECHOSLOVAKIA
TEL: 724525
TLX: 121579
COM: 07

LAMPTON MICHAEL
SPACE SCIENCES LABORATORY
UNIVERSITY OF CALIFORNIA
BERKELEY CA 94729
U.S.A.
TEL: 415-642-3576
TLX: 9103667945
COM: 48

LANDI-DESSY J DR
OBSERVATORIO NACIONAL
5000 CORDOBA
ARGENTINA
TEL:
TLX:
COM:

LAGO MARIA TERESA V T MS
GRUPO DE MATEM. APLICADA
UNIVERSIDADE DO PORTO
RUA DAS TAIPAS 135
4000 PORTO
PORTUGAL
TEL: 380313
TLX: 28109
COM: 27,46

LAMB DONALD QUINCY JR DR
UNIVERSITY OF CHICAGO
5801 ELLIS AVE
CHICAGO IL 60637
U.S.A.
TEL: 312-962-8203
TLX:
COM: 35,42,48

LAMY PHILIPPE DR
MPI FUER KERNPHYSIK
POSTFACH 103980
D-6900 HEIDELBERG 1
GERMANY, F.R.
TEL: 6221-51-612
TLX: 461666
COM: 15,21C,22

LANDINI MASSIMO PROF
OSSERVATORIO ASTROFISICO
DI ARCETRI
LARGO E. FERMI 5
I-50125 FIRENZE
ITALY
TEL: 22-00-34
TLX:
COM:

LAHIRI N C
INDIAN ASTRON EPHEM UNIT
INDIAN METEOROLOGIC. DEPT
P-546 BLOCK N
NEW ALIPORE CALCUTTA
INDIA
TEL:
TLX:
COM: 04

LAMB FREDERICK K PROF
PHYSICS DEPT
UNIVERSITY OF ILLINOIS
1110 W GREEN STREET
URBANA IL 61801
U.S.A.
TEL: 217-333-6363
TLX: 6502272050 MCI
COM: 48

LANCASTER BROWN PETER
10A ST PETER'S ROAD
ALDEBURGH, SUFFOLK
U.K.
TEL:
TLX:
COM: 15

LANDMAN DONALD ALAN DR
INSTITUTE FOR ASTRONOMY
UNIVERSITY OF HAWAII
2680 WOODLAWN DRIVE
HONOLULU HI 96822
U.S.A.
TEL:
TLX:
COM: 10,12,14

LAHULLA J FORNIES DR
OBSERVATORIO ASTRONOMICO
ALFONSO XII-3
28014 MADRID
SPAIN
TEL: 2270107
TLX: 22465 IGC E
COM:

LAMB SUSAN ANN DR
DEPARTMENT OF PHYSICS
UNIVERSITY OF MISSOURI
8001 NATURAL BRG RD
ST LOUIS MO 63121
U.S.A.
TEL:
TLX:
COM: 35,48

LANDE KENNETH PROF
PHYSICS DEPT
UNIV OF PENNSYLVANIA
PHILADELPHIA PA 19104
U.S.A.
TEL: 215-898-8177
TLX:
COM:

LANDOLFI MARCO
OSSERVATORIO ASTROFISICO
DI ARCETRI
LARGO E. FERMI 5
I-50125 FIRENZE
ITALY
TEL: 39-55-22-00-34
TLX: 572268 ARCETR
COM: 12

LAI SEBASTIANA
ISTITUTO DI ASTRONOMIA
VIA OSPEDALE 72
I-09100 CAGLIARI
ITALY
TEL:
TLX:
COM:

LAMBECK KURT PROF
AUSTRALIA NAT UNIVERSITY
RESEARCH SCHOOL,EARTH SCI
GPO BOX 4
CANBERRA 2600
AUSTRALIA
TEL: 49-2487
TLX: 62693
COM: 19

LANDECKER PETER BRUCE DR
HUGHES AIRCRAFT CO
SPACE & COMM GR/BLDG S41
MS B322, PO BOX 92919
LOS ANGELES CA 90009
U.S.A.
TEL: 213-648-0815
TLX: 664480
COM: 10

LANDOLT ARLO U PROF
DEPT PHYSICS & ASTRONOMY
LOUISIANA STATE UNIV
BATON ROUGE LA 70803
U.S.A.
TEL: 504-388-8276
TLX: 559184
COM: 25,27,42

LANDSTREET JOHN D PROF
DEPT OF ASTRONOMY
UNIV OF WESTERN ONTARIO
LONDON ONT N6A 3K7
CANADA
TEL: 519-679-3186
TLX:
COM: 25

LANE ADAIR P
BOSTON UNIVERSITY
ASTRONOMY DEPT
725 COMMONWEALTH AVE.
BOSTON MA 02215
U.S.A.
TEL: 617-353-2633
TLX:
COM:

LANE ARTHUR LONNE DR
JET PROPULSION LABORATORY
4800 OAK GROVE DRIVE
PASADENA CA 91109
U.S.A.
TEL: 818-345-2725
TLX:
COM: 16

LANEY CLIFTON
S A A O
P O BOX 9
OBSERVATORY 7935
SOUTH AFRICA
TEL:
TLX:
COM: 27

LANG JAMES DR
RUTHERFORD APPLETON LAB
CHILTON, DIDCOT OX11 0QX
U.K.
TEL: 0235-21900
TLX: 83159
COM: 14

LANG KENNETH R ASST PROF
DEPT OF PHYSICS
TUFTS UNIVERSITY
ROBINSON HALL
MEDFORD MA 02155
U.S.A.
TEL: 617-381-3390
TLX:
COM: 10,40,41

LANGER GEORGE EDWARD DR
DEPT OF PHYSICS
COLORADO COLLEGE
COLORADO SPRINGS CO 80903
U.S.A.
TEL: 303-4732233x578
TLX:
COM: 29

LANGER WILLIAM DAVID DR
PLASMA PHYSICS LAB
PRINCETON UNIVERSITY
PO BOX 451
PRINCETON NJ 08544
U.S.A.
TEL: 609-683-2262
TLX:
COM: 34,40

LANTOS PIERRE DR
OBSERVATOIRE DE PARIS
SECTION DE MEUDON
DASOP
F-92195 MEUDON PL CEDEX
FRANCE
TEL: 1-45-34-75-30
TLX:
COM: 05C,40

LAPASSET EMILIO DR
OBSERVATORIO ASTRONOMICO
LAPRIDA 854
5000 CORDOBA
ARGENTINA
TEL: 051-36876
TLX: 51822 BUCOR
COM: 37, 42

LAPOINTE S M DR
UNIVERSITE DU QUEBEC
2875 BOUL LAURIER
STE-FOY PQ G1V 2M3
CANADA
TEL: 418-657-3551
TLX: 051-31-623
COM:

LAPUSHKA K K DR
LATVIAN STATE UNIVERSITY
ASTRONOMICAL OBSERVATORY
226098 RIGA
U.S.S.R.
TEL: 223149/611984
TLX:
COM: 24

LAQUES PIERRE DR
OBS DU PIC DU MIDI
R DU PONT DE LA MOULETTE
F-65200 BAGNERES-DE-B.
FRANCE
TEL: 62-95-19-69
TLX: 531625
COM: 09,51

LARGE MICHAEL I DR
SCHOOL OF PHYSICS
UNIVERSITY OF SYDNEY
SYDNEY NSW 2006
AUSTRALIA
TEL: 2-692-2222
TLX: 26169 UNISYD AA
COM: 40

LARI C DR
LABORATORIO NAZIONALE DI
RADIOASTRONOMIA
VIA IRNERIO 46
I-40126 BOLOGNA
ITALY
TEL:
TLX:
COM:

LARINK JOHANNES PROF DR
HAMBURGER STERNWARTE
BERGEDORF
GOJENBERGSWEG 78
D-2050 HAMBURG 80
GERMANY, F.R.
TEL:
TLX:
COM:

LARSON HAROLD P DR
LUNAR AND PLANETARY LAB
UNIVERSITY OF ARIZONA
TUCSON AZ 85721
U.S.A.
TEL: 602-621-6943
TLX:
COM: 15,16

LARSON RICHARD B PROF
ASTRONOMY DEPARTMENT
YALE UNIVERSITY
BOX 6666
NEW HAVEN CT 06511
U.S.A.
TEL: 203-436-8318
TLX:
COM: 28,33,35

LARSON STEPHEN M
DEPT PLANETARY SCIENCES
UNIVERSITY OF ARIZONA
TUCSON AZ 85721
U.S.A.
TEL: 602-621-4973
TLX: 910-952-1143
COM: 15,16

LARSSON-LEANDER G PROF
LUND OBSERVATORY
BOX 43
S-221 00 LUND
SWEDEN
TEL: 46-10-70-00
TLX: 331990BSNOT S
COM: 29,37,42

LASENBY ANTHONY
MULLARD RADIO ASTRON OBS
CAVENDISH LABORATORY
MADINGLEY ROAD
CAMBRIDGE CBJ 0HE
U.K.
TEL: 223-66477
TLX: 81292 CAVLAB G
COM: 40

LASHER GORDON JEWETT DR
IBM TJ WATSON RES CTR
YORKTOWN HEIGHTS NY 10598
U.S.A.
TEL: 914-945-1901
TLX: 137456
COM: 48

LASKARIDES PAUL G ASSPROF
DEPT OF ASTRONOMY
UNIVERSITY OF ATHENS
PANEPISTIMIOPOLIS
GR-15771 ZOGRAFOS
GREECE
TEL: 01-7243211
TLX:
COM: 25,27,35

LASKER BARRY M DR
SPACE TELESCOPE SC INST
HOMEWOOD CAMPUS
3700 SAN MARTIN DRIVE
BALTIMORE MD 21218
U.S.A.
TEL: 301-338-4840
TLX: 6849191 STSI
COM: 09,25,28,34

LASOTA JEAN-PIERRE DR
OBSERVATOIRE DE PARIS
SECTION DE MEUDON
F-92195 MEUDON PL CEDEX
FRANCE
TEL: 1-45-34-75-30
TLX: 201571
COM: 35,47

LATHAM DAVID W DR
CENTER FOR ASTROPHYSICS
60 GARDEN STREET
CAMBRIDGE MA 02138
U.S.A.
TEL: 617-495-7215
TLX: 921428 SATELLITE CAM
COM: 30V

LATOUR JEAN J
OBSERVATOIRE PIC-DU-MIDI
ET TOULOUSE
14 AVENUE EDOUARD BELIN
F-31400 TOULOUSE
FRANCE
TEL: 61-25-21-01
TLX:
COM: 35

LATTIMER JAMES M DR
ASTROPHYSICS PROGRAM
DEPT EARTH & SPACE SCS
SUNY AT STONY BROOK
STONY BROOK NY 11794
U.S.A.
TEL: 516-246-8223
TLX:
COM: 48

LATYPOV A A DR
ASTRONOMICAL INSTITUTE
UZBEKIAN ACADEMY OF SCI
700052 TASHKENT
U.S.S.R.
TEL: 358102
TLX: 116012 VREMJA
COM: 24

LAUBERTS ANDRIS DR
ASTRONOMICAL OBSERVATORY
BOX 515
S-751 20 UPPSALA
SWEDEN
TEL: 018-115208
TLX:
COM: 28

LAUNAY JEAN-MICHEL DR
OBSERVATOIRE DE PARIS
SECTION DE MEUDON
F-92195 MEUDON PL CEDEX
FRANCE
TEL: 1-45-34-75-70
TLX:
COM: 14

LAURENT BERTEL E PROF
INST FOR THEORETICAL PHYS
VANADISVAEGEN 9
S-113 46 STOCKHOLM
SWEDEN
TEL: 468-16-45-00
TLX: 15433 FYSTO S
COM:

LAURENT CLAUDINE DR
INSU
77 AVE DENFERT-ROCHEREAU

FRANCE
TEL: 1-43-20-13-30
TLX: 270070
COM: 34

LAUSBERG ANDRE DR
INSTITUT D'ASTROPHYSIQUE
UNIVERSITE DE LIEGE
AVENUE DE COINTE 5
B-4200 COINTE-OUGREE
BELGIUM
TEL: 41-52-99-80
TLX:
COM: 28-47

LAUSTSEN SVEND DR
ASTRONOMICAL INSTITUTE
UNIVERSITY OF AARHUS
DK-8000 AARHUS C
DENMARK
TEL: 06-128899
TLX: 64767 AAUSCI DK
COM: 08

LAUTMAN D A DR
SMITHSONIAN ASTROPHYS CTR
60 GARDEN STREET
CAMBRIDGE MA 02138
U.S.A.
TEL:
TLX:
COM:

LAVAL ANNIE DR
OBSERVATOIRE DE MARSEILLE
2 PLACE LE VERRIER
F-13248 MARSEILLE CDX 04
FRANCE
TEL: 91-95-90-88
TLX: 420241 F
COM: 37

LAVROV M I PROF
ENGELHARDT ASTR OBS
UNIVERSITY
420008 KAZAN
U.S.S.R.
TEL:
TLX:
COM: 42

LAVRUKHINA A K PROF DR
INSTITUTE OF GEOCHEMISTRY
USSR ACADEMY OF SCIENCES
117334 MOSCOW
U.S.S.R.
TEL: 137-75-38
TLX:
COM:

LAWRENCE G M DR
LASP
UNIVERSITY OF COLORADO
CAMPUS BOX 392
BOULDER CO 80309
U.S.A.
TEL:
TLX:
COM: 14

LAWRIE DAVID G
THE AEROSPACE CORPORATION
PO BOX 92957
MS M4/041
LOS ANGELES CA 90009
U.S.A.
TEL: 213-648-6142
TLX:
COM:

LAYZER DAVID PROF
HARVARD COLLEGE OBS
MAIL STOP 31
60 GARDEN STREET
CAMBRIDGE MA 02138
U.S.A.
TEL:
TLX:
COM: 14,28,47

LAZOVIC JOVAN P PROF
DEPT OF ASTRONOMY
FACULTY OF SCIENCES
STUDENTSKI TRG 16
11000 BEOGRAD
YUGOSLAVIA
TEL: 11-638-715
TLX:
COM: 07

LE CONTEL JEAN-MICHEL
OBSERVATOIRE DE NICE
B P 139
F-06003 NICE CEDEX
FRANCE
TEL: 93-89-04-20
TLX: 460004
COM:

LE DOURNEUF MARYVONNE
OBSERVATOIRE DE PARIS
SECTION DE MEUDON
F-92190 MEUDON PL CEDEX
FRANCE
TEL: 1-45-34-75-70
TLX: 201571 F
COM: 14

LE POOLE RUDOLF S DR
STERREWACHT
POSTBUS 9513
NL-2300 RA LEIDEN
NETHERLANDS
TEL: 071-148333
TLX: 39058 ASTRO NL
COM: 24

LE SERGEANT D'HENDECOURT
GR. PHYSIQUE DES SOLIDES
T23, UNIVERSITE PARIS VII
4 PLACE JUSSIEU
F-75251 PARIS CEDEX 05
FRANCE
TEL:
TLX:
COM: 34

LE SQUEREN ANNE-MARIE DR
OBSERVATOIRE DE PARIS
SECTION DE MEUDON
F-92195 MEUDON PL CEDEX
FRANCE
TEL: 1-45-34-75-30
TLX:
COM: 34,40

LEA SUSAN MAUREEN DR
PHYSICS & ASTRONOMY DEPT
SAN FRANSISCO STATE UNIV
1600 HOLLOWAY AVE
SAN FRANSISCO CA 94132
U.S.A.
TEL: 405-469-1880
TLX:
COM: 48

LEACOCK ROBERT JAY
DEPT OF ASTRONOMY
UNIVERSITY OF FLORIDA
211 SSRB
GAINESVILLE FL 32611
U.S.A.
TEL: 904-392-2052
TLX:
COM: 28

LEBEDINETS V N DR
ASTRONOMICAL COUNCIL
USSR ACADEMY OF SCIENCES
PYATNITSKAYA UL 48
109017 MOSCOW
U.S.S.R.
TEL:
TLX:
COM: 22

LEBLANC YOLANDE DR
OBSERVATOIRE DE PARIS
SECTION DE MEUDON
F-92195 MEUDON PL CEDEX
FRANCE
TEL: 1-45-34-75-30
TLX:
COM: 40

LEBOFSKY LARRY ALLEN
LUNAR & PLANETARY LAB
UNIVERSITY OF ARIZONA
TUCSON AZ 85721
U.S.A.
TEL: 602-621-6947
TLX:
COM: 15

LEBOVITZ NORMAN R PROF
MATHEMATICS DEPT
UNIVERSITY OF CHICAGO
5734 S. UNIVERSITY AVE
CHICAGO IL 60637
U.S.A.
TEL: 312-753-8074
TLX:
COM: 35

LECAR MYRON DR
CENTER FOR ASTROPHYSICS
60 GARDEN STREET
CAMBRIDGE MA 02138
U.S.A.
TEL: 617-495-7251
TLX: 921428 SATELLITE CAM
COM: 33

LECKRONE DAVID S DR
GODDARD SPACE FLIGHT CTR
CODE 681
GREENBELT MD 20771
U.S.A.
TEL: 301-344-8904
TLX:
COM: 29,44

LEDERLE TRUDPERT DR
ASTRONOMISCHES-RECHEN
INSTITUT
MOENCHHOFSTR 12-14
D-6900 HEIDELBERG 1
GERMANY, F.R.
TEL: 6221-49026
TLX: 461336 ARIHD D
COM: 04C,05,08,19,31

LEDOUX P J PROF
RUE DE LA FAILLE 55
B-4000 LIEGE
BELGIUM
TEL: 41-52-12-45
TLX:
COM: 27,35

LEE PAUL D DR
LOUISIANA STATE UNIV
BATON ROUGE LA 70803
U.S.A.
TEL:
TLX:
COM:

LEE SEE-WOO DR
DEPT OF ASTRONOMY
SEOUL NATIONAL UNIVERSITY
SEOUL CITY
KOREA, REPUBLIC
TEL: 877-2131-9x3308
TLX:
COM:

LEE TERENCE J DR
HEAD OF TECHNOLOGY
ROYAL OBSERVATORY
BLACKFORD HILL
EDINBURGH EH9 3HJ
U.K.
TEL: 031-667-3321
TLX: 72383 RGEDING UK
COM: 34

LEER EGIL PROF
AURORAL OBSERVATORY
UNIVERSITY OF TROMSO
P O BOX 953
N-9001 TROMSO
NORWAY
TEL: 83-86060
TLX: 64124 AUROB N
COM:

LEFEBVRE MICHEL DR
CNES/GRGS
18 AVENUE EDOUARD BELIN
F-31055 TOULOUSE CEDEX
FRANCE
TEL:
TLX:
COM: 19

LEFEVRE JEAN DR
OBSERVATOIRE DE NICE
BP 139
F-06003 NICE CEDEX
FRANCE
TEL: 93-89-04-20
TLX:
COM:

LIST OF MEMBERS

LEGG THOMAS H DR
HERZBERG INST ASTROPHYS
NATL RESEARCH COUNCIL
OTTAWA ONT K1A 0R6
CANADA
TEL: 613-593-6060
TLX: 053-3715
COM: 40

LEHNERT B P PROF
DEPT PLASMA PHYS. AND
FUSION RESEARCH
ROYAL INST. TECHNOLOGY
S-100 44 STOCKHOLM 70
SWEDEN
TEL: 7877763
TLX: 10389 KTHB S
COM:

LEIBACHER JOHN DR
NATL SOLAR OBSERVATORY
PO BOX 26732
TUCSON AZ 85726-6732
U.S.A.
TEL: 602-325-9302
TLX: 0666-484
COM: 10,12,36

LEIBOWITZ ELIA M DR
DEPT PHYSICS & ASTRONOMY
TEL-AVIV UNIVERSITY
TEL-AVIV 69978
ISRAEL
TEL: 03-413788
TLX: 343171 VERSY IL
COM: 50

LEIGHTON R B PROF
CALTECH
1201 E. CALIFORNIA BLVD
PASADENA CA 91125
U.S.A.
TEL: 818-356-4286
TLX:
COM: 12

LEIKIN G A DR
ASTRONOMICAL COUNCIL
USSR ACADEMY OF SCIENCES
PYATNITSKAYA UL 48
109017 MOSCOW
U.S.S.R.
TEL: 231-54-61
TLX: 412623 SCSTP SU
COM: 16

LEINERT CHRISTOPH DR
MPI FUER ASTRONOMIE
KOENIGSTUHL
D-6900 HEIDELBERG 1
GERMANY, F.R.
TEL: 06221-528-264
TLX: 461789 MPIAD
COM: 21

LEITE SCHEID PAULO DR
OBSERVATORIO NACIONAL
RUA GENERAL BRUCE 586
20000 RIO DE JANEIRO
BRAZIL
TEL:
TLX:
COM: 27

LELIEVRE GERARD DR
CANADA-FRANCE-HAWAII
TELESCOPE CORP
PO BOX 1597
KAMUELA HI 96743
U.S.A.
TEL: 808-885-7944
TLX: 633147 CFHT
COM: 09,28

LEMAIRE PHILIPPE DR
LPSP CNRS
B P 10
F-91371 VERRIERES-LE-B
FRANCE
TEL: 1-69-20-10-60
TLX: 600252
COM: 44

LEMAITRE ANNE DR
FAC UNIV N.D. DE LA PAIX
DEPT DE MATHEMATIQUES
REMPART DE LA VIERGE 8
B-5000 NAMUR
BELGIUM
TEL: 081-22-90-61
TLX: 59222 FACNAM B
COM: 07

LEMAITRE GERARD R DR
OBSERVATOIRE DE MARSEILLE
2 PLACE LE VERRIER
F-13004 MARSEILLE
FRANCE
TEL: 91-95-90-88
TLX:
COM: 09

LEMKE DIETRICH DR
MPI FUER ASTRONOMIE
KOENIGSTUHL
D-6900 HEIDELBERG 1
GERMANY, F.R.
TEL: 49-6221-528259
TLX: 461789 IMPIA-D
COM:

LENA PIERRE J PROF
OBSERVATOIRE DE PARIS
SECTION DE MEUDON
F-92195 MEUDON PL CEDEX
FRANCE
TEL: 1-45-34-75-70
TLX: 201571F
COM:

LENZEN RAINER DR
MPI FUER ASTRONOMIE
KOENIGSTUHL
D-6900 HEIDELBERG 1
GERMANY, F.R.
TEL:
TLX:
COM: 25

LEORAT JACQUES DR
OBSERVATOIRE DE MEUDON
SECTION DE MEUDON
F-91370 MEUDON PL CEDEX
FRANCE
TEL: 1-45-34-75-70
TLX: 201-571 LAM F
COM:

LEPINE JACQUES R D DR
DEPTO DE ASTRONOMIA
IAG/USP
AV MIGUEL STEFANO 4200
04301 SAO PAULO SP
BRAZIL
TEL: 275-37-20
TLX: 1136221
COM: 35,40

LEQUEUX JAMES DR
OBSERVATOIRE DE MARSEILLE
2 PLACE LE VERRIER
F-13248 MARSEILLE CEDEX 4
FRANCE
TEL: 91-95-90-88
TLX: 420241 F
COM: 05,28C,34P,40,47

LEROY JEAN-LOUIS
OBSERVATOIRE PIC-DU-MIDI
ET TOULOUSE
14 AVENUE EDOUARD BELIN
F-31400 TOULOUSE
FRANCE
TEL: 61-25-21-01
TLX: 530776 F
COM: 10,12

LESTER JOHN B DR
ASTRONOMY DEPT
ERINDALE COLLEGE
UNIVERSITY OF TORONTO
MISSISSAUGA L5L 1C6
CANADA
TEL: 416-828-5356
TLX:
COM: 29

LETFUS VOJTECH DR
ASTRONOMICAL INSTITUTE
CZECH. ACAD. OF SCIENCES
251 65 ONDREJOV
CZECHOSLOVAKIA
TEL:
TLX:
COM:

LEUNG CHUN MING DR
DEPT OF PHYSICS
RENSSELAER POLYTECH INST
TROY NY 12180-3590
U.S.A.
TEL: 518-266-6318
TLX:
COM: 34,40

LEUNG KAM CHING PROF
BEHLEN OBSERVATORY
DEPT PHYSICS & ASTRONOMY
UNIVERSITY OF NEBRASKA
LINCOLN NB 68588
U.S.A.
TEL: 402-472-2770
TLX: 484340 UNL
COM: 27,38C,42C

LEVASSEUR-REGOURD A.C. PR
SERVICE AERONOMIE CNRS
BP 3
F-91370 VERRIERES-LE-B.
FRANCE
TEL: 1-69-20-10-60
TLX: 692400
COM: 15,21V,22

LEVATO ORLANDO HUGO DR
COMPLEJO ASTRONOMICO
EL LEONCITO
CASILLA DE CORREO 467
5400 SAN JUAN
ARGENTINA
TEL: 064-22-5718
TLX: 59134 ENTOP AR
COM: 29C,45

LEVIN BORIS J DR
ASTRONOMICAL COUNCIL
USSR ACADEMY OF SCIENCES
PYATNITSKAYA UL 48
109017 MOSCOW
U.S.S.R.
TEL:
TLX:
COM: 15,16,22

LEVINE RANDOLPH H DR
50 CARVER ROAD
NEWTON MA 02161
U.S.A.
TEL: 617-965-5953
TLX:
COM:

LEVY EUGENE H DR
DEPT PLANETARY SCIENCES
LUNAR & PLANETARY LAB
UNIVERSITY OF ARIZONA
TUCSON AZ 85721
U.S.A.
TEL: 602-621-6962
TLX: 9109521143
COM: 49

LEVY JACQUES R DR
OBSERVATOIRE DE PARIS
61 AVE DE L'OBSERVATOIRE
F-75014 PARIS
FRANCE
TEL: 1-43-20-12-10
TLX:
COM: 41

LEWIN WALTER H G PROF
PHYSICS DEPT
MIT 37-627
CAMBRIDGE MA 02139
U.S.A.
TEL: 617-253-4282
TLX:
COM: 44

LEWIS BRIAN MURRAY DR
ARECIBO OBSERVATORY
NAIC
PO BOX 95
ARECIBO PR 00612
U.S.A.
TEL: 809-878-2612
TLX:
COM:

LEWIS J S
DEPT PLANETARY SCIENCES
UNIVERSITY OF ARIZONA
TUCSON AZ 85721
U.S.A.
TEL: 602-621-4972
TLX:
COM: 16

LI CHUN-SHENG
DEPT OF ASTRONOMY
NANJING UNIVERSITY
NANJING 0909
CHINA, PEOPLE'S REP.
TEL: 34651-2882
TLX: 34151 PRCNU CN
COM: 10,40

LI J Y
PYONGYANG ASTRON OBS
ACADEMY OF SCIENCES DPRK
TAESONG DISTRICT
PYONGYANG
KOREA DPR
TEL:
TLX:
COM:

LI TIPEI
INSTITUTE OF HIGH ENERGY
PHYSICS
BEIJING
CHINA, PEOPLE'S REP.
TEL: 812971-464
TLX: 22082 IHEP CN
COM: 44,48

LI ZONG-WEI
DEPT OF ASTRONOMY
BEIJING NORMAL UNIVERSITY
BEIJING
CHINA, PEOPLE'S REP.
TEL: 65-6531 x 683
TLX:
COM: 35,48

LI DEPEI
NANJING ASTRONOMICAL
INSTRUMENT FACTORY
NANJING
CHINA, PEOPLE'S REP.
TEL: 56191
TLX: 1131
COM: 09

LI JING
BEIJING OBSERVATORY
ACADEMIA SINICA
BEIJING
CHINA, PEOPLE'S REP.
TEL: 22040 BAOAS CN
TLX:
COM: 28,33

LI WEI BAO
YUNNAN OBSERVATORY
P.O. BOX 110
KUNMING
CHINA, PEOPLE'S REP.
TEL:
TLX:
COM: 10

LI ZONG-YUN
DEPT OF ASTRONOMY
NANJING UNIVERSITY
NANJING
CHINA, PEOPLE'S REP.
TEL:
TLX: 34151 PRCNU CN
COM:

LI DONG-MING
PURPLE MOUNTAIN
OBSERVATORY
NANJING
CHINA, PEOPLE'S REP.
TEL:
TLX:
COM: 08

LI NED C DR
CALIFORNIA UNIVERSITY
6531 WITHWORTH RD
LOS ANGELES CA 90035
U.S.A.
TEL:
TLX:
COM:

LI XIAO-QING
PURPLE MOUNTAIN
OBSERVATORY
NANJING
CHINA, PEOPLE'S REP.
TEL: 31096
TLX: 34144 PMO NJ CN
COM: 28,49

LIANG SHI-GUANG
SHANGHAI OBSERVATORY
ACADEMIA SINICA
SHANGHAI
CHINA, PEOPLE'S REP.
TEL: 386191
TLX: 33164 SHAO CN
COM: 40

LI GI MAN
PYONGYANG ASTRON OBS
ACADEMY OF SCIENCES DPRK
TAESONG DISTRICT
PYONGYANG
KOREA DPR
TEL:
TLX:
COM: 04

LI NENG-YAO
PURPLE MOUNTAIN
OBSERVATORY
NANJING
CHINA, PEOPLE'S REP.
TEL: 37609
TLX: 34144 PMONJ CN
COM: 04,08

LI ZHENG-XIN
INSTITUTE OF GEODESY
AND CARTOGRAPHY
BEIJING
CHINA, PEOPLE'S REP.
TEL: 386191
TLX: 33164 SHAO CN
COM: 19

LIANG ZHONG-HUAN
P.O. BOX 18
LINTONG
XIAN
CHINA, PEOPLE'S REP.
TEL:
TLX:
COM: 31

LI GYONG WON
PYONGYANG ASTRON OBS
ACADEMY OF SCIENCES DPRK
TAESONG DISTRICT
PYONGYANG
KOREA DPR
TEL:
TLX:
COM: 40

LI QI-BIN
BEIJING OBSERVATORY
ACADEMIA SINICA
BEIJING
CHINA, PEOPLE'S REP.
TEL: 281968
TLX: 22040 BADAS CN
COM: 28C,48

LI ZHI-FANG
SHANGHAI OBSERVATORY
ACADEMIA SINICA
SHANGHAI
CHINA, PEOPLE'S REP.
TEL: 386191
TLX: 33164 SHAO CN
COM: 08,47C

LIDDELL U MR
NASA LUNAR & PLANET PRG
OFFICE OF SPACE SCIENCES
SPACE SCI & APPLICATIONS
WASHINGTON DC 20546
U.S.A.
TEL:
TLX:
COM:

LI HEN
SHANGHAI OBSERVATORY
SHANGHAI
CHINA, PEOPLE'S REP.
TEL: 386191
TLX: 33164 SHAO CN
COM: 35

LI SIN HYONG
PYONGYANG ASTRON OBS
ACADEMY OF SCIENCES DPRK
TAESONG DISTRICT
PYONGYANG
KOREA DPR
TEL:
TLX:
COM: 25

LI ZHI-SEN
BEIJING ASTRONOMICAL OBS
ACADEMIA SINICA
BEIJING
CHINA, PEOPLE'S REP.
TEL: 28-1698
TLX: 9053
COM: 41

LIEBSCHER DIERCK-E DR
ZNTRLINST. F. ASTROPHYSIK
STERNWARTE BABELSBERG
ROSA-LUXEMBURG-STR 17A
DDR-1502 POTSDAM
GERMANY, D.R.
TEL:
TLX:
COM: 47

LI HONG-WEI
DEPT OF ASTRONOMY
NANJING UNIVERSITY
NANJING
CHINA, PEOPLE'S REP.
TEL: 34651, 34751
TLX: 34151 PRCNU CN
COM: 40

LI SON JAE
PYONGYANG ASTRON OBS
ACADEMY OF SCIENCES DPRK
TAESONG DISTRICT
PYONGYANG
KOREA DPR
TEL:
TLX:
COM: 10

LI ZHIGANG
SHAANXI ASTRONOMICAL OBS
P.O. BOX 18
LINTONG
XIAN
CHINA, PEOPLE'S REP.
TEL: 32255 XIAN
TLX: 70121 CSAO CN
COM: 08

LIESKE JAY H DR
JPL/CALTECH
MS 264-664
4800 OAK GROVE DRIVE
PASADENA CA 91109
U.S.A.
TEL: 818-354-3642
TLX: 675429
COM: 04C,07,19,20,31

LI HYOK HO
PYONGYANG ASTRON OBS
ACADEMY OF SCIENCES DPRK
TAESONG DISTRICT
PYONGYANG
KOREA DPR
TEL:
TLX:
COM: 04

LI TING
NANJING ASTRONOMICAL
INSTRUMENT FACTORY
NANJING
CHINA, PEOPLE'S REP.
TEL:
TLX:
COM: 09

LI ZHONGYUAN
DEPT EARTH & SPACE SCI.
UNIV SCIENCE & TECHNOLOGY
HEFEI, ANHUI
CHINA, PEOPLE'S REP.
TEL: 63300
TLX: 90028 USTC CN
COM: 44

LILLER WILLIAM DR
INSTITUTO ISAAC NEWTON
CASILLA 437
VINA DEL MAR
CHILE
TEL: 03-970864
TLX:
COM: 15,34

LILLEY EDWARD A PROF
HARVARD COLLEGE
OBSERVATORY
60 GARDEN STREET
CAMBRIDGE MA 02138
U.S.A.
TEL: 617-495-3971
TLX: 921428 SATELLITE CAM
COM: 40,51

LILLIE CHARLES F DR
TRW ELECTRONICS & DEFENSE
R5/2261
1 SPACE PARK
REDONDO BEACH, CA 90278
U.S.A.
TEL: 213-536-4080
TLX: 910-325-6611
COM: 15,21

LIN CHIA C PROF
MIT
DEPT OF MATHEMATICS
77 MASSACHUSETTS AVE
CAMBRIDGE MA 02139
U.S.A.
TEL: 617-253-1796
TLX: 921473 MIT CAM
COM: 28,33,34

LIN DOUGLAS N. C. DR
LICK OBSERVATORY
UNIVERSITY OF CALIFORNIA
SANTA CRUZ CA 95064
U.S.A.
TEL: 408-429-2732
TLX:
COM:

LIN YUANZHANG
BEIJING ASTRONOMICAL OBS
ACADEMIA SINICA
BEIJING
CHINA, PEOPLE'S REP.
TEL: 281698
TLX: 22040 BAOAS CN
COM: 10,12

LINCOLN J VIRGINIA MISS
2005 ALPINE DRIVE
BOULDER CO 80302
U.S.A.
TEL: 303-442-6757
TLX:
COM:

LINDBLAD BERTIL A DR
LUND OBSERVATORY
BOX 43
S-221 00 LUND
SWEDEN
TEL: 46-10-70-00
TLX: 33199 OBSNOT S
COM: 20,22,44

LINDBLAD PER OLOF PROF
STOCKHOLM OBSERVATORY
S-133 00 SALTSJOEBADEN
SWEDEN
TEL: 87-170195
TLX: 12972 SOBSERV S
COM: 28,33

LINDEGREN LENNART DR
LUND OBSERVATORY
BOX 43
S-221 00 LUND
SWEDEN
TEL: 46-10-70-00
TLX: 33199 OBSNOT S
COM: 08C

LINDSEY CHARLES ALLAN
UNIVERSITY OF HAWAII
AT MANOA
2680 WOODLAWN DRIVE
HONOLULU HI 96822
U.S.A.
TEL: 808-948-6526
TLX:
COM: 15

LING CHIH-BING DR
INSTITUTE OF MATHEMATICS
ACADEMIA SINICA
P O BOX NO 143
TAIPEI
CHINA, TAIWAN
TEL:
TLX:
COM:

LINGENFELTER RICHARD E
UNIVERSITY OF CALIFORNIA
CASS C-011
LA JOLLA CA 92093
U.S.A.
TEL: 619-452-2464
TLX: 9103371271 SIOCCAN
COM:

LINKE RICHARD ALAN DR
BELL LABORATORIESS
CRAWFORD HILL LAB
HOLMDEL NJ 07733
U.S.A.
TEL:
TLX:
COM: 34,40

LINNELL ALBERT P PROF
DEPT PHYSICS & ASTRONOMY
MICHIGAN STATE UNIVERSITY
EAST LANSING MI 48824
U.S.A.
TEL: 517-353-6670
TLX:
COM: 35,36,42

LINNIK V P PROF DR
MAIN ASTRONOMICAL OBS
PULKOVO
196140 LENINGRAD
U.S.S.R.
TEL:
TLX:
COM:

LINSKY JEFFREY L DR
JILA
UNIVERSITY OF COLORADO
CAMPUS BOX 440
BOULDER CO 80309
U.S.A.
TEL: 303-492-7838
TLX: 755842 JILA
COM: 12,36,44

LINSLEY JOHN
DEPT PHYSICS & ASTRONOMY
UNIVERSITY OF NEW MEXICO
ALBUQUERQUE NM 87131
U.S.A.
TEL: 505-243-1924
TLX: 910989
COM: 44,48

LIPOVETSKY V A
SPECIAL ASTROPHYSICAL OBS
NIZHNIJ ARKHYZ
357147 STAVROPOLSKIJ KRAJ
U.S.S.R.
TEL: 93-2-42
TLX:
COM: 28

LIPPINCOTT SARAH LEE DR
SPROUL OBSERVATORY
SWARTHMORE COLLEGE
507 CEDAR LANE
SWARTHMORE PA 19081
U.S.A.
TEL: 215-543-9058
TLX:
COM: 24,26,51

LISZT HARVEY STEVEN
NRAO
EDGEMONT ROAD
CHARLOTTESVILLE VA 22901
U.S.A.
TEL: 804-296-0344
TLX: 910-997-0714
COM: 34

LITTLE LESLIE T DR
ELECTRONICS LAB
UNIVERSITY OF KENT
CANTERBURY, KENT CT2 7NT
U.K.
TEL: 0227-66822
TLX: 965449 UKCLIB
COM: 40

LITTLETON JOHN E
DEPT OF PHYSICS
WEST VIRGINIA UNIVERSITY
PO BOX 6023
MORGANTOWN WV 26506-6023
U.S.A.
TEL: 304-293-3498
TLX: 710-921-0309
COM: 35

LITVAK MARVIN M DR
TRW INC.
MS 01/1070
ONE SPACE PARK
REDONDO BEACH CA 90278
U.S.A.
TEL: 213-536-4770
TLX:
COM:

LIU BAO-LIN
PURPLE MOUNTAIN
OBSERVATORY
NANJING
CHINA, PEOPLE'S REP.
TEL: 42817 / 46700
TLX: 34144 PMONJ CN
COM: 04

LIU CAIPIN
PURPLE MOUNTAIN
OBSERVATORY
NANJING
CHINA, PEOPLE'S REP.
TEL: 42817 NANJING
TLX: 34144 PMONJ CN
COM: 36

LIU JINYI
INST OF HISTORY OF NAT SC
GONG YUAN WEST STREET 1
BEIJING
CHINA, PEOPLE'S REP.
TEL: 557180 BEIJING
TLX:
COM: 41,48

LIU LIAO
DEPT OF PHYSICS
BEIJING NORMAL UNIVERSITY
BEIJING
CHINA, PEOPLE'S REP.
TEL:
TLX:
COM: 47

LIU LIN
DEPT OF ASTRONOMY
NANJING UNIVERSITY
NANJING
CHINA, PEOPLE'S REP.
TEL: 34651 x 2882
TLX: 34151 PRCNU CN
COM:

LIU LIN-ZHONG
PURPLE MOUNTAIN
OBSERVATORY
NANJING
CHINA, PEOPLE'S REP.
TEL: 46700
TLX: 34144 PMONT CN
COM: 15

LIU RU-LIANG
PURPLE MOUNTAIN
OBSERVATORY
NANJING
CHINA, PEOPLE'S REP.
TEL: 42817 / 46700
TLX: 34144 PMONJ CN
COM: 28,48

LIU SOU-YANG DR
COMPUTER SCIENCES CORP
SYSTEM SCIENCES DIVISION
8728 COLESVILLE ROAD
SILVER SPRING MD 20910
U.S.A.
TEL: 301-589-1545
TLX:
COM:

LIU XUEFU
DEPT OF ASTRONOMY
BEIJING NORMAL UNIVERSITY
BEIJING
CHINA, PEOPLE'S REP.
TEL: 656531-6285
TLX: 8511
COM: 42

LIU YONG-ZHEN
GRADUATE SCHOOL
UNIV SCIENCE & TECHNOLOGY
P.O. BOX 3908
BEIJING
CHINA, PEOPLE'S REP.
TEL: 817031
TLX:
COM: 28,47

LIVINGSTON WILLIAM C
NOAO/NSO
PO BOX 26732
TUCSON AZ 85726
U.S.A.
TEL: 602-327-5511
TLX: 0666484 AURA NOAO TU
COM: 09C,12

LIVIO MARIO
DEPARTMENT OF PHYSICS
TECHNION
HAIFA 32000
ISRAEL
TEL: 04-293549
TLX: 46650 TECLI IL
COM: 35

LIVSHITS M A DR
INSTITUTE OF TERRESTRIAL
MAGNETISM & IONOSPHERE
IZMIRAN
142092 TROITSK MOSCOW REG
U.S.S.R.
TEL:
TLX: 412623 SCSTP SU
COM: 10

LLOYD-EVANS THOMAS DR
S A A O
P O BOX 9
OBSERVATORY 7935
SOUTH AFRICA
TEL: 021-47-0026
TLX: 5720309 SA
COM: 37,45C

LO KWOK-YUNG DR
OWENS VALLEY RADIO OBS
DEPARTMENT OF ASTRONOMY
CALTECH 105-24
PASADENA CA 91125
U.S.A.
TEL: 818-356-4415
TLX: 675425 CALTECH PSD
COM: 28,34,40

LOCANTHI DOROTHY DAVIS DR
2180 PINECREST DRIVE
ALTADENA CA 91001
U.S.A.
TEL: 213-797-0629
TLX:
COM: 29

LOCHMAN JAN
ASTRONOMICAL INSTITUTE
CZECH. ACAD. OF SCIENCES
DVORAKOVA 298
511 01 TURNOV
CZECHOSLOVAKIA
TEL: 0436-22622
TLX:
COM: 09

LOCHTE-HOLTGREVEN W PROF
INSTITUT FUR EXPERIMENTAL
PHYSIK DER UNIVERSITAET
LEIBNIZSTR
D-2300 KIEL
GERMANY, F.R.
TEL: 0431-332420
TLX:
COM: 14

LOCKE JACK L DR
250 BRAESIDE AVENUE
OTTAWA ONT K1H 7J5
CANADA
TEL: 613-523-0812
TLX:
COM: 12,40

LOCKMAN FELIX J
NRAO
EDGEMONT ROAD
CHARLOTTESVILLE VA 22903
U.S.A.
TEL: 804-296-0211
TLX: 910-997-0174
COM: 33,34,40

LOCKWOOD G WESLEY DR
LOWELL OBSERVATORY
1400 W. MARS HILL RD
FLAGSTAFF AZ 86001
U.S.A.
TEL: 607-774-3358
TLX:
COM: 16,25,27

LODEN KERSTIN R DR
STOCKHOLM OBSERVATORY
S-133 00 SALTSJOEBADEN
SWEDEN
TEL: 08-7170195
TLX: 12972 SOBSERV S
COM: 33,45

LODEN LARS OLOF PROF
ASTRONOMICAL OBSERVATORY
BOX 515
S-751 20 UPPSALA
SWEDEN
TEL: 018-11-44-90
TLX: 76024
COM: 33,37,51

LOISEAU NORA DR
INSTITUTO ARGENTINO DE
RADIOASTRONOMIA
CASILLA DE CORREO No. 5
1894 VILLA ELISA (Bs.As.)
ARGENTINA
TEL:
TLX:
COM:

LOMB NICHOLAS RALPH DR
SYDNEY OBSERVATORY
MUSEUM APPLIED ARTS & SC
P.O.BOX K346
HAYMARKET NSW 2000
AUSTRALIA
TEL:
TLX:
COM: 20,46

LONGAIR M S PROF
ASTRONOMER ROYAL FOR
SCOTLAND
ROYAL OBSERVATORY
EDINBURGH EH9 3JH
U.K.
TEL: 031-667-3321
TLX: 72383
COM: 40,47C,48

LONGMORE ANDREW J
UK INFRARED TELESCOPE
UNIT
665 KOMOHANA STREET
HILO HI 96720
U.S.A.
TEL: 808-961-3756
TLX: 633135
COM:

LOPEZ GARCIA ZUELMA L DR
OBSERVATORIO ASTRONOMICO
FELIX AGUILAR
AV BENAVIDEZ 8175 OESTE
5407 MARQUESADO (S.J.)
ARGENTINA
TEL:
TLX:
COM:

LOPEZ JOSE A ING
OBSERVATORIO ASTRONOMICO
FELIX AGUILAR
AV BENAVIDEZ 8175 OESTE
5407 MARQUESADO (S.J.)
ARGENTINA
TEL: 064-231494
TLX: 59100 UNSJA AR
COM: 08

LOPEZ-ARROYO M
OBSERVATORIO ASTRONOMICO
ALFONSO XII-5
28014 MADRID
SPAIN
TEL:
TLX:
COM: 12

LOPEZ-GARCIA FRANCISCO DR
OBSERVATORIO ASTRONOMICO
FELIX AGUILAR
AV. BENAVIDEZ 8175 OESTE
5407 MARQUESADO (S.J.)
ARGENTINA
TEL:
TLX:
COM:

LOPEZ-MORENO JOSE JUAN
INST DE ASTROFISICA DE
ANDALUCIA, APDO 2144
PROFESOR ALBAREDA 1
18080 GRANADA
SPAIN
TEL: 58-12-13-00
TLX: 78573 IAAG E
COM: 16,21

LOPEZ-PUERTAS MANUEL
INST DE ASTROFISICA DE
ANDALUCIA, APDO 2144
PROFESOR ALBAREDA 1
18080 GRANADA
SPAIN
TEL: 58-12-13-00
TLX: 78573 IAAG E
COM: 16,21

LOREN ROBERT BRUCE DR
DEPT OF ASTRONOMY
UNIVERSITY OF TEXAS
AUSTIN TX 78712
U.S.A.
TEL: 512-892-5280
TLX:
COM: 34,40

LORENZ HILMAR
ZNTRLINST. F. ASTROPHYSIK
AKAD. WISSENSCHAFTEN DDR
ROSA-LUXEMBURG-STR 17A
DDR-1502 POTSDAM-BABELSB.
GERMANY, D.R.
TEL:
TLX:
COM: 28,40

LORTET MARIE CLAIRE
OBSERVATOIRE DE PARIS
SECTION DE MEUDON
DAPHE
F-92195 MEUDON PL CEDEX
FRANCE
TEL: 1-45-34-75-70
TLX: 201-571 LAM F
COM: 05,34

LOSCO LUCETTE DR
FACULTE DES SCIENCES
F-25030 BESANCON CEDEX
FRANCE
TEL:
TLX:
COM:

LOTOVA N A DR
IZMIRAN
AKADEMGORODOK
142092 MOSCOW REGION
U.S.S.R.
TEL:
TLX:
COM: 49

LOUGHHEAD RALPH E DR
CSIRO
DIV OF APPLIED PHYSICS
P.O.BOX 218
LINDFIELD NSW 2070
AUSTRALIA
TEL: 02-467-6355
TLX: 26296
COM: 10,12

LOUISE RAYMOND PROF
FACULTE DES SCIENCES
DEPT DE PHYSIQUE
33 RUE ST-LEU
F-80039 AMIENS
FRANCE
TEL:
TLX:
COM: 34

LOULERGUE MICHELLE DR
OBSERVATOIRE DE PARIS
SECTION DE MEUDON
F-92195 MEUDON PL CEDEX
FRANCE
TEL: 1-45-34-75-70
TLX: 270912 OBSASTR
COM: 14

LOVAS FRANCIS JOHN DR
MOLECULAR SPECTROSCOPIC
DIV 545
NATL BUREAU OF STANDARDS
WASHINGTON DC 20234
U.S.A.
TEL: 301-921-2023
TLX: 898993
COM: 14C,34

LOVAS MIKLOS
KONKOLY OBSERVATORY
BOX 67
H-1525 BUDAPEST
HUNGARY
TEL: 366621 BUDAPEST
TLX: 227460 KONOB
COM: 20

LOVELACE RICHARD V E DR
SPACE SCIENCES BLDG
CORNELL UNIVERSITY
ITHACA NY 14853
U.S.A.
TEL: 607-256-3968
TLX:
COM: 28,48

LOVELL SIR BERNARD PROF
NUFFIELD RADIO ASTR LABS
JODRELL BANK
MACCLESFIELD SK11 9PL
U.K.
TEL: 0477-71321
TLX: 36149
COM: 22,40,44,51

LOW BOON CHYE
HIGH ALTITUDE OBSERVATORY
NCAR
PO BOX 3000
BOULDER CO 80307
U.S.A.
TEL: 303-497-1553
TLX: 45694
COM: 10

LOW FRANK J DR
4940 CALLE BARRIL
TUCSON AZ 85718
U.S.A.
TEL: 602-621-2779
TLX:
COM: 28,34,45

LOWE ROBERT P DR
DEPT OF PHYSICS
UNIV OF WESTERN ONTARIO
LONDON ONT N6A 3K7
CANADA
TEL: 519-679-2917
TLX:
COM:

LOYOLA PATRICIO DR
OBS ASTRONOMICO NACIONAL
UNIVERSIDAD DE CHILE
CASILLA 36-D
SANTIAGO
CHILE
TEL:
TLX:
COM: 08

LOZINSKAYA TAT'YANA A DR
STERNBERG STATE
ASTRONOMICAL INSTITUTE
119899 MOSCOW B-234
U.S.S.R.
TEL: 139-10-30
TLX:
COM: 34,40

LOZINSKIJ A M DR
ASTRONOMICAL COUNCIL
USSR ACADEMY OF SCIENCES
PYATNITSKAYA UL 48
109017 MOSCOW
U.S.S.R.
TEL: 231-54-61
TLX: 412623 SCSTP SU
COM: 24,31

LU BEN-KUI
PURPLE MOUNTAIN
OBSERVATORY
NANJING
CHINA, PEOPLE'S REP.
TEL: 32893
TLX: 34144 PMONJ CN
COM: 07

LU CHUN-LIN
PURPLE MOUNTAIN
OBSERVATORY
NANJING
CHINA, PEOPLE'S REP.
TEL: 42700
TLX: 34144 PMONJ CN
COM: 08

LU PHILLIP K DR
DEPT PHYSICS & ASTRONOMY
WESTERN CONN STATE UNIV
181 WHITE ST
DANBURY CT 06810
U.S.A.
TEL: 203-797-4218
TLX:
COM: 24,37

LU TAN
DEPT OF ASTRONOMY
NANJING UNIVERSITY
NANJING
CHINA, PEOPLE'S REP.
TEL: 34651-2882
TLX: 34151 PRCNU CN
COM: 47,48

LU YANG
DEPT OF ASTRONOMY
NANJING UNIVERSITY
NANJING
CHINA, PEOPLE'S REP.
TEL: 34651-2882
TLX: 34151 PRCNU CN
COM: 40

LUB JAN DR
STERREWACHT
HUYGENS LABORATORIUM
POSTBUS 9513
NL-2300 RA LEIDEN
NETHERLANDS
TEL: 071-148333
TLX: 39068 ASTRO NL
COM: 25,27

LUCAS ROBERT DR
GROUPE D'ASTROPHYSIQUE
UNIV SCIENT & MEDICALE
CERMO BP 68
F-38402 ST-MARTIN-D'HERES
FRANCE
TEL: 76-51-46-00
TLX:
COM: 34

LUCCHIN FRANCESCO
ISTITUTO DI FISICA
G. GALILEI
VIA MARZOLO 8
I-35100 PADOVA
ITALY
TEL: 049-844333
TLX: 430308 DF GGPDI
COM:

LUCKE PETER B DR
DEPT PHYSICS & ASTRONOMY
MOUNT UNION COLLEGE
ALLIANCE OH 44601
U.S.A.
TEL: 216-821-5320
TLX:
COM:

LUCY LEON B PROF
ESO
KARL-SCHWARZSCHILD-STR 2
D-8046 GARCHING B MUNCHEN
GERMANY, F.R.
TEL: 89-32006-249
TLX: 0528282-0
COM: 42

LUEST REIMAR PROF
EUROPEAN SPACE AGENCY
8-10 RUE MARIO NIKIS
F-75738 PARIS
FRANCE
TEL: 1-42-73-74-04
TLX: 202746 ESA
COM: 12,44,48,49

LUEST RHEA DR
MPI FUER PHYSIK UND
ASTROPHYSIK
K-SCHWARZSCHILDSTR 1
D-8046 GARCHING B MUNCHEN
GERMANY, F.R.
TEL: 89-320-32990
TLX: 524629 ASTROD
COM: 15

LUGGER PHYLLIS M
INDIANA UNIVERSITY
DEPT OF ASTRONOMY
SWAIN WEST 319
BLOOMINGTON IN 47405
U.S.A.
TEL: 812-335-6929
TLX:
COM: 28

LUKACEVIC ILIJA S DR
FACULTY OF SCIENCES
DEPT OF MECHANICS
STUDENTSKI TRG 16
11000 BEOGRAD
YUGOSLAVIA
TEL:
TLX:
COM:

LUMINET JEAN-PIERRE
OBSERVATOIRE DE PARIS
SECTION DE MEUDON
F-92195 MEUDON PL CEDEX
FRANCE
TEL: 1-45-34-75-70
TLX: 201571 F
COM: 28,47,48

LUMME KARI A DR
OBSERVATORY
TAHTITORNINMAKI
SF-00130 HELSINKI 13
FINLAND
TEL: 1912910
TLX:
COM: 16

LUNA HOMERO G. DR
INSTITUTO ARGENTINO DE
RADIOASTRONOMIA
CASILLA DE CORREO No. 5
1894 VILLA ELISA (Bs.As.)
ARGENTINA
TEL:
TLX:
COM: 25

LUNDQUIST CHARLES A DR
RESEARCH INSTITUTE
THE UNIVERSITY OF ALABAMA
BOX 209
HUNTSVILLE AL 35899
U.S.A.
TEL: 205-895-6100
TLX:
COM: 07

LUNEL MADELEINE DR
OBSERVATOIRE DE LYON
AVENUE CHARLES ANDRE
F-69230 ST-GENIS-LAVAL
FRANCE
TEL: 78-56-07-05
TLX: 310-926
COM: 33

LUNGU NICOLAIE DR
INSTITUTUL POLITEHNIC
CATEDRA DE MATEMATICA
STR EMIL ISAC 15
3400 CLUJ NAPOCA
RUMANIA
TEL: 951-17229
TLX:
COM:

LUO BAO-RONG
YUNNAN OBSERVATORY
KUNMING
CHINA, PEOPLE'S REP.
TEL:
TLX:
COM: 10

LUO DING-JIANG
BEIJING OBSERVATORY
ZHONG-GUAN-CUN
WESTERN SUBURB
BEIJING
CHINA, PEOPLE'S REP.
TEL: 281698
TLX: 22040 BAO ASCN
COM: 08C,19

LUO DINGCHANG
BEIJING ASTRONOMICAL OBS
ACADEMIA SINICA
WESTERN SUBURB
BEIJING
CHINA, PEOPLE'S REP.
TEL: 275580
TLX: 22040
COM: 31

LUO SHI-FANG
SHANGHAI OBSERVATORY
ACADEMIA SINICA
SHANGHAI
CHINA, PEOPLE'S REP.
TEL: 386191
TLX: 33164 SHAO CN
COM: 19,31

LUO XIANHAN
DEPT OF GEOPHYSICS
BEIJING UNIVERSITY
BEIJING
CHINA, PEOPLE'S REP.
TEL: 22239
TLX:
COM: 10,40

LUPISHKO DMITRIJ F
ASTRONOMICAL OBSERVATORY
SUMSKAYA STR 35
310022 KHARKOV
U.S.S.R.
TEL:
TLX:
COM: 15

LUTZ BARRY L DR
LOWELL OBSERVATORY
1400 W. MARS HILL RD
FLAGSTAFF AZ 86001
U.S.A.
TEL: 602-774-3358
TLX: 6502358958 MCI
COM: 14,16,34

LUTZ JULIE H DR
PROGRAM IN ASTRONOMY
WASHINGTON STATE UNIV
PULLMAN WA 99164-2930
U.S.A.
TEL: 509-335-3136
TLX: 5107741091 WSUOIPPMA
COM: 45

LUTZ THOMAS E DR
PROGRAM IN ASTRONOMY
WASHINGTON STATE UNIV
PULLMAN WA 99164-2930
U.S.A.
TEL: 509-335-3141
TLX: 5107741091 WSUOIPPMA
COM: 24C

LUUD LAURI DR
TARTU ASTROPHYSICAL OBS
TARTU 41-258
TARTU KOMETA
202444 TORAVERE, ESTONIA
U.S.S.R.
TEL:
TLX:
COM: 27,29

LUYTEN WILLEM J PROF
SPACE SCIENCE CENTER
UNIVERSITY OF MINNESOTA
MINNEAPOLIS MN 55455
U.S.A.
TEL: 612-373-3366
TLX:
COM: 24,26,33

LYNAS-GRAY ANTHONY E
DEPT PHYSICS & ASTRONOMY
UNIVERSITY COLLEGE LONDON
GOWER STREET
LONDON WC1E 6BT
U.K.
TEL:
TLX:
COM: 29

LYNCH DAVID K
AEROSPACE CORPORATION
SPACE PHYSICS LABS
PO BOX 92957, MS M2-226
LOS ANGELES CA 90009
U.S.A.
TEL: 213-648-6686
TLX: 664460
COM: 09

LYNDEN-BELL DONALD PROF
INSTITUTE OF ASTRONOMY
MADINGLEY ROAD
CAMBRIDGE CB3 0HA
U.K.
TEL: 0223-62204
TLX: 817297 ASTRON G
COM: 28C,33,37,48

LYNDS BEVERLY T DR
KITT PEAK NATL OBS
PO BOX 26732
TUCSON AZ 85726
U.S.A.
TEL: 602-325-9396
TLX: 0666-484 AURA NOAO
COM: 28,34

LYNDS ROGER C DR
KITT PEAK NATL OBS
PO BOX 26732
TUCSON AZ 85726
U.S.A.
TEL: 602-327-5511
TLX:
COM: 28

LYNE ANDREW G DR
NRAL
JODRELL BANK
MACCLESFIELD SK11 9PL
U.K.
TEL: 0477-71321
TLX: 36149
COM: 40

LYNGA GOSTA DR
LUND OBSERVATORY
BOX 43
S-221 00 LUND
SWEDEN
TEL: 46-10-72-98
TLX: 33199 OBSNOT S
COM: 05,33C,37,45

LYTTKENS EJNAR DR
SKOLGATAN 33 B
S-752 21 UPPSALA
SWEDEN
TEL:
TLX:
COM:

LYTTLETON RAYMOND A PROF
INSTITUTE OF ASTRONOMY
CAMBRIDGE
U.K.
TEL: 0223-62204
TLX: 817297 ASTRON G
COM: 15

LYUTY VICTOR M DR
CRIMEAN STATION OF
STERNBERG INSTITUTE
NAUCHNYJ
334413 CRIMEA
U.S.S.R.
TEL:
TLX:
COM: 28,42

MA ER
BEIJING ASTRONOMICAL OBS
BEIJING
CHINA, PEOPLE'S REP.
TEL: 28-16-98
TLX: 22040 BAOAS CN
COM: 28

MA XING-YUAN
DEPARTMENT OF GEOGRAPHY
BEIJING TEACHERS COLLEGE
BALIZHUANG
BEIJING
CHINA, PEOPLE'S REP.
TEL:
TLX:
COM:

MA YU-QIAN
INSTITUTE OF HIGH ENERGY
PHYSICS
PO BOX 918-3
BEIJING
CHINA, PEOPLE'S REP.
TEL: 812-971 x 464
TLX: 22082 IHEP CN
COM: 44,48

MACALPINE GORDON M
UNIVERSITY OF MICHIGAN
DEPT OF ASTRONOMY
ANN ARBOR MI 48109
U.S.A.
TEL: 313-764-3433
TLX: 810-223-6056
COM: 28

MACCACARO TOMMASO DR
HARVARD-SMITHSONIAN CTR
FOR ASTROPHYSICS
60 GARDEN STREET
CAMBRIDGE MA 02138
U.S.A.
TEL: 617-495-7253
TLX: 921428 SATELLITE CAM
COM: 48

MACCAGNI DARIO
IST. DI FISICA COSMICA
CNR
VIA BASSINI 15
I-20133 MILANO
ITALY
TEL: 02-298-237
TLX: 313839 MUACNR I
COM: 48

MACCALLUM MALCOLM A H
SCHOOL OF MATH. SCIENCES
QUEEN MARY COLLEGE
MILE END ROAD
LONDON E1 4NS
U.K.
TEL: 01-980-4811
TLX: 893750 QMCUOL
COM: 47

MACCHETTO FERDINANDO DR
SPACE TELESCOPE SCI INST
HOMEWOOD CAMPUS
3700 SAN MARTIN DRIVE
BALTIMORE MD 21218
U.S.A.
TEL: 301-338-4790
TLX: 6849101
COM: 28,40,44C,48

MACCONNELL DARRELL J DR
DEPT PHYS & ASTRONOMY
PHYSICS-ASTRO BLDG
MICHIGAN STATE UNIVERSITY
EAST LANSING MI 48824
U.S.A.
TEL: 517-353-4541
TLX:
COM: 33,45C

MACDONALD GEOFFREY H DR
ELECTRONICS LABORATORY
UNIVERSITY OF KENT
CANTERBURY, KENT CT2 7NT
U.K.
TEL: 0227-66822 X258
TLX: 965449 UKCLIB
COM: 40

MACDONALD JAMES
DEPARTMENT OF PHYSICS
UNIVERSITY OF DELAWARE
NEWARK DE 19716
U.S.A.
TEL: 302-451-2661
TLX:
COM: 40,42

MACERONI CARLA
OSSERVATORIO ASTRONOMICO
DI ROMA
VIA DEL PARCO MELLINI 84
I-00136 ROMA
ITALY
TEL:
TLX:
COM: 42

MACGILLIVRAY HARVEY T DR
ROYAL OBSERVATORY
BLACKFORD HILL
EDINBURGH EH9 3HJ
U.K.
TEL: 31-667-3321
TLX: 72383 ROEDIN G
COM: 28

MACHADO JOSE M A B DR
TECHNICAL UNIVERSITY
OF LISBON
AV DA IGREJA 17 1 D
1700 LISBOA
PORTUGAL
TEL: 892225
TLX:
COM:

MACHADO LUIZ E. DA SILVA
OBSERVATORIO DO VALONGO
UNIV. FEDERAL RIO DE J.
LAD. PEDRO ANTONIO. 43
20080 RIO DE JANEIRO
BRAZIL
TEL: 021-263-0685
TLX: 2122924 UFRJ BR
COM: 20

MACHALSKI JERZY DR
ASTRONOMICAL OBSERVATORY
JAGIELLONIAN UNIVERSITY
UL. MAZOWIECKA 36/33
30-019 KRAKOW
POLAND
TEL:
TLX: 0322297 UJ PL
COM: 40

MACIEL WALTER J DR
UNIVERSIDADE DE SAO PAULO
INST ASTRON. E GEOFISICO
CAIXA POSTAL 30627
01051 SAO PAULO SP
BRAZIL
TEL:
TLX:
COM: 34

MACK PETER
S A A O
P O BOX 9
OBSERVATORY 7935
SOUTH AFRICA
TEL: 47-0025
TLX:
COM: 09

MACKAY CRAIG D DR
INSTITUTE OF ASTRONOMY
UNIVERSITY OF CAMBRIDGE
MADINGLEY RD
CAMBRIDGE CB3 0HA
U.K.
TEL: 44-223-62204
TLX: 817297 ASTRON G
COM: 28

MACKINNON ALEXANDER L
DEPT OF ASTRONOMY
UNIVERSITY OF GLASGOW
GLASGOW G12 8QW
U.K.
TEL: 41-339-8855
TLX: 777070 UNIGLA
COM: 10

MACLEOD JOHN M DR
HERZBERG INST ASTROPHYS
NATL RESEARCH COUNCIL
OTTAWA ONT K1A 0R6
CANADA
TEL: 613-593-6060
TLX:
COM: 34,40

MACQUEEN ROBERT M DR
NCAR
HIGH ALTITUDE OBSERVATORY
PO BOX 3000
BOULDER CO 80307
U.S.A.
TEL: 303-497-1500
TLX: 45694
COM: 10,49

MACRAE DONALD A PROF
DAVID DUNLAP OBSERVATORY
P O BOX 360
RICHMOND HILL ONT L4C 4Y6
CANADA
TEL: 416-884-9562
TLX:
COM: 33,38,40

MACRIS CONSTANTIN J PROF
RSAAM
ACADEMY OF ATHENS
ANAGNOSTOPOULOU 14
GR-10673 ATHENS
GREECE
TEL: 3613589
TLX:
COM: 10

MACY WILLIAM WRAY DR
LOCKHEED RESEARCH LAB
BLDG 202, ORG 91-30
3251 HANOVER STREET
PALO ALTO CA 94304
U.S.A.
TEL: 415-424-2836
TLX:
COM:

MADDISON RONALD CH DR
UNIVERSITY OF KEELE
2 CHURCH PLANTATION
KEELE PARK
KEELE, STAFFS
U.K.
TEL: 0782-621111
TLX:
COM: 46

MADEJ JERZY
WARSAW UNIVERSITY OBS
AL. UJAZDOWSKIE 4
00-478 WARSAW
POLAND
TEL: 4822-29-40-11
TLX:
COM: 36

MADORE BARRY FRANCIS DR
DAVID DUNLAP OBSERVATORY
UNIVERSITY OF TORONTO
RICHMOND HL ONT L4C 4Y6
CANADA
TEL: 416-884-9562
TLX:
COM: 27,28

MADSEN JES
ASTRONOMISK INSTITUT
UNIVERSITY OF AARHUS
DK-8000 AARHUS C
DENMARK
TEL: 06-12-88-99
TLX: 64767 AAUSCI DK
COM:

MAEDA KOITIRO
DEPT OF PHYSICS
HYOGO COLL OF MEDICINE
NISHINOMIYA
HYOGO 663
JAPAN
TEL: 798-45-6111
TLX:
COM:

MAEDER ANDRE PROF
OBSERVATOIRE DE GENEVE
CH-1290 SAUVERNY
SWITZERLAND
TEL: 22-552611
TLX: 27720 OBSG CH
COM: 27,35V,37

MAEHARA HIDEO DR
TOKYO ASTRONOMICAL
OBSERVATORY
OSAWA MITAKA
TOKYO 181
JAPAN
TEL: 0422-32-5111
TLX: 2822307 TAONTK J
COM: 45

MAETZLER CHRISTIAN DR
INSTITUTE APPLIED PHYSICS
UNIVERSITY OF BERN
SIDLERSTRASSE 5
CH-3012 BERN
SWITZERLAND
TEL: 031-65-89-11
TLX: 32320 PHYBE CH
COM:

MAFFEI PAOLO PROF
UNIVERSITA DI PERUGIA
CATTEDRA DI ASTROFISICA
VIA DELL'ELCE DI SOTTO
I-06100 PERUGIA
ITALY
TEL: 075-45647
TLX:
COM: 27,51

MAGAIN PIERRE DR
INSTITUT D'ASTROPHYSIQUE
UNIVERSITE DE LIEGE
AVENUE DE COINTE 5
B-4200 COINTE-OUGREE
BELGIUM
TEL:
TLX:
COM:

MAGALASHVILI N L DR
ABASTUMANI ASTROPHYSICAL
OBSERVATORY
383762 ABASTUMANI,GEORGIA
U.S.S.R.
TEL:
TLX:
COM: 26,42

MAGALHAES ANTONIO A S ENG
OBSERVATORIO ASTRONOMICO
UNIVERSIDADE DO PORTO
MONTE DA VIRGEM
4400 VILA NOVA GAIA
PORTUGAL
TEL: 782-0404
TLX:
COM:

MAGALHAES ANTONIO MARIO
INST ASTRON. E GEOFISICO
UNIVERSIDADE DE SAO PAULO
CAIXA POSTAL 30627
01051 SAO PAULO
BRAZIL
TEL: 55-11-275-3720
TLX: 1136221 IAGM BR
COM:

MAGNAN CHRISTIAN DR
LAT
98 BIS BOULEVARD ARAGO
F-75014 PARIS
FRANCE
TEL: 1-43-20-14-25
TLX:
COM:

MAGNARADZE N G DR
STATE UNIVERSITY
380043 TBILISI
U.S.S.R.
TEL:
TLX:
COM: 07

MAGNI GIANFRANCO
IST. ASTROFISICA SPAZIALE
VIALE DELL'UNIVERSITA 11
I-00185 ROMA
ITALY
TEL:
TLX:
COM:

MAGUN ANDREAS DR
INST APPLIED PHYSICS
SIDLERSTRASSE 5
CH-3012 BERN
SWITZERLAND
TEL: 031-658923
TLX:
COM:

MAHDY HAMED A DR
HELWAN OBSERVATORY
HELWAN-CAIRO
EGYPT
TEL:
TLX:
COM:

MAHESWARAN MURUGESAPILLAI
INST OF FUNDAMENT STUDIES
380/72 BAUDDHALOKA
MAWATHA
COLOMBO 7
SRI LANKA
TEL: 01-597538
TLX: 21700 IFS CE
COM: 35

MAHMOUD FAROUK M A B DR
HELWAN OBSERVATORY
HELWAN-CAIRO
EGYPT
TEL:
TLX:
COM: 27

MAHRA H S DR
UTTAR PRADESH STATE OBS
MANORA PEAK
NAINITAL 263 129
INDIA
TEL: 2136, 2583
TLX:
COM: 09,16,20,27,50

MAKISHIMA KAZUO
INST SPACE & ASTRON SCI
4-6-1 KOMABA
MEGURO-KU
TOKYO 153
JAPAN
TEL: 03-467-1111x303
TLX: 34757 ISASTRO J
COM:

MALIN STUART
NATIONAL MARITIME MUSEUM
GREENWICH
LONDON SE10 9NF
U.K.
TEL: 01-858-1167
TLX:
COM: 41

MANARA ALESSANDRO A DR
OSSERVATORIO ASTRONOMICO
DI MILANO
VIA BRERA 28
I-20121 MILANO
ITALY
TEL: 02-87-4444
TLX:
COM: 44

MAIHARA TOSHINORI DR
DEPT OF PHYSICS
KYOTO UNIVERSITY
SAKYOKU
KYOTO 606
JAPAN
TEL: 075-751-2111
TLX: 5422693 LIBKYU J
COM: 34

MAKITA MITSUGU DR
TOKYO ASTRONOMICAL
OBSERVATORY
OSAWA MITAKA
TOKYO 181
JAPAN
TEL: 0422-32-5111
TLX: 2822307 TAOMTK J
COM: 10,12

MALITSON HARRIET H MS
13315 MAGELLAN AVE
ROCKVILLE MD 20853
U.S.A.
TEL: 301-946-0496
TLX:
COM: 10,44

MANCHANDA R K DR
TATA INSTITUTE OF
FUNDAMENTAL RESEARCH
HOMI BHABHA RD
BOMBAY 400 005
INDIA
TEL: 219-111 x 336
TLX: 113009 TIFR IN
COM:

MAILLARD JEAN-PIERRE DR
UNIVERSITE DE MONTREAL
DEPT DE PHYSIQUE
CP 6128 SUCC A
MONTREAL PQ H3C 3J6
CANADA
TEL:
TLX:
COM: 09,14

MALACARA DANIEL
CENTRO DE INVESTIGACIONES
EN OPTICA
APDO POSTAL 948
37000 LEON, GTO
MEXICO
TEL: 758-23
TLX:
COM:

MALLIA EDWARD A DR
DEPT OF ASTROPHYSICS
SOUTH PARKS ROAD
OXFORD OX1 3RQ
U.K.
TEL:
TLX:
COM:

MANCHESTER RICHARD N DR
CSIRO
DIVISION OF RADIOPHYSICS
P.O. BOX 76
EPPING NSW 2121
AUSTRALIA
TEL: 02-868-0225
TLX: 26320 ASTRO
COM: 33,34,40

MAITZEN HANS M DR
INSTITUT FUER ASTRONOMIE
TUERKENSCHANZSTR 17
A-1180 WIEN
AUSTRIA
TEL: 0222-345360-94
TLX: 116222 PHYSI A
COM: 29

MALAGNINI MARIA LUCIA
OSSERVATORIO ASTRONOMICO
VIA TIEPOLO 11
PO BOX SUCC TRIESTE 5
I-34131 TRIESTE
ITALY
TEL: 040-793921
TLX: 461137 OAT I
COM: 28

MALLIK D C V DR
INDIAN INSTITUTE OF
ASTROPHYSICS
BANGALORE 560 034
INDIA
TEL: 566585/566497
TLX: 845-763 IIAB IN
COM: 34,35

MANCUSO SANTI PROF
CAPODIMONTE ASTR OBS
VIA MOIARIELLO 16
I-80131 NAPOLI
ITALY
TEL: 44-01-01
TLX:
COM:

MAKARENKO EKATERINA N DR
ASTRONOMICAL OBSERVATORY
PARK SHEVCHENKO
270014 ODESSA
U.S.S.R.
TEL:
TLX:
COM: 27

MALAISE DANIEL J DR
INSTITUT D'ASTROPHYSIQUE
UNIVERSITE DE LIEGE
AVENUE DE COINTE 5
B-4200 COINTE-OUGREE
BELGIUM
TEL:
TLX:
COM: 15,44

MALTBY PER PROF
INST THEORET ASTROPHYSICS
UNIVERSITY OF OSLO
P O BOX 1029
N-0315 BLINDERN, OSLO 3
NORWAY
TEL: 2-456509
TLX:
COM: 10

MANDELSTAM S L PROF
INST OF SPECTROSCOPY
USSR ACADEMY OF SCIENCES
142092 AKADEMGORODOK
U.S.S.R.
TEL: 334-55-79
TLX:
COM: 10,14C,44

MAKAROV VALENTINE I
KISLOVODSK STATION OF THE
PULKOVO OBSERVATORY
357741 KISLOVODSK
U.S.S.R.
TEL:
TLX:
COM: 10,12

MALAKPUR IRADJ DR
INSTITUTE OF GEOPHYSICS
TEHRAN UNIVERSITY
KARGAR SHOMALI
TEHRAN 14394
IRAN
TEL: 631081-3
TLX: 215319 UTIG
COM:

MALVILLE J MCKIM PROF
DEPT OF ASTROGEOPHYSICS
UNIVERSITY OF COLORADO
BOULDER CO 80302
U.S.A.
TEL: 303-492-8788
TLX:
COM: 10

MANDOLESI NAZZARENO
ISTITUTO TE.S.R.E.-C.N.R.
VIA DE CASTAGNOLI 1
I-40126 BOLOGNA
ITALY
TEL:
TLX:
COM: 40,44,47

MAKAROVA ELENA A DR
STERNBERG ASTRONOMICAL
INSTITUTE
117234 MOSCOW
U.S.S.R.
TEL: 139-1973
TLX: 113037 JAPET
COM: 12

MALARODA STELLA M DR
COMPLEJO ASTRONOMICO
EL LEONCITO
CASILLA DE CORREO 467
5400 SAN JUAN
ARGENTINA
TEL: 64-22-5718
TLX: 59134 ENTOP AR
COM: 29,45

MAMMANO AUGUSTO DR
OSSERVATORIO ASTROFISICO
I-36012 ASIAGO, VICENZA
ITALY
TEL:
TLX:
COM: 42

MANFROID JEAN DR
INSTITUT D'ASTROPHYSIQUE
UNIVERSITE DE LIEGE
AVENUE DE COINTE 5
B-4200 COINTE-OUGREE
BELGIUM
TEL: 41-559980
TLX: 41264 ASTRLG
COM: 34

MAKINO FUMIYOSHI DR
INST SPACE & ASTRONAUT SC
6-1 KOMABA 4-CHOME
MEGURO-KU
TOKYO 153
JAPAN
TEL: 03-467-1111
TLX: 24550 SPACETKY J
COM:

MALIN DAVID F MR
ANGLO-AUSTRALIAN OBS
P.O.BOX 296
EPPING NSW 2121
AUSTRALIA
TEL: 02-868-1666
TLX: 23999 AA
COM: 09C

MANABE SEIJI DR
INTL LATITUDE OBSERVATORY
MIZUSAWA
IWATE 023
JAPAN
TEL:
TLX:
COM: 19

MANGENEY ANDRE DR
OBSERVATOIRE DE MEUDON
SECTION DE MEUDON
F-92195 MEUDON PL CEDEX
FRANCE
TEL: 1-45-34-75-30
TLX:
COM: 49

MANNINO GIUSEPPE PROF
ISTITUTO MATEMATICO
VIA CAMPI 181
I-41100 MODENA
ITALY
TEL:
TLX:
COM: 27

MARANO BRUNO
DIPT. DI ASTRONOMIA
UNIVERSITA DI BOLOGNA
C P 596
I-40100 BOLOGNA
ITALY
TEL: 222956
TLX: 211664 INFNBO I
COM: 47

MARGONI RINO
OSSERVATORIO ASTROFISICO
DI ASIAGO
I-36012 ASIAGO, VICENZA
ITALY
TEL: 0424-62665
TLX: 430110 SETOUR
COM:

MARKELLOS VASSILIS V DR
UNIVERSITY OF PATRAS
DEPT ENGINEERING SCIENCE
260 00 RION
GREECE
TEL: 061-991-465
TLX:
COM: 07

MANRIQUE WALTER T PROF
OBSERVATORIO ASTRONOMICO
FELIX AGUILAR
AV BENAVIDEZ 8175 OESTE
5407 MARQUESADO (S.J.)
ARGENTINA
TEL:
TLX:
COM: 08

MARAR T M K
TECHNICAL PHYSICS DIV
ISRO SATELLITE CENTRE
AIRPORT RD, VIMANAPURA
BANGALORE 560 017
INDIA
TEL: 566 251
TLX:
COM: 44

MARGRAVE THOMAS EWING JR
400 JOHNSON STREET
VIENNA VA 22180
U.S.A.
TEL:
TLX:
COM: 27,51

MARKKANEN TAPIO DR
OBSERVATORY
TAHTITORNINMAKI
SF-00130 HELSINKI 13
FINLAND
TEL: 90-8391
TLX:
COM: 25,37,50

MANSFIELD VICTOR N PROF
COLGATE UNIVERSITY
HAMILTON NY 13346
U.S.A.
TEL: 315-824-1000
TLX:
COM:

MARASCHI LAURA DR
ISTITUTO DI FISICA
VIA CELORIA 16
I-20133 MILANO
ITALY
TEL:
TLX:
COM:

MARIE M A DR
DEPT OF ASTRONOMY
FACULTY OF SCIENCE
CAIRO UNIVERSITY
CAIRO
EGYPT
TEL:
TLX:
COM:

MARKOWITZ WILLIAM DR
APT 15-B
2800 E. SUNRISE BLVD
FORT LAUDERDALE FL 33304
U.S.A.
TEL: 305-563-2859
TLX:
COM: 19,31

MANTEGAZZA LUCIANO
OSSERVATORIO ASTRONOMICO
DI MERATE
VIA BIANCHI 46
I-22055 MERATE
ITALY
TEL: 039-592035
TLX:
COM: 27

MARCELIN MICHEL
OBSERVATOIRE DE MARSEILLE
2 PLACE LE VERRIER
F-13248 MARSEILLE CEDEX 4
FRANCE
TEL: 91-95-90-88
TLX: 420241 F
COM: 28

MARIK MIKLOS DR.
ASTRONOMICAL DEPT
L. EOTVOS UNIVERSITY
KUN BELA TER 2
H-1083 BUDAPEST
HUNGARY
TEL: 141-019
TLX:
COM: 12,38,46

MARLBOROUGH J M PROF
DEPT OF ASTRONOMY
UNIV OF WESTERN ONTARIO
LONDON ONT N6A 3K7
CANADA
TEL: 519-679-3184
TLX: 064-7134
COM: 36

MANTOVANI FRANCO
IST. DI RADIOASTRONOMIA
VIA IRNERIO 46
I-40126 BOLOGNA
ITALY
TEL: 51-23-2856
TLX: 211664 INFNBO I
COM:

MARCHAL CHRISTIAN DR
DEPT ETUDES DE SYNTHESE
ONERA
F-92320 CHATILLON
FRANCE
TEL: 1-46-57-11-60
TLX: 260907 F
COM: 07

MARILLI ETTORE DR
OSSERVATORIO ASTROFISICO
CITTA UNIVERSITARIA
I-95125 CATANIA
ITALY
TEL: 095-33-05-33
TLX: 970359 ASTRCT I
COM: 12,42

MARMOLINO CIRO
DIPARTIMENTO DI FISICA
MOSTRA D'OLTRAMARE
PAD. 19
I-80125 NAPOLI
ITALY
TEL: 081-7253428
TLX: 720320 INFNNA I
COM: 12

MAO WEI
YUNNAN OBSERVATORY
KUNMING, YUNNAN PROVINCE
CHINA, PEOPLE'S REP.
TEL:
TLX:
COM: 08

MARDIROSSIAN FABIO
DIPT. DI ASTRONOMIA
UNIVERSITA DEGLI STUDI
VIA G.B. TIEPOLO 11
I-34131 TRIESTE
ITALY
TEL: 040-793921/221
TLX: 461137 OAT I
COM: 42,47

MARINO BRIAN F ENG
AUCKLAND OBSERVATORY
BOX 72009
NORTHCOTE
AUCKLAND 9
NEW ZEALAND
TEL:
TLX:
COM: 42

MAROCHNIK L S PROF DR
USSR ACADEMY OF SCIENCES
SPACE RESEARCH INSTITUTE
PROFSOJOSNAJ A 84/32
117810 MOSCOW
U.S.S.R.
TEL: 333-31-22
TLX: 411498 STAR SU
COM: 33

MARABINI RODOLFO JOSE
UNIVERSIDAD NACIONAL
FACULTAD DE CIENCIAS
ASTRON Y GEOFISICAS
1900 LA PLATA
ARGENTINA
TEL: 021-217-308
TLX:
COM:

MAREK JOHN
44 PERCY ROAD
WREXHAM, CLWYD
U.K.
TEL:
TLX:
COM: 47

MARISKA JOHN THOMAS
NAVAL RESEARCH LABORATORY
CODE 4175M
WASHINGTON DC 20375
U.S.A.
TEL: 202-767-2605
TLX:
COM: 12

MAROV MIKHAIL YA PROF
INST OF APPLIED MATHS
USSR ACADEMY OF SCIENCES
MIUSSKAYA SQ 4
125047 MOSCOW
U.S.S.R.
TEL:
TLX:
COM: 16C,44,51,W

MARAN STEPHEN P DR
NASA/GSFC
CODE 680
GREENBELT MD 20771
U.S.A.
TEL: 301-344-8607
TLX: 89675
COM: 15,40,44

MARGON BRUCE H PROF
ASTRONOMY DEPT FM-20
UNIVERSITY OF WASHINGTON
SEATTLE WA 98195
U.S.A.
TEL: 206-543-0089
TLX: 4740096
COM:

MARK JAMES WAI-KEE DR
PHYSICS DEPT
L.LIVERMORE NATL LAB L477
UNIVERSITY OF CALIFORNIA
LIVERMORE CA 94550
U.S.A.
TEL: 415-422-5931
TLX: 910-386-8339 UCCLLL
COM: 28,33

MARQUES DOS SANTOS P PROF
INST ASTRON. E GEOFISICO
UNIVERSIDADE DE SAO PAULO
CAIXA POSTAL 30627
01051 SAO PAULO SP
BRAZIL
TEL: 11-276-3941
TLX: 36221 IAGM BR
COM: 28,40

MARQUES MANUEL N DR
OBSERVATORIO ASTRONOMICO
TAPADA DA AJUDA
1300 LISBOA 3
PORTUGAL
TEL:
TLX:
COM:

MARTEL MARIE-THERESE DR
CARCADIS
F-04870 ST-MICHEL-L'OBS.
FRANCE
TEL:
TLX:
COM: 15,34

MARTIN WILLIAM C DR
NATL BUREAU OF STANDARDS
A167 PHYSICS BLDG
GAITHERSBURG MD 20899
U.S.A.
TEL: 301-921-2011
TLX:
COM: 14

MARTRES MARIE-JOSEPHE
OBSERVATOIRE DE PARIS
SECTION DE MEUDON
F-92195 MEUDON PL CEDEX
FRANCE
TEL: 1-45-34-75-30
TLX:
COM: 10

MARRACO HUGO G DR
UNIV NACIONAL DE LA PLATA
FACULTAD DE CIENCIAS
ASTRONOMICAS Y GEOFISICAS
1900 LA PLATA
ARGENTINA
TEL: 54-21-21-7308
TLX: 31151 BULAP AR
COM: 25,37

MARTIN ANTHONY R DR
UK CULHAM LABORATORY
RM F4/135
ABINGDON OX14 3DB
U.K.
TEL: 0235-21840
TLX: 83189
COM: 51

MARTIN WILLIAM L DR
ROYAL GREENWICH OBS
HERSTMONCEUX CASTLE
HAILSHAM BN27 1RP
U.K.
TEL: 0323-833171
TLX: 87451 RGOBSY G
COM: 27

MARTYNOV D YA PROF DR
STERNBERG STATE
ASTRONOMICAL INSTITUTE
117234 MOSCOW
U.S.S.R.
TEL:
TLX:
COM: 05,06,16,42

MARSCHALL LAURENCE A
GETTYSBURG COLLEGE
DEPT OF PHYSICS
GETTYSBURG PA 17325
U.S.A.
TEL: 717-337-1865
TLX:
COM: 24

MARTIN DEREK H PROF
QUEEN MARY COLLEGE
MILE END ROAD
LONDON E1 4NS
U.K.
TEL:
TLX:
COM:

MARTIN-LORON M DR
HERMANOS MIRALLES 14
MADRID 1
SPAIN
TEL:
TLX:
COM:

MARVIN URSULA B DR
CENTER FOR ASTROPHYSICS
60 GARDEN STREET
CAMBRIDGE MA 02138
U.S.A.
TEL: 617-495-7270
TLX: 921428 SATELLITE CAM
COM: 22

MARSCHER ALAN PATRICK
ASTRONOMY DEPT
BOSTON UNIVERSITY
725 COMMONWEALTH AVE
BOSTON MA 02215
U.S.A.
TEL: 617-353-5029
TLX: 951289 BOS UNIV BSN
COM: 40

MARTIN FRANCOIS DR
DEPT D'ASTROPHYSIQUE
UNIVERSITE DE NICE
PARC VALROSE
F-06034 NICE CEDEX
FRANCE
TEL: 93-51-91-00
TLX: 970281
COM:

MARTIN-PINTADO JESUS
CENTRO ASTRON DE YEBES
O A N
APARTADO CORREOS 148
19080 GUADALAJARA
SPAIN
TEL: 911-223358
TLX:
COM:

MARX GYORGY PROF.
L. EOTVOS UNIVERSITY
PUSHKIN U 5-7
H-1088 BUDAPEST
HUNGARY
TEL:
TLX:
COM: 35,51V

MARSDEN BRIAN G DR
SMITHSONIAN ASTROPHYSICAL
OBSERVATORY
60 GARDEN STREET
CAMBRIDGE MA 02138
U.S.A.
TEL: 617-495-7244
TLX: 7103206842 ASTROGRAM
COM: 06C,07,15,20C

MARTIN INACIO MALMONGE DR
UNIV DE CAMPINAS-UNICAMP
INSTITUTO DE FISICA
DEPTO RAIOS COSMICOS
13100 CAMPINAS SP
BRAZIL
TEL: 0192-391301
TLX: 019-1150
COM: 48

MARTINET LOUIS PROF
OBSERVATOIRE DE GENEVE
CH-1290 SAUVERNY
SWITZERLAND
TEL: 55-26-11
TLX: 27720
COM: 28,33

MARX SIEGFRIED DR
ZNTRLINST. F. ASTROPHYSIK
KARL-SCHWARZSCHILD OBS
DDR-6901 TAUTENBURG
GERMANY, D.R.
TEL: JENA 23530
TLX: 5886284 KSOT DD
COM: 50

MARSDEN PHILIP L PROF
DEPT OF PHYSICS
UNIVERSITY OF LEEDS
LEEDS LS2 9JT
U.K.
TEL: 0532-431751
TLX: 556473 UNIDS
COM:

MARTIN NICOLE DR
OBSERVATOIRE DE MARSEILLE
2 PLACE LE VERRIER
F-13248 MARSEILLE CEDEX 4
FRANCE
TEL: 91-95-90-88
TLX: 420241
COM: 30

MARTINEZ MARIO DR
CIESE
DEPTO DE GEOFISICA
APDO POSTAL 2732
22860 ENSENADA, B. CALIF.
MEXICO
TEL:
TLX:
COM:

MASANI A PROF
OSSERVATORIO ASTRONOMICO
DI BRERA
VIA BRERA 28
I-20100 MILANO
ITALY
TEL:
TLX:
COM: 25,27,35

MARSH JULIAN C D
HATFIELD POLYTECHNIC
OBSERVATORY
BAYFORDBURY
HERTFORD, HERTS SG13 8LD
U.K.
TEL: 0992-558451
TLX: 262413
COM: 46

MARTIN PETER G PROF
INST FOR THEOR ASTROPHYS
UNIVERSITY OF TORONTO
TORONTO ONT M5S 1A7
CANADA
TEL: 416-978-6840
TLX:
COM:

MARTINI ALDO DR
ASTROFISICA SPAZIALE
CNR
C P 67
I-00044 FRASCATI, ROMA
ITALY
TEL:
TLX:
COM:

MASLOWSKI JOZEF DR
ASTRONOMICAL OBSERVATORY
UL. ORLA 171
30-244 KRAKOW
POLAND
TEL: 34-10-41
TLX: 0322297 UJ PL
COM: 40

MARSOGLU A DR
UNIVERSITY OBSERVATORY
UNIVERSITE
ISTANBUL
TURKEY
TEL: 11-522-35-97
TLX:
COM:

MARTIN ROBERT N DR
STEWARD OBSERVATORY
UNIVERSITY OF ARIZONA
TUCSON AZ 85721
U.S.A.
TEL: 602-621-1539
TLX: 467175
COM: 34,40

MARTINS DONALD HENRY DR
DEPT PHYSICS & ASTRONOMY
UNIVERSITY OF ALASKA
3221 UAA DRIVE
ANCHORAGE AK 99508
U.S.A.
TEL: 907-786-1238
TLX:
COM: 09,37

MASNOU FRANCOISE DR
28 ALLEE GAMBAUDERIE
F-91190 GIF-S-YVETTE
FRANCE
TEL:
TLX:
COM:

MASNOU J L DR
OBSERVATOIRE DE PARIS
SECTION DE MEUDON
LA 173
F-92195 MEUDON PL CEDEX
FRANCE
TEL: 1-45-34-75-70
TLX:
COM:

MASON GLENN M
UNIVERSITY OF MARYLAND
DEPT PHYSICS & ASTRONOMY
COLLEGE PARK MD 20742
U.S.A.
TEL: 301-454-2616
TLX: 71-8261125
COM: 10,49

MASON HELEN E DR
DEPT APPLIED MATHS
SILVER STREET
CAMBRIDGE CB3 9EW
U.K.
TEL: 0223 351645
TLX: 81240
COM: 14

MASON KEITH OWEN
MULLARD SPACE SCIENCE LAB
HOLMBURY ST MARY
DORKING, SURREY RH5 6NT
U.K.
TEL: 0306-70292
TLX: 859185
COM: 48

MASSAGLIA SILVANO
IST. DI FISICA GENERALE
CORSO M. D'AZEGLIO 46
I-10125 TORINO
ITALY
TEL: 011-657694
TLX: 211041
COM: 36

MASSEVICH ALLA G DR
ASTRONOMICAL COUNCIL
USSR ACADEMY OF SCIENCES
PYATNITSKAYA UL 48
109017 MOSCOW
U.S.S.R.
TEL: 231-54-61
TLX: 412623 SCSTP SU
COM: 35

MASSEY PHILIP L
KITT PEAK NATIONAL OBS
PO BOX 26732
TUCSON AZ 85726-6732
U.S.A.
TEL: 602-327-5511
TLX:
COM:

MASSON COLIN R
CALTECH
MS 405-47
PASADENA CA 91125
U.S.A.
TEL: 213-356-6229
TLX: 675425
COM: 34,40

MASURSKY HAROLD DR
US GEOLOGICAL SURVEY
BRANCH OF ASTRO GEOLOGY
2255 NORTH GEMINI DR
FLAGSTAFF AZ 86001
U.S.A.
TEL: 601-527-2003
TLX:
COM: 16C,WP

MATAS VLADIMIR R DR
ASTRONOMISCHES-RECHEN
INSTITUT
MOENCHHOFSTR 12-14
D-6900 HEIDELBERG
GERMANY, F.R.
TEL:
TLX:
COM: 07

MATERNE JUERGEN DR
ARETINSTR 27
D-8000 MUNCHEN 90
GERMANY, F.R.
TEL:
TLX:
COM: 28,47

MATHER JOHN CROMWELL
LAB FOR EXTRATERR PHYSICS
NASA/GSFC, CODE 697
GREENBELT MD 20771
U.S.A.
TEL: 301-344-8720
TLX: 89675
COM: 47

MATHESON DAVID NICHOLAS
SERC
RUTHERFORD APPLETON LAB
CHILTON, DIDCOT OX11 0QX
U.K.
TEL: 0235-21900
TLX: 83159
COM: 40

MATHEWS WILLIAM G PROF
LICK OBSERVATORY
UNIVERSITY OF CALIFORNIA
SANTA CRUZ CA 95064
U.S.A.
TEL: 408-429-2074
TLX:
COM: 34

MATHEWSON DONALD S PROF
MT STROMLO & SIDING
SPRING OBSERVATORIES
PRIVATE BAG
WODEN P.O. ACT 2606
AUSTRALIA
TEL: 062-881111
TLX: 62270 AA
COM: 28,33,34

MATHEZ GUY
OBSERVATOIRE PIC-DU-MIDI
ET DE TOULOUSE
14 AVENUE EDOUARD BELIN
F-31400 TOULOUSE
FRANCE
TEL: 61-25-21-01
TLX: 530776 OBSTLSE F
COM:

MATHIS JOHN S PROF
DEPT OF ASTRONOMY
UNIVERSITY OF WISCONSIN
475 N. CHARTER ST
MADISON WI 53706
U.S.A.
TEL: 608-262-5994
TLX: 265452 UOFWISC MDS
COM: 34V

MATHUR B S DR
NATL PHYSICAL LABORATORY
HILLSIDE ROAD
NEW DELHI 110 012
INDIA
TEL: 586168
TLX: 31-62454 RSD IN
COM: 31

MATHYS GAUTIER DR
OBSERVATOIRE DE GENEVE
CHEMIN DES MAILLETTES 51
CH-1290 SAUVERNY
SWITZERLAND
TEL: 55-26-11
TLX: 27720 OBSG CH
COM: 29

MATIAGIN VALERY S DR
ASTROPHYSICAL INSTITUTE
480068 ALMA-ATA
U.S.S.R.
TEL:
TLX:
COM: 24

MATSAKIS DEMETRIOS N
US NAVAL OBSERVATORY
34 & MASSACHUSETTS AVE NW
WASHINGTON DC 20390
U.S.A.
TEL: 202-653-1873
TLX:
COM: 19,31,40,51

MATSON DENNIS L DR
JPL 183-501
4800 OAK GROVE DRIVE
PASADENA CA 91103
U.S.A.
TEL: 213-354-2984
TLX:
COM: 15,16

MATSUDA TAKUYA PROF
DEPT AERONAUTIC ENGINEERG
KYOTO UNIVERSITY
YOSHIDAHONMACHI SAKYOKU
KYOTO 606
JAPAN
TEL: 075-751-2111
TLX: 05422693 LIBKYUJ
COM: 36

MATSUMOTO MASAMICHI PROF
FACULTY OF ENGINEERING
GIFU UNIVERSITY
501-11 GIFU
JAPAN
TEL:
TLX:
COM: 36

MATSUMOTO TOSHIO DR
DEPT OF PHYSICS
NAGOYA UNIVERSITY
FUROCHO CHIKUSAKU
NAGOYA 464
JAPAN
TEL: 052-781-5111
TLX: 4477323 SCUNAG J
COM: 21,47

MATSUOKA MASARU DR
INST SPACE & ASTRONAUT.
SCIENCES
KOMABA MEGURO
TOKYO 153
JAPAN
TEL: 03-467-1111
TLX: 24550 SPACEKY J
COM: 44,48

MATSUSHIMA SATOSHI DR
DEPT ASTRONOMY
PENNSYLVANIA STATE UNIV
525 DAVEY LAB
UNIVERSITY PARK PA 16802
U.S.A.
TEL: 814-865-0418
TLX:
COM: 12,36

MATSUURA OSCAR T DR
DEPTO DE ASTRONOMIA
IAG-USP
CAIXA POSTAL 30627
01051 SAO PAULO SP
BRAZIL
TEL: 011-275-3720
TLX: 1136221 IAGM BR
COM: 10,15,49

MATTEI JANET AKYUZ DR
AAVSO
25 BIRCH STREET
CAMBRIDGE MA 02138
U.S.A.
TEL: 617-354-0484
TLX:
COM: 27,42

MATTEUCCI FRANCESCA
IST. ASTROFISICA SPAZIALE
C P 67
I-00044 FRASCATI
ITALY
TEL:
TLX:
COM:

MATTHEWS THOMAS A DR
ASTRONOMY PROGRAM
UNIVERSITY OF MARYLAND
COLLEGE PK MD 20742
U.S.A.
TEL:
TLX:
COM:

MATTIG W PROF DR
KIEPENHEUER INSTITUT
FUER SONNENPHYSIK
SCHOENECKSTRASSE 6
D-7800 FREIBURG-IM-BR.
GERMANY, F.R.
TEL: 761-32864
TLX: 7721552 KIS D
COM: 10,12,50

MATTILA KALEVI DR
OBSERVATORY
TAHTITORNINMAKI
SF-00130 HELSINKI 13
FINLAND
TEL: 90-1912947
TLX: 124690 UNIH SF
COM: 21P,34,40

MATVEYENKO L I DR
SPACE RESEARCH INSTITUTE
USSR ACADEMY OF SCIENCES
117810 MOSCOW
U.S.S.R.
TEL: 333-31-22
TLX: 411498 STAR SU
COM: 40C

MATZNER RICHARD A PROF
PHYSICS DEPT
UNIVERSITY OF TEXAS
AUSTIN TX 78712
U.S.A.
TEL: 512-471-5062
TLX:
COM:

MAUDER HORST PROF DR
ASTRONOMISCHES INSTITUT
WALDHAUSER-STR 64
D-7400 TUEBINGEN
GERMANY, F.R.
TEL:
TLX:
COM: 42

MAURICE ERIC N
OBSERVATOIRE DE MARSEILLE
2 PLACE LE VERRIER
F-13248 MARSEILLE CDX 04
FRANCE
TEL: 91-95-90-88
TLX: 420241 F
COM: 28,30C

MAVRAGANIS A G PROF
DEPT OF ENG SECT OF MECH
NATL TECHN UNIV/5 HEROES
POLYTECH AVE
GR-15773 ATHENS
GREECE
TEL: 6433170
TLX:
COM: 07

MAVRIDES STAMATIA DR
OBSERVATOIRE DE PARIS
SECTION DE MEUDON
DEPT RADIOASTRONOMIE
F-92195 MEUDON PL CEDEX
FRANCE
TEL: 1-45-34-75-30
TLX:
COM: 28,47

MAVRIDIS L N PROF
DEPT GEODETIC ASTRONOMY
UNIVERSITY THESSALONIKI
GR-54006 THESSALONIKI
GREECE
TEL:
TLX:
COM: 08,27C,33,46,50,51

MAVROMICHALAKI HELEN DR
UNIVERSITY / PHYSICS DEPT
NUCLEAR PHYSICS SECTION
104 SOLONOS ST
GR-10680 ATHENS
GREECE
TEL: 3639439
TLX:
COM: 49

MAX CLAIRE E DR
LAWRENCE LIVERMORE LAB
L-413
PO BOX 808
LIVERMORE CA 94550
U.S.A.
TEL: 415-422-5442
TLX: 9103868339 UCLLLLVMR
COM:

MAXWELL ALAN DR
HARVARD SMITHSONIAN CTR
FOR ASTROPHYSICS
60 GARDEN STREET
CAMBRIDGE MA 02138
U.S.A.
TEL: 617-495-9059
TLX:
COM: 10,40

MAY ANDREW
DEPT OF THEORETICAL PHYS
1 KEBLE ROAD
OXFORD OX1 3NP
U.K.
TEL: 0865-53281
TLX: 83295 NUCCOX
COM: 28

MAY J
OBS RADIOASTR DE MAIPU
UNIVERSIDAD DE CHILE
CASILLA 68
MAIPU
CHILE
TEL: 2294101
TLX:
COM:

MAYALL MARGARET W
5 SPARKS STREET
CAMBRIDGE MA 02138
U.S.A.
TEL: 617-876-1563
TLX:
COM: 27

MAYALL NICHOLAS U ASTRON
7206 E. CAMINO VECINO
TUCSON AZ 85715
U.S.A.
TEL: 602-886-2423
TLX:
COM: 28

MAYER CORNELL H
SPACE SCIENCE DIVISION
NAVAL RESEARCH LAB
CODE 4130M
WASHINGTON DC 20375
U.S.A.
TEL: 202-767-2495
TLX:
COM: 16,40

MAYER PAVEL DR
DEPT OF ASTRONOMY
CHARLES UNIVERSITY
SVEDSKA 8
150 00 PRAHA 5
CZECHOSLOVAKIA
TEL: 540395
TLX:
COM: 25,42

MAYFIELD EARLE B DR
THE AEROSPACE CORPORATION
5536 MICHELLE DRIVE
TORRANCE CA 90503
U.S.A.
TEL: 213-648-7088
TLX: 664460
COM:

MAYOR MICHEL DR
OBSERVATOIRE DE GENEVE
CHEMIN DES MAILLETTES 51
CH-1290 SAUVERNY
SWITZERLAND
TEL: 22-55-26-11
TLX: 27720 OBSG CH
COM: 30C,33V,37

MAZA JOSE
DEPTO DE ASTRONOMIA
UNIVERSIDAD DE CHILE
CASILLA 36-D
SANTIAGO
CHILE
TEL:
TLX:
COM:

MAZURE ALAIN DR
OBSERVATOIRE DE PARIS
SECTION DE MEUDON
DAF
F-92195 MEUDON PL CEDEX
FRANCE
TEL: 1-45-34-75-70
TLX:
COM:

MAZUREK THADDEUS JOHN DR
MISSION RESEARCH CORP
PO DRAWER 719
SANTA BARBARA CA 93102
U.S.A.
TEL: 805-963-8761
TLX:
COM: 35,48

MAZZITELLI ITALO DR
IST. ASTROFISICA SPAZIALE
C P 67
I-00044 FRASCATI
ITALY
TEL: 06-9421483
TLX: 610261 CNRFRA
COM: 35

MAZZUCCONI FABRIZIO DR
OSSERVATORIO ASTROFISICO
DI ARCETRI
LARGO E. FERMI 5
I-50125 FIRENZE
ITALY
TEL:
TLX:
COM:

MCADAM W BRUCE DR
SCHOOL OF PHYSICS
UNIVERSITY OF SYDNEY
SYDNEY NSW 2006
AUSTRALIA
TEL: 692-2222
TLX: 26169 UNISYD
COM: 40

MCALISTER HAROLD A DR
DEPT PHYSICS & ASTRONOMY
GEORGIA STATE UNIVERSITY
ATLANTA GA 30303
U.S.A.
TEL: 404-658-2932
TLX:
COM: 24,26V,51

MCBREEN BRIAN PHILIP DR
PHYSICS DEPT
UNIVERSITY COLLEGE
BELFIELD
DUBLIN 4
IRELAND
TEL: 693244 671218
TLX: 36293
COM: 28,48

MCCABE MARIE K MS
INSTITUTE FOR ASTRONOMY
2680 WOODLAWN DRIVE
HONOLULU HI 96822
U.S.A.
TEL: 808-948-8306
TLX: 7238459 UHAST HR
COM: 10

MCCALL MARSHALL LESTER DR
DAVID DUNLAP OBSERVATORY
UNIVERSITY OF TORONTO
P O BOX 360
RICHMOND HILL ONT L4C 4Y6
CANADA
TEL: 416-978-4165
TLX:
COM: 34

MCCAMMON DAN
UNIVERSITY OF WISCONSIN
PHYSICS DEPT
1150 UNIVERSITY AVE
MADISON WI 53706
U.S.A.
TEL: 608-262-5916
TLX: 265452 UOFWISC MDS
COM:

MCCARROLL RONALD PROF
LAB D'ASTROPHYSIQUE
UNIVERSITE DE BORDEAUX I
F-33405 TALENCE
FRANCE
TEL:
TLX:
COM:

MCCARTHY DENNIS D DR
US NAVAL OBSERVATORY
34 & MASSACHUSETTS AVE NW
WASHINGTON DC 20390
U.S.A.
TEL: 202-653-0066
TLX: 7108221970
COM: 19,31P

LIST OF MEMBERS

MCCARTHY MARTIN F DR
SPECOLA VATICANA
00120 CITTA DEL VATICANO
VATICAN CITY STATE
TEL: 698-3411
TLX: 5042020 VAT OBS VA
COM: 25,33,45,46,50

MCCLAIN EDWARD F
4133 MAPLE ROAD
MORNINGSIDE MD 20746
U.S.A.
TEL: 301-736-8933
TLX:
COM:

MCCLINTOCK JEFFREY E DR
CENTER FOR ASTROPHYSICS
SMITHSONIAN ASTROPHYS OBS
60 GARDEN STREET
CAMBRIDGE MA 02138
U.S.A.
TEL: 617-495-7136
TLX:
COM:

MCCLURE ROBERT D PROF
DOMINION ASTROPHYS OBS
5071 W SAANICH ROAD
RR 5
VICTORIA BC V8X 4M6
CANADA
TEL: 604-388-0230
TLX:
COM: 28,30C,45

MCCLUSKEY GEORGE E JR DR
DIV OF ASTRONOMY
DEPT OF MATHEMATICS
LEHIGH UNIVERSITY
BETHLEHEM PA 18015
U.S.A.
TEL: 215-861-3721
TLX:
COM: 42,44

MCCORD THOMAS B DR
PLANETARY GEOSCIENCES DIV
HAWAII INST OF GEOPHYSICS
2525 CORREA RD
HONOLULU HI 96822
U.S.A.
TEL: 808-948 6488
TLX:
COM: 15,16

MCCRACKEN KENNETH G DR
CSIRO - COSSA
LIMESTONE AVENUE
P.O. BOX 225
DICKSON ACT 2602
AUSTRALIA
TEL: 062-484-595
TLX: 62003 AA
COM: 44

MCCRAY RICHARD DR
JILA
UNIVERSITY OF COLORADO
BOULDER CO 80309
U.S.A.
TEL: 303-492-7835
TLX:
COM: 34,48

MCCREA J DERMOTT
DEPT OF MATH PHYSICS
UNIVERSITY COLLEGE
BELFIELD
DUBLIN 4
IRELAND
TEL:
TLX:
COM: 34,35,47

MCCREA WILLIAM SIR
ASTRONOMY CENTRE
SUSSEX UNIVERSITY
BRIGHTON BN1 9QH
U.K.
TEL: 0273-606755
TLX: 877259 UNISEX G
COM: 28,47

MCCROSKY RICHARD E DR
CENTER FOR ASTROPHYSICS
60 GARDEN STREET
CAMBRIDGE MA 02138
U.S.A.
TEL: 617-495-7212
TLX:
COM: 15,20,22

MCCULLOCH PETER M DR
DEPT OF PHYSICS
UNIVERSITY OF TASMANIA
HOBART, TASMANIA
AUSTRALIA
TEL: 002-20-24-20
TLX: 58150
COM: 40

MCCUTCHEON WILLIAM H PROF
DEPT OF PHYSICS
UNIV OF BRITISH COLUMBIA
2075 WESBROOK MALL
VANCOUVER BC V6T 2A6
CANADA
TEL: 604-228-3853
TLX: 04508576 UBCPHYSICS
COM:

MCDONALD FRANK B DR
NASA/GSFC
CODE 660
GREENBELT MD 20771
U.S.A.
TEL:
TLX:
COM:

MCDONALD J K PETRIE DR
2195 CUBBON DRIVE
VICTORIA BC V8R 1R4
CANADA
TEL: 604-592-6880
TLX:
COM:

MCDONNELL J A M PROF
UNIT FOR SPACE SCIENCES
UNIVERSITY OF KENT
CANTERBURY, KENT CT2 7NR
U.K.
TEL: 0227-459616
TLX: 965449 UKCLIB G
COM: 22C

MCDONOUGH THOMAS R DR
CALIF INST OF TECHNOLOGY
500 S. OAK KNOLL, NO.46
PASADENA CA 91101
U.S.A.
TEL: 818-795-0147
TLX:
COM: 51

MCELROY M B DR
KITT PEAK NAT OBSERVATORY
950 N. CHERRY AVE
TUCSON AZ 85726
U.S.A.
TEL:
TLX:
COM: 16

MCGEE JAMES D PROF
10 HILLTOP CRESCENT
APT 3E
FAIRLIGHT
SYDNEY NSW 2094
AUSTRALIA
TEL: 02-949-3723
TLX:
COM: 09,40

MCGEE RICHARD X DR
CSIRO
DIVISION OF RADIOPHYSICS
P.O.BOX 76
EPPING NSW 2121
AUSTRALIA
TEL: 02-868-0222
TLX: 26230 ASTRO
COM: 34

MCGIMSEY BEN Q JR DR
DEPT OF PHYSICS & ASTR
GEORGIA STATE UNIVERSITY
UNIVERSITY PLAZA
ATLANTA GA 30303
U.S.A.
TEL: 404-658-2279
TLX:
COM: 28

MCGREGOR PETER JOHN DR
MT STROMLO & SIDING
SPRING OBSERVATORIES
PRIVATE BAG
WODEN P.O. ACT 2606
AUSTRALIA
TEL: 062-88-1111
TLX: 62270 CANOPUS AA
COM: 33

MCINTOSH BRUCE A DR
HERZBERG INST ASTROPHYS
NATL RESEARCH COUNCIL
OTTAWA ONT K1A 0R6
CANADA
TEL:
TLX:
COM: 22

MCKEE CHRISTOPHER F PROF
PHYSICS DEPT
UNIVERSITY OF CALIFORNIA
BERKELEY CA 94720
U.S.A.
TEL: 415-642-0805
TLX: 820181 UCB AST RALUD
COM: 34

MCKEITH CONAL D DR
QUEEN'S UNIVERSITY
PURE AND APPLIED PHYSICS
BELFAST BT7 1NN
U.K.
TEL: 0232-245133
TLX: 74487 QUB ADM
COM: 34

MCKEITH NIALL ENDA DR
ST PATRICK'S COLLEGE
MAYNOOTH CO KILDARE
IRELAND
TEL:
TLX:
COM:

MCKENNA LAWLOR SUSAN
ST PATRICK'S COLLEGE
DEPT OF EXPERIMENTAL
PHYSICS
MAYNOOTH CO KILDARE
IRELAND
TEL: 285222
TLX: 31493 SPCM EI
COM: 10,12,15,40,41

MCLAREN ROBERT A DR
CANADA-FRANCE-HAWAII
TELESCOPE CORPORATION
PO BOX 1597
KAMUELA HI 96743
U.S.A.
TEL: 808-885-7944
TLX: 633147 CFHT
COM:

MCLEAN BRIAN JOHN
SPACE TELESCOPE SCI INST
HOMEWOOD CAMPUS
BALTIMORE MD 21218
U.S.A.
TEL: 301-333-9101
TLX: 6849101 STSCI
COM:

MCLEAN DONALD J DR
CSIRO
DIVISION OF RADIOPHYSICS
P.O.BOX 76
EPPING NSW 2121
AUSTRALIA
TEL: 02-868-0222
TLX: 26230 ASTRO AA
COM: 10,40

MCLEAN IAN S DR
ROYAL OBSERVATORY
BLACKFORD HILL
EDINBURGH EH9 3HJ
U.K.
TEL: 31-667-3321x290
TLX: 72383 ROEDIN G
COM: 25V

MCMULLAN DENNIS DR
CAVENDISH LABORATORY
MADINGLEY ROAD
CAMBRIDGE CB3 0HE
U.K.
TEL:
TLX:
COM: 09

LIST OF MEMBERS

MCNALLY DEREK DR
UNIVERSITY OF LONDON OBS
MILL HILL PARK
LONDON NW7 2QS
U.K.
TEL: 01-959-7367
TLX: 28722 UCPHYS G
COM: 34,46

MCNAMARA DELBERT H DR
DEPT PHYSICS & ASTRONOMY
BRIGHAM YOUNG UNIVERSITY
PROVO UT 84602
U.S.A.
TEL: 801-378-2298
TLX:
COM: 05,27,29,45

MCVITTIE GEORGE C PROF
74 OLD DOVER ROAD
CANTERBURY, KENT CT1 3AY
U.K.
TEL: 0227-60704
TLX:
COM: 28,47

MCWHIRTER R W PETER DR
SPACE & ASTROPHYSICS DIV
RUTHERFORD APPLETON LAB
CHILTON, DIDCOT OX11 0QX
U.K.
TEL: 0235-446424
TLX: 83159 RUTHLB G
COM: 14,44

MEABURN J DR
DEPT OF ASTRONOMY
THE UNIVERSITY
MANCHESTER M13 9PL
U.K.
TEL:
TLX:
COM: 34

MEAD JAYLEE MONTAGUE DR
NASA/GSFC
CODE 680
GREENBELT MD 20771
U.S.A.
TEL: 301-344-8543
TLX:
COM: 05,45

MEADOWS A JACK PROF
ASTRONOMY & HISTORY
OF SCIENCE DEPT
UNIVERSITY OF LEICESTER
LEICESTER LE1 7RH
U.K.
TEL: 0533-54455
TLX: 341198 LEICUL
COM: 05,16,41

MEBOLD ULRICH DR PROF
RADIOASTRONOMISCHES INST
DER UNIVERSITAT BONN
AUF DEM HUEGEL 71
D-5300 BONN 1
GERMANY, F.R.
TEL:
TLX:
COM: 34,40

MEDVEDEV YURI A DR
ASTRONOMICAL OBSERVATORY
ODESSA STATE UNIVERSITY
270014 ODESSA
U.S.S.R.
TEL: 22-84-42
TLX:
COM:

MEEKS M LITTLETON DR
MEEKS ASSOCIATES,INC.
P O BOX 643
LINCOLN MA 01773
U.S.A.
TEL: 617-259-0093
TLX:
COM: 40

MEGESSIER CLAUDE DR
OBSERVATOIRE DE PARIS
SECTION DE MEUDON
LAM
F-92195 MEUDON PL CEDEX
FRANCE
TEL: 1-45-34-75-30
TLX: 201571 LAM
COM: 29

MEGRELISHVILI T G PROF
ABASTUMANI ASTROPHYSICAL
OBSERVATORY
383762 ABASTUMANI,GEORGIA
U.S.S.R.
TEL: 22-66
TLX: 327409 TERMIT
COM: 21

MEIER DAVID L
JET PROPULSION LABORATORY
CODE 264-700
4800 OAK GROVE DRIVE
PASADENA CA 91109
U.S.A.
TEL: 213-354-5062
TLX: 675429
COM: 28,40

MEIER ROBERT R
NAVAL RESEARCH LAB
CODE 4140
WASHINGTON DC 20375
U.S.A.
TEL: 202-767-2773
TLX:
COM: 34

MEIKLE WILLIAM P S
ASTROPHYSICS GROUP
IMPERIAL COLLEGE
PRINCE CONSORT ROAD
LONDON SW7 2BZ
U.K.
TEL: 589-5111
TLX: 261503
COM: 28

MEIN NICOLE DR
OBSERVATOIRE DE PARIS
SECTION DE MEUDON
DASOP
F-92195 MEUDON PL CEDEX
FRANCE
TEL: 1-45-34-75-30
TLX:
COM:

MEIN PIERRE
OBSERVATOIRE DE PARIS
SECTION DE MEUDON
5 PLACE JANSSEN
F-92195 MEUDON PL CEDEX
FRANCE
TEL: 1-45-34-75-30
TLX: 270912 OBSASTR
COM: 05,10,12

MEINEL ADEN B PROF
JPL
MS 186-134
4800 OAK GROVE DRIVE
PASADENA CA 91109
U.S.A.
TEL: 818-354-6827
TLX:
COM: 09

MEINIG MANFRED DR
ZENTRALINSTITUT FUR
PHYSIK DER ERDE
TELEGRAFENBERG A17
DDR-1500 POTSDAM
GERMANY, D.R.
TEL: 4551
TLX: 15305 VDE PDM DD
COM: 19

MEIRE RAPHAEL
ASTRONOMICAL OBSERVATORY
GHENT STATE UNIVERSITY
WEIDESTRAAT 11
B-9050 EVERGEM
BELGIUM
TEL: 091-53-87-55
TLX:
COM: 07

MEISEL DAVID D DR
DEPT PHYSICS & ASTRONOMY
STATE UNIVERSITY COLLEGE
SUNY
GENESEO NY 14454
U.S.A.
TEL: 716-245-5284
TLX:
COM: 15,22

MEKLER YURI PROF
DEPT GEOPHYS & PLANET SCI
TEL-AVIV UNIVERSITY
TEL-AVIV 69978
ISRAEL
TEL: 3-413-505
TLX:
COM:

MELBOURNE WILLIAM G DR
JPL - MAILSTOP 238-540
MAILSTOP 238-540
4800 OAK GROVE DRIVE
PASADENA CA 91109
U.S.A.
TEL:
TLX:
COM: 07,19,31

MELCHIOR PAUL J PROF DIR
OBS ROYAL DE BELGIQUE
AVENUE CIRCULAIRE 3
B-1180 BRUXELLES
BELGIUM
TEL: 32-2-374-38-01
TLX: 21565 OBSBEL B
COM: 08,19,31

MELIK-ALAVERDIAN YU DR
BYURAKAN ASTROPHYSICAL
OBSERVATORY
378433 ARMENIA
U.S.S.R.
TEL:
TLX:
COM: 35

MELNICK GARY J
HARVARD-SMITHSONIAN CTR
FOR ASTROPHYSICS
60 GARDEN STREET
CAMBRIDGE MA 02138
U.S.A.
TEL: 617-495-7388
TLX:
COM: 34,44

MELNICK JORGE
DEPTO DE ASTRONOMIA
UNIVERSIDAD DE CHILE
CASILLA 36-D
SANTIAGO
CHILE
TEL: 2294101
TLX:
COM:

MELROSE DONALD B PROF
DEPT OF THEORETICAL
PHYSICS
UNIVERSITY OF SYDNEY
SYDNEY NSW 2006
AUSTRALIA
TEL:
TLX:
COM: 10,48

MEN' A V DR
INST RADIOPHYS & ELECTRON
UKRAINIAN ACADEMY OF SCI
310085 KHARKOV
U.S.S.R.
TEL:
TLX:
COM:

MENDEZ MANUEL DR
INSTITUTO DE ASTRONOMIA
UNAM
APDO POSTAL 70-264
04510 MEXICO DF
MEXICO
TEL:
TLX:
COM:

MENDEZ ROBERTO H DR
INST ASTRONOMIA Y FISICA
DEL ESPACIO
CASILLA 67 SUCURSAL 28
1428 BUENOS AIRES
ARGENTINA
TEL: 781-6755
TLX: 22414 CEDOC AR
COM: 34

MENDIS DEVANITTA ASOKA DR
EECS
UNIVERSITY OF CALIFORNIA
AT SAN DIEGO
LA JOLLA CA 92093
U.S.A.
TEL: 619-452-2719
TLX:
COM: 15,49

MENDOZA CLAUDIO
IBM VENEZUELA SCIENT. CTR
P.O. BOX 388
CARACAS 1010A
VENEZUELA
TEL: 02-9088697
TLX: 23283 IBMVE VC
COM:

MENDOZA V EUGENIO E DR
CIDA
APARTADO 264
MERIDA
VENEZUELA
TEL:
TLX:
COM: 25,45,51

MENEGUZZI MAURICE M DR
SERVICE D'ASTROPHYSIQUE
CEN SACLAY
F-91191 GIF-S-YVETTE
FRANCE
TEL: 1-69-08-44-38
TLX: 690860
COM:

MENG XINMIN
YUNNAN OBSERVATORY
P.O. BOX 110
KUNMING
CHINA, PEOPLE'S REP.
TEL: 72946
TLX: 64040 YUOBS CN
COM: 09

MENNESSIER MARIE-ODILE DR
LABORATOIRE D'ASTRONOMIE
U S T L
F-34060 MONTPELLIER
FRANCE
TEL: 67-63-91-44
TLX: 490944 USTMONT F
COM: 24,27,33

MENON T K PROF
DEPT GEOPHYS & ASTRONOMY
UNIV OF BRITISH COLUMBIA
VANCOUVER BC V6T 1W5
CANADA
TEL: 604-228-2082
TLX: 04542245
COM: 28,34,37,40

MENTESE HUSEYIN DR
UNIVERSITY OBSERVATORY
UNIVERSITY OF ISTANBUL
ISTANBUL
TURKEY
TEL: 522-35-97
TLX:
COM:

MENZIES JOHN W DR
S A A O
P O BOX 9
OBSERVATORY 7935
SOUTH AFRICA
TEL: 47-0025
TLX: 5720309
COM: 25,37,50

MERAT PARVIZ
INSTITUT D'ASTROPHYSIQUE
98 BIS BOULEVARD ARAGO
F-75014 PARIS
FRANCE
TEL: 1-43-20-14-25
TLX:
COM: 47

MERCIER CLAUDE DR
OBSERVATOIRE DE PARIS
SECTION DE MEUDON
DASOP
F-92195 MEUDON PL CEDEX
FRANCE
TEL: 1-45-34-75-30
TLX: 201571 F
COM:

MERGENTALER JAN PROF
ASTRONOMICAL INSTITUTE
UL. KOPERNIKA 19
51-617 WROCLAW
POLAND
TEL: 48-23-29
TLX:
COM: 10,12

MERLEAU-PONTY J PROF
5 R. GENERAL DE CASTELNAU
F-75015 PARIS
FRANCE
TEL:
TLX:
COM: 41

MERMAN G A DR
INSTITUTE OF THEORETICAL
ASTRONOMY
10 KUTUZOV QUAY
192187 LENINGRAD
U.S.S.R.
TEL:
TLX:
COM: 07

MERMAN NATALIA V DR
PULKOVO OBSERVATORY
196140 LENINGRAD
U.S.S.R.
TEL:
TLX:
COM:

MERMILLIOD JEAN-CLAUDE DR
INSTITUT D'ASTRONOMIE
UNIVERSITE DE LAUSANNE
CH-1290 CHAVANNES-DES-B.
SWITZERLAND
TEL: 22-55-26-11
TLX: 27720 OBSG CH
COM: 05,37

MERRIAM JAMES B
DEPT GEOLOGICAL SCIENCES
UNIV OF SASKATCHEWAN
SASKATOON SA S7N 0W0
CANADA
TEL: 306-966-5716
TLX:
COM: 19

MERRILL JOHN E DR
DEPT OF ASTRONOMY
UNIVERSITY OF FLORIDA
SSRB 211
GAINESVILLE FL 32611
U.S.A.
TEL: 904-392-2052
TLX:
COM: 42

MERTZ LAWRENCE N DR
287 FAIRFIELD COURT
PALO ALTO CA 94306
U.S.A.
TEL:
TLX:
COM: 09

MERZANIDES CONSTANTINOS
DEPT OF ASTRONOMY
UNIVERSITY THESSALONIKI
GR-54006 THESSALONIKI
GREECE
TEL:
TLX:
COM:

MESSAGE PHILIP J DR
DEPT APPLIED MATHS AND
THEORETICAL PHYSICS
THE UNIVERSITY
LIVERPOOL L69 3BX
U.K.
TEL: 0510709-6022
TLX: 627095
COM: 07

MESSINA ANTONIO
DIPT. DI ASTRONOMIA
C P 596
I-40100 BOLOGNA
ITALY
TEL:
TLX:
COM:

MESTEL LEON PROF
ASTRONOMY CENTRE
UNIVERSITY OF SUSSEX
BRIGHTON BN1 9QH
U.K.
TEL: 273-60-6755
TLX: 877159 UNISEX G
COM: 35,48,49

MESZAROS PETER DR
PENNSYLVANIA STATE UNIV
525 DAVEY LABORATORY
DEPT OF ASTRONOMY
UNIVERSITY PARK PA 16802
U.S.A.
TEL: 814-865-0418
TLX: 842510
COM: 34,47,48

METZ KLAUS DR
INSTITUT F ASTRONOMIE
SCHEINERSTR 1
D-8000 MUENCHEN 80
GERMANY, F.R.
TEL: 089-989021
TLX:
COM: 27

MEURERS JOSEPH PROF DR
SCHLECHINGER-STR 7
8211 SCHLECHING-ETTENHAUS
GERMANY, F.R.
TEL: 6-08649-483
TLX:
COM: 24,37

MEWE R DR
LAB VOOR RUIMTEONDERZOEK
BENELUXLAAN 21
NL-3527 HS UTRECHT
NETHERLANDS
TEL: 31-30-937145
TLX: 47224
COM: 12,14,44

MEYER CLAUDE DR
C E R G A
AVENUE COPERNIC
F-06130 GRASSE
FRANCE
TEL: 93-36-58-49
TLX: 470865
COM: 26

MEYER FRIEDRICH DR
MPI FUER PHYSIK UND
ASTROPHYSIK
KARL-SCHWARZSCHILD-ST 1
D-8046 GARCHING B MUNCHEN
GERMANY, F.R.
TEL: 89-32990
TLX: 524629 ASTRO D
COM: 12,48

MEYER JEAN-PAUL DR
SERVICE D'ASTROPHYSIQUE
CEN SACLAY
BP NO 2
F-91191 GIF/YVETTE CDX
FRANCE
TEL: 1-69-08-50-25
TLX: 690860 PHYSPAC
COM: 48

MEYER-HOFMEISTER E DR
MPI F PHYSIK & ASTROPHYS
KARL-SCHWARZSCHILD-STR 1
D-8046 GARCHING B MUNCHEN
GERMANY, F.R.
TEL: 089-32990
TLX: 524629 ASTRO D
COM: 35,42

MEYERS KARIE ANN
OHIO STATE UNIVERSITY
ASTRONOMY DEPT
174 WEST 18TH AVENUE
COLUMBUS OH 43210
U.S.A.
TEL: 614-422-7871
TLX:
COM:

MEZGER PETER G PROF
MPI FUER RADIOASTRONOMIE
AUF DEM HUEGEL 69
D-5300 BONN 1
GERMANY, F.R.
TEL: 0228-525297
TLX: 0886440 MPIFRD
COM: 33,34,40V

MEZZETTI MARINO
OSSERVATORIO ASTRONOMICO
DI TRIESTE
VIA G.B. TIEPOLO 11
I-34131 TRIESTE
ITALY
TEL: 040-793221
TLX: 461137 OAT I
COM: 42,47

MIANES PIERRE DR
OBSERVATOIRE DE TOULOUSE
14 AVENUE EDOUARD BELIN
F-31400 TOULOUSE
FRANCE
TEL: 61-25-21-01
TLX:
COM: 25

MIAO YONG-KUAN
DEPT OF ASTRONOMY
NANJING UNIVERSITY
NANJING 210008
CHINA, PEOPLE'S REP.
TEL:
TLX:
COM: 08,46C

MIAO YONG-RUI
SHAANXI OBSERVATORY
P.O. BOX 10
LINTONG XIAN
SHAANXI
CHINA, PEOPLE'S REP.
TEL: XIAN 32255
TLX: 70121 CSAO CN
COM: 31C

MICHALEC ADAM
ASTRONOMICAL OBSERVATORY
JAGIELLONIAN UNIVERSITY
UL. ORLA 171
30-244 KRAKOW
POLAND
TEL: 221817/3856
TLX: 0322297 UJ PL
COM: 40

MICHALITSIANOS ANDREW
NASA/GSFC
CODE 684.1
GREENBELT MD 20771
U.S.A.
TEL: 301-344-6177
TLX:
COM: 10,34,44

MICHARD RAYMOND DR
OBSERVATOIRE DE NICE
BP 139
F-06003 NICE CEDEX
FRANCE
TEL: 93-89-04-20
TLX: 460004
COM: 10,12

MICHAUD GEORGES J DR
250 DU FINISTERE
ST-LAMBERT J4S 1P5
CANADA
TEL: 514-343-6672
TLX: 055-62425 UDEMPHYSAS
COM: 35,36

MICHEL F CURTIS PROF
RICE UNIVERSITY
HOUSTON TX 77251
U.S.A.
TEL: 713-527-4925
TLX: 556457
COM: 48,49

MICZAIKA G R DR
TRW SYSTEMS R5-2291
1 SPACE PARK
REDONDO BEACH CA 90278
U.S.A.
TEL:
TLX:
COM:

MIDDLEHURST BARBARA M MS
LUNAR & PLANETARY INST
3303 NASA ROAD 1
HOUSTON TX 77058
U.S.A.
TEL:
TLX:
COM: 16

MIETELSKI JAN S DR
ASTRONOMICAL OBSERVATORY
JAGIELLONIAN UNIVERSITY
UL. ORLA 171
30-244 KRAKOW
POLAND
TEL: 48-12-22-38-56
TLX: 0322297 UJ PL
COM: 19

MIGEOTTE MARCEL V PROF
INSTITUT D'ASTROPHYSIQUE
UNIVERSITE DE LIEGE
AVENUE DE COINTE 5
B-4200 COINTE-OUGREE
BELGIUM
TEL: 41-52-9980
TLX: 41264 ASTRLG
COM: 12.14

MIGNARD FRANCOIS DR
CERGA
AVENUE COPERNIC
F-06130 GRASSE
FRANCE
TEL: 93-36-58-49
TLX: 470865
COM: 07

MIHAILA IERONIM PROF
BUCHAREST UNIVERSITY
ACADEMIEI 14
70109 BUCAREST
RUMANIA
TEL: 230819
TLX:
COM:

MIHALAS BARBARA R WEIBEL
HIGH ALTITUDE OBSERVATORY
NCAR
PO BOX 3000
BOULDER CO 80307
U.S.A.
TEL: 304-494-5151
TLX:
COM:

MIHALAS DIMITRI DR
DEPARTMENT OF ASTRONOMY
UNIVERSITY OF ILLINOIS
1011 W. SPRINGFIELD AVE.
URBANA IL 61801
U.S.A.
TEL: 217-333-3090
TLX:
COM: 12,36C

MIKAMI TAKAO DR
OSAKA GAKUIN UNIVERSITY
2-36-1, KISHIBE-MINAMI
SUITA-SHI
OSAKA 564
JAPAN
TEL: 06-381-8434
TLX:
COM:

MIKESELL ALFRED H MR
2509 N. CAMPBELL #69
TUCSON AZ 85719
U.S.A.
TEL: 602-327-2381
TLX:
COM:

MIKHAIL FAHMY I PROF DR
AIN SHAMS UNIVERSITY
FACULTY OF SCIENCE
CAIRO
EGYPT
TEL: 575887
TLX: 94070 USHMS UN
COM:

MIKHAIL JOSEPH SIDKY DR
HELWAN OBSERVATORY
CAIRO
EGYPT
TEL:
TLX:
COM: 16

MIKHELSON NIKOLAJ N DR
MAIN ASTRON OBSERVATORY
PULKOVO
196140 LENINGRAD
U.S.S.R.
TEL: 2-979-465
TLX:
COM: 09

MIKKOLA SEPPO DR
TURKU UNIVERSITY OBS
ITAINEN PITKAKATU 1
SF-20520 TURKU
FINLAND
TEL: 921-645976
TLX:
COM: 07,26,33,37

MILANI ANDREA
ISTITUTO DI MATEMATICA
UNIVERSITA' DI PISA
VIA BUONARROTI 2
I-56100 PISA
ITALY
TEL:
TLX:
COM: 07

MILANO LEOPOLDO DR
OSSERVATORIO ASTRONOMICO
DI CAPODIMONTE
VIA MOIARIELLO 16
I-80131 NAPOLI
ITALY
TEL:
TLX:
COM:

MILES HOWARD G MR
LANE PARK
PITYNE, ST MINVER
WADEBRIDGE PL27 6PN
U.K.
TEL: 020-886-3153
TLX:
COM: 22,46

MILET BERNARD L DR
OBSERVATOIRE DE NICE
BP 139
F-06003 NICE CEDEX
FRANCE
TEL: 93-89-04-20
TLX:
COM: 15,20,51

MILEY G K DR
SPACE TELESCOPE SCI INST
HOMEWOOD CAMPUS
3700 SAN MARTIN DRIVE
BALTIMORE MD 21218
U.S.A.
TEL: 301-338-4760
TLX:
COM: 40

MILKEY ROBERT W DR
SPACE TEL SCIENCE INST
HOMEWOOD CAMPUS
3700 SAN MARTIN DRIVE
BALTIMORE MD 21218
U.S.A.
TEL: 301-338-4720
TLX: 6849101
COM: 12

MILLAR THOMAS J DR
MATHEMATICS DEPT
UMIST
P O BOX 88
MANCHESTER M60 1QD
U.K.
TEL: 061-236-3311
TLX: 666094
COM: 34

MILLER FREEMAN D PROF
UNIVERSITY OF MICHIGAN
DEPT OF ASTRONOMY
DENNISON BLDG
ANN ARBOR MI 48109
U.S.A.
TEL: 313-764-3447
TLX:
COM: 15

MILLER HUGH R PROF
DEPT OF PHYSICS
GEORGIA STATE UNIVERSITY
ATLANTA GA 30303
U.S.A.
TEL: 404-658-2279
TLX:
COM: 28

MILLER JOHN C DR
DEPT OF ASTROPHYSICS
UNIVERSITY OF OXFORD
SOUTH PARKS ROAD
OXFORD OX1 3RQ
U.K.
TEL: 44-865-511336
TLX: 83295 NUCLOX G
COM: 48

MILLER JOSEPH S PROF
LICK OBSERVATORY
UNIVERSITY OF CALIFORNIA
SANTA CRUZ CA 95060
U.S.A.
TEL: 408-429-2135
TLX:
COM: 25C,28,34

MILLER RICHARD H DR
ASTRONOMY DEPT
UNIVERSITY OF CHICAGO
5640 ELLIS AVENUE
CHICAGO IL 60637
U.S.A.
TEL: 312-962-8201
TLX: 6871133
COM: 28,33

MILLET JEAN DR
LAS
TRAVERSE DU SIPHON
LES TROIS LUCS
F-13012 MARSEILLE
FRANCE
TEL: 91-66-08-32
TLX: 420584 ASTROSP F
COM:

MILLIARD BRUNO
LAB ASTRONOMIE SPATIALE
TRAVERSE DU SIPHON
LES TROIS LUCS
F-13012 MARSEILLE
FRANCE
TEL: 91-05-59-00
TLX: 420584 ASTROSP F
COM:

MILLIGAN J E
ASTROPHYSICS BRANCH
GODDARD SPACE FLIGHT CTR
GREENBELT MD 20771
U.S.A.
TEL:
TLX:
COM: 29

MILLIKAN ALLAN G MR
RESEARCH LABS B-59
EASTMAN KODAK CO
343 STATE STREET
ROCHESTER NY 14650
U.S.A.
TEL: 716-722-0277
TLX:
COM: 09

MILLIS ROBERT L DR
LOWELL OBSERVATORY
1400 W. MARS HILL ROAD
FLAGSTAFF AZ 86001
U.S.A.
TEL: 602-774-3358
TLX:
COM: 16,20C

MILLMAN PETER M DR
HERZBERG INST ASTROPHYS
NATL RES COUNCIL CANADA
OTTAWA ONT K1A 0R6
CANADA
TEL: 613-990-0705
TLX: 053-3715
COM: 15,16,22,W

MILLS ALLAN A DR
DEPT OF ASTRONOMY
THE UNIVERSITY
LEICESTER LE1 7RH
U.K.
TEL: 0533-554455
TLX:
COM:

MILLS BERNARD Y PROF
SCHOOL OF PHYSICS
UNIVERSITY OF SYDNEY
SYDNEY NSW 2006
AUSTRALIA
TEL: 02-692-2544
TLX: 26169 UNISYD
COM: 28,40

MILNE DOUGLAS K DR
CSIRO
DIVISION OF RADIOPHYSICS
P.O.BOX 76
EPPING NSW 2121
AUSTRALIA
TEL: 02-868-0222
TLX: 26230 ASTRO
COM: 34,40

MILOGRADOV-TURIN JELENA
INSTITUTE OF ASTRONOMY
UNIVERSITY OF BEOGRAD
STUDENTSKI TRG 16
YU-11000 BEOGRAD
YUGOSLAVIA
TEL: 638-715
TLX:
COM: 40

MILONE EUGENE F PROF
PHYSICS DEPT
UNIVERSITY OF CALGARY
2500 UNIVERSITY DR NW
CALGARY AB T2N 1N4
CANADA
TEL: 403-220-5412
TLX: 03821545
COM: 25C,27,42

MILONE LUIS A DR
OBSERVATORIO ASTRONOMICO
LAPRIDA 854
5000 CORDOBA
ARGENTINA
TEL: 36876 - 40629
TLX: 51822 BUCOR
COM: 27

MILOVANOVIC VLADETA DR
INSTITUT ZA GEODEZIJU
BULEVAR REVOLUCIJE 73
11000 BEOGRAD
YUGOSLAVIA
TEL:
TLX:
COM: 19

MINAROVJECH MILAN
ASTRONOMICAL INSTITUTE
SLOVAK ACAD. OF SCIENCES
059 60 TATRANSKA LOMNICA
CZECHOSLOVAKIA
TEL: 967-8668x0969
TLX: 78277
COM: 09

MININ I N PROF
LENINGRAD STATE UNIV
ASTRONOMICAL OBSERVATORY
BIBLIOTECHNAJA PL. 2
198904 LENINGRAD-PETRODV.
U.S.S.R.
TEL: 257-94-89
TLX:
COM: 34

MINN YOUNG KEY DR
KOREAN NATIONAL
OBSERVATORY
KANGNAM-KU
SEOUL 134-03
KOREA, REPUBLIC
TEL: 567-0751
TLX: 24230 MIOST K
COM: 34,51

MINNET HARRY C MR
CSIRO
DIVISION OF RADIOPHYSICS
P.O.BOX 76
EPPING NSW 2121
AUSTRALIA
TEL:
TLX:
COM:

MINTZ BLANCO BETTY MRS
CERRO TOLOLO
INTERAMERICAN OBSERVATORY
CASILLA 603
LA SERENA
CHILE
TEL: 213352
TLX:
COM: 20,25

MIRABEL IGOR FELIX DR
IAFE
CC 67, SUC. 28
1428 BUENOS AIRES
U.S.A.
TEL:
TLX:
COM: 28,33,40,51

MIRONOV NIKOLAY T
MAIN ASTRONOMICAL OBS
UKRAINIAN ACADEMY OF SCI
252127 KIEV
U.S.S.R.
TEL: 66-47-59
TLX: 131406 SKY
COM: 19

MIRZOYAN L V DR PROF
BYURAKAN ASTROPHYSICAL
OBSERVATORY
378433 ARMENIA
U.S.S.R.
TEL:
TLX:
COM: 27C,33

MISCONI NEBIL YOUSIF DR
SPACE ASTRONOMY LAB
UNIVERSITY OF FLORIDA
1810 N.W. 6TH STREET
GAINESVILLE FL 32609
U.S.A.
TEL:
TLX:
COM: 21,22

MISNER CHARLES W PROF
DEPT PHYSICS & ASTRONOMY
UNIVERSITY OF MARYLAND
COLLEGE PARK MD 20742
U.S.A.
TEL:
TLX:
COM: 47

MISSANA MARCO DR
OSSERVATORIO ASTRONOMICO
DI BRERA
VIA CREMAGNANI 13/11
I-20059 VIMERCATE
ITALY
TEL:
TLX:
COM:

MISSANA NATALE PROF
VIA PUCCINI 2
I-20025 PINO TORINESE
ITALY
TEL:
TLX:
COM:

MITALAS ROMAS ASSOC PROF
DEPT OF ASTRONOMY
UNIV OF WESTERN ONTARIO
LONDON ONT N6A 5B9
CANADA
TEL: 519-679-3184
TLX:
COM: 35

MITCHELL GEORGE F DR
DEPT OF ASTRONOMY
SAINT MARY'S UNIVERSITY
HALIFAX NS B3H 3C3
CANADA
TEL: 902-429-9780
TLX:
COM:

MITCHELL RICHARD MR
12704 LA CUEVA N.E.
ALBUQUERQUE NM 87123
U.S.A.
TEL: 505-292-0309
TLX:
COM: 25

MITCHELL WALTER E JR
ASTRONOMY DEPT
OHIO STATE UNIVERSITY
174 W. 18TH AVE
COLUMBUS OH 43210
U.S.A.
TEL: 614-422-5554
TLX:
COM:

MITIC LJUBISA A DR
OBSERVATOIRE DE BELGRADE
VOLGINA 7
11050 BELGRADE
YUGOSLAVIA
TEL: 011-419-357
TLX:
COM: 08

MITRA A P DR
NATIONAL PHYSICAL LAB
NEW DELHI 110 012
INDIA
TEL: 585298/581440
TLX: 3162454 RSD IN
COM:

MITROFANOVA LYUDMILA A DR
PULKOVO OBSERVATORY
196140 LENINGRAD
U.S.S.R.
TEL:
TLX:
COM:

MITTON SIMON DR
CAMBRIDGE UNIV PRESS
SHAFTSBURY ROAD
CAMBRIDGE CB2 2RU
U.K.
TEL: 0-223-312-393
TLX: 817256 CUPCAM UK
COM: 05C

MIYADI MASASI DR
SENGAWA 3-6-11
CHOFU-SHI 182
JAPAN
TEL: 03-300-4632
TLX:
COM: 19

MIYAJI SHIGEKI DR
DEPT NATURAL HISTORY
COLLEGE ARTS & SCS
1-33 YAYOICHO
CHIBA 260
JAPAN
TEL: 472-51-1111
TLX:
COM: 35,42,48

MIYAMA SYOKEN
DEPT OF PHYSICS
KYOTO UNIVERSITY
KITASHIRAKAWA OIWAKECHO
KYOTO 606
JAPAN
TEL: 075-751-2111
TLX:
COM: 34

MIYAMOTO MASANORI DR
TOKYO ASTRONOMICAL
OBSERVATORY
OSAWA MITAKA
TOKYO 181
JAPAN
TEL: 0422-32-5111
TLX: 2822307 TAOMTK J
COM: 08V,33

MIYAMOTO SIGENORI PROF
DEPT OF PHYSICS/FAC SCI
OSAKA UNIVERSITY
MACHIKANEYAMA-CHO
TOYONAKA OSAKA 560
JAPAN
TEL: 06-844-1151
TLX:
COM: 16,36,44,48

MIYAMOTO SYOTARO PROF DR
KASAN OBSERVATORY
YAMASHINA
KYOTO 607
JAPAN
TEL: 075-581-1235
TLX:
COM:

MIZUNO SHUN
KANAZAWA INST TECHNOLOGY
7-1 OGIGAOKA
NONOICHIMACHI
ISHIKAWA 921
JAPAN
TEL: 0762-48-1100
TLX: 5122456 KITLC J
COM: 34

MNATSAKANIAN MAMIKON A DR
BYURAKAN ASTROPHYSICAL
OBSERVATORY
378433 ARMENIA
U.S.S.R.
TEL: 56-34-53
TLX:
COM: 36

MO JING-ER
PURPLE MOUNTAIN
OBSERVATORY
NANJING
CHINA. PEOPLE'S REP.
TEL: 36967
TLX: 34144 PMONJ CN
COM:

MOCHNACKI STEPHAN W DR
DAVID DUNLAP OBSERVATORY
DEPT OF ASTRONOMY
UNIVERSITY OF TORONTO
TORONTO ONT M5S 1A7
CANADA
TEL:
TLX:
COM: 42

MOCZKO JANUSZ DR
BOROWIEC
63-120 KORNIK
POLAND
TEL:
TLX:
COM: 19

MODALI SARMA B DR
QSM SYSTEMS & RESEARCH
8401 CORPORATION DR.
LANDOVER MD 20785
U.S.A.
TEL: 301-459-3322
TLX:
COM:

MODISETTE JERRY L PROF
HOUSTON BAPTIST UNIV
7502 FONDREN AVE
HOUSTON TX 77074
U.S.A.
TEL: 375-774-7661
TLX:
COM: 44

MOEHLMANN DIEDRICH
INST. F. KOSMOSFORSCHUNG
RUDOWER CHAUSSEE 5
DDR-1199 BERLIN
GERMANY. D.R.
TEL: 6743485
TLX: 113132 IKF DD
COM: 15,16

MOELLENHOFF CLAUS DR
LANDESSTERNWARTE
KOENIGSTUHL
D-6900 HEIDELBERG 1
GERMANY, F.R.
TEL: 06221-10036
TLX: 461789 MPIA D
COM: 35

MOERDIJK WILLY G DR
ASTRONOMISCH OBS R.U.G.
ST PIETERSAALSTSTRAAT 171
B-9000 GENT
BELGIUM
TEL: 091-221233
TLX:
COM:

MOESGAARD KRISTIAN P
HISTORY OF SC DEPT
UNIVERSITY OF AARHUS
BYGADAN 1 / TORRILD
DK-8300 ODDER
DENMARK
TEL: 06-53-1004
TLX:
COM: 41

MOFFAT ANTHONY F J DR
DEPT DE PHYSIQUE
UNIVERSITE DE MONTREAL
CP 6128 - SUCC A
MONTREAL PQ H3C 3J7
CANADA
TEL: 514-343-6682
TLX: 05562425 UDEMPHYSAS
COM: 29,33,37

MOFFAT JOHN W. DR
DEPT OF PHYSICS
UNIVERSITY OF TORONTO
TORONTO ONT M5S 1A7
CANADA
TEL: 416-978-2949
TLX:
COM:

MOFFATT HENRY KEITH PROF
DEPT APPLIED MATHS
SILVER STREET
CAMBRIDGE CB3 9EW
U.K.
TEL: 0223-351645
TLX: 81249 CAMSPL G
COM:

MOFFET ALAN T PROF
OWENS VALLEY RADIO OBS
CALTECH 105 24
PASADENA CA 91125
U.S.A.
TEL: 818-356-4977
TLX: 675425
COM: 40

MOFFETT THOMAS J PROF
DEPT OF PHYSICS
PURDUE UNIVERSITY
WEST LAFAYETTE IN 47907
U.S.A.
TEL: 317-494-5508
TLX:
COM: 25,27

MOGILEVSKIJ EH I DR
INSTITUTE OF TERRESTRIAL
MAGNETISM, IONOSPHERE AND
RADIO WAVE PROPAGATION
142092 TROITSK. MOSCOW R.
U.S.S.R.
TEL: 232-19-31
TLX: 412623 SCSTP SU
COM: 10

MOHLER ORREN C PROF
405 AWIXA RD
ANN ARBOR MI 48104
U.S.A.
TEL: 313-662-2770
TLX:
COM: 07,10,14

MOISEEV I G DR
CRIMEAN ASTROPHYS OBS
USSR ACADEMY OF SCIENCES
USSR ACADEMY OF SCIENCES
334413 CRIMEA
U.S.S.R.
TEL:
TLX:
COM: 10,40

MOLCHANOV A P PROF
LENINGRAD STATE UNIV
ASTRONOMICAL OBSERVATORY
199178 LENINGRAD
U.S.S.R.
TEL:
TLX:
COM: 40

MOLES MARIANO J DR
INST DE ASTROFISICA DE
ANDALUCIA, APDO 2144
PROFESOR ALBAREDA 1
18080 GRANADA
SPAIN
TEL: 58-12-13-11
TLX: 78573 IAAG E
COM: 28

MOLINA ANTONIO
INST DE ASTROFISICA DE
ANDALUCIA, APDO 2144
PROFESOR ALBAREDA 1
18080 GRANADA
SPAIN
TEL: 58-12-13-00
TLX: 78573 IAAG E
COM: 16,21

MOLNAR MICHAEL R PROF
AT&T BELL LABORATORIES
ROOM 2B-422
CRAWFORDS CORNER ROAD
HOLMDEL NJ 07733
U.S.A.
TEL:
TLX:
COM:

MOMCHEV GOSPODIN
ASTRONOMICAL OBSERVATORY
P.O. BOX 7
8800 SLIVEN
BULGARIA
TEL: 2-72-04
TLX:
COM:

MONAGHAN JOSEPH J DR
MATHEMATICS DEPT
MONASH UNIVERSITY
CLAYTON VIC 3168
AUSTRALIA
TEL: 03-541-2563
TLX: MONASH AA 32691
COM: 35

MONET DAVID G
US NAVAL OBSERVATORY
PO BOX 1149
FLAGSTAFF AZ 86001
U.S.A.
TEL:
TLX:
COM: 24,33,44

MONFILS ANDRE G PROF
IAL SPACE
UNIVERSITE DE LIEGE
AVENUE DU PRE AILY
B-4900 ANGLEUR-LIEGE
BELGIUM
TEL: 041-67-66-68
TLX: 41320 IAL SP B
COM: 14,44

MONNET GUY J DR
OBSERVATOIRE DE LYON
AVENUE CHARLES ANDRE
F-69230 ST GENIS-LAVAL
FRANCE
TEL: 78-56-07-05
TLX: 310926
COM: 33

MONSIGNORI FOSSI BRUNA DR
OSSERVATORIO ASTROFISICO
DI ARCETRI
LARGO E. FERMI 5
I-50125 FIRENZE
ITALY
TEL:
TLX:
COM:

MONTES CARLOS DR
OBSERVATOIRE DE NICE
BP 139
F-06003 NICE CEDEX
FRANCE
TEL: 93-89-04-20
TLX: 460004
COM:

MONTMERLE THIERRY DR
SERVICE D'ASTROPHYSIQUE
CEN SACLAY
F-91191 GIF/YVETTE CDX 1
FRANCE
TEL: 1-69-08-57-22
TLX: 690860
COM:

MOOK DELO E PROF
DEPT PHYSICS & ASTRONOMY
DARTMOUTH COLLEGE
HANOVER NH 03755
U.S.A.
TEL: 603-646-2972
TLX:
COM:

MOONS MICHELE B M M
DEPT DE MATHEMATIQUE
FAC UNIV N.D. DE LA PAIX
REMPART DE LA VIERGE 8
B-5000 NAMUR
BELGIUM
TEL: 081-229061x2438
TLX: 59222 FACNAM B
COM: 07

MOORE DANIEL R DR
DEPT APPLIED MATHS
SILVER STREET
CAMBRIDGE CB3 9EW
U.K.
TEL:
TLX:
COM: 35

MOORE ELLIOTT P PROF
JOINT OBSERVATORY
FOR COMETARY RESEARCH
CAMPUS STATION
SOCORRO NM 87801
U.S.A.
TEL: 505-835-5431
TLX:
COM: 15

MOORE PATRICK DR
FARTHINGS
39 WEST STREET
SELSEY, SUSSEX
U.K.
TEL: 0243-603-668
TLX:
COM: 16

MOORE RONALD L DR
NASA/MARSHALL SPACE
FLIGHT CENTER ES 52
SPACE SCIENCE LABORATORY
HUNTSVILLE AL 35812
U.S.A.
TEL: 205-453-0118
TLX: 59-4416 NASA MSFC HT
COM:

MOORHEAD JAMES M DR
ASTRONOMY DEPT
UNIVERSITY OF W. ONTARIO
LONDON ONT N6A 3K7
CANADA
TEL: 519-679-3186
TLX:
COM:

MOORWOOD ALAN F M
ESO
KARL-SCHWARZSCHILD-STR 2
D-8046 GARCHING B MUNCHEN
GERMANY, F.R.
TEL: 089-320-06-294
TLX: 05 28 282 24 EO D
COM: 28

MOOS HENRY WARREN DR
DEPT PHYSICS & ASTRONOMY
JOHNS HOPKINS UNIVERSITY
BALTIMORE MD 21218
U.S.A.
TEL: 301-338-7337
TLX: 7102341090
COM: 29,44

MORAN JAMES M DR
CENTER FOR ASTROPHYSICS
60 GARDEN STREET
CAMBRIDGE MA 02138
U.S.A.
TEL: 617-495-7477
TLX: 921428 SATELLITE CAM
COM: 40

MORANDO BRUNO L DR
BUREAU DES LONGITUDES
77 AVE DENFERT-ROCHEREAU
F-75014 PARIS
FRANCE
TEL: 1-43-20-12-10
TLX:
COM: 04P,07,20

MORBEY CHRISTOPHER L
DOMINION ASTROPHYS OBS
5071 W SAANICH ROAD
RR 5
VICTORIA BC V8X 4M6
CANADA
TEL: 604-388-0220
TLX: 0497295
COM: 26,28,30

MOREELS GUY DR
OBSERVATOIRE DE BESANCON
41B AVE DE L'OBSERVATOIRE
F-25000 BESANCON
FRANCE
TEL: 81-50-22-66
TLX: 361144 F
COM:

MOREL PIERRE JACQUES DR
OBSERVATOIRE DE NICE
BP 139
F-06003 NICE CEDEX
FRANCE
TEL: 93-89-04-20
TLX: 460004
COM: 26

MORENO HUGO PROF
DEPTO DE ASTRONOMIA
UNIVERSIDAD DE CHILE
CASILLA 36-D
SANTIAGO
CHILE
TEL: 229-4101/4002
TLX: 440001 PBCZ OBSERNAL
COM: 25

MORENO-INSERTIS FERNANDO
INST DE ASTROFISICA DE
CANARIAS
38071 TENERIFE
SPAIN
TEL: 922-262211
TLX: 92640
COM: 10,12

MORETON G E
15-5 THE ESPLANADE
BALMORAL BEACH NSW 2088
AUSTRALIA
TEL:
TLX:
COM: 10

MORGAN BRIAN LEALAN
BLACKETT LABORATORY
IMPERIAL COLLEGE
LONDON SW7 2BZ
U.K.
TEL: 01-589-5111
TLX: 261503
COM: 09

MORGAN DAVID H DR
ROYAL OBSERVATORY
BLACKFORD HILL
EDINBURGH EH9 3HJ
U.K.
TEL: 031-667-3321
TLX: 72383 ROEDIN G
COM: 21,34

MORGAN JOHN ADRIAN
THE AEROSPACE CORPORATION
MS M4/041
PO BOX 92957
LOS ANGELES CA 90009
U.S.A.
TEL:
TLX:
COM: 35

MORGAN PETER DR
CANBERRA COLL ADV EDUC
SCHOOL OF APPLIED SCIENCE
P.O.BOX 1
BELCONNEN ACT 2616
AUSTRALIA
TEL: 062-52-2111
TLX: 62267 CANCOL AA
COM: 19,31

MORGAN THOMAS H DR
JOHNSON SPACE CENTER
CODE SN3
HOUSTON TX 77058
U.S.A.
TEL: 713-483-5039
TLX: 762931
COM: 42,44

MORGAN WILLIAM W PROF
YERKES OBSERVATORY
PO BOX 258
WILLIAMS WI 53191
U.S.A.
TEL: 414-245-5555
TLX:
COM: 28,45

MORGULEFF NINA ING
INSTITUT D'ASTROPHYSIQUE
98 BIS BOULEVARD ARAGO
F-75014 PARIS
FRANCE
TEL: 1-43-20-14-25
TLX:
COM: 27,29,45

MOROZHENKO N N DR
MAIN ASTRONICAL OBS
UKRAINIAN ACADEMY OF SCI
GOLOSEEVO
252127 KIEV
U.S.S.R.
TEL: 663110
TLX: 131406 SKY
COM: 10

MORRISON PHILIP PROF
DEPT OF PHYSICS
MIT 6-205
CAMBRIDGE MA 02139
U.S.A.
TEL: 617-253-5086
TLX:
COM: 47,48,51

MOURADIAN ZADIG M DR
OBSERVATOIRE DE PARIS
SECTION DE MEUDON
DASOP -LA326
F-92195 MEUDON PL CEDEX
FRANCE
TEL: 1-45-34-75-30
TLX:
COM: 12

MORIMOTO MASAKI DR
TOKYO ASTRONOMICAL
OBSERVATORY
OSAWA MITAKA
TOKYO 181
JAPAN
TEL:
TLX:
COM: 34,40,51

MORRAS RICARDO DR.
INSTITUTO ARGENTINO
DE RADIOASTRONOMIA
CASILLA DE CORREO 5
1894 VILLA ELISA (Bs.As.)
ARGENTINA
TEL: 021-43793
TLX: 18052 CICYR-AR
COM: 40

MORTON DONALD C DR
HERZBERG INST ASTROPHYS
NATIONAL RESEARCH COUNCIL
100 SUSSEX DRIVE
OTTAWA ONT K1A 0R6
CANADA
TEL:
TLX:
COM: 09,29,34,44

MOUSCHOVIAS TELEMACHOS CH
DEPT OF ASTRONOMY
UNIVERSITY OF ILLINOIS
1011 W. SPRINGFIELD AVE
URBANA IL 61801
U.S.A.
TEL: 217-333-3090
TLX:
COM: 34

MORISON IAN MR
NRAL
JODRELL BANK
MACCLESFIELD SK11 9DL
U.K.
TEL: 0477-71321
TLX: 36149
COM: 40

MORRIS DAVID DR
IRAM
VOIE 10
DOMAINE UNIVERSITAIRE
F-38406 ST-MARTIN-D'HERES
FRANCE
TEL: 75-42-33-83
TLX: 950753
COM: 40

MORTON G A DR
1122 SKYCREST DR. APT 6
WALNUT CREEK CA 94595
U.S.A.
TEL:
TLX:
COM:

MOUSSAS XENOPHON PH D
NATIONAL UNIVERSITY
ASTROPHYSICS LABET
20 SKYLITSI STREET
GR-11473 ATHENS
GREECE
TEL: 7235122/8843877
TLX:
COM: 49

MORITA KAZUHIKO
DEPT OF PHYSICS
HOKKAIDO UNIVERSITY
NISHI 8-CHOME KITA 10-JYO
SAPPORO, HOKKAIDO 060
JAPAN
TEL: 11-711-2111
TLX:
COM: 40

MORRIS MARK ROOT DR
DEPT OF ASTRONOMY
MATH-SCIENCES BLDG
UCLA
LOS ANGELES CA 90024
U.S.A.
TEL: 213-825-3320
TLX:
COM: 33,34,40,51

MOSS CHRISTOPHER DR
INSTITUTE OF ASTRONOMY
MADINGLEY ROAD
CAMBRIDGE CB3 0HA
U.K.
TEL: 0223-337548
TLX: 817297 ASTRON G
COM: 28

MOUTSOULAS MICHAEL PROF
DEPT OF EARTH SCIENCES
UNIVERSITY OF ATHENS
GR-15784 ATHENS
GREECE
TEL: 7247569
TLX: 215255 GR
COM: 16,51

MORIYAMA FUMIO PROF
OSAKA GAKUIN UNIVERSITY
2-36-1 KISHIBE MINAMI
SUITA-SHI
OSAKA-FU 564
JAPAN
TEL: 06-381-8434
TLX:
COM: 10,12,40

MORRIS STEPHEN C DR
DOMINION ASTROPHYS OBS
5071 W SAANICH ROAD
RR 5
VICTORIA BC V8X 4M6
CANADA
TEL: 604-388-3976
TLX: 0497295
COM: 25,35

MOSS DAVID L DR
MATHEMATICS DEPT
MANCHESTER UNIVERSITY
MANCHESTER M13 9PL
U.K.
TEL:
TLX:
COM: 35

MOVAHED REZA DR
P O BOX 6
BABOLSAR
IRAN
TEL:
TLX:
COM:

MOROSSI CARLO
OSSERVATORIO ASTRONOMICO
VIA G.B. TIEPOLO
I-34131 TRIESTE
ITALY
TEL:
TLX:
COM: 25,29,45

MORRISON DAVID PROF
INSTITUTE FOR ASTRONOMY
UNIVERSITY OF HAWAII
HONOLULU HI 96822
U.S.A.
TEL: 808-948-8531
TLX:
COM: 15,16V,W

MOTTA SANTO DR
DIPT. DI MATIMATICA
CITTA UNIVERSITARIA
VIALE A. DORIA 6
I-95125 CATANIA
ITALY
TEL: 095-330533x668
TLX: 970359 ASTRCT-I
COM: 10

MRKOS ANTONIN DR
DEPT OF ASTRONOMY
CHARLES UNIVERSITY
SVEDSKA 8
150 00 PRAHA 5
CZECHOSLOVAKIA
TEL:
TLX: 144307 KLET CZ
COM: 06P,15,20

MOROZ V I PROF DR
SPACE RESEARCH INSTITUTE
USSR ACADEMY OF SCIENCES
117810 MOSCOW
U.S.S.R.
TEL: 333-31-22
TLX: 411498 CTAP CY
COM: 16,51

MORRISON LESLIE V
ROYAL GREENWICH OBS
HERSTMONCEUX CASTLE
HAILSHAM BN27 1RP
U.K.
TEL: 032-181-3171
TLX: 87451 RGOBSY G
COM: 04,08,19

MOTZ LLOYD PROF
DEPT OF ASTRONOMY
COLUMBIA UNIVERSITY
PUPIN HALL BOX 57
NEW YORK NY 10027
U.S.A.
TEL: 212-280-3279
TLX:
COM:

MUELLER EDITH A PROF
RENNWEG 15
CH-4052 BASEL
SWITZERLAND
TEL: 061-42-31-68
TLX:
COM: 12,36,38P,44,46

MOROZHENKO A V DR
MAIN ASTRONOMICAL OBS
UKRAINIAN ACADEMY OF SCI
GOLOSEEVO
252127 KIEV
U.S.S.R.
TEL: 663110
TLX: 131406 SKY
COM: 16C

MORRISON NANCY DUNLAP DR
DEPT PHYSICS & ASTRONOMY
UNIVERSITY OF TOLEDO
2801 W BANCROFT ST
TOLEDO OH 43606
U.S.A.
TEL: 419-537-2659
TLX:
COM: 27,29

MOULD JEREMY R
DIV PHYSICS & ASTRONOMY
CALTECH 105-24
PASADENA CA 91125
U.S.A.
TEL: 818-356-4168
TLX:
COM: 28,37

MUELLER EWALD
MPI F PHYS & ASTROPHYSIK
INSTITUT F. ASTROPHYSIK
KARL-SCHWARZSCHILD-STR. 1
D-8046 GARCHING B MUNCHEN
GERMANY, F.R.
TEL: 089-3299-0
TLX: 524629 ASTRO D
COM: 35

MUELLER HELMUT O PROF DR
HERZOGENMUEHLESTR. 4
CH-8051 ZUERICH
SWITZERLAND
TEL: 01-41-11-47
TLX:
COM:

MUELLER IVAN I PROF
GEODETIC SCI & SURVEYING
OHIO STATE UNIVERSITY
1958 NEIL AVENUE
COLUMBUS OH 43210-1247
U.S.A.
TEL: 614-422-2269
TLX:
COM: 04,19C,31C

MUENCH GUIDO PROF
MPI FUER ASTRONOMIE
KOENIGSTUHL
D-6900 HEIDELBERG 1
GERMANY, F.R.
TEL: 06221-528210
TLX: 461 789 MPIA D
COM: 33,34,36

MUFSON STUART LEE DR
ASTRONOMY DEPT
INDIANA UNIVERSITY
319 SWAIN WEST
BLOOMINGTON IN 47401
U.S.A.
TEL: 812-335-6927
TLX:
COM: 34

MUKAI SONOYO DR
KANAZAWA TECHNOLOGIC INST
7-1 OGIGAOKA
NONOICHIMACHI
ISHIKAWA 921
JAPAN
TEL: 0762-48-1100
TLX: 5122456 KITLCJ
COM: 21,36

MUKAI TADASHI DR
KANAZAWA TECHNOLOGIC INST
7-1 OGIGAOKA
NONOICHIMACHI
ISHIKAWA 921
JAPAN
TEL: 0762-48-1100
TLX: 5122456 KITLCJ
COM: 21C

MULDERS GERARD F W
4519 EVERETT STREET
KENSINGTON MD 20895
U.S.A.
TEL: 301-564-0090
TLX:
COM:

MULHOLLAND J DERRAL DR
SPACE ASTRONOMY LAB
UNIVERSITY OF FLORIDA
1810 NW 6TH ST
GAINSVILLE FL 32609
U.S.A.
TEL: 904-392-5450
TLX: 810-825-2308
COM: 07,16,20

MULLALY RICHARD F DR
SCHOOL OF ELECTRICAL
ENGINEERING
UNIVERSITY OF SYDNEY
SYDNEY NSW 2006
AUSTRALIA
TEL:
TLX:
COM:

MULLAN DERMOTT J DR
BARTOL RES FOUNDATION
UNIVERSITY OF DELAWARE
NEWARK DE 19716
U.S.A.
TEL: 301-398-3368
TLX: 5106660805
COM:

MULLER A B DR
THOMASLAAN 40
NL-5631 GM EINDHOVEN
NETHERLANDS
TEL: 040-430322
TLX:
COM: 25

MULLER C A PROF JR
ODINKSVELD 8
NL-7491 HD DELDEN
NETHERLANDS
TEL: 05407-2428
TLX:
COM: 40

MULLER PAUL
LES GENETS
CHEMIN DES GABRES
F-06520 MAGAGNOSC
FRANCE
TEL:
TLX:
COM: 26

MULLER RICHARD A
LAWRENCE BERKELEY LAB
BLDG 50, RM 238
BERKELEY CA 94720
U.S.A.
TEL: 415-486-5235
TLX:
COM: 47,51

MULLER RICHARD DR
OBSERVATOIRE PIC-DU-MIDI
F-65200 BAGNERES-DE-B.
FRANCE
TEL: 62-95-00-69
TLX:
COM: 10,12

MUMFORD GEORGE S PROF
DEPARTMENT OF EDUCATION
TUFTS UNIVERSITY
FILENE CENTER
MEDFORD MA 02155
U.S.A.
TEL: 617-653-8923
TLX:
COM: 25,27,42

MUMMA MICHAEL JON
NASA/GSFC
CODE 693
GREENBELT MD 20771
U.S.A.
TEL: 301-344-6994
TLX:
COM: 14,15,16

MUNRO RICHARD H DR
HIGH ALTITUDE OBSERVATORY
PO BOX 3000
BOULDER CO 80307
U.S.A.
TEL: 303-497-1564
TLX: 45694
COM: 12

MURAKAMI TOSHIO
INST SPACE & ASTRON SCI
4-6-1 KOMABA
MEGORU-KU
TOKYO 153
JAPAN
TEL: 03-467-1111x303
TLX: 34757 ISASTRO J
COM:

MURDIN PAUL G DR
ROYAL GREENWICH
OBSERVATORY
HERSTMONCEUX CASTLE
HAILSHAM BN27 1RP
U.K.
TEL:
TLX:
COM: 27,50C

MURDOCH HUGH S DR
ASTROPHYSICS DEPT
UNIVERSITY OF SYDNEY
SYDNEY NSW 2006
AUSTRALIA
TEL: 02-692-2222
TLX: 26169 UNISYD
COM: 28,40

MURDOCK THOMAS LEE
GENERAL RESEARCH CORP.
1891 PROFESSIONAL BLDG
LIBERTY SQUARE
DANVERS MA 01923
U.S.A.
TEL: 617-771-6584
TLX:
COM: 44

MURPHY ROBERT E DR
NASA HEADQUARTERS
CODE EEL
WASHINGTON DC 20546
U.S.A.
TEL: 202-453-1720
TLX: 89530 NASA WSH
COM: 16

MURRAY C ANDREW
ROYAL GREENWICH
OBSERVATORY
HERSTMONCEUX CASTLE
HAILSHAM BN27 1RP
U.K.
TEL: 0323-833171
TLX: 87451
COM: 08,24,33,37

MURRAY CARL D
SCHOOL OF MATHEMAT. SCI.
QUEEN MARY COLLEGE
MILE END ROAD
LONDON E1 4NS
U.K.
TEL: 01-980-4811
TLX: 893750
COM: 07,20

MURRAY JOHN B DR
UNIVERSITY OF LONDON
OBSERVATORY
MILL HILL PARK
LONDON NW7 2QS
U.K.
TEL:
TLX:
COM:

MURRAY STEPHEN S DR
HARVARD-SMITHSONIAN
CENTER FOR ASTROPHYSICS
60 GARDEN STREET
CAMBRIDGE MA 02138
U.S.A.
TEL: 617-495-7205
TLX: 921428 SATELLITE CAM
COM: 09,28

MUSEN PETER DR
8804 ORBIT LANE
LANHAM MD 20801
U.S.A.
TEL: 301-552-3848
TLX:
COM: 07

MUSMAN STEVEN DR
NATIONAL GEODETIC SURVEY
CHARTING & GEOD. SURVEY
NOS/NOAA - N/CG112
ROCKVILLE MD 20852
U.S.A.
TEL:
TLX:
COM:

MUSTEL E R PROF DR
ASTRONOMICAL COUNCIL
USSR ACADEMY OF SCIENCES
PYATNITSKAYA UL 48
109017 MOSCOW
U.S.S.R.
TEL: 231-54-61
TLX: 412623 SCSTP SU
COM: 10,29,36

MUTEL ROBERT LUCIEN
DEPT PHYSICS & ASTRONOMY
UNIVERSITY OF IOWA
IOWA CITY IA 52242
U.S.A.
TEL: 319-353-7205
TLX:
COM: 40

MUTSCHLECNER J PAUL DR
ASTRONOMY DEPT
INDIANA UNIVERSITY
SWAIN HALL WEST
BLOOMINGTON IN 47405
U.S.A.
TEL:
TLX:
COM: 36

LIST OF MEMBERS

MUXLOW THOMAS
UNIVERSITY OF MANCHESTER
NUFFIELD RADIO ASTR LABS
JODRELL BANK
MACCLESFIELD SK11 9DL
U.K.
TEL: 0477-71321
TLX: 36149
COM: 40

MUZZIO JUAN C PROF
OBSERVATORIO ASTRONOMICO
FAC CIENCIAS ASTRONOMICAS
PASEO DEL BOSQUE
1900 LA PLATA
ARGENTINA
TEL: 21-7308/3-8810
TLX: 31151 BULAP
COM: 28,33

MYACHIN V F DR
INSTITUTE OF THEORETICAL
ASTRONOMY
10 KUTUZOV QAY
192187 LENINGRAD
U.S.S.R.
TEL:
TLX:
COM: 07

MYERS PHILIP C
HARVARD-SMITHSONIAN CTR
FOR ASTROPHYSICS, MS 42
60 GARDEN STREET
CAMBRIDGE MA 02138
U.S.A.
TEL:
TLX:
COM: 34,40

NACOZY PAUL E DR
FEDEREAL SPACE SYSTEMS
PO BOX 50205
AUSTIN TX 78763
U.S.A.
TEL: 512-467-6659
TLX:
COM: 07,20

NADEAU DANIEL DR
DEPARTEMENT DE PHYSIQUE
UNIVERSITE DE MONREAL
C P 6128 SUCC A
MONTREAL PQ H3C 3J7
CANADA
TEL: 514-343-6676
TLX:
COM: 40

NADOLSCHI V PROF DR
COM.ARDEOANI OF TESCANI
JUD
BACAU
RUMANIA
TEL:
TLX:
COM:

NADYOZHIN D K DR
INSTITUTE OF THEORETICAL
CHEREMUSHKINSKAJA 25
117259 MOSCOW
U.S.S.R.
TEL: 123-02-92
TLX: 411059 CERII SU
COM: 35

NAGASAWA SHINGO PROF
TOKYO SCIENCE UNIVERSITY
2641 YAMAZAKI
HIGASHI KAMEYAMA
NODASHI 278
JAPAN
TEL: 0471-14-1501
TLX:
COM: 10

NAGASE FUMIAKI DR
DEPT OF ASTROPHYSICS
NAGOYA UNIVERSITY
CHIKUSA-KU, FURO-CHO
NAGOYA 464
JAPAN
TEL: 052-781-5111
TLX: 4477323 SCUNAG J
COM:

NAGIRNER DMITRIJ I DR
LENINGRAD UNIVERSITY
ASTRONOMICAL OBSERVATORY
BIBLIOTECHNAJA PL. 2
198904 LENINGRAD-PETRODV.
U.S.S.R.
TEL: 257-94-89
TLX:
COM: 36

NAGNIBEDA VALERY G DR
LENINGRAD UNIVERSITY
ASTRONOMICAL OBSERVATORY
BIBLIOTECHNAJA PL 2
198904 LENINGRAD
U.S.S.R.
TEL: 257-94-91
TLX:
COM: 40

NAHON FERNAND PROF
25 AVENUE DE L'EUROPE
F-92310 SEVRES
FRANCE
TEL: 1-45-34-18-05
TLX:
COM: 07,33

NAIDENOV VICTOR O
A.F. IOFFE PHYS TECH INST
USSR ACADEMY OF SCIENCES
POLYTECHNICHESKAYA 26
194021 LENINGRAD
U.S.S.R.
TEL:
TLX:
COM:

NAKADA YOSHIKAZU DR
DEPT OF ASTRONOMY
FACULTY OF SCIENCE
UNIV TOKYO, BUNKYO-KU
TOKYO 113
JAPAN
TEL: 03-812-2111
TLX:
COM: 34

NAKAGAWA NAOYA DR
UNIVERSITY OF ELECTRO-
COMMUNICATIONS
CHOFU-SHI
TOKYO 182
JAPAN
TEL: 0424-83-2161
TLX: 2822446 UEC J
COM:

NAKAGAWA YOSHINARI DR
CHIBA INST OF TECHNOLOGY
NARASHINO 275
JAPAN
TEL: 0474-75-2111
TLX:
COM: 10,49,51

NAKAGAWA YOSHITSUGU DR
GEOPHYSICAL INSTITUTE
FACULTY OF SCIENCE
UNIVERSITY OF TOKYO
TOKYO 113
JAPAN
TEL: 3-812-2111x4311
TLX:
COM: 16

NAKAI YOSHIHIRO
KWASAN AND HIDA
OBSERVATORIES
KYOTO UNIVERSITY
KYOTO 607
JAPAN
TEL: 075-581-1235
TLX:
COM: 09

NAKAJIMA HIROSHI
NOBEYAMA SOLAR RADIO OBS
TOKYO ASTRON OBSERVATORY
MINAMIMAKI-MURA
NAGANO 384-13
JAPAN
TEL: 267-98-2034
TLX: 3329005 TAONRO J
COM: 10

NAKAMURA TAKASHI DR
DEPT OF PHYSICS
KYOTO UNIVERSITY
KYOTO 606
JAPAN
TEL:
TLX:
COM: 35

NAKAMURA TSUKO DR
TOKYO ASTRONOMICAL OBS
OSAWA, MITAKA
TOKYO 181
JAPAN
TEL: 03-812-2111
TLX:
COM: 15,20

NAKAMURA YASUHISA
KOMABA HIGH SCHOOL
OHASHI 2-18-1
MEGURO-KU
TOKYO 153
JAPAN
TEL: 03-466-2481
TLX:
COM: 42

NAKANO SABURO DR
KOBINATO 1-21-7
BUNKYO-KU
TOKYO 112
JAPAN
TEL:
TLX:
COM:

NAKANO SYUICHI
COMPUTING CENTER, SUMOTO
P.O. BOX 32
SUMOTO P.O.
HYOGO-KEN 656-91
JAPAN
TEL: 3-919-1485
TLX:
COM: 06,20

NAKANO TAKENORI DR
DEPT OF PHYSICS
KYOTO UNIVERSITY
KITASHIRAKAWA OIWAKE
SAKYOKU KYOTO 606
JAPAN
TEL: 075-751-2111
TLX:
COM: 35

NAKAYAMA SHIGERU DR
FACULTY GENERAL EDUC
UNIVERSITY OF TOKYO
KOMABA MEGURO
TOKYO 153
JAPAN
TEL: 03-467-1171
TLX:
COM: 41

NAKAZAWA KIYOSHI DR
TOKYO GEOPHYSICAL INST
YAYOI 2-11-16
BUNKYOKU TOKYO 113
JAPAN
TEL: 3-812-2111x4304
TLX:
COM: 22,35

NAMBA OSAMU DR
SONNENBORGH OBSERVATORY
UNIVERSITY OF UTRECHT
ZONNENBURG 2
NL-3512 NL UTRECHT
NETHERLANDS
TEL: 030-312841
TLX: 47224
COM: 10,12

NAN REN-DONG
BEIJING ASTRONOMICAL OBS
ACADEMIA SINICA
BEIJING
CHINA, PEOPLE'S REP.
TEL: 28 1698
TLX: 22040 BAOBS CN
COM: 40

NANDY KASHINATH DR
ROYAL OBSERVATORY
BLACKFORD HILL
EDINBURGH EH9 3HJ
U.K.
TEL: 31-667-3321
TLX: 72383 ROEDING
COM: 34,45

NAPIER WILLIAM M DR
ROYAL OBSERVATORY
BLACKFORD HILL
EDINBURGH EH9 3HJ
U.K.
TEL: 031-667-3321
TLX: 72383 ROEDIN G
COM:

LIST OF MEMBERS

NAQVI S I H PROF
DEPT PHYSICS & ASTRONOMY
UNIVERSITY OF REGINA
REGINA S4S 0A2
CANADA
TEL: 306-584-4262
TLX: 071-2683 U R REG
COM:

NATALI GIULIANO DR
LAB. ASTROFISICA SPAZIALE
C P 67
I-00044 FRASCATI
ITALY
TEL:
TLX:
COM:

NEEMAN YUVAL PROF
DEPT PHYSICS & ASTRONOMY
TEL-AVIV UNIVERSITY
TEL-AVIV 69978
ISRAEL
TEL: 03-425411
TLX: 342171 VERSY IL
COM: 47.48

NEMEC JAMES
DEPT GEOPHYSICS & ASTRON
UNIV OF BRITISH COLUMBIA
VANCOUVER BC V6T 1W5
CANADA
TEL: 604-652-4517
TLX:
COM: 37

NARANAN S PROF
TATA INSTITUTE OF
FUNDAMENTAL RESEARCH
HOMI BHABHA RD
BOMBAY 400 005
INDIA
TEL: 219-111
TLX: 0113009
COM:

NATHER R EDWARD
DEPT OF ASTRONOMY
UNIVERSITY OF TEXAS
AUSTIN TX 78712
U.S.A.
TEL:
TLX:
COM: 27.42

NEFEDEVA ANTONINA I PROF
ENGELHARDT ASTRONOMICAL
OBSERVATORY
OBSERVATORY STATION
422526 KAZAN
U.S.S.R.
TEL: 324827
TLX:
COM: 08

NEMIRO ANDREJ A DR PROF
PULKOVO OBSERVATORY
196140 LENINGRAD
U.S.S.R.
TEL: 2992242
TLX:
COM: 08

NARAYAN RAMESH DR
STEWARD OBSERVATORY
UNIVERSITY OF ARIZONA
TUCSON AZ 85721
U.S.A.
TEL: 602-621-2560
TLX: 467175
COM:

NATTA ANTONELLA DR
CENTRO PER ASTRONOMIA IR
LAPGC E FERMI 5
I-50125 FIRENZE
ITALY
TEL: 220034
TLX: 572268 ARCETRI
COM:

NEFF JOHN S
UNIVERSITY OF IOWA
605 BROOKLAND PARK DRIVE
IOWA CITY IA 52240
U.S.A.
TEL: 319-353-4340
TLX:
COM: 15.27.36

NESCI ROBERTO
ISTITUTO ASTRONOMICO
UNIVERSITA DI ROMA
VIA LANCISI 29
I-00161 ROMA
ITALY
TEL: 39-6-867-525
TLX: 613255 INFNRO
COM: 37

NARAYANA J V
REGIONAL METEOROLOGICAL
OFFICE
4 COLLEGE ROAD
MADRAS 600 006
INDIA
TEL:
TLX:
COM:

NAUMOV VITALIJ A DR
PULKOVO OBSERVATORY
196140 LENINGRAD
U.S.S.R.
TEL:
TLX:
COM: 19.31

NELSON ALISTAIR H DR
DEPT OF APPLIED MATHS
UNIVERSITY COLLEGE
CARDIFF
U.K.
TEL: 0222-44211x2668
TLX: 498635 ULIBCFG
COM: 33

NESS NORMAN F DR
NASA/GSFC
CODE 690
GREENBELT MD 20771
U.S.A.
TEL: 301-344-8112
TLX:
COM: 16.44

NARIAI HIDEKAZU
RESEARCH INSTITUTE FOR
THEORETICAL PHYSICS
HIROSHIMA UNIVERSITY
725 TAKEHARA
JAPAN
TEL: 08462-2-2362
TLX:
COM: 35.47

NAWAR SAMIR DR
HELWAN OBSERVATORY
HELWAN
EGYPT
TEL:
TLX:
COM: 21

NELSON BURT DR
DEPT OF ASTRONOMY
SAN DIEGO STATE UNIV
SAN DIEGO CA 92182
U.S.A.
TEL: 619-265-6175
TLX:
COM: 42.50

NEUGEBAUER GERRY DR
PHYSICS DEPARTMENT
CALIF INST OF TECHNOLOGY
320 DOWNS
PASADENA CA 91125
U.S.A.
TEL: 818-356-4284
TLX: 675425
COM: 28.34

NARIAI KYOJI DR
TOKYO ASTRONOMICAL
OBSERVATORY
MITAKA
TOKYO 181
JAPAN
TEL: 0422-32-5111
TLX:
COM: 36.42

NECKEL HEINZ DR
HAMBURGER STERNWARTE
GOJENSBERGSWEG 112
D-2050 HAMBURG 80
GERMANY, F.R.
TEL: 49-40-7252-4130
TLX: 217884 HAMST D
COM: 12

NELSON GRAHAM JOHN DR
CSIRO
DIVISION OF RADIOPHYSICS
P.O.BOX 76
EPPING NSW 2121
AUSTRALIA
TEL: 02-868-0222
TLX: 26230
COM: 10

NEUKUM G DR
D F V L R
NE-OE-PE
D-8031 WESSLING
GERMANY, F.R.
TEL: 8153-28731
TLX: 0526419 DVLOP D
COM: 16

NARITA SHINJI DR
DOSHISHA UNIVERSITY
KYOTO 602
JAPAN
TEL:
TLX:
COM: 35

NECKEL TH DR
MPI FUER ASTRONOMIE
KOENIGSTUHL
D-6900 HEIDELBERG
GERMANY, F.R.
TEL: 06221-528288
TLX: 461789 MPIA D
COM: 33

NELSON JERRY
LAWRENCE BERKELEY LAB
UNIVERSITY OF CALIFORNIA
BLDG 50, RM 351
BERKELEY CA 94720
U.S.A.
TEL: 415-486-5913
TLX:
COM: 09

NEUPERT WERNER M DR
NASA/GSFC
CODE 680
GREENBELT MD 20771
U.S.A.
TEL: 301-344-8169
TLX:
COM: 10.44

NARLIKAR JAYANT V PROF
TATA INSTITUTE OF
FUNDAMENTAL RESEARCH
HOMI BHABHA RD
BOMBAY 400 005
INDIA
TEL: BOMBAY 219111
TLX: 0113009 TIFR IN
COM: 28,47

NEE TSU-WEI DR
INST PHYSICS & ASTRONOMY
NATL CENTRAL UNIVERSITY
CHUNG-LI
CHINA, TAIWAN
TEL:
TLX:
COM:

NELSON ROBERT M
183/501 JPL
4800 OAK GROVE DRIVE
PASADENA CA 91109
U.S.A.
TEL: 213-354-68939
TLX:
COM:

NEUZIL LUDEK DR
ASTRONOMICAL INSTITUTE
CZECH. ACAD. OF SCIENCES
OBSERVATORY
251 65 ONDREJOV
CZECHOSLOVAKIA
TEL:
TLX:
COM: 21

NEVEN LUC DR
OBS ROYAL DE BELGIQUE
AVENUE CIRCULAIRE 3
B-1180 BRUXELLES
BELGIUM
TEL:
TLX:
COM: 12,36

NGUYEN-QUANG RIEU DR
OBSERVATOIRE DE PARIS
SECTION DE MEUDON
F-92195 MEUDON PL CEDEX
FRANCE
TEL: 1-45-34-75-30
TLX: 270912
COM: 34,40

NICOLET MARCEL PROF
INST D'AERONOMIE SPATIALE
30 AVE DEN DOORN
B-1180 BRUXELLES
BELGIUM
TEL: 322-3742949
TLX: 21563 ESPACE B
COM: 12,21

NIIMI HIDEYUKI DR
FACULTY ENGINEERING
UNIVERSITY OF KYOTO
YOSHIDAHONMACHI
SAKYOKU KYOTO 606
JAPAN
TEL:
TLX:
COM: 33

NEVIN THOMAS E PROF
HUNTERS MOON
SYDENHAM VILLAS
DUBLIN 14
IRELAND
TEL: DUBLIN 984 341
TLX:
COM: 14

NHA IL-SEONG DR
YONSEI UNIVERSITY OBS
134 SINCHON-DONG
SEODAEMUN-KU
SEOUL 120
KOREA, REPUBLIC
TEL: 392-0131
TLX:
COM: 38,42

NICOLOV NIKOLAI S DR
UNIVERSITY OF SOFIA
DEPT OF ASTRONOMY
ANTON IVANOV STR 5
1126 SOFIA
BULGARIA
TEL: 51-24-05
TLX:
COM: 46

NIIMI YUKIO
TOKYO ASTRONOMICAL OBS
2-21-1 OSAWA
MITAKA
TOKYO 181
JAPAN
TEL: 0422-32-5111
TLX: 2822307 TAOMTK J
COM: 31

NEWBURN RAY L JR
3226 EMERALD ISLE DRIVE
GLENDALE CA 91206
U.S.A.
TEL:
TLX:
COM: 15,22

NIARCHOS PANAYIOTIS PH D
DEPT OF ASTRONOMY
UNIVERSITY OF ATHENS
PANEPISTIMIOPOLIS
GR-15771 ATHENS
GREECE
TEL:
TLX:
COM: 27,51

NICOLSON GEORGE D DR
RADIO ASTRONOMY
OBSERVATORY
P O BOX 3718
JOHANNESBURG 2000
SOUTH AFRICA
TEL: 11-642 4692
TLX: 321006 SA
COM: 40C

NIKITIN A A DR
LENINGRAD STATE UNIV
ASTRONOMICAL OBSERVATORY
198904 LENINGRAD
U.S.S.R.
TEL: 293-22-62
TLX:
COM: 29

NEWELL EDWARD B DR
MT STROMLO & SIDING
SPRING OBSERVATORIES
PRIVATE BAG
WODEN ACT 2606
AUSTRALIA
TEL: 062-881111
TLX: AA 62270 CANOPUS
COM: 37

NICHOLLS RALPH W PROF
CRESS, DEPT OF PHYSICS
YORK UNIVERSITY
4700 KEELE STREET
NORTH YORK ONT M3J1P3
CANADA
TEL: 416-667-383
TLX: 06524736
COM: 14P,29

NIELL ARTHUR E DR
HAYSTACK OBSERVATORY
WESTFORD MA 01886
U.S.A.
TEL: 617-692-4764
TLX: 948149
COM:

NIKOLOFF IVAN DR
PERTH OBSERVATORY
6 PEOPLES AVENUE
GOOSEBERRY HILL 6076
AUSTRALIA
TEL: 2931865
TLX:
COM: 08

NEWMAN MICHAEL JOHN DR
LOS ALAMOS NATIONAL LAB
X-2 MS B220
PO BOX 1663
LOS ALAMOS NM 87545
U.S.A.
TEL: 505-667-7698
TLX:
COM: 35

NICHOLSON WILLIAM
ROYAL GREENWICH OBS
HERSTMONCEUX CASTLE
HAILSHAM BN27 1RP
U.K.
TEL:
TLX:
COM: 24

NIEMELA VIRPI S DR
CALLE 51 ESQ 11
1894 VILLA ELISA (Bs.As.)
ARGENTINA
TEL:
TLX:
COM:

NIKOLOV ANDREJ DR
UNIVERSITY OF SOFIA
DEPT OF ASTRONOMY
ANTON IVANOV STR 5
1126 SOFIA
BULGARIA
TEL: 62561x375
TLX:
COM: 27

NEWSOM GERALD H PROF
ASTRONOMY DEPT
OHIO STATE UNIVERSITY
174 W. 18TH AVENUE
COLUMBUS OH 43210
U.S.A.
TEL: 614-422-7082
TLX:
COM: 14

NICOLACI DA COSTA LUIZ-A.
OBSERVATORIO NACIONAL
RUA GENERAL BRUCE 586
SAO CRISTOVAO
20921 RIO DE JANEIRO
BRAZIL
TEL: 021-580-7313
TLX: 021-21288
COM:

NIEMI AIMO
TURUN YLIOPISTO
ITAINEN PITKARATU 1
SF-20520 TURKU 52
FINLAND
TEL: 921-431863
TLX:
COM: 09,19

NIKONOV V B DR
CRIMEAN ASTROPHYSICAL OBS
USSR ACADEMY OF SCIENCES
NAUCHNIY 3-3
334413 CRIMEA
U.S.S.R.
TEL:
TLX:
COM: 09,25,45

NEWTON ROBERT R DR
JOHNS HOPKINS UNIVERSITY
APPLIED PHYSICS LAB
JOHNS HOPKINS RD
LAUREL MD 20707
U.S.A.
TEL: 301-953-7100
TLX:
COM:

NICOLAS KENNETH ROBERT
NAVAL RESEARCH LABORATORY
CODE 4163
OVERLOOK AVENUE
WASHINGTON DC 20375
U.S.A.
TEL: 202-767-2517
TLX:
COM: 12

NIETO JEAN-LUC
OBSERVATOIRE DE TOULOUSE
14 AVENUE EDOUARD BELIN
F-31400 TOULOUSE
FRANCE
TEL: 61-25-21-01
TLX: 530776 F
COM: 28

NILSON PETER DR
ASTRONOMICAL OBSERVATORY
BOX 515
S-751 20 UPPSALA
SWEDEN
TEL:
TLX:
COM:

NEY EDWARD P PROF
DEPT OF ASTRONOMY
TATE LAB OF PHYSICS
UNIVERSITY OF MINNESOTA
MINNEAPOLIS MN 55455
U.S.A.
TEL: 612-373-4687
TLX:
COM: 21

NICOLET BERNARD
OBSERVATOIRE DE GENEVE
CH-1290 SAUVERNY
SWITZERLAND
TEL: 22-552611
TLX: 27720 OBSG CH
COM: 25,45

NIEUWENHUIJZEN HANS DR
ASTRONOMICAL INSTITUTE
ZONNENBURG 2
NL-3512 NL UTRECHT
NETHERLANDS
TEL: 030-312841
TLX:
COM:

NILSSON CARL DR
SMITHSONIAN ASTROPHYSICAL
OBSERVATORY
60 GARDEN STREET
CAMBRIDGE MA 02138
U.S.A.
TEL:
TLX:
COM:

LIST OF MEMBERS

NINKOVIC SLOBODAN
ASTRONOMICAL OBSERVATORY
VOLGINA 7
YU-11050 BEOGRAD
YUGOSLAVIA
TEL: 011-419-357
TLX:
COM: 33,38

NISHI KEIZO DR
TOKYO ASTRONOMICAL
OBSERVATORY
OSAWA MITAKA
TOKYO 181
JAPAN
TEL: 0422-32-5111
TLX: 02822307 TAOMTK J
COM: 10,12

NISHIDA MINORU PROF
DEPT OF PHYSICS
KYOTO UNIVERSITY
KITASHIRAFAWA OIWAKE
SAKYOKU KYOTO 606
JAPAN
TEL:
TLX: 5422693 LIBKYU J
COM: 33,35,47

NISHIDA MITSUGU
DEPT OF LITERATURE
KOBE WOMEN'S UNIVERSITY
SUMA-KU
KOBE 654
JAPAN
TEL: 078-731-4416
TLX:
COM: 33

NISHIMURA JUN DR
INST SPACE & AERON SCI
6-1 KOMABA 4-CHOME
MEGURO-KU
TOKYO 153
JAPAN
TEL: 03-467-1111x388
TLX: J 24550 SPACETKY
COM:

NISHIMURA MASAKI
DEPT OF PHYSICS
HOKKAIDO UNIVERSITY
NISHI 8-CHOME KITA 10-JYO
SAPPORO, HOKKAIDO 060
JAPAN
TEL: 11-71--2111
TLX:
COM: 28

NISHIMURA SHIRO DR
TOKYO ASTRONOMICAL
OBSERVATORY
OSAWA MITAKA
TOKYO 181
JAPAN
TEL: 422-32-5111
TLX: 2822307 TAOMTK J
COM: 05,09,29

NISHIMURA TETSUO DR
STEWARD OBSERVATORY
UNIVERSITY OF ARIZONA
TUCSON AZ 85721
U.S.A.
TEL: 602-621-2054
TLX: 467175
COM: 21

NISSEN POUL E PROF
INSTITUTE OF ASTRONOMY
UNIVERSITY OF AARHUS
LANGELANDSGADE
DK-8000 AARHUS C
DENMARK
TEL: 06-128899
TLX: 64767 AAUSCI DK
COM: 37C

NITTMAN JOHANN
ETUDES ET FABRICATION
DOWELL SCHLUMBERGER
BP 90
F-42003 ST-ETIENNE
FRANCE
TEL: 77-32-64-23
TLX:
COM: 34

NITYANANDA R DR
RAMAN RESEARCH INSTITUTE
BANGALORE 560 080
INDIA
TEL: 0812-360-126
TLX: 845671
COM:

NOBILI ANNA M
ISTITUTO DI MATEMATICA
UNIVERSITA' DI PISA
VIA BUONARROTI 2
I-56100 PISA
ITALY
TEL:
TLX:
COM: 07,19,20

NOBILI L DR
DIPT. DI FISICA
G. GALILEI
VIA MARZOLO 8
I-35131 PADOVA
ITALY
TEL: 49-844205/111
TLX: 430308 DFGGPDI
COM:

NOCI GIANCARLO PROF
ISTITUTO DI ASTRONOMIA
UNIVERSITA DI FIRENZE
I-50125 FIRENZE
ITALY
TEL: 55-22-0034
TLX: 572268
COM:

NOEL FERNANDO
DEPTO DE ASTRONOMIA
UNIVERSIDAD DE CHILE
CASILLA 36-D
SANTIAGO
CHILE
TEL: 229-4101
TLX: 40853
COM: 08C,31

NOELS ARLETTE DR
50 AVENUE DE LA PAIX
BOITE 063
B-4030 GRIVEGNEE
BELGIUM
TEL: 41-52-9980/7517
TLX:
COM: 35

NOERDLINGER PETER D PROF
JET PROPULSION LABORATORY
MS 264-748
4800 OAK GROVE DRIVE
PASADENA CA 91109
U.S.A.
TEL: 818-354-8324
TLX:
COM: 47

NOLLEZ GERARD DR
INSTITUT D'ASTROPHYSIQUE
98 BIS BOULEVARD ARAGO
F-75014 PARIS
FRANCE
TEL: 1-43-20-14-25
TLX:
COM: 14

NOMOTO KEN'ICHI DR
DEPT EARTH SCI/ASTRONOMY
COLLEGE ARTS & SCIENCES
UNIVERSITY OF TOKYO
MEGURO-KU TOKYO 153
JAPAN
TEL: 03-467-1171
TLX: 25510 UNITOKYO
COM: 35

NOONAN THOMAS W PROF
PHYSICS DEPT
SUNY
BROCKPORT NY 14420
U.S.A.
TEL: 716-395-5581
TLX:
COM: 28,47

NORDH H LENNART DR
STOCKHOLM OBSERVATORY
S-133 00 SALTSJOEBADEN
SWEDEN
TEL: 08-7170195
TLX: 12972
COM: 34,44

NORDLUND AKE DR
UNIVERSITY OBSERVATORY
OESTER VOLDGADE 3
DK-1350 COPENHAGEN K
DENMARK
TEL: 1-14-17-90
TLX: 44155 DANAST
COM: 12,36

NORDSTROM BIRGITTA DR
COPENHAGEN UNIVERSITY OBS
BRORFELDEVEJ 23
DK-4340 TOLLOSE
DENMARK
TEL: 03-488195
TLX: 44155 DANAST DK
COM: 30,42

NORGAARD-NIELSEN HANS U
UNIVERSITY OBSERVATORY
OSTER VOLDGADE 3
DK-1350 COPENHAGEN K
DENMARK
TEL:
TLX:
COM:

NORMAN COLIN A PROF
SPACE TELESCOPE SCI INST
HOMEWOOD CAMPUS
3700 SAN MARTIN DRIVE
BALTIMORE MD 21218
U.S.A.
TEL: 301-338-4895
TLX:
COM:

NORRIS JOHN DR
MT STROMLO OBSERVATORY
PRIVATE BAG
WODEN P.O. ACT 2606
AUSTRALIA
TEL: 062-88-1111
TLX: AA 62270 CANOPUS
COM:

NORRIS RAYMOND PAUL
CSIRO
DIVISION OF RADIOPHYSICS
P.O.BOX 76
EPPING NSW 2121
AUSTRALIA
TEL: 02-868 0222
TLX: 26230 ASTRO
COM:

NORTH JOHN DAVID PROF
FILOSOFISCH INSTITUUT
RIJKSUNIVERSITEIT
GRONINGEN
NETHERLANDS
TEL: 05907-1846
TLX:
COM: 41V

NORTH PIERRE
INSTITUT D'ASTRONOMIE
UNIVERSITE DE LAUSANNE
CH-1290 CHAVANNES-DES-B.
SWITZERLAND
TEL: 022-55-26-11
TLX: 27720 OBSG CH
COM: 45

NOSKOV BORIS N DR
STERNBERG STATE
ASTRONOMICAL INSTITUTE
119899 MOSCOW
U.S.S.R.
TEL:
TLX:
COM: 07

NOTNI P DR
ZNTRLINST. F. ASTROPHYSIK
STERNWARTE BABELSBERG
ROSA-LUXEMBURG-STR 17A
DDR-1502 POTSDAM
GERMANY, D.R.
TEL:
TLX:
COM: 25,45

NOTTALE LAURENT
OBSERVATOIRE DE PARIS
SECTION DE MEUDON
DAF
F-92195 MEUDON PL CEDEX
FRANCE
TEL: 1-45-34-75-70
TLX: 201571
COM: 47

NOVICK ROBERT
DEPT OF PHYSICS
COLUMBIA UNIVERSITY
538 W. 120 STREET
NEW YORK NY 10027
U.S.A.
TEL: 212-280-3293
TLX: 22094 COLU UR
COM: 44,48

NOVIKOV I D DR
SPACE RESEARCH INSTITUTE
USSR ACADEMY OF SCIENCES
117810 MOSCOW
U.S.S.R.
TEL:
TLX:
COM: 47C

NOVIKOV SERGEJ B DR
STERNBERG STATE
ASTRONOMICAL INSTITUTE.
117234 MOSCOW
U.S.S.R.
TEL:
TLX:
COM:

NOVOSELOV V S PROF DR
LENINGRAD STATE UNIV.
ASTRONOMICAL OBSERVATORY
BIBLIOTECHNAJA PL. 2
198904 LENINGRAD
U.S.S.R.
TEL: 257-94-91
TLX:
COM: 07

NOVOTNY VACLAV
ASTRONOMICAL INSTITUTE
CZECH. ACAD. OF SCIENCES
ONDREJOV OBSERVATORY
251 65 ONDREJOV
CZECHOSLOVAKIA
TEL: 72-45-25
TLX: 121579
COM: 44

NOYES ROBERT W PROF
CENTER FOR ASTROPHYSICS
60 GARDEN STREET
CAMBRIDGE MA 02138
U.S.A.
TEL: 617-495-7424
TLX: 921428 SATELLITE CAM
COM: 10,12C,44

NUGIS TIIT
W.STRUVE ASTROPHYS OBS
ESTONIAN SSR
TORAVERE
202444 TARTU
U.S.S.R.
TEL:
TLX:
COM: 27

NULSEN PAUL DR
MT STROMLO & SIDING
SPRING OBSERVATORIES
PRIVATE BAG
WODEN P.O. ACT 2606
AUSTRALIA
TEL: 062-88-1111
TLX: 62270 AA
COM:

NUNES ROGERIO S DE SOUSA
GRUPO DE MATEM. APLICADA
UNIVERSIDADE DO PORTO
RUA DAS TAIPAS 135
4000 PORTO
PORTUGAL
TEL: 380313/769
TLX:
COM: 09

NUNEZ JORGE DR
OBSERVATORIO FABRA
TIBIDADO
08022 BARCELONA
SPAIN
TEL: 2475736
TLX:
COM:

NUSSBAUMER HARRY PROF
INSTITUT FUER ASTRONOMIE
ETH-ZENTRUM
CH-8092 ZUERICH
SWITZERLAND
TEL: 1-256-3631
TLX: 53178 ETHBI CH
COM: 10,14C,34

NUTH JOSEPH A III
NASA/GSFC, CODE 691
LAB FOR EXTERR PHYSICS
GREENBELT MD 20771
U.S.A.
TEL: 301-344-6364
TLX:
COM: 22,34

O'CONNELL ROBERT F PROF
LOUISIANA STATE UNIV
BATON ROUGE LA 70803
U.S.A.
TEL: 504-388-6848
TLX:
COM: 48

O'CONNELL ROBERT WEST DR
ASTRONOMY DEPT
UNIVERSITY OF VIRGINIA
PO BOX 3818, UNIV STATION
CHARLOTTESVILLE VA 22903
U.S.A.
TEL: 804-924-7494
TLX: 510-587-5453
COM: 28,47

O'CONNOR SEAMUS L DR
PHYSICS DEPT
UNIVERSITY COLLEGE
BELFIELD
DUBLIN 4
IRELAND
TEL: 353-1-693244
TLX: 32693 UCD EI
COM:

O'DELL CHARLES R DR
DEPT SPACE PHYS & ASTRON
RICE UNIVERSITY
PO BOX 1892
HOUSTON TX 77251
U.S.A.
TEL: 713-527-8101
TLX: 556457
COM: 09,15,34

O'DELL STEPHEN L
PHYSICS DEPT
VIRGINIA TECH
ROBESON HALL
BLACKSBURG VA 24061
U.S.A.
TEL: 703-961-5206
TLX:
COM: 34

O'DONOGHUE DARRAGH
DEPT OF ASTRONOMY
UNIVERSITY OF CAPE TOWN
RONDEBOSCH 7700
SOUTH AFRICA
TEL:
TLX:
COM: 27

O'HANDLEY DOUGLAS A DR
JET PROPULSION LAB
4800 OAK GROVE DRIVE
PASADENA CA 91109
U.S.A.
TEL: 818-577-6967
TLX: 675429
COM: 04,07

O'HORA NATHY P J
ROYAL GREENWICH OBS
HERSTMONCEUX CASTLE
HAILSHAM BN27 1RP
U.K.
TEL:
TLX:
COM: 19

O'KEEFE JOHN A DR
GODDARD SPACE FLIGHT CTR
CODE 681
GREENBELT MD 20771
U.S.A.
TEL: 301-344-8445
TLX: 89675
COM: 15,16,22

O'LEARY BRIAN T
SAIC
2615 PACIFIC COAST HWY
SUITE 300
HERMOSA BEACH CA 90254
U.S.A.
TEL: 213-318-2611
TLX:
COM:

O'MARA BERNARD J PROF
DEPT OF PHYSICS
UNIVERSITY OF QUEENSLAND
ST LUCIA
BRISBANE QLD 4067
AUSTRALIA
TEL:
TLX:
COM: 36

O'MONGAIN EON
PHYSICS DEPT
UNIVERSITY COLLEGE
BELFIELD
DUBLIN 4
IRELAND
TEL: 01-693244
TLX: 32693 UCD
COM: 44

O'SULLIVAN DENIS F
DUBLIN INSTITUTE FOR
ADVANCED STUDIES
5 MERRION SQUARE
DUBLIN 2
IRELAND
TEL: 353-1-774321
TLX: 31687 DIAS EI
COM: 48

O'SULLIVAN JOHN DAVID DR
DIVISION OF RADIOPHYSICS
CSIRO
P.O.BOX 76
EPPING NSW 2121
AUSTRALIA
TEL:
TLX: 26230 ASTRO
COM: 40

OBASHEV SAKEN O DR
ASTROPHYSICAL INSTITUTE
480068 ALMA-ATA
U.S.S.R.
TEL:
TLX: 275
COM:

OBI SHINYA PROF
FAC OF GENERAL EDUCATION
UNIVERSITY OF TOKYO
KOMABA MEGURO
TOKYO 153
JAPAN
TEL:
TLX:
COM: 14

OBLAK EDOUARD
OBSERVATOIRE DE BESANCON
41B AVE DE L'OBSERVATOIRE
F-25044 BESANCON CEDEX
FRANCE
TEL: 81-50-30-88
TLX: 361144
COM:

OBREGON DIAZ OCTAVIO J DR
DEPART. FISICA
UNIDAD IZTAPALAPA. UAM
PO BOX 55-534
09340 MEXICO DF
MEXICO
TEL: 6860322
TLX: 1764296 UAM ME
COM:

OBRIDKO VLADIMIR N DR
IZMIRAN
ACADEMGORODOK
142092 MOSCOW REGION
U.S.S.R.
TEL: 232-1921
TLX: 412623 SCSTP
COM: 10

OCCHIONERO FRANCO PROF
ISTITUTO ASTRONOMICO
VIA G.M. LANCISI 29
I-00161 ROMA
ITALY
TEL: 867525/8442977
TLX: 613255
COM:

LIST OF MEMBERS

OCHSENBEIN FRANCOIS DR
E S O
KARL-SCHWARZSCHILD-STR 2
D-8046 GARCHING B MUNCHEN
GERMANY, F.R.
TEL:
TLX:
COM: 05

ODA MINORU PROF
INST SPACE & ASTRONAUT SC
KOMABA MEGURO-KU
TOKYO 153
JAPAN
TEL: 03-467-1111
TLX: 24550 SPACE TKY
COM: 44C,48,51

ODA NAOKI
MICRO-ELECTRONICS RES LAB
NIPPON ELECTRIC COMPANY
4-1-1 MIYAZAKI MIYAMAE-KU
KAWASAKI, KANAGAWA 213
JAPAN
TEL:
TLX:
COM: 44

ODELL ANDREW P
INSTITUT FUER ASTRONOMIE
TUERKENSCHANZSTR 17
A-1180 WIEN
AUSTRIA
TEL: 0222-34-53-60
TLX:
COM: 35

ODGERS GRAHAM J DR
DOMINION ASTROPH OBS
5071 W SAANICH ROAD
RR 5
VICTORIA BC V8X 4M6
CANADA
TEL: 604-388-3977
TLX: 049-7295
COM: 09,27

OEGERLE WILLIAM R
5924 BERWYN ROAD
COLLEGE PARK MD20740
U.S.A.
TEL:
TLX:
COM:

OEHMAN YNGVE PROF
THULELEM 53
S-223 67 LUND
SWEDEN
TEL: 046-143362
TLX:
COM:

OEMLER AUGUSTUS JR DR
YALE UNIVERSITY OBS
PO BOX 6666
NEW HAVEN CT 06511
U.S.A.
TEL: 203-436-3460
TLX:
COM: 28,47

OESTERWINTER CLAUS
COMMANDER
NAVAL SURFACE WEAPONS CTR
K10
DAHLGREN VA 22448
U.S.A.
TEL: 703-663-7426
TLX:
COM: 04,07

OESTGAARD ERLEND
DEPARTMENT OF PHYSICS
UNIVERSITY OF TRONDHEIM
AVH
N-7055 DRAGVOLLM
NORWAY
TEL: 07-920411 x 117
TLX:
COM:

OESTREICHER ROLAND
LANDESSTERNWARTE
KOENIGSTUHL
D-6900 HEIDELBERG
GERMANY, F.R.
TEL: 06221-10036
TLX:
COM: 25

OETKEN L DR
ZNTRLINST. F. ASTROPHYSIK
ASTROPHYSIK. OBSERVATOR.
TELEGRAFENBERG
DDR-1500 POTSDAM
GERMANY, D.R.
TEL:
TLX:
COM: 14,29,30

OGAWARA YOSHIAKI
INST SPACE & ASTRON SCI
4-6-1 KOMABA
MEGURO-KU
TOKYO 153
JAPAN
TEL: 3-467-1111
TLX: 34757 ISASTRO J
COM: 44

OGORODNIKOV KYRILL P PROF
DEPT OF ASTRONOMY
LENINGRAD UNIVERSITY
199164 LENINGRAD
U.S.S.R.
TEL:
TLX:
COM: 05,33

OGURA KATSUO DR
KOKUGAKUIN UNIVERSITY
COLLEGE OF LITERATURE
HIGASHI 4-10-28
SHIBUYAKU TOKYO 150
JAPAN
TEL: 298-42-6913
TLX: 28899 SIBINBTH J
COM: 37

OHKI KENICHIRO DR
TOKYO ASTRONOMICAL
OBSERVATORY
OSAWA
MITAKA TOKYO 181
JAPAN
TEL: 0422-32-5111
TLX: 2822307
COM: 10

OHRING GEORGE PROF
DEPT GEOPHYS & PLANET SCI
TEL-AVIV UNIVERSITY
TEL-AVIV 69978
ISRAEL
TEL:
TLX:
COM:

OHTANI HIROSHI DR
DEPT OF ASTRONOMY
UNIVERSITY OF KYOTO
KYOTO 606
JAPAN
TEL: 075-751-2111
TLX: 5422693 LIBKYU J
COM: 34

OHYAMA NOBORU PROF
FACULTY OF ENGINEERING
SHIZUOKA UNIVERSITY
3 CHOME JYOHOKU
HAMAMATSU 432
JAPAN
TEL:
TLX:
COM: 35

OJA HEIKKI DR
OBS & ASTROPHYSICS LAB
UNIVERSITY OF HELSINKI
TAHTITORNINMAKI
SF-00130 HELSINKI 13
FINLAND
TEL: 358-0-1912942
TLX:
COM: 46

OJA TARMO PROF
KVISTABERG OBSERVATORY
S-197 00 BRO
SWEDEN
TEL: 0758-40157
TLX:
COM: 24,33,45

OKA TAKESHI DR
CHEMISTRY DEPT
UNIVERSITY OF CHICAGO
5735 S. ELLIS AVE
CHICAGO IL 60637
U.S.A.
TEL: 312-962-7070
TLX:
COM: 14

OKAMOTO ISAO DR
INTERNATIONAL LATITUDE
OBSERVATORY
HOSHIGAOKA MIZUSAWA
IWATE 023
JAPAN
TEL: 0197-24-7111
TLX: 8376-28 ILSMIZ J
COM: 19,35

OKAMURA SADANORI DR
KISO BRANCH OF THE
TOKYO ASTRON OBSERVATORY
MITAKEMURA KISOGUN
NAGANOKEN 397-01
JAPAN
TEL: 0264-52-3360
TLX: 3347577 KSOOBS J
COM: 28

OKAZAKI AKIRA DR
TSUDA COLLEGE
KODAIRA
TOKYO 187
JAPAN
TEL: 0423-41-2441
TLX:
COM: 42

OKAZAKI SEICHI DR
2-4-4 OSAWA, MITAKA
TOKYO 181
JAPAN
TEL: 0422-31-6770
TLX:
COM: 19

OKE J BEVERLEY PROF
CALTECH 105-24
PASADENA CA 91125
U.S.A.
TEL: 818-356-4007
TLX:
COM: 28,29

OKI TOSIO PROF DR
DEPT EARTH SC FAC OF EDUC
FUKUSHIMA UNIVERSITY
MATSUKAWA MACHI
FUKUSHIMA 960-12
JAPAN
TEL: 0245-48-5151
TLX:
COM:

OKOYE SAMUEL E PROF
DEPT OF PHYSICS & ASTR
UNIVERSITY OF NIGERIA
NSUKKA
NIGERIA
TEL: 042-770752
TLX:
COM: 38,40,46,47,48

OKUDA HARUYUKI DR PROF
INST F.SPACE & ASTRONAUT.
SCIENCE
4-6-1 KOMABA, MEGURO-KU
TOKYO 153
JAPAN
TEL: 03-467-1111
TLX: 24550 SPACETKY J
COM: 33,34

OKUDA TORU
INST OF EARTH SCIENCE
HOKKAIDO UNIV OF EDUCAT
1-2 HACHIMAN-CHO
HAKODATE 040
JAPAN
TEL: 0138-41-1121
TLX:
COM: 44

OLAH KATALIN DR
KONKOLY OBSERVATORY
HUNGARIAN ACADEMY OF SCI
BOX 67
H-1525 BUDAPEST
HUNGARY
TEL: 166-426/366-621
TLX: 227460 KONOB H
COM: 27

OLANO CARLOS ALBERTO DR
INSTITUTO ARGENTINO DE
RADIOASTRONOMIA
CASILLA DE CORREO No. 5
1894 VILLA ELISA (Bs.As.)
ARGENTINA
TEL:
TLX:
COM: 33

OLSEN ERIK H
COPENHAGEN UNIVERSITY OBS
BRORFELDEVEJ 23
DK-4340 TOLLOSE
DENMARK
TEL: 03-488195
TLX: 44155 DANAST DK
COM: 45C

ONAKA TAKASHI
DEPT OF ASTRONOMY
UNIVERSITY OF TOKYO
2-11-16 YAYOI, BUNKYO-KU
TOKYO 113
JAPAN
TEL: 3-812-2111
TLX: 33659 UTYOSCI J
COM: 34

OPROIU TIBERIU DR
ASTRONOMICAL OBSERVATORY
BLOC P-5 SC I & II AP 10
STR BUCIUM 25
3400 CLUJ NAPOCA
RUMANIA
TEL: 951-62616
TLX:
COM:

OLEAK H DR
ZNTRLINST. F. ASTROPHYSIK
STERNWARTE BABELSBERG
ROSA-LUXEMBURG-STR 17A
DDR-1502 POTSDAM
GERMANY, D.R.
TEL:
TLX:
COM: 28

OLSEN KENNETH H DR
LOS ALAMOS NAT LABORATORY
BOX 1663, MS C335
LOS ALAMOS NM 87545
U.S.A.
TEL: 505-667-1007
TLX:
COM:

ONDERLICKA BEDRICH DR
DEPT OF ASTRONOMY
PURKYNE UNIVERSITY
KOTLARSKA 2
611 37 BRNO
CZECHOSLOVAKIA
TEL:
TLX:
COM:

ORCHISTON WAYNE DR
FACULTY OF APPL. SCIENCE
VICTORIA COLLEGE
662 BLACKBURN ROAD
CLAYTON, VICTORIA 3168
AUSTRALIA
TEL: 542-7353
TLX:
COM: 41

OLIVER BERNARD M PROF
CHIEF SETI PROGRAM OFFICE
NASA-AMES RESEARCH CENTER
MOFFETT FIELD CA 94035
U.S.A.
TEL: 415-694-5166
TLX:
COM: 51

OLSON EDWARD C PROF
OBSERVATORY
UNIVERSITY OF ILLINOIS
1011 W. SPRINGFIELD AVENUE
URBANA IL 62801
U.S.A.
TEL: 217-333-5531
TLX:
COM: 42

ONEGINA A B DR
MAIN ASTRONOMICAL OBS
UKRAINIAN ACADEMY OF SCI
GOLOSEEVO
252127 KIEV
U.S.S.R.
TEL: 66-37-44
TLX:
COM: 24

ORLIN HYMAN DR
NATL ACADEMY OF SCIENCES
2101 CONSTITUTION AVE NW
WASHINGTON DC 20418
U.S.A.
TEL:
TLX:
COM:

OLIVER JOHN PARKER DR
DEPTARTMENT OF ASTRONOMY
UNIVERSITY OF FLORIDA
GAINESVILLE FL 32611
U.S.A.
TEL:
TLX:
COM: 42

OLTHOF HINDERICUS DR
ESA
8-10 RUE MARIO NIKIS
F-75738 PARIS CEDEX 15
FRANCE
TEL: 1-42-73-71-03
TLX: 202746
COM: 44

ONO YORO PROF
DEPT OF PHYSICS
UNIVERSITY HOKKAIDO
KITAHACHIJYO NISHI 8
SAPPORO HOKKAIDO 063
JAPAN
TEL:
TLX:
COM:

ORLOV MIKHAIL DR
MAIN ASTRONOMICAL OBS
252127 KIEV
U.S.S.R.
TEL: 66-31-10
TLX: 131406 SKY
COM: 29

OLLONGREN A PROF DR
DEPT MATHS & COMPUT SCI
WASSENAARSEWEG 80
POSTBOX 9512
NL-2300 RA LEIDEN
NETHERLANDS
TEL: 071-148333-5006
TLX: 39058 ASTRO NL
COM: 33.51

OMAROV TUKEN B PROF
ASTROPHYSICAL INSTITUTE
480068 ALMA ATA
U.S.S.R.
TEL: 64-40-40
TLX:
COM: 07

OOE MASATSUGU DR
INTERNATIONAL LATITUDE
OBSERVATORY
HOSHIGACKA MIZUSAWA
IWATE 023
JAPAN
TEL: 0197-24-7111
TLX: 837628 MIZ J
COM: 19

ORRALL FRANK Q PROF
INSTITUTE FOR ASTRONOMY
2680 WOODLAWN DRIVE
HONOLULU HI 96822
U.S.A.
TEL: 808-948-8667
TLX: 723-8459
COM: 10,12,36

OLNON FRISO
RADIOSTERRENWACHT
STICHTING RZM
POSTBUS 2
NL-7990 AA DWINGELOO
NETHERLANDS
TEL: 05219-7244
TLX: 42043 SRZM NL
COM:

OMER GUY C JR PROF
1080 S.W. 11TH TERRACE
GAINESVILLE FL 32601
U.S.A.
TEL: 904-378-4627
TLX:
COM: 28,41,47

OORT JAN H PROF
PRES. KENNEDYLAAN 169
NL-2343 GZ OEGSTGEEST
NETHERLANDS
TEL:
TLX:
COM: 28,33,34,40,47

ORTE ALBERTO
CECILLO PUJAZON 22-3 A
SAN FERNANDO (CADIZ)
SPAIN
TEL: 89-54-41
TLX:
COM: 19,31

OLOFSSON HANS
ONSALA SPACE OBSERVATORY
S-439 00 ONSALA
SWEDEN
TEL: 0300-60650
TLX: 8542400 ONSPACE
COM: 34

OMNES ROLAND PROF
LPTHE BAT 211
UNIVERSITE DE PARIS-SUD
F-91405 ORSAY
FRANCE
TEL: 1-69-41-77-44
TLX: 692166 FACORS
COM: 47

OPHER REUVEN PROF
ASTRONOMY DEPT
IAG/USP
CAIXA POSTAL 30627
01051 SAO PAULO SP
BRAZIL
TEL: 275-3720
TLX: 1136221 IAGM BR
COM:

ORTOLANI SERGIO
OSSERVATORIO ASTRONOMICO
DI ASIAGO
I-36012 ASIAGO
ITALY
TEL: 0-424-65457
TLX: 430110 SETURIST
COM:

OLOFSSON S GOERAN DR
STOCKHOLM OBSERVATORY
S-133 00 SALTSJOEBADEN
SWEDEN
TEL: 8-7172639
TLX: 12972
COM:

OMONT ALAIN PROF
CERMO-USMG
BP 68
F-38041 ST MARTIN D'HERES
FRANCE
TEL: 76-51-47-90
TLX: 980753 F
COM: 14

OPOLSKI ANTONI PROF
ASTRONOMICAL OBSERVATORY
UL. KOPERNIKA 11
51-622 WROCLAW
POLAND
TEL:
TLX:
COM: 27,38

ORTON GLENN S DR
JPL
MS 183-301
4800 OAK GROVE DRIVE
PASADENA CA 91109
U.S.A.
TEL: 818-354-2460
TLX:
COM: 14

ORUS JUAN J PROF
DPTO FIS TIERRA & COSMOS
UNIVERSIDAD DE BARCELONA
08028 BARCELONA
SPAIN
TEL:
TLX:
COM: 07

OSAKI TORU DR
RYUKOKU UNIVERSITY
FUKAKUSA TSUKAMOTO
FUSHIMIKU
KYOTO 612
JAPAN
TEL: 075-642-1111
TLX:
COM: 34

OSAKI YOJI DR
DEPT OF ASTRONOMY
UNIVERSITY OF TOKYO
YAYOI BUNKYO
TOKYO 113
JAPAN
TEL: 03-812-2111
TLX: 33659 UTYOSCI J
COM: 35C,42

OSAWA KIYOTERU DR
TOKYO ASTRONOMICAL
OBSERVATORY
OSAWA
MITAKA TOKYO 181
JAPAN
TEL: 0422-32-5111
TLX:
COM: 25,27,29,45

OSBORN WAYNE DR
PHYSICS DEPT
CENTRAL MICHIGAN UNIV
MT PLEASANT MI 48859
U.S.A.
TEL: 517-774-3321
TLX:
COM: 37,45,46

OSBORNE JOHN L DR
DEPT OF PHYSICS
UNIVERSITY OF DURHAM
SOUTH ROAD
DURHAM DH1 3LE
U.K.
TEL: 0385-64971
TLX: 537351 DURLIB G
COM: 34

OSKANYAN V S DR
BYURAKAN ASTROPHYSICAL
OBSERVATORY
378433 ARMENIA
U.S.S.R.
TEL:
TLX:
COM: 25,27

OSMAN ANAS MOHAMED DR
HELWAN OBSERVATORY
HELWAN-CAIRO
EGYPT
TEL: 780645 HIAG UN
TLX:
COM: 28,37

OSMER PATRICK S DR
NOAO/KPNO
PO BOX 26732
TUCSON AZ 85718
U.S.A.
TEL: 602-327-5511
TLX: 666484
COM:

OSORIO JOSE J S P PROF
OBSERVATORIO ASTRONOMICO
UNIVERSIDADE DO PORTO
MONTE DA VIRGEM
4400 VILA NOVA DE GAIA
PORTUGAL
TEL: 7820404
TLX: 23121 UNIPOR P
COM: 07,08,46,50

OSTER LUDWIG F PROF DR
NATL SCIENCE FOUNDATION
1800 G STREET N.W.
WASHINGTON DC 20550
U.S.A.
TEL: 202-357-9857
TLX:
COM: 12

OSTERBROCK DONALD E PROF
LICK OBSERVATORY
UNIVERSITY OF CALIFORNIA
SANTA CRUZ CA 95064
U.S.A.
TEL: 408-429-2605
TLX:
COM: 28,34,40,41

OSTRIKER JEREMIAH P PROF
PRINCETON UNIVERSITY
OBSERVATORY
PEYTON HALL
PRINCETON NJ 08544
U.S.A.
TEL: 609-452-3800
TLX: 322409
COM: 33,35,48,51

OTERMA LIISI PROF
SIRKKALANKATU 31
SF-20700 TURKU
FINLAND
TEL: 358-21-332081
TLX:
COM: 19,20

OTHMAN MAZLAN
DEPT OF PHYSICS
UNIVERSITI KEBANGSAAN
MALAYSIA
43600 BANGI SELANGOR
MALAYSIA
TEL: 8250001
TLX: 31496 UNIKEB MA
COM: 46

OTTELET I J DR
INSTITUT D'ASTROPHYSIQUE
UNIVERSITE DE LIEGE
AVENUE DE COINTE 5
B-4200 COINTE-OUGREE
BELGIUM
TEL: 041-529980
TLX: 41264
COM: 16

OVENDEN MICHAEL W PROF
DEPT GEOPHYS & ASTRONOMY
UNIV OF BRITISH COLUMBIA
VANCOUVER BC V6T 1W5
CANADA
TEL: 604-228-2138
TLX:
COM: 33,42

OVERBEEK MICHIEL DANIEL
P O BOX 212
EDENVALE 1610
SOUTH AFRICA
TEL: 11-53-5447
TLX:
COM:

OWAKI NAOAKI DR
TOKYO GAKUGEI UNIVERSITY
NUKUIKITAMACHI
KOGANEI TOKYO 184
JAPAN
TEL:
TLX:
COM: 46

OWEN FRAZER NELSON DR
VLA NRAO
1000 BULLOCK BLVD
PO BOX 0
SOCORRO NM 87801
U.S.A.
TEL: 505-772-4011
TLX: 910-988-1710
COM: 28,40

OWEN TOBIAS C PROF
DEPT OF EARTH & SPACE SCI
STATE UNIVERSITY NEW YORK
STONY BROOK NY 11794
U.S.A.
TEL: 516-246-6705
TLX: 5102287767
COM: 16C,44,51,W

OWREN LEIF DR
FYSISK INSTITUTT
UNIVERSITETET I BERGEN
ALLEGATEN 55
N-5000 BERGEN
NORWAY
TEL: 47-05-213050
TLX:
COM:

OXENIUS JOACHIM DR
UNIV LIBRE BRUXELLES
CP 231
CAMPUS PLAINE ULB
B-1050 BRUSSELS
BELGIUM
TEL: 32-02-640-00-15
TLX: 23069 UNILIB B
COM: 36

OZERNOY LEONID M PROF
PHYSICAL INSTITUTE
USSR ACADEMY OF SCIENCES
117924 MOSCOW
U.S.S.R.
TEL:
TLX:
COM: 34,47,48

OZGUC ATILA
BOGAZICI UNIVERSITY
KANDILLI OBSERVATORY
CENGELKOY
ISTANBUL 81220
TURKEY
TEL: 90-1-332-02-40
TLX:
COM:

OZSVATH I PROF
UNIVERSITY OF TEXAS
PROGRAMS IN MATHEMAT. SCI
PO BOX 830688
RICHARDSON TX 75083-0688
U.S.A.
TEL: 214-690-2174
TLX:
COM: 47

PAAL GYORGY DR
KONKOLY OBSERVATORY
BOX 67
H-1525 BUDAPEST
HUNGARY
TEL: 166-506
TLX: 227460 KONOB H
COM: 47

PACHNER JAROSLAV PROF
390 WOODSWORTH RD 11
TORONTO ONT M2L 2T9
CANADA
TEL: 416-447-1015
TLX:
COM: 28,47

PACHOLCZYK ANDRZEJ G PROF
STEWARD OBSERVATORY
UNIVERSITY OF ARIZONA
TUCSON AZ 85721
U.S.A.
TEL: 602-621-6928
TLX: 467175
COM: 28,40,48

PACINI FRANCO PROF
ISTITUTO DI ASTRONOMIA
LARGO E FERMI 5
LARGO E. FERMI 5
I-50125 FIRENZE
ITALY
TEL: 055-220-034
TLX: 572268 ARCETR-I
COM: 44,48C,51C

PACZYNSKI BOHDAN PROF
COPERNICUS ASTRON CENTER
UL. BARTYCKA
00-716 WARSZAWA
POLAND
TEL:
TLX:
COM: 35,42

PADALIA T D DR
UTTAR PRADESH STATE OBS
MANORA PARK
NAINITAL 263 129
INDIA
TEL: 2136
TLX:
COM: 42

LIST OF MEMBERS

PADEVET VLADIMIR DR
ASTRONOMICAL INSTITUTE
CZECH. ACAD. OF SCIENCES
OBSERVATORY
251 65 ONDREJOV
CZECHOSLOVAKIA
TEL: 724525 PRAHA
TLX: 121579
COM: 22

PADMAN RACHAEL
MULLARD RADIO ASTRON OBS
CAVENDISH LABORATORY
MADINGLEY ROAD
CAMBRIDGE CB3 0HE
U.K.
TEL: 223-66477
TLX: 81292
COM: 40

PADMANABHAN T DR
TATA INSTITUTE OF
FUNDAMENTAL RESEARCH
BOMBAY 400 005
INDIA
TEL: 219111
TLX: 0113009 TIFR IN
COM: 47

PADRIELLI LUCIA
IST. DI RADIOASTRONOMIA
VIA IRNERIO 46
I-40126 BOLOGNA
ITALY
TEL: 51-232856
TLX: 211664 INF BO I
COM:

PAGE ARTHUR MR
MT TAMBORINE OBSERVATORY
PO BOX 44
ASPLEY, QLD 4034
AUSTRALIA
TEL: 61-7-263-4813
TLX:
COM: 25

PAGE CLIVE G DR
PHYSICS DEPT
UNIVERSITY OF LEICESTER
LEICESTER LE1 7RH
U.K.
TEL: 533-554455 x 23
TLX: 341664 LUXRAY G
COM: 48

PAGE DON NELSON
104 DAVEY LABORATORY
PENNSYLVANIA STATE UNIV.
UNIVERSITY PARK PA 16802
U.S.A.
TEL: 814-863-0163
TLX: 842510
COM: 47

PAGE THORNTON L DR
NASA JOHNSON SPACE CENTER
18639 POINT LOOKOUT DR
HOUSTON TX 77058
U.S.A.
TEL: 713-483-3728
TLX:
COM: 28,51

PAGEL BERNARD E J PROF
ROYAL GREENWICH OBS
HERSTMONCEUX CASTLE
HAILSHAM BN27 1RP
U.K.
TEL: 0323833171
TLX: 87451 RGOBSY G
COM: 29,36

PAL ARPAD PROF DR
UNIV OF CLUJ-NAPOCA
FACULTY OF MATHEMATICS
STR RAKOCZI 72
3400 CLUJ NAPOCA
RUMANIA
TEL: 951-16101/11592
TLX:
COM: 07

PALLA FRANCESCO
OSSERVATORIO ASTROFISICO
DI ARCETRI
LARGO E. FERMI 5
I-50125 FIRENZE
ITALY
TEL: 055-220034
TLX: 572268 ARCETR I
COM: 34

PALLAVICINI ROBERTO DR
OSSERVATORIO ASTROFISICO
DI ARCETRI
I-50125 FIRENZE
ITALY
TEL: 55-220034
TLX: 572268 ARCETR I
COM: 10,36

PALMEIRA RICARDO A R DR
INST PESQUISAS ESPACIAIS
CAIXA POSTAL 515
12200 S. JOSE DOS CAMPOS
BRAZIL
TEL:
TLX:
COM:

PALMER PATRICK E PROF
DEPT OF ASTRONOMY
UNIVERSITY OF CHICAGO
5640 S. ELLIS AVE
CHICAGO IL 60637
U.S.A.
TEL: 312-962-7972
TLX: 6871133
COM: 33,34,40

PALOUS JAN DR
ASTRONOMICAL INSTITUTE
CZECH. ACAD. OF SCIENCES
BUDECSKA 6
120 23 PRAGUE 2
CZECHOSLOVAKIA
TEL: 25-87-57
TLX: 122486
COM: 33

PALUMBO GIORGIO G C DR
TE-S R E LAB CNR
VIA DE CASTAGNOLI 1
I-40100 BOLOGNA
ITALY
TEL: 051-51-95-93
TLX: 511350 CNR-80
COM: 48

PAMYATNIKH A A DR
ASTRONOMICAL COUNCIL
USSR ACADEMY OF SCIENCES
48 PYATNITSKAYA ST
109017 MOSCOW
U.S.S.R.
TEL: 231-54-61
TLX: 412623 SCSTP SU
COM: 05,35

PAN JUN-HUA
NANJING ASTRONOMICAL
INSTRUMENTS FACTORY
NANJING
CHINA. PEOPLE'S REP.
TEL: 46191
TLX: 34136 GLYNJ CN
COM:

PAN LIANDE
SHAANXI OBSERVATORY
P.O. BOX 18
LINTONG
XIAN
CHINA. PEOPLE'S REP.
TEL: 3-2255
TLX: 70121 CSAO CN
COM: 10

PAN NING-BAO
BEIJING ASTRONOMICAL OBS
ACADEMIA SINICA
BEIJING
CHINA. PEOPLE'S REP.
TEL: 281698
TLX:
COM: 28

PAN RONG-SHI
SHANGHAI OBSERVATORY
ACADEMIA SINICA
SHANGHAI
CHINA. PEOPLE'S REP.
TEL: 386191
TLX: 33164 SHAO CN
COM: 24,28,47

PAN XIAO-PEI
SHAANXI ASTRONOMICAL OBS
P.O. BOX 18
LINTONG
SHAANXI
CHINA. PEOPLE'S REP.
TEL: XIAN 32255x406
TLX: 70121 CSAO CN
COM: 19

PANAGIA NINO DR
SPACE TELESCOPE SCI INST
3700 SAN MARTIN DRIVE
BALTIMORE MD 21218
U.S.A.
TEL: 301-338-4916
TLX: 6849101 ST SCI
COM: 34

PANDE GIRISH CHANDRA PROF
DEPT OF MATHEMATICS
UNIVERSITY OF PATRAS
GR-26110 PATRAS
GREECE
TEL: 061-991-889/991
TLX: 312239 EFAP GR
COM: 35

PANDE MAHESH CHANDRA DR
UP STATE OBSERVATORY
MANORA PEAK
NAINI TAL 263 129
INDIA
TEL: 2136
TLX:
COM: 12

PANEK ROBERT J DR
DEPT OF ASTRONOMY
525 DAVEY LAB
UNIV OF PENNSYLVANIA
UNIVERSITY PARK PA 16802
U.S.A.
TEL:
TLX:
COM: 36

PANG KEVIN
JET PROPULSION LAB
MS 183-601
4800 OAK GROVE DRIVE
PASADENA CA 91109
U.S.A.
TEL: 818-354-5392
TLX: 675429
COM: 16

PANKONIN VERNON LEE DR
DIV ASTRONOMICAL SCIENCES
NATL SCIENCE FOUNDATION
1800 G STREET N.W.
WASHINGTON DC 20550
U.S.A.
TEL: 202-357-9696
TLX:
COM: 34,40,50C

PANNUNZIO RENATO
OSSERVATORIO ASTRONOMICO
STRADA OSSERVATORIO 20
I-10025 PINO TORINESE
ITALY
TEL: 011-841067
TLX: 213236 TOASTR I
COM: 26

PANOV KIRIL DR
DEPT OF ASTRONOMY
BULGARIAN ACAD SCIENCES
7TH NOVEMBER STR 1
1000 SOFIA
BULGARIA
TEL: 7341
TLX: 23561 ECF BAN BG

PAOLICCHI PAOLO DR
ISTITUTO DI ASTRONOMIA
UNIVERSITA' DI PISA
PIAZZA TORRICELLI 2
I-56100 PISA
ITALY
TEL: 50-43343
TLX:
COM: 15,16

PAP JUDIT
DEPT OF ASTRONOMY
L. EOTVOS UNIVERSITY
KUN BELA TER 2
H-1083 BUDAPEST
HUNGARY
TEL: 36-1-141-019
TLX: 225467
COM:

PAPAGIANNIS MICHAEL D PRO
DEPT OF ASTRONOMY
BOSTON UNIVERSITY
725 COMMONWEALTH AVE
BOSTON MA 02215
U.S.A.
TEL: 617-353-2626
TLX:
COM: 40,44,51P

PAPALIOLIOS COSTAS DR
SMITHSONIAN ASTROPHYS OBS
60 GARDEN STREET
CAMBRIDGE MA 02138
U.S.A.
TEL:
TLX:
COM:

PAPALOIZOU JOHN C B DR
INSTITUTE OF ASTRONOMY
MADINGLEY ROAD
CAMBRIDGE CB3 0HA
U.K.
TEL:
TLX:
COM: 27,35

PAPARO MARGIT DR
KONKOLY OBSERVATORY
HUNGARIAN ACADEMY OF SCI
BOX 67
H-1525 BUDAPEST
HUNGARY
TEL: 166-426
TLX: 227460
COM:

PAPATHANASOGLOU D DR
DEPT OF ASTRONOMY
UNIVERSITY OF ATHENS
PANEPISTIMIOPOLIS
GR-15771 ATHENS
GREECE
TEL: 7243414
TLX:
COM: 12

PAPAYANNOPOULOS TH DR
DEPT OF ASTRONOMY
UNIVERSITY OF ATHENS
PANEPISTIMIOPOLIS
GR-15771 ZOGRAFOS
GREECE
TEL: 01 7243414
TLX:
COM: 33

PAPOUSEK JIRI
DEPT OF ASTROPHYSICS KFTA
FACULTY OF SCIENCE UJEP
KOTLARSKA 2
611 37 BRNO
CZECHOSLOVAKIA
TEL: 51112
TLX:
COM: 27

PAQUET PAUL EG DR
OBS ROYAL DE BELGIQUE
AVENUE CIRCULAIRE 3
B-1180 BRUXELLES
BELGIUM
TEL: 32-2-374-38-01
TLX: 21565 OBSBEL
COM: 19,31V

PARCELIER PIERRE DR
OBSERVATOIRE DE PARIS
61 AVE DE L'OBSERVATOIRE
F-75014 PARIS
FRANCE
TEL: 1-43-20-12-10
TLX: 270776
COM: 31

PARESCE FRANCESCO DR
SPACE SCIENCES LABORATORY
UNIVERSITY OF CALIFORNIA
BERKELEY CA 94720
U.S.A.
TEL:
TLX:
COM: 21,49C

PARIJSKIJ N N PROF
INST OF PHYSICS OF EARTH
USSR ACADEMY OF SCIENCES
123810 MOSCOW
U.S.S.R.
TEL: 252-07-21
TLX: 411196 IFZAN US
COM: 19

PARIJSKIJ YU N DR
SPECIAL ASTROPHYSICAL OBS
USSR ACADEMY OF SCIENCES
LENINGRAD BRANCH
196140 LENINGRAD
U.S.S.R.
TEL: 2979452
TLX:
COM: 40,51

PARISOT JEAN-PAUL
OBSERVATOIRE DE BESANCON
41B AVE DE L'OBSERVATOIRE
F-25044 BESANCON CEDEX
FRANCE
TEL: 81-50-30-88
TLX: 361144 OBS
COM:

PARKER EDWARD A DR
ELECTRONICS LABORATORIES
THE UNIVERSITY
CANTERBURY CT2 7NT
U.K.
TEL: 0227-66822
TLX: 965449
COM: 40

PARKER EUGENE N
LAB ASTROPHYS SPACE RES
UNIVERSITY OF CHICAGO
933 E 56TH STREET
CHICAGO IL 60637
U.S.A.
TEL: 312-962-7847
TLX: 910-221-5617
COM: 34,48,49

PARKER ROBERT A R
NASA/JOHNSON SPACE CENTER
CODE CB
HOUSTON TX 77058
U.S.A.
TEL: 713-483-2221
TLX:
COM:

PARKINSON JOHN H DR
MULLARD SPACE SCIENCE LAB
UNIVERSITY COLLEGE LONDON
HOLMBURY ST MARY
DORKING. SURREY RH56NS
U.K.
TEL: 030-670-292
TLX: 859185
COM: 10,44,48

PARKINSON TRUMAN DR
KITT PEAK NATL OBS
950 N. CHERRY AVE
TUCSON AZ 85726
U.S.A.
TEL:
TLX:
COM:

PARKINSON WILLIAM H
HARVARD COLLEGE OBS
60 GARDEN STREET
CAMBRIDGE MA 02138
U.S.A.
TEL: 617-495-4865
TLX: 921428
COM: 10,12,14C,44

PARMA PAOLA
IST. DI RADIOASTRONOMIA
CNR
VIA IRNERIO 46
I-40126 BOLOGNA
ITALY
TEL: 51-232856
TLX:
COM:

PARRISH ALLAN DR.
STATE UNIVERSITY OF N.Y.
C/O 6191 GRC U MASS.
AMHERST MA 01003
U.S.A.
TEL:
TLX:
COM: 40

PARSAMYAN ELMA S DR
BYURAKAN ASTROPHYSICAL
OBSERVATORY
378433 ARMENIA
U.S.S.R.
TEL: 56-34-53
TLX:
COM: 27,37

PARSONS SIDNEY B DR
SPACE TELESCOPE SCI INST
HOMEWOOD CAMPUS
3700 SAN MARTIN DRIVE
BALTIMORE MD 21218
U.S.A.
TEL: 301-338-4807
TLX:
COM: 29,45

PARTHASARATHY M DR
INDIAN INSTITUTE OF
ASTROPHYSICS
BANGALORE 560 034
INDIA
TEL: 566585/566497
TLX: 845763 IIAB IN
COM: 27,29,42

PARTRIDGE ROBERT B PROF
HAVERFORD COLLEGE
HAVERFORD PA 19041
U.S.A.
TEL: 215-896-1144
TLX:
COM:

PASACHOFF JAY M PROF
WILLIAMS COLLEGE
HOPKINS OBSERVATORY
WILLIAMSTOWN MA 01267
U.S.A.
TEL: 413-597-2105
TLX:
COM: 12,40,46

PASCOAL ANTONIO J B SCI
OBSERVATORIO ASTRONOMICO
PROF MANUEL DE BARROS
MONTE DA VIRGEM
4400 VILA NOVA DE GAIA
PORTUGAL
TEL: 7820404
TLX:
COM:

PASCU DAN DR
US NAVAL OBSERVATORY
WASHINGTON DC 20390
U.S.A.
TEL: 202-653-1178
TLX:
COM: 20

PASIAN FABIO
OSSERVATORIO ASTRONOMICO
DI TRIESTE
VIA G.B. TIEPOLO 11
I-34131 TRIESTE
ITALY
TEL: 39-40-768005
TLX: 461137 OAT I
COM: 09

PASINETTI LAURA E PROF
DIPARTIMENTO DI FISICA
UNIVERSITA DI MILANO
VIA CELORIA 16
I-20133 MILANO
ITALY
TEL: 2-2392275/272
TLX: 334687 INFN MI
COM: 05,29,36,45

PASTORI LIVIO
OSSERVATORIO ASTRONOMICO
VIA BIANCHI 46
I-22055 MERATE
ITALY
TEL: 039-592035
TLX:
COM:

PASTORIZA MIRIANI G DR
INSTITUTO DE FISICA
UNIVERSIDADE FEDERAL
DO RIO GRANDE DO SUL
90000 PORTO ALEGRE
BRAZIL
TEL: 512-71-666
TLX: 051-1055 UFRS BR
COM: 28

PATERNO LUCIO PROF
OSSERVATORIO ASTROFISICO
CITTA UNIVERSITARIA
I-95125 CATANIA
ITALY
TEL: 39-95-33-0533
TLX: 970359 ASTRCT I
COM: 10,27

PAVLOVSKAYA E D DR
STERNBERG STATE
ASTRONOMICAL INSTITUTE
117234 MOSCOW
U.S.S.R.
TEL:
TLX:
COM: 33

PEARSON TIMOTHY J
OWENS VALLEY RADIO OBS
CALTECH 105-24
PASADENA CA 91125
U.S.A.
TEL: 818-356-4980
TLX: 675425
COM: 40

PEDOUSSAUT ANDRE
OBSERVATOIRE PIC-DU-MIDI
ET TOULOUSE
14 AVENUE EDOUARD BELIN
F-31400 TOULOUSE CEDEX
FRANCE
TEL: 61-25-21-01
TLX: 530776 OBSTLSE
COM: 29,30

PATERSON-BEECKMANS F
VINCENT VAN GOGHLAAN 19
NL-2343 RH OEGSTGEEST
NETHERLANDS
TEL: 31-071-170829
TLX:
COM: 29

PAXTON HAROLD J B R
ROYAL GREENWICH OBS
HERSTMONCEUX CASTLE
HAILSHAM BN27 1RP
U.K.
TEL:
TLX:
COM:

PEAT D W DR
COLLEGE OF RIPON
LORD MAYOR'S WALK
YORK YO3 7EX
U.K.
TEL:
TLX:
COM: 29

PEDREROS MARIO DR
DEPTO DE ASTRONOMIA
UNIVERSIDAD DE CHILE
CASILLA 36-D
SANTIAGO
CHILE
TEL: 52-2-2294101
TLX:
COM: 25,37

PATHRIA RAJ K PROF
DEPT OF PHYSICS
UNIVERSITY OF WATERLOO
WATERLOO ONT N2L 3G1
CANADA
TEL: 519-885-1211
TLX: 069-55259
COM:

PAYNE DAVID G
JET PROPULSION LABORATORY
MS 264-748
4800 OAK GROVE DRIVE
PASADENA CA 91190
U.S.A.
TEL: 818-354-4630
TLX:
COM: 28,40

PECINA PETR
ASTRONOMICAL INSTITUTE
CZECH. ACAD. OF SCIENCES
251 65 ONDREJOV
CZECHOSLOVAKIA
TEL: 25-87-57
TLX: 122486
COM: 22

PEEBLES P JAMES E
JOSEPH HENRY LABS
JADWIN HALL
PRINCETON NJ 08544
U.S.A.
TEL: 609-452-4386
TLX: 499-3512
COM: 47

PATKOS LASZLO DR
KONKOLY OBSERVATORY
HUNGARIAN ACADEMY OF SCI
BOX 67
H-1525 BUDAPEST
HUNGARY
TEL:
TLX: 227460
COM: 42

PEACH GILLIAN DR
DEPT PHYSICS & ASTRONOMY
UNIVERSITY COLLEGE
GOWER STREET
LONDON WC1E 6BT
U.K.
TEL: 01-387-7050
TLX: 28722
COM: 14

PECKER JEAN-CLAUDE PROF
IAT
98 BIS BOULEVARD ARAGO
F-75014 PARIS
FRANCE
TEL: 1-43-20-14-25
TLX:
COM: 05,12,34,36,47

PEERY BENJAMIN F PROF
DEPT PHYSICS & ASTRONOMY
HOWARD UNIVERSITY
WASHINGTON DC 20059
U.S.A.
TEL: 202-636-6267
TLX:
COM: 29,46

PATRIARCHI PATRIZIO DR
OSSERVATORIO ASTROFISICO
DI ARCETRI
LARGO E. FERMI 5
I-50125 FIRENZE
ITALY
TEL: 39-55-22-00-34
TLX: 572268 ARCETR I
COM:

PEACH JOHN V DR
DEPT OF ASTROPHYSICS
SOUTH PARKS ROAD
OXFORD OX1 3RQ
U.K.
TEL: 0865-511336
TLX:
COM:

PEDERSEN BENT M DR
OBSERVATOIRE DE PARIS
SECTION DE MEUDON
F-92195 MEUDON PL CEDEX
FRANCE
TEL: 1-45-34-75-30
TLX:
COM: 10

PEIMBERT MANUEL DR
INSTITUTO DE ASTRONOMIA
UNAM
APDO POSTAL 70-264
04510 MEXICO DF
MEXICO
TEL: 905-548-5306
TLX: 01760155 CIMCE
COM: 28,33,34C

PATUREL GEORGES
OBSERVATOIRE DE LYON
F-69230 ST-GENIS-LAVAL
FRANCE
TEL: 78-56-07-05
TLX: 310926
COM: 28

PEACOCK JOHN ANDREW
ROYAL OBSERVATORY
BLACKFORD HILL
EDINBURGH EH9 3HJ
U.K.
TEL: 031-667-3321
TLX: 72383 ROEDIN G
COM: 47

PEDERSEN HOLGER DR
E S O
CASILLA 19001
SANTIAGO 19
CHILE
TEL: 056-2-88757
TLX: 240881 ESOGO CL
COM:

PEKERIS CHAIM LEIB PROF
DEPT APPLIED MATHS
WEIZMANN INST OF SCIENCE
REHOVOT 76100
ISRAEL
TEL: 08-483292
TLX: 361900
COM:

PAULINY TOTH IVAN K K DR
MPI FUER RADIOASTRONOMIE
AUF DEM HUEGEL 69
D-5300 BONN
GERMANY, F.R.
TEL: 228-525-243
TLX: 886440
COM: 40,48

PEALE STANTON J PROF
DEPT OF PHYSICS
UNIVERSITY OF CALIFORNIA
SANTA BARBARA CA 93106
U.S.A.
TEL: 805-961-2977
TLX:
COM: 07

PEDERSEN OLAF PROF
HISTORY OF SCIENCE INST
UNIVERSITY OF AARHUS
NY MUNKEGADE
DK-8000 AARHUS C
DENMARK
TEL: 06-127188
TLX:
COM: 41C

PEKUENLUE E RENNAN DR
EGE UNIVERSITY
FACULTY OF SCIENCE
BORNOVA-IZMIR
TURKEY
TEL: 222295
TLX:
COM:

PAULS THOMAS ALBERT DR
NAVAL RESEARCH LABORATORY
CODE 4130
WASHINGTON DC 20375-5000
U.S.A.
TEL:
TLX:
COM: 33,34,40

PEARSE REGINALD W B DR
60 GREENLANDS ROAD
STAINES TW18 4LR
U.K.
TEL:
TLX:
COM: 21

PEDLAR ALAN DR
VLA PROJECT
BOX 0
SOCORRO NM 87801
U.S.A.
TEL:
TLX:
COM: 40

PEL JAN WILLEM DR
KAPTEIJN STERREWACHT
MENSINGHEWEG 20
NL-9301 KA RODEN
NETHERLANDS
TEL:
TLX: 53767 KSW RO NL
COM: 25

PELLAS PAUL DR
LABORATOIRE MINERALOGIE
61 RUE BUFFON
F-75005 PARIS
FRANCE
TEL: 1-47-07-28-24
TLX:
COM: 15

PELLET ANDRE
OBSERVATOIRE DE MARSEILLE
2 PLACE LE VERRIER
F-13248 MARSEILLE CEDEX 4
FRANCE
TEL: 91-95-90-88
TLX: 420241 F
COM:

PELSENEER JEAN KON PROF
UNIV LIBRE DE BRUXELLES
76 AV DES GRENADIERS
BOITE 6
B-1150 BRUXELLES
BELGIUM
TEL:
TLX:
COM: 41

PELTIER LESLIE C
327 S. BREDEICK STREET
DELPHOS OH 45833
U.S.A.
TEL:
TLX:
COM: 27

PENG QIU-HE
DEPT OF ASTRONOMY
NANJING UNIVERSITY
NANJING
CHINA, PEOPLE'S REP.
TEL: 34651 - 2882
TLX: 34151 PRCNU CN
COM: 28,48

PENG YUN-LOU
DEPT OF ASTRONOMY
NANJING UNIVERSITY
NANJING
CHINA, PEOPLE'S REP.
TEL: 37551, 2882
TLX: 34151 PRCNU CN
COM: 40

PENNY ALAN JOHN DR
RUTHERFORD APPLETON LAB
CHILTON, DIDCOT OX11 0QX
U.K.
TEL: 0235-21900
TLX: 83159 RUTHBL G
COM: 09,25C,37

PENSADO JOSE DR
OBSERVATORIO ASTRONOMICO
ALFONSO XII-5
28014 MADRID
SPAIN
TEL: 2270107
TLX:
COM:

PENSTON MARGARET
ROYAL GREENWICH OBS
HERSTMONCEUX CASTLE
HAILSHAM BN27 1RP
U.K.
TEL:
TLX:
COM:

PENSTON MICHAEL V DR
ROYAL GREENWICH OBS
MADINGLEY ROAD
HAILSHAM BN27 1RP
U.K.
TEL: 323-833171
TLX: 87451 RGOBSY G
COM: 34

PENZIAS ARNO A DR
AT&T BELL LABORATORIES
ROOM 6A-409
600 MOUNTAIN AVENUE
MURRAY HILL NJ 07974
U.S.A.
TEL: 201-582-3361
TLX: 13-8650 OR 219348
COM: 34,40,47

PEQUIGNOT DANIEL
OBSERVATOIRE DE PARIS
SECTION DE MEUDON
DAF
F-92195 MEUDON PL CEDEX
FRANCE
TEL: 1-45-34-75-70
TLX: 201571
COM:

PERAIAH ANNAMANENI DR
INDIAN INSTITUTE OF
ASTROPHYSICS
BANGALORE 560 034
INDIA
TEL: 566585, 566497
TLX: 845763 IIAB IN
COM: 36

PERCY JOHN R PROF
DEPARTMENT OF ASTRONOMY
UNIVERSITY OF TORONTO
TORONTO ONT M5S 1A1
CANADA
TEL: 416-978-4971
TLX:
COM: 27C,46C

PERDANG JEAN M DR
INSTITUT D'ASTROPHYSIQUE
UNIVERSITE DE LIEGE
AVENUE DE COINTE 5
B-4200 COINTE-OUGREE
BELGIUM
TEL: 041-52-99-80
TLX: 41264 ASTRLG B
COM:

PERDOMO RAUL
FACULTAD DE CIENCIAS
ASTRON. y GEOFISICAS
PASEO DEL BOSQUE
1900 LA PLATA
ARGENTINA
TEL: 213-8810
TLX: 31151 BULAP
COM: 19

PEREK LUBOS DR
ASTRONOMICAL INSTITUTE
CZECH. ACAD. OF SCIENCES
BUDECSKA 6
120 23 PRAHA 2
CZECHOSLOVAKIA
TEL: 254234
TLX: 122486
COM: 33,51

PEREZ-DE-TEJADA H A DR
INSTITUTO DE GEOFISICA
UNAM
22860 ENSENADA, B. CALIF.
MEXICO
TEL: 706-674-0601
TLX:
COM: 15

PEREZ-PERAZA JORGE DR
INAOE
AP POSTALES 216 y 51
72000 PUEBLA, PUE.
MEXICO
TEL: 47-04-19
TLX:
COM:

PERINOTTO MARIO PROF
ISTITUTO DI ASTRONOMIA
LARGO E. FERMI 5
I-50125 FIRENZE
ITALY
TEL: 55-22-00-34
TLX: 572268 ARCETR
COM: 34

PERKINS FRANCIS W DR
PLASMA PHYSICS LAB
PRINCETON UNIVERSITY
PO BOX 451
PRINCETON NJ 08540
U.S.A.
TEL: 609-683-2603
TLX: 5106852399
COM: 49

PERLEY RICHARD ALAN
NRAO
PO BOX 0
SOCORRO NM 87801
U.S.A.
TEL: 505-772-4011
TLX: 910-988-1710
COM: 40

PEROLA G C DR
ISTITUTO ASTRONOMICO
VIA LANCISI 29
I-00161 ROMA
ITALY
TEL: 06-867525
TLX: 613255 INFNRO
COM: 48

PERRIN MARIE-NOEL DR
OBSERVATOIRE DE PARIS
61 AVE DE L'OBSERVATOIRE
F-75014 PARIS
FRANCE
TEL: 1-43-20-12-10
TLX:
COM: 29

PERRY CHARLES L DR
DEPT PHYSICS & ASTRONOMY
LOUISIANA STATE UNIV
BATON ROUGE LA 70803
U.S.A.
TEL: 504-388-8287
TLX: 559184
COM: 25,30,33,45

PERRY JUDITH J DR
INSTITUTE OF ASTRONOMY
CAMBRIDGE CB3 0HA
U.K.
TEL: 0223-62204
TLX: 817297
COM:

PERRYMAN MICHAEL A C
ASTROPHYSICS DIVISION
SPACE SCIENCE DEPT ESA
ESTEC, POSTBUS 299
NL-2200 AG NOORDWIJK
NETHERLANDS
TEL: 1719-83615
TLX: 39098
COM: 08,09,24,47

PERSI PAOLO
IST. ASTROFISICA SPAZIALE
C P 67
I-00044 FRASCATI
ITALY
TEL: 396-9425655
TLX: 610261 CNR FRA
COM: 34

PERSIDES SOTIRIOS C
DEPT OF ASTRONOMY
UNIVERSITY THESSALONIKI
GR-54006 THESSALONIKI
GREECE
TEL: 991357
TLX:
COM: 47

PESCH PETER DR
DIV ASTRONOMICAL SCIENCES
NATL SCIENCE FOUNDATION
1800 G STREET, N.W.
WASHINGTON DC 20550
U.S.A.
TEL: 202-357-7622
TLX:
COM: 33

PESEK RUDOLPH PROF
CZECH. ACAD. OF SCIENCES
PRAGUE
CZECHOSLOVAKIA
TEL:
TLX:
COM: 51

PETERS GERALDINE JOAN DR
DEPT OF ASTRONOMY
UNIV SOUTHERN CALIFORNIA
UNIVERSITY PARK
LOS ANGELES CA 90007
U.S.A.
TEL: 213-743-6962
TLX:
COM: 29,36,42,44

PETERS WILLIAM L III DR
ASTRONOMY DEPT
UNIVERSITY OF TEXAS
AUSTIN TX 78712
U.S.A.
TEL:
TLX:
COM: 28,34,40

PETON ALAIN DR
OBSERVATOIRE DE MARSEILLE
2 PLACE LE VERRIER
F-13248 MARSEILLE CEDEX 4
FRANCE
TEL: 91-95-90-88
TLX:
COM:

PETROV G M DR
NIKOLAEV DEPT OF THE
MAIN ASTRONOMICAL OBS
327000 NIKOLAEV
U.S.S.R.
TEL: 36-18-24
TLX:
COM: 08

PETTINI MAX
ROYAL GREENWICH OBS
HERSTMONCEUX CASTLE
HAILSHAM BN27 1RP
U.K.
TEL: 44-323-833171
TLX: 87451 RGOBSY G
COM:

PETERSEN J O DR
UNIVERSITY OBSERVATORY
OESTER VOLDGADE 3
DK-1350 COPENHAGEN K
DENMARK
TEL: 1-14-17-90
TLX:
COM: 27

PETRI WINFRIED PROF DR
UNTERLEITEN 2
POSTFACH 106
D-8162 SCHLIERSEE
GERMANY, F.R.
TEL: 08026-6428
TLX:
COM: 41

PETROV GENNADIJ M
INST RADIOTECH & ELECTRON
USSR ACADEMY OF SCIENCES
MARKS AVENJU 18
103907 MOSCOW GSP-3
U.S.S.R.
TEL:
TLX:
COM:

PEYTREMANN ERIC DR
50 WATTEN VIEW
SINGAPORE 1128
SINGAPORE
TEL:
TLX:
COM:

PETERSON BRADLEY MICHAEL
DEPT OF ASTRONOMY
OHIO STATE UNIVERSITY
174 W 18TH AVENUE
COLUMBUS OH 43210
U.S.A.
TEL: 614-422-7886
TLX:
COM: 28

PETRINI DANIEL DR
OBSERVATOIRE DE NICE
BP 139
F-06003 NICE CEDEX
FRANCE
TEL: 93-89-04-20
TLX: 460004
COM: 14

PETROV GEORGY TRENDAFILOV
DEPT OF ASTRONOMY
BULGARIAN ACAD SCIENCES
72 LENIN BLVD
1784 SOFIA
BULGARIA
TEL: 75-89-27
TLX: 23561 ECF BAN BG
COM:

PEYTURAUX ROGER H PROF
INSTITUT D'ASTROPHYSIQUE
98 BIS BOULEVARD ARAGO
F-75014 PARIS
FRANCE
TEL: 1-43-20-14-25
TLX:
COM: 12

PETERSON BRUCE A DR
MT STROMLO & SIDING
SPRING OBSERVATORIES
AUSTRALIAN NAT UNIVERSITY
WODEN P.O. ACT 2606
AUSTRALIA
TEL: 61-62-88-1111
TLX: 62270
COM: 47,48

PETRO LARRY DAVID
SPACE TELESCOPE SCI INST
HOMEWOOD CAMPUS
3700 SAN MARTIN DRIVE
BALTIMORE MD 21218
U.S.A.
TEL: 301-338-4501
TLX: 7102341090
COM:

PETROV NIKOLAI
ASTRONOMICAL OBSERVATORY
PO BOX 120
9000 VARNA
BULGARIA
TEL: 22-28-90
TLX:
COM:

PFAU WERNER
UNIVERSITY OBSERVATORY
SCHILLERGAESSCHEN 2
DDR-6900 JENA
GERMANY, D.R.
TEL: 058861347
TLX:
COM: 25

PETERSON CHARLES JOHN DR
DEPT PHYSICS & ASTRONOMY
UNIVERSITY OF MISSOURI
223 PHYSICS BLDG
COLUMBIA MO 65211
U.S.A.
TEL: 314-882-3217
TLX:
COM: 28,37,41

PETROPOULOS BASIL CH DR
RES CTR ASTRON APPL MATHS
ACADEMY OF ATHENS
14 ANAGNOSTOPOULOU
GR-10673 ATHENS
GREECE
TEL: 3613589
TLX:
COM: 14,16

PETROVSKAYA M S DR
INST OF THEORET ASTRONOMY
10 KUTUZOV QUAY
191187 LENINGRAD
U.S.S.R.
TEL: 121578 ITA SU
TLX:
COM: 07,37

PFEIFFER RAYMOND J
8 BARBARA LANE
TITUSVILLE NJ 08560
U.S.A.
TEL: 609-883-4612
TLX:
COM: 25

PETERSON LAURENCE E PROF
CASS C-011
UNIVERSITY OF CALIFORNIA
AT SAN DIEGO
LA JOLLA CA 92093
U.S.A.
TEL: 619-452-3461
TLX: 910-337-1271 SIOCEAN
COM: 44,48

PETROSIAN VAHE PROF
CTR FOR SPACE SCIENCE &
ASTROPHYSICS, ERL RM 304
STANFORD UNIVERSITY
STANFORD CA 94306
U.S.A.
TEL: 415-497-1435
TLX:
COM: 10,34,47,48

PETTENGILL GORDON H PROF
MIT
RM 37-241
CAMBRIDGE MA 02139
U.S.A.
TEL: 617-253-7501
TLX: 92-1473
COM: 16,40

PFENNIG HANS H DR
MPI F PHYSIK & ASTROPHYS
KARL-SCHWARZSCHILD-STR 1
D-8046 GARCHING B MUNCHEN
GERMANY, F.R.
TEL: 89-32-99-94-35
TLX: 524629 ASTRO D
COM: 14

PETERSON RUTH CAROL DR
607 MARION PLACE
PALO ALTO CA 94301
U.S.A.
TEL: 415-321-1281
TLX:
COM: 29

PETROU MARIA DR
UNIVERSITY OF OXFORD
DEPT THEORETICAL PHYSICS
1 KEBLE ROAD
OXFORD OX1 3NP
U.K.
TEL: 0865-53281
TLX:
COM: 28

PETTERSEN BJOERN RAGNVALD
INST THEORET ASTROPHYSICS
UNIVERSITY OF OSLO
P O BOX 1029
N-0315 BLINDERN OSLO 3
NORWAY
TEL: 02-45-65-01
TLX: 72705 ASTRO N
COM:

PFLEIDERER JORG PROF
INSTITUT FUER ASTRONOMIE
UNIVERSITAETSSTR 4
A-6020 INNSBRUCK
AUSTRIA
TEL: 52-22-724-6610
TLX:
COM: 21

PETFORD A DAVID DR
DEPT OF ASTROPHYSICS
SOUTH PARKS ROAD
OXFORD OX1 3RQ
U.K.
TEL: 865-511336
TLX:
COM: 09

PETROV G I PROF DR
SPACE RESEARCH INSTITUTE
USSR ACADEMY OF SCIENCES
117810 MOSCOW
U.S.S.R.
TEL:
TLX:
COM:

PETTINI MARCO
OSSERVATORIO ASTROFISICO
DI ARCETRI
LARGO E. FERMI 5
I-50125 FIRENZE
ITALY
TEL: 055-220034
TLX: 572268 ARCETR
COM: 14

PFLUG KLAUS DR
ZNTRLINST. F. ASTROPHYSIK
SONNENOBSERVATORIUM
EINSTEINTURM
DDR-1500 POTSDAM
GERMANY, D.R.
TEL:
TLX:
COM: 10,12,49

PHAM-VAN JACQUELINE MME
CERGA
AVENUE COPERNIC
F-06130 GRASSE
FRANCE
TEL: 93-36-58-49
TLX: 470865
COM: 08

PHILIP A G DAVIS
1125 OXFORD PLACE
SCHENECTADY NY 12308
U.S.A.
TEL: 518-374-5636
TLX:
COM: 05,25,30C,33,37,45

PHILLIPS ANTHONY PAUL
23 CROUCH HALL RD, FLAT 5
CROUCH END
LONDON N8 8HT
U.K.
TEL:
TLX:
COM: 34

PHILLIPS JOHN G PROF
ASTRONOMY DEPT
UNIVERSITY OF CALIFORNIA
601 CAMPBELL HALL
BERKELEY CA 94720
U.S.A.
TEL: 415-642-5275
TLX:
COM: 14,36

PHILLIPS JOHN PETER
PHYSICS DEPT
QUEEN MARY COLLEGE
MILE END ROAD
LONDON E1 4NS
U.K.
TEL: 01-980-4811
TLX: 893750 QMEUOL G
COM:

PHILLIPS KENNETH J H
SPACE & ASTROPHYSICS DIV
RUTHERFORD APPLETON LAB
CHILTON, DIDCOT OX11 0QX
U.K.
TEL: 0235-21900
TLX:
COM: 10,12,44

PHILLIPS MARK M DR
CERRO TOLOLO INTER-
AMERICAN OBSERVATORY NOAO
CASILLA 603
LA SERENA
CHILE
TEL: 51-213352
TLX: 620301 AURA CT
COM: 28,35

PHILLIPS THOMAS GOULD DR
CALTECH
320-47
PASADENA CA 91125
U.S.A.
TEL: 818-356-4278
TLX:
COM: 34,40

PIAZZA LILIANA RIZZO
INST. PESQUISAS ESPACIAIS
C.P. 515
12200 S. JOSE DOS CAMPOS
BRAZIL
TEL: 011-825365
TLX: 11-34061 INPE BR
COM:

PICAT JEAN-PIERRE DR
OBSERVATOIRE PIC-DU-MIDI
14 AVENUE EDOUARD BELIN
F-31400 TOULOUSE
FRANCE
TEL: 61-25-21-01
TLX:
COM: 09

PICCIONI ADALBERTO
ISTITUTO ASTRONOMICO
UNIVERSITARIO
C P 596
I-40100 BOLOGNA
ITALY
TEL: 051-222956
TLX: 211664 INFNBO I
COM: 42

PICK MONIQUE DR
OBSERVATOIRE DE PARIS
SECTION DE MEUDON
DASOP
F-92195 MEUDON PL CEDEX
FRANCE
TEL: 1-45-34-75-30
TLX: 200590
COM: 10P,40

PIDDINGTON JACK H RES FEL
NATIONAL MEASUREMENT LAB
CSIRO, P O B 218
LINDFIELD
SYDNEY NSW 2070
AUSTRALIA
TEL: 467 6211
TLX: 26296 AA
COM: 10,48

PIERCE A KEITH DR
NATL SOLAR OBSERVATORY
PO 26732
TUCSON AZ 85726
U.S.A.
TEL: 602-327-5511
TLX:
COM: 07,12

PIERCE DAVID ALLEN
SCIENCE & MATH DIVISION
EL CAMINO COLLEGE
TORRANCE CA 90506
U.S.A.
TEL:
TLX:
COM: 20

PIIROLA VILPPU E DR
OBSERVATORY
TAHTITORNINMAKI
SF-00130 HELSINKI 13
FINLAND
TEL: 90-1912801
TLX: 124690 UNIH SF
COM: 25C,27,42

PIKE CHRISTOPHER DAVID
ROYAL GREENWICH OBS
HERSTMONCEUX CASTLE
HAILSHAM BN27 1RP
U.K.
TEL:
TLX:
COM:

PILACHOWSKI CATHERINE DR
KITT PEAK NATL OBS
NTL OPTICAL ASTR OBS
PO BOX 26732
TUCSON AZ 85726
U.S.A.
TEL: 602-327-5511
TLX: 0666484 AURA NOACTUC
COM: 29,37C

PILCHER CARL BERNARD DR
INSTITUTE FOR ASTRONOMY
UNIVERSITY OF HAWAII
2680 WOODLAWN DRIVE
HONOLULU HI 96822
U.S.A.
TEL: 808-948-7954
TLX: 723-8459 UHAST
COM: 15

PILKINGTON JOHN D H DR
ROYAL GREENWICH OBS
HERSTMONCEUX CASTLE
HAILSHAM BN27 1RP
U.K.
TEL: 44 323 833171
TLX: 87451 RGOBSY G
COM: 19,31C

PILOWSKI K PROF DR
GEODAETISCHES INSTITUT
TECHNISCHE UNIVERSITAET
NIENBURGER STR 1
D-3000 HANNOVER
GERMANY, F.R.
TEL:
TLX:
COM: 08,33

PINEAULT SERGE DR
DEPT DE PHYSIQUE
UNIVERSITE LAVAL
SAINTE-FOY PQ G1K 7P4
CANADA
TEL: 418-656-3901
TLX:
COM:

PINES DAVID PROF
DEPT OF PHYSICS
UNIVERSITY OF ILLINOIS
URBANA IL 61801
U.S.A.
TEL: 217-333-0115
TLX: 9103806599 PHYSICS S
COM: 35

PINGREE DAVID PROF
BROWN UNIVERSITY
PO BOX 1900
PROVIDENCE RI 02912
U.S.A.
TEL: 401-863-2101
TLX:
COM: 41

PINIGIN GENNADIJ I DR
NIKOLAYEV BRANCH
CENTRAL ASTRONOMICAL OBS
OBSERVATORNAYA 1
327001 NIKOLAYEV REGIONAL
U.S.S.R.
TEL:
TLX:
COM: 08C

PINKAU K PROF
MPI FUER PLASMAPHYSIK
D-8046 GARCHING B MUNCHEN
GERMANY, F.R.
TEL: 89-3299-342
TLX: 05-215-808
COM: 44,48

PINOTSIS ANTONIS D DR
DEPT OF ASTRONOMY
UNIVERSITY OF ATHENS
PANEPISTIMIOPOLIS
ATHENS 621
GREECE
TEL:
TLX:
COM: 35

PINTO GIROLAMO PROF
OSSERVATORIO ASTRONOMICO
I-35100 PADOVA
ITALY
TEL:
TLX:
COM:

PIPHER JUDITH L
PHYSICS & ASTRONOMY DEPT
UNIVERSITY OF ROCHESTER
ROCHESTER NY 14627
U.S.A.
TEL: 716-275-4402
TLX:
COM:

PIRRONELLO VALERIO
OSSERVATORIO ASTROFISICO
CITTA UNIVERSITARIA
VIALE A. DORIA
I-95125 CATANIA
ITALY
TEL: 095-330533
TLX: 970359 ASTRCT I
COM:

PISKUNOV ANATOLY E
ASTRONOMICAL COUNCIL
USSR ACADEMY OF SCIENCES
PYATNITSKAYA 48
109017 MOSCOW
U.S.S.R.
TEL: 231-54-61
TLX: 412623 SCSTP SU
COM: 37

PISMIS DE RECILLAS PARIS
INSTITUTO DE ASTRONOMIA
UNAM
APDO POSTAL 70-264
04510 MEXICO DF
MEXICO
TEL: 905-548-5306
TLX: 1760155 CICME
COM: 28,33,34

PITTICH EDUARD M DR
ASTRONOMICAL INSTITUTE
SLOVAK ACAD. OF SCIENCES
DUBRAVSKA CESTA 9
842 28 BRATISLAVA
CZECHOSLOVAKIA
TEL: 427-375157
TLX: 93373 SEIS
COM: 15,20

PITZ ECKHART DR
MPI FUER ASTRONOMIE
KOENIGSTUHL
D-6900 HEIDELBERG
GERMANY, F.R.
TEL: 06221-5281
TLX: 461789 MPIA D
COM: 21

PIZZELLA G DR
DIPARTIMENTO DI FISICA
UNIVERSITA DI ROMA
PIAZZALE ALDO MORO 2
I-00185 ROMA
ITALY
TEL: 6-4940156
TLX: 613255 INFNRO
COM:

PIZZICHINI GRAZIELLA
ISTITUTO TESRE/CNR
VIA DE CASTAGNOLI 1
I-40126 BOLOGNA
ITALY
TEL: 051-519593
TLX: 511350 CNR BO
COM: 28

PLAKIDIS STAVROS PROF
EVRYTANIAS 16
KATO HALANDRI
GR-15231 ATHENS
GREECE
TEL: 6721770
TLX:
COM:

PLANESAS PERE
CENTRO ASTRON DE YEBES
O A N
APARTADO CORREOS 148
19080 GUADALAJARA
SPAIN
TEL: 11-22-33-58
TLX:
COM:

PLASSARD J DR
KSARA OBSERVATORY
KSARA
LEBANON
TEL:
TLX:
COM:

PLAVEC MIREK J PROF
DEPT OF ASTRONOMY
UNIVERSITY OF CALIFORNIA
MS 8979
LOS ANGELES CA 90024
U.S.A.
TEL: 213-825-1672
TLX:
COM: 29,35,42

PLAVEC ZDENKA DR
DEPT OF ASTRONOMY
UCLA
405 HILGARD AVE
LOS ANGELES CA 90024
U.S.A.
TEL: 213-206-8596
TLX:
COM: 22

PNEUMAN GERALD W
HIGH ALTITUDE OBSERVATORY
PO BOX 3000
BOULDER CO 80302
U.S.A.
TEL: 303-497-1000
TLX: 45694
COM: 10,49

PODOBED V V DR
STERNBERG STATE
ASTRONOMICAL INSTITUTE
117234 MOSCOW
U.S.S.R.
TEL:
TLX:
COM: 08,24

POECKERT ROLAND H DR
DEFENCE RESEARCH
ESTABLISHMENT PACIFIC
FMO CFB ESQUIMALT
VICTORIA BC V0S 1B0
CANADA
TEL:
TLX:
COM: 29

POEPPEL WOLFGANG G L DR
INSTITUTO ARGENTINO
DE RADIOASTRONOMIA
CASILLA DE CORREO 5
1894 VILLA ELISA (Bs.As.)
ARGENTINA
TEL: 021-43793
TLX: 18052 CICYT-AR
COM: 34

POGO ALEXANDER DR
MT WILSON & LAS CAMPANAS
OBSERVATORIES
813 SANTA BARBARA ST
PASADENA CA 91101
U.S.A.
TEL: 213-577-1122
TLX:
COM: 41

POHL ECKHARD DR
STERNWARTE NUERNBERG
REGIOMONTANUSWEG 1
D-8500 NUERNBERG 20
GERMANY, F.R.
TEL: 0911-593540
TLX:
COM:

POLAND ARTHUR I DR
NASA/GSFC
CODE 682
GREENBELT MD 20771
U.S.A.
TEL: 301-344-7334
TLX: 89675
COM: 10

POLCARO V F
IST. ASTROFISICA SPAZIALE
C P 67
I-00044 FRASCATI
ITALY
TEL: 9425651
TLX: 610261
COM:

POLETTO GIANNINA PROF
OSSERVATORIO ASTROFISICO
DI ARCETRI
LARGO E. FERMI 5
I-50125 FIRENZE
ITALY
TEL: 55-220034
TLX: 572268 ARCETR I
COM: 10

POLIDAN RONALD S
UNIVERSITY OF ARIZONA
LUNAR & PLANETARY LAB
3625 E. AJO WAY
TUCSON AZ 85713
U.S.A.
TEL:
TLX:
COM: 42,44

POLLACK JAMES B DR
SPACE SCIENCE DIVISION
NASA-AMES RESEARCH CTR
MS 245-3
MOFFETT FIELD CA 94035
U.S.A.
TEL: 415-694-5530
TLX:
COM: 16,51

POLNITZKY GERHARD DR
INSTITUT FUER ASTRONOMIE
UNIVERSITAET WIEN
TUERKENSCHANZSTR 17
A-1180 WIEN
AUSTRIA
TEL: 0222-345360-90
TLX: 116222 PHYSI A
COM: 08,22

POLOZHENTSEV DIMITRIJ DR
PULKOVO OBSERVATORY
196140 LENINGRAD
U.S.S.R.
TEL: 298-22-42
TLX:
COM: 08,24

POLUPAN P N DR
KIEV STATE UNIVERSITY
ASTRONOMICAL OBSERVATORY
252053 KIEV
U.S.S.R.
TEL: 26-09-08
TLX: 132201
COM: 10

POMA ANGELO DR
INTL ASTRONOMICAL STATION
VIA OSPEDALE 72
I-09100 CAGLIARI
ITALY
TEL: 070-66-35-44
TLX: 790326 OSSAST
COM: 08,19

PONNAMPERUMA CYRIL PROF
DEPT OF CHEMISTRY
UNIVERSITY OF MARYLAND
COLLEGE PARK MD 20472
U.S.A.
TEL:
TLX:
COM: 51

PONSONBY JOHN E B DR
NRAL
JODRELL BANK
MACCLESFIELD SK11 9DL
U.K.
TEL: 0477-71321
TLX: 36149
COM: 40,51

POOLEY GUY DR
CAVENDISH LABORATORY
MADINGLEY ROAD
CAMBRIDGE CB3 0HE
U.K.
TEL: 223-66477
TLX: 81292
COM: 40

POPELAR JOSEF DR
GRAVITY & GEODYNAMICS DIV
EARTH PHYSICS BRANCH
3 OBSERVATORY CRESCENT
OTTAWA ONT K1A 0E4
CANADA
TEL: 613-9925419
TLX: 0533117 EMAR-OTT
COM: 19,31

POPOV VASIL NIKOLOV
DEPT OF ASTRONOMY
BULGARIAN ACAD SCIENCES
72 LENIN BLVD
1784 SOFIA
BULGARIA
TEL: 449-477
TLX:
COM: 28

POPOV VICTOR S DR
MAIN ASTRONOMICAL OBS
USSR ACADEMY OF SCIENCES
PULKOVO M-140
196140 LENINGRAD
U.S.S.R.
TEL:
TLX:
COM: 30

POPOVA MALINA D PROF DR
DEPT OF ASTRONOMY
BULGARIAN ACAD SCIENCES
72 LENIN BLVD
1784 SOFIA
BULGARIA
TEL: 449-477
TLX:
COM: 27,35,37

POPOVIC BOZIDAR PROF DR
OGNJENA PRICE 80
11000 BEOGRAD
YUGOSLAVIA
TEL:
TLX:
COM: 07,20

LIST OF MEMBERS

POPOVIC GEORGIJE DR
ASTRONOMICAL OBSERVATORY
VOLGINA 7
11050 BEOGRAD
YUGOSLAVIA
TEL: 38-11-419357
TLX:
COM: 26

POPPER DANIEL M PROF
DEPT OF ASTRONOMY
UNIVERSITY OF CALIFORNIA
LOS ANGELES CA 90024
U.S.A.
TEL: 213-825-3622
TLX: 9103427597
COM: 42

POQUERUSSE MICHEL
OBSERVATOIRE DE PARIS
SECTION DE MEUDON
DESPA
F-92195 MEUDON PL CEDEX
FRANCE
TEL: 1-45-34-75-30
TLX: 204464
COM: 10,12

PORCAS RICHARD DR
MPI FUER RADIOASTRONOMIE
AUF DEM HUEGEL 69
D-5300 BONN
GERMANY. F.R.
TEL: 0228-525-282
TLX: 0886440 MPIFR D
COM: 40

PORFIR'EV V V DR
KRUPSKAJA PEDAGOGOC INST
107846 MOSCOW
U.S.S.R.
TEL:
TLX:
COM: 35

PORTER NEIL A PROF
PHYSICS DEPT
UNIVERSITY COLLEGE
BELFIELD
DUBLIN 4
IRELAND
TEL: 1-693-244x211
TLX: 32693 UCDEI
COM: 41,48

PORUBCAN VLADIMIR DR
ASTRONOMICAL INSTITUTE
SLOVAK ACAD. OF SCIENCES
DUBRAVSKA 9
842 28 BRATISLAVA
CZECHOSLOVAKIA
TEL: 427-375157
TLX: 93373 SEIS
COM: 22

POTTASCH STUART R PROF
KAPTEYN LABORATORIUM
POSTBUS 800
NL-9700 AV GRONINGEN
NETHERLANDS
TEL: 50-1166641
TLX: 53572 STARS NL
COM: 34,36

POTTER HEINO I DR
PULKOVO OBSERVATORY
196140 LENINGRAD
U.S.S.R.
TEL: 298-22-42
TLX:
COM: 24

POULAKOS CONSTANTINE DR
RESEARCH CTR F. ASTRONOMY
AND APPLIED MATHS
ACADEMY OF ATHENS
GR-10673 ATHENS
GREECE
TEL:
TLX:
COM:

POULLE EMMANUEL PROF
ECOLE NATLE DES CHARTES
19 RUE DE LA SORBONNE
F-75005 PARIS
FRANCE
TEL: 1-45-89-48-57
TLX:
COM: 41

POUMEYROL FERNAND MR
OBSERVATOIRE DE BORDEAUX
AVENUE P. SEMIROT
F-33270 FLOIRAC
FRANCE
TEL: 56-86-43-30
TLX:
COM:

POUNDS KENNETH A PROF
DEPT OF PHYSICS
THE UNIVERSITY
UNIVERSITY ROAD
LEICESTER LE1 7RH
U.K.
TEL: 0533-954455x151
TLX: 341664 LUXRAYG
COM: 06,44V,48

POUQUET ANNICK DR
OBSERVATOIRE DE NICE
BP 139
F-06003 NICE CEDEX
FRANCE
TEL: 93-89-04-20
TLX: 460004
COM:

POVEDA ARCADIO DR
INSTITUTO DE ASTRONOMIA
UNAM
APDO POSTAL 70-264
04510 MEXICO DF
MEXICO
TEL: 550-5805
TLX: 1760155 CICME
COM: 26,28,35,37

POYET JEAN-PIERRE DR
OBSERVATOIRE PIC-DU-MIDI
ET DE TOULOUSE
14 AVENUE EDOUARD BELIN
F-31400 TOULOUSE
FRANCE
TEL: 61-25-21-01
TLX:
COM:

PRABHU TUSHAR P
INDIAN INSTITUTE OF
ASTROPHYSICS
BANGALORE 560 034
INDIA
TEL:
TLX: 845763 IIAB IN
COM: 28

PRADERIE FRANCOISE DR
OBSERVATOIRE DE PARIS
SECTION DE MEUDON
DEPT RECHERCHE SPATIALE
F-92195 MEUDON PL CEDEX
FRANCE
TEL: 1-45-34-75-30
TLX: 204464
COM: 36

PRADHAN DR
JILA QPD 525
UNIVERSITY OF COLORADO
BOULDER CO 80309
U.S.A.
TEL: 303-492-7812
TLX: 755842
COM:

PRASAD SHEO S
JET PROPULSION LABORATORY
MS 183-601
4800 OAK GROVE DRIVE
PASADENA CA 91109
U.S.A.
TEL: 213-354-6423
TLX: 675429
COM: 34

PRASANNA A R DR
PHYSICAL RESEARCH LAB
NAVRANGPURA
AHMEDABAD 380 009
INDIA
TEL: 462129
TLX: 021-397 PRL IN
COM:

PRATAP R DR
INSTITUTE OF APPLIED
SCIENCES
COCHIN 682 317
INDIA
TEL:
TLX:
COM:

PRAVDO STEVEN H
JET PROPULSION LABORATORY
MS 168-222
4800 OAK GROVE DRIVE
PASADENA CA 91109
U.S.A.
TEL: 818-354-4134
TLX: 910-588-3294
COM:

PREITE-MARTINEZ ANDREA DR
IST. ASTROFISICA SPAZIALE
C P 67
I-00044 FRASCATI
ITALY
TEL:
TLX: 610261
COM: 34

PRENDERGAST KEVIN H PROF
DEPT OF ASTRONOMY
COLUMBIA UNIVERSITY
538 W. 120TH STREET
NEW YORK NY 10027
U.S.A.
TEL: 212-280-3280
TLX:
COM: 28

PRENTICE ANDREW J R DR
DEPT OF MATHEMATICS
MONASH UNIVERSITY
CLAYTON VIC 3168
AUSTRALIA
TEL:
TLX:
COM: 35

PRESS WILLIAM H DR
HARVARD COLLEGE OBS
60 GARDEN STREET
CAMBRIDGE MA 02138
U.S.A.
TEL: 617-495-4908
TLX: 921428 SATELLITE CAM
COM: 28,47

PRESTON GEORGE W DR
MT WILSON & LAS CAMPANAS
OBSERVATORIES
813 SANTA BARBARA STREET
PASADENA CA 91101
U.S.A.
TEL: 818-577-1122
TLX:
COM: 29,30,45

PRESTON ROBERT ARTHUR
138-307 JPL
4800 OAK GROVE DRIVE
PASADENA CA 91109
U.S.A.
TEL: 213-354-6895
TLX: 675429
COM: 40

PREUSS EUGEN DR
MPI FUER RADIOASTRONOMIE
AUF DEM HUEGEL 69
D-5300 BONN 1
GERMANY. F.R.
TEL: 228-5251
TLX: 886440 MPIFR D
COM: 40,48

PREVOT LOUIS DR
OBSERVATOIRE DE MARSEILLE
2 PLACE LE VERRIER
F-13248 MARSEILLE CDX 04
FRANCE
TEL: 91-95-90-88
TLX: 420241
COM: 30

PREVOT-BURNICHON M.L. DR
OBSERVATOIRE DE MARSEILLE
2 PLACE LE VERRIER
F-13248 MARSEILLE CDX 04
FRANCE
TEL: 91-95-90-88
TLX: 420241
COM: 28

PRICE MICHAEL J. DR.
SCIENCE APPLICATIONS
5151 E BROADWAY
SUITE 1100
TUCSON AZ 85711
U.S.A.
TEL: 602-748-7400
TLX:
COM:

PRICE R MARCUS DR
DEPT PHYSICS & ASTRONOMY
UNIVERSITY OF NEW MEXICO
ALBUQUERQUE NM 87131
U.S.A.
TEL: 505-277-2616
TLX:
COM: 33,34,40

PRICE STEPHAN DONALD
2 POLLEY ROAD
WESTFORD MA 01886
U.S.A.
TEL: 617-861-4552
TLX:
COM: 44

PRIEST ERIC R PROF
APPLIED MATHS DEPT
THE UNIVERSITY
ST ANDREWS, FIFE KY16 9SS
U.K.
TEL: 0334-76161
TLX: 76213 SAULIB
COM: 10V,12

PRIESTER WOLFGANG PROF
INSTITUT F ASTROPHYSIK
AUF DEM HUEGEL 71
D-5300 BONN
GERMANY, F.R.
TEL: 0228-73-3671
TLX: 886440
COM: 33,40

PRIETO MERCEDES
INST DE ASTROFISICA DE
CANARIAS
LA LAGUNA
38071 TENERIFE
SPAIN
TEL: 922-262211
TLX: 92640
COM:

PRINCE HELEN DODSON PROF
4800 FILLMORE AVE
ALEXANDRIA VA 22311
U.S.A.
TEL: 703-578-1000
TLX:
COM:

PRINGLE JAMES E DR
INSTITUTE OF ASTRONOMY
MADINGLEY ROAD
CAMBRIDGE CB3 0HA
U.K.
TEL: 0223-62204
TLX: 817297 ASTRON G
COM: 27,42

PRITCHET CHRISTOPHER J DR
PHYSICS DEPT
UNIVERSITY OF VICTORIA
P O BOX 1700
VICTORIA BC V8W 2Y2
CANADA
TEL: 604-721-7704
TLX: 049-7222
COM: 09.28.37

PROBSTEIN R F DR
DEPT MECHANICAL ENGINEERG
MIT
CAMBRIDGE MA 02139
U.S.A.
TEL: 617-253-2240
TLX: 921473 MIT CAM
COM:

PROCHAZKA FRANZ V DR
SONNENOBSERVATORIUM
KANZELHOEHE
A-9520 SATTENDORF
AUSTRIA
TEL: 0-42-48-27-17
TLX: 45699
COM: 24,51

PRODAN Y I DR
STERNBERG STATE
ASTRONOMICAL INSTITUTE
119899 MOSCOW
U.S.S.R.
TEL: 139-55-43
TLX:
COM: 19

PROISY PAUL E DR
OBSERVATOIRE DE LYON
F-69230 ST-GENIS-LAVAL
FRANCE
TEL: 78-56-07-05
TLX:
COM: 15

PROKAKIS THEODORE J DR
ASTRONOMICAL INSTITUTE
NATL OBSERVATORY OF ATHEN
P C BOX 20048
GR-11810 ATHENS
GREECE
TEL: 8040619-3461191
TLX: 21 5530
COM: 10,12,41

PROKOF'EV VLADIMIR K PROF
CRIMEAN ASTROPHYSICAL
OBSERVATORY
PO NAUCHNY
334413 CRIMEA
U.S.S.R.
TEL:
TLX:
COM: 14,44

PROKOF'EVA IRINA A DR
PULKOVO OBSERVATORY
196140 LENINGRAD
U.S.S.R.
TEL:
TLX:
COM:

PROKOF'EVA VALENTINA V DR
CRIMEAN ASTROPHYSICAL
OBSERVATORY
NAUCHNYJ
334413 CRIMEA
U.S.S.R.
TEL: 1-24
TLX:
COM: 09

PRONIK I I DR
CRIMEAN ASTROPHYS OBS
USSR ACADEMY OF SCIENCES
PO NAUCHNY
334413 CRIMEA
U.S.S.R.
TEL: 569
TLX:
COM: 28.34

PRONIK V I DR
CRIMEAN ASTROPHYS OBS
USSR ACADEMY OF SCIENCES
PO NAUCHNY
334413 CRIMEA
U.S.S.R.
TEL: 569
TLX:
COM: 28

PROSZYNSKI MIECZYSLAW
COPERNICUS ASTRON CENTER
UL. BARTYCKA 18
00-716 WARSAW
POLAND
TEL:
TLX:
COM: 44

PROTHEROE RAYMOND J DR
DEPT OF PHYSICS
UNIVERSITY OF ADELAIDE
ADELAIDE, S. AUSTR. 5001
AUSTRALIA
TEL: 08-228-5996
TLX: 89141 UNIVAD AA
COM:

PROTHEROE WILLIAM M PROF
DEPT OF ASTRONOMY
OHIO STATE UNIVERSITY
174 W. 18TH AVENUE
COLUMBUS OH 43210
U.S.A.
TEL: 614-422-7891
TLX:
COM:

PROTICH MILORAD B
ASTRONOMICAL OBSERVATORY
VOLGINA 7
11050 BELGRADE
YUGOSLAVIA
TEL: 011-402-365
TLX:
COM: 20

PROUST DOMINIQUE
OBSERVATOIRE DE PARIS
SECTION DE MEUDON
DAPHE
F-92195 MEUDON PL CEDEX
FRANCE
TEL: 1-45-34-75-70
TLX:
COM: 28

PROVERBIO EDOARDO PROF
ISTITUTO DI ASTRONOMIA
VIA OSPEDALE 72
I-09100 CAGLIARI
ITALY
TEL: 070-657657
TLX: 790326 OSSAST I
COM: 08,19,31C.46.51

PROVOST JANINE DR
OBSERVATOIRE DE NICE
BP 139
F-06003 NICE CEDEX
FRANCE
TEL: 93-89-04-20
TLX: 460004
COM: 27,35

PRYCE MAURICE H L DR
DEPT OF PHYSICS
UNIV OF BRITISH COLUMBIA
2075 WESBROOK MALL
VANCOUVER BC V6T 1W5
CANADA
TEL:
TLX:
COM:

PSKOVSKIJ JU P DR
STERNBERG STATE
ASTRONOMICAL INSTITUTE
119899 MOSCOW
U.S.S.R.
TEL: 139-37-21
TLX:
COM: 27,34

PUCILLO MAURO DR
OSSERVATORIO ASTRONOMICO
VIA TIEPOLO 11
I-34131 TRIESTE
ITALY
TEL: 040-793921
TLX: 461137 OAT I
COM:

PUGET JEAN-LOUP DR
RADIOASTRONOMIE
LAB. DE PHYSIQUE E.N.S.
24 RUE LHOMOND
F-75005 PARIS
FRANCE
TEL: 1-43-29-12-25
TLX: 270912
COM: 34

PUNETHA LALIT MOHAN DR
UP STATE OBSERVATORY
MANORA PEAK
NAINI TAL 263 129
INDIA
TEL: 2136
TLX:
COM:

PURCELL EDWARD M PROF
DEPT OF PHYSICS
HARVARD UNIVERSITY
CAMBRIDGE MA 02138
U.S.A.
TEL: 617-495-2860
TLX:
COM: 51

PURTON CHRISTOPHER R DR
DOMINION RADIO ASTROPHYS
OBSERVATORY, NRC
P O BOX 248
PENTICTON BC V2A 6K3
CANADA
TEL: 604-497-5321
TLX: 048-88127
COM:

PUSCHELL JEFFERY JOHN
THE TITAN CORPORATION
PO BOX 12139
LA JOLLA CA 92037
U.S.A.
TEL: 619-453-9500
TLX:
COM: 40

PUSHKIN SERGEY B DR
TIME & FREQUENCY SERVICE
GOSSTANDARD USSR
117049 MOSCOW
U.S.S.R.
TEL:
TLX:
COM: 31

PYE JOHN P DR
PHYSICS DEPT
UNIVERSITY OF LEICESTER
UNIVERSITY ROAD
LEICESTER LE1 7RH
U.K.
TEL: 533-554455-23
TLX: 341664 LUXRAY G
COM:

PYPER SMITH DIANE M DR
PHYSICS DEPARTMENT
TEL-AVIV UNIVERSITY
UNIVERSITY OF NEVADA
LAS VEGAS NV 89154
U.S.A.
TEL:
TLX:
COM:

QIAN BO-CHEN
SHANGHAI OBSERVATORY
ACADEMIA SINICA
SHANGHAI
CHINA, PEOPLE'S REP.
TEL: 386191
TLX: 33164 SHAO CN
COM: 37

QIAN JING-KUI
DEPT OF GEOPHYSICS
PEKING UNIVERSITY
BEIJING
CHINA, PEOPLE'S REP.
TEL: 282471-3888
TLX: 22239 PKUNI
COM: 10

QIAN SHAN-JIE
BEIJING ASTRONOMICAL OBS
ACADEMIA SINICA
BEIJING
CHINA, PEOPLE'S REP.
TEL: 28-2194
TLX: 22040 BAOAS CN
COM: 40

QIAN ZHI-HAN DR
SHANGHAI OBSERVATORY
ACADEMIA SINICA
SHANGHAI
CHINA, PEOPLE'S REP.
TEL: 386191
TLX: 33164 SHAO CN
COM: 08

QIAN ZHONG-YU
BEIJING ASTRONOMICAL OBS
ACADEMIA SINICA
BEIJING
CHINA, PEOPLE'S REP.
TEL:
TLX: 22040 BAOAS CN
COM: 33

QIAO GUOJUN
BEIJING UNIVERSITY
ROOM 214, BLDG 39
BEIJING
CHINA, PEOPLE'S REP.
TEL:
TLX:
COM: 42

QIN DAO
PURPLE MOUNTAIN
OBSERVATORY
NANJING
CHINA, PEOPLE'S REP.
TEL: 46700
TLX: 34144 PMONJ CN
COM: 24

QIN SONG-NIAN
YUNNAN OBSERVATORY
P.O. BOX 110
KUNMING
CHINA, PEOPLE'S REP.
TEL:
TLX:
COM:

QIN ZHI-HAI
DEPT OF ASTRONOMY
NANJING UNIVERSITY
NANJING
CHINA, PEOPLE'S REP.
TEL: 34651-2882
TLX: 0909
COM: 34

QIU YU-HAI
BEIJING ASTRONOMICAL OBS.
ACADEMIA SINICA
BEIJING 100080
CHINA, PEOPLE'S REP.
TEL:
TLX: 22040 BAOAS CN
COM: 40

QU QIN-YUE
DEPT OF ASTRONOMY
NANJING UNIVERSITY
NANJING
CHINA, PEOPLE'S REP.
TEL: 37551 EXT2741
TLX: 34151 PRCNU CN
COM: 35,47,48C

QUAMAR JAWAID
D-19, STAFF TOWN
UNIVERSITY OF KARACHI
KARACHI 3201
PAKISTAN
TEL: 46 54 91
TLX:
COM:

QUAN HEJUN
SHANGHAI OBSERVATORY
ACADEMIA SINICA
SHANGHAI
CHINA, PEOPLE'S REP.
TEL: 386191
TLX: 33164 SHAO CN
COM: 41

QUAST GERMANO RODRIGO
OBSERVATORIO NACIONAL
RUA CORONEL RENNO 07
CAIXA POSTAL 21
37500 ITAJUBA MG
BRAZIL
TEL: 035-6220788
TLX: 031 2603
COM:

QUENBY JOHN J DR
BLACKETT LABORATORY
IMPERIAL COLLEGE
PRINCE CONSORT ROAD
LONDON SW7 2BZ
U.K.
TEL: 1-589-5111x6661
TLX: 261503
COM:

QUERCI FRANCOIS R DR
OBS DU PIC-DU-MIDI
ET DE TOULOUSE
14 AVENUE EDOUARD BELIN
F-31400 TOULOUSE
FRANCE
TEL: 61-25-21-01
TLX: 530776 F
COM: 14,29,36

QUERCI MONIQUE DR
OBS DU PIC-DU-MIDI
ET DE TOULOUSE
14 AVENUE EDOUARD BELIN
F-31400 TOULOUSE
FRANCE
TEL: 61-25-21-01
TLX: 530776 F
COM: 29,36

QUIJANO LUIS
INSTITUTO Y OBSERVATORIO
DE MARINA
SAN FERNANDO (CADIZ)
SPAIN
TEL: 956-883-548
TLX: 76108 IOM E
COM: 08C,20,24

QUINTANA HERNAN DR
DEPTO DE ASTRONOMIA
UNIVERSIDAD CATOLICA
CASILLA 114-D
SANTIAGO
CHILE
TEL: 775474
TLX: 240395 PUCVA CL
COM: 28C,48,51

QUINTANA JOSE M DR
INST DE ASTROFISICA DE
ANDALUCIA, APDO 2144
PROFESOR ALBAREDA 1
18080 GRANADA
SPAIN
TEL: 58-121300
TLX: 78573 IAAG E
COM: 51

QUIRK WILLIAM J DR
LAWRENCE LIVERMORE NATL
LABORATORY L 35
BOX 808
LIVERMORE CA 94550
U.S.A.
TEL: 415-422-1852
TLX:
COM:

QVIST BERTIL PROF
MATHEMATICS DEPT
ABO AKADEMI
SF-20500 ABO 50
FINLAND
TEL:
TLX:
COM:

RAADU MICHAEL A DR
DEPT PLASMA PHYSICS
ROYAL INST OF TECHNOLOGY
S-100 44 STOCKHOLM 70
SWEDEN
TEL: 08-78-77000
TLX: 10389 KTHB STOCKHOLM
COM: 10,49

RABIN DOUGLAS MARK
NATIONAL SOLAR OBS
NATL OPTICAL ASTR OBS
PO BOX 26732
TUCSON AZ 85726-6732
U.S.A.
TEL: 602-325-9331
TLX: 0666484 AURA NOAOTUC
COM: 10,12

RACHKOVSKY D N DR
CRIMEAN ASTROPHYSICAL OBS
USSR ACADEMY OF SCIENCES
NAUCHNIY
334413 CRIMEA
U.S.S.R.
TEL: 1-03
TLX: 192
COM: 8

RACINE RENE DR
DEPT DE PHYSIQUE
UNIVERSITE DE MONTREAL
BP 6128
MONTREAL PQ H3C 3J7
CANADA
TEL: 514-343-6718
TLX: 055-62425 UDEMPHYSAS
COM: 09

RACKHAM THOMAS W DR
39 MEADOW AVENUE
GOOSTREY
CREWE CW4 8LS
U.K.
TEL: 0477-33004
TLX:
COM:

LIST OF MEMBERS

RADHAKRISHNAN V PROF
RAMAN RESEARCH INSTITUTE
SADASHIVANAGAR
BANGALORE 560 080
INDIA
TEL: 360522 360122
TLX: 845 671 RRI IN
COM: 34,40,48

RAHE JURGEN PROF
NASA HEADQUARTERS
CODE EL
WASHINGTON DC 20546
U.S.A.
TEL: 202-453-1590
TLX:
COM: 15V,42C,44C

RAJU P K DR
INDIAN INSTITUTE OF
ASTROPHYSICS
BANGALORE 560 034
INDIA
TEL: 566-585
TLX: 845763 IIAB IN
COM:

RAMELLA MASSIMO
OSSERVATORIO ASTRONOMICO
VIA G.B. TIEPOLO 11
I-34131 TRIESTE
ITALY
TEL: 040-76-85-06
TLX: 461137 CAT I
COM: 29,47

RADIMAN IRATIUS
BOSSCHA OBSERVATORY
LAMBANG, JAVA
INDONESIA
TEL:
TLX:
COM:

RAHUNEN TIMO
TAMPERE SAERKAENNIEMI OY
SAERKAENNIEMI
SF-33410 TAMPERE
FINLAND
TEL: 931-31333
TLX:
COM: 42

RAKAVY GIDEON PROF
EINSTEIN INST OF PHYSICS
HEBREW UNIV. OF JERUSALEM
JERUSALEM 91904
ISRAEL
TEL:
TLX:
COM:

RAMSEY LAWRENCE W DR
DEPT OF ASTRONOMY
PENNSYLVANIA STATE UNIV
525 DAVEY LAB
UNIVERSITY PARK PA 16802
U.S.A.
TEL: 814-865-3418
TLX:
COM: 09.36

RADLOVA L N DR
INSTITUTE OF SCIENCES
AND TECHNICS INFORMATION
DEPT OF ASTRONOMY
125219 MOSCOW
U.S.S.R.
TEL:
TLX:
COM: 05

RAIKOVA DONKA DR
DEPT OF ASTRONOMY
BULGARIAN ACAD SCIENCES
7TH NOVEMBER STR 1
1000 SOFIA
BULGARIA
TEL: 7341
TLX: 23561 ECF BAN BG
COM:

RAKOS KARL D PROF
INSTITUT FUER ASTRONOMIE
UNIVERSITAET WIEN
TUERKENSCHANZSTR 17
A-1180 WIEN
AUSTRIA
TEL: 0222-345360-95
TLX: 133099 VIAST A
COM: 09,26P,27,42

RANDIC LEO PROF DR
GEODETICAL FACULTY
GUNDULICEVA 54
ZAGREB 41000
YUGOSLAVIA
TEL: 041-44-66-75
TLX:
COM: 19,31

RADOSLAVOVA TSVETANKA
DEPT OF ASTRONOMY
BULGARIAN ACAD SCIENCES
72 LENIN BLVD
1784 SOFIA
BULGARIA
TEL:
TLX:
COM:

RAIMOND ERNST DR
NETHERLAND FOUNDATION
FOR RADIOASTRONOMY
POST BUS 2
NL-7990 AA DWINGELOO
NETHERLANDS
TEL: 05219-7244
TLX: 42043
COM: 08,34,40

RAKSHIT H PROF
BENGAL ENGINEERG COLLEGE
SIBPORE
HEWRAH
INDIA
TEL:
TLX:
COM:

RANIERI MARCELLO
IST. ASTROFISICA SPAZIALE
C P 67
I-00044 FRASCATI
ITALY
TEL:
TLX:
COM:

RAEDLER K H DR
ZNTRLINST. F. ASTROPHYSIK
ROSA-LUXEMBURG-STR 17A
DDR-1502 POTSDAM
GERMANY, D.R.
TEL:
TLX:
COM: 35

RAINE DEREK J DR
DEPT OF ASTRONOMY
UNIVERSITY OF LEICESTER
LEICESTER LEI 7RH
U.K.
TEL: 533-554455
TLX: 341198 LEICUL
COM:

RAM SAGAR DR
INDIAN INST OF ASTROPHYS
BANGALORE 560 034
INDIA
TEL: 566585/566497
TLX: 845763 IIAB IN
COM: 37

RANK DAVID M PROF
LICK OBSERVATORY
UNIVERSITY OF CALIFORNIA
SANTA CRUZ CA 95064
U.S.A.
TEL: 408-429-2277
TLX:
COM:

RAFANELLI PIERO DR
OSSERVATORIO ASTRONOMICO
VICOLO DELL'OSSERVATORIO
I-35100 PADOVA
ITALY
TEL: 49-661499
TLX:
COM:

RAITALA JOUKO T
DEPT OF ASTRONOMY
UNIVERSITY OF OULU
SF-90570 OULU 57
FINLAND
TEL: 81-35-21-06
TLX: 32375
COM:

RAMADURAI SOURIRAJA DR
ASTRONOMY GROUP
DEPARTMENT OF PHYSICS
INDIAN INST OF SCIENCE
BANGALORE 560 012
INDIA
TEL: BGL 364411x314
TLX: 08458349 BG
COM: 35,46,48

RANKIN JOANNA M DR
DEPT OF PHYSICS
UNIVERSITY OF VERMONT
A405 COOK BUILDING
BURLINGTON VT 05405
U.S.A.
TEL: 802-656-2644
TLX: 510-299-0021
COM:

RAFERT JAMES BRUCE
DEPT PHYSICS & ASTRONOMY
APPALACHIAN STATE UNIV
BOONE NC 28608
U.S.A.
TEL:
TLX:
COM: 42

RAJAMOHAN R DR
INDIAN INSTITUTE OF
ASTROPHYSICS
BANGALORE 560 034
INDIA
TEL: 566497/585
TLX: 845763 IIAB IN
COM: 37,51

RAMATY REUVEN DR
LAB HIGH ENERGY ASTROPHYS
NASA/GSFC, CODE 665
GREENBELT MD 20771
U.S.A.
TEL: 301-344-8715
TLX:
COM: 40

RAO A PRAMESH DR
RADIO ASTRONOMY CENTER
P O BOX NO 8
UDHAGAMANDALAM 643 001
INDIA
TEL: 2651 - 2032
TLX: 8458488 TIFR
COM: 10,40

RAGHAVAN NIRUPAMA DR
2133 INDIAN INSTITUTE
OF TECHNOLOGY
CAMPUS
NEW DELHI 110 029
INDIA
TEL:
TLX:
COM:

RAJCHL JAROSLAV DR
ASTRONOMICAL INSTITUTE
CZECH. ACAD. OF SCIENCES
OBSERVATORY
251 65 ONDREJOV
CZECHOSLOVAKIA
TEL: 724525 PRAHA
TLX: 121579
COM: 22

RAMBERG JOERAN M PROF
GENVAEGEN 4
S-133 00 SALTSJOEBADEN
SWEDEN
TEL: 46-8-717-1926
TLX:
COM: 33

RAO K NARAHARI
OHIO STATE UNIVERSITY
DEPT OF PHYSICS
174 W. 18TH AVENUE
COLUMBUS OH 43210
U.S.A.
TEL: 614-422-6505
TLX:
COM: 14

RAO K RAMANUJA DR
DEPT OF ASTRONOMY
OSMANIA UNIVERSITY
HYDERABAD 500 007
INDIA
TEL:
TLX:
COM:

RAO M N DR
PHYS. RESEARCH LABORATORY
NAVRANGPURA
AHMEDABAD 380 009
INDIA
TEL: 462129
TLX: 121397
COM: 16

RAO N KAMESWARA
INDIAN INSTITUTE OF
ASTROPHYSICS
BANGALORE 560 034
INDIA
TEL:
TLX:
COM: 27,29

RAO P VIVEKANANDA DR
DEPT OF ASTRONOMY
OSMANIA UNIVERSITY
HYDERABAD 500 007
INDIA
TEL:
TLX:
COM: 25

RAO RAMACHANDRA V PROF
ISRO SATELLITE CENTER
PEENYA
BANGALORE 560 058
INDIA
TEL:
TLX:
COM: 44

RAPAPORT MICHEL DR
OBSERVATOIRE DE BORDEAUX
AVENUE PIERRE SEMIROT
F-33270 FLOIRAC
FRANCE
TEL: 56-86-43-30
TLX:
COM: 20,21

RAPLEY CHRISTOPHER G DR
MULLARD SPACE SCIENCE LAB
UNIVERSITY COLLEGE LONDON
LONDON
U.K.
TEL: 030-670-292
TLX: 859185
COM:

RATNATUNGA KAVAN U.
INST FOR ADVANCED STUDIES
PRINCETON NJ 08540
U.S.A.
TEL: 609-734-8020
TLX: 229734 IAS UR
COM:

RAUBENHEIMER BAREND C
COSMIC RAY RESEARCH UNIT
POTCHEFSTROOM UNIVERSITY
POTCHEFSTROOM 2520
SOUTH AFRICA
TEL: 01481-27511
TLX: 421363
COM: 48

RAUTELA B S DR
UTTAR PRADESH STATE OBS
MANORA PEAK
NAINITAL 263 129
INDIA
TEL:
TLX:
COM: 45

RAY ALAK DR
TATA INSTITUTE OF
FUNDAMENTAL RESEARCH
BOMBAY 400 005
INDIA
TEL:
TLX:
COM:

RAY THOMAS P
DUNSINK OBSERVATORY
CASTLEKNOCK
DUBLIN 15
IRELAND
TEL: 387958
TLX:
COM: 40

RAYMOND JOHN CHARLES
CENTER FOR ASTROPHYSICS
60 GARDEN STREET
CAMBRIDGE MA 02138
U.S.A.
TEL:
TLX:
COM: 34

RAYROLE JEAN R DR
OBSERVATOIRE DE PARIS
SECTION DE MEUDON
F-92195 MEUDON PL CEDEX
FRANCE
TEL: 1-45-34-75-30
TLX:
COM: 10

RAZDAN HIRALAL
BHABHA ATOMIC RES CTR
ZAKURA SRINIGAR
KASHMIR 190 006
INDIA
TEL:
TLX:
COM: 48

RAZIN V A DR
RADIOPHYSICAL RESEARCH
INSTITUTE
603600 GORKIJ
U.S.S.R.
TEL: 36-72-94
TLX:
COM: 40

READHEAD ANTHONY C S DR
RADIO ASTRONOMY DEPT
CALTECH
ROBINSON BLDG
PASADENA CA 91125
U.S.A.
TEL: 213-356-4972
TLX: 675425 CALTECH PSD
COM: 40,49

REASENBERG ROBERT D DR
CENTER FOR ASTROPHYSICS
ROOM B 217
60 GARDEN STREET
CAMBRIDGE MA 02138
U.S.A.
TEL: 617-495-7108
TLX: 921428 SATELITE CAM
COM: 04

REAVES GIBSON PROF
DEPT OF ASTRONOMY
UNIV OF S. CALIFIFORNIA
LOS ANGELES CA 90089-1342
U.S.A.
TEL: 213-743-2039
TLX:
COM: 28

REAY NEWRICK K DR
ASTROPHYSICS GROUP
BLACKETT LABORATORY
IMPERIAL COLLEGE
LONDON SW7 2BZ
U.K.
TEL: 1-589-5111x6669
TLX: 261503 IMPCOL
COM: 09,49,51

REBEIROT EDITH DR
OBSERVATOIRE DE MARSEILLE
2 PLACE LE VERRIER
F-13248 MARSEILLE CDX 04
FRANCE
TEL: 91-95-90-88
TLX: 320241
COM: 30

REBER GROTE DR
GENERAL DELIVERY
BOTHWELL TASM 7411
AUSTRALIA
TEL: 002-237371
TLX:
COM: 40

RECILLAS-CRUZ ELSA DR
INSTITUTO DE ASTRONOMIA
UNAM
APDO POSTAL 70-264
04510 MEXICO DF
MEXICO
TEL:
TLX:
COM:

REES DAVID ELWYN DR
DEPT APPLIED MATHS
UNIVERSITY OF SYDNEY
SYDNEY NSW 2006
AUSTRALIA
TEL: 02-692-3724
TLX: UNISYD AA 26169
COM: 10,12

REES MARTIN J PROF
INSTITUTE OF ASTRONOMY
MADINGLEY RD
CAMBRIDGE CB3 0HA
U.K.
TEL: 223-62204
TLX: 817297 ASTRON G
COM: 44,47,48,51C

REEVES EDMOND M DR
NASA HEADQUARTERS
CODE EM
600 INDEPENDENCE AVE
WASHINGTON DC 20546
U.S.A.
TEL: 202-453-1571
TLX: 89530
COM: 10,12,44

REEVES HUBERT PROF
SEP-SES BAT 28
CEN SACLAY
BP 2
F-91190 GIF S/YVETTE
FRANCE
TEL: 1-69-08-51-59
TLX:
COM: 10,35,47,48

REFSDAL S PROF DR
HAMBURGER STERNWARTE
GOJENBERGSWEG 112
D-2050 HAMBURG 80
GERMANY, F.R.
TEL: 49-40-72524124
TLX: 217884 HAMST
COM: 42

REGO FERNANDEZ M DR
ASTROFISICA
FACULTA FISICA
UNIVERSIDAD COMPLUTENSE
28040 MADRID
SPAIN
TEL: 449-53-16
TLX: 47273 FF UC
COM: 29,46

REICH WOLFGANG
MPI FUER RADIOASTRONOMIE
AUF DEM HUEGEL 69
D-5300 BONN
GERMANY, F.R.
TEL:
TLX: 886440 MPIFR D
COM: 40

REID MARK JONATHAN DR
CENTER FOR ASTROPHYSICS
60 GARDEN STREET
CAMBRIDGE MA 02138
U.S.A.
TEL: 617-495-7470
TLX: 921428 SATELLITE CAM
COM: 40

REID NEILL
ROYAL GREENWICH OBS
HERSTMONCEUX CASTLE
HAILSHAM BN27 1RP
U.K.
TEL: 0323 833171
TLX: 87451 RGOBSY G
COM: 33

REIMERS DIETER PROF
HAMBURGER STERNWARTE
UNIVERSITAET HAMBURG
GOJENBERGSWEG 112
D-2050 HAMBURG 80
GERMANY, F.R.
TEL: 0407252 4112
TLX:
COM: 29,36

REINISCH GILBERT DR
OBSERVATOIRE DE NICE
BP 139
F-06003 NICE CEDEX
FRANCE
TEL: 93-89-04-20
TLX:
COM:

REIPURTH BO
E.S.C.
CASILLA 19001
SANTIAGO 19
CHILE
TEL: 6988757 SANTIAG
TLX: 240881
COM:

REITSEMA HAROLD J
BALL AEROSPACE SYSTEMS
DIVISION
PO BOX 1062
BOULDER CO 80306
U.S.A.
TEL: 303-441-5026
TLX:
COM: 20

REIZ ANDERS PROF
LOVSPRINGSVEJ 3 B
DK-2920 CHARLOTTENLUND
DENMARK
TEL: 1-63-25-36
TLX:
COM: 08,35,38

REMY BATTIAU LILIANE G A
CONSEIL DE LA RECHERCHE
UNIVERSITE DE LIEGE
7 PLACE DU XX AOUT
B-4000 LIEGE
BELGIUM
TEL: 41-42-00-80
TLX: 41397 UNIV ULG
COM: 05,15

REN JIANG-PING
DEPT OF ASTRONOMY
NANJING UNIVERSITY
NANJING
CHINA, PEOPLE'S REP.
TEL: 34651-2882
TLX: 34151 PRCNU CN
COM: 19

RENGARAJAN T N DR
IR ASTRONOMY
TIFR
HOMI BHABHA ROAD
BOMBAY 400 005
INDIA
TEL: 219111
TLX: 011-3009 TIFR IN
COM: 34,48

RENSE WILLIAM A DR
DEPT PHYSICS & ASTROPHYS
UNIVERSITY OF COLORADO
DUANE PHYSICAL LABS
BOULDER CO 80302
U.S.A.
TEL: 303-492-0111
TLX:
COM: 44

RENSON P F M DR
INSTITUT D'ASTROPHYSIQUE
UNIVERSITE DE LIEGE
AVENUE DE COINTE 5
B-4200 COINTE-OUGREE
BELGIUM
TEL: 41-52-99-80
TLX:
COM: 05,27

RENZINI ALVIO PROF
DIPT. DI ASTRONOMIA
VIA ZAMBONI 33
I-40126 BOLOGNA
ITALY
TEL: 51-222956
TLX: 211664 INFNBC
COM: 35

REQUIEME YVES DR
OBSERVATOIRE DE BORDEAUX
B.P. 21
F-33270 FLOIRAC
FRANCE
TEL: 56-86-43-30
TLX:
COM: 08P,24

REUNING ERNEST G DR
DEPT PHYSICS & ASTRONOMY
UNIVERSITY OF GEORGIA
ATHENS GA 30602
U.S.A.
TEL: 404-542-2485
TLX:
COM: 42

REVELLE DOUGLAS ORSON DR
METEOROLOGY PROGRAM
DEPT OF GEOGRAPHY
815 DAVIS HALL
DEKALB IL 60115
U.S.A.
TEL: 815-753-0631
TLX:
COM: 15,22

REYES FRANCISCO DR
DEPT OF ASTRONOMY
UNIVERSITY OF FLORIDA
GAINESVILLE FL 32611
U.S.A.
TEL: 904-392-2361
TLX:
COM: 40

REYNOLDS JOHN H PROF
DEPT OF PHYSICS
UNIVERSITY OF CALIFORNIA
BERKELEY CA 94720
U.S.A.
TEL: 415-642-4863
TLX: 9103667114
COM:

REYNOLDS RONALD J DR
SPACE PHYSICS GROUP
UNIVERSITY OF WISCONSIN
1150 UNIVERSITY AVENUE
MADISON WI 53706
U.S.A.
TEL: 608-262-5916
TLX:
COM: 34

REYNOLDS STEPHEN P
DEPARTMENT OF PHYSICS
NORTH CAROLINA STATE UNIV
BOX 8202
RALEIGH NC 27695-8202
U.S.A.
TEL: 919-737-7751
TLX:
COM:

RHODES EDWARD J JR
11801 KILLIMORE AVE
NORTHRIDGE CA 91326
U.S.A.
TEL:
TLX:
COM:

RIBES ELIZABETH DR
OBSERVATOIRE DE PARIS
SECTION DE MEUDON
F-92195 MEUDON PL CEDEX
FRANCE
TEL: 1-45-34-75-30
TLX:
COM:

RIBES JEAN-CLAUDE DR
INSU
77 AVE DENFERT-ROCHEREAU
F-75014 PARIS
FRANCE
TEL: 1-43-20-13-30
TLX: 270070
COM: 40

RICE JOHN B DR
DEPT PHYSICS & ASTRONOMY
BRANDON UNIVERSITY
BRANDON MAN R7A 6A9
CANADA
TEL: 204-727-9693
TLX: 07-502721
COM:

RICHARDSON E HARVEY DR
HERZBERG INST ASTROPHYS
DOMINION ASTROPHYS OBS
5071 W SAANICH RD
VICTORIA BC V8X 4M6
CANADA
TEL:
TLX:
COM: 09

RICHARDSON R S
GRIFFITH OBSERVATORY
PO BOX 27787
LOS FELIX STATION
LOS ANGELES CA 90027
U.S.A.
TEL:
TLX:
COM:

RICHER HARVEY B DR
DEPT GEOPHYS & ASTRONOMY
UNIV OF BRITISH COLUMBIA
2075 WESBROOK PLACE
VANCOUVER BC V6T 1W5
CANADA
TEL: 604-228-4134
TLX:
COM: 28,37

RICHTER G A DR
ZNTRLINST. F. ASTROPHYSIK
STERNWARTE BABELSBERG
ROSA-LUXEMBURG-STR 17A
DDR-6400 SONNEBERG
GERMANY, D.R.
TEL: 2287
TLX: 6288180 STEW DD
COM: 27

RICHTER JOHANNES PROF
INST F EXPERIMENT. PHYSIK
PHYSIKZENTRUM
OLSHAUSENSTRASSE
D-2300 KIEL
GERMANY, F.R.
TEL: 0431-880-3835
TLX: 292706 IAPKID
COM: 14

RICKARD JAMES JOSEPH DR
PO BOX 777
BORREGO SPRINGS CO 92004
U.S.A.
TEL: 714-767-5462
TLX:
COM:

RICKARD LEE J DR
NAVAL RESEARCH LABORATORY
CODE4138RRD
WASHINGTON DC 20375-5000
U.S.A.
TEL: 202-767-2495
TLX:
COM: 28,34,40

RICKER GEORGE R DR
CENTER FOR SPACE RESEARCH
MIT RM 37-527
77 MASSACHUSSETS AVENUE
CAMBRIDGE MA 02139
U.S.A.
TEL: 617-253-7532
TLX: 92-14-73
COM:

RICKETT BARNABY JAMES DR
DEPT OF ELECTRICAL ENG
AND COMPUTER SCIENCE
UNIV CALIF AT SAN DIEGO
LA JOLLA CA 92093
U.S.A.
TEL: 619-452-2731
TLX:
COM: 40,49

RICKMAN HANS DR
ASTRONOMISKA OBS
BOX 515
S-751 20 UPPSALA
SWEDEN
TEL: 46-18113522
TLX: 76024 UNIVUPS
COM: 15,20

LIST OF MEMBERS

RICORT GILBERT DR
DEPT D'ASTROPHYSIQUE
UNIVERSITE DE NICE
PARC VALROSE
F-06034 NICE
FRANCE
TEL:
TLX:
COM:

RIDDLE ANTHONY C DR
700 GRANT PL
BOULDER CO 80302
U.S.A.
TEL: 303-447-8127
TLX:
COM: 49

RIEGEL KURT W DR
NATL SCIENCE FOUNDATION
1800 G STREET NW
WASHINGTON DC 20550
U.S.A.
TEL: 202-357-9450
TLX:
COM: 33

RIGHINI ALBERTO PROF
ISTITUTO DI ASTRONOMIA
UNIV DEGLI STUDI FIRENZE
LARGO E. FERMI 5
I-50125 FIRENZE
ITALY
TEL: 055-220034
TLX: 572268 ARCETRI I
COM:

RIGHINI-COHEN GIOVANNA DR
DEPT EARTH & SPACE SCI
SUNY
STONY BROOK NY 11794
U.S.A.
TEL:
TLX:
COM: 12,34,44

RIGUTTI MARIO PROF
OSSERVATORIO ASTRONOMICO
DI CAPODIMONTE
MOIARIELLO 16
I-80131 NAPOLI
ITALY
TEL: 440101
TLX:
COM: 12,46

RIIHIMAA JORMA J DR
AARNE KARJALAINEN OBS
UNIVERSITY OF OULU
SF-90570 OULU
FINLAND
TEL:
TLX:
COM: 40,51

RILEY JULIA M DR
MULLARD RADIO ASTRON OBS
CAVENDISH LABORATORY
MADINGLEY ROAD
CAMBRIDGE CB3 OHE
U.K.
TEL: 0223-66477
TLX: 81292
COM: 40

RINDLER WOLFGANG PROF
UNIV OF TEXAS AT DALLAS
U.T.D.
BOX 830688
RICHARDSON TX 75083-0688
U.S.A.
TEL: 214-690 2885
TLX: 791-880
COM: 28,47

RING JAMES PROF
THE BLACKETT LABORATORY
IMPERIAL COLLEGE
PRINCE CONSORT ROAD
LONDON SW7 2BZ
U.K.
TEL: 01-589-5111
TLX: 261503
COM: 09

RINGNES TRULS S DR
INST THEORET ASTROPHYSICS
UNIVERSITY OF OSLO
P O BOX 1029
N-0315 BLINDERN, OSLO 3
NORWAY
TEL: 472-456-503
TLX: 72425 UNIOS N
COM:

RINGUELET ADELA E DR
49 342
1900 LA PLATA
ARGENTINA
TEL:
TLX:
COM: 29

RIPKEN HARTMUT W DR
CELSIUSSTR 24
D-5300 BONN 1
GERMANY, F.R.
TEL: 0228-25-18-16
TLX:
COM: 21,22,49

RITTER HANS DR
UNIVERSITAETS STERNWARTE
SCHEINERSTR 1
D-8000 MUENCHEN
GERMANY, F.R.
TEL: 89-98-90-21
TLX: 529815 UNIVM D
COM: 42

RIVOLO ARTHUR REX
DEPT OF EARTH & SPACE SCI
SUNY
STONY BROOK NY 11794
U.S.A.
TEL: 516-246-8223
TLX:
COM: 47

RIZVANOV NAUFAL G DR
ENGELHARDT OBSERVATORY
422526 KASAN
U.S.S.R.
TEL: 324827
TLX:
COM: 24

ROACH FRANKLIN E
2969 KALAKAVA AVE,APT 605
HONOLULU HI 96815
U.S.A.
TEL: 808-923-1405
TLX:
COM: 21,49

ROARK TERRY P PROF
KENT STATE UNIVERSITY
171 MAJORS LANE
KENT OH 44240
U.S.A.
TEL: 216-672-2220
TLX:
COM:

ROBBINS R ROBERT PROF
UNIVERSITY OF TEXAS
ASTRONOMY DEPT
AUSTIN TX 78712
U.S.A.
TEL: 512-471-7312
TLX:
COM: 34,46C

ROBE H A G DR
INSTITUT D'ASTROPHYSIQUE
UNIVERSITE DE LIEGE
AVENUE DE COINTE 5
B-4200 COINTE-OUGREE
BELGIUM
TEL: 41-52-9980
TLX: 41264 ASTRLG
COM:

ROBERTI GIUSEPPE DR
ISTITUTO DI FISICA
PAD 19
MOSTRA D'OLTREMARE
I-80125 NAPOLI
ITALY
TEL:
TLX:
COM: 12

ROBERTS BERNARD DR
DEPT OF APPLIED MATHS
UNIVERSITY OF ST ANDREWS
ST ANDREWS, FIFE KY16 9SS
U.K.
TEL: 0334-76161
TLX: 76213
COM: 10,12

ROBERTS DAVID HALL DR
PHYSICS DEPARTMENT
BRANDEIS UNIVERSITY
WALTHAM MA 02254
U.S.A.
TEL: 617-647-2846
TLX: 703013
COM: 40,47

ROBERTS JAMES A DR
CSIRO
DIVISION OF RADIOPHYSICS
P.O.BOX 76
EPPING NSW 2121
AUSTRALIA
TEL: 61-2-868-0313
TLX: 26230 ASTRO
COM: 40

ROBERTS MORTON S DR
NRAC
EDGEMONT ROAD
CHARLOTTESVILLE VA 22903
U.S.A.
TEL: 804-296-0252
TLX: 910 997-0174
COM: 28,33,40

ROBERTS WALTER ORR DR
UNIVERSITY CORP FOR
ATMOSPHERIC RESEARCH
P O BOX 3000
BOULDER CO 80307
U.S.A.
TEL: 303-497-1610
TLX: 45694
COM:

ROBERTS WILLIAM W JR PROF
DEPT OF APPLIED MATHS
UNIVERSITY OF VIRGINIA
THORNTON HALL
CHARLOTTESVILLE VA 22901
U.S.A.
TEL: 804-924-1038
TLX:
COM: 28,33,34

ROBERTSON DOUGLAS S
NOAA NGS N/CG114
11400 ROCKVILLE PIKE
ROCKVILLE MD 20852
U.S.A.
TEL: 301-443-8423
TLX:
COM: 19,31,40

ROBERTSON JAMES GORDON DR
ANGLO-AUSTRALIAN
OBSERVATORY
P.O.BOX 296
EPPING NSW 2121
AUSTRALIA
TEL: 02-868-1666
TLX: 23999 AAOSYD AA
COM: 28,40

ROBERTSON JOHN ALISTAIR
DEPT OF APPLIED MATHS
UNIVERSITY OF ST ANDREWS
NORTH HAUGH
ST ANDREWS, FIFE KY16 9SS
U.K.
TEL: 0334-76161
TLX: 76213
COM: 42

ROBERTSON WILLIAM H
1 DUNMORE ROAD
EPPING NSW 2121
AUSTRALIA
TEL: 02-869-8713
TLX:
COM: 08,20,24

ROBINSON BRIAN J DR
CSIRO
DIVISION OF RADIOPHYSICS
P.O.BOX 76
EPPING NSW 2121
AUSTRALIA
TEL: 02-868-0222
TLX: 26230 ASTRO AA
COM: 33,34,40

ROBINSON EDWARD LEWIS DR
DEPT OF ASTRONOMY
UNIVERSITY OF TEXAS
AUSTIN TX 78712
U.S.A.
TEL: 512-471-3401
TLX:
COM: 25,27,42

ROBINSON I PROF
UNIVERSITY OF TEXAS
BOX 688, MS BE 32
RICHARDSON TX 75080
U.S.A.
TEL:
TLX:
COM: 28,47

ROBINSON JR RICHARD D DR
ANGLO-AUSTRALIAN
OBSERVATORY
P.O.BOX 296
EPPING NSW 2121
AUSTRALIA
TEL: 02-868-1666
TLX: 23999
COM: 10,40

ROBINSON LEIF J
SKY & TELESCOPE
49 BAY STATE RD
CAMBRIDGE MA 02238
U.S.A.
TEL: 617-864-7360
TLX:
COM: 46,51

ROBINSON LLOYD B DR
LICK OBSERVATORY
UNIVERSITY OF CALIFORNIA
SANTA CRUZ CA 95064
U.S.A.
TEL: 408-429-2437
TLX:
COM: 09

ROBINSON WILLIAM J DR
DEPT OF MATHEMATICS
THE UNIVERSITY
BRADFORD BD7 1DP
U.K.
TEL: 733466
TLX:
COM: 07

ROBLEY R DR
9 ALLEES FR. VERDIER
F-31000 TOULOUSE
FRANCE
TEL: 61-52-22-73
TLX:
COM: 21

ROBSON IAN E DR
SCHOOL OF PHYSICS & ASTR
LANCASHIRE POLYTECHNIC
PRESTON PR1 2TQ
U.K.
TEL: 772-22141x2188
TLX: 677409 LANPOL
COM:

ROCA CORTES TEODORO
INST DE ASTROFISICA DE
CANARIAS
LA LAGUNA
38071 TENERIFE
SPAIN
TEL: 922-26-22-11
TLX: 92640
COM: 10,12

ROCCA-VOLMERANGE BRIGITTE
INSTITUT D'ASTROPHYSIQUE
98 BIS BOULEVARD ARAGO
F-75014 PARIS
FRANCE
TEL: 1-43-20-14-25
TLX: 270070 INSU
COM:

ROCHESTER MICHAEL G PROF
DEPT OF EARTH SCIENCES
MEMORIAL UNIVERSITY
OF NEWFOUNDLAND
ST JOHNS, NFLD A1B 3X7
CANADA
TEL: 709-737-7565
TLX: 0164101
COM: 19C

RODDIER CLAUDE DR
UER DE MATHEMATIQUES
UNIVERSITE DE PROVENCE
PLACE VICTOR HUGO
F-13331 MARSEILLE
FRANCE
TEL: 91-95-90-71
TLX: 402014
COM: 09

RODDIER FRANCOIS PROF
NOAO/ADP DIVISION
PO BOX 26732
950 N. CHERRY AVENUE
TUCSON AZ 85726
U.S.A.
TEL: 602-325-9220
TLX: 0666484 AURA NOAO TU
COM: 09,12

RODGERS ALEX W DR
MT STROMLO OBSERVATORY
WODEN P.O. ACT 2606
AUSTRALIA
TEL: 062-881111
TLX: 62270 CANOPUS AA
COM: 27,29,46

RODMAN RICHARD B DR
65 LOCUST AVE
LEXINGTON MA 02173
U.S.A.
TEL: 617-861-8149
TLX:
COM:

RODONO MARCELLO DR
INSTITUTE OF ASTRONOMY
UNIVERSITY OF CATANIA
VIALE ANDREA DORIA 6
I-95125 CATANIA
ITALY
TEL: 33-07-34
TLX: 970359 ASTRCT I
COM: 27,42C

RODRIGO RAFAEL
INST DE ASTROFISICA DE
ANDALUCIA, APDO 2144
PROFESOR ALBAREDA 1
18080 GRANADA
SPAIN
TEL: 58-12-13-00
TLX: 78573 IAAG E
COM: 16,21

RODRIGUEZ LUIS F
INSTITUTO DE ASTRONOMIA
UNAM
APDO POSTAL 70-264
04510 MEXICO DF
MEXICO
TEL: 905-548-5306
TLX: 1760155 CICME
COM: 34,40,46

ROEDER ROBERT C PROF
DEPT OF PHYSICS
SOUTHWESTERN UNIVERSITY
UNIVERSITY AVENUE
GEORGETOWN TX 78626
U.S.A.
TEL: 512-863-1633
TLX: 910-350-1677
COM: 40,47

ROEMER ELIZABETH PROF
LUNAR AND PLANETARY LAB
UNIVERSITY OF ARIZONA
TUCSON AZ 85721
U.S.A.
TEL: 602-621-2897
TLX: 467175
COM: 06V,15,20C,24

ROEMER MAX PROF
INST F ASTROPHYSIK &
EXTRATERR FORSCHUNG
AUF DEM HUEGEL 71
D-5300 BONN 1
GERMANY, F.R.
TEL: 228-733670
TLX:
COM: 10

ROENNAENG BERNT O DR
ONSALA SPACE OBSERVATORY
S-439 00 ONSALA
SWEDEN
TEL: 300 62637
TLX: 2400
COM: 40

ROESER HANS-PETER
MPI FUER RADIOASTRONOMIE
AUF DEM HUEGEL 69
D-5300 BONN 1
GERMANY, F.R.
TEL: 0228/525265
TLX: 886440 MPIFR D
COM: 34,40

ROESER HERMANN-JOSEF DR
MPI FUER ASTRONOMIE
KOENIGSTUHL
D-6900 HEIDELBERG 1
GERMANY, F.R.
TEL: 06221-528(1)206
TLX: 461789 MPIA D
COM: 28

ROESER SIEGFRIED DR
ASTRONOMISCHES RECHEN-
INSTITUT
MOENCHHOFSTR 12-14
D-6900 HEIDELBERG 1
GERMANY, F.R.
TEL: 06221-49026
TLX: 461336 ARIHD D
COM: 08,24

ROGER ROBERT S DR
DOMINION RADIO ASTRO-
PHYSICAL OBSERVATORY
P O BOX 248
PENTICTON BC V2A 6K3
CANADA
TEL: 604-497-5321
TLX: 048-88127
COM: 34,40

ROGERS ALAN E E DR
HAYSTACK OBSERVATORY
WESTFORD MA 01886
U.S.A.
TEL: 617-692-4764
TLX: 948149 HAYSTACK WFRD
COM: 34,40

ROGERS CHRISTOPHER DR
DEPT OF ASTRONOMY
UNIVERSITY OF TORONTO
TORONTO ONT M5S 1A1
CANADA
TEL: 416-978-4833
TLX: 06-986766
COM:

ROGERSON JOHN B PROF
PRINCETON UNIVERSITY
DEPT ASTROPHYS SCIENCES
PEYTON HALL
PRINCETON NJ 08540
U.S.A.
TEL: 609-452-3806
TLX: 322409
COM:

ROGSTAD DAVID H DR
MAIL CODE 264-748
JET PROPULSION LAB
4800 OAK GROVE DRIVE
PASADENA CA 91109
U.S.A.
TEL: 818-354-3573
TLX: 67529
COM: 40

ROHLFS K PROF DR
RUHR UNIVERSITAET BOCHUM
INSTITUT FUR ASTROPHYSIK
POSTFACH 102 148
D-4630 BOCHUM 1
GERMANY, F.R.
TEL: 0234-700-5802
TLX: 0825860
COM: 33,34,40

ROLAND GINETTE DR
INSTITUT D'ASTROPHYSIQUE
UNIVERSITE DE LIEGE
AVENUE DE COINTE 5
B-4200 COINTE-OUGREE
BELGIUM
TEL: 41-52-99-80
TLX: 41254 ASTRLG B
COM: 12

LIST OF MEMBERS

ROLLAND ANGEL DR
INSTITUTO DE ASTROFISICA
DE ANDALUCIA
APDO 2144
18080 GRANADA
SPAIN
TEL: 958-121-300
TLX: 78573
COM:

ROMAN NANCY G DR
APT 306W
4260 NORTH PARK AVE
CHEVY CHASE MD 20815
U.S.A.
TEL: 301-656-6092
TLX:
COM: 44,45

ROMANCHUK P R DR
KIEV STATE UNIVERSITY
ASTRONOMICAL OBSERVATORY
252053 KIEV
U.S.S.R.
TEL:
TLX:
COM: 10

ROMANO GIULIANO PROF
V. S.ANTONIO DA PADOVA 7
I-31100 TREVISO
ITALY
TEL:
TLX:
COM: 27

ROMANOV YURI S DR
ODESSA ASTRONOMICAL
OBSERVATORY
SHEVCHENKO PARK
270014 ODESSA
U.S.S.R.
TEL: 22-03-96
TLX:
COM: 27,30

ROMERO PEREZ M PILAR
INST ASTRON & GEODESIA
FAC DE CIENCIAS MATEMAT.
UNIVERSIDAD COMPLUTENSE
28040 MADRID
SPAIN
TEL: 2442501
TLX:
COM: 04

ROMNEY JONATHAN D DR
NRAO
EDGEMONT ROAD
CHARLOTTESVILLE VA 22903
U.S.A.
TEL: 804-296-0242
TLX: 910-997-0174
COM: 40

ROMPOLT BOGDAN DR
ASTRONOMICAL INSTITUTE
UL. KOPERNIKA 11
51-622 WROCLAW
POLAND
TEL: 071-48-24-34
TLX: 0712791 UWR PL
COM: 10

RONAN COLIN A
13 ACORN AVENUE
BAR HILL
CAMBRIDGE CB3 8DT
U.K.
TEL: 0954-81058
TLX:
COM: 41

RONG JIAN-XIANG
DEPT OF ASTRONOMY
NANJING UNIVERSITY
NANJING
CHINA. PEOPLE'S REP.
TEL: 34651 - 2882
TLX: 34151 PRCNU CN
COM: 33

ROOD HERBERT J
SCHOOL OF NATURAL SCS
INST FOR ADVANCED STUDY
PRINCETON NJ 08540
U.S.A.
TEL:
TLX:
COM: 28

ROOD ROBERT T DR
UNIVERSITY OF VIRGINIA
BOX 3818
UNIVERSITY STATION
CHARLOTTESVILLE VA 22903
U.S.A.
TEL: 804-924-4904
TLX:
COM: 35,51

ROOSEN ROBERT G DR
RAINBOW OBSERVATORY
RR1
P.O. BOX 5068
PAHOA HI 96778
U.S.A.
TEL:
TLX:
COM: 21,22

ROSA DOROTHEA DR
EMIL-KURZ-STR 4
D-8045 ISMANING
GERMANY, F.R.
TEL: 89-96-42-99
TLX:
COM:

ROSA MICHAEL RICHARD DR
ST/ECF
C/O ESO
KARL-SCHWARZSCHILD-STR 2
D-8046 GARCHING B MÜNCHEN
GERMANY, F.R.
TEL: 49-89-32006-0
TLX: 528-282-22-EO D
COM: 28,34

ROSADO MARGARITA DR
INSTITUTO DE ASTRONOMIA
UNAM
APDO POSTAL 70-264
04510 MEXICO DF
MEXICO
TEL: 905-548-5306
TLX:
COM:

ROSCH JEAN PROF
OBSERVATOIRES PIC-DU MIDI
ET TOULOUSE
F-65200 BAGNERES-DE-B.
FRANCE
TEL: 62-95-19-69
TLX: 531625 F
COM: 09,10,16

ROSE JAMES ANTHONY
INST FOR ASTRONOMY
UNIVERSITY OF HAWAII
2680 WOODLAWN DRIVE
HONOLULU HI 96822
U.S.A.
TEL: 808-948-6837
TLX:
COM: 28

ROSE WILLIAM K DR
ASTRONOMY PROGRAM
UNIVERSITY OF MARYLAND
COLLEGE PK MD 20742
U.S.A.
TEL: 301-299-2777
TLX:
COM: 34

ROSEN E DR
DEPT OF HISTORY
CITY COLLEGE OF THE
CITY UNIVERSITY NEW YORK
NEW YORK NY 10031
U.S.A.
TEL:
TLX:
COM:

ROSENBERG J DR
STATE UNIV OF UTRECHT
HEIDELBERLAAN 8
NL-3584 CS UTRECHT
NETHERLANDS
TEL: 030-535124
TLX:
COM:

ROSENDHAL JEFFREY D DR
NASA HEADQUARTERS
CODE E
WASHINGTON DC 20546
U.S.A.
TEL: 202-453-1410
TLX:
COM: 44

ROSINO LEONIDA PROF
OSSERVATORIO ASTRONOMICO
VICOLO DELL'OSSERVATORIO
I-35100 PADOVA
ITALY
TEL: 049-661499
TLX: 430176 UNPADU
COM: 06,27,28,34,37

ROSLUND CURT DR
DEPT OF ASTRONOMY
CHALMERS UNIV TECHNOLOGY
S-412 96 GOTHENBURG
SWEDEN
TEL: 46-31-810-100
TLX:
COM: 25,46

ROSNER ROBERT
HARVARD-SMITHSONIAN
CENTER FOR ASTROPHYSICS
60 GARDEN STREET
CAMBRIDGE MA 02138
U.S.A.
TEL: 617-495-5879
TLX: 921428 SATELLITE CAM
COM: 48,49

ROSQUIST KJELL
INST OF THEORETICAL PHYS
VANADISVAEGEN 9
S-113 46 STOCKHOLM
SWEDEN
TEL: 46-8-228160x225
TLX: 15433 FYSTO S
COM:

ROSS DENNIS K PROF
PHYSICS DEPT
IOWA STATE UNIVERSITY
AMES IA 50011
U.S.A.
TEL: 515-294-6010
TLX:
COM:

ROSS JOHN E R DR
PHYSICS DEPT
UNIVERSITY OF QUEENSLAND
ST LUCIA
BRISBANE QLD 4067
AUSTRALIA
TEL: 07-377-3429
TLX: 40315 UNIVQLD AA
COM: 14,36

ROSSELLO GASPAR
DEPTO FIS TIERRA & COSMOS
UNIVERSIDAD DE BARCELONA
DIAGONAL 645
08028 BARCELONA
SPAIN
TEL:
TLX:
COM: 04

ROSSI BRUNO B PROF
MIT
RM 37-667
CAMBRIDGE MA 02139
U.S.A.
TEL: 617-253-4283
TLX: 92-1473
COM: 48

ROSSI LUCIO
IST. ASTROFISICA SPAZIALE
C P 67
I-00044 FRASCATI
ITALY
TEL: 06-9425651/2/3
TLX: 610261 CNR FRA
COM: 29

ROSTAS FRANCOIS DR
OBSERVATOIRE DE PARIS
SECTION DE MEUDON
F-92195 MEUDON PL CEDEX
FRANCE
TEL: 1-45-34-75-70
TLX:
COM:

ROTH MARIA LUISE PH D
HAMBURGER STERNWARTE
GOJENSBERGSWEG
D-2050 HAMBURG 80
GERMANY, F.R.
TEL: 040-72524112
TLX: 217884 HAMST D
COM: 37

ROUSSEAU JEANINE DR
OBSERVATOIRE DE LYON
F-69230 ST GENIS-LAVAL
FRANCE
TEL: 78-56-07-05
TLX: 310926
COM:

ROXBURGH IAN W PROF
SCHOOL OF MATHEMATICAL SC
QUEEN MARY COLLEGE
MILE END ROAD
LONDON E1 4NS
U.K.
TEL: 01-980-4811
TLX:
COM: 10,34,35,42,47,49C

RUBIN ROBERT HOWARD
NASA AMES RESEARCH CENTER
MS 245-6
MOFFETT FIELD CA 94035
U.S.A.
TEL: 415-965-5528
TLX: 348408
COM: 34,40,51

ROTH MIGUEL R DR
INSTITUTO DE ASTRONOMIA
UNAM
APDO POSTAL 877
22860 ENSENADA, B. CALIF.
MEXICO
TEL: 667-40887
TLX: 56539 CICEME
COM:

ROUTLEDGE DAVID DR
ELECTRICAL ENGR DEPT
UNIVERSITY OF ALBERTA
EDMONTON AB T6G 2G7
CANADA
TEL: 403-432-5668
TLX:
COM:

ROY ARCHIE E PROF
DEPT OF ASTRONOMY
GLASGOW UNIVERSITY
GLASGOW G12 8QQ
U.K.
TEL: 41-339-8855x502
TLX: 778421 GLASUL
COM: 07,46

RUBIN VERA C DR
DEPT TERRESTR. MAGNETISM
CARNEGIE INST. WASHINGTON
5241 BROAD BRANCH RD N.W.
WASHINGTON DC 20015
U.S.A.
TEL: 202-966-0863
TLX: 440427 MAGN UI
COM: 28C,30,33,34,47

ROTS ARNOLD H DR
NRAO
PO BOX 0
SOCORRO NM 87801
U.S.A.
TEL: 505-772-4259
TLX: 910-988-1710
COM: 28

ROUTLY PAUL M DR
US NAVAL OBSERVATORY
34 & MASSACHUSETTS AVE NW
WASHINGTON DC 20390
U.S.A.
TEL: 202-653-1532
TLX:
COM: 38

ROY JEAN-RENE
DEPT DE PHYSIQUE
UNIVERSITE LAVAL
CITE UNIVERSITAIRE
QUEBEC G1K 7P4
CANADA
TEL: 418-656-5816
TLX: 5131621 UNILAVAL
COM:

RUBIO MONICA DR
DEPTO DE ASTRONOMIA
UNIVERSIDAD DE CHILE
CASILLA 36-D
SANTIAGO
CHILE
TEL: 2294101
TLX: 440005 ATTN OBSERNAL
COM: 40

ROTTENBERG J A DR
2911 BAYVIEW AVE
SUITE 110C
WILLOWDALE ONT M2K 1E8
CANADA
TEL:
TLX:
COM:

ROVIRA MARTA GRACIELA
INST. DE ASTRONOMIA y
FISICA DEL ESPACIO
C.C. 67, SUC. 28
1428 CAPITAL
ARGENTINA
TEL:
TLX:
COM: 12,36

ROZELOT JEAN P
CERGA
AVENUE COPERNIC
F-06130 GRASSE
FRANCE
TEL: 93-36-58-49
TLX: 470865
COM: 10

RUCINSKI SLAWOMIR M DR
DAVID DUNLAP OBSERVATORY
P O BOX 360
RICHMOND HILL ONT L4C 4Y6
CANADA
TEL: 416-884-9562
TLX: 06-986766
COM: 42

ROUEFF EVELYNE M A DR
OBSERVATOIRE DE PARIS
SECTION DE MEUDON
DAF
F-92195 MEUDON PL CEDEX
FRANCE
TEL: 1-45-34-75-70
TLX:
COM: 14

ROVITHIS PETER DR
NATL OBSERVATORY ATHENS
P.O. BOX 20048
ATHENS 306
GREECE
TEL: 01-3461191
TLX: 215530 OBSA GR
COM: 42

ROZHKOVSKIJ DIMITRIJ A
ASTROPHYSICAL INSTITUE
480068 ALMA-ATA
U.S.S.R.
TEL: 62-40-40
TLX:
COM: 21,34

RUDAK BRONISLAW
COPERNICUS ASTRON CENTER
UL. CHOPINA 12/18
87-100 TORUN
POLAND
TEL: 26037 x 10
TLX: 813978 ZAPAN PL
COM:

ROUNTREE JANET DR
AFE/TECHN SERVICES STAFF
BOLLING AFB
WASHINGTON DC 20332
U.S.A.
TEL: 202-767-3968
TLX:
COM: 27,37,45

ROVITHIS-LIVANIOU HELEN
SECTION OF ASTROPHYSICS
ASTRONOMY AND MECHANICS
DEPT OF PHYSICS
GR-15771 ZOGRAFOS
GREECE
TEL: 01-724-3414
TLX:
COM: 42

ROZYCZKA MICHAL
WARSAW UNIVERSITY OBS
AL. UJAZDOWSKIE 4
00-478 WARSAW
POLAND
TEL:
TLX: 813978 ZAPAN PL
COM: 34

RUDDY VINCENT P DR
REGIONAL TECHNICAL COLL.
ROSSA AVENUE
CORK
IRELAND
TEL:
TLX:
COM: 47

ROUSE CARL A DR
627 15TH STREET
DEL MAR CA 92014
U.S.A.
TEL: 619-455-4015
TLX: 695065
COM: 35

ROWAN-ROBINSON MICHAEL DR
DEPT OF APPLIED MATHS
QUEEN MARY COLLEGE
MILE END ROAD
LONDON E1 4NS
U.K.
TEL:
TLX:
COM: 51

RUBASHEV BORIS M DR
PULKOVO OBSERVATORY
196140 LENINGRAD
U.S.S.R.
TEL:
TLX:
COM: 10

RUDER HANNS
LEHRSTUHL F THEORET ASTRO
PHYSIK DER UNIV TUEBINGEN
AUF DER MORGENSTELLE 12,C
D-7400 TUEBINGEN
GERMANY, F.R.
TEL: 07071/292487
TLX:
COM: 09,14,19,24,44

ROUSSEAU JEAN-MICHEL MR
OBSERVATOIRE DE BORDEAUX
AVENUE P. SEMIROT
F-33270 FLOIRAC
FRANCE
TEL: 56-86-43-30
TLX:
COM: 08

ROWSON BARRIE DR
NRAL
JODRELL BANK
MACCLESFIELD SK11 9DL
U.K.
TEL: 047-77-1321
TLX: 36149
COM: 40

RUBEN G PROF DR
ZNTRLINST. F. ASTROPHYSIK
ROSA-LUXEMBURG-STR 17A
DDR-1502 POTSDAM
GERMANY, D.R.
TEL:
TLX:
COM: 35,38,44

RUDERMAN MALVIN A
PHYSICS DEPT
COLUMBIA UNIVERSITY
NEW YORK NY 10027
U.S.A.
TEL: 212-280 3317
TLX:
COM:

RUDKJOBING MOGENS PROF
INSTITUTE OF ASTRONOMY
UNIVERSITY OF AARHUS
LANGELANDSGADE
DK-8000 AARHUS C
DENMARK
TEL: 06-12-88-99
TLX: 64767 AAUSCI DK
COM: 45

RUDNICK LAWRENCE DR
UNIVERSITY OF MINNESOTA
116 CHURCH STREET S.E.
MINNEAPOLIS MN 55455
U.S.A.
TEL: 612-373-5457
TLX:
COM: 40,47

RUDNICKI KONRAD PROF
OBSERVATORY FORT SKALA
UL. ORLA 171
30-244 KRAKOW
POLAND
TEL: 22-18-77
TLX:
COM: 28

RUDZIKAS ZENONAS B
INSTITUTE OF PHYSICS
LITHUANIAN SSR
K. POZELOS 54
232600 VILNIUS
U.S.S.R.
TEL: 612610
TLX:
COM: 14C

RUFENER FREDY G PROF
OBSERVATOIRE DE GENEVE
CH-1290 SAUVERNY
SWITZERLAND
TEL: 41-22-552611
TLX: 27720 OBSG CH
COM: 25P

RUFFINI REMO
DIPARTIMENTO DI FISICA
UNIVERSITA DI ROMA
PIAZZALE ALDO MORO 2
I-00185 ROMA
ITALY
TEL: 4976304
TLX: 613255 INFNRO I
COM:

RUGGE HUGO R DR
SPACE SCIENCES LABORATORY
AEROSPACE CORPORATION
PO BOX 92957
LOS ANGELES CA 90009
U.S.A.
TEL: 213-648-7086
TLX:
COM:

RUIZ MARIA TERESA DR
OBS ASTRONOMICO NACIONAL
UNIVERSIDAD DE CHILE
CASILLA 36-D
SANTIAGO
CHILE
TEL: 2294101
TLX:
COM: 33

RULE BRUCE H
HALE OBSERVATORIES
2205 MONTE VISTA STREET
PASADENA CA 91107
U.S.A.
TEL: 818-794-6593
TLX:
COM:

RUMSEY NORMAN J
PHYSICS AND ENGINEERING
LABORATORY
D S I R
LOWER HUTT
NEW ZEALAND
TEL:
TLX:
COM:

RUNCORN S K PROF
SCHOOL OF PHYSICS
THE UNIVERSITY
NEWCASTLE/TYNE NE1 7RU
U.K.
TEL: 0632-32511
TLX: 53654 UNINEW G
COM: 16,19

RUPRECHT JAROSLAV DR
ASTRONOMICAL INSTITUTE
CZECH. ACAD. OF SCIENCES
BUDECSKA 6
120 23 PRAHA 2
CZECHOSLOVAKIA
TEL: 258757
TLX: 122 486
COM: 37

RUSCONI LUIGIA DR
DIPT. DI ASTRONOMIA
UNIVERSITA DI TRIESTE
VIA TIEPOLO 11
I-34131 TRIESTE
ITALY
TEL: 40-794863
TLX: 461137 OAOTI
COM: 09

RUSIN VOJTECH
ASTRONOMICAL INSTITUTE
SLOVAK ACAD. OF SCIENCES
059 60 TATRANSKA LOMNICA
CZECHOSLOVAKIA
TEL: 0969-967866/7/8
TLX: 80-78277 AUSSAV C
COM: 10,12

RUSKOL EUGENIA L DR
OJSCHMIDT INSTITUTE
OF PHYSICS OF THE EARTH
USSR ACADEMY OF SCIENCES
123810 MOSCOW
U.S.S.R.
TEL: 252-07-26
TLX: 411196 IFZAN SU
COM: 16

RUSSELL CHRISTOPHER T
INST OF GEOPHYSICS
UNIVERSITY OF CALIFORNIA
LOS ANGELES CA 90024
U.S.A.
TEL: 213-825-3188
TLX: 910-342-6981
COM: 49

RUSSELL JANE L DR
SPACE TELESCOPE SCI INST
HOMEWOOD CAMPUS
3700 SAN MARTIN DRIVE
BALTIMORE MD 21218
U.S.A.
TEL: 301-338-4843
TLX: 6849101 STSCI
COM: 08,24,26.51

RUSSELL JOHN A PROF
DEPT OF ASTRONOMY
UNIV SOUTHERN CALIFORNIA
UNIVERSITY PARK
LOS ANGELES CA 90089
U.S.A.
TEL: 213-743-0231
TLX:
COM: 22

RUSSEV RUSCHO DR
UNIVERSITY OF SOFIA
DEPT OF ASTRONOMY
ANTON IVANOV STR 5
1126 SOFIA
BULGARIA
TEL: 6-25-61
TLX:
COM: 27

RUSSEVA TATJANA
DEPT OF ASTRONOMY AND
NATL ASTRON OBSERVATORY
72 LENIN BLVD
1784 SOFIA
BULGARIA
TEL: 73-41-559
TLX: 23561 ECF BAN BG
COM: 37

RUSSO GUIDO DR
ST-ECF
ESO
KARL-SCHWARZSCHILD-STR 2
D-8046 GARCHING B MUNCHEN
GERMANY, F.R.
TEL: 89-320-06346
TLX: 52828222
COM:

RUST DAVID M DR
APPLIED PHYSICS LAB
JOHNS HOPKINS UNIVERSITY
JOHNS HOPKINS ROAD
LAUREL MD 20707
U.S.A.
TEL: 301-953-5414
TLX: 89-548 APL JHU LAUR
COM: 10

RUSU I DR
ASTRONOMICAL OBSERVATORY
CUTITUL DE ARGINT 5
75212 BUCAREST 28
RUMANIA
TEL: 23-63-01
TLX:
COM: 08,19

RUSU L DR
ASTRONOMICAL OBSERVATORY
CUTITUL DE ARGINT 5
75212 BUCAREST 28
RUMANIA
TEL: 23-63-01
TLX:
COM:

RUTTEN ROBERT J. DR
STERREWACHT SONNENBORGH
ZONNENBURG 2
NL-3512 NL UTRECHT
NETHERLANDS
TEL: 30-312841
TLX: 47224 ASTRO NL
COM: 12,36

RUZDJAK VLADIMIR DR
INSTITUTE OF PHYSICS
UNIVERSITY OF ZAGREB
P O BOX 304
41001 ZAGREB
YUGOSLAVIA
TEL:
TLX:
COM: 10

RUZICKOVA-TOPOLOVA B DR
ASTRONOMICAL INSTITUTE
CZECH. ACAD. OF SCIENCES
OBSERVATORY
251 65 ONDREJOV
CZECHOSLOVAKIA
TEL: 724525 PRAHA
TLX: 121579
COM: 10

RYABOV YU A PROF DR
MATHEMATICS DEPT OF MADI
LENINGRADSKY PROSP. 64
125319 MOSCOW
U.S.S.R.
TEL: 1550326
TLX:
COM: 07

RYBANSKY MILAN
ASTRONOMICAL INSTITUTE
SLOVAK ACAD. OF SCIENCES
059 60 TATRANSKA LOMNICA
CZECHOSLOVAKIA
TEL: 0969-967-866
TLX: 80-78277 AUSAV C
COM: 10.12

RYBICKI GEORGE B DR
HARVARD SMITHSONIAN
CENTER FOR ASTROPHYSICS
60 GARDEN STREET
CAMBRIDGE MA 02138
U.S.A.
TEL: 617-495-7452
TLX: 92-1428
COM: 33,36

RYBKA EUGENIUSZ PROF DR
UL. PIATOWSKA 48/3
50-361 WROCLAW
POLAND
TEL: 22-44-02
TLX:
COM: 25,41

RYBKA PRZEMYSLAW DR
INSTITUTE OF HISTORY
OF SCIENCE
NOWY SWIAT 72
00-330 WARSZAWA
POLAND
TEL:
TLX:
COM: 41

RYDBECK GUSTAF H B DR
ONSALA SPACE OBSERVATORY
S-439 00 ONSALA
SWEDEN
TEL: 0300-6208
TLX:
COM: 28,40

RYDBECK OLOF E H PROF
ONSALA SPACE OBSERVATORY
S-439 00 ONSALA
SWEDEN
TEL: 0300-62081
TLX: 8542400 ONSPACE
COM: 40

RYDGREN ALFRED ERIC JR DR
COMPUTER SCIENCE CORP.
SPACE TELESCOPE SCI INST
HOMEWOOD CAMPUS
BALTIMORE MD 21218
U.S.A.
TEL: 301-338-4902
TLX:
COM: 25

RYKHLOVA LIDIJA V DR
ASTRONOMICAL COUNCIL
USSR ACADEMY OF SCIENCES
PYATNITSKAYA 48
109017 MOSCOW
U.S.S.R.
TEL: 231-54-61
TLX: 412623 SCSTP SU
COM: 19

RYLOV VALERIJ S DR
SPECIAL ASTROPHYSICAL OBS
USSR ACADEMY OF SCIENCES
357140 N ARKHYZ
U.S.S.R.
TEL:
TLX:
COM: 09

RYTER CHARLES E DR
CEN SACLAY
DPhG/SAp
BAT 28
F-91191 GIF/YVETTE CEDEX
FRANCE
TEL: 1-69-08-39-12
TLX:
COM:

RYZHKOV N F DR
SPECIAL ASTROPHYSICAL OBS
LENINGRAD BRANCH
196140 LENINGRAD
U.S.S.R.
TEL:
TLX:
COM: 40

RZHIGA O N DR
INST OF RADIO & ELECTRON
USSR ACADEMY OF SCIENCES
103907 MOSCOW
U.S.S.R.
TEL:
TLX:
COM:

SAAR ENN DR
TARTU ASTROPHYSICAL
OBSERVATORY
202444 TORAVERE, ESTONIA
U.S.S.R.
TEL:
TLX:
COM: 33,47

SABANO YUTAKA DR
ASTRONOMICAL INSTITUTE
TOHOKU UNIVERSITY
ARAMAKI
SENDAI 980
JAPAN
TEL: 0222-22-1800
TLX:
COM: 34

SABBADIN FRANCO DR
OSSERVATORIO ASTROFISICO
I-36012 ASIAGO
ITALY
TEL: 0424-62665
TLX: SETUR 430110
COM: 34

SACK NOAM DR
DEPT OF THEORETICAL PHYS
HEBREW UNIV. OF JERUSALEM
JERUSALEM 91904
ISRAEL
TEL:
TLX:
COM:

SACKMANN I JULIANA DR
KELLOGG RADIATION LAB
CALTECH
PASADENA CA 91125
U.S.A.
TEL: 818-356-4256
TLX:
COM: 35

SADAKANE KOZO DR
ASTRONOMICAL INSTITUTE
OSAKA KYOIKU UNIVERSITY
TENNOJI-KU
OSAKA 543
JAPAN
TEL:
TLX:
COM: 29

SADEH D DR
DEPT PHYSICS & ASTRONOMY
TEL-AVIV UNIVERSITY
TEL-AVIV 69978
ISRAEL
TEL: 3-420-553
TLX: 34271 VERSY
COM:

SADIK AZIZ R DR
ASTRONOMY & SPACE RES CTR
COUNCIL FOR SCI RESEARCH
P O BOX 2441
JADIRIYAH, BAGHDAD
IRAQ
TEL: 01-7765127
TLX: 213976 SRC
COM: 27,42

SADLER DONALD H DR
8 COLLINGTON RISE
BEXHILL-ON-SEA TN39 3RT
U.K.
TEL: 042-43-3572
TLX:
COM:

SADLER ELAINE MARGARET
KITT PEAK NATIONAL OBS
PO BOX 26732
TUCSON AZ 85726
U.S.A.
TEL: 602-327-5511
TLX: 0666484 AURA NOAO TU
COM: 28

SADZAKOV SOFIJA DR
ASTRONOMICAL OBSERVATORY
VOLGINA 7
11050 BEOGRAD
YUGOSLAVIA
TEL: 419-357/421-875
TLX:
COM: 08C,19

SAEMUNDSON THORSTEINN
RAUNVISINDASTOFNUN
HASKOLANS
DUNHAGA 3
IS-107 REYKJAVIK
ICELAND
TEL: 354-1-21340
TLX: 2307 ISINFO
COM: 10

SAFKO JOHN L
DEPT PHYSICS & ASTRONOMY
UNIVERSITY OF S. CAROLINA
COLUMBIA SC 29208
U.S.A.
TEL: 803-777-6466
TLX: UNIVSCAROL CLB
COM: 46

SAFRONOV VICTOR S DR
INSTITUTE OF PHYSICS
OF THE EARTH
B GRUZINSKAYA 10
123242 MOSCOW
U.S.S.R.
TEL: 252-07-26
TLX: 411196 IFZAN SU
COM: 16

SAGAN CARL DR
CORNELL UNIVERSITY
302 SPACE SCIENCE BLDG
ITHACA NY 14853
U.S.A.
TEL: 607-256-4971
TLX: 937478
COM: 16,51

SAGGION ANTONIO PROF
ISTITUTO DI FISICA
G. GALILEI
VIA MARZOLO 8
I-35100 PADOVA
ITALY
TEL: 049-844254
TLX: 430308 DFGGPDI
COM:

SAGITOV M U DR
STERNBERG STATE
ASTRONOMICAL INSTITUTE
119859 MOSCOW V-234
U.S.S.R.
TEL:
TLX:
COM: 16

SAGNIER JEAN-LOUIS DR
UNIVERSIDADE DE SAO PAULO
DEPT ASTRONOMIA
CAIXA POSTAL 30627
01051 SAO PAULO SP
BRAZIL
TEL:
TLX:
COM: 07,20

SAHADE JORGE PROF
INST ARGENTINO DE
RADIOASTRONOMIA
CASILLA DE CORREO NO. 5
1894 VILLA ELISA (Bs.As.)
ARGENTINA
TEL: 54-021-43793
TLX: 31216 CESLA AR
COM: 29,38,42

SAHAL-BRECHOT SYLVIE DR
OBSERVATOIRE DE PARIS
SECTION DE MEUDON
F-92195 MEUDON PL CEDEX
FRANCE
TEL: 1-45-34-75-70
TLX: 201 571
COM: 14V

SAIJO KEIICHI
DEPT OF PHYSICAL SCIENCES
NATIONAL SCIENCE MUSEUM
7-20 UENO PARK, TAITO-KU
TOKYO 110
JAPAN
TEL: 3-822-0111
TLX:
COM: 42

SAIKIA D J DR
RADIO ASTRONOMY CENTRE
TIFR
POST BOX 1234
BANGALORE 560 012
INDIA
TEL: 362816 / 364062
TLX: 8458488
COM: 40

SAIO HIDEYUKI DR
DEPT OF ASTRONOMY
UNIVERSITY OF TOKYO
BUNKYO-KU
TOKYO 113
JAPAN
TEL: 3-812-2111
TLX: UTYOSCI J33659
COM: 35

SAISSAC JOSEPH DR
OBSERVATOIRE PIC-DU-MIDI
ET TOULOUSE
14 AVENUE EDOUARD BELIN
F 65200 BAGNERES-DE-B.
FRANCE
TEL: 62-95-19-69
TLX: 531625 S
COM: 16

SAITO KUNIJI PROF
TOKYO ASTRONOMICAL
OBSERVATORY
OSAWA MITAKA
TOKYO 181
JAPAN
TEL:
TLX:
COM: 10,36

SAITO MAMORU DR
DEPT OF ASTRONOMY
UNIVERSITY OF KYOTO
SAKYOKU
KYOTO 606
JAPAN
TEL: 0757512111x3904
TLX: 5422693 LIBKYU J
COM:

SAITO SUMISABURO DR
KWASAN OBSERVATORY
YAMASHINA
KYOTO 607
JAPAN
TEL: 075-581-1235
TLX: 5422693 LIBKYU J
COM:

SAKAI JUN-ICHI
HIGH ALTITUDE OBSERVATORY
BOULDER CO 80307
U.S.A.
TEL:
TLX:
COM: 12

SAKASHITA SHIRO PROF
DEPT OF PHYSICS
HOKKAIDO UNIVERSITY
KITA 10 NISHI 8
SAPPORO HOKKAIDO 060
JAPAN
TEL: 011-716-2111
TLX:
COM: 35

SAKHAROV VLADIMIR I DR
PULKOVO OBSERVATORY
196140 LENINGRAD
U.S.S.R.
TEL:
TLX:
COM: 19

SAKHIBULLIN NAIL A DR
DEPT OF ASTRONOMY
KAZAN STATE UNIVERSITY
420008 KAZAN
U.S.S.R.
TEL: 32-36-41
TLX:
COM: 36

SAKURAI KUNITOMO PROF
DEPT OF PHYSICS
KANAGAWA UNIVERSITY
KANAGAWAKU
YOKOHAMA 221
JAPAN
TEL: 045-481-5661
TLX:
COM: 10,51

SAKURAI TAKASHI DR
DEPT OF ASTRONOMY
UNIVERSITY OF TOKYO
BUNKYO-KU
TOKYO 113
JAPAN
TEL: 03-812-2111
TLX: UTYOSCI J 33659
COM: 10,12

SAKURAI TAKEO T PROF
FACULTY OF ENGINEERING
KYOTO UNIVERSITY
SAKYO-KU
KYOTO 606
JAPAN
TEL: 75-7512111x5792
TLX: 05422693 LIBKYUJ
COM:

SALETIC DUSAN
ASTRONOMICAL OBSERVATORY
VOLGINA 7
11050 BEOGRAD
YUGOSLAVIA
TEL: 157-022
TLX:
COM: 08

SALINARI PIERO
OSSERVATORIO ASTROFISICO
DI ARCETRI
LARGO E. FERMI 5
I-50125 FIRENZE
ITALY
TEL: 55-220034
TLX: 572268
COM: 34

SALISBURY J W DR
US GEOLOGICAL SURVEY
927 NATIONAL CENTER
RESTON VA 22092
U.S.A.
TEL: 703-860-6668
TLX: 92178
COM:

SALO HEIKKI
UNIVERSITY OF OULU
DEPT OF ASTRONOMY
SF-90570 OULU 57
FINLAND
TEL: 981-352028
TLX:
COM:

SALOMONOVICH A E DR
PHYSICAL INSTITUTE
USSR ACADEMY OF SCIENCES
LENINSKY PROSPECT 53
117924 MOSCOW
U.S.S.R.
TEL: 135-22-50
TLX:
COM: 40,44

SALPETER EDWIN E PROF
NEWMAN LAB OF NUCLEAR STU
CORNELL UNIVERSITY
ITHACA NY 14853
U.S.A.
TEL: 607-256-3302
TLX: 937478
COM: 34,35,40,48C

SALUKVADZE G N DR
ABASTUMANY ASTROPHYSICAL
OBSERVATORY
383762 ABASTUMANY
U.S.S.R.
TEL:
TLX:
COM: 26,37C

SALVADOR-SOLE EDUARDO
DEPTO FIS TIERRA & COSMOS
UNIVERSIDAD DE BARCELONA
DIAGONAL 645
08028 BARCELONA
SPAIN
TEL: 33007311
TLX:
COM: 28,47

SALVATI MARCO
IST ASTROFISICA SPAZIALE
C P 67
I-00044 FRASCATI
ITALY
TEL: 39-6-942-5655
TLX: 610261 CNRFRA I
COM: 48

SAMAIN DENYS DR
LPSP
BP 10
F-91370 VERRIERES-LE-B.
FRANCE
TEL:
TLX:
COM: 12

SAMPSON DOUGLAS H PROF
DEPT OF ASTRONOMY
PENNSYLVANIA STATE UNIV
525 DAVEY LAB
UNIVERSITY PARK PA 16802
U.S.A.
TEL: 814-865-0261
TLX: 842510
COM:

SAMUS NIKOLAI N DR
ASTRONOMICAL COUNCIL
USSR ACADEMY OF SCIENCES
PYATNITSKAYA 48
109017 MOSCOW
U.S.S.R.
TEL: 231-54-61
TLX: 412623 SCSTP SU
COM: 27,37

SANAHUJA BLAS
DEPTO FIS TIERRA & COSMOS
UNIVERSIDAD DE BARCELONA
DIAGONAL 645
08028 BARCELONA
SPAIN
TEL: 3307311 x298
TLX:
COM: 34

SANAMIAN V A DR
BYURAKAN ASTROPHYSICAL
OBSERVATORY
378433 ARMENIA
U.S.S.R.
TEL: 563453
TLX:
COM: 40

SANCHEZ FRANCISCO PROF
INSTITUTO DE ASTROFISICA
DE CANARIAS
LA LAGUNA
38071 TENERIFE
SPAIN
TEL: 922-262211
TLX: 92640 IAC E
COM: 21,50

SANCHEZ MAGRO C DR
INSTITUTO DE ASTROFISICA
DE CANARIAS
LA LAGUNA
38071 TENERIFE
SPAIN
TEL: 262211
TLX: 92640 IACE E
COM: 09,21

SANCHEZ MANUEL
INSTITUTO Y OBSERVATORIO
DE MARINA
SAN FERNANDO (CADIZ)
SPAIN
TEL: 956-883548
TLX: 76108
COM: 08,19

SANCHEZ-SAAVEDRA M LUISA
FACULDAD DE CIENCIAS
UNIVERSIDAD DE GRANADA
18080 GRANADA
SPAIN
TEL: 958-20-22-12
TLX:
COM: 21,33,34

SANCISI RENZO DR
KAPTEYN LABORATORIUM
POSTBUS 800
NL-9700 AV GRONINGEN
NETHERLANDS
TEL: 050-116695
TLX: 53572 STARS NL
COM: 28,34,51

SANDAGE ALLAN
MT WILSON & LAS CAMPANAS
OBSERVATORIES
813 SANTA BARBARA ST
PASADENA CA 91101
U.S.A.
TEL: 818-577-1122
TLX:
COM:

SANDAKOVA E V DR
KIEV STATE UNIVERSITY
ASTRONOMICAL OBSERVATORY
252053 KIEV
U.S.S.R.
TEL:
TLX:
COM:

SANDELL GORAN HANS L DR
OBS & ASTROPHYSICS LAB
UNIVERSITY OF HELSINKI
KOPERNIKUKSENTIE 1
SF-00130 HELSINKI 13
FINLAND
TEL: 358-0-1912943
TLX:
COM: 34,40

SANDERS ROBERT DR
KAPTEYN LABORATORY
POSTBUS 800
NL-9700 AV GRONINGEN
NETHERLANDS
TEL: 50-11-66-95
TLX: 53572 STARS NL
COM: 28

SANDERS W L PROF
NEW MEXICO STATE UNIV
BOX 4500
LAS CRUCES NM 88003
U.S.A.
TEL: 505-646-4914
TLX:
COM: 24,37

SANDERS WILTON TURNER III
DEPT OF PHYSICS
UNIVERSITY OF WISCONSIN
MADISON WI 53706
U.S.A.
TEL: 608-262-5916
TLX:
COM: 48

SANDFORD MAXWELL T II
LOS ALAMOS SCIENTIFIC LAB
LOS ALAMOS NM 87545
U.S.A.
TEL: 505-667-6384
TLX:
COM:

SANDMANN WILLIAM HENRY
PHYSICS DEPT
HARVEY MUDD COLLEGE
CLAREMONT CA 91711
U.S.A.
TEL: 714-621-8024
TLX:
COM: 27

SANDQVIST AAGE DR
STOCKHOLM OBSERVATORY
S-133 00 SALTSJOEBADEN
SWEDEN
TEL: 08-717-2149
TLX: 12972
COM: 33,34,46V

SANDULEAK NICHOLAS DR
WARNER & SWASEY OBS
CASE WESTERN RESERVE UNIV
CLEVELAND OH 44106
U.S.A.
TEL: 216-368-6696
TLX:
COM: 33,45

SANFORD PETER WILLIAM MR
DEPT PHYSICS & ASTRONOMY
UNIVERSITY COLLEGE LONDON
GOWER STREET
LONDON WC1E 6BT
U.K.
TEL:
TLX:
COM:

SANG GAK LEE
DEPT OF ASTRONOMY
COLLEGE NATURAL SCIENCES
SEOUL NATIONAL UNIVERSITY
SEOUL 151-00
KOREA, REPUBLIC
TEL: 877-0101-3315
TLX:
COM: 33,45,51

SANTIN PAOLO DR
OSSERVATORIO ASTRONOMICO
VIA G.B. TIEPOLO 11
CP SUCC T5 5
I-34131 TRIESTE
ITALY
TEL: 040-793921
TLX: 461137 OAT I
COM: 28

SANWAL N B DR
DEPT OF ASTRONOMY
OSMANIA UNIVERSITY
HYDERABAD 500 007
INDIA
TEL: 71951 x247
TLX:
COM: 30,42,45,46

SANYAL ASHIT DR
4618 OLYMPIA AVE
BELTSVILLE MD 20705
U.S.A.
TEL: 301-937-8943
TLX:
COM: 27,42

SAPAR ARVED DR
IAPHA ESTONIAN ACADEMY
OF SCIENCES
TARTU OBSERVATORY
202444 TORAVERE
U.S.S.R.
TEL:
TLX:
COM: 36,47

SARAZIN CRAIG L DR
DEPT OF ASTRONOMY
UNIVERSITY OF VIRGINIA
PO BOX 3818 UNIV STATION
CHARLOTTESVILLE VA 22903
U.S.A.
TEL: 804-924-4903
TLX:
COM: 28,34

SAREYAN JEAN-PIERRE DR
OBSERVATOIRE DE NICE
BP 139
F-06003 NICE CEDEX
FRANCE
TEL: 93-89-04-20
TLX: 460004
COM: 27

SARGENT ANNEILA I
CALTECH 320-47
DOWNS LAB. OF PHYSICS
PASADENA CA 91125
U.S.A.
TEL: 818-356-6622
TLX: 675425
COM:

SARGENT WALLACE L W DR
ASTRONOMY DEPT
CALTECH 105-24
1201 E CALIFORNIA ST
PASADENA CA 91125
U.S.A.
TEL: 818-356-4055
TLX: 675425 CALTECH PSD
COM: 28,47

SARMA M B K PROF
DEPT OF ASTRONOMY
OSMANIA UNIVERSITY
HYDERABAD 500 007
INDIA
TEL: 65228
TLX:
COM: 25,27

SARMA N V G PROF
RAMAN RESEARCH INSTITUTE
BANGALORE 560 080
INDIA
TEL: 360122
TLX: 845671 RRI IN
COM: 40

SARRIS ELEFTHERIOS PH D
NATL OBSERVATORY ATHENS
ASTRONOMICAL INSTITUTE
GR-11810 ATHENS
GREECE
TEL:
TLX: 215530 OBSA GR
COM:

SARRIS EMMANUEL T PH D
DEPT OF ELECT ENGINEERING
DEMOCRITOS UNIV OF THRACE
GR-67100 XANTHI
GREECE
TEL: 0541-26948
TLX: 452312 POLX GR
COM: 49

SARTORI LEO PROF
BEHLEN LAB OF PHYSICS
UNIVERSITY OF NEBRASKA
LINCOLN NB 68588
U.S.A.
TEL:
TLX:
COM: 48

SASAKI MISAO
RES INST FOR THEORET PHYS
HIROSHIMA UNIVERSITY
TAKEHARA-CHO
TAKEHARA 725
JAPAN
TEL: 08462-2-2362
TLX:
COM: 47

SASAO TETSUO DR
INTERNATIONAL LATITUDE
OBSERVATORY OF MIZUSAWA
MIZUSAWA-SHI
IWATE 023
JAPAN
TEL: 197-24-7111
TLX: 837628 ILSMIZJ
COM: 19

SASLAW WILLIAM C PROF
ASTRONOMY DEPT
UNIVERSITY OF VIRGINIA
BOX 3818 UNIV STATION
CHARLOTTESVILLE VA 22903
U.S.A.
TEL: 304-924-4892
TLX:
COM: 28,48

SASTRI HANUMATH J DR
INDIAN INSTITUTE OF
ASTROPHYSICS
BANGALORE 560 034
INDIA
TEL: 566-585
TLX: 845763 IIAB IN
COM: 49

SASTRY CH V
INDIAN INSTITUTE OF
ASTROPHYSICS
BANGALORE 560 034
INDIA
TEL:
TLX:
COM: 40

SASTRY SHANKARA K
DEPT OF ASTRONOMY
OSMANIA UNIVERSITY
HYDERABAD 500 007
INDIA
TEL:
TLX:
COM: 28

SATO FUMIO DR
HYOGO UNIVERSITY OF
TEACHER EDUCATION
YASHIRO
HYOGO 673-14
JAPAN
TEL: 07954-4-1101
TLX:
COM: 34,40

SATO HUMITAKA PROF
DEPARTMENT OF PHYSICS
UNIVERSITY OF KYOTO
SAKYO-KU
KYOTO 606
JAPAN
TEL:
TLX:
COM: 47V

SATO KATSUHIKO PROF
DEPT OF PHYSICS
FACULTY OF SCI UNIV TOKYO
BUNKYO-KU
TOKYO 113
JAPAN
TEL: 03-812-2111
TLX: 23472 UTPHYSIC
COM: 35,44,47

SATO KOICHI DR
INTERNATIONAL LATITUDE
OBSERVATORY OF MIZUSAWA
MIZUSAWA
IWATE 023
JAPAN
TEL: 0197-24-7111
TLX: 837628 ILSMIZJ
COM: 08,19

SATO NAONOBU PROF
AKITA UNIVERSITY
1-1 TEGATA GAKUENCHO
AKITA 010
JAPAN
TEL: 0188-33-5261
TLX:
COM: 27,41

SATO SHUJI DR
DEPT OF PHYSICS
UNIVERSITY OF KYOTO
KITASHIRAKAWA
SAKYOKU KYOTO 606
JAPAN
TEL: 075-701-5377
TLX: 5422693 LIBKYU J
COM: 34

SATO YUZO DR
OSAWA 4-8-19, MITAKA
TOKYO 181
JAPAN
TEL:
TLX:
COM:

SAUNDERS RICHARD D.E.
RADIO ASTRONOMY GROUP
CAVENDISH LABORATORY
MADINGLEY ROAD
CAMBRIDGE CB3 OHE
U.K.
TEL: 223-66477
TLX: 81292
COM: 40

SAUVAL A JACQUES DR
OBS ROYAL DE BELGIQUE
AVENUE CIRCULAIRE 3
B-1180 BRUXELLES
BELGIUM
TEL: 02-375-2484
TLX: 21565 OBSBEL B
COM:

SAVAGE ANN DR
UK SCHMIDT TELESCOPE
PRIVATE BAG
COONABARABRAN NSW 2357
AUSTRALIA
TEL:
TLX:
COM: 28,40,47

SAVAGE BLAIR D DR
DEPT OF ASTRONOMY
UNIVERSITY OF WISCONSIN
475 N.CHARTER STR
MADISON WI 53706
U.S.A.
TEL: 608-262-3072
TLX: 265452 UOFWISC-MDS
COM: 34,44

SAVEDOFF MALCOLM P PROF
DEPT PHYSICS & ASTRONOMY
UNIVERSITY OF ROCHESTER
BAUSCH AND LOMB BLDG
ROCHESTER NY 14627
U.S.A.
TEL: 716-275-4357
TLX: 978374 UNIBOOK ROC
COM: 34,35,48

SAVONIJE GERRIT JAN DR
ASTRONOMICAL INSTITUTE
UNIVERSITY OF AMSTERDAM
ROETERSSTRAAT 15
NL-1018 WB AMSTERDAM
NETHERLANDS
TEL: 020-5223004
TLX: 16460 FACWN NL
COM: 35,42

SAWYER CONSTANCE B DR
850 20TH STREET # 705
BOULDER CO 803023
U.S.A.
TEL:
TLX:
COM: 10,49

SAWYER-HOGG HELEN B DR
DAVID DUNLAP OBSERVATORY
P O BOX 360
RICHMOND HILL ONT L4C 4Y6
CANADA
TEL:
TLX:
COM: 27,37

SAXENA A K DR
INDIAN INSTITUTE OF
ASTROPHYSICS
BANGALORE 560 034
INDIA
TEL: 566585, 566497
TLX: 845763 IIAB IN
COM: 09

SAXENA P P DR
DEPT OF MATHS & ASTRONOMY
LUCKNOW UNIVERSITY
LUCKNOW
INDIA
TEL:
TLX:
COM: 21,46

SCALISE JR EUGENIO DR
INPE/CRAAM
AV DOS ASTRONAUTAS 1758
CAIXA POSTAL 515
12200 S. JOSE DOS CAMPOS
BRAZIL
TEL: 55-011-8266588
TLX: 34061 INPE BR
COM: 40

SCALO JOHN MICHAEL
DEPT OF ASTRONOMY
UNIVERSITY OF TEXAS
AUSTIN TX 78712
U.S.A.
TEL: 512-471-4461
TLX:
COM: 34,35

SCALTRITI FRANCO DR
OSSERVATORIO ASTRONOMICO
DI TORINO
STRADA OSSERVATORIO 20
I-10025 PINO TORINESE
ITALY
TEL: 011-841067
TLX: 213236 TOASTR I
COM: 15,42

SCARDIA MARCO
OSSERVATORIO ASTRONOMICO
DI BRERA
VIA E. BIANCHI 46
I-22055 MERATE
ITALY
TEL:
TLX:
COM: 26

SCARFE COLIN D DR
DEPT OF PHYSICS
UNIVERSITY OF VICTORIA
P O BOX 1700
VICTORIA BC V8W 2Y2
CANADA
TEL: 604-721-7740
TLX: 049-7222
COM: 26,30,42

SCARGLE JEFFREY D DR
NASA-AMES RESEARCH CENTER
MS 245-3
MOFFETT FIELD CA 94035
U.S.A.
TEL: 415-694-6330
TLX:
COM: 48,51

SCARIA K K DR
INDIAN INSTITUTE OF
ASTROPHYSICS
BANGALORE 560 034
INDIA
TEL:
TLX:
COM: 37

SCARROTT STANLEY M DR
PHYSICS DEPT
UNIVERSITY OF DURHAM
SOUTH ROAD
DURHAM DH1 3LE
U.K.
TEL:
TLX:
COM:

SCHADEE AERT DR
STERREWACHT SONNENBORGH
ZONNENBURG 2
NL-3512 NL UTRECHT
NETHERLANDS
TEL: 030-312841
TLX:
COM: 14

SCHAIFERS KARL DR
STEINBACHWEG 37
D-6900 HEIDELBERG
GERMANY, F.R.
TEL: 06221-801511
TLX:
COM:

SCHALEN CARL PROF
LUND OBSERVATORY
BOX 43
S-221 00 LUND
SWEDEN
TEL: 46-10 70 00
TLX: 33199 OBSNOT S
COM: 34

SCHANDA ERWIN PROF
INST OF APPLIED PHYSICS
SIDLERSTRASSE 5
CH-3012 BERN
SWITZERLAND
TEL: 031-65-89-10
TLX: 32320
COM:

SCHARMER GOERAN BJARNE
STOCKHOLM OBSERVATORY
S-133 00 SALTSJOEBADEN
SWEDEN
TEL: 8-717-0195
TLX: 12972
COM: 36

SCHATTEN KENNETH H DR
PHYSICS DEPT
VICTORIA UNIVERSITY
PRIVATE BAG
WELLINGTON
NEW ZEALAND
TEL:
TLX:
COM: 10,35,48

SCHATZMAN EVRY PROF
OBSERVATOIRE DE NICE
BP 139
F-06003 NICE CEDEX
FRANCE
TEL: 93-89-04-20
TLX: 460004
COM: 34,35,47,48,49,51

SCHEEPMAKER ANTON DR
COSMIC RAY WORKING GROUP
HUYGENS LABORATORY
WASSENAARSEWEG 78
NL-2300 RA LEIDEN
NETHERLANDS
TEL:
TLX:
COM:

SCHEFFLER HELMUT PROF
LANDESSTERNWARTE
KOENIGSTUHL
D-6900 HEIDELBERG 1
GERMANY, F.R.
TEL: 06221-10036
TLX:
COM:

SCHEIDECKER JEAN-PAUL DR
OBSERVATOIRE DE NICE
BP 139
F-06003 NICE CEDEX
FRANCE
TEL: 93-89-04-20
TLX: 460004
COM:

SCHERB FRANK PROF
PHYSICS DEPT
UNIVERSITY OF WISCONSIN
MADISON WI 53706
U.S.A.
TEL: 608-262-6879
TLX:
COM: 34,49

SCHERRER PHILIP H DR
CTR FOR SPACE SCIENCES &
ASTROPHYSICS
STANFORD UNIVERSITY, ERL
STANFORD CA 94305
U.S.A.
TEL: 415-497-1505
TLX: 348402 STANFRD STNU
COM:

SCHLICKEISER REINHARD DR
MPI FUER RADIOASTRONOMIE
AUF DEM HUEGEL 69
D-5300 BONN
GERMANY, F.R.
TEL: 0228-5251
TLX: 0886440 MPIFR D
COM: 40

SCHMALBERGER DONALD C DR
DUDLEY OBSERVATORY
1202 TROY-SCHENECTADY RD
LATHAM NY 12110
U.S.A.
TEL:
TLX:
COM: 36

SCHMIDT THOMAS DR
RUDOLF-STEINER-SCHULE
AN DER STIFTSKIRCHE 13
D-4800 BIELEFELD 1
GERMANY, F.R.
TEL: 0521-880407
TLX:
COM: 34,46

SCHEUER PETER A G DR
CAVENDISH LABORATORY
MADINGLEY RD
CAMBRIDGE CB3 0HE
U.K.
TEL: 0223-66477x344
TLX: 81292
COM: 34,40,47,48C

SCHLOERB F. PETER
UNIV. OF MASSACHUSETTS
DEPT PHYSICS & ASTRONOMY
AMHERST MA 01003
U.S.A.
TEL: 413-545-4303
TLX: 955491
COM: 16

SCHMEIDLER F PROF DR
INST ASTRON & ASTROPHYSIK
SCHEINERSTR 1
D-8000 MUENCHEN 80
GERMANY, F.R.
TEL: 089-98-90-21
TLX:
COM: 08

SCHMIDT-KALER TH PROF
ASTRONISCHES INSTITUT
RUHR-UNIVERSITAET BOCHUM
STEINHUEGEL 105
D-5810 WITTEN
GERMANY, F.R.
TEL: 0234-7003454
TLX: 0825860
COM: 33,34,45

SCHILD HANSRUEDI
ROYAL GREENWICH OBS
HERSTMONCEUX CASTLE
HAILSHAM BN27 1RP
U.K.
TEL: 0323-83-31-71
TLX: 87451
COM: 35,37

SCHLOSSER WOLFHARD PROF
ASTRONOMISCHES INSTITUT
POSTFACH 102148
D-4630 BOCHUM
GERMANY, F.R.
TEL: 0234-700-3454
TLX: 0825860
COM: 46

SCHMID-BURGK J DR PROF
MPI FUER RADIOASTRONOMIE
AUF DEM HUEGEL 69
D-5300 BONN 1
GERMANY, F.R.
TEL: 0449-228-525271
TLX: 0886440 MPIFR D
COM: 34,36

SCHMIEDER BRIGITTE DR
OBSERVATOIRE DE PARIS
SECTION DE MEUDON
F-92195 MEUDON PL CEDEX
FRANCE
TEL: 1-45-34-75-30
TLX:
COM: 10

SCHILD RUDOLPH E DR
CENTER FOR ASTROPHYSICS
60 GARDEN STREET
CAMBRIDGE MA 02138
U.S.A.
TEL: 617-495-7426
TLX: 921428 SATELLITE CAM
COM: 29,45,51

SCHLUETER A PROF DR
MPI FUER PLASMAPHYSIK
D-8046 GARCHING B MUNCHEN
GERMANY, F.R.
TEL: 089-3299-347
TLX: 05-215808 IPP D
COM: 10

SCHMIDT EDWARD G
DEPT OF PHYSICS & ASTRON
UNIVERSITY OF NEBRASKA
LINCOLN NB 68588-0111
U.S.A.
TEL: 402-472-2788
TLX:
COM: 25,27

SCHMITTER EDWARD F DR
DEPT OF PHYSICS
UNIVERSITY OF LAGOS
AKOKA
LAGOS
NIGERIA
TEL: 01-83-78-64
TLX:
COM:

SCHILIZZI RICHARD T DR
RADIOSTERRENWACHT
POSTBUS 2
NL-7990 AA DWINGELOO
NETHERLANDS
TEL: 5219-7244
TLX: 42043 SRZM NL
COM: 40,48,50

SCHLUETER DIETER PROF
INST THEOR PHYS & STERNW
NEUE UNIV PHYSIK ZENTRUM
OLSHAUSENST GEB N 61C
D-2300 KIEL 1
GERMANY, F.R.
TEL: 880-4109
TLX:
COM:

SCHMIDT H U DR
MPI FUER PHYSIK UND
ASTROPHYSIK
KARL-SCHWARZSCHILD-STR 1
D-8046 GARCHING B MUNCHEN
GERMANY, F.R.
TEL: 89-32999413/4
TLX: 524629 ASTRO D
COM: 10,15,49

SCHMUTZ WERNER
INST F. THEORET. PHYSIK
OLSHAUSENSTRASSE
D-2300 KIEL 1
GERMANY, F.R.
TEL: 0431-880 4102
TLX: 292706 IAPKI D
COM: 36

SCHILLER KARL PROF DR
PIRSCHWEG 6
6072 DREIEICH-BUCHSCHLAG
GERMANY, F.R.
TEL:
TLX:
COM:

SCHMADEL LUTZ D DR
ASTRONOMISCHES RECHEN
INSTITUT
MOENCHHOFSTR 12-14
D-6900 HEIDELBERG 1
GERMANY, F.R.
TEL: 06221-49026
TLX: 461336 ARIHD D
COM: 05,20

SCHMIDT HANS PROF
UNIVERSITAETSSTERNWARTE
AUF DEM HUEGEL 71
D-5300 BONN 1
GERMANY, F.R.
TEL:
TLX:
COM: 25,33,42

SCHNEIDER JEAN
OBSERVATOIRE DE PARIS
SECTION DE MEUDON
F-92195 MEUDON PL CEDEX
FRANCE
TEL: 1-45-34-75-70
TLX:
COM: 47

SCHINDLER KARL PROF DR
INST FUER THEORET PHYSIK
RUHR-UNIVERSITAET BOCHUM
D-4630 BOCHUM
GERMANY, F.R.
TEL:
TLX:
COM: 49

SCHMAHL EDWARD J DR
ASTRONOMY PROGRAM
UNIVERSITY OF MARYLAND
COLLEGE PARK MD 20742
U.S.A.
TEL: 301-454-6074
TLX:
COM: 10,12

SCHMIDT K H DR
ZNTRLINST. F. ASTROPHYSIK
STERNWARTE BABELSBERG
ROSA-LUXEMBURG-STR 17A
DDR-1502 POTSDAM
GERMANY, D.R.
TEL:
TLX:
COM: 05,33

SCHNELL ANNELIESE DR
INSTITUT FUER ASTRONOMIE
UNIVERSITAT WIEN
TUERKENSCHANZSTR 17
A-1180 WIEN
AUSTRIA
TEL: 222-34-53-60-93
TLX:
COM:

SCHLESINGER BARRY M DR
SASC TECHNOLOGIES INC.
5809 ANNAPOLIS ROAD
HYATTSVILLE MD 20784
U.S.A.
TEL: 301-699-6171
TLX: 317630
COM:

SCHMAHL GUENTER PROF
UNIVERSITAETSSTERNWARTE
GEISMARLANDSTR 11
D-3400 GOETTINGEN
GERMANY, F.R.
TEL: 0551-395061
TLX: 96753 USTERN
COM:

SCHMIDT MAARTEN PROF
CALTECH
ASTRONOMY 105-24
PASADENA CA 91125
U.S.A.
TEL: 818-356-4204
TLX: 675425
COM: 15,28,33,40,47

SCHNEPS MATTHEW H
HARVARD-SMITHSONIAN CTR
FOR ASTROPHYSICS
60 GARDEN STREET
CAMBRIDGE MA 02138
U.S.A.
TEL: 617-495-7472
TLX: 921428 SATELLITE CAM
COM:

SCHNOPPER HERBERT W DR
DANISH SPACE RESEARCH INS
LUNDTOFTEVEJ 7
DK-2800 LYNGBY
DENMARK
TEL: 02-88 22 77
TLX: 37198 DANRU
COM:

SCHNUR GERHARD F O
ASTRONOMISCHES INSTITUT
RUHR UNIVERSITAET
POSTFACH 102148
D-4630 BOCHUM
GERMANY, F.R.
TEL:
TLX:
COM:

SCHOBER HANS J DR
INSTITUT FUER ASTRONOMIE
UNIVERSITAETSPLATZ 5
A-8010 GRAZ
AUSTRIA
TEL: 0316-380-5273
TLX: 31078 OBSLGZ
COM: 10,12,15,20,42,51

SCHOEFFEL EBERHARD F DR
MERIANERSTR 42
D-8600 BAMBERG
GERMANY, F.R.
TEL:
TLX:
COM: 42

SCHOEMBS ROLF DR
INSTITUT FUER ASTRONOMIE
UND ASTROPHYSIK
SCHEINERSTR 1
D-8000 MUENCHEN 80
GERMANY, F.R.
TEL: 98-90-21
TLX:
COM: 27

SCHOENBERNER DETLEF PROF
INST THEOR PHYS & STERNW.
OLSHAUSENSTRASSE
D-2300 KIEL
GERMANY, F.R.
TEL: 0431-8804100
TLX: 292706 IAPKI D
COM: 36

SCHOENEICH W DR
ZNTRLINST. F. ASTROPHYSIK
ROSA-LUXEMBURG-STR 17A
DDR-1502 POTSDAM
GERMANY, D.R.
TEL:
TLX:
COM: 25,44

SCHOENFELDER VOLKER DR
MPI F EXTRATERRESTRISCHE
PHYSIK
D-8046 GARCHING B MUNCHEN
GERMANY, F.R.
TEL: 49-89-3299-578
TLX: 5215845 XTER D
COM:

SCHOLL HANS DR
ASTRONOMISCHES RECHEN
INSTITUT
MOENCHHOFSTR 12-14
D-6900 HEIDELBERG
GERMANY, F.R.
TEL:
TLX:
COM: 07C,20

SCHOLZ GERHARD DR
ZNTRLINST. F. ASTROPHYSIK
AKAD. WISSENSCHAFTEN DDR
ROSA-LUXEMBURG-STR 17A
DDR-1502 POTSDAM-BABELSB.
GERMANY, D.R.
TEL:
TLX:
COM: 29

SCHOLZ M PROF
INST F THEORETISCHE
ASTROPHYS DER UNIVERSITAT
NEUENHEIMER FELD 561
D-6900 HEIDELBERG
GERMANY, F.R.
TEL:
TLX:
COM: 36

SCHOOLMAN STEPHEN A DR
LOCKHEED PA RES LAB
3251 HANOVER STREET
PALO ALTO CA 94304
U.S.A.
TEL:
TLX:
COM:

SCHRAMM DAVID N PROF
UNIVERSITY CHICAGO
ASTRON & ASTROPHYS CENTER
5640 SO ELLIS AVENUE
CHICAGO IL 60637
U.S.A.
TEL: 312-962-8202
TLX: 6871133 UNCGO UW
COM: 35,47,48C

SCHREIBER ROMAN
COPERNICUS ASTRON CENTER
ASTROPHYSICS LAB
UL. CHOPINA 12/18
87-100 TORUN
POLAND
TEL: 48-5626017
TLX: 0552234 ASTR PL
COM: 49

SCHRIJVER JOHANNES DR
LABORATORIUM VOOR
RUIMTEONDERZOEK
BENELUXLAAN 21
NL-3527 HS UTRECHT
NETHERLANDS
TEL: 30-93-71-45
TLX: 47224 ASTRO NL
COM: 14

SCHROEDER DANIEL J PROF
DEPT PHYSICS & ASTRONOMY
BELOIT COLLEGE
BELOIT WI 53511
U.S.A.
TEL: 608-365-3391
TLX:
COM: 09

SCHROEDER ROLF DR
MOEOERKENWEG 3712
D-2050 HAMBURG 80
GERMANY, F.R.
TEL:
TLX:
COM:

SCHROETER EGON H PROF
KIEPENHEUER INSTITUT
FUER SONNENPHYSIK
SCHOENECKSTRASSE 6
D-7800 FREIBURG I BR.
GERMANY, F.R.
TEL: 0761-32864
TLX: 7721552 KIS D
COM: 10

SCHROLL ALFRED DR
SONNENOBSERVATORIUM
KANZELHOEHE
A-9520 SATTENDORF
AUSTRIA
TEL: 04248-2717
TLX: 45699 SOLOBS A
COM:

SCHRUEFER EBERHARD DR
INSTITUT FUER ASTROPHYSIK
UNIVERSITAET BONN
AUF DEM HUEGEL 71
D-5300 BONN 1
GERMANY, F.R.
TEL: 228-73-33-90
TLX: 886440 MPIFR D
COM:

SCHRUTKA-RECHTENSTAMM PR.
WILLERGASSE 27/4/7
A-1238 WIEN
AUSTRIA
TEL: 0222-8848132
TLX:
COM: 20

SCHUBART JOACHIM DR
ASTRONOMISCHES RECHEN-
INSTITUT
MOENCHHOFSTR 12-14
D-6900 HEIDELBERG
GERMANY, F.R.
TEL: 49-6221-4-90-26
TLX: 461336 ARIHD D
COM: 07,20

SCHUCH NELSON JORGE
OBSERVATORIO NACIONAL
UFSM/CTRO TECNOLOGIA
CIDADE UNIVERSITARIA
97100 SANTA MARIA
BRAZIL
TEL: 055-226-1616
TLX: 0552230 UFSM
COM:

SCHUECKING E L DR
DEPT OF PHYSICS
NEW YORK UNIVERSITY
NEW YORK NY 10012
U.S.A.
TEL:
TLX:
COM: 28,47

SCHUESSLER MANFRED DR
KIEPENHEUER-INSTITUT
FUER SONNENPHYSIK
SCHOENECKSTR 6
D-7800 FREIBURG
GERMANY, F.R.
TEL: 761-32864
TLX: 7721552 KIS D
COM: 12

SCHULER WALTER DR
STERNWARTE DER
KANTONSSCHULE
CH-4500 SOLOTHURN
SWITZERLAND
TEL: 065-23-20-55
TLX:
COM: 31

SCHULTE D H DR
ITEK CORPORATION
10 MAGUIRE ROAD
LEXINGTON MA 02173
U.S.A.
TEL:
TLX:
COM:

SCHULTZ G V DR
MPI FUER RADIOASTRONOMIE
AUF DEM HUEGEL 69
D-5300 BONN
GERMANY, F.R.
TEL: 0228-52-52-91
TLX: 0886440 ASTROD
COM: 09,28,34,40,44,47

SCHULZ HARTMUT DR
ASTRONOMISCHES INSTITUT
UNIVERSITAT BOCHUM
POSTFACH 10 21 48
D-4630 BOCHUM 1
GERMANY, F.R.
TEL: 234-700-3454
TLX: 0825860
COM: 28

SCHULZ ROLF ANDREAS
MPI FUER RADIOASTRONOMIE
AUF DEM HUEGEL 69
D-5300 BONN 1
GERMANY, F.R.
TEL: 228-525-232
TLX: 886440 MPIFR D
COM: 34,40

SCHUMANN JOERG DIETER DR
OBSERVATORIUM HOHER LIST
UNIV STERNWARTE BONN
D-5568 DAUN
GERMANY, F.R.
TEL: 06592-2937
TLX:
COM: 09

SCHUSTER HANS-EMIL
E S O
CASILLA 19001
SANTIAGO 19
CHILE
TEL: 6988757
TLX: 240881
COM: 09,20,28,37

SCHUSTER WILLIAM JOHN DR
INSTITUTO DE ASTRONOMIA
UNAM
APDO POSTAL 877
22860 ENSENADA, B. CALIF.
MEXICO
TEL: 706-67-83093
TLX: 56539 CICE ME
COM:

SCHUTZ BERNARD F PROF
APPLIED MATHS & ASTRONOMY
UNIVERSITY COLLEGE
CARDIFF CF1 1XL
U.K.
TEL: 0222-44211
TLX: 448635 ULIBCF
COM: 35

SCHUTZ BOB EWALD
CENTER FOR SPACE RESEARCH
UNIVERSITY OF TEXAS
AUSTIN TX 78712
U.S.A.
TEL: 512-471-1356
TLX: 704265 CSRUTX UD
COM: 19C

SCHWAN HEINER DR
ASTRONOMISCHES RECHEN-
INSTITUT
MOENCHHOFSTR 12-14
D-6900 HEIDELBERG
GERMANY, F.R.
TEL: 06221-49026
TLX: 461336 ARIHD D
COM: 04.08C

SCHWARTZ DANIEL A DR
CENTER FOR ASTROPHYSICS
60 GARDEN STREET
CAMBRIDGE MA 02138
U.S.A.
TEL: 617-495-7232
TLX:
CCM: 44.48

SCHWARTZ PHILIP R DR
NAVAL RESEARCH LABORATORY
CODE 4138
WASHINGTON DC 20375
U.S.A.
TEL: 202-767-3391
TLX:
COM: 27,34,40

SCHWARTZ RICHARD D
PHYSICS DEPT
UNIVERSITY OF MISSOURI
8001 NATURAL BRIDGE RD
ST LOUIS MO 63121
U.S.A.
TEL: 314-553-5025
TLX: 447658 UMSL BOOKSTOR
COM: 34

SCHWARTZ ROLF PH D
MPI FUER RADIOASTRONOMIE
AUF DEM HUEGEL 69
D-5300 BONN 1
GERMANY, F.R.
TEL: 228-525-303
TLX:
COM:

SCHWARTZ STEVEN JAY
THEORET ASTR UNIT/MATH SC
QUEEN MARY COLLEGE
MILE END ROAD
LONDON E1 4NS
U.K.
TEL: 1-980-4811x3849
TLX:
COM: 12,44,49

SCHWARZ ULRICH J DR
KAPTEYN LABORATORIUM
POSTBUS 800
NL-9700 AV GRONINGEN
NETHERLANDS
TEL: 050-116695
TLX: 53572 STARS NL
COM: 28,34,40

SCHWARZENBERG-CZERNY A
WARSAW UNIVERSITY OBS
AL. UJAZDOWSKIE 4
00-478 WARSAW
POLAND
TEL: 29 40 11
TLX: 813978 ZAPAN PL
COM: 27

SCHWARZSCHILD MARTIN PROF
PRINCETON UNIVERSITY
OBSERVATORY
PEYTON HALL
PRINCETON NJ 08544
U.S.A.
TEL: 609-452-3812
TLX:
COM: 35

SCHWEIZER FRANCOIS DR
DEPT TERREST. MAGNETISM
CARNEGIE INST. WASHINGTON
5241 BROAD BRANCH RD N.W.
WASHINGTON DC 20015
U.S.A.
TEL: 202-966-0863
TLX: 440427 MAGN UI
COM: 28

SCHWERDTFEGER HANS-M. DR
ASTRONOMISCHES RECHEN-
INSTITUT
MOENCHHOFSTR 12-14
D-6900 HEIDELBERG 1
GERMANY, F.R.
TEL: 0049-6221-49026
TLX: 461336 ARIHD D
COM: 33

SCIAMA DENNIS W DR
DEPT OF ASTROPHYSICS
SOUTH PARKS ROAD
OXFORD OX1 3RQ
U.K.
TEL: 0865-511336
TLX:
COM: 28,47,48

SCONZO PASQUALE DR
29 OLD MYSTIC STREET
ARLINGTON MA 02174
U.S.A.
TEL: 617-646-9315
TLX:
COM: 07

SCOTT ELIZABETH L PROF
STATISTICS
UNIVERSITY OF CALIFORNIA
367 EVANS HALL
BERKELEY CA 94720
U.S.A.
TEL: 415-642-2777
TLX: 910-366-7114 UC BERK
COM: 28,47

SCOTT EUGENE HOWARD
GODDARD SPACE FLIGHT CTR
CODE 684.9
GREENBELT MD 20771
U.S.A.
TEL: 301-344-8746
TLX:
COM: 34

SCOTT JOHN S DR
STEWARD OBSERVATORY
UNIVERSITY OF ARIZONA
TUCSON AZ 85721
U.S.A.
TEL:
TLX:
COM: 40,48

SCOTT PAUL F DR
CAVENDISH LABORATORY
MADINGLEY ROAD
CAMBRIDGE CB3 0HE
U.K.
TEL: 0223-66477
TLX: 81292
COM: 40

SCOVILLE NICHOLAS Z
ASTRONOMY & PHYSICS DEPT
UNIV OF MASSACHUSETTS
GRADUATE RESEARCH CENTER
AMHERST MA 01003
U.S.A.
TEL: 413-545-0789
TLX:
COM: 28

SCRIMGER J. NORMAN DR
DEPT OF ASTRONOMY
ST MARY'S UNIVERSITY
HALIFAX NS B3H 3C3
CANADA
TEL:
TLX:
COM:

SCUFLAIRE RICHARD DR
INSTITUT D'ASTROPHYSIQUE
UNIVERSITE DE LIEGE
AVENUE DE COINTE 5
B-4200 COINTE-OUGREE
BELGIUM
TEL: 041-52-99-80
TLX: 41264 ASTRLG B
COM: 27,35

SEAQUIST ERNEST R PROF
DEPT OF ASTRONOMY
UNIVERSITY OF TORONTO
TORONTO ONT M5S 1A7
CANADA
TEL: 416-978-3146
TLX: 06-986766
COM: 40C

SEARLE LEONARD DR
HALE OBSERVATORIES
813 SANTA BARBARA ST
PASADENA CA 91101
U.S.A.
TEL: 818-304-0220
TLX:
COM: 28

SEARS RICHARD LANGLEY DR
DEPT OF ASTRONOMY
UNIVERSITY OF MICHIGAN
ANN ARBOR MI 48109
U.S.A.
TEL: 313-763-3295
TLX:
COM: 35

SEATON MICHAEL J PROF
DEPT PHYSICS & ASTRONOMY
UNIVERSITY COLLEGE LONDON
GOWER STREET
LONDON WC1E 6BT
U.K.
TEL: 01-387-7050
TLX: 28722
COM: 12,14,34,36C

SEDLMAYER ERWIN DR
INST FUER ASTRONOMIE &
ASTROPHYSIK DER TECHN UNI
ERNST-REUTER-PLATZ 7
D-1000 BERLIN 10
GERMANY, F.R.
TEL:
TLX:
COM: 36

SEDMAK GIORGIO PROF
DIPT. DI ASTRONOMIA
UNIVERSITA DI TRIESTE
VIA TIEPOLO 11
I-34131 TRIESTE
ITALY
TEL: 40-79-4863
TLX: 461137
COM: 05,09

SEEGER CHARLES LOUIS III
SAN FRANCISCO STATE UNIV
473 JAMES ROAD
PALO ALTO CA 94306
U.S.A.
TEL: 415-493-6005
TLX:
COM: 51

SEEGER PHILIP A DR
LOS ALAMOS NATIONAL LAB
MS/H805
PO BOX 1663
LOS ALAMOS NM 87545
U.S.A.
TEL: 505-667-8843
TLX:
COM:

SEGAL IRVING E DR
MIT
2-224
CAMBRIDGE MA 02139
U.S.A.
TEL: 617-253-4985
TLX:
COM: 28,47

LIST OF MEMBERS

SEGALUVITZ ALEXANDER DR
11 HABANIM ST
KEFAR SAVA
ISRAEL
TEL:
TLX:
COM:

SEGGEWISS WILHELM PROF
OBSERVATORIUM HOHER LIST
UNIVERSITAETS-STERNWARTE
D-5568 DAUN EIFEL
GERMANY, F.R.
TEL: 06592-2150
TLX:
COM: 29,33,42

SEHNAL LADISLAV DR
ASTRONOMICAL INSTITUTE
CZECH. ACAD. OF SCIENCES
OBSERVATORY
251 65 ONDREJOV
CZECHOSLOVAKIA
TEL: 258757
TLX: 121579
COM: 07

SEIDELMANN P KENNETH DR
US NAVAL OBSERVATORY
34 & MASSACHUSETTS AVE NW
WASHINGTON DC 20390
U.S.A.
TEL: 202-653-1545
TLX: 710-822-1970
COM: 04V,07C,20

SEIDEN PHILIP E
IBM RESEARCH CENTER
PO BOX 218
YORKTOWN HEIGHTS NY 10598
U.S.A.
TEL: 914-945-1424
TLX: 137456
COM: 28,47

SEIDOV ZAKIR F DR
SHEMAKHA ASTROPHYSICAL
OBSERVATORY
373243 SHEMAKHA, AZERB.
U.S.S.R.
TEL:
TLX:
COM: 35

SEIELSTAD GEORGE A
NRAO
PO BOX 2
GREEN BANK WV 24944
U.S.A.
TEL: 304-456-2301
TLX: 710-938-1530
COM: 40,47,48,51

SEIRADAKIS JOHN HUGH DR
DEPT OF ASTRONOMY
UNIVERSITY THESSALONIKI
GR-54006 THESSALONIKI
GREECE
TEL: 031-991357
TLX:
COM: 40,51

SEITTER WALTRAUT C PROF
ASTRONOMISCHES INSTITUT
DOMAGKSTR 75
D-4400 MUENSTER/W
GERMANY, F.R.
TEL: 251-83-3561
TLX: 892529 UNI MS D
COM: 45

SEKANINA ZDENEK DR
EARTH & SPACE SCI DIV
JPL
4800 OAK GROVE DRIVE
PASADENA CA 91103
U.S.A.
TEL: 818-354-7589
TLX:
COM: 15,20,22

SEKI MUNEZO DR
DEPT OF EARTH SCIENCES
COLL OF GENERAL EDUCATION
TOHOKU UNIV, KAWAUCHI
SENDAI 980
JAPAN
TEL: 0222-12-1800
TLX:
COM: 34

SEKIGUCHI NAOSUKE PROF
TOKYO ASTRONOMICAL
OBSERVATORY
OSAWA, MITAKA
TOKYO 181
JAPAN
TEL: 0422-32-5111
TLX: 02822307 TAOMTK J
COM: 19

SELLWOOD JEREMY ARTHUR
THE UNIVERSITY
DEPT OF ASTRONOMY
MANCHESTER M13 9PL
U.K.
TEL: 061-273-7121
TLX:
COM: 28,33

SELVELLI PIERLUIGI DR
OSSERVATORIO ASTRONOMICO
VIA TIEPOLO 11
I-34131 TRIESTE
ITALY
TEL: 40-793221
TLX: 461137 OAT I
COM: 44

SEMEL MEIR DR
OBSERVATOIRE DE PARIS
SECTION DE MEUDON
F-92195 MEUDON PL CEDEX
FRANCE
TEL: 1-45-34-75-30
TLX:
COM: 10,12

SEMENIUK IRENA DR
WARSAW UNIVERSITY OBS
AL. UJAZDOWSKIE 4
00-478 WARSZAWA
POLAND
TEL: 29-40-11/12
TLX: 815548 OAUW
COM: 42

SEMENZATO ROBERTO
UNIVERSITA DI PADOVA
PHYSICS DEPT
VIA MARZOLO 8
I-35131 PADOVA
ITALY
TEL: 049-844-247
TLX: 430308 DF GGPD I
COM:

SEN S N DR
INDIAN ASSOCIATION FOR
THE CULTIVATION OF SCI
JADAVPUR
INDIA
TEL:
TLX:
COM:

SENGBUSCH KURT V DR
MPI FUER PHYSIK UND
ASTROPHYSIK
KARL-SCHWARZSCHILD-STR 1
D-8046 GARCHING B MUNCHEN
GERMANY, F.R.
TEL:
TLX:
COM: 35

SERAFIN RICHARD AUGUST
ASTRONOMICAL OBSERVATORY
A. MICKIEWICZ UNIVERSITY
SLONECZNA 36
60-286 POZNAN
POLAND
TEL: 679 670
TLX:
COM:

SERRANO ALFONSO DR
INSTITUTO DE ASTRONOMIA
UNAM
APDO POSTAL 70-264
04510 MEXICO DF
MEXICO
TEL:
TLX:
COM: 28

SERSIC J L DR
OBSERVATORIO ASTRONOMICO
LAPRIDA 854
5000 CORDOBA
ARGENTINA
TEL: 051 25072
TLX: 51-822 BUCOR OBSASTR
COM: 28,47

SERVAN BERNARD
OBSERVATOIRE DE PARIS
61 AVE DE L'OBSERVATOIRE
F-75014 PARIS
FRANCE
TEL: 1-43-20-12-10
TLX: 270776
COM: 09

SESSIN WAGNER DR
INST TECN. DE AERONAUTICA
DEPTO DE ASTRONOMIA
12200 S. JOSE DOS CAMPOS
BRAZIL
TEL: 0123-22-9088
TLX: 011 73437 ZWO-24-73
COM: 07

SETTI GIANCARLO PROF
ESO
KARL-SCHWARZSCHILD-STR 2
D-8046 GARCHING B MUNCHEN
GERMANY,F.R.
TEL: 049-89-320-06-0
TLX: 52828222 EO D
COM: 28,40C,47P,48,49

SEVARLIC BRANISLAV M PROF
DEPT OF ASTRONOMY
UNIVERSITY BEOGRAD
VOLGINA 7
11050 BEOGRAD
YUGOSLAVIA
TEL:
TLX:
COM: 08,19

SEVERINO GIUSEPPE
OSSERVATORIO ASTRONOMICO
DI CAPODIMONTE
VIA MOIARIELLO 16
I-80131 NAPOLI
ITALY
TEL: 081-440101
TLX:
COM: 12

SEVERNYJ A B PROF DR
CRIMEAN ASTROPHYSICAL OBS
USSR ACADEMY OF SCIENCES
NAUCHNIY
334413 CRIMEA
U.S.S.R.
TEL:
TLX:
COM: 10,12,44,49

SEVILLA MIGUEL J DR
INST DE ASTRON Y GEODESIA
ALMANSA 76
28040 MADRID
SPAIN
TEL: 2 442501
TLX:
COM: 19

SEWARD FREDERICK D
HARVARD-SMITHSONIAN CTR
FOR ASTROPHYSICS
60 GARDEN STREET
CAMBRIDGE MA 02138
U.S.A.
TEL: 617-495-7282
TLX:
COM:

SEYMOUR P A H
57 HERMITAGE ROAD
PLYMOUTH, DEVON
U.K.
TEL:
TLX:
COM: 46

SEZER CENGIZ DR
EGE UNIVERSITY
FACULTY OF SCIENCE
CAMPUS P K 21
BORNOVA-IZMIR
TURKEY
TEL: 90-51-180110
TLX:
COM:

SHAFFER DAVID B DR
NASA/GSFC
CODE 621.9
GREENBELT MD 20771
U.S.A.
TEL: 301-344-6434
TLX:
COM: 40

SHAH GHANSHYAM A DR
INDIAN INST ASTROPHYSICS
SARJAPUR RD
KORAMANGALA
BANGALORE 560 034
INDIA
TEL: 566585 & 566497
TLX: 845-763 IIAB IN
COM: 34

SHAHAM JACOB PROF
COLUMBIA UNIVERSITY
PHYSICS DEPARTMENT
NEW YORK NY 10027
U.S.A.
TEL: 212-280-3349
TLX: 220094 COLU UR
COM: 48

SHAHBAZIAN ROMELIA K DR
BYURAKAN ASTROPHYSICAL
OBSERVATORY
378433 BYURAKAN, ARMENIA
U.S.S.R.
TEL: 56-34-53
TLX:
COM: 28

SHAKESHAFT JOHN R DR
CAVENDISH LABORATORY
MADINGLEY ROAD
CAMBRIDGE CB3 0HE
U.K.
TEL: 223-66477
TLX: 81292 CAVLAB G
COM: 05,28,40

SHAKHBAZYAN YURIJ L DR
BYURAKAN ASTROPHYSICAL
OBSERVATORY
378433 ARMENIA
U.S.S.R.
TEL:
TLX:
COM: 09,

SHAKHOVSKOJ N M DR
CRIMEAN ASTROPHYSICAL OBS
USSR ACADEMY OF SCIENCES
NAUCHNIY
334413 CRIMEA
U.S.S.R.
TEL:
TLX:
COM: 25C

SHAKURA NICHOLAJ I DR
STERNBERG STATE
ASTRONOMICAL INSTITUTE
117234 MOSCOW
U.S.S.R.
TEL:
TLX:
COM: 42,48

SHALLIS MICHAEL J DR
DEPT OF ASTROPHYSICS
SOUTH PARKS ROAD
OXFORD OX1 3RQ
U.K.
TEL:
TLX:
COM: 12

SHALTOUT MESALAM A M DR
HELWAN OBSERVATORY
HELWAN
EGYPT
TEL:
TLX:
COM:

SHANE WILLIAM W DR
STERRENKUNDIG INSTITUUT
KATHOLIEKE UNIVERSITEIT
TOERNOOIVELD
NL-6525 ED NIJMEGEN
NETHERLANDS
TEL: 31-080-558833
TLX: 48228 WINAT NL
COM: 33,34

SHAO CHENG-YUAN
HARVARD-SMITHSONIAN CTR
FOR ASTROPHYSICS
60 GARDEN STREET
CAMBRIDGE MA 02138
U.S.A.
TEL: 617-495-7212
TLX:
COM: 22,34

SHAPIRO IRWIN I PROF
CENTER FOR ASTROPHYSICS
ROOM P 209
60 GARDEN STREET
CAMBRIDGE MA 02138
U.S.A.
TEL: 617-495-7100
TLX: 921428 SATELLITE CAM
COM: 04,07,16,19

SHAPIRO MAURICE M PROF
205 YOAKUM PKWY 2-1720
ALEXANDRIA VA 22304
U.S.A.
TEL:
TLX:
COM: 48,51

SHAPIRO STUART L
CTR RADIOPHYS & SPACE RES
CORNELL UNIVERSITY
ITHACA NY 14853
U.S.A.
TEL: 607-256-4936
TLX:
COM: 34

SHAPLEY ALAN H
NOAA
BOULDER CO 80302
U.S.A.
TEL:
TLX:
COM: 10

SHARA MICHAEL DR
SPACE TELESCOPE SCI INST
HOMEWOOD CAMPUS
3700 SAN MARTIN DRIVE
BALTIMORE MD 21218
U.S.A.
TEL: 301-338-4743
TLX: 6849101 STSCI UW
COM: 27

SHARAF MOHAMED ADEL DR
DEPT OF ASTRONOMY
CAIRO UNIVERSITY
CAIRO
EGYPT
TEL:
TLX:
COM:

SHARAF SH G DR
INSTITUTE OF THEORETICAL
ASTRONOMY
10 KUTUZOV QUAY
192187 LENINGRAD
U.S.S.R.
TEL:
TLX:
COM: 07

SHAROV A S DR
STERNBERG STATE
ASTRONOMICAL INSTITUTE
119899 MOSCOW
U.S.S.R.
TEL: 139-26-57
TLX:
COM: 21,33,37

SHARPLESS STEWART PROF
DEPT PHYSICS & ASTRONOMY
UNIVERSITY OF ROCHESTER
ROCHESTER NY 14627
U.S.A.
TEL: 716-275-4389
TLX:
COM: 34,45

SHAVER PETER A DR
ESO
KARL-SCHWARTZSCHILD-STR 2
D-8046 GARCHING B MUNCHEN
GERMANY, F.R.
TEL: 089-320060
TLX: 52828222 EOD
COM: 28,34C,40,47,48

SHAVIV GIORA DR
DEPT OF PHYSICS
ISRAEL INST OF TECHNOLOGY
TECHNION
HAIFA 32000
ISRAEL
TEL:
TLX:
COM: 35,42C,47,48

SHAW JAMES SCOTT DR
DEPT PHYSICS & ASTRONOMY
UNIVERSITY OF GEORGIA
ATHENS GA 30602
U.S.A.
TEL: 404-542-2485
TLX:
COM:

SHAW JOHN H PROF
OHIO STATE UNIVERSITY
174 W. 18TH AVENUE
COLUMBUS OH 43210
U.S.A.
TEL: 614-422 7968
TLX:
COM:

SHAW R WILLIAM PROF
105 HALCYON HILL
ITHACA NY 14850
U.S.A.
TEL: 607-257-1948
TLX:
COM:

SHAWHAN STANLEY D DR
DEPT PHYSICS & ASTRONOMY
UNIVERSITY OF IOWA
IOWA CITY IA 52242
U.S.A.
TEL: 319-353-3294
TLX:
COM: 49

SHAWL STEPHEN J DR
ASTRONOMY DEPARTMENT
UNIVERSITY OF CALIFORNIA
BERKELEY CA 94720
U.S.A.
TEL:
TLX:
COM: 25,34,37

SHCHEGLOV P V DR
STERNBERG STATE
ASTRONOMICAL INSTITUTE
119899 MOSCOW
U.S.S.R.
TEL: 139-19-73
TLX:
COM: 09,34,50C

SHCHEGOLEV DIMITRIJ E DR
PULKOVO OBSERVATORY
196140 LENINGRAD
U.S.S.R.
TEL:
TLX:
COM: 29

SHCHERBINA-SAMOJLOVA I DR
INSTITUTE OF SCIENCE
AND TECHNICS INFORMATION
DEPARTMENT OF ASTRONOMY
125219 MOSCOW
U.S.S.R.
TEL: 1554237
TLX:
COM: 05

SHEA MARGARET A DR
AIR FORCE GEOPHYSICS LAB
HANSCOM AFB
BEDFORD MA 01732
U.S.A.
TEL:
TLX:
COM: 10,49

SHEELEY NEIL R DR
NAVAL RESEARCH LABORATORY
CODE 4172
WASHINGTON DC 20375
U.S.A.
TEL: 202-767-2777
TLX:
COM: 10,12

SHEFFER EUGENE K DR
STERNBERG STATE
ASTRONOMICAL INSTITUTE
119899 MOSCOW V-234
U.S.S.R.
TEL: 1392046
TLX:
COM:

SHEFOV NICOLAI N
INST OF PHYSICS OF THE
ATMOSPHERE
PYZHEVSKY 3
109017 MOSCOW
U.S.S.R.
TEL:
TLX:
COM: 21

SHELUS PETER J DR
ASTRONOMY DEPT
UNIVERSITY OF TEXAS
RLM 15-316
AUSTIN TX 78712
U.S.A.
TEL: 512-471-3339
TLX: 910-874-1351
COM: 20

SHEN BENJAMIN S P PROF
DEPT ASTRONOMY E1
UNIV OF PENNSYLVANIA
PHILADELPHIA PA 19104
U.S.A.
TEL: 215-898-8176
TLX:
COM: 28

SHEN CHANGJUN
PURPLE MOUNTAIN
OBSERVATORY
NANJING
CHINA, PEOPLE'S REP.
TEL:
TLX: 34144 PMONJ CN
COM: 09

SHEN CHUN-SHAN
ASTRONOMICAL STY OF CHINA
NATL TSING-HUA UNIVERSITY
HSIN-CHU 300
CHINA, TAIWAN
TEL: 886-35-719039
TLX:
COM: 46,51

SHEN KAIXIAN
SHAANXI OBSERVATORY
P.O. BOX 18
LINTONG
XIAN
CHINA, PEOPLE'S REP.
TEL: 3-2155
TLX: 70121 CSAO CN
COM: 08

SHEN LIANG-ZHAO
BEIJING ASTRONOMICAL OBS
BEIJING 100080
CHINA, PEOPLE'S REP.
TEL:
TLX: 22040 BAOAS CN
COM: 42

SHEN LONG-XIANG
BEIJING OBSERVATORY
ACADEMIA SINICA
BEIJING
CHINA, PEOPLE'S REP.
TEL: 28-1968
TLX: 22040 BAOAS CN
COM: 12

SHEN PARN-AN
NANJING ASTRONOMICAL
INSTRUMENT FACTORY
NANJING
CHINA, PEOPLE'S REP.
TEL: 46191
TLX: 34136 GLYNJ c/o NAIF
COM: 09

SHER DAVID DR
2837 MINTO DR, APT 2
CINCINNATI OH 45208
U.S.A.
TEL: 513-871-8850
TLX:
COM: 33,37

SHERIDAN K V DR
17B/23 THORNTON STREET
DARLING POINT NSW 2027
AUSTRALIA
TEL:
TLX:
COM: 40

SHERWOOD WILLIAM A DR
MPI FUER RADIOASTRONOMIE
AUF DEM HUEGEL 69
D-5300 BONN
GERMANY, F.R.
TEL: 0228-525-362
TLX: 0886440 ASTROD
COM: 27,28,34

SHEVCHENKO VLADISLAV V DR
STERNBERG STATE
ASTRONOMICAL INSTITUTE
UNIVERSITETSKY PROSP. 13
119899 MOSCOW
U.S.S.R.
TEL:
TLX:
COM: 16C,W

SHI GUANG-CHEN
PURPLE MOUNTAIN
OBSERVATORY
NANJING
CHINA, PEOPLE'S REP.
TEL: 33921
TLX: 34144 PMOAS CN
COM: 08,19,24

SHI ZHONG-XIAN
BEIJING ASTRONOMICAL OBS
ACADEMIA SINICA
BEIJING
CHINA, PEOPLE'S REP.
TEL: 28-1698
TLX: 9053
COM: 10

SHIBAHASHI HIROMOTO DR
DEPT OF ASTRONOMY
UNIVERSITY OF TOKYO
BUNKYO-KU
TOKYO 113
JAPAN
TEL: 03-812-2111
TLX: 33659 UTYOSCI J
COM: 35

SHIBASAKI KIYOTO
RES INST OF ATMOSPHERICS
NAGOYA UNIVERSITY
3-13 HONOHRA, TOYOKAWA
AICHI 442
JAPAN
TEL: 5338-6-3154
TLX: 4322310 TYKW J
COM: 10

SHIBATA YUKIO DR
RESEARCH INSTITUTE FOR
SCIENTIFIC MEASUREMENTS
TOHOKU UNIVERSITY
SENDAI 980
JAPAN
TEL:
TLX:
COM: 35

SHIELDS GREGORY A DR
DEPT OF ASTRONOMY
UNIVERSITY OF TEXAS
RLM 15.212
AUSTIN TX 78712
U.S.A.
TEL: 512-471-4461
TLX: 910-874-1351
COM: 28,34,48

SHIM WOON-TAIK PROF
236-53 SINDAN-DONG
JOONG-KU
SEOUL 100
KOREA, REPUBLIC
TEL:
TLX:
COM:

SHIMIZU MIKIO PROF
INST SPACE & ASTRONAUT.
SCIENCES
KOMABA MEGURO KU
TOKYO 153
JAPAN
TEL: 03-467-1111
TLX:
COM: 16,51

SHIMIZU TSUTOMU PROF EMER
26-16 TERADA OHTANI
JYOYO 610-01
JAPAN
TEL: 07745-2-7298
TLX:
COM: 16,33

SHIMMINS ALBERT JOHN
18 PAGE STREET
ALBERT PARK VIC 3206
AUSTRALIA
TEL: 03-6903803
TLX:
COM: 40

SHINE RICHARD A DR
DEPARTMENT 91-30
LOCKHEED P/ALTO RES LAB
3170 PORTER DRIVE
PALO ALTO CA 94304-1211
U.S.A.
TEL: 415-858-4135
TLX:
COM: 10,12,36

SHIPMAN HENRY L DR
PHYSICS DEPT
UNIVERSITY OF DELAWARE
NEWARK DE 19711
U.S.A.
TEL: 302-451-2986
TLX:
COM: 36,46

SHIRYAEV A V DR
LENINGRAD STATE UNIV
ASTRONOMICAL OBSERVATORY
199178 LENINGRAD
U.S.S.R.
TEL:
TLX:
COM:

SHIVANANDAN KANDIAH DR
NAVAL RESEARCH LABORATORY
CODE 4138-S
WASHINGTON DC 20375-5000
U.S.A.
TEL: 202-767-2749
TLX: 202-767-6473
COM: 09,44,47

SHKODROV V G DR
DEPT OF ASTRONOMY
BULGARIAN ACAD SCIENCES
72 LENIN BLVD
1784 SOFIA
BULGARIA
TEL: 7341 x 559
TLX: 23761 ECF BAN BG
COM: 20

SHOBBROOK ROBERT R DR
CHATTERTON ASTRONOMY DEPT
UNIVERSITY OF SYDNEY
SYDNEY NSW 2006
AUSTRALIA
TEL: 61-2-692-3604
TLX: 26169 UNISYD AA
COM: 27,37

SHOEMAKER EUGENE M
BRANCH OF ASTROGEOLOGY
US GEOLOGICAL SURVEY
2255 N. GEMINI DRIVE
FLAGSTAFF AZ 86001
U.S.A.
TEL: 602-527-7181
TLX:
COM: 16

SHOLOMITSKY G B DR
SPACE RESEARCH INSTITUTE
USSR ACADEMY OF SCIENCES
117810 MOSCOW
U.S.S.R.
TEL: 333-31-22
TLX: 411498 STAR SU
COM: 40,44C

SHOR VIKTOR A DR
INST F. THEORET ASTRONOMY
NABEREZHNAYA KUTOZOVA 10
191187 LENINGRAD
U.S.S.R.
TEL:
TLX: 121578
COM: 20

SHORE BRUCE W
LAWRENCE LIVERMORE LAB
LIVERMORE CA 94550
U.S.A.
TEL: 415-447-1100
TLX:
COM: 14

SHORE STEVEN N
NEW MEXICO INSTITUTE OF
MINING & TECHNOLOGY
SOCORRO NM 87801
U.S.A.
TEL: 505-835-5328
TLX:
COM: 28,29

SHOSTAK G SETH DR
KAPTEYN LABORATORIUM
POSTBUS 800
NL-9700 AV GRONINGEN
NETHERLANDS
TEL: 050-116655
TLX: 53572 STARS NL
COM: 28,51

SHTEINS K A DR
LATVIAN STATE UNIVERSITY
ASTRONOMICAL OBSERVATORY
226098 RIGA
U.S.S.R.
TEL:
TLX:
COM: 07,20

SHU FRANK H PROF
ASTRONOMY DEPT
UNIVERSITY OF CALIFORNIA
CAMPBELL HALL
BERKELEY CA 94720
U.S.A.
TEL: 415-642-2529
TLX:
COM: 33,34,42

SHUKLA K
DEPT MATHS & ASTRONOMY
LUCKNOW UNIVERSITY
LUCKNOW U P
INDIA
TEL:
TLX:
COM: 41

SHUKRE C S DR
RAMAN RESEARCH INSTITUTE
SADASHIVANAGAR
BANGALORE 560 080
INDIA
TEL: 360122
TLX: 845671
COM: 48

SHUL'BERG A M DR
ODESSA STATE UNIVERSITY
ASTRONOMICAL OBSERVATORY
270014 ODESSA
U.S.S.R.
TEL: 250356
TLX:
COM: 26,42

SHUL'MAN L M DR
MAIN ASTRONOMICAL OBS
UKRAINIAN ACADEMY OF SCI
252127 KIEV
U.S.S.R.
TEL:
TLX: 131406 SKY SU
COM: 15C

SHULL JOHN MICHAEL
UNIVERSITY OF COLORADO
JILA
BOULDER CO 80309
U.S.A.
TEL: 303-492-7827
TLX:
COM: 34

SHULOV O S DR
LENINGRAD STATE UNIV
ASTRONOMICAL OBSERVATORY
199178 LENINGRAD
U.S.S.R.
TEL:
TLX:
COM:

SHUSTOV BORIS M DR
ASTRONOMICAL COUNCIL
USSR ACADEMY OF SCIENCES
PYATNITSKAYA 48
109017 MOSCOW
U.S.S.R.
TEL: 231-54-61
TLX: 412623 SCSTP SU
COM: 34C,35

SHUTER WILLIAM L H DR
DEPT OF PHYSICS
UNIV OF BRITISH COLUMBIA
6224 AGRICULTURE ROAD
VANCOUVER BC V6T 2A6
CANADA
TEL: 604-228-4269
TLX: 04508576
COM: 33,34,40,51

SIBILLE FRANCOIS
OBSERVATOIRE DE LYON
F-69230 ST-GENIS-LAVAL
FRANCE
TEL: 78-56-07-05
TLX: 310926
COM:

SIDLICHOVSKY MILOS DR
ASTRONOMICAL INSTITUTE
CZECH. ACAD. OF SCIENCES
BUDECSKA 6
120 23 PRAHA 2
CZECHOSLOVAKIA
TEL: PRAHA 258757
TLX: 122486
COM: 07

SIDORENKOV NIKOLAY S
HYDROMETCENTRE OF USSR
123376 MOSCOW
U.S.S.R.
TEL:
TLX:
COM: 19

SIEBER WOLFGANG PH D
MPI FUER RADIOASTRONOMIE
AUF DEM HUEGEL 69
D-5300 BONN 1
GERMANY, F.R.
TEL: 0228-525-317
TLX: 0886440 MPIFR D
COM: 40

SIENKIEWICZ RYSZARD DR
COPERNICUS ASTRON CENTER
UL. BARTYCKA 18
00-716 WARSAW
POLAND
TEL: 411086
TLX: 813878 ZAPAN PL
COM: 35

SIGNORE MONIQUE DR
ECOLE NORMALE SUPERIEURE
RADIOASTRONOMIE
24 RUE LHOMOND
F-75231 PARIS CEDEX 05X
FRANCE
TEL: 1-45-29-12-25
TLX:
COM: 35,41,47,48

SIKORA MAREK
COPERNICUS ASTRON CENTER
UL. BARTYCKA 18
00-716 WARSAW
POLAND
TEL:
TLX:
COM:

SILBERBERG REIN DR
NAVAL RESEARCH LABORATORY
CODE 4154
WASHINGTON DC 20375
U.S.A.
TEL: 202-767-2803
TLX:
COM: 10,34,48

SILK JOSEPH I PROF
ASTRONOMY DEPT
UNIVERSITY OF CALIFORNIA
BERKELEY CA 94720
U.S.A.
TEL: 415-642-2113
TLX: 820181 UCB AST
COM: 34

SILVERBERG ERIC C DR
MCDONALD OBSERVATORY
UNIVERSITY OF TEXAS
PO BOX 1337
FORT DAVIS TX 79734
U.S.A.
TEL:
TLX:
COM: 19

SILVESTRO GIOVANNI
ISTITUTO DI FISICA
UNIVERSITA DI TORINO
CORSO M. D'AZEGLIO 46
I-10125 TORINO
ITALY
TEL: 11-650-8623
TLX: 211041 INFNTO
COM: 34,35,44

SIM MARY E MISS
ROYAL OBSERVATORY
BLACKFORD HILL
EDINBURGH EH9 3HJ
U.K.
TEL: 031-667-3321
TLX: 72383 ROEDIN G
COM: 09

SIMA ZDISLAV DR
ASTRONOMICAL INSTITUTE
CZECH. ACAD. OF SCIENCES
BUDECSKA 6
120 23 PRAHA 2
CZECHOSLOVAKIA
TEL: 42-2-258757
TLX: 66-122486
COM: 07,42

SIMEK MILOS DR
ASTRONOMICAL INSTITUTE
CZECH. ACAD. OF SCIENCES
251 65 ONDREJOV
CZECHOSLOVAKIA
TEL: PRAHA 724525
TLX: 121579
COM: 22

SIMIEN FRANCOIS DR
OBSERVATOIRE DE LYON
AVENUE CHARLES ANDRE
F-69230 ST-GENIS-LAVAL
FRANCE
TEL: 78-56-07-05
TLX: 310926
COM: 28

SINKIN SUSAN M DR
MICHIGAN STATE UNIVERSITY
DEPT PHYSICS & ASTRONOMY
EAST LANSING MI 48824
U.S.A.
TEL: 517-353-4540
TLX:
COM: 28

SIMMONS JOHN FRANCIS L
31 HAVELOCK STREET
GLASGOW G11 5HA
U.K.
TEL:
TLX:
COM: 42

LIST OF MEMBERS

SIMNETT GEORGE M
DEPT OF SPACE RESEARCH
UNIVERSITY OF BIRMINGHAM
BIRMINGHAM B15 2TT
U.K.
TEL:
TLX:
COM: 10

SIMON NORMAN R PROF
BEHLEN LAB OF PHYSICS
UNIVERSITY OF NEBRASKA
LINCOLN NE 68588-0111
U.S.A.
TEL: 402-472-2788
TLX:
COM:

SIMOVLJEVITCH JOVAN L DR
DEPT OF ASTRONOMY
FACULTY OF SCIENCES
STUDENTSKI TRG 16
11000 BEOGRAD
YUGOSLAVIA
TEL: 011-638-715
TLX:
COM:

SINNERSTAD ULF E PROF
STOCKHOLM OBSERVATORY
S-133 00 SALTSJOEBADEN
SWEDEN
TEL: 08-7170195
TLX:
COM: 29,45

SINO CHARLES DR
UNIVERSIDAD DE BARCELONA
FACULDAD DE MATEMATICAS
AV JOSE ANTONIO 585
BARCELONA 7
SPAIN
TEL:
TLX:
COM:

SIMON PAUL A DR
OBSERVATOIRE DE PARIS
SECTION DE MEUDON
F-92195 MEUDON PL CEDEX
FRANCE
TEL: 1-45-34-75-30
TLX:
COM: 40,44

SIMS KENNETH P DR
SYDNEY OBSERVATORY
OBSERVATORY PARK
SYDNEY NSW 2000
AUSTRALIA
TEL:
TLX:
COM: 08,24

SINTON WILLIAM M
UNIVERSITY OF HAWAII
INSTITUTE FOR ASTRONOMY
2680 WOODLAWN DRIVE
HONOLULU HI 96822
U.S.A.
TEL: 808-948-8007
TLX:
COM: 16

SIMODA MAHIRO PROF
DEPT ASTRON/EARTH SCI
TOKYO GAKUGEI UNIVERSITY
KOGANEI
TOKYO 184
JAPAN
TEL: 0423-25-2111
TLX:
COM: 37

SIMON PAUL C DR
INSTITUT AERONOMIE SPAT.
AVENUE CIRCULAIRE 3
B-1180 BRUXELLES
BELGIUM
TEL: 2-375-15-79
TLX: 21563
COM: 44

SINCLAIR ANDREW T DR
ROYAL GREENWICH OBS
HERSTMONCEUX CASTLE
HAILSHAM BN27 1RP
U.K.
TEL: 0323-833171
TLX: 87451 RGOBSY G
COM: 07C,20

SINVHAL SHAMBHU DAYAL DR.
55/2 AMOD PATH
UNIV. OF ROORKEE
ROORKEE 247 667
INDIA
TEL:
TLX:
COM: 27,42

SIMON GEORGE W DR
AFGL-PHS
NATL SOLAR OBSERVATORY
SUNSPOT NM 88349
U.S.A.
TEL: 505-434-1390
TLX:
COM: 12

SIMON RENE L E PROF
INSTITUT D'ASTROPHYSIQUE
UNIVERSITE DE LIEGE
AVENUE DE COINTE 5
B-4200 COINTE-OUGREE
BELGIUM
TEL: 041-52-99-80
TLX: 41264 ASTRLG B
COM: 47

SINGH H P
DEPT OF PHYSICS AND
ASTROPHYSICS
DELHI UNIVERSITY
DELHI 110 007
INDIA
TEL: 324853
TLX:
COM: 51

SINZI AKIRA M DR
HYDROGRAPHIC DEPT
TSUKIJI 5 CHUO KU
TOKYO 104
JAPAN
TEL:
TLX:
COM: 04,31

SIMON GUY
OBSERVATOIRE DE PARIS
SECTION DE MEUDON
F-92195 MEUDON PL CEDEX
FRANCE
TEL: 1-45-34-75-30
TLX:
COM: 10,12

SIMON THEODORE
INSTITUTE FOR ASTRONOMY
UNIVERSITY OF HAWAII
2680 WOODLAWN DRIVE
HONOLULU HI 96822
U.S.A.
TEL: 808-948-8968
TLX: 723-8459 UHAST HR
COM: 36

SINGH JAGDEV DR
INDIAN INSTITUTE OF
ASTROPHYSICS
BANGALORE 560 034
INDIA
TEL: 566-585/566-497
TLX: 0845763 IIAB IN
COM: 12

SION EDWARD MICHAEL
DEPT OF ASTRONOMY
VILLANOVA UNIVERSITY
VILLANOVA PA 19085
U.S.A.
TEL: 215-645-4822
TLX:
COM: 35

SIMON JEAN-LOUIS MR
BUREAU DES LONGITUDES
77 AVE DENFERT-ROCHEREAU
F-75014 PARIS
FRANCE
TEL: 1-43-20-12-10
TLX:
COM: 04,07

SIMONNEAU EDUARDO DR
INSTITUT D'ASTROPHYSIQUE
98 BIS BOULEVARD ARAGO
F-75014 PARIS
FRANCE
TEL: 1-43-20-14-25
TLX:
COM: 36

SINGH PATAN DEEN DR
UNIVERSIDADE DE SAO PAULO
INST ASTRON E GEOFISICO
CAIXA POSTAL 30627
01051 SAO PAULO SP
BRAZIL
TEL: 011-275-3720
TLX: 011-36221 IAGM BR
COM: 34

SIROKY JAROMIR DR
PALACKY UNIVERSITY
DEPT PHYSICS & ASTRONOMY
LENIN STR 26
771 46 OLOMOUC
CZECHOSLOVAKIA
TEL: 22451
TLX:
COM:

SIMON KLAUS PETER
INST F ASTRON & ASTROPHYS
DER UNIVERSITAET MUENCHEN
SCHEINERSTR 1
D-8000 MUENCHEN 80
GERMANY, F.R.
TEL: 089-98-90-21
TLX: 529815 UNIVM D
COM: 36

SIMONS STUART DR
SCHOOL OF MATHEMAT. SCI.
QUEEN MARY COLLEGE
MILE END ROAD
LONDON E1 4NSY
U.K.
TEL: 01-980-4811
TLX:
COM: 34

SINHA K DR
UTTAR PRADESH STATE OBS
MANORA PEAK
NAINITAL 263 129
INDIA
TEL: 2136
TLX: CABLE : ASTRONOMY
COM: 12

SIRY JOSEPH W
4438 42ND STREET N.W.
WASHINGTON DC 20016
U.S.A.
TEL:
TLX:
COM: 07

SIMON MICHAL PROF
DEPT EARTH SPACE SCIENCES
SUNY
STONY BROOK NY 11794
U.S.A.
TEL: 516-246-7672
TLX: 510-228-7767
COM:

SIMONSON S CHRISTIAN DR
1061 RUSSELL AVENUE
LOS ALTOS CA 94022
U.S.A.
TEL: 415-968-0473
TLX:
COM: 33

SINHA RAMESHWAR P
SYSTEMS & APPL SCI CORP
4400 FORBES BLVD
LANHAM MD 20706
U.S.A.
TEL: 301-743-5203
TLX: 317630
COM: 40

SISSON GEORGE M MR
PLANETREES
WALL
HEXHAM NE46 4EQ
U.K.
TEL: 0434-81-434
TLX:
COM:

SISTERO ROBERTO F DR
OBSERVATORIO ASTRONOMICO
LAPRIDA 854
5000 CORDOBA
ARGENTINA
TEL: 40613 - 36876
TLX: 51822 BUCOR
COM: 42,47

SITARSKI GRZEGORZ PROF
CENTER FOR SPACE RESEARCH
UL. BARTYCKA 18
00-716 WARSAW
POLAND
TEL: 410041
TLX: 815670 CBK PL
COM: 20

SITKO MICHAEL L
KITT PEAK NAT OBSERVATORY
PO BOX 26732
950 N. CHERRY AVENUE
TUCSON AZ 85726
U.S.A.
TEL: 602-327-5511
TLX:
COM: 28,34

SITNIK G F PROF
STERNBERG STATE
ASTRONOMICAL INSTITUTE
UNIVERSITETSKIJ PROSP. 13
119899 MOSCOW
U.S.S.R.
TEL: 139-19-73
TLX:
COM: 10,12,36

SITTERLY CHARLOTTE M DR
3711 BRANDYWINE ST N.W.
WASHINGTON DC 20016
U.S.A.
TEL: 202-966-9044
TLX:
COM: 12,14

SIVAN JEAN-PIERRE DR
LAB D'ASTRONOMIE SPATIALE
TRAVERSE DU SIPHON
LES TROIS LUCS
F-13012 MARSEILLE
FRANCE
TEL: 91-66-08-32
TLX: 420584 ASTROSP
COM: 34

SIVARAM C DR
INDIAN INSTITUTE OF
ASTROPHYSICS
BANGALORE 560 034
INDIA
TEL: 566585, 566497
TLX: 845763 IIAB IN
COM: 47,51

SIVARAMAN K R DR
INDIAN INSTITUTE OF
ASTROPHYSICS
BANGALORE 560 034
INDIA
TEL: 566585
TLX: 845-763 IIAB IN
COM: 12C,15

SJOGREN WILLIAM L MR
JPL
MS 264-664
4800 OAK GROVE DRIVE
PASADENA CA 91109
U.S.A.
TEL: 818-354-4868
TLX: 675421
COM: 16

SKALAFURIS ANGELO J
NAVAL RESEARCH LABORATORY
C-5307
WASHINGTON DC 20375-5000
U.S.A.
TEL: 302-767-3227
TLX:
COM:

SKILLING JOHN DR
DEPT APPLIED MATHS
SILVER STREET
CAMBRIDGE CB3 9EW
U.K.
TEL:
TLX:
COM: 34,48

SKUMANICH ANDRE PROF
HIGH ALTITUDE OBSERVATORY
PO BOX 3000
BOULDER CO 80307
U.S.A.
TEL: 303-497-1528
TLX: 45694
COM: 12,36

SLADE MARTIN A III DR
JPL
264-737
4800 OAK GROVE DRIVE
PASADENA CA 91109
U.S.A.
TEL: 818-354-6538
TLX:
COM: 19,40

SLEBARSKI TADEUSZ DR
UNIVERSITY OBSERVATORY
BUCHANAN GARDENS
ST ANDREWS, FIFE KY16 9LZ
U.K.
TEL:
TLX:
COM:

SLEE O B DR
CSIRO
DIVISION OF RADIOPHYSICS
P.O.BOX 76
EPPING NSW 2121
AUSTRALIA
TEL: 868-0222
TLX: 26230 ASTRO
COM: 40

SLETTEBAK ARNE PROF
PERKINS OBSERVATORY
PO BOX 449
DELAWARE OH 43015
U.S.A.
TEL: 614-363-1257
TLX:
COM: 29,33,45

SLONIM E M DR
ASTRONOMICAL INSTITUTE
UZBEKIAN ACADEMY OF SCI
700000 TASHKENT
U.S.S.R.
TEL:
TLX:
COM: 10

SLYSH V I DR
SPACE RESEARCH INSTITUTE
USSR ACADEMY OF SCIENCES
117810 MOSCOW
U.S.S.R.
TEL:
TLX:
COM: 40,51

SMAK JOSEPH I PROF
COPERNICUS ASTRON CENTER
UL. BARTYCKA 18
00-716 WARSAW
POLAND
TEL: 41-00-41
TLX: 813978 ZAPAN PL
COM: 27,29,42P

SMALDONE LUIGI ANTONIO
DIPARTIMENTO DI FISICA
MOSTRA D'OLTREMARE
PAD 19
I-80125 NAPOLI
ITALY
TEL: 81-7253428
TLX: 720320 INFNNA I
COM: 10

SMEYERS PAUL PROF
ASTRONOMISCH INSTITUUT
KATHOLIEKE UNIV LEUVEN
CELESTIJNENLAAN 200B
B-3000 LEUVEN
BELGIUM
TEL: 016-20-06-56
TLX: 25715 KULBI B
COM: 27,35

SMIT J A PROF
PRINCETONPLEIN 5
NL-3584 CC UTRECHT
NETHERLANDS
TEL:
TLX:
COM:

SMITH ALEX G PROF
DEPT OF ASTRONOMY
UNIVERSITY OF FLORIDA
211 SPACE SCI BLDG
GAINESVILLE FL 32611
U.S.A.
TEL: 904-392-6135
TLX:
COM: 40

SMITH ANDREW M DR
CODE 681
GODDARD SPACE FLIGHT CTR
GREENBELT MD 20771
U.S.A.
TEL: 301-344-8648
TLX:
COM:

SMITH BARHAM W DR
LOS ALAMOS NATIONAL LAB
MS D-436
LOS ALAMOS NM 87545
U.S.A.
TEL: 505-667-1585
TLX:
COM: 34,48

SMITH BRADFORD A PROF
DEPT OF PLANETARY SCI
UNIVERSITY OF ARIZONA
TUCSON AZ 85721
U.S.A.
TEL: 602-621-6930
TLX: 910-952-1143
COM: 16C,44,W

SMITH BRUCE F DR
THEORETICAL STUDIES BR.
NASA-AMES RESEARCH CTR
245-3
MOFFETT FIELD CA 94035
U.S.A.
TEL: 415-694-5515
TLX:
COM: 28

SMITH CHARLES DITTO
MUSEUM OF SCE & INDUSTRY
4801 EAST FOWLER AVENUE
TAMPA FL 22617
U.S.A.
TEL: 813-985-5531
TLX:
COM: 09,25

SMITH CLAYTON A JR DR
US NAVAL OBSERVATORY
WASHINGTON DC 20390
U.S.A.
TEL: 202-653-1511
TLX: 710-822-1970
COM: 08C,24

SMITH DEAN F DR
BERKELEY RESEARCH ASSOC.
290 GREEN ROCK DRIVE
BOULDER CO 80302
U.S.A.
TEL: 303-444-1922
TLX:
COM: 10,40,49

SMITH ELSKE V P DR
COLL HUMANITIES/SCIENCES
VIRGINIA COMMONW UNIV
900 PARK AVENUE
RICHMOND VA 23284
U.S.A.
TEL: 804-257-1674
TLX:
COM: 38

SMITH F GRAHAM PROF
DIRECTOR NRAL
JODRELL BANK
MACCLESFIELD SK11 9DL
U.K.
TEL: 0477-71321
TLX: 36149
COM: 19,38V,40,48,50,51

SMITH GEOFFREY DR
DEPT OF ASTROPHYSICS
SOUTH PARKS ROAD
OXFORD OX1 3RQ
U.K.
TEL: 0865-511336
TLX:
COM: 14

SMITH HARDING E JR DR
CASS C-011
UNIVERSITY OF CALIFORNIA
LA JOLLA CA 92093
U.S.A.
TEL: 419-542-4558
TLX:
COM: 28,47

SMITH HARLAN J PROF
ASTRONOMY DEPT
UNIVERSITY OF TEXAS
R.L. MOORE HALL 15.206
AUSTIN TX 78712
U.S.A.
TEL: 512-471-4461
TLX:
COM: 16,27,44,51

SMITH HAYWOOD C DR
DEPT OF ASTRONOMY
211 SPACE SC RES BLDG
UNIVERSITY OF FLORIDA
GAINESVILLE FL 32611
U.S.A.
TEL: 904-392-1079
TLX:
COM: 28

SMITH HORACE A
MICHIGAN STATE UNIVERSITY
DEPT PHYSICS & ASTRONOMY
EAST LANSING MI 48824
U.S.A.
TEL: 517-353-6784
TLX:
COM:

SMITH HOWARD ALAN
SPACE SCIENCE DIVISION
CODE 4138SM
NAVAL RESEARCH LABORATORY
WASHINGTON DC 20375-5000
U.S.A.
TEL: 202-767-3058
TLX:
COM: 34,44,51

SMITH HUMPHRY M
23 NORMANDALE
BEXHILL-ON-SEA TN39 3LU
U.K.
TEL: 0424-214288
TLX:
COM: 19,31

SMITH LINDA J
DEPT PHYSICS & ASTRONOMY
UNIVERSITY COLLEGE LONDON
GOWER STREET
LONDON WC1E 6BT
U.K.
TEL: 01-387-7050x788
TLX: 28722 UCPHYS G
COM: 44

SMITH LINDSEY F DR
DEPT OF PHYSICS
UNIVERSITY OF WOLLONGONG
P.O.BOX 1144
WOLLONGONG NSW 2500
AUSTRALIA
TEL: 042-270-555
TLX: 29022
COM:

SMITH MALCOLM G DR
UK INFRARED TELESCOPE
665 KOMOHANA STREET
HILO HI 96720
U.S.A.
TEL: 808-961-3756
TLX: 708-633-135
COM: 28

SMITH MYRON A ASST PROF
NATL SOLAR OBSERVATORY
BOX 26732
TUCSON AZ 85726-6732
U.S.A.
TEL: 602-327-5511
TLX: 0666484 UURA/NOAO TU
COM: 27C,29C,30

SMITH PETER L DR
CENTER FOR ASTROPHYSICS
MS-50
60 GARDEN STREET
CAMBRIDGE MA 02138
U.S.A.
TEL: 617-495-4984
TLX: 921428 SATELLITE CAM
COM: 14,34

SMITH ROBERT CONNON DR
ASTRONOMICAL CENTRE
UNIVERSITY OF SUSSEX
PHYSICS BLDG, FALMER
BRIGHTON BN1 9QH
U.K.
TEL: 273-606755x3101
TLX: 877159 UNISEX G
COM: 35,42

SMITH WM HAYDEN PROF
MCDONNELL CENTER FOR
SPACE SCIENCES
WASHINGTON UNIVERSITY
ST LOUIS MO 63130
U.S.A.
TEL: 314-889-6574
TLX:
COM: 14

SMOL'KOV GENNADIJ YA DR
SIBIZMIR
P B 4
664697 IRKUTSK 33
U.S.S.R.
TEL:
TLX:
COM: 10,40

SMOLINSKI JAN DR
COPERNICUS ASTRON CENTER
UL. CHOPINA 12/18
87-100 TORUN
POLAND
TEL:
TLX:
COM: 29C

SMOLUCHOWSKI ROMAN PROF
ASTRONOMY DEPT
UNIVERSITY OF TEXAS
RLM HALL 15.314
AUSTIN TX 78712
U.S.A.
TEL: 512-471-1305
TLX: 910-874-1351
COM: 15,16

SMOOT III GEORGE F.
LAWRENCE BERKELEY LAB
UNIVERSITY OF CALIFORNIA
BLDG 50-230
BERKELEY CA 94720
U.S.A.
TEL: 415-486-5237
TLX:
COM: 47

SMRIGLIO FILIPPO PROF
IST. DI ASTRONOMIA
UNIVERSITA LA SAPIENZA
VIA LANCISI 29
I-00161 ROMA
ITALY
TEL: 867525-8442977
TLX:
COM:

SMYLIE DOUGLAS E DR
DEPT EARTH & ATMOSPH SC
YORK UNIVERSITY
4700 KEELE STREET
DOWNSVIEW ONT M3J1P3
CANADA
TEL: 416-667-6430
TLX:
COM: 19,31

SMYTH MICHAEL J DR
DEPT OF ASTRONOMY
ROYAL OBSERVATORY
EDINBURGH EH9 3HJ
U.K.
TEL: 031-667-3321
TLX: 72383
COM: 09,25

SNEDEN CHRISTOPHER A
UNIVERSITY OF TEXAS
DEPT OF ASTRONOMY
AUSTIN TX 78712
U.S.A.
TEL: 512-471-4461
TLX:
COM:

SNELL RONALD L
FIVE COLLEGE RADIO ASTRON
OBS, GRC TOWER B
UNIV OF MASSACHUSETTS
AMHERST MA 01003
U.S.A.
TEL: 413-545-1949
TLX: 955491
COM: 34

SNEZHKO LEONID I
SPECIAL ASTROPHYS OBS
NIZHNIJ ARKHYZ
357147 STAVROPOLSKIJ KRAJ
U.S.S.R.
TEL: 93513
TLX: 297140 ZENIT
COM: 09,36

SNIJDERS MATTHEUS A J
DEPT PHYSICS & ASTRONOMY
UNIVERSITY COLLEGE LONDON
LONDON WC1E 6BT
U.K.
TEL:
TLX:
COM: 36

SNOW THEODORE P PROF
CASS CB 391
UNIVERSITY OF COLORADO
BOULDER CO 80309
U.S.A.
TEL: 303-492-6857
TLX:
COM: 29,34,44

SNYDER LEWIS E
DEPT OF ASTRONOMY
UNIVERSITY OF ILLINOIS
URBANA IL 61801
U.S.A.
TEL: 217-333-5530
TLX: 910-245-2434
COM: 15,51

SOBERMAN ROBERT K DR
FRANKLIN RESEARCH CENTER
ARVIN-CALSPAN
20TH & RACE STREETS
PHILADELPHIA PA 19103
U.S.A.
TEL: 215-448-1058
TLX: 710-670-1889
COM: 21,22

SOBIESKI STANLEY DR
CODE 673
GODDARD SPACE FLIGHT CTR
GREENBELT MD 20771
U.S.A.
TEL:
TLX:
COM: 42

SOBOLEV V V DR
ASTRONOMICAL OBSERVATORY
LENINGRAD UNIVERSITY
199178 LENINGRAD
U.S.S.R.
TEL:
TLX:
COM: 34,36

SOBOLEV VLADISLAV M DR
MAIN ASTRONOMICAL OBS
PULKOVO
196140 LENINGRAD
U.S.S.R.
TEL: 298-22-42
TLX:
COM: 12

SOBOLEVA N S DR
SPECIAL ASTROPHYSICAL OBS
USSR ACADEMY OF SCIENCES
LENINGRAD BRANCH
196140 LENINGRAD
U.S.S.R.
TEL:
TLX:
COM: 40

SOBOUTI YOUSEF PROF
DEPARTMENT OF PHYSICS
SHIRAZ UNIVERSITY
SHIRAZ
IRAN
TEL: 1198-71-57339
TLX:
COM: 28,35,36

SOKOLOWSKI LECH
ASTRONOMICAL OBSERVATORY
JAGIELLONIAN UNIVERSITY
UL. ORLA 171
30-244 KRAKOW
POLAND
TEL: 012-22-38-56
TLX: 0322723 UJ PL
COM: 47

SOLOMON PHILIP M DR
ASTROPHYSICS PROGRAM
DEPT EARTH & SPACE SCI
SUNY AT STONY BROOK
STONY BROOK NY 11794
U.S.A.
TEL: 516-246-8383
TLX: 510-228-7767
COM: 33.34

SORENSEN GUNNAR DR
INSTITUTE OF PHYSICS
LANGELANDSGADE
DK-8000 AARHUS C
DENMARK
TEL:
TLX:
COM: 14

SOCHILINA ALLA S DR
INST TO THEORET ASTRONOMY
USSR ACADEMY OF SCIENCES
10 KUTUZOV QUAY
191187 LENINGRAD
U.S.S.R.
TEL: 278-88-98
TLX: 121578 ITA SU
COM: 04

SOLARIC NIKOLA
GEODETSKI FAKULTET
UNIVERSITY OF ZAGREB
KACICEVA 26
41000 ZAGREB
YUGOSLAVIA
TEL: 041-521-548
TLX:
COM: 08

SOLTAN ANDRZEJ MARIA DR
COPERNICUS ASTRON CENTER
UL. BARTYCKA 18
00-716 WARSAW
POLAND
TEL:
TLX:
COM: 28

SORENSEN SOREN-AKSEL DR
DEPT COMPUTER SCIENCE
UNIVERSITY COLLEGE LONDON
LONDON WC1E 6BT
U.K.
TEL:
TLX:
COM:

SODERBLOM DAVID R
SPACE TELESCOPE SCI INST
3700 SAN MARTIN DRIVE
BALTIMORE MD 21218
U.S.A.
TEL: 301-338-4830
TLX: 6849101 STSCI
COM:

SOLC IVAN DR
ASTRONOMICAL INSTITUTE
GROUP OF OPTICS
DVORAKOWA 298
511 01 TURNOV
CZECHOSLOVAKIA
TEL: 42-2-540395
TLX: 121673 MFF
COM:

SOMERVILLE WILLIAM B DR
DEPT PHYSICS & ASTRONOMY
UNIVERSITY COLLEGE LONDON
GOWER STREET
LONDON WC1E 6BT
U.K.
TEL: 01-382-7050
TLX: 28722
COM: 14.34

SOROCHENKO R L DR
PHYSICAL INSTITUTE
USSR ACADEMY OF SCIENCES
LENINSKY PROSPECT 53
117924 MOSCOW
U.S.S.R.
TEL: 135-01-71
TLX: 411479 NEOD SU
COM: 40

SODERBLOM LARRY DR
BRANCH OF ASTROGEOLOGIC
US GEOLOGICAL SURVEY
2555 NORTH GEMINI DRIVE
FLAGSTAFF AZ 86001
U.S.A.
TEL:
TLX:
COM: 16

SOLC MARTIN
DEPT ASTRONOMY/ASTROPHYS
CHARLES UNIVERSITY PRAGUE
SVEDSKA 8
150 00 PRAHA 5
CZECHOSLOVAKIA
TEL: 02-540395
TLX:
COM: 15,34,41

SONETT CHARLES P PROF
DEPT PLANETARY SCIENCES
UNIVERSITY OF ARIZONA
TUCSON AZ 85721
U.S.A.
TEL: 602-621-6935
TLX: 9109521143
COM: 16,49

SORU-ESCAUT IRINA MRS
OBSERVATOIRE DE PARIS
SECTION DE MEUDON
F-92195 MEUDON PL CEDEX
FRANCE
TEL: 1-45-34-75-30
TLX:
COM:

SOEDERHJELM STAFFAN DR
LUND OBSERVATORY
BOX 43
S-221 00 LUND
SWEDEN
TEL: 46-10 73 03
TLX: 33199 OBSNOT S
COM: 08,42

SOLF JOSEF DR
MPI FUER ASTRONOMIE
KOENIGSTUHL
D-6900 HEIDELBERG 1
GERMANY, F.R.
TEL: 6221-528-226
TLX: 461789 MPIA D
COM:

SONG GUO-XUAN
SHANGHAI OBSERVATORY
ACADEMIA SINICA
SHANGHAI
CHINA. PEOPLE'S REP.
TEL: 386191
TLX: 33164 SHAO CN
COM: 28,33

SOTIROVSKI PASCAL DR
OBSERVATOIRE DE PARIS
SECTION DE MEUDON
F-92195 MEUDON PL CEDEX
FRANCE
TEL: 1-45-34-75-30
TLX: 270912
COM: 10,12

SOFIA SABATINO PROF
YALE UNIVERSITY OBS
PO BOX 6666
NEW HAVEN CT 06511
U.S.A.
TEL: 203-436-3460
TLX: 710-465-3041
COM: 34,35,44,48

SOLHEIM JAN ERIK
INST F. MATEMAT REALFAG
P O BOX 953
N-9001 TROMSO
NORWAY
TEL: 083-86060
TLX: 64124
COM: 42

SONG JIN-AN
P.C. BOX 18
LINTONG
XIAN
CHINA, PEOPLE'S REP.
TEL:
TLX:
COM: 31

SOUFFRIN PIERRE B DR
OBSERVATOIRE DE NICE
BP 139
F-06003 NICE CEDEX
FRANCE
TEL: 93-89-04-20
TLX:
COM: 12,35,36

SOFUE YOSHIAKI PROF
NOBEYAMA RADIO ASTRONOMY
OBSERVATORY
MINAMISAKU-GUN
NAGANO-KEN 384-13
JAPAN
TEL: 267-98-2831
TLX: 3329005 TAONROJ
COM: 40,51

SOLIMAN MOHAMED AHMED
HELWAN OBSERVATORY
HELWAN
EGYPT
TEL:
TLX:
COM:

SONG MU-TAO
PURPLE MOUNTAIN
OBSERVATORY
NANJING
CHINA, PEOPLE'S REP.
TEL:
TLX:
COM: 12

SOULIE GUY
OBSERVATOIRE DE BORDEAUX
F-33270 FLOIRAC
FRANCE
TEL: 56-86-43-30
TLX:
COM:

SOIFER BARUCH T DR
PHYSICS DEPT
CALTECH
DOWNES LAB 320-47
PASADENA CA 91125
U.S.A.
TEL: 818-356-6626
TLX: 675425
COM:

SOLLAZZO CLAUDIO
EUROPEAN SPACE OPERATIONS
CENTER
ROBERT-BOSCH-STR 5
D-6100 DARMSTADT
GERMANY, F.R.
TEL: 06151-8861
TLX: 419453 ESOC D
COM:

SONGSATHAPORN RUANGSAK DR
PHYSICS DEPARTMENT
CHIANG MAI UNIVERSITY
CHIANG MAI 50002
THAILAND
TEL: 221934 X 135
TLX: 43553 UNICHIM TH
COM:

SOUTHWORTH R B DR
HARVARD COLLEGE OBS
60 GARDEN STREET
CAMBRIDGE MA 02138
U.S.A.
TEL:
TLX:
COM: 22

LIST OF MEMBERS

SPADA GIANFRANCO DR
T E S R E
CNR
VIA DE CASTAGNOLI 1
I-40126 BOLOGNA
ITALY
TEL: 51-95-93
TLX: 511350 CNR BO
COM: 44,48

SPAENHAUER ANDREAS MARTIN
ASTRONOMISCHES INSTITUT
UNIVERSITAET BASEL
VENUSSTRASSE 7
CH-4102 BINNINGEN
SWITZERLAND
TEL: 061-227711
TLX:
COM:

SPARKE LINDA
KAPTEYN LABORATORIUM
GRONINGEN UNIVERSITY
POSTBUS 800
NL-9700 AV GRONINGEN
NETHERLANDS
TEL: 050-634056
TLX: 53572 STARS NL
COM: 33

SPARKS WARREN M DR
LOS ALAMOS NATL LAB
MS-F669
LOS ALAMOS NM 87545
U.S.A.
TEL: 505-667-4922
TLX:
COM: 35,42

SPARKS WILLIAM BRIAN
ROYAL GREENWICH OBS
HERSTMONCEUX CASTLE
HAILSHAM BN27 1RP
U.K.
TEL: 03230833171
TLX: 87451
COM: 28

SPARROW JAMES G DR
AERONAUTICAL RESEARCH
LABORATORIES
BOX 4331
MELBOURNE 3001
AUSTRALIA
TEL: 03-647 7623
TLX: 39391 ARL AA
COM: 21

SPASOVA NEDKA MARINOVA
DEPT OF ASTRONOMY
BULGARIAN ACAD SCIENCES
72 LENIN BLVD
1784 SOFIA
BULGARIA
TEL: 7341 x379
TLX:
COM:

SPEER R J DR
DEPT OF PHYSICS
IMPERIAL COLLEGE
PRINCE CONSORT ROAD
LONDON SW7 2BZ
U.K.
TEL: 01-589-5111
TLX: 261503 IMPCOL
COM: 44

SPENCER JOHN HOWARD
NAVAL RESEARCH LABORATORY
CODE 4134
WASHINGTON DC 20375
U.S.A.
TEL: 202-767-3050
TLX:
COM: 40

SPENCER RALPH E DR
NUFFIELD RADIO ASTROMY
LABORATORIES
JODRELL BANK
MACCLESFIELD SK11 9DL
U.K.
TEL: 0477-71-321
TLX: 36149 JODREL G
COM: 40

SPICER DANIEL SHIELDS DR
LAB F.ASTRON & SOLAR PHYS
NASA/GSFC CODE 682
GREENBELT MD 20771
U.S.A.
TEL: 301-344-7334
TLX:
COM: 10,12

SPIEGEL E DR
ASTRONOMY DEPARTMENT
COLUMBIA UNIVERSITY
NEW YORK NY 10027
U.S.A.
TEL:
TLX:
COM: 33,35,36

SPINRAD HYRON PROF
DEPT OF ASTRONOMY
UNIVERSITY OF CALIFORNIA
BERKELEY CA 94720
U.S.A.
TEL: 415-642-2078
TLX:
COM: 15,28

SPITE FRANCOIS M DR
OBSERVATOIRE DE PARIS
SECTION DE MEUDON
F-92195 MEUDON PL CEDEX
FRANCE
TEL: 1-45-34-75-30
TLX: 270 912
COM: 05C,29,36

SPITE MONIQUE DR
OBSERVATOIRE DE PARIS
SECTION DE MEUDON
F-92195 MEUDON PL CEDEX
FRANCE
TEL: 1-45-34-75-30
TLX: 270 912
COM: 29C,36

SPITHAS ELEFTERIOS N DR
DEPT OF ASTRONOMY
UNIVERSITY OF ATHENS
PANEPISTIMIOPOLIS
GR-15771 ATHENS
GREECE
TEL:
TLX:
COM:

SPITZER LYMAN JR DR
PRINCETON UNIVERSITY
OBSERVATORY
PEYTON HALL
PRINCETON NJ 08544
U.S.A.
TEL: 609-452-3809
TLX: 322409
COM: 34,44

SPOELSTRA T A TH DR
NETHERLANDS FOUNDATION
FOR RADIO ASTRONOMY
OUDE HOOGEVEENSEDIJK 4
NL-7991 PD DWINGELOO
NETHERLANDS
TEL: 05219-7244
TLX: 42043 SRZM NL
COM:

SPRUIT HENK C DR
MPI FUER ASTROPHYSIK
KARL-SCHWARZSCHILD-STR 1
D-8046 GARCHING B MUNCHEN
GERMANY, F.R.
TEL: 089-32999420
TLX: 524629 ASTRO D
COM: 10,36

SPYROU NICOLAOS PROF
DEPT OF ASTRONOMY
UNIVERSITY THESSALONIKI
GR-54006 THESSALONIKI
GREECE
TEL: 031-992658
TLX: 412181
COM: 47

SRAMEK RICHARD A DR
NRAO
PO BOX 0
SOCORRO NM 87801
U.S.A.
TEL: 505-772-4011
TLX: 9109881710
COM: 40

SREEKANTAN B V DR
TATA INSTITUTE OF
FUNDAMENTAL RESEARCH
HOMI BHABHA RD
BOMBAY 400 005
INDIA
TEL: 219111
TLX: 011-3009
COM:

SREENIVASAN S RANGA PROF
PHYSICS DEPARTMENT
UNIVERSITY OF CALGARY
2500 UNIVERSITY DR,NW
CALGARY AB T2N 1N4
CANADA
TEL: 403-284-5385
TLX:
COM: 35

SRINIVASAN G
RAMAN RESEARCH INSTITUTE
BANGALORE 560 080
INDIA
TEL:
TLX:
COM:

SRIVASTAVA J B DR
UTTAR PRADESH STATE OBS
MANORA PEAK
NAINITAL 263 129
INDIA
TEL: 2136
TLX:
COM: 42

STABELL ROLF DR
INST THEORET ASTROPHYSICS
UNIVERSITY OF OSLO
P C BOX 1029
N-0315 BLINDERN, OSLO 3
NORWAY
TEL: 2-456-530
TLX: 72705 ASTRO N
COM:

STACHNIK ROBERT V
HARVARD-SMITHSONIAN CTR
FOR ASTROPHYSICS
60 GARDEN STREET
CAMBRIDGE MA 02138
U.S.A.
TEL: 617-495-2829
TLX: 921428 SATELLITE CAM
COM: 44

STAGNI RUGGERO
OSSERVATORIO ASTROFISICO
DI ASIAGO
I-36012 ASIAGO
ITALY
TEL: 0424-62665
TLX: 430110 SETOUR I
COM:

STAHR-CARPENTER M DR
1101 HILL TOP ROAD
CHARLOTTESVILLE VA 22903
U.S.A.
TEL: 804-293-7063
TLX:
COM: 40

STALIO ROBERTO DR
DIPT. DI ASTRONOMIA
UNIVERSITA DI TRIESTE
VIA TIEPOLO 11
I-34131 TRIESTE
ITALY
TEL: 40-793921/221
TLX: 461137 OAT I
COM: 29,36,51

STANDISH E MYLES DR
JET PROPULSION LAB
JPL 264-664
PASADENA CA 91109
U.S.A.
TEL: 818-354-3959
TLX:
COM: 04,07

STANGA RUGGERO
ESO
KARL-SCHWARZSCHILD-STR 2
D-8046 GARCHING B.MUNCHEN
GERMANY, F.R.
TEL:
TLX:
COM: 34

STANGE LOTHAR
TECHNICAL UNIVERSITY
DRESDEN
MOMMSENSTR.13
DDR-8027 DRESDEN
GERMANY, D.R.
TEL: 463-4652
TLX: 02278
COM: 08,24

STAUBERT RUDIGER PROF DR
ASTRONOMISCHES INSTITUT
UNIVERSITAET TUEBINGEN
WALDHAUSERSTR 64
D-7400 TUEBINGEN
GERMANY, F.R.
TEL: 7071-294980
TLX: 7262714 AIT D
COM: 44,48

STEFANOVITCH-GOMEZ A E DR
OBSERVATOIRE DE PARIS
SECTION DE MEUDON
F-92195 MEUDON PL CEDEX
FRANCE
TEL: 1-45-34-75-30
TLX: 201571 F
COM: 33

STEINER JOAO E DR
INST PESQUISAS ESPACIAIS
CAIXA POSTAL 515
AV. DOS ASTRONAUTAS 1758
12200 S. JOSE DOS CAMPOS
BRAZIL
TEL: 0123-22-9977
TLX:
COM:

STANILA GEORGE DR
ASTRONOMICAL OBSERVATORY
CUTITUL DE ARGINT 5
75212 BUCAREST 28
RUMANIA
TEL: 23-68-92
TLX:
COM: 19,31

STAUDE HANS JAKOB PH D
MPI FUER ASTRONOMIE
KOENIGSTUHL
D-6900 HEIDELBERG 1
GERMANY, F.R.
TEL: 06221-528229
TLX: 461739 MPIA D
COM: 21

STEFL VLADIMIR
DEPT THEORET PHYS & ASTRO
FACULTY NATURAL SCIENCES
J.E. PURKYNE UNIVERSITY
611 37 BRNO
CZECHOSLOVAKIA
TEL: 51112
TLX:
COM:

STEINITZ RAPHAEL PROF
PHYSICS DEPT
BEN GOURION UNIVERSITY OF NEGEV
BEERSHEVA 84105
ISRAEL
TEL: 57-70985
TLX:
COM:

STANKEVICH KAZIMIR S DR
RADIOPHYSICAL RESEARCH
INSTITUTE
603600 GORKIJ
U.S.S.R.
TEL: 38-90-91
TLX:
COM:

STAUDE JUERGEN DR
ZNTRLINST. F. ASTROPHYSIK
SONNENOBSERVATORIUM
EINSTEINTURM
DDR-1500 POTSDAM
GERMANY, D.R.
TEL:
TLX:
COM: 12

STEIGER W R PROF
BERNICE P. BISHOP MUSEUM
1525 BERNICE STREET
HONOLULU HI 96817
U.S.A.
TEL: 808-847-3511
TLX:
COM:

STEINLIN ULI PROF
ASTRONOMISCHES INSTITUT
UNIVERSITAET BASEL
VENUSSTRASSE 7
CH-4102 BINNINGEN
SWITZERLAND
TEL: 061-227711
TLX:
COM: 25.33.45

STANLEY G J
PO BOX 1348
CARMEL VALLEY CA 93924
U.S.A.
TEL: 408-659-2940
TLX:
COM: 40

STAWIKOWSKI ANTONI DR
ASTRONOMICAL CENTER
UL. CHOPINA 12/18
87-100 TORUN
POLAND
TEL:
TLX:
COM: 29

STEIGMAN GARY PROF
BARTOL RESEARCH
FOUNDATION
UNIVERSITY OF DELAWARE
NEWARK DE 19716
U.S.A.
TEL: 510-666-3805
TLX:
COM: 47.48

STELLINGWERF ROBERT F DR
MISSION RESEARCH CORP.
1720 RANDOLPH RD SE
ALBUQUERQUE NM 87106
U.S.A.
TEL: 505-843-7200
TLX:
COM: 27,35

STANNARD DAVID DR
NRAL
JODRELL BANK
MACCLESFIELD SK11 9DL
U.K.
TEL: 0477-71321
TLX: 36149 JODREL G
COM: 40

STEBBINS ROBIN
NATL SOLAR OBSERVATORY
SUNSPOT NM 88349
U.S.A.
TEL: 505-434-1394
TLX:
COM: 12

STEIN JOHN WILLIAM
555 HILL STREET
SEWICKLEY PA 15143
U.S.A.
TEL: 412-741-4182
TLX:
COM: 24.51

STELLMACHER GOETZ
INSTITUT D'ASTROPHYSIQUE
98 BIS BOULEVARD ARAGO
F-75014 PARIS
FRANCE
TEL: 1-43-20-14-25
TLX:
COM: 10

STARK ANTONY A
AT & T BELL LABORATORIES
HOH L-231
HOLMDELL NJ 07733
U.S.A.
TEL: 201-949-4842
TLX:
COM:

STECHER THEODORE P
GODDARD SPACE FLIGHT CTR
CODE 680
GREENBELT MD 20771
U.S.A.
TEL: 301-344-8718
TLX:
COM: 29,44,34

STEIN ROBERT F ASSOC PROF
PHYSICS-ASTRONOMY DEPT
MICHIGAN STATE UNIVERSITY
EAST LANSING MI 48824
U.S.A.
TEL: 517-353-8661
TLX:
COM: 36

STELLMACHER IRENE DR
BUREAU DES LONGITUDES
77 AVE DENFERT-ROCHEREAU
F-75014 PARIS
FRANCE
TEL: 1-43-20-12-10
TLX:
COM: 07,20

STARRFIELD SUMNER PROF
DEPT OF PHYSICS
ARIZONA STATE UNIVERSITY
TEMPE AZ 85281
U.S.A.
TEL: 602-965-3561
TLX: 667391 ARIZ ST U TMP
COM: 27,35,42

STECKER FLOYD W DR
HIGH ENERGY ASTROPHYS LAB
NASA/GSFC CODE 660
GREENBELT MD 20771
U.S.A.
TEL: 301-344-6057
TLX:
COM: 33,47

STEIN WAYNE A PROF
SCHOOL OF PHYS & ASTRON
UNIVERSITY OF MINNESOTA
MINNEAPOLIS MN 55455
U.S.A.
TEL: 612-373-9963
TLX:
COM: 28

STENCEL ROBERT EDWARD
CASA
UNIVERSITY OF COLORADO
CAMPUS BOX 391
BOULDER CO 80309
U.S.A.
TEL: 303-492-7178
TLX:
COM: 29,42,44

STASINSKA GRAZYNA DR
OBSERVATOIRE DE PARIS
SECTION DE MEUDON
F-92195 MEUDON PL CEDEX
FRANCE
TEL: 1-45-34-75-70
TLX:
COM:

STEENMAN-CLARK LOIS DR
OBSERVATOIRE DE NICE
BP 139
F-06003 NICE CEDEX
FRANCE
TEL: 93-89-04-20
TLX:
COM: 14

STEINBERG JEAN-LOUIS DR
OBSERVATOIRE DE PARIS
SECTION DE MEUDON
F-92195 MEUDON PL CEDEX
FRANCE
TEL: 1-45-34-75-30
TLX: 204464 F
COM: 40,44

STENFLO JAN O DR
INSTITUT FUER ASTRONOMIE
ETH-ZENTRUM
CH-8092 ZUERICH
SWITZERLAND
TEL: 01-256-3813
TLX: 53178 ETHBI CH
COM: 10,12C

STENHOLM LARS
STOCKHOLM OBSERVATORY
S-133 00 SALTSJOEBADEN
SWEDEN
TEL: 08-7170195
TLX: 12972 SOBSERV S
COM:

STEPANIAN A A DR
CRIMEAN ASTROPHYSICAL OBS
USSR ACADEMY OF SCIENCES
NAUCHNIY
334413 CRIMEA
U.S.S.R.
TEL:
TLX:
COM: 48

STEPANIAN N N DR
CRIMEAN ASTROPHYSICAL OBS
USSR ACADEMY OF SCIENCES
NAUCHNIY
334413 CRIMEA
U.S.S.R.
TEL: 1-86, 5-55
TLX:
COM: 10

STEPANOV V E PROF
SIBIZMIR
P B 4
664697 IRKUTSK 33
U.S.S.R.
TEL:
TLX:
COM: 10,12

STEPHENS S A DR
TATA INSTITUTE OF
FUNDAMENTAL RESEARCH
HOMI BHABHA RD
BOMBAY 400 005
INDIA
TEL: 219111
TLX: 011-3009 TIFR IN
COM:

STEPHENSON C BRUCE PROF
WARNER & SWASEY OBS
CASE WESTERN RESERVE UNIV
CLEVELAND OH 44106
U.S.A.
TEL: 216-368-3728
TLX:
COM: 33,45

STEPHENSON F RICHARD DR
DEPARTMENT OF PHYSICS
UNIVERSITY OF DURHAM
DURHAM DH1 3LE
U.K.
TEL: 0385-64971 x208
TLX: 537351 DURLIB G
COM: 19,41

STEPIEN KAZIMIERZ DR
ASTRONOMICAL OBSERVATORY
AL. UJAZDOWSKIE 4
00-478 WARSAW
POLAND
TEL: 29-40-11
TLX:
COM: 27,36

STEPPE HANS DR
IRAM
AV. DIVINA PASTORA 7
BLOQUE 6/2B
18012 GRANADA
SPAIN
TEL:
TLX:
COM:

STERKEN CHRISTIAAN LEO DR
ASTROPHYSICAL INSTITUT
VRIJE UNIV BRUSSEL
PLEINLAAN 2
B-1050 BRUSSELS
BELGIUM
TEL: 0032-2-6413469
TLX: 61051 VUBCO
COM: 27

STERN ROBERT ALLAN
LOCKHEED P. ALTO RES LAB
DEPT 91-20 / BLDG 255
3251 HANOVER ST
PALO ALTO CA 94304
U.S.A.
TEL: 415-858-4072
TLX:
COM: 44

STESHENKO N V DR
CRIMEAN ASTROPHYSICAL OBS
USSR ACADEMY OF SCIENCES
NAUCHNIY
334413 CRIMEA
U.S.S.R.
TEL: 065-54-32-945
TLX:
COM: 09C,10

STETSON PETER B. DR
49-4061 LARCHWOOD DRIVE
VICTORIA BC V8N 4P1
CANADA
TEL:
TLX:
COM: 37

STEVENS GERARD A DR
LABORATORIUM VOOR
RUIMTEONDERZOEK
BENELUXLAAN 21
NL-3527 HS UTRECHT
NETHERLANDS
TEL:
TLX:
COM:

STEWART JOHN MALCOLM DR
DEPT OF APPLIED MATH &
THEORETICAL PHYSICS
SILVER STREET
CAMBRIDGE CB3 9EW
U.K.
TEL: 223-351-645
TLX:
COM: 47

STEWART PAUL DR
MATHEMATICS DEPT
THE UNIVERSITY
MANCHESTER M13 9PL
U.K.
TEL:
TLX:
COM: 40

STEWART RONALD T MR
CSIRO
DIVISION OF RADIOPHYSICS
P.O.BOX 76
EPPING NSW 2121
AUSTRALIA
TEL: 868-0222
TLX: 26230
COM: 10,40

STEYAERT HERMAN PROF DR
STERRENKUNDIG INSTITUT
KRYGSLAAN 271 S9
B-9000 GENT
BELGIUM
TEL: 91-22-5715x2572
TLX:
COM:

STIBBS DOUGLAS W N PROF
UNIVERSITY OBSERVATORY
BUCHANAN GARDENS
ST ANDREWS, FIFE KY16 9LZ
U.K.
TEL: 0334-76161
TLX: 72613 SAULIB GB
COM: 28,33,35,36

STICKLAND DAVID J DR
SPACE & ASTROPHYS DIV
RUTHERFORD APPLETON LAB
CHILTON, DIDCOT OX1 OQX
U.K.
TEL: 0235-21900
TLX: 83159
COM:

STIER MARK T
PERKIN-ELMER CORPORATION
SPACE SCIENCE DIVISION
MS-897
DANBURY CT 06810
U.S.A.
TEL: 203-797-5708
TLX: 965954
COM: 44

STIFT MARTIN JOHANNES DR
INSTITUT FUER ASTRONOMIE
TUERKENSCHANZSTR 17
A-1180 WIEN
AUSTRIA
TEL: 0222-345360/96
TLX:
COM: 27

STINEBRING DANIEL R
DEPARTMENT OF PHYSICS
PRINCETON UNIVERSITY
PRINCETON NJ 08544
U.S.A.
TEL: 609-452-5578
TLX:
COM:

STIX MICHAEL DR
KIEPENHEUER-INSTITUT
FUER SONNENPHYSIK
SCHOENECKSTR 6
D-7800 FREIBURG IM BR.
GERMANY, F.R.
TEL:
TLX: 7721552 KIS D
COM: 10,12

STOBIE ROBERT S DR
ROYAL OBSERVATORY
BLACKFORD HILL
EDINBURGH EH9 3HJ
U.K.
TEL: 031-667-3321
TLX: 72383 ROEDIN G
COM: 27

STOCK JURGEN D
CENTRO DE INVESTIGACION
DE ASTRONOMIA
APARTADO 264
MERIDA
VENEZUELA
TEL: 074-639930
TLX: 74174
COM: 24C,25,30,45

STOCKMAN HERVEY S JR DR
SPACE TELESCOPE SCI INST
HOMEWOOD CAMPUS
3700 SAN MARTIN DRIVE
BALTIMORE MD 21218
U.S.A.
TEL: 301-338-4820
TLX: 6849101 STSCI UW
COM: 25,44,48

STOCKTON ALAN N DR
INSTITUTE FOR ASTRONOMY
2680 WOODLAWN DR
HONOLULU HI 96822
U.S.A.
TEL:
TLX:
COM:

STODOLKIEWICZ JERZY S DR
COPERNICUS ASTRON CENTER
UL. BARTYCKA 18
00-716 WARSAW
POLAND
TEL: 41-00-41
TLX: 813978
COM:

STOEGER WILLIAM R DR
SPECOLA VATICANA
00120 CITTA DEL VATICANO
VATICAN CITY STATE
TEL: 06-698-3411
TLX: 504-2020 VATOBS VA
COM: 47

STOHL JAN DR
ASTRONOMICAL INSTITUTE
SLOVAK ACAD. OF SCIENCES
842 28 BRATISLAVA
CZECHOSLOVAKIA
TEL: 427-375157
TLX: 093355
COM: 22C

STOKER PIETER H
COSMIC RAY RESEARCH UNIT
POTCHEFSTROOM UNIVERSITY
POTCHEFSTROOM 2520
SOUTH AFRICA
TEL: 27-1481-25360
TLX: 421363
COM: 49

STONE EDWARD C DR
CALTECH 103-33
PASADENA CA 91125
U.S.A.
TEL: 213-356-4516
TLX: 675425
COM: 16

STRAND KAJ AA DR
3200 ROWLAND PL N.W.
WASHINGTON DC 20008
U.S.A.
TEL: 202-966-0495
TLX:
COM: 24,26

STROHMEIER WOLFGANG PROF
VOLKFELDSTR 5
D-8600 BAMBERG
GERMANY, F.R.
TEL: 0951-55394
TLX:
COM: 25,27,42

STRUBLE MITCHELL F
DEPT ASTRON & ASTROPHYS
UNIV OF PENNSYLVANIA
PHILADELPHIA PA 19104
U.S.A.
TEL: 215-243-8176
TLX:
COM: 47

STONE R G DR
LAB FOR EXTRATERR PHYSICS
NASA/GSFC CODE 690
GREENBELT MD 20771
U.S.A.
TEL: 301-344-8631
TLX: 710-82089716
COM: 40,44,49

STRASSL HANS L PROF
ASTRON INST UNIV MUENSTER
DOMAGKSTR 75
D-4400 MUENSTER
GERMANY, F.R.
TEL: 0251-86-24-63
TLX:
COM:

STROM KAREN M
ASTRONOMY PROGRAM
UNIV OF MASSACHUSETTS
GRC 518 B6732
AMHERST MA 07003
U.S.A.
TEL: 413-545-2290
TLX:
COM:

STRUCK-MARCELL CURTIS J
IOWA STATE UNIVERSITY
PHYSICS DEPARTMENT
AMES IA 50011
U.S.A.
TEL: 515-294-5440
TLX:
COM:

STONE REMINGTON P S DR
LICK OBSERVATORY
MOUNT HAMILTON CA 95140
U.S.A.
TEL: 408-274-1809
TLX:
COM: 25,28

STRAZZULLA GIOVANNI
OSSERVATORIO ASTROFISICO
CITTA UNIVERSITARIA
I-95125 CATANIA
ITALY
TEL: 95-330533
TLX: 970359 ASTRCT I
COM:

STROM RICHARD G DR
RADIOSTERREWACHT
POSTBUS 2
NL-7990 AA DWINGELOO
NETHERLANDS
TEL: 5219-7244
TLX: 42043 SRZM NL
COM: 34,40C

STRYKER LINDA L
LICK OBSERVATORY
UNIVERSITY OF CALIFORNIA
SANTA CRUZ CA 95064
U.S.A.
TEL: 408-429-2844
TLX:
COM:

STONE RONALD CECIL
BLACK BIRCH ASTROMETRIC
OBSERVATORY
P O BOX 770
BLEINHEIM
NEW ZEALAND
TEL: 64-057-87164
TLX:
COM: 08,24

STREL'NITSKIJ VLADIMIR DR
ASTRONOMICAL COUNCIL
USSR ACADEMY OF SCIENCES
PYATNITSKAYA UL 48
109017 MOSCOW
U.S.S.R.
TEL: 231-54-61
TLX: 412623 SCSTP SU
COM: 14

STROM ROBERT G PROF
DEPT OF PLANETARY SCI
UNIVERSITY OF ARIZONA
TUCSON AZ 85721
U.S.A.
TEL: 602-621-2720
TLX: 9109521143
COM: 16,28

STUMPFF PETER PROF DR
MPI FUER RADIOASTRONOMIE
AUF DEM HUEGEL 69
D-5300 BONN
GERMANY, F.R.
TEL: 0228-525360
TLX: 886440 MPIFR D
COM:

STOREY JOHN W V DR
DEPT OF PHYSICS
UNIV NEW SOUTH WALES
P.O.BOX 1
KENSINGTON NSW 2033
AUSTRALIA
TEL: 61-2-6974591
TLX: 26054 AA
COM: 09

STRITTMATTER PETER A PROF
STEWARD OBSERVATORY
TUCSON AZ 85721
U.S.A.
TEL: 602-621-6532
TLX:
COM: 35

STROM STEPHEN E
ASTRONOMY PROGRAM
UNIV OF MASSACHUSETTS
GRC 518 B
AMHERST MA 01003
U.S.A.
TEL: 418-545-2290
TLX:
COM: 36

STURCH CONRAD R DR
COMPUTER SCIENCES CORP
SPACE TEL SCIENCE INST
3700 SAN MARTIN DRIVE
BALTIMORE MD 21218
U.S.A.
TEL: 301-338-4856
TLX:
COM: 33

STOYKO ANNA
11 RUE ERNEST CRESSON
F-75014 PARIS
FRANCE
TEL: 1-45-39-56-35
TLX:
COM: 19,31

STROBEL ANDRZEJ DR
INSTITUTE OF ASTRONOMY
N COPERNICUS UNIVERSITY
UL. CHOPINA 12/18
87-100 TORUN
POLAND
TEL: 260-18
TLX: 0552234 ASTR PL
COM: 33,45

STROMGREN BENGT PROF
GAMLE CARLSBERGVEG 15
DK-2500 COPENHAGEN VALBY
DENMARK
TEL: 01-31-12 X25
TLX:
COM: 35,45

STURROCK PETER A PROF
CTR FOR SPACE SCIENCE &
ASTROPHYSICS
STANFORD UNIV. ERL 306
STANFORD CA 94305
U.S.A.
TEL: 415-497-1438
TLX:
COM: 10,48,49,51

STRAFELLA FRANCESCO
DIPARTIMENTO DI FISICA
UNIVERSITA DI LECCE
I-73100 LECCE
ITALY
TEL: 832-627-247
TLX: 860830 UNSTLE I
COM:

STROBEL DARRELL F
DEPT OF EARTH & PLANETARY
JOHNS HOPKINS UNIVERSITY
BALTIMORE MD 21218
U.S.A.
TEL:
TLX:
COM: 16

STRONG IAN B DR
LOS ALAMOS NATIONAL LAB
MS 436
P O BOX 1663
LOS ALAMOS NM 87545
U.S.A.
TEL: 505-667-4823
TLX:
COM: 48

SU BUMEI
YUNNAN OBSERVATORY
P.O. BOX 110
KUNMING
CHINA, PEOPLE'S REP.
TEL: 72946
TLX: 64040 YUOBS CN
COM: 34

STRAIZYS V PROF DR
ASTROPHYSICAL DEPARTMENT
INSTITUTE OF PHYSICS
POZELOS 54
232600 VILNIUS, LITHUANIA
U.S.S.R.
TEL: 73-12-27
TLX:
COM: 25,45C,51

STROBEL WILLI DR
ASTRONOMISCHES
RECHEN-INSTITUT
MOENCHHOFSTR 12-14
D-6900 HEIDELBERG 1
GERMANY, F.R.
TEL: 06-221-49026
TLX:
COM: 20

STRONG JOHN D PROF
ASTRON RESEARCH FACILITY
UNIVERSITY MASSACHUSETTS
AMHERST MA 01003
U.S.A.
TEL:
TLX:
COM: 16

SU DING-QIANG
NANJING ASTRONOMICAL
INSTRUMENT FACTORY
JIANGSU PROVINCE
CHINA, PEOPLE'S REP.
TEL: 41191
TLX:
COM: 09

SU HONG-JUN
PURPLE MOUNTAIN
OBSERVATORY
NANJING
CHINA, PEOPLE'S REP.
TEL: 025-36967
TLX: 34144 PMONT CN
COM: 28

SU WAN-ZHEN
PURPLE MOUNTAIN
OBSERVATORY
NANJING
CHINA, PEOPLE'S REP.
TEL: 33583
TLX: 34144 PMONJ CN
COM: 44

SUBRAHMANYA C R
SCHOOL OF PHYSICS
UNIVERSITY OF SYDNEY
SYDNEY NSW 2006
AUSTRALIA
TEL: 692-2622
TLX: 26169 UNISYD
COM: 47

SUBRAHMANYAM P V DR
DEPT OF ASTRONOMY
OSMANIA UNIVERSITY
HYDERABAD 500 007
INDIA
TEL: 71951 x 247
TLX:
COM: 28

SUDA JAN
ASTRONOMICAL INSTITUTE
CZECH. ACAD. OF SCIENCES
ONDREJOV OBSERVATORY
251 65 ONDREJOV
CZECHOSLOVAKIA
TEL: 724525
TLX: 121579
COM: 10

SUDA KAZUO PROF
ASTRONOMICAL INSTITUTE
TOHOKU UNIVERSITY
ARAMAKI SENDAI 980
JAPAN
TEL: 0222-22-1800
TLX:
COM: 35

SUEMOTO ZENZABURO PROF DR
TOKYO ASTRONOMICAL
OBSERVATORY
OSAWA MITAKA
TOKYO 181
JAPAN
TEL:
TLX:
COM: 10,12

SUESS STEVEN T DR
SPACE SCIENCES LAB
CODE ES 52
NASA/MARSHALL SFC
HUNTSVILLE AL 35812
U.S.A.
TEL: 205-453-2824
TLX:
COM: 49C

SUGAWA CHIKARA DR
HANANCI 1586-25
KASHIWA-SHI
CHIBA-KEN 277
JAPAN
TEL: 0471-33-3825
TLX:
COM: 19

SUGIMOTO DAIICHIRO PROF
DEPT EARTH SCI & ASTRCN
COLL ARTS & SCIENCES
UNIV OF TOKYO, KOMABA
MEGURO-KU TOKYO 153
JAPAN
TEL: 03-467-1171
TLX: 25510 UNITOKYO J
COM: 35P,37,42

SULENTIC JACK W DR
DEPT PHYSICS & ASTRONOMY
UNIVERSITY OF ALABAMA
P O BOX 1921
UNIVERSITY AL 35486
U.S.A.
TEL: 205-348-5050
TLX: 810-729-5845
COM: 28

SULLIVAN DENIS JOHN DR
PHYSICS DEPARTMENT
VICTORIA UNIVERSITY
PRIVATE BAG
WELLINGTON
NEW ZEALAND
TEL: 721000
TLX:
COM: 25

SULLIVAN WOODRUFF T III
DEPT OF ASTRONOMY FM 20
UNIVERSITY OF WASHINGTON
SEATTLE WA 98195
U.S.A.
TEL: 206-543-2888
TLX: 9104240096 WWUI
COM: 28,40,41,51

SULTANOV G F ACAD
SHEMAKA ASTROPHYSICAL OBS
373243 AZERBAIDZAN
U.S.S.R.
TEL:
TLX:
COM: 07,20

SUMMERS HUGH P DR
JET JOINT UNDERTAKING
CULHAM LABORATORY
ABINGDON OX14 3EA
U.K.
TEL: 0235-28822
TLX: 837505 JETEUR G
COM: 14

SUN JIN
DEPT OF ASTRONOMY
BEIJING NORMAL UNIVERSITY
BEIJING
CHINA, PEOPLE'S REP.
TEL: 65-6531, 6285
TLX:
COM: 34

SUN KAI
ASTROPHYSICS DIVISION
GEOPHYSICS DEPARTMENT
BEIJING UNIVERSITY
BEIJING
CHINA, PEOPLE'S REP.
TEL:
TLX: 22239 PKUNI CN
COM: 10

SUN YI-SUI
DEPT OF ASTRONOMY
NANJING UNIVERSITY
NANJING
CHINA, PEOPLE'S REP.
TEL: 37551
TLX: 34151 PRCNU CN
COM: 07

SUN YONGXIANG
INSTITUTE OF GEODESY &
GEOPHYSICS
XU DONG LU
WUHAN
CHINA, PEOPLE'S REP.
TEL:
TLX:
COM: 19

SUNDMAN ANITA DR
STOCKHOLM OBSERVATORY
S-133 00 SALTSJOEBADEN
SWEDEN
TEL: 08-717-06-34
TLX: 12972 SWEDEN
COM: 41,42

SUNTZEFF NICHOLAS B
NATL OPTICAL ASTR OBS
CTIAO
CASILLA 603
LA SERENA
CHILE
TEL: 565-121-3352
TLX: 620301 AURA CT
COM: 29

SUNYAEV R A DR
SPACE RESEARCH INSTITUE
USSR ACADEMY OF SCIENCES
117810 MOSCOW
U.S.S.R.
TEL:
TLX:
COM: 47,48V

SURDEJ JEAN M G
INSTITUT D'ASTROPHYSIQUE
UNIVERSITE DE LIEGE
AVENUE DE COINTE 5
B-4200 COINTE-OUGREE
BELGIUM
TEL: 32-41-529980
TLX: 41264 ASTRLG B
COM: 47

SUTANTYO WINARDI
BOSSCHA OBSERVATORY
LEMBANG, JAVA
INDONESIA
TEL: 6001 LEMBANG
TLX:
COM:

SUTHERLAND PETER G DR
PHYSICS DEPT
MCMASTER UNIVERSITY
HAMILTON ONT L8S 4M1
CANADA
TEL: 416-525-9140
TLX:
COM:

SUZUKI HIROKO
NOBEYAMA RADIO OBS OF
TOKYO ASTRON OBSERVATORY
NOBEYAMA, MINAMISAKU
NAGANO 384-13
JAPAN
TEL: 267-98-2831
TLX: 3329005
COM: 40

SUZUKI YOSHIMASA PROF
23-1 NAKAJIMA
HIRONOMACHI
UJI SHI 611
JAPAN
TEL:
TLX:
COM:

SVALGAARD LEIF DR
WAVERSEBAAN 111
AUD-HEVERLEE
B-3036 LEUVEN
BELGIUM
TEL:
TLX:
COM:

SVECHNIKOV M A DR
ASTRONCMICAL DEPT OF
URALSKIJ STATE UNIV
620083 SVERDLOVSK
U.S.S.R.
TEL:
TLX:
COM: 42

SVENSSON ROLAND
NORDITA
BLEGDAMSVEJ 17
DK-2100 COPENHAGEN
DENMARK
TEL: 00945-1-421616
TLX: 15216 NBI DK
COM:

SVESTKA ZDENEK DR
SPACE RESEARCH LABORATORY
BENELUXLAAN 21
NL-3527 HS UTRECHT
NETHERLANDS
TEL: 030-937145
TLX: 47224 ASTRO NL
COM: 10,12

SVOLOPOULOS SOTIRIOS PROF
DEPT OF ASTROPHYSICS
UNIVERSITY OF ATHENS
PANEPISTIMIOPOLIS
GR-15771 ATHENS
GREECE
TEL:
TLX:
COM: 29,33,41

LIST OF MEMBERS

SVOREN JAN
ASTRONOMICAL INSTITUTE
SLOVAK ACAD. OF SCIENCES
059 60 TATRANSKA LOMNICA
CZECHOSLOVAKIA
TEL: 42-969-967866
TLX: 78277 AU SAV CS
COM: 15,20

SWANENBURG B N DR
SPACE RESEARCH LABORATORY
P.O. BOX 9504
WASSENAARSEWEG 78
NL-2300 RA LEIDEN
NETHERLANDS
TEL: 071-148333
TLX: 39058 ASTRO NL
COM:

SWANK JEAN HEBB
NASA/GSFC
CODE 661
GREENBELT MD 20771
U.S.A.
TEL: 301-344-6188
TLX: 89675 NASCOM GBLT
COM: 48

SWARUP GOVIND PROF
RADIO ASTRONOMY CENTRE
TIFR
POST BOX 8
OOTACAMUND 643 001
INDIA
TEL: 0423-2651, 2032
TLX: 8458488 TIFR IN
COM: 38C,40

SWEET PETER A PROF
DEPT OF ASTRONOMY
THE UNIVERSITY
GLASGOW G12 8QW
U.K.
TEL: 041-339-8855
TLX:
COM: 35

SWEIGART ALLEN V DR
NASA/GSFC
CODE 681
GREENBELT MD 20771
U.S.A.
TEL: 301-344-6274
TLX: 89675
COM: 35

SWENSON GEORGE W JR PROF
ELECT & COMPUTER ENG DEPT
UNIVERSITY OF ILLINOIS
1406 WEST GREEN ST
URBANA IL 61801
U.S.A.
TEL: 217-333-4498
TLX:
COM: 40

SWENSSON JOHN W DR
DEPT OF THEORETICAL PHYS
SOELVEGATAN 14 A
S-223 62 LUND
SWEDEN
TEL: 40-10969686
TLX:
COM: 12,29

SWERDLOW NOEL PROF
UNIVERSITY OF CHICAGO
5640 S. ELLIS AVENUE
CHICAGO IL 60637
U.S.A.
TEL: 312-962-7969
TLX:
COM: 41

SWIHART THOMAS L DR
STEWARD OBSERVATORY
UNIVERSITY OF ARIZONA
TUCSON AZ 85721
U.S.A.
TEL: 602-621-6525
TLX:
COM: 36

SWINGS JEAN-PIERRE DR
INSTITUT D'ASTROPHYSIQUE
UNIVERSITE DE LIEGE
AVENUE DE COINTE 5
B-4200 COINTE-OUGREE
BELGIUM
TEL: 41-52-9980
TLX: 41264 ASTRLG B
COM: 09.14.29

SYKES-HART AVRIL B DR
DEPT OF ASTROPHYSICS
SOUTH PARKS ROAD
OXFORD OX1 3RQ
U.K.
TEL:
TLX:
COM:

SYKORA JULIUS DR
ASTRONOMICAL INSTITUTE
SLOVAK ACAD. OF SCIENCES
SKALNATE PLESO OBS
059 60 TATRANSKA LOMNICA
CZECHOSLOVAKIA
TEL: 0969-967866
TLX: 78277 AUSAV CZ
COM: 10

SYLWESTER JANUSZ
SPACE RESEARCH CENTER
POLISH ACAD OF SCIENCES
UL. KOPERNIKA 11
51-622 WROCLAW
POLAND
TEL: 48 18 01
TLX:
COM: 10

SYNNOTT STEPHEN P
JET PROPULSION LABORATORY
MS 264-686
4800 OAK GROVE DRIVE
PASADENA CA 91109
U.S.A.
TEL: 818-354-6933
TLX:
COM: 16,20

SZABADOS LASZLO PH D
KONKOLY OBSERVATORY
BOX 67
H-1525 BUDAPEST
HUNGARY
TEL: 1-166-426
TLX: 227460 KONOBH
COM: 27

SZAFRANIEC ROZALIA DR
UL. KOPERNIKA 27
31-501 KRAKOW
POLAND
TEL:
TLX:
COM: 42

SZEBEHELY VICTOR G PROF
DEPT AEROSPACE ENGINEERG
UNIVERSITY OF TEXAS
WRW 414
AUSTIN TX 78712
U.S.A.
TEL: 512-471-4239
TLX: 9108741305
COM: 07.33

SZECSENYI-NAGY GABOR DR
DEPT OF ASTRONOMY
LORAND EOTVOS UNIVERSITY
KUN BELA TER 2
H-1083 BUDAPEST
HUNGARY
TEL: 1141019
TLX:
COM: 27.37

SZEIDL BELA DR
KONKOLY OBSERVATORY
BOX 67
H-1525 BUDAPEST
HUNGARY
TEL: 1-366-621
TLX: 227460 KONOB
COM: 27P

SZKODY PAULA DR
DEPT OF ASTRONOMY
UNIVERSITY OF WASHINGTON
SEATTLE WA 98195
U.S.A.
TEL: 206-543-1988
TLX:
COM: 25,27,42

TABARA HIROTO DR
FACULTY OF EDUCATION
UTSUNOMIYA UNIVERSITY
MINEMACHI
UTSUNOMIYA 321
JAPAN
TEL: 0286-36-1515
TLX:
COM: 40

TADEMARU EUGENE DR
UNIV OF MASSACHUSETTS
AMHERST MA 01002
U.S.A.
TEL:
TLX:
COM:

TAFFARA SALVATORE PROF
VIA CALZA 5BIS
I-35128 PADOVA
ITALY
TEL: 049-8071-624
TLX:
COM: 29

TAGLIAFERRI GIUSEPPE PROF
OSSERVATORIO ASTROFISICO
DI ARCETRI
LARGO E. FERMI 5
I-50125 FIRENZE
ITALY
TEL:
TLX:
COM:

TAKADA-HIDAI MASAHIDE DR
RES INST OF CIVILIZATION
TOKAI UNIVERSITY
1117 KITAKANAME
KANAGAWA 259-12
JAPAN
TEL: 0463-58-1211
TLX: 2423402 UNITOK J
COM: 29.51

TAKAGI KOJIRO PROF
DEPT OF PHYSICS
TOYAMA UNIVERSITY
3190 GOFUKU
TOYAMA 930
JAPAN
TEL: 0764-234716
TLX:
COM: 40

TAKAGI SHIGETSUGU DR
DEPT. DE FISICA-CCE-UFRN
CAMPUS UNIVERSITARIO
59000 NATAL-RN
BRAZIL
TEL:
TLX:
COM: 19,31

TAKAHARA FUMIO DR
NOBEYAMA RADIO OBS
TOKYO ASTR OBS/UNIV TOKYO
NOBEYAMA MINAMISAKU
NAGANO 384-13
JAPAN
TEL: 0267-98-2831
TLX: 3329005 TAONRO J
COM: 48

TAKAHARA MARIKO
DEPT OF ASTRONOMY
UNIVERSITY OF TOKYO
2-11-16 YAYOI, BUNKYO-KU
TOKYO 113
JAPAN
TEL: 03-812-2111
TLX: 33659 UTYOSCI
COM: 35

TAKAKUBO KEIYA PROF
ASTRONOMICAL INSTITUTE
TOHOKU UNIVERSITY
ARAMAKI AZA AOBA
SENDAI 980
JAPAN
TEL: 222-22-1800
TLX: 852246 THUCOM J
COM: 34,40

TAKAKURA TATSUO PROF EMER
DEPT OF ASTRONOMY
UNIVERSITY OF TOKYO
BUNKYO-KU
TOKYO 113
JAPAN
TEL: 03-812-2111
TLX: 33659 UTYOSCI J
COM: 10,40,44

TAKARADA KATSUO DR
KYOTO INST OF TECHNOLOGY
MATSUGASAKI SAKYOKU
KYOTO 606
JAPAN
TEL: 075-791-3211
TLX:
COM: 28

TAKASE BUNSHIRO PROF
TOKYO ASTRONOMICAL
OBSERVATORY
OSAWA, MITAKA
TOKYO 181
JAPAN
TEL: 0422-32-5111
TLX: 2822307 TAOMTK
COM: 28

TAKAYANAGI KAZUO PROF
INSTITUTE OF SPACE AND
ASTRONAUTICAL SCIENCES
KOMABA MEGURO KU
TOKYO 153
JAPAN
TEL: 03-467-1111
TLX: 24550 SPACETRY J
COM: 14

TAKEDA HIDENORI DR
DEPT OF AERONAUT. ENGINRG
KYOTO UNIVERSITY
SAKYOKU
KYOTO 606
JAPAN
TEL:
TLX:
COM: 15

TAKENOUCHI TADAO DR
1-28-30 KICHIJYOJI
KITA-MACHI, MUSASHINO
TOKYO 180
JAPAN
TEL:
TLX:
COM:

TAKENS ROELF JAN DR
ASTRONOMICAL INSTITUTE
ROETERSSTRAAT 15
NL-1018 WB AMSTERDAM
NETHERLANDS
TEL: 31-205223009
TLX: 16460 FAC WN
COM:

TAKEUTI MINE DR
ASTRONOMICAL INSTITUTE
TOHOKU UNIVERSITY
ARAMAKI AZA AOBA
SENDAI 980
JAPAN
TEL: 222-22-1800
TLX: 852246 THUCOM J
COM: 27

TALBOT RAYMOND J JR DR
THE AEROSPACE CORPORATION
1927 CURTIS AVENUE
REDONDO BEACH CA 90278
U.S.A.
TEL: 213-379-9927
TLX:
COM: 28

TALON RAOUL DR
CESR
9 AVENUE DU COLONEL ROCHE
BP 4346
F-31029 TOULOUSE CEDEX
FRANCE
TEL:
TLX:
COM: 10

TALWAR SATYA P DR
PHYSICS DEPT
DELHI UNIVERSITY
DELHI 110 007
INDIA
TEL:
TLX:
COM:

TAMENAGA TATSUO DR
FACULTY OF EDUCATION
MIE UNIVERSITY
TSU-SHI
MIE 514
JAPAN
TEL:
TLX:
COM: 10

TAMMANN G ANDREAS PROF DR
ASTRONOMISCHES INSTITUT
UNIVERSITAET BASEL
VENUSSTRASSE 7
CH-4102 BINNINGEN
SWITZERLAND
TEL: 061-227711
TLX:
COM: 27,28V,33,47C

TAMURA SHINICHI DR
DEPT OF ASTRONOMY
TOHOKU UNIVERSITY
ARAMAKI
SENDAI 980
JAPAN
TEL: 222-22-1800
TLX: 852246 THUCOM J
COM: 34

TAN HUISONG
YUNNAN OBSERVATORY
P.O. BOX 110
KUNMING
CHINA. PEOPLE'S REP.
TEL: 72946
TLX: 64040 YUOBS CN
COM: 42

TANABE HIROYOSHI DR
TOKYO ASTRONOMICAL
OBSERVATORY
OSAWA, MITAKA
TOKYO 181
JAPAN
TEL: 0422-32-5111
TLX: 02822307 TAOMTK J
COM: 15,21C

TANAKA KATSUO DR
TOKYO ASTRONOMICAL
OBSERVATORY
OSAWA, MITAKA
TOKYO 181
JAPAN
TEL: 0422-32-5111
TLX: 2822307 TAOMTK J
COM: 10,12

TANAKA RIICHIRO PROF
FACULTY GENERAL EDUCATION
NIIGATA UNIVERSITY
ASAHIMACHIDORI
NIIGATA 951
JAPAN
TEL:
TLX:
COM: 40

TANAKA WATARU DR
DEPT OF ASTRONOMY
UNIVERSITY OF TOKYO
BUNKYO-KU
TOKYO 113
JAPAN
TEL: 03-812-2111
TLX:
COM:

TANAKA YASUO DR
FACULTY OF EDUCATION
IBARAKI UNIVERSITY
BUNKYO
MITO 310
JAPAN
TEL: 292-26-1621x372
TLX:
COM:

TANAKA YASUO PROF
INST SPACE & ASTRONAUT
SCIENCES
4-6-1 KOMABA,MEGURO-KU
TOKYO 153
JAPAN
TEL: 03-467-1111
TLX: J24550 SPACE TKY
COM: 44C

TANDBERG-HANSSEN EINAR A
NASA/MSFC
ES01
HUNTSVILLE AL 35812
U.S.A.
TEL: 205-544-7578
TLX:
COM: 10,12

TANDON JAGDISH NARAIN DR
DEPT OF PHYSICS AND
ASTROPHYSICS
UNIVERSITY OF DELHI
DELHI 110 007
INDIA
TEL:
TLX:
COM: 10

TANDON S N PROF
TATA INSTITUTE OF
FUNDAMENTAL RESEARCH
HOMI BHABHA RD
BOMBAY 400 005
INDIA
TEL: 219111 x 339
TLX: 0113009 TIFR IN
COM: 25

TANG YU-HUA
DEPT OF ASTRONOMY
NANJING UNIVERSITY
NANJING
CHINA. PEOPLE'S REP.
TEL: 37651
TLX: 0909
COM: 10

TANGO WILLIAM J. DR
SCHOOL OF PHYSICS
UNIVERSITY OF SYDNEY
SYDNEY NSW 2006
AUSTRALIA
TEL: 02-692-3953
TLX: 26169 UNISYD AA
COM: 09

TANZELLA-NITTI GIUSEPPE
OSSERVATORIO ASTRONOMICO
DI TORINO
I-10025 PINO TORINESE
ITALY
TEL: 011-841067
TLX: 213236 TO ASTR I
COM:

TANZI ENRICO G
IST. DI FISICA COSMICA
CNR
VIA BASSANI 15
I-20133 MILANO
ITALY
TEL:
TLX:
COM:

TAPIA MAURICIO DR
INSTITUTO DE ASTRONOMIA
UNAM
APDO POSTAL 877
22860 ENSENADA, B. CALIF.
MEXICO
TEL: 4-08-80/8-30-93
TLX: 56739 CICEME
COM: 26

TAPIA-PEREZ SANTIAGO
STEWARD OBSERVATORY
UNIVERSITY OF ARIZONA
TUCSON AZ 85721
U.S.A.
TEL: 602-621-2876
TLX:
COM: 25

TAPLEY BYRON D DR
DEPT AEROSPACE ENGR
AND ENGR MECHANICS
UNIV OF TEXAS, WRW 402
AUSTIN TX 78712
U.S.A.
TEL: 512-471-1356
TLX:
COM: 19

TARADY VLADIMIR K DR
MAIN ASTRON OBSERVATORY
UKRAINIAN ACADEMY OF SCI
GOLOSEEVO
252127 KIEV
U.S.S.R.
TEL: 662286
TLX: 131406 SKY SU
COM: 19

TARAFDAR SHANKAR P DR
TATA INSTITUTE OF
FUNDAMENTAL RESEARCH
HOMI BHABHA RD
BOMBAY 400 005
INDIA
TEL: 219111
TLX: 011-3009 TIFR IN
COM: 34,36

LIST OF MEMBERS

TARENGHI MASSIMO DR
ESO
KARL-SCHWARZSCHILD-STR 2
D-8046 GARCHING B MUNCHEN
GERMANY, F.R.
TEL: 089-32006236
TLX: 52828223 EO D
COM:

TATON RENE PROF
CENTRE ALEXANDRE KOYRE
12 RUE COLBERT
F-75002 PARIS
FRANCE
TEL: 1-42-97-52-45
TLX:
COM: 41

TAYLOR DONALD J DR
DEPT PHYSICS & ASTRONOMY
UNIVERSITY OF NEBRASKA
LINCOLN NB 68588
U.S.A.
TEL: 402-472-3686
TLX:
COM:

TEJFEL VIKTOR G DR
LAB OF LUNAR & PLANETARY
PHYSICS
ASTROPHYSICAL INSTITUTE
480068 ALMA-ATA
U.S.S.R.
TEL: 68-30-53
TLX:
COM: 16C.51,W

TARNSTROM GUY DR
MIT LINCOLN LABORATORY
PO BOX 73
LEXINGTON MA 02173
U.S.A.
TEL: 617-863-5500
TLX: 923355
COM:

TATUM JEREMY B DR
CLIMENHOGA OBSERVATORY
UNIVERSITY OF VICTORIA
VICTORIA BC V8W 2Y2
CANADA
TEL:
TLX:
COM: 14

TAYLOR GORDON E
FIVE FIRS
CINDERFORD LANE, COWBEECH
HAILSHAM BN27 1RP
U.K.
TEL: 0323-833255
TLX:
COM: 04.20

TEKTUNALI H GOKMEN DR
UNIVERSITY OBSERVATORY
UNIVERSITY OF ISTANBUL
ISTANBUL
TURKEY
TEL: 9015223597
TLX:
COM:

TARRAB IRENE
INSTITUT D'ASTROPHYSIQUE
98 BIS BOULEVARD ARAGO
F-75014 PARIS
FRANCE
TEL: 1-43-20-14-25
TLX:
COM:

TAUBER GERALD E PROF
DEPT OF PHYSICS
TEL-AVIV UNIVERSITY
TEL-AVIV 69978
ISRAEL
TEL: 3-420692
TLX: 342171 VERSY IL
COM: 47

TAYLOR JOSEPH H PROF
PRINCETON UNIVERSITY
PHYSICS DEPARTMENT
PRINCETON NJ 08544
U.S.A.
TEL: 609-452-4368
TLX: 4993512
COM:

TELEKI GEORGE DR
ASTRONOMSKA OPSERVATORIJA
VOLGINA 7
11050 BEOGRAD
YUGOSLAVIA
TEL: 11-419-357
TLX:
COM: 08,19

TARTER C BRUCE DR
LAWRENCE LIVERMORE LAB
L-295
UNIVERSITY OF CALIFORNIA
LIVERMORE CA 94550
U.S.A.
TEL: 415-422-4169
TLX:
COM:

TAVAKOL REZA
SCHOOL OF MATHEMAT. SCI.
QUEEN MARY COLLEGE
MILE END ROAD
LONDON E1 4NS
U.K.
TEL:
TLX:
COM: 51

TAYLOR KEITH DR
ROYAL GREENWICH OBS
HERSTMONCEUX CASTLE
HAILSHAM BN27 1RP
U.K.
TEL: 0373-833171
TLX: 87451
COM:

TEMPESTI PIERO PROF
ISTITUTO DI ASTRONOMIA
DELL'UNIVERSITA
VIA G.M. LANCISI 29
I-00161 ROMA
ITALY
TEL: 06-8442977
TLX:
COM: 27

TARTER JILL C DR
NASA-AMES RESEARCH CTR
MS 229-8
MOFFETT FIELD CA 94035
U.S.A.
TEL: 415-694-5727
TLX: 820181
COM: 40,47,51

TAVARES J T L DR
AV DIAS DA SILVA
173 R/C ESQ
3000 COIMBRA
PORTUGAL
TEL:
TLX:
COM:

TAYLOR KENNETH N R PROF
105A COPELAND ROAD
BEECROFT NSW 2119
AUSTRALIA
TEL:
TLX:
COM: 34,46

TENORIO-TAGLE G DR
MPI F PHYSIK & ASTROPHYS
KARL-SCHWARZSCHILD-STR 1
D-8046 GARCHING B MUNCHEN
GERMANY, F.R.
TEL: 089-32990
TLX: 524629 ASTRO D
COM: 34

TASSOUL JEAN-LOUIS PROF
DEPT DE PHYSIQUE
UNIVERSITE DE MONTREAL
CP 6128
MONTREAL PQ H3C 3J7
CANADA
TEL: 514-343-7274
TLX:
COM: 35C

TAWADROS MAHET JACOUB DR
HELWAN OBSERVATORY
HELWAN-CAIRO
EGYPT
TEL: 780 645
TLX: 93070 HIAG
COM: 07

TEDESCO EDWARD F
JET PROPULSION LAB
MS 183-501
4800 OAK GROVE DRIVE
PASADENA CA 91109
U.S.A.
TEL: 818-354-4739
TLX:
COM: 15,22,51

TEPLITSKAYA R B DR
SIBIZMIR
P B 4
664697 IRKUTSK 33
U.S.S.R.
TEL: 6-23-65
TLX:
COM: 12

TASSOUL MONIQUE DR
C/O DEPT DE PHYSIQUE
UNIVERSITE DE MONTREAL
C P 6128
MONTREAL P Q H3C 3J7
CANADA
TEL: 514-343-7274
TLX:
COM: 35

TAYLER ROGER J PROF
ASTRONOMY CENTRE
UNIVERSITY OF SUSSEX
BRIGHTON BN1 9QH
U.K.
TEL: 273-606755
TLX: 877159 UNISEX G
COM: 35,47,48

TEERIKORPI VELI PEKKA DR
TURKU UNIVERSITY OBS
TUORLA
SF-21500 PIIKKIO
FINLAND
TEL: 921-431863
TLX:
COM:

TER HAAR DIRK
MAGDALEN COLLEGE
OXFORD OX1 4AU
U.K.
TEL:
TLX:
COM:

TATEVYAN S K DR
ASTRONOMICAL COUNCIL
USSR ACADEMY OF SCIENCES
PYATNITSKAYA UL 48
109017 MOSCOW
U.S.S.R.
TEL: 231-54-61
TLX: 412623 SCSTP SU
COM: 07

TAYLOR DONALD BOGGIA DR
ROYAL GREENWICH OBS
HERSTMONCEUX CASTLE
HAILSHAM BN27 1RP
U.K.
TEL: 0323-833272
TLX: 87451 RGOBSY G
COM: 07,20

TEHERANY D
83 AVENUE REY
TEHERAN
IRAN
TEL:
TLX:
COM:

TERASHITA YOICHI PROF
KANAZAWA INST TECHNOLOGY
NONOICHI CHO
MINAMI KYOKU
KANAZAWA 921
JAPAN
TEL:
TLX:
COM: 05

TEREBIZH VALERY YU DR
CRIMEAN STATION OF
STERNBERG ASTRON INST
NAUCHNY
334413 CRIMEA
U.S.S.R.
TEL: SIMPHEROPOL 382
TLX:
COM: 28

TERENTJEVA ALEXANDRA K DR
ASTRONOMICAL COUNCIL
USSR ACADEMY OF SCIENCES
PYATNITSKAYA 48
109017 MOSCOW
U.S.S.R.
TEL: 231-54-61
TLX: 412623 SCSTP SU
COM: 22

TERLEVICH ROBERTO JUAN
ROYAL GREENWICH OBS
HERSTMONCEUX CASTLE
HAILSHAM BN27 1RP
U.K.
TEL:
TLX:
COM: 28

TERRAZAS MR L.R.
INAOE
AP POSTALES 216 y 51
72000 PUEBLA,PUE.
MEXICO
TEL:
TLX:
COM:

TERRELL NELSON JAMES JR
ESS-9 MS/D436
LOS ALAMOS NATL LAB
BOX 1663
LOS ALAMOS NM 87545
U.S.A.
TEL: 505-667-2044
TLX:
COM: 48

TERRIEN JEAN
103, RUE DE VERSAILLES
F-92410 VILLE D'AVRAY
FRANCE
TEL: 1-47-09-10-34
TLX:
COM:

TERRILE RICHARD JOHN
JET PROPULSION LABORATORY
MS 183-30
PASADENA CA 91109
U.S.A.
TEL:
TLX:
COM: 16

TERZAN AGOP DR
OBSERVATOIRE DE LYON
F-69230 ST GENIS-LAVAL
FRANCE
TEL: 78-56-07-05
TLX: 310926
COM: 27,37

TERZIAN YERVANT PROF
CORNELL UNIVERSITY
SPACE SCIENCES BLDG
ITHACA NY 14853
U.S.A.
TEL: 607-256-4935
TLX: 932454
COM: 28,34,40,51

TERZIDES CHARALAMBOS DR
DEPT OF ASTRONOMY
UNIVERSITY THESSALONIKI
GR-54006 THESSALONIKI
GREECE
TEL:
TLX:
COM: 33

TESKE RICHARD G PROF
DEPT OF ASTRONOMY
UNIVERSITY OF MICHIGAN
ANN ARBOR MI 48109
U.S.A.
TEL: 313-764-3398
TLX:
COM: 10

TEXEREAU JEAN M
CERGA
AVENUE COPERNIC
F-06130 GRASSE
FRANCE
TEL: 93-36-58-49
TLX:
COM:

THADDEUS PATRICK PROF
INST FOR SPACE STUDIES
2880 BROADWAY
NEW YORK NY 10025
U.S.A.
TEL: 212-678-5621
TLX:
COM: 34,51

THE PIK-SIN PROF
ASTRONOMICAL INSTITUTE
ANTON PANNEKOEK
ROETERSSTRAAT 15
NL-1018 WB AMSTERDAM
NETHERLANDS
TEL: 20-522-3004
TLX: 16460 FACWN NL
COM: 33,34,37

THERNOE KARL-AUGUST
VINKELVEJ 36
DK-2800 LYNGBY
DENMARK
TEL:
TLX:
COM:

THIELEMANN FRIEDRICH-KARL
MPI F PHYS & ASTROPHYSIK
INSTITUT F. ASTROPHYSIK
KARL-SCHWARZSCHILD-STR 1
D-8046 GARCHING B MUNCHEN
GERMANY, F.R.
TEL: 089-3299-0
TLX: 524629 ASTRO D
COM: 35

THIELHEIM KLAUS O DR
ABTEILUNG MATHEM. PHYSIK
UNIVERSITAET KIEL
OLSHAUSENSTR 40/60
D-2300 KIEL
GERMANY, F.R.
TEL: 0431-880-3216
TLX: 292979 IFKKI
COM: 33

THIRY YVES R PROF
UNIVERSITE DE PARIS VI
TOUR 66
4 PLACE JUSSIEU
F-75230 PARIS CEDEX 05
FRANCE
TEL: 1-43-36-25-25
TLX:
COM: 07

THOBURN CHRISTINE
ROYAL GREENWICH OBS
HERSTMONCEUX CASTLE
HAILSHAM BN27 1RP
U.K.
TEL: 0323-833171
TLX: 87451
COM: 08

THOMAS DAVID V DR
SCIENCE RESEARCH COUNCIL
CENTRAL OFFICE
P O BOX 18
SWINDON SN2 1ET WILT
U.K.
TEL:
TLX:
COM: 08,19,24

THOMAS HANS-CHRISTOPH DR
MPI FUER PHYSIK UND
ASTROPHYSIK
KARL-SCHWARZSCHILD-STR 1
D-8046 GARCHING B MUNCHEN
GERMANY, F.R.
TEL:
TLX:
COM: 35

THOMAS JOHN A PROF
PHYSICS (RAAF) DEPT
UNIVERSITY OF MELBOURNE
PARKVILLE VIC 3052
AUSTRALIA
TEL: 3446821
TLX: AA 35185 UNIMEL
COM: 25

THOMAS JOHN H PROF
DEPT MECH & AEROSPACE SCI
UNIVERSITY OF ROCHESTER
ROCHESTER NY 14627
U.S.A.
TEL: 716-275-4083
TLX:
COM: 10,12

THOMAS RICHARD N DR
1155 TIMBERLANE
PINEBROOK HILLS
BOULDER CO 80302
U.S.A.
TEL: 303-443-9290
TLX:
COM: 12,36

THOMAS ROGER J DR
NASA/GSFC
CODE 682
GREENBELT MD 20771
U.S.A.
TEL: 301-344-7921
TLX:
COM: 10,44

THOMASSON PETER DR
NRAL
JODRELL BANK
MACCLESFIELD SK11 9DL
U.K.
TEL: 0477-71321
TLX: 36149
COM: 40

THOMPSON A RICHARD DR
NRAO
VLBA PROJECT
2015 IVY ROAD
CHARLOTTESVILLE VA 22903
U.S.A.
TEL: 804-296-0211
TLX:
COM: 34,40

THOMPSON G I DR
ROYAL OBSERVATORY
EDINBURGH EH9 3HJ
U.K.
TEL:
TLX:
COM: 29

THOMPSON KEITH DR
PHYSICS DEPT
MONASH UNIVERSITY
WELLINGTON ROAD
CLAYTON, VICT. 3168
AUSTRALIA
TEL: 03-541-3639
TLX: 32691 AA
COM: 42

THOMPSON LAIRD A DR
INSTITUTE FOR ASTRONOMY
2680 WOODLAWN DRIVE
HONOLULU HI 96822
U.S.A.
TEL: 808-948-8102
TLX: 723-8459 UHAST HR
COM: 28,47

THOMPSON RODGER I PROF
STEWARD OBSERVATORY
UNIVERSITY OF ARIZONA
TUCSON AZ 85721
U.S.A.
TEL: 602-621-6527
TLX: 467175
COM:

THOMPSON THOMAS WILLIAM
2043 CLOUD CREST AVE
LA CRESCENTA CA 91214
U.S.A.
TEL:
TLX:
COM: 16

THOMSEN BJARNE B LECT
INSTITUTE OF ASTRONOMY
UNIVERSITY OF AARHUS
DK-8000 AARHUS C
DENMARK
TEL:
TLX:
COM:

THONNARD NORBERT DR
DEPT TERREST. MAGNETISM
CARNEGIE INST. WASHINGTON
5241 BROAD BRANCH RD N.W.
WASHINGTON DC 20015
U.S.A.
TEL: 202-966-0863
TLX: 440427 MAGN UI
COM: 28,34

THOREN VICTOR E PROF
130 GOODBODY HALL
INDIANA UNIVERSITY
BLOOMINGTON IN 47401
U.S.A.
TEL: 812-825-5970
TLX:
COM: 41

THORNE KIP S PROF
CALTECH 130-33
PASADENA CA 91125
U.S.A.
TEL: 213-356-4598
TLX: 675425
COM: 48

THORSTENSEN JOHN R
DARTMOUTH COLLEGE
DEPT PHYSICS & ASTRONOMY
HANOVER NH 03755
U.S.A.
TEL: 603-646-2869
TLX:
COM:

THRONSON HARLEY ANDREW JR
DEPT PHYSICS & ASTRONOMY
UNIVERSITY OF WYOMING
LARAMIE WY 82071
U.S.A.
TEL: 307-766-6150
TLX:
COM: 34

THUAN TRINH XUAN DR
DEPT OF ASTRONOMY
UNIVERSITY OF VIRGINIA
BOX 3818 UNIV STATION
CHARLOTTESVILLE VA 22903
U.S.A.
TEL: 804-924-4894
TLX:
COM: 28,47

THUM CLEMENS DR
IRAM
AV. DIVINA PASTORA 7
BLOQUE 6/2B
18012 GRANADA
SPAIN
TEL: 958-480413
TLX: 78521 IRAM E
COM: 40

TIFFT WILLIAM G PROF
STEWARD OBSERVATORY
UNIVERSITY OF ARIZONA
TUCSON AZ 85721
U.S.A.
TEL: 602-621-6532
TLX: 467175
COM: 28,47

TIFREA EMILIA DR
OBSERVATOIRE DE BUCAREST
CUTITUL DE ARGINT 5
75212 BUCAREST 28
RUMANIA
TEL: 23-60-10
TLX: 09-26-29
COM: 10

TIMOTHY J GETHYN DR
CTR FOR SPACE SCIENCE &
ASTROPHYSICS
STANFORD UNIV., ERL 314
STANFORD CA 94305
U.S.A.
TEL: 415-497-0059
TLX: 348402 STANFRD STNU
COM:

TINBERGEN JAAP DR
KAPTEYN STERREWACHT
WERKGROEP
MENSINGHEWEG 20
NL-9301 KA RODEN
NETHERLANDS
TEL: 31-5908-19631
TLX: 53767 KSWRO NL
COM: 25C

TING YEOU-TSWEN
ASTRONOMY SECTION
CENTRAL WEATHER BUREAU
64 KUNG YUEN ROAD
TAIPEI 100
CHINA, TAIWAN
TEL: 3713181-281
TLX:
COM: 04

TIURI MARTTI PROF
HELSINKI UNIV TECHNOLOGY
RADIO LABORATORY
OTAKAARI 5 A
SF-02150 ESPOO 15
FINLAND
TEL: 358-0-451-2545
TLX: 122771 RORTA SF
COM:

TJIN-A-DJIE HERMAN R E DR
KOEKOELAAN 106
NL-1403 EJ BUSSUM
NETHERLANDS
TEL: 31-2159-17076
TLX: 16460 FACWN NL
COM: 27,35

TLAMICHA ANTONIN DR
ASTRONOMICAL OBSERVATORY
CZECH. ACAD. OF SCIENCES
251 65 ONDREJOV
CZECHOSLOVAKIA
TEL: 72-45-25
TLX: 121579 ASTR CZ
COM: 10,40

TOBIN WILLIAM
OBSERVATOIRE DE MARSEILLE
2 PLACE LE VERRIER
F-13248 MARSEILLE CEDEX
FRANCE
TEL: 91-95-90-88
TLX: 420241 F
COM: 33

TODORAN IOAN DR
ASTRONOMICAL OBSERVATORY
STR CIRESILOR 19
3400 CLUJ NAPOCA
RUMANIA
TEL:
TLX:
COM: 25,42

TOFANI GIANNI PROF
OSSERVATORIO ASTROFISICO
DI ARCETRI
LARGO E. FERMI 5
I-50125 FIRENZE
ITALY
TEL: 35-22-00-34
TLX: 572268 ARCETRI
COM: 40

TOHLINE JOEL EDWARD
DEPT PHYSICS & ASTRONOMY
LOUISIANA STATE UNIV.
BATON ROUGE LA 70803
U.S.A.
TEL: 504-388-6851
TLX: 559184
COM: 35

TOLBERT CHARLES R DR
LEANDER MCCORMICK OBS
BOX 3818
UNIVERSITY STATION
CHARLOTTESVILLE VA 22903
U.S.A.
TEL: 804-924-7494
TLX:
COM: 25,38C,40,51

TOMASI PAOLO DR
LAB. DI RADIOASTRONOMIA
VIA IRNERIO 46
I-40126 BOLOGNA
ITALY
TEL:
TLX:
COM: 40

TOMASKO MARTIN G DR
LUNAR & PLANETARY LAB
UNIVERSITY OF ARIZONA
SPACE SCIENCES BLDG
TUCSON AZ 85721
U.S.A.
TEL: 602-621-6969
TLX:
COM:

TOMBAUGH CLYDE W PROF
DEPT OF ASTRONOMY
NEW MEXICO STATE UNIV
BOX 4500
LAS CRUCES NM 88003
U.S.A.
TEL: 505-646-2107
TLX:
COM: 16

TOMIMATSU AKIRA DR
DEPT OF PHYSICS
NAGOYA UNIVERSITY
NAGOYA 464
JAPAN
TEL:
TLX:
COM: 47,48

TOMITA KENJI PROF
RES INST FOR THEORET PHYS
HIROSHIMA UNIVERSITY
TAKEHARA 725
JAPAN
TEL: 08462-2-2362
TLX:
COM: 28,47

TOMITA KOICHIRO MR
11-20 4 CHOME YOGA
SETAGAYAKU
TOKYO 158
JAPAN
TEL: 037000066
TLX:
COM: 15,20,22C

TOMOV ALEXANDER NIKOLOV
DEPT OF ASTRONOMY
BULGARIAN ACAD SCIENCES
72 LENIN BLVD
1784 SOFIA
BULGARIA
TEL:
TLX:
COM:

TONG FU
PURPLE MOUNTAIN
OBSERVATORY
NANJING
CHINA, PEOPLE'S REP.
TEL: 33921
TLX: 34144 PMOAS CN
COM: 07

TONG YI
DEPARTMENT OF ASTRONOMY
BEIJING NORMAL UNIVERSITY
19 XINJISKOW OUT-STREET
BEIJING
CHINA, PEOPLE'S REP.
TEL: 656531-6285
TLX:
COM: 28,33

TOOMRE ALAR DR
MIT
ROOM 2-371
77 MASSACHUSETTS AVE
CAMBRIDGE MA 02139
U.S.A.
TEL: 617-253-4326
TLX:
COM: 28C,33

TOOMRE JURI
DEPT ASTRO-GEOPHYSICS
JILA
UNIVERSITY OF COLORADO
BOULDER CO 80309
U.S.A.
TEL: 303-492-7854
TLX:
COM: 33,35,36

LIST OF MEMBERS

TOPAKTAS LATIF A DR
KING SAUD UNIVERSITYRY
COLLEGE OF SCIENCE
P.C. BOX 2455
RIYADH 11453
SAUDI ARABIA
TEL:
TLX:
COM:

TORROJA J PROF
CATEDRA DE ASTRONOMIA
FACULTAD DE CIENCIAS
UNIVERSIDAD COMPLUTENSE
28040 MADRID
SPAIN
TEL:
TLX:
COM:

TRAFTON LAURENCE M DR
ASTRONOMY DEPT
UNIVERSITY OF TEXAS
AT AUSTIN
AUSTIN TX 78712
U.S.A.
TEL: 512-471-1476
TLX:
COM: 16

TREHAN SURINDAR K PROF
DEPT OF MATHEMATICS
PANJAB UNIVERSITY
CHANDIGARH 160 014
INDIA
TEL: 29938
TLX:
COM:

TORAO MASAHISA
5-7-9-401 YOYOGI
SHIBUYA
TOKYO 151
JAPAN
TEL:
TLX:
COM: 19

TOSA MAKOTO DR
ASTRONOMICAL INSTITUTE
TOHOKU UNIVERSITY
SENDAI 980
JAPAN
TEL: 0222-22-1800
TLX: 852246 THUCOM J
COM: 33C

TRAN MINH NGUYET DR
OBSERVATOIRE DE PARIS
SECTION DE MEUDON
F-92195 MEUDON PL CEDEX
FRANCE
TEL: 1-45-34-75-70
TLX: 270912 OBSASTR
COM:

TRELLIS MICHEL DR
OBSERVATOIRE DE NICE
BP 139
F-06003 NICE CEDEX
FRANCE
TEL: 93-89-04-20
TLX:
COM: 10

TORELLI M DR
OSSERVATORIO ASTRONOMICO
VIA DEL PARCO MELLINI 84
I-00136 ROMA
ITALY
TEL: 347056
TLX:
COM: 12

TOSI MONICA
OSSERVATORIO ASTRONOMICO
UNIVERSITARIO
C P 596
I-40100 BOLOGNA
ITALY
TEL: 51-222956
TLX: 211664 INFN 30 I
COM: 34

TRAUB WESLEY ARTHUR
CENTER FOR ASTROPHYSICS
60 GARDEN STREET
CAMBRIDGE MA 02138
U.S.A.
TEL: 617-495-7406
TLX: 921428 SATELLITE CAM
COM: 09,44

TREMAINE SCOTT DUNCAN
CITA, MC LENNAN LABSO
UNIVERSITY OF TORONTO
60 ST GEORGE STREET
TORONTO M5S 1A1
CANADA
TEL: 416-978-6879
TLX:
COM: 28,47

TORNAMBE AMEDEO
IST. ASTROFISICA SPAZIALE
C P 67
I-00044 FRASCATI
ITALY
TEL:
TLX:
COM:

TOUSEY RICHARD DR
NAVAL RESEARCH LABORATORY
CODE 7140
WASHINGTON DC 20375
U.S.A.
TEL: 202-767-3441
TLX:
COM: 12,14

TRAVING GERHARD PROF
INSTITUT FUER
THEORETISCHE ASTROPHYSIK
NEUENHEIMER FELD 561
D-6900 HEIDELBERG
GERMANY, F.R.
TEL: 06221-562815
TLX: 461515
COM: 36

TREMKO JOZEF DR
ASTRONOMICAL INSTITUTE
SLOVAK ACAD. OF SCIENCES
SKALNATE PLESO OBS
059 60 TATRANSKA LOMNICA
CZECHOSLOVAKIA
TEL: 967866
TLX: 78277 AUSAV CZ
COM: 27,42,50C

TOROSHLIDZE TEIMURAZ I DR
ABASTUMANI ASTROPHYSICAL
OBSERVATORY
383762 ABASTUMANI,GEORGIA
U.S.S.R.
TEL:
TLX:
COM: 21

TOVMASSIAN H M DR
BYURAKAN ASTROPHYSICAL
OBSERVATORY
375433 ARMENIA
U.S.S.R.
TEL: 56-34-53
TLX:
COM: 28,40,44,51

TREDER H J PROF DR
ZNTRLINST. F. ASTROPHYSIK
STERNWARTE BABELSBERG
ROSA-LUXEMBURG-STR 17A
DDR-1502 POTSDAM
GERMANY, D.R.
TEL: 762225
TLX: 15471 ADW RZB DO
COM: 28,47

TREUMANN RUDOLF A. DR
MPI F. PHYS & ASTROPHYSIK
INST F EXTRATERR PHYSIK
D-8046 GARCHING B MUNCHEN
GERMANY, F.R.
TEL: 89-3299831
TLX: 5215845 XTER D
COM: 10

TORRES CARLOS ALBERTO DR
OBSERVATORIO NACIONAL/LNA
RUA CORONEL RENNO 07
CAIXA POSTAL 21
37500 ITAJUBA MG
BRAZIL
TEL: 035-622-0788
TLX: 031-2603
COM: 27,50C

TOWNES CHARLES HARD DR
DEPT OF PHYSICS
UNIVERSITY OF CALIFORNIA
RM 557 BIRGE HALL
BERKELEY CA 94720
U.S.A.
TEL: 415-642-1128
TLX:
COM: 34,40,51

TREFFERS RICHARD R
ASTRONOMY DEPT
UNIVERSITY OF CALIFORNIA
BERKELEY CA 94720
U.S.A.
TEL: 415-642-4223
TLX:
COM: 34

TREVESE DARIO
OSSERVATORIO ASTRONOMICO
VIA DEL PARCO MELLINI 84
I-00136 ROMA
ITALY
TEL: 6-347-056
TLX:
COM: 47

TORRES CARLOS DR
OBS ASTRONOMICO NACIONAL
UNIVERSIDAD DE CHILE
CASILLA 36-D
SANTIAGO
CHILE
TEL: 56-2-229-4101
TLX: 440001 ITT BOOTH FOR
COM: 20,50

TOZER DAVID C DR
SCHOOL OF PHYSICS
UNIVERSITY OF NEWCASTLE
NEWCASTLE/TYNE NE1 7RU
U.K.
TEL:
TLX:
COM:

TREFFTZ ELEONORE E DR
MPI F PHYSIK UND
ASTROPHYSIK
KARL-SCHWARZSCHILD-STR 1
D-8046 GARCHING B MUNCHEN
GERMANY, F.R.
TEL: 89-32990
TLX: 524629 ASTRO D
COM: 14

TREXLER JAMES H MR
NAVAL RESEARCH LAB
OXON HILL MD 20745
U.S.A.
TEL: 202-767-3305
TLX:
COM:

TORRES-PEIMBERT SILVIA DR
INSTITUTO DE ASTRONOMIA
UNAM
APDO POSTAL 70-264
04510 MEXICO DF
MEXICO
TEL: 905-548-5306
TLX: 1760155 CIC ME
COM: 34

TOZZI GIAN PAOLO
OSSERVATORIO ASTROFISICO
DI ARCETRI
LARGO E. FERMI 5
I-50125 FIRENZE
ITALY
TEL: 55-220034
TLX: 572268 ARCETR I
COM: 14

TREFZGER CHARLES F DR
ASTRONOMISCHES INSTITUT
UNIVERSITAET BASEL
VENUSSTRASSE 7
CH-4102 BINNINGEN
SWITZERLAND
TEL: 061-22-77-11
TLX:
COM: 33

TRIMBLE VIRGINIA L DR
DEPT OF PHYSICS
UNIVERSITY OF CALIFORNIA
IRVINE CA 92717
U.S.A.
TEL: 714-856-6948
TLX:
COM: 28,35,42,47C,48C,51

TRINCHIERI GINEVRA
OSSERVATORIO ASTROFISICO
DI ARCETRI
LARGO E. FERMI 5
I-50125 FIRENZE
ITALY
TEL: 39-55-220034
TLX: 572268 ARCETR I
COM: 28

TROLAND THOMAS HUGH
PHYSICS-ASTRONOMY DEPT
UNIVERSITY OF KENTUCKY
LEXINGTON KY 40506
U.S.A.
TEL: 606-257-8620
TLX:
COM: 40

TSAO MO PROF
NO 47 SEC 3
HSIN-I ROAD
TAIPEI 106
CHINA, TAIWAN
TEL: 02-7047795
TLX:
COM: 19,31

TSUBOKAWA IETSUNE DR
INTERNATIONAL LATITUDE
OBSERVATORY
MIZUSAWA
IWATE 023
JAPAN
TEL: 0197247111
TLX: 837628
COM: 19

TRIPATHI B M DR
UTTAR PRADESH STATE OBS
MANORA PEAK
NAINITAL 263 129
INDIA
TEL: 2136
TLX: CABLE : ASTRONOMY
COM: 12

TROTTET GERARD DR
OBSERVATOIRE DE PARIS
SECTION DE MEUDON
DASOP
F-92195 MEUDON PL CEDEX
FRANCE
TEL: 1-45-34-75-30
TLX:
COM: 10

TSAP T T DR
CRIMEAN ASTROPHYS OBS
USSR ACADEMY OF SCIENCES
NAUCHNIY
334413 CRIMEA
U.S.S.R.
TEL: 132
TLX:
COM: 12

TSUCHIYA ATSUSHI DR PROF
TOKYO ASTRONOMICAL OBS
OSAWA
MITAKA
TOKYO 181
JAPAN
TEL: 0422-32-5111
TLX: 02822307 TAOMTK J
COM: 31

TRITAKIS BASIL P DR
ASTRONOMY & APPL MATHS
ACADEMY OF ATHENS
14 ANAGNOSTOPOULOU
GR-10673 ATHENS
GREECE
TEL: 1-3613589
TLX:
COM: 10,49

TRUEMPER JOACHIM PROF
MPI F EXTRATERRESTRISCHE
PHYSIK
D-8046 GARCHING B MUNCHEN
GERMANY, F.R.
TEL: 089-3299559
TLX: 5215845 XTER D
COM: 44C,48C

TSCHARNUTER WERNER M DR
UNIVERSITAETS-STERNWARTE
TUERKENSCHANZSTR 17
A-1180 WIEN
AUSTRIA
TEL: 0222-345360x4
TLX:
COM:

TSUJI TAKASHI
TOKYO ASTRONOMICAL
OBSERVATORY
MITAKA
TOKYO 181
JAPAN
TEL: 0422-32-5111
TLX: 02822307 TAOMTK J
COM: 29,36

TRITTON KEITH P DR
APARTADO DE CORREOS 321
SANTA CRUZ DE LA PALMA
38071 TENERIFE
SPAIN
TEL: 3422-414148
TLX: 92757 URLOE
COM: 40,46

TRULSEN JAN K PROF
UNIVERSITY OF TROMSO
P O BOX 953
N-9001 TROMSO
NORWAY
TEL:
TLX:
COM:

TSESEVICH V P PROF DR
ODESSA STATE UNIVERSITY
ASTRONOMICAL OBSERVATORY
270014 ODESSA
U.S.S.R.
TEL:
TLX:
COM: 27,42

TSURUTA SACHIKO DR
DEPT OF PHYSICS
UNIVERSITY OF TOKYO
C/O S.ISHIMARU
TOKYO
JAPAN
TEL:
TLX:
COM: 40,48

TRITTON SUSAN BARBARA
ROYAL OBSERVATORY
BLACKFORD HILL
EDINBURGH EH9 3HJ
U.K.
TEL: 031-667-3321
TLX: 72383 ROEDIN G
COM: 05

TRURAN JAMES W JR
DEPT OF ASTRONOMY
UNIVERSITY OF ILLINOIS
URBANA IL 61801
U.S.A.
TEL: 217-333-3090
TLX:
COM: 35C,48

TSEYTLIN NAUM M
RADIOPHYSICAL RES INST
LYADOV STR. 25/14
603600 GORKY
U.S.S.R.
TEL: 36-01-29
TLX: 1113 LUNA
COM:

TSVETANOV ZLATAN IVANOV
DEPT OF ASTRONOMY
BULGARIAN ACAD SCIENCES
72 LENIN BLVD
1784 SOFIA
BULGARIA
TEL: 7341 x379
TLX: 23561 ECF BAN BG
COM:

TROCHE-BOGGINO A E DR
INST DE CIENCIAS BASICAS
UNIV NACIONAL DE ASUNCION
C.CORREO 1039-1804
ASUNCION
PARAGUAY
TEL:
TLX:
COM: 46

TRUSSONI EDOARDO
IST. DI COSMOGEOFISICA
DEL CNR
CORSO FIUME 4
I-10133 TORINO
ITALY
TEL: 011-657694/8979
TLX: 211041 INFNTO
COM:

TSIKOUDI VASSILIKI PH D
DEPT OF PHYSICS
DIV OF ASTRO-GEOPHYSICS
UNIVERSITY OF IOANNINA
GR-45332 IOANNINA
GREECE
TEL: 0651-91084
TLX:
COM:

TSVETKOV MILCHO K DR
DEPT OF ASTRONOMY
BULGARIAN ACAD SCIENCES
72 LENIN BLVD
1784 SOFIA
BULGARIA
TEL: 758927
TLX: 23561 ECF BAN BG
COM: 27

TRODAHL HARRY JOSEPH DR
VICTORIA UNIVERSITY
PRIVATE BAG
WELLINGTON
NEW ZEALAND
TEL: 721-000
TLX:
COM: 25

TRUTSE YU L DR
INST PHYSICS OF ATMOSPH
USSR ACADEMY OF SCIENCES
109017 MOSCOW
U.S.S.R.
TEL:
TLX:
COM: 21

TSIOUMIS ALEXANDROS DR
DEPT GEODETIC ASTRONOMY
UNIVERSITY THESSALONIKI
GR-54006 THESSALONIKI
GREECE
TEL:
TLX:
COM: 27,33

TSVETKOV TSVETAN DR
DEPT OF ASTRONOMY
FACULTY OF PHYSICS
ANTON IVANOV STR 5
1126 SOFIA
BULGARIA
TEL:
TLX:
COM:

TROITSKY V S PROF DR
RADIOPHYSICAL RESEARCH
INSTITUTE
LYADOV STREET 25/14
603600 GORKIJ
U.S.S.R.
TEL: 36-04-40
TLX:
COM: 16,40,51C

TSAI CHANG-HSIEN DIRECTOR
ASTRONOMICAL STY OF CHINA
TAIPEI OBSERVATORY
TAIPEI 104
CHINA, TAIWAN
TEL:
TLX:
COM:

TSUBAKI TOKIO PROF
DEPT OF EARTH SCIENCE
SHIGA UNIVERSITY
2-5-1 HIRATSU
OHTSU 520
JAPAN
TEL: 0775-37-0081
TLX:
COM: 10,12

TSVETKOVA KATIA
DEPT OF ASTRONOMY AND
NATL ASTRON OBSERVATORY
72 LENIN BLVD
1784 SOFIA
BULGARIA
TEL: 73-41-379
TLX: 23561 ECF BAN BG
COM: 37

LIST OF MEMBERS

TUCHMAN YTZHAK
RACAH INST. OF PHYSICS
HEBREW UNIV. OF JERUSALEM
JERUSALEM 91904
ISRAEL
TEL: 02-584417
TLX: 25391 HUIL
COM:

TUCKER WALLACE H DR
PO BOX 266
BONSALL CA 92003
U.S.A.
TEL: 619-728-7103
TLX:
COM:

TUEG HELMUT DR
ALFRED-WEGENER INSTITUT
FUR POLARFORSCHUNG
COLUMBUS CENTER
D-2850 BREMERHAVEN
GERMANY, F.R.
TEL:
TLX:
COM: 09

TUFEKCIOGLU ZEKI DR
ASTRONOMY DEPT
UNIVERSITY OF ANKARA
FEN FAKULTESI
ANKARA
TURKEY
TEL:
TLX:
COM:

TULL ROBERT G
DEPT OF ASTRONOMY
UNIVERSITY OF TEXAS
AT AUSTIN RLM 15 308
AUSTIN TX 78712
U.S.A.
TEL: 512-471-3337
TLX: 910-874-1351
COM: 09C

TULLY JOHN A DR
OBSERVATOIRE DE NICE
BP 139
F-06003 NICE CEDEX
FRANCE
TEL: 93-89-04-20
TLX: 460004
COM:

TULLY RICHARD BRENT DR
INSTITUTE FOR ASTRONOMY
UNIVERSITY OF HAWAII
2680 WOODLAWN DR
HONOLULU HI 96822
U.S.A.
TEL: 808-948-8606
TLX: 723-8459 UHAST HR
COM: 28,47

TUNCA ZEYNEL DR
EGE UNIVERSITY
FACULTY OF SCIENCE
DEPARTMENT OF ASTRONOMY
BORNOVA-IZMIR
TURKEY
TEL: 180110-2332
TLX:
COM:

TUOHY IAN R DR
MT STROMLO & SIDING
SPRING OBSERVATORIES
WODEN P.O. ACT 2606
AUSTRALIA
TEL: 062-88-1111
TLX: 62270 AA
COM:

TUOMINEN ILKKA V DR
OBSERVATORY
UNIVERSITY OF HELSINKI
TAHTITORNINMAKI
SF-00130 HELSINKI 13
FINLAND
TEL: 3580-1912946
TLX:
COM: 29,35

TUOMINEN JAAKKO V PROF
PIHLAJATIE 49 B 20
SF-00270 HELSINKI 27
FINLAND
TEL: 0-484352
TLX:
COM: 10,35

TURLO ZYGMUNT DR
ASTRONOMICAL CENTER
UL. CHOPINA 12/18
87-100 TORUN
POLAND
TEL:
TLX:
COM: 40

TURNER BARRY E DR
NRAO
EDGEMONT ROAD
CHARLOTTESVILLE VA 22901
U.S.A.
TEL: 804-296-0337
TLX: 910-997-0174
COM: 34,40

TURNER DAVID G DR
DEPT OF ASTRONOMY
ST MARY'S UNIVERSITY
HALIFAX NS B3H 3C3
CANADA
TEL: 9024299780x2254
TLX:
COM: 37

TURNER EDWIN L DR
PRINCETON UNIVERSITY OBS
PEYTON HALL
PRINCETON NJ 08544
U.S.A.
TEL: 609-452-3577
TLX:
COM: 28,47,51

TURNER KENNETH C DR
ARECIBO OBSERVATORY
PO BOX 995
ARECIBO PR 00613
U.S.A.
TEL: 809-878-2612
TLX: 385638
COM: 34,40

TURNER MARTIN J L DR
X-RAY ASTRONOMY GROUP
PHYSICS DEPT
UNIV OF LEICESTER
LEICESTER LE1 7RH
U.K.
TEL: 533-554455
TLX: 341664 LUXRAY G
COM:

TURNER MICHAEL S
ASTRON & ASTROPHYS CENTER
UNIVERSITY OF CHICAGO
5460 S. ELLIS AVE
CHICAGO IL 60637
U.S.A.
TEL: 312-962-7974
TLX: 6871133 UNCGO VW
COM: 47

TURON PIERRE
40 RUE DE LUZARCHES
F-95270 SEUGY
FRANCE
TEL:
TLX:
COM:

TURON-LACARRIEU C DR
OBSERVATOIRE DE PARIS
SECTION DE MEUDON
DEPEG
F-92195 MEUDON PL CEDEX
FRANCE
TEL: 1-45-34-75-30
TLX:
COM: 24,33

TURTLE A J DR
DEPT OF PHYSICS
UNIVERSITY OF SYDNEY
SYDNEY NSW 2006
AUSTRALIA
TEL: 02-692-2222
TLX: 26169 UNISYD
COM: 40

TUTUKOV A V DR
ASTRONOMICAL COUNCIL
USSR ACADEMY OF SCIENCES
PYATNITSKAYA UL 48
109017 MOSCOW
U.S.S.R.
TEL: 231-54-61
TLX: 412623 SCSTP SU
COM: 27,35,42

TUZI KONOSUKE DR
2-2-2- OSAWA, MITAKA
TOKYO 181
JAPAN
TEL:
TLX:
COM:

TWAROG BRUCE A
UNIVERSITY OF KANSAS
DEPT PHYSICS & ASTRONOMY
LAWRENCE KS 66045
U.S.A.
TEL: 913-864-5163
TLX:
COM:

TWISS R Q DR
C/O A.R. BOSCHI
96A HOLLAND ROAD
LONDON W14 8BD
U.K.
TEL:
TLX:
COM:

TWORKOWSKI ANDRZEJ S
SCHOOL OF MATHEMAT. SCI.
QUEEN MARY COLLEGE
MILE END ROAD
LONDON E1 4NS
U.K.
TEL: 01-980-4822
TLX: 893750
COM:

TYLENDA ROMUALD DR
COPERNICUS ASTRON CENTER
UL. CHOPINA 12/18
87-100 TORUN
POLAND
TEL:
TLX:
COM: 27

TYLER JR G LEONARD DR
CTR/RADAR ASTRONOMY - SEL
STANFORD UNIVERSITY
STANFORD CA 94305
U.S.A.
TEL: 415-497-3535
TLX:
COM: 16

TYPHOON LEE
INST EARTH SCIENCES
ACADEMIA SINICA
P O BOX 23-59
TAIPEI 107
CHINA, TAIWAN
TEL: 02-396-3211
TLX:
COM: 15,35

TYSON JOHN A DR
BELL LABS
RM 1D-316
600 MOUNTAIN AVE
MURRAY HILL NJ 07974
U.S.A.
TEL: 201-582-6028
TLX: 138650 BELL LABS MUH
COM: 21,28

UBERTINI PIETRO
IST. ASTROFISICA SPAZIALE
C P 67
I-00044 FRASCATI
ITALY
TEL: 06-9425132
TLX: 610261 CNRFRA
COM:

UCHIDA JUICHI DR
TOHOKU GAKUEN UNIVERSITY
TAGAJYO CITY
MIYAGI 985
JAPAN
TEL:
TLX:
COM: 35

UCHIDA YUTAKA PROF
TOKYO ASTRONOMICAL
OBSERVATORY
OSAWA. MITAKA
TOKYO 181
JAPAN
TEL:
TLX:
COM: 12

UDAL'TSOV V A DR
PHYSICAL INSTITUTE
USSR ACADEMY OF SCIENCES
LENINSKI PROSPECT 53
117924 MOSCOW
U.S.S.R.
TEL: SERPUHOV 32889
TLX: 411479 NEOD SU
COM: 40

UENO SUEO PROF
KANAZAWA INSTITUTE
OF TECHNOLOGY
NONOICHIMACHI
ISHIKAWA 921
JAPAN
TEL: 0762-48-1100
TLX: 5122456 KIY LCJ
COM: 36

UESUGI AKIRA DR
DEPT OF ASTRONOMY
UNIVERSITY OF KYOTO
SAKYOKU
KYOTO 606
JAPAN
TEL:
TLX:
COM: 05,36

UKITA NOBUHARU
NOBEYAMA RADIO OBS
TOKYO ASTRON OBSERVATORY
NOBEYAMA, MINAMISAKU
NAGANO 384-13
JAPAN
TEL: 267-98-2831
TLX: 3329005
COM: 40

ULFBECK OLE DR
NIELS BOHR INSTITUTE
BLEGDAMSVEJ 17
DK-2100 COPENHAGEN
DENMARK
TEL:
TLX:
COM:

ULICH BOBBY LEE
STEWARD OBSERVATORY
UNIVERSITY OF ARIZONA
TUCSON AZ 85721
U.S.A.
TEL: 602-621-1537
TLX:
COM: 09

ULMER MELVILLE P PROF
DEARBORN OBSERVATORY
NORTHWESTERN UNIVERSITY
EVANSTON IL 60201
U.S.A.
TEL: 312-491-5633
TLX: 9102310040
COM:

ULMSCHNEIDER PETER PROF
INST FUER THEORETISCHE
ASTROPHYSIK
IM NEUENHEIMER FELD 561
D-6900 HEIDELBERG
GERMANY, F.R.
TEL: 6221-562837
TLX: 461515 UNIHD D
COM: 36

ULRICH BRUCE T PROF
HILTENSPERGER STR 93
D-8000 MUNCHEN 40
GERMANY, F.R.
TEL:
TLX:
COM: 25,40

ULRICH MARIE-HELENE D DR
ESO
KARL-SCHWARSZCHILD-STR 2
D-8046 GARCHING B MUNCHEN
GERMANY, F.R.
TEL: 89-320-06-229
TLX: 52828222 EOD
COM: 28,34,40

ULRICH ROGER K PROF
DEPT OF ASTRONOMY
UNIVERSITY OF CALIFORNIA
8931 MSB
LOS ANGELES CA 90024
U.S.A.
TEL: 213-825-4270
TLX:
COM: 35

UMLENSKI VASIL
DEPT OF ASTRONOMY AND
NATL ASTRON OBSERVATORY
72 LENIN BLVD
1784 SOFIA
BULGARIA
TEL: 73-41-559
TLX: 23561 ECF BAN BG
COM:

UNDERHILL ANNE B DR
4696 WEST 10TH AVE #301
VANCOUVER BC V6R 2J5
CANADA
TEL: 604-224-3552
TLX:
COM: 29,36C,44

UNDERWOOD JAMES H DR
CTR F X-RAY OPTICS 80-101
LAWRENCE BERKELEY LAB
UNIVERSITY OF CALIFORNIA
BERKELEY CA 94720
U.S.A.
TEL: 415-486-4958
TLX: 910-366-2037
COM: 10,44

UNNO WASABURO PROF
RES INST SCI & TECHNOLOGY
KINKI UNIVERSITY
3-4-1 KOWAKAE, HIGASHI
OSAKA 577
JAPAN
TEL: 06-721-2332
TLX:
COM: 12,34,35,36

UNSOELD ALBRECHT PROF
INST THEOR PHYS & STERNW
NEUE UNIV PHYSIK ZENTRUM
OLSHAUSENST GEB N 61C
D-2300 KIEL 1
GERMANY, F.R.
TEL: 431-84205
TLX:
COM:

UNWIN STEPHEN C
CALTECH
105-24
PASADENA CA 91125
U.S.A.
TEL: 213-356-4973
TLX: 675425
COM: 40

UPGREN ARTHUR R DR
VAN VLECK OBSERVATORY
WESLEYAN UNIVERSITY
MIDDLETOWN CT 06457
U.S.A.
TEL: 203-347-9411
TLX:
COM: 24P,33,37,45

UPSON WALTER L II DR
PRINCETON UNIVERSITY OBS
PEYTON HALL
PRINCETON NJ 08540
U.S.A.
TEL:
TLX:
COM: 44

UPTON E K L DR
ASTRONOMY DEPT
UCLA
LOS ANGELES CA 90024
U.S.A.
TEL:
TLX:
COM:

URAS SILVANO DR
ISTITUTO DI ASTRONOMIA
VIA OSPEDALE 72
I-09100 CAGLIARI
ITALY
TEL: 070-711246
TLX: 790326 OSSAST I
COM:

URASIN LIRIK A DR
ENGELHARDT ASTRONOMICAL
OBSERVATORY
422526 KAZAN
U.S.S.R.
TEL: 32-48-27
TLX:
COM: 34

URBANIK MAREK DR
ASTRONOMICAL OBSERVATORY
JAGIELLONIAN UNIVERSITY
UL. ORLA 171
30-244 KRAKOW
POLAND
TEL: 4812-221877
TLX: 0322297 UJ PL
COM: 28

URBARZ H DR
ASTRONOMISCHES INSTITUT
DER UNIVERSITAET
AUSSENSTELLE WEISSENAU
7980 RAVENSBURG/RASTHALDE
GERMANY, F.R.
TEL: 0751-61621
TLX:
COM:

URECHE VASILE DR
UNIVERSITY OF CLUJ-NAPOCA
FACULTY OF MATHEMATICS
STR M KOGALNICEANU 1
3400 CLUJ NAPOCA
RUMANIA
TEL: 951-16101/11592
TLX:
COM: 25,42

URPO SEPPO I
HELSINKI UNIV OF TECHNOL.
RADIO LABORATORY
OTAKAARI 5 A
SF-02150 ESPOO 15
FINLAND
TEL: 358-0-4512548
TLX: 122771 RORTA SF
COM: 10,40

USHER PETER D DR
DEPT OF ASTRONOMY
PENNSYLVANIA STATE UNIV
507 DAVEY LAB
UNIVERSITY PARK PA 16802
U.S.A.
TEL: 814-865-3509
TLX: 842510 PENNSTBSTRCG
COM: 27

UTSUMI KAZUHIKO DR
DEPT OF ASTRONOMY
HIROSHIMA UNIVERSITY
NAKA-KU, HIGASHI-SENDA
HIROSHIMA 730
JAPAN
TEL: 082-241-1221
TLX:
COM: 29

UUS UNDO DR
TARTU ASTROPHYSICAL
OBSERVATORY
202444 TORAVERE, ESTONIA
U.S.S.R.
TEL:
TLX:
COM: 12,35

VAGER ZEEV DR
DEPT OF PHYSICS
WEIZMANN INSTITUTE
REHOVOT
ISRAEL
TEL:
TLX:
COM:

VAGHI SERGIO DR
ESTEC/PHA
POSTBUS 299
NL-2200 AG NOORDWIJK
NETHERLANDS
TEL: 31-1719-83453
TLX: 39098
COM: 20

VAGNETTI FAUSTO
DIPARTIMENTO DI FISICA
II UNIVERSITA' DI ROMA
VIA ORAZIO RAIMONDO
I-00173 ROMA
ITALY
TEL: 6-7979-2323
TLX: 611462 UNIVRM
COM:

VALSECCHI GIOVANNI B
IST. ASTROFISICA SPAZIALE
REPARTO PLANETOLOGIA
VIALE UNIVERSITA 11
I-00185 ROMA
ITALY
TEL: 39-6-4956951
TLX: 610261 CNR FRA
COM: 15,20

VAN BEEK FRANK DR
SPACE RESEARCH LABORATORY
BENELUXLAAN 21
NL-3527 HS UTRECHT
NETHERLANDS
TEL: 030-937145
TLX: 47224
COM: 44

VAN DE STADT HERMAN DR
ASTRONOMICAL INSTITUTE
ZONNENBURG 2
NL-3512 NL UTRECHT
NETHERLANDS
TEL: 030-312841
TLX:
COM:

VAIANA GIUSEPPE S DR
HARVARD COLLEGE OBS
HARVARD UNIVERSITY
60 GARDEN STREET
CAMBRIDGE MA 02138
U.S.A.
TEL: 617-495-4311
TLX:
COM:

VALTAOJA ESKO
DEPT OF PHYSICAL SCIENCES
UNIVERSITY OF TURKU
SF-20500 TURKU 50
FINLAND
TEL: 921-645651
TLX: 62683 TYF
COM: 40

VAN BLERKOM DAVID J PROF
ASTRONOMY DEPT
UNIV OF MASSACHUSETTS
AMHERST MA 01002
U.S.A.
TEL:
TLX:
COM:

VAN DEN BERGH SIDNEY PROF
HERZBERG INST ASTROPHYS
DOMINION ASTROPHYS OBS
5071 W SAANICH ROAD
VICTORIA BC V8X 4M6
CANADA
TEL: 604-388-3924
TLX: 0497295
COM: 28,37,50P

VAIDYA P C PROF
34 SHARDA NAGAR
PALDI
AHMEDABAD 380 007
INDIA
TEL: 413322
TLX:
COM: 47

VALTIER JEAN-CLAUDE DR
OBSERVATOIRE DE NICE
BP 139
F-06003 NICE CEDEX
FRANCE
TEL: 93-89-04-20
TLX: 460004
COM: 27

VAN BREDA IAN G DR
ROYAL GREENWICH OBS
HERSTMONCEUX CASTLE
HAILSHAM BN27 1RP
U.K.
TEL: 0323-833171
TLX: 87451
COM:

VAN DEN HEUVEL EDWARD P J
ASTRONOMICAL INSTITUT
ROETERSSTRAAT 15
NL-1018 WB AMSTERDAM
NETHERLANDS
TEL: 020-5223004
TLX: 16460 FACWN NL
COM: 35,38,42

VAINSTEIN L A DR
PHYSICAL INSTITUTE
USSR ACADEMY OF SCIENCES
LENINSKY PROSP. 53
117924 MOSCOW
U.S.S.R.
TEL: 135-22-50
TLX:
COM: 49

VALTONEN MAURI J PROF
DEPT OF PHYSICAL SCIENCES
UNIVERSITY OF TURKU
IT. PITKAKATU 1
SF-20520 TURKU 52
FINLAND
TEL: 358-21-645973
TLX: 62683 TYF SF
COM: 07

VAN BREUGEL WIL
RADIOASTRONOMY LABORATORY
UNIVERSITY OF CALIFORNIA
601 CAMPBELL HALL
BERKELEY CA 94720
U.S.A.
TEL: 415-642-5275
TLX:
COM:

VAN DER BORGHT RENE PROF
DEPT OF MATHEMATICS
MONASH UNIVERSITY
CLAYTON VIC 3168
AUSTRALIA
TEL: 541-2580
TLX: 32691 MONASH AA
COM: 35

VALBOUSQUET ARMAND DR
OBS DE STRASBOURG
11 RUE DE L'UNIVERSITE
F-67000 STRASBOURG
FRANCE
TEL: 88-35-43-00
TLX:
COM: 24,26,51

VAN AGT S L TH J DR
STERRENKUNDIG INSTITUUT
TOERNOOIVELD
NL-6525 ED NIJMEGEN
NETHERLANDS
TEL: 80-558-833
TLX: 48228 WINAT NL
COM: 27

VAN BUEREN HENDRIK G PROF
ADVISORY COUNCIL FOR
SCIENCE POLICY
P O BOX 18524
NL-2502 EM THE HAGUE
NETHERLANDS
TEL: 070-639922
TLX:
COM:

VAN DER HUCHT KAREL A DR
SRON SPACE RESEARCH LAB.
RESEARCH UTRECHT
BENELUXLAAN 21
NL-3527 HS UTRECHT
NETHERLANDS
TEL: 030-937145
TLX: 47224
COM: 26,29,44C

VALENTIJN EDWIN A DR
KAPTEYN LABORATORIUM
POSTBUS 800
NL-9700 AV GRONONGEN
NETHERLANDS
TEL: 50-11-66-95
TLX:
COM:

VAN ALBADA TJEERD S DR
KAPTEYN LABORATORIUM
POSTBUS 800
NL-9700 AV GRONINGEN
NETHERLANDS
TEL: 050-116695
TLX: 53572 STARS NL
COM: 28

VAN CITTERS GORDON W DR
DIV OF ASTRONOMICAL SCI
NTL SCIENCE FOUNDATION
WASHINGTON DC 20550
U.S.A.
TEL:
TLX:
COM: 09

VAN DER HULST JAN M DR
NETHERLANDS FOUNDATION OF
RADIOASTRONOMY
POSTBUS 2
NL-7990 AA DWINGELOO
NETHERLANDS
TEL: 31-5219-7244
TLX: 42043 SRZM NL
COM: 28,40

VALLEE JACQUES P DR
HERZBERG INSTITUTE
CNRC
100 SUSSEX DRIVE
OTTAWA ONT K1A 0R6
CANADA
TEL: 613-993-6060
TLX:
COM: 40,51

VAN ALLEN JAMES A PROF
DEPT PHYSICS & ASTRONOMY
UNIVERSITY OF IOWA
IOWA CITY IA 52242
U.S.A.
TEL: 319-353-4531
TLX:
COM: 10,16,21,49

VAN DE HULST H C PROF DR
STERREWACHT
POSTBUS 9513
NL-2300 RA LEIDEN
NETHERLANDS
TEL: 071-148333
TLX: 39058
COM: 21,34,40,44

VAN DER KRUIT PIETER C DR
KAPTEYN LABORATORIUM
POSTBUS 800
NL-9700 AV GRONINGEN
NETHERLANDS
TEL: 50-63 40 73
TLX: 53572 STARS NL
COM: 28P,33,40

VALNICEK BORIS DR
ASTRONOMICAL INSTITUTE
CZECH. ACAD. OF SCIENCES
OBSERVATORY
251 65 ONDREJOV
CZECHOSLOVAKIA
TEL: 204-999-202
TLX: 121579
COM: 09,10C,44

VAN ALTENA WILLIAM F PROF
YALE UNIVERSITY OBS
PO BOX 6666
NEW HAVEN CT 06511
U.S.A.
TEL: 203-436-8318
TLX:
COM: 24V,37

VAN DE KAMP PETER
AMSTEL 244
NL-1017 AK AMSTERDAM
NETHERLANDS
TEL: 020-22-33-77
TLX:
COM: 24,26,51

VAN DER LAAN H PROF DR
STERREWACHT
POSTBUS 9513
NL-2300 RA LEIDEN
NETHERLANDS
TEL: 71-148333-5848
TLX: 39058 ASTRO NL
COM: 28,34,40,47

VAN DER RAAY HERMAN B
DEPT OF PHYSICS
UNIVERSITY OF BIRMINGHAM
P O BOX 363
BIRMINGHAM B15 2TT
U.K.
TEL: 021-472-1301
TLX: 228938 SPAPHY G
COM: 35

VAN HAMME WALTER
UNIV OF SOUTH CAROLINA
COASTAL CAROLINA COLLEGE
PO BOX 19541 (S9)
CONWAY SC 29526
U.S.A.
TEL:
TLX:
COM: 42

VAN NIEUWKOOP J DR IR
STERREWACHT SONNENBORGH
ZONNENBURG 2
NL-3512 WK UTRECHT
NETHERLANDS
TEL:
TLX: 47224 ASTRO
COM: 40

VAN'T VEER-MENNERET CL DR
INSTITUT D'ASTROPHYSIQUE
98 BIS BOULEVARD ARAGO
F-75014 PARIS
FRANCE
TEL: 1-320-14-25
TLX:
COM: 29,36

VAN DESSEL EDWIN LUDO DR
KONINKLIJKE STERRENWACHT
RINGLAAN 3
B-1180 BRUSSELS
BELGIUM
TEL: 32-26-735366
TLX: 21565 OBSBEL
COM: 26,30

VAN HERK G
STERREWACHT LEIDEN
POSTBUS 9513
NL-2300 RA LEIDEN
NETHERLANDS
TEL:
TLX:
COM:

VAN PARADIJS JOHANNES DR
ASTRONOMICAL INSTITUTE
ROETERSTRAAT 15
NL-1018 WB AMSTERDAM
NETHERLANDS
TEL: 31-20-522003/4
TLX: 16460 FACWN NL
COM: 42

VANDEN BOUT PAUL A
NRAO
EDGEMONT ROAD
CHARLOTTESVILLE VA 22903
U.S.A.
TEL: 804-296-0241
TLX: 910-997-0174
COM: 34,40

VAN DIGGELEN J DR
STERREWACHT SONNENBORGH
ZONNENBURG 2
NL-3512 NL UTRECHT
NETHERLANDS
TEL: 02940-12996
TLX:
COM:

VAN HOOF A PROF EM
ASTRONOMISCH INSTITUUT
KATHOLIEKE UNIV LEUVEN
CELESTIJNENLAAN 200B
B-3000 LEUVEN
BELGIUM
TEL:
TLX:
COM: 27,33,42

VAN REGEMORTER HENRI DR
OBSERVATOIRE DE PARIS
SECTION DE MEUDON
F-92195 MEUDON PL CEDEX
FRANCE
TEL: 1-534-75-70
TLX: 201571 F
COM: 14,36

VANDENBERG DON DR
C/O PHYSICS DEPT
UNIVERSITY OF VICTORIA
P O BOX 1700
VICTORIA BC V8W 2Y2
CANADA
TEL: 604-721-7739
TLX: 0497222
COM: 35,37

VAN DORN BRADT HALE DR
MIT RM 37-581
CENTER FOR SPACE RESEARCH
CAMBRIDGE MA 02139
U.S.A.
TEL: 617-253-7550
TLX: 921473 MITCAM
COM:

VAN HORN HUGH M PROF
DEPT PHYSICS & ASTRONOMY
UNIVERSITY OF ROCHESTER
ROCHESTER NY 14627
U.S.A.
TEL: 716-275-4344
TLX:
COM: 35

VAN RENSBERGEN WALTER DR
ASTROPHYSISCH INSTITUUT
VRIJE UNIVERSITEIT
PLEINLAAN 2
B-1050 BRUSSEL
BELGIUM
TEL: 02-641-34-97
TLX:
COM: 14

VANDERVOORT PETER O DR
DEPT ASTRON & ASTROPHYS
5640 ELLIS AVENUE
CHICAGO IL 60637
U.S.A.
TEL: 312-962-8209
TLX:
COM: 33

VAN DUINEN R J DR
FOKKER BV
P O BOX 7600
NL-1117 ZJ SCHIPHOL
NETHERLANDS
TEL: 020-5442030
TLX: 12227
COM: 44

VAN HOUTEN C J DR
STERREWACHT LEIDEN
POSTBUS 9513
NL-2300 RA LEIDEN
NETHERLANDS
TEL: 71-148333
TLX: 39058 ASTRO NL
COM: 20

VAN RIPER KENNETH A DR
LOS ALAMOS NATIONAL LAB
MS B 226 X-6
PO BOX 1663
LOS ALAMOS NM 87545
U.S.A.
TEL: 505-667-8104
TLX:
COM: 35,48

VANYSEK VLADIMIR PROF
DEPT OF ASTRONOMY
CHARLES UNIVERSITY
SVEDSKA 8
150 00 PRAHA 5
CZECHOSLOVAKIA
TEL: 00422-540395
TLX: 121673 MFF
COM: 15,34,47

VAN FLANDERN THOMAS DR
V.F. ASSOCIATES
6327 WESTERN AVE NW
WASHINGTON DC 20015
U.S.A.
TEL: 202-363-3860
TLX:
COM: 04,16,20,51

VAN HOUTEN-GROENEVELD I
STERREWACHT LEIDEN
POSTBUS 9513
NL-2300 RA LEIDEN
NETHERLANDS
TEL: 71-148333x5881
TLX: 39058 ASTRO NL
COM: 20

VAN SPEYBROECK LEON P DR
CENTER FOR ASTROPHYSICS
60 GARDEN STREET
CAMBRIDGE MA 02138
U.S.A.
TEL: 617-495-7233
TLX:
COM: 44

VAPILLON LOIC J DR
OBSERVATOIRE DE PARIS
SECTION DE MEUDON
F-92195 MEUDON PL CEDEX
FRANCE
TEL: 1-45-34-75-30
TLX: 20571
COM:

VAN GENDEREN A M DR
STERREWACHT LEIDEN
POSTBUS 9513
NL-2300 RA LEIDEN
NETHERLANDS
TEL: 071-148333
TLX: 31476 ASTRO NL
COM: 27C

VAN HOVEN GERARD DR
DEPT OF PHYSICS
UNIVERSITY OF CALIFORNIA
IRVINE CA 92717
U.S.A.
TEL: 714-856-5145
TLX: 683322 IRIN
COM: 10,12

VAN WOERDEN HUGO PROF DR
KAPTEYN LABORATORIUM
POSTBUS 800
NL-9700 AV GRONINGEN
NETHERLANDS
TEL: 50-116-695
TLX: 53572 STARS NL
COM: 28,33,34,40

VARDANIAN R A DR
BYURAKAN ASTROPPHYSICAL
OBSERVATORY
378433 ARMENIA
U.S.S.R.
TEL: 28-41-42
TLX:
COM: 25

VAN GORKOM JACQUELINE H
NRAO
PO BOX 0
SOCORRO NM 87801
U.S.A.
TEL: 505-772-4302
TLX: 910-997-0174
COM: 40

VAN LEER B DR
DELFT UNIV OF TECHNOLOGY
DEPT MATHS & INFORMATICS
P.O. BOX 356
NL-2600 AJ DELFT
NETHERLANDS
TEL: 015-783634
TLX: 38151 BHTHD
COM:

VAN'T VEER FRANS DR
INSTITUT D'ASTROPHYSIQUE
98 BIS BOULEVARD ARAGO
F-75014 PARIS
FRANCE
TEL: 1-43-20-14-25
TLX: 270776 OBSASTR
COM: 10,36,42

VARDAVAS ILIAS MIHAIL
ALLIGATOR RIVERS RES INST
P.O. BOX 387
BONDI JUNCTION NSW 2022
AUSTRALIA
TEL: 02-3870697
TLX: 23984 ARRIS
COM: 36

VARDYA M S DR
TATA INSTITUTE OF
FUNDAMENTAL RESEARCH
HOMI BHABHA RD
BOMBAY 400 005
INDIA
TEL: 219111 x221
TLX: 011-3009 TIFR IN
COM: 35.36

VARMA RAM KUMAR PROF
PHYSICAL RESEARCH LAB
AHMEDABAD 380 009
INDIA
TEL: 272-462-129
TLX: 0121397 PRL IN
COM: 28

VARSAVSKY C M DR
LAS HERAS 1975 (6°A)
BUENOS AIRES
ARGENTINA
TEL:
TLX:
COM: 14,33,40

VARSHALOVICH DIMITRIJ PR
PHYSIKO-TECHNICAL INST
USSR ACADEMY OF SCIENCES
194021 LENINGRAD
U.S.S.R.
TEL: 247-22-55
TLX:
COM: 14,34,51

VARVOGLIS H DR
DEPT OF ASTRONOMY
UNIVERSITY THESSALONIKI
GR-54006 THESSALONIKI
GREECE
TEL: 30-31-991357
TLX: 412181
COM: 07

VASHKOV'YAK SOF'YA N DR
STERNBERG STATE
ASTRONOMICAL INSTITUTE
UNIVERSITETSKIJ PROSP 13
119889 MOSCOW
U.S.S.R.
TEL: 139-37-64
TLX: 113037 JAPET
COM: 07

VASILEVA GALINA J DR
PULKOVO OBSERVATORY
196140 LENINGRAD
U.S.S.R.
TEL:
TLX:
COM: 12

VASILEVSKIS STANISLAUS
455 GRANT AVE 16
PALO ALTO CA 94306
U.S.A.
TEL: 415-325-8567
TLX:
COM: 24

VAUCLAIR GERARD P DR
OBS DU PIC-DU-MIDI
ET DE TOULOUSE
14 AVENUE EDOUARD BELIN
F-31400 TOULOUSE
FRANCE
TEL: 61-25-21-01
TLX: 530776
COM: 35

VAUCLAIR SYLVIE D DR
OBS DU PIC-DU-MIDI
ET DE TOULOUSE
14 AVENUE EDOUARD BELIN
F-31400 TOULOUSE
FRANCE
TEL: 61-25-21-01
TLX:
COM:

VAUGHAN ARTHUR H DR
PERKIN-ELMER CORP
7421 ORANGEWOOD AVE
GARDEN GROVE CA 92641
U.S.A.
TEL: 714-895-1667
TLX:
COM: 10,12,25

VAZ LUIZ PAULO RIBEIRO
OBSERVATORIO ASTRONOMICO
DEPTO DE FISICA-ICEX-UFMG
CAIXA POSTAL 702
30161 BELO HORIZONTE - MG
BRAZIL
TEL: 55-31-4412541
TLX: 312308 UFMG BR
COM: 42

VAZQUEZ MANUEL DR
INSTITUTO DE ASTROFISICA
DE CANARIAS
LA LAGUNA
38071 TENERIFE
SPAIN
TEL:
TLX:
COM: 51

VEEDER GLENN J DR
JPL
MS 183-501
4800 OAK GROVE DRIVE
PASADENA CA 91109
U.S.A.
TEL: 213-354-7388
TLX:
COM: 15

VEGA E. IRENE DR
OBSERVATORIO ASTRONOMICO
DE LA UNLP
PASEO DEL BOSQUE s/n
1900 LA PLATA
ARGENTINA
TEL: 021-21-7308
TLX: 31151 BULAP AR
COM: 33

VEILLET CHRISTIAN
CERGA
AVENUE COPERNIC
F-06130 GRASSE
FRANCE
TEL: 93-36-58-49
TLX: 470865
COM: 07,19,20

VEIS GEORGE PH D
GEODESY LABORATORY
NATL TECHNICAL UNIVERSITY
ATHENS
GREECE
TEL:
TLX:
COM: 19

VEISMANN UNO DR
TARTU OBSERVATORY
202444 ESTONIA
U.S.S.R.
TEL:
TLX:
COM:

VELGHE ALBERT G PROF DR
AHORNLAAN 29
B-9910 GENT
BELGIUM
TEL:
TLX:
COM: 05,25,33

VELKOV KIRIL
DEPT OF ASTRONOMY AND
NATL ASTRON OBSERVATORY
72 LENIN BLVD
1784 SOFIA
BULGARIA
TEL: 73-41-614
TLX: 23561 ECF BQN BG
COM: 09,10

VELUSAMY T DR
RADIO ASTRONOMY CTR
P O BOX NO 8
UDHAGAMANDALAM 643 001
INDIA
TEL: 2651 & 2032
TLX: 8458488 TIFR IN
COM: 40

VENKATAKRISHNAN P DR
INDIAN INSTITUTE OF
ASTROPHYSICS
BANGALORE 560 034
INDIA
TEL:
TLX:
COM: 12

VENKATESAN DORASWAMY DR
PHYSICS DEPT
UNIVERSITY OF CALGARY
CALGARY ALB T2N 1N4
CANADA
TEL:
TLX:
COM: 10

VENUGOPAL V R DR
RADIO ASTRONOMY
CENTRE OF TIFR
POST BOX 8
UDHAGAMANDALAM 643 001
INDIA
TEL: 2651 & 2032
TLX: 0853-241 RAC IN
COM: 33,40,51

VERBEEK PAUL DR
GEORGE MINNELAAN 50
B-9830 S MARTENS-LATEM
BELGIUM
TEL: 09-82-61-19
TLX:
COM:

VERDET JEAN-PIERRE DR
OBSERVATOIRE DE PARIS
61 AVE DE L'OBSERVATOIRE
F-75014
FRANCE
TEL: 1-329-12-10
TLX:
COM: 41

VERES FERENC
BAJA OBSERVATORY
ASTRONOMICAL INSTITUTE
BOX 110
H-6500 BAJA
HUNGARY
TEL: 11064, 12170
TLX: 281303
COM:

VERGNANO A PROF
OSSERVATORIO ASTRONOMICO
I-10025 PINO TORINESE
ITALY
TEL:
TLX:
COM:

VERHEEST FRANK ASSOC PROF
INST THEORET. MECHANIKA
RIJKSUNIVERSITEIT GENT
KRIJGSLAAN 281 (S9)
B-9000 GENT
BELGIUM
TEL: 091-22 57 15
TLX: 12 754 RUGENT B
COM: 49

VERMA R P DR
TATA INSTITUTE OF
FUNDAMENTAL RESEARCH
BOMBAY 400 005
INDIA
TEL:
TLX:
COM: 25

VERNIANI FRANCO PROF
DIPARTIMENTO DI FISICA
VIA IRNERIO 46
I-40126 BOLOGNA
ITALY
TEL: 26-09-91
TLX: 211664
COM: 22

VERON MARIE-PAULE DR
OBS DE HAUTE-PROVENCE
F-04870 ST-MICHEL L'OBS.
FRANCE
TEL: 92-76-63-68
TLX:
COM: 28

VERON PHILIPPE DR
OBS DE HAUTE-PROVENCE
F-04870 ST-MICHEL L'OBS.
FRANCE
TEL: 92-76-63-68
TLX:
COM: 28,40

VIDAL JEAN-LOUIS DR
OBS DU PIC-DU-MIDI
F-65200 BAGNERES-DE-B.
FRANCE
TEL: 62-95-19-69
TLX: 531625
COM: 34

VILHENA DE MORAES R DR
DEPTO DE ASTRONOMIA
ITA-CTA
12200 S. JOSE DOS CAMPOS
BRAZIL
TEL: 55-123-229088
TLX: 01173437 ZWO-24-73
COM: 07

VISHNIAC ETHAN T
UNIVERSITY OF TEXAS
DEPT OF ASTRONOMY
AUSTIN TX 78712
U.S.A.
TEL: 512-471-1429
TLX:
COM: 47

VERSCHUUR GERRIT L PROF
ARECIBO OBSERVATORY
PO BOX 995
ARECIBO PR 00613
U.S.A.
TEL:
TLX:
COM: 33,34,40,51

VIDAL NISSIM V DR
DEPT PHYSICS & ASTRONOMY
TEL-AVIV UNIVERSITY
TEL-AVIV 69978
ISRAEL
TEL:
TLX:
COM: 48

VILHU OSMI DR
OBS AND ASTROPHYS LAB
UNIVERSITY OF HELSINKI
TAHTITORNINMAKI
SF-00130 HELSINKI 13
FINLAND
TEL:
TLX:
COM: 29,35,42

VISHVESHWARA C V PROF
RAMAN RESEARCH INSTITUTE
BANGALORE 560 080
INDIA
TEL: 360122
TLX: 845671
COM: 47

VESECKY J F DR
DEPT OF ASTRONOMY
THE UNIVERSITY
LEICESTER LE1 7RH
U.K.
TEL:
TLX:
COM:

VIDAL-MADJAR ALFRED DR
INSTITUT D'ASTROPHYSIQUE
98 BIS BOULEVARD ARAGO
F-75014 PARIS
FRANCE
TEL: 1-43-20-14-25
TLX:
COM: 34,44

VINER MELVYN R DR
DEPT OF METALLURGICAL
ENGINEERING
QUEEN'S UNIVERSITY
KINGSTON ONT K7L 3N6
CANADA
TEL: 613-547-2634
TLX:
COM: 34,40

VISVANATHAN NATARAJAN DR
MT STROMLO OBSERVATORY
WODEN P.O. ACT 2606
AUSTRALIA
TEL: 062-881111
TLX: 62270 TLG CANOPUS AA
COM: 25,28,34

VETESNIK MIROSLAV DR
GARYOUNIS UNIVERSITY
FACULTY OF SCIENCE
POB 9480 2
BENGHAZI
LIBYA
TEL:
TLX:
COM: 33,42

VIEIRA MARTINS ROBERTO DR
OBSERVATORIO NACIONAL
RUA GENERAL BRUCE 586
SAO CRISTOVAO
20921 RIO DE JANEIRO
BRAZIL
TEL: 021-580-7313
TLX: 021-21288
COM: 20

VINLUAN RENATO
UNIVERSITY OF SOUTHERN
PHILIPPINES
OBRERO DAVAO CITY 9501
PHILIPPINES
TEL:
TLX:
COM: 10

VITINSKIJ YURIJ I DR
PULKOVO OBSERVATORY
196140 LENINGRAD
U.S.S.R.
TEL: 298-22-42
TLX:
COM: 10,12

VETTOLANI GIAMPAOLO
IST. DI RADIOASTRONOMIA
CNR
VIA IRNERIO 46
I-40126 BOLOGNA
ITALY
TEL: 39-51-232856
TLX: 211664 INFN BO I
COM:

VIGIER JEAN-PIERRE DR
INSTITUT H. POINCARE
11 RUE P.& M. CURIE
F-75005 PARIS
FRANCE
TEL:
TLX:
COM:

VINOD S KRISHAN MRS DR
INDIAN INSTITUTE OF
ASTROPHYSICS
BANGALORE 560 034
INDIA
TEL: 566585/566497
TLX: 0845-763 IIAB IN
COM: 10C,49

VITON MAURICE DR
L A S
TRAVERSE DU SIPHON
LES TROIS LUCS
F-13012 MARSEILLE
FRANCE
TEL:
TLX:
COM:

VEVERKA JOSEPH DR
CORNELL UNIVERSITY
312 SPACE SCI BLDG
ITHACA NY 14853
U.S.A.
TEL: 607-256-3507
TLX: 937478
COM: 15,16

VIGOTTI MARIO
LAB. DI RADIOASTRONOMIA
VIA IRNERIO 46
I-40126 BOLOGNA
ITALY
TEL: 51-232856
TLX:
COM:

VINTI JOHN P DR
MIT
MEASUREMENT SYSTEMS LAB
RM W59-216
CAMBRIDGE MA 02139
U.S.A.
TEL: 417-782-2470
TLX:
COM: 07

VITTONE ALBERTO ANGELO
OSSERVATORIO ASTRONOMICO
DI CAPODIMONTE
VIA MOIARIELLO 16
I-80131 NAPOLI
ITALY
TEL: 81-440101
TLX:
COM:

VIAL JEAN-CLAUDE
LPSP
BP 10
F-91371 VERRIERES-LE-B.
FRANCE
TEL: 1-69-20-10-60
TLX: 600252
COM: 10,12,44

VIIK TONU DR
TORAVERE OBSERVATORY
202444 TARTU RAJ. ESTONIA
U.S.S.R.
TEL: 4-11-81 TARTU
TLX:
COM: 36

VIOTTI ROBERTO DR
CNR ASTROFISICA SPAZIALE
C P 67
I-00044 ROMA
ITALY
TEL: 06-942-5655
TLX: 610261
COM: 27,29,44

VITTORIO NICOLA
ISTITUTO ASTRONOMICO
UNIVERSITA DI ROMA
VIA LANCISI 29
I-00161 ROMA
ITALY
TEL:
TLX:
COM:

VICENTE RAIMUNDO O PROF
FACULDADE CIENCIAS LISBOA
RUA MESTRE AVIZ 30 R C
1495 LISBOA
PORTUGAL
TEL: 2112888
TLX:
COM: 19,31

VILA SAMUEL C PROF
DEPT OF ASTRONOMY
UNIV OF PENNSYLVANIA
33RD & WALNUT STREETS
PHILADELPHIA PA 19104
U.S.A.
TEL: 215-898-5994
TLX:
COM: 35

VIRGOPIA NICOLA PROF
DIPT. DI MATEMATICA
UNIV DI ROMA LA SAPIENZA
CITTA UNIVERSITARIA
I-00185 ROMA
ITALY
TEL:
TLX:
COM:

VIVEKANAND M DR
RAMAN RESEARCH INSTITUTE
BANGALORE 560 080
INDIA
TEL: 360122
TLX: 845671 RRI IN
COM: 40

LIST OF MEMBERS

VIVES TEODORO JOSE DR
CTR ASTRON HISPANO ALEMAN
REINA 66 9°B
CORREOS 511
04002 ALMERIA
SPAIN
TEL: 23-09-88
TLX: 78812 DSAZ E
COM:

VLACHOS DEMETRIUS G PROF
UNIV OF THESSALONIKI
DEPT GEODESY & SURVEYING
FACULTY OF ENGINEERING
GR-54006 THESSALONIKI
GREECE
TEL: 031-991520
TLX: 412181 AUTH GR
COM:

VLADIMIROV SIMEON
ASTRONOMICAL OBSERVATORY
72 LENIN BLVD
POST BOX 15
SOFIA 1309
BULGARIA
TEL: 23-13-97
TLX:
COM: 09

VOELK HEINRICH J PROF
MPI FUER KERNPHYSIK
POSTFACH 103 980
D-6900 HEIDELBERG
GERMANY, F.R.
TEL: 6221-516-295
TLX: 461666
COM: 14,48

VOGT NIKOLAUS DR
DEPTO DE ASTRONOMIA
UNIVERSIDAD CATOLICA
CASILLA 6014
SANTIAGO
CHILE
TEL: 775474
TLX: 240395 PUCVA CL
COM: 27,37,51

VOGT STEVEN SCOTT
LICK OBSERVATORY
UNIVERSITY OF CALIFORNIA
SANTA CRUZ CA 95064
U.S.A.
TEL: 408-429-2844
TLX: 910-598-4408
COM: 29

VOIGT HANS H PROF
UNIVERSITAETS-STERNWARTE
GEISMARLANDSTR 11
D-3400 GOETTINGEN
GERMANY, F.R.
TEL: 0551-395041
TLX: 96753 USTERN D
COM: 05,29,38C

VOLLAND H DR
ASTRONOMISCHES INSTITUT
DER UNIVERSITAET
AUF DEM HUEGEL 71
D-5300 BONN
GERMANY, F.R.
TEL: 0228-733674
TLX: 0886440
COM:

VOLONTE SERGE DR
DEPT D'ASTROPHYSIQUE
UNIVERSITE DE MONS
19 AVENUE MAISTRIAU
B-7020 MONS
BELGIUM
TEL: 065-31-51-71
TLX: 57764 UEMONS B
COM: 12,14

VON BORZESZKOWSKI H H DR
EINSTEIN-LABORATORIUM
AKAD. WISSENSCHAFTEN DDR
ROSA-LUXEMBURG-STR 17A
DDR-1502 POTSDAM
GERMANY, D.R.
TEL: 762225
TLX:
COM: 47

VON DER HEIDE JOHANN DR
ALARDUSSTR 12
D-2000 HAMBURG 20
GERMANY, F.R.
TEL: 40-491-4016
TLX:
COM: 8

VON HOERNER SEBASTIAN DR
KRUMMENACKER-STR 186
D-7300 ESSLINGEN
GERMANY, F.R.
TEL:
TLX:
COM: 51

VON SOCHER HERMANN DR
INSTITUT FUER ASTRONOMIE
DER UNIVERSITAET WIEN
TUERKENSCHANZSTR 17
A-1180 WIEN
AUSTRIA
TEL:
TLX:
COM:

VON WEIZSAECKER C F PROF
MAX-PLANCK INSTITUT
RIEMERSCHMID-STR 7
D-8130 STARNBERG
GERMANY, F.R.
TEL:
TLX:
COM:

VONDRAK JAN DR
ASTRONOMICAL INSTITUTE
CZECH. ACAD. OF SCIENCES
BUDECSKA 6
120 23 PRAHA 2
CZECHOSLOVAKIA
TEL: 42-2-258757
TLX: 66-122486
COM: 19C

VORONTSOV-VEL'YAMINOV B A
STERNBERG STATE ASTRONOM.
INSTITUTE
117234 MOSCOW
U.S.S.R.
TEL:
TLX:
COM: 28,34

VOROSHILOV V I DR
MAIN ASTRONOMICAL OBS
UKRAINIAN ACADEMY OF SCI
252127 KIEV
U.S.S.R.
TEL: 66-31-10
TLX: 131406 SKY SU
COM: 33

VORPAHL JOAN A DR
AEROSPACE CORPORATION
PO BOX 92957
LOS ANGELES CA 90009
U.S.A.
TEL:
TLX:
COM:

VRBA FREDERICK J DR
US NAVAL OBSERVATORY
PO BOX 1149
FLAGSTAFF AZ 86002
U.S.A.
TEL: 602-779-5132
TLX:
COM: 09,25C,34

VREUX JEAN MARIE DR
INSTITUT D'ASTROPHYSIQUE
UNIVERSITE DE LIEGE
AVENUE DE COINTE 5
B-4200 COINTE-OUGREE
BELGIUM
TEL: 41-529980
TLX: 41264
COM: 29

VU DUONG TUYEN DR
BUREAU DES LONGITUDES
77 AVE DENFERT-ROCHEREAU
F-75014 PARIS
FRANCE
TEL: 1-43-20-12-10
TLX:
COM: 20

VUJNOVIC VLADIS DR
INSTITUTE OF PHYSICS
OF THE UNIVERSITY
P O B 304
41001 ZAGREB
YUGOSLAVIA
TEL: 041-271211
TLX: 22203 IFS YU
COM: 14,46

VUKICEVIC K M PROF DR
DEPT OF ASTRONOMY
FACULTY OF SCIENCES
STUDENTSKI TRG 16
11000 BEOGRAD
YUGOSLAVIA
TEL:
TLX:
COM: 12

VYALSHIN GENNADIJ F DR
MAIN ASTRONOMICAL OBS
PULKOVO
196140 LENINGRAD
U.S.S.R.
TEL:
TLX:
COM: 10

WACHMANN A A PROF DR
SCHNIEDERSBERG 2B
D-2057 REINBEK
GERMANY, F.R.
TEL:
TLX:
COM: 27

WACKERNAGEL H BEAT DR
51 BROADMOOR HILLS DRIVE
COLORADO SPRINGS CO 80906
U.S.A.
TEL: 303-576-2382
TLX:
COM: 04,31

WADDINGTON C JAKE PROF
PHYSICS DEPT
UNIVERSITY OF MINNESOTA
116 CHURCH ST S.E.
MINNEAPOLIS MN 55455
U.S.A.
TEL: 612-624-2566
TLX: 910-576-2955
COM:

WADE CAMPBELL M DR
NRAO
PO BOX 0
SOCORRO NM 87801
U.S.A.
TEL: 505-835-5351
TLX: 910-988-1710
COM: 40

WAELKENS CHRISTOFFEL
ASTRONOMISCH INSTITUUT
KATHOLIEKE UNIV. LEUVEN
CELESTIJNENLAAN 200B
B-3030 HEVERLEE
BELGIUM
TEL: 016-20-06-56
TLX: 25715
COM: 27

WAGNER RAYMOND L DR
FORD AEROSPACE
F-10
2880 E. FOUNTAIN BLVD
COLORADO SPRINGS CO 80910
U.S.A.
TEL: 303-635-8911
TLX:
COM:

WAGNER WILLIAM J DR
NOAA/SPACE ENVIR. LAB
325 BROADWAY
PO BOX 3000
BOULDER CO 80303
U.S.A.
TEL: 303-497-3274
TLX:
COM:

WAGONER ROBERT V PROF
STANFORD UNIVERSITY
VARIAN PHYSICS BLDG
STANFORD CA 94305
U.S.A.
TEL: 415-723-4561
TLX: 348402
COM: 47

WAKAMATSU KEN-ICHI DR
JUNIOR DEPARTMENT
COLLEGE OF TECHNOLOGY
GIFU UNIVERSITY
GIFU 501-11
JAPAN
TEL: 582-30-1111
TLX:
COM: 28C

WAKO KOJIRO DR
INTERNATIONAL LATITUDE
OBSERVATORY
MIZUSAWA
IWATE 023
JAPAN
TEL:
TLX:
COM: 19

WALBORN NOLAN R DR
SPACE TELESCOPE SCI INST
3700 SAN MARTIN DRIVE
BALTIMORE MD 21218
U.S.A.
TEL: 301-338-4915
TLX: 6849101 STSCI UW
COM: 45

WALCH JEAN-JACQUES
CERGA
AVENUE COPERNIC
F-06130 GRASSE
FRANCE
TEL: 93-36-58-49
TLX: 470865 CERGA F
COM: 07

WALDMEIER MAX PROF DR
SWISS FEDERAL OBSERVATORY
WIRZENWEID 15
CH-8053 ZUERICH
SWITZERLAND
TEL:
TLX:
COM: 10,12

WALKER ALISTAIR ROBIN DR
S A A O
P O BOX 9
OBSERVATORY 7935
SOUTH AFRICA
TEL: 021-470025
TLX: 5-20309 SA
COM: 09,25

WALKER ARTHUR B C JR PROF
CENTER FOR SPACE SCIENCE
STANFORD UNIV, ERL 310
STANFORD CA 94305
U.S.A.
TEL: 415-497-1486
TLX:
COM:

WALKER EDWARD N MR
ROYAL GREENWICH OBS
HERSTMONCEUX CASTLE
HAILSHAM BN27 1RP
U.K.
TEL: 0323-833171
TLX: 87451
COM: 27

WALKER GORDON A H PROF
DEPT GEOPHYS & ASTRONOMY
UNIV OF BRITISH COLUMBIA
2075 WESBROOK PLACE
VANCOUVER BC V6T 1W5
CANADA
TEL: 604-228-4133
TLX: 0454245
COM: 09C,34,37,45

WALKER HELEN
STERREWACHT LEIDEN
POSTBUS 9513
NL-2300 RA LEIDEN
NETHERLANDS
TEL:
TLX:
COM:

WALKER IAN WALTER
DEPT OF ASTRONOMY
THE UNIVERSITY
GLASGOW G12 8QQ
U.K.
TEL:
TLX:
COM: 07

WALKER MERLE F PROF
LICK OBSERVATORY
UNIVERSITY OF CALIFORNIA
SANTA CRUZ CA 95064
U.S.A.
TEL: 408-429-2526
TLX:
COM: 09,27,37,50C

WALKER RICHARD L
US NAVAL OBSERVATORY
FLAGSTAFF STATION
BOX 1149
FLAGSTAFF AZ 86002
U.S.A.
TEL: 602-774-6623
TLX:
COM: 26,42

WALKER ROBERT M A PROF
PHYSICS DEPT
WASHINGTON UNIVERSITY
BOX 1105
ST LOUIS MO 63130
U.S.A.
TEL: 314-889-6225
TLX:
COM: 16

WALKER WILLIAM S G
14 APPLEYARD CRESC.
AUCKLAND 5
NEW ZEALAND
TEL:
TLX:
COM: 42

WALL JASPER V DR
ROYAL GREENWICH OBS
HERSTMONCEUX CASTLE
HAILSHAM BN27 1RP
U.K.
TEL: 323-833-171
TLX: 87451
COM: 40

WALLACE LLOYD V DR
KITT PEAK NATIONAL
OBSERVATORY
PO BOX 26732
TUCSON AZ 85726
U.S.A.
TEL: 602-327-5511
TLX:
COM: 16,21

WALLACE PATRICK T MR
STARLINK RUTHERFORD
APPLETON LAB
CHILTON, DIDCOT OX11 0QX
U.K.
TEL: 44-235-445-472
TLX: 83159
COM: 05,08,09

WALLACE RICHARD K
LOS ALAMOS NAT LABORATORY
X-7, MS B257
LOS ALAMOS NM 87545
U.S.A.
TEL: 505-667-5000
TLX:
COM:

WALLENQUIST AAKE A E PROF
ASTRONOMICAL OBSERVATORY
NORRLANDSGATAN 34 D
S-752 29 UPPSALA
SWEDEN
TEL: 18-13 56 85
TLX:
COM: 25,37

WALLERSTEIN GEORGE PROF
ASTRONOMY DEPT FM 20
UNIVERSITY OF WASHINGTON
SEATTLE WA 98195
U.S.A.
TEL: 206-543-2888
TLX:
COM: 27,29

WALLIS MAX K DR
DEPT APPL MATHS & ASTRON
UNIVERSITY COLLEGE
P O BOX 78
CARDIFF CF1 1XL, WALES
U.K.
TEL: 222-44211
TLX: 488635
COM: 15,49,51

WALMSLEY C MALCOLM DR
MPI FUER RADIOASTRONOMIE
AUF DEM HUEGEL 69
D-5300 BONN 1
GERMANY, F.R.
TEL: 0228-525305
TLX: 0886440 MPIFR D
COM: 40

WALRAVEN TH DR
P O BOX 98
CORNELIA
ORANGE FREESTATE 9850
SOUTH AFRICA
TEL:
TLX:
COM: 25,27

WALSH DENNIS DR
NRAL
JODRELL BANK
MACCLESFIELD SK11 9DL
U.K.
TEL: 0477-71321
TLX: 36149
COM: 40,44

WALTER FREDERICK M
UNIVERSITY OF COLORADO
CASA
PO BOX 391
BOULDER CO 80309
U.S.A.
TEL: 303-492-7606
TLX: 9109403441
COM:

WALTER HANS G DR
ASTRONOMISCHES RECHEN
INSTITUT
MCENCHHOFSTR 12-14
D-6900 HEIDELBERG 1
GERMANY, F.R.
TEL: 49026
TLX: 461336 ARIHD D
COM: 08,24

WALTER KURT PROF DR
ASTRONOMISCHES INSTITUT
DER UNIVERSITAET
WALDHAUSERSTR 64
D-7400 TUEBINGEN
GERMANY, F.R.
TEL: 07071-296126
TLX: 7262714 AIT D
COM: 42

WAMPLER E JOSEPH PROF
ESO
KARL-SCHWARZSCHILD-STR 2
D-8046 GARCHING B MUNCHEN
GERMANY, F.R.
TEL: 49-89-320-06297
TLX: 52828222 EO D
COM: 09

WAMSTEKER WILLEM DR
ESA IUE GROUND STATION
VILLAFRANCA DE CASTILLO
P O BOX 54065
28080 MADRID
SPAIN
TEL: 34-1-401-9661
TLX: 42555
COM: 44

WAN FOOK SUN
DEPT OF MATHEMATICS
NATL UNIVERSITY SINGAPORE
KENT RIDGE
SINGAPORE 0511
SINGAPORE
TEL: 772-2742
TLX:
COM:

WAN LAI
ZO-SE SECTION
SHANGHAI OBSERVATORY
SHANGHAI
CHINA, PEOPLE'S REP.
TEL: 380696
TLX: 33164 SHAO CN
COM: 24,37

WAN TONG-SHAN
SHANGHAI OBSERVATORY
80 NAN DAN ROAD
SHANGHAI
CHINA, PEOPLE'S REP.
TEL: 386191
TLX: 33164 SHAO CN
COM: 19,40

WANAS M I DR
DEPT OF ASTRONOMY
CAIRO UNIVERSITY
GEZA ORMAN
EGYPT
TEL:
TLX:
COM:

WANG CHUAN-JIN
PURPLE MOUNTAIN
OBSERVATORY
NANJING
CHINA, PEOPLE'S REP.
TEL: 46700
TLX: 34144 PMONJ CN
COM: 25

WANG DEYU
PURPLE MOUNTAIN
OBSERVATORY
NANJING
CHINA, PEOPLE'S REP.
TEL: 42817, 46700
TLX: 34144 PMONJ CN
COM:

WANG JIA-JI
SHANGHAI OBSERVATORY
ACADEMIA SINICA
SHANGHAI
CHINA, PEOPLE'S REP.
TEL: 386191
TLX: 33164 SHAO CN
COM: 24

WANG JIA-LONG
BEIJING ASTRONOMICAL OBS
ACADEMIA SINICA
BEIJING
CHINA, PEOPLE'S REP.
TEL:
TLX: 22040 BAOBS CN
COM: 10

WANG JING-SHENG
YUNNAN OBSERVATORY
P.O. BOX 110
KUNMING
CHINA, PEOPLE'S REP.
TEL: 72946
TLX: 64040 YUOBS CN
COM: 40

WANG JING-XIU
BEIJING ASTRONOMICAL OBS
ACADEMIA SINICA
BEIJING
CHINA, PEOPLE'S REP.
TEL: 28 1698
TLX: 22040 BAOBS CN
COM: 10,12

WANG LAN-JUAN
SHANGHAI OBSERVATORY
ACADEMIA SINICA
SHANGHAI
CHINA, PEOPLE'S REP.
TEL: 386191
TLX: 33164 SHAO CN
COM: 09

WANG RENCHUAN
CENTER FOR ASTROPHYSICS
UNIV SCIENCE & TECHNOLOGY
HEFEI, ANHUI
CHINA, PEOPLE'S REP.
TEL:
TLX: 90028 USTC CN
COM: 47

WANG SHOU-GUAN
BEIJING OBSERVATORY
ACADEMIA SINICA
BEIJING
CHINA, PEOPLE'S REP.
TEL: 281261
TLX: 22040 BADAS CN
COM: 40C,48

WANG SHUI
DEPT EARTH & SPACE SCI.
UNIV SCIENCE & TECHNOLOGY
HEFEI, ANHUI
CHINA, PEOPLE'S REP.
TEL: 63300-209
TLX: 4430
COM: 44

WANG SI-CHAO
PURPLE MOUNTAIN
OBSERVATORY
NANJING
CHINA, PEOPLE'S REP.
TEL: 44205
TLX: 34144 PMONJ CN
COM: 15

WANG YANAN
NANJING ASTRONOMICAL
INSTRUMENT FACTORY
NANJING
CHINA, PEOPLE'S REP.
TEL: 46191
TLX: 34136 GLYNJ c/o NAIF
COM: 09

WANG YIMING
YUNNAN OBSERVATORY
P.O. BOX 110
KUNMING, YUNNAN PROVINCE
CHINA, PEOPLE'S REP.
TEL:
TLX:
COM: 09

WANG ZHEN-RU
DEPT OF ASTRONOMY
NANJING UNIVERSITY
NANJING
CHINA, PEOPLE'S REP.
TEL: 37551 x 2685
TLX: 34151 PRCNU CN
COM: 48

WANG ZHEN-YI
PURPLE MOUNTAIN
OBSERVATORY
NANJING
CHINA, PEOPLE'S REP.
TEL: 46700
TLX: 34144 PMONJ CN
COM: 12

WANG ZHENG MING
SHAANXI ASTRONOMICAL OBS
P.O. BOX 18
LINTONG
XIAN
CHINA, PEOPLE'S REP.
TEL: 32255 XIAN
TLX: 70121 CSAO CN
COM: 19

WANNIER PETER GREGORY DR
JET PROPULSION LAB 169506
CALIF INST OF TECHNOLOGY
4800 OAK GROVE DRIVE
PASADENA CA 91109
U.S.A.
TEL: 8188-354-3347
TLX: 67-5429
COM: 34C,40

WARD MARTIN JOHN
INSTITUTE OF ASTRONOMY
MADINGLEY ROAD
CAMBRIDGE CB3 0HA
U.K.
TEL:
TLX:
COM: 28,42

WARD RICHARD A DR
MAIL CODE L-23
LAWRENCE LIVERMORE LAB
PO BOX 808
LIVERMORE CA 94550
U.S.A.
TEL: 415-423-2679
TLX: 910-386-8339 UCLLL
COM: 35

WARD WILLIAM R DR
JPL
4800 OAK GROVE DRIVE
PASADENA CA 91103
U.S.A.
TEL:
TLX:
COM: 19

WARDLE JOHN F C PROF
PHYSICS DEPT
BRANDEIS UNIVERSITY
WALTHAM MA 02154
U.S.A.
TEL: 617-647-2889
TLX:
COM: 40

WARES GORDON W DR
73 PERKINS STREET
WEST NEWTON MA 02165
U.S.A.
TEL:
TLX:
COM: 14

WARGAU WALTER FRIEDRICH
UNIV OF SOUTH AFRICA
DEPT MATHS, APPL MATHS &
ASTRONOMY, P.O. BOX 392
PRETORIA 0001
SOUTH AFRICA
TEL: 27-12-440-2133
TLX: 350068 TA UNISA TTX
COM: 42

WARMAN J DR
INSTITUTO DI ASTRONOMIA
UNAM
APDO POSTAL 70-264
04510 MEXICO DF
MEXICO
TEL:
TLX:
COM:

WARNER BRIAN PROF
DEPT OF ASTRONOMY
DARTMOUTH COLLEGE
HANOVER NH 03755
U.S.A.
TEL:
TLX:
COM: 27C,42,45

WARNER JOHN W DR
PERKIN-ELMER CORP
M/S 892
100 WOOSTER HEIGHTS RD
DANBURY CT 86810
U.S.A.
TEL: 203-796-7919
TLX:
COM: 28,44

WARNER PETER J DR
CAVENDISH LABORATORY
MADINGLEY ROAD
CAMBRIDGE CB3 0HE
U.K.
TEL: 223-66477
TLX: 81292 CAVLAB G
COM: 40

WARREN WAYNE H JR DR
NASA/GSFC
CODE 633.8
GREENBELT MD 20771
U.S.A.
TEL: 301-344-8310
TLX: 89675 NASCOM GBLT
COM: 05,25,37,45

WARWICK JAMES W DR
DEPT OF ASTROPHYS,PLANET
UNIVERSITY OF COLORADO
BOULDER CO 80309
U.S.A.
TEL: 303-447-9524
TLX:
COM: 12,40

WARWICK ROBERT S DR
PHYSICS DEPT
UNIVERSITY OF LEICESTER
LEICESTER LE1 7RH
U.K.
TEL: 533-554455
TLX: 341664 LUXRAY G
COM:

WARZEE J DR
115 RUE MATTOT
B-1410 WATERLOO
BELGIUM
TEL:
TLX:
COM:

WATSON FREDERICK GARNETT
UK SCHMIDT TELESCOPE
PRIVATE BAG
COONABARABRAN NSW 2357
AUSTRALIA
TEL: 068-421622
TLX: 63945 CANOPUS AA
COM: 09,51

WEBBINK RONALD F DR
DEPT OF ASTRONOMY
UNIVERSITY OF ILLINOIS
1011 W. SPRINGFIELD AVE
URBANA IL 61801
U.S.A.
TEL: 217-333-9582
TLX: 910-245-2434 AST
COM: 27,35,42

WEHINGER PETER A DR
PHYSICS DEPARTMENT
ASTRONOMY GROUP
ARIZONA STATE UNIVERSITY
TEMPE AZ 85287
U.S.A.
TEL: 602-965-4063
TLX: 140289 HALLEY ASU UT
COM: 15,28,29

WASHIMI HARUICHI DR
RESEARCH INSTITUTE
OF ATMOSPHERICS
NAGOYA UNIVERSITY
TOYOKAWA AICHI 442
JAPAN
TEL: 05338-6-3154
TLX: 4322311
COM:

WATSON WILLIAM D PROF
PHYSICS DEPT
UNIVERSITY OF ILLINOIS
URBANA IL 61801
U.S.A.
TEL: 217-33-7240
TLX:
COM:

WEBER STEPHEN VANCE
L-477
LAWRENCE LIVERMORE LAB
PO BOX 808
LIVERMORE CA 94550
U.S.A.
TEL: 415-422-5433
TLX: 910-386-8339
COM: 36

WEHLAU AMELIA DR
ASTRONOMY DEPT
UNIV OF WESTERN ONTARIO
LONDON ONT N6A 3K7
CANADA
TEL: 519-679-3186
TLX: 064-7134
COM: 27,29,37

WASSERMAN LAWRENCE H DR
LOWELL OBSERVATORY
1400 W. MARS HILL ROAD
FLAGSTAFF AZ 86001
U.S.A.
TEL: 602-774-3358
TLX:
COM: 16,20,24

WATT GRAEME DAVID
RADIOSTERRENWACHT
POSTBUS 2
NL-7990 AA DWINGELOO
NETHERLANDS
TEL: 05219-7244
TLX: 42043 SRZM NL
COM: 34

WEBROVA LUDMILA DR
ASTRONOMICAL INSTITUTE
CZECH. ACAD. OF SCIENCES
BUDECSKA 6
120 23 PRAHA
CZECHOSLOVAKIA
TEL:
TLX:
COM: 31

WEHLAU WILLIAM H PROF
ASTRONOMY DEPT
UNIV OF WESTERN ONTARIO
LONDON ONT N6A 3K7
CANADA
TEL: 519-679-3183
TLX: 064-7134
COM: 27,29,42

WASSON JOHN T
INSTITUTE OF GEOPHYSICS
UNIVERSITY OF CALIFORNIA
LOS ANGELES CA 90024
U.S.A.
TEL: 213-825-1986
TLX:
COM: 15C,16

WATTENBERG D PROF
LINDERHOFSTR 57
DDR-1147 BERLIN
GERMANY, D.R.
TEL: 527-77-72
TLX:
COM: 41

WEBSTER ADRIAN S DR
ROYAL OBSERVATORY
BLACKFORD HILL
EDINBURGH EH9 3HJ
U.K.
TEL: 031-667-3321
TLX: 72383 ROEDIN G
COM: 47,48

WEHRSE RAINER DR
INST F THEOR ASTROPHYSIK
IM NEUENHEIMER FELD 561
D-6900 HEIDELBERG
GERMANY, F.R.
TEL: 06221-562837
TLX: 461515 UNIHD D
COM: 36

WATANABE TAKASHI DR
RESEARCH INSTITUTE
OF ATMOSPHERICS
NAGOYA UNIVERSITY
TOYOKAWA 442
JAPAN
TEL: 0532-54-3067
TLX: 2822307 TAOMTK J
COM: 49

WAYMAN PATRICK A PROF
DUNSINK OBSERVATORY
DUBLIN 15
IRELAND
TEL: 353-1-387911
TLX: 31687 DIAS
COM: 05C,33,50

WEBSTER B LOUISE DR
ANGLO AUSTRALIAN
OBSERVATORY
P.O. BOX 296
EPPING NSW 2121
AUSTRALIA
TEL:
TLX:
COM: 34

WEI MINGZHI
BEIJING ASTRONOMICAL OBS
ACADEMIA SINICA
BEIJING
CHINA, PEOPLE'S REP.
TEL: 28 1698
TLX: 22040 BAOBS CN
COM: 40

WATANABE TETSUYA
TOKYO ASTRONOMICAL
OBSERVATORY
2-21-1 OSAWA, MITAKA
TOKYO 181
JAPAN
TEL: 422-32-5111
TLX:
COM: 36

WEAVER HAROLD F PROF
DEPT OF ASTRONOMY
UNIVERSITY OF CALIFORNIA
BERKELEY CA 94720
U.S.A.
TEL:
TLX: 820181 UCS AST
COM: 33,34,37,40

WEEDMAN DANIEL W PROF
PENNSYLVANIA STATE UNIV
ASTRONOMY DEPARTMENT
525 DAVEY LABORATORY
UNIVERSITY PARK PA 16802
U.S.A.
TEL: 814-865-0418
TLX: 842510
COM: 28

WEIDEMANN VOLKER PROF
INST THEOR PHYS & STERNW
NEUE UNIV PHYSIK ZENTRUM
OLSHAUSENST GEB N 61C
D-2300 KIEL 1
GERMANY, F.R.
TEL: 0431-880-4110
TLX: 2922706 IAPKI D
COM: 05,35C,36

WATERFIELD REGINALD L DR
WOOLSTON OBSERVATORY
NORTH CADBURY
YEOVIL, SOMERSET
U.K.
TEL: 40457
TLX:
COM: 15

WEAVER THOMAS A DR
PHYSICS DEPT L-17
LAWRENCE LIVERMORE LAB
PO BOX 808
LIVERMORE CA 94550
U.S.A.
TEL: 415-423-1850
TLX:
COM: 35,48

WEEKES TREVOR C DR
FRED LAWRENCE WHIPPLE OBS
HARVARD-SMITHSONIAN CTR
PO BOX 97
AMADO AZ 85645-0097
U.S.A.
TEL: 602-629-6741
TLX:
COM:

WEIGERT ALFRED PROF
HAMBURGER STERNWARTE
GOJENSBERGWEG 112
D-2050 HAMBURG 80
GERMANY, F.R.
TEL: 72244112
TLX: 217884 HAMST D
COM: 35,42

WATERWORTH MICHAEL DR
UNIVERSITY OF TASMANIA
GPO BOX 252C
HOBART, TASMANIA 7001
AUSTRALIA
TEL: 61-102-202418
TLX: 58150 AA
COM: 29

WEBBER JOHN C DR
MIT HAYSTACK OBSERVATORY
WESTFORD MA 01886
U.S.A.
TEL: 617-692-4764
TLX:
COM:

WEGNER GARY ALAN
DEPT PHYSICS & ASTRONOMY
WILDER LABORATORY
DARTMOUTH COLLEGE
HANOVER NH 03755
U.S.A.
TEL: 603-646-2359
TLX:
COM: 29

WEILER EDWARD J DR
PRINCETON UNIVERSITY OBS
PEYTON HALL
PRINCETON NJ 08540
U.S.A.
TEL:
TLX:
COM: 42,44

WEILER KURT W DR
HULBURT CTR F. SPACE RES
NAVAL RESEARCH LABORATORY
WASHINGTON DC 20375-5000
U.S.A.
TEL: 202-767-3010
TLX:
COM: 34

WEISS NIGEL O DR
UNIVERSITY OF CAMBRIDGE
DEPT APPL MATH/THEO PHYS
SILVER STREET
CAMBRIDGE CB3 9EW
U.K.
TEL: 0223-351645
TLX: 81240
COM: 12,35

WELLER CHARLES S DR
NAVAL RESEARCH LABORATORY
CODE 5382
WASHINGTON DC 20375
U.S.A.
TEL: 202-767-2003
TLX:
COM: 49

WENZEL W DR
ZNTRLINST. F. ASTROPHYSIK
STERNWARTE SONNEBERG
DDR-6400 SONNEBERG
GERMANY, D.R.
TEL:
TLX:
COM: 27

WEILL GILBERT M DR
SPOT IMAGE CORPORATION
1897 PRESTON WHITE DRIVE
RESTON VA 22091-4326
U.S.A.
TEL: 620-22-00
TLX: 4993073
COM: 21

WEISS WERNER W DR
INSTITUT FUER ASTRONOMIE
DER UNIVERSITAET WIEN
TUERKENSCHANZSTR 17
A-1130 WIEN
AUSTRIA
TEL: 0-22-2-34-53-60
TLX: 116222 PHYSI A
COM: 29

WELLGATE G BERNARD MR
CANEHEATH HOUSE
ARLINGTON
POLEGATE BN26 6SJ
U.K.
TEL:
TLX:
COM:

WESEMAEL FRANCOIS DR
DEPT DE PHYSIQUE
UNIVERSITE DE MONTREAL
C P 6128 SUCC A
MONTREAL PQ H3C 3J7
CANADA
TEL: 514-343-7355
TLX: 05562425 UDEMPHYSAS
COM:

WEIMER THEOPHILE P F DR
OBSERVATOIRE DE PARIS
61 AVE DE L'OBSERVATOIRE
F-75014 PARIS
FRANCE
TEL: 1-43-20-12-10
TLX:
COM: 16

WEISSKOPF MARTIN CH DR
NASA MSFC
CODE ES-65
HUNTSVILLE AL 35812
U.S.A.
TEL: 205-453-3238
TLX:
COM: 48

WELLINGTON KELVIN DR
CSIRO
DIVISION OF RADIOPHYSICS
P.O.BOX 76
EPPING NSW 2121
AUSTRALIA
TEL: 02-8680222
TLX: 26230 AA
COM: 40

WESSELINK ADRIAAN J DR
143 FALLS ROAD
BETHANY CT 06525
U.S.A.
TEL: 203-393-3297
TLX:
COM: 24,25,27,42

WEINBERG J L DR
SPACE ASTRONOMY LAB
UNIVERSITY OF FLORIDA
1810 N.W. 6TH STREET
GAINESVILLE FL 32609
U.S.A.
TEL: 904-392-5450
TLX: 810-825-2308 SPACELA
COM: 21C,22,44

WEISSMAN PAUL ROBERT
JET PROPULSION LABORATORY
MS 183-301
4800 OAK GROVE DRIVE
PASADENA CA 91109
U.S.A.
TEL: 818-354-2636
TLX: 675429
COM: 15,20

WELLMANN PETER PROF DR
INST FUER ASTRONOMIE &
ASTROPHYSIK
SCHEINER-STR 1
D-8000 MUENCHEN 80
GERMANY, F.R.
TEL:
TLX:
COM: 29,36,42

WESSELIUS PAUL R DR
SPACE RESEARCH LABORATORY
POSTBUS 800
NL-9700 AV GRONINGEN
NETHERLANDS
TEL:
TLX:
COM: 25,34,44,45

WEINBERG STEVEN DR
DEPT OF PHYSICS
UNIVERSITY OF TEXAS
AUSTIN TX 78712
U.S.A.
TEL: 512-471-4394
TLX: 910-874-1305
COM: 47

WEISTROP DONNA DR
LAB ASTRON & SOLAR PHYS
NASA/GSFC CODE 681
GREENBELT MD 20771
U.S.A.
TEL: 301-344-5781
TLX:
COM: 25,33

WELLS DONALD C III DR
NRAO
EDGEMONT ROAD
CHARLOTTESVILLE VA 22901
U.S.A.
TEL: 804-296-0211
TLX:
COM: 05

WESSON PAUL S DR
DEPT OF PHYSICS
UNIVERSITY OF WATERLOO
WATERLOO ONT N2L 3G1
CANADA
TEL: 519-885-1211
TLX:
COM: 47

WEIS EDWARD W DR
VAN VLECK OBSERVATORY
WESLEYAN UNIVERSITY
MIDDLETOWN CT 06457
U.S.A.
TEL: 203-347-9411
TLX:
COM: 26

WELCH GARY A DR
DEPT OF ASTRONOMY
ST MARY'S UNIVERSITY
HALIFAX NS B3H 3C3
CANADA
TEL: 902-429-9780
TLX:
COM: 28

WENDKER HEINRICH J PROF
HAMBURGER STERNWARTE
GOJENBERGSWEG 112
D-2050 HAMBURG 80
GERMANY, F.R.
TEL: 040-7252-4112
TLX: 217884 HAMST D
COM: 34,40

WEST RICHARD M DR
ESO
KARL-SCHWARZSCHILD-STR 2
D-8046 GARCHING B MUNCHEN
GERMANY, F.R.
TEL: 89-32006275
TLX: 52828220 ESO D
COM: 09,20,45,46

WEISBERG JOEL MARK
CARLETON COLLEGE
DEPT PHYSICS & ASTRONOMY
NORTHFIELD MN 55057
U.S.A.
TEL: 507-663-4367
TLX:
COM:

WELCH WILLIAM J PROF
RADIO ASTRONOMY LAB
UNIVERSITY OF CALIFORNIA
601 CAMPBELL HALL
BERKELEY CA 94720
U.S.A.
TEL: 415-642-6679
TLX: 820181 UCB AST RAL
COM: 40,51

WENIGER SCHAME DR
OBSERVATOIRE DE PARIS
SECTION DE MEUDON
F-92195 MEUDON PL CEDEX
FRANCE
TEL: 1-45-34-75-30
TLX:
COM: 14,21,29

WEST ROBERT ALAN
JET PROPULSION LABORATORY
MS 183-301
4800 OAK GROVE DRIVE
PASADENA CA 91109
U.S.A.
TEL: 818-354-0479
TLX: 675429
COM: 16

WEISHEIT JON C DR
PHYSICS DEPARTMENT L-297
L. LIVERMORE NTL LAB
PO BOX 808
LIVERMORE CA 94550
U.S.A.
TEL: 415-423-4254
TLX:
COM: 34,48

WELIACHEW LEONID DR
IRAM
VOIE 10
DOMAINE UNIVERSITAIRE
F-38406 ST MARTIN D'HERES
FRANCE
TEL: 76-42-33-83
TLX: 980 753 F
COM: 40

WENTZEL DONAT G DR
ASTRONOMY PROGRAM
UNIVERSITY OF MARYLAND
COLLEGE PARK MD 20742
U.S.A.
TEL: 301-454-5969
TLX: 710-826-0352
COM: 10,12,46C,48

WESTERHOUT GART DR
SCIENTIFIC DIRECTOR
US NAVAL OBSERVATORY
WASHINGTON DC 20390
U.S.A.
TEL: 202-653-1513
TLX: 710-822-1970
COM: 05C,24,33,40

WESTERLUND BENGT E PROF
ASTRONOMICAL OBSERVATORY
BOX 515
S-751 20 UPPSALA
SWEDEN
TEL: 46-18-135157
TLX: 76024 UNIV UPPSS
COM: 28,33,45

WHITAKER EWEN A
LUNAR & PLANETARY LAB
UNIVERSITY OF ARIZONA
TUCSON AZ 85711
U.S.A.
TEL: 602-621-2888
TLX: 910-952-1143
COM: 16,41

WHITE RICHARD L
SPACE TELESCOPE SCI INST
HOMEWOOD CAMPUS
3700 SAN MARTIN DRIVE
BALTIMORE MD 21218
U.S.A.
TEL: 301-338-4797
TLX:
COM: 34,36

WHITROW GERALD JAMES PROF
41 HOME PARK RD
WIMBLEDON
LONDON SW19 7HS
U.K.
TEL: 1-947-34-3467
TLX:
COM: 41,47

WESTFOLD KEVIN C PROF
MONASH UNIVERSITY
CLAYTON VIC 3168
AUSTRALIA
TEL: 03-5413080
TLX: 32691 AA
COM: 40,48

WHITE GLENN J
PHYSICS DEPT
QUEEN MARY COLLEGE
MILE END ROAD
LONDON E1 4NS
U.K.
TEL: 980-4811 x4045
TLX: 893750
COM: 34

WHITE SIMON DAVID MANION
STEWARD OBSERVATORY
UNIVERSITY OF ARIZONA
TUCSON AZ 85721
U.S.A.
TEL: 602-621-6530
TLX:
COM: 28,47

WHITTET DOUGLAS C B DR
SCHOOL OF PHYSICS & ASTR
LANCASHIRE POLYTECHNIC
CORPORATION ST
PRESTON PR1 2TQ
U.K.
TEL: 772-22141
TLX:
COM: 33,34

WESTPHAL JAMES A PROF
CALTECH
170-25
1201 E. CALIFORNIA
PASADENA CA 91125
U.S.A.
TEL: 213-356-4900
TLX:
COM: 09,44

WHITE GRAEME LINDSAY DR
CSIRO
DIVISION OF RADIOPHYSICS
P.O.BOX 76
EPPING NSW 2121
AUSTRALIA
TEL: 868-0222 x420
TLX: 26230 ASTRO AA
COM: 24,41

WHITELOCK PATRICIA ANN DR
SAAO
P O BOX 9
OBSERVATORY
CAPE 7935
SOUTH AFRICA
TEL: 470025
TLX: 57-20309
COM: 27,34

WHITWORTH ANTHONY PETER
DEPT OF APPLIED MATHS
UNIVERSITY COLLEGE
CARDIFF CF1 1XL
U.K.
TEL: 0222-44211
TLX: 498635 ULIBCFG
COM: 34

WETHERILL GEORGE W
DEPT TERRESTR. MAGNETISM
CARNEGIE INST. WASHINGTON
5241 BROAD BRANCH RD N.W.
WASHINGTON DC 20015
U.S.A.
TEL: 202-966-0863
TLX: 440427 MAGN UI
COM: 15,16,22,51

WHITE NATHANIEL M DR
LOWELL OBSERVATORY
1400 W. MARS HILL ROAD
FLAGSTAFF AZ 86001
U.S.A.
TEL: 602-774-3358
TLX:
COM: 25

WHITEOAK J B DR
CSIRO
DIVISION OF RADIOPHYSICS
P.O.BOX 76
EPPING NSW 2121
AUSTRALIA
TEL: 612868-0226
TLX: 26230
COM: 33,34,40

WICKRAMASINGHE D T DR
AUSTRALIAN NAT UNIVERSITY
DEPT OF APPLIED MATHS
P.O.BOX 4
CANBERRA ACT 2600
AUSTRALIA
TEL:
TLX:
COM:

WEYMANN RAY J PROF
STEWARD OBSERVATORY
UNIVERSITY OF ARIZONA
TUCSON AZ 85721
U.S.A.
TEL: 602-621-2375
TLX: 467175
COM: 34

WHITE ORAN R DR
7590 ROAD 39
MANCOS CO 80307
U.S.A.
TEL: 303-533-7318
TLX:
COM:

WHITFORD ALBERT E PROF
LICK OBSERVATORY
UNIVERSITY OF CALIFORNIA
SANTA CRUZ CA 95064
U.S.A.
TEL: 408-429-2149
TLX:
COM: 28

WICKRAMASINGHE N C PROF
UNIVERSITY COLLEGE
DEPT OF APPLIED MATHS
CARDIFF CF1 1XL
U.K.
TEL: 222-44211
TLX: 498635 ULIBCFG
COM: 34,36,40

WHEELER J CRAIG PROF
ASTRONOMY DEPT
UNIVERSITY OF TEXAS
AUSTIN TX 78712
U.S.A.
TEL: 512-471-4461
TLX: 910-874-1351
COM: 35C

WHITE R STEPHEN PROF
IGPP
UNIVERSITY OF CALIFORNIA
RIVERSIDE CA 92521
U.S.A.
TEL: 714-787-4503
TLX:
COM:

WHITMORE BRADLEY C
SPACE TELESCOPE SCI INST
HOMEWOOD CAMPUS
3700 SAN MARTIN DRIVE
BALTIMORE MD 21218
U.S.A.
TEL: 301-338-4713
TLX:
COM: 28

WIDING KENNETH G DR
US NAVAL RESEARCH LAB
CODE 7144
WASHINGTON DC 20375
U.S.A.
TEL: 202-767-2605
TLX:
COM:

WHEELER JOHN A DR
DEPT OF PHYSICS
UNIVERSITY OF TEXAS
AUSTIN TX 78712
U.S.A.
TEL: 512-471-3751
TLX: 910-874-1351
COM: 47,48

WHITE RAYMOND E DR
STEWARD OBSERVATORY
UNIVERSITY OF ARIZONA
TUCSON AZ 85721
U.S.A.
TEL: 602-621-6528
TLX: 467175
COM: 33,37

WHITNEY BALFOUR S
1102 E. MISSOURI
NORMAN OK 73071
U.S.A.
TEL: 405-321-3547
TLX:
COM:

WIEDLING TOR DR
STUDSVIK SCIENCE
RESEARCH LABORATORY
S-611 82 NYKOEPING
SWEDEN
TEL: 0155-22100
TLX:
COM:

WHIPPLE FRED L DR
CENTER FOR ASTROPHYSICS
60 GARDEN STREET
CAMBRIDGE MA 02138
U.S.A.
TEL: 617-495-7200
TLX:
COM: 15,20,22

WHITE RICHARD E
SMITH COLLEGE
ASTRONOMY DEPARTMENT
CLARK SCIENCE CENTER
NORTHAMPTON MA 01063
U.S.A.
TEL: 413-584-2700
TLX:
COM:

WHITNEY CHARLES A PROF
CENTER FOR ASTROPHYSICS
60 GARDEN STREET
CAMBRIDGE MA 02138
U.S.A.
TEL: 617-495-7451
TLX:
COM:

WIEHR EBERHARD DR
UNIVERSITAETS STERNWARTE
GEISMARLANDSTR 11
D-3400 GOETTINGEN
GERMANY, F.R.
TEL: 0551-395053
TLX:
COM: 10

WIELEBINSKI RICHARD PROF
MPI FUER RADIOASTRONOMIE
AUF DEM HUEGEL 69
D-5300 BONN
GERMANY, F.R.
TEL: 0228-525-300
TLX: 886440 MPIFR D
COM: 25C,28,33,40,51

WIELEN ROLAND PROF DR
ASTRONOMISCH. RECHEN-INST
MOENCHHOFSTR. 12-14
D-6900 HEIDELBERG 1
GERMANY, F.R.
TEL: 06221-490264
TLX: 461 336 ARIHD D
COM: 08,28,33C,37

WIESE WOLFGANG L DR
NATL BUREAU OF STANDARDS
DIVISION 531
ROOM A267, BLDG 221
GAITHERSBURG MD 20899
U.S.A.
TEL: 301-921-2071
TLX: WU 898493
COM: 14C

WIETH-KNUDSEN NIELS P DR
SVEND TROSTSVEJ 12
DK-1912 FREDERIKSBERG C
DENMARK
TEL: 45-1-249131
TLX:
COM: 26,31

WIITA PAUL JOSEPH
DEPT ASTRON & ASTROPHYS
UNIV OF PENNSYLVANIA. E1
PHILADELPHIA PA19104
U.S.A.
TEL: 215-898-8176
TLX:
COM: 28

WIJNBERGEN JAN DR
LAB VOOR RUIMTEONDERZOEK
HOOGBOUW WSN
POSTBUS 800
NL-9700 AV GRONINGEN
NETHERLANDS
TEL: 50-116660
TLX: 53572 STARS NL
COM:

WILCOCK WILLIAM L PROF
SCHOOL OF PHYSICAL AND
MOLECULAR SCIENCES
UNIV COLLEGE OF N. WALES
BANGOR GWYNEDD LL57 2UW
U.K.
TEL: 0248-351151
TLX: 61100
COM: 09C

WILD JOHN PAUL DR
CSIRO
P.O.BOX 225
DICKSON ACT 2602
AUSTRALIA
TEL:
TLX:
COM: 10,38,40,49

WILD PAUL PROF
ASTRONOMISCHES INSTITUT
UNIVERSITAET BERN
SIDLERSTRASSE 5
CH-3012 BERN
SWITZERLAND
TEL: 31-65-85-96
TLX: 32320 PHYBE CH
COM: 20,28

WILDEY ROBERT L PROF DR
NORTHERN ARIZONA UNIV
ASTROPHYSICAL OBSERVATORY
FLAGSTAFF AZ 86011
U.S.A.
TEL: 602-523-2661
TLX:
COM: 16

WILKENING LAUREL L DR
UNIVERSITY OF ARIZONA
ADMIN BLDG 601
TUCSON AZ 85721
U.S.A.
TEL: 602-626-3513
TLX:
COM: 15

WILKES BELINDA J
SMITHSONIAN ASTROPHYSICAL
OBSERVATORY
60 GARDEN STREET
CAMBRIDGE MA 02138
U.S.A.
TEL: 617-495-7268
TLX: 921428
COM:

WILKINS GEORGE A DR
ROYAL GREENWICH OBS
HERSTMONCEUX CASTLE
HAILSHAM BN27 1RP
U.K.
TEL: 323-833171
TLX: 87451 RGOBSY G
COM: 04,05P,19C,31

WILKINSON ALTHEA
ASTRONOMY DEPT
UNIVERSITY OF MANCHESTER
MANCHESTER M13 9PL
U.K.
TEL: 061-273-7121
TLX:
COM: 28

WILKINSON DAVID T
PRINCETON UNIVERSITY
JADWIN HALL
PO BOX 708
PRINCETON NJ 08540
U.S.A.
TEL: 609-452-4406
TLX:
COM: 47

WILKINSON PETER N DR
NRAL
JODRELL BANK
MACCLESFIELD SK11 9DL
U.K.
TEL: 0-477-71321
TLX: 36149
COM: 40

WILL CLIFFORD M DR
DEPT OF PHYSICS
WASHINGTON UNIVERSITY
ST LOUIS MO 63130
U.S.A.
TEL: 314-889-6244
TLX:
COM: 47,48

WILLIAMON RICHARD M
FERNBANK SCIENCE CENTER
156 HEATON PARK DRIVE
ATLANTA GA 30307
U.S.A.
TEL: 404-378-4313
TLX:
COM: 27,42

WILLIAMS BARBARA A
UNIVERSITY OF DELAWARE
PHYSICS DEPT
SHARP LAB.
NEWARK DE 19716
U.S.A.
TEL: 302-451-738-266
TLX:
COM: 28

WILLIAMS CAROL A
DEPT OF MATHEMATICS
UNIVERSITY OF S. FLORIDA
TAMPA FL 33620
U.S.A.
TEL: 813-974-2643
TLX:
COM: 07

WILLIAMS D DR
RADIO-ASTRONOMY LAB
UNIVERSITY OF CALIFORNIA
BERKELEY CA 94720
U.S.A.
TEL: 415-524-5631
TLX: 820181
COM: 34,40

WILLIAMS DAVID A PROF
MATHEMATICS DEPT
UMIST
P O BOX 88
MANCHESTER M60 1QD
U.K.
TEL: 061-236-3311
TLX: 666094
COM: 34

WILLIAMS IWAN P DR
THEORET. ASTRONOMY UNIT
QUEEN MARY COLLEGE
MILE END ROAD
LONDON E1 4NS
U.K.
TEL: 01-980-4811
TLX: 893750
COM: 16,20,22,34,51

WILLIAMS JAMES G DR
JPL 264-700
4800 OAK GROVE DRIVE
PASADENA CA 91109
U.S.A.
TEL: 818-354-6466
TLX: 910-588-3269 JPL
COM: 04,16,19,20

WILLIAMS JOHN A DR
PHYSICS DEPT
ALBION COLLEGE
ALBION MI 49224
U.S.A.
TEL: 517-629-5511
TLX:
COM: 45

WILLIAMS PEREDUR M DR
ROYAL OBSERVATORY
BLACKFORD HILL
EDINBURGH EH9 3HJ
U.K.
TEL: 31-667-3321
TLX: 72383 ROEDIN G
COM: 29

WILLIAMS ROBERT E DR
CERRO TOLOLO
INTERAMERICAN OBSERVATORY
CASILLA 603
LA SERENA
CHILE
TEL: 56-51-213-352
TLX: 645227 AURA CT
COM: 28,34,42

WILLIAMS THEODORE B DR
DET PHYSICS & ASTRONOMY
RUTGERS UNIVERSITY
PO BOX 849
PISCATAWAY NJ 08854
U.S.A.
TEL: 201-932-2516
TLX:
COM: 28

WILLIS ALLAN J DR
DEPT PHYSICS & ASTRONOMY
UNIVERSITY COLLEGE LONDON
GOWER STREET
LONDON WC1E 6BT
U.K.
TEL: 01-387-7050
TLX: 28722
COM: 34,44C

WILLIS ANTHONY GORDON DR
DEPT OF PHYSICS
BRANDEIS UNIVERSITY
WALTHAM MA 02154
U.S.A.
TEL:
TLX:
COM: 40

WILLMORE A PETER PROF
SPACE RESEARCH DEPT
UNIVERSITY OF BIRMINGHAM
PO BOX 363
BIRMINGHAM B15 2TT
U.K.
TEL: 021-472-1301
TLX: 338938 SPAPHY G
COM:

WILLNER STEVEN PAUL DR
CENTER FOR ASTROPHYSICS
60 GARDEN STREET
CAMBRIDGE MA 02138
U.S.A.
TEL: 617-495-7123
TLX: 921428 SATELLITE CAM
COM: 34,44

WILLS BEVERLEY J DR
ASTRONOMY DEPT RLM 15 308
UNIVERSITY OF TEXAS
AUSTIN TX 78712
U.S.A.
TEL: 512-471-3424
TLX: 910-874-1351
COM: 28,40

WILLS DEREK DR
ASTRONOMY DEPT RLM 15 308
UNIVERSITY OF TEXAS
AUSTIN TX 78712
U.S.A.
TEL: 512-471-4461
TLX: 910-874-1351
COM: 28,40

WILLSON LEE ANNE DR
ASTRONOMY PROGRAM
PHYSICS DEPT
IOWA STATE UNIVERSITY
AMES IA 50011
U.S.A.
TEL: 515-294-6765
TLX: 910-520-1157
COM: 27,35,36

WILLSON ROBERT FREDERICK
DEPT OF PHYSICS
TUFTS UNIVERSITY
MEDFORD MD 02155
U.S.A.
TEL: 617-628--5000
TLX:
COM: 40,51

WILLSTROP RODERICK V DR
INSTITUTE OF ASTRONOMY
MADINGLEY ROAD
CAMBRIDGE CB3 0HA
U.K.
TEL: 0223-62204
TLX: 817297 ASTRON G
COM: 25,30

WILSON ALBERT G DR
RESEARCH PROGRAM STUDIES
PO BOX 113
TOPANGA CA 90290
U.S.A.
TEL: 818-716-6332
TLX:
COM: 28,47

WILSON ANDREW S DR
ASTRONOMY PROGRAM
UNIVERSITY OF MARYLAND
COLLEGE PARK MD 20742
U.S.A.
TEL: 301-454-6061
TLX: 7108260352
COM: 40

WILSON BRIAN G PROF
UNIVERSITY OF QUEENSLAND
55 WALCOTT STREET
ST LUCIA, QUEENSLAND 4067
AUSTRALIA
TEL: 61-7-3772200
TLX: 40315 UNIQLD AA
COM:

WILSON CURTIS A
ST JOHN'S COLLEGE
PO BOX 1671
ANNAPOLIS MD 21404
U.S.A.
TEL: 301-263-2371
TLX:
COM: 41

WILSON JAMES R DR
LAWRENCE LIVERMORE LAB
L-35
LIVERMORE CA 94550
U.S.A.
TEL: 415-422-1659
TLX: 910-386-8339 LLNL
COM: 48

WILSON LIONEL DR
ENV. SCIENCE DEPT
LANCASTER UNIVERSITY
LANCASTER LA1 4YQ
U.K.
TEL: 524-65201x4075
TLX: 65111 LQNCULG
COM: 27

WILSON P DR
INST ANGEWANDTE GEODAESIE
RICHARD-STRAUSS-ALLEE 11
D-6000 FRANKFURT/MAIN 70
GERMANY, F.R.
TEL: 49-69-6333260
TLX: 413592
COM: 19

WILSON PETER R PROF
DEPT OF APPLIED MATHS
UNIVERSITY OF SYDNEY
SYDNEY NSW 2007
AUSTRALIA
TEL:
TLX:
COM: 10,12,36

WILSON RAYMOND H DR
5325 GAINSBOROUGH DR
FAIRFAX VA 22032
U.S.A.
TEL: 703-978-3889
TLX:
COM: 26

WILSON RAYMOND N DR
ESO
KARL-SCHWARZSCHILD-STR 2
D-8046 GARCHING B MUNCHEN
GERMANY, F.R.
TEL: 089-320-06-274
TLX: 528282-0 EO D
COM:

WILSON ROBERT E PROF
DEPT OF ASTRONOMY
UNIVERSITY OF FLORIDA
GAINESVILLE FL 32611
U.S.A.
TEL: 904-392-1182
TLX:
COM: 35,42

WILSON ROBERT PROF
DEPT PHYSICS & ASTRONOMY
UNIVERSITY COLLEGE LONDON
GOWER STREET
LONDON WC1E 6BT
U.K.
TEL: 01-387-70-50
TLX: 28722
COM: 14,29,44

WILSON ROBERT W DR
AT & BELL LABORATORIES
HOH L239
BOX 400
HOLMDEL NJ 07733
U.S.A.
TEL: 201-949-3803
TLX:
COM: 34,40

WILSON S J
DEPT OF MATHEMATICS
NATL UNIV OF SINGAPORE
KENT RIDGE
SINGAPORE 0511
SINGAPORE
TEL:
TLX:
COM: 36

WILSON THOMAS L DR
MPI FUER RADIOASTRONOMIE
AUF DEM HUEGEL 69
D-5300 BONN
GERMANY, F.R.
TEL: 0228-525-378
TLX: 886440 MPIFR D
COM: 33,40C

WILSON WILLIAM J DR
JPL
BLDG 168-327
4800 OAK GROVE DRIVE
PASADENA CA 91109
U.S.A.
TEL: 818-354-5699
TLX:
COM: 40

WINCKLER JOHN R PROF
SCHOOL OF PHYS & ASTRON
UNIVERSITY OF MINNESOTA
MINNEAPOLIS MN 55455
U.S.A.
TEL: 612-373-4688
TLX:
COM:

WING ROBERT F PROF
ASTRONOMY DEPT
OHIO STATE UNIVERSITY
174 W. 18TH AVENUE
COLUMBUS OH 43210
U.S.A.
TEL: 614-422-7876
TLX:
COM: 27,29,45

WINGET DONALD E
UNIV. OF TEXAS AT AUSTIN
ASTRONOMY DEPT
AUSTIN TX 78712
U.S.A.
TEL: 512-471-4461
TLX:
COM:

WINIARSKI MACIEJ
CRACOW ASTRONOMICAL OBS
UL. ORLA 171
30-244 KRAKOW
POLAND
TEL:
TLX:
COM: 25

WINK JOERN ERHARD DR
MPI FUER RADIOASTRONOMIE
AUF DEM HUEGEL 69
D-5300 BONN
GERMANY, F.R.
TEL:
TLX: 886440 MPIFR D
COM: 40

WINKLER GERNOT M R DR
US NAVAL OBSERVATORY
TIME SERVICE DEPT
34TH & MASSACHUSETTS AVE
WASHINGTON DC 20390-5100
U.S.A.
TEL: 202-653-1520
TLX: 710-822-1970 NAVOBSY
COM: 04,19,31

WINKLER KARL-HEINZ A DR
LOS ALAMOS NATIONAL LAB
X-DOT, MS-B218
PO BOX 1663
LOS ALAMOS NM 87545
U.S.A.
TEL:
TLX:
COM: 28,35

WINKLER PAUL FRANK DR
DEPT OF PHYSICS
MIDDLEBURY COLLEGE
MIDDLEBURY VT 05753
U.S.A.
TEL: 802-388-3711
TLX: 353249
COM:

WINNBERG ANDERS DR
ONSALA SPACE OBSERVATORY
S-439 00 ONSALA
SWEDEN
TEL: 46-300-60653
TLX: 2400 ONSPACE S
COM: 34,40

WINNEWISSER GISBERT DR
UNIVERSITAT ZU KOLN
I. PHYSIKALISCHES INST
UNIVERSITATSSTRASSE 14
D-5000 KOLN 41
GERMANY, F.R.
TEL: 211-470-3567
TLX:
COM: 14,40

WISNIEWSKI WIESLAW Z
UNIVERSITY OF ARIZONA
LUNAR & PLANETARY LAB
TUCSON AZ 85721
U.S.A.
TEL: 602-621-6956
TLX: 910-952-1143
COM: 15,25,27

WITHBROE GEORGE L DR
CENTER FOR ASTROPHYSICS
HARVARD COLLEGE OBS
60 GARDEN STREET
CAMBRIDGE MA 02138
U.S.A.
TEL: 617-495-7438
TLX:
COM:

WITT ADOLF N DR
RITTER ASTROPHYSICAL
RESEARCH CENTER
UNIVERSITY OF TOLEDO
TOLEDO OH 43606
U.S.A.
TEL: 419-537-2709
TLX: 810 442 1633
COM: 21,34

WITTEN LOUIS PROF
DEPT OF PHYSICS
UNIVERSITY OF CINCINNATI
CINCINNATI OH 45221-0111
U.S.A.
TEL: 513-475-6492
TLX:
COM:

WITTMANN AXEL D. PH D
UNIVERSITAETS-STERNWARTE
GEISMARLANDSR 11
D-3400 GOETTINGEN
GERMANY. F.R.
TEL: 0551-395042
TLX: 96753 USTERN D
COM: 10,12

WITZEL ARNO DR
MPI FUER RADIOASTRONOMIE
AUF DEM HUEGEL 69
D-5300 BONN
GERMANY. F.R.
TEL: 525211
TLX: 886440 MPIFR D
COM: 40

WLERICK GERARD DR
OBSERVATOIRE DE PARIS
61 AVE DE L'OBSERVATOIRE
F-75014 PARIS
FRANCE
TEL: 1-43-20-12-10
TLX: 27776 OBS PARIS
COM: 09C,28

WNUK EDWIN
ASTRONOMICAL OBSERVATORY
A. MICKIEWICZ UNIVERSITY
UL. SLONECZNA 36
60-286 POZNAN
POLAND
TEL:
TLX:
COM: 07

WOEHL HUBERTUS DR
KIEPENHEUER INSTITUT
FUER SONNENPHYSIK
SCHOENECKSTRASSE 6
D-7800 FREIBURG
GERMANY. F.R.
TEL: 0049-761-32864
TLX: 7721552 KIS D
COM: 09,10,12,36

WOLF BERNHARD PH D
LANDESSTERNWARTE
KOENIGSTUHL
D-6900 HEIDELBERG 1
GERMANY. F.R.
TEL: 06221-10036
TLX:
COM: 29

WOLF RAINER E A DR
MPI FUER ASTRONOMIE
KOENIGSTUHL
D-6900 HEIDELBERG 1
GERMANY. F.R.
TEL: 06221-528-1
TLX: 461789 MPIA D
COM:

WOLFE ARTHUR M PROF
DEPT PHYSICS & ASTRONOMY
UNIVERSITY OF PITTSBURGH
PITTSBURGH PA 15260
U.S.A.
TEL: 412-624-4318
TLX:
COM:

WOLFENDALE ARNOLD W PROF
PHYSICS DEPT
THE UNIVERSITY
SOUTH ROAD
DURHAM DH1 3LE
U.K.
TEL: 0385-64971
TLX: 537351 DURLIB G
COM: 48C

WOLFF SIDNEY C DR
KITT PEAK NAT OBSERVATORY
PO BOX 26732
TUCSON AZ 85726-6732
U.S.A.
TEL: 602-327-5511
TLX: 666-484 AURA NOAO
COM: 29C

WOLFSON C JACOB
LOCKHEED RESEARCH LABS
DEPT 91-30
BLDG 202
PALO ALTO CA 94304
U.S.A.
TEL: 415-424-2855
TLX:
COM:

WOLSTENCROFT RAMON D DR
ROYAL OBSERVATORY
BLACKFORD HILL
EDINBURGH EH9 3HJ
U.K.
TEL: 031-667-3321
TLX: 72383
COM: 21,25,34,48,51

WOLSZCZAN ALEKSANDER DR
COPERNICUS ASTRON CENTER
UL. CHOPINA 12/18
87-100 TORUN
POLAND
TEL:
TLX:
COM:

WOLTJER LODEWIJK PROF
ESO
KARL-SCHWARZSCHILD-STR 2
D-8046 GARCHING B MUNCHEN
GERMANY. F.R.
TEL: 89-320-06-227
TLX: 5282820 EO D
COM: 10,33,34,40,47,48C

WOO JONG OK
FAC. OF NATURAL SCIENCES
KOREA NATL UNIV OF EDUC
CHUNGWON-GUN
CHUNGBUK 320-23
KOREA, REPUBLIC
TEL: 431-60-3712
TLX:
COM: 25,46

WOOD DAVID B DR
6 TURNING MILL ROAD
LEXINGTON MA 02173
U.S.A.
TEL:
TLX:
COM: 42

WOOD F BRADSHAW PROF
DEPT OF ASTRONOMY
UNIVERSITY OF FLORIDA
SSRB 211
GAINESVILLE FL 32611
U.S.A.
TEL: 904-392-2059
TLX:
COM: 38C,42

WOOD III H J DR
DEPT OF ASTRONOMY
INDIANA UNIVERSITY
SWAIN HALL W 319
BLOOMINGTON IN 47405
U.S.A.
TEL:
TLX:
COM: 29

WOOD JOHN A DR
CENTER FOR ASTROPHYSICS
60 GARDEN STREET
CAMBRIDGE MA 02138
U.S.A.
TEL: 617-495-7278
TLX: 921428 SATELLITE CAM
COM: 15,16,22

WOOD PETER R DR
MT STROMLO & SIDING
SPRING OBSERVATORIES
PRIVATE BAG
WODEN P.O. ACT 2606
AUSTRALIA
TEL: 62-881111
TLX: 62270 CANOPUS AA
COM: 27,35

WOOD ROGER DR
ROYAL GREENWICH OBS
HERSTMONCEUX CASTLE
HAILSHAM BN27 1RP
U.K.
TEL: 323-833171x3391
TLX: 87451 RGOBSY G
COM: 28

WOODSWORTH ANDREW W.DR
HERZBERG INST ASTROPHYS
NATL RESEARCH COUNCIL
OTTAWA ONT K1A 0R6
CANADA
TEL: 613-993-1906
TLX: 053-3145
COM: 40

WOODWARD PAUL R DR
DEPARTMENT OF ASTRONOMY
UNIVERSITY OF MINNESOTA
116 CHURCH STREET S.E.8
MINNEAPOLIS MS 55455
U.S.A.
TEL:
TLX:
COM: 33,34

WOOLF NEVILLE J
STEWARD OBSERVATORY
UNIVERSITY OF ARIZONA
TUCSON AZ 85721
U.S.A.
TEL:
TLX:
COM: 34,50

WOOLFSON MICHAEL M PROF
DEPT OF PHYSICS
UNIVERSITY OF YORK
HESLINGTON, YORK YO1 5DD
U.K.
TEL: 904-59861
TLX: 57933 YORKULG
COM: 15,16,21,22

WOOLSEY E G
1909 LAUDER DRIVE
OTTAWA ONT K2A 1A9
CANADA
TEL:
TLX:
COM:

WOOSLEY S E PROF
LICK OBSERVATORY
UNIVERSITY OF CALIFORNIA
SANTA CRUZ CA 95064
U.S.A.
TEL: 408-429-2976
TLX:
COM: 28

WOOTTEN HENRY ALWYN
NRAO
EDGEMONT ROAD
CHARLOTTESVILLE VA 22901
U.S.A.
TEL: 804-296-0211
TLX: 510-587-5482
COM: 34,40

WORDEN SIMON P DR
STRATEGIC DEFENSE INITIA-
TIVE ORGANIZATION
OFF SECRETARY OF DEFENSE
WASHINGTON DC 20301-7100
U.S.A.
TEL:
TLX:
COM: 09,12

WORLEY CHARLES E
ASTROM & ASTROPHYS DIV.
US NAVAL OBSERVATORY
WASHINGTON DC 20390
U.S.A.
TEL: 202-653-1588
TLX:
COM: 05C,26

WRIGHT EDWARD L DR
DEPT OF ASTRONOMY
UCLA
LOS ANGELES CA 90024
U.S.A.
TEL: 213-825-5755
TLX:
COM: 34,47

WROBLEWSKI HERBERT DR
DEPTO DE ASTRONOMIA
UNIVERSIDAD DE CHILE
CASILLA 36-D
SANTIAGO
CHILE
TEL: 229-4101
TLX:
COM: 20,24

WU SHENGYIN
BEIJING ASTR OBSERVATORY
ACADEMIA SINICA
BEIJING
CHINA, PEOPLE'S REP.
TEL:
TLX: 22040 BAOAS CN
COM: 40

WORRALL DIANA MARY
HARVARD-SMITHSONIAN CTR
FOR ASTROPHYSICS
60 GARDEN STREET
CAMBRIDGE MA 02138
U.S.A.
TEL: 617-495-7139
TLX: 921428 SATELLITE CAM
COM: 28,48

WRIGHT FRANCES W DR
ASTRONOMY DEPT
HARVARD COLLEGE OBS
60 GARDEN ST
CAMBRIDGE MA 02138
U.S.A.
TEL: 617-495-2647
TLX:
COM: 27

WU CHI CHAO DR
COMPUTER SCIENCES CORP
SPACE TELESCOPE SC INST
HOMEWOOD CAMPUS
BALTIMORE MD 21218
U.S.A.
TEL: 301-338-4770
TLX: U.S.A.
COM: 34,44

WU SHI TSAN DR
SCHOOL OF ENGINEERING
UNIVERSITY OF ALABAMA
IN HUNTSVILLE
HUNTSVILLE AL 35899
U.S.A.
TEL: 205-895-6413
TLX:
COM: 10,49

WORRALL GORDON DR
BIRDSWOOD
EARDISLEY, HEREFDS
U.K.
TEL:
TLX:
COM:

WRIGHT GEOFFREY A E DR
8 BEECHWAYS DRIVE
NESTON S. WIRRAL C64 6TF
U.K.
TEL: 51-336-4208
TLX:
COM: 35

WU FEI
BEIJING OBSERVATORY
ACADEMIA SINICA
BEIJING
CHINA, PEOPLE'S REP.
TEL: 28-1698
TLX: 22040 BAOBS CN
COM: 10,12

WU SHOU-XIAN
SHAANXI OBSERVATORY
P.O. BOX 18
LINTONG XIAN
SHAANXI
CHINA, PEOPLE'S REP.
TEL: 51696 XIAN
TLX: 70121 CSAO CN
COM: 19,31

WORSWICK SUSAN
ROYAL GREENWICH OBS
HERSTMONCEUX CASTLE
HAILSHAM BN27 1RP
U.K.
TEL: 0323-833171
TLX: 87451
COM: 09

WRIGHT HELEN
173 FOREST LAKE DRIVE
ANDOVER NJ 07821
U.S.A.
TEL: 201-786-6566
TLX:
COM: 41

WU HSIN-HENG DR
DEPARTMENT OF PHYSICS
NATL CENTRAL UNIVERSITY
CHUNG-LI
CHINA, TAIWAN
TEL:
TLX:
COM: 37,45

WU XINJI
DEPT OF GEOPHYSICS
PEKING UNIVERSITY
BEIJING
CHINA, PEOPLE'S REP.
TEL: 282471-3929
TLX: 22239 PKUNI CN
COM: 40

WOSZCZYK ANDRZEJ PROF
INSTITUTE OF ASTRONOMY
UL. CHOPINA 12/18
87-100 TORUN
POLAND
TEL: 2-60-18
TLX: 00552234 ASTR PL
COM: 15,16,50

WRIGHT JAMES P DR
DIV ASTRONOMICAL SCIENCES
NAT SC FOUNDATION RM 615
1800 G STREET N.W.
WASHINGTON DC 20550
U.S.A.
TEL: 202-357-7639
TLX:
COM:

WU HUAI-WEI
SHANGHAI OBSERVATORY
ACADEMIA SINICA
80 NAN DAN ROAD
SHANGHAI
CHINA, PEOPLE'S REP.
TEL: 386191
TLX: 33164 SHAO CN
COM: 40

WU XUE-LIN
NANJING ASTRONOMICAL
INSTRUMENT FACTORY
JIANGSU PROVINCE
CHINA, PEOPLE'S REP.
TEL:
TLX:
COM:

WRAMDEMARK STIG S O DR
LUND OBSERVATORY
BOX 43
S-221 00 LUND
SWEDEN
TEL: 46-10 7303
TLX: 33199 OBSNOT S
COM: 25,33,37

WRIGHT KENNETH O DR
DOMINION ASTROPHYS OBS
5071 W SAANICH ROAD
RR 5
VICTORIA BC V8X 4M6
CANADA
TEL: 604-388-3157
TLX:
COM: 29,36,42

WU LIAN-DA
PURPLE MOUNTAIN
OBSERVATORY
NANJING
CHINA, PEOPLE'S REP.
TEL: 32893
TLX: 34144 PMONJ CN
COM: 07

WUNNER GUENTER
LEHRSTUHL F THEORET ASTRO
PHYSIK DER UNIV TUEBINGEN
AUF DER MORGENSTELLE 12,C
D-7400 TUEBINGEN
GERMANY, F.R.
TEL: 7071-292487
TLX: 7262714 AIT D
COM: 14,44

WRAY JAMES D DR
DEPT OF ASTRONOMY
UNIVERSITY OF TEXAS
RL MOORE HALL 15-212
AUSTIN TX 78712
U.S.A.
TEL:
TLX:
COM: 28,44

WRIGHT MELVYN C H DR
RADIO ASTRONOMY LAB
UNIVERSITY OF CALIFORNIA
BERKELEY CA 94720
U.S.A.
TEL: 415-642-0420
TLX:
COM:

WU LIN-XIANG
DEPT OF GEOPHYSICS
BEIJING UNIVERSITY
BEIJING
CHINA, PEOPLE'S REP.
TEL:
TLX:
COM: 09,12

WYCKOFF SUSAN DR
PHYSICS DEPT/ASTRON GROUP
ARIZONA STATE UNIVERSITY
TEMPE AZ 85287
U.S.A.
TEL: 602-965-3561
TLX: 140289 HALLEU ASU UT
COM: 15C,29,45

WRIGHT ALAN E DR
AUSTRALIAN NATIONAL
RADIO ASTRONOMY OBS
P.O.BOX 276
PARKES NSW 2870
AUSTRALIA
TEL: 068-62-3677
TLX: 63999 QASAR
COM: 40

WRIXON GERARD T DR
NATL MICROELECTRONICS
RESEARCH CENTER
UNIVERSITY COLLEGE
CORK
IRELAND
TEL: 353-21-508375
TLX: 26050
COM:

WU MING-CHAN
YUNNAN OBSERVATORY
P.O. BOX 110
KUNMING, YUNNAN PROVINCE
CHINA, PEOPLE'S REP.
TEL: 72946
TLX: 64040 YUOBS CN
COM: 50

WYLLER ARNE A PROF
GRUPO SUECO
APARTADO 66
SANTA CRUZ DE LA PALMA
38071 TENERIFE
SPAIN
TEL: 34-22-40-00-16
TLX:
COM: 09,12,29,36

LIST OF MEMBERS

WYNN-WILLIAMS C G DR
INSTITUTE FOR ASTRONOMY
UNIVERSITY OF HAWAII
2680 WOODLAWN DRIVE
HONOLULU HI 96822
U.S.A.
TEL: 808-948-8807
TLX:
COM: 28,34

WYNNE CHARLES G PROF
ROYAL GREENWICH OBS
HERSTMONCEUX CASTLE
HAILSHAM BN27 1RP
U.K.
TEL: 0323-533-171
TLX: 87451
COM: 09

XANTHAKIS JOHN N PROF
R C A A M
ACADEMY OF ATHENS
14 ANAGNOSTOPOULOU
GR-10673 ATHENS
GREECE
TEL: 3613589
TLX:
COM: 10

XANTHOPOULOS B C DR
DEPT OF PHYSICS
UNIVERSITY OF CRETE
GR-71110 IRAKLION
GREECE
TEL: 081-235576
TLX: 262728
COM: 47

XI ZE-ZONG
INSTITUTE OF THE HISTORY
OF NATURAL SCIENCE
BEIJING
CHINA, PEOPLE'S REP.
TEL:
TLX:
COM: 41C

XIA JIONGYU
INSTITUTE OF GEODESY &
GEOPHYSICS
XU DONG LU
WUCHANG, HUBEI
CHINA, PEOPLE'S REP.
TEL:
TLX:
COM: 19

XIA YI-FEI
DEPT OF ASTRONOMY
NANJING UNIVERSITY
NANJING
CHINA, PEOPLE'S REP.
TEL: 34651-2882
TLX: 34151 PRCNU CN
COM: 08

XIAN DING-ZHANG
PURPLE MOUNTAIN
OBSERVATORY
NANJING
CHINA, PEOPLE'S REP.
TEL: 37609
TLX: 34144 PMONJ CN
COM: 04

XIANG DELIN
PURPLE MOUNTAIN
OBSERVATORY
NANJING
CHINA, PEOPLE'S REP.
TEL: 33738
TLX: 34144 PMONJ CN
COM: 34

XIANG SHOUPING
UNIV SCIENCE & TECHNOLOGY
HEFEI, ANHUI
CHINA, PEOPLE'S REP.
TEL:
TLX:
COM: 47

XIAO NAI-YUAN
DEPT OF ASTRONOMY
NANJING UNIVERSITY
NANJING
CHINA, PEOPLE'S REP.
TEL: 34651-2882
TLX: 34151 PRCNU CN
COM: 19

XIAO XING HUA
DEPT OF ASTRONOMY
BEIJING NORMAL UNIVERSITY
BEIJING
CHINA, PEOPLE'S REP.
TEL: 656531/1367
TLX:
COM: 47

XIE GUANG-ZHONG
YUNNAN OBSERVATORY
ACADEMIA SINICA
PO BOX 110
KUNMING
CHINA, PEOPLE'S REP.
TEL: 72946
TLX: 64040 YUOBS CN
COM: 06

XIE LIANGYUN
INSTITUTE OF GEODEDY &
GEOPHYSICS
XU DONG LU
WUCHANG, HUBEI
CHINA, PEOPLE'S REP.
TEL:
TLX:
COM: 08

XING JUN
DEPT OF GEOPHYSICS
BEIJING UNIVERSITY
BEIJING
CHINA, PEOPLE'S REP.
TEL:
TLX:
COM: 34

XIONG DA-RUN
PURPLE MOUNTAIN
OBSERVATORY
NANJING
CHINA, PEOPLE'S REP.
TEL: 42817
TLX: 34144 PMONJ CN
COM: 27,35

XU AO-AO
DEPT OF ASTRONOMY
NANJING UNIVERSITY
NANJING
CHINA, PEOPLE'S REP.
TEL:
TLX: 34151 PRC NU CN
COM: 10

XU BANG-XIN
DEPT OF ASTRONOMY
NANJING UNIVERSITY
NANJING
CHINA, PEOPLE'S REP.
TEL:
TLX:
COM: 08

XU JIA-YAN
P.O. BOX 18
LINTONG
SHAANXI
CHINA, PEOPLE'S REP.
TEL:
TLX:
COM:

XU PEI-YUAN
INST OF ELECTRON. PHYSICS
UNIV SCIENCE & TECHNOLOGY
SHANGHAI
CHINA, PEOPLE'S REP.
TEL: 951602
TLX:
COM: 40

XU PINXIN
PURPLE MOUNTAIN
OBSERVATORY
NANJING
CHINA, PEOPLE'S REP.
TEL: 32893
TLX: 34144 PMONJ CN
COM: 07

XU TONG-QI
SHANGHAI OBSERVATORY
80 NAN DAN ROAD
SHANGHAI
CHINA, PEOPLE'S REP.
TEL: 386191
TLX: 33164 SHAO CN
COM: 08,19

XU ZHENTAO
PURPLE MOUNTAIN OBS.
2 WEST BEIJING ROAD
NANJING
CHINA, PEOPLE'S REP.
TEL: 31096
TLX: 34144 PMONJ CN
COM: 10,41

XU ZHI-CAI
2 WEST BEIJING STREET
NANJING
CHINA, PEOPLE'S REP.
TEL:
TLX:
COM: 40

YABUSHITA SHIN A PROF
DEPT APPLIED MATHS & PHYS
KYOTO UNIVERSITY
SAKYOKU
KYOTO 606
JAPAN
TEL: 075-751-2111
TLX:
COM: 20,34

YABUUTI KIYOSHI PROF
20 TANAKA HIGASKI
HINCKUCH MACHI
SAKYOKU 606 KYOTO
JAPAN
TEL:
TLX:
COM: 41

YAHIL AMOS DR.
ASTRONOMY PROGRAM
SUNY AT STONY BROOK
ESS BLDG
STONY BROOK NY 11794-2100
U.S.A.
TEL: 516-246-6545
TLX: 510-228-7767
COM:

YAKOVKIN N A DR
KIEV STATE UNIVERSITY
ASTRONOMICAL OBSERVATORY
252053 KIEV
U.S.S.R.
TEL:
TLX:
COM: 10

YALLOP BERNARD D DR
ROYAL GREENWICH OBS
HERSTMONCEUX CASTLE
HAILSHAM BN27 1RP
U.K.
TEL: 0323-833171
TLX: 87451
COM: 04C

YAMAGUCHI SHICHIRO
DEPT OF APPLIED MATHS
FAC ENGG, GIFU UNIVERSITY
YANAGIDO
GIFU 501-11
JAPAN
TEL: 582-30-1111
TLX:
COM:

YAMAKOSHI KAZUO
COSMIC MATTER DIVISION
COSMIC RAY RES/TOKYO UNIV
3 CHOME, 2-1 TANASHI
TOKYO 188
JAPAN
TEL: 0424-61-4131
TLX: 02822371 ICRTU J
COM: 22

YAMAMOTO TETSUO DR
INST OF SPACE &
ASTRONAUTICAL SCIENCES
4-6-1 KOMABA MEGURO-KU
TOKYO 153
JAPAN
TEL: 03-467-1111
TLX: 24550 SPACETKY J
COM:

YAMASAKI ATSUMA DR
DEPT EARTH SC & ASTRONOMY
UNIVERSITY OF TOKYO
KOMABA MEGURO-KU
TOKYO 153
JAPAN
TEL:
TLX:
COM: 42

YANG KE-JUN
P.O. BOX 18
LINTONG
XIAN
CHINA, PEOPLE'S REP.
TEL:
TLX:
COM: 31

YAROV-YAROVOJ M S DR
MATHEMATICS DEPT
MVTU
VTORAYA BAUMANSKAYA 5
107005 MOSCOW
U.S.S.R.
TEL: 267-03-92
TLX: 111572
COM: 07

YEIVIN Y PROF
TEL-AVIV UNIVERSITY
TEL-AVIV 69978
ISRAEL
TEL:
TLX:
COM:

YAMASHITA KOJUN DR
DEPT OF PHYSICS
OSAKA UNIVERSITY
MACHIKANEYAMACHO 1-1
TOYONAKA OSAKA 560
JAPAN
TEL: 06-844-1151
TLX:
COM: 44

YANG LAN-TIAN
DEPT OF PHYSICS
HUAZHONG NORMAL UNIV.
WUHAN
CHINA, PEOPLE'S REP.
TEL: 75601x300 OR401
TLX:
COM: 47,48

YASUDA HARUO PROF DR
TOKYO ASTRONOMICAL
OBSERVATORY
OSAWA, MITAKA
TOKYO 181
JAPAN
TEL:
TLX:
COM: 08

YEN JUI-LIN PROF
DEPT OF ELECTRICAL
ENGINEERING
UNIVERSITY TORONTO
TORONTO ONT M5S 1A4
CANADA
TEL: 416-978-8756
TLX:
COM:

YAMASHITA YASUMASA PROF
TOKYO ASTRONOMICAL
OBSERVATORY
OSAWA, MITAKA
TOKYO 181
JAPAN
TEL:
TLX: 2822307 TAOMTK J
COM: 25,29

YANG SHI JIE
PURPLE MOUNTAIN
OBSERVATORY
NANJING
CHINA, PEOPLE'S REP.
TEL:
TLX: 34144 PMO NJ CN
COM: 09

YATSKIV YA S DR
MAIN ASTRONOMICAL OBS
UKRAINIAN ACADEMY OF SCI
252127 KIEV
U.S.S.R.
TEL: 663110
TLX: 131406 SKY SU
COM: 08,19C

YEOMANS DONALD K DR
JET PROPULSION LAB
MS 264-6
4800 OAK GROVE DRIVE
PASADENA CA 91109
U.S.A.
TEL: 818-354-2127
TLX: 675429 JPL CCMM PSD
COM: 15,20C,22,41

YAMAZAKI AKIRA DR
HYDROGRAPHIC DEPT
TSUKIJI 5 / CHUO-KU
TOKYO 104
JAPAN
TEL: 03-541-3811
TLX: 0 252 2222 JAHYD J
COM: 04C,08

YANKULOVA IVANKA DR
DEPT OF ASTRONOMY
FACULTY OF PHYSICS
ANTON IVANCV STR 5
1126 SOFIA
BULGARIA
TEL:
TLX:
COM:

YAVNEL ALEXANDER A DR
METEORITE COMMITTEE
USSR ACADEMY OF SCIENCES
UL M ULIANCVOJ 3 K 1
117313 MOSCOW
U.S.S.R.
TEL: 1377538
TLX:
COM: 15,22

YI ZHAO-HUA
DEPT OF ASTRONOMY
NANJING UNIVERSITY
NANJING
CHINA, PEOPLE'S REP.
TEL:
TLX:
COM: 07C,46

YAN LIN-SHAN
SHANGHAI OBSERVATORY
ACADEMIA SINICA
SHANGHAI
CHINA, PEOPLE'S REP.
TEL: 386191
TLX: 33164 SHAO CN
COM: 26

YANOVITSKIJ EDGARD G DR
MAIN ASTRONOMICAL OBS
UKRAINIAN ACADEMY OF SCI
252127 KIEV
U.S.S.R.
TEL: 66-31-10
TLX: 131406 SKY SU
COM: 36

YE BINXUN
YUNNAN OBSERVATORY
P.O. BOX 110
KUNMING
CHINA, PEOPLE'S REP.
TEL: 72946
TLX: 64040 YUOBS CN
COM: 09

YILMAZ FATMA DR
UNIVERSITY OBSERVATORY
ISTANBUL
TURKEY
TEL:
TLX:
COM:

YANG FUMIN
SHANGHAI OBSERVATORY
ACADEMIA SINICA
SHANGHAI
CHINA, PEOPLE'S REP.
TEL: 386191
TLX: 33164 SHAO CN
COM: 19

YAO BAO-AN
SHANGHAI OBSERVATORY
ACADEMIA SINICA
SHANGHAI
CHINA, PEOPLE'S REP.
TEL: 386191
TLX: 33164 SHAO CN
COM: 27

YE SHI-HUI
PURPLE MOUNTAIN
OBSERVATORY
NANJING
CHINA, PEOPLE'S REP.
TEL: 46700
TLX: 34144 PMONTJ CN
COM: 10

YILMAZ NIHAL DR
DEPT OF ASTRONOMY
FEN FAKULTESI
UNIVERSITY OF ANKARA
BESEVLER, ANKARA
TURKEY
TEL: 236-550
TLX:
COM:

YANG HAI SHOU
ASTROPHYSICS DIVISION
GEOPHYSICS DEPT
BEIJING UNIVERSITY
BEIJING
CHINA, PEOPLE'S REP.
TEL: 282471-3888
TLX: 22239 TKUNI CN
COM: 10,48

YAO ZHENG-QIU
NANJING ASTRONOMICAL
INSTRUMENT FACTORY
NANJING
CHINA, PEOPLE'S REP.
TEL: 46191
TLX: 34136 GLYNJ c/o NAIF
COM: 09

YE SHU-HUA
SHANGHAI OBSERVATORY
SHANGHAI
CHINA, PEOPLE'S REP.
TEL: 386191
TLX: 33164 SHAO CN
COM: 08,19C,31C,38C

YIN JI-SHENG
BEIJING ASTRONOMICAL OBS
ACADEMIA SINICA
BEIJING
CHINA, PEOPLE'S REP.
TEL: 28-1203
TLX: 22040 BAOAS CN
COM: 25

YANG JIAN
PURPLE MOUNTAIN
OBSERVATORY
NANJING
CHINA, PEOPLE'S REP.
TEL:
TLX: 34144 PMONJ CN
COM: 40

YAPLEE B S
8 CREST VIEW COURT
ROCKVILLE MD 20854
U.S.A.
TEL: 301-762-0935
TLX:
COM:

YEE HOWARD K.C. DR
DEPARTEMENT DE PHYSIQUE
UNIVERSITE DE MONTREAL
C P 6128 SUCC A
MONTREAL PQ H3C 3J7
CANADA
TEL: 514-343-7274
TLX:
COM:

YIN QI-FENG
DEPT OF GEOPHYSICS
BEIJING UNIVERSITY
BEIJING
CHINA, PEOPLE'S REP.
TEL: 28-2471 x3888
TLX: 22239 PKUNI CN
COM: 40

YODER CHARLES F
JET PROPULSION LAB
MS 1831501
4800 OAK GROVE DRIVE
PASADENA CA 91109
U.S.A.
TEL: 818-354-2444
TLX: 617-5429
COM: 16

YOKOYAMA KOICHI DR
INTERNATIONAL LATITUDE
OBSERVATORY
MIZUSAWA
IWATE 023
JAPAN
TEL: 0197-24-7111
TLX: 837628
COM: 19C

YONEYAMA TADAOKI DR
2-1-16 HIBARIGAOKA-KITA
HOYA-SHI
TOKYO 202
JAPAN
TEL:
TLX:
COM:

YORK DONALD G DR
AAC
UNIVERSITY OF CHICAGO
5640 S. ELLIS AVENUE
CHICAGO IL 60637
U.S.A.
TEL: 312-962-8930
TLX: 910-221-5617
COM: 34C

YORKE HAROLD W DR
UNIVERSITAETS STERNWARTE
GEISMARLANDSTR 11
D-3400 GOETTINGEN
GERMANY, F.R.
TEL: 395042
TLX: 96753
COM: 34,35,36

YOSHIDA HARUO
DEPT OF ASTRONOMY
UNIVERSITY OF TOKYO
2-11-16 YAYOI, BUNKYO-KU
TOKYO 113
JAPAN
TEL: 3-812-2111
TLX:
COM: 07

YOSHIDA JUNZO PROF
DEPT OF PHYSICS
KYOTO SANGYO UNIVERSITY
KAMIGAMO KITA-KU
KYOTO 603
JAPAN
TEL: 075-701-2151
TLX: 5422661 KSUJ
COM: 07

YOSHII YUZURU DR
TOKYO ASTRON OBS
UNIVERSITY OF TOKYO
MITAKA 181
JAPAN
TEL: 0422-32-5111
TLX: 2822307 TAOMTK J
COM: 33

YOSHIMURA HIROKAZU DR
FACULTY OF SCIENCE
DEPT OF ASTRONOMY
UNIVERSITY OF TOKYO
TOKYO 113
JAPAN
TEL: 03-812-2111
TLX: 33659 UTYOSCI
COM: 10,12

YOSHIZAWA MASANORI DR
TOKYO ASTRONOMICAL
OBSERVATORY
OSAWA 2-21-1, MITAKA
TOKYO 181
JAPAN
TEL: 0422-32-5111
TLX:
COM: 08C

YOSS KENNETH M DR
ASTRONOMY DEPT
UNIVERSITY OF ILLINOIS
1011 W. SPRINGFIELD AVE
URBANA IL 61801
U.S.A.
TEL: 217-333-3295
TLX:
COM: 30,45

YOU JIAN-QI
PURPLE MOUNTAIN
OBSERVATORY
NANJING
CHINA, PEOPLE'S REP.
TEL: 46700
TLX: 34144 PMONJ CN
COM: 10,12

YOU JUNHAN
ASTROPHYS RES DIV
UNIV OF SCI & TECHNOLOGY
HEIFI, ANHUI PROVINCE
CHINA, PEOPLE'S REP.
TEL: 63300
TLX: 90028 USTC CN
COM: 48

YOUNG ANDREW T DR
DEPT OF PHYSICS
SAN DIEGO STATE UNIV
SAN DIEGO CA 92182-0334
U.S.A.
TEL: 619-265-5817
TLX:
COM: 16,25C

YOUNG ARTHUR DR
ASTRONOMY DEPT
SAN DIEGO STATE UNIV
SAN DIEGO CA 92182
U.S.A.
TEL: 619-265-6167
TLX:
COM:

YOUNG JUDITH SHARN
FIVE COLLEGE RADIOASTR
OBSERVATORY
UNIV OF MASSACHUSETTS
AMHERST MA 01003
U.S.A.
TEL: 413-545-0789
TLX: 95-5491
COM: 28

YOUNG LOUISE GRAY DR
DEPT OF ASTRONOMY
SAN DIEGO STATE UNIV
SAN DIEGO CA 92182
U.S.A.
TEL: 619-287-8890
TLX:
COM: 14,16

YOUNIS SAAD M
ASTRONOMY & SPACE RES CTR
COUNCIL FOR SCI RESEARCH
P O BOX 2441
JADIRIYAH, BAGHDAD
IRAQ
TEL: 7765127
TLX: 2187 BATHILMI IK
COM: 24,28,33,34,40

YOUSEF SHAHINAZ M DR
DEPT OF ASTRONOMY
FACULTY OF SCIENCE
CAIRO UNIVERSITY
CAIRO
EGYPT
TEL:
TLX:
COM:

YOUSSEF NAHED H DR
DEPT OF ASTRONOMY
FACULTY OF SCIENCE
CAIRO UNIVERSITY
CAIRO
EGYPT
TEL: 586041
TLX:
COM: 12

YU KYUNG-LOH PROF
SEOUL NATL UNIVERSITY
SINLIM-DONG
KWANAK-KU
SEOUL 151
KOREA, REPUBLIC
TEL:
TLX:
COM: 08

YUAN CHI PROF
DEPT OF PHYSICS
CITY COLLEGE OF N Y
138 ST CONVENE AVE
NEW YORK NY 10031
U.S.A.
TEL: 212-690-6823
TLX:
COM: 33

YUASA MANABU DR
DEPT OF MATH & PHYSICS
KINKI UNIVERSITY
HIGASHI-OSAKA
OSAKA 577
JAPAN
TEL:
TLX:
COM: 07

YUE ZENG-YUAN
DEPT OF GEOPHYSICS
PEKING UNIVERSITY
PEKING
CHINA, PEOPLE'S REP.
TEL:
TLX:
COM:

YULDASHBAEV TAIMAS S
ASTRONOMICAL INSTITUTE
USSR ACADEMY OF SCIENCES
700052 TASHKENT
U.S.S.R.
TEL:
TLX:
COM:

YUMI S PROF DR
TOKYO UNIVERSITY
HAKUSAN 5-28-20
BUNKYO-KU
TOKYO 112
JAPAN
TEL: 03-945-7392
TLX:
COM: 19,31

YUN HONG-SIK PROF
SEOUL NAT UNIVERSITY
SINLIM-DONG
KWANAK-KU
SEOUL 151
KOREA, REPUBLIC
TEL: 877-2130 x2542
TLX:
COM: 10

YUNGELSON LEV R
ASTRONOMICAL COUNCIL
USSR ACADEMY OF SCIENCES
PYATNITSKAYA 48
109017 MOSCOW
U.S.S.R.
TEL: 231-54-61
TLX: 412623 SCSTP SU
COM: 35

ZABRISKIE F R PROF
RD 1
ALEXANDRIA PA 16611
U.S.A.
TEL: 814-669-4483
TLX:
COM:

ZACHAROV IGOR DR
ASTRONOMICAL INSTITUTE
CZECH. ACAD. OF SCIENCES
OBSERVATORY
251 65 ONDREJOV
CZECHOSLOVAKIA
TEL: 724525
TLX: 121579
COM: 09

ZADUNAISKY PEDRO E PROF
CNIE - CENTRO ESPACIAL
SAN MIGUEL
AV MITRE 3100
1663 SAN MIGUEL (Bs.As.)
ARGENTINA
TEL: 667-2371
TLX:
COM: 20

ZAHN JEAN-PAUL DR
OBS DU PIC-DU-MIDI
ET DE TOULOUSE
14 AVENUE EDOUARD BELIN
F-31400 TOULOUSE
FRANCE
TEL: 61-25-21-01
TLX: 530776 F
COM: 35,36

ZAITSEV VALERII V DR
INST OF APPLIED PHYSICS
ULYANOVA ST. 46
603600 GORKY
U.S.S.R.
TEL:
TLX:
COM: 40

ZAMBON GIULIO
MT STROMLO OBSERVATORY
WODEN P.O. ACT 2606
AUSTRALIA
TEL: 062-881111
TLX: 62270 AA
COM: 09

ZAMORANI GIOVANNI
IST. DI RADIOASTRONOMIA
CNR
VIA IRNERIO 46
I-40126 BOLOGNA
ITALY
TEL: 39-51-232856
TLX: 211664 INFN BO I
COM:

ZANDER RODOLPHE DR
INSTITUT D'ASTROPHYSIQUE
UNIVERSITE DE LIEGE
AVENUE DE COINTE 5
B-4200 COINTE-OUGREE
BELGIUM
TEL: 041-529980
TLX: 41264 ASTRLG B
COM:

ZANINETTI LORENZO
IST. DI FISICA GENERALE
CORSO M. D'AZEGLIO 46
I-10125 TORINO
ITALY
TEL: 011-657694
TLX: 211041 INFNTO I
COM:

ZAPPALA ROSARIO ALDO DR
ISTITUTO DI ASTRONOMIA
CITTA UNIVERSITARIA
I-95125 CATANIA
ITALY
TEL: 33-05-33 x 493
TLX: 970359 ASTRCT I
COM: 10

ZAPPALA VINCENZO PROF
OSSERVATORIO ASTRONOMICO
I-10025 PINO TORINESE
ITALY
TEL: 11-841067
TLX: 213236 TO ASTR I
COM: 15C,20

ZARE KHALIL DR
1180 AWALT DRIVE
MOUNTAIN VIEW CA 94040
U.S.A.
TEL: 415-940-1881
TLX:
COM: 07

ZARNECKI JAN CHARLES DR
UNIT FOR SPACE SCIENCES
PHYSICS LABORATORIES
UNIV KENT AT CANTERBURY
CANTERBURY CT2 7NR
U.K.
TEL: 0227-66822
TLX: 965449 UKLIB
COM: 44

ZASOV ANATOLE V DR
STERNBERG ASTRONOMICAL
INSTITUTE
119899 MOSCOW V-234
U.S.S.R.
TEL:
TLX:
COM: 28

ZAVATTI FRANCO
DIPT. DI ASTRONOMIA
VIA ZAMBONI 33
I-40126 BOLOGNA
ITALY
TEL: 051-222956
TLX: 211664 INFNBO I
COM:

ZDANAVICIUS KAZIMERAS DR
ASTRONOMIJOS OBS
VILNIUS 31
CIURLIONIO 29, LITHUNIA
U.S.S.R.
TEL:
TLX:
COM: 45

ZEALEY WILLIAM J DR
UNIVERSITY OF WOLLONGONG
PHYSICS DEPT
BOX 1144
WOLLONGONG NSW 2500
AUSTRALIA
TEL: 042-270-555
TLX: 29022 AA
COM: 09,34,46

ZEILIK MICHAEL II DR
DEPT PHYSICS & ASTRONOMY
UNIVERSITY OF NEW MEXICO
800 YALE BLVD NE
ALBUQUERQUE NM 87131
U.S.A.
TEL: 505-277-4442
TLX:
COM: 34,46

ZEIPPEN CLAUDE DR
OBSERVATOIRE DE PARIS
SECTION DE MEUDON
F-92195 MEUDON PL CEDEX
FRANCE
TEL: 1-45-34-75-70
TLX: 201571 LAM F
COM: 14

ZEKL HANS WILHELM
TON BELLER GMBH
BURGSTRASSE 22
D-6140 BENSHEIM 3
GERMANY, F.R.
TEL: 06251-73001
TLX: 468352 TBELL D
COM:

ZEL'DOVICH YA B ACAD
SPACE RESEARCH INSTITUTE
USSR ACADEMY OF SCIENCES
117810 MOSCOW
U.S.S.R.
TEL:
TLX:
COM: 28,47,48

ZEL'MANOV A L DR
STERNBERG STATE
ASTRONOMICAL INSTITUTE
119899 MOSCOW
U.S.S.R.
TEL:
TLX:
COM: 47

ZELENKA ANTOINE DR
DACHSLENBERGSTR. 56
CH-8180 BUELACH
SWITZERLAND
TEL:
TLX:
COM: 10,12

ZELLNER BENJAMIN H DR
SPACE TELESCOPE INSTITUTE
HOMEWOOD CAMPUS
3700 SAN MARTIN DRIVE
BALTIMORE MD 21218
U.S.A.
TEL:
TLX:
COM: 15

ZERULL REINER H DR
RUHR-UNIVERSITAET BOCHUM
BEREICH EXTRATERR. PHYSIK
D-4630 BOCHUM
GERMANY, F.R.
TEL: 234-700-4576
TLX: 0825860
COM: 21

ZHAI DI-SHENG
BEIJING ASTRONOMICAL OBS
ACADEMIA SINICA
BEIJING
CHINA, PEOPLE'S REP.
TEL: 28-1698
TLX: 22040 BAOAS CN
COM: 42

ZHAI ZAOCHENG
SHANGHAI OBSERVATORY
ACADEMIA SINICA
80 NAN DAN ROAD
SHANGHAI
CHINA, PEOPLE'S REP.
TEL: 386191
TLX: 33164 SHAO CN
COM: 31

ZHANG BAI-RONG
YUNNAN OBSERVATORY
KUNMING, YUNNAN PROVINCE
CHINA, PEOPLE'S REP.
TEL: 72946
TLX: 64040 YUOBS CN
COM: 10C,50

ZHANG BIN
DEPT OF GEOPHYSICS
BEIJING UNIVERSITY
BEIJING
CHINA, PEOPLE'S REP.
TEL:
TLX:
COM: 33

ZHANG CHENG-YUE
PURPLE MOUNTAIN
OBSERVATORY
NANJING
CHINA, PEOPLE'S REP.
TEL:
TLX: 34144 PMONJ CN
COM: 34

ZHANG GUO-DONG
BEIJING OBSERVATORY
ACADEMIA SINICA
BEIJING
CHINA, PEOPLE'S REP.
TEL:
TLX:
COM: 19

ZHANG HE-QI
PURPLE MOUNTAIN
OBSERVATORY
NANJING, JIANGSU PROVINCE
CHINA, PEOPLE'S REP.
TEL:
TLX: 34144 PMONJ CN
COM: 10,48

ZHANG HUI
P.O. BOX 18
LINTONG
XIAN
CHINA, PEOPLE'S REP.
TEL: XIAN 32255
TLX: 70121 CSAO CN
COM: 08

ZHANG JIA-LU
ASTROPHYSICS RESEARCH DIV
UNIV SCIENCE & TECHNOLOGY
HEFEI, ANHUI PROVINCE
CHINA, PEOPLE'S REP.
TEL: 63300 HEFEI
TLX: 90028 USTC CN
COM: 47,48

ZHANG JIA-XIANG
PURPLE MOUNTAIN
OBSERVATORY
ACADEMIA SINICA
NANJING
CHINA, PEOPLE'S REP.
TEL:
TLX: 34144 PMONJ CN
COM: 20

ZHANG JINTONG
INSTITUTE OF GEODESY &
GEOPHYSICS
XU DONG LU
WUHAN
CHINA, PEOPLE'S REP.
TEL:
TLX:
COM: 31

ZHANG MING-CHANG
DEPT OF ASTRONOMY
NANJING UNIVERSITY
NANJING
CHINA, PEOPLE'S REP.
TEL: 37651
TLX: 0909
COM: 16

ZHANG PEIYU
PURPLE MOUNTAIN
OBSERVATORY
NANJING
CHINA, PEOPLE'S REP.
TEL: 37521
TLX: 34144 PMONJ CN
COM: 41

ZHANG SHENG-PAN
DEPT OF ASTRONOMY
NANJING UNIVERSITY
NANJING
CHINA, PEOPLE'S REP.
TEL: 34651 - 2882
TLX: 34151 PRCNU CN
COM: 07

ZHANG YOUYI
PURPLE MOUNTAIN
OBSERVATORY
NANJING
CHINA, PEOPLE'S REP.
TEL:
TLX: 34144 PMONJ CN
COM: 09

ZHANG YU-ZHE
PURPLE MOUNTAIN
OBSERVATORY
NANJING
CHINA, PEOPLE'S REP.
TEL:
TLX:
COM: 20C

ZHANG ZHEN-DA
DEPT OF ASTRONOMY
NANJING UNIVERSITY
NANJING
CHINA, PEOPLE'S REP.
TEL: 34651-2882
TLX: 34151 PRCNU CN
COM: 10

ZHANG ZHEN-JIU
DEPT OF PHYSICS
HUAZHONG NORMAL UNIV.
WUHAN
CHINA, PEOPLE'S REP.
TEL: WUHAN 75601
TLX: 6908
COM: 47,48

ZHAO GANG
SHANGHAI OBSERVATORY
ACADEMIA SINICA
SHANGHAI
CHINA, PEOPLE'S REP.
TEL: 386191
TLX: 33164 SHAO CN
COM: 31

ZHAO JUN-LIANG
SHANGHAI OBSERVATORY
SHANGHAI
CHINA, PEOPLE'S REP.
TEL: 386191
TLX: 33164 SHAO CN
COM: 33,37

ZHAO MING
SHANGHAI OBSERVATORY
ACADEMIA SINICA
SHANGHAI
CHINA, PEOPLE'S REP.
TEL: 386191
TLX: 33164 SHAO CN
COM: 19

ZHAO REN-YANG
BEIJING ASTRONOMICAL OBS
ACADEMIA SINICA
BEIJING
CHINA, PEOPLE'S REP.
TEL:
TLX:
COM: 40

ZHAO XIAN-ZI
PURPLE MOUNTAIN
OBSERVATORY
NANJING
CHINA, PEOPLE'S REP.
TEL: 31337
TLX: 085 34144 PMONJ CN
COM: 04,07

ZHELEZNIAKOV VLADIMIR V
APPLIED PHYSICS INSTITUTE
USSR ACADEMY OF SCIENCES
ULYANOV STREET 46
603600 GORKIJ
U.S.S.R.
TEL:
TLX:
COM: 40

ZHENG DA-WEI
SHANGHAI OBSERVATORY
ACADEMIA SINICA
80 NAN DAN ROAD
SHANGHAI
CHINA, PEOPLE'S REP.
TEL: 386191
TLX: 33164 SHAO CN
COM: 19

ZHENG JIA-QING
PURPLE MOUNTAIN
OBSERVATORY
NANJING
CHINA, PEOPLE'S REP.
TEL: 46700
TLX: 34144 PMONJ CN
COM: 07

ZHENG WENGUANG
BEIJING ASTRONOMICAL OBS
BEIJING
CHINA, PEOPLE'S REP.
TEL:
TLX:
COM:

ZHENG XUE-TANG
DEPT OF ASTRONOMY
BEIJING NORMAL UNIVERSITY
BEIJING
CHINA, PEOPLE'S REP.
TEL: 633531-6285
TLX: 8511
COM: 07

ZHENG YI-JIA
BEIJING ASTRONOMICAL OBS
BEIJING
CHINA, PEOPLE'S REP.
TEL:
TLX: 22040 BAOBS CN
COM: 40

ZHENG YING
PURPLE MOUNTAIN
OBSERVATORY
NANJING
CHINA, PEOPLE'S REP.
TEL:
TLX: 34144 PMONJ CN
COM: 31

ZHEVAKIN S A PROF DR
RADIOPHYSICAL RESEARCH
INSTITUTE
LYADOV STREET 25/14
603600 GORKIJ
U.S.S.R.
TEL: 36-67-51
TLX:
COM: 35

ZHOU DAOQI
DEPT OF GEOPHYSICS
PEKING UNIVERSITY
BEIJING
CHINA, PEOPLE'S REP.
TEL: 282471-3888
TLX: 22239 PKUNI
COM: 10,12

ZHOU HONG-NAN
DEPT OF ASTRONOMY
NANJING UNIVERSITY
NANJING
CHINA, PEOPLE'S REP.
TEL: 34651-2882
TLX: 34151 PRCNU CN
COM: 07

ZHOU TI-JIAN
DEPT OF GEOPHYSICS
PEKING UNIVERSITY
BEIJING
CHINA, PEOPLE'S REP.
TEL: 28-2471-3888
TLX: 22239 PKUNI
COM: 40

ZHOU XING-HAI
PURPLE MOUNTAIN
OBSERVATORY
NANJING
CHINA, PEOPLE'S REP.
TEL: 46700
TLX: 34144 PMONJ CN
COM: 15,24

ZHOU YOU-YUAN
CENTER FOR ASTROPHYSICS
UNIV SCIENCE & TECHNOLOGY
OF CHINA
HEFEI, ANHUI
CHINA, PEOPLE'S REP.
TEL: 63300 HEFEI
TLX: 90028 USTC CN
COM: 28,47

ZHOU ZHEN-PU
PURPLE MOUNTAIN
OBSERVATORY
NANJING
CHINA, PEOPLE'S REP.
TEL: 33738
TLX: 34114 PMONTJ CN
COM: 34

ZHU CI-SHENG
DEPT OF ASTRONOMY
NANJING UNIVERSITY
NANJING
CHINA, PEOPLE'S REP.
TEL: 37551/2882
TLX: 34151 PRCNU CN
COM: 42

ZHU NENGHONG
SHANGHAI OBSERVATORY
ACADEMIA SINICA
SHANGHAI
CHINA, PEOPLE'S REP.
TEL: 386191
TLX: 33164 SHAO CN
COM: 09

ZHU SHI-CHANG
DEPT OF PHYSICS
SHANGHAI TEACHERS UNIV
SHANGHAI
CHINA, PEOPLE'S REP.
TEL: 384-301
TLX: 9016
COM: 47

ZHU WEN-YAO
SHANGHAI OBSERVATORY
ACADEMIA SINICA
SHANGHAI
CHINA, PEOPLE'S REP.
TEL: 386191
TLX: 33164 SHAO CN
COM: 07

ZHU XINGFENG
CENTER FOR ASTROPHYSICS
UNIV SCIENCE & TECHNOLOGY
HEFEI, ANHUI
CHINA, PEOPLE'S REP.
TEL:
TLX: 90028 USTC CN
COM: 47

ZHU YONG-HE
BEIJING ASTRONOMICAL OBS
BEIJING
CHINA, PEOPLE'S REP.
TEL: 281-1698
TLX: 22040 BAOBS CN
COM: 19

ZHUANG QIXIANG
US NAVAL OBSERVATORY
WASHINGTON DC 20390
U.S.A.
TEL: 202-653-1205
TLX:
COM: 31

ZHUANG WEIFENG
BEIJING OBSERVATORY
ACADEMIA SINICA
BEIJING
CHINA, PEOPLE'S REP.
TEL:
TLX:
COM: 41

ZHUGZHDA YUZEF D DR
INSTITUTE OF TERRESTRIAL
MAGNETISM & IONOSPHERE
142092 ACADEMGORODOK
U.S.S.R.
TEL:
TLX:
COM: 10,12

ZIEBA ANDRZEJ PROF
ASTRONOMICAL OBSERVATORY
UL. ORLA 171
30-244 KRAKOW
POLAND
TEL:
TLX:
COM: 47

ZIEBA STANISLAW DR
OBSERVATORIUM ASTRON.
JAGIELLONIAN UNIVERSITY
UL. ORLA 171
30-244 KRAKOW
POLAND
TEL: 223856, 221877
TLX: 0322297 UJ PL
COM: 40,47

ZIKIDES MICHAEL C DR
DEPT OF ASTRONOMY
PANEPISTIMIOPOLIS
GR-15771 ATHENS
GREECE
TEL:
TLX:
COM:

ZIMMERMANN HELMUT DR
UNIVERSITAETS-STERNWARTE
SCHILLERGAESSCHEN 2
DDR-6900 JENA
GERMANY, D.R.
TEL: 27122
TLX:
COM: 34,46

ZINN ROBERT J DR
DEPT OF ASTRONOMY
YALE UNIVERSITY
260 WHITNEY AVE BOX 6666
NEW HAVEN CT 06511
U.S.A.
TEL: 203-436-3460
TLX:
COM: 28,37

ZINNECKER HANS
ROYAL OBSERVATORY
BLACKFORD HILL
EDINBURGH EH9 3HJ
U.K.
TEL: 031-667-3321
TLX: 72383 ROEDIN G
COM:

ZIOLKOWSKI JANUSZ DR
COPERNICUS ASTRON CENTER
UL. BARTYCKA 18
00-716 WARSAW
POLAND
TEL:
TLX:
COM: 35,42

ZIOLKOWSKI KRZYSZTOF DR
SPACE RESEARCH CENTER
UL. BARTYCKA 18
00-716 WARSAW
POLAND
TEL: 0-22-410041
TLX:
COM: 20

ZIRIN HAROLD DR
CALTECH 264-33
PASADENA CA 91125
U.S.A.
TEL: 818-356-3857
TLX:
COM: 10,12,14

ZIRKER JACK B DR
NATL SOLAR OBSERVATORY
SUNSPOT NM 88349
U.S.A.
TEL: 505-434-1390
TLX:
COM: 12

ZLATEV SLAVEY
ASTRONOMICAL OBSERVATORY
OF KARDGALI
6600 KARDGALI
BULGARIA
TEL: 25-95
TLX: 47421
COM:

ZLOBEC PAOLO DR
OSSERVATORIO ASTRONOMICO
G.B. TIEPOLO 11
P.O.B. SUCC. TRIESTE 5
I-34131 TRIESTE
ITALY
TEL: 40-793921
TLX: 461137 OAT I
COM: 10,40

ZOMBECK MARTIN V DR
CENTER FOR ASTROPHYSICS
60 GARDEN STREET
CAMBRIDGE MA 02138
U.S.A.
TEL: 617-495-7227
TLX: 921428 SATELLITE CAM
COM: 44,48

ZOSIMOVICH IRINA D
INSTITUTE OF HISTORY
ACAD SCI UKRAINIAN SSR
KIROV STR 4
252001 KIEV
U.S.S.R.
TEL: 29-02-72
TLX:
COM: 41

ZOU HUI-CHENG
SHANGHAI OBSERVATORY
80 NAN DAN ROAD
SHANGHAI
CHINA, PEOPLE'S REP.
TEL: 386191
TLX: 33164 SHAO CNCN
COM: 44

ZOU YI-XIN
BEIJING OBSERVATORY
CHUNGKUANTSEN
W. SUBURB
BEIJING
CHINA, PEOPLE'S REP.
TEL: 281261
TLX: 22040
COM: 10

ZOU ZHEN-LONG
BEIJING OBSERVATORY
ACADEMIA SINICA
BEIJING
CHINA, PEOPLE'S REP.
TEL:
TLX: 22040 BAOAS CN
COM: 28,47

ZUCCARELLO FRANCESCA
ISTITUTO DI ASTRONOMIA
VIALE A. DORIA 6
I-95100 CATANIA
ITALY
TEL: 330-533
TLX: 970359 ASTRCT I
COM:

ZUCKERMAN BEN M DR
ASTRONOMY DEPT
UCLA
LOS ANGELES CA 90024
U.S.A.
TEL: 213-825-9338
TLX: 910 342 7597
COM: 27,34,40,51

ZUIDERWIJK EDWARDUS J
BEECHES
JOES LANE
WINDMILL HILL
HAILSHAM E SUSSEX
U.K.
TEL:
TLX:
COM: 42,47

ZVEREV MITROFAN S PROF DR
PULKOVO OBSERVATORY
196140 LENINGRAD
U.S.S.R.
TEL:
TLX:
COM: 08

ZVOLANKOVA JUDITA
ASTRONOMICAL INSTITUTE
SLOVAK ACAD. OF SCIENCES
842 28 BRATISLAVA
CZECHOSLOVAKIA
TEL: 427-375157
TLX: 93373 SEIS
COM: 22

ZWAAN CORNELIS PROF DR
OBSERVATORY
ZONNENBURG 2
NL-3512 NL UTRECHT
NETHERLANDS
TEL: 30-312841
TLX: 47224
COM: 10,12,36

RAYMOND H. FOGLER LIBRARY
DATE DUE

BOOKS ARE SUBJECT TO
RECALL AFTER TWO WEEKS